The Digital Signal Processing Handbook, Second Edition

Digital Signal Processing Fundamentals
Video, Speech, and Audio Signal Processing and Associated Standards
Wireless, Networking, Radar, Sensor Array Processing, and Nonlinear Signal Processing

CRC Press
Taylor & Francis Group
6000 Broken Sound Parkway NW, Suite 300
Boca Raton, FL 33487-2742

Printed in the United States of America on acid-free paper
10 9 8 7 6 5 4 3 2 1

International Standard Book Number: 978-1-4200-4604-5 (Hardback)

Library of Congress Cataloging-in-Publication Data

Wireless, networking, radar, sensor array processing, and nonlinear signal processing / Vijay K. Madisetti.
p. cm.
"Second edition of the DSP Handbook has been divided into three parts."
Includes bibliographical references and index.
ISBN 978-1-4200-4604-5 (alk. paper)
1. Signal processing--Digital techniques. 2. Wireless communication systems. 3. Array processors. 4. Computer networks. 5. Radar. I. Madisetti, V. (Vijay) II. Digital signal processing handbook. III. Title.

TK5102.9.W555 2009
621.382'2--dc22

2009022597

Visit the Taylor & Francis Web site at
http://www.taylorandfrancis.com

and the CRC Press Web site at
http://www.crcpress.com

The Digital Signal Processing Handbook

SECOND EDITION

Wireless, Networking, Radar, Sensor Array Processing, and Nonlinear Signal Processing

EDITOR-IN-CHIEF

Vijay K. Madisetti

CRC Press
Taylor & Francis Group
Boca Raton London New York

CRC Press is an imprint of the
Taylor & Francis Group, an **informa** business

The Electrical Engineering Handbook Series

Series Editor
Richard C. Dorf
University of California, Davis

Titles Included in the Series

The Handbook of Ad Hoc Wireless Networks, Mohammad Ilyas
The Avionics Handbook, Second Edition, Cary R. Spitzer
The Biomedical Engineering Handbook, Third Edition, Joseph D. Bronzino
The Circuits and Filters Handbook, Second Edition, Wai-Kai Chen
The Communications Handbook, Second Edition, Jerry Gibson
The Computer Engineering Handbook, Vojin G. Oklobdzija
The Control Handbook, William S. Levine
The CRC Handbook of Engineering Tables, Richard C. Dorf
The Digital Avionics Handbook, Second Edition Cary R. Spitzer
The Digital Signal Processing Handbook, Second Edition, Vijay K. Madisetti
The Electrical Engineering Handbook, Second Edition, Richard C. Dorf
The Electric Power Engineering Handbook, Second Edition, Leonard L. Grigsby
The Electronics Handbook, Second Edition, Jerry C. Whitaker
The Engineering Handbook, Third Edition, Richard C. Dorf
The Handbook of Formulas and Tables for Signal Processing, Alexander D. Poularikas
The Handbook of Nanoscience, Engineering, and Technology, Second Edition William A. Goddard, III, Donald W. Brenner, Sergey E. Lyshevski, and Gerald J. Iafrate
The Handbook of Optical Communication Networks, Mohammad Ilyas and Hussein T. Mouftah
The Industrial Electronics Handbook, J. David Irwin
The Measurement, Instrumentation, and Sensors Handbook, John G. Webster
The Mechanical Systems Design Handbook, Osita D.I. Nwokah and Yidirim Hurmuzlu
The Mechatronics Handbook, Second Edition, Robert H. Bishop
The Mobile Communications Handbook, Second Edition, Jerry D. Gibson
The Ocean Engineering Handbook, Ferial El-Hawary
The RF and Microwave Handbook, Second Edition, Mike Golio
The Technology Management Handbook, Richard C. Dorf
The Transforms and Applications Handbook, Second Edition, Alexander D. Poularikas
The VLSI Handbook, Second Edition, Wai-Kai Chen

Contents

Preface

Digital signal processing (DSP) is concerned with the theoretical and practical aspects of representing information-bearing signals in a digital form and with using computers, special-purpose hardware and software, or similar platforms to extract information, process it, or transform it in useful ways. Areas where DSP has made a significant impact include telecommunications, wireless and mobile communications, multimedia applications, user interfaces, medical technology, digital entertainment, radar and sonar, seismic signal processing, and remote sensing, to name just a few.

Given the widespread use of DSP, a need developed for an authoritative reference, written by the top experts in the world, that would provide information on both theoretical and practical aspects in a manner that was suitable for a broad audience—ranging from professionals in electrical engineering, computer science, and related engineering and scientific professions to managers involved in technical marketing, and to graduate students and scholars in the field. Given the abundance of basic and introductory texts on DSP, it was important to focus on topics that were useful to engineers and scholars without overemphasizing those topics that were already widely accessible. In short, the DSP handbook was created to be relevant to the needs of the engineering community.

A task of this magnitude could only be possible through the cooperation of some of the foremost DSP researchers and practitioners. That collaboration, over 10 years ago, produced the first edition of the successful DSP handbook that contained a comprehensive range of DSP topics presented with a clarity of vision and a depth of coverage to inform, educate, and guide the reader. Indeed, many of the chapters, written by leaders in their field, have guided readers through a unique vision and perception garnered by the authors through years of experience.

The second edition of the DSP handbook consists of *Digital Signal Processing Fundamentals*; *Video, Speech, and Audio Signal Processing and Associated Standards*; and *Wireless, Networking, Radar, Sensor Array Processing, and Nonlinear Signal Processing* to ensure that each part is dealt with in adequate detail, and that each part is then able to develop its own individual identity and role in terms of its educational mission and audience. I expect each part to be frequently updated with chapters that reflect the changes and new developments in the technology and in the field. The distribution model for the DSP handbook also reflects the increasing need by professionals to access content in electronic form anywhere and at anytime.

Wireless, Networking, Radar, Sensor Array Processing, and Nonlinear Signal Processing, as the name implies, provides a comprehensive coverage of the foundations of signal processing related to wireless, radar, space–time coding, and mobile communications, together with associated applications to networking, storage, and communications.

This book needs to be continuously updated to include newer aspects of these technologies, and I look forward to suggestions on how this handbook can be improved to serve you better.

MATLAB® is a registered trademark of The MathWorks, Inc. For product information, please contact:

The MathWorks, Inc.
3 Apple Hill Drive
Natick, MA 01760-2098 USA
Tel: 508 647 7000
Fax: 508-647-7001
E-mail: info@mathworks.com
Web: www.mathworks.com

Editor

Vijay K. Madisetti is a professor in the School of Electrical and Computer Engineering at the Georgia Institute of Technology in Atlanta. He teaches graduate and undergraduate courses in digital signal processing and computer engineering, and leads a strong research program in digital signal processing, telecommunications, and computer engineering.

Dr. Madisetti received his BTech (Hons) in electronics and electrical communications engineering in 1984 from the Indian Institute of Technology, Kharagpur, India, and his PhD in electrical engineering and computer sciences in 1989 from the University of California at Berkeley.

He has authored or edited several books in the areas of digital signal processing, computer engineering, and software systems, and has served extensively as a consultant to industry and the government. He is a fellow of the IEEE and received the 2006 Frederick Emmons Terman Medal from the American Society of Engineering Education for his contributions to electrical engineering.

Contributors

Naofal Al-Dhahir
Department of Electrical Engineering
The University of Texas at Dallas
Richardson, Texas

Mishal Al-Gharabally
Electrical Engineering Department
College of Engineering and Petroleum
Safat, Kuwait

K. Balemarthy
Department of Electrical and Computer Engineering
Georgia Institute of Technology
Atlanta, Georgia

Victor A. N. Barroso
Department of Electrical and Computer Engineering
Instituo Superior Tecnico
Instituto de Sistemas e Robótica
Lisbon, Portugal

John R. Barry
School of Electrical and Computer Engineering
Georgia Institute of Technology
Atlanta, Georgia

Robert J. Baxley
Georgia Tech Research Institute
Atlanta, Georgia

Kevin M. Buckley
Department of Electrical and Computer Engineering
Villanova University
Villanova, Pennsylvania

Robert Calderbank
Department of Electrical Engineering
Princeton University
Princeton, New Jersey

Chaitali Chakrabarti
School of Electrical, Computer and Energy Engineering
Arizona State University
Tempe, Arizona

Jean-Yves Chouinard
Department of Electronic Engineering and Computer Science
Laval University
Quebec, Quebec, Canada

Jimmy Chui
Department of Electrical Engineering
Princeton University
Princeton, New Jersey

Kevin M. Cuomo
Lincoln Laboratory
Massachusetts Institute of Technology
Lexington, Massachusetts

Pankaj Das
Department of Electrical and Computer Engineering
University of California
San Diego, California

Sushanta Das
Phillips Research N.A.
New York, New York

J. Debardelaben
Department of Electrical and Computer Engineering
Georgia Institute of Technology
Atlanta, Georgia

R. D. DeGroat
Broadcom Corporation
Denver, Colorado

N. Desai
Department of Electrical and Computer Engineering
Georgia Institute of Technology
Atlanta, Georgia

Suhas Diggavi
Ecole Polytechnique
Lausanne, Switzerland

E. M. Dowling
Department of Electrical Engineering
The University of Texas at Dallas
Richardson, Texas

Lan-Rong Dung
Department of Electrical and Computer Engineering
Georgia Institute of Technology
Atlanta, Georgia

T. Egolf
Department of Electrical and Computer Engineering
Georgia Institute of Technology
Atlanta, Georgia

Atef Z. Elsherbeni
Department of Electrical Engineering
University of Mississippi
Oxford, Mississippi

S. Famorzadeh
Department of Electrical and Computer Engineering
Georgia Institute of Technology
Atlanta, Georgia

Daniel R. Fuhrmann
Department of Electrical and System Engineering
Washington University
St. Louis, Missouri

Egemen Gönen
Globalstar
San Jose, California

Martin Haardt
Communication Research Laboratory
Ilmenau University of Technology
Ilmenau, Germany

Emile A. Hendriks
Information and Communication Theory Group
Delft University of Technology
Delft, the Netherlands

R. Hezar
Department of Electrical and Computer Engineering
Georgia Institute of Technology
Atlanta, Georgia

Steven H. Isabelle
Department of Electrical Engineering and Computer Science
Massachusetts Institute of Technology
Cambridge, Massachusetts

Yong-kyu Jung
Department of Electrical and Computer Engineering
Georgia Institute of Technology
Atlanta, Georgia

D. I. Kaklamani
Department of Electrical and Computer Engineering
National Technical University of Athens
Athens, Greece

D. A. Kateros
Department of Electrical and Computer Engineering
National Technical University of Athens
Athens, Greece

Mostafa Kaveh
Department of Electrical and Computer Engineering
University of Minnesota
Minneapolis, Minnesota

A. Kavipurapu
Department of Electrical and Computer Engineering
Georgia Institute of Technology
Atlanta, Georgia

Brian Kelley
Department of Electrical and Computer Engineering
The University of Texas at San Antonio
San Antonio, Texas

M. Khan
Department of Electrical and Computer Engineering
Georgia Institute of Technology
Atlanta, Georgia

Yuan Lin
Advanced Computer Architecture Laboratory
University of Michigan at Ann Arbor
Ann Arbor, Michigan

D. A. Linebarger
Department of Electrical Engineering
The University of Texas at Dallas
Richardson, Texas

K. J. Ray Liu
Department of Electrical and Computer Engineering
University of Maryland
College Park, Maryland

Xiaoli Ma
School of Electrical and Computer Engineering
Georgia Institute of Technology
Atlanta, Georgia

Vijay K. Madisetti
School of Electrical and Computer Engineering
Georgia Institute of Technology
Atlanta, Georgia

Bassem R. Mahafza
Deceibel Research, Inc.
Huntsville, Alabama

Scott Mahlke
Advanced Computer Architecture Laboratory
University of Michigan at Ann Arbor
Ann Arbor, Michigan

Petros Maragos
Department of Electrical and Computer Engineering
National Technical University of Athens
Athens, Greece

Cherian P. Mathews
Department of Electrical and Computer Engineering
University of the Pacific
Stockton, California

Jerry M. Mendel
Department of Electrical Engineering
University of Southern California
Los Angeles, California

David L. Milliner
School of Electrical and Computer Engineering
Georgia Institute of Technology
Atlanta, Georgia

José M. F. Moura
Department of Electrical and Computer Engineering
Carnegie Mellon University
Pittsburgh, Pennsylvania

Trevor Mudge
Advanced Computer Architecture Laboratory
University of Michigan at Ann Arbor
Ann Arbor, Michigan

Arye Nehorai
Department of Electrical and Computer Engineering
The University of Illinois at Chicago
Chicago, Illinois

Alan V. Oppenheim
Department of Electrical Engineering and Computer Science
Massachusetts Institute of Technology
Cambridge, Massachusetts

Eytan Paldi
Department of Mathematics
Israel Institute of Technology
Technion City, Haifa, Israel

C. B. Papadias
Broadband Wireless
Athens Information Technology
Peania Attikis, Greece

Panos Papamichalis
Texas Instruments
Dallas, Texas

A. Paulraj
Department of Electrical Engineering
Stanford University
Stanford, California

Athina P. Petropulu
Department of Electrical and Computer Engineering
Drexel University
Philadelphia, Pennsylvania

M. Pettigrew
Department of Electrical and Computer Engineering
Georgia Institute of Technology
Atlanta, Georgia

G. N. Prezerakos
Department of Electrical and Computer Engineering
National Technical University of Athens
Athens, Greece

and

Technological Education Institute of Piraeus
Athens, Greece

Javier Ramos
Department of Signal Processing and Communications
Universidad Rey Juan Carlos
Madrid, Spain

André Redert
Philips Research Europe
Eindhoven, the Netherlands

Zoltan Safar
Department of Innovation
IT University of Copenhagen
Copenhagen, Denmark

Sangwon Seo
Advanced Computer Architecture Laboratory
University of Michigan at Ann Arbor
Ann Arbor, Michigan

Andrew C. Singer
Sanders (A Lockhead Martin Company)
Manchester, New Hampshire

Mohanned O. Sinnokrot
Department of Electrical and Computer Engineering
Georgia Institute of Technology
Atlanta, Georgia

Brian J. Smith
U.S. Army Aviation and Missile Command
Redstone Arsenal, Alabama

P. Stoica
Information Technology Department
Uppsala University
Uppsala, Sweden

Weifeng Su
Department of Electrical Engineering
State University of New York at Buffalo
Buffalo, New York

Barry Van Veen
Department of Electrical and Computer Engineering
University of Wisconsin
Madison, Wisconsin

Mats Viberg
Department of Signal and Systems
Chalmers University of Technology
Goteborg, Sweden

Hong Wang
Department of Electrical and Computer Engineering
Syracuse University
Syracuse, New York

Xianbin Wang
Department of Electrical and Computer Engineering
University of Western Ontario
London, Ontario, Canada

Douglas B. Williams
School of Electrical and Computer Engineering
Georgia Institute of Technology
Atlanta, Georgia

Mark Woh
Advanced Computer Architecture Laboratory
University of Michigan at Ann Arbor
Ann Arbor, Michigan

M. Wong
Department of Electrical and Computer Engineering
McMaster University
Hamilton, Ontario, Canada

Gregory W. Wornell
Department of Electrical Engineering and Computer Science
Massachusetts Institute of Technology
Cambridge, Massachusetts

Q. Wu
CELWAVE
Claremont, North Carolina

Yiyan Wu
Communications Research Centre
Ottawa, Ontario, Canada

D. A. Zarbouti
Department of Electrical and Computer Engineering
National Technical University of Athens
Athens, Greece

Wei Zhang
School of Electrical and Computer Engineering
Georgia Institute of Technology
Atlanta, Georgia

G. Tong Zhou
Department of Electrical and Computer Engineering
Georgia Institute of Technology
Atlanta, Georgia

Michael D. Zoltowski
School of Electrical and Computer Engineering
Purdue University
West Lafayette, Indiana

I

Sensor Array Processing

Mostafa Kaveh
University of Minnesota

A SENSOR ARRAY SYSTEM CONSISTS OF a number of spatially distributed elements, such as dipoles, hydrophones, geophones or microphones, followed by receivers and a processor. The array samples propagate wavefields in time and space. The receivers and the processor vary in mode of implementation and complexity according to the types of signals encountered, the desired operation, and the adaptability of the array. For example, the array may be narrowband or wideband and the processor may be for determining the directions of the sources of signals or for beamforming to reject interfering signals and to enhance the quality of the desired signal in a communication system. The broad range of applications and the multifaceted nature of technical challenges for modern array signal processing have provided a fertile ground for contributions by and collaborations among researchers and practitioners from many disciplines, particularly those from the signal processing, statistics, and numerical linear algebra communities.

The following chapters present a sampling of the latest theory, algorithms, and applications related to array signal processing. The range of topics and algorithms include some which have been in use for more than a decade as well as some which are results of active current research. The sections on applications give examples of current areas of significant research and development.

Modern array signal processing often requires the use of the formalism of complex variables in modeling received signals and noise. Chapter 1 provides an introduction to complex random processes which are useful for bandpass communication systems and arrays. A classical use for arrays of sensors is to exploit the differences in the location (direction) of sources of transmitted signals to perform spatial filtering. Such techniques are reviewed in Chapter 2.

Another common use of arrays is the estimation of informative parameters about the wavefields impinging on the sensors. The most common parameter of interest is the direction of arrival (DOA) of a wave. Subspace techniques have been advanced as a means of estimating the DOAs of sources, which are very close to each other, with high accuracy. The large number of developments in such techniques is reflected in the topics covered in Chapters 3 through 7. Chapter 3 gives a general overview of subspace processing for direction finding, while Chapter 4 discusses a particular type of subspace algorithm that is extended to sensing of azimuth and elevation angles with planar arrays. Most estimators assume knowledge of the needed statistical characteristics of the measurement noise. This requirement is relaxed in the approach given in Chapter 5. Chapter 6 extends the capabilities of traditional sensors to those which can measure the complete electric and magnetic field components and provides estimators which exploit such information. When signal sources move, or when computational requirements for real-time processing prohibit batch estimation of the subspaces, computationally efficient adaptive subspace updating techniques are called for. Chapter 7 presents many of the recent techniques that have been developed for this purpose. Before subspace methods are used for estimating the parameters of the waves received by an array, it is necessary to determine the number of sources which generate the waves. This aspect of the problem, often termed detection, is discussed in Chapter 8.

An important area of application for arrays is in the field of communications, particularly as it pertains to emerging mobile and cellular systems. Chapter 9 gives an overview of a number of techniques for improving the reception of signals in mobile systems, while Chapter 10 considers problems that arise in beamforming in the presence of multipath signals—a common occurrence in mobile communications. Chapter 12 discusses radar systems that employ sensor arrays, thereby providing the opportunity for space–time signal processing for improved resolution and target detection.

1

Complex Random Variables and Stochastic Processes

Daniel R. Fuhrmann
Washington University

1.1 Introduction

Much of modern digital signal processing is concerned with the extraction of information from signals which are noisy, or which behave randomly while still revealing some attribute or parameter of a system or environment under observation. The term in popular use now for this kind of computation is "statistical signal processing," and much of this handbook is devoted to this very subject. Statistical signal processing is classical statistical inference applied to problems of interest to electrical engineers, with the added twist that answers are often required in "real time," perhaps seconds or less. Thus, computational algorithms are often studied hand-in-hand with statistics.

One thing that separates the phenomena electrical engineers study from that of agronomists, economists, or biologists, is that the data they process are very often complex; that is, the data points come in pairs of the form $x + jy$, where x is called the real part, y the imaginary part, and $j = \sqrt{-1}$. Complex numbers are entirely a human intellectual creation: there are no complex physical measurable quantities such as time, voltage, current, money, employment, crop yield, drug efficacy, or anything else. However, it is possible to attribute to physical phenomena an underlying mathematical model that associates complex causes with real results. Paradoxically, the introduction of a complex-number-based theory can often simplify mathematical models.

Beyond their use in the development of analytical models, complex numbers often appear as actual data in some information processing systems. For representation and computation purposes, a complex number is nothing more than an ordered pair of real numbers. One just mentally attaches the "j" to one of the two numbers, then carries out the arithmetic or signal processing that this interpretation of the data implies.

One of the most well-known systems in electrical engineering that generates complex data from real measurements is the quadrature, or IQ, demodulator, shown in Figure 1.1. The theory behind this system is as follows. A real bandpass signal, with bandwidth small compared to its center frequency, has the form

$$s(t) = A(t)\cos(\omega_c t + \phi(t)), \tag{1.1}$$

where

ω_c is the center frequency
$A(t)$ and $\phi(t)$ are the amplitude and angle modulation, respectively

By viewing $A(t)$ and $\phi(t)$ together as the polar coordinates for a complex function $g(t)$, i.e.,

$$g(t) = A(t)e^{j\phi(t)}, \tag{1.2}$$

we imagine that there is an underlying "complex modulation" driving the generation of $s(t)$, and thus

$$s(t) = \mathrm{Re}\{g(t)e^{j\omega_c t}\}. \tag{1.3}$$

Again, $s(t)$ is physically measurable, while $g(t)$ is a mathematical creation. However, the introduction of $g(t)$ does much to simplify and unify the theory of bandpass communication. It is often the case that information to be transmitted via an electronic communication channel can be mapped directly into the magnitude and phase, or the real and imaginary parts, of $g(t)$. Likewise, it is possible to demodulate $s(t)$, and thus "retrieve" the complex function $g(t)$ and the information it represents. This is the purpose of the quadrature demodulator shown in Figure 1.1. In Section 1.2, we will examine in some detail the operation of this demodulator, but for now note that it has one real input and two real outputs, which are interpreted as the real and imaginary parts of an information-bearing complex signal.

Any application of statistical inference requires the development of a probabilistic model for the received or measured data. This means that we imagine the data to be a "realization" of a multivariate random variable, or a stochastic process, which is governed by some underlying probability space of which we have incomplete knowledge. Thus, the purpose of this section is to give an introduction to probabilistic models for complex data. The topics covered are second-order stochastic processes and their

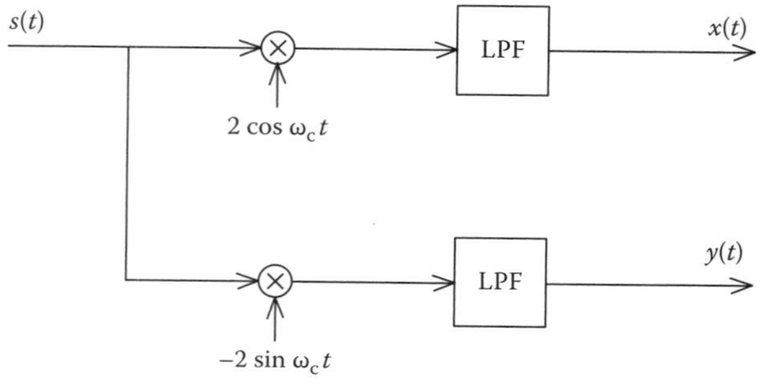

FIGURE 1.1 Quadrature demodulator.

complex representations, the multivariate complex Gaussian distribution, and related distributions which appear in statistical tests. Special attention will be paid to a particular class of random variables, called "circular" complex random variables. Circularity is a type of symmetry in the distributions of the real and imaginary parts of complex random variables and stochastic processes, which can be physically motivated in many applications and is almost always assumed in the statistical signal processing literature. Complex representations for signals and the assumption of circularity are particularly useful in the processing of data or signals from an array of sensors, such as radar antennas. The reader will find them used throughout this chapter of the handbook.

1.2 Complex Envelope Representations of Real Bandpass Stochastic Processes

1.2.1 Representations of Deterministic Signals

The motivation for using complex numbers to represent real phenomena, such as radar or communication signals, may be best understood by first considering the complex envelope of a real deterministic finite-energy signal.

Let $s(t)$ be a real signal with a well-defined Fourier transform $S(\omega)$. We say that $s(t)$ is bandlimited if the support of $S(\omega)$ is finite, that is,

$$S(\omega)\begin{cases} = 0 & \omega \notin B \\ \neq 0 & \omega \notin B \end{cases}, \tag{1.4}$$

where B is the frequency band of the signal, usually a finite union of intervals on the ω-axis such as

$$B = [-\omega_2, -\omega_1] \cup [\omega_1, \omega_2]. \tag{1.5}$$

The Fourier transform of such a signal is illustrated in Figure 1.2.

Since $s(t)$ is real, the Fourier transform $S(\omega)$ exhibits conjugate symmetry, i.e., $S(-\omega) = S^*(\omega)$. This implies that knowledge of $S(\omega)$, for $\omega \geq 0$ only, is sufficient to uniquely identify $s(t)$.

The complex envelope of $s(t)$, which we denote $g(t)$, is a frequency-shifted version of the complex signal whose Fourier transform is $S(\omega)$ for positive ω, and 0 for negative ω. It is found by the operation indicated graphically by the diagram in Figure 1.3, which could be written

$$g(t) = \text{LPF}\{2s(t)e^{-j\omega_c t}\}. \tag{1.6}$$

ω_c is the center frequency of the band B

"LPF" represents an ideal lowpass filter whose bandwidth is greater than half the bandwidth of $s(t)$, but much less than $2\omega_c$

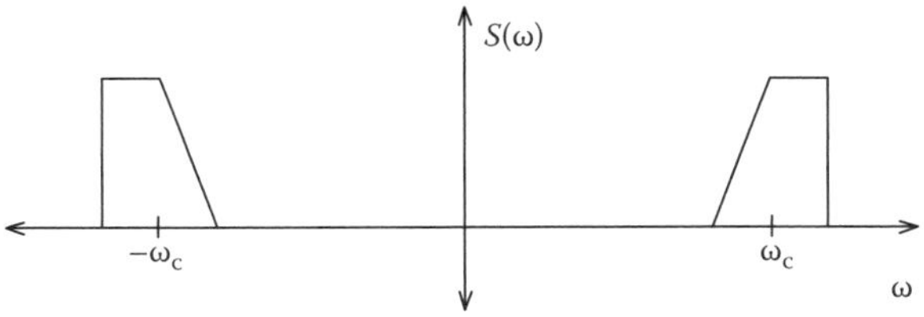

FIGURE 1.2 Fourier transform of a bandpass signal.

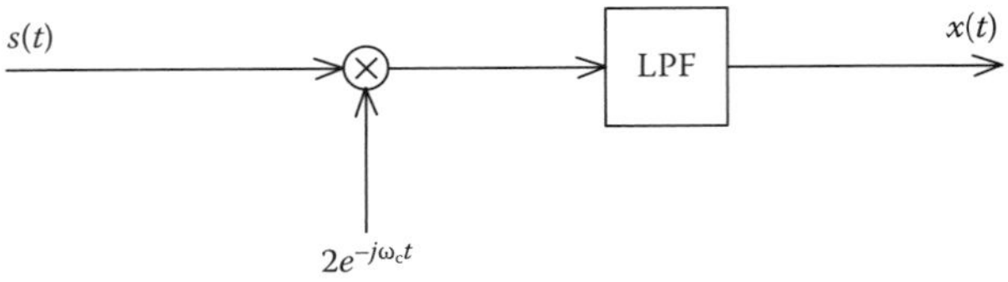

FIGURE 1.3 Quadrature demodulator.

The Fourier transform of $g(t)$ is given by

$$G(\omega) = \begin{cases} 2S(\omega - \omega_c) & |\omega| < BW \\ 0 & \text{otherwise} \end{cases}. \tag{1.7}$$

The Fourier transform of $g(t)$, for $s(t)$ as given in Figure 1.2, is shown in Figure 1.4.

The inverse operation which gives $s(t)$ from $g(t)$ is

$$s(t) = \text{Re}\{g(t)e^{j\omega_c t}\}. \tag{1.8}$$

Our interest in $g(t)$ stems from the information it represents. Real bandpass processes can be written in the form

$$s(t) = A(t)\cos(\omega_c t + \phi(t)), \tag{1.9}$$

where $A(t)$ and $\phi(t)$ are slowly varying functions relative to the unmodulated carrier $\cos(\omega_c t)$, and carry information about the signal source. From the complex envelope representation (Equation 1.3), we know that

$$g(t) = A(t)e^{j\phi(t)} \tag{1.10}$$

and hence $g(t)$, in its polar form, is a direct representation of the information-bearing part of the signal.

In what follows we will outline a basic theory of complex representations for real stochastic processes, instead of the deterministic signals discussed above. We will consider representations of second-order stochastic processes, those with finite variances and correlations and well-defined spectral properties. Two classes of signals will be treated separately: those with finite energy (such as radar signals) and those with finite power (such as radio communication signals).

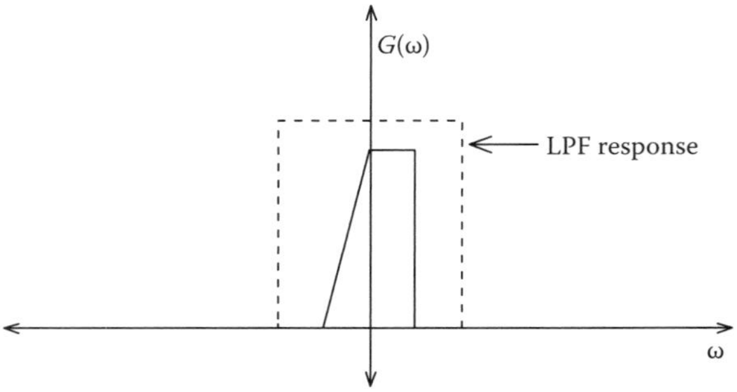

FIGURE 1.4 Fourier transform of the complex representation.

1.2.2 Finite-Energy Second-Order Stochastic Processes

Let $\mathbf{x}(t)$ be a real, second-order stochastic process, with the defining property

$$E\{\mathbf{x}^2(t)\} < \infty, \quad \text{all } t. \tag{1.11}$$

Furthermore, let $\mathbf{x}(t)$ be finite-energy, by which we mean

$$\int_{-\infty}^{\infty} E\{\mathbf{x}^2(t)\}dt < \infty. \tag{1.12}$$

The autocorrelation function for $\mathbf{x}(t)$ is defined as

$$R_{xx}(t_1, t_2) = E\{\mathbf{x}(t_1)\mathbf{x}(t_2)\}, \tag{1.13}$$

and from Equation 1.11 and the Cauchy–Schwartz inequality we know that R_{xx} is finite for all t_1, t_2.

The bifrequency energy spectral density function is

$$S_{xx}(\omega_1, \omega_2) = \int_{-\infty}^{\infty}\int_{-\infty}^{\infty} R_{xx}(t_1, t_2)e^{-j\omega_1 t_1}e^{+j\omega_2 t_2}dt_1 dt_2. \tag{1.14}$$

It is assumed that $S_{xx}(\omega_1, \omega_2)$ exists and is well defined. In an advanced treatment of stochastic processes (e.g., Loeve [1]) it can be shown that $S_{xx}(\omega_1, \omega_2)$ exists if and only if the Fourier transform of $\mathbf{x}(t)$ exists with probability 1; in this case, the process is said to be "harmonizable."

If $\mathbf{x}(t)$ is the input to a linear time-invariant (LTI) system $\mathbf{H}$, and $\mathbf{y}(t)$ is the output process, as shown in Figure 1.5, then $\mathbf{y}(t)$ is also a second-order finite-energy stochastic process. The bifrequency energy spectral density of $\mathbf{y}(t)$ is

$$S_{yy}(\omega_1, \omega_2) = H(\omega_1)H^*(\omega_2)S_{xx}(\omega_1, \omega_2). \tag{1.15}$$

This last result aids in a natural interpretation of the function $S_{xx}(\omega, \omega)$, which we denote as the "energy spectral density." For any process, the total energy E_x is given by

$$E_x = \frac{1}{2\pi}\int_{-\infty}^{\infty} S_{xx}(\omega, \omega)d\omega. \tag{1.16}$$

If we pass $\mathbf{x}(t)$ through an ideal filter whose frequency response is 1 in the band B and 0 elsewhere, then the total energy in the output process is

$$E_y = \frac{1}{2\pi}\int_{B} S_{xx}(\omega, \omega)d\omega. \tag{1.17}$$

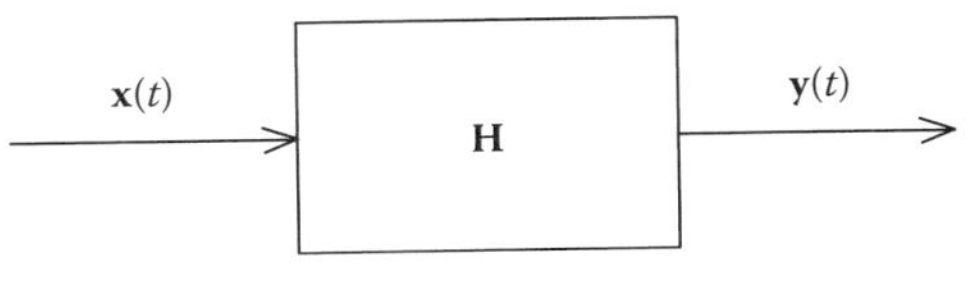

FIGURE 1.5 LTI system with stochastic input and output.

This says that the energy in the stochastic process $\mathbf{x}(t)$ can be partitioned into different frequency bands, and the energy in each band is found by integrating $S_{xx}(\omega, \omega)$ over the band.

We can define a "bandpass" stochastic process, with band B, as one that passes undistorted through an ideal filter $\mathbf{H}$ whose frequency response is 1 within the frequency band and 0 elsewhere. More precisely, if $\mathbf{x}(t)$ is the input to an ideal filter $\mathbf{H}$, and the output process $\mathbf{y}(t)$ is equivalent to $\mathbf{x}(t)$ in the mean-square sense, that is,

$$E\{(\mathbf{x}(t) - \mathbf{y}(t))^2\} = 0, \quad \text{all } t, \tag{1.18}$$

then we say that $\mathbf{x}(t)$ is a bandpass process with frequency band equal to the passband of $\mathbf{H}$. This is equivalent to saying that the integral of $S_{xx}(\omega_1, \omega_2)$ outside of the region $\omega_1, \omega_2 \in B$ is 0.

1.2.3 Second-Order Complex Stochastic Processes

A "complex" stochastic process $\mathbf{z}(t)$ is one given by

$$\mathbf{z}(t) = \mathbf{x}(t) + j\mathbf{y}(t) \tag{1.19}$$

where the real and imaginary parts, $\mathbf{x}(t)$ and $\mathbf{y}(t)$, respectively, are any two stochastic processes defined on a common probability space. A finite-energy, second-order complex stochastic process is one in which $\mathbf{x}(t)$ and $\mathbf{y}(t)$ are both finite-energy, second-order processes, and thus have all the properties given above. Furthermore, because the two processes have a joint distribution, we can define the "cross-correlation function"

$$R_{xy}(t_1, t_2) = E\{\mathbf{x}(t_1)\mathbf{y}(t_2)\}. \tag{1.20}$$

By far the most widely used class of second-order complex processes in signal processing is the class of "circular" complex processes. A circular complex stochastic process is one with the following two defining properties:

$$R_{xx}(t_1, t_2) = R_{yy}(t_1, t_2) \tag{1.21}$$

and

$$R_{xy}(t_1, t_2) = -R_{yx}(t_1, t_2), \quad \text{all } t_1, t_2. \tag{1.22}$$

From Equations 1.21 and 1.22 we have that

$$E\{\mathbf{z}(t_1)\mathbf{z}^*(t_2)\} = 2R_{xx}(t_1, t_2) + 2jR_{yx}(t_1, t_2) \tag{1.23}$$

and furthermore

$$E\{\mathbf{z}(t_1)\mathbf{z}(t_2)\} = 0, \quad \text{all } t_1, t_2. \tag{1.24}$$

This implies that all of the joint second-order statistics for the complex process $\mathbf{z}(t)$ are represented in the function

$$R_{zz}(t_1, t_2) = E\{\mathbf{z}(t_1)\mathbf{z}^*(t_2)\} \tag{1.25}$$

which we define unambiguously as the autocorrelation function for $\mathbf{z}(t)$. Likewise, the bifrequency spectral density function for $\mathbf{z}(t)$ is given by

$$S_{zz}(\omega_1, \omega_2) = \int_{-\infty}^{\infty}\int_{-\infty}^{\infty} R_{zz}(t_1, t_2)e^{-j\omega_1 t_1}e^{+j\omega_2 t_2}dt_1 dt_2. \tag{1.26}$$

The functions $R_{zz}(t_1, t_2)$ and $S_{zz}(\omega_1, \omega_2)$ exhibit Hermitian symmetry, i.e.,

$$R_{zz}(t_1, t_2) = R_{zz}^*(t_2, t_1) \tag{1.27}$$

and

$$S_{zz}(\omega_1, \omega_2) = S_{zz}^*(\omega_2, \omega_1). \tag{1.28}$$

However, there is no requirement that $S_{zz}(\omega_1, \omega_2)$ exhibit the conjugate symmetry for positive and negative frequencies, given in Equation 1.6, as is the case for real stochastic processes.

Other properties of real second-order stochastic processes given above carry over to complex processes. Namely, if $\mathbf{H}$ is a LTI system with arbitrary complex impulse response $h(t)$, frequency response $H(\omega)$, and complex input $\mathbf{z}(t)$, then the complex output $\mathbf{w}(t)$ satisfies

$$S_{ww}(\omega_1, \omega_2) = H(\omega_1)H^*(\omega_2)S_{zz}(\omega_1, \omega_2). \tag{1.29}$$

A bandpass circular complex stochastic process is one with finite spectral support in some arbitrary frequency band B.

Complex stochastic processes undergo a frequency translation when multiplied by a deterministic complex exponential. If $\mathbf{z}(t)$ is circular, then

$$\mathbf{w}(t) = e^{j\omega_c t}\mathbf{z}(t) \tag{1.30}$$

is also circular, and has bifrequency energy spectral density function

$$S_{ww}(\omega_1, \omega_2) = S_{zz}(\omega_1 - \omega_c, \omega_2 - \omega_c). \tag{1.31}$$

1.2.4 Complex Representations of Finite-Energy Second-Order Stochastic Processes

Let $\mathbf{s}(t)$ be a bandpass finite-energy second-order stochastic process, as defined in Section 1.2.2. The complex representation of $\mathbf{s}(t)$ is found by the same down-conversion and filtering operation described for deterministic signals:

$$\mathbf{g}(t) = \text{LPF}\{2\mathbf{s}(t)e^{-j\omega_c t}\}. \tag{1.32}$$

The lowpass filter (LPF) in Equation 1.32 is an ideal filter that passes the baseband components of the frequency-shifted signal, and attenuates the components centered at frequency $-2\omega_c$.

The inverse operation for Equation 1.32 is given by

$$\hat{\mathbf{s}}(t) = \text{Re}\{\mathbf{g}(t)e^{j\omega_c t}\}. \tag{1.33}$$

Because the operation in Equation 1.32 involves the integral of a stochastic process, which we define using mean-square stochastic convergence, we cannot say that $\mathbf{s}(t)$ is identically equal to $\hat{s}(t)$ in the manner that we do for deterministic signals. However, it can be shown that $\mathbf{s}(t)$ and $\hat{s}(t)$ are equivalent in the mean-square sense, that is,

$$E\{(\mathbf{s}(t) - \hat{\mathbf{s}}(t))^2\} = 0, \quad \text{all } t. \tag{1.34}$$

With this interpretation, we say that $\mathbf{g}(t)$ is the unique complex envelope representation for $\mathbf{s}(t)$.

The assumption of circularity of the complex representation is widespread in many signal processing applications. There is an equivalent condition which can be placed on the real bandpass signal that guarantees its complex representation has this circularity property. This condition can be found indirectly by starting with a circular $\mathbf{g}(t)$ and looking at the $\mathbf{s}(t)$ which results.

Let $\mathbf{g}(t)$ be an arbitrary lowpass circular complex finite-energy second-order stochastic process. The frequency-shifted version of this process is

$$\mathbf{p}(t) = \mathbf{g}(t)e^{+j\omega_c t} \tag{1.35}$$

and the real part of this is

$$\mathbf{s}(t) = \frac{1}{2}(\mathbf{p}(t) + \mathbf{p}^{\star}(t)). \tag{1.36}$$

By the definition of circularity, $\mathbf{p}(t)$ and $\mathbf{p}^{\star}(t)$ are orthogonal processes ($E\{\mathbf{p}(t_1)(\mathbf{p}^{\star}(t_2))^{\star} = 0\}$) and from this we have

$$\begin{aligned} S_{ss}(\omega_1, \omega_2) &= \frac{1}{4}(S_{pp}(\omega_1, \omega_2) + S_{p^{\star}p^{\star}}(\omega_1, \omega_2)) \\ &= \frac{1}{4}(S_{gg}(\omega_1 - \omega_c, \omega_2 - \omega_c) + S^{\star}_{gg}(-\omega_1 - \omega_c, -\omega_2 - \omega_c)). \end{aligned} \tag{1.37}$$

Since $\mathbf{g}(t)$ is a baseband signal, the first term in Equation 1.37 has spectral support in the first quadrant in the (ω_1, ω_2) plane, where both ω_1 and ω_2 are positive, and the second term has spectral support only for both frequencies negative. This situation is illustrated in Figure 1.6.

It has been shown that a necessary condition for $\mathbf{s}(t)$ to have a circular complex envelope representation is that it have spectral support only in the first and third quadrants of the (ω_1, ω_2) plane. This condition is also sufficient: if $\mathbf{g}(t)$ is not circular, then the $\mathbf{s}(t)$ which results from the operation in Equation 1.33 will have nonzero spectral components in the second and fourth quadrants of the (ω_1, ω_2) plane, and this contradicts the mean-square equivalence of $\mathbf{s}(t)$ and $\hat{\mathbf{s}}(t)$.

An interesting class of processes with spectral support only in the first and third quadrants is the class of processes whose autocorrelation function is separable in the following way:

$$R_{ss}(t_1, t_2) = R_1(t_1 - t_2)R_2\left(\frac{t_1 + t_2}{2}\right). \tag{1.38}$$

For these processes, the bifrequency energy spectral density separates in a like manner:

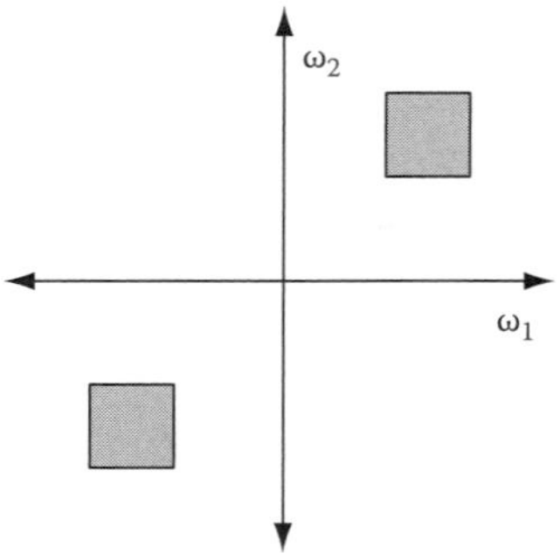

FIGURE 1.6 Spectral support for bandpass process with circular complex representation.

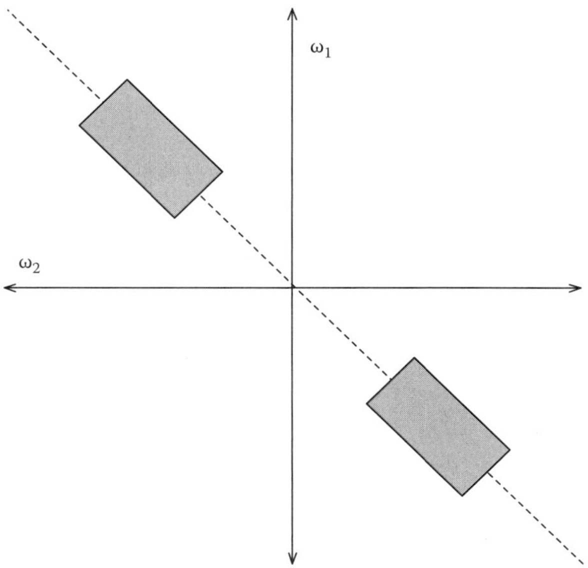

FIGURE 1.7 Spectral support for bandpass process with separable autocorrelation.

$$S_{ss}(\omega_1, \omega_2) = S_1(\omega_1 - \omega_2)S_2\left(\frac{\omega_1 + \omega_2}{2}\right). \tag{1.39}$$

In fact, S_1 is the Fourier transform of R_2 and vice versa. If S_1 is a lowpass function, and S_2 is a bandpass function, then the resulting product has spectral support illustrated in Figure 1.7.

The assumption of circularity in the complex representation can often be physically motivated. For example, in a radar system, if the reflected electromagnetic wave undergoes a phase shift, or if the reflector position cannot be resolved to less than a wavelength, or if the reflection is due to a sum of reflections at slightly different path lengths, then the absolute phase of the return signal is considered random and uniformly distributed. Usually it is not the absolute phase of the received signal which is of interest; rather, it is the "relative phase" of the signal value at two different points in time, or of two different signals at the same instance in time. In many radar systems, particularly those used for direction-of-arrival estimation or delay-Doppler imaging, this relative phase is central to the signal processing objective.

1.2.5 Finite-Power Stochastic Processes

The second major class of second-order processes we wish to consider is the class of finite-power signals. A finite-power signal $\mathbf{x}(t)$ as one whose mean-square value exists, as in Equation 1.4, but whose total energy, as defined in Equation 1.12, is infinite. Furthermore, we require that the time-averaged mean-square value, given by

$$P_x = \lim_{T\to\infty} \frac{1}{2T} \int_{-T}^{T} R_{xx}(t, t)\mathrm{d}t, \tag{1.40}$$

exist and be finite. P_x is called the "power" of the process $\mathbf{x}(t)$.

The most commonly invoked stochastic process of this type in communications and signal processing is the "wide-sense-stationary" (w.s.s.) process, one whose autocorrelation function $R_{xx}(t_1, t_2)$

is a function of the time difference $t_1 - t_2$ only. In this case, the mean-square value is constant and is equal to the average power. Such a process is used to model a communication signal that transmits for a long period of time, and for which the beginning and end of transmission are considered unimportant.

A w.s.s. process may be considered to be the limiting case of a particular type of finite-energy process, namely a process with separable autocorrelation as described by Equations 1.38 and 1.39. If in Equation 1.38 the function $R_2\left(\frac{t_1+t_2}{2}\right)$ is equal to a constant, then the process is w.s.s. with second-order properties determined by the function $R_1(t_1 - t_2)$. The bifrequency energy spectral density function is

$$S_{xx}(\omega_1, \omega_2) = 2\pi\delta(\omega_1 - \omega_2)S_2\left(\frac{\omega_1 + \omega_2}{2}\right) \tag{1.41}$$

where

$$S_2(\omega) = \int_{-\infty}^{\infty} R_1(\tau)e^{-j\omega\tau}d\tau. \tag{1.42}$$

This last pair of equations motivates us to describe the second-order properties of $\mathbf{x}(t)$ with functions of one argument instead of two, namely the autocorrelation function $R_{xx}(\tau)$ and its Fourier transform $S_{xx}(\omega)$, known as the power spectral density. From basic Fourier transform properties we have

$$P_x = \frac{1}{2\pi}\int_{-\infty}^{\infty} S_{xx}(\omega)d\omega. \tag{1.43}$$

If w.s.s. $\mathbf{x}(t)$ is the input to a LTI system with frequency response $H(\omega)$ and output $\mathbf{y}(t)$, then it is not difficult to show that

1. $\mathbf{y}(t)$ is w.s.s.
2. $S_{yy}(\omega) = |H(\omega)|^2 S_{xx}(\omega)$.

These last results, combined with Equation 1.43, lead to a natural interpretation of the power spectral density function. If $\mathbf{x}(t)$ is the input to an ideal bandpass filter with passband B, then the total power of the filter output is

$$P_y = \frac{1}{2\pi}\int_{B} S_x(\omega)d\omega. \tag{1.44}$$

This shows how the total power in the process $\mathbf{x}(t)$ can be attributed to components in different spectral bands.

1.2.6 Complex Wide-Sense-Stationary Processes

Two real stochastic processes $\mathbf{x}(t)$ and $\mathbf{y}(t)$, defined on a common probability space, are said to be jointly w.s.s. if:

1. Both $\mathbf{x}(t)$ and $\mathbf{y}(t)$ are w.s.s.
2. The cross-correlation $R_{xy}(t_1, t_2) = E\{\mathbf{x}(t_1)\mathbf{y}(t_2)\}$ is a function of $t_1 - t_2$ only.

For jointly w.s.s. processes, the cross-correlation function is normally written with a single argument, e.g., $R_{xy}(\tau)$, with $\tau = t_1 - t_2$. From the definition we see that

$$R_{xy}(\tau) = R_{yx}(-\tau). \tag{1.45}$$

A complex w.s.s. stochastic process $\mathbf{z}(t)$ is one that can be written

$$\mathbf{z}(t) = \mathbf{x}(t) + j\mathbf{y}(t) \tag{1.46}$$

where $\mathbf{x}(t)$ and $\mathbf{y}(t)$ are jointly w.s.s. A "circular" complex w.s.s. process is one in which

$$R_{xx}(\tau) = R_{yy}(\tau) \tag{1.47}$$

and

$$R_{xy}(\tau) = -R_{yx}(\tau), \quad \text{all } \tau. \tag{1.48}$$

The reader is cautioned not to confuse the meanings of Equations 1.45 and 1.48.

For circular complex w.s.s. processes, it is easy to show that

$$E\{\mathbf{z}(t_1)\mathbf{z}(t_2)\} = 0, \quad \text{all } t_1, t_2, \tag{1.49}$$

and therefore the function

$$\begin{aligned} R_{zz}(t_1, t_2) &= E\{\mathbf{z}(t_1)\mathbf{z}^*(t_2)\} \\ &= 2R_{xx}(t_1, t_2) + 2jR_{yx}(t_1, t_2) \end{aligned} \tag{1.50}$$

defines all the second-order properties of $\mathbf{z}(t)$. All the quantities involved in Equation 1.50 are functions of $\tau = t_1 - t_2$ only, and thus the single-argument function $R_{zz}(\tau)$ is defined as the autocorrelation function for $\mathbf{z}(t)$.

The power spectral density for $\mathbf{z}(t)$ is

$$S_{zz}(\omega) = \int_{-\infty}^{\infty} R_{zz}(\tau)e^{-j\omega\tau}d\tau. \tag{1.51}$$

$R_{zz}(\tau)$ exhibits conjugate symmetry ($R_{zz}(\tau) = R_{zz}^*(-\tau)$); $S_{zz}(\omega)$ is nonnegative but otherwise has no symmetry constraints.

If $\mathbf{z}(t)$ is the input to a complex LTI system with frequency response $H(\omega)$, then the output process $\mathbf{w}(t)$ is wide-sense-stationarity with power spectral density

$$S_{ww}(\omega) = |H(\omega)|^2 S_{zz}(\omega). \tag{1.52}$$

A bandpass w.s.s. process is one with finite (possible asymmetric) support in frequency.

If $\mathbf{z}(t)$ is a circular w.s.s. process, then

$$\mathbf{w}(t) = e^{j\omega_c t}\mathbf{z}(t) \tag{1.53}$$

is also circular, and has power spectral density

$$S_{ww}(\omega) = S_{zz}(\omega - \omega_c). \tag{1.54}$$

1.2.7 Complex Representations of Real Wide-Sense-Stationary Signals

Let $\mathbf{s}(t)$ be a real bandpass w.s.s. stochastic process. The complex representation for $\mathbf{s}(t)$ is given by the now-familiar expression

$$\mathbf{g}(t) = \text{LPF}\{2\mathbf{s}(t)e^{-j\omega_c t}\} \tag{1.55}$$

with inverse relationship

$$\hat{\mathbf{s}}(t) = \text{Re}\{\mathbf{g}(t)e^{j\omega_c t}\}. \tag{1.56}$$

In Equations 1.55 and 1.56, ω_c is the center frequency for the passband of $\mathbf{s}(t)$, and the LPF has bandwidth greater than that of $\mathbf{s}(t)$ but much less than $2\omega_c$. $\mathbf{s}(t)$ and $\hat{\mathbf{s}}(t)$ are equivalent in the mean-square sense, implying that $\mathbf{g}(t)$ is the unique complex envelope representation for $\mathbf{s}(t)$.

For arbitrary real w.s.s. $\mathbf{s}(t)$, the circularity of the complex representation comes without any additional conditions like the ones imposed for finite-energy signals. If w.s.s. $\mathbf{s}(t)$ is the input to a quadrature demodulator, then the output signals $\mathbf{x}(t)$ and $\mathbf{y}(t)$ are jointly w.s.s., and the complex process

$$\mathbf{g}(t) = \mathbf{x}(t) + j\mathbf{y}(t) \tag{1.57}$$

is circular. There are various ways of showing this, with the simplest probably being a proof by contradiction. If $\mathbf{g}(t)$ is a complex process that is not circular, then the process $\text{Re}\{\mathbf{g}(t)e^{j\omega_c t}\}$ can be shown to have an autocorrelation function with nonzero terms which are a function of $t_1 + t_2$, and thus it cannot be w.s.s.

Communication signals are often modeled as w.s.s. stochastic processes. The stationarity results from the fact that the carrier phase, as seen at the receiver, is unknown and considered random, due to lack of knowledge about the transmitter and path length. This in turn leads to a circularity assumption on the complex modulation.

In many communication and surveillance systems, the quadrature demodulator is an actual electronic subsystem which generates a pair of signals interpreted directly as a complex representation of a bandpass signal. Often these signals are sampled, providing complex digital data for further digital signal processing. In array signal processing, there are multiple such receivers, one behind each sensor or antenna in a multisensor system. Data from an array of receivers is then modeled as a "vector" of complex random variables. In the next section, we consider multivariate distributions for such complex data.

1.3 The Multivariate Complex Gaussian Density Function

The discussions of Section 1.2 centered on the second-order (correlation) properties of real and complex stochastic processes, but to this point nothing has been said about joint probability distributions for these processes. In this section, we consider the distribution of samples from a complex process in which the real and imaginary parts are Gaussian distributed. The key concept of this section is that the assumption of circularity on a complex stochastic process (or any collection of complex random variables) leads to a compact form of the density function which can be written directly as a function of a complex argument z rather than its real and imaginary parts.

From a data processing point-of-view, a collection of N complex numbers is simply a collection of $2N$ real numbers, with a certain mathematical significance attached to the N numbers we call the "real parts" and the other N numbers we call the "imaginary parts." Likewise, a collection of N complex random variables is really just a collection of $2N$ real random variables with some joint distribution in R^{2N}. Because these random variables have an interpretation as real and imaginary parts of some complex numbers, and because the $2N$-dimensional distribution may have certain symmetries such as those resulting from circularity, it is often natural and intuitive to express joint densities and distributions using a notation which makes explicit the complex nature of the quantities involved. In this section we develop such a density for the case where the random variables have a Gaussian distribution and are samples of a circular complex stochastic process.

Let $\mathbf{z}_i, i = 1, \ldots, N$ be a collection of complex numbers that we wish to model probabilistically. Write

$$\mathbf{z}_i = \mathbf{x}_i + j\mathbf{y}_i \tag{1.58}$$

and consider the vector of numbers $[\mathbf{x}_1, \mathbf{y}_1, \ldots, \mathbf{x}_N, \mathbf{y}_N]^{\mathrm{T}}$ as a set of $2N$ random variables with a distribution over $\mathbb{R}^{2N}$. Suppose further that the vector $[\mathbf{x}_1, \mathbf{y}_1, \ldots, \mathbf{x}_N, \mathbf{y}_N]^{\mathrm{T}}$ is subject to the usual multivariate Gaussian distribution with $2N \times 1$ mean vector μ and $2N \times 2N$ covariance matrix $\mathbf{R}$. For compactness, denote the entire random vector with the symbol $\mathbf{x}$. The density function is

$$f_{\mathbf{x}}(x) = (2\pi)^{\frac{-2N}{2}}(\det \mathbf{R})^{\frac{-1}{2}} e^{-\frac{x^{\mathrm{T}}\mathbf{R}^{-1}x}{2}}. \tag{1.59}$$

We seek a way of expressing the density function of Equation 1.59 directly in terms of the complex variable z, i.e., a density of the form $f_{\mathbf{z}}(z)$. In so doing it is important to keep in mind what such a density represents. $f_{\mathbf{z}}(z)$ will be a nonnegative real-valued function $f : \mathbb{C}^N \to \mathbb{R}^+$, with the property that

$$\int\limits_{\mathbb{C}^N} f_{\mathbf{z}}(z)\mathrm{d}z = 1. \tag{1.60}$$

The probability that $\mathbf{z} \in A$, where A is some subset of $\mathbb{C}^N$, is given by

$$P(A) = \int\limits_A f_{\mathbf{z}}(z)\mathrm{d}z. \tag{1.61}$$

The differential element $\mathrm{d}z$ is understood to be

$$\mathrm{d}z = \mathrm{d}x_1\mathrm{d}y_1\mathrm{d}x_2\mathrm{d}y_2 \ldots \mathrm{d}x_N\mathrm{d}y_N. \tag{1.62}$$

The most general form of the complex multivariate Gaussian density is in fact given by Equation 1.59, and further simplification requires further assumptions. Circularity of the underlying complex process is one such key assumption, and it is now imposed. To keep the following development simple, it is assumed that the mean vector μ is 0. The results for nonzero μ are not difficult to obtain by extension.

Consider the four real random variables $\mathbf{x}_i, \mathbf{y}_i, \mathbf{x}_k, \mathbf{y}_k$. If these numbers represent the samples of a circular complex stochastic process, then we can express the 4×4 covariance as

$$E\left\{\begin{bmatrix}\mathbf{x}_i \\ \mathbf{y}_i \\ \mathbf{x}_k \\ \mathbf{y}_k\end{bmatrix}[\mathbf{x}_i\mathbf{y}_i\mathbf{x}_k\mathbf{y}_k]\right\} = \frac{1}{2}\left[\begin{array}{cc|cc}\alpha_{ii} & 0 & \alpha_{ik} & -\beta_{ik} \\ 0 & \alpha_{ii} & \beta_{ik} & \alpha_{ik} \\ \hline \alpha_{ki} & -\beta_{ki} & \alpha_{kk} & 0 \\ \beta_{ki} & \alpha_{ki} & 0 & \alpha_{kk}\end{array}\right], \tag{1.63}$$

where

$$\alpha_{ik} = 2E\{\mathbf{x}_i\mathbf{x}_k\} = 2E\{\mathbf{y}_i\mathbf{y}_k\} \tag{1.64}$$

and

$$\beta_{ik} = -2E\{\mathbf{x}_i\mathbf{y}_k\} = +2E\{\mathbf{x}_k\mathbf{y}_i\}. \tag{1.65}$$

Extending this to the full $2N \times 2N$ covariance matrix $\mathbf{R}$, we have

$$\mathbf{R} = \frac{1}{2}\left[\begin{array}{cc|cc|c|cc} \alpha_{11} & 0 & \alpha_{12} & -\beta_{12} & \cdots & \alpha_{1N} & -\beta_{1N} \\ 0 & \alpha_{11} & \beta_{12} & \alpha_{12} & \cdots & \beta_{1N} & \alpha_{1N} \\ \hline \alpha_{21} & -\beta_{21} & \alpha_{22} & 0 & \cdots & \alpha_{2N} & -\beta_{2N} \\ \beta_{21} & \alpha_{21} & 0 & \alpha_{22} & \cdots & \beta_{2N} & \alpha_{2N} \\ \hline \vdots & \vdots & \vdots & \vdots & \ddots & \vdots & \vdots \\ \hline \alpha_{N1} & -\beta_{N1} & \alpha_{N2} & -\beta_{N2} & \cdots & \alpha_{NN} & 0 \\ \beta_{N1} & \alpha_{N1} & \beta_{N2} & \alpha_{N2} & \cdots & 0 & \alpha_{NN} \end{array}\right]. \tag{1.66}$$

The key thing to notice about the matrix in Equation 1.66 is that, because of its special structure, it is completely specified by N^2 real quantities: one for each of the 2×2 diagonal blocks, and two for each of the 2×2 upper off-diagonal blocks. This is in contrast to the $N(2N + 1)$ free parameters one finds in an unconstrained $2N \times 2N$ real Hermitian matrix.

Consider now the complex random variables $\mathbf{z}_i$ and $\mathbf{z}_k$. We have that

$$\begin{aligned} E\{\mathbf{z}_i\mathbf{z}_i^*\} &= E\{(\mathbf{x}_i + j\mathbf{y}_i)(\mathbf{x}_i - j\mathbf{y}_i)\} \\ &= E\{\mathbf{x}_i^2 + \mathbf{y}_i^2\} = \alpha_{ii} \end{aligned} \tag{1.67}$$

and

$$\begin{aligned} E\{\mathbf{z}_i\mathbf{z}_k^*\} &= E\{(\mathbf{x}_i + j\mathbf{y}_i)(\mathbf{x}_k - j\mathbf{y}_k)\} \\ &= E\{\mathbf{x}_i\mathbf{x}_k + \mathbf{y}_i\mathbf{y}_k - j\mathbf{x}_k\mathbf{y}_i + j\mathbf{x}_i\mathbf{y}_k\} \\ &= \alpha_{ik} + j\beta_{ik}. \end{aligned} \tag{1.68}$$

Similarly

$$E\{\mathbf{z}_k\mathbf{z}_i^*\} = \alpha_{ik} - j\beta_{ik} \tag{1.69}$$

and

$$E\{\mathbf{z}_k\mathbf{z}_k^*\} = \alpha_{kk}. \tag{1.70}$$

Using Equations 1.66 through 1.70, it is possible to write the following $N \times N$ complex Hermitian matrix:

$$E\{\mathbf{z}\mathbf{z}^{\mathrm{H}}\} = \begin{bmatrix} \alpha_{11} & \alpha_{12} + j\beta_{12} & \cdots & \alpha_{1N} + j\beta_{1N} \\ \alpha_{21} + j\beta_{21} & \alpha_{22} & \cdots & \alpha_{2N} + j\beta_{2N} \\ \vdots & \vdots & \vdots & \vdots \\ \alpha_{N1} + j\beta_{N1} & \alpha_{N2} + j\beta_{N2} & \cdots & \alpha_{NN} \end{bmatrix}. \tag{1.71}$$

Note that this complex matrix has exactly the same N^2 free parameters as did the $2N \times 2N$ real matrix $\mathbf{R}$ in Equation 1.66, and thus it tells us everything there is to know about the joint distribution of the real and imaginary components of $\mathbf{z}$. Under the symmetry constraints imposed on $\mathbf{R}$, we can define

$$\mathbf{C} = E\{\mathbf{z}\mathbf{z}^{\mathrm{H}}\} \tag{1.72}$$

and call this matrix the covariance matrix for $\mathbf{z}$. In the 0-mean Gaussian case, this matrix parameter uniquely identifies the multivariate distribution for $\mathbf{z}$.

The derivation of the density function $f_z(z)$ rests on a set of relationships between the $2N \times 1$ real vector $\mathbf{x}$, and its $N \times 1$ complex counterpart $\mathbf{z}$. We say that $\mathbf{x}$ and $\mathbf{z}$ are "isomorphic" to one another, and denote this with the symbol

$$\mathbf{z} \approx \mathbf{x}. \tag{1.73}$$

Likewise we say that the $2N \times 2N$ real matrix $\mathbf{R}$, given in Equation 1.66, and the $N \times N$ complex matrix $\mathbf{C}$, given in Equation 1.71 are isomorphic to one another, or

$$\mathbf{C} \approx \mathbf{R}. \tag{1.74}$$

The development of the complex Gaussian density function $f_z(z)$ is based on three claims based on these isomorphisms.

Proposition 1.1. If $\mathbf{z} \approx \mathbf{x}$, and $\mathbf{R} \approx \mathbf{C}$, then

$$\mathbf{x}^{\mathrm{T}}(2\mathbf{R})\mathbf{x} = \mathbf{z}^{\mathrm{H}}\mathbf{C}\mathbf{z}. \tag{1.75}$$

Proposition 1.2. If $\mathbf{R} \approx \mathbf{C}$, then

$$\frac{1}{4}\mathbf{R}^{-1} \approx \mathbf{C}^{-1}. \tag{1.76}$$

Proposition 1.3. If $\mathbf{R} \approx \mathbf{C}$, then

$$\det \mathbf{R} = |\det \mathbf{C}|^2 \left(\frac{1}{2}\right)^{2N}. \tag{1.77}$$

The density function $f_z(z)$ is found by substituting the results from Propositions 1.1 through 1.3 directly into the density function $f_x(x)$. This is possible because the mapping from **z** to **x** is one-to-one and onto, and the Jacobian is 1 [see Equation 1.62]. We have

$$f_z(z) = (2\pi)^{\frac{-2N}{2}}(\det \mathbf{R})^{\frac{-1}{2}} e^{-\frac{x^T \mathbf{R}^{-1} x}{2}}$$

$$= \left(\frac{1}{2}\right)^{-N}(2\pi)^{-N}(\det \mathbf{C})^{-1} e^{-z^H \mathbf{C}^{-1} z} \tag{1.78}$$

$$= \pi^{-N}(\det \mathbf{C})^{-1} e^{-z^H \mathbf{C}^{-1} z}. \tag{1.79}$$

At this point it is straightforward to introduce a nonzero mean μ, which is the complex vector isomorphic to the mean of the real random vector **x**. The resulting density is

$$f_z(z) = \pi^{-N}(\det \mathbf{C})^{-1} e^{-(z-\mu)^H \mathbf{C}^{-1}(z-\mu)}. \tag{1.80}$$

The density function in Equation 1.80 is commonly referred to as the "complex Gaussian density function," although in truth one could be more general and have an arbitrary $2N$-dimension Gaussian distribution on the real and imaginary components of **z**. It is important to recognize that the use of Equation 1.80 implies those symmetries in the real covariance of **x** implied by circularity of the underlying complex process. This symmetry is expressed by some authors in the equation

$$E\{\mathbf{z}\mathbf{z}^T\} = 0 \tag{1.81}$$

where the superscript "T" indicates transposition without complex conjugation. This comes directly from Equations 1.24 and 1.49.

For many, the functional form of the complex Gaussian density in Equation 1.80 is actually simpler and cleaner than its N-dimensional real counterpart, due to elimination of the various factors of 2 which complicate it. This density is the starting point for virtually all of the multivariate analysis of complex data seen in the current signal and array processing literature.

1.4 Related Distributions

In many problems of interest in statistical signal processing, the raw data may be complex and subject to a complex Gaussian distribution described in the density function in Equation 1.80. The processing may take the form of the computation of a test statistic for use in a hypothesis test. The density functions for these test statistics are then used to determine probabilities of false alarm and/or detection. Thus, it is worthwhile to study certain distributions that are closely related to the complex Gaussian in this way.

In this section we will describe and give the functional form for four densities related to the complex Gaussian: the complex χ^2, the complex F, the complex β, and the complex t. Only the "central" versions of these distributions will be given, i.e., those based on 0-mean Gaussian data. The central distributions are usually associated with the null hypothesis in a detection problem and are used to compute probabilities of false alarm. The noncentral densities, used in computing probabilities of detection, do not exist in closed form but can be easily tabulated.

1.4.1 Complex Chi-Squared Distribution

One very common type of detection problem in radar problems is the "signal present" vs. "signal absent" decision problem. Often under the "signal absent" hypothesis, the data is zero-mean complex Gaussian, with known covariance, whereas under the "signal present" hypothesis the mean is nonzero, but perhaps unknown or subject to some uncertainty. A common test under these circumstances is to compute the sum of squared magnitudes of the data points (after prewhitening, if appropriate) and compare this to a threshold. The resulting test statistic has a χ^2-squared distribution.

Let $\mathbf{z}_1 \ldots \mathbf{z}_N$ be N complex Gaussian random variables, independent and identically distributed with mean 0 and variance 1 (meaning that the covariance matrix for the $\mathbf{z}$ vector is $\mathbf{I}$). Define the real nonnegative random variable $\mathbf{q}$ according to

$$\mathbf{q} = \sum_i^N |\mathbf{z}_i|^2. \tag{1.82}$$

Then the density function for $\mathbf{q}$ is given by

$$f_{\mathrm{q}}(q) = \frac{1}{(N-1)!} q^{N-1} \mathrm{e}^{-q} U(q). \tag{1.83}$$

To establish this result, show that the density function for $|\mathbf{z}_i|^2$ is a simple exponential. Equation 1.83 is the N-fold convolution of this exponential density function with itself.

We often say that $\mathbf{q}$ is χ^2 with N complex degrees of freedom. A "complex degree of freedom" is like two real degrees of freedom. Note, however, that Equation 1.83 is not the usual χ^2 density function with $2N$ degrees of freedom. Each of the real variables going into the computation of $\mathbf{q}$ has variance $\frac{1}{2}$, not 1. $f_{\mathrm{q}}(q)$ is a gamma density with an integer parameter N, and, like the complex Gaussian density in Equation 1.60, it is cleaner and simpler than its real counterpart.

1.4.2 Complex *F*-Distribution

In some "signal present" vs. "signal absent" problems, the variance or covariance of the noise is not known under the null hypothesis, and must be estimated from some auxiliary data. Then the test statistic becomes the ratio of the sum of square magnitudes of the test data to the sum of square magnitudes of the auxiliary data. The resulting test statistic is subject to a particular form of the *F*-distribution.

Let $\mathbf{q}_1$ and $\mathbf{q}_2$ be two independent random variables subject to the χ^2 distribution with N and M complex degrees of freedom, respectively. Define the real, nonnegative random variable $\mathbf{f}$ according to

$$\mathbf{f} = \frac{\mathbf{q}_1}{\mathbf{q}_2}. \tag{1.84}$$

The density function for $\mathbf{f}$ is

$$f_{\mathrm{f}}(f) = \frac{(N+M-1)!}{(N-1)!(M-1)!} \frac{f^{N-1}}{(1+f)^{N+M}} U(f). \tag{1.85}$$

We say that $\mathbf{f}$ is subject to an *F*-distribution with N and M complex degrees of freedom.

1.4.3 Complex Beta Distribution

An *F*-distributed random variable can be transformed in such a way that the resulting density has finite support. The random variable $\mathbf{b}$, defined by

$$\mathbf{b} = \frac{1}{(1+\mathbf{f})}, \tag{1.86}$$

where **f** is an F-distributed random variable, has this property. The density function is given by

$$f_b(b) = \frac{(N+M-1)!}{(N-1)!(M-1)!} b^{M-1}(1-b)^{N-1} \tag{1.87}$$

on the interval $0 \leq b \leq 1$, and is 0 elsewhere.

The random variable **b** is said to be beta-distributed, with N and M complex degrees of freedom.

1.4.4 Complex Student-*t* Distribution

In the "signal present" vs. "signal absent" problem, if the signal is known exactly (including phase) then the optimal detector is a prewhitener followed by a matched filter. The resulting test statistic is complex Gaussian, and the detector partitions the complex plane into two half-planes which become the decision regions for the two hypotheses. Now it may be that the signal is known, but the variance of the noise is not. In this case, the Gaussian test statistic must be scaled by an estimate of the standard deviation, obtained as before from zero-mean auxiliary data. In this case the test statistic is said to have a complex t (or Student-t) distribution. Of the four distributions discussed in this section, this is the only one in which the random variables themselves are complex: the χ^2, F, and β distributions all describe real random variables functionally dependent on complex Gaussians.

Let **z** and **q** be independent scalar random variables. **z** is complex Gaussian with mean 0 and variance 1, and **q** is χ^2 with N complex degrees of freedom. Define the random variable **t** according to

$$\mathbf{t} = \frac{\mathbf{z}}{\sqrt{\mathbf{q}/N}}. \tag{1.88}$$

The density of **t** is then given by

$$f_t(t) = \frac{1}{\pi\left(1 + \frac{|t|^2}{N}\right)^{N+1}}. \tag{1.89}$$

This density is said to be "heavy-tailed" relative to the Gaussian, and this is a result in the uncertainty in the estimate of the standard deviation. Note that as $N \rightarrow \infty$, the denominator Equation 1.88 approaches 1 (i.e., the estimate of the standard deviation approaches truth) and thus $f_t(t)$ approaches the Gaussian density $\pi^{-1}e^{-|t|^2}$ as expected.

1.5 Conclusion

In this chapter, we have outlined a basic theory of complex random variables and stochastic processes as they most often appear in statistical signal and array processing problems. The properties of complex representations for real bandpass signals were emphasized, since this is the most common application in electrical engineering where complex data appear. Models for both finite-energy signals, such as radar pulses, and finite-power signals, such as communication signals, were developed. The key notion of circularity of complex stochastic processes was explored, along with the conditions that a real stochastic process must satisfy in order for it to have a circular complex representation. The complex multivariate Gaussian distribution was developed, again building on the circularity of the underlying complex stochastic process. Finally, related distributions which often appear in statistical inference problems with complex Gaussian data were introduced.

The general topic of random variables and stochastic processes is fundamental to modern signal processing, and many good textbooks are available. Those by Papoulis [2], Leon-Garcia [3], and Melsa and Sage [4] are recommended. The original short paper deriving the complex multivariate Gaussian density function is by Wooding [5]; another derivation and related statistical analysis is given in Goodman [6], whose name is more often cited in connection with complex random variables. The monograph by Miller [7] has a mathematical flavor, and covers complex stochastic processes, stochastic differential equations, parameter estimation, and least-squares problems. The paper by Neeser and Massey [8] treats circular (which they call "proper") complex stochastic processes and their application in information theory. There is a good discussion of complex random variables in Kay [9], which includes Cramer–Rao lower bounds and optimization of functions of complex variables. Kelly and Forsythe [10] is an advanced treatment of inference problems for complex multivariate data, and contains a number of appendices with valuable background information, including one on distributions related to the complex Gaussian.

References

1. Loeve, M., *Probability Theory*, D. Van Nostrand Company, New York, 1963.
2. Papoulis, A., *Probability, Random Variables, and Stochastic Processes*, 3rd edn., McGraw-Hill, New York, 1991.
3. Leon-Garcia, A., *Probability and Random Processes for Electrical Engineering*, 2nd edn., Addison-Wesley, Reading, MA, 1994.
4. Melsa, J. and Sage, A., *An Introduction to Probability and Stochastic Processes*, Prentice-Hall, Englewood Cliffs, NJ, 1973.
5. Wooding, R., The multivariate distribution of complex normal variables, *Biometrika*, 43, 212–215, 1956.
6. Goodman, N., Statistical analysis based on a certain multivariate complex Gaussian distribution, *Ann. Math. Stat.*, 34, 152–177, 1963.
7. Miller, K., *Complex Stochastic Processes*, Addison-Wesley, Reading, MA, 1974.
8. Neeser, F. and Massey, J., Proper complex random processes with applications to information theory, *IEEE Trans. Inform. Theory*, 39(4), 1293–1302, July 1993.
9. Kay, S., *Fundamentals of Statistical Signal Processing: Estimation Theory*, Prentice-Hall, Englewood Cliffs, NJ, 1993.
10. Kelly, E. and Forsythe, K., Adaptive detection and parameter estimation for multidimensional signal models, MIT Lincoln Laboratory Technical Report 848, April 1989.

2

Beamforming Techniques for Spatial Filtering

Barry Van Veen
University of Wisconsin

Kevin M. Buckley
Villanova University

2.1 Introduction

Systems designed to receive spatially propagating signals often encounter the presence of interference signals. If the desired signal and interferers occupy the same temporal frequency band, then temporal filtering cannot be used to separate signal from interference. However, desired and interfering signals often originate from different spatial locations. This spatial separation can be exploited to separate signal from interference using a spatial filter at the receiver.

A beamformer is a processor used in conjunction with an array of sensors to provide a versatile form of spatial filtering. The term "beamforming" derives from the fact that early spatial filters were designed to form pencil beams (see polar plot in Figure 2.5c) in order to receive a signal radiating from a specific location and attenuate signals from other locations. "Forming beams" seems to indicate radiation of energy; however, beamforming is applicable to either radiation or reception of energy. In this section we discuss the formation of beams for reception, providing an overview of beamforming from a signal processing perspective. Data-independent, statistically optimum, adaptive, and partially adaptive beamforming are discussed.

Implementing a temporal filter requires processing of data collected over a temporal aperture. Similarly, implementing a spatial filter requires processing of data collected over a spatial aperture. A single sensor such as an antenna, sonar transducer, or microphone collects impinging energy over a continuous aperture, providing spatial filtering by summing coherently waves that are in phase across the aperture while destructively combining waves that are not. An array of sensors provides a discrete sampling across its aperture. When the spatial sampling is discrete, the processor that performs the spatial filtering is termed a beamformer. Typically a beamformer linearly combines the spatially sampled time series from each sensor to obtain a scalar output time series in the same manner that an FIR filter linearly combines temporally sampled data. Two principal advantages of spatial sampling with an array of sensors are discussed in the following.

Spatial discrimination capability depends on the size of the spatial aperture; as the aperture increases, discrimination improves. The absolute aperture size is not important, rather its size in wavelengths is the critical parameter. A single physical antenna (continuous spatial aperture) capable of providing the requisite discrimination is often practical for high-frequency signals because the wavelength is short. However, when low-frequency signals are of interest, an array of sensors can often synthesize a much larger spatial aperture than that practical with a single physical antenna.

A second very significant advantage of using an array of sensors, relevant at any wavelength, is the spatial filtering versatility offered by discrete sampling. In many application areas, it is necessary to change the spatial filtering function in real time to maintain effective suppression of interfering signals. This change is easily implemented in a discretely sampled system by changing the way in which the beamformer linearly combines the sensor data. Changing the spatial filtering function of a continuous aperture antenna is impractical.

This section begins with the definition of basic terminology, notation, and concepts. Succeeding sections cover data-independent, statistically optimum, adaptive, and partially adaptive beamforming. We then conclude with a summary.

Throughout this section we use methods and techniques from FIR filtering to provide insight into various aspects of spatial filtering with beamformer. However, in some ways beamforming differs significantly from FIR filtering. For example, in beamforming a source of energy has several parameters that can be of interest: range, azimuth and elevation angles, polarization, and temporal frequency content. Different signals are often mutually correlated as a result of multipath propagation. The spatial sampling is often nonuniform and multidimensional. Uncertainty must often be included in characterization of individual sensor response and location, motivating development of robust beamforming techniques. These differences indicate that beamforming represents a more general problem than FIR filtering and, as a result, more general design procedures and processing structures are common.

2.2 Basic Terminology and Concepts

In this section we introduce terminology and concepts employed throughout. We begin by defining the beamforming operation and discussing spatial filtering. Next we introduce second-order statistics of the array data, developing representations for the covariance of the data received at the array and discussing distinctions between narrowband and broadband beamforming. Last, we define various types of beamformers.

2.2.1 Beamforming and Spatial Filtering

Figure 2.1 depicts two beamformers. The first, which samples the propagating wave field in space, is typically used for processing narrowband signals. The output at time k, $y(k)$, is given by a linear combination of the data at the J sensors at time k:

$$y(k) = \sum_{l=1}^{J} w_l^* x_l(k), \tag{2.1}$$

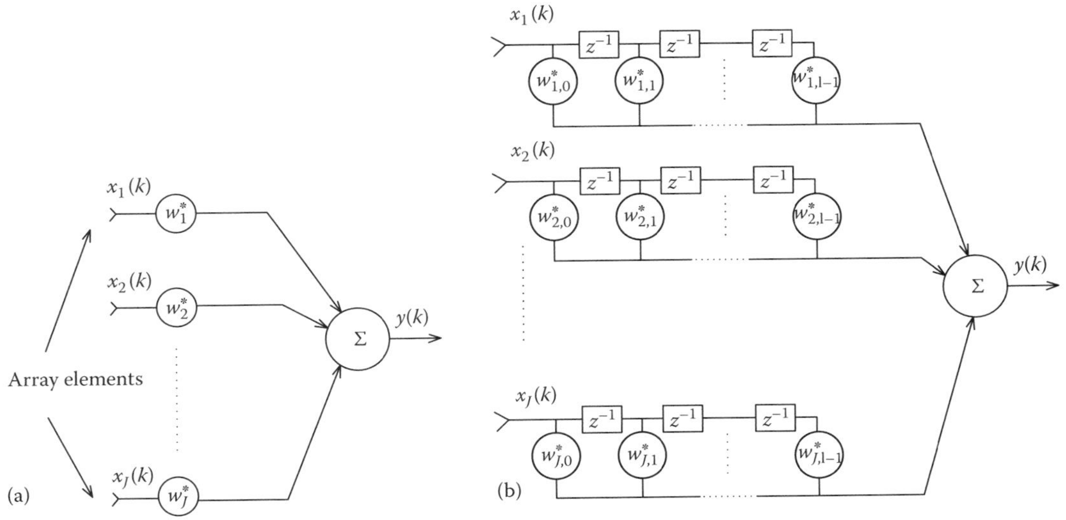

FIGURE 2.1 A beamformer forms a linear combination of the sensor outputs. In (a), sensor outputs are multiplied by complex weights and summed. This beamformer is typically used with narrowband signals. A common broadband beamformer is illustrated in (b).

where $*$ represents complex conjugate. It is conventional to multiply the data by conjugates of the weights to simplify notation. We assume throughout that the data and weights are complex since in many applications a quadrature receiver is used at each sensor to generate in phase and quadrature (I and Q) data. Each sensor is assumed to have any necessary receiver electronics and an A/D converter if beamforming is performed digitally.

The second beamformer in Figure 2.1 samples the propagating wave field in both space and time and is often used when signals of significant frequency extent (broadband) are of interest. The output in this case can be expressed as

$$y(k) = \sum_{l=1}^{J} \sum_{p=0}^{K-1} w_{l,p}^{*} x_l(k-p), \tag{2.2}$$

where $K-1$ is the number of delays in each of the J sensor channels. If the signal at each sensor is viewed as an input, then a beamformer represents a multi-input single output system.

It is convenient to develop notation that permits us to treat both beamformers in Figure 2.1 simultaneously. Note that Equations 2.1 and 2.2 can be written as

$$y(k) = \mathbf{w}^{\mathrm{H}}\mathbf{x}(k), \tag{2.3}$$

by appropriately defining a weight vector $\mathbf{w}$ and data vector $\mathbf{x}(k)$. We use lower and uppercase boldface to denote vector and matrix quantities, respectively, and let superscript represent Hermitian (complex conjugate) transpose. Vectors are assumed to be column vectors. Assume that $\mathbf{w}$ and $\mathbf{x}(k)$ are N dimensional; this implies that $N=KJ$ when referring to Equation 2.2 and $N=J$ when referring to Equation 2.1. Except for Section 2.5 on adaptive algorithms, we will drop the time index and assume that its presence is understood throughout the remainder of this chapter. Thus, Equation 2.3 is written as $y=\mathbf{w}^{\mathrm{H}}\mathbf{x}$. Many of the techniques described in this section are applicable to continuous time as well as discrete time beamforming.

The frequency response of an FIR filter with tap weights w_p^*, $1 \leq p \leq J$ and a tap delay of T seconds is given by

$$r(\omega) = \sum_{p=1}^{J} w_p^* e^{-j\omega T(p-1)}. \tag{2.4}$$

Alternatively

$$r(\omega) = \mathbf{w}^H \mathbf{d}(\omega), \tag{2.5}$$

where

$\mathbf{w}^H = [w_1^* \; w_2^* \ldots w_J^*]$

$\mathbf{d}(\omega) = [1 \; e^{j\omega T} \; e^{j\omega 2T} \ldots e^{j\omega(J-1)T}]^H$

$r(\omega)$ represents the response of the filter* to a complex sinusoid of frequency ω

$\mathbf{d}(\omega)$ is a vector describing the phase of the complex sinusoid at each tap in the FIR filter relative to the tap associated with w_1

Similarly, beamformer response is defined as the amplitude and phase presented to a complex plane wave as a function of location and frequency. Location is, in general, a three-dimensional quantity, but often we are only concerned with one- or two-dimensional direction of arrival (DOA). Throughout the remainder of the section we do not consider range. Figure 2.2 illustrates the manner in which an array of sensors samples a spatially propagating signal. Assume that the signal is a complex plane wave with DOA θ and frequency ω. For convenience let the phase be zero at the first sensor. This implies $x_1(k) = e^{j\omega k}$ and $x_l(k) = e^{j\omega[k - \Delta_l(\theta)]}$, $2 \leq l \leq J$. $\Delta_l(\theta)$ represents the time delay due to propagation from the first to the lth sensor. Substitution into Equation 2.2 results in the beamformer output

$$y(k) = e^{j\omega k} \sum_{l=1}^{J} \sum_{p=0}^{K-1} w_{l,p}^* e^{-j\omega[\Delta_l(\theta)+p]} = e^{j\omega k} \; r(\theta\omega), \tag{2.6}$$

where $\Delta_1(\theta) = 0$. $r(\theta, \omega)$ is the beamformer response and can be expressed in vector form as

$$r(\theta, \omega) = \mathbf{w}^H \mathbf{d}(\theta, \omega). \tag{2.7}$$

The elements of $\mathbf{d}(\theta, \omega)$ correspond to the complex exponentials $e^{j\omega[\Delta_l(\theta)+p]}$. In general it can be expressed as

$$\mathbf{d}(\theta, \omega) = [1 \; e^{j\omega\tau_2(\theta)} \; e^{j\omega\tau_3(\theta)} \ldots e^{j\omega\tau_N(\theta)}]^H \tag{2.8}$$

where the $\tau_i(\theta)$, $2 \leq i \leq N$ are the time delays due to propagation and any tap delays from the zero phase reference to the point at which the ith weight is applied. We refer to $\mathbf{d}(\theta, \omega)$ as the array response vector. It is also known as the steering vector, direction vector, or array manifold vector. Nonideal sensor characteristics can be incorporated into $\mathbf{d}(\theta, \omega)$ by multiplying each phase shift by a function $a_i(\theta, \omega)$, which describes the associated sensor response as a function of frequency and direction.

* An FIR filter is by definition linear, so an input sinusoid produces at the output a sinusoid of the same frequency. The magnitude and argument of $r(\omega)$ are, respectively, the magnitude and phase responses.

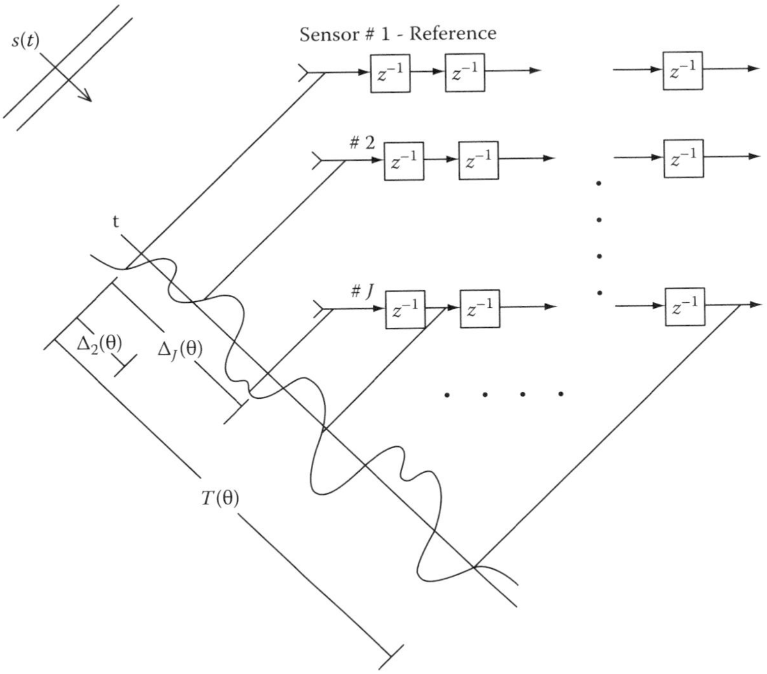

FIGURE 2.2 An array with attached delay lines provides a spatial/temporal sampling of propagating sources. This figure illustrates this sampling of a signal propagating in plane waves from a source located at DOA θ. With J sensors and K samples per sensor, at any instant in time the propagating source signal is sampled at JK nonuniformly spaced points. $T(\theta)$, the time duration from the first sample of the first sensor to the last sample of the last sensor, is termed the temporal aperture of the observation of the source at θ. As notation suggests, temporal aperture will be a function of DOA θ. Plane wave propagation implies that at any time k a propagating signal, received anywhere on a planar front perpendicular to a line drawn from the source to a point on the plane, has equal intensity. Propagation of the signal between two points in space is then characterized as pure delay. In this figure, $\Delta_l(\theta)$ represents the time delay due to plane wave propagation from the first (reference) to the lth sensor.

The "beampattern" is defined as the magnitude squared of $r(\theta, \omega)$. Note that each weight in $\mathbf{w}$ affects both the temporal and spatial responses of the beamformer. Historically, use of FIR filters has been viewed as providing frequency dependent weights in each channel. This interpretation is somewhat incomplete since the coefficients in each filter also influence the spatial filtering characteristics of the beamformer. As a multi-input single output system, the spatial and temporal filtering that occurs is a result of mutual interaction between spatial and temporal sampling.

The correspondence between FIR filtering and beamforming is closest when the beamformer operates at a single temporal frequency ω_o and the array geometry is linear and equispaced as illustrate in Figure 2.3. Letting the sensor spacing be d, propagation velocity be c, and θ represent DOA relative to broadside (perpendicular to the array), we have $\tau_i(\theta) = (i-1)(d/c)\sin\theta$. In this case we identify the relationship between temporal frequency ω in $\mathbf{d}(\omega)$ (FIR filter) and direction θ in $\mathbf{d}(\theta, \omega_o)$ (beamformer) as $\omega = \omega_o(d/c)$ sin θ. Thus, temporal frequency in an FIR filter corresponds to the sine of direction in a narrowband linear equispaced beamformer. Complete interchange of beamforming and FIR filtering methods is possible for this special case provided the mapping between frequency and direction is accounted for.

The vector notation introduced in Equation 2.3 suggests a vector space interpretation of beamforming. This point of view is useful both in beamformer design and analysis. We use it here in consideration of spatial sampling and array geometry. The weight vector $\mathbf{w}$ and the array response vectors $\mathbf{d}(\theta, \omega)$ are vectors in an

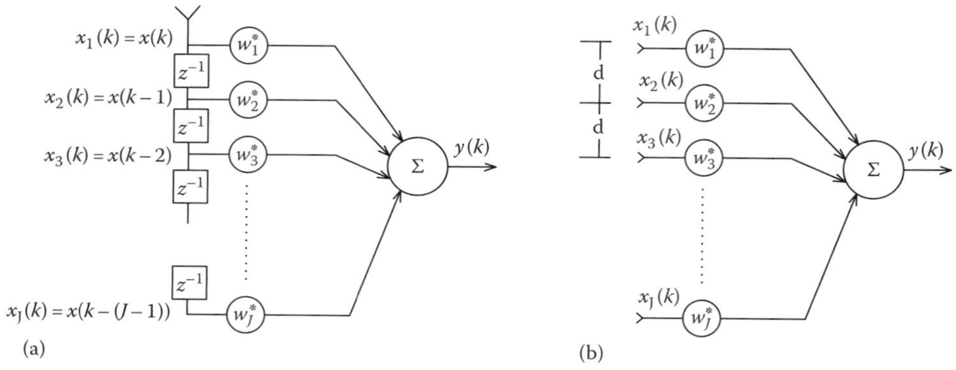

FIGURE 2.3 The analogy between (a) an equispaced omnidirectional narrowband line array and (b) a single-channel FIR filter is illustrated in this figure.

N-dimensional vector space. The angles between $\mathbf{w}$ and $\mathbf{d}(\theta, \omega)$ determine the response $r(\theta, \omega)$. For example, if for some (θ, ω) the angle between $\mathbf{w}$ and $\mathbf{d}(\theta, \omega)$ 90° (i.e., if $\mathbf{w}$ is orthogonal to $\mathbf{d}(\theta, \omega)$), then the response is zero. If the angle is close to 0°, then the response magnitude will be relatively large. The ability to discriminate between sources at different locations and/or frequencies, say (θ_1, ω_1) and (θ_2, ω_2), is determined by the angle between their array response vectors, $\mathbf{d}(\theta_1, \omega_1)$ and $\mathbf{d}(\theta_2, \omega_2)$.

The general effects of spatial sampling are similar to temporal sampling. Spatial aliasing corresponds to an ambiguity in source locations. The implication is that sources at different locations have the same array response vector, e.g., for narrowband sources $\mathbf{d}(\theta_1, \omega_o)$ and $\mathbf{d}(\theta_2, \omega_o)$. This can occur if the sensors are spaced too far apart. If the sensors are too close together, spatial discrimination suffers as a result of the smaller than necessary aperture; array response vectors are not well dispersed in the N-dimensional vector space. Another type of ambiguity occurs with broadband signals when a source at one location and frequency cannot be distinguished from a source at a different location and frequency, i.e., $\mathbf{d}(\theta_1, \omega_1) = \mathbf{d}(\theta_2, \omega_2)$. For example, this occurs in a linear equispaced array whenever $\omega_1 \sin\, \theta_1 = \omega_2 \sin\theta_2$. (The addition of temporal samples at one sensor prevents this particular ambiguity.)

A primary focus of this section is on designing response via weight selection; however, Equation 2.7 indicates that response is also a function of array geometry (and sensor characteristics if the ideal omnidirectional sensor model is invalid). In contrast with single channel filtering where A/D converters provide a uniform sampling in time, there is no compelling reason to space sensors regularly. Sensor locations provide additional degrees of freedom in designing a desired response and can be selected so that over the range of (θ, ω) of interest the array response vectors are unambiguous and well dispersed in the N-dimensional vector space. Utilization of these degrees of freedom can become very complicated due to the multidimensional nature of spatial sampling and the nonlinear relationship between $r(\theta, \omega)$ and sensor locations.

2.2.2 Second-Order Statistics

Evaluation of beamformer performance usually involves power or variance, so the second-order statistics of the data play an important role. We assume the data received at the sensors are zero mean throughout this section. The variance or expected power of the beamformer output is given by $E\{|y|^2\} = \mathbf{w}^H E\{\mathbf{x}\,\mathbf{x}^H\}\mathbf{w}$. If the data are wide sense stationary, then $\mathbf{R}_x = E\{\mathbf{x}\,\mathbf{x}^H\}$, the data covariance matrix, is independent of time. Although we often encounter nonstationary data, the wide sense stationary assumption is used in developing statistically optimal beamformers and in evaluating steady state performance.

Suppose **x** represents samples from a uniformly sampled time series having a power spectral density $S(\omega)$ and no energy outside of the spectral band $[\omega_a, \omega_b]$. $\mathbf{R}_x$ can be expressed in terms of the power spectral density of the data using the Fourier transform relationship as

$$\mathbf{R}_x = \frac{1}{2\pi} \int_{\omega_a}^{\omega_b} S(\omega)\ \mathbf{d}(\omega)\ \mathbf{d}^H(\omega) d\omega, \tag{2.9}$$

with $\mathbf{d}(\omega)$ as defined for Equation 2.5. Now assume the array data **x** is due to a source located at direction θ. In like manner to the time series case we can obtain the covariance matrix of the array data as

$$\mathbf{R}_x = \frac{1}{2\pi} \int_{\omega_a}^{\omega_b} S(\omega)\ \mathbf{d}(\theta, \omega)\ \mathbf{d}^H(\theta, \omega) d\omega. \tag{2.10}$$

A source is said to be narrowband of frequency ω_o if $\mathbf{R}_x$ can be represented as the rank one outer product

$$\mathbf{R}_x = \sigma_s^2\ \mathbf{d}(\theta, \omega_o)\mathbf{d}^H(\theta, \omega_o), \tag{2.11}$$

where σ_s^2 is the source variance or power.

The conditions under which a source can be considered narrowband depend on both the source bandwidth and the time over which the source is observed. To illustrate this, consider observing an amplitude modulated sinusoid or the output of a narrowband filter driven by white noise on an oscilloscope. If the signal bandwidth is small relative to the center frequency (i.e., if it has small fractional bandwidth), and the time intervals over which the signal is observed are short relative to the inverse of the signal bandwidth, then each observed waveform has the shape of a sinusoid. Note that as the observation time interval is increased, the bandwidth must decrease for the signal to remain sinusoidal in appearance. It turns out, based on statistical arguments, that the observation time bandwidth product (TBWP) is the fundamental parameter that determines whether a source can be viewed as narrowband (see Buckley [2]).

An array provides an effective temporal aperture over which a source is observed. Figure 2.2 illustrates this temporal aperture $T(\theta)$ for a source arriving from direction θ. Clearly the TBWP is dependent on the source DOA. An array is considered narrowband if the observation TBWP is much less than one for all possible source directions.

Narrowband beamforming is conceptually simpler than broadband since one can ignore the temporal frequency variable. This fact, coupled with interest in temporal frequency analysis for some applications, has motivated implementation of broadband beamformers with a narrowband decomposition structure, as illustrated in Figure 2.4. The narrowband decomposition is often performed by taking a discrete Fourier transform (DFT) of the data in each sensor channel using an fast Fourier transform (FFT) algorithm. The data across the array at each frequency of interest are processed by their own beamformer. This is usually termed frequency domain beamforming. The frequency domain beamformer outputs can be made equivalent to the DFT of the broadband beamformer output depicted in Figure 2.1b with proper selection of beamformer weights and careful data partitioning.

2.2.3 Beamformer Classification

Beamformers can be classified as either data independent or statistically optimum, depending on how the weights are chosen. The weights in a data-independent beamformer do not depend on the array data and are chosen to present a specified response for all signal/interference scenarios. The weights in a statistically optimum beamformer are chosen based on the statistics of the array data to "optimize" the array response.

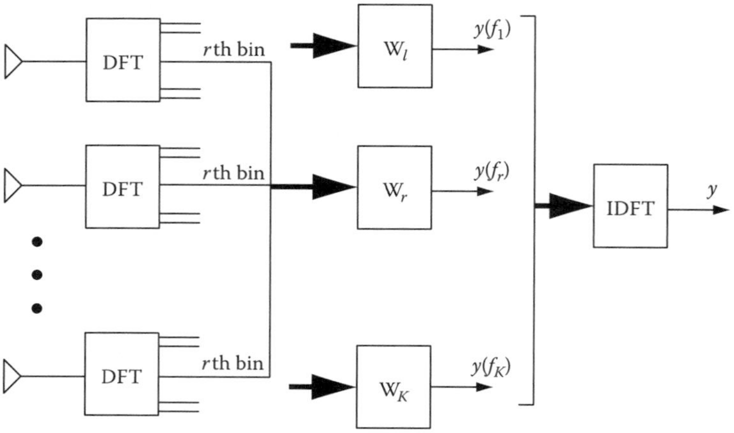

FIGURE 2.4 Beamforming is sometimes performed in the frequency domain when broadband signals are of interest. This figure illustrates transformation of the data at each sensor into the frequency domain. Weighted combinations of data at each frequency (bin) are performed. An inverse discrete Fourier transform produces the output time series.

In general, the statistically optimum beamformer places nulls in the directions of interfering sources in an attempt to maximize the signal-to-noise ratio (SNR) at the beamformer output. A comparison between data-independent and statistically optimum beamformers is illustrated in Figure 2.5.

Sections 2.3 through 2.6 cover data-independent, statistically optimum, adaptive, and partially adaptive beamforming. Data-independent beamformer design techniques are often used in statistically optimum beamforming (e.g., constraint design in linearly constrained minimum variance (LCMV) beamforming). The statistics of the array data are not usually known and may change over time so adaptive algorithms are typically employed to determine the weights. The adaptive algorithm is designed so the beamformer response converges to a statistically optimum solution. Partially adaptive beamformers reduce the adaptive algorithm computational load at the expense of a loss (designed to be small) in statistical optimality.

2.3 Data-Independent Beamforming

The weights in a data-independent beamformer are designed so the beamformer response approximates a desired response independent of the array data or data statistics. This design objective—approximating a desired response—is the same as that for classical finite impulse response (FIR) filter design (see, e.g., Parks and Burrus [8]). We shall exploit the analogies between beamforming and FIR filtering where possible in developing an understanding of the design problem. We also discuss aspects of the design problem specific to beamforming.

The first part of this section discusses forming beams in a classical sense, i.e., approximating a desired response of unity at a point of direction and zero elsewhere. Methods for designing beamformers having more general forms of desired response are presented in the second part.

2.3.1 Classical Beamforming

Consider the problem of separating a single complex frequency component from other frequency components using the J tap FIR filter illustrated in Figure 2.3. If frequency ω_o is of interest, then the desired frequency response is unity at ω_o and zero elsewhere. A common solution to this problem is to choose $\mathbf{w}$ as the vector $\mathbf{d}(\omega_o)$. This choice can be shown to be optimal in terms of minimizing the squared error between the actual response and desired response. The actual response is characterized by a main lobe (or beam) and many sidelobes. Since $\mathbf{w} = \mathbf{d}(\omega_o)$, each element of $\mathbf{w}$ has unit magnitude. Tapering or windowing the amplitudes of the elements of $\mathbf{w}$ permits trading of main lobe or beam width against sidelobe levels to form the response into a desired shape. Let $\mathbf{T}$ be a J by J diagonal matrix

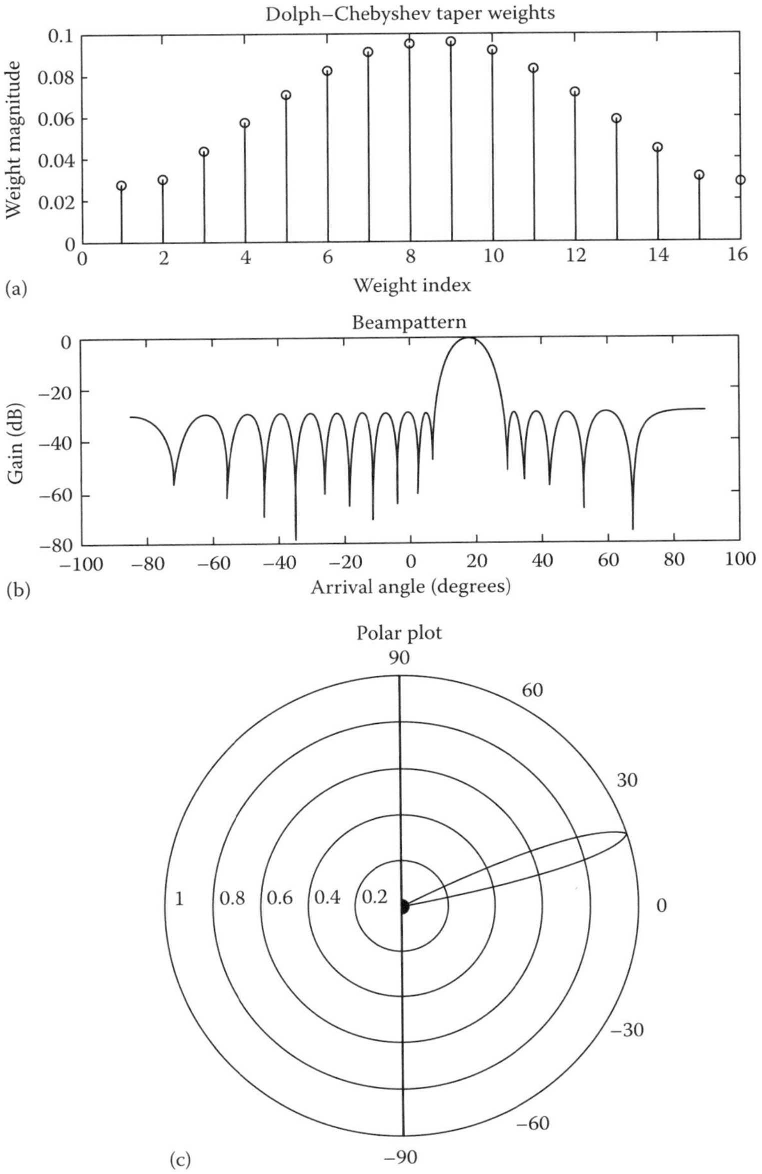

FIGURE 2.5 Beamformers come in both data-independent and statistically optimum varieties. In (a) through (e) we consider an equispaced narrowband array of 16 sensors spaced at one-half wavelength. In (a), (b), and (c) the magnitude of the weights, the beampattern, and the beampattern, in polar coordinates are shown, respectively, for a Dolph–Chebyshev beamformer with −30 dB sidelobes.

(continued)

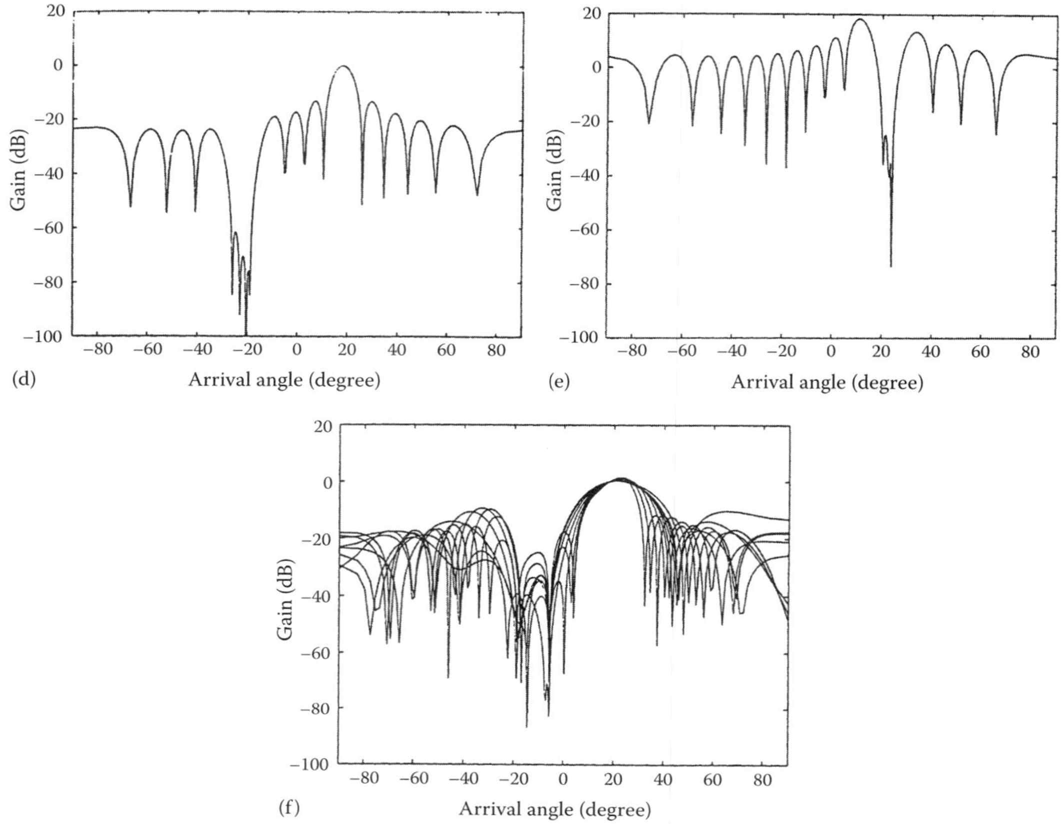

FIGURE 2.5 (continued) In (d) and (e) beampatterns are shown of statistically optimum beamformers which were designed to minimize output power subject to a constraint that the response be unity for an arrival angle of 18°. Energy is assumed to arrive at the array from several interference sources. In (d) several interferers are located between −20° and −23°, each with power of 30 dB relative to the uncorrelated noise power at a single sensor. Deep nulls are formed in the interferer directions. The interferers in (e) are located between 20° and 23°, again with relative power of 30 dB. Again deep nulls are formed at the interferer directions; however, the sidelobe levels are significantly higher at other directions. (f) depicts the broadband LCMV beamformer magnitude response at eight frequencies on the normalized frequency interval [2π/5, 4π/5] when two interferers arrive from directions −5.75° and −17.5° in the presence of white noise. The interferers have a white spectrum on [2π/5, 4π/5] and have powers of 40 and 30 dB relative to the white noise, respectively. The constraints are designed to present a unit gain and linear phase over [2π/5, 4π/5] at a DOA of 18°. The array is linear equispaced with 16 sensors spaced at one-half wavelength for frequency 4π/5 and five tap FIR filters are used in each sensor channel.

with the real-valued taper weights as diagonal elements. The tapered FIR filter weight vector is given by $\mathbf{Td}(\omega_o)$. A detailed comparison of a large number of tapering functions is given in [5].

In spatial filtering one is often interested in receiving a signal arriving from a known location point θ_o. Assuming the signal is narrowband (frequency ω_o), a common choice for the beamformer weight vector is the array response vector $\mathbf{d}(\theta_o, \omega_o)$. The resulting array and beamformer is termed a phased array because the output of each sensor is phase shifted prior to summation. Figure 2.5b depicts the magnitude of the actual response when $\mathbf{w} = \mathbf{Td}(\theta_o, \omega_o)$, where $\mathbf{T}$ implements a common Dolph–Chebyshev tapering function. As in the FIR filter discussed above, beam width and sidelobe levels are the important characteristics of the response. Amplitude tapering can be used to control the shape of the response, i.e., to form the beam. The equivalence of the narrowband linear equispaced array and FIR filter (see Figure 2.3) implies that the same techniques for choosing taper functions are applicable to either problem. Methods for choosing tapering weights also exist for more general array configurations.

2.3.2 General Data-Independent Response Design

The methods discussed in this section apply to design of beamformers that approximate an arbitrary desired response. This is of interest in several different applications. For example, we may wish to receive any signal arriving from a range of directions, in which case the desired response is unity over the entire range. As another example, we may know that there is a strong source of interference arriving from a certain range of directions, in which case the desired response is zero in this range. These two examples are analogous to bandpass and bandstop FIR filtering. Although we are no longer "forming beams," it is conventional to refer to this type of spatial filter as a beamformer.

Consider choosing $\mathbf{w}$ so the actual response $r(\theta,\omega)=\mathbf{w}^H\mathbf{d}(\theta,\omega)$ approximates desired response $r_d(\theta,\omega)$. Ad hoc techniques similar to those employed in FIR filter design can be used for selecting $\mathbf{w}$. Alternatively, formal optimization design methods can be employed (see, e.g., Parks and Burrus [8]). Here, to illustrate the general optimization design approach, we only consider choosing $\mathbf{w}$ to minimize the weighted averaged square of the difference between desired and actual response.

Consider minimizing the squared error between the actual and desired response at P points (θ_i,ω_i), $1<i<P$. If $P>N$, then we obtain the overdetermined least squares problem

$$\min_{\mathbf{w}} |\mathbf{A}^H\mathbf{w} - \mathbf{r}_d|^2, \tag{2.12}$$

where

$$\mathbf{A} = [\mathbf{d}(\theta_1,\omega_1),\ \mathbf{d}(\theta_2,\omega_2)\ldots\mathbf{d}(\theta_P,\omega_P)]; \tag{2.13}$$

$$\mathbf{r}_d = [r_d(\theta_1,\omega_1),\ r_d(\theta_2,\omega_2)\ldots r_d(\theta_P,\omega_P)]^H. \tag{2.14}$$

Provided $\mathbf{AA}^H$ is invertible (i.e., $\mathbf{A}$ is full rank), then the solution to Equation 2.12 is given as

$$\mathbf{w} = \mathbf{A}^+\mathbf{r}_d, \tag{2.15}$$

where $\mathbf{A}^+=(\mathbf{AA}^H)^{-1}\,\mathbf{A}$ is the pseudoinverse of $\mathbf{A}$.

A note of caution is in order at this point. The white noise gain of a beamformer is defined as the output power due to unit variance white noise at the sensors. Thus, the norm squared of the weight vector, $\mathbf{w}^H\mathbf{w}$, represents the white noise gain. If the white noise gain is large, then the accuracy by which $\mathbf{w}$ approximates the desired response is a moot point because the beamformer output will have a poor SNR due to white noise contributions. If $\mathbf{A}$ is ill-conditioned, then $\mathbf{w}$ can have a very large norm and still approximate the desired response. The matrix $\mathbf{A}$ is ill-conditioned when the effective numerical dimension of the space spanned by the $\mathbf{d}(\theta_i,\omega_i)$, $1\le i\le P$, is less than N. For example, if only one source direction is sampled, then the numerical rank of $\mathbf{A}$ is approximately given by the TBWP for that direction. Low rank approximates of $\mathbf{A}$ and $\mathbf{A}^+$ should be used whenever the numerical rank is less than N. This ensures that the norm of $\mathbf{w}$ will not be unnecessarily large.

Specific directions and frequencies can be emphasized in Equation 2.12 by selection of the sample points (θ_i,ω_i) and/or unequally weighting of the error at each (θ_i,ω_i). Parks and Burrus [8] discuss this in the context of FIR filtering.

2.4 Statistically Optimum Beamforming

In statistically optimum beamforming, the weights are chosen based on the statistics of the data received at the array. Loosely speaking, the goal is to "optimize" the beamformer response so the output contains minimal contributions due to noise and interfering signals. We discuss several different criteria for choosing statistically optimum beamformer weights. Table 2.1 summarizes these different approaches. Where possible, equations describing the criteria and weights are confined to Table 2.1. Throughout the section we assume that the data is wide-sense stationary and that its second-order statistics are known. Determination of weights when the data statistics are unknown or time varying is discussed in the following section on adaptive algorithms.

2.4.1 Multiple Sidelobe Canceller

The multiple sidelobe canceller (MSC) is perhaps the earliest statistically optimum beamformer. An MSC consists of a "main channel" and one or more "auxiliary channels" as depicted in Figure 2.6a. The main channel can be either a single high gain antenna or a data-independent beamformer (see Section 2.3). It has a highly directional response, which is pointed in the desired signal direction. Interfering signals are assumed to enter through the main channel sidelobes. The auxiliary channels also receive the interfering signals. The goal is to choose the auxiliary channel weights to cancel the main channel interference component. This implies that the responses to interferers of the main channel and linear combination of auxiliary channels must be identical. The overall system then has a response of zero as illustrated in Figure 2.6b. In general, requiring zero response to all interfering signals is either not possible or can result in significant white noise gain. Thus, the weights are usually chosen to trade off interference suppression for white noise gain by minimizing the expected value of the total output power as indicated in Table 2.1.

Choosing the weights to minimize output power can cause cancellation of the desired signal because it also contributes to total output power. In fact, as the desired signal gets stronger it contributes to a larger fraction of the total output power and the percentage cancellation increases. Clearly this is an undesirable effect. The MSC is very effective in applications where the desired signal is very weak (relative to the interference), since the optimum weights will not pay any attention to it, or when the desired signal is known to be absent during certain time periods. The weights can then be adapted in the absence of the desired signal and frozen when it is present.

2.4.2 Use of a Reference Signal

If the desired signal were known, then the weights could be chosen to minimize the error between the beamformer output and the desired signal. Of course, knowledge of the desired signal eliminates the need for beamforming. However, for some applications, enough may be known about the desired signal to generate a signal that closely represents it. This signal is called a reference signal. As indicated in Table 2.1, the weights are chosen to minimize the mean square error between the beamformer output and the reference signal.

The weight vector depends on the cross covariance between the unknown desired signal present in $\mathbf{x}$ and the reference signal. Acceptable performance is obtained provided this approximates the covariance of the unknown desired signal with itself. For example, if the desired signal is amplitude modulated, then acceptable performance is often obtained by setting the reference signal equal to the carrier. It is also assumed that the reference signal is uncorrelated with interfering signals in $\mathbf{x}$. The fact that the direction of the desired signal does not need to be known is a distinguishing feature of the reference signal approach. For this reason it is sometimes termed "blind" beamforming. Other closely related blind beamforming techniques choose weights by exploiting properties of the desired signal such as constant modulus, cyclostationarity, or third and higher order statistics.

TABLE 2.1 Summary of Optimum Beamformers

Type	MSC	Reference Signal	Max SNR	LCMV
Definitions	$\mathbf{x}_a$—auxiliary data y_m—primary data $\mathbf{r}_{ma} = E\{\mathbf{x}_a \mathbf{y}_m^*\}$ $\mathbf{R}_a = E\{\mathbf{x}_a \mathbf{x}_a^H\}$ Output: $\mathbf{y} = \mathbf{y}_m - \mathbf{w}_a^H \mathbf{x}_a$	$\mathbf{x}$—array data y_d—desired signal $\mathbf{r}_{xd} = E\{\mathbf{x}\mathbf{y}_d^*\}$ $\mathbf{R}_x = E\{\mathbf{x}\mathbf{x}^H\}$ Output: $\mathbf{y} = \mathbf{w}^H\mathbf{x}$	$\mathbf{x} = \mathbf{s} + \mathbf{x}$—array data $\mathbf{s}$—signal component $\mathbf{n}$—noise component $\mathbf{R}_s = E\{\mathbf{s}\mathbf{s}^H\}$ $\mathbf{R}_n = E\,\{\mathbf{n}\mathbf{n}^H\}$ Output: $\mathbf{y} = \mathbf{w}^H\mathbf{x}$	$\mathbf{x}$—array data $\mathbf{C}$—constraint matrix $\mathbf{f}$—response vector $\mathbf{R}_x = E\{\mathbf{x}\mathbf{x}^H\}$ Output: $\mathbf{y} = \mathbf{w}^H\mathbf{x}$
Criterion	$\min_{\mathbf{w}_a} E\left\{\left\|\mathbf{y}_m - \mathbf{w}_a^H \mathbf{x}_a\right\|^2\right\}$	$\min_{\mathbf{w}} E\left\{\left\|\mathbf{y} - \mathbf{y}_d\right\|^2\right\}$	$\max_{\mathbf{w}} \frac{\mathbf{w}^H \mathbf{R}_s \mathbf{w}}{\mathbf{w}^H \mathbf{R}_n \mathbf{w}}$	$\min_{\mathbf{w}} \{\mathbf{w}^H \mathbf{R}_x \mathbf{w}\}$ s.t. $\mathbf{C}^H \mathbf{w} = \mathbf{f}$
Optimum weights	$\mathbf{w}_a = \mathbf{R}_a^{-1} \mathbf{r}_{ma}$	$\mathbf{w}_a = \mathbf{R}_x^{-1} \mathbf{r}_{rd}$	$\mathbf{R}_n^{-1} \mathbf{R}_s \mathbf{w} = \lambda_{max} \mathbf{w}$	$\mathbf{w} = \mathbf{R}_x^{-1} \mathbf{C}\left[\mathbf{C}^H \mathbf{R}_x^{-1} \mathbf{C}\right]^{-1} f$
Advantages	Simple	Direction of desired signal can be unknown	True maximization of SNR	Flexible and general constraints
Disadvantages	Requires absence of desired signal from auxiliary channels for weight determination	Must generate reference signal	Must know $\mathbf{R}_s$ and $\mathbf{R}_n$ Solve generalized eigenproblem for weights	Computation of constrained weight vector
References	Applebaum (1976)	Widrow (1967)	Monzingo and Miller (1980)	Frost (1972)

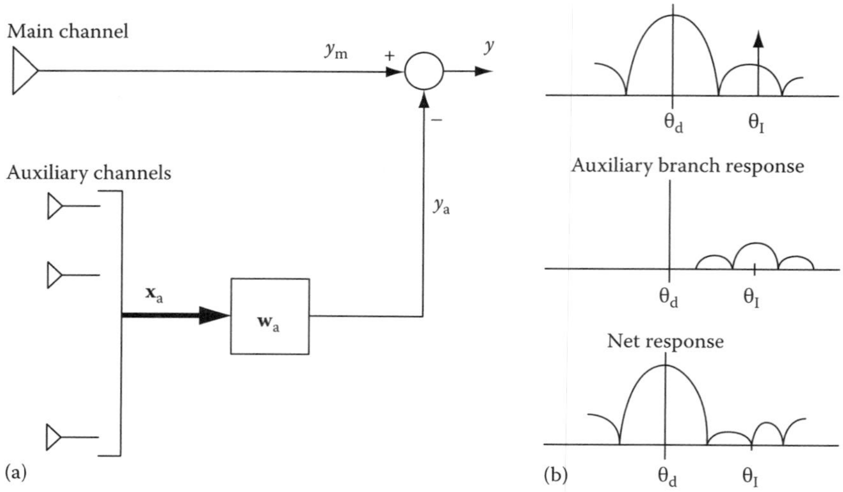

FIGURE 2.6 The multiple sidelobe canceller consists of a main channel and several auxiliary channels as illustrated in (a). The auxiliary channel weights are chosen to "cancel" interference entering through sidelobes of the main channel. (b) Depicts the main channel, auxiliary branch, and overall system response when an interferer arrives from direction θ_I.

2.4.3 Maximization of Signal-to-Noise Ratio

Here the weights are chosen to directly maximize the SNR as indicated in Table 2.1. A general solution for the weights requires knowledge of both the desired signal, $\mathbf{R}_s$, and noise, $\mathbf{R}_n$, covariance matrices. The attainability of this knowledge depends on the application. For example, in an active radar system $\mathbf{R}_n$ can be estimated during the time that no signal is being transmitted and $\mathbf{R}_s$ can be obtained from knowledge of the transmitted pulse and direction of interest. If the signal component is narrowband, of frequency ω, and direction θ, then $\mathbf{R}_s = \sigma^2\mathbf{d}(\theta,\omega)\mathbf{d}^H(\theta,\omega)$ from the results in Section 2.2. In this case, the weights are obtained as

$$\mathbf{w} = \alpha\mathbf{R}_n^{-1}\mathbf{d}(\theta,\omega), \tag{2.16}$$

where the α is some nonzero complex constant. Substitution of Equation 2.16 into the SNR expression shows that the SNR is independent of the value chosen for α.

2.4.4 Linearly Constrained Minimum Variance Beamforming

In many applications none of the above approaches are satisfactory. The desired signal may be of unknown strength and may always be present, resulting in signal cancellation with the MSC and preventing estimation of signal and noise covariance matrices in the maximum SNR processor. Lack of knowledge about the desired signal may prevent utilization of the reference signal approach. These limitations can be overcome through the application of linear constraints to the weight vector. Use of linear constraints is a very general approach that permits extensive control over the adapted response of the beamformer. In this section we illustrate how linear constraints can be employed to control beamformer response, discuss the optimum linearly constrained beamforming problem, and present the generalized sidelobe canceller (GSC) structure.

The basic idea behind LCMV beamforming is to constrain the response of the beamformer so signals from the direction of interest are passed with specified gain and phase. The weights are chosen to minimize output variance or power subject to the response constraint. This has the effect of preserving the desired signal while minimizing contributions to the output due to interfering signals and noise arriving from directions other than the direction of interest. The analogous FIR filter has the weights chosen to minimize the filter output power subject to the constraint that the filter response to signals of frequency ω_o be unity.

In Section 2.2, we saw that the beamformer response to a source at angle θ and temporal frequency ω is given by $\mathbf{w}^H\mathbf{d}(\theta,\omega)$. Thus, by linearly constraining the weights to satisfy $\mathbf{w}^H\mathbf{d}(\theta,\omega)=g$ where g is a complex constant, we ensure that any signal from angle θ and frequency ω is passed to the output with response g. Minimization of contributions to the output from interference (signals not arriving from θ with frequency ω) is accomplished by choosing the weights to minimize the output power or variance $E\{|y|^2\}=\mathbf{w}^H\mathbf{R}_x\mathbf{w}$. The LCMV problem for choosing the weights is thus written

$$\min_{\mathbf{w}} \mathbf{w}^H\mathbf{R}_x\mathbf{w} \quad \text{subject to} \quad \mathbf{d}^H(\theta,\omega)\mathbf{w}=g^{\star}. \tag{2.17}$$

The method of Lagrange multipliers can be used to solve Equation 2.17 resulting in

$$\mathbf{w}=g^{\star}\frac{\mathbf{R}_x^{-1}\mathbf{d}(\theta,\omega)}{\mathbf{d}^H(\theta,\omega)\mathbf{R}_x^{-1}\mathbf{d}(\theta,\omega)}. \tag{2.18}$$

Note that, in practice, the presence of uncorrelated noise will ensure that $\mathbf{R}_x$ is invertible. If $g=1$, then Equation 2.18 is often termed the minimum variance distortionless response (MVDR) beamformer. It can be shown that Equation 2.18 is equivalent to the maximum SNR solution given in Equation 2.16 by substituting $\sigma^2\mathbf{d}(\theta,\omega)\mathbf{d}^H(\theta,\omega)+\mathbf{R}_n$ for $\mathbf{R}_x$ in Equation 2.18 and applying the matrix inversion lemma.

The single linear constraint in Equation 2.17 is easily generalized to multiple linear constraints for added control over the beampattern. For example, if there is fixed interference source at a known direction ϕ, then it may be desirable to force zero gain in that direction in addition to maintaining the response g to the desired signal. This is expressed as

$$\begin{bmatrix}\mathbf{d}^H(\theta,\omega)\\ \mathbf{d}^H(\phi,\omega)\end{bmatrix}\mathbf{w}=\begin{bmatrix}g^{\star}\\ 0\end{bmatrix}. \tag{2.19}$$

If there are $L<N$ linear constraints on $\mathbf{w}$, we write them in the form $\mathbf{C}^H\mathbf{w}=\mathbf{f}$ where the N by L matrix $\mathbf{C}$ and L-dimensional vector $\mathbf{f}$ are termed the constraint matrix and response vector. The constraints are assumed to be linearly independent so $\mathbf{C}$ has rank L. The LCMV problem and solution with this more general constraint equation are given in Table 2.1.

Several different philosophies can be employed for choosing the constraint matrix and response vector. Specifically point, derivative, and eigenvector constraint approaches are popular. Each linear constraint uses one degree of freedom in the weight vector so with L constraints there are only $N-L$ degrees of freedom available for minimizing variance. See Van Veen and Buckley [11] or Van Veen [12] for a more in-depth discussion on this topic.

Generalized sidelobe canceller. The GSC represents an alternative formulation of the LCMV problem, which provides insight, is useful for analysis, and can simplify LCMV beamformer implementation. It also illustrates the relationship between MSC and LCMV beamforming. Essentially, the GSC is a mechanism for changing a constrained minimization problem into unconstrained form.

Suppose we decompose the weight vector $\mathbf{w}$ into two orthogonal components $\mathbf{w}_o$ and $-\mathbf{v}$ (i.e., $\mathbf{w}=\mathbf{w}_o-\mathbf{v}$) that lie in the range and null spaces of $\mathbf{C}$, respectively. The range and null spaces of a matrix span the entire space so this decomposition can be used to represent any $\mathbf{w}$. Since $\mathbf{C}^H\mathbf{v}=0$, we must have

$$\mathbf{w}_o=\mathbf{C}(\mathbf{C}^H\mathbf{C})^{-1}\mathbf{f}, \tag{2.20}$$

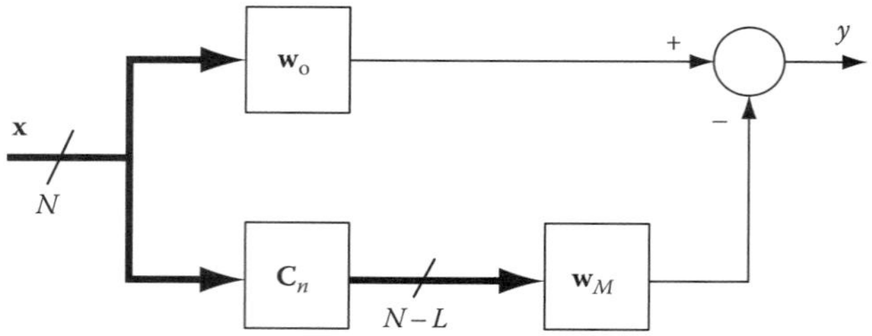

FIGURE 2.7 The generalized sidelobe canceller represents an implementation of the LCMV beamformer in which the adaptive weights are unconstrained. It consists of a preprocessor composed of a fixed beamformer $\mathbf{w}_o$ and a blocking matrix $\mathbf{C}_n$, and a standard adaptive filter with unconstrained weight vector $\mathbf{w}_M$.

if $\mathbf{w}$ is to satisfy the constraints. Equation 2.20 is the minimum L_2 norm solution to the underdetermined equivalent of Equation 2.12. The vector $\mathbf{v}$ is a linear combination of the columns of an N by M ($M = N - L$) matrix $\mathbf{C}_n$ (i.e., $\mathbf{v} = \mathbf{C}_n \mathbf{w}_M$) provided the columns of $\mathbf{C}_n$ form a basis for the null space of $\mathbf{C}$. $\mathbf{C}_n$ can be obtained from $\mathbf{C}$ using any of several orthogonalization procedures such as Gram–Schmidt, QR decomposition, or singular value decomposition. The weight vector $\mathbf{w} = \mathbf{w}_o - \mathbf{C}_n \mathbf{w}_M$ is depicted in block diagram form in Figure 2.7. The choice for $\mathbf{w}_o$ and $\mathbf{C}_n$ implies that $\mathbf{w}$ satisfies the constraints independent of $\mathbf{w}_M$ and reduces the LCMV problem to the unconstrained problem

$$\min_{\mathbf{w}_M} [\mathbf{w}_o - \mathbf{C}_n \mathbf{w}_M]^{\mathrm{H}} \mathbf{R}_x [\mathbf{w}_o - \mathbf{C}_n \mathbf{w}_M]. \tag{2.21}$$

The solution is

$$\mathbf{w}_M = \left(\mathbf{C}_n^{\mathrm{H}} \mathbf{R}_x \mathbf{C}_n\right)^{-1} \mathbf{C}_n^{\mathrm{H}} \mathbf{R}_x \mathbf{w}_o. \tag{2.22}$$

The primary implementation advantages of this alternate but equivalent formulation stem from the facts that the weights $\mathbf{w}_M$ are unconstrained and a data-independent beamformer $\mathbf{w}_o$ is implemented as an integral part of the optimum beamformer. The unconstrained nature of the adaptive weights permits much simpler adaptive algorithms to be employed and the data-independent beamformer is useful in situations where adaptive signal cancellation occurs (see Section 2.4.5).

As an example, assume the constraints are as given in Equation 2.17. Equation 2.20 implies $\mathbf{w}_o = g^* \mathbf{d}(\theta, \omega) / \left[\mathbf{d}^{\mathrm{H}}(\theta, \omega) \mathbf{d}(\theta, \omega)\right]$. $\mathbf{C}_n$ satisfies $\mathbf{d}^{\mathrm{H}}(\theta, \omega)\mathbf{C}_n = 0$ so each column $[\mathbf{C}_n]_i$; $1 < i < N - L$, can be viewed as a data-independent beamformer with a null in direction θ at frequency ω: $\mathbf{d}^{\mathrm{H}}(\theta, \omega)[\mathbf{C}_n]_j = 0$. Thus, a signal of frequency ω and direction θ arriving at the array will be blocked or nulled by the matrix $\mathbf{C}_n$. In general, if the constraints are designed to present a specified response to signals from a set of directions and frequencies, then the columns of $\mathbf{C}_n$ will block those directions and frequencies. This characteristic has led to the term "blocking matrix" for describing $\mathbf{C}_n$. These signals are only processed by $\mathbf{w}_o$ and since $\mathbf{w}_o$ satisfies the constraints, they are presented with the desired response independent of $\mathbf{w}_M$. Signals from directions and frequencies over which the response is not constrained will pass through the upper branch in Figure 2.7 with some response determined by $\mathbf{w}_o$. The lower branch chooses $\mathbf{w}_M$ to estimate the signals at the output of $\mathbf{w}_o$ as a linear combination of the data at the output of the blocking matrix. This is similar to the operation of the MSC, in which weights are applied to the output of auxiliary sensors in order to estimate the primary channel output (see Figure 2.6).

2.4.5 Signal Cancellation in Statistically Optimum Beamforming

Optimum beamforming requires some knowledge of the desired signal characteristics, either its statistics (for maximum SNR or reference signal methods), its direction (for the MSC), or its response vector $\mathbf{d}(\theta, \omega)$ (for the LCMV beamformer). If the required knowledge is inaccurate, the optimum beamformer will attenuate the desired signal as if it were interference. Cancellation of the desired signal is often significant, especially if the SNR of the desired signal is large. Several approaches have been suggested to reduce this degradation (e.g., Cox et al. [3]).

A second cause of signal cancellation is correlation between the desired signal and one or more interference signals. This can result either from multipath propagation of a desired signal or from smart (correlated) jamming. When interference and desired signals are uncorrelated, the beamformer attenuates interferers to minimize output power. However, with a correlated interferer the beamformer minimizes output power by processing the interfering signal in such a way as to cancel the desired signal. If the interferer is partially correlated with the desired signal, then the beamformer will cancel the portion of the desired signal that is correlated with the interferer. Methods for reducing signal cancellation due to correlated interference have been suggested (e.g., Widrow et al. [13], Shan and Kailath [10]).

2.5 Adaptive Algorithms for Beamforming

The optimum beamformer weight vector equations listed in Table 2.1 require knowledge of second-order statistics. These statistics are usually not known, but with the assumption of ergodicity, they (and therefore the optimum weights) can be estimated from available data. Statistics may also change over time, e.g., due to moving interferers. To solve these problems, weights are typically determined by adaptive algorithms.

There are two basic adaptive approaches: (1) block adaptation, where statistics are estimated from a temporal block of array data and used in an optimum weight equation; and (2) continuous adaptation, where the weights are adjusted as the data is sampled such that the resulting weight vector sequence converges to the optimum solution. If a nonstationary environment is anticipated, block adaptation can be used, provided that the weights are recomputed periodically. Continuous adaptation is usually preferred when statistics are time-varying or, for computational reasons, when the number of adaptive weights M is moderate to large; values of $M > 50$ are common.

Among notable adaptive algorithms proposed for beamforming are the Howells–Applebaum adaptive loop developed in the late 1950s and reported by Howells [7] and Applebaum [1], and the Frost LCMV algorithm [4]. Rather than recapitulating adaptive algorithms for each optimum beamformer listed in Table 2.1, we take a unifying approach using the standard adaptive filter configuration illustrated on the right side of Figure 2.7.

In Figure 2.7 the weight vector $\mathbf{w}_M$ is chosen to estimate the desired signal y_d as linear combination of the elements of the data vector $\mathbf{u}$. We select $\mathbf{w}_M$ to minimize the mean squared error (MSE)

$$J(\mathbf{w}_M) = E\left\{\left|y_d - \mathbf{w}_M^H \mathbf{u}\right|^2\right\} = \sigma_d^2 - \mathbf{w}_M^H \mathbf{r}_{ud} - \mathbf{r}_{ud}^H \mathbf{w}_M + \mathbf{w}_M^H \mathbf{R}_u \mathbf{w}_M, \tag{2.23}$$

where

$$\sigma_d^2 = E\left\{|y_d|^2\right\}$$
$$\mathbf{r}_{ud} = E\{\mathbf{u}\, y_d^*\}$$
$$\mathbf{R}_u = E\{\mathbf{u}\, \mathbf{u}^H\}$$

$J(\mathbf{w}_M)$ is minimized by

$$\mathbf{w}_{opt} = \mathbf{R}_u^{-1} \mathbf{r}_{ud}. \tag{2.24}$$

Comparison of Equation 2.23 and the criteria listed in Table 2.1 indicates that this standard adaptive filter problem is equivalent to both the MSC beamformer problem (with $y_d = y_m$ and $\mathbf{u} = \mathbf{x}_a$) and the reference signal beamformer problem (with $\mathbf{u} = \mathbf{x}$). The LCMV problem is apparently different. However closer examination of Figure 2.7 and Equations 2.22 and 2.24 reveals that the standard adaptive filter problem is equivalent to the LCMV problem implemented with the GSC structure. Setting $\mathbf{u} = \mathbf{C}_n^{\mathrm{H}}\mathbf{x}$ and $y_{\mathrm{d}} = \mathbf{w}_{\mathrm{o}}^{\mathrm{H}}\mathbf{x}$ implies $\mathbf{R}_{\mathrm{u}} = \mathbf{C}_n^{\mathrm{H}}\mathbf{R}_x\mathbf{C}_n$ and $\mathbf{r}_{\mathrm{ud}} = \mathbf{C}_n^{\mathrm{H}}\mathbf{R}_x\mathbf{w}_{\mathrm{o}}$. The maximum SNR beamformer cannot in general be represented by Figure 2.7 and Equation 2.24. However, it was noted after Equation 2.18 that if the desired signal is narrowband, then the maximum SNR and the LCMV beamformers are equivalent.

The block adaptation approach solves Equation 2.24 using estimates of $\mathbf{R}_{\mathrm{u}}$ and $\mathbf{r}_{\mathrm{ud}}$ formed from K samples of $\mathbf{u}$ and y_{d}: $\mathbf{u}(k)$, $y_{\mathrm{d}}(k)$; $0 < k < K - 1$. The most common are the sample covariance matrix

$$\hat{\mathbf{R}}_{\mathrm{u}} = \frac{1}{K}\sum_{k=0}^{K-1}\mathbf{u}(k)\mathbf{u}^{\mathrm{H}}(k), \tag{2.25}$$

and sample cross-covariance vector

$$\hat{\mathbf{r}}_{\mathrm{ud}} = \frac{1}{K}\sum_{k=0}^{K-1}\mathbf{u}(k)y_{\mathrm{d}}^{\star}(k). \tag{2.26}$$

Performance analysis and guidelines for selecting the block size K are provided in Reed et al. [9].

Continuous adaptation algorithms are easily developed in terms of Figure 2.7 and Equation 2.23. Note that $J(\mathbf{w}_M)$ is a quadratic error surface. Since the quadratic surface's "Hessian" $\mathbf{R}_{\mathrm{u}}$ is the covariance matrix of noisy data, it is positive definite. This implies that the error surface is a "bowl." The shape of the bowl is determined by the eigenstructure of $\mathbf{R}_{\mathrm{u}}$. The optimum weight vector $\mathbf{w}_{\mathrm{opt}}$ corresponds to the bottom of the bowl.

One approach to adaptive filtering is to envision a point on the error surface that corresponds to the present weight vector $\mathbf{w}_M(k)$. We select a new weight vector $\mathbf{w}_M(k+1)$ so as to descend on the error surface. The gradient vector

$$\nabla_{\mathbf{w}_M(k)} = \frac{\partial}{\partial \mathbf{w}_M} J(\mathbf{w}_M)\bigg|_{\mathbf{w}_M = \mathbf{w}_m(k)} = -2\mathbf{r}_{\mathrm{ud}} + 2\mathbf{R}_{\mathrm{u}}\mathbf{w}_M(k), \tag{2.27}$$

tells us the direction in which to adjust the weight vector. Steepest descent, i.e., adjustment in the negative gradient direction, leads to the popular least mean-square (LMS) adaptive algorithm. The LMS algorithm replaces $\nabla_{\mathbf{w}_M(k)}$ with the instantaneous gradient estimate $\hat{\nabla}_{\mathbf{w}_M(k)} = -2[\mathbf{u}(k)y_{\mathrm{d}}^{\star}(k) - \mathbf{u}(k)\mathbf{u}^{\mathrm{H}}(k)\mathbf{w}_M(k)]$. Denoting $y(k) = y_{\mathrm{d}}(k) - \mathbf{w}_M^{\mathrm{H}}\mathbf{u}(k)$, we have

$$\mathbf{w}_M(k+1) = \mathbf{w}_M(k) + \mu\mathbf{u}(k)y^{\star}(k). \tag{2.28}$$

The gain constant μ controls convergence characteristics of the random vector sequence $\mathbf{w}_M(k)$. Table 2.2 provides guidelines for its selection.

The primary virtue of the LMS algorithm is its simplicity. Its performance is acceptable in many applications; however, its convergence characteristics depend on the shape of the error surface and therefore the eigenstructure of $\mathbf{R}_{\mathrm{u}}$. When the eigenvalues are widely spread, convergence can be slow and other adaptive algorithms with better convergence characteristics should be considered. Alternative procedures for searching the error surface have been proposed in addition to algorithms based on least squares and Kalman filtering. Roughly speaking, these algorithms trade off computational requirements with speed of convergence to $\mathbf{w}_{\mathrm{opt}}$. We refer you to texts on adaptive filtering for detailed descriptions and analysis (Widrow and Stearns [14], Haykin [6], and others).

One alternative to LMS is the exponentially weighted recursive least squares (RLS) algorithm. At the Kth time step, $\mathbf{w}_M(K)$ is chosen to minimize a weighted sum of past squared errors

$$\min_{\mathbf{w}_M(K)} \sum_{k=0}^{K} \lambda^{K-k}\left|y_{\mathrm{d}}(k) - \mathbf{w}_M^{\mathrm{H}}(K)\mathbf{u}(k)\right|^2. \tag{2.29}$$

TABLE 2.2 Comparison of the LMS and RLS Weight Adaptation Algorithms

Algorithm	LMS	RLS
Initialization	$\mathbf{w}_M(0) = 0$ $\mathbf{y}(0) = \mathbf{y}_\mathrm{d}(0)$ $0 < \mu < \frac{1}{\mathrm{Trace}[\mathbf{R}_\mathrm{u}]}$	$\mathbf{w}_M(0) = 0$ $\mathbf{P}(0) = \delta^{-1}\mathbf{I}$ δ small, $\mathbf{I}$ identity matrix
Update Equations	$\mathbf{w}_M(k) = \mathbf{w}_M(k-1) + \mu\mathbf{u}(k-1)y^*(k-1)$ $y(k) = y_\mathrm{d}(k) - \mathbf{w}_M^\mathrm{H}(k)\mathbf{u}(k)$	$\mathbf{v}(k) = \mathbf{P}(k-1)\mathbf{u}(k)$ $\mathbf{k}(k) = \frac{\lambda^{-1}\mathbf{v}(k)}{1+\lambda^{-1}\mathbf{u}^\mathrm{H}(k)\mathbf{v}(k)}$ $\alpha(k) = y_\mathrm{d}(k) - \mathbf{w}_M^\mathrm{H}(k-1)\mathbf{u}(k)$ $\mathbf{w}_M(k) = \mathbf{w}_M(k-1) + \mathbf{k}(k)\alpha^*(k)$ $\mathbf{P}(k) = \lambda^{-1}\mathbf{P}(k-1) - \lambda^{-1}\mathbf{k}(k)\mathbf{v}^\mathrm{H}(k)$
Multiplies per update	$2M$	$4M^2 + 4M + 2$
Performance Characteristics	Under certain conditions, convergence of $\mathbf{w}_M(k)$ to the statistically optimum weight vector $\mathbf{w}_\mathrm{opt}$ in the mean-square sense is guaranteed if μ is chosen as indicated above. The convergence rate is governed by the eigenvalue spread of $\mathbf{R}_\mathrm{u}$. For large eigenvalue spread, convergence can be very slow.	The $\mathbf{w}_M(k)$ represents the least squares solution at each instant k and are optimum in a deterministic sense. Convergence to the statistically optimum weight vector $\mathbf{w}_\mathrm{opt}$ is often faster than that obtained using the LMS algorithm because it is independent of the eigenvalue spread of $\mathbf{R}_\mathrm{u}$.

λ is a positive constant less than one which determines how quickly previous data are de-emphasized. The RLS algorithm is obtained from Equation 2.29 by expanding the magnitude squared and applying the matrix inversion lemma. Table 2.2 summarizes both the LMS and RLS algorithms.

2.6 Interference Cancellation and Partially Adaptive Beamforming

The computational requirements of each update in adaptive algorithms are proportional to either the weight vector dimension M (e.g., LMS) or dimension squared M^2 (e.g., RLS). If M is large, this requirement is quite severe and for practical real time implementation it is often necessary to reduce M. Furthermore, the rate at which an adaptive algorithm converges to the optimum solution may be very slow for large M. Adaptive algorithm convergence properties can be improved by reducing M.

The concept of "degrees of freedom" is much more relevant to this discussion than the number of weights. The expression degrees of freedom refers to the number of unconstrained or "free" weights in an implementation. For example, an LCMV beamformer with L constraints on N weights has $N - L$ degrees of freedom; the GSC implementation separates these as the unconstrained weight vector $\mathbf{w}_M$. There are M degrees of freedom in the structure of Figure 2.7. A fully adaptive beamformer uses all available degrees of freedom and a partially adaptive beamformer uses a reduced set of degrees of freedom. Reducing degrees of freedom lowers computational requirements and often improves adaptive response time. However, there is a performance penalty associated with reducing degrees of freedom. A partially adaptive beamformer cannot generally converge to the same optimum solution as the fully adaptive beamformer. The goal of partially adaptive beamformer design is to reduce degrees of freedom without significant degradation in performance.

The discussion in this section is general, applying to different types of beamformers although we borrow much of the notation from the GSC. We assume the beamformer is described by the adaptive structure of Figure 2.7 where the desired signal y_d is obtained as $y_\mathrm{d} = \mathbf{w}_\mathrm{o}^\mathrm{H}\mathbf{x}$ and the data vector $\mathbf{u}$ as $\mathbf{u} = \mathbf{T}^\mathrm{H}\mathbf{x}$. Thus, the beamformer output is $y = \mathbf{w}^\mathrm{H}\mathbf{x}$ where $\mathbf{w} = \mathbf{w}_\mathrm{o} - \mathbf{T}\mathbf{w}_M$. In order to distinguish between fully and partially adaptive implementations, we decompose $\mathbf{T}$ into a product of two matrices $\mathbf{C}_n\mathbf{T}_M$.

The definition of $\mathbf{C}_n$ depends on the particular beamformer and $\mathbf{T}_M$ represents the mapping which reduces degrees of freedom. The MSC and GSC are obtained as special cases of this representation. In the MSC $\mathbf{w}_o$ is an N vector that selects the primary sensor, $\mathbf{C}_n$ is an N by $N-1$ matrix that selects the $N-1$ possible auxiliary sensors from the complete set of N sensors, and $\mathbf{T}_M$ is an $N-1$ by M matrix that selects the M auxiliary sensors actually utilized. In terms of the GSC, $\mathbf{w}_o$ and $\mathbf{C}_n$ are defined as in Section 2.4.4 and $\mathbf{T}_M$ is an $N-L$ by M matrix that reduces degrees of freedom ($M < N - L$).

The goal of partially adaptive beamformer design is to choose $\mathbf{T}_M$ (or $\mathbf{T}$) such that good interference cancellation properties are retained even though M is small. To see that this is possible in principle, consider the problem of simultaneously canceling two narrowband sources from direction θ_1 and θ_2 at frequency ω_o. Perfect cancellation of these sources requires $\mathbf{w}^H\mathbf{d}(\theta_1, \omega_o) = 0$ and $\mathbf{w}^H\mathbf{d}(\theta_2, \omega_o) = 0$ so we must choose $\mathbf{w}_M$ to satisfy

$$\mathbf{w}_M^H\left[\mathbf{T}^H\mathbf{d}(\theta_1, \omega_o)\mathbf{T}^H\mathbf{d}(\theta_2, \omega_o)\right] = [g_1, g_2], \tag{2.30}$$

where $g_i = \mathbf{w}_o^H\mathbf{d}(\theta_i, \omega_o)$ is the response of the $\mathbf{w}_o$ branch to the ith interferer. Assuming $\mathbf{T}^H\mathbf{d}(\theta_1, \omega_o)$ and $\mathbf{T}^H\mathbf{d}(\theta_2, \omega_o)$ are linearly independent and nonzero, and provided $M \geq 2$, then at least one $\mathbf{w}_M$ exists that satisfies Equation 2.30. Extending this reasoning, we see that $\mathbf{w}_M$ can be chosen to cancel M narrowband interferers (assuming the $\mathbf{T}^H\mathbf{d}(\theta_i, \omega_o)$ are linearly independent and nonzero), independent of $\mathbf{T}$. Total cancellation occurs if $\mathbf{w}_M$ is chosen so the response of $\mathbf{T}\,\mathbf{w}_M$ perfectly matches the $\mathbf{w}_o$ branch response to the interferers. In general, M narrowband interferers can be canceled using M adaptive degrees of freedom with relatively mild restrictions on $\mathbf{T}$.

No such rule exists in the broadband case. Here complete cancellation of a single interferer requires choosing $\mathbf{T}\mathbf{w}_M$ so that the response of the adaptive branch, $\mathbf{w}_M^H\mathbf{T}^H\mathbf{d}(\theta_1, \omega)$, matches the response of the $\mathbf{w}_o$ branch, $\mathbf{w}_o^H\mathbf{d}(\theta_1, \omega)$, over the entire frequency band of the interferer. In this case, the degree of cancellation depends on how well these two responses match and is critically dependent on the interferer direction, frequency content, and $\mathbf{T}$. Good cancellation can be obtained in some situations when $M = 1$, while in others even large values of M result in poor cancellation.

A variety of intuitive and optimization-based techniques have been proposed for designing $\mathbf{T}_M$ that achieve good interference cancellation with relatively small degrees of freedom. See Van Veen and Buckley [11] and Van Veen [12] for further review and discussion.

2.7 Summary

A beamformer forms a scalar output signal as a weighted combination of the data received at an array of sensors. The weights determine the spatial filtering characteristics of the beamformer and enable separation of signals having overlapping frequency content if they originate from different locations. The weights in a data-independent beamformer are chosen to provide a fixed response independent to the received data. Statistically optimum beamformers select the weights to optimize the beamformer response based on the statistics of the data. The data statistics are often unknown and may change with time so adaptive algorithms are used to obtain weights that converge to the statistically optimum solution. Computational and response time considerations dictate the use of partially adaptive beamformers with arrays composed of large numbers of sensors.

Defining Terms

Array response vector: Vector describing the amplitude and phase relationships between propagating wave components at each sensor as a function of spatial direction and temporal frequency. Forms the basis for determining the beamformer response.

Beamformer: A device used in conjunction with an array of sensors to separate signals and interference on the basis of their spatial characteristics. The beamformer output is usually given by a weighted combination of the sensor outputs.

Beampattern: The magnitude squared of the beamformer's spatial filtering response as a function of spatial direction and possibly temporal frequency.

Data-independent, statistically optimum, adaptive, and partially adaptive beamformers: The weights in a data-independent beamformer are chosen independent of the statistics of the data. A statistically optimum beamformer chooses its weights to optimize some statistical function of the beamformer output, such as SNR. An adaptive beamformer adjusts its weights in response to the data to accommodate unknown or time varying statistics. A partially adaptive beamformer uses only a subset of the available adaptive degrees of freedom to reduce the computational burden or improve the adaptive convergence rate.

Generalized sidelobe canceller: Structure for implementing LCMV beamformers that separates the constrained and unconstrained components of the adaptive weight vector. The unconstrained components adaptively cancel interference that leaks through the sidelobes of a data-independent beamformer designed to satisfy the constraints.

LCMV beamformer: Beamformer in which the weights are chosen to minimize the output power subject to a linear response constraint. The constraint preserves the signal of interest while power minimization optimally attenuates noise and interference.

Multiple sidelobe canceller: Adaptive beamformer structure in which the data received at low gain auxiliary sensors is used to adaptively cancel the interference arriving in the mainlobe or sidelobes of a spatially high gain sensor.

MVDR beamformer: A form of LCMV beamformer employing a single constraint designed to pass a signal of given direction and frequency with unit gain.

References

1. Applebaum, S.P., Adaptive arrays, Syracuse University Research Corp., Report SURC SPL TR 66-001, August 1966 (reprinted in *IEEE Trans. AP,* AP-24, 585–598, September 1976).
2. Buckley, K.M., Spatial/spectral filtering with linearly-constrained minimum variance beamformers, *IEEE Trans. ASSP*, ASSP-35, 249–266, March 1987.
3. Cox, H., Zeskind, R.M., and Owen, M.M., Robust adaptive beamforming, *IEEE Trans. ASSP*, ASSP-35, 1365–1375, October 1987.
4. Frost III, O.L., An algorithm for linearly constrained adaptive array processing, *Proc. IEEE*, 60, 926–935, August 1972.
5. Harris, F.J., On the use of windows for harmonic analysis with the discrete Fourier transform, *Proc. IEEE*, 66, 51–83, January 1978.
6. Haykin, S., *Adaptive Filter Theory*, 3rd edn., Prentice-Hall, Englewood Cliffs, NJ, 1996.
7. Howells, P.W., Explorations in fixed and adaptive resolution at GE and SURC, *IEEE Trans. AP*, AP-24, 575–584, September 1976.
8. Parks, T.W. and Burrus, C.S., *Digital Filter Design*, Wiley-Interscience, New York, 1987.
9. Reed, I.S., Mallett, J.D., and Brennen, L.E., Rapid convergence rate in adaptive arrays, *IEEE Trans. AES*, AES-10, 853–863, November 1974.
10. Shan, T. and Kailath, T., Adaptive beamforming for coherent signals and interference, *IEEE Trans. ASSP*, ASSP-33, 527–536, June 1985.
11. Van Veen, B. and Buckley, K., Beamforming: a versatile approach to spatial filtering, *IEEE ASSP Mag.*, 5(2), 4–24, April 1988.
12. Van Veen, B., Minimum variance beamforming, in *Adaptive Radar Detection and Estimation*, Haykin, S. and Steinhardt, A., eds., John Wiley & Sons, New York, Chap. 4, pp. 161–236, 1992.

13. Widrow, B., Duvall, K.M., Gooch, R.P., and Newman, W.C., Signal cancellation phenomena in adaptive arrays: Causes and cures, *IEEE Trans. AP*, AP-30, 469–478, May 1982.
14. Widrow, B. and Stearns, S., *Adaptive Signal Processing*, Prentice-Hall, Englewood Cliffs, NJ, 1985.

Further Readings

For further information, we refer the reader to the following books

Compton, R.T., Jr. *Adaptive Antennas: Concepts and Performance*, Prentice-Hall, Englewood Cliffs, NJ, 1988.

Haykin, S., ed. *Array Signal Processing*, Prentice-Hall, Englewood Cliffs, NJ, 1985.

Johnson, D. and Dudgeon, D., *Array Signal Processing: Concepts and Techniques*, Prentice-Hall, Englewood Cliffs, NJ, 1993.

Monzingo, R. and Miller, T., *Introduction to Adaptive Arrays*, John Wiley & Sons, New York, 1980.

Widrow, P.E., Mantey, P.E., Griffiths, L.J., and Goode, B.B., Adaptive Antenna Systems, *Proc. IEEE*, 55(12), 2143–2159, December 1967.

Tutorial Articles

Gabriel, W.F., Adaptive arrays: An introduction, *Proc. IEEE*, 64, 239–272, August 1976 and bibliography.

Marr, J., A selected bibliography on adaptive antenna arrays, *IEEE Trans. AES*, AES-22, 781–798, November 1986.

Several special journal issues have been devoted to beamforming—*IEEE Transactions on Antennas and Propagation*, September 1976 and March 1986, and the *Journal of Ocean Engineering*, 1987. Papers devoted to beamforming are often found in the *IEEE Transactions on: Antennas and Propagation, Signal Processing, Aerospace and Electronic Systems*, and in the *Journal of the Acoustical Society of America*.

3

Subspace-Based Direction-Finding Methods

Egemen Gönen
Globalstar

Jerry M. Mendel
University of Southern California

3.1 Introduction

Estimating bearings of multiple narrowband signals from measurements collected by an array of sensors has been a very active research problem for the last two decades. Typical applications of this problem are radar, communication, and underwater acoustics. Many algorithms have been proposed to solve the bearing estimation problem. One of the first techniques that appeared was beamforming which has a resolution limited by the array structure. Spectral estimation techniques were also applied to the problem. However, these techniques fail to resolve closely spaced arrival angles for low signal-to-noise ratios (SNRs). Another approach is the maximum-likelihood (ML) solution. This approach has been well documented in the literature. In the stochastic ML method [29], the sbgv signals are assumed to be Gaussian whereas they are regarded as arbitrary and deterministic in the deterministic ML method [37]. The sensor noise is modeled as Gaussian in both methods, which is a reasonable assumption due to the central limit theorem. The stochastic ML estimates of the bearings achieve the Cramer–Rao bound (CRB). On the other hand, this does not hold for deterministic ML estimates [32]. The common problem with the ML methods in general is the necessity of solving a nonlinear multidimensional (MD) optimization problem which has a high computational cost and for which there is no guarantee of global convergence. "Subspace-based" (or, super-resolution) approaches have attracted much attention, after the work of Schmidt [29], due to their computational simplicity as compared to the ML approach, and their possibility of overcoming the Rayleigh bound on the resolution power of classical direction-finding methods. Subspace-based direction-finding methods are summarized in this section.

3.2 Formulation of the Problem

Consider an array of M antenna elements receiving a set of plane waves emitted by $P(P < M)$ sources in the far field of the array. We assume a narrowband propagation model, i.e., the signal envelopes do not change during the time it takes for the wave fronts to travel from one sensor to another. Suppose that the signals have a common frequency of f_0; then, the wavelength $\lambda = c/f_0$ where c is the speed of propagation. The received M-vector $\mathbf{r}(t)$ at time t is

$$\mathbf{r}(t) = \mathbf{A}\mathbf{s}(t) + \mathbf{n}(t), \tag{3.1}$$

where

$\mathbf{s}(t) = [s_1(t), \ldots, s_P(t)]^T$ is the P-vector of sources

$\mathbf{A} = [\mathbf{a}(\theta_1), \ldots, \mathbf{a}(\theta_P)]$ is the $M \times P$ steering matrix in which $\mathbf{a}(\theta_i)$, the ith steering vector, is the response of the array to the ith source arriving from θ_i

$\mathbf{n}(t) = [n_1(t), \ldots, n_M(t)]^T$ is an additive noise process

We assume (1) the source signals may be statistically independent, partially correlated, or completely correlated (i.e., coherent); the distributions are unknown; (2) the array may have an arbitrary shape and response; and (3) the noise process is independent of the sources, zero-mean, and it may be either partially white or colored; its distribution is unknown. These assumptions will be relaxed, as required by specific methods, as we proceed.

The direction finding problem is to estimate the bearings [i.e., directions of arrival (DOA)] $\{\theta_i\}_{i=1}^P$ of the sources from the snapshots $\mathbf{r}(t)$, $t = 1, \ldots, N$.

In applications, the Rayleigh criterion sets a bound on the resolution power of classical direction-finding methods. In the next sections we summarize some of the so-called super-resolution direction-finding methods which may overcome the Rayleigh bound. We divide these methods into two classes, those that use second-order and those that use second- and higher-order statistics.

3.3 Second-Order Statistics-Based Methods

The second-order methods use the sample estimate of the array spatial covariance matrix $\mathbf{R} = E\{\mathbf{r}(t)\mathbf{r}(t)^H\} = \mathbf{A}\mathbf{R}_s\mathbf{A}^H + \mathbf{R}_n$, where $\mathbf{R}_s = E\{\mathbf{s}(t)\mathbf{s}(t)^H\}$ is the $P \times P$ signal covariance matrix and $\mathbf{R}_n = E\{\mathbf{n}(t)\mathbf{n}(t)^H\}$ is the $M \times M$ noise covariance matrix. For the time being, let us assume that the noise is spatially white, i.e., $\mathbf{R}_n = \sigma^2\mathbf{I}$. If the noise is colored and its covariance matrix is known or can be estimated, the measurements can be "whitened" by multiplying the measurements from the left by the matrix $\Lambda^{-1/2}\mathbf{E}_n^H$ obtained by the orthogonal eigendecomposition $\mathbf{R}_n = \mathbf{E}_n\Lambda\mathbf{E}_n^H$. The array spatial covariance matrix is estimated as $\hat{\mathbf{R}} = \sum_{t=1}^N \mathbf{r}(t)\mathbf{r}(t)^H/N$.

Some spectral estimation approaches to the direction finding problem are based on optimization. Consider the "minimum variance" (MV) algorithm, for example. The received signal is processed by a beamforming vector $\mathbf{w}_o$ which is designed such that the output power is minimized subject to the constraint that a signal from a desired direction is passed to the output with unit gain. Solving this optimization problem, we obtain the array output power as a function of the arrival angle θ as

$$P_{mv}(\theta) = \frac{1}{\mathbf{a}^H(\theta)\mathbf{R}^{-1}\mathbf{a}(\theta)}.$$

The arrival angles are obtained by scanning the range $[-90°, 90°]$ of θ and locating the peaks of $P_{mv}(\theta)$. At low SNRs the conventional methods, such as MV, fail to resolve closely spaced arrival angles. The resolution of conventional methods are limited by SNR even if exact $\mathbf{R}$ is used, whereas in subspace methods, there is no resolution limit; hence, the latter are also referred to as "super-resolution" methods. The limit comes from the sample estimate of $\mathbf{R}$.

The subspace-based methods exploit the eigendecomposition of the estimated array covariance matrix $\hat{\mathbf{R}}$. To see the implications of the eigendecomposition of $\hat{\mathbf{R}}$, let us first state the properties of $\mathbf{R}$: (1) If the source signals are independent or partially correlated, $\text{rank}(\mathbf{R}_s) = P$. If there are coherent sources, $\text{rank}(\mathbf{R}_s) < P$. In the methods explained in Sections 3.3.1 and 3.3.2, except for the weighted subspace fitting (WSF) method (see Section 3.3.1.1), it will be assumed that there are no coherent sources. The coherent signals case is described in Section 3.3.2. (2) If the columns of $\mathbf{A}$ are independent, which is generally true when the source bearings are different, then $\mathbf{A}$ is of full-rank P. (3) Properties 1 and 2 imply $\text{rank}(\mathbf{A}\mathbf{R}_s\mathbf{A}^H) = P$; therefore, $\mathbf{A}\mathbf{R}_s\mathbf{A}^H$ must have P nonzero eigenvalues and $M - P$ zero eigenvalues. Let the eigendecomposition of $\mathbf{A}\mathbf{R}_s\mathbf{A}^H$ be $\mathbf{A}\mathbf{R}_s\mathbf{A}^H = \sum_{i=1}^{M} \alpha_i \mathbf{e}_i \mathbf{e}_i^H$; then $\alpha_1 \geq \alpha_2 \geq \cdots \geq \alpha_P \geq \alpha_{P+1} = \cdots = \alpha_M = 0$ are the rank-ordered eigenvalues, and $\{\mathbf{e}_i\}_{i=1}^{M}$ are the corresponding eigenvectors. (4) Because $\mathbf{R}_n = \sigma^2\mathbf{I}$, the eigenvectors of $\mathbf{R}$ are the same as those of $\mathbf{A}\mathbf{R}_s\mathbf{A}^H$, and its eigenvalues are $\lambda_i = \alpha_i + \sigma^2$, if $1 \leq i \leq P$, or $\lambda_i = \sigma^2$, if $P + 1 \leq i \leq M$. The eigenvectors can be partitioned into two sets: $\mathbf{E}_s \triangleq [\mathbf{e}_1, \ldots, \mathbf{e}_P]$ forms the "signal subspace," whereas $\mathbf{E}_n \triangleq [\mathbf{e}_{P+1}, \ldots, \mathbf{e}_M]$ forms the "noise subspace." These subspaces are orthogonal. The signal eigenvalues $\Lambda_s \triangleq \text{diag}\{\lambda_1, \ldots, \lambda_P\}$, and the noise eigenvalues $\Lambda_n \triangleq \text{diag}\{\lambda_{P+1}, \ldots, \lambda_M\}$. (5) The eigenvectors corresponding to zero eigenvalues satisfy $\mathbf{A}\mathbf{R}_s\mathbf{A}^H\mathbf{e}_i = 0$, $i = P + 1, \ldots, M$; hence, $\mathbf{A}^H\mathbf{e}_i = 0$, $i = P + 1, \ldots, M$, because $\mathbf{A}$ and $\mathbf{R}_s$ are full rank. This last equation means that steering vectors are orthogonal to noise subspace eigenvectors. It further implies that because of the orthogonality of signal and noise subspaces, spans of signal eigenvectors and steering vectors are equal. Consequently there exists a nonsingular $P \times P$ matrix $\mathbf{T}$ such that $\mathbf{E}_s = \mathbf{A}\mathbf{T}$.

Alternatively, the signal and noise subspaces can also be obtained by performing a singular value decomposition (SVD) directly on the received data without having to calculate the array covariance matrix. Li and Vaccaro [17] state that the properties of the bearing estimates do not depend on which method is used; however, SVD must then deal with a data matrix that increases in size as the new snapshots are received. In the sequel, we assume that the array covariance matrix is estimated from the data and an eigendecomposition is performed on the estimated covariance matrix.

The eigenvalue decomposition of the spatial array covariance matrix, and the eigenvector partitionment into signal and noise subspaces, leads to a number of subspace-based direction-finding methods. The signal subspace contains information about where the signals are whereas the noise subspace informs us where they are not. Use of either subspace results in better resolution performance than conventional methods. In practice, the performance of the subspace-based methods is limited fundamentally by the accuracy of separating the two subspaces when the measurements are noisy [18]. These methods can be broadly classified into signal subspace and noise subspace methods. A summary of direction-finding methods based on both approaches is discussed in the following.

3.3.1 Signal Subspace Methods

In these methods, only the signal subspace information is retained. Their rationale is that by discarding the noise subspace we effectively enhance the SNR because the contribution of the noise power to the covariance matrix is eliminated. Signal subspace methods are divided into search-based and algebraic methods, which are explained in Sections 3.3.1.1 and 3.3.1.2.

3.3.1.1 Search-Based Methods

In search-based methods, it is assumed that the response of the array to a single source, "the array manifold" $\mathbf{a}(\theta)$, is either known analytically as a function of arrival angle, or is obtained through the calibration of the array. For example, for an M-element uniform linear array, the array response to a signal from angle θ is analytically known and is given by

$$\mathbf{a}(\theta) = \left[1, e^{-j2\pi\frac{d}{\lambda}\sin(\theta)}, \ldots, e^{-j2\pi(M-1)\frac{d}{\lambda}\sin(\theta)}\right]^T,$$

where
d is the separation between the elements
λ is the wavelength

In search-based methods to follow (except for the subspace fitting [SSF] methods), which are spatial versions of widely known power spectral density estimators, the estimated array covariance matrix is approximated by its signal subspace eigenvectors, or its "principal components," as $\hat{\mathbf{R}} \approx \sum_{i=1}^{P} \lambda_i \mathbf{e}_i \mathbf{e}_i^{\mathrm{H}}$. Then the arrival angles are estimated by locating the peaks of a function, $S(\theta)$ $(-90^\circ \leq \theta \leq 90^\circ)$, which depends on the particular method. Some of these methods and the associated function $S(\theta)$ are summarized in the following [13,18,20]:

Correlogram method: In this method, $S(\theta) = \mathbf{a}(\theta)^{\mathrm{H}}\hat{\mathbf{R}}\mathbf{a}(\theta)$. The resolution obtained from the Correlogram method is lower than that obtained from the MV and autoregressive (AR) methods.

Minimum variance [1] *method*: In this method, $S(\theta) = 1/\mathbf{a}(\theta)^{\mathrm{H}}\hat{\mathbf{R}}^{-1}\mathbf{a}(\theta)$. The MV method is known to have a higher resolution than the correlogram method, but lower resolution and variance than the AR method.

Autoregressive method: In this method, $S(\theta) = 1/|\mathbf{u}^{\mathrm{T}}\hat{\mathbf{R}}^{-1}\mathbf{a}(\theta)|^2$ where $\mathbf{u} = [1, 0, \ldots, 0]^{\mathrm{T}}$. This method is known to have a better resolution than the previous ones.

Subspace fitting and weighted subspace fitting methods: In Section 3.2 we saw that the spans of signal eigenvectors and steering vectors are equal; therefore, bearings can be solved from the best least-squares (LS) fit of the two spanning sets when the array is calibrated [35]. In the SSF method the criterion $[\hat{\theta}, \hat{\mathbf{T}}] = \operatorname{argmin}\|\mathbf{E}_s\mathbf{W}^{1/2} - \mathbf{A}(\theta)\mathbf{T}\|^2$ is used, where $\|.\|$ denotes the Frobenius norm, $\mathbf{W}$ is a positive definite weighting matrix, $\mathbf{E}_s$ is the matrix of signal subspace eigenvectors, and the notation for the steering matrix is changed to show its dependence on the bearing vector θ. This criterion can be minimized directly with respect to $\mathbf{T}$, and the result for $\mathbf{T}$ can then be substituted back into it, so that

$$\hat{\theta} = \operatorname{argmin} \operatorname{Tr}\left\{(\mathbf{I} - \mathbf{A}(\theta)\mathbf{A}(\theta)^{\#})\mathbf{E}_s\mathbf{W}\mathbf{E}_s^{\mathrm{H}}\right\},$$

where $\mathbf{A}^{\#} = (\mathbf{A}^{\mathrm{H}}\mathbf{A})^{-1}\mathbf{A}^{\mathrm{H}}$.

Viberg and Ottersten have shown that a class of direction finding algorithms can be approximated by this SSF formulation for appropriate choices of the weighting matrix $\mathbf{W}$. For example, for the deterministic ML method $\mathbf{W} = \Lambda_s - \sigma^2\mathbf{I}$, which is implemented using the empirical values of the signal eigenvalues, Λ_s, and the noise eigenvalue σ^2. Total least square (TLS)-estimation of signal parameters via rotational invariance techniques (ESPRIT), which is explained in the next section, can also be formulated in a similar but more involved way. Viberg and Ottersten have also derived an optimal WSF method, which yields the smallest estimation error variance among the class of SSF methods. In WSF, $\mathbf{W} = (\Lambda_s - \sigma^2\mathbf{I})^2\Lambda_s^{-1}$. The WSF method works regardless of the source covariance (including coherence) and has been shown to have the same asymptotic properties as the stochastic ML method; hence, it is asymptotically efficient for Gaussian signals (i.e., it achieves the stochastic CRB). Its behavior in the finite sample case may be different from the asymptotic case [34]. Viberg and Ottersten have also shown that the asymptotic properties of the WSF estimates are identical for both cases of Gaussian and non-Gaussian sources. They have also developed a consistent detection method for arbitrary signal correlation, and an algorithm for minimizing the WSF criterion. They do point out several practical implementation problems of their method, such as the need for accurate calibrations of the array manifold and knowledge of the derivative of the steering vectors w.r.t. θ. For nonlinear and nonuniform arrays, MD search methods are required for SSF, hence it is computationally expensive.

3.3.1.2 Algebraic Methods

Algebraic methods do not require a search procedure and yield DOA estimates directly.

ESPRIT [23]: The ESPRIT algorithm requires "translationally invariant" arrays, i.e., an array with its "identical copy" displaced in space. The geometry and response of the arrays do not have to be known; only the measurements from these arrays and the displacement between the identical arrays are required. The computational complexity of ESPRIT is less than that of the search-based methods.

Let $\mathbf{r}^1(t)$ and $\mathbf{r}^2(t)$ be the measurements from these arrays. Due to the displacement of the arrays the following holds

$$\mathbf{r}^1(t) = \mathbf{A}\mathbf{s}(t) + \mathbf{n}_1(t) \quad \text{and} \quad \mathbf{r}^2(t) = \mathbf{A}\Phi\mathbf{s}(t) + \mathbf{n}_2(t),$$

where $\Phi = \text{diag}\left\{e^{-j2\pi\frac{d}{\lambda}\sin\theta_1}, \ldots, e^{-j2\pi\frac{d}{\lambda}\sin\theta_P}\right\}$ in which d is the separation between the identical arrays, and the angles $\{\theta_i\}_{i=1}^{P}$ are measured with respect to the normal to the displacement vector between the identical arrays. Note that the auto covariance of $\mathbf{r}^1(t)$, $\mathbf{R}^{11}$, and the cross-covariance between $\mathbf{r}^1(t)$ and $\mathbf{r}^2(t)$, $\mathbf{R}^{21}$, are given by

$$\mathbf{R}^{11} = \mathbf{A}\mathbf{D}\mathbf{A}^{\mathrm{H}} + \mathbf{R}_{\mathrm{n}_1},$$

and

$$\mathbf{R}^{21} = \mathbf{A}\Phi\mathbf{D}\mathbf{A}^{\mathrm{H}} + \mathbf{R}_{\mathrm{n}_2\mathrm{n}_1},$$

where

$\mathbf{D}$ is the covariance matrix of the sources

$\mathbf{R}_{\mathrm{n}_1}$ and $\mathbf{R}_{\mathrm{n}_2\mathrm{n}_1}$ are the noise auto- and cross-covariance matrices

The ESPRIT algorithm solves for Φ, which then gives the bearing estimates. Although the subspace separation concept is not used in ESPRIT, its LS and TLS versions are based on a signal subspace formulation. The LS and TLS versions are more complicated, but are more accurate than the original ESPRIT, and are summarized in the next subsection. Here we summarize the original ESPRIT:

1. Estimate the autocovariance of $\mathbf{r}^1(t)$ and cross covariance between $\mathbf{r}^1(t)$ and $\mathbf{r}^2(t)$, as

$$\mathbf{R}^{11} = \frac{1}{N}\sum_{t=1}^{N}\mathbf{r}^1(t)\mathbf{r}^1(t)^{\mathrm{H}},$$

and

$$\mathbf{R}^{21} = \frac{1}{N}\sum_{t=1}^{N}\mathbf{r}^2(t)\mathbf{r}^1(t)^{\mathrm{H}}.$$

2. Calculate $\hat{\mathbf{R}}^{11} = \mathbf{R}^{11} - \mathbf{R}_{\mathrm{n}_1}$ and $\hat{\mathbf{R}}^{21} = \mathbf{R}^{21} - \mathbf{R}_{\mathrm{n}_2\mathrm{n}_1}$, where $\mathbf{R}_{\mathrm{n}_1}$ and $\mathbf{R}_{\mathrm{n}_2\mathrm{n}_1}$ are the estimated noise covariance matrices.
3. Find the singular values λ_i of the matrix pencil $\hat{\mathbf{R}}^{11} - \lambda_i\hat{\mathbf{R}}^{21}$, $i = 1, \ldots, P$.
4. The bearings, θ_i $(i = 1, \ldots, P)$, are readily obtained by solving the equation

$$\lambda_i = e^{j2\pi\frac{d}{\lambda}\sin\theta_i},$$

for θ_i. In the above steps, it is assumed that the noise is spatially and temporally white or the covariance matrices $\mathbf{R}_{\mathrm{n}_1}$ and $\mathbf{R}_{\mathrm{n}_2\mathrm{n}_1}$ are known.

LS and TLS-ESPRIT [28]:

1. Follow Steps 1 and 2 of ESPRIT.
2. Stack $\hat{\mathbf{R}}^{11}$ and $\hat{\mathbf{R}}^{21}$ into a $2M \times M$ matrix $\mathbf{R}$, as $\mathbf{R} \triangleq [\hat{\mathbf{R}}^{11T}\hat{\mathbf{R}}^{21T}]^{\mathrm{T}}$, and perform an SVD of $\mathbf{R}$, keeping the first $2M \times P$ submatrix of the left singular vectors of $\mathbf{R}$. Let this submatrix be $\mathbf{E}_s$.
3. Partition $\mathbf{E}_s$ into two $M \times P$ matrices $\mathbf{E}_{s1}$ and $\mathbf{E}_{s2}$ such that

$$\mathbf{E}_s = \left[\mathbf{E}_{s1}^{\mathrm{T}}\mathbf{E}_{s2}^{\mathrm{T}}\right]^{\mathrm{T}}.$$

4. For LS-ESPRIT, calculate the eigendecomposition of $\left(\mathbf{E}_{s1}^{\mathrm{H}}\mathbf{E}_{s1}\right)^{-1}\mathbf{E}_{s1}^{\mathrm{H}}\mathbf{E}_{s2}$. The eigenvalue matrix gives

$$\Phi = \mathrm{diag}\left\{\mathrm{e}^{-j2\pi\frac{d}{\lambda}\sin\theta_1}, \ldots, \mathrm{e}^{-j2\pi\frac{d}{\lambda}\sin\theta_P}\right\},$$

from which the arrival angles are readily obtained. For TLS-ESPRIT, proceed as follows.

5. Perform an SVD of the $M \times 2P$ matrix $[\mathbf{E}_{s1}, \mathbf{E}_{s2}]$, and stack the last P right singular vectors of $[\mathbf{E}_{s1}, \mathbf{E}_{s2}]$ into a $2P \times P$ matrix denoted $\mathbf{F}$.
6. Partition $\mathbf{F}$ as

$$\mathbf{F} \triangleq \left[\mathbf{F}_x^{\mathrm{T}}\mathbf{F}_y^{\mathrm{T}}\right]^{\mathrm{T}},$$

where $\mathbf{F}_x$ and $\mathbf{F}_y$ are $P \times P$.

7. Perform the eigendecomposition of $-\mathbf{F}_x\mathbf{F}_y^{-1}$. The eigenvalue matrix gives

$$\Phi = \mathrm{diag}\left\{\mathrm{e}^{-j2\pi\frac{d}{\lambda}\sin\theta_1}, \ldots, \mathrm{e}^{-j2\pi\frac{d}{\lambda}\sin\theta_P}\right\},$$

from which the arrival angles are readily obtained.

Different versions of ESPRIT have different statistical properties. The Toeplitz approximation method (TAM) [16], in which the array measurement model is represented as a state-variable model, although different in implementation from LS-ESPRIT, is equivalent to LS-ESPRIT; hence, it has the same error variance as LS-ESPRIT.

Generalized eigenvalues utilizing signal subspace eigenvectors (GEESE) [24]

1. Follow Steps 1 through 3 of TLS-ESPRIT.
2. Find the singular values λ_i of the pencil

$$\mathbf{E}_{s1} - \lambda_i\mathbf{E}_{s2}, \quad i = 1, \ldots, P.$$

3. The bearings, θ_i ($i = 1, \ldots, P$), are readily obtained from

$$\lambda_i = \mathrm{e}^{j2\pi\frac{d}{\lambda}\sin\theta_i}.$$

The GEESE method is claimed to be better than ESPRIT [24].

3.3.2 Noise Subspace Methods

These methods, in which only the noise subspace information is retained, are based on the property that the steering vectors are orthogonal to any linear combination of the noise subspace eigenvectors. Noise subspace methods are also divided into search-based and algebraic methods, which are explained next.

3.3.2.1 Search-Based Methods

In search-based methods, the array manifold is assumed to be known, and the arrival angles are estimated by locating the peaks of the function $S(\theta) = 1/\mathbf{a}(\theta)^{\mathrm{H}}\mathbf{N}\mathbf{a}(\theta)$, where $\mathbf{N}$ is a matrix formed using the noise space eigenvectors.

Pisarenko method: In this method, $\mathbf{N} = \mathbf{e}_M\mathbf{e}_M^{\mathrm{H}}$, where $\mathbf{e}_M$ is the eigenvector corresponding to the minimum eigenvalue of $\mathbf{R}$. If the minimum eigenvalue is repeated, any unit-norm vector which is a linear combination of the eigenvectors corresponding to the minimum eigenvalue can be used as $\mathbf{e}_M$. The basis of this method is that when the search angle θ corresponds to an actual arrival angle, the denominator of $S(\theta)$ in the Pisarenko method, $|\mathbf{a}(\theta)^{\mathrm{H}}\mathbf{e}_M|^2$, becomes small due to orthogonality of steering vectors and noise subspace eigenvectors; hence, $S(\theta)$ will peak at an arrival angle.

Multiple signal classification (MUSIC) [29] *method*: In this method, $\mathbf{N} = \sum_{i=P+1}^{M} \mathbf{e}_i\mathbf{e}_i^{\mathrm{H}}$. The idea is similar to that of the Pisarenko method; the inner product $|\mathbf{a}(\theta)^{\mathrm{H}} \sum_{i=P+1}^{M} \mathbf{e}_i|^2$ is small when θ is an actual arrival angle. An obvious signal-subspace formulation of MUSIC is also possible. The MUSIC spectrum is equivalent to the MV method using the exact covariance matrix when SNR is infinite, and therefore performs better than the MV method.

Asymptotic properties of MUSIC are well established [32,33], e.g., MUSIC is known to have the same asymptotic variance as the deterministic ML method for uncorrelated sources. It is shown by Xu and Buckley [38] that although, asymptotically, bias is insignificant compared to standard deviation, it is an important factor limiting the performance for resolving closely spaced sources when they are correlated.

In order to overcome the problems due to finite sample effects and source correlation, a MD version of MUSIC has been proposed [28,29]; however, this approach involves a computationally involved search, as in the ML method. MD MUSIC can be interpreted as a norm minimization problem, as shown in Ephraim et al. [8]; using this interpretation, strong consistency of MD MUSIC has been demonstrated. An optimally weighted version of MD MUSIC, which outperforms the deterministic ML method, has also been proposed in Viberg and Ottersten [35].

Eigenvector (EV) method: In this method,

$$\mathbf{N} = \sum_{i=P+1}^{M} \frac{1}{\lambda_i} \mathbf{e}_i\mathbf{e}_i^{\mathrm{H}}.$$

The only difference between the EV method and MUSIC is the use of inverse eigenvalue (the λ_i are the noise subspace eigenvalues of $\mathbf{R}$) weighting in eigenvector and unity weighting in MUSIC, which causes eigenvector to yield fewer spurious peaks than MUSIC [13]. The EV method is also claimed to shape the noise spectrum better than MUSIC.

Method of direction estimation (MODE): MODE is equivalent to WSF when there are no coherent sources. Viberg and Ottersten [35] claim that, for coherent sources, only WSF is asymptotically efficient. A minimum-norm interpretation and proof of strong consistency of MODE for ergodic and stationary signals, has also been reported [8]. The norm measure used in that work involves the source covariance matrix. By contrasting this norm with the Frobenius norm that is used in MD MUSIC, Ephraim et al. relate MODE and MD MUSIC.

Minimum-norm [15] *method*: In this method, the matrix $\mathbf{N}$ is obtained as follows [12]:

1. Form $\mathbf{E}_n = [\mathbf{e}_{P+1}, \ldots, \mathbf{e}_M]$.
2. Partition $\mathbf{E}_n$ as $\mathbf{E}_n = [\mathbf{c}\mathbf{C}^{\mathrm{T}}]^{\mathrm{T}}$, to establish $\mathbf{c}$ and $\mathbf{C}$.
3. Compute $\mathbf{d} = [1((\mathbf{c}^{\mathrm{H}}\mathbf{c})^{-1}\mathbf{C}^{*}\mathbf{c})^{\mathrm{T}}]^{\mathrm{T}}$, and, finally, $\mathbf{N} = \mathbf{d}\mathbf{d}^{\mathrm{H}}$.

For two closely spaced, equal power signals, the minimum-norm method has been shown to have a lower SNR threshold (i.e., the minimum SNR required to separate the two sources) than MUSIC [14].

Li and Vaccaro [17] derive and compare the mean-squared errors of the DOA estimates from minimum-norm and MUSIC algorithms due to finite sample effects, calibration errors, and noise modeling errors for the case of finite samples and high SNR. They show that mean-squared errors for DOA estimates produced by the MUSIC algorithm are always lower than the corresponding mean-squared errors for the minimum-norm algorithm.

3.3.2.2 Algebraic Methods

When the array is uniform linear, so that

$$\mathbf{a}(\theta) = \left[1, e^{-j2\pi\frac{d}{\lambda}\sin(\theta)}, \ldots, e^{-j2\pi(M-1)\frac{d}{\lambda}\sin(\theta)}\right]^{\mathrm{T}},$$

the search in $S(\theta) = 1/\mathbf{a}(\theta)^{\mathrm{H}}\mathbf{N}\mathbf{a}(\theta)$ for the peaks can be replaced by a root-finding procedure which yields the arrival angles. So doing results in better resolution than the search-based alternative because the root-finding procedure can give distinct roots corresponding to each source whereas the search function may not have distinct maxima for closely spaced sources. In addition, the computational complexity of algebraic methods is lower than that of the search-based ones. The algebraic version of MUSIC (root-MUSIC) is given next; for algebraic versions of Pisarenko, EV, and minimum-norm, the matrix $\mathbf{N}$ in root-MUSIC is replaced by the corresponding $\mathbf{N}$ in each of these methods.

Root-MUSIC method: In root-MUSIC, the array is required to be uniform linear, and the search procedure in MUSIC is converted into the following root-finding approach:

1. Form the $M \times M$ matrix $\mathbf{N} = \sum_{i=P+1}^{M} \mathbf{e}_i\mathbf{e}_i^{\mathrm{H}}$.
2. Form a polynomial $p(z)$ of degree $2M-1$ which has for its ith coefficient $c_i = \mathrm{tr}_i[\mathbf{N}]$, where tr_i denotes the trace of the ith diagonal, and $i = -(M-1), \ldots, 0, \ldots, M-1$. Note that tr_0 denotes the main diagonal, tr_1 denotes the first super-diagonal, and tr_{-1} denotes the first sub-diagonal.
3. The roots of $p(z)$ exhibit inverse symmetry with respect to the unit circle in the z-plane. Express $p(z)$ as the product of two polynomials $p(z) = h(z)h^*(z^{-1})$.
4. Find the roots z_i $(i = 1, \ldots, M)$ of $h(z)$. The angles of roots that are very close to (or, ideally on) the unit circle yield the DOA estimates, as

$$\theta_i = \sin^{-1}\left(\frac{\lambda}{2\pi d}\angle z_i\right), \quad \text{where } i = 1, \ldots, P.$$

The root-MUSIC algorithm has been shown to have better resolution power than MUSIC [27]; however, as mentioned previously, root-MUSIC is restricted to uniform linear arrays (ULA). Steps 2 through 4 make use of this knowledge. Li and Vaccaro show that algebraic versions of the MUSIC and minimum-norm algorithms have the same mean-squared errors as their search-based versions for finite samples and high SNR case. The advantages of root-MUSIC over search-based MUSIC is increased resolution of closely spaced sources and reduced computations.

3.3.3 Spatial Smoothing

When there are coherent (completely correlated) sources, $\mathrm{rank}(\mathbf{R}_s)$, and consequently $\mathrm{rank}(\mathbf{R})$, is less than P, and hence the above described subspace methods fail. If the array is uniform linear, then by applying the spatial smoothing method, described below, a new rank-P matrix is obtained which can be used in place of $\mathbf{R}$ in any of the subspace methods described earlier.

Spatial smoothing [9,31] starts by dividing the M-vector $\mathbf{r}(t)$ of the ULA into $K = M - S + 1$ overlapping subvectors of size S, $\mathbf{r}^{\mathrm{f}}_{S,k}(k = 1, \ldots, K)$, with elements $\{r_k, \ldots, r_{k+S-1}\}$, and $\mathbf{r}^{\mathrm{b}}_{S,k}(k = 1, \ldots, K)$, with elements $\{r^{\star}_{M-k+1}, \ldots, r^{\star}_{M-S-k+2}\}$. Then, a forward and backward spatially smoothed matrix $\mathbf{R}^{\mathrm{fb}}$ is calculated as

$$\mathbf{R}^{\mathrm{fb}} = \sum_{t=1}^{N} \sum_{k=1}^{K} \left(\mathbf{r}^{\mathrm{f}}_{S,k}(t)\mathbf{r}^{\mathrm{f}\,\mathrm{H}}_{S,k}(t) + \mathbf{r}^{\mathrm{b}}_{S,k}(t)\mathbf{r}^{\mathrm{b}\,\mathrm{H}}_{S,k}(t)\right)/KN.$$

The rank of $\mathbf{R}^{\mathrm{fb}}$ is P if there are at most $2M/3$ coherent sources. S must be selected such that

$$P_c + 1 \leq S \leq M - P_c/2 + 1,$$

in which P_c is the number of coherent sources. Then, any subspace-based method can be applied to $\mathbf{R}^{\mathrm{fb}}$ to determine the DOA. It is also possible to do spatial smoothing based only on $\mathbf{r}^{\mathrm{f}}_{S,k}$ or $\mathbf{r}^{\mathrm{b}}_{S,k}$, but in this case at most $M/2$ coherent sources can be handled.

3.3.4 Discussion

The application of all the subspace-based methods requires exact knowledge of the number of signals, in order to separate the signal and noise subspaces. The number of signals can be estimated from the data using either the Akaike information criterion (AIC) [36] or minimum descriptive length (MDL) [37] methods. The effect of underestimating the number of sources is analyzed by Radich and Buckley [26], whereas the case of overestimating the number of signals can be treated as a special case of the analysis in Stoica and Nehorai [32].

The second-order methods described above have the following disadvantages:

1. Except for ESPRIT (which requires a special array structure), all of the above methods require calibration of the array which means that the response of the array for every possible combination of the source parameters should be measured and stored; or, analytical knowledge of the array response is required. However, at any time, the antenna response can be different from when it was last calibrated due to environmental effects such as weather conditions for radar, or water waves for sonar. Even if the analytical response of the array elements is known, it may be impossible to know or track the precise locations of the elements in some applications (e.g., towed array). Consequently, these methods are sensitive to errors and perturbations in the array response. In addition, physically identical sensors may not respond identically in practice due to lack of synchronization or imbalances in the associated electronic circuitry.
2. In deriving the above methods, it was assumed that the noise covariance structure is known; however, it is often unrealistic to assume that the noise statistics are known due to several reasons. In practice, the noise is not isolated; it is often observed along with the signals. Moreover, as Swindlehurst and Kailath [33] state, there are noise phenomena effects that cannot be modeled accurately, e.g., channel crosstalk, reverberation, near-field, wideband, and distributed sources.
3. None of the methods in Sections 3.3.1 and 3.3.2, except for the WSF method and other MD search-based approaches, which are computationally very expensive, work when there are coherent (completely correlated) sources. Only if the array is uniform linear, can the spatial smoothing method in Section 3.3.2 be used. On the other hand, higher-order statistics of the received signals can be exploited to develop direction-finding methods which have less restrictive requirements.

3.4 Higher-Order Statistics-Based Methods

The higher-order statistical direction-finding methods use the spatial cumulant matrices of the array. They require that the source signals be non-Gaussian so that their higher than second-order statistics convey extra information. Most communication signals (e.g., Quadrature Amplitude Modulation (QAM)) are "complex circular" (a signal is complex circular if its real and imaginary parts are independent and symmetrically distributed with equal variances) and hence their third-order cumulants vanish; therefore, even-order cumulants are used, and usually fourth-order cumulants are employed. The fourth-order cumulant of the source signals must be nonzero in order to use these methods. One important feature of cumulant-based methods is that they can suppress Gaussian noise regardless of its coloring. Consequently, the requirement of having to estimate the noise covariance, as in second-order statistical processing methods, is avoided in cumulant-based methods. It is also possible to suppress non-Gaussian noise [6], and, when properly applied, cumulants extend the aperture of an array [5,30], which means that more sources than sensors can be detected. As in the second-order statistics-based methods, it is assumed that the number of sources is known or is estimated from the data.

The fourth-order moments of the signal $\mathbf{s}(t)$ are

$$E\{s_i s_j^\star s_k s_l^\star\}, \quad 1 \leq i, j, k, l \leq P,$$

and the fourth-order cumulants are defined as

$$\begin{aligned} c_{4,s}(i,j,k,l) &= \mathrm{cum}(s_i, s_j^\star, s_k, s_l^\star) \\ &= E\{s_i s_j^\star s_k s_l^\star\} - E\{s_i s_j^\star\}E\{s_k s_l^\star\} - E\{s_i s_l^\star\}E\{s_k s_j^\star\} - E\{s_i s_j\}E\{s_k^\star s_l^\star\}, \end{aligned}$$

where $1 \leq i, j, k, l \leq P$. Note that two arguments in the above fourth-order moments and cumulants are conjugated and the other two are unconjugated. For circularly symmetric signals, which is often the case in communication applications, the last term in $c_{4,s}(i,j,k,l)$ is zero.

In practice, sample estimates of the cumulants are used in place of the theoretical cumulants, and these sample estimates are obtained from the received signal vector $\mathbf{r}(t)$ $(t = 1, \ldots, N)$, as

$$\begin{aligned} \hat{c}_{4,r}(i,j,k,l) = {} & \sum_{t=1}^{N} r_i(t) r_j^\star(t) r_k(t) r_l^\star(t)/N - \sum_{t=1}^{N} r_i(t) r_j^\star(t) \sum_{t=1}^{N} r_k(t) r_l^\star(t)/N^2 \\ & - \sum_{t=1}^{N} r_i(t) r_l^\star(t) \sum_{t=1}^{N} r_k(t) r_j^\star(t)/N^2, \end{aligned}$$

where $1 \leq i, j, k, l \leq M$. Note that the last term in $c_{4,r}(i,j,k,l)$ is zero and, therefore, it is omitted.

Higher-order statistical subspace methods use fourth-order spatial cumulant matrices of the array output, which can be obtained in a number of ways by suitably selecting the arguments i, j, k, l of $c_{4,r}(i,j,k,l)$. Existing methods for the selection of the cumulant matrix, and their associated processing schemes are summarized next.

Pan–Nikias [22] *and Cardoso–Moulines* [2] *method*: In this method, the array needs to be calibrated, or its response must be known in analytical form. The source signals are assumed to be independent or partially correlated (i.e., there are no coherent signals). The method is as follows:

1. An estimate of an $M \times M$ fourth-order cumulant matrix $\mathbf{C}$ is obtained from the data. The following two selections for $\mathbf{C}$ are possible [2,22]:

$$c_{ij} = c_{4,r}(i,j,j,j), \quad 1 \leq i, j \leq M,$$

or

$$c_{ij} = \sum_{m=1}^{M} c_{4,r}(i,j,m,m), \quad 1 \leq i,j \leq M.$$

Using cumulant properties [19], and Equation 3.1, and a_{ij} for the ijth element of **A**, it is easy to verify that

$$c_{4,r}(i,j,j,j) = \sum_{p=1}^{P} a_{ip} \sum_{q,r,s=1}^{P} a_{jq}^{\star} a_{jr} a_{js}^{\star} c_{4,s}(p,q,r,s),$$

which, in matrix format, is $\mathbf{C} = \mathbf{AB}$ where **A** is the steering matrix and **B** is a $P \times M$ matrix with elements

$$b_{ij} = \sum_{q,r,s=1}^{P} a_{iq}^{\star} a_{jr} a_{js}^{\star} c_{4,s}(i,q,r,s).$$

Similarly,

$$\sum_{m=1}^{M} c_{4,r}(i,j,m,m) = \sum_{p,q=1}^{P} a_{ip} \left(\sum_{r,s=1}^{P} \sum_{m=1}^{M} a_{mr} a_{ms}^{\star} c_{4,s}(p,q,r,s) \right) a_{jq}^{\star}, \quad 1 \leq i,j \leq M,$$

which, in matrix form, can be expressed as $\mathbf{C} = \mathbf{ADA}^{\mathrm{H}}$, where **D** is a $P \times P$ matrix with elements

$$d_{ij} = \sum_{r,s=1}^{P} \sum_{m=1}^{M} a_{mr} a_{ms}^{\star} c_{4,s}(i,j,r,s).$$

Note that additive Gaussian noise is suppressed in both C matrices because higher than second-order statistics of a Gaussian process are zero.

2. The P left singular vectors of $\mathbf{C} = \mathbf{AB}$, corresponding to nonzero singular values or the P eigenvectors of $\mathbf{C} = \mathbf{ADA}^{\mathrm{H}}$ corresponding to nonzero eigenvalues form the signal subspace. The orthogonal complement of the signal subspace gives the noise subspace. Any of the Section 3.3 covariance-based search and algebraic direction finding (DF) methods (except for the EV method and ESPRIT) can now be applied (in exactly the same way as described in Section 3.3) either by replacing the signal and noise subspace eigenvectors and eigenvalues of the array covariance matrix by the corresponding subspace eigenvectors and eigenvalues of $\mathbf{ADA}^{\mathrm{H}}$, or by the corresponding subspace singular vectors and singular values of **AB**. A cumulant-based analog of the EV method does not exist because the eigenvalues and singular values of $\mathbf{ADA}^{\mathrm{H}}$ and **AB** corresponding to the noise subspace are theoretically zero. The cumulant-based analog of ESPRIT is explained later.

The same assumptions and restrictions for the covariance-based methods apply to their analogs in the cumulant domain. The advantage of using the cumulant-based analogs of these methods is that there is no need to know or estimate the noise-covariance matrix.

The asymptotic covariance of the DOA estimates obtained by MUSIC based on the above fourth-order cumulant matrices are derived in Cardoso and Moulines [2] for the case of Gaussian measurement noise with arbitrary spatial covariance, and are compared to the asymptotic covariance of the DOA estimates from the covariance-based MUSIC algorithm. Cardoso and Moulines show that covariance- and fourth-order cumulant-based MUSIC have similar performance for the high SNR case, and as SNR decreases below a certain SNR threshold, the variances of the fourth-order cumulant-based MUSIC DOA estimates increase with the fourth power of the reciprocal of the SNR, whereas the variances of covariance-based MUSIC DOA estimates increase with the square of the reciprocal of the SNR. They also observe that for high SNR and uncorrelated sources, the covariance-based MUSIC DOA estimates are uncorrelated, and the asymptotic variance of any particular source depends only on the power of that source (i.e., it is independent of the powers of the other sources). They observe, on the other hand, that DOA estimates from cumulant-based MUSIC, for the same case, are correlated, and the variance of the DOA estimate of a weak source increases in the presence of strong sources. This observation limits the use of cumulant-based MUSIC when the sources have a high dynamic range, even for the case of high SNR. Cardoso and Moulines state that this problem may be alleviated when the source of interest has a large fourth-order cumulant.

Porat and Friedlander [25] *method*: In this method, the array also needs to be calibrated, or its response is required in analytical form. The model used in this method divides the sources into groups that are partially correlated (but not coherent) within each group, but are statistically independent across the groups, i.e.,

$$\mathbf{r}(t) = \sum_{g=1}^{G} \mathbf{A}_g \mathbf{s}_g + \mathbf{n}(t),$$

where G is the number of groups each having p_g sources $\left(\sum_{g=1}^{G} p_g = P\right)$. In this model, the p_g sources in the gth group are partially correlated, and they are received from different directions. The method is as follows:

1. Estimate the fourth-order cumulant matrix, $\mathbf{C}_r$, of $\mathbf{r}(t) \otimes \mathbf{r}(t)^*$, where $\otimes$ denotes the Kronecker product. It can be verified that

$$\mathbf{C}_r = \sum_{g=1}^{G} (\mathbf{A}_g \otimes \mathbf{A}_g^*)\mathbf{C}_{s_g}(\mathbf{A}_g \otimes \mathbf{A}_g^*)^{\mathrm{H}},$$

   where $\mathbf{C}_{s_g}$ is the fourth-order cumulant matrix of $\mathbf{s}_g$. The rank of $\mathbf{C}_r$ is $\sum_{g=1}^{G} p_g^2$, and since $\mathbf{C}_r$ is $M^2 \times M^2$, it has $M^2 - \sum_{g=1}^{G} p_g^2$ zero eigenvalues which correspond to the noise subspace. The other eigenvalues correspond to the signal subspace.
2. Compute the SVD of $\mathbf{C}_r$ and identify the signal and noise subspace singular vectors. Now, second-order subspace-based search methods can be applied, using the signal or noise subspaces, by replacing the array response vector $\mathbf{a}(\theta)$ by $\mathbf{a}(\theta) \otimes \mathbf{a}^*(\theta)$.

The eigendecomposition in this method has computational complexity $O(M^6)$ due to the Kronecker product, whereas the second-order statistics-based methods (e.g., MUSIC) have complexity $O(M^3)$.

Chiang–Nikias [4] *method*: This method uses the ESPRIT algorithm and requires an array with its entire identical copy displaced in space by distance d; however, no calibration of the array is required. The signals

$$\mathbf{r}^1(t) = \mathbf{A}\mathbf{s}(t) + \mathbf{n}_1(t),$$

and

$$\mathbf{r}^2(t) = \mathbf{A}\boldsymbol{\Phi}\mathbf{s}(t) + \mathbf{n}_2(t).$$

Two $M \times M$ matrices $\mathbf{C}^1$ and $\mathbf{C}^2$ are generated as follows:

$$c_{ij}^1 = \text{cum}\left(r_i^1, r_j^{1*}, r_k^1, r_k^{1*}\right), \quad 1 \le i,j,k \le M,$$

and

$$c_{ij}^2 = \text{cum}\left(r_i^2, r_j^{1*}, r_k^1, r_k^{1*}\right), \quad 1 \le i,j,k \le M.$$

It can be shown that $\mathbf{C}^1 = \mathbf{AEA}^{\mathrm{H}}$ and $\mathbf{C}^2 = \mathbf{A\Phi EA}^{\mathrm{H}}$, where

$$\Phi = \text{diag}\left\{ \mathrm{e}^{-j2\pi\frac{d}{\lambda}\sin\theta_1}, \ldots, \mathrm{e}^{-j2\pi\frac{d}{\lambda}\sin\theta_P} \right\},$$

in which d is the separation between the identical arrays, and $\mathbf{E}$ is a $P \times P$ matrix with elements

$$e_{ij} = \sum_{q,r=1}^{P} a_{kq} a_{kr}^{*} c_{4,s}(i,q,r,j).$$

Note that these equations are in the same form as those for covariance-based ESPRIT (the noise cumulants do not appear in $\mathbf{C}^1$ and $\mathbf{C}^2$ because the fourth-order cumulants of Gaussian noises are zero); therefore, any version of ESPRIT or GEESE can be used to solve for Φ by replacing $\mathbf{R}^{11}$ and $\mathbf{R}^{21}$ by $\mathbf{C}^1$ and $\mathbf{C}^2$, respectively.

Virtual cross-correlation computer (VC^3) [5]: In VC^3, the source signals are assumed to be statistically independent. The idea of VC^3 can be demonstrated as follows: Suppose we have three identical sensors as in Figure 3.1, where $r_1(t), r_2(t)$, and $r_3(t)$ are measurements, and $\vec{d}_1$, $\vec{d}_2$, and $\vec{d}_3$ ($\vec{d}_3 = \vec{d}_1 + \vec{d}_2$) are the vectors joining these sensors. Let the response of each sensor to a signal from θ be $a(\theta)$. A "virtual" sensor is one at which no measurement is actually made. Suppose that we wish to compute the correlation between the virtual sensor $v_1(t)$ and $r_2(t)$, which (using the plane wave assumption) is

$$E\{r_2^*(t)v_1(t)\} = \sum_{p=1}^{P} |a(\theta_p)|^2 \sigma_p^2 e^{-j\vec{k}_p \cdot \vec{d}_3}.$$

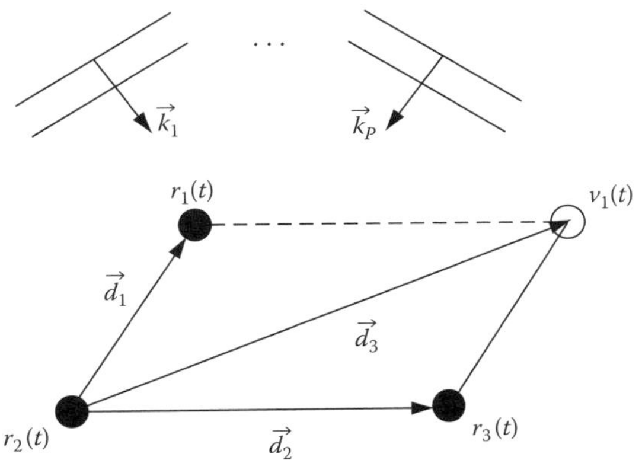

FIGURE 3.1 Demonstration of VC^3.

Consider the following cumulant

$$\begin{aligned}\text{cum}(r_2^\star(t), r_1(t), r_2^\star(t), r_3(t)) &= \sum_{p=1}^{P} |a(\theta_p)|^4 \gamma_p e^{-j\vec{k}_p \cdot \vec{d}_1} e^{-j\vec{k}_p \cdot \vec{d}_2} \\ &= \sum_{p=1}^{P} |a(\theta_p)|^4 \gamma_p e^{-j\vec{k}_p \cdot \vec{d}_3}.\end{aligned}$$

This cumulant carries the same angular information as the cross correlation $E\{r_2^\star(t)v_1(t)\}$, but for sources having different powers.

The fact that we are interested only in the directional information carried by correlations between the sensors therefore let us interpret a cross correlation as a vector (e.g., $\vec{d}_3$), and a fourth-order cumulant as the addition of two vectors (e.g., $\vec{d}_1 + \vec{d}_2$). This interpretation leads to the idea of decomposing the computation of a cross correlation into that of computing a cumulant. Doing this means that the directional information that would be obtained from the cross correlation between nonexisting sensors (or between an actual sensor and a nonexisting sensor) at certain virtual locations in the space can be obtained from a suitably defined cumulant that uses the real sensor measurements.

One advantage of virtual cross-correlation computation is that it is possible to obtain a larger aperture than would be obtained by using only second-order statistics. This means that more sources than sensors can be detected using cumulants. For example, given an M element ULA, VC^3 lets its aperture be extended from M to $2M - 1$ sensors, so that $2M - 2$ targets can be detected (rather than $M - 1$) just by using the array covariance matrix obtained by VC^3 in any of the subspace-based search methods explained earlier. This use of VC^3 requires the array to be calibrated. Another advantage of VC^3 is a fault tolerance capability. If sensors at certain locations in a given array fail to operate properly, these sensors can be replaced using VC^3.

Virtual ESPRIT (VESPA) [5]: For VESPA, the array only needs two identical sensors; the rest of the array may have arbitrary and unknown geometry and response. The sources are assumed to be statistically independent. VESPA uses the ESPRIT solution applied to cumulant matrices. By choosing a suitable pair of cumulants in VESPA, the need for a copy of the entire array, as required in ESPRIT, is totally eliminated. VESPA preserves the computational advantage of ESPRIT over search-based algorithms. An example array configuration is given in Figure 3.2.

Without loss of generality, let the signals received by the identical sensor pair be r_1 and r_2. The sensors r_1 and r_2 are collectively referred to as the "guiding sensor pair." The VESPA algorithm is

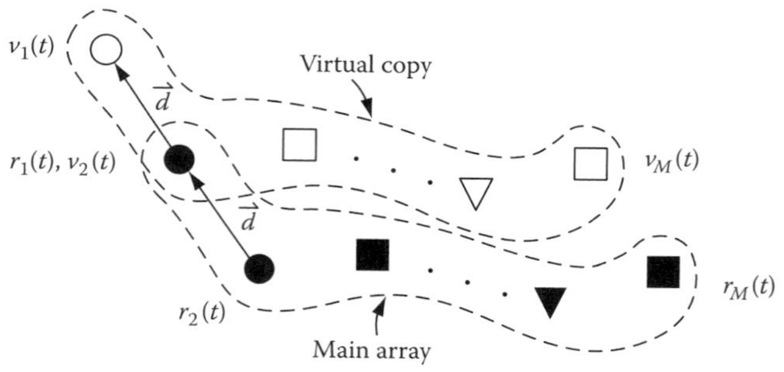

FIGURE 3.2 The main array and its virtual copy.

1. Two $M \times M$ matrices, $\mathbf{C}^1$ and $\mathbf{C}^2$, are generated as follows:

$$c_{ij}^1 = \text{cum}(r_1, r_1^*, r_i, r_j^*), \quad 1 \le i,j \le M,$$
$$c_{ij}^2 = \text{cum}(r_2, r_1^*, r_i, r_j^*), \quad 1 \le i,j \le M.$$

It can be shown that these relations can be expressed as $\mathbf{C}^1 = \mathbf{AFA}^{\mathrm{H}}$ and $\mathbf{C}^2 = \mathbf{A\Phi FA}^{\mathrm{H}}$, where the $P \times P$ matrix

$$\mathbf{F} = \text{diag}\{\gamma_{4,s_1}|a_{11}|^2, \ldots, \gamma_{4,s_P}|a_{1P}|^2\}, \{\gamma_{4,s_P}\}_{p=1}^{P},$$

and Φ has been defined before.

2. Note that these equations are in the same form as ESPRIT and Chiang and Nikias's ESPRIT-like method; however, as opposed to these methods, there is no need for an identical copy of the array; only an identical response sensor pair is necessary for VESPA. Consequently, any version of ESPRIT or GEESE can be used to solve for Φ by replacing $\mathbf{R}^{11}$ and $\mathbf{R}^{21}$ by $\mathbf{C}^1$ and $\mathbf{C}^2$, respectively.

Note, also, that there exists a very close link between VC^3 and VESPA. Although the way we chose $\mathbf{C}^1$ and $\mathbf{C}^2$ above seems to be not very obvious, there is a unique geometric interpretation to it. According to VC^3, as far as the bearing information is concerned, $\mathbf{C}^1$ is equivalent to the autocorrelation matrix of the array, and $\mathbf{C}^2$ is equivalent to the cross-correlation matrix between the array and its virtual copy (which is created by displacing the array by the vector that connects the second and the first sensors).

If the noise component of the signal received by one of the guiding sensor pair elements is independent of the noises at the other sensors, VESPA suppresses the noise regardless of its distribution [6]. In practice, the noise does affect the standard deviations of results obtained from VESPA.

An iterative version of VESPA has also been developed for cases where the source powers have a high dynamic range [11]. Iterative VESPA has the same hardware requirements and assumptions as in VESPA.

Extended VESPA [10]: When there are coherent (or completely correlated) sources, all of the above second- and higher-order statistics methods, except for the WSF method and other MD search-based approaches, fail. For the WSF and other MD methods, however, the array must be calibrated accurately and the computational load is expensive. The coherent signals case arises in practice when there are multipaths. Porat and Friedlander present a modified version of their algorithm to handle the case of coherent signals; however, their method is not practical because it requires selection of a highly redundant subset of fourth-order cumulants that contains $O(N^4)$ elements, and no guidelines exist for its selection and second-, fourth-, sixth-, and eighth-order moments of the data are required. If the array is "uniform linear," coherence can be handled using spatial smoothing as a preprocessor to the usual second- or higher-order [3,39] methods; however, the array aperture is reduced. Extended VESPA can handle coherence and provides increased aperture. Additionally, the array does not have to be completely uniform linear or calibrated; however, a uniform linear subarray is still needed. An example array configuration is shown in Figure 3.3.

Consider a scenario in which there are G statistically independent narrowband sources, $\{u_g(t)\}_{i=1}^{G}$. These source signals undergo multipath propagation, and each produces p_i coherent wave fronts

$$\{s_{1,1}, \ldots, s_{1,p_1}, \ldots, s_{G,1}, \ldots, s_{G,p_G}\} \left(\sum_{i=1}^{G} p_i = P \right),$$

that impinge on an M element sensor array from directions

$$\{\theta_{1,1}, \ldots, \theta_{1,p_1}, \ldots, \theta_{G,1}, \ldots, \theta_{G,p_G}\},$$

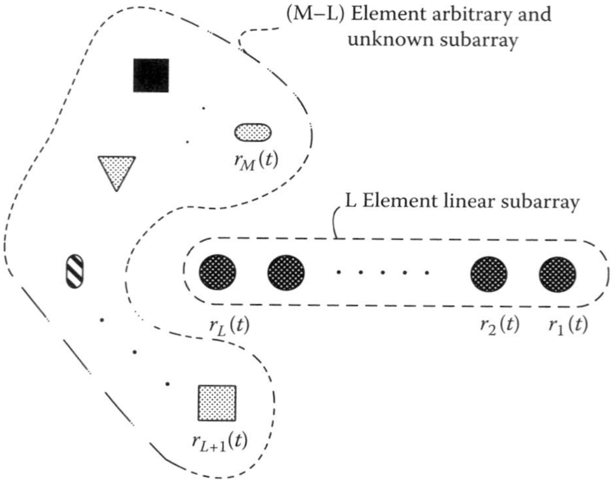

FIGURE 3.3 An example array configuration. There are M sensors, L of which are uniform linearly positioned; $r_1(t)$ and $r_2(t)$ are identical guiding sensors. Linear subarray elements are separated by Δ.

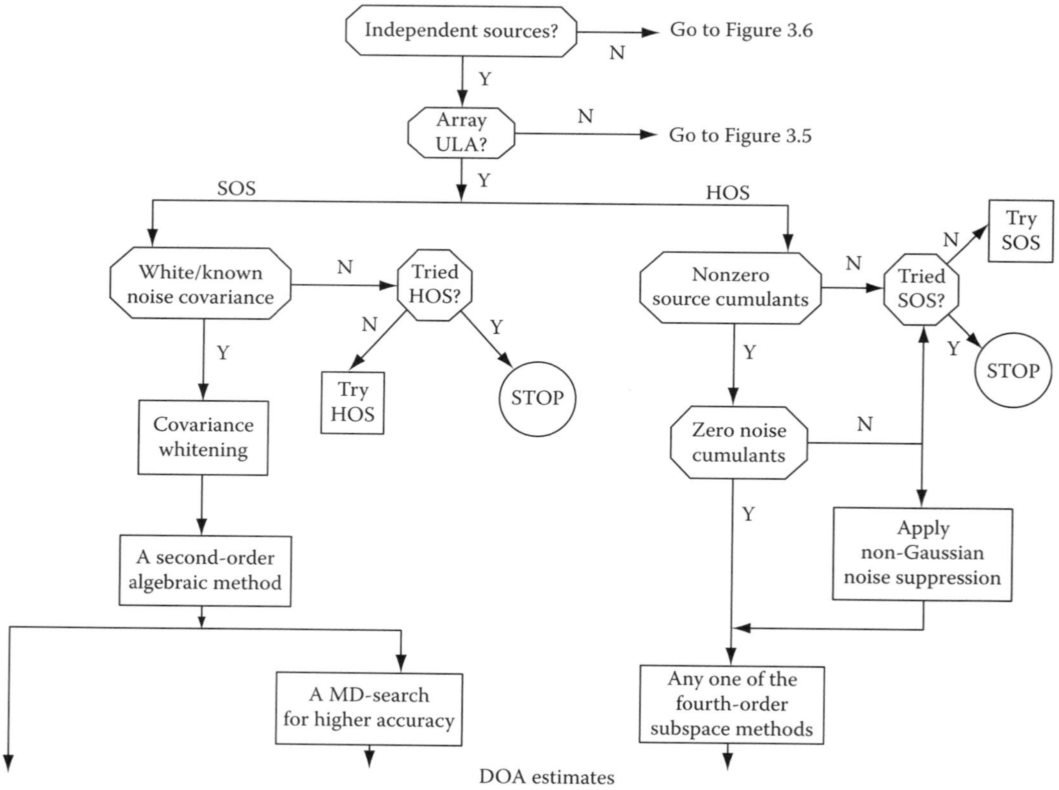

FIGURE 3.4 Second- or higher-order statistics-based subspace DF algorithm. Independent sources and ULA.

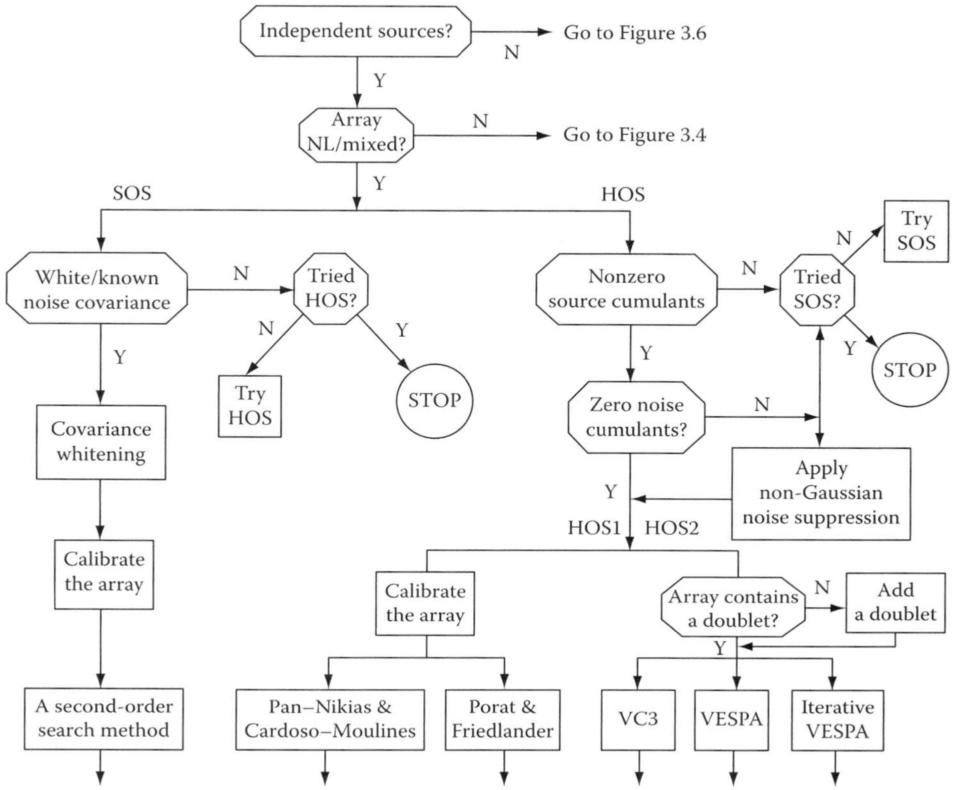

FIGURE 3.5 Second- or higher-order statistics-based subspace DF algorithm. Independent sources and NL/mixed array.

where $\theta_{m,p}$ represents the angle-of-arrival of the wave front $s_{g,p}$ that is the pth coherent signal in the gth group. The collection of p_i coherent wave fronts, which are scaled and delayed replicas of the ith source, are referred to as the ith group. The wave fronts are represented by the P-vector $\mathbf{s}(t)$. The problem is to estimate the DOAs $\{\theta_{1,1}, \ldots, \theta_{1,p_1}, \ldots, \theta_{G,1}, \ldots, \theta_{G,p_G}\}$.

When the multipath delays are insignificant compared to the bit durations of signals, then the signals received from different paths differ by only amplitude and phase shifts, thus the coherence among the received wave fronts can be expressed by the following equation:

$$\mathbf{s}(t) = \begin{bmatrix} \mathbf{s}_1(t) \\ \mathbf{s}_2(t) \\ \vdots \\ \mathbf{s}_G(t) \end{bmatrix} = \begin{bmatrix} \mathbf{c}_1 & 0 & \cdots & 0 \\ 0 & \mathbf{c}_2 & \cdots & 0 \\ \vdots & \vdots & \ddots & \vdots \\ 0 & 0 & \cdots & \mathbf{c}_G \end{bmatrix} \begin{bmatrix} u_1(t) \\ u_2(t) \\ \vdots \\ u_G(t) \end{bmatrix} = \mathbf{Q}\mathbf{u}(t), \tag{3.2}$$

where

$\mathbf{s}_i(t)$ is a $p_i \times 1$ signal vector representing the coherent wave fronts from the ith independent source $u_i(t)$

$\mathbf{c}_i$ is a $p_i \times 1$ complex attenuation vector for the ith source $(1 \leq i \leq G)$

$\mathbf{Q}$ is $P \times G$

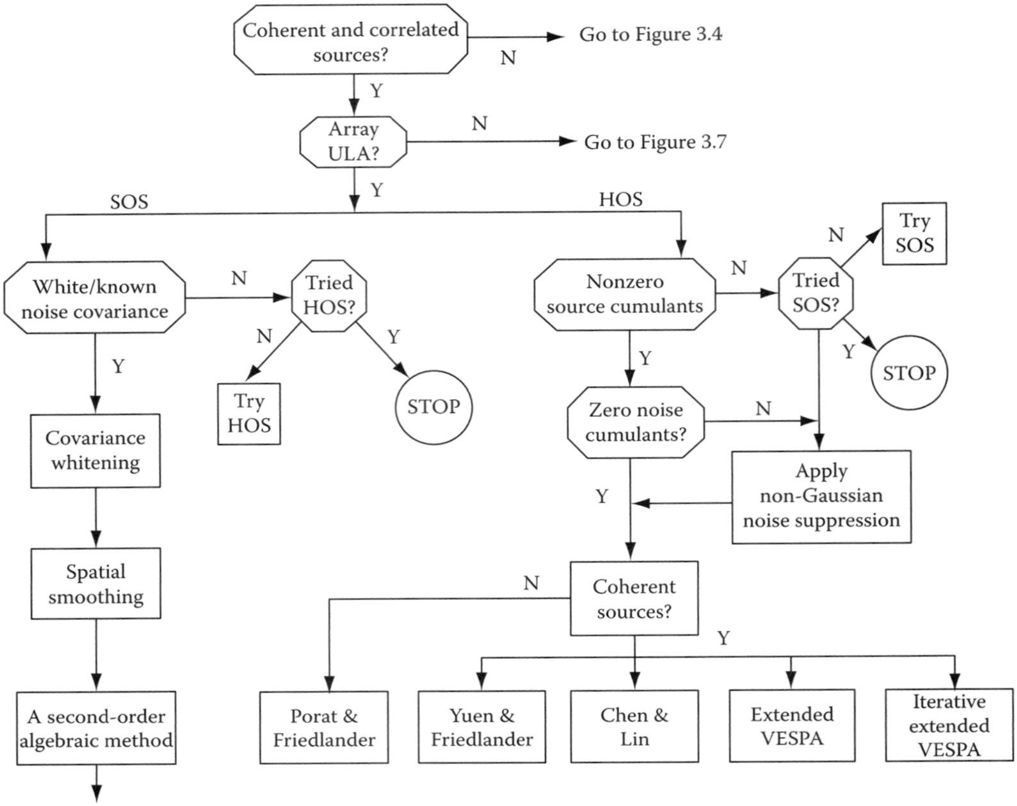

FIGURE 3.6 Second- or higher-order statistics-based subspace DF algorithms. Coherent and correlated sources and ULA.

The elements of $\mathbf{c}_i$ account for the attenuation and phase differences among the multipaths due to different arrival times. The received signal can then be written in terms of the independent sources as follows:

$$\mathbf{r}(t) = \mathbf{A}\mathbf{s}(t) + \mathbf{n}(\mathbf{t}) = \mathbf{A}\mathbf{Q}\mathbf{u}(t) + \mathbf{n}(\mathbf{t}) = \mathbf{B}\mathbf{u}(t) + \mathbf{n}(\mathbf{t}), \tag{3.3}$$

where $\mathbf{B} \triangleq \mathbf{AQ}$. The columns of $M \times G$ matrix $\mathbf{B}$ are known as the "generalized steering vectors."

Extended VESPA has three major steps:

Step 1: Use Step (1) of VESPA by choosing $r_1(t)$ and $r_2(t)$ as any two sensor measurements. In this case $\mathbf{C}^1 = \mathbf{BGB}^H$ and $\mathbf{C}^2 = \mathbf{BCGB}^H$, where

$$\mathbf{G} = \mathrm{diag}\left(\gamma_{4,u_1}|b_{11}|^2, \ldots, \gamma_{4,u_G}|b_{1G}|^2\right), \quad \left\{\gamma_{4,u_g}\right\}_{g=1}^{G},$$

$$\mathbf{C} = \mathrm{diag}\left(\frac{b_{21}}{b_{11}}, \ldots, \frac{b_{2G}}{b_{1G}}\right).$$

Due to the coherence, the DOAs cannot be obtained at this step from just $\mathbf{C}^1$ and $\mathbf{C}^2$ because the columns of $\mathbf{B}$ depend on a vector of DOAs (all those within a group). In the independent sources case, the columns of $\mathbf{A}$ depend only on a single DOA. Fortunately, the columns of $\mathbf{B}$ can be solved for as follows:

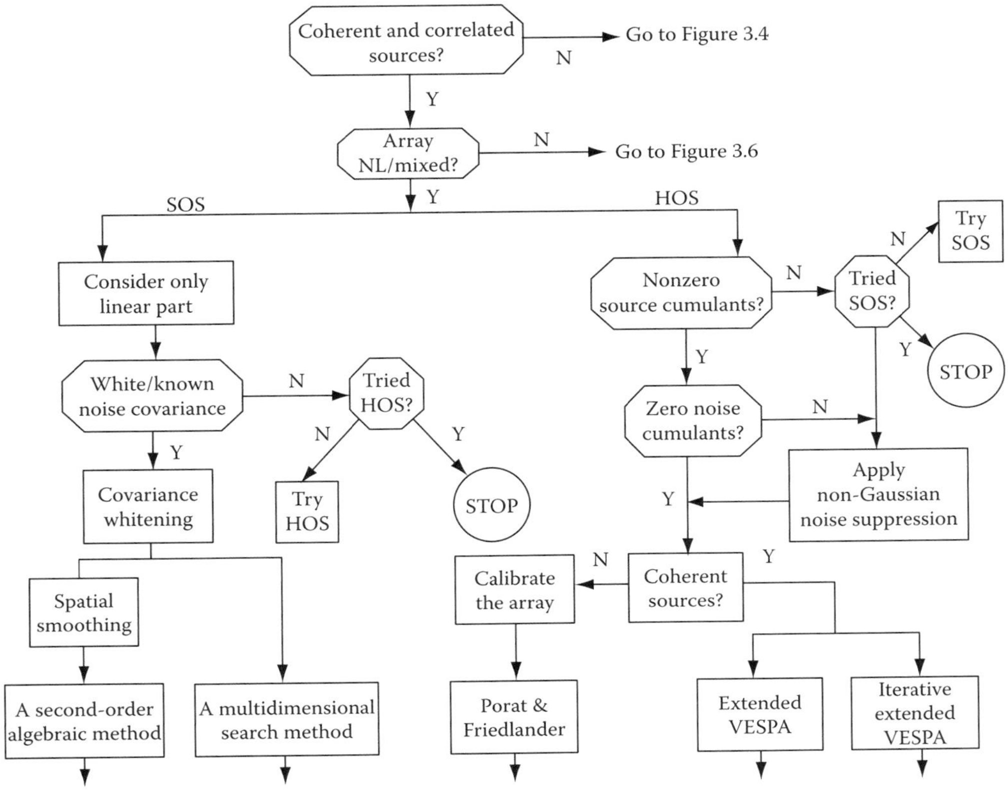

FIGURE 3.7 Second- or higher-order statistics-based subspace DF algorithms. Coherent and correlated sources and NL/mixed array.

(1) follow Steps 2 through 5 of TLS-ESPRIT by replacing $\mathbf{R}^{11}$ and $\mathbf{R}^{21}$ by $\mathbf{C}^1$ and $\mathbf{C}^2$, respectively, and using appropriate matrix dimensions; (2) determine the eigenvectors and eigenvalues of $-\mathbf{F}_x\mathbf{F}_y^{-1}$; let the eigenvector and eigenvalue matrices of $-\mathbf{F}_x\mathbf{F}_y^{-1}$ be $\mathbf{E}$ and $\mathbf{D}$, respectively; and (3) obtain an estimate of $\mathbf{B}$ to within a diagonal matrix, as $\mathbf{B} = (\mathbf{U}_{11}\mathbf{E} + \mathbf{U}_{12}\mathbf{E}\mathbf{D}^{-1})/2$, for use in Step 2.

Step 2: Partition the matrices $\mathbf{B}$ and $\mathbf{A}$ as $\mathbf{B} = [\mathbf{b}_1, \ldots, \mathbf{b}_G]$ and $\mathbf{A} = [\mathbf{A}_1, \ldots, \mathbf{A}_G]$, where the steering vector for the ith group $\mathbf{b}_i$ is $M \times 1$, $\mathbf{A}_i \stackrel{\Delta}{=} [\mathbf{a}(\theta_{i,1}), \ldots, \mathbf{a}(\theta_{i,p_i})]$ is $M \times p_i$, and $\theta_{i,m}$ is the angle-of-arrival of the mth source in the ith coherent group ($1 \le m \le p_i$). Using the fact that the ith column of $\mathbf{Q}$ has p_i nonzero elements, express $\mathbf{B}$ as $\mathbf{B} = \mathbf{A}\mathbf{Q} = [\mathbf{A}_1\mathbf{c}_1, \ldots, \mathbf{A}_G\mathbf{c}_G]$; therefore, the ith column of $\mathbf{B}$, $\mathbf{b}_i$ is $\mathbf{b}_i = \mathbf{A}_i\mathbf{c}_i$ where $i = 1, \ldots, G$. Now, the problem of solving for the steering vectors is transformed into the problem of solving for the steering vectors from each coherent group separately. To solve this new problem, each generalized steering vector $\mathbf{b}_i$ can be interpreted as a received signal for an array illuminated by p_i coherent signals having a steering matrix $\mathbf{A}_i$, and covariance matrix $\mathbf{c}_i\mathbf{c}_i^H$. The DOAs could then be solved for by using a second-order-statistics-based high-resolution method such as MUSIC, if the array was calibrated, and the rank of $\mathbf{c}_i\mathbf{c}_i^H$ was p_i; however, the array is not calibrated and $\text{rank}(\mathbf{c}_i\mathbf{c}_i^H) = 1$. The solution is to keep the portion of each $\mathbf{b}_i$ that corresponds to the uniform linear part of the array, $\mathbf{b}_{L,i}$, and to then apply the Section 3.3.3 spatial smoothing technique to a pseudo-covariance matrix $\mathbf{b}_{L,i}\mathbf{b}_{L,i}^H$ for $i = 1, \ldots, G$. Doing this "restores" the rank of $\mathbf{c}_i\mathbf{c}_i^H$ to p_i. In Section 3.3.3, we must replace $\mathbf{r}(t)$ by $\mathbf{b}_{L,i}$ and set $N = 1$.

The conditions on the length of the linear subarray and the parameter S under which the rank of $\mathbf{b}_{S,i}\mathbf{b}_{S,i}^H$ is restored to p_i are [11]: (a) $L \ge 3p_i/2$, which means that the linear subarray must have at least

Second-Order Statistics based Subspace Methods for Direction Finding

Signal Subspace Methods

> *SNR is enhanced effectively by retaining the signal subspace only*

Search Based Methods

> *Select if array is calibrated or response is known analytically*

Correlogram
> *Lower resolution than MV and AR*

Minimum Variance (MV)
> *Narrower mainlobe and smoother sidelobes than conventional beamformers*
> *Higher resolution than Correlogram*
> *Lower resolution than AR*
> *Lower variance than AR*

Autoregressive (AR)
> *Higher resolution than MV and Correlogram*

Subspace Fitting (SF)
> *Weighted SF works regardless of source correlation, and has the same asymptotic properties as the stochastic ML method, i.e., it achieves CRB.*
> *Requires accurate calibration of the manifold and its derivative with respect to arrival angle*

Algebraic Methods

> *Select if the array is ULA or its identical copy exists*
> *Computationally simpler than search-based methods.*

ESPRIT
> *Select if the array has an identical copy*
> *Computationally simple as compared to search-based methods*
> *Sensitive to perturbations in the sensor response and array geometry*
> *LS and TLS versions are best. They have the same asymptotic performance, but TLS converges faster and is better than LS for low SNR and short data lengths*

Toeplitz Approximation Method (TAM)
> *Equivalent to LS-ESPRIT*

GEESE
> *Better than ESPRIT*

Noise Subspace Methods

> *Methods are based on the orthogonality of steering vectors and noise subspace eigenvectors*

Search Based Methods

> *Select if array is calibrated or response is known analytically*

Eigenvector (EV)
> *Produces fewer spurious peaks than MUSIC*
> *Shapes the noise spectrum better than MUSIC*

Pisarenko
> *Performance with short data is poor*

MUSIC
> *Better than MV*
> *Same asymptotic performance as the deterministic ML for uncorrelated sources*

Minimum Norm
> *Select if the array is ULA*
> *Lower SNR threshold than MUSIC for resolution of closely spaced sources*

Method of Direction Estimation (MODE)
> *Consistent for ergodic and stationary signals*

Algebraic Methods

> *Select if the array is ULA*
> *Algebraic versions of EV, Pisarenko, MUSIC, and Minimum Norm are possible*
> *Better resolution than search-based versions*

Root MUSIC
> *Lower SNR threshold than MUSIC for resolution of closely spaced sources*
> *Simple root-finding procedure*

FIGURE 3.8 Pros and cons of all the methods considered.

$3p_{\max}/2$ elements, where $p_{\max}$ is the maximum number of multipaths in anyone of the G groups; and (b) given L and $p_{\max}$, the parameter S must be selected such that $p_{\max}+1 \leq S \leq L - p_{\max}/2 + 1$.

Step 3: Apply any second-order-statistics-based subspace technique (e.g., root-MUSIC, etc.) to R_i^{fb} $(i = 1, \ldots, G)$ to estimate DOAs of up to $2L/3$ coherent signals in each group.

Note that the matrices **C** and **G** in $\mathbf{C}^1$ and $\mathbf{C}^2$ are not used; however, if the received signals are independent, choosing $r_1(t)$ and $r_2(t)$ from the linear subarray lets DOA estimates be obtained from **C** in Step 1 because, in that case,

$$\mathbf{C} = \mathrm{diag}\left\{ e^{-j2\pi\frac{d}{\lambda}\sin\theta_1}, \ldots, e^{-j2\pi\frac{d}{\lambda}\sin\theta_P} \right\};$$

hence, extended VESPA can also be applied to the case of independent sources.

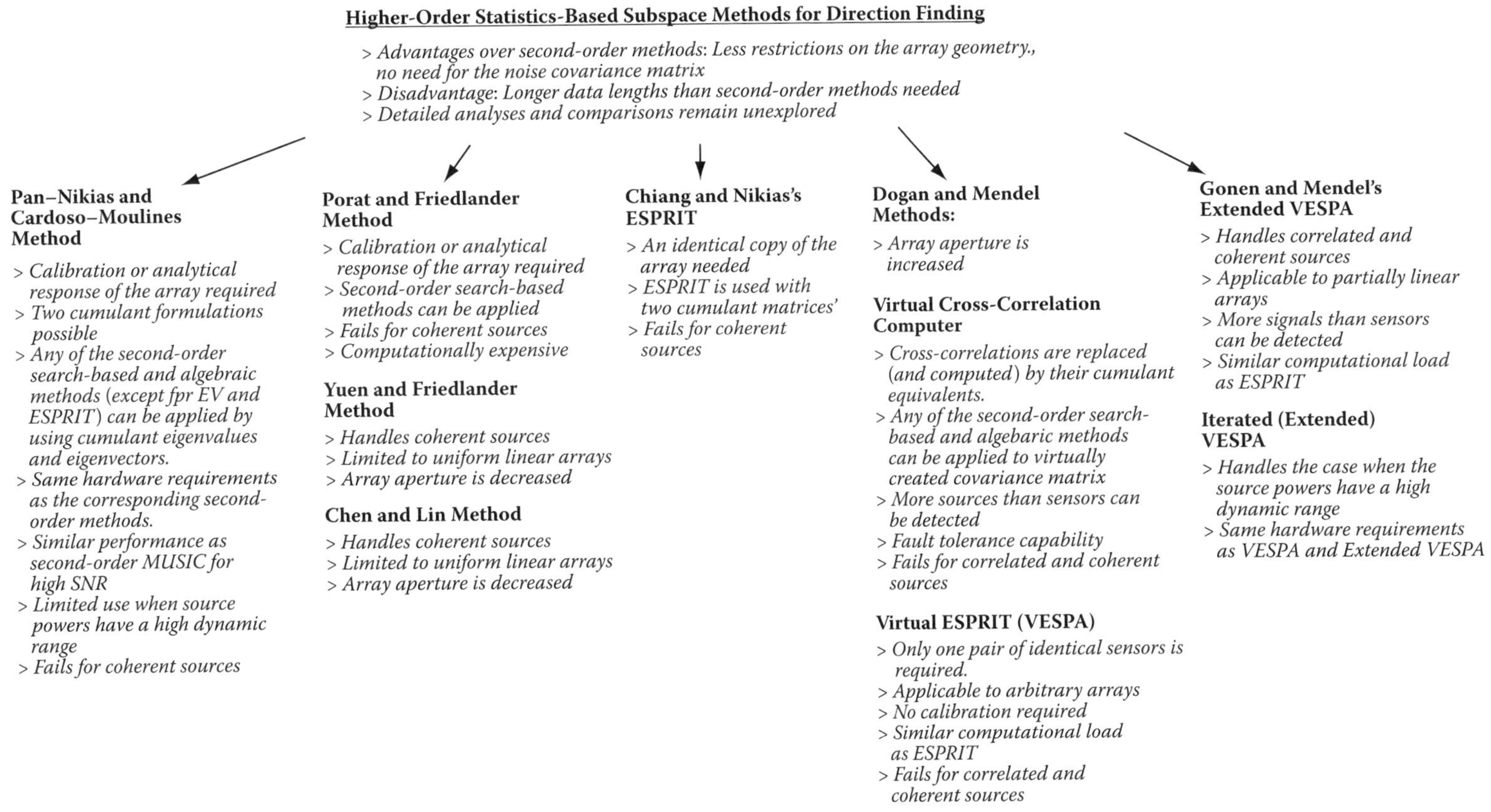
Higher-Order Statistics-Based Subspace Methods for Direction Finding
> Advantages over second-order methods: Less restrictions on the array geometry., no need for the noise covariance matrix
> Disadvantage: Longer data lengths than second-order methods needed
> Detailed analyses and comparisons remain unexplored
Pan–Nikias and Cardoso–Moulines Method
> Calibration or analytical response of the array required
> Two cumulant formulations possible
> Any of the second-order search-based and algebraic methods (except fpr EV and ESPRIT) can be applied by using cumulant eigenvalues and eigenvectors.
> Same hardware requirements as the corresponding second-order methods.
> Similar performance as second-order MUSIC for high SNR
> Limited use when source powers have a high dynamic range
> Fails for coherent sources
Porat and Friedlander Method
> Calibration or analytical response of the array required
> Second-order search-based methods can be applied
> Fails for coherent sources
> Computationally expensive
Yuen and Friedlander Method
> Handles coherent sources
> Limited to uniform linear arrays
> Array aperture is decreased
Chen and Lin Method
> Handles coherent sources
> Limited to uniform linear arrays
> Array aperture is decreased
Chiang and Nikias's ESPRIT
> An identical copy of the array needed
> ESPRIT is used with two cumulant matrices'
> Fails for coherent sources
Dogan and Mendel Methods:
> Array aperture is increased
Virtual Cross-Correlation Computer
> Cross-correlations are replaced (and computed) by their cumulant equivalents.
> Any of the second-order search-based and algebaric methods can be applied to virtually created covariance matrix
> More sources than sensors can be detected
> Fault tolerance capability
> Fails for correlated and coherent sources
Virtual ESPRIT (VESPA)
> Only one pair of identical sensors is required.
> Applicable to arbitrary arrays
> No calibration required
> Similar computational load as ESPRIT
> Fails for correlated and coherent sources
Gonen and Mendel's Extended VESPA
> Handles correlated and coherent sources
> Applicable to partially linear arrays
> More signals than sensors can be detected
> Similar computational load as ESPRIT
Iterated (Extended) VESPA
> Handles the case when the source powers have a high dynamic range
> Same hardware requirements as VESPA and Extended VESPA

FIGURE 3.9 Pros and cons of all the methods considered.

3.4.1 Discussion

One advantage of using higher-order statistics-based methods over second-order methods is that the covariance matrix of the noise is not needed when the noise is Gaussian. The fact that higher-order statistics have more arguments than covariances leads to more practical algorithms that have less restrictions on the array structure (for instance, the requirement of maintaining identical arrays for ESPRIT is reduced to only maintaining two identical sensors for VESPA). Another advantage is more sources than sensors can be detected, i.e., the array aperture is increased when higher-order statistics are properly applied; or, depending on the array geometry, unreliable sensor measurements can be replaced by using the VC^3 idea. One disadvantage of using higher-order statistics-based methods is that sample estimates of higher-order statistics require longer data lengths than covariances; hence, computational complexity is increased. In their recent study, Cardoso and Moulines [2] present a comparative performance analysis of second- and fourth-order statistics-based MUSIC methods. Their results indicate that dynamic range of the sources may be a factor limiting the performance of the fourth-order statistics-based MUSIC. A comprehensive performance analysis of the above higher-order statistical methods is still lacking; therefore, a detailed comparison of these methods remains as a very important research topic.

3.5 Flowchart Comparison of Subspace-Based Methods

Clearly, there are many subspace-based direction-finding methods. In order to see the forest from the trees, to know when to use a second-order or a higher-order statistics-based method, we present Figures 3.4 through 3.9. These figures provide a comprehensive summary of the existing subspace-based methods for direction finding and constitute guidelines to selection of a proper direction-finding method for a given application.

Note that: Figure 3.4 depicts independent sources and ULA, Figure 3.5 depicts independent sources and NL/mixed array, Figure 3.6 depicts coherent and correlated sources and ULA, and Figure 3.7 depicts coherent and correlated sources and NL/mixed array.

All four figures show two paths: SOS (second-order statistics) and HOS (higher-order statistics). Each path terminates in one or more method boxes, each of which may contain a multitude of methods. Figures 3.8 and 3.9 summarize the pros and cons of all the methods we have considered in this chapter.

Using Figures 3.4 through 3.9, it is possible for a potential user of a subspace-based direction-finding method to decide which method(s) is (are) most likely to give best results for his/her application.

Acknowledgments

The authors would like to thank Profs. A. Paulraj, V.U. Reddy, and M. Kaveh for reviewing the manuscript.

References

1. Capon, J., High-resolution frequency-wavenumber spectral analysis, *Proc. IEEE*, 57(8), 1408–1418, August 1969.
2. Cardoso, J.-F. and Moulines, E., Asymptotic performance analysis of direction-finding algorithms based on fourth-order cumulants, *IEEE Trans. Signal Process.*, 43(1), 214–224, January 1995.
3. Chen, Y.H. and Lin, Y.S., A modified cumulant matrix for DOA estimation, *IEEE Trans. Signal Process.*, 42, 3287–3291, November 1994.
4. Chiang, H.H. and Nikias, C.L., The ESPRIT algorithm with higher-order statistics, in *Proceedings of the Workshop on Higher-Order Spectral Analysis*, Vail, CO, pp. 163–168, June 28–30, 1989.

5. Dogan, M.C. and Mendel, J.M., Applications of cumulants to array processing, Part I: Aperture extension and array calibration, *IEEE Trans. Signal Process.*, 43(5), 1200–1216, May 1995.
6. Dogan, M.C. and Mendel, J.M., Applications of cumulants to array processing, Part II: Non-Gaussian noise suppression, *IEEE Trans. Signal Process.*, 43(7), 1661–1676, July 1995.
7. Dogan, M.C. and Mendel, J.M., Method and apparatus for signal analysis employing a virtual cross-correlation computer, U.S. Patent No. 5,459,668, October 17, 1995.
8. Ephraim, T., Merhav, N., and Van Trees, H.L., Min-norm interpretations and consistency of MUSIC, MODE and ML, *IEEE Trans. Signal Process.*, 43(12), 2937–2941, December 1995.
9. Evans, J.E., Johnson, J.R., and Sun, D.F., High resolution angular spectrum estimation techniques for terrain scattering analysis and angle of arrival estimation, in *Proceedings of the First ASSP Workshop Spectral Estimation*, Communication Research Laboratory, McMaster University, Hamilton, Ontario, Canada, August 1981.
10. Gönen, E., Dogan, M.C., and Mendel, J.M., Applications of cumulants to array processing: Direction finding in coherent signal environment, in *Proceedings of 28th Asilomar Conference on Signals, Systems, and Computers*, Asilomar, CA, pp. 633–637, 1994.
11. Gönen, E., Cumulants and subspace techniques for array signal processing, PhD thesis, University of Southern California, Los Angeles, CA, December 1996.
12. Haykin, S.S., *Adaptive Filter Theory*, Prentice-Hall, Englewood Cliffs, NJ, 1991.
13. Johnson, D.H. and Dudgeon, D.E., *Array Signal Processing: Concepts and Techniques*, Prentice-Hall, Englewood Cliffs, NJ, 1993.
14. Kaveh, M. and Barabell, A.J., The statistical performance of the MUSIC and the Minimum-Norm algorithms in resolving plane waves in noise, *IEEE Trans. Acoust. Speech Signal Process.*, 34, 331–341, April 1986.
15. Kumaresan, R. and Tufts, D.W., Estimating the angles of arrival multiple plane waves, *IEEE Trans. Aerosp. Electron. Syst.*, AES-19, 134–139, January 1983.
16. Kung, S.Y., Lo, C.K., and Foka, R., A Toeplitz approximation approach to coherent source direction finding, *Proceedings of the ICASSP*, Tokyo, Japan, 1986.
17. Li, F. and Vaccaro, R.J., Unified analysis for DOA estimation algorithms in array signal processing, *Signal Process.*, 25(2), 147–169, November 1991.
18. Marple, S.L., *Digital Spectral Analysis with Applications*, Prentice-Hall, Englewood Cliffs, NJ, 1987.
19. Mendel, J.M., Tutorial on higher-order statistics (spectra) in signal processing and system theory: Theoretical results and some applications, *Proc. IEEE*, 79(3), 278–305, March 1991.
20. Nikias, C.L. and Petropulu, A.P., *Higher-Order Spectra Analysis: A Nonlinear Signal Processing Framework*, Prentice-Hall, Englewood Cliffs, NJ, 1993.
21. Ottersten, B., Viberg, M., and Kailath, T., Performance analysis of total least squares ESPRIT algorithm, *IEEE Trans. Signal Process.*, 39(5), 1122–1135, May 1991.
22. Pan, R. and Nikias, C.L., Harmonic decomposition methods in cumulant domains, in *Proceedings of the ICASSP'88*, New York, pp. 2356–2359, 1988.
23. Paulraj, A., Roy, R., and Kailath, T., Estimation of signal parameters via rotational invariance techniques-ESPRIT, in *Proceedings of the 19th Asilomar Conference on Signals, Systems, and Computers*, Asilomar, CA, November 1985.
24. Pillai, S.U., *Array Signal Processing*, Springer-Verlag, New York, 1989.
25. Porat, B. and Friedlander, B., Direction finding algorithms based on high-order statistics, *IEEE Trans. Signal Process.*, 39(9), 2016–2023, September 1991.
26. Radich, B.M. and Buckley, K., The effect of source number underestimation on MUSIC location estimates, *IEEE Trans. Signal Process.*, 42(1), 233–235, January 1994.
27. Rao, D.V.B. and Hari, K.V.S., Performance analysis of Root-MUSIC, *IEEE Trans. Acoust. Speech Signal Process.*, ASSP-37, 1939–1949, December 1989.
28. Roy, R.H., ESPRIT-estimation of signal parameters via rotational invariance techniques, PhD dissertation, Stanford University, Stanford, CA, 1987.

29. Schmidt, R.O., A signal subspace approach to multiple emitter location and spectral estimation, PhD dissertation, Stanford University, Stanford, CA, November 1981.
30. Shamsunder, S. and Giannakis, G.B., Detection and parameter estimation of multiple nonGaussian sources via higher order statistics, *IEEE Trans. Signal Process.*, 42, 1145–1155, May 1994.
31. Shan, T.J., Wax, M., and Kailath, T., On spatial smoothing for direction-of-arrival estimation of coherent signals, *IEEE Trans. Acoust. Speech Signal Process.*, ASSP-33(2), 806–811, August 1985.
32. Stoica, P. and Nehorai, A., MUSIC, maximum likelihood and Cramer–Rao bound: Further results and comparisons, *IEEE Trans. Signal Process.*, 38, 2140–2150, December 1990.
33. Swindlehurst, A.L. and Kailath, T., A performance analysis of subspace-based methods in the presence of model errors, Part 1: The MUSIC algorithm, *IEEE Trans. Signal Process.*, 40(7), 1758–1774, July 1992.
34. Viberg, M., Ottersten, B., and Kailath, T., Detection and estimation in sensor arrays using weighted subspace fitting, *IEEE Trans. Signal Process.*, 39(11), 2436–2448, November 1991.
35. Viberg, M. and Ottersten, B., Sensor array processing based on subspace fitting, *IEEE Trans. Signal Process.*, 39(5), 1110–1120, May 1991.
36. Wax, M. and Kailath, T., Detection of signals by information theoretic criteria, *IEEE Trans. Acoust. Speech Signal Process.*, ASSP-33(2), 387–392, April 1985.
37. Wax, M., Detection and estimation of superimposed signals, PhD dissertation, Stanford University, Stanford, CA, March 1985.
38. Xu, X.-L. and Buckley, K., Bias and variance of direction-of-arrival estimates from MUSIC, MIN-NORM and FINE, *IEEE Trans. Signal Process.*, 42(7), 1812–1816, July 1994.
39. Yuen, N. and Friedlander, B., DOA estimation in multipath based on fourth-order cumulants, in *Proceedings of the IEEE Signal Processing ATHOS Workshop on Higher-Order Statistics*, Aiguablava, Spain, pp. 71–75, June 1995.

4

ESPRIT and Closed-Form 2-D Angle Estimation with Planar Arrays

Martin Haardt
Ilmenau University of Technology

Michael D. Zoltowski
Purdue University

Cherian P. Mathews
University of the Pacific

Javier Ramos
Universidad Rey Juan Carlos

4.1 Introduction ... 4-1
Notation
4.2 The Standard ESPRIT Algorithm ... 4-3
4.3 1-D Unitary ESPRIT ... 4-6
1-D Unitary ESPRIT in Element Space • 1-D Unitary ESPRIT in DFT Beamspace
4.4 UCA-ESPRIT for Circular Ring Arrays ... 4-11
Results of Computer Simulations
4.5 FCA-ESPRIT for Filled Circular Arrays ... 4-14
Computer Simulation
4.6 2-D Unitary ESPRIT ... 4-16
2-D Array Geometry • 2-D Unitary ESPRIT in Element Space • Automatic Pairing of the 2-D Frequency Estimates • 2-D Unitary ESPRIT in DFT Beamspace • Simulation Results
References ... 4-27

4.1 Introduction

Estimating the directions of arrival (DOAs) of propagating plane waves is a requirement in a variety of applications including radar, mobile communications, sonar, and seismology. Due to its simplicity and high-resolution capability, estimation of signal parameters via rotational invariance techniques (ESPRIT) [23] has become one of the most popular signal subspace-based DOA or spatial frequency estimation schemes. ESPRIT is explicitly premised on a point source model for the sources and is restricted to use with array geometries that exhibit so-called invariances [23]. However, this requirement is not very restrictive as many of the common array geometries used in practice exhibit these invariances, or their output may be transformed to effect these invariances.

ESPRIT may be viewed as a complement to the multiple signal classification (MUSIC) algorithm, the forerunner of all signal subspace-based DOA methods, in that it is based on properties of the signal eigenvectors, whereas MUSIC is based on properties of the noise eigenvectors. This chapter concentrates solely on the use of ESPRIT to estimate the DOAs of plane waves incident upon an antenna array. It should be noted, though, that ESPRIT may be used in the dual problem of estimating the frequencies of sinusoids embedded in a time series [23]. In this application, ESPRIT is more generally applicable than MUSIC as it can handle damped sinusoids and provide estimates of the damping factors as well as the constituent frequencies. The standard ESPRIT algorithm for one-dimensional (1-D) arrays is reviewed in Section 4.2. There are three primary steps in any ESPRIT-type algorithm:

1. *Signal subspace estimation*: Computation of a basis for the estimated signal subspace
2. *Solution of the invariance equation*: Solution of an (in general) overdetermined system of equations, the so-called invariance equation, derived from the basis matrix estimated in step 1
3. *Spatial frequency estimation*: Computation of the eigenvalues of the solution of the invariance equation formed in step 2

Many antenna arrays used in practice have geometries that possess some form of symmetry. For example, a linear array of equispaced identical antennas is symmetric about the center of the linear aperture it occupies. In Section 4.3.1, an efficient implementation of ESPRIT is presented that exploits the symmetry present in so-called centrosymmetric arrays to formulate the three steps of ESPRIT in terms of real-valued computations, despite the fact that the input to the algorithm needs to be the complex analytic signal output from each antenna. This reduces the computational complexity significantly. A reduced dimension beamspace version of ESPRIT is developed in Section 4.3.2. Advantages to working in beamspace include reduced computational complexity [3], decreased sensitivity to array imperfections [1], and lower signal-to-noise ratio (SNR) resolution thresholds [14].

With a 1-D array, one can only estimate the angle of each incident plane wave relative to the array axis. For source localization purposes, this only places the source on a cone whose axis of symmetry is the array axis. The use of a two-dimensional 2-D or planar array enables one to passively estimate the 2-D arrival angles of each emitting source. The remainder of the chapter presents ESPRIT-based techniques for use in conjunction with circular and rectangular arrays that provide estimates of the azimuth and elevation angle of each incident signal. As in the 1-D case, the symmetries present in these array geometries are exploited to formulate the three primary steps of ESPRIT in terms of real-valued computations.

If the transmitted source signals are real-valued (or noncircular, where the magnitude of the noncircularity rate is one), e.g., by using binary phase shift keying (BPSK) or M-ary amplitude shift keying (M-ASK) modulation schemes, this a priori knowledge can be exploited to increase the parameter estimation accuracy even further and estimate the parameters of twice as many sources as for complex-valued source signals, as explained in [10]. Also minimum shift keying (MSK) and offset quadrature phase shift keying (O-QPSK) can be mapped onto noncircular constellations by a derotation step.

4.1.1 Notation

Throughout this chapter, column vectors and matrices are denoted by lowercase and uppercase boldfaced letters, respectively. For any positive integer p, $\mathbf{I}_p$ is the $p \times p$ identity matrix and Π_p is the $p \times p$ exchange matrix with ones on its antidiagonal and zeros elsewhere:

$$\mathbf{\Pi}_p = \begin{bmatrix} & & 1 \\ & 1 & \\ \cdot & & \\ 1 & & \end{bmatrix} \in \mathbb{R}^{p \times p}. \tag{4.1}$$

Premultiplication of a matrix by Π_p will reverse the order of its rows, while postmultiplication of a matrix by Π_p reverses the order of its columns. Furthermore, the superscripts $(\cdot)^{\mathrm{H}}$ and $(\cdot)^{\mathrm{T}}$ denote complex conjugate transposition and transposition without complex conjugation, respectively. Complex conjugation by itself is denoted by an overbar $(\bar{\cdot})$, such that $\boldsymbol{X}^{\mathrm{H}} = \bar{\boldsymbol{X}}^{\mathrm{T}}$. A diagonal matrix Φ with the diagonal elements $\phi_1, \phi_2, \ldots, \phi_d$ may be written as

$$\Phi = \mathrm{diag}\{\phi_i\}_{i=1}^{d} = \begin{bmatrix} \phi_1 & & & \\ & \phi_2 & & \\ & & \cdot & \\ & & & \phi_d \end{bmatrix} \in \mathbb{C}^{d \times d}.$$

Moreover, matrices $\mathbf{Q} \in \mathbb{C}^{p\times q}$ satisfying

$$\mathbf{\Pi}_p \bar{\mathbf{Q}} = \mathbf{Q} \tag{4.2}$$

will be called left Π-real [13]. Often left Π-real matrices are also called conjugate centrosymmetric [27].

4.2 The Standard ESPRIT Algorithm

The algorithm ESPRIT [23] must be used in conjunction with an M-element sensor array composed of m pairs of pairwise identical, but displaced, sensors (doublets) as depicted in Figure 4.1. If the subarrays do not overlap, i.e., if they do not share any elements, $M = 2m$, but in general $M \leq 2\ m$ since overlapping subarrays are allowed; cf. Figure 4.2. Let Δ denote the distance between the two subarrays. Incident on both subarrays are d narrowband noncoherent* planar wave fronts with distinct DOAs θ_i, $1 \leq i \leq d$, relative to the displacement between the two subarrays.† Their complex envelope at an arbitrary reference point may be expressed as $s_i(t) = \alpha_i(t)e^{j(2\pi f_c t + \beta_i(t))}$, where f_c denotes the common carrier frequency of the d wave fronts. Without loss of generality, we assume that the reference point is the array centroid. The signals are called "narrowband" if their amplitudes $\alpha_i(t)$ and phases $\beta_i(t)$ vary slowly with respect to the propagation time across the array τ, i.e., if

$$\alpha_i(t-\tau) \cong \alpha_i(t) \quad \text{and} \quad \beta_i(t-\tau) \cong \beta_i(t). \tag{4.3}$$

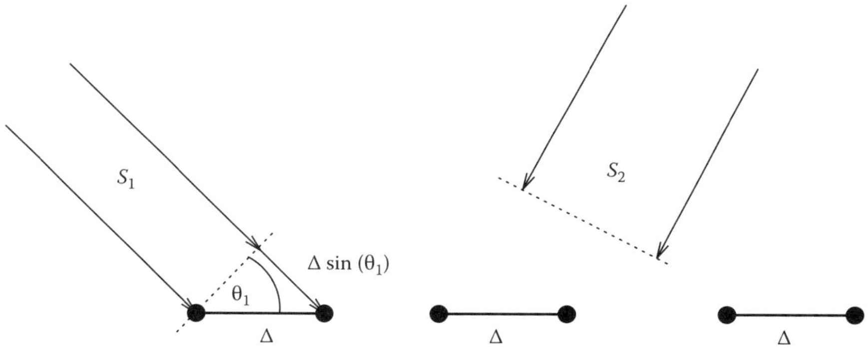

FIGURE 4.1 Planar array composed of $m = 3$ pairwise identical, but displaced, sensors (doublets).

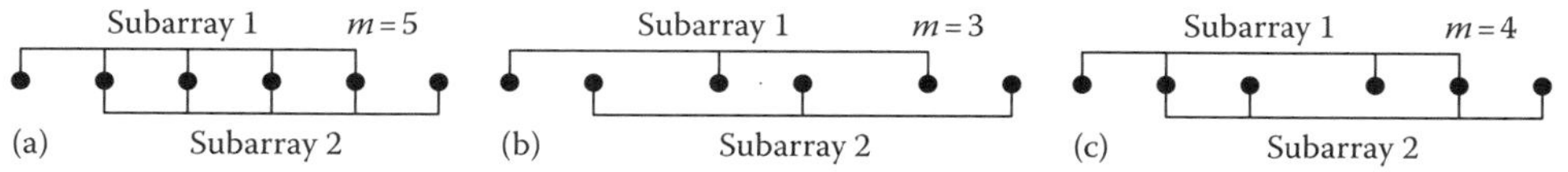

FIGURE 4.2 Three centrosymmetric line arrays of $M = 6$ identical sensors and the corresponding subarrays required for ESPRIT-type algorithms.

* This restriction can be modified later as Unitary ESPRIT can estimate the DOAs of two coherent wave fronts due to an inherent forward–backward averaging effect. Two wave fronts are called "coherent" if their cross-correlation coefficient has magnitude one. The directions of arrival of "more than two" coherent wave fronts can be estimated by using spatial smoothing as a preprocessing step.

† $\theta_k = 0$ corresponds to the direction perpendicular to Δ.

In other words, the narrowband assumption allows the time-delay of the signals across the array τ to be modeled as a simple phase shift of the carrier frequency, such that

$$s_i(t-\tau) \approx \alpha_i(t)e^{j(2\pi f_c(t-\tau)+\beta_i(t))} = e^{-j2\pi f_c \tau} s_i(t).$$

Figure 4.1 shows that the propagation delay of a plane wave signal between the two identical sensors of a doublet equals $\tau_i = \frac{\Delta \sin\theta_i}{c}$, where c denotes the signal propagation velocity. Due to the narrowband assumption (Equation 4.3), this propagation delay τ_i corresponds to the multiplication of the complex envelope signal by the complex exponential $e^{j\mu_i}$, referred to as the phase factor, such that

$$s_i(t-\tau_i) = e^{-j\frac{2\pi f_c}{c}\Delta \sin\theta_i} s_i(t) = e^{j\mu_i} s_i(t), \tag{4.4}$$

where the "spatial frequencies" μ_i are given by $\mu_i = -\frac{2\pi}{\lambda}\Delta \sin\theta_i$. Here, $\lambda = \frac{c}{f_c}$ denotes the common wavelength of the signals. We also assume that there is a one-to-one correspondence between the spatial frequencies $-\pi < \mu_i < \pi$ and the range of possible DOAs. Thus, the maximum range is achieved for $\Delta \leq \lambda/2$. In this case, the DOAs are restricted to the interval $-90° < \theta_i < 90°$ to avoid ambiguities.

In the sequel, the d impinging signals $s_i(t)$, $1 \leq i \leq d$, are combined to a column vector $\mathbf{s}(t)$. Then the noise-corrupted measurements taken at the M sensors at time t obey the linear model

$$\boldsymbol{x}(t) = [\boldsymbol{a}(\mu_1) \quad \boldsymbol{a}(\mu_2) \quad \cdots \quad \boldsymbol{a}(\mu_d)] \begin{bmatrix} s_1(t) \\ s_2(t) \\ \vdots \\ s_d(t) \end{bmatrix} + \boldsymbol{n}(t) = \boldsymbol{A}\boldsymbol{s}(t) + \boldsymbol{n}(t) \in \mathbb{C}^M, \tag{4.5}$$

where the columns of the array steering matrix $\boldsymbol{A} \in \mathbb{C}^{M\times d}$ and the array response or array steering vectors $\boldsymbol{a}(\mu_i)$ are functions of the unknown spatial frequencies μ_i, $1 \leq i \leq d$. For example, for a uniform linear array (ULA) of M identical omnidirectional antennas,

$$\boldsymbol{a}(\mu_i) = e^{-j\left(\frac{M-1}{2}\right)\mu_i}\begin{bmatrix} 1 & e^{j\mu_i} & e^{j2\mu_i} & \cdots & e^{j(M-1)\mu_i} \end{bmatrix}^{\mathrm{T}}, \quad 1 \leq i \leq d.$$

Moreover, the additive noise vector $\boldsymbol{n}(t)$ is taken from a zero-mean, spatially uncorrelated random process with variance σ_N^2, which is also uncorrelated with the signals. Since every row of $\mathbf{A}$ corresponds to an element of the sensor array, a particular subarray configuration can be described by two selection matrices, each choosing m elements of $\boldsymbol{x}(t) \in \mathbb{C}^M$, where m, $d \leq m < M$, is the number of elements in each subarray. Figure 4.2, for example, displays the appropriate subarray choices for three centrosymmetric arrays of $M = 6$ identical sensors. In case of a ULA with maximum overlap, cf. Figure 4.2a, $\boldsymbol{J}_1$ picks the first $m = M-1$ rows of $\boldsymbol{A}$, while $\boldsymbol{J}_2$ selects the last $m = M-1$ rows of the array steering matrix. In this case, the corresponding selection matrices are given by

$$\boldsymbol{J}_1 = \begin{bmatrix} 1 & 0 & 0 & \cdots & 0 & 0 \\ 0 & 1 & 0 & \cdots & 0 & 0 \\ \vdots & \vdots & \vdots & \ddots & \vdots & \vdots \\ 0 & 0 & 0 & \cdots & 1 & 0 \end{bmatrix} \in \mathbb{R}^{m\times M} \quad \text{and} \quad \boldsymbol{J}_2 = \begin{bmatrix} 0 & 1 & 0 & \cdots & 0 & 0 \\ 0 & 0 & 1 & \cdots & 0 & 0 \\ \vdots & \vdots & \vdots & \ddots & \vdots & \vdots \\ 0 & 0 & 0 & \cdots & 0 & 1 \end{bmatrix} \in \mathbb{R}^{m\times M}.$$

Notice that $\boldsymbol{J}_1$ and $\boldsymbol{J}_2$ are centrosymmetric with respect to one another, i.e., they obey $\boldsymbol{J}_2 = \Pi_m \boldsymbol{J}_1 \Pi_M$. This property holds for all centrosymmetric arrays and plays a key role in the derivation of Unitary ESPRIT [8]. Since we have two identical, but physically displaced, subarrays, Equation 4.4 indicates that an array steering vector of the "second" subarray $J_2 a(\mu_i)$ is just a scaled version of the corresponding array steering vector of the "first" subarray $J_1 a(\mu_i)$, namely,

$$J_1 a(\mu_i) e^{j\mu_i} = J_2 a(\mu_i), \quad 1 \leq i \leq d. \tag{4.6}$$

This "shift invariance property" of all d array steering vectors $a(\mu_i)$ may be expressed in compact form as

$$J_1 A \Phi = J_2 A,$$

where

$$\Phi = \mathrm{diag}\{e^{j\mu_i}\}_{i=1}^{d} \tag{4.7}$$

is the unitary diagonal $d \times d$ matrix of the phase factors. All ESPRIT-type algorithms are based on this invariance property of the array steering matrix A, where A is assumed to have full column rank d.

Let X denote an $M \times N$ complex data matrix composed of N snapshots $x(t_n)$, $1 \leq n \leq N$;

$$\begin{aligned} X &= [x(t_1)\, x(t_2) \ldots x(t_N)] \\ &= A[s(t_1)\, s(t_2) \ldots s(t_N)] + [n(t_1)\, n(t_2) \ldots n(t_N)] \\ &= A \cdot S + N \in \mathbb{C}^{M \times N}. \end{aligned} \tag{4.8}$$

The starting point is a singular value decomposition (SVD) of the noise-corrupted data matrix X (direct data approach). Assume that $U_s \in \mathbb{C}^{M \times d}$ contains the d left singular vectors corresponding to the d largest singular values of X. Alternatively, U_s can be obtained via an eigendecomposition of the (scaled) sample covariance matrix XX^{H} (covariance approach). Then $U_s \in \mathbb{C}^{M \times d}$ contains the d eigenvectors corresponding to the d largest eigenvalues of XX^{H}.

Asymptotically, i.e., as the number of snapshots N becomes infinitely large, the range space of U_s is the d-dimensional range space of the array steering matrix A referred to as the "signal subspace." Therefore, there exists a nonsingular $d \times d$ matrix T, such that $A \approx U_s T$. Let us express the shift invariance property (Equation 4.7) in terms of the matrix U_s that spans the estimated signal subspace;

$$J_1 U_s T \Phi \approx J_2 U_s T \Leftrightarrow J_1 U_s \Psi \approx J_2 U_s,$$

where

$$\Psi = T \Phi T^{-1}$$

is a nonsingular $d \times d$ matrix. Since Φ in Equation 4.7 is diagonal, $T\Phi T^{-1}$ is in the form of an eigenvalue decomposition. This implies that $e^{j\mu_i}$, $1 \leq i \leq d$, is the eigenvalue of Ψ. These observations form the basis for the subsequent steps of the algorithm. By applying the two selection matrices to the signal subspace matrix, the following (in general) overdetermined set of equations is formed;

$$J_1 U_s \Psi \approx J_2 U_s \in \mathbb{C}^{m \times d}. \tag{4.9}$$

This set of equations, the so-called invariance equation, is usually solved in the least-squares (LS) or total least-squares (TLS) sense. Notice, however, that Equation 4.9 is highly structured if overlapping subarray configurations are used. Structured least squares (SLS) solves the invariance equation by preserving its structure [7]. Formally, SLS was derived as a linearized iterative solution of a nonlinear optimization problem. If SLS is initialized with the LS solution of the invariance equation, only one "iteration," i.e., the

TABLE 4.1 Summary of the Standard ESPRIT Algorithm

1. *Signal subspace estimation*: Compute $U_s \in \mathbb{C}^{M\times d}$ as the d dominant left singular vectors of $X \in \mathbb{C}^{M\times N}$.
2. *Solution of the invariance equation*: Solve
$$\underbrace{J_1 U_s}_{\mathbb{C}^{m\times d}} \Psi \approx \underbrace{J_2 U_s}_{\mathbb{C}^{m\times d}}$$
by means of LS, TLS, or SLS.
3. *Spatial frequency estimation*: Calculate the eigenvalues of the resulting complex-valued solution
$$\Psi = T\Phi T^{-1} \in \mathbb{C}^{d\times d} \quad \text{with } \Phi = \operatorname{diag}\{\phi_i\}_{i=1}^{d}$$
- $\mu_i = \arg(\phi_i),\ 1 \le i \le d$.

solution of one linear system of equations, is required to achieve a significant improvement of the estimation accuracy [7].

Then an eigendecomposition of the resulting solution $\Psi \in \mathbb{C}^{d\times d}$ may be expressed as

$$\Psi = T\Phi T^{-1} \quad \text{with } \Phi = \operatorname{diag}\{\phi_i\}_{i=1}^{d}. \tag{4.10}$$

The eigenvalues ϕ_i, i.e., the diagonal elements of Φ, represent estimates of the phase factors $e^{j\mu_i}$. Notice that the ϕ_i are not guaranteed to be on the unit circle. Notwithstanding, estimates of the spatial frequencies μ_i and the corresponding DOAs θ_i are obtained via the relationships,

$$\mu_i = \arg(\phi_i) \quad \text{and} \quad \theta_i = \arcsin\left(-\frac{\lambda}{2\pi\Delta}\cdot\mu_i\right), \quad 1 \le i \le d. \tag{4.11}$$

To end this section, a brief summary of the standard ESPRIT algorithm is given in Table 4.1.

4.3 1-D Unitary ESPRIT

In contrast to the standard ESPRIT algorithm, Unitary ESPRIT is efficiently formulated in terms of real-valued computations throughout [8]. It is applicable to centrosymmetric array configurations that possess the discussed invariance structure; cf. Figures 4.1 and 4.2. A sensor array is called "centrosymmetric" [26] if its element locations are symmetric with respect to the centroid. Assuming that the sensor elements have identical radiation characteristics, the array steering matrix of a centrosymmetric array satisfies

$$\Pi_M \bar{A} = A, \tag{4.12}$$

if the array centroid is chosen as the phase reference.

4.3.1 1-D Unitary ESPRIT in Element Space

Before presenting an efficient element space implementation of Unitary ESPRIT, let us define the sparse unitary matrices

$$Q_{2n} = \frac{1}{\sqrt{2}}\begin{bmatrix} I_n & jI_n \\ \Pi_n & -j\Pi_n \end{bmatrix} \quad \text{and} \quad Q_{2n+1} = \frac{1}{\sqrt{2}}\begin{bmatrix} I_n & 0 & jI_n \\ 0^T & \sqrt{2} & 0^T \\ \Pi_n & 0 & -j\Pi_n \end{bmatrix}. \tag{4.13}$$

They are left Π-real matrices of even and odd order, respectively.

Since Unitary ESPRIT involves forward–backward averaging, it can efficiently be formulated in terms of real-valued computations throughout, due to a one-to-one mapping between centro-Hermitian and real matrices [13]. The forward–backward averaged sample covariance matrix is centro-Hermitian and can, therefore, be transformed into a real-valued matrix of the same size; cf. [15] and [8]. A real-valued square root factor of this transformed sample covariance matrix is given by

$$\mathcal{T}(\boldsymbol{X}) = \boldsymbol{Q}_M^{\mathrm{H}}[\boldsymbol{X}\ \boldsymbol{\Pi}_M\ \bar{\boldsymbol{X}}\ \boldsymbol{\Pi}_N]\boldsymbol{Q}_{2N} \in \mathbb{R}^{M\times 2N}, \tag{4.14}$$

where $\boldsymbol{Q}_M$ and $\boldsymbol{Q}_{2N}$ were defined in Equation 4.13.* If M is even, an efficient computation of $\mathcal{T}(\boldsymbol{X})$ from the complex-valued data matrix $\boldsymbol{X}$ only requires $M \times 2N$ real additions and no multiplication [8]. Instead of computing a complex-valued SVD as in the standard ESPRIT case, the signal subspace estimate is obtained via a real-valued SVD of $\mathcal{T}(\boldsymbol{X})$ (direct data approach). Let $\boldsymbol{E}_s \in \mathbb{R}^{M\times d}$ contain the d left singular vectors corresponding to the d largest singular values of $\mathcal{T}(\boldsymbol{X})$.† Then the columns of

$$\boldsymbol{U}_s = \boldsymbol{Q}_M \boldsymbol{E}_s \tag{4.15}$$

span the estimated signal subspace, and spatial frequency estimates could be obtained from the eigenvalues of the complex-valued matrix $\boldsymbol{\Psi}$ that solves Equation 4.9. These complex-valued computations, however, are not required, since the transformed array steering matrix

$$\boldsymbol{D} = \boldsymbol{Q}_M^{\mathrm{H}}\boldsymbol{A} = [\,\boldsymbol{d}(\mu_1)\quad \boldsymbol{d}(\mu_2)\quad \cdots\quad \boldsymbol{d}(\mu_d)\,] \in \mathbb{R}^{M\times d} \tag{4.16}$$

satisfies the following shift invariance property

$$\boldsymbol{K}_1\boldsymbol{D}\boldsymbol{\Omega} = \boldsymbol{K}_2\boldsymbol{D}, \quad \text{where } \boldsymbol{\Omega} = \operatorname{diag}\left\{\tan\left(\frac{\mu_i}{2}\right)\right\}_{i=1}^{d} \tag{4.17}$$

and the transformed selection matrices $\boldsymbol{K}_1$ and $\boldsymbol{K}_2$ are given by

$$\boldsymbol{K}_1 = 2\cdot \operatorname{Re}\left\{\boldsymbol{Q}_m^{\mathrm{H}}\boldsymbol{J}_2\ \boldsymbol{Q}_M\right\} \quad \text{and} \quad \boldsymbol{K}_2 = 2\cdot \operatorname{Im}\left\{\boldsymbol{Q}_m^{\mathrm{H}}\boldsymbol{J}_2\boldsymbol{Q}_M\right\}. \tag{4.18}$$

Here, Re $\{\cdot\}$ and Im $\{\cdot\}$ denote the real and the imaginary part, respectively. Notice that Equation 4.17 is similar to Equation 4.7 except for the fact that all matrices in Equation 4.17 are real-valued.

Let us take a closer look at the transformed selection matrices defined in Equation 4.18. If $\boldsymbol{J}_2$ is sparse, $\boldsymbol{K}_1$ and $\boldsymbol{K}_2$ are also sparse. This is illustrated by the following example. For the ULA with $M = 6$ sensors and maximum overlap sketched in Figure 4.2a, $\boldsymbol{J}_2$ is given by

$$\boldsymbol{J}_2 = \begin{bmatrix} 0 & 1 & 0 & 0 & 0 & 0 \\ 0 & 0 & 1 & 0 & 0 & 0 \\ 0 & 0 & 0 & 1 & 0 & 0 \\ 0 & 0 & 0 & 0 & 1 & 0 \\ 0 & 0 & 0 & 0 & 0 & 1 \end{bmatrix} \in \mathbb{R}^{5\times 6}.$$

According to Equation 4.18, straightforward calculations yield the transformed selection matrices

$$\boldsymbol{K}_1 = \begin{bmatrix} 1 & 1 & 0 & 0 & 0 & 0 \\ 0 & 1 & 1 & 0 & 0 & 0 \\ 0 & 0 & \sqrt{2} & 0 & 0 & 0 \\ 0 & 0 & 0 & 1 & 1 & 0 \\ 0 & 0 & 0 & 0 & 1 & 1 \end{bmatrix} \quad \text{and} \quad \boldsymbol{K}_2 = \begin{bmatrix} 0 & 0 & 0 & -1 & 1 & 0 \\ 0 & 0 & 0 & 0 & -1 & 1 \\ 0 & 0 & 0 & 0 & 0 & -\sqrt{2} \\ 1 & -1 & 0 & 0 & 0 & 0 \\ 0 & 1 & -1 & 0 & 0 & 0 \end{bmatrix}.$$

In this case, applying $\boldsymbol{K}_1$ or $\boldsymbol{K}_2$ to $\boldsymbol{E}_s$ only requires $(m-1)d$ real additions and d real multiplications.

* The results of this chapter also hold if $\boldsymbol{Q}_M$ and $\boldsymbol{Q}_{2N}$ denote arbitrary left Π-real matrices that are also unitary.

† Alternatively, $\boldsymbol{E}_s$ can be obtained through a real-valued eigendecomposition of $\mathcal{T}(X)\mathcal{T}(X)^{\mathrm{H}}$ (covariance approach).

TABLE 4.2 Summary of 1-D Unitary ESPRIT in Element Space

1. *Signal subspace estimation*: Compute $\boldsymbol{E}_s \in \mathbb{R}^{M\times d}$ as the d dominant left singular vectors of $\mathcal{T}(\boldsymbol{X}) \in \mathbb{R}^{M\times 2N}$.
2. *Solution of the invariance equation*: Then solve

$$\underbrace{\boldsymbol{K}_1\boldsymbol{E}_s}_{\mathbb{R}^{m\times d}}\Upsilon \approx \underbrace{\boldsymbol{K}_2\boldsymbol{E}_s}_{\mathbb{R}^{m\times d}}$$

by means of LS, TLS, or SLS.

3. *Spatial frequency estimation*: Calculate the eigenvalues of the resulting real-valued solution

$$\Upsilon = \boldsymbol{T}\Omega\boldsymbol{T}^{-1} \in \mathbb{R}^{d\times d} \quad \text{with } \Omega = \text{diag}\{\omega_i\}_{i=1}^{d}$$

- $\mu_i = 2\arctan(\omega_i)$, $1 \le i \le d$.

Asymptotically, the real-valued matrices $\boldsymbol{E}_s$ and $\boldsymbol{D}$ span the same d-dimensional subspace, i.e., there is a nonsingular matrix $\boldsymbol{T} \in \mathbb{R}^{d\times d}$ such that $\boldsymbol{D} \approx \boldsymbol{E}_s\boldsymbol{T}$. Substituting this into Equation 4.17, yields the real-valued invariance equation

$$\boldsymbol{K}_1\boldsymbol{E}_s\Upsilon \approx \boldsymbol{K}_2\boldsymbol{E}_s \in \mathbb{R}^{m\times d}, \quad \text{where } \Upsilon = \boldsymbol{T}\Omega\boldsymbol{T}^{-1}. \tag{4.19}$$

Thus, the eigenvalues of the solution $\Upsilon \in \mathbb{R}^{d\times d}$ to the matrix equation above are

$$\omega_i = \tan\left(\frac{\mu_i}{2}\right) = \frac{1}{j}\frac{e^{j\mu_i}-1}{e^{j\mu_i}+1}, \quad 1 \le i \le d. \tag{4.20}$$

This reveals a spatial frequency warping identical to the temporal frequency warping incurred in designing a digital filter from an analog filter via the bilinear transformation. Consider $\Delta = \frac{\lambda}{2}$ so that $\mu_i = -\frac{2\pi}{\lambda}\Delta\sin\theta_i = -\pi\sin\theta_i$. In this case, there is a one-to-one mapping between $-1 < \sin\theta_i < 1$, corresponding to the range of possible values for the DOAs $-90° < \theta_i < 90°$, and $-\infty < \omega_i < \infty$.

Note that the fact that the eigenvalues of a real matrix have to either be real-valued or occur in complex conjugate pairs gives rise to an ad hoc "reliability test." That is, if the final step of the algorithm yields a complex conjugate pair of eigenvalues, then either the SNR is too low, not enough snapshots have been averaged, or two corresponding signal arrivals have not been resolved. In the latter case, taking the tangent inverse of the real part of the eigenvalues can sometimes provide a rough estimate of the DOA of the two closely spaced signals. In general, though, if the algorithm yields one or more complex conjugate pairs of eigenvalues in the final stage, the estimates should be viewed as unreliable. In this case, pseudonoise resampling may be used to improve the estimation accuracy [5].

The element space implementation of 1-D Unitary ESPRIT is summarized in Table 4.2.

4.3.2 1-D Unitary ESPRIT in DFT Beamspace

Reduced dimension processing in beamspace, yielding reduced computational complexity, is an option when one has a priori information on the general angular locations of the incident signals, as in a radar application, for example. In the case of a ULA, transformation from element space to discrete Fourier transform (DFT) beamspace may be effected by premultiplying the data by those rows of the DFT matrix forming beams encompassing the sector of interest. (Each row of the DFT matrix forms a beam pointed to a different angle.) If there is no a priori information, one may examine the DFT spectrum and apply Unitary ESPRIT in DFT beamspace to a small set of DFT values around each spectral peak above a particular threshold. In a more general setting, Unitary ESPRIT in DFT beamspace can simply be applied via parallel processing to each of a number of sets of successive DFT values corresponding to overlapping sectors.

Note, though, that in the development to follow, we will initially employ all M DFT beams for the sake of notational simplicity. Without loss of generality, we consider an omnidirectional ULA. Let $\boldsymbol{W}_M^{\mathrm{H}} \in \mathbb{C}^{M\times M}$ be the scaled M-point DFT matrix with its M rows given by

$$\boldsymbol{w}_k^{\mathrm{H}} = \mathrm{e}^{j\left(\frac{M-1}{2}\right)k\frac{2\pi}{M}}\left[1 \quad \mathrm{e}^{-jk\frac{2\pi}{M}} \quad \mathrm{e}^{-j2k\frac{2\pi}{M}} \quad \cdots \quad \mathrm{e}^{-j(M-1)k\frac{2\pi}{M}}\right], \quad 0 \le k \le (M-1). \tag{4.21}$$

Notice that $\boldsymbol{W}_M$ is left $\boldsymbol{\Pi}$-real or column conjugate symmetric, i.e., $\boldsymbol{\Pi}_M \bar{\boldsymbol{W}}_M = \boldsymbol{W}_M$. Thus, as pointed out for $\boldsymbol{D}$ in Equation 4.16, the transformed steering matrix of the ULA

$$\boldsymbol{B} = \boldsymbol{W}_M^{\mathrm{H}}\boldsymbol{A} = \left[\boldsymbol{b}(\mu_1) \quad \boldsymbol{b}(\mu_2) \quad \cdots \quad \boldsymbol{b}(\mu_d)\right] \in \mathbb{R}^{M\times d} \tag{4.22}$$

is real-valued. It has been shown in [27] that $\boldsymbol{B}$ satisfies a shift invariance property which is similar to Equation 4.17, namely

$$\boldsymbol{\Gamma}_1\boldsymbol{B}\boldsymbol{\Omega} = \boldsymbol{\Gamma}_2\boldsymbol{B}, \quad \text{where } \boldsymbol{\Omega} = \operatorname{diag}\left\{\tan\left(\frac{\mu_i}{2}\right)\right\}_{i=1}^{d}. \tag{4.23}$$

Here, the selection matrices $\boldsymbol{\Gamma}_1$ and $\boldsymbol{\Gamma}_2$ of size $M \times M$ are defined as

$$\boldsymbol{\Gamma}_1 = \begin{bmatrix} 1 & \cos\left(\frac{\pi}{M}\right) & 0 & 0 & \cdots & 0 & 0 \\ 0 & \cos\left(\frac{\pi}{M}\right) & \cos\left(\frac{2\pi}{M}\right) & 0 & \cdots & 0 & 0 \\ 0 & 0 & \cos\left(\frac{2\pi}{M}\right) & \cos\left(\frac{3\pi}{M}\right) & \cdots & 0 & 0 \\ \vdots & \vdots & \vdots & \vdots & \cdot & \vdots & \vdots \\ 0 & 0 & 0 & 0 & \cdots & \cos\left((M-2)\frac{\pi}{M}\right) & \cos\left((M-1)\frac{\pi}{M}\right) \\ (-1)^M & 0 & 0 & 0 & \cdots & 0 & \cos\left((M-1)\frac{\pi}{M}\right) \end{bmatrix}, \tag{4.24}$$

$$\boldsymbol{\Gamma}_2 = \begin{bmatrix} 0 & \sin\left(\frac{\pi}{M}\right) & 0 & 0 & \cdots & 0 & 0 \\ 0 & \sin\left(\frac{\pi}{M}\right) & \sin\left(\frac{2\pi}{M}\right) & 0 & \cdots & 0 & 0 \\ 0 & 0 & \sin\left(\frac{2\pi}{M}\right) & \sin\left(\frac{3\pi}{M}\right) & \cdots & 0 & 0 \\ \vdots & \vdots & \vdots & \vdots & \cdot & \vdots & \vdots \\ 0 & 0 & 0 & 0 & \cdots & \sin\left((M-2)\frac{\pi}{M}\right) & \sin\left((M-1)\frac{\pi}{M}\right) \\ 0 & 0 & 0 & 0 & \cdots & 0 & \sin\left((M-1)\frac{\pi}{M}\right) \end{bmatrix}. \tag{4.25}$$

As an alternative to Equation 4.14, another real-valued square root factor of the transformed sample covariance matrix is given by

$$[\operatorname{Re}\{\boldsymbol{Y}\} \quad \operatorname{Im}\{\boldsymbol{Y}\}] \in \mathbb{R}^{M\times 2N}, \quad \text{where } \boldsymbol{Y} = \boldsymbol{W}_M^{\mathrm{H}}\boldsymbol{X} \in \mathbb{C}^{M\times N}. \tag{4.26}$$

Here, $\boldsymbol{Y}$ can efficiently be computed via an fast Fourier transform (FFT), which exploits the Vandermonde form of the rows of the DFT matrix, followed by an appropriate scaling, cf. Equation 4.21. Let the columns of $\boldsymbol{E}_s \in \mathbb{R}^{M\times d}$ contain the d left singular vectors corresponding to the d largest singular values of Equation 4.26. Asymptotically, the real-valued matrices $\boldsymbol{E}_s$ and $\boldsymbol{B}$ span the same d-dimensional subspace, i.e., there is a nonsingular matrix $\boldsymbol{T} \in \mathbb{R}^{d\times d}$, such that $\boldsymbol{B} \approx \boldsymbol{E}_s\boldsymbol{T}$. Substituting this into Equation 4.23 yields the real-valued invariance equation

$$\boldsymbol{\Gamma}_1\boldsymbol{E}_s\Upsilon \approx \boldsymbol{\Gamma}_2\boldsymbol{E}_s \in \mathbb{R}^{M\times d}, \quad \text{where } \Upsilon = \boldsymbol{T\Omega T}^{-1}. \tag{4.27}$$

Thus, the eigenvalues of the solution $\Upsilon \in \mathbb{R}^{d\times d}$ to the matrix equation above are also given by Equation 4.20.

It is a crucial observation that one row of the matrix Equation 4.23 relates to "two successive components" of the transformed array steering vectors $\boldsymbol{b}(\mu_i)$, cf. Equations 4.24 and 4.25. This insight enables us to apply only $B \ll M$ successive rows of $\boldsymbol{W}_M^{\mathrm{H}}$ (instead of all M rows) to the data matrix $\boldsymbol{X}$ in Equation 4.26. To stress the reduced number of rows, we call the resulting beamforming matrix $\boldsymbol{W}_B^{\mathrm{H}} \in \mathbb{C}^{B\times M}$. The number of its rows B depends on the width of the sector of interest and may be substantially less than the number of sensors M. Thereby, the SVD of Equation 4.26 and, therefore, also $\boldsymbol{E}_s \in \mathbb{R}^{B\times d}$ and the invariance Equation 4.27 will have a reduced dimensionality. Employing the appropriate subblocks of Γ_1 and Γ_2 as selection matrices, the algorithm is the same as the one described previously except for its reduced dimensionality. In the sequel, the resulting selection matrices of size $(B-1)\times B$ will be called $\boldsymbol{\Gamma}_1^{(B)}$ and $\boldsymbol{\Gamma}_2^{(B)}$. The whole algorithm, that operates in a B-dimensional DFT beamspace, is summarized in Table 4.3.

Consider, for example, a ULA of $M=8$ sensors. The structure of the corresponding selection matrices Γ_1 and Γ_2 is sketched in Figure 4.3. Here, the symbol $\times$ denotes entries of both selection matrices that might be nonzero, cf. Equations 4.24 and 4.25. If one employed rows 4, 5, and 6 of $\boldsymbol{W}_8^{\mathrm{H}}$ to form $B=3$ beams in estimating the DOAs of two closely spaced signal arrivals, as in the low-angle radar tracking scheme described by Zoltowski and Lee [29], the corresponding 2×3 subblock of the selection matrices Γ_1 and Γ_2 is shaded in Figure 4.3a.* Notice that the first and the last (Mth) row of $\boldsymbol{W}_M^{\mathrm{H}}$ steer beams that are also physically adjacent to one another (the wraparound property of the DFT). If, for example, one employed rows 8, 1, and 2 of $\boldsymbol{W}_8^{\mathrm{H}}$ to form $B=3$ beams in estimating the DOAs of two closely spaced signal arrivals, the corresponding subblocks of the selection matrices Γ_1 and Γ_2 are shaded in Figure 4.3b.†

TABLE 4.3 Summary of 1-D Unitary ESPRIT in DFT Beamspace

0. *Transformation to beamspace*: $\boldsymbol{Y} = \boldsymbol{W}_B^{\mathrm{H}}\boldsymbol{X} \in \mathbb{C}^{B\times N}$
1. *Signal subspace estimation*: Compute $\boldsymbol{E}_s \in \mathbb{R}^{B\times d}$ as the d dominant left singular vectors of $[\mathrm{Re}\{\boldsymbol{Y}\} \quad \mathrm{Im}\{\boldsymbol{Y}\}] \in \mathbb{R}^{B\times 2N}$.
2. *Solution of the invariance equation*: Solve

$$\underbrace{\boldsymbol{\Gamma}_1^{(B)}\boldsymbol{E}_s}_{\mathbb{R}^{(B-1)\times d}}\Upsilon \approx \underbrace{\boldsymbol{\Gamma}_2^{(B)}\boldsymbol{E}_s}_{\mathbb{R}^{(B-1)\times d}}$$

by means of LS, TLS, or SLS.

3. *Spatial frequency estimation*: Calculate the eigenvalues of the resulting real-valued solution

$$\Upsilon = \boldsymbol{T\Omega T}^{-1} \in \mathbb{R}^{d\times d} \quad \text{with } \boldsymbol{\Omega} = \mathrm{diag}\{\omega_i\}_{i=1}^{d}$$

- $\mu_i = 2\arctan(\omega_i)$, $1 \le i \le d$.

* Here, the first row of $\Gamma_1^{(3)}$ and $\Gamma_2^{(3)}$ combines beams 4 and 5, while the second row of $\Gamma_1^{(3)}$ and $\Gamma_2^{(3)}$ combines beams 5 and 6.

† Here, the first row of $\boldsymbol{\Gamma}_1^{(3)}$ and $\boldsymbol{\Gamma}_2^{(3)}$ combines beams 1 and 2, while the second row of $\boldsymbol{\Gamma}_1^{(3)}$ and $\boldsymbol{\Gamma}_2^{(3)}$ combines beams 1 and 8.

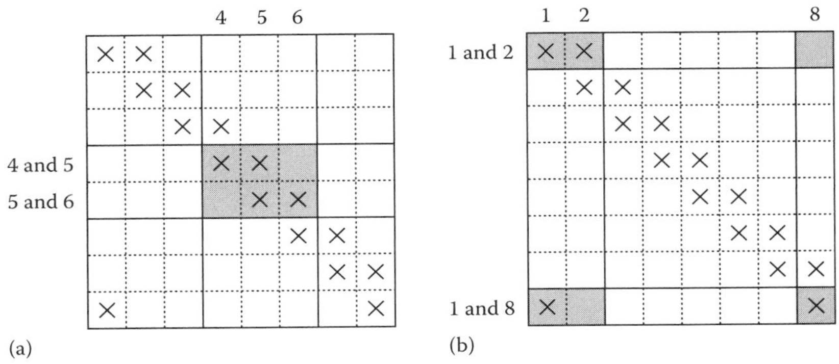

FIGURE 4.3 Structure of the selection matrices Γ_1 and Γ_2 for a ULA of $M = 8$ sensors. The symbol $\times$ denotes entries of both selection matrices that might be nonzero. The shaded areas illustrate how to choose the appropriate subblocks of the selection matrices for reduced dimension processing, i.e., how to form $\boldsymbol{\Gamma}_1^{(B)}$ and $\boldsymbol{\Gamma}_2^{(B)}$, if only $B = 3$ successive rows of $\boldsymbol{W}_8^{\mathrm{H}}$ are applied to the data matrix $\boldsymbol{X}$. Here, the following two examples are used: (a) rows 4, 5, and 6 and (b) rows 8, 1, and 2.

Cophasal beamforming using the DFT weight vectors defined in Equation 4.21 yield array patterns having peak sidelobe levels of −13.5 dB relative to the main lobe. Thus when scanning a sector of space (for example, the sector spanned by beams 4, 5, and 6 as outlined above), strong sources lying outside the sector can leak in due to the relatively high sidelobe levels. Reference [21] describes how the algorithm can be modified to employ cosine or Hanning windows, providing peak sidelobe levels of −23.5 and −32.3 dB, respectively. The improved suppression of out-of-band sources allows for more effective parallel searches for sources in different spatial sectors.

4.4 UCA-ESPRIT for Circular Ring Arrays

Uniform circular array (UCA)-ESPRIT [18–20] is a 2-D angle estimation algorithm developed for use with UCAs. The algorithm provides automatically paired azimuth and elevation angle estimates of far-field signals incident on the UCA via a closed-form procedure. The rotational symmetry of the UCA makes it desirable for a variety of applications where one needs to discriminate in both azimuth and elevation, as opposed to just conical angle of arrival which is all the ULA can discriminate upon. For example, UCAs are commonly employed as part of an antijam spatial filter for global positioning system (GPS) receivers. Some experimental UCA-based systems are described in [4]. The development of a closed-form 2-D angle estimation technique for a UCA provides further motivation for the use of a UCA in a given application.

Consider an M-element UCA in which the array elements are uniformly distributed over the circumference of a circle of radius R. We will assume that the array is located in the x–y plane, with its center at the origin of the coordinate system. The elevation angles θ_i and azimuth angles ϕ_i of the d impinging sources are defined in Figure 4.4, as are the direction cosines u_i and v_i, $1 \leq i \leq d$. UCA-ESPRIT is premised on phase-mode-excitation-based beamforming. The maximum phase-mode (integer-valued) excitable by a given UCA is

$$K \approx \frac{2\pi R}{\lambda},$$

where λ is the common (carrier) wavelength of the incident signals. Phase-mode-excitation-based beamforming requires $M > 2K$ array elements ($M = 2K + 3$ is usually adequate). UCA-ESPRIT can resolve a maximum of $d_{\max} = K - 1$ sources. As an example, if the array radius is $R = \lambda$, $K = 6$ (the largest integer

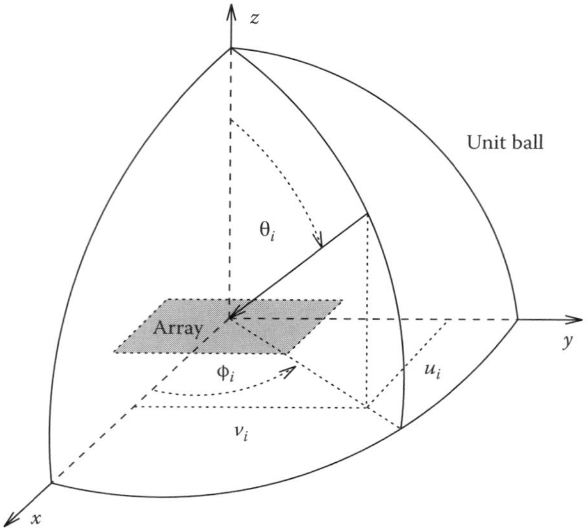

FIGURE 4.4 Definitions of azimuth ($-180° < \phi_i \leq 180°$) and elevation ($0° \leq \theta_i \leq 90°$). The direction cosines u_i and v_i are the rectangular coordinates of the projection of the corresponding point on the unit ball onto the equatorial plane.

smaller than 2π) and at least $M = 15$ array elements are needed. UCA-ESPRIT can resolve five sources in conjunction with this UCA.

UCA-ESPRIT operates in a $K' = 2K + 1$-dimensional beamspace. It employs a $K' \times M$ beamforming matrix to transform from element space to beamspace. After this transformation, the algorithm has the same three basic steps of any ESPRIT-type algorithm: (1) the computation of a basis for the signal

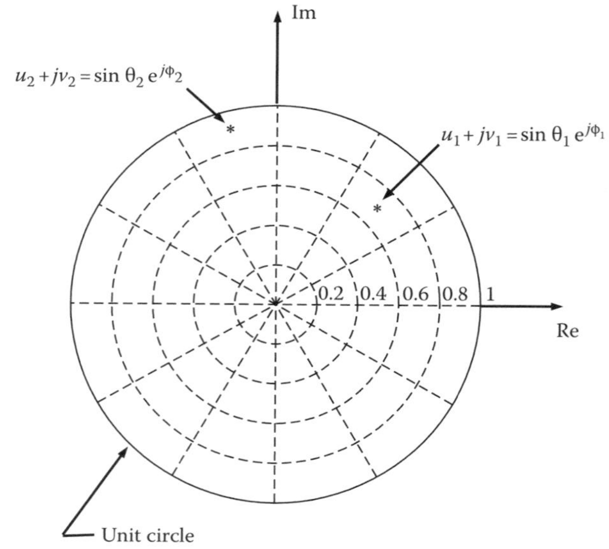

FIGURE 4.5 Illustrating the form of signal roots (eigenvalues) obtained with UCA-ESPRIT or FCA-ESPRIT.

subspace, (2) the solution to an (in general) overdetermined system of equations derived from the matrix of vectors spanning the signal subspace, and (3) the computation of the eigenvalues of the solution to the system of equations formed in step (2). As illustrated in Figure 4.5, the ith eigenvalue obtained in the final step is ideally of the form $\xi_i = \sin\theta_i e^{j\phi_i}$, where ϕ_i and θ_i are the azimuth and elevation angles of the ith source. Note that

$$\xi_i = \sin\theta_i e^{j\phi_i} = u_i + jv_i, \quad 1 \leq i \leq d,$$

where u_i and v_i are the direction cosines of the ith source relative to the x- and y-axis, respectively, as indicated in Figure 4.4.

The formulation of UCA-ESPRIT is based on the special structure of the resulting K'-dimensional beamspace manifold. The following vector and matrix definitions are needed to summarize the algorithm in Table 4.4.

$$\boldsymbol{v}_k^{\mathrm{H}} = \frac{1}{M}\left[1 \quad e^{jk\frac{2\pi}{M}} \quad e^{j2k\frac{2\pi}{M}} \quad \ldots \quad e^{j(M-1)k\frac{2\pi}{M}}\right], \tag{4.28}$$

$$\begin{aligned}
\boldsymbol{V} &= \sqrt{M}[\boldsymbol{v}_{-K} \quad \ldots \quad \boldsymbol{v}_{-1} \quad \boldsymbol{v}_0 \quad \boldsymbol{v}_1 \quad \ldots \quad \boldsymbol{v}_K] \in \mathbb{C}^{M\times K'},\\
\boldsymbol{C}_v &= \mathrm{diag}\{j^{-|k|}\}_{k=-K}^{K} \in \mathbb{C}^{K'\times K'},\\
\boldsymbol{F}_r^{\mathrm{H}} &= \boldsymbol{Q}_{K'}^{\mathrm{T}}\boldsymbol{C}_v\boldsymbol{V}^{\mathrm{H}} \in \mathbb{C}^{K'\times M},\\
\boldsymbol{C}_o &= \mathrm{diag}\{\mathrm{sign}(k)^{-k}\}_{k=-K}^{K} \in \mathbb{R}^{K'\times K'},\\
\boldsymbol{D} &= \mathrm{diag}\{(-1)^{|k|}\}_{k=-(K-2)}^{K} \in \mathbb{R}^{(K'-2)\times(K'-2)},\\
\boldsymbol{\Gamma} &= \frac{\lambda}{\pi R}\cdot \mathrm{diag}\{k\}_{k=-(K-1)}^{(K-1)} \in \mathbb{R}^{(K'-2)\times(K'-2)}.
\end{aligned} \tag{4.29}$$

Note that the columns of the matrix $\boldsymbol{V}$ consist of the DFT weight vectors $\boldsymbol{v}_k$ defined in Equation 4.28. The beamforming matrix $\boldsymbol{F}_r^{\mathrm{H}}$ in Equation 4.29 synthesizes a real-valued beamspace manifold and facilitates signal subspace estimation via a real-valued SVD or eigendecomposition. Recall that sparse left Π-real

TABLE 4.4 Summary of UCA-ESPRIT

0. *Transformation to beamspace*: $\boldsymbol{Y} = \boldsymbol{F}_r^{\mathrm{H}}\boldsymbol{X} \in \mathbb{C}^{K'\times N}$
1. *Signal subspace estimation*: Compute $\boldsymbol{E}_s \in \mathbb{R}^{K'\times d}$ as the d dominant left singular vectors of $[\mathrm{Re}\{\boldsymbol{Y}\} \quad \mathrm{Im}\{\boldsymbol{Y}\}] \in \mathbb{R}^{K'\times 2N}$.
2. *Solution of the invariance equation*:
   - Compute $\boldsymbol{E}_u = \boldsymbol{C}_o\bar{\boldsymbol{Q}}_{K'}\boldsymbol{E}_s$. Form the matrix $\boldsymbol{E}_{-1}$ that consists of all but the last two rows of $\boldsymbol{E}_u$. Similarly form the matrix $\boldsymbol{E}_0$ that consists of all but the first and last rows of $\boldsymbol{E}_u$.
   - Compute $\underline{\boldsymbol{\Psi}} \in \mathbb{C}^{2d\times d}$, the LS solution to the system
   $$[\boldsymbol{E}_{-1} \quad \boldsymbol{D\Pi}_{(K'-2)}\bar{\boldsymbol{E}}_{-1}]\underline{\boldsymbol{\Psi}} = \boldsymbol{\Gamma}\boldsymbol{E}_0 \in \mathbb{C}^{(K'-2)\times d}.$$
   Recall that the overbar denotes complex conjugation. Form $\boldsymbol{\Psi}$ by extracting the upper $d\times d$ block from $\underline{\boldsymbol{\Psi}}$. Note that Ψ can be efficiently computed by solving a "real-valued" system of $2d$ equations (see [20]).
3. *Spatial frequency estimation*: Compute the eigenvalues ξ_i, $1 \leq i \leq d$, of $\boldsymbol{\Psi} \in \mathbb{C}^{d\times d}$. The estimates of the elevation and azimuth angles of the ith source are
   $$\theta_i = \arcsin(|\xi_i|) \quad \text{and} \quad \phi_i = \arg(\xi_i),$$
   respectively. If direction cosine estimates are desired, we have
   $$u_i = \mathrm{Re}\{\xi_i\} \quad \text{and} \quad v_i = \mathrm{Im}\{\xi_i\}.$$
   Again, ξ_i can be efficiently computed via a "real-valued" eigenvalue decomposition (EVD) (see [20]).

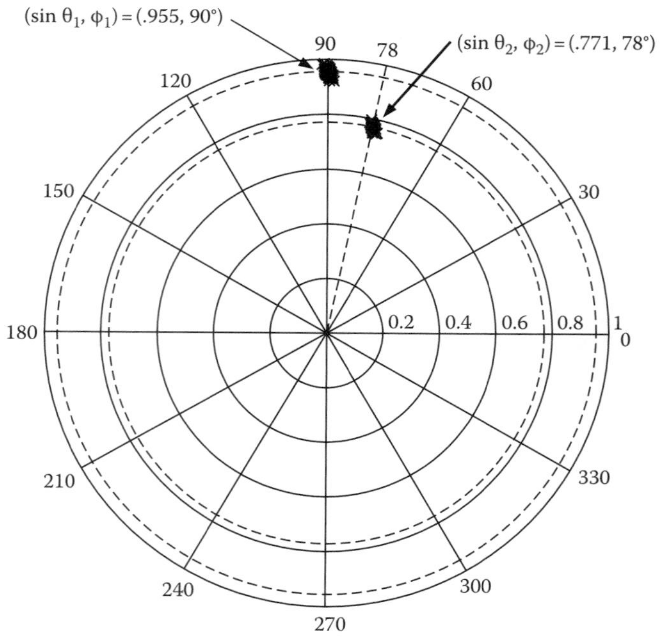

FIGURE 4.6 Plot of the UCA-ESPRIT eigenvalues $\xi_1 = \sin\theta_1 e^{j\phi_1}$ and $\xi_2 = \sin\theta_2 e^{j\phi_2}$ for 200 trials.

matrix $\boldsymbol{Q}_{K'} \in \mathbb{C}^{K' \times K'}$ has been defined in Equation 4.13. The complete UCA-ESPRIT algorithm is summarized in Table 4.4.

4.4.1 Results of Computer Simulations

Simulations were conducted with a UCA of radius $R = \lambda$, with $K = 6$ and $M = 19$ (performance close to that reported below can be expected even if $M = 15$ elements are employed). The simulation employed two sources with arrival angles given by $(\theta_1, \phi_1) = (72.73°, 90°)$ and $(\theta_2, \phi_2) = (50.44°, 78°)$. The sources were highly correlated, with the correlation coefficient referred to the center of the array being $0.9e^{j\frac{\pi}{4}}$. The SNR was 10 dB (per array element) for each source. The number of snapshots was $N = 64$ and arrival angle estimates were obtained for 200 independent trials. Figure 4.6 depicts the results of the simulation. Here, the UCA-ESPRIT eigenvalues ξ_i are denoted by the symbol $\times$.* The results from all 200 trials are superimposed in the figure. The eigenvalues are seen to be clustered around the expected locations (the dashed circles indicate the true elevation angles).

4.5 FCA-ESPRIT for Filled Circular Arrays

The use of a circular ring array and the attendant use of UCA-ESPRIT are ideal for applications where the array aperture is not very large as on the top of a mobile communications unit. For much larger array apertures as in phased array surveillance radars, too much of the aperture is devoid of elements so that a lot of the signal energy impinging upon the aperture is not intercepted. As an example, each of the four

* The horizontal axis represents $\text{Re}\{\xi_i\}$, and the vertical axis represents $\text{Im}\{\xi_i\}$.

panels comprising either the SPY-1A or SPY-1B radars of the AEGIS series is composed of 4400 identical elements regularly spaced on a flat panel over a circular aperture [24]. The sampling lattice is hexagonal. Recent prototype arrays for satellite-based communications have also employed the filled circular array (FCA) geometry [2].

This section presents an algorithm similar to UCA-ESPRIT that provides the same closed-form 2-D angle estimation capability for a "FCA." Similar to UCA-ESPRIT, the far-field pattern arising from the sampled excitation is approximated by the far-field pattern arising from the continuous excitation from which the sampled excitation is derived through sampling. (Note, Steinberg [25] shows that the array pattern for a ULA of N elements with interelement spacing d is nearly identical to the far-field pattern for a continuous linear aperture of length $(N+1)d$, except near the fringes of the visible region.) That is, it is assumed that the interelement spacings have been chosen so that aliasing effects are negligible as in the generation of phase modes with a single ring array. It can be shown that this is the case for any sampling lattice as long as the intersensor spacings are roughly half a wavelength or less on the average and that the sources of interest are at least 20° in elevation above the plane of the array, i.e., we require that the elevation angle of the ith source satisfies $0^\circ \leq \theta_i \leq 70^\circ$. In practice, many phased arrays only provide reliable coverage for $0^\circ \leq \theta_i \leq 60^\circ$ (plus or minus 60° away from boresite) due to a reduced aperture effect and the fact that the gain of each individual antenna has a significant roll-off at elevation angles near the horizon, i.e., the plane of the array. FCA-ESPRIT has been successfully applied to rectangular, hexagonal, polar raster, and random-sampling lattices.

The key to the development of UCA-ESPRIT was phase-mode (DFT) excitation and exploitation of a recurrence relationship that Bessel functions satisfy. In the case of a FCA, the same type of processing is facilitated by the use of a phase-mode-dependent aperture taper derived from an integral relationship that Bessel functions satisfy.

Consider an M-element FCA where the array elements are distributed over a circular aperture of radius R. We assume that the array is centered at the origin of the coordinate system and contained in the x–y plane. The ith element is located at a radial distance r_i from the origin and at an angle γ_i relative to the x-axis measured counterclockwise in the x–y plane. In contrast to a UCA, $0 \leq r_i \leq R$. The beamforming weight vectors employed in FCA-ESPRIT are

$$\boldsymbol{w}_m = \frac{1}{M}\begin{bmatrix} A_1\left(\frac{r_1}{R}\right)^{|m|} e^{-jm\gamma_1} \\ \vdots \\ A_i\left(\frac{r_i}{R}\right)^{|m|} e^{-jm\gamma_i} \\ \vdots \\ A_M\left(\frac{r_M}{R}\right)^{|m|} e^{-jm\gamma_M} \end{bmatrix}, \tag{4.30}$$

where m ranges from $-K$ to K with $K = \frac{2\pi R}{\lambda}$. Here A_i is proportional to the area surrounding the ith array element. A_i is a constant (and can be omitted) for hexagonal and rectangular lattices and proportional to the radius ($A_i = r_i$) for a polar raster. The transformation from element space to beamspace is effected through premultiplication by the beamforming matrix

$$\boldsymbol{W} = \sqrt{M}[\boldsymbol{w}_{-K} \quad \ldots \quad \boldsymbol{w}_{-1} \quad \boldsymbol{w}_0 \quad \boldsymbol{w}_1 \quad \ldots \quad \boldsymbol{w}_K] \in \mathbb{C}^{M\times K'} (K' = 2K+1). \tag{4.31}$$

The following matrix definitions are needed to summarize FCA-ESPRIT.

TABLE 4.5 Summary of FCA-ESPRIT

0. *Transformation to beamspace*: $Y = B_r^H X$
1. *Signal subspace estimation*: Compute $E_s \in \mathbb{R}^{K' \times d}$ as the d dominant left singular vectors of $[\mathrm{Re}\{Y\} \quad \mathrm{Im}\{Y\}] \in \mathbb{R}^{K' \times 2N}$.
2. *Solution of the invariance equation*:
   - Compute $E_u = FQ_{K'}E_s$. Form the matrices E_{-1}, E_0, and E_1 that consist of all but the last two, first and last, and first two rows, respectively.
   - Compute $\underline{\Psi} \in \mathbb{C}^{2d \times d}$, the LS solution to the system

$$[E_{-1} \; C_1 \; E_1] \; \underline{\Psi} = \Gamma E_0 \in \mathbb{C}^{(K'-2) \times d}.$$

   Form Ψ by extracting the upper $d \times d$ block from $\underline{\Psi}$.
3. *Spatial frequency estimation*: Compute the eigenvalues ξ_i, $1 \le i \le d$, of $\Psi \in \mathbb{C}^{d \times d}$. The estimates of the elevation and azimuth angles of the ith source are

$$\theta_i = \arcsin(|\xi_i|) \quad \text{and} \quad \phi_i = \arg(\xi_i),$$

respectively.

$$\begin{aligned}
B &= WC \in \mathbb{C}^{M \times K'}, \\
C &= \mathrm{diag}\{\mathrm{sign}(k) \cdot j^k\}_{k=-K}^{K} \in \mathbb{C}^{K' \times K'}, \\
B_r &= BFQ_{K'} \in \mathbb{C}^M \times K', \\
F &= \mathrm{diag}\big([(-1)^{-M-1}, \ldots, (-1)^{-2}, 1, 1, \ldots, 1]\big) \in \mathbb{R}^{K' \times K'}, \\
\Gamma &= \frac{\lambda}{\pi R} \mathrm{diag}([\overbrace{-M, \ldots, -3, -2}^{M-1}, 0, \overbrace{2, \ldots, M}^{M-1}]) \in \mathbb{R}^{(K'-2) \times (K'-2)}, \\
C_1 &= \mathrm{diag}([\overbrace{1, \ldots, 1}^{M-1}, -1, -1, \overbrace{1, \ldots, 1}^{M-2}]) \in \mathbb{R}^{(K'-2) \times (K'-2)}.
\end{aligned} \tag{4.32}$$

The whole algorithm is summarized in Table 4.5. B_r synthesizes a real-valued manifold that facilitates signal subspace estimation via a real-valued SVD or eigenvalue decomposition in the first step. As in UCA-ESPRIT, the eigenvalues of Ψ computed in the final step are asymptotically of the form $\sin(\theta_i)e^{j\phi_i}$, where θ_i and ϕ_i are the elevation and azimuth angles of the ith source, respectively.

4.5.1 Computer Simulation

As an example, a simulation involving a "random-filled array" is presented. The element locations are depicted in Figure 4.7. The outer radius is $R = 5\lambda$ and the average distance between elements is $\lambda/4$. Two plane waves of equal power were incident upon the array. The SNR per antenna per signal was 0 dB. One signal arrived at 10° elevation and 40° azimuth, while the other arrived at 30° elevation and 60° azimuth. Figure 4.8 shows the results of 32 independent trials of FCA-ESPRIT overlaid; each execution of the algorithm (with a different realization of the noise) produced two eigenvalues. The eigenvalues are observed to be clustered around the expected locations (the dashed circles indicate the true elevation angles).

4.6 2-D Unitary ESPRIT

For UCAs and FCAs, UCA-ESPRIT and FCA-ESPRIT provide closed-form, automatically paired 2-D angle estimates as long as the direction cosine pair of each signal arrival is unique. In this section, we develop 2-D Unitary ESPRIT, a closed-form 2-D angle estimation algorithm that achieves automatic

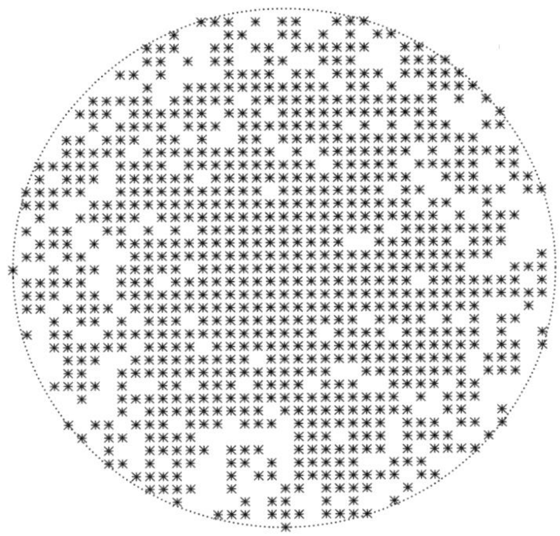

FIGURE 4.7 Random-filled array.

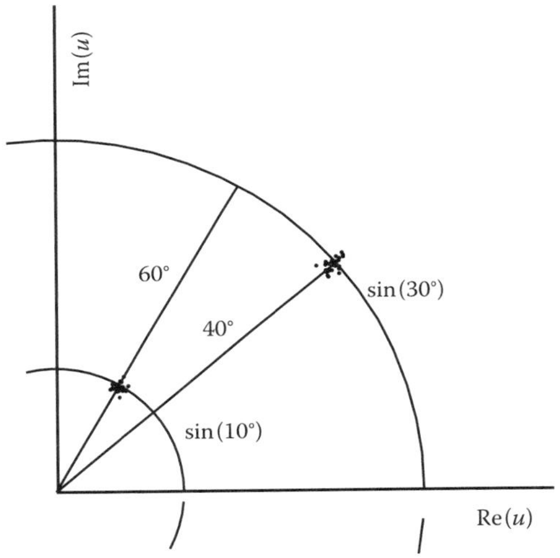

FIGURE 4.8 Plot of the FCA-ESPRIT eigenvalues from 32 independent trials.

pairing in a similar fashion. It is applicable to 2-D centrosymmetric array configurations with a dual invariance structure such as uniform rectangular arrays (URAs). In the derivations of UCA-ESPRIT and FCA-ESPRIT, it was necessary to approximate the sampled aperture pattern by the continuous aperture pattern. Such an approximation is not required in the development of 2-D Unitary ESPRIT.

Apart from the 2-D extension presented here, Unitary ESPRIT has also been extended to the general R-dimensional case to solve the R-dimensional harmonic retrieval problem, where $R \geq 3$. R-D Unitary ESPRIT is a closed-form algorithm to estimate several undamped R-dimensional modes (or frequencies) along with their correct pairing. In [9], automatic pairing of the R-dimensional frequency estimates is achieved through a simultaneous Schur decomposition (SSD) of R real-valued, nonsymmetric matrices that reveals their "average eigenstructure." Like its 1-D and 2-D counterparts, R-D Unitary ESPRIT inherently includes forward–backward averaging and is efficiently formulated in terms of real-valued computations throughout.

In channel sounding applications, for example, an $R = 6$-dimensional extension of Unitary ESPRIT can be used to estimate the 2-D departure angles at the transmit array (e.g., in terms of azimuth and elevation), the 2-D arrival angles at the receive array (e.g., in terms of azimuth and elevation), the Doppler shifts, and the propagation delays of the dominant multipath components jointly [11].

4.6.1 2-D Array Geometry

Consider a two-dimensional (2-D) centrosymmetric sensor array of M elements lying in the x–y plane (Figure 4.4). Assume that the array also exhibits a "dual" invariance, i.e., two identical subarrays of m_x elements are displaced by Δ_x along the x-axis, and another pair of identical subarrays, consisting of m_y elements each, is displaced by Δ_y along the y-axis. Notice that the four subarrays can overlap and m_x is not required to equal m_y. Such array configurations include URAs, uniform rectangular frame arrays (URFAs), i.e., URAs without some of their center elements, and cross-arrays consisting of two orthogonal linear arrays with a common phase center as shown in Figure 4.9.* Extensions to more general hexagonal array geometries have been developed in [22].

Incident on the array are d narrowband planar wave fronts with wavelength λ, azimuth ϕ_i, and elevation θ_i, $1 \leq i \leq d$. Let

$$u_i = \cos\phi_i \sin\theta_i \quad \text{and} \quad v_i = \sin\phi_i \sin\theta_i, \quad 1 \leq i \leq d,$$

denote the direction cosines of the ith source relative to the x- and y-axis, respectively. These definitions are illustrated in Figure 4.4. The fact that $\xi_i = u_i + jv_i = \sin\theta_i e^{j\phi_i}$ yields a simple formula to determine azimuth ϕ_i and elevation θ_i from the corresponding direction cosines u_i and v_i, namely

$$\phi_i = \arg(\xi_i) \quad \text{and} \quad \theta_i = \arcsin(|\xi_i|), \quad \text{with } \xi_i = u_i + jv_i, \quad 1 \leq i \leq d. \tag{4.33}$$

Similar to the 1-D case, the data matrix $\boldsymbol{X}$ is an $M \times N$ matrix composed of N snapshots $\boldsymbol{x}(t_n)$, $1 \leq n \leq N$, of data as columns. Referring to Figure 4.10 for a URA of $M = 4 \times 4 = 16$ sensors as an illustrative example, the antenna element outputs are stacked columnwise. Specifically, the first element of $\boldsymbol{x}(t_n)$ is the output of the antenna in the upper left corner. Then sequentially progress downwards along the positive x-axis such that the fourth element of $\boldsymbol{x}(t_n)$ is the output of the antenna in the bottom left corner. The fifth element of $\boldsymbol{x}(t_n)$ is the output of the antenna at the top of the second column; the eighth element of $\boldsymbol{x}(t_n)$ is the output of the antenna at the bottom of the second column, etc. This forms a 16×1 vector at each sampling instant t_n.

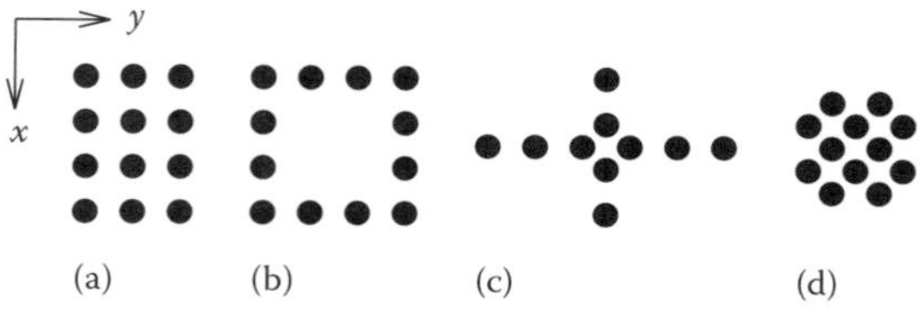

FIGURE 4.9 Centrosymmetric array configurations with a dual invariance structure: (a) URA with $M = 12$, $m_x = 9$, $m_y = 8$. (b) URFA with $M = 12$, $m_x = m_y = 6$. (c) Cross-array with $M = 10$, $m_x = 3$, $m_y = 5$. (d) $M = 12$, $m_x = m_y = 7$.

* In the examples of Figure 4.9, all values of m_x and m_y correspond to selection matrices with maximum overlap in both directions. For a URA of $M = M_x \cdot M_y$ elements, cf. Figure 4.9a, this assumption implies $m_x = (M_x - 1)M_y$ and $m_y = M_x(M_y - 1)$.

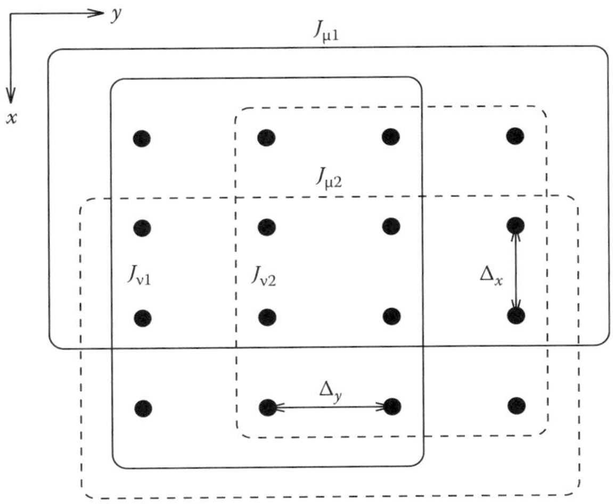

FIGURE 4.10 Subarray selection for a URA of $M = 4 \cdot 4 = 16$ sensor elements (maximum overlap in both directions: $m_x = m_y = 12$).

Similar to the 1-D case, the array measurements may be expressed as $\boldsymbol{x}(t) = \boldsymbol{A}\boldsymbol{s}(t) + \boldsymbol{n}(t) \in \mathbb{C}^M$. Due to the centrosymmetry of the array, the steering matrix $\boldsymbol{A} \in \mathbb{C}^{M \times d}$ satisfies Equation 4.12. The goal is to construct two pairs of selection matrices that are centrosymmetric with respect to each other, i.e.,

$$\boldsymbol{J}_{\mu 2} = \boldsymbol{\Pi}_{m_x} \boldsymbol{J}_{\mu 1} \boldsymbol{\Pi}_M \quad \text{and} \quad \boldsymbol{J}_{\nu 2} = \boldsymbol{\Pi}_{m_y} \boldsymbol{J}_{\nu 1} \boldsymbol{\Pi}_M \tag{4.34}$$

and cause the array steering matrix $\boldsymbol{A}$ to satisfy the following two invariance properties,

$$\boldsymbol{J}_{\mu 1} \boldsymbol{A} \boldsymbol{\Phi}_\mu = \boldsymbol{J}_{\mu 2} \boldsymbol{A} \quad \text{and} \quad \boldsymbol{J}_{\nu 1} \boldsymbol{A} \boldsymbol{\Phi}_\nu = \boldsymbol{J}_{\nu 2} \boldsymbol{A}, \tag{4.35}$$

where the diagonal matrices

$$\boldsymbol{\Phi}_\mu = \operatorname{diag}\{e^{j\mu_i}\}_{i=1}^d \quad \text{and} \quad \boldsymbol{\Phi}_\nu = \operatorname{diag}\{e^{j\nu_i}\}_{i=1}^d \tag{4.36}$$

are unitary and contain the desired 2-D angle information. Here $\mu_i = \frac{2\pi}{\lambda} \Delta_x u_i$ and $\nu_i = \frac{2\pi}{\lambda} \Delta_y v_i$ are the spatial frequencies in x- and y-direction, respectively.

Figure 4.10 visualizes a possible choice of the selection matrices for a URA of $M = 4 \times 4 = 16$ sensor elements. Given the stacking procedure described above and the 1-D selection matrices for a ULA of four elements

$$\boldsymbol{J}_1^{(4)} = \begin{bmatrix} 1 & 0 & 0 & 0 \\ 0 & 1 & 0 & 0 \\ 0 & 0 & 1 & 0 \end{bmatrix} \quad \text{and} \quad \boldsymbol{J}_2^{(4)} = \begin{bmatrix} 0 & 1 & 0 & 0 \\ 0 & 0 & 1 & 0 \\ 0 & 0 & 0 & 1 \end{bmatrix},$$

the appropriate selection matrices corresponding to maximum overlap are

$$
\boldsymbol{J}_{\mu 1} = \boldsymbol{I}_{M_y} \otimes \boldsymbol{J}_1^{(M_x)} = \begin{bmatrix}
1 & 0 & 0 & 0 & 0 & 0 & 0 & 0 & 0 & 0 & 0 & 0 & 0 & 0 & 0 & 0 \\
0 & 1 & 0 & 0 & 0 & 0 & 0 & 0 & 0 & 0 & 0 & 0 & 0 & 0 & 0 & 0 \\
0 & 0 & 1 & 0 & 0 & 0 & 0 & 0 & 0 & 0 & 0 & 0 & 0 & 0 & 0 & 0 \\
0 & 0 & 0 & 0 & 1 & 0 & 0 & 0 & 0 & 0 & 0 & 0 & 0 & 0 & 0 & 0 \\
0 & 0 & 0 & 0 & 0 & 1 & 0 & 0 & 0 & 0 & 0 & 0 & 0 & 0 & 0 & 0 \\
0 & 0 & 0 & 0 & 0 & 0 & 1 & 0 & 0 & 0 & 0 & 0 & 0 & 0 & 0 & 0 \\
0 & 0 & 0 & 0 & 0 & 0 & 0 & 0 & 1 & 0 & 0 & 0 & 0 & 0 & 0 & 0 \\
0 & 0 & 0 & 0 & 0 & 0 & 0 & 0 & 0 & 1 & 0 & 0 & 0 & 0 & 0 & 0 \\
0 & 0 & 0 & 0 & 0 & 0 & 0 & 0 & 0 & 0 & 1 & 0 & 0 & 0 & 0 & 0 \\
0 & 0 & 0 & 0 & 0 & 0 & 0 & 0 & 0 & 0 & 0 & 0 & 1 & 0 & 0 & 0 \\
0 & 0 & 0 & 0 & 0 & 0 & 0 & 0 & 0 & 0 & 0 & 0 & 0 & 1 & 0 & 0 \\
0 & 0 & 0 & 0 & 0 & 0 & 0 & 0 & 0 & 0 & 0 & 0 & 0 & 0 & 1 & 0
\end{bmatrix} \in \mathbb{R}^{12 \times 16},
$$

$$
\boldsymbol{J}_{\mu 2} = \boldsymbol{I}_{M_y} \otimes \boldsymbol{J}_2^{(M_x)} = \begin{bmatrix}
0 & 1 & 0 & 0 & 0 & 0 & 0 & 0 & 0 & 0 & 0 & 0 & 0 & 0 & 0 & 0 \\
0 & 0 & 1 & 0 & 0 & 0 & 0 & 0 & 0 & 0 & 0 & 0 & 0 & 0 & 0 & 0 \\
0 & 0 & 0 & 1 & 0 & 0 & 0 & 0 & 0 & 0 & 0 & 0 & 0 & 0 & 0 & 0 \\
0 & 0 & 0 & 0 & 0 & 1 & 0 & 0 & 0 & 0 & 0 & 0 & 0 & 0 & 0 & 0 \\
0 & 0 & 0 & 0 & 0 & 0 & 1 & 0 & 0 & 0 & 0 & 0 & 0 & 0 & 0 & 0 \\
0 & 0 & 0 & 0 & 0 & 0 & 0 & 1 & 0 & 0 & 0 & 0 & 0 & 0 & 0 & 0 \\
0 & 0 & 0 & 0 & 0 & 0 & 0 & 0 & 0 & 1 & 0 & 0 & 0 & 0 & 0 & 0 \\
0 & 0 & 0 & 0 & 0 & 0 & 0 & 0 & 0 & 0 & 1 & 0 & 0 & 0 & 0 & 0 \\
0 & 0 & 0 & 0 & 0 & 0 & 0 & 0 & 0 & 0 & 0 & 1 & 0 & 0 & 0 & 0 \\
0 & 0 & 0 & 0 & 0 & 0 & 0 & 0 & 0 & 0 & 0 & 0 & 0 & 1 & 0 & 0 \\
0 & 0 & 0 & 0 & 0 & 0 & 0 & 0 & 0 & 0 & 0 & 0 & 0 & 0 & 1 & 0 \\
0 & 0 & 0 & 0 & 0 & 0 & 0 & 0 & 0 & 0 & 0 & 0 & 0 & 0 & 0 & 1
\end{bmatrix} \in \mathbb{R}^{12 \times 16},
$$

$$
\boldsymbol{J}_{\nu 1} = \boldsymbol{J}_1^{(M_y)} \otimes \boldsymbol{I}_{M_x} = \begin{bmatrix}
1 & 0 & 0 & 0 & 0 & 0 & 0 & 0 & 0 & 0 & 0 & 0 & 0 & 0 & 0 & 0 \\
0 & 1 & 0 & 0 & 0 & 0 & 0 & 0 & 0 & 0 & 0 & 0 & 0 & 0 & 0 & 0 \\
0 & 0 & 1 & 0 & 0 & 0 & 0 & 0 & 0 & 0 & 0 & 0 & 0 & 0 & 0 & 0 \\
0 & 0 & 0 & 1 & 0 & 0 & 0 & 0 & 0 & 0 & 0 & 0 & 0 & 0 & 0 & 0 \\
0 & 0 & 0 & 0 & 1 & 0 & 0 & 0 & 0 & 0 & 0 & 0 & 0 & 0 & 0 & 0 \\
0 & 0 & 0 & 0 & 0 & 1 & 0 & 0 & 0 & 0 & 0 & 0 & 0 & 0 & 0 & 0 \\
0 & 0 & 0 & 0 & 0 & 0 & 1 & 0 & 0 & 0 & 0 & 0 & 0 & 0 & 0 & 0 \\
0 & 0 & 0 & 0 & 0 & 0 & 0 & 1 & 0 & 0 & 0 & 0 & 0 & 0 & 0 & 0 \\
0 & 0 & 0 & 0 & 0 & 0 & 0 & 0 & 1 & 0 & 0 & 0 & 0 & 0 & 0 & 0 \\
0 & 0 & 0 & 0 & 0 & 0 & 0 & 0 & 0 & 1 & 0 & 0 & 0 & 0 & 0 & 0 \\
0 & 0 & 0 & 0 & 0 & 0 & 0 & 0 & 0 & 0 & 1 & 0 & 0 & 0 & 0 & 0 \\
0 & 0 & 0 & 0 & 0 & 0 & 0 & 0 & 0 & 0 & 0 & 1 & 0 & 0 & 0 & 0
\end{bmatrix} \in \mathbb{R}^{12 \times 16},
$$

$$
\boldsymbol{J}_{\nu 2} = \boldsymbol{J}_2^{(M_y)} \otimes \boldsymbol{I}_{M_x} = \begin{bmatrix}
0 & 0 & 0 & 0 & 1 & 0 & 0 & 0 & 0 & 0 & 0 & 0 & 0 & 0 & 0 & 0 \\
0 & 0 & 0 & 0 & 0 & 1 & 0 & 0 & 0 & 0 & 0 & 0 & 0 & 0 & 0 & 0 \\
0 & 0 & 0 & 0 & 0 & 0 & 1 & 0 & 0 & 0 & 0 & 0 & 0 & 0 & 0 & 0 \\
0 & 0 & 0 & 0 & 0 & 0 & 0 & 1 & 0 & 0 & 0 & 0 & 0 & 0 & 0 & 0 \\
0 & 0 & 0 & 0 & 0 & 0 & 0 & 0 & 1 & 0 & 0 & 0 & 0 & 0 & 0 & 0 \\
0 & 0 & 0 & 0 & 0 & 0 & 0 & 0 & 0 & 1 & 0 & 0 & 0 & 0 & 0 & 0 \\
0 & 0 & 0 & 0 & 0 & 0 & 0 & 0 & 0 & 0 & 1 & 0 & 0 & 0 & 0 & 0 \\
0 & 0 & 0 & 0 & 0 & 0 & 0 & 0 & 0 & 0 & 0 & 1 & 0 & 0 & 0 & 0 \\
0 & 0 & 0 & 0 & 0 & 0 & 0 & 0 & 0 & 0 & 0 & 0 & 1 & 0 & 0 & 0 \\
0 & 0 & 0 & 0 & 0 & 0 & 0 & 0 & 0 & 0 & 0 & 0 & 0 & 1 & 0 & 0 \\
0 & 0 & 0 & 0 & 0 & 0 & 0 & 0 & 0 & 0 & 0 & 0 & 0 & 0 & 1 & 0 \\
0 & 0 & 0 & 0 & 0 & 0 & 0 & 0 & 0 & 0 & 0 & 0 & 0 & 0 & 0 & 1
\end{bmatrix} \in \mathbb{R}^{12 \times 16},
$$

where $M_x = M_y = 4$. Notice, however, that it is not required to compute all four selection matrices explicitly, since they are related via Equation 4.34. In fact, to be able to compute the four transformed selection matrices for 2-D Unitary ESPRIT, it is sufficient to specify $\boldsymbol{J}_{\mu 2}$ and $\boldsymbol{J}_{\nu 2}$, cf. Equations 4.38 and 4.39.

4.6.2 2-D Unitary ESPRIT in Element Space

Similar to Equation 4.16 in the 1-D case, let us define the transformed 2-D array steering matrix as $\boldsymbol{D} = \boldsymbol{Q}_M^{\mathrm{H}} \boldsymbol{A}$. Based on the two invariance properties of the 2-D array steering matrix $\boldsymbol{A}$ in Equation 4.35, it is a straightforward 2-D extension of the derivation of 1-D Unitary ESPRIT to show that the transformed array steering matrix $\boldsymbol{D}$ satisfies

$$\boldsymbol{K}_{\mu 1} \boldsymbol{D} \cdot \boldsymbol{\Omega}_{\mu} = \boldsymbol{K}_{\mu 2} \boldsymbol{D} \quad \text{and} \quad \boldsymbol{K}_{\nu 1} \boldsymbol{D} \cdot \boldsymbol{\Omega}_{\nu} = \boldsymbol{K}_{\nu 2} \boldsymbol{D}, \tag{4.37}$$

where the two pairs of transformed selection matrices are defined as

$$\boldsymbol{K}_{\mu 1} = 2 \cdot \mathrm{Re}\left\{\boldsymbol{Q}_{m_x}^{\mathrm{H}} \boldsymbol{J}_{\mu 2}\, \boldsymbol{Q}_M\right\} \quad \boldsymbol{K}_{\mu 2} = 2 \cdot \mathrm{Im}\left\{\boldsymbol{Q}_{m_x}^{\mathrm{H}} \boldsymbol{J}_{\mu 2}\, \boldsymbol{Q}_M\right\}, \tag{4.38}$$

$$\boldsymbol{K}_{\nu 1} = 2 \cdot \mathrm{Re}\left\{\boldsymbol{Q}_{m_y}^{\mathrm{H}} \boldsymbol{J}_{\nu 2}\, \boldsymbol{Q}_M\right\} \quad \boldsymbol{K}_{\nu 2} = 2 \cdot \mathrm{Im}\left\{\boldsymbol{Q}_{m_y}^{\mathrm{H}} \boldsymbol{J}_{\nu 2}\, \boldsymbol{Q}_M\right\}, \tag{4.39}$$

and the real-valued diagonal matrices

$$\boldsymbol{\Omega}_{\mu} = \mathrm{diag}\left\{\tan\left(\frac{\mu_i}{2}\right)\right\}_{i=1}^{d} \quad \text{and} \quad \boldsymbol{\Omega}_{\nu} = \mathrm{diag}\left\{\tan\left(\frac{\nu_i}{2}\right)\right\}_{i=1}^{d} \tag{4.40}$$

contain the desired (spatial) frequency information.

Given the noise-corrupted data matrix $\boldsymbol{X}$, a real-valued matrix $\boldsymbol{E}_s$, spanning the dominant subspace of $\mathcal{T}(\boldsymbol{X})$, is obtained as described in Section 4.3.1 for the 1-D case. Asymptotically or without additive noise, $\boldsymbol{E}_s$ and $\boldsymbol{D}$ span the same d-dimensional subspace, i.e., there is a nonsingular matrix $\boldsymbol{T}$ of size $d \times d$ such that $\boldsymbol{D} \approx \boldsymbol{E}_s \boldsymbol{T}$. Substituting this relationship into Equation 4.37 yields two "real-valued" invariance equations

$$\boldsymbol{K}_{\mu 1} \boldsymbol{E}_s \Upsilon_{\mu} \approx \boldsymbol{K}_{\mu 2} \boldsymbol{E}_s \in \mathbb{R}^{m_x \times d} \quad \text{and} \quad \boldsymbol{K}_{\nu 1} \boldsymbol{E}_s \Upsilon_{\nu} \approx \boldsymbol{K}_{\nu 2} \boldsymbol{E}_s \in \mathbb{R}^{m_y \times d}, \tag{4.41}$$

where $\Upsilon_{\mu} = \boldsymbol{T} \boldsymbol{\Omega}_{\mu} \boldsymbol{T}^{-1} \in \mathbb{R}^{d \times d}$ and $\Upsilon_{\nu} = \boldsymbol{T} \boldsymbol{\Omega}_{\nu} \boldsymbol{T}^{-1} \in \mathbb{R}^{d \times d}$. Thus, Υ_{μ} and Υ_{ν} are related with the diagonal matrices $\boldsymbol{\Omega}_{\mu}$ and $\boldsymbol{\Omega}_{\nu}$ via eigenvalue-preserving similarity transformations. Moreover, the real-valued matrices Υ_{μ} and Υ_{ν} share the "same set of eigenvectors." As in the 1-D case, the two real-valued invariance Equation 4.41 can be solved independently via LS or TLS [12]. As an alternative, they may be solved jointly via 2-D SLS, which is a 2-D extension of SLS [7].

4.6.3 Automatic Pairing of the 2-D Frequency Estimates

Asymptotically or without additive noise, the real-valued eigenvalues of the solutions $\Upsilon_{\mu} \in \mathbb{R}^{d \times d}$ and $\Upsilon_{\nu} \in \mathbb{R}^{d \times d}$ to the invariance equations above are given by $\tan(\mu_i/2)$ and $\tan(\nu_i/2)$, respectively. If these eigenvalues were calculated independently, it would be quite difficult to pair the resulting two distinct sets of frequency estimates. Notice that one can choose a real-valued eigenvector matrix $\boldsymbol{T}$ such that all matrices that appear in the spectral decompositions of $\Upsilon_{\mu} = \boldsymbol{T} \boldsymbol{\Omega}_{\mu} \boldsymbol{T}^{-1}$ and $\Upsilon_{\nu} = \boldsymbol{T} \boldsymbol{\Omega}_{\nu} \boldsymbol{T}^{-1}$ are real-valued. Moreover, the subspace spanned by the columns of $\boldsymbol{T} \in \mathbb{R}^{d \times d}$ is unique. These observations are critical to achieve automatic pairing of the spatial frequencies μ_i and ν_i, and $1 \leq i \leq d$.

With additive noise and a finite number of snapshots N, however, the real-valued matrices Υ_μ and Υ_ν do not exactly share the same set of eigenvectors. To determine an approximation of the set of common eigenvectors from one of these matrices is, obviously, not the best solution, since this strategy would rely on an arbitrary choice and would also discard information contained in the other matrix. Moreover, Υ_μ and Υ_ν might have some degenerate (multiple) eigenvalues, while both of them have well-determined common eigenvectors $\boldsymbol{T}$ (for $N \to \infty$ or $\sigma_N^2 \to 0$). 2-D Unitary ESPRIT circumvents these difficulties and achieves automatic pairing of the spatial frequency estimates μ_i and ν_i by computing the eigenvalues of the "complexified" matrix $\Upsilon_\mu + j\Upsilon_\nu$ since this complex-valued matrix may be spectrally decomposed as

$$\Upsilon_\mu + j\Upsilon_\nu = \boldsymbol{T}(\boldsymbol{\Omega}_\mu + j\boldsymbol{\Omega}_\nu)\boldsymbol{T}^{-1}. \tag{4.42}$$

Here, automatically paired estimates of Ω_μ and Ω_ν in Equation 4.40 are given by the real and imaginary parts of the complex eigenvalues of $\Upsilon_\mu + j\Upsilon_\nu$. The maximum number of sources 2-D Unitary ESPRIT can handle is the minimum of m_x and m_y, assuming that at least $d/2$ snapshots are available. If only a single snapshot is available (or more than two sources are highly correlated), one can extract $d/2$ or more identical subarrays out of the overall array to get the effect of multiple snapshots (spatial smoothing), thereby decreasing the maximum number of sources that can be handled. A brief summary of the described element space implementation of 2-D Unitary ESPRIT is given in Table 4.6.

It is instructive to examine a very simple numerical example. Consider a URA of $M = 2 \times 2 = 4$ sensor elements, i.e., $M_x = M_y = 2$. Effecting maximum overlap we have $m_x = m_y = 2$. For the sake of simplicity, assume that the true covariance matrix of the noise-corrupted measurements

$$\boldsymbol{R}_{xx} = \mathrm{E}\{\boldsymbol{x}(t)\boldsymbol{x}^{\mathrm{H}}(t)\} = \boldsymbol{A}\boldsymbol{R}_{ss}\boldsymbol{A}^{\mathrm{H}} + \sigma_N^2\boldsymbol{I}_4 = \begin{bmatrix} 3 & 0 & 1-j & -1+j \\ 0 & 3 & 1-j & 1-j \\ 1+j & 1+j & 3 & 0 \\ -1-j & 1+j & 0 & 3 \end{bmatrix},$$

is known. Here, $\boldsymbol{R}_{ss} = \mathrm{E}\{\boldsymbol{s}(t)\boldsymbol{s}^{\mathrm{H}}(t)\} \in \mathbb{C}^{d \times d}$ denotes the unknown signal covariance matrix. Furthermore, the measurement vector $\boldsymbol{x}\ (t)$ is defined as

$$\boldsymbol{x}(t) = [x_{11}(t) \quad x_{12}(t) \quad x_{21}(t) \quad x_{22}(t)]^{\mathrm{T}}. \tag{4.43}$$

In this example, we have to use a covariance approach instead of the direct data approach summarized in Table 4.6, since the array measurements $\boldsymbol{x}(t)$ themselves are not known. To this end, we will compute the eigendecomposition of the real part of the transformed covariance matrix as, for instance, discussed

TABLE 4.6 Summary of 2-D Unitary ESPRIT in Element Space

1. *Signal subspace estimation*: Compute $\boldsymbol{E}_s \in \mathbb{R}^{M \times d}$ as the d dominant left singular vectors of $\mathcal{T}(\boldsymbol{X}) \in \mathbb{R}^{M \times 2N}$.
2. *Solution of the invariance equations*: Solve

$$\underbrace{\boldsymbol{K}_{\mu 1}\boldsymbol{E}_s}_{\mathbb{R}^{m_x \times d}} \Upsilon_\mu \approx \underbrace{\boldsymbol{K}_{\mu 2}\boldsymbol{E}_s}_{\mathbb{R}^{m_x \times d}} \quad \text{and} \quad \underbrace{\boldsymbol{K}_{\nu 1}\boldsymbol{E}_s}_{\mathbb{R}^{m_y \times d}} \Upsilon_\nu \approx \underbrace{\boldsymbol{K}_{\nu 2}\boldsymbol{E}_s}_{\mathbb{R}^{m_y \times d}}$$

by means of LS, TLS, or 2-D SLS.

3. *Spatial frequency estimation*: Calculate the eigenvalues of the complex-valued $d \times d$ matrix

$$\Upsilon_\mu + j\Upsilon_\nu = \boldsymbol{T}\boldsymbol{\Lambda}\boldsymbol{T}^{-1} \quad \text{with } \boldsymbol{\Lambda} = \mathrm{diag}\{\lambda_i\}_{i=1}^d$$

- $\mu_i = 2$ arctan (Re $\{\lambda_i\}$), $1 \leq i \leq d$.
- $\nu_i = 2$ arctan (Im $\{\lambda_i\}$), $1 \leq i \leq d$.

in [28]. According to Equation 4.13, the left Π-real transformation matrices $\boldsymbol{Q}_M$ and $\boldsymbol{Q}_{m_x} = \boldsymbol{Q}_{m_y}$ take the form

$$\boldsymbol{Q}_4 = \frac{1}{\sqrt{2}} \begin{bmatrix} 1 & 0 & j & 0 \\ 0 & 1 & 0 & j \\ 0 & 1 & 0 & -j \\ 1 & 0 & -j & 0 \end{bmatrix} \quad \text{and} \quad \boldsymbol{Q}_2 = \frac{1}{\sqrt{2}} \begin{bmatrix} 1 & j \\ 1 & -j \end{bmatrix},$$

respectively. Therefore, we have

$$\boldsymbol{R}_Q = \operatorname{Re}\{\boldsymbol{Q}_4^{\mathrm{H}} \boldsymbol{R}_{xx} \boldsymbol{Q}_4\} = \boldsymbol{Q}_4^{\mathrm{H}} \boldsymbol{R}_{xx} \boldsymbol{Q}_4 = \begin{bmatrix} 2 & 1 & 1 & -1 \\ 1 & 4 & -1 & -1 \\ 1 & -1 & 4 & -1 \\ -1 & -1 & -1 & 2 \end{bmatrix}. \tag{4.44}$$

The eigenvalues of $\boldsymbol{R}_Q$ are given by $\rho_1 = 5, \rho_2 = 5, \rho_3 = 1$, and $\rho_4 = 1$. Clearly, ρ_1 and ρ_2 are the dominant eigenvalues, and the variance of the additive noise is identified as $\sigma_N^2 = \rho_3 = \rho_4 = 1$. Therefore, there are $d = 2$ impinging wave fronts. The columns of

$$\boldsymbol{E}_s = \begin{bmatrix} 1 & 0 \\ 1 & 1 \\ 1 & -1 \\ -1 & 0 \end{bmatrix},$$

contain eigenvectors of $\boldsymbol{R}_Q$ corresponding to the $d = 2$ largest eigenvalues ρ_1 and ρ_2. The four selection matrices

$$\boldsymbol{J}_{\mu 1} = \begin{bmatrix} 1 & 0 & 0 & 0 \\ 0 & 0 & 1 & 0 \end{bmatrix}, \quad \boldsymbol{J}_{\mu 2} = \begin{bmatrix} 0 & 1 & 0 & 0 \\ 0 & 0 & 0 & 1 \end{bmatrix}, \quad \boldsymbol{J}_{\nu 1} = \begin{bmatrix} 1 & 0 & 0 & 0 \\ 0 & 1 & 0 & 0 \end{bmatrix}, \quad \boldsymbol{J}_{\nu 2} = \begin{bmatrix} 0 & 0 & 1 & 0 \\ 0 & 0 & 0 & 1 \end{bmatrix},$$

are constructed in accordance with Equation 4.43, cf. Figure 4.10, yielding

$$\boldsymbol{K}_{\mu 1} = \begin{bmatrix} 1 & 1 & 0 & 0 \\ 0 & 0 & 1 & 1 \end{bmatrix}, \quad \boldsymbol{K}_{\mu 2} = \begin{bmatrix} 0 & 0 & -1 & 1 \\ 1 & -1 & 0 & 0 \end{bmatrix},$$

$$\boldsymbol{K}_{\nu 1} = \begin{bmatrix} 1 & 1 & 0 & 0 \\ 0 & 0 & 0 & -1 \end{bmatrix}, \quad \boldsymbol{K}_{\nu 2} = \begin{bmatrix} 0 & 0 & -1 & -1 \\ 1 & -1 & 0 & 0 \end{bmatrix},$$

according to Equations 4.38 and 4.39. With these definitions, the invariance Equation 4.41 turns out to be

$$\begin{bmatrix} 2 & 1 \\ 0 & -1 \end{bmatrix} \Upsilon_\mu \approx \begin{bmatrix} -2 & 1 \\ 0 & -1 \end{bmatrix} \quad \text{and} \quad \begin{bmatrix} 2 & 1 \\ 2 & -1 \end{bmatrix} \Upsilon_\nu \approx \begin{bmatrix} 0 & 1 \\ 0 & -1 \end{bmatrix}.$$

Solving these matrix equations, we get

$$\Upsilon_\mu = \begin{bmatrix} -1 & 0 \\ 0 & 1 \end{bmatrix} \quad \text{and} \quad \Upsilon_\nu = \begin{bmatrix} 0 & 0 \\ 0 & 1 \end{bmatrix}.$$

Finally, the eigenvalues of the "complexified" 2×2 matrix $\Upsilon_\mu + j\Upsilon_\nu$ are observed to be $\lambda_1 = -1$ and $\lambda_2 = 1 + j$, corresponding to the spatial frequencies

$$\mu_1 = -\frac{\pi}{2}, \quad \nu_1 = 0 \quad \text{and} \quad \mu_2 = \frac{\pi}{2}, \quad \nu_2 = \frac{\pi}{2}.$$

If we assume that $\Delta_x = \Delta_y = \lambda/2$, the direction cosines are given by $u_i = \mu_i/\pi$ and $v_i = \nu_i/\pi$, $i = 1, 2$. According to Equation 4.33, the corresponding azimuth and elevation angles can be calculated as

$$\phi_1 = 180^\circ, \quad \theta_1 = 30^\circ, \quad \text{and} \quad \phi_2 = 45^\circ, \quad \theta_2 = 45^\circ.$$

4.6.4 2-D Unitary ESPRIT in DFT Beamspace

Here, we will restrict the presentation of 2-D Unitary ESPRIT in DFT beamspace to URAs of $M = M_x \cdot M_y$ identical sensors, cf. Figure 4.10.* Without loss of generality, assume that the M sensors are omnidirectional and that the centroid of the URA is chosen as the phase reference.

Let us form B_x out of M_x beams in x-direction and B_y out of M_y beams in y-direction, yielding a total of $B = B_x \cdot B_y$ beams. Then the corresponding scaled DFT-matrices $\boldsymbol{W}_{B_x}^{\mathrm{H}} \in \mathbb{C}^{B_x \times M_x}$ and $\boldsymbol{W}_{B_y}^{\mathrm{H}} \in \mathbb{C}^{B_y \times M_y}$ are formed as discussed in Section 4.3.2. Now, viewing the array output at a given snapshot as an $M_x \times M_y$ matrix, premultiply this matrix by $\boldsymbol{W}_{B_x}^{\mathrm{H}}$ and postmultiply it by $\bar{\boldsymbol{W}}_{B_y}$.† Then apply the vec $\{\cdot\}$-operator, and place the resulting $B \times 1$ vector ($B = B_x \cdot B_y$) as a column of a matrix $\boldsymbol{Y} \in \mathbb{C}^{B \times N}$. The vec$\{\cdot\}$-operator maps a $B_x \times B_y$ matrix to a $B \times 1$ vector by stacking the columns of the matrix. Note that if $\boldsymbol{X}$ denotes the $M \times N$ complex-valued element space data matrix, it is easy to show that the relationship between $\boldsymbol{Y}$ and $\boldsymbol{X}$ may be expressed as $\boldsymbol{Y} = (\boldsymbol{W}_{B_y}^{\mathrm{H}} \otimes \boldsymbol{W}_{B_x}^{\mathrm{H}})\boldsymbol{X}$ [27]. Here, the symbol $\otimes$ denotes the Kronecker matrix product [6].

Let the columns of $\boldsymbol{E}_s \in \mathbb{R}^{B \times d}$ contain the d left singular vectors of

$$[\,\mathrm{Re}\{\boldsymbol{Y}\} \quad \mathrm{Im}\{\boldsymbol{Y}\}\,] \in \mathbb{R}^{B \times 2N}, \tag{4.45}$$

corresponding to its d largest singular values. To set up two overdetermined sets of equations similar to Equation 4.41, but with a reduced dimensionality, let us define the selection matrices

$$\boldsymbol{\Gamma}_{\mu 1} = \boldsymbol{I}_{B_y} \otimes \boldsymbol{\Gamma}_1^{(B_x)} \quad \text{and} \quad \boldsymbol{\Gamma}_{\mu 2} = \boldsymbol{I}_{B_y} \otimes \boldsymbol{\Gamma}_2^{(B_x)}, \tag{4.46}$$

of size $b_x \times B$ for the x-direction ($b_x = (B_x - 1) \cdot B_y$) and

$$\boldsymbol{\Gamma}_{\nu 1} = \boldsymbol{\Gamma}_1^{(B_y)} \otimes \boldsymbol{I}_{B_x} \quad \text{and} \quad \boldsymbol{\Gamma}_{\nu 2} = \boldsymbol{\Gamma}_2^{(B_y)} \otimes \boldsymbol{I}_{B_x}, \tag{4.47}$$

of size $b_y \times B$ for the y-direction ($b_y = B_x \cdot (B_y - 1)$). Then $\Upsilon_\mu \in \mathbb{R}^{d \times d}$ and $\Upsilon_\nu \in \mathbb{R}^{d \times d}$ can be calculated as the LS, TLS, or 2-D SLS solution of

$$\boldsymbol{\Gamma}_{\mu 1} \boldsymbol{E}_s \Upsilon_\mu \approx \boldsymbol{\Gamma}_{\mu 2} \boldsymbol{E}_s \in \mathbb{R}^{b_x \times d} \quad \text{and} \quad \boldsymbol{\Gamma}_{\nu 1} \boldsymbol{E}_s \Upsilon_\nu \approx \boldsymbol{\Gamma}_{\nu 2} \boldsymbol{E}_s \in \mathbb{R}^{b_y \times d}, \tag{4.48}$$

respectively. Finally, the desired "automatically paired" spatial frequency estimates μ_i and ν_i, $1 \leq i \leq d$, are obtained from the real and imaginary part of the eigenvalues of the "complexified" matrix $\Upsilon_\mu + j\Upsilon_\nu$

* In [27], we have also described how to use 2-D Unitary ESPRIT in DFT beamspace for cross-arrays as depicted in Figure 4.9c.

† This can be achieved via a 2-D FFT with appropriate scaling.

TABLE 4.7 Summary of 2-D Unitary ESPRIT in DFT Beamspace

0. *Transformation to beamspace*: Compute a 2-D DFT (with appropriate scaling) of the $M_x \times M_y$ matrix of array outputs at each snapshot, apply the vec{·}-operator, and place the result as a column of $\boldsymbol{Y} \Rightarrow \boldsymbol{Y} = (\boldsymbol{W}_{B_y}^{\mathrm{H}} \otimes \boldsymbol{W}_{B_x}^{\mathrm{H}})\boldsymbol{X} \in \mathbb{C}^{B\times N} (B = B_x \cdot B_y)$.
1. *Signal subspace estimation*: Compute $\boldsymbol{E}_s \in \mathbb{R}^{B\times d}$ as the d dominant left singular vectors of $[\mathrm{Re}\{\boldsymbol{Y}\} \quad \mathrm{Im}\{\boldsymbol{Y}\}] \in \mathbb{R}^{B\times 2N}$.
2. *Solution of the invariance equations*: Solve

$$\underbrace{\boldsymbol{\Gamma}_{\mu 1}\boldsymbol{E}_s}_{\mathbb{R}^{b_x\times d}} \Upsilon_\mu \approx \underbrace{\boldsymbol{\Gamma}_{\mu 2}\boldsymbol{E}_s}_{\mathbb{R}^{b_x\times d}} \quad \text{and} \quad \underbrace{\boldsymbol{\Gamma}_{\nu 1}\boldsymbol{E}_s}_{\mathbb{R}^{b_y\times d}} \Upsilon_\nu \approx \underbrace{\boldsymbol{\Gamma}_{\nu 2}\boldsymbol{E}_s}_{\mathbb{R}^{b_y\times d}}$$

$$b_x = (B_x - 1)\cdot B_y \qquad\qquad b_y = B_x \cdot (B_y - 1)$$

by means of LS, TLS, or 2-D SLS.

3. *Spatial frequency estimation*: Calculate the eigenvalues of the complex-valued $d \times d$ matrix

$$\Upsilon_\mu + j\Upsilon_\nu = \boldsymbol{T}\boldsymbol{\Lambda}\boldsymbol{T}^{-1} \quad \text{with } \boldsymbol{\Lambda} = \mathrm{diag}\{\lambda_i\}_{i=1}^d$$

- $\mu_i = 2 \arctan(\mathrm{Re}\{\lambda_i\}), 1 \le i \le d$
- $\nu_i = 2 \arctan(\mathrm{Im}\{\lambda_i\}), 1 \le i \le d$

as discussed in Section 4.6.1. Here, the maximum number of sources we can handle is given by the minimum of b_x and b_y, assuming that at least $d/2$ snapshots are available. A summary of 2-D Unitary ESPRIT in DFT beamspace is presented in Table 4.7.

4.6.5 Simulation Results

Simulations were conducted employing a URA of 8×8 elements, i.e., $M_x = M_y = 8$, with $\Delta_x = \Delta_y = \lambda/2$. The source scenario consisted of $d = 3$ equipowered, uncorrelated sources located at $(u_1, v_1) = (0, 0), (u_2, v_2) = (1/8, 0)$, and $(u_3, v_3) = (0, 1/8)$, where u_i and v_i are the direction cosines of the ith source relative to the x- and y-axis, respectively. Notice that sources 1 and 2 have the same v-coordinates, while sources 2 and 3 have the same u-coordinates. A given trial run at a given SNR level (per source per element) involved $N = 64$ snapshots. The noise was independent and identically distributed (*i.i.d.*) from element to element and from snapshot to snapshot. The RMS error defined as

$$\mathrm{RMSE}_i = \sqrt{\mathrm{E}\{(\hat{u}_i - u_i)^2\} + \mathrm{E}\{(\hat{v}_i - v_i)^2\}}, \quad i = 1, 2, 3, \tag{4.49}$$

was employed as the performance metric. Let $(\hat{u}_{i_k}, \hat{v}_{i_k})$ denote the coordinate estimates of the ith source obtained at the kth run. Sample performance statistics were computed from $K = 1000$ independent trials as

$$\widehat{\mathrm{RMSE}}_i = \sqrt{\frac{1}{K}\sum_{k=1}^{T}\{(\hat{u}_{i_k} - u_i)^2 + (\hat{v}_{i_k} - v_i)^2\}}, \quad i = 1, 2, 3. \tag{4.50}$$

2-D Unitary ESPRIT in DFT beamspace was implemented with a set of $B = 9$ beams centered at $(u, v) = (0, 0)$, using $B_x = 3$ out of $M_x = 8$ in x-direction (rows 8, 1, and 2 of $\boldsymbol{W}_8^{\mathrm{H}}$) and also $B_y = 3$ out of $M_y = 8$ in y-direction (again, rows 8, 1, and 2 of $\boldsymbol{W}_8^{\mathrm{H}}$). Thus, the corresponding subblocks of the selection matrices $\boldsymbol{\Gamma}_1 \in \mathbb{R}^{8\times 8}$ and $\boldsymbol{\Gamma}_2 \in \mathbb{R}^{8\times 8}$, used to form $\boldsymbol{\Gamma}_1^{(B_x)}$ and $\boldsymbol{\Gamma}_2^{(B_x)}$ in Equation 4.46 and also used to form $\boldsymbol{\Gamma}_1^{(B_y)}$ and $\boldsymbol{\Gamma}_2^{(B_y)}$ in Equation 4.47, are shaded in Figure 4.3b. The bias of 2-D Unitary ESPRIT in element space and DFT beamspace was found to be negligible, facilitating comparison with the Cramér–Rao (CR) lower bound [18]. The resulting performance curves are plotted in Figures 4.11 through 4.13. We have also included theoretical performance predictions of both implementations based on an asymptotic

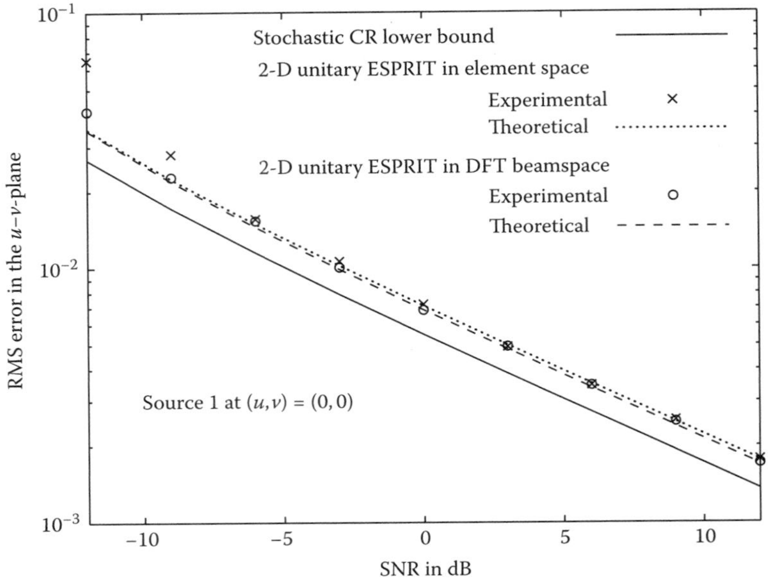

FIGURE 4.11 RMS error of source 1 at $(u_1, v_1) = (0, 0)$ in the u–v plane as a function of the SNR (8×8 sensors, $N = 64$, 1000 trial runs).

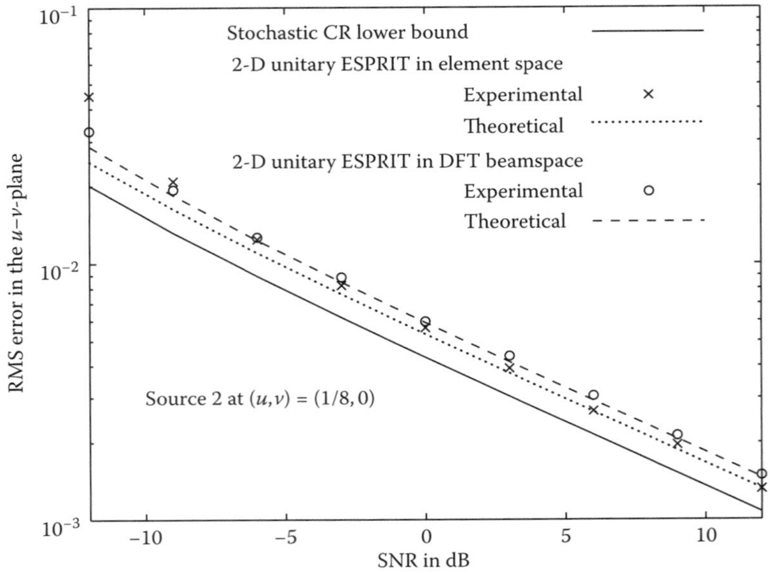

FIGURE 4.12 RMS error of source 2 at $(u_2, v_2) = (1/8, 0)$ in the u–v plane as a function of the SNR (8×8 sensors, $N = 64$, 1000 trial runs).

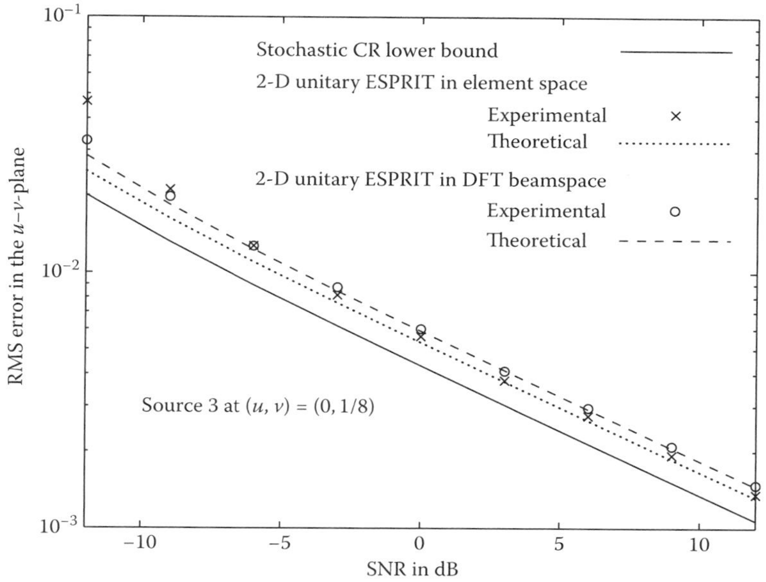

FIGURE 4.13 RMS error of source 3 at $(u_3, v_3) = (0, 1/8)$ in the u–v plane as a function of the SNR (8×8 sensors, $N = 64$, 1000 trial runs).

performance analysis [16,17]. Observe that the empirical root mean squared errors (RMSEs) closely follow the theoretical predictions, except for deviations at low SNRs. The performance of the DFT beamspace implementation is comparable to that of the element space implementation. However, the former requires significantly less computations than the latter, since it operates in a $B = B_x \cdot B_y =$ 9-dimensional beamspace as opposed to an $M = M_x \cdot M_y = 64$-dimensional element space.

For SNRs lower than −9 dB, the DFT beamspace version outperformed the element space version of 2-D Unitary ESPRIT. This is due to the fact that the DFT beamspace version exploits a priori information on the source locations by forming beams pointed in the general directions of the sources.

References

1. G. Bienvenu and L. Kopp, Decreasing high resolution method sensitivity by conventional beamforming preprocessing, in *Proceedings of the IEEE International Conference on Acoustics, Speech, Signal Processing*, pp. 33.2.1–33.2.4, San Diego, CA, March 1984.
2. P. V. Brennan, A low cost phased array antenna for land-mobile satcom applications, *IEE Proc. H*, 138, 131–136, April 1991.
3. K. M. Buckley and X. L. Xu, Spatial-spectrum estimation in a location sector, *IEEE Trans. Acoust. Speech Signal Process.*, ASSP-38, 1842–1852, November 1990.
4. D. E. N. Davies, *The Handbook of Antenna Design*, vol. 2, Chapter 12, Peter Peregrinus, London, U.K., 1983.
5. A. B. Gershman and M. Haardt, Improving the performance of Unitary ESPRIT via pseudo-noise resampling, *IEEE Trans. Signal Process.*, 47, 2305–2308, August 1999.
6. A. Graham, *Kronecker Products and Matrix Calculus: With Applications*, Ellis Horwood Ltd., Chichester, U.K., 1981.

7. M. Haardt, Structured least squares to improve the performance of ESPRIT-type algorithms, *IEEE Trans. Signal Process.*, 45, 792–799, March 1997.
8. M. Haardt and J. A. Nossek, Unitary ESPRIT: How to obtain increased estimation accuracy with a reduced computational burden, *IEEE Trans. Signal Process.*, 43, 1232–1242, May 1995.
9. M. Haardt and J. A. Nossek, Simultaneous Schur decomposition of several non-symmetric matrices to achieve automatic pairing in multidimensional harmonic retrieval problems, *IEEE Trans. Signal Process.*, 46, 161–169, January 1998.
10. M. Haardt and F. Römer, Enhancements of Unitary ESPRIT for non-circular sources, in *Proceedings of the IEEE International Conference on Acoustics, Speech, and Signal Processing (ICASSP)*, vol. II, pp. 101–104, Montreal, Quebec, Canada, May 2004.
11. M. Haardt, R. S. Thomä, and A. Richter, Multidimensional high-resolution parameter estimation with applications to channel sounding, in *High-Resolution and Robust Signal Processing*, Chapter 5, Y. Hua, A. Gershman, and Q. Chen, eds., pp. 255–338, Marcel Dekker, New York, 2004.
12. M. Haardt, M. D. Zoltowski, C. P. Mathews, and J. A. Nossek, 2D unitary ESPRIT for efficient 2D parameter estimation, in *Proceedings of the IEEE International Conference on Acoustics, Speech, and Signal Processing (ICASSP)*, vol. 3, pp. 2096–2099, Detroit, MI, May 1995.
13. A. Lee, Centrohermitian and skew-centrohermitian matrices, *Linear Algebra Appl.*, 29, 205–210, 1980.
14. H. B. Lee and M. S. Wengrovitz, Resolution threshold of beamspace MUSIC for two closely spaced emitters, *IEEE Trans. Acoustics Speech Signal Process.*, 38, 1545–1559, September 1990.
15. D. A. Linebarger, R. D. DeGroat, and E. M. Dowling, Efficient direction finding methods employing forward/backward averaging, *IEEE Trans. Signal Process.*, 42, 2136–2145, August 1994.
16. C. P. Mathews, M. Haardt, and M. D. Zoltowski, Implementation and performance analysis of 2D DFT Beamspace ESPRIT, in *Proceedings of the 29th Asilomar Conference on Signals, Systems, and Computers*, vol. 1, pp. 726–730, Pacific Grove, CA, November 1995. IEEE Computer Society Press, Los Alamitos, CA.
17. C. P. Mathews, M. Haardt, and M. D. Zoltowski, Performance analysis of closed-form, ESPRIT based 2-D angle estimator for rectangular arrays, *IEEE Signal Process. Lett.*, 3, 124–126, April 1996.
18. C. P. Mathews and M. D. Zoltowski, Eigenstructure techniques for 2-D angle estimation with uniform circular arrays, *IEEE Trans. Signal Process.*, 42, 2395–2407, September 1994.
19. C. P. Mathews and M. D. Zoltowski, Performance analysis of the UCA-ESPRIT algorithm for circular ring arrays, *IEEE Trans. Signal Process.*, 42, 2535–2539, September 1994.
20. C. P. Mathews and M. D. Zoltowski, Closed-form 2D angle estimation with circular arrays/apertures via phase mode exitation and ESPRIT, in *Advances in Spectrum Analysis and Array Processing*, S. Haykin, ed., vol. III, pp. 171–218, Prentice Hall, Englewood Cliffs, NJ, 1995.
21. C. P. Mathews and M. D. Zoltowski, Beamspace ESPRIT for multi-source arrival angle estimation employing tapered windows, in *Proceedings of the IEEE International Conference on Acoustics, Speech, Signal Processing*, vol. 3, pp. III–3009–III–3012, Orlando, FL, May 2002.
22. F. Römer and M. Haardt, Using 3-D Unitary ESPRIT on a hexagonal shaped ESPAR antenna for 1-D and 2-D direction of arrival estimation, in *Proceedings of the IEEE/ITG Workshop on Smart Antennas*, Duisburg, Germany, April 2005.
23. R. Roy and T. Kailath, ESPRIT—Estimation of signal parameters via rotational invariance techniques, *IEEE Trans. Acoustics Speech Signal Process.*, 37, 984–995, July 1989.
24. A. Sensi, *Aspects of Modern Radar*, Artech House, Norwood MA, 1988.
25. B. D. Steinberg, Introduction to periodic array synthesis, in *Principle of Aperture and Array System Design*, Chapter 6, pp. 98–99, John Wiley & Sons, New York, 1976.
26. G. Xu, R. H. Roy, and T. Kailath, Detection of number of sources via exploitation of centro-symmetry property, *IEEE Trans. Signal Process.*, 42, 102–112, January 1994.

27. M. D. Zoltowski, M. Haardt, and C. P. Mathews, Closed-form 2D angle estimation with rectangular arrays in element space or beamspace via Unitary ESPRIT, *IEEE Trans. Signal Process.*, 44, 316–328, February 1996.
28. M. D. Zoltowski, G. M. Kautz, and S. D. Silverstein, Beamspace root-MUSIC, *IEEE Trans. Signal Process.*, 41, 344–364, January 1993.
29. M. D. Zoltowski and T. Lee, Maximum likelihood based sensor array signal processing in the beamspace domain for low-angle radar tracking, *IEEE Trans. Signal Process.*, 39, 656–671, March 1991.

5

A Unified Instrumental Variable Approach to Direction Finding in Colored Noise Fields

P. Stoica
Uppsala University

Mats Viberg
Chalmers University of Technology

M. Wong
McMaster University

Q. Wu
CELWAVE

5.1 Introduction 5-2
5.2 Problem Formulation 5-3
5.3 The IV-SSF Approach 5-5
5.4 The Optimal IV-SSF Method 5-6
Optimal Selection of $\hat{W}$ • Optimal Selection of $\hat{W}_R$ and $\hat{W}_L$ • Optimal IV-SSF Criteria
5.5 Algorithm Summary 5-10
5.6 Numerical Examples 5-11
5.7 Concluding Remarks 5-15
Acknowledgment 5-15
Appendix A: Introduction to IV Methods 5-15
References 5-17

The main goal herein is to describe and analyze, in a unifying manner, the "spatial" and "temporal" IV-SSF approaches recently proposed for array signal processing in colored noise fields. (The acronym IV-SSF stands for "Instrumental Variable–Signal Subspace Fitting"). Despite the generality of the approach taken herein, our analysis technique is simpler than those used in previous more specialized publications. We derive a general, optimally weighted (optimal, for short), IV-SSF direction estimator and show that this estimator encompasses the UNCLE estimator of Wong and Wu, which is a spatial IV-SSF method, and the temporal IV-SSF estimator of Viberg, Stoica, and Ottersten. The later two estimators have seemingly different forms (among others, the first of them makes use of four weights, whereas the second one uses three weights only), and hence their asymptotic equivalence shown in this chapter comes as a surprising unifying result. We hope that the present chapter, along with the original works aforementioned, will stimulate the interest in the IV-SSF approach to array signal processing, which is sufficiently flexible to handle colored noise fields, coherent signals and indeed also situations were only some of the sensors in the array are calibrated.

5.1 Introduction

Most parametric methods for direction-of-arrival (DOA) estimation require knowledge of the spatial (sensor-to-sensor) color of the background noise. If this information is unavailable, a serious degradation of the quality of the estimates can result, particularly at low signal-to-noise ratio (SNR) [1–3]. A number of methods have been proposed over the recent years to alleviate the sensitivity to the noise color. If a parametric model of the covariance matrix of the noise is available, the parameters of the noise model can be estimated along with those of the interesting signals [4–7]. Such an approach is expected to perform well in situations where the noise can be accurately modeled with relatively few parameters. An alternative approach, which does not require a precise model of the noise, is based on the principle of instrumental variables (IVs). See Söderström and Stoica [8,9] for thorough treatments of IV methods (IVMs) in the context of identification of linear time-invariant dynamical systems. A brief introduction is given in the appendix. Computationally simple IVMs for array signal processing appeared in [10,11]. These methods perform poorly in difficult scenarios involving closely spaced DOAs and correlated signals.

More recently, the combined instrumental variable–signal subspace fitting (IV-SSF) technique has been proposed as a promising alternative to array signal processing in spatially colored noise fields [12–15]. The IV-SSF approach has a number of appealing advantages over other DOA estimation methods. These advantages include

- IV-SSF can handle noises with arbitrary spatial correlation, under minor restrictions on the signals or the array. In addition, estimation of a noise model is avoided, which leads to statistical robustness and computational simplicity.
- The IV-SSF approach is applicable to both noncoherent and coherent signal scenarios.
- The spatial IV-SSF technique can make use of the information contained in the output of a completely uncalibrated subarray under certain weak conditions, which other methods cannot.

Depending on the type of "IVs" used, two classes of IVMs have appeared in the literature:

1. *Spatial IVM*, for which the IVs are derived from the output of a (possibly uncalibrated) subarray the noise of which is uncorrelated with the noise in the main calibrated subarray under consideration (see [12,13]).
2. *Temporal IVM*, which obtains IVs from the delayed versions of the array output, under the assumption that the temporal-correlation length of the noise field is shorter than that of the signals (see [11,14]).

The previous literature on IV-SSF has treated and analyzed the above two classes of spatial and temporal methods separately, ignoring their common basis. In this contribution, we reveal the common roots of these two classes of DOA estimation methods and study them under the same umbrella. Additionally, we establish the statistical properties of a general (either spatial or temporal) weighted IV-SSF method and present the optimal weights that minimize the variance of the DOA estimation errors. In particular, we point out that the optimal four-weight spatial IV-SSF of Wu and Wong [12,13] (called UNCLE there, and arrived at by using canonical correlation decomposition ideas) and the optimal three-weight temporal IV-SSF of Viberg et al. [14] are asymptotically equivalent when used under the same conditions. This asymptotic equivalence property, which is a main result of the present section, is believed to be important as it shows the close ties that exist between two seemingly different DOA estimators.

This section is organized as follows. In Section 5.2, the data model and technical assumptions are introduced. Next, in Section 5.3, the IV-SSF method is presented in a fairly general setting. In Section 5.4, the statistical performance of the method is presented along with the optimal choices of certain user-specified quantities. The data requirements and the optimal IV-SSF (UNCLE) algorithm are summarized in Section 5.5. The anxious reader may wish to jump directly to this point to investigate the usefulness of the algorithm in a specific application. In Section 5.6, some numerical examples and computer simulations are presented to illustrate the performance. The conclusions are given in Section 5.7. In Appendix A we give a brief introduction to IVMs. The reader who is not familiar with IV might be helped by reading the appendix before the rest of the chapter. Background material on the subspace-based approach to DOA estimation can be found in Chapter 3.

5.2 Problem Formulation

Consider a scenario in which n narrowband plane waves, generated by point sources, impinge on an array comprising m calibrated sensors. Assume, for simplicity, that the n sources and the array are situated in the same plane. Let $\boldsymbol{a}(\theta)$ denote the complex array response to a unit-amplitude signal with DOA parameter equal to θ. Under these assumptions, the output of the array, $\boldsymbol{y}(t) \in C^{m\times 1}$, can be described by the following well-known equation [16,17]:

$$\boldsymbol{y}(t) = \boldsymbol{A}\boldsymbol{x}(t) + \boldsymbol{e}(t) \tag{5.1}$$

where
- $\boldsymbol{x}(t) \in C^{n\times 1}$ denotes the signal vector
- $\boldsymbol{e}(t) \in C^{m\times 1}$ is a noise term

$$\boldsymbol{A} = [\boldsymbol{a}(\theta_1) \dots \boldsymbol{a}(\theta_n)] \tag{5.2}$$

Hereafter, θ_k denotes the kth DOA parameter.

The following assumptions on the quantities in the array equation (Equation 5.1) are considered to hold throughout this section:

A1. The signal vector $x(t)$ is a normally distributed random variable with zero mean and a possibly singular covariance. The signals may be temporally correlated; in fact the temporal IV-SSF approach relies on the assumption that the signals exhibit some form of temporal correlation (see below for details).

A2. The noise $\boldsymbol{e}(t)$ is a random vector that is temporally white, uncorrelated with the signals and circularly symmetric normally distributed with zero mean and unknown covariance matrix* $\boldsymbol{Q} > \boldsymbol{O}$,

$$\mathrm{E}[\boldsymbol{e}(t)\boldsymbol{e}^*(s)] = \boldsymbol{Q}\delta_{t,s}; \quad \mathrm{E}[\boldsymbol{e}(t)\boldsymbol{e}^{\mathrm{T}}(s)] = \boldsymbol{O} \tag{5.3}$$

A3. The manifold vectors $\{\boldsymbol{a}(\theta)\}$, corresponding to any set of m different values of θ, are linearly independent.

Note that Assumption A1 above allows for coherent signals, and that in Assumption A2 the noise field is allowed to be arbitrarily spatially correlated with an unknown covariance matrix. Assumption A3 is a well-known condition that, under a weak restriction on m, guarantees DOA parameter identifiability in the case $\boldsymbol{Q}$ is known (to within a multiplicative constant) [18]. When $\boldsymbol{Q}$ is completely unknown, DOA identifiability can only be achieved if further assumptions are made on the scenario under consideration. The following assumption is typical of the IV-SSF approach:

A4. There exists a vector $z(t) \in C^{\bar{m}\times 1}$, which is normally distributed and satisfies

$$\mathrm{E}[z(t)e^*(s)] = \boldsymbol{O} \quad \text{for } t \leq s \tag{5.4}$$

$$\mathrm{E}[z(t)e^{\mathrm{T}}(s)] = \boldsymbol{O} \quad \text{for all } t, s \tag{5.5}$$

* Henceforth, the superscript "*" denotes the conjugate transpose; whereas the transpose is designated by a superscript "T." The notation $A \geq B$, for two Hermitian matrices A and B, is used to mean that $(A-B)$ is a nonnegative definite matrix. Also, O denotes a zero matrix of suitable dimension.

Furthermore, denote

$$\mathbf{\Gamma} = \mathrm{E}[z(t)\boldsymbol{x}^*(t)] \quad (\bar{m} \times n) \tag{5.6}$$

$$\bar{n} = \mathrm{rank}(\mathbf{\Gamma}) \leq \bar{m} \tag{5.7}$$

It is assumed that no row of $\mathbf{\Gamma}$ is identically zero and that the inequality

$$\bar{n} > 2n - m \tag{5.8}$$

holds (note that a rank-one $\mathbf{\Gamma}$ matrix can satisfy the condition 5.8 if m is large enough, and hence the condition in question is rather weak). Owing to its (partial) uncorrelatedness with $\{\boldsymbol{e}(t)\}$, the vector $\{\boldsymbol{z}(t)\}$ can be used to eliminate the noise from the array output equation (Equation 5.1), and for this reason $\{z(t)\}$ is called an IV vector. In Examples 5.1 through 5.3, we briefly describe three possible ways to derive an IV vector from the available data measured with an array of sensors (for more details on this aspect, the reader should consult [12–14]).

Example 5.1: Spatial IV

Assume that the n signals, which impinge on the main (sub)array under consideration, are also received by another (sub)array that is sufficiently distanced from the main one so that the noise vectors in the two subarrays are uncorrelated with one another. Then $z(t)$ can be made from the outputs of the sensors in the second subarray (note that those sensors need not be calibrated) [12,13,15].

Example 5.2: Temporal IV

When a second subarray, as described above, is not available but the signals are temporally correlated, one can obtain an IV vector by delaying the output vector: $\boldsymbol{z}(t) = [\boldsymbol{y}^{\mathrm{T}}(t-1)\boldsymbol{y}^{\mathrm{T}}(t-2)\ldots]^{\mathrm{T}}$. Clearly, such a vector $z(t)$satisfies Equations 5.4 and 5.5, and it also satisfies Equation 5.8 under weak conditions on the signal temporal correlation. This construction of an IV vector can be readily extended to cases where $e(t)$ is temporally correlated, provided that the signal temporal-correlation length is longer than that corresponding to the noise [11,14].

In a sense, the above examples are both special cases of the following more general situation.

Example 5.3: Reference Signal

In many systems a reference or pilot signal [19,20] $z(t)$ (scalar or vector) is available. If the reference signal is sufficiently correlated with all signals of interest (in the sense of Equation 5.8) and uncorrelated with the noise, it can be used as an IV. Note that all signals that are not correlated with the reference will be treated as noise. Reference signals are commonly available in communication applications, for example, a PN-code in spread spectrum communication [20] or a training signal used for synchronization and/or equalizer training [21]. A closely related possibility is utilization of cyclostationarity (or self-coherence), a property that is exhibited by many man-made signals. The reference signal(s) can then consist, for example, of sinusoids of different frequencies [22,23]. In these techniques, the data are usually preprocessed by computing the autocovariance function (or a higher-order statistic) before correlating with the reference signal.

The problem considered in this section concerns the estimation of the DOA vector

$$\boldsymbol{\theta} = [\theta_1, \ldots, \theta_n]^{\mathrm{T}} \tag{5.9}$$

given N snapshots of the array output and of the IV vector, $\{y(t), z(t)\}_{t=1}^{N}$. The number of signals, n, and the rank of the covariance matrix Γ, $\bar{n}$, are assumed to be given (for the estimation of these integer-valued parameters by means of IV/SSF-based methods, we refer to [24,25]).

5.3 The IV-SSF Approach

Let

$$\hat{R} = \hat{W}_{\mathrm{L}} \left[\frac{1}{N} \sum_{t=1}^{N} z(t) y^*(t) \right] \hat{W}_{\mathrm{R}} \quad (\bar{m} \times m) \tag{5.10}$$

where $\hat{W}_{\mathrm{L}}$ and $\hat{W}_{\mathrm{R}}$ are two nonsingular Hermitian weighting matrices which are possibly data-dependent (as indicated by the fact that they are roofed). Under the assumptions made, as $N \to \infty$, $\hat{R}$ converges to the matrix:

$$R = W_{\mathrm{L}} \mathrm{E}[z(t) y^*(t)] W_{\mathrm{R}} = W_{\mathrm{L}} \Gamma A^* W_{\mathrm{R}} \tag{5.11}$$

where W_{L} and W_{R} are the limiting weighting matrices (assumed to be bounded and nonsingular). Owing to Assumptions A2 and A3,

$$\mathrm{rank}(R) = \bar{n} \tag{5.12}$$

Hence, the singular value decomposition (SVD) [26] of R can be written as

$$R = [U \ ?] \begin{bmatrix} \Lambda & O \\ O & O \end{bmatrix} \begin{bmatrix} S^* \\ ? \end{bmatrix} = U \Lambda S^* \tag{5.13}$$

where $U^* U = S^* S = I$, $\Lambda \in \mathcal{R}^{\bar{n} \times \bar{n}}$ is diagonal and nonsingular, and the question marks stand for blocks that are of no importance for the present discussion.

The following key equality is obtained by comparing the two expressions for R in Equations 5.11 and 5.13

$$S = W_{\mathrm{R}} A C \tag{5.14}$$

where $C \triangleq \Gamma^* W_{\mathrm{L}} U \Lambda^{-1} \in \mathcal{C}^{n \times \bar{n}}$ has full column rank. For a given S, the true DOA vector can be obtained as the unique solution to Equation 5.14 under the parameter identifiability condition (Equation 5.8) (see, e.g., [18]). In the more realistic case when S is unknown, one can make use of Equation 5.14 to estimate the DOA vector in the following steps.

The IV step—Compute the pre- and postweighted sample covariance matrix $\hat{R}$ in Equation 5.10, along with its SVD:

$$\hat{R} = \begin{bmatrix} \hat{U} & ? \end{bmatrix} \begin{bmatrix} \hat{\Lambda} & O \\ O & ? \end{bmatrix} \begin{bmatrix} \hat{S}^* \\ ? \end{bmatrix} \tag{5.15}$$

where $\hat{\Lambda}$ contains the $\bar{n}$ largest singular values. Note that $\hat{U}, \hat{\Lambda}$, and $\hat{S}$ are consistent estimates of U, Λ, and S in the SVD of R.

The signal subspace fitting (SSF) step—Compute the DOA estimate as the minimizing argument of the following SSF criterion:

$$\min_{\theta} \left\{ \min_{C} \left[\mathrm{vec}(\hat{S} - \hat{W}_{\mathrm{R}} A C) \right]^* \hat{V} \left[\mathrm{vec}(\hat{S} - \hat{W}_{\mathrm{R}} A C) \right] \right\} \tag{5.16}$$

where
$\hat{V}$ is a positive definite weighting matrix
"vec" is the vectorization operator*

Alternatively, one can estimate the DOA instead by minimizing the following criterion:

$$\min_{\boldsymbol{\theta}} \left\{ \left[\mathrm{vec}\left(\boldsymbol{B}^* \hat{\boldsymbol{W}}_{\mathrm{R}}^{-1} \hat{\boldsymbol{S}}\right)\right]^* \hat{\boldsymbol{W}} \left[\mathrm{vec}\left(\boldsymbol{B}^* \hat{\boldsymbol{W}}_{\mathrm{R}}^{-1} \hat{\boldsymbol{S}}\right)\right] \right\} \tag{5.17}$$

where
$\hat{\boldsymbol{W}}$ is a positive definite weight
$\boldsymbol{B} \in C^{m\times(m-n)}$ is a matrix whose columns form a basis of the null-space of A^* (hence, $\boldsymbol{B}^*\boldsymbol{A} = 0$ and rank$(\boldsymbol{B}) = m - n$)

The alternative fitting criterion above is obtained from the simple observation that Equation 5.14 along with the definition of $\boldsymbol{B}$ imply that

$$\boldsymbol{B}^* \boldsymbol{W}_{\mathrm{R}}^{-1} \boldsymbol{S} = 0 \tag{5.18}$$

It can be shown [27] that the classes of DOA estimates derived from Equations 5.16 and 5.17, respectively, are asymptotically equivalent. More exactly, for any $\hat{V}$ in Equation 5.16 one can choose $\hat{\boldsymbol{W}}$ in Equation 5.17 so that the DOA estimates obtained by minimizing Equation 5.16 and, respectively, Equation 5.17 have the same asymptotic distribution and vice-versa.

In view of the previous result, in an asymptotical analysis it suffices to consider only one of the two criteria above. In the following, we focus on Equation 5.17. Compared with Equation 5.16, the criterion (Equation 5.17) has the advantage that it depends on the DOA only. On the other hand, for a general array there is no known closed-form parameterization of $\boldsymbol{B}$ in terms of $\boldsymbol{\theta}$. However, as shown in the following, this is no drawback because the optimally weighted criterion (which is the one to be used in applications) is an explicit function of $\boldsymbol{\theta}$.

5.4 The Optimal IV-SSF Method

In what follows, we deal with the essential problem of choosing the weights $\hat{\boldsymbol{W}}, \hat{\boldsymbol{W}}_{\mathrm{R}}$, and $\hat{\boldsymbol{W}}_{\mathrm{L}}$ in the IV-SSF criterion (Equation 5.17) so as to maximize the DOA estimation accuracy. First, we optimize the accuracy with respect to $\hat{\boldsymbol{W}}$, and then with respect to $\hat{\boldsymbol{W}}_{\mathrm{R}}$ and $\hat{\boldsymbol{W}}_{\mathrm{L}}$.

5.4.1 Optimal Selection of $\hat{W}$

Define

$$\boldsymbol{g}(\boldsymbol{\theta}) = \mathrm{vec}\left(\boldsymbol{B}^* \hat{\boldsymbol{W}}_{\mathrm{R}}^{-1} \hat{\boldsymbol{S}}\right) \tag{5.19}$$

and observe that the criterion function in Equation 5.17 can be written as

$$\boldsymbol{g}^*(\boldsymbol{\theta}) \hat{\boldsymbol{W}} \boldsymbol{g}(\boldsymbol{\theta}) \tag{5.20}$$

In Stoica et al. [27] it is shown that $\boldsymbol{g}(\boldsymbol{\theta})$ (evaluated at the true DOA vector) has, asymptotically in N, a circularly symmetric normal distribution with zero mean and the following covariance:

$$\boldsymbol{G}(\boldsymbol{\theta}) = \frac{1}{N}\left[(\boldsymbol{W}_{\mathrm{L}} \boldsymbol{U} \boldsymbol{\Lambda}^{-1})^* \boldsymbol{R}_z (\boldsymbol{W}_{\mathrm{L}} \boldsymbol{U} \boldsymbol{\Lambda}^{-1})\right]^{\mathrm{T}} \otimes [\boldsymbol{B}^* \boldsymbol{R}_y \boldsymbol{B}] \tag{5.21}$$

* If x_k is the kth column of a matrix X, then vec $(X) = \left[x_1^{\mathrm{T}}\ x_2^{\mathrm{T}}\ \cdots\right]^{\mathrm{T}}$.

where $\otimes$ denotes the Kronecker matrix product [28]; and where, for a stationary signal $\boldsymbol{s}(t)$, we use the notation

$$\boldsymbol{R}_s = \mathrm{E}[\boldsymbol{s}(t)\boldsymbol{s}^*(t)] \tag{5.22}$$

Then, it follows from the asymptotically best consistent (ABC) theory of parameter estimation* to that the minimum variance estimate, in the class of estimates under discussion, is given by the minimizing argument of the criterion in Equation 5.20 with $\hat{\boldsymbol{W}} = \hat{\boldsymbol{G}}^{-1}(\boldsymbol{\theta})$, that is

$$f(\boldsymbol{\theta}) = \boldsymbol{g}^*(\boldsymbol{\theta})\hat{\boldsymbol{G}}^{-1}(\boldsymbol{\theta})\boldsymbol{g}(\boldsymbol{\theta}) \tag{5.23}$$

where

$$\hat{\boldsymbol{G}}(\boldsymbol{\theta}) = \frac{1}{N}\left[(\hat{\boldsymbol{W}}_{\mathrm{L}}\hat{\boldsymbol{U}}\hat{\boldsymbol{\Lambda}}^{-1})^*\hat{\boldsymbol{R}}_z(\hat{\boldsymbol{W}}_{\mathrm{L}}\hat{\boldsymbol{U}}\hat{\boldsymbol{\Lambda}}^{-1})\right]^{\mathrm{T}} \otimes [\boldsymbol{B}^*\hat{\boldsymbol{R}}_y\boldsymbol{B}] \tag{5.24}$$

and where $\hat{\boldsymbol{R}}_z$ and $\hat{\boldsymbol{R}}_y$ are the usual sample estimates of $\boldsymbol{R}_z$ and $\boldsymbol{R}_y$. Furthermore, it is easily shown that the minimum variance estimate, obtained by minimizing Equation 5.23, is asymptotically normally distributed with mean equal to the true parameter vector and the following covariance matrix:

$$\boldsymbol{H} = \tfrac{1}{2}\{\mathrm{Re}[\boldsymbol{J}^*\boldsymbol{G}^{-1}(\theta)\boldsymbol{J}]\}^{-1} \tag{5.25}$$

where

$$\boldsymbol{J} = \lim_{N\to\infty} \frac{\partial \boldsymbol{g}(\boldsymbol{\theta})}{\partial \boldsymbol{\theta}} \tag{5.26}$$

The following more explicit formula for $\boldsymbol{H}$ is derived in Stoica et al. [27]:

$$\boldsymbol{H} = \frac{1}{2N}\left(\mathrm{Re}\left\{\left[\boldsymbol{D}^*\boldsymbol{R}_y^{-1/2}\boldsymbol{\Pi}^{\perp}_{R_y^{-1/2}A}\boldsymbol{R}_y^{-1/2}\boldsymbol{D}\right] \odot \boldsymbol{\Omega}^{\mathrm{T}}\right\}\right)^{-1} \tag{5.27}$$

where $\odot$ denotes the Hadamard–Schur matrix product (element-wise multiplication) and

$$\boldsymbol{\Omega} = \boldsymbol{\Gamma}^*\boldsymbol{W}_{\mathrm{L}}\boldsymbol{U}(\boldsymbol{U}^*\boldsymbol{W}_{\mathrm{L}}\boldsymbol{R}_z\boldsymbol{W}_{\mathrm{L}}\boldsymbol{U})^{-1}\boldsymbol{U}^*\boldsymbol{W}_{\mathrm{L}}\boldsymbol{\Gamma} \tag{5.28}$$

Furthermore, the notation $\boldsymbol{Y}^{-1/2}$ is used for a Hermitian (for notational convenience) square root of the inverse of a positive definite matrix $\boldsymbol{Y}$, the matrix $\boldsymbol{D}$ is made from the direction vector derivatives,

$$D = [d_1 \ldots d_n]; \quad d_k = \frac{\partial a(\theta_k)}{\partial \theta_k}$$

and, for a full column-rank matrix $\boldsymbol{X}$, $\boldsymbol{\Pi}_X^{\perp}$ defines the orthogonal projection onto the nullspace of $\boldsymbol{X}^*$ as

$$\boldsymbol{\Pi}_X^{\perp} = I - \boldsymbol{\Pi}_X; \quad \boldsymbol{\Pi}_X = \boldsymbol{X}(\boldsymbol{X}^*\boldsymbol{X})^{-1}\boldsymbol{X}^* \tag{5.29}$$

* For details on the ABC theory, which is an extension of the classical best linear unbiased estimation (BLUE) Markov theory of linear regression to a class of nonlinear regressions with asymptotically vanishing residuals, the reader is referred to [9,29].

To summarize, for fixed $\hat{W}_R$ and $\hat{W}_L$, the statistically optimal selection of $\hat{W}$ leads to DOA estimates with an asymptotic normal distribution with mean equal to the true DOA vector and covariance matrix given by Equation 5.27.

5.4.2 Optimal Selection of $\hat{W}_R$ and $\hat{W}_L$

The optimal weights $\hat{W}_R$ and $\hat{W}_L$ are, by definition, those that minimize the limiting covariance matrix $\boldsymbol{H}$ of the DOA estimation errors. In the expression (Equation 5.27) of $\boldsymbol{H}$, only $\boldsymbol{\Omega}$ depends on $\boldsymbol{W}_R$ and $\boldsymbol{W}_L$ (the dependence on W_R is implicit, via U). Since the matrix Γ has rank $\bar{n}$, it can be factorized as follows:

$$\boldsymbol{\Gamma} = \boldsymbol{\Gamma}_1\boldsymbol{\Gamma}_2^* \tag{5.30}$$

where both $\boldsymbol{\Gamma}_1 \in C^{\bar{m}\times\bar{n}}$ and $\boldsymbol{\Gamma}_2 \in C^{n\times\bar{n}}$ have full column rank. Insertion of Equation 5.30 into the equality $\boldsymbol{W}_L\boldsymbol{\Gamma}\boldsymbol{A}^*\boldsymbol{W}_R = \boldsymbol{U}\boldsymbol{\Lambda}\boldsymbol{S}^*$ yields the following equation, after a simple manipulation,

$$\boldsymbol{W}_L\boldsymbol{\Gamma}_1\boldsymbol{T} = \boldsymbol{U} \tag{5.31}$$

where $\boldsymbol{T} = \boldsymbol{\Gamma}_2^*A^*\boldsymbol{W}_R\boldsymbol{S}\boldsymbol{\Lambda}^{-1} \in C^{\bar{n}\times\bar{n}}$ is a nonsingular transformation matrix. By using Equation 5.31 in Equation 5.28, we obtain

$$\boldsymbol{\Omega} = \boldsymbol{\Gamma}_2\left(\boldsymbol{\Gamma}_1^*W_L^2\boldsymbol{\Gamma}_1\right)\left(\boldsymbol{\Gamma}_1^*\boldsymbol{W}_L^2\boldsymbol{R}_z\boldsymbol{W}_L^2\boldsymbol{\Gamma}_1\right)^{-1}\left(\boldsymbol{\Gamma}_1^*W_L^2\boldsymbol{\Gamma}_1\right)\boldsymbol{\Gamma}_2^* \tag{5.32}$$

Observe that $\boldsymbol{\Omega}$ does not actually depend on $\boldsymbol{W}_R$. Hence, $\hat{\boldsymbol{W}}_R$ can be arbitrarily selected, as any nonsingular Hermitian matrix, without affecting the asymptotics of the DOA parameter estimates!

Concerning the choice of $\hat{\boldsymbol{W}}_L$, it is easily verified that

$$\boldsymbol{\Omega} \le \boldsymbol{\Omega}\big|_{\boldsymbol{W}_L=\boldsymbol{R}_z^{-1/2}} = \boldsymbol{\Gamma}_2\left(\boldsymbol{\Gamma}_1^*\boldsymbol{R}_z^{-1}\boldsymbol{\Gamma}_1\right)\boldsymbol{\Gamma}_2^* = \boldsymbol{\Gamma}^*\boldsymbol{R}_z^{-1}\boldsymbol{\Gamma} \tag{5.33}$$

Indeed,

$$\begin{aligned}\boldsymbol{\Gamma}^*\boldsymbol{R}_z^{-1}\boldsymbol{\Gamma} - \boldsymbol{\Omega} &= \boldsymbol{\Gamma}_2\left[\boldsymbol{\Gamma}_1^*\boldsymbol{R}_z^{-1}\boldsymbol{\Gamma}_1 - \left(\boldsymbol{\Gamma}_1^*W_L^2\boldsymbol{\Gamma}_1\right)\left(\boldsymbol{\Gamma}_1^*\boldsymbol{W}_L^2\boldsymbol{R}_z\boldsymbol{W}_L^2\boldsymbol{\Gamma}_1\right)^{-1}\left(\boldsymbol{\Gamma}_1^*\boldsymbol{W}_L^2\boldsymbol{\Gamma}_1\right)\right]\boldsymbol{\Gamma}_2^* \\ &= \boldsymbol{\Gamma}^*\boldsymbol{R}_z^{-1/2}\boldsymbol{\Pi}^{\perp}_{R_z^{1/2}W_L^2\Gamma_1}\boldsymbol{R}_z^{-1/2}\boldsymbol{\Gamma}\end{aligned} \tag{5.34}$$

which is obviously a nonnegative definite matrix. Hence, $\boldsymbol{W}_L = \boldsymbol{R}_z^{-1/2}$ maximizes $\boldsymbol{\Omega}$. Then, it follows from the expression of the matrix $\boldsymbol{H}$ and the properties of the Hadamard–Schur product that this same choice of $\boldsymbol{W}_L$ minimizes $\boldsymbol{H}$. The conclusion is that the optimal weight $\hat{\boldsymbol{W}}_L$, which yields the best limiting accuracy, is

$$\hat{\boldsymbol{W}}_L = \hat{\boldsymbol{R}}_z^{-1/2} \tag{5.35}$$

The (minimum) covariance matrix $\mathbf{H}$, corresponding to the above choice, is given by

$$\boldsymbol{H}_o = \frac{1}{2N}\left\{\mathrm{Re}\left[\left(\boldsymbol{D}^*\boldsymbol{R}_y^{-1/2}\boldsymbol{\Pi}^{\perp}_{R_y^{-1/2}A}\boldsymbol{R}_y^{-1/2}\boldsymbol{D}\right)\odot\left(\boldsymbol{\Gamma}^*\boldsymbol{R}_z^{-1}\boldsymbol{\Gamma}\right)^T\right]\right\}^{-1} \tag{5.36}$$

Remark—It is worth noting that $\boldsymbol{H}_o$ monotonically decreases as $\bar{m}$ (the dimension of $z(t)$) increases. The proof of this claim is similar to the proof of the corresponding result in Söderström and Stoica [9]. Hence, as could be intuitively expected, one should use all available instruments (spatial and/or temporal) to obtain maximal theoretical accuracy. However, practice has shown that too large a dimension of the IV vector may in fact decrease the empirically observed accuracy. This phenomenon can be explained by the fact that increasing $\bar{m}$ means that a longer data set is necessary for the asymptotic results to be valid.

5.4.3 Optimal IV-SSF Criteria

Fortunately, the criterion (Equations 5.23 and 5.24) can be expressed in a functional form that depends on the indeterminate $\boldsymbol{\theta}$ in an explicit way (recall that, for most cases, the dependence of $\boldsymbol{B}$ in Equation 5.23 on θ is not available in explicit form). By using the following readily verified equality [28],

$$\mathrm{tr}(\boldsymbol{A}\boldsymbol{X}^{*}\boldsymbol{B}\boldsymbol{Y}) = [\mathrm{vec}(\boldsymbol{X})]^{*}[\boldsymbol{A}^{\mathrm{T}} \otimes \boldsymbol{B}]\,[\mathrm{vec}(\mathbf{Y})] \tag{5.37}$$

which holds for any conformable matrices $\boldsymbol{A}, \boldsymbol{X}, \boldsymbol{B}$, and $\boldsymbol{Y}$, one can write Equation 5.23 as*

$$f(\boldsymbol{\theta}) = \mathrm{tr}\left\{[(\hat{\boldsymbol{W}}_{\mathrm{L}}\hat{\boldsymbol{U}}\hat{\boldsymbol{\Lambda}}^{-1})^{*}\hat{\boldsymbol{R}}_{z}(\hat{\boldsymbol{W}}_{\mathrm{L}}\hat{\boldsymbol{U}}\hat{\boldsymbol{\Lambda}}^{-1})]^{-1}\hat{\boldsymbol{S}}^{*}\hat{\boldsymbol{W}}_{\mathrm{R}}^{-1}\boldsymbol{B}(\boldsymbol{B}^{*}\hat{\boldsymbol{R}}_{y}\boldsymbol{B})^{-1}\boldsymbol{B}^{*}\hat{\boldsymbol{W}}_{\mathrm{R}}^{-1}\hat{\boldsymbol{S}}\right\} \tag{5.38}$$

However, observe that

$$\boldsymbol{B}(\boldsymbol{B}^{*}\hat{\boldsymbol{R}}_{y}\boldsymbol{B})^{-1}\boldsymbol{B}^{*} = \hat{\boldsymbol{R}}_{y}^{-1/2}\boldsymbol{\Pi}_{\hat{R}_{y}^{1/2}B}\hat{\boldsymbol{R}}_{y}^{-1/2} = \hat{\boldsymbol{R}}_{y}^{-1/2}\boldsymbol{\Pi}^{\perp}_{\hat{R}_{y}^{-1/2}A}\hat{\boldsymbol{R}}_{y}^{-1/2} \tag{5.39}$$

Inserting Equation 5.39 into Equation 5.38 yields

$$f(\boldsymbol{\theta}) = \mathrm{tr}\left[\hat{\boldsymbol{\Lambda}}(\hat{\boldsymbol{U}}^{*}\hat{\boldsymbol{W}}_{\mathrm{L}}\hat{\boldsymbol{R}}_{z}\hat{\boldsymbol{W}}_{\mathrm{L}}\hat{U})^{-1}\hat{\boldsymbol{\Lambda}}\hat{\boldsymbol{S}}^{*}\hat{\boldsymbol{W}}_{\mathrm{R}}^{-1}\hat{\boldsymbol{R}}_{y}^{-1/2}\boldsymbol{\Pi}^{\perp}_{\hat{R}_{y}^{-1/2}A}\hat{R}_{y}^{-1/2}\hat{\boldsymbol{W}}_{\mathrm{R}}^{-1}\hat{\boldsymbol{S}}\right] \tag{5.40}$$

which is an explicit function of $\boldsymbol{\theta}$. Insertion of the optimal choice of W_{L} into Equation 5.40 leads to a further simplification of the criterion as seen below.

Owing to the arbitrariness in the choice of $\hat{\boldsymbol{W}}_{\mathrm{R}}$, there exists an infinite class of optimal IV-SSF criteria. In what follows, we consider two members of this class.

Let

$$\hat{\boldsymbol{W}}_{\mathrm{R}} = \hat{\boldsymbol{R}}_{y}^{-1/2} \tag{5.41}$$

Insertion of Equation 5.41, along with Equation 5.35, into Equation 5.40 yields the following criterion function:

$$f_{WW}(\boldsymbol{\theta}) = \mathrm{tr}\left(\boldsymbol{\Pi}^{\perp}_{\hat{R}_{y}^{-1/2}A}\tilde{\mathbf{S}}\tilde{\boldsymbol{\Lambda}}^{2}\tilde{\boldsymbol{S}}^{*}\right) \tag{5.42}$$

where $\tilde{\mathbf{S}}$ and $\tilde{\boldsymbol{\Lambda}}$ are made from the principal singular right vectors and singular values of the matrix

$$\tilde{\boldsymbol{R}} = \tilde{\boldsymbol{R}}_{z}^{-1/2}\tilde{\boldsymbol{R}}_{zy}\tilde{\boldsymbol{R}}_{y}^{-1/2} \tag{5.43}$$

(with $\hat{\boldsymbol{R}}_{zy}$ defined in an obvious way). The function (Equation 5.42) is the UNCLE (spatial IV-SSF) criterion of Wong and Wu [12,13].

Next, choose $\hat{W}_{\mathrm{R}}$ as

$$\hat{\boldsymbol{W}}_{\mathrm{R}} = \boldsymbol{I} \tag{5.44}$$

The corresponding criterion function is

$$f_{VSO}(\theta) = \mathrm{tr}\left(\boldsymbol{\Pi}^{\perp}_{\hat{R}_{y}^{-1/2}A}\hat{\boldsymbol{R}}_{y}^{-1/2}\bar{\mathbf{S}}\bar{\boldsymbol{\Lambda}}^{2}\bar{\boldsymbol{S}}^{*}\hat{\boldsymbol{R}}_{y}^{-1/2}\right) \tag{5.45}$$

* To within a multiplicative constant.

where $\hat{\mathbf{S}}$ and $\hat{\boldsymbol{\Lambda}}$ are made from the principal singular pairs of

$$\bar{\boldsymbol{R}} = \hat{\boldsymbol{R}}_z^{-1/2}\hat{\boldsymbol{R}}_{zy} \tag{5.46}$$

The function (Equation 5.45) above is recognized as the optimal (temporal) IV-SSF criterion of Viberg et al. [14].

An important consequence of the previous discussion is that the DOA estimation methods of [12,13] and [14], respectively, which were derived in seemingly unrelated contexts and by means of somewhat different approaches, are in fact asymptotically equivalent when used under the same conditions. These two methods have very similar computational burdens, which can be seen by comparing Equations 5.42 and 5.43 with Equations 5.45 and 5.46. Also, their finite sample properties appear to be rather similar, as demonstrated in the simulation examples. Numerical algorithms for the minimization of the type of criterion function associated with the optimal IV-SSF methods are discussed in Ottersten et al. [17]. Some suggestions are also given in the summary below.

5.5 Algorithm Summary

The estimation method presented in this section is useful for direction finding in the presence of noise of unknown spatial color. The underlying assumptions and the algorithm can be summarized as follows:

Assumptions—A batch of N samples of the array output $y(t)$, that can accurately be described by the model (Equation 5.1 and 5.2) is available. The array is calibrated in the sense that $a(\theta)$ is a known function of its argument θ. In addition, N samples of the IV-vector $z(t)$, fulfilling Equations 5.4 through 5.8, are given. In words, the IV vector is uncorrelated with the noise but well correlated with the signal. In practice, $z(t)$ may be taken from a second subarray, a delayed version of $y(t)$, or a reference (pilot) signal. In the former case, the second subarray need not be calibrated.

Algorithm—In the following we summarize the UNCLE version (Equation 5.42) of the algorithm. First, compute $\tilde{R}$ from the sample statistics of $y(t)$ and $z(t)$, according to

$$\tilde{\boldsymbol{R}} = \hat{\boldsymbol{R}}_z^{-1/2}\hat{\boldsymbol{R}}_{zy}\hat{\boldsymbol{R}}_y^{-1/2}$$

From a numerical point of view, this is best done using QR factorization. Next, partition the SVD of $\tilde{\boldsymbol{R}}$ according to

$$\tilde{\boldsymbol{R}} = [\,\tilde{\boldsymbol{U}} \quad ?\,]\begin{bmatrix}\tilde{\boldsymbol{\Lambda}} & \boldsymbol{O}\\ \boldsymbol{O} & ?\end{bmatrix}\begin{bmatrix}\tilde{\mathbf{S}}^*\\ ?\end{bmatrix}$$

where $\tilde{\mathbf{S}}$ contains the $\bar{n}$ principal right singular vectors and the diagonal matrix $\tilde{\boldsymbol{\Lambda}}$ the corresponding singular values. If $\bar{n}$ is unknown, it can be estimated as the number of significant singular values. Finally, compute the DOA estimates as the minimizing arguments of the criterion function

$$f_{WW}(\boldsymbol{\theta}) = \mathrm{tr}\left(\boldsymbol{\Pi}^{\perp}_{\hat{R}_y^{-1/2}A}\tilde{\mathbf{S}}\tilde{\boldsymbol{\Lambda}}^2\tilde{\mathbf{S}}^*\right)$$

using $n = \bar{n}$. If the minimum value of the criterion is "large," it is an indication that more than $\bar{n}$ sources are present. In the general case, a numerical search must be performed to find the minimum. The **leastsq** implementation in Matlab®, which uses the Levenberg–Marquardt or Gauss–Newton techniques [30], is a possible choice. To initialize the search, one can use the alternating projection procedure [31]. In short, a grid search over $f_{WW}(\theta)$ is first performed assuming $n = 1$, i.e., using $f_{WW}(\theta_1)$. The resulting DOA estimate $\hat{\theta}_1$ is then "projected out" from the data, and a grid search for the second DOA is performed using the modified criterion $f_2(\theta_2)$. The procedure is repeated until initial estimates are available for all DOAs. The kth modified criterion can be expressed as

$$f_k(\theta_k) = -\frac{\boldsymbol{a}^*(\theta_k)\boldsymbol{\Pi}^{\perp}_{\hat{R}_y^{-1/2}\hat{A}_{k-1}}\tilde{\boldsymbol{S}}\tilde{\boldsymbol{\Lambda}}^2\tilde{\boldsymbol{S}}^*\boldsymbol{\Pi}^{\perp}_{\hat{R}_y^{-1/2}\hat{A}_{k-1}}a(\theta_k)}{\boldsymbol{a}^*(\theta_k)\boldsymbol{\Pi}^{\perp}_{\hat{R}_y^{-1/2}\hat{A}_{k-1}}\boldsymbol{a}(\theta_k)}$$

where

$$\hat{\boldsymbol{A}}_k = \boldsymbol{A}(\hat{\boldsymbol{\theta}}_k)$$
$$\hat{\boldsymbol{\theta}}_k = [\hat{\theta}_1, \ldots, \hat{\theta}_k]^{\mathrm{T}}$$

The initial estimate of θ_k is taken as the minimizing argument of $f_k(\theta_k)$. Once all DOAs have been initialized one can, in principle, continue the alternating projection minimization in the same way. However, the procedure usually converges rather slowly and therefore it is recommended instead to switch to a Newton-type search as indicated above. Empirical investigations in [17,32] using similar subspace fitting criteria, have indicated that this indeed leads to the global minimum with high probability.

5.6 Numerical Examples

This section reports the results of a comparative performance study based on Monte–Carlo simulations. The scenarios are identical to those presented in Stoica et al. [33] (spatial IV-SSF) and Viberg et al. [14] (temporal IV-SSF). The plots presented below contain theoretical standard deviations of the DOA estimates along with empirically observed root-mean-square (RMS) errors. The former are obtained from Equation 5.36, whereas the latter are based on 512 independent noise and signal realizations. The minimizers of Equation 5.42 (UNCLE) and Equation 5.45 (IV-SSF) are computed using a modified Gauss–Newton search initialized at the true DOAs (since here we are interested only in the quality of the global optimum). DOA estimates that are more than 5° off the true value are declared failures, and not included in the empirical RMS calculation. If the number of failures exceeds 30%, no RMS value is calculated.

In all scenarios, two planar wave fronts arrive from DOAs 0° and 5° relative to the array broadside. Unless otherwise stated, the emitter signals are zero-mean Gaussian with signal covariance matrix $\boldsymbol{P} = \boldsymbol{I}$. Only the estimation statistics for $\theta_1 = 0°$ are shown in the plots below, the ones for θ_2 being similar.

The array output (both subarrays in the spatial IV scenario) is corrupted by additive zero-mean temporally white Gaussian noise. The noise covariance matrix has klth element

$$\boldsymbol{Q}_{kl} = \sigma^2 0.9^{|k-l|} e^{j\frac{\pi}{2}(k-l)} \tag{5.47}$$

The noise level σ^2 is adjusted to give a desired SNR, defined as $\boldsymbol{P}_{11}/\sigma^2 = \boldsymbol{P}_{22}/\sigma^2$. This noise is reminiscent of a strong signal cluster at the location $\theta = 30°$.

Example 5.4: Spatial IVM

In the first example, a uniform linear array (ULA) of 16 elements and half-wavelength separation is employed. The first $m = 8$ contiguous sensors form a calibrated subarray, whereas the outputs of the last $\bar{m} = 8$ sensors are used as IVs, and these sensors could therefore be uncalibrated. Letting $y(t)$ denote the 16-element array output, we thus take

$$y(t) = \tilde{y}_{1:8}(t), \quad z(t) = \tilde{y}_{9:16}(t)$$

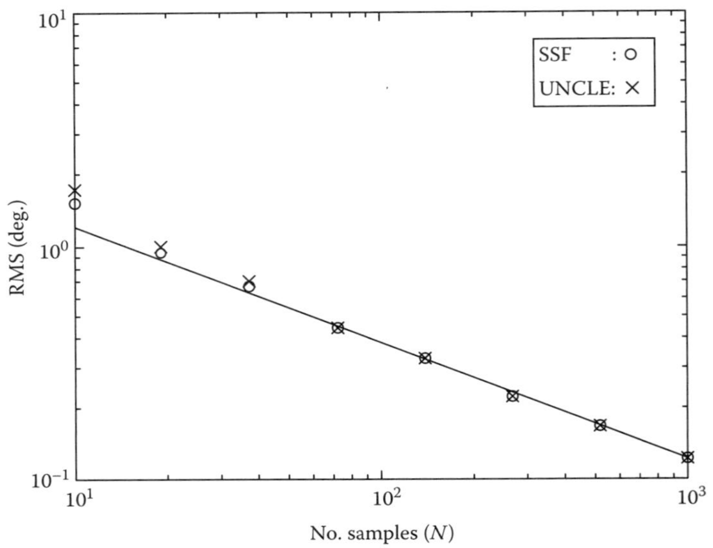

FIGURE 5.1 RMS error of DOA estimate vs. number of snapshots. Spatial IVM. The solid line is the theoretical standard deviation.

Both subarray outputs are perturbed by independent additive noise vectors, both having 8×8 covariance matrices given by Equation 5.47. In this example, the emitter signals are assumed to be temporally white.

In Figure 5.1, the theoretical and empirical RMS errors are displayed vs. the number of samples. The SNR is fixed at 6 dB.

Figure 5.2 shows the theoretical and empirical RMS errors vs. the SNR. The number of snapshots is here fixed to $N = 100$.

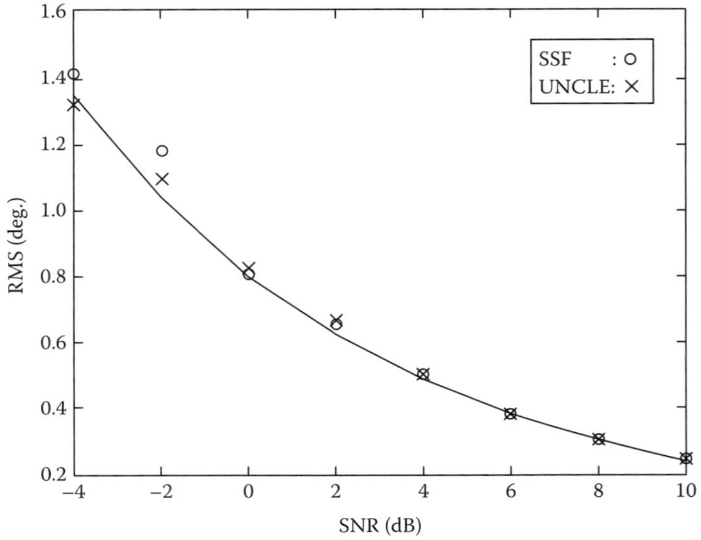

FIGURE 5.2 RMS error of DOA estimate vs. SNR. Spatial IVM. The solid line is the theoretical standard deviation.

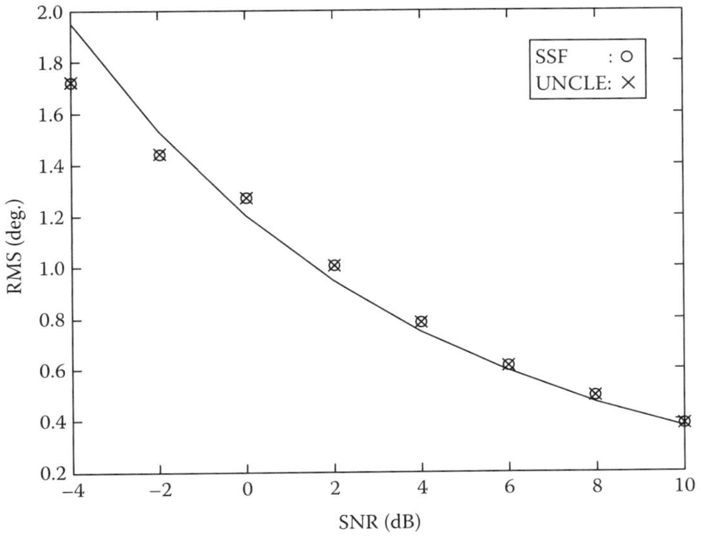

FIGURE 5.3 RMS error of DOA estimate vs. SNR. Spatial IVM. Coherent signals. The solid line is the theoretical standard deviation.

To demonstrate the applicability to situations involving highly correlated signals, Figure 5.2 is repeated but using the signal covariance

$$\boldsymbol{P} = \begin{bmatrix} 1 & 1 \\ 1 & 1 \end{bmatrix}$$

The resulting RMS errors are plotted with their theoretical values in Figure 5.3. By comparing Figures 5.2 and 5.3, we see that the methods are not insensitive to the signal correlation. However, the observed RMS errors agree well with the theoretically predicted values, and in spatial scenarios this is the best possible RMS performance (the empirical RMS error appears to be lower than the Cramer–Rao bound (CRB) for low SNR; however this is at the price of a notable bias).

In conclusion, no significant performance difference is observed between the two IV-SSF versions. The observed RMS errors of both methods follow the theoretical curves quite closely, even in fairly difficult scenarios involving closely spaced DOAs and highly correlated signals.

Example 5.5: Temporal IVM

In this example, the temporal IV approach is investigated. The array is a 6-element ULA of half wavelength interelement spacing. The real and imaginary parts of both signals are generated as uncorrelated first-order complex autoregressive (AR) processes with identical spectra. The poles of the driving AR-processes are 0.6. In this case, $y(t)$ is the array output, whereas the IV vector is chosen as $\mathbf{z}(t) = \mathbf{y}(t-1)$.

In Figure 5.4, we show the theoretical and empirical RMS errors vs. the number of snapshots. The SNR is fixed at 10 dB. Figure 5.5 displays the theoretical and empirical RMS errors vs. the SNR. The number of snapshots is here fixed at $N = 100$.

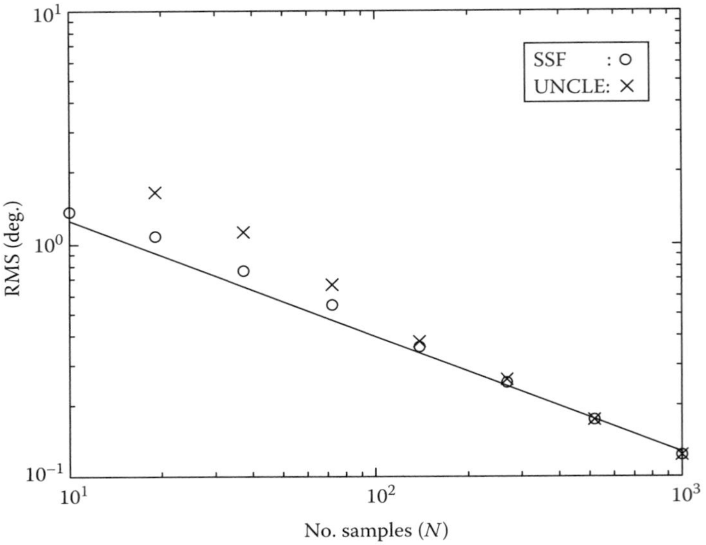

FIGURE 5.4 RMS error of DOA estimate vs. number of snapshots. Temporal IVM. The solid line is the theoretical standard deviation.

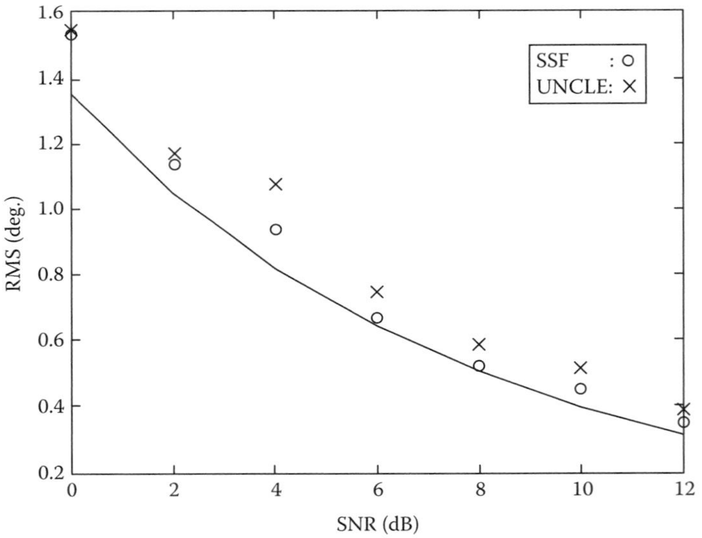

FIGURE 5.5 RMS error of DOA estimate vs. SNR. Temporal IVM. The solid line is the theoretical standard deviation.

The figures indicate a slight performance difference among the methods in temporal scenarios, namely when the number of samples is small but the SNR is relatively high. However, no definite conclusions can be drawn regarding this somewhat unexpected phenomenon from our limited simulation study.

5.7 Concluding Remarks

The main points made by the present contribution can be summarized as follows:

1. The spatial and temporal IV-SSF approaches can be treated in a unified manner under general conditions. In fact, a general IV-SSF approach using both spatial and temporal instruments is also possible.
2. The optimization of the DOA parameter estimation accuracy, for fixed weights $\hat{W}_L$ and $\hat{W}_R$, can be most conveniently carried out using the ABC theory. The resulting derivations are more concise than those based on other analysis techniques.
3. The column (or post)weight $\hat{W}_R$ has no effect on the asymptotics.
4. An important corollary of the above-mentioned result is that the optimal IV-SSF methods of Wu and Wong [12,13] and, respectively, Viberg et al. [14] are asymptotically equivalent when used on the same data.

In closing this section, we reiterate the fact that the IV-SSF approaches can deal with coherent signals, handle noise fields with general (unknown) spatial correlations, and, in their spatial versions, can make use of outputs from completely uncalibrated sensors. They are also comparatively simple from a computational standpoint, since no noise modeling is required. Additionally, the optimal IV-SSF methods provide highly accurate DOA estimates. More exactly, in spatial IV scenarios these DOA estimation methods can be shown to be asymptotically statistically efficient under weak conditions [33]. In temporal scenarios, they are no longer exactly statistically efficient, yet their accuracy is quite close to the best possible one [14]. All these features and properties should make the optimal IV-SSF approach appealing for practical array signal processing applications. The IV-SSF approach can also be applied, with some modifications, to system identification problems [34] and is hence expected to play a role in that type of application as well.

Acknowledgment

This work was supported in part by the Swedish Research Council for Engineering Sciences (TFR).

Appendix A: Introduction to IV Methods

In this appendix, we give a brief introduction to IVMs in their original context, which is time series analysis. Let $y(t)$ be a real-valued scalar time series, modeled by the autoregressive moving average (ARMA) equation

$$y(t) + a_1 y(t-1) + \cdots + a_p y(t-p) = e(t) + b_1 e(t-1) + \cdots + b_q e(t-q) \tag{5.A.1}$$

Here, $e(t)$ is assumed to be a stationary white noise. Suppose we are given measurements of $y(t)$ for $t = 1, \ldots, N$ and wish to estimate the AR parameters $a_1, \ldots, a_p$. The roots of the AR polynomial $z^p + a_1 z^{p-1} + \cdots + a_p$ are the system poles, and their estimation is of importance, for instance, for stability monitoring. Also, the first step of any "linear" method for ARMA modeling involves finding the AR parameters as the first step. The optimal way to approach the problem requires a nonlinear search over the entire parameter set $\{a_k\}_{k=1}^p, \{b_k\}_{k=1}^q$; using a maximum likelihood or a prediction error criterion [9,35]. However, in many cases this is computationally prohibitive, and in addition the "noise model" (the MA parameters) is sometimes of less interest per se. In contrast, the IV approach produces estimates of the AR part from a solution of a (possibly overdetermined) linear system of equations as follows: Rewrite Equation 5.A.1 as

$$y(t) = \boldsymbol{\varphi}^T(t)\boldsymbol{\theta} + v(t) \tag{5.A.2}$$

where

$$\boldsymbol{\varphi}(t) = [-y(t-1), \ldots, -y(t-p)]^{\mathrm{T}} \tag{5.A.3}$$

$$\boldsymbol{\theta} = [a_1, \ldots, a_p]^{\mathrm{T}} \tag{5.A.4}$$

$$v(t) = e(t) + b_1 e(t-1) + \cdots + b_q e(t-q) \tag{5.A.5}$$

Note that Equation 5.A.2 is a linear regression in the unknown parameter $\boldsymbol{\theta}$. A standard least squares (LS) estimate is obtained by minimizing the LS criterion

$$V_{\mathrm{LS}}(\boldsymbol{\theta}) = \mathrm{E}[(y(t) - \boldsymbol{\varphi}^{\mathrm{T}}(t)\boldsymbol{\theta})^2] \tag{5.A.6}$$

Equating the derivative of Equation 5.A.6 (w.r.t. θ) to zero gives the so-called normal equations

$$\mathrm{E}[\boldsymbol{\varphi}(t)\boldsymbol{\varphi}^{\mathrm{T}}(t)]\hat{\boldsymbol{\theta}} = \mathrm{E}[\boldsymbol{\varphi}(t)y(t)] \tag{5.A.7}$$

resulting in

$$\hat{\boldsymbol{\theta}} = \boldsymbol{R}_{\varphi\varphi}^{-1}\boldsymbol{R}_{\varphi y} = (\mathrm{E}[\boldsymbol{\varphi}(t)\boldsymbol{\varphi}^{\mathrm{T}}(t)])^{-1}\mathrm{E}[\boldsymbol{\varphi}(t)y(t)] \tag{5.A.8}$$

Inserting Equation 5.A.2 into Equation 5.A.8 shows that

$$\hat{\boldsymbol{\theta}} = \boldsymbol{\theta} + \boldsymbol{R}_{\varphi\varphi}^{-1}\boldsymbol{R}_{\varphi v} \tag{5.A.9}$$

In case $q = 0$ (i.e., $y(t)$ is an AR process), we have $v(t) = e(t)$. Because $\boldsymbol{\varphi}(t)$ and $e(t)$ are uncorrelated, Equation 5.A.9 shows that the LS method produces a consistent estimate of $\boldsymbol{\theta}$. However, when $q > 0$, $\boldsymbol{\varphi}(t)$ and $v(t)$ are in general correlated, implying that the LS method gives biased estimates.

From the above we conclude that the problem with the LS estimate in the ARMA case is that the regression vector $\boldsymbol{\varphi}(t)$ is correlated with the "equation error noise" $v(t)$. An IV vector $\boldsymbol{\zeta}(t)$ is one that is uncorrelated with $v(t)$, while still "sufficiently correlated" with $\boldsymbol{\varphi}(t)$. The most natural choice in the ARMA case (provided the model orders are known) is

$$\boldsymbol{\zeta}(t) = \boldsymbol{\varphi}(t-q) \tag{5.A.10}$$

which clearly fulfills both requirements. Now, multiply both sides of the linear regression model (Equation 5.A.2) by $\boldsymbol{\zeta}(t)$ and take expectation, resulting in the "IV normal equations"

$$\mathrm{E}[\boldsymbol{\zeta}(t)y(t)] = \mathrm{E}[\boldsymbol{\zeta}(t)\boldsymbol{\varphi}^{\mathrm{T}}(t)]\boldsymbol{\theta} \tag{5.A.11}$$

The IV estimate is obtained simply by solving the linear system of Equation 5.A.11, but with the unknown cross-covariance matrices $\boldsymbol{R}_{\zeta\varphi}$ and $\boldsymbol{R}_{\zeta y}$ replaced by their corresponding estimates using time averaging. Since the latter are consistent, so are the IV estimates of $\boldsymbol{\theta}$. The method is also referred to as the *extended Yule–Walker* approach in the literature. Its finite sample properties may often be improved upon by increasing the dimension of the IV vector, which means that Equation 5.A.11 must be solved in an LS sense, and also by appropriately prefiltering the IV-vector. This is quite similar to the optimal weighting proposed herein.

In order to make the connection to the IV-SSF method more clear, a slightly modified version of Equation 5.A.11 is presented. Let us rewrite Equation 5.A.11 as follows

$$\boldsymbol{R}_{\zeta\phi}\begin{bmatrix}1\\ \theta\end{bmatrix} = 0 \tag{5.A.12}$$

where

$$\boldsymbol{R}_{\zeta\phi} = \mathrm{E}\{\zeta(t)[y(t), -\boldsymbol{\varphi}^{\mathrm{T}}(t)]\} \tag{5.A.13}$$

The relation (Equation 5.A.12) shows that $\boldsymbol{R}_{\zeta\phi}$ is singular, and that $\boldsymbol{\theta}$ can be computed from a suitably normalized vector in its one-dimensional nullspace. However, when $\boldsymbol{R}_{\zeta\phi}$ is estimated using a finite number of data, it will with probability one have full rank. The best (in a LS sense) low-rank approximation of $\boldsymbol{R}_{\zeta\phi}$ is obtained by truncating its SVD. A natural estimate of $\boldsymbol{\theta}$ can therefore be obtained from the right singular vector of $\boldsymbol{R}_{\zeta\phi}$ that corresponds to the minimum singular value. The proposed modification is essentially an IV-SSF version of the extended Yule–Walker method, although the SSF step is trivial because the parameter vector of interest can be computed directly from the estimated subspace.

Turning to the array processing problem, the counterpart of Equation 5.A.2 is the (Hermitian transposed) data model (Equation 5.1)

$$\boldsymbol{y}^{*}(t) = \boldsymbol{x}^{*}(t)\boldsymbol{A}^{*} + e^{*}(t)$$

Note that this is a nonlinear regression model, owing to the nonlinear dependence of $\boldsymbol{A}$ on $\boldsymbol{\theta}$. Also observe that $y(t)$ is a complex vector as opposed to the real scalar $y(t)$ in Equation 5.A.2. Similar to Equation 5.A.11, the IV normal equations are given by

$$\mathrm{E}[z(t)\boldsymbol{y}^{*}(t)] = \mathrm{E}[z(t)\boldsymbol{x}^{*}(t)]\boldsymbol{A}^{*} \tag{5.A.14}$$

under the assumption that the IV-vector $\boldsymbol{z}(t)$ is uncorrelated with the noise $\boldsymbol{e}(t)$. Unlike the standard IV problem, the "regressor" $\boldsymbol{x}(t)$ [corresponding to $\varphi(t)$ in Equation 5.A.2] cannot be measured. Thus, it is not possible to get a direct estimate of the "regression variable" $\boldsymbol{A}$. However, its range space, or at least a subset thereof, can be computed from the principal right singular vectors. In the finite sample case, the performance can be improved by using row and column weighting, which leads to the weighted IV normal equation (Equation 5.11). The exact relation involving the principal right singular vectors is Equation 5.14, and two SSF formulations for revealing $\boldsymbol{\theta}$ from the computed signal subspace are given in Equations 5.16 and 5.17.

References

1. Li, F. and Vaccaro, R.J., Performance degradation of DOA estimators due to unknown noise fields, *IEEE Trans. SP*, SP-40(3), 686–689, March 1992.
2. Viberg, M., Sensitivity of parametric direction finding to colored noise fields and undermodeling, *Signal Process.*, 34(2), 207–222, November 1993.
3. Swindlehurst, A. and Kailath, T., A performance analysis of subspace-based methods in the presence of model errors: Part 2—Multidimensional algorithms, *IEEE Trans. SP*, SP-41, 2882–2890, September 1993.
4. Böhme, J.F. and Kraus, D., On least squares methods for direction of arrival estimation in the presence of unknown noise fields, in *Proceedings of the ICASSP 88*, pp. 2833–2836, New York, 1988.
5. Le Cadre, J.P., Parametric methods for spatial signal processing in the presence of unknown colored noise fields, *IEEE Trans. ASSP*, ASSP-37(7), 965–983, July 1989.

6. Nagesha, V. and Kay, S., Maximum likelihood estimation for array processing in colored noise, in *Proceedings of the ICASSP 93*, 4, pp. 240–243, Minneapolis, MN, 1993.
7. Ye, H. and DeGroat, R., Maximum likelihood DOA and unknown colored noise estimation with asymptotic Cramér-Rao bounds, in *Proceedings of the 27th Asilomar Conference on Signals, Systems, and Computers*, pp. 1391–1395, Pacific Grove, CA, November 1993.
8. Söderström, T. and Stoica, P., *Instrumental Variable Methods for System Identification*, Springer-Verlag, Berlin, 1983.
9. Söderström, T. and Stoica, P., *System Identification*, Prentice-Hall, London, U.K., 1989.
10. Moses, R.L. and Beex, A.A., Instrumental variable adaptive array processing, *IEEE Trans. AES*, AES-24, 192–202, March 1988.
11. Stoica, P., Viberg, M., and Ottersten, B., Instrumental variable approach to array processing in spatially correlated noise fields, *IEEE Trans. SP*, SP-42, 121–133, January 1994.
12. Wu, Q. and Wong, K.M., UN-MUSIC and UN-CLE: An application of generalized canonical correlation analysis to the estimation of the directions of arrival of signals in unknown correlated noise, *IEEE Trans. SP*, 42, 2331–2341, September 1994.
13. Wu, Q. and Wong, K.M., Estimation of DOA in unknown noise: Performance analysis of UN-MUSIC and UN-CLE, and the optimality of CCD, *IEEE Trans. SP*, 43, 454–468, February 1995.
14. Viberg, M., Stoica, P., and Ottersten, B., Array processing in correlated noise fields based on instrumental variables and subspace fitting, *IEEE Trans. SP*, 43, 1187–1199, May 1995.
15. Stoica, P., Viberg, M., Wong, M., and Wu, Q., Maximum-likelihood bearing estimation with partly calibrated arrays in spatially correlated noise fields, *IEEE Trans. SP*, 44, 888–899, April 1996.
16. Schmidt, R.O., Multiple emitter location and signal parameter estimation, *IEEE Trans. AP*, 34, 276–280, March 1986.
17. Ottersten, B., Viberg, M., Stoica, P., and Nehorai, A., Exact and large sample ML techniques for parameter estimation and detection in array processing, in *Radar Array Processing*, Haykin, S., Litva, J., and Shepherd, T.J. eds., Springer-Verlag, Berlin, 1993, pp. 99–151.
18. Wax, M. and Ziskind, I., On unique localization of multiple sources by passive sensor arrays, *IEEE Trans. ASSP*, ASSP-37(7), 996–1000, July 1989.
19. Hudson, J.E., *Adaptive Array Principles*, Peter Peregrinus, London, U.K., 1981.
20. Compton, R.T., Jr., *Adaptive Antennas*, Prentice-Hall, Englewood Cliffs, NJ, 1988.
21. Lee, W.C.Y., *Mobile Communications Design Fundamentals*, 2nd edn., John Wiley & Sons, New York, 1993.
22. Agee, B.G., Schell, A.V., and Gardner, W.A., Spectral self-coherence restoral: A new approach to blind adaptive signal extraction using antenna arrays, *Proc. IEEE*, 78, 753–767, April 1990.
23. Shamsunder, S. and Giannakis, G., Signal selective localization of nonGaussian cyclostationary sources, *IEEE Trans. SP*, 42, 2860–2864, October 1994.
24. Zhang, Q.T. and Wong, K.M., Information theoretic criteria for the determination of the number of signals in spatially correlated noise, *IEEE Trans. SP*, SP-41(4), 1652–1663, April 1993.
25. Wu, Q. and Wong, K.M., Determination of the number of signals in unknown noise environments, *IEEE Trans. SP*, 43, 362–365, January 1995.
26. Golub, G.H. and VanLoan, C.F., *Matrix Computations*, 2nd edn., Johns Hopkins University Press, Baltimore, MD, 1989.
27. Stoica, P., Viberg, M., Wong, M., and Wu, Q., A unified instrumental variable approach to direction finding in colored noise fields: Report version, Technical Report CTH-TE-32, Chalmers University of Technology, Gothenburg, Sweden, July 1995.
28. Brewer, J.W., Kronecker products and matrix calculus in system theory, *IEEE Trans. CAS*, 25(9), 772–781, September 1978.
29. Porat, B., *Digital Processing of Random Signals*, Prentice-Hall, Englewood Cliffs, NJ, 1993.
30. Gill, P.E., Murray, W., and Wright, M.H., *Practical Optimization*, Academic Press, London, 1981.

31. Ziskind, I. and Wax, M., Maximum likelihood localization of multiple sources by alternating projection, *IEEE Trans. ASSP*, ASSP-36, 1553–1560, October 1988.
32. Viberg, M., Ottersten, B., and Kailath, T., Detection and estimation in sensor arrays using weighted subspace fitting, *IEEE Trans. SP*, SP-39(11), 2436–2449, November 1991.
33. Stoica, P., Viberg, M., Wong, M., and Wu, Q., Optimal direction finding with partly calibrated arrays in spatially correlated noise fields, in *Proceedings of the 28th Asilomar Conference on Signals, Systems, and Computers*, Pacific Grove, CA, October 1994.
34. Cedervall, M. and Stoica, P., System identification from noisy measurements by using instrumental variables and subspace fitting, in *Proceedings of the ICASSP 95*, pp. 1713–1716, Detroit, MI, May 1995.
35. Ljung, L., *System Identification: Theory for the User*, Prentice-Hall, Englewood Cliffs, NJ, 1987.

6

Electromagnetic Vector-Sensor Array Processing*

6.1 Introduction 6-1
6.2 Measurement Models 6-3
Single-Source Single-Vector Sensor Model • Multisource Multivector Sensor Model
6.3 Cramér–Rao Bound for a Vector-Sensor Array 6-10
Statistical Model • The Cramér–Rao Bound
6.4 MSAE, CVAE, and Single-Source Single-Vector Sensor Analysis 6-12
The MSAE • DST Source Analysis • SST Source (DST Model) Analysis • SST Source (SST Model) Analysis • CVAE and SST Source Analysis in the Wave Frame • A Cross-Product-Based DOA Estimator
6.5 Multisource Multivector Sensor Analysis 6-21
Results for Multiple Sources, Single-Vector Sensor
6.6 Concluding Remarks 6-24
Acknowledgments 6-25
Appendix A: Definitions of Some Block Matrix Operators 6-25
References 6-26

Arye Nehorai
The University of Illinois at Chicago

Eytan Paldi
Israel Institute of Technology

6.1 Introduction

This chapter (see also [1,2]) considers new methods for multiple electromagnetic source localization using sensors whose output is a "vector" corresponding to the complete electric and magnetic fields at the sensor. These sensors, which will be called "vector sensors," can consist, for example, of two orthogonal triads of scalar sensors that measure the electric and magnetic field components. Our approach is in contrast to other chapters that employ sensor arrays in which the output of each sensor is a scalar corresponding, for example, to a scalar function of the electric field. The main advantage of the vector sensors is that they make use of all available electromagnetic information and hence should outperform the scalar sensor arrays in accuracy of direction of arrival (DOA) estimation. Vector sensors should also allow the use of smaller array apertures while improving performance. (Note that we use the term "vector sensor" for a device that measures a complete physical vector quantity.)

Section 6.2 derives the measurement model. The electromagnetic sources considered can originate from two types of transmissions: (1) single signal transmission (SST), in which a single signal message is

* Dedicated to the memory of our physics teacher, Isaac Paldi.

transmitted, and (2) dual signal transmission (DST), in which two separate signal messages are transmitted simultaneously (from the same source), see for example [3,4]. The interest in DST is due to the fact that it makes full use of the two spatial degrees of freedom present in a transverse electromagnetic plane wave. This is particularly important in the wake of increasing demand for economical spectrum usage by existing and emerging modern communication technologies.

Section 6.3 analyzes the minimum attainable variance of unbiased DOA estimators for a general vector-sensor array model and multielectromagnetic sources that are assumed to be stochastic and stationary. A compact expression for the corresponding Cramér–Rao bound (CRB) on the DOA estimation error that extends previous results for the scalar sensor array case in [5] (see also [6]) is presented.

A significant property of the vector sensors is that they enable DOA (azimuth and elevation) estimation of an electromagnetic source with a single-vector sensor and a single snapshot. This result is explicitly shown by using the CRB expression for this problem in Section 6.4. A bound on the associated normalized mean-square angular error (MSAE, to be defined later) which is invariant to the reference coordinate system is used for an in-depth performance study. Compact expressions for this MSAE bound provide physical insight into the SST and DST source localization problems with a single-vector sensor.

The CRB matrix for an SST source in the sensor coordinate frame exhibits some nonintrinsic singularities (i.e., singularities that are not inherent in the physical model while being dependent on the choice of the reference coordinate system) and has complicated entry expressions. Therefore, we introduce a new vector angular error defined in terms of the incoming wave frame. A bound on the normalized asymptotic covariance of the vector angular error (CVAE) is derived. The relationship between the CVAE and MSAE and their bounds is presented. The CVAE matrix bound for the SST source case is shown to be diagonal, easy to interpret, and to have only intrinsic singularities.

We propose a simple algorithm for estimating the source DOA with a single-vector sensor, motivated by the Poynting vector. The algorithm is applicable to various types of sources (e.g., wideband and non-Gaussian); it does not require a minimization of a cost function and can be applied in real time. Statistical performance analysis evaluates the variance of the estimator under mild assumptions and compares it with the MSAE lower bound.

Section 6.5 extends these results to the multisource multivector sensor case, with special attention to the two-source single-vector sensor case. Section 6.5 summarizes the main results and gives some ideas of possible extensions.

The main difference between the topics of this chapter and other chapters on source direction estimation is in our use of vector sensors with complete electric and magnetic data. Most papers have dealt with scalar sensors. Other papers that considered estimation of the polarization state and source direction are [7–12]. Reference [7] discussed the use of subspace methods to solve this problem using diversely polarized electric sensors. References [8–10] devised algorithms for arrays with two-dimensional electric measurements. Reference [11] provided performance analysis for arrays with two types of electric sensor polarizations (diversely polarized). An earlier reference, [12], proposed an estimation method using a three-dimensional vector sensor and implemented it with magnetic sensors. All these references used only part of the electromagnetic information at the sensors, thereby reducing the observability of DOAs. In most of them, time delays between distributed sensors played an essential role in the estimation process.

For a plane wave (typically associated with a single source in the far-field) the magnitude of the electric and magnetic fields can be found from each other. Hence, it may be felt that one (complete) field is deducible from the other. However, this is not true when the source direction is unknown. Additionally, the electric and magnetic fields are orthogonal to each other and to the source DOA vector, hence measuring both fields increases significantly the accuracy of the source DOA estimation. This is true in particular for an incoming wave which is nearly linearly polarized, as will be explicitly shown by the CRB (see Table 6.1).

TABLE 6.1 MSAE Bounds for a SST Source

	Elliptical	Circular	Linear
General $\text{MSAE}_{\text{CR}}^{\text{S}}$	Equation 6.34	$\frac{2(1+\#)}{\#^2}$	$\frac{(1+\#)(\sigma_E^2+\sigma_H^2)}{2\#\,\sigma_s^2}$
Precise electric measurement	0	0	$\frac{\sigma_H^2}{2\sigma_s^2}$
Electric measurement only	$\frac{\sigma_E^2(\sigma_E^2+\sigma_s^2)}{2\sigma_s^4\sin^2\theta_4\cos^2\theta_4}$	$\frac{2\sigma_E^2(\sigma_E^2+\sigma_s^2)}{\sigma_s^4}$	∞

The use of the complete electromagnetic vector data enables source parameter estimation with a single sensor (even with a single snapshot) where time delays are not used at all. In fact, this is shown to be possible for at least two sources. As a result, the derived CRB expressions for this problem are applicable to wideband sources. The source DOA parameters considered include azimuth and elevation. This section also considers direction estimation to DST sources, as well as the CRB on wave ellipticity and orientation angles (to be defined later) for SST sources using vector sensors, which were first presented in [1,2]. This is true also for the MSAE and CVAE quality measures and the associated bounds. Their application is not limited to electromagnetic vector-sensor processing.

We comment that electromagnetic vector sensors as measuring devices are commercially available and actively researched. EMC Baden Ltd. in Baden, Switzerland, is a company that manufactures them for signals in the 75 Hz–30 MHz frequency range, and Flam and Russell, Inc. in Horsham, Pennsylvania, makes them for the 2–30 MHz frequency band. Lincoln Labs at MIT has performed some preliminary localization tests with vector sensors [13]. Some examples of recent research on sensor development are [14,15].

Following the recent impressive progress in the performance of DSP processors, there is a trend to fuse as much data as possible using smart sensors. Vector sensors, which belong to this category of sensors, are expected to find larger use and provide important contribution in improving the performance of DSP in the near future.

6.2 Measurement Models

This section presents the measurement models for the estimation problems that are considered in the latter parts of the chapter.

6.2.1 Single-Source Single-Vector Sensor Model

6.2.1.1 Basic Assumptions

Throughout the chapter it will be assumed that the wave is traveling in a nonconductive, homogeneous, and isotropic medium. Additionally, the following will be assumed:

A1: Plane wave at the sensor: This is equivalent to a far-field assumption (or maximum wavelength much smaller than the source to sensor distance), a point source assumption (i.e., the source size is much smaller than the source to sensor distance) and a point-like sensor (i.e., the sensor's dimensions are small compared to the minimum wavelength).

A2: Band-limited spectrum: The signal has a spectrum including only frequencies ω satisfying $\omega_{\min} \leq |\omega| \leq \omega_{\max}$ where $0 < \omega_{\min} < \omega_{\max} < \infty$. This assumption is satisfied in practice. The lower and upper limits on ω are also needed, respectively, for the far-field and point-like sensor assumptions.

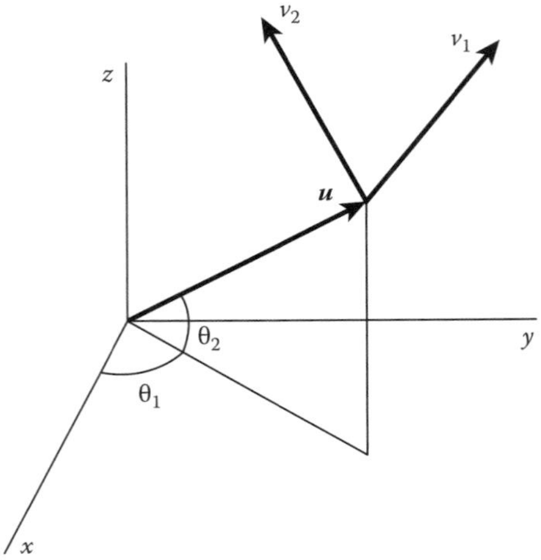

FIGURE 6.1 The orthonormal vector triad ($\boldsymbol{u}, v_1, v_2$).

Let $E(t)$ and $H(t)$ be the vector phasor representations (or complex envelopes, see, e.g., [1,16,17]) of the electric and magnetic fields at the sensor. Also, let $\boldsymbol{u}$ be the unit vector at the sensor pointing toward the source, i.e.,

$$\boldsymbol{u} = \begin{bmatrix} \cos\theta_1 \cos\theta_2 \\ \sin\theta_1 \cos\theta_2 \\ \sin\theta_2 \end{bmatrix} \tag{6.1}$$

where θ_1 and θ_2 denote, respectively, the azimuth and elevation angles of $\boldsymbol{u}$, see Figure 6.1. Thus, $\theta_1 \in [0, 2\pi)$ and $|\theta_2| \leq \pi/2$.

In [1, Appendix A] it is shown that for plane waves Maxwell's equations can be reduced to an equivalent set of two equations without any loss of information. Under the additional assumption of a band-limited signal, these two equations can be written in terms of phasors. The results are summarized in the following theorem.

THEOREM 6.1

Under Assumption A1, Maxwell's equations can be reduced to an equivalent set of two equations. With the additional band-limited spectrum Assumption A2, they can be written as

$$\boldsymbol{u} \times \varepsilon(t) = -\mathcal{H}(t) \tag{6.2a}$$

$$\boldsymbol{u} \cdot \varepsilon(t) = 0 \tag{6.2b}$$

where

η is the intrinsic impedance of the medium

"$\times$" and "$\cdot$" are the cross and inner products of $\mathbb{R}^3$ applied to vectors in $\mathbb{C}^3$

(That is, if $\mathbf{v}, \mathbf{w} \in \mathbb{C}^3$ then $\boldsymbol{v} \cdot \boldsymbol{w} = \sum_i v_i w_i$. This is different than the usual inner product of $\mathbb{C}^3$.)

Proof 6.1 See [1, Appendix A]. (Note that $u = -\kappa$, where κ is the unit vector in the direction of the wave propagation.)

Thus, under the plane and band-limited wave assumptions, the vector phasor equations (Equation 6.2) provide all the information contained in the original Maxwell equations. This result will be used in the following to construct measurement models in which the Maxwell equations are incorporated entirely.

6.2.1.2 The Measurement Model

Suppose that a vector sensor measures all six components of the electric and magnetic fields. (It is assumed that the sensor does not influence the electric and magnetic fields.) The measurement model is based on the phasor representation of the measured electromagnetic data (with respect to a reference frame) at the sensor. Let $y_E(t)$ be the measured electric field phasor vector at the sensor at time t and $e_E(t)$ its noise component. Then the electric part of the measurement will be

$$y_E(t) = \varepsilon(t) + e_E(t) \tag{6.3}$$

Similarly, from Equation 6.2a, after appropriate scaling, the magnetic part of the measurement will be taken as

$$\mathbf{y}_H(t) = \mathbf{u} \times \varepsilon(t) + \mathbf{e}_H(t) \tag{6.4}$$

In addition to Equations 6.3 and 6.4, we have the constraint 6.2b.

Define the matrix cross-product operator that maps a vector $\mathbf{v} \in \mathbb{R}^{3\times 1}$ to $(\boldsymbol{u} \times \boldsymbol{v}) \in \mathbb{R}^{3\times 1}$ by

$$(\mathbf{u}\times) \stackrel{\Delta}{=} \begin{bmatrix} 0 & -u_z & u_y \\ u_z & 0 & -u_x \\ -u_y & u_x & 0 \end{bmatrix} \tag{6.5}$$

where u_x, u_y, u_z are the x, y, z components of the vector $\boldsymbol{u}$. With this definition, Equations 6.3 and 6.5 can be combined to

$$\begin{bmatrix} \mathbf{y}_E(t) \\ \mathbf{y}_H(t) \end{bmatrix} = \begin{bmatrix} I_3 \\ (\mathbf{u}\times) \end{bmatrix} \varepsilon(t) + \begin{bmatrix} \mathbf{e}_E(t) \\ \mathbf{e}_H(t) \end{bmatrix} \tag{6.6}$$

where I_3 denotes the 3×3 identity matrix. For notational convenience the dimension subscript of the identity matrix will be omitted whenever its value is clear from the context.

The constraint 6.2b implies that the electric phasor $\varepsilon(t)$ can be written

$$\varepsilon(t) = V\xi(t) \tag{6.7}$$

where V is a 3×2 matrix whose columns span the orthogonal complement of $\boldsymbol{u}$ and $\xi(t) \in \mathcal{C}^{2\times 1}$. It is easy to check that the matrix

$$V = \begin{bmatrix} -\sin\theta_1 & -\cos\theta_1 \sin\theta_2 \\ \cos\theta_1 & -\sin\theta_1 \sin\theta_2 \\ 0 & \cos\theta_2 \end{bmatrix} \tag{6.8}$$

whose columns are orthonormal, satisfies this requirement. We note that since $\|\mathbf{u}\|^2 = 1$ the columns of V, denoted by $\boldsymbol{v}_1$ and $\boldsymbol{v}_2$, can be constructed, for example, from the partial derivatives of $\boldsymbol{u}$ with respect to θ_1 and θ_2 and postnormalization when needed. Thus,

$$\boldsymbol{v}_1 = \frac{1}{\cos\theta_2}\frac{\partial \boldsymbol{u}}{\partial \theta_2} \tag{6.9a}$$

$$\boldsymbol{v}_2 = \boldsymbol{u} \times \boldsymbol{v}_1 = \frac{\partial \boldsymbol{u}}{\partial \theta_2} \tag{6.9b}$$

and $(\mathbf{u}, \mathbf{v}_1, \mathbf{v}_2)$ is a right orthonormal triad, see Figure 6.1. (Observe that the two coordinate systems shown in the figure actually have the same origin). The signal $\xi(t)$ fully determines the components of $E(t)$ in the plane where it lies, namely the plane orthogonal to $\boldsymbol{u}$ spanned by $\mathbf{v}_1, \mathbf{v}_2$. This implies that there are two degrees of freedom present in the spatial domain (or the wave's plane), or two independent signals can be transmitted simultaneously.

Combining Equations 6.6 and 6.7 we now have

$$\begin{bmatrix} y_{\mathrm{E}}(t) \\ y_{\mathrm{H}}(t) \end{bmatrix} = \begin{bmatrix} I \\ (\boldsymbol{u}\times) \end{bmatrix} V\xi(t) + \begin{bmatrix} e_{\mathrm{E}}(t) \\ e_{\mathrm{H}}(t) \end{bmatrix} \tag{6.10}$$

This system is equivalent to Equation 6.6 with Equation 6.2b.

The measured signals in the sensor reference frame can be further related to the original source signal at the transmitter using the following lemma.

LEMMA 6.1

Every vector $\xi = [\xi_1, \xi_2]^{\mathrm{T}} \in \mathbb{C}^{2\times 1}$ has the representation

$$\xi = \|\xi\| e^{i\varphi} \boldsymbol{Q}\boldsymbol{w} \tag{6.11}$$

where

$$\boldsymbol{Q} = \begin{bmatrix} \cos\theta_3 & \sin\theta_3 \\ -\sin\theta_3 & \cos\theta_3 \end{bmatrix} \tag{6.12a}$$

$$\boldsymbol{w} = \begin{bmatrix} \cos\theta_4 \\ i\sin\theta_4 \end{bmatrix} \tag{6.12b}$$

and where $\varphi \in (-\pi, \pi]$, $\theta_3 \in (-\pi/2, \pi/2]$, $\theta_4 \in [-\pi/4, \pi/4]$. Moreover, $\|\xi\|, \varphi, \theta_3, \theta_4$ in Equation 6.11 are uniquely determined if and only if $\xi_1^2 + \xi_2^2 \neq 0$.

Proof 6.2 See [1, Appendix B].

The equality $\xi_1^2 + \xi_2^2 = 0$ holds if and only if $|\theta_4| = \pi/4$, corresponding to circular polarization (defined below). Hence, from Lemma 6.1 the representation (Equations 6.11 and 6.12) is not unique in this case as should be expected, since the orientation angle θ_3 is ambiguous. It should be noted that the representation (Equations 6.11 and 6.12) is known and was used (see, e.g., [18]) without a proof. However, Lemma 6.1 of existence and uniqueness appears to be new. The existence and uniqueness properties are important to guarantee identifiability of parameters.

The physical interpretations of the quantities in the representation (Equations 6.11 and 6.12) are as follows. $\|\xi\|e^{i\varphi}$: Complex envelope of the source signal (including amplitude and phase).

w: Normalized overall transfer vector of the source's antenna and medium, i.e., from the source complex envelope signal to the principal axes of the received electric wave.

$\mathbf{Q}$: A rotation matrix that performs the rotation from the principal axes of the incoming electric wave to the $(\mathbf{v}_1, \mathbf{v}_2)$ coordinates.

Let ω_c be the reference frequency of the signal phasor representation, see [1, Appendix A]. In the narrow-band SST case, the incoming electric wave signal $\mathrm{Re}\{e^{i\omega_c t}\|\xi(t)\|e^{i\varphi(t)}Qw\}$ moves on a quasistationary ellipse whose semimajor and semiminor axes' lengths are proportional, respectively, to $\cos\theta_4$ and $\sin\theta_4$, see Figure 6.2 and [19]. The ellipse's eccentricity is thus determined by the magnitude of θ_4. The sign of θ_4 determines the spin sign or direction. More precisely, a positive (negative) θ_4 corresponds to a positive (negative) spin with right (left)-handed rotation with respect to the wave propagation vector $\kappa = -\mathbf{u}$. As shown in Figure 6.2, θ_3 is the rotation angle between the $(\mathbf{v}_1, \mathbf{v}_2)$ coordinates and the electric ellipse axes $\tilde{\mathbf{v}}_1, \tilde{\mathbf{v}}_2)$. The angles θ_3 and θ_4 will be referred to, respectively, as the orientation and ellipticity angles of the received electric wave ellipse. In addition to the electric ellipse, there is also a similar but perpendicular magnetic ellipse.

It should be noted that if the transfer matrix from the source to the sensor is time invariant, then so are θ_3 and θ_4.

The signal $\xi(t)$ can carry information coded in various forms. In the following we discuss briefly both existing forms and some motivated by the above representation.

6.2.1.3 Single Signal Transmission Model

Suppose that a single modulated signal is transmitted. Then, using Equation 6.11, this is a special case of Equation 6.10 with

$$\xi(t) = \mathbf{Q}ws(t) \tag{6.13}$$

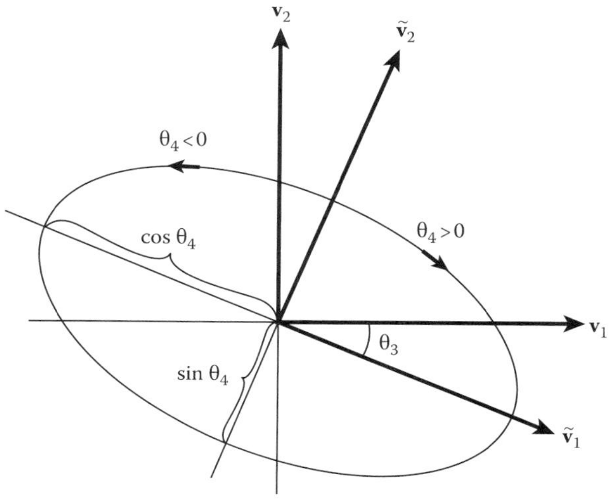

FIGURE 6.2 The electric polarization ellipse.

where $s(t)$ denotes the complex envelope of the (scalar) transmitted signal. Thus, the measurement model is

$$\begin{bmatrix} y_E(t) \\ y_H(t) \end{bmatrix} = \begin{bmatrix} I \\ (\boldsymbol{u}\times) \end{bmatrix} V\boldsymbol{Q}\boldsymbol{w}s(t) + \begin{bmatrix} \mathbf{e}_E(t) \\ \mathbf{e}_H(t) \end{bmatrix} \tag{6.14}$$

Special cases of this transmission are linear polarization with $\theta_4 = 0$ and circular polarization with $|\theta_4| = \pi/4$.

Recall that since there are two spatial degrees of freedom in a transverse electromagnetic plane wave, one could, in principle, transmit two separate signals simultaneously. Thus, the SST method does not make full use of the two spatial degrees of freedom present in a transverse electromagnetic plane wave.

6.2.1.4 Dual Signal Transmission Model

Methods of transmission in which two separate signals are transmitted simultaneously from the same source will be called "dual signal transmissions." Various DST forms exist, and all of them can be modeled by Equation 6.10 with $\xi(t)$ being a linear transformation of the two-dimensional source signal vector.

One DST form uses two linearly polarized signals that are spatially and temporally orthogonal with an amplitude or phase modulation (see, e.g., [3,4]). This is a special case of Equation 6.10, where the signal $\xi(t)$ is written in the form

$$\xi(t) = Q\begin{bmatrix} s_1(t) \\ is_2(t) \end{bmatrix} \tag{6.15}$$

where $s_1(t)$ and $s_2(t)$ represent the complex envelopes of the transmitted signals. To guarantee unique decoding of the two signals (when θ_3 is unknown) using Lemma 6.1, they have to satisfy $s_1(t) \neq 0$, $s_2(t)/s_1(t) \in (-1, 1)$. (Practically this can be achieved by using a proper electronic antenna adapter that yields a desirable overall transfer matrix.)

Another DST form uses two circularly polarized signals with opposite spins. In this case

$$\xi(t) = \boldsymbol{Q}[\boldsymbol{w}\tilde{s}_1(t) + \bar{\boldsymbol{w}}\tilde{s}_2(t)] \tag{6.16a}$$

$$\boldsymbol{w} = (1/\sqrt{2})[1, i]^{\mathrm{T}} \tag{6.16b}$$

where $\bar{\boldsymbol{w}}$ denotes the complex conjugate of $\boldsymbol{w}$. The signals $\tilde{s}_1(t), \tilde{s}_2(t)$ represent the complex envelopes of the transmitted signals. The first term on the r.h.s. of Equations 6.16 corresponds to a signal with positive spin and circular polarization ($\theta_4 = \pi/4$), while the second term corresponds to a signal with negative spin and circular polarization ($\theta_4 = -\pi/4$). The uniqueness of Equations 6.16 is guaranteed without the conditions needed for the uniqueness of Equation 6.15.

The above-mentioned DST models can be applied to communication problems. Assuming that u is given, it is possible to measure the signal $\xi(t)$ and recover the original messages as follows. For Equation 6.15, an existing method resolves the two messages using mechanical orientation of the receiver's antenna (see, e.g., [4]). Alternatively, this can be done electronically using the representation of Lemma 6.1, without the need to know the orientation angle. For Equations 6.16, note that $\xi(t) = we^{i\theta_3}\tilde{s}_1(t) + \bar{w}e^{-i\theta_3}\tilde{s}_2(t)$, which implies the uniqueness of Equations 6.16 and indicates that the orientation angle has been converted into a phase angle whose sign depends on the spin sign. The original signals can be directly recovered from $\xi(t)$ up to an additive constant phase without knowledge of the orientation angle. In some cases, it is of interest to estimate the orientation angle. Let W be a matrix whose columns are $w, \bar{w}$. For Equations 6.16 this can be done using equal calibrating signals and then premultiplying the measurement by W^{-1} and measuring the phase difference between the two components of the result. This can also be used for real time estimation of the angular velocity $\mathrm{d}\theta_3/\mathrm{d}t$.

In general it can be stated that the advantage of the DST method is that it makes full use of the spatial degrees of freedom of transmission. However, the above DST methods need the knowledge of u and, in addition, may suffer from possible cross polarizations (see, e.g., [3]), multipath effects, and other unknown distortions from the source to the sensor.

The use of the proposed vector sensor can motivate the design of new improved transmission forms. Here we suggest a new DST method that uses on line electronic calibration in order to resolve the above problems. Similar to the previous methods it also makes full use of the spatial degrees of freedom in the system. However, it overcomes the need to know $\boldsymbol{u}$ and the overall transfer matrix from source to sensor.

Suppose the transmitted signal is $z(t) \in \mathbb{C}^{2\times 1}$ (this signal is as it appears before reaching the source's antenna). The measured signal is

$$\begin{bmatrix} y_{\mathrm{E}}(t) \\ y_{\mathrm{H}}(t) \end{bmatrix} = C(t)z(t) + \begin{bmatrix} \boldsymbol{e}_{\mathrm{E}}(t) \\ \boldsymbol{e}_{\mathrm{H}}(t) \end{bmatrix} \tag{6.17}$$

where $C(t) \in \mathbb{C}^{6\times 2}$ is the unknown source to sensor transfer matrix that may be slowly varying due to, for example, the source dynamics. To facilitate the identification of $z(t)$, the transmitter can send calibrating signals, for instance, transmit $z_1(t) = [1, 0]^{\mathrm{T}}$ and $z_2(t) = [0, 1]^{\mathrm{T}}$ separately. Since these inputs are in phasor form, this means that actually constant carrier waves are transmitted. Obviously, one can then estimate the columns of $C(t)$ by averaging the received signals, which can be used later for finding the original signal $z(t)$ by using, for example, least-squares estimation. Better estimation performance can be achieved by taking into account *a priori* information about the model.

The use of vector sensors is attractive in communication systems as it doubles the channel capacity (compared with scalar sensors) by making full use of the electromagnetic wave properties. This spatial multiplexing has vast potential for performance improvement in cellular communications.

In future research it would be of interest to develop optimal coding methods (modulation forms) for maximum channel capacity while maintaining acceptable distortions of the decoded signals despite unknown varying channel characteristics. It would also be of interest to design communication systems that utilize entire arrays of vector sensors.

Observe that actually any combination of the variables $\|\xi\|$, φ, θ_3, and θ_4 can be modulated to carry information. A binary signal can be transmitted using the spin sign of the polarization ellipse (sign of θ_4). Lemma 6.1 guarantees the identifiability of these signals from $\xi(t)$.

6.2.2 Multisource Multivector Sensor Model

Suppose that waves from n distant electromagnetic sources are impinging on an array of m vector sensors and that Assumptions A1 and A2 hold for each source. To extend the Model 6.10 to this scenario we need the following additional assumptions, which imply that Assumptions A1, A2 hold uniformly on the array:

A3: Plane wave across the array: In addition to Assumption A1, for each source the array size d_{A} has to be much smaller than the source to array distance, so that the vector u is approximately independent of the individual sensor positions.

A4: Narrow-band signal assumption: The maximum frequency of $E(t)$, denoted by ω_m, satisfies $\omega_m d_{\mathrm{A}}/c \ll 1$, where c is the velocity of wave propagation (i.e., the minimum modulating wavelength is much larger than the array size). This implies that $\varepsilon(t-\tau) \simeq \varepsilon(t)$ for all differential delays τ of the source signals between the sensors.

Note that (under the assumption $\omega_m < \omega_c$) since $\omega_m = \max\{|\omega_{\min} - \omega_c|, |\omega_{\max} - \omega_c|\}$, it follows that Assumption A4 is satisfied if $(\omega_{\max} - \omega_{\min})d_{\mathrm{A}}/2c \ll 1$ and ω_c is chosen to be close enough to $(\omega_{\max} + \omega_{\min})/2$.

Let $y_{\text{EH}}(t)$ and $e_{\text{EH}}(t)$ be the $6m \times 1$ dimensional electromagnetic sensor phasor measurement and noise vectors,

$$y_{\text{EH}}(t) \triangleq \left[\left(y_{\text{E}}^{(1)}(t)\right)^{\text{T}}, \left(y_{\text{H}}^{(1)}(t)\right)^{\text{T}}, \ldots, \left(y_{\text{E}}^{(m)}(t)\right)^{\text{T}}, \left(y_{\text{H}}^{(m)}(t)\right)^{\text{T}}\right]^{\text{T}} \tag{6.18a}$$

$$e_{\text{EH}}(t) \triangleq \left[\left(e_{\text{E}}^{(1)}(t)\right)^{\text{T}}, \left(e_{\text{H}}^{(1)}(t)\right)^{\text{T}}, \ldots, \left(e_{\text{E}}^{(m)}(t)\right)^{\text{T}}, \left(e_{\text{H}}^{(m)}(t)\right)^{\text{T}}\right]^{\text{T}} \tag{6.18b}$$

where
- $y_{\text{E}}^{(j)}(t)$ and $y_{\text{H}}^{(j)}(t)$ are, respectively, the measured phasor electric and magnetic vector fields at the jth sensor
- $e_{\text{E}}^{(j)}(t)$ and $e_{\text{H}}^{(j)}(t)$ are the noise components

Then, under Assumptions A3 and A4 and from Equation 6.10, we find that the array measured phasor signal can be written as

$$y_{\text{EH}}(t) = \sum_{k=1}^{n} e_k \otimes \begin{bmatrix} I_3 \\ (u_k \times) \end{bmatrix} V_k \xi_k(t) + e_{\text{EH}}(t) \tag{6.19}$$

where $\otimes$ is the Kronecker product, e_k denotes the kth column of the matrix $E \in \mathbb{C}^{m \times n}$ whose (j, k) entry is

$$E_{jk} = \text{e}^{-i\omega_c \tau_{jk}} \tag{6.20}$$

where τ_{jk} is the differential delay of the kth source signal between the jth sensor and the origin of some fixed reference coordinate system (e.g., at one of the sensors). Thus, $\tau_{jk} = -(u_k \cdot r_j)/c$, where $\mathbf{u}_k$ is the unit vector in the direction from the array to the kth source and $\mathbf{r}_j$ is the position vector of the jth sensor in the reference frame. The rest of the notation in Equation 6.19 is similar to the single source case, cf. Equations 6.1, 6.8, and 6.10. The vector $\xi_k(t)$ can have either the SST or the DST form described above.

Observe that the signal manifold matrix in Equation 6.19 can be written as the Khatri–Rao product (see, e.g., [20,21]) of E and a second matrix whose form depends on the source transmission type (i.e., SST or DST), see also later.

6.3 Cramér–Rao Bound for a Vector-Sensor Array

6.3.1 Statistical Model

Consider the problem of finding the parameter vector θ in the following discrete-time vector-sensor array model associated with n vector sources and m vector sensors:

$$y(t) = A(\theta)x(t) + \mathbf{e}(t) \quad t = 1, 2, \ldots \tag{6.21}$$

where
- $y(t) \in \mathbb{C}^{\bar{\mu} \times 1}$ is the vectors of observed sensor outputs (or snapshots)
- $x(t) \in \mathbb{C}^{\bar{\nu} \times 1}$ is the unknown source signals
- $e(t) \in \mathbb{C}^{\bar{\mu} \times 1}$ is the additive noise vectors

The transfer matrix $A(\theta) \in C^{\bar{\mu}\times\bar{\nu}}$ and the parameter vector $\theta \in R^{\bar{q}\times 1}$ are given by

$$\boldsymbol{A}(\theta)\left[A_1(\theta^{(1)})\cdots A_n(\theta^{(n)})\right] \tag{6.22a}$$

$$\theta = \left[(\theta^{(1)})^{\mathrm{T}}, \ldots, (\theta^{(n)})^{\mathrm{T}}\right]^{\mathrm{T}} \tag{6.22b}$$

where $A_k(\theta^{(k)}) \in^{\bar{\mu}\times\nu_k}$ and the parameter vector of the kth source $\theta^{(k)} \in \mathbb{R}^{q_k\times 1}$, thus $\bar{\nu} = \sum_{k=1}^{n} \nu_k$ and $\bar{q} = \sum_{k=1}^{n} q_k$.

The following notation will also be used:

$$\boldsymbol{y}(t) = \left[(\boldsymbol{y}^{(1)}(t))^{\mathrm{T}}, \ldots, (\boldsymbol{y}^{(m)}(t))^{\mathrm{T}}\right]^{\mathrm{T}} \tag{6.23a}$$

$$\boldsymbol{x}(t) = \left[(\boldsymbol{x}^{(1)}(t))^{\mathrm{T}}, \ldots, (\boldsymbol{x}^{(n)}(t))^{\mathrm{T}}\right]^{\mathrm{T}} \tag{6.23b}$$

where $\boldsymbol{y}^{(j)}(t) \in \mathbb{C}^{\mu_j\times 1}$ is the vector measurement of the jth sensor, implying $\bar{\mu} = \sum_{j=1}^{m} \mu_j$, and $\boldsymbol{x}^{(k)}(t) \in \mathbb{C}^{\nu_k\times 1}$ $C^{\nu_k\times 1}$ is the vector signal of the kth source. Clearly $\bar{\mu}$ and $\bar{\nu}$ correspond, respectively, to the total number of sensor components and source signal components.

The Model 6.21 generalizes the commonly used multiscalar source multiscalar sensor one (see, e.g., [7,22]). It will be shown later that the electromagnetic multivector source multivector sensor data models are special cases of Equation 6.21 with appropriate choices of matrices.

For notational simplicity, the explicit dependence on θ and t will be occasionally omitted.

We make the following commonly used assumptions on the Model 6.21:

A5: The source signal sequence $\{\mathbf{x}(1), \mathbf{x}(2), \ldots\}$ is a sample from a temporally uncorrelated stationary (complex) Gaussian process with zero mean and

$$Ex(t)x^{*}(s) = P\delta_{t,s}$$

$$Ex(t)x^{\mathrm{T}}(s) = 0 \quad \text{(for all } t \text{ and } s)$$

where

E is the expectation operator
"$*$" denotes the conjugate transpose
$\delta_{t,s}$ is the Kronecker delta

A6: The noise $\mathbf{e}(t)$ is (complex) Gaussian distributed with zero mean and

$$E\mathbf{e}(t)\mathbf{e}^{*}(s) = \sigma^2 I\delta_{t,s}$$

$$E\mathbf{e}(t)\mathbf{e}^{\mathrm{T}}(s) = 0 \quad \text{(for all } t \text{ and } s)$$

It is also assumed that the signals $x(t)$ and the noise $\mathbf{e}(s)$ are independent for all t and s.

A7: The matrix A has full rank $\bar{\nu} < \bar{\mu}$ (thus $A^{*}A$ is p.d.) and a continuous Jacobian $\partial A/\partial\theta$ in some neighborhood of the true θ. The matrix $APA^{*} + \sigma^2 I$ is assumed to be positive definite, which implies that the probability density functions of the model are well defined in some neighborhood of the true θ, P, σ^2. Additionally, the matrix in braces in Equation 6.24 below is assumed to be nonsingular.

The unknown parameters in the Model 6.21 include the vector θ, the signal covariance matrix P, and the noise variance σ^2. The problem of estimating θ in Equation 6.21 from N snapshots $y(1), \ldots, y(N)$ and the statistical performance of estimation methods are the main concerns of this chapter.

6.3.2 The Cramér–Rao Bound

Consider the estimation of θ in the Model 6.21 under the above assumptions and with θ, P, σ^2 unknown. We have the following theorem.

THEOREM 6.2

The Cramér–Rao lower bound on the covariance matrix of any (locally) unbiased estimator of the vector θ in the Model 6.21, under Assumptions A5 through A7 with $\boldsymbol{\theta}, \boldsymbol{P}, \sigma^2$ unknown and $\nu_k = \nu$ for all k, is a positive definite matrix given by

$$\mathrm{CRB}(\theta) = \frac{\sigma^2}{2N}\left\{\mathrm{Re}\left[\mathrm{btr}\left((1 \times U) \times (D^{*}\Pi_c D)^{bT}\right)\right]\right\}^{-1} \tag{6.24}$$

where

$$U = (A^{*}AP + \sigma^2 I)^{-1}A^{*}AP \tag{6.25a}$$

$$\Pi_c = I - \Pi \tag{6.25b}$$

$$\Pi = A(A^{*}A)^{-1}A^{*} \tag{6.25c}$$

$$D = \left[D_1^{(1)} \cdots D_{q_1}^{(1)} \cdots D_1^{(n)} \cdots D_{q_n}^{(n)}\right] \tag{6.25d}$$

$$D_\ell^{(k)} = \frac{\partial A_k}{\partial \theta_\ell^{(k)}} \tag{6.25e}$$

and where 1 denotes a $\bar{q} \times \bar{q}$ matrix with all entries equal to one, and the block trace operator btr($\cdot$), the block Kronecker product $\times$, the block Schur–Hadamard product $\bullet$, and the block transpose operator bT are as defined in the Appendix with blocks of dimensions $\nu \times \nu$, except for the matrix 1 that has blocks of dimensions $q_i \times q_j$.

Furthermore, the CRB in Equation 6.24 remains the same independently of whether σ^2 is known or unknown.

Proof 6.3 See [1, Appendix C].

Theorem 6.2 can be extended to include a larger class of unknown sensor noise covariance matrices (see [1, Appendix D]).

6.4 MSAE, CVAE, and Single-Source Single-Vector Sensor Analysis

This section introduces the MSAE and CVAE quality measures and their bounds for source direction and orientation estimation in three-dimensional space. The bounds are applied to analyze the statistical performance of parameter estimation of an electromagnetic source whose covariance is unknown using a single-vector sensor. Note that single-vector sensor analysis is valid for *wideband* sources, as Assumptions A3 and A4 are not needed.

6.4.1 The MSAE

We define the MSAE which is a quality measure that is useful for gaining physical insight into DOA (azimuth and elevation) estimation and for performance comparisons. The analysis of this subsection is not limited to electromagnetic measurements or to Gaussian data.

The angular error, say δ, corresponding to a direction error $\Delta\boldsymbol{u}$ in $\boldsymbol{u}$, can be shown to be $\delta = 2 \arcsin(\|\Delta\boldsymbol{u}\|/2)$. Hence, $\delta^2 = \|\Delta\boldsymbol{u}\|^2 + O(\|\Delta\boldsymbol{u}\|^4)$. Since $\Delta\boldsymbol{u} = \left(\frac{\partial \boldsymbol{u}}{\partial\theta_1}\right)\Delta\theta_1 + \left(\frac{\partial \boldsymbol{u}}{\partial\theta_2}\right)\Delta\theta_2 + O((\Delta\theta_1)^2 + (\Delta\theta_2)^2)$ where $\Delta\theta_1$, $\Delta\theta_2$ are the errors in θ_1 and θ_2, we have

$$\delta^2 = (\cos\theta_2 \cdot \Delta\theta_1)^2 + (\Delta\theta_2)^2 + O\left(|\Delta\theta_1|^3 + |\Delta\theta_2|^3\right) \tag{6.26}$$

We introduce the following definitions.

Definition 6.1: A model will be called "regular" if it satisfies any set of sufficient conditions for the CRB to hold (see, e.g., [23,24]).

Definition 6.2: The "normalized asymptotic" MSAE of a direction estimator will be defined as

$$\text{MSAE} \stackrel{\Delta}{=} \lim_{N\to\infty} \{NE(\delta^2)\} \tag{6.27}$$

whenever this limit exists.

Definition 6.3: A direction estimator will be called regular if its errors satisfy $E\left[|\Delta\theta_1|^3 + |\Delta\theta_2|^3\right] = o(1/N)$, the gradient of its bias with respect to θ_1,θ_2 exists and is $o(1)$ as $N \to \infty$, and its MSAE exists. (If $|\theta_2| = \pi/2$ then θ_1 is undefined and we can use the equivalent condition $E\left[\|\Delta\boldsymbol{u}\|^3\right] = o(1/N)$).

Equation 6.26 shows that under the assumptions that the model and estimator are regular we have

$$E(\delta)^2 \geq \left[\cos^2\theta_2 \cdot \text{CRB}(\theta_1) + \text{CRB}(\theta_2)\right] + o(1/N) \quad \text{as } N \to \infty \tag{6.28}$$

where $\text{CRB}(\theta_1)$ and $\text{CRB}(\theta_2)$ are, respectively, the CRBs for the azimuth and elevation.
Using Equation 6.28 we have the following theorem.

THEOREM 6.3

For a regular model MSAE of any regular direction estimator is bounded from below by

$$\text{MSAE}_{\text{CR}} \stackrel{\Delta}{=} N\left[\cos^2\theta_2 \cdot \text{CRB}(\theta_1) + \text{CRB}(\theta_2)\right] \tag{6.29}$$

Observe that MSAE_{CR} is not a function of N. Additionally, MSAE_{CR} is a tight bound if it is attained by some second order efficient regular estimator (usually the maximum likelihood [ML] estimator, see, e.g., [25]). For vector sensor measurements this bound has the desirable property of being invariant to the choice of reference coordinate frame, since the information content in the data is invariant under rotational transformations. This invariance property also holds for the MSAE of an estimator if the estimate is independent of known rotational transformations of the data.

For a regular model, the bound (Equation 6.29) can be used for performance analysis of any regular direction (azimuth and elevation) finding algorithm.

It is of interest to note that the bound (Equation 6.29) actually holds for finite data, when the estimators of u are unbiased and constrained to be of unit norm, see [26].

6.4.2 DST Source Analysis

Assume that it is desired to estimate the direction to a DST source whose covariance is unknown using a vector sensor. We will first present a statistical model for this problem as a special case of Equation 6.21 and then investigate in detail the resulting CRB and MSAE.

The measurement model for the DST case is given in Equation 6.10. Suppose the noise vector of Equation 6.10 is (Complex) Gaussian with zero mean and the following covariances:

$$E\begin{bmatrix} \boldsymbol{e}_{\mathrm{E}}(t) \\ \boldsymbol{e}_{\mathrm{H}}(t) \end{bmatrix}\left[\boldsymbol{e}_{\mathrm{E}}^{*}(s), \quad \boldsymbol{e}_{\mathrm{H}}^{*}(s)\right] = \begin{bmatrix} \sigma_{\mathrm{E}}^2 I_3 & 0 \\ 0 & \sigma_{\mathrm{H}}^2 I_3 \end{bmatrix}\delta_{t,s}$$

$$E\begin{bmatrix} \boldsymbol{e}_{\mathrm{E}}(t) \\ \boldsymbol{e}_{\mathrm{H}}(t) \end{bmatrix}\left[\boldsymbol{e}_{\mathrm{E}}^{\mathrm{T}}(s), \quad \boldsymbol{e}_{\mathrm{H}}^{\mathrm{T}}(s)\right] = 0 \quad \text{(for all } t \text{ and } s\text{)}$$

Our assumption that the noise components are statistically independent stems from the fact that they are created separately at different sensor components (even if the sensor components belong to a vector sensor). Note that under Assumption A1 the measurement includes a source plane wave component and sensor self-noise.

To relate the Model 6.10 through 6.21, define a scaled measurement $\boldsymbol{y}(t) \triangleq \left[r\boldsymbol{y}_{\mathrm{E}}^{\mathrm{T}}(t), \boldsymbol{y}_{\mathrm{H}}^{\mathrm{T}}(t)\right]^{\mathrm{T}}$ where $r = \sigma_{\mathrm{H}}/\sigma_{\mathrm{E}}$ is assumed to be known. (The results of this section actually hold also when r is unknown as is explained in [1]). The resulting scaled noise vector $\boldsymbol{e}(t) = \left[r\boldsymbol{e}_{\mathrm{E}}^{\mathrm{T}}(t), \boldsymbol{e}_{\mathrm{H}}^{\mathrm{T}}(t)\right]^{\mathrm{T}}$ then satisfies Assumption A6 with $\sigma = \sigma_{\mathrm{H}}$. Assume further that the signal $\xi(t)$ satisfies Assumption A5 with $x(t) = \xi(t)$. Then, under these assumptions, the scaled version of the DST source 6.10 can be viewed as a special case of Equation 6.21 with $m = n = 1$ and

$$\begin{aligned} &\boldsymbol{A}\begin{bmatrix} rV(22) \\ (\boldsymbol{u}\times)V \end{bmatrix} \quad \boldsymbol{x}(t) = \xi(t) \quad \sigma^2 = \sigma_{\mathrm{H}}^2 \\ &\theta = [\theta_1, \quad \theta_2]^{\mathrm{T}} \end{aligned} \tag{6.30}$$

where the unknown parameters are θ, P, σ^2. The parameter vector of interest is θ while P and σ^2 are the so-called nuisance parameters.

The above discussion shows that the CRB expression 6.24 is applicable to the present problem with the special choice of variables in Equation 6.30, thus $n = 1$ and $\bar{q} = 2$. The computation of the CRB is given in [1]. The result is independent of whether r is known or unknown.

Using the CRB results of [1] we find that $\mathrm{MSAE_{CR}}$ for the present DST problem is

$$\mathrm{MSAE}_{\mathrm{CR}}^{D} = \frac{(\sigma_{\mathrm{E}}^2 + \sigma_{\mathrm{H}}^2)\sigma_{\mathrm{E}}^2\sigma_{\mathrm{H}}^2 \operatorname{tr} U}{2\left[\sigma_{\mathrm{E}}^2\sigma_{\mathrm{H}}^2(\operatorname{tr} U)^2 + (\sigma_{\mathrm{E}}^2 - \sigma_{\mathrm{H}}^2)^2 \det(\operatorname{Re} U)\right]} \tag{6.31}$$

Observe that $\mathrm{MSAE}_{\mathrm{CR}}^{D}$ is symmetric with respect to $\sigma_{\mathrm{E}}, \sigma_{\mathrm{H}}$, as should be expected from the Maxwell equations. $\mathrm{MSAE}_{\mathrm{CR}}^{D}$ is not a function of $\theta_1, \theta_2, \theta_3$, as should be expected since for vector sensor measurements the MSAE bound is by definition invariant to the choice of coordinate system. Note that $\mathrm{MSAE}_{\mathrm{CR}}^{D}$ is independent of whether σ_{E} and σ_{H} are known or unknown.

6.4.3 SST Source (DST Model) Analysis

Consider the MSAE for a SST source when the estimation is done under the assumption that the source is of a DST type. In this case, the Model 6.10 has to be used but with a signal in the form of Equation 6.13. The signal covariance is then

$$P = \sigma_s^2 Q\boldsymbol{w}(Q\boldsymbol{w})^* \tag{6.32}$$

where $\sigma_s^2 = \mathrm{E}\, s^2(t)$ and Q and $\boldsymbol{w}$ are defined in Equation 6.12. Thus, rank $P = 1$ and P has a unit norm eigenvector $Q\boldsymbol{w}$ with an eigenvalue σ_s^2.

Let

$$\sigma_{\|}^2 \triangleq \frac{\sigma_{\mathrm{E}}^2 \cdot \sigma_{\mathrm{H}}^2}{\sigma_{\mathrm{E}}^2 + \sigma_{\mathrm{H}}^2} \tag{6.33}$$

The variance $\sigma_{\|}^2$ can be viewed as an equivalent noise variance of two measurements with independent noise variances σ_{E}^2 and σ_{H}^2. Define $\varrho, \sigma_s^2/\sigma_{\|}^2$, which is an effective SNR.

Using the analysis of U in [1] and expression 6.31 we find that

$$\begin{aligned}\mathrm{MSAE}_{\mathrm{CR}}^{\mathrm{S}} &= \frac{(1+\varrho)\left(\sigma_{\mathrm{E}}^2 + \sigma_{\mathrm{H}}^2\right)^2}{2\varrho^2\left[\sigma_{\mathrm{E}}^2\sigma_{\mathrm{H}}^2 + \left(\sigma_{\mathrm{E}}^2 - \sigma_{\mathrm{H}}^2\right)^2 \sin^2\theta_4 \cos^2\theta_4\right]} \\ &= \frac{(1+\varrho)(1+r^2)^2}{2\varrho^2\left[r^2 + (1-r^2)^2 \sin^2\theta_4 \cos^2\theta_4\right]}\end{aligned} \tag{6.34}$$

where $\mathrm{MSAE}_{\mathrm{CR}}^{\mathrm{S}}$ denotes the $\mathrm{MSAE}_{\mathrm{CR}}$ bound for the SST problem under the DST model. (It will be shown later that the same result also holds under the SST model.) Observe that $\mathrm{MSAE}_{\mathrm{CR}}^{\mathrm{S}}$ is symmetric with respect to $\sigma_{\mathrm{E}}, \sigma_{\mathrm{H}}$. It is also independent of whether σ_{H} and σ_{E} are known or unknown, as can be shown from Theorem 6.2 and [1, Appendix]. Also, $\mathrm{MSAE}_{\mathrm{CR}}^{\mathrm{S}}$ is not a function of θ_1, θ_2, θ_3, since for vector sensor measurements the MSAE bound is invariant under rotational transformations of the reference coordinate system. On the other hand, $\mathrm{MSAE}_{\mathrm{CR}}^{\mathrm{S}}$ is influenced by the ellipticity angle θ_4 through the difference in the electric and magnetic noise variances.

Table 6.1 summarizes several special cases of the expression 6.34 for $\mathrm{MSAE}_{\mathrm{CR}}^{\mathrm{S}}$. The elliptical polarization column corresponds to an arbitrary polarization angle $\theta_4 \in [-\pi/4, \pi/4]$. The circular and linear polarization columns are obtained, respectively, as special cases of Equation 6.34 with $|\theta_4| = \pi/4$ and $\theta_4 = 0$. The row of precise (noise-free) electric measurement (with noisy magnetic measurements) is obtained by substituting $\sigma_{\mathrm{E}}^2 = 0$ in Equation 6.34. The row of electric measurement only is obtained by deriving the corresponding CRB and $\mathrm{MSAE}_{\mathrm{CR}}^{\mathrm{S}}$. Alternatively, $\mathrm{MSAE}_{\mathrm{CR}}^{\mathrm{S}}$ can be found for this case by taking the limit of Equation 6.34 as $\sigma_{\mathrm{H}}^2 \to \infty$.

Observe from Equation 6.34 that when $\sigma_{\mathrm{H}}^2 \neq \sigma_{\mathrm{E}}^2$, $\mathrm{MSAE}_{\mathrm{CR}}^{\mathrm{S}}$ is minimized for circular polarization and maximized for linear polarization. This result is illustrated in Figure 6.3, which shows the square root of $\mathrm{MSAE}_{\mathrm{CR}}^{\mathrm{S}}$ as a function of $r = \sigma_{\mathrm{H}}/\sigma_{\mathrm{E}}$ for three types of polarizations ($\theta_4 = 0, \pi/12, \pi/4$). The equivalent signal-to-noise ratio $\mathrm{SNR} = \sigma_s^2/\sigma_{\|}^2$ is kept at one, while the individual electric and magnetic noise variances are varied to give the desired value of r. As r becomes larger or smaller than one, $\mathrm{MSAE}_{\mathrm{CR}}^{\mathrm{S}}$ increases more significantly for sources with polarization closer to linear.

When the electric (or magnetic) field is measured precisely and the source polarization is circular or elliptical, the $\mathrm{MSAE}_{\mathrm{CR}}^{\mathrm{S}}$ is zero (i.e., no angular error), while for linearly polarized sources it remains positive. In the latter case, the contribution to $\mathrm{MSAE}_{\mathrm{CR}}^{\mathrm{S}}$ stems from the magnetic (or electric) noisy measurement. When only the electric (or magnetic) field is measured, $\mathrm{MSAE}_{\mathrm{CR}}^{\mathrm{S}}$ increases as the

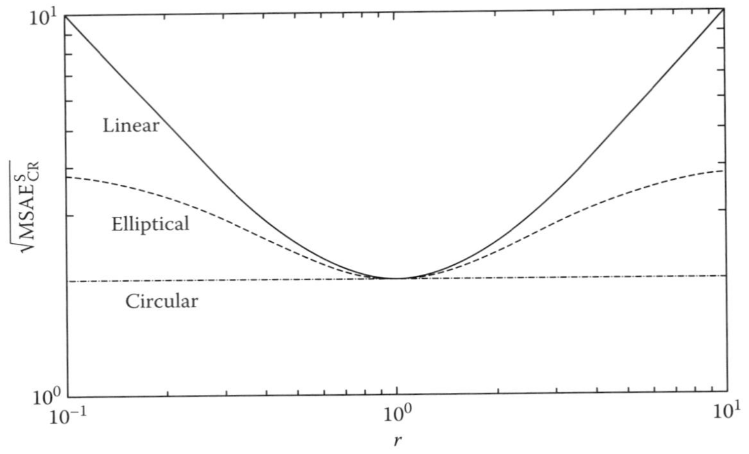

FIGURE 6.3 Effect of change in $r = \sigma_H/\sigma_E$ on $\text{MSAE}_{\text{CR}}^{\text{S}}$ for three types of polarizations ($\theta_4 = 0, \pi/12, \pi/4$). A single SST source, $\text{SNR} = \sigma_S^2/\sigma_{\|}^2 = 1$.

polarization changes from circular to linear. In the linear polarization case, $\text{MSAE}_{\text{CR}}^{\text{S}}$ tends to infinity. In this case, it is impossible to uniquely identify the source direction $\boldsymbol{u}$ from the electric field only, since $\boldsymbol{u}$ can then be anywhere in the plane orthogonal to the electric field vector.

The immediate conclusion is that as the source becomes closer to being linearly polarized it becomes more important to measure both the electric and magnetic fields to get good direction estimates using a single-vector sensor.

These results are illustrated in Figure 6.4, which shows the square root of $\text{MSAE}_{\text{CR}}^{\text{S}}$ as a function of σ_H^2 and three polarization types ($\theta_4 = 0, \pi/12, \pi/4$). The standard deviations of the signal and electric noise are $\sigma_S = \sigma_E = 1$. The left side of the figure corresponds to (nearly) precise magnetic measurement, while the right side to (nearly) electric measurement only.

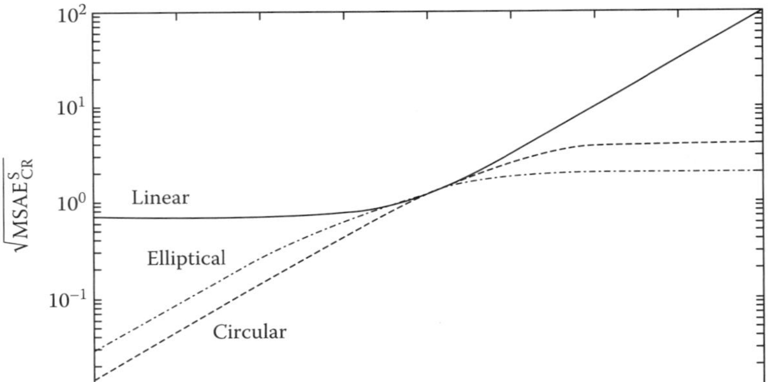

FIGURE 6.4 Effect of change in magnitude of σ_H^2 on $\text{MSAE}_{\text{CR}}^{\text{S}}$ for three types of polarizations ($\theta_4 = 0, \pi/12, \pi/4$). A single SST source, $\sigma_S = \sigma_E = 1$.

6.4.4 SST Source (SST Model) Analysis

Suppose that it is desired to estimate the direction to an SST source whose variance is unknown using a single-vector sensor, and the estimation is done under the correct model of an SST source. In the following, the CRB for this problem will be derived and it will be shown that the resulting MSAE bound remains the same as when the estimation was done under the assumption of a DST source. That is, knowledge of the source type does not improve the accuracy of its direction estimate.

To get a statistical model for the SST measurement Model 6.14 as a special case of Equation 6.21, we will make the same assumptions on the noise and use a similar data scaling as in the above DST source case. That will give again equal noise variances in all the sensor coordinates. Assume also that the signal envelope $s(t)$ satisfies Assumption A5 with $x(t) = s(t)$ in Equation 6.21. Then the resulting statistical model becomes a special case of Equation 6.21 with

$$\begin{aligned} \boldsymbol{A} &= \begin{bmatrix} rV(28) \\ (u\times)V \end{bmatrix} Q\boldsymbol{w} \quad x(t) = s(t) \quad \sigma^2 = \sigma_{\mathrm{H}}^2 \\ \boldsymbol{\theta} &= [\theta_1, \quad \theta_2, \quad \theta_3, \quad \theta_4]^{\mathrm{T}} \end{aligned} \tag{6.35}$$

The unknown parameters are θ, P, σ^2.

The matrix expression of CRB(θ) was calculated and its entries are presented in [1, Appendix F]. The results show that the ellipticity angle θ_4 is decoupled from the rest of the parameters and that its variance is not a function of these parameters. Additionally, the parameter vector θ is decoupled from σ_{E} and σ_{H}.

The MSAE bound for an SST source under the SST model was calculated using the analysis of [1]. The result coincides with Equation 6.34. That is, the MSAE bound for an SST source is the same under both the SST and the DST models.

The CRB expression in [1, Appendix F] implies that the CRB variance of the orientation angle θ_3 tends to infinity as the elevation angle θ_2 approaches $\pi/2$ or $-\pi/2$. This singularity is explained by the fact that the orientation angle is a function of the azimuth (through v_1, v_2), and the latter becomes increasingly sensitive to measurement errors as the elevation angle approaches the zenith or nadir. (Note that the azimuth is undefined in the zenith and nadir elevations). However, this singularity is not an intrinsic one, as it depends on the chosen reference system, while the information in the vector measurement does not.

6.4.5 CVAE and SST Source Analysis in the Wave Frame

In order to get performance results intrinsic to the SST estimation problem and thereby solve the singularity problems associated with the above model, we choose an alternative error vector that is invariant under known rotational transformations of the coordinate system. The details of the following analysis appear in [1, Appendix G].

Denote by W the wave frame whose coordinate axes are $(\boldsymbol{u}, \tilde{\boldsymbol{v}}_1, \tilde{\boldsymbol{v}}_2)$ where $\tilde{\boldsymbol{v}}_1$ and $\tilde{\boldsymbol{v}}_2$ correspond, respectively, to the major and minor axes of the source's electric wave ellipse (see Figure 6.2). For any estimator $\widehat{\theta}_i, i = 1, 2, 3$ there is an associated estimated wave frame $\widehat{W}$. Define the vector angular error $\boldsymbol{\phi}_{W\widehat{W}}$ which is the vector angle by which $\widehat{W}$ is (right-handed) rotated about W, and by $[\boldsymbol{\phi}_{W\widehat{W}}]_W$ the representation of $\boldsymbol{\phi}_{W\widehat{W}}$ in the coordinate system W (see [1, Appendix G]). The proposed vector angular error will be $[\boldsymbol{\phi}_{W\widehat{W}}]_W$.

Observe that $[\boldsymbol{\phi}_{W\widehat{W}}]_W$ depends, by definition, only on the frames W, $\widehat{W}$. Thus, for an estimator that is independent of known rotations of the data, the estimated wave frame $\widehat{W}$, the vector angular error and its covariance are independent of the sensor frame. We introduce the following definitions.

Definition 6.4: The "normalized asymptotic CVAE in the wave frame" is defined as

$$\mathrm{CVAE} \stackrel{\Delta}{=} \lim_{N\to\infty} \left\{ NE\left(\left[\boldsymbol{\phi}_{W\widehat{W}} \right]_W \left[\boldsymbol{\phi}_{W\widehat{W}} \right]_W^{\mathrm{T}} \right) \right\} \tag{6.36}$$

whenever this limit exists.

Definition 6.5: A direction and orientation estimator will be called regular if its errors satisfy $E\sum_{i=1}^{3} |\Delta\theta_i|^3 = o(1/N)$ and the gradient of its bias with respect to $\theta_1, \theta_2, \theta_3$ is $o(1)$ as $N \to \infty$.

Then we have the following theorems.

THEOREM 6.4

For a regular model the CVAE of any regular direction and orientation estimator, whenever it exists, is bounded from below by

$$\mathrm{CVAE}_{\mathrm{CR}} \stackrel{\Delta}{=} N \cdot K\mathrm{CRB}(\theta_1, \theta_2, \theta_3)K^{\mathrm{T}} \tag{6.37}$$

where

$$K = \begin{bmatrix} \sin\theta_2 & 0 & -1 \\ -\cos\theta_2 \sin\theta_3 & -\cos\theta_3 & 0 \\ \cos\theta_2 \cos\theta_3 & -\sin\theta_3 & 0 \end{bmatrix} \tag{6.38}$$

and $CRB\ (\theta_1, \theta_2, \theta_3)$ is the Cramér–Rao submatrix bound for the azimuth, elevation, and orientation angles for the particular model used.

Proof 6.4 See [1, Appendix G]

Observe that the result of Theorem 6.4 is obtained using geometrical considerations only. Hence, it is applicable to general direction and orientation estimation problems and is not limited to the SST problem only. It is dependent only on the ability to define a wave frame. For example, one can apply this theorem to a DST source with a wave frame defined by the orientation angle that diagonalizes the source signal covariance matrix. A generalization of this theorem to estimating nonunit vector systems is given in [26].

For vector sensor measurements, $\mathrm{CVAE}_{\mathrm{CR}}$ has the desirable property of being invariant to the choice of reference coordinate frame. This invariance property also holds for the CVAE of an estimator if the estimate is independent of deterministic rotational transformations of the data. Note that $\mathrm{CVAE}_{\mathrm{CR}}$ is not a function of N.

THEOREM 6.5

The MSAE and CVAE of any regular estimator are related through

$$\mathrm{MSAE} = [\mathrm{CVAE}]_{2,2} + [\mathrm{CVAE}]_{3,3} \tag{6.39}$$

Furthermore, a similar equality holds for a regular model where the MSAE and CVAE in Equation 6.39 are replaced by their lower bounds MSAE_{CR} and CVAE_{CR}.

Proof 6.5 See [1, Appendix G]

In our case, $\text{CRB}(\theta_1,\theta_2,\theta_3)$ is the 3×3 upper left block entry of the CRB matrix in the sensor frame given in [1, Appendix F]. Substituting this block entry into Equation 6.37 and denoting the CVAE matrix bound for the SST problem by $\text{CVAE}_{\text{CR}}^{\text{S}}$, we have that this matrix is diagonal with nonzero entries given by

$$\left[\text{CVAE}_{\text{CR}}^{\text{S}}\right]_{1,1} = \frac{(1+\varrho)}{2\varrho^2 \cos^2 2\theta_4} \tag{6.40a}$$

$$\left[\text{CVAE}_{\text{CR}}^{\text{S}}\right]_{2,2} = \frac{(1+\varrho)\left(\sigma_{\text{E}}^2 + \sigma_{\text{H}}^2\right)}{2\varrho^2\left[\sigma_{\text{H}}^2 \sin^2\theta_4 + \sigma_{\text{E}}^2 \cos^2\theta_4\right]} \tag{6.40b}$$

$$\left[\text{CVAE}_{\text{CR}}^{\text{S}}\right]_{3,3} = \frac{(1+\varrho)\left(\sigma_{\text{E}}^2 + \sigma_{\text{H}}^2\right)}{2\varrho^2\left[\sigma_{\text{E}}^2 \sin^2\theta_4 + \sigma_{\text{H}}^2 \cos^2\theta_4\right]} \tag{6.40c}$$

Some observations in Equations 6.40 are summarized in the following:

- Rotation around $\boldsymbol{u}$: Singular only for a circularly polarized signal.
- Rotation around $\tilde{\boldsymbol{v}}_1$ (electric ellipse's major axis): Singular only for a linearly polarized signal and no magnetic measurement.
- Rotation around $\tilde{\boldsymbol{v}}_2$ (electric ellipse's minor axis): Singular only for a linearly polarized signal and no electric measurement.
- The rotation variances around $\tilde{\boldsymbol{v}}_1$ and $\tilde{\boldsymbol{v}}_2$ are symmetric with respect to the electric and magnetic measurements.
- All the three variances in Equation 6.40 are bounded from below by $(1+\varrho)/2\varrho^2$ (independent of the wave parameters).

The singular cases above are found by checking when their variances in $\text{CVAE}_{\text{CR}}^{\text{S}}$ tend to infinity (see, e.g., [25, Theorem 6.3]). The three singular cases above should be expected as the corresponding rotations are unobservable. These singularities are intrinsic to the SST estimation problem and are independent of the reference coordinate system. The symmetry of the variances of the rotations around the major and minor axes of the ellipse with respect to the magnetic and electric measurements should be expected as their axes have a spatial angle difference of $\pi/2$.

The fact that the resulting singularities in the rotational errors are intrinsic (independent of the reference coordinate system) as well as the diagonality of the $\text{CVAE}_{\text{CR}}^{\text{S}}$ bound matrix with its simple entry expressions indicate that the wave frame is a natural system in which to do the analysis.

6.4.6 A Cross-Product-Based DOA Estimator

We propose a simple algorithm for estimating the DOA of a single electromagnetic source using the measurements of a single-vector sensor. The motivation for this algorithm stems from the average cross-product Poynting vector. Observe that $-\boldsymbol{u}$ is the unit vector in the direction of the Poynting vector given by [27],

$$\begin{aligned} S(t) &= E(t) \times H(t) = \text{Re}\left\{e^{i\omega_c t}E(t)\right\} \times \text{Re}\left\{e^{i\omega_c t}H(t)\right\} \\ &= \frac{1}{2}\text{Re}\{E(t) \times \bar{H}(t)\} + \frac{1}{2}\text{Re}\left\{e^{i2\omega_c t}E(t) \times H(t)\right\} \end{aligned}$$

where $\bar{\mathcal{H}}$ denotes the complex conjugate of $\mathcal{H}$. The carrier time average of the Poynting vector is defined as $\langle S\rangle_t = \frac{1}{2}\mathrm{Re}\{\varepsilon(t) \times \bar{\mathcal{H}}(t)\}$. Note that unlike $\varepsilon(t)$ and $\mathcal{H}(t)$ this average is not a function of ω_c. Thus, it has an intrinsic physical meaning.

At this point we can see two possible ways for estimating u:

1. Phasor time averaging of $\langle S\rangle_t$ yielding a vector denoted by $\langle S\rangle$ with the estimated u taken as the unit vector in the direction of $-\langle S\rangle$.
2. Estimation of u by phasor time averaging of the unit vectors in the direction of $\mathrm{Re}\{\varepsilon(t) \times \bar{H}(t)\}$.

Clearly, the first way is preferable, since then u is estimated after the measurement noise is reduced by the averaging process, while the estimated u in the second way is more sensitive to the measurement noises which may be magnified considerably.

Thus, the proposed algorithm computes

$$\hat{\boldsymbol{s}} = \frac{1}{N}\sum_{t=1}^{N}\mathrm{Re}\{\boldsymbol{y}_{\mathrm{E}}(t) \times \bar{\boldsymbol{y}}_{\mathrm{H}}(t)\} \tag{6.41a}$$

$$\hat{\boldsymbol{u}} = \hat{s}/\|\hat{s}\| \tag{6.41b}$$

This algorithm and some of its variables have been patented [28].

The statistical performance of this estimator $\hat{\boldsymbol{u}}$ is analyzed in [1] under the previous assumptions on $\xi(t), e_{\mathrm{E}}(t), e_{\mathrm{H}}(t)$, except that the Gaussian assumption is omitted. The results are summarized by the following theorem.

THEOREM 6.6

The estimator $\hat{u}$ has the following properties (for both DST and SST sources):

(a) If $\|\xi(t)\|^2, \|\boldsymbol{e}_{\mathrm{E}}(t)\|, \|\boldsymbol{e}_{\mathrm{H}}(t)\|$ have finite first order moments, then $\hat{\boldsymbol{u}} \to \boldsymbol{u}$ almost surely.

(b) If $\|\xi(t)\|^2, \|\boldsymbol{e}_{\mathrm{E}}(t)\|, \|\boldsymbol{e}_{\mathrm{H}}$ have finite second order moments, then $\sqrt{N}(\hat{u} - u)$ is asymptotically normal.

(c) If $\|\xi(t)\|^2, \|\boldsymbol{e}_{\mathrm{E}}(t)\|, \|\boldsymbol{e}_{\mathrm{H}}(t)\|$ have finite fourth order moments, then the MSAE is

$$\mathrm{MSAE} = \tfrac{1}{2}\varrho^{-1}\left(1 + 4\varrho^{-1}\right)\left(r + r^{-1}\right)^2 \tag{6.42}$$

where $\varrho = \mathrm{tr}(P)/\sigma_{\|}^2 = \mathrm{SNR}$.

(d) Under the conditions of (c), $N\delta^2$ is asymptotically χ^2 distributed with two degrees of freedom.

Proof 6.6 See [1, Appendix H]

For the Gaussian SST case, the ratio between the MSAE of this estimator to $\mathrm{MSAE}_{\mathrm{CR}}^{\mathrm{S}}$ in Equation 6.34 is

$$\mathrm{eff} \triangleq \frac{\mathrm{MSAE}}{\mathrm{MSAE}_{\mathrm{CR}}^{\mathrm{S}}} = \frac{\varrho + 4}{\varrho + 1}\left[1 + (r - r^{-1})^2 \sin^2\theta_4 \cos^2\theta_4\right] \tag{6.43}$$

Hence, this estimator is nearly efficient if the following two conditions are met:

$$\varrho \gg 1 \tag{6.44a}$$

$$r \simeq 1 \quad \text{or} \quad \theta_4 \simeq 0 \tag{6.44b}$$

Figure 6.5 illustrates these results using plots of the efficiency factor 6.43 as a function of the ellipticity angle θ_4 for $\mathrm{SNR} = \varrho = 10$ and three different values of r.

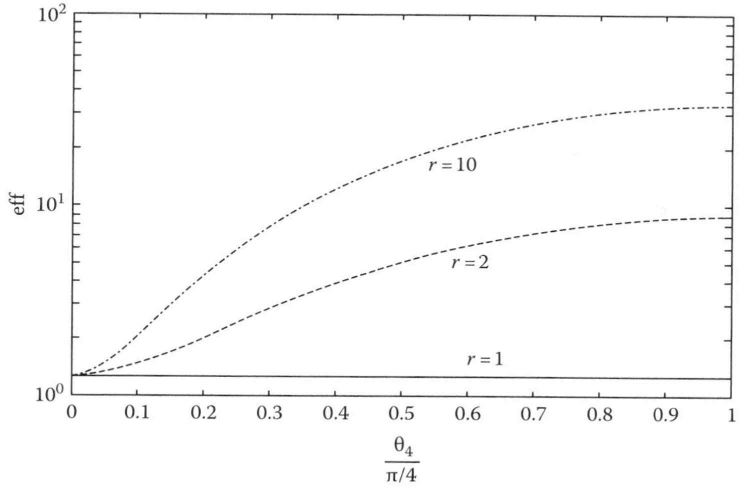

FIGURE 6.5 The efficiency factor (35) of the cross-product-based direction estimator as a function of the normalized ellipticity angle for three values of $r = \sigma_H/\sigma_E$. A single source, SNR = 10.

The estimator 6.41 can be improved using a weighted average of cross products between all possible pairs of real and imaginary parts of $y_E(t)$ and $y_H(s)$ taken at arbitrary times t and s. (Note that these cross products have directions nearly parallel to the basic estimator $\hat{u}$ in Equation 6.41; however, before averaging, these cross products should be premultiplied by $+1$ or -1 in accordance with the direction of the basic estimator $\hat{\boldsymbol{u}}$.) A similar algorithm suitable for real time applications can also be developed in the time domain without preprocessing needed for phasor representation. It can be extended to non-stationary inputs by using a moving average window on the data. It is of interest to find the optimal weights and the performances of these estimators.

The main advantages of the proposed cross-product-based algorithm 6.41 or one of its variants above are

- It can give a direction estimate instantly, i.e., with one time sample.
- It is simple to implement (does not require minimization of a cost function) and can be applied in real time.
- It is equally applicable to sources of various types, including SST, DST, wideband, and non-Gaussian.
- Its MSAE is nearly optimal in the Gaussian SST case under Equation 6.44.
- It does not depend on time delays and therefore does not require data synchronization among different sensor components.

6.5 Multisource Multivector Sensor Analysis

Consider the case in which it is desired to estimate the directions to multiple electromagnetic sources whose covariance is unknown using an array of vector sensors. The MSAE_{CR} and CVAE_{CR} bound expressions in Equations 6.29 and 6.37 are applicable to each of the sources in the multisource multi-vector sensor scenario. Suppose that the noise vector $e_{EH}(t)$ in Equation 6.19 is complex white Gaussian with zero mean and diagonal covariance matrix (i.e., noises from different sensors are uncorrelated) and with electric and magnetic variances σ_E^2 and σ_H^2, respectively. Suppose also that $r = \sigma_H/\sigma_E$ is known. Similarly to the single sensor case, multiply the electric measurements in Equation 6.19 by r to obtain equal noise variances in all the sensor coordinates. The resulting models then become special cases of Equation 6.21 as follows.

For DST signals, the block columns $A_k \in C^{6m\times 2}$ and the signals $x(t) \in C^{2n\times 1}$ are

$$A_k = \boldsymbol{e}_k \otimes \begin{bmatrix} rI_3 \\ (\boldsymbol{u}_k\times) \end{bmatrix} V_k \tag{6.45a}$$

$$\boldsymbol{x}(t) = \left[\xi_1^{\mathrm{T}}(t), \ldots, \xi_n^{\mathrm{T}}(t)\right]^{\mathrm{T}} \tag{6.45b}$$

The parameter vector of the kth source includes here its azimuth and elevation.

For the SST case, the columns $A_k \in \mathbb{C}^{6m\times 1}$ and the signals $x(t) \in \mathbb{C}^{n\times 1}$ are

$$A_k = e_k \otimes \begin{bmatrix} rI_3 \\ (\boldsymbol{u}_k\times) \end{bmatrix} V_k Q_k \boldsymbol{w}_k \tag{6.46a}$$

$$\boldsymbol{x}(t) = [s_1(t), \ldots, s_n(t)]^{\mathrm{T}} \tag{6.46b}$$

The parameter vector of the kth source includes here its azimuth, elevation, orientation, and ellipticity angles.

The matrices A whose (block) columns are given in Equations 6.45a and 6.46a are the Khatri–Rao products (see, e.g., [20,21]) of the two matrices whose (block) columns are the arguments of the Kronecker products in these equations.

Mixed single and DSTs are also special cases of Equation 6.21 with appropriate combinations of the above expressions.

6.5.1 Results for Multiple Sources, Single-Vector Sensor

We present several results for the multiple-source model and a single-vector sensor. It is assumed that the signal and noise vectors satisfy, respectively, Assumptions A5 and A6. The results are applicable to wideband sources since a single-vector sensor is used and thus Assumptions A3 and A4 are not needed.

We first present results obtained by numerical evaluation concerning the localization of two uncorrelated sources, assuming r is known:

1. When only the electric field is measured, the information matrix is singular.
2. When the electric measurement is precise, the CRB variances are generally nonzero.
3. The $\mathrm{MSAE}_{\mathrm{CR}}^{\mathrm{S}}$ can increase without bound with decreasing source angular separation for sources with the same ellipticity and spin direction, but remarkably it remains bounded for sources with different ellipticities or opposite spin directions.

Properties 1 and 2 are, in general, different from the single source case. Property 1 shows that it is necessary to include both the electric and magnetic measurements to estimate the direction to more than one source. Property 3 demonstrates the great advantage of using the electromagnetic vector sensor, in that it allows high resolution of sources with different ellipticities or opposite spins. Note that this generally requires a very large aperture using a scalar sensor array.

The above result on the ability to resolve two sources that are different only in their ellipticity or spin direction appears to be new. Note also the analogy to Pauli's "exclusion principle," as in our case two narrow-band SST sources are distinguishable if and only if they have different sets of parameters. The set in our case includes wavelength, direction, ellipticity, and spin sign.

Now we present conditions for identifiability of multiple SST (or polarized) sources and a single-vector sensor, which are analytically proven in [29,30], assuming the noise variances are known:

1. A single source is always identifiable.
2. Two sources that are not fully correlated are identifiable if they have different DOAs.

3. Two fully correlated sources are identifiable if they have different DOAs and ellipticities.
4. Three sources that are not fully correlated are identifiable if they have different DOAs and ellipticities.

Note that by identifiability we refer to both the DOA and polarization parameters.

Figures 6.6 and 6.7 illustrate the resolution of two uncorrelated equal power SST sources with a single electromagnetic vector sensor. The figures show the square root of the $\mathrm{MSAE}_{\mathrm{CR}}^{\mathrm{S}}$ of one of the sources for a variety of spin directions, ellipticities, and orientation angles, as a function of the separation

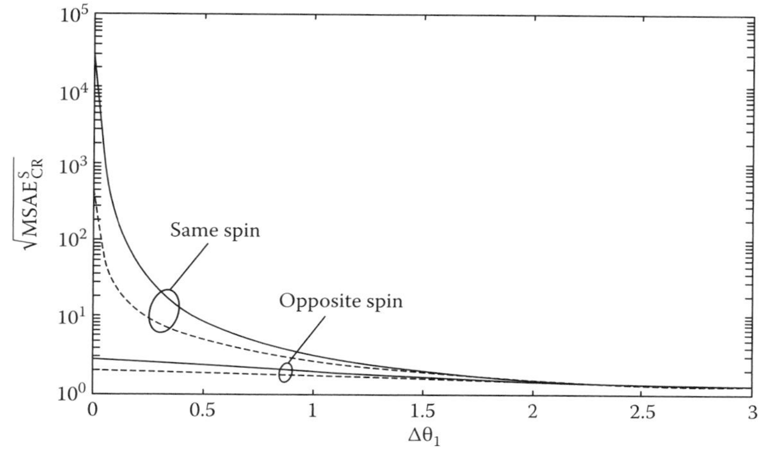

FIGURE 6.6 $\mathrm{MSAE}_{\mathrm{CR}}^{\mathrm{S}}$ for two uncorrelated equal power SST sources and a single-vector sensor as a function of the source angular separation. Upper two curves: same spin directions ($\theta_4^{(1)} = \theta_4^{(2)} = \pi/12$). Lower two curves: opposite spin directions ($\theta_4^{(1)} = -\theta_4^{(2)} = \pi/12$). Solid curves: same orientation angles ($\theta_3^{(1)} = \theta_3^{(2)} = \pi/4$). Dashed curves: different orientation angles ($\theta_3^{(1)} = -\theta_3^{(2)} = \pi/4$). Remaining parameters are $\theta_1^{(1)} = \theta_2^{(1)} = \theta_2^{(2)} = 0$, $\Delta\theta_1 \triangleq \theta_1^{(2)}$, $P = I_2, \sigma_E = \sigma_H = 1$.

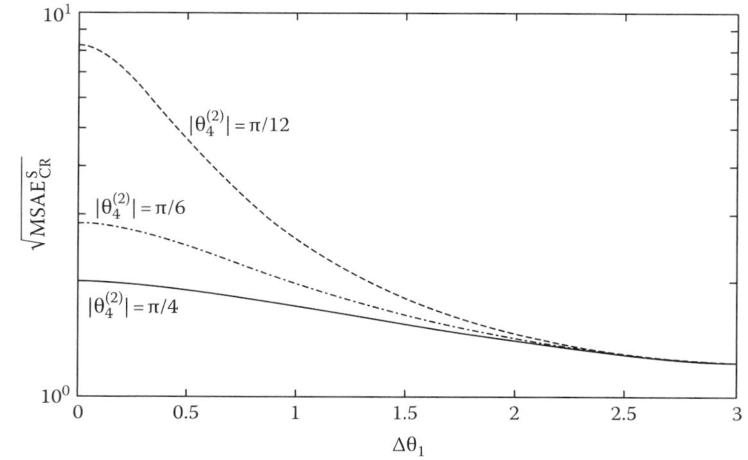

FIGURE 6.7 $\mathrm{MSAE}_{\mathrm{CR}}^{\mathrm{S}}$ for two uncorrelated equal power SST sources and a single-vector sensor as a function of the source angular separation. Sources are with the same orientation angles ($\theta_3^{(1)} = \theta_3^{(2)} = \pi/4$) and different ellipticity angles ($\theta_4^{(1)} = 0$ and $\theta_4^{(2)}$ as shown in the figure). Remaining parameters are as in Figure 6.6.

angle between the sources. (The $\text{MSAE}_{\text{CR}}^{\text{S}}$ values of the two sources are found to be equal in all the following cases.) The covariances of the signals and noise are normalized such that $P = I_2, \sigma_{\text{E}} = \sigma_{\text{H}} = 1$. The azimuth angle of the first source and the elevation angles of the two sources are kept constant $(\theta_1^{(1)} = \theta_2^{(1)} = \theta_2^{(2)} = 0)$. The second source's azimuth is varied to give the desired separation angle $\Delta\theta_1, \theta_1^{(2)}$. In Figure 6.5, the cases shown are of same spin directions $(\theta_4^{(1)} = \theta_4^{(2)} = \pi/12)$ and opposite spin directions $(\theta_4^{(1)} = -\theta_4^{(2)} = \pi/12)$, same orientation angles $(\theta_3^{(1)} = \theta_3^{(2)} = \pi/4)$ and different orientation angles $(\theta_3^{(1)} = -\theta_3^{(2)} = \pi/4)$. The figure shows that the resolution of the two sources with a single-vector sensor is remarkably good when the sources have opposite spin directions. In particular, the $\text{MSAE}_{\text{CR}}^{\text{S}}$ remains bounded even for zero separation angle and equal orientation angles! On the other hand, the resolution is not so significant when the two sources have different orientation angles but equal ellipticity angles (then, for example, the $\text{MSAE}_{\text{CR}}^{\text{S}}$ tends to infinity for zero separation angle). In Figure 6.7, the orientation angles of the sources is the same $(\theta_3^{(1)} = \theta_3^{(2)} = \pi/4)$, the polarization of the first source is kept linear $(\theta_4^{(1)} = 0)$ while the ellipticity angle of the second source is varied $(|\theta_4^{(2)}| = \pi/12, \pi/6, \pi/4)$ to illustrate the remarkable resolvability due to different ellipticities. It can be seen that the $\text{MSAE}_{\text{CR}}^{\text{S}}$ remains bounded here even for zero separation angle.

Thus, Figures 6.6 and 6.7 show that with one vector sensor it is possible to resolve extremely well two uncorrelated SST sources that have only different spin directions or different ellipticities (these sources can have the same DOA and the same orientation angle). This demonstrates a great advantage of the vector sensor over scalar sensor arrays, in that the latter require large array apertures to resolve sources with small separation angle.

6.6 Concluding Remarks

An approach has been presented for the localization of electromagnetic sources using vector sensors. We summarize some of the main results of this chapter and give an outlook to their possible extensions.

Models: New models that include the complete electromagnetic data at each sensor have been introduced. Furthermore, new signal models and vector angular error models in the wave frame have been proposed. The wave frame model provides simple performance expressions that are easy to interpret and have only intrinsic singularities. Extensions of the proposed models may include additional structures for specific applications.

CRBs and quality measures: A compact expression for the CRB for multivector source multivector sensor processing has been derived. The derivation gave rise to new block matrix operators. New quality measures in three-dimensional space, such as the MSAE for direction estimation and CVAE for direction and orientation estimation, have been defined. Explicit bounds on the MSAE and CVAE, having the desirable property of being invariant to the choice of the reference coordinate frame, have been derived and can be used for performance analysis. Some generalizations of the bounds appear in [26]. These bounds are not limited to electromagnetic vector-sensor processing. Performance comparisons of vector-sensor processing with scalar sensor counterparts are of interest.

Identifiability: The derived bounds and the identifiability analysis of [29,30] were used to show that the fusion of magnetic and electric data at a single-vector sensor increases the number of identifiable sources (or resolution capacity) in three-dimensional space from one source in the electric data case to up to three sources in the electromagnetic case. For a SST source, in order to get good direction estimates, the fusion of the complete data becomes more important as the polarization gets closer to linear. Finding the number of identifiable sources per sensor in a general vector-sensor array is of interest. Preliminary results on this issue can be found in [29,31].

Resolution: Source resolution using vector sensors is inherently different from scalar sensors, where the latter case is characterized by the classical Rayleigh principle. For example, it was shown that a single-vector sensor can be used to resolve two sources in three-dimensional space. In particular, a vector sensor exhibits remarkable resolvability when the sources have opposite spin directions or different ellipticity angles. This is very different from the scalar sensor array case in which a plane array with large aperture is required to achieve the same goal. Analytical results on source resolution using vector-sensor arrays and comparisons with their scalar counterparts are of interest.

Algorithms: A simple algorithm has been proposed and analyzed for finding the direction to a single source using a single-vector sensor based on the cross-product operation. It is of interest to analyze the performance of the aforementioned variants of this algorithm and to extend them to more general source scenarios (e.g., larger number of sources). It is also of interest to develop new algorithms for the vector-sensor array case.

Communication: The main considerations in communication are transmission of signals over channels with limited bandwidth and their recovery at the sensor. Vector sensors naturally fit these considerations as they have maximum observability to incoming signals and they double the channel capacity (compared with scalar sensors) with DST signals. This has vast potential for performance improvement in cellular communications. Future goals will include development of optimum signal estimation algorithms, communication forms, and coding design with vector-sensor arrays.

Implementations: The proposed methods should be implemented and tested with real data.

Sensor development: The use of complete electromagnetic data seems to be virtually nonexistent in the literature on source localization. It is hoped that the results of this research will motivate the systematic development of high quality electromagnetic sensors that can operate over a broad range of frequencies. Recent references on this topic can be found in [14,15].

Extensions: The vector sensor concept can be extended to other areas and open new possibilities. An example of this can be found in [32,33] for the acoustic case.

Acknowledgments

This work was supported by the U.S. Air Force Office of Scientific Research under Grant no. F49620-97-1-0481, the Office of Naval Research under Grant no. N00014-96-1-1078, the National Science Foundation under Grant No. MIP-9615590, and the HTI fellowship. The authors are grateful to Professor I.Y. Bar-Itzhack from the Department of Aeronautical Engineering, Technion, Israel, for bringing reference [34] to their attention.

Appendix A: Definitions of Some Block Matrix Operators

This appendix defines several block matrix operators that are found to be useful in this chapter. The following notation will be used for a blockwise partitioned matrix A:

$$A = \begin{bmatrix} A_{<11>} & \cdots & A_{<1n>} \\ \vdots & & \vdots \\ A_{<m1>} & \cdots & A_{<mn>} \end{bmatrix} \triangleq [A_{<ij>}] \tag{6.A.1}$$

with the block entries $A_{<ij>}$ of dimensions $\mu_i \times \nu_j$. Define $\bar{\mu} \triangleq \sum_{i=1}^{m} \mu_i, \bar{\nu} \triangleq \sum_{j=1}^{n} \nu_j$, so A is a $\bar{\mu} \times \bar{\nu}$ matrix. Since the block entries may not be of the same size, this is sometimes called an unbalanced partitioning. The following definitions will be considered.

Definition 6.6: *Block transpose*. Let A be an $m\mu \times n\nu$ blockwise partitioned matrix, with blocks $A_{<ij>}$ of equal dimensions $\mu \times \nu$. Then the block transpose A^{bT} is an $n\mu \times m\nu$ matrix defined through

$$(A^{bT})_{<ij>} = A_{<ij>} \tag{6.A.2}$$

Definition 6.7: *Block Kronecker product.* Let A be a blockwise partitioned matrix of dimension $\bar{\mu} \times \bar{\nu}$, with block entries $A_{<ij>}$ of dimensions $\mu_i \times \nu_j$, and let B be a blockwise partitioned matrix of dimensions $\bar{\eta} \times \bar{\rho}$, with block entries $B_{<ij>}$ of dimensions $\eta_i \times \rho_j$. Also $\bar{\mu} = \sum_{i=1}^{m} \mu_i, \bar{\nu} = \sum_{j=1}^{n} \nu_j$, $\bar{\eta} = \sum_{i=1}^{m} \eta_i$, $\bar{\rho} = \sum_{j=1}^{n} \rho_j$. Then the block Kronecker product $A \times B$ is an $\left(\sum_{i=1}^{m} \mu_i \eta_i \times \sum_{j=1}^{n} \nu_j \rho_j\right)$ matrix defined through

$$(A \oplus B)_{<ij>} = A_{<ij>} \otimes B_{<ij>} \qquad (6.A.3)$$

i.e., the (i, j) block entry of $A \times B$ *is* $A_{<ij>} \otimes B_{<ij>}$ of dimension $\mu_i \eta_i \times \nu_j \rho_j$.

Definition 6.8: *Block Schur–Hadamard product.* Let A be an $m\mu \times n\nu$ matrix consisting of blocks $A_{<ij>}$ of dimensions $\mu \times \nu$, and let B be an $m\nu \times n\eta$ matrix consisting of blocks $B_{<ij>}$ of dimensions $\nu \times \eta$. Then the block Schur–Hadamard product $|A \bullet B|$ is an $m\mu \times n\eta$ matrix defined through

$$(A \odot B)_{<ij>} = A_{<ij>} B_{<ij>} \qquad (6.A.4)$$

Thus, each block of the product is a *usual* product of a pair of blocks and is of dimension $\mu \times \eta$.

Definition 6.9: *Block trace operator.* Let A be an $m\mu \times n\mu$ matrix consisting of blocks $A_{<ij>}$ of dimensions $\mu \times \mu$. Then the block trace matrix operator $\text{btr}[A]$ is an $m \times n$ matrix defined by

$$(\text{btr}[A])_{ij} = \text{tr } A_{<ij>} \qquad (6.A.5)$$

References

1. Nehorai, A. and Paldi, E., Vector-sensor array processing for electromagnetic source localization, *IEEE Trans. Signal Process.*, SP-42, 376–398, Feb. 1994.
2. Nehorai, A. and Paldi, E., Vector sensor processing for electromagnetic source localization, *Proceedings of the 25th Asilomar Conference on Signals, Systems and Computers*, Pacific Grove, CA, Nov. 1991, pp. 566–572.
3. Schwartz, M., Bennett, W.R., and Stein, S., *Communication Systems and Techniques*, McGraw-Hill, New York, 1966.
4. Keiser, B.E., *Broadband Coding, Modulation, and Transmission Engineering*, Prentice-Hall, Englewood Cliffs, NJ, 1989.
5. Stoica, P. and Nehorai, A., Performance study of conditional and unconditional direction-of-arrival estimation, *IEEE Trans. Acoust. Speech Signal Process.*, ASSP-38, 1783–1795, Oct. 1990.
6. Ottersten, B., Viberg, M., and Kailath, T., Analysis of subspace fitting and ML techniques for parameter estimation from sensor array data, *IEEE Trans. Signal Process.*, SP-40, 590–600, Mar. 1992.
7. Schmidt, R.O., A signal subspace approach to multiple emitter location and spectral estimation, PhD dissertation, Stanford University, Stanford, CA, Nov. 1981.
8. Ferrara, E.R. Jr. and Parks, T.M., Direction finding with an array of antennas having diverse polarization, *IEEE Trans. Antennas Propagat.*, AP-31, 231–236, Mar. 1983.
9. Ziskind, I. and Wax, M., Maximum likelihood localization of diversely polarized sources by simulated annealing, *IEEE Trans. Antennas Propagat.*, AP-38, 1111–1114, July 1990.
10. Li, J. and Compton, R.T. Jr., Angle and polarization estimation using ESPRIT with a polarization sensitive array, *IEEE Trans. Antennas Propagat.*, AP-39, 1376–1383, Sept. 1991.

11. Weiss, A.J. and Friedlander, B., Performance analysis of diversely polarized antenna arrays, *IEEE Trans. Signal Process.*, SP-39, 1589–1603, July 1991.
12. Means, J.D., Use of three-dimensional covariance matrix in analyzing the polarization properties of plane waves, *J. Geophys. Res.*, 77, 5551–5559, Oct. 1972.
13. Hatke, G.F., Performance Analysis of the SuperCART Antenna Array, Project Report No. AST-22, Lincoln Laboratory, Massachusetts Institute of Technology, Lexington, MA, Mar. 1992.
14. Kanda, M., An electromagnetic near-field sensor for simultaneous electric and magnetic-field measurements, *IEEE Trans. Electromag. Compatibility*, 26(1), 102–110, Aug. 1984.
15. Kanda, M. and Hill, D., A three-loop method for determining the radiation characteristics of an electrically small source, *IEEE Trans. Electromag. Compatibility*, 34(1), 1–3, Feb. 1992.
16. Dugundji, J., Envelopes and pre-envelopes of real waveforms, *IRE Trans. Inf. Theory*, IT-4, 53–57, Mar. 1958.
17. Rice, S.O., Envelopes of narrow-band signals, *Proc. IEEE*, 70, 692–699, July 1982.
18. Giuli, D., Polarization diversity in radars, *Proc. IEEE*, 74, 245–269, Feb. 1986.
19. Born, M. and Wolf, E., Eds., *Principles of Optics*, 6th edn., Pergamon Press, Oxford, 1980 [1st edn., 1959].
20. Khatri, C.G. and Rao, C.R., Solution to some functional equations and their applications to characterization of probability distribution, *Sankhyā Ser. A*, 30, 167–180, 1968.
21. Rao, C.R. and Mitra, S.K., *Generalized Inverse of Matrices and Its Applications*, John Wiley & Sons, New York, 1971.
22. Stoica, P. and Nehorai, A., MUSIC, maximum likelihood and Cramér-Rao bound, *IEEE Trans. Acoust. Speech Signal Process.*, ASSP-37, 720–741, May 1989.
23. Ibragimov, I.A. and Has'minskii, R.Z., *Statistical Estimation: Asymptotic Theory*, Springer-Verlag, New York, 1981.
24. Paldi, E. and Nehorai, A., A generalized Cramér-Rao bound, in preparation.
25. Caines, P.E., *Linear Stochastic Systems*, John Wiley & Sons, New York, 1988.
26. Nehorai, A. and Hawkes, M., Performance bounds on estimating vector systems, in preparation.
27. Jackson, J.D., *Classical Electrodynamics*, 2nd edn., John Wiley & Sons, New York, 1975 [1st edn., 1962].
28. Nehorai, A. and Paldi, E., Method for electromagnetic source localization, U.S. Patent No. 5,315,308, May 24, 1994.
29. Hochwald, B. and Nehorai, A., Identifiability in array processing models with vector-sensor applications, *IEEE Trans. Signal Process.*, SP-44, 83–95, Jan. 1996.
30. Ho, K.-C., Tan, K.-C., and Ser, W., An investigation on number of signals whose direction-of-arrival are uniquely determinable with an electromagnetic vector sensor, *Signal Process.*, 47, 41–54, Nov. 1995.
31. Tan, K.-C., Ho, K.-C., and Nehorai, A., Uniqueness study of measurements obtainable with arrays of electromagnetic vector sensors, *IEEE Trans. Signal Process.*, SP-44, 1036–1039, Apr. 1996.
32. Nehorai, A. and Paldi, E., Acoustic vector-sensor array processing, *IEEE Trans. Signal Process.*, SP-42, 2481–2491, Sept. 1994; A short version appeared in *Proceedings of the 26th Asilomar Conference Signals, Systems and Computers*, Pacific Grove, CA, Oct. 1992, pp. 192–198.
33. Hawkes, M. and Nehorai, A., Acoustic vector-sensor beamforming and capon direction estimation, *IEEE International Conference on Acoustic Speech Signal Processing*, Detroit, MI, May 1995, pp. 1673–1676.
34. Shuster, M.D., A survey of attitude representations, *J. Astronaut. Sci.*, 41(4), 439–517, Oct.–Dec. 1993.

7 Subspace Tracking

7.1 Introduction 7-1
7.2 Background 7-2
Eigenvalue Decomposition vs. Singular Value Decomposition • Short Memory Windows for Time Varying Estimation • Classification of Subspace Methods • Historical Overview of MEP Methods • Historical Overview of Adaptive, Non-MEP Methods
7.3 Issues Relevant to Subspace and Eigen Tracking Methods 7-5
Bias due to Time Varying Nature of Data Model • Controlling Roundoff Error Accumulation and Orthogonality Errors • Forward–Backward Averaging • Frequency vs. Subspace Estimation Performance • The Difficulty of Testing and Comparing Subspace Tracking Methods • Spherical Subspace Updating—A General Framework for Simplified Updating • Initialization of Subspace and Eigen Tracking Algorithms • Detection Schemes for Subspace Tracking
7.4 Summary of Subspace Tracking Methods Developed Since 1990 7-11
Modified Eigenproblems • Gradient-Based Eigen Tracking • The URV and Rank Revealing QR Updates • Miscellaneous Methods
References 7-14

R. D. DeGroat
Broadcom Corporation

E. M. Dowling
The University of Texas at Dallas

D. A. Linebarger
The University of Texas at Dallas

7.1 Introduction

Most high resolution direction-of-arrival (DOA) estimation methods rely on subspace or eigen-based information which can be obtained from the eigenvalue decomposition (EVD) of an estimated correlation matrix, or from the singular value decomposition (SVD) of the corresponding data matrix. However, the expense of directly computing these decompositions is usually prohibitive for real-time processing. Also, because the DOA angles are typically time-varying, repeated computation is necessary to track the angles. This has motivated researchers in recent years to develop low cost eigen and subspace tracking methods. Four basic strategies have been pursued to reduce computation: (1) computing only a few eigencomponents, (2) computing a subspace basis instead of individual eigencomponents, (3) approximating the eigencomponents or basis, and (4) recursively updating the eigencomponents or basis. The most efficient methods usually employ several of these strategies.

In 1990, an extensive survey of SVD tracking methods was published by Comon and Golub [7]. They classified the various algorithms according to complexity and basically two categories emerge: $O(n^2r)$ and $O(nr^2)$ methods, where n is the snapshot vector size and r is the number of extreme eigenpairs to be tracked. Typically, $r < n$ or $r \ll n$, so the $O(nr^2)$ methods involve significantly fewer computations than the $O(n^2r)$ algorithms. However, since 1990, a number of $O(nr)$ algorithms have been developed.

This article will primarily focus on recursive subspace and eigen updating methods developed since 1990, especially, the $O(nr^2)$ and $O(nr)$ algorithms.

7.2 Background

7.2.1 Eigenvalue Decomposition vs. Singular Value Decomposition

Let $X = [x_1|x_2|\ldots|x_N]$ be an $n \times N$ data matrix where the kth column corresponds to the kth snapshot vector, $x_k \in C^n$. With block processing, the correlation matrix for a zero mean, stationary, ergodic vector process is typically estimated as $R = \frac{1}{N}XX^{\mathrm{H}}$ where the true correlation matrix, $\Phi = E[x_k x_k^{\mathrm{H}}] = E[R]$.

The EVD of the estimated correlation matrix is closely related to the SVD of the corresponding data matrix. The SVD of X is given by $X = USV^{\mathrm{H}}$ where $U \in C^{n\times n}$ and $V \in C^{N\times N}$ are unitary matrices and $S \in C^{n\times N}$ is a diagonal matrix whose nonzero entries are positive. It is easy to see that the left singular vectors of X are the eigenvectors of $XX^{\mathrm{H}} = USS^{\mathrm{T}}U^{\mathrm{H}}$, and the right singular vectors of X are the eigenvectors of $X^{\mathrm{H}}X = VS^{\mathrm{T}}SV^{\mathrm{H}}$. This is so because XX^{H} and $X^{\mathrm{H}}X$ are positive definite Hermitian matrices (which have orthogonal eigenvectors and real, positive eigenvalues). Also note that the nonzero singular values of X are the positive square roots of the nonzero eigenvalues of XX^{H} and $X^{\mathrm{H}}X$. Mathematically, the eigen information contained in the SVD of X or the EVD of XX^{H} (or $X^{\mathrm{H}}X$) is equivalent, but the dynamic range of the eigenvalues is twice that of the corresponding singular values. With finite precision arithmetic, the greater dynamic range can result in a loss of information. For example, in rank determination, suppose the smallest singular value is ε where ε is machine precision. The corresponding eigenvalue, ε^2, would be considered a machine precision zero and the EVD of XX^{H} (or $X^{\mathrm{H}}X$) would incorrectly indicate a rank deficiency. Because of the dynamic range issue, it is generally recommended to use the SVD of X (or a square root factor of R). However, because additive sensor noise usually dominates numerical errors, this choice may not be critical in most signal processing applications.

7.2.2 Short Memory Windows for Time Varying Estimation

Ultimately, we are interested in tracking some aspect of the eigenstructure of a time varying correlation (or data) matrix. For simplicity we will focus on time varying estimation of the correlation matrix, realizing that the EVD of R is trivially related to the SVD of X. A time varying estimator must have a short term memory in order to track changes. An example of long memory estimation is an estimator that involves a growing rectangular data window. As time goes on, the estimated quantities depend more and more on the old data, and less and less on the new data. The two most popular short memory approaches to estimating a time varying correlation matrix involve (1) a moving rectangular window and (2) an exponentially faded window. Unfortunately, an unbiased, causal estimate of the true instantaneous correlation matrix at time k, $\Phi_k = E[x_k x_k^{\mathrm{H}}]$, is not possible if averaging is used and the vector process is truly time varying. However, it is usually assumed that the process is varying slowly enough within the effective observation window that the process is approximately stationary and some averaging is desirable. In any event, at time k, a length N moving rectangular data window results in a rank two modification of the correlation matrix estimate, i.e.,

$$R_k^{(\mathrm{rect})} = R_{k-1}^{(\mathrm{rect})} + \frac{1}{N}\left(x_k x_k^{\mathrm{H}} - x_{k-N}x_{k-N}^{\mathrm{H}}\right) \tag{7.1}$$

where

x_k is the new snapshot vector

x_{k-N} is the oldest vector which is being removed from the estimate

The corresponding data matrix is given by $X_k^{(\mathrm{rect})} = [x_k|x_{k-1}|\cdots|x_{k-N+1}]$ and $R_k^{(\mathrm{rect})} = \frac{1}{N}X_k^{(\mathrm{rect})}\left(X_k^{(\mathrm{rect})}\right)^{\mathrm{H}}$. Subtracting the rank-one matrix from the correlation estimate is referred to as a rank-one downdate.

Downdating moves all the eigenvalues down (or unchanged). Updating, on the other hand, moves all eigenvalues up (or unchanged). Downdating is potentially ill-conditioned because the smallest eigenvalue can move toward zero.

An exponentially faded data window produces a rank-one modification in

$$\boldsymbol{R}_k^{(\text{fade})} = \alpha \boldsymbol{R}_{k-1}^{(\text{fade})} + (1-\alpha) x_k x_k^{\mathrm{H}} \tag{7.2}$$

where α is the fading factor with $0 \leq \alpha \leq 1$. In this case, the data matrix is growing in size, but the older data is deemphasized with a diagonal weighting matrix, $X_k^{(\text{fade})} = [x_k | x_{k-1} | \cdots | x_1]$ $\text{sqrt}(\text{diag}(1, \alpha, \alpha^2, \ldots, \alpha^{k-1}))$ and $R_k^{(\text{fade})} = (1-\alpha) X_k^{(\text{fade})} \left(X_k^{(\text{fade})}\right)^{\mathrm{H}}$.

Of course, the two windows could be combined to produce an exponentially faded moving rectangular window, but this kind of hybrid short memory window has not been the subject of much study in the signal processing literature. Similarly, not much attention has been paid to which short memory windowing scheme is most appropriate for a given data model. Since downdating is potentially ill-conditioned, and since two rank-one modifications usually involve more computation than one, the exponentially faded window has some advantages over the moving rectangular window. The main advantage of a (short) rectangular window is in tracking sudden changes. Assuming stationarity within the effective observation window, the power in a rectangular window will be equal to the power in an exponentially faded window when

$$N \approx \frac{1}{(1-\alpha)} \quad \text{or equivalently} \quad \alpha \approx 1 - \frac{1}{N} = \frac{N-1}{N}. \tag{7.3}$$

Based on a Fourier analysis of linearly varying frequencies, equal frequency lags occur when [14]

$$N \approx \frac{(1+\alpha)}{(1-\alpha)} \quad \text{or equivalently} \quad \alpha \approx \frac{N-1}{N+1}. \tag{7.4}$$

Either one of these relationships could be used as a rule of thumb for relating the effective observation window of the two most popular short memory windowing schemes.

7.2.3 Classification of Subspace Methods

Eigenstructure estimation can be classified as (1) block or (2) recursive. Block methods simply compute an EVD, SVD, or related decomposition based on a block of data. Recursive methods update the previously computed eigen information using new data as it arrives. We focus on recursive subspace updating methods in this article.

Most subspace tracking algorithms can also be broadly categorized as (1) modified eigenproblem (MEP) methods or (2) adaptive (or non-MEP) methods. With short memory windowing, MEP methods are adaptive in the sense that they can track time varying eigen information. However, when we use the word adaptive, we mean that exact eigen information is not computed at each update, but rather, an adaptive method tends to move toward an EVD (or some aspect of an EVD) at each update. For example, gradient-based, perturbation-based, and neural network-based methods are classified as adaptive because on average they move towards an EVD at each update. On the other hand, rank one, rank k, and sphericalized EVD and SVD updates are, by definition, MEP methods because exact eigen information associated with an explicit matrix is computed at each update. Both MEP and adaptive methods are supposed to track the eigen information of the instantaneous, time varying correlation matrix.

7.2.4 Historical Overview of MEP Methods

Many researchers have studied SVD and EVD tracking problems. Golub [19] introduced one of the first eigen-updating schemes, and his ideas were developed and expanded by Bunch and coworkers in [3,4]. The basic idea is to update the EVD of a symmetric (or Hermitian) matrix when modified by a rank-one matrix. The rank-one eigen update was simplified in [37], when Schreiber introduced a transformation that makes the core eigenproblem real. Based on an additive white noise model, Karasalo [21] and Schreiber [37] suggested that the noise subspace be "sphericalized," i.e., replace the noise eigenvalues by their average value so that deflation [4] could be used to significantly reduce computation. By deflating the noise subspace and only tracking the r dominant eigenvectors, the computation is reduced from $O(n^3)$ to $O(nr^2)$ per update. DeGroat reduced computation further by extending this concept to the signal subspace [8]. By sphericalizing and deflating both the signal and the noise subspaces, the cost of tracking the r dimensional signal (or noise) subspace is $O(nr)$ and no iteration is involved. To make eigen updating more practical, DeGroat and Roberts developed stabilization schemes to control the loss of orthogonality due to the buildup of roundoff error [10]. Further work related to eigenvector stabilization is reported in [15,28–30]. Recently, a more stable version of Bunch's algorithm was developed by Gu and Eisenstat [20]. In [46], Yu extended rank-one eigen updating to rank k updating.

DeGroat showed in [8] that forcing certain subspaces of the correlation matrix to be spherical, i.e., replacing the associated eigenvalues with a fixed or average value, is an easy way to deflate the size of the updating problem and reduce computation. Basically, a spherical subspace (SS) update is a rank-one EVD update of a sphericalized correlation matrix. Asymptotic convergence analysis of SS updating is found in [11,13]. A four level SS update capable of automatic signal subspace rank and size adjustment is described in [9,11]. The four level and the two level SS updates are the only MEP updates to date that are $O(nr)$ and noniterative. For more details on SS updating, see Section 7.3.6.

In [42], Xu et al. present a Lanczos based subspace tracking method with an associated detection scheme to track the number of sources. A reference list for systolic implementations of SVD based subspace trackers is contained in [12].

7.2.5 Historical Overview of Adaptive, Non-MEP Methods

Owsley pioneered orthogonal iteration and stochastic-based subspace trackers in [32]. Yang and Kaveh extended Owsley's work in [44] by devising a family of constrained gradient-based algorithms. A highly parallel algorithm, denoted the inflation method, is introduced for the estimation of the noise subspace. The computational complexity of this family of gradient-based methods varies from (approximately) n^2r to $\frac{7}{2}nr$ for the adaptation equation. However, since the eigenvectors are only approximately orthogonal, an additional nr^2 flops may be needed if Gram Schmidt orthogonalization is used. It may be that a partial orthogonalization scheme (see Section 7.3.2) can be combined with Yang and Kaveh's methods to improve orthogonality enough to eliminate the $O(nr^2)$ Gram Schmidt computation. Karhunen [22] also extended Owsley's work by developing a stochastic approximation method for subspace computation. Bin Yang [43] used recursive least squares (RLS) methods with a projection approximation approach to develop the projection approximation subspace tracker (PAST) which tracks an arbitrary basis for the signal subspace, and PASTd which uses deflation to track the individual eigencomponents. A multi-vector eigen tracker based on the conjugate gradient method is developed in [18]. Previous conjugate gradient-based methods tracked a single eigenvector only. Orthogonal iteration, lossless adaptive filter, and perturbation-based subspace trackers appear in [40], [36], and [5] respectively. A family of non-EVD subspace trackers is given in [16]. An adaptive subspace method that uses a linear operator, referred to as the Propagator, is given in [26]. Approximate SVD methods that are based on a QR update step followed by a single (or partial) Jacobi sweep to move the triangular factor towards a diagonal form appear in [12,17,30]. These methods can be described as approximate SVD methods because they will converge to an SVD if the Jacobi sweeps are repeated.

Subspace estimation methods based on URV or rank revealing QR (RRQR) decompositions are referenced in [6]. These rank revealing decompositions can divide a set of orthonormal vectors into sets that span the signal and noise subspaces. However, a threshold (noise power) level that lies between the largest noise eigenvalue and the smallest signal eigenvalue must be known in advance. In some ways, the URV decomposition (URVD) can be viewed as an approximate SVD. For example, the transposed QR (TQR) iteration [12] can be used to compute the SVD of a matrix, but if the iteration is stopped before convergence, the resulting decomposition is URV-like.

Artificial neural networks (ANN) have also been used to estimate eigen information [35]. In 1982, Oja [31] was one of the first to develop an eigenvector estimating ANN. Using a Hebbian type learning rule, this ANN adaptively extracts the first principal eigenvector. Much research has been done in this area since 1982. For an overview and a list of references, see [35].

7.3 Issues Relevant to Subspace and Eigen Tracking Methods

7.3.1 Bias due to Time Varying Nature of Data Model

Because DOA angles are typically time varying, a range of spatial frequencies is usually included in the effective observation window. Most spatial frequency estimation methods yield frequency estimates that are approximately equal to the effective frequency average in the window. Consequently, the estimates lag the true instantaneous frequency. If the frequency variation is assumed to be linear within the effective observation window, this lag (or bias) can be easily estimated and compensated [14].

7.3.2 Controlling Roundoff Error Accumulation and Orthogonality Errors

Numerical algorithms are generally defined as stable if the roundoff error accumulates in a linear fashion. However, recursive updating algorithms cannot tolerate even a linear buildup of error if large (possibly unbounded) numbers of updates are to be performed. For real time processing, periodic reinitialization is undesirable. Most of the subspace tracking algorithms involve the product of at least k orthogonal matrices by the time the kth update is computed. According to Parlett [33], the error propagated by a product of orthogonal matrices is bounded as

$$|U_k U_k^{\mathrm{H}} - I|_E \leq (k+1)n^{1.5}\varepsilon \tag{7.5}$$

where the $n \times n$ matrix $U_k = U_{k-1}Q_k = Q_k Q_{k-1} \cdots Q_1$ is a product of k matrices that are each orthogonal to working accuracy, ε is machine precision, and $|\ .\ |_E$ denotes the Euclidean matrix norm. Clearly, if k is large enough, the roundoff error accumulation can be significant.

There are really only two sources of error in updating a symmetric or Hermitian EVD: (1) the eigenvalues and (2) the eigenvectors. Of course, the eigenvectors and eigenvalues are interrelated. Errors in one tend to produce errors in the other. At each update, small errors may occur in the EVD update so that the eigenvalues become slowly perturbed and the eigenvectors become slowly nonorthonormal. The solution is to prevent significant errors from ever accumulating in either.

We do not expect the main source of error to be from the eigenvalues. According to Stewart [38], the eigenvalues of a Hermitian matrix are perfectly conditioned, having condition numbers of one. Moreover, it is easy to show that when exponential weighting is used, the accumulated roundoff error is bounded by a constant, assuming no significant errors are introduced by the eigenvectors. By contrast, if exponential windowing is not used, the bound for the accumulated error builds up in a linear fashion. Thus, the fading factor not only fades out old data, but also old roundoff errors that accumulate in the eigenvalues.

Unfortunately, the eigenvectors of a Hermitian matrix are not guaranteed to be well conditioned. An eigenvector will be ill-conditioned if its eigenvalue is closely spaced with other eigenvalues. In this case, small roundoff perturbations to the matrix may cause relatively large errors in the eigenvectors.

The greatest potential for nonorthogonality then is between eigenvectors with adjacent (closely spaced) eigenvalues. This observation led to the development of a partial orthogonalization scheme known as pairwise Gram Schmidt (PGS) [10] which attacks the roundoff error buildup problem at the point of greatest numerical instability—nonorthogonality of adjacent eigenvectors. If the intervening rotations (orthogonal matrix products) inherent in the eigen update are random enough, the adjacent vector PGS can be viewed as a full orthogonalization spread out over time. When PGS is combined with exponential fading, the roundoff accumulation in both the eigenvectors and the eigenvalues is controlled. Although PGS was originally designed to stabilize Bunch's EVD update, it is generally applicable to any EVD, SVD, URV, QR, or orthogonal vector update. Moonen et al. [29] suggested that the bulk of the eigenvector stabilization in the PGS scheme is due to the normalization of the eigenvectors. Simulations seem to indicate that normalization alone stabilizes the eigenvectors almost as well as the PGS scheme, but not to working precision orthogonality. Edelman and Stewart provide some insight into the normalization only approach to maintaining orthogonality [15]. For additional analysis and variations on the basic idea of spreading orthogonalization out over time, see [30] and especially [28].

Many of the $O(nr)$ adaptive subspace methods produce eigenvector estimates that are only approximately orthogonal and normalization alone does not always provide enough stabilization to keep the orthogonality and other error measures small enough. We have found that PGS stabilization can noticeably improve both the subspace estimation performance as well as the DOA (or spatial frequency) estimation performance. For example, without PGS (but with normalization only), we found that Champagne's $O(nr)$ perturbation-based eigen tracker (method PC) [5] sometimes gives spurious MUSIC-based frequency estimates. On the other hand, with PGS, Champagne's PC method produced improved subspace and frequency estimates. The orthogonality error was also significantly reduced. Similar performance boosts could be expected for any subspace or eigen tracking method (especially those that produce eigenvector estimates that are only approximately orthogonal, e.g., PAST and PASTd [43] or Yang and Kaveh's family of gradient-based methods [44,45]). Unfortunately, normalization only and PGS are $O(nr)$. Adding this kind of stabilization to an $O(nr)$ subspace tracking method could double its overall computation.

Other variations on the original PGS idea involve symmetrizing the 2×2 transformation and making the pairwise orthogonalization cyclic [28]. The symmetric transformation assumes that the vector pairs are almost orthogonal so that higher order error terms can be ignored. If this is the case, the symmetric version can provide slightly better results at a somewhat higher computational cost. For methods that involve working precision orthogonal vectors, the original PGS scheme is overkill. Instead of doing PGS orthogonalization on each adjacent vector pair, cyclic PGS orthogonalizes only one pair of vectors per update, but cycles through all possible combinations over time. Thus, cyclic PGS covers all vector pairs without relying on the randomness of intervening rotations. Cyclic PGS spreads the orthogonalization process out in time even more than the adjacent vector PGS method. Moreover, cyclic PGS (or cyclic normalization) involves $O(n)$ flops per update, but there is a small overhead associated with keeping track of the vector pair cycle.

In summary, we can say that stabilization may not be needed for a small number of updates. On the other hand, if an unbounded number of updates is to be performed, some kind of stabilization is recommended. For methods that yield nearly orthogonal vectors at each update, only a small amount of orthogonalization is needed to control the error buildup. In these cases, cyclic PGS may be best. However, for methods that produce vectors that are only approximately orthogonal, a more complete orthogonalization scheme may be appropriate, e.g., a cyclic scheme with two or three vector pairs orthogonalized per update will produce better results than a single pair scheme.

7.3.3 Forward–Backward Averaging

In many subspace tracking problems, forward–backward (FB) averaging can improve subspace as well as DOA (or frequency) estimation performance. Although FB averaging is generally not appropriate for nonstationary processes, it does appear to improve spatial frequency estimation performance if the frequencies vary linearly within the effective observation window. Based on Fourier analysis of linearly varying frequencies, we infer that this is probably due to the fact that the average frequency in the window is identical for both the forward and the backward cases [14]. Consequently, the frequency estimates are reinforced by FB averaging. Besides improved estimation performance, FB averaging can be exploited to reduce computation by as much as 75% [24]. FB averaging can also reduce computer memory requirements because (conjugate symmetric or antisymmetric) symmetries in the complex eigenvectors of an FB averaged correlation matrix (or the singular vectors of an FB data matrix) can be exposed through appropriate normalization.

7.3.4 Frequency vs. Subspace Estimation Performance

It has recently been shown with asymptotic analysis that a better subspace estimate does not necessarily result in a better MUSIC-based frequency estimate [23]. In subspace tracking simulations, we have also observed that some methods produce better subspace estimates, but the associated MUSIC-based frequency estimates are not always better. Consequently, if DOA estimation is the ultimate goal, subspace estimation performance may not be the best criterion for evaluating subspace tracking methods.

7.3.5 The Difficulty of Testing and Comparing Subspace Tracking Methods

A significant amount of research has been done on subspace and eigen tracking algorithms in the past few years, and much progress has been made in making subspace tracking more efficient. Not surprisingly, all of the methods developed to date have different strengths and weaknesses. Unfortunately, there has not been enough time to thoroughly analyze, study, and evaluate all of the new methods. Over the years, several tests have been devised to "experimentally" compare various methods, e.g., convergence tests [44], response to sudden changes [7], and crossing frequency tracks (where the signal subspace temporarily collapses) [8]. Some methods do well on one test, but not so well on another. It is difficult to objectively compare different subspace tracking methods because optimal operating parameters are usually unknown and therefore unused, and the performance criteria may be ill-defined or contradictory.

7.3.6 Spherical Subspace Updating—A General Framework for Simplified Updating

Most eigen and subspace tracking algorithms are based directly or indirectly on tracking some aspect of the EVD of a time varying correlation matrix estimate that is recursively updated according to Equation 7.1 or 7.2. Since Equations 7.1 and 7.2 involve rank-one and rank-two modifications to the correlation matrix, most subspace tracking algorithms explicitly or implicitly involve rank-one (or two) modification of the correlation matrix. Since rank-two modifications can be computed as two rank-one modifications, we will focus on rank-one updating.

Basically, SS updates are simplified rank-one EVD updates. The simplification involves sphericalizing subsets of eigenvalues (i.e., forcing each subset to have the same eigenlevel) so that the sphericalized subspaces can be deflated.

Based on an additive white noise signal model, Karasalo [21] and Schreiber [37] first suggested that the "noise" eigenvalues be replaced by their average value in order to reduce computation by deflation. Using Ljung's ODE-based method for analyzing stochastic recursive algorithms [25], it has recently been shown that, if the noise subspace is sphericalized, the dominant eigenstructure of a correlation matrix asymptotically converges to the true eigenstructure with probability one (under any noise assumption) [11]. It is important to realize that averaging the noise eigenvalues yields a SS in which the eigenvectors can be arbitrarily oriented as long as they form an orthonormal basis for the subspace. A rank-one modification affects only one component of the sphericalized subspace. Thus, only one of the multiple noise eigenvalues is changed by a rank-one modification. Consequently, making the noise subspace spherical (by averaging the noise eigenvalues, or replacing them with a constant eigenlevel) deflates theeigenproblem to an $(r+1)\times(r+1)$ problem, which corresponds to a signal subspace of dimension r, and the single noise component whose power is changed. For details on deflation, see [4].

The analysis in [11] shows that any number of sphericalized eigenlevels can be used to track various subspace spans associated with the correlation matrix. For example, if both the noise and the signal subspaces are sphericalized (i.e., the dominant and subdominant set of eigenvalues is replaced by their respective averages), the problem deflates to a 2×2 eigenproblem that can be solved in closed form, noniteratively. We will call this doubly deflated SS update, SA2 (signal averaged, two eigenlevels) [8]. In [13] we derived the SA2 algorithm ODE and used a Lyapunov function to show asymptotic convergence to the true subspaces w.p. 1 under a diminishing gain assumption. In fact, the SA2 subspace trajectories can be described with Lie bracket notation and follow an isospectral flow as described by Brockett's ODE [2]. A four level SS update (called SA4) was introduced in [9] to allow for information theoretic source detection (based on the eigenvalues at the boundary of the signal and noise subspaces) and automatic subspace size adjustment. A detailed analysis of SA4 and an SA4 minimum description length (SA4-MDL) detection scheme can be found in [11,41]. SA4 sphericalizes all the signal eigenvalues except the smallest one, and all the noise eigenvalues except the largest one, resulting in a 4×4 deflated eigenproblem. By tracking the eigenvalues that are on the boundary of the signal and noise subspaces, information theoretic detection schemes can be used to decide if the signal subspace dimension should be increased, decreased, or remain unchanged. Both SA2 and SA4 are $O(nr)$ and noniterative.

The deflated core problem in SS updating can involve any EVD or SVD method that is desired. It can also involve other decompositions, e.g., the URVD [34]. To illustrate the basic idea of SS updating, we will explicitly show how an update is accomplished when only the smallest $(n-r)$ "noise" eigenvalues are sphericalized. This particular SS update is called a signal eigenstructure (SE) update because only the dominant r "signal" eigencomponents are tracked. This case is equivalent to that described by Schreiber [37] and an SVD version is given by Karasalo [21].

To simplify and more clearly illustrate the idea SS updating, we drop the normalization factor, $(1-\alpha)$, and the k subscripts from Equation 7.2 and use the eigendecomposition of $\boldsymbol{R}=\boldsymbol{UDU}^{\mathrm{H}}$ to expose a simpler underlying structure for a single rank-one update

$$\tilde{\boldsymbol{R}} = \alpha\boldsymbol{R} + xx^{\mathrm{H}} \tag{7.6}$$

$$= \alpha\boldsymbol{UDU}^{\mathrm{H}} + xx^{\mathrm{H}} \tag{7.7}$$

$$= \boldsymbol{U}(\alpha\boldsymbol{D} + \beta\beta^{\mathrm{H}})\boldsymbol{U}^{\mathrm{H}}, \quad \beta = U^{\mathrm{H}}x \tag{7.8}$$

$$= \boldsymbol{UG}(\alpha\boldsymbol{D} + \gamma\gamma^{\mathrm{T}})\boldsymbol{G}^{\mathrm{H}}\boldsymbol{U}^{\mathrm{H}}, \quad \gamma = \boldsymbol{G}^{\mathrm{H}}\beta \tag{7.9}$$

$$= \boldsymbol{UGH}(\alpha\boldsymbol{D} + \zeta\zeta^{\mathrm{T}})\boldsymbol{H}^{\mathrm{T}}\boldsymbol{G}^{\mathrm{H}}\boldsymbol{U}^{\mathrm{H}}, \quad \zeta = \boldsymbol{H}^{\mathrm{T}}\gamma \tag{7.10}$$

$$= \boldsymbol{UGH}(\boldsymbol{Q}\tilde{\boldsymbol{D}}\boldsymbol{Q}^{\mathrm{T}})\boldsymbol{H}^{\mathrm{T}}\boldsymbol{G}^{\mathrm{H}}\boldsymbol{U}^{\mathrm{H}} \tag{7.11}$$

$$= \tilde{U}\tilde{D}\tilde{U}^{H}, \quad \tilde{U} = UGHQ \tag{7.12}$$

where

$G = \mathrm{diag}\,(\beta_1/|\beta_1|,\ldots,\beta_n/|\beta_n|$ is a diagonal unitary transformation that has the effect of making the matrix inside the parenthesis real [37]

$\boldsymbol{H}$ is an embedded Householder transformation that deflates the core problem by zeroing out certain elements of ζ (see the SE case below)

$\boldsymbol{Q}\tilde{\boldsymbol{D}}\boldsymbol{Q}^{\mathrm{T}}$ is the EVD of the simplified, deflated core matrix, $(\alpha\boldsymbol{D} + \zeta\zeta^{\mathrm{T}})$

In general, $\boldsymbol{H}$ and $\boldsymbol{Q}$ will involve smaller matrices embedded in an $n\times n$ identity matrix. In order to more clearly see the details of deflation, we must concentrate on finding the eigendecomposition of the completely real matrix, $S = (\alpha \boldsymbol{D} + \gamma\gamma^{\mathrm{T}})$ for a specific case. Let us consider the SE update and assume that the noise eigenvalues contained in the diagonal matrix have been replaced by their average values, $d^{(n)}$, to produce a sphericalized noise subspace. We must then apply block Householder transformations to concentrate all of the power in the new data vector into a single component of the noise subspace. The update is thus deflated to an $(r+1) \times (r+1)$ embedded eigenproblem as shown below,

$$S = (\alpha \boldsymbol{D} + \gamma\gamma^{\mathrm{T}}) \tag{7.13}$$

$$= H(\alpha \boldsymbol{D} + \zeta\zeta^{\mathrm{T}})H^{\mathrm{T}}, \quad \zeta = \boldsymbol{H}^{\mathrm{T}}\gamma \tag{7.14}$$

$$= \begin{bmatrix} \boldsymbol{I}_r & 0(15) \\ 0 & \boldsymbol{H}_{n-r}^{(n)} \end{bmatrix} \left(\alpha \begin{bmatrix} \boldsymbol{D}_r^{(s)} & 0(16) \\ 0 & d^{(n)} I_{n-r} \end{bmatrix} + \zeta\zeta^{\mathrm{T}} \right) \begin{bmatrix} \boldsymbol{I}_r & 0 \\ 0 & \boldsymbol{H}_{n-r}^{(n)} \end{bmatrix}^{\mathrm{T}} \tag{7.15}$$

$$= \begin{bmatrix} \boldsymbol{I}_r & 0 \\ 0 & \boldsymbol{H}_{n-r}^{(n)} \end{bmatrix} \left(\begin{bmatrix} \boldsymbol{Q}_{r+1} & 0 \\ 0 & I_{n-r-1} \end{bmatrix} \begin{bmatrix} \tilde{D}_r^{(s)} & 0 & 0 \\ 0 & \tilde{d}^{(n)} & 0 \\ 0 & 0 & \alpha d^{(n)} I_{n-r-1} \end{bmatrix} \begin{bmatrix} \boldsymbol{Q}_{r+1} & 0 \\ 0 & \boldsymbol{I}_{n-r-1} \end{bmatrix}^{\mathrm{T}} \begin{bmatrix} \boldsymbol{I}_r & 0 \\ 0 & \boldsymbol{H}_{n-r}^{(n)} \end{bmatrix}^{\mathrm{T}} \right) \tag{7.16}$$

$$= H(Q\tilde{D}Q^{\mathrm{T}})H^{\mathrm{T}} \tag{7.17}$$

where

$$\zeta^{\mathrm{T}} = (\boldsymbol{H}^{\mathrm{T}}\gamma)^{\mathrm{T}} = [\gamma^{(s)}, |\gamma^{(n)}|, 0_{(n-r-1)\times 1}]^{\mathrm{T}} \tag{7.18}$$

$$\boldsymbol{H}_{n-r}^{(n)} = I_{n-r} - 2\frac{\nu^{(n)}(\nu^{(n)})^{\mathrm{T}}}{(\nu^{(n)})^{\mathrm{T}}\nu^{(n)}} \tag{7.19}$$

$$\boldsymbol{H} = \begin{bmatrix} \boldsymbol{I}_r & 0 \\ 0 & \boldsymbol{H}_{n-r}^{(n)} \end{bmatrix} \tag{7.20}$$

$$\gamma = \begin{bmatrix} \gamma^{(s)} \\ \gamma^{(n)} \end{bmatrix} \begin{matrix} \}r \\ \}n-r \end{matrix} \tag{7.21}$$

$$\nu^{(n)} = \gamma^{(n)} + |\gamma^{(n)}| \begin{bmatrix} 1 \\ 0_{(n-r-1)\times 1} \end{bmatrix} \tag{7.22}$$

The superscripts (s) and (n) denote signal and noise subspace, respectively, and the subscripts denote the size of the various block matrices. In the actual implementation of the SE algorithm, the Householder transformations are not explicitly computed, as we will see below. Moreover, it should be stressed that the Householder transformation does not change the span of the noise subspace, but merely "aligns" the subspace so that all of the new data vector, x, that projects into the noise subspace lies in a single component of the noise subspace.

The embedded (deflated) $(r+1) \times (r+1)$ eigenproblem,

$$\boldsymbol{E} = \left(\begin{bmatrix} D^{(s)} & 0 \\ 0 & d^{(n)} \end{bmatrix} + \begin{bmatrix} \gamma^{(s)} \\ |\gamma^{(n)}| \end{bmatrix} \begin{bmatrix} \gamma^{(s)} \\ |\gamma^{(n)}| \end{bmatrix}^{\mathrm{T}} \right)_{(r+1)\times(r+1)} = \boldsymbol{Q}_{r+1}\tilde{D}_{r+1}Q_{r+1}^{\mathrm{T}} \tag{7.23}$$

can be solved using any EVD algorithm. Or, an SVD (square root) version can be computed by finding the SVD of

$$F = \begin{bmatrix} \Sigma^{(s)} & 0 & \gamma^{(s)} \\ 0 & \sigma^{(n)} & |\gamma^{(n)}| \end{bmatrix}_{(r+1)\times(r+2)} = Q_{r+1}\tilde{\Sigma}_{r+1}P_{r+1}^{\mathrm{T}} \tag{7.24}$$

where

$E = FF^{\mathrm{T}}$

$\Sigma^{(s)} = \mathrm{sqrt}(D^{(s)})$

$\sigma^{(n)} = \sqrt{d^{(n)}}$

$\widetilde{\Sigma}_{r+1} = \mathrm{sqrt}(\tilde{D}_{r+1})$

The right singular vectors, P_{r+1}, are generally not needed or explicitly computed in most subspace tracking problems. The new signal and noise subspaces are thus given by

$$\tilde{U} = [\tilde{U}^{(s)}, \tilde{U}^{(n)}] \tag{7.25}$$

$$= UGHQ \tag{7.26}$$

$$= \left[\underbrace{U^{(s)}G^{(s)}}_{n\times r}, \underbrace{U^{(n)}G^{(n)}H^{(n)}}_{n\times(n-r)}\right]\begin{bmatrix} Q_{r+1} & 0 \\ 0 & I_{n-r-1} \end{bmatrix} \tag{7.27}$$

where

$U^{(s)}$ and $U^{(n)}$ are the old signal and noise subspaces

G represents the diagonal unitary transformation that makes the rest of the problem real

H is the block Householder transformation that rotates (or more precisely, reflects) the SSs so that all of the noise power contained in the new data vector can be concentrated into a single component of noise subspace

Q represents the evolution and interaction of the two subspaces induced by the new data vector

Basically, this update partitions the data space into two subspaces: the signal subspace is not sphericalized and all of its eigencomponents are explicitly tracked whereas the noise subspace is sphericalized and not explicitly tracked (to save computation). Using the properties of the Householder transformation, it can be shown that the single component of the noise subspace that mixes with the signal subspace via Q_{r+1} is given by

$$U^{(n)} = \text{the first column of } U^{(n)}G^{(n)}H^{(n)} \tag{7.28}$$

$$= \frac{U^{(n)}(U^{(n)})^{\mathrm{H}}x}{|U^{(n)}(U^{(n)})^{\mathrm{H}}x|} \tag{7.29}$$

$$= \frac{1}{|\gamma^{(n)}|}(I - U^{(s)}(U^{(s)})^{\mathrm{H}})x \tag{7.30}$$

$$= \frac{1}{|\gamma^{(n)}|}(x - U^{(s)}\gamma^{(s)}) \tag{7.31}$$

where

$U^{(n)}$ is the projection of x into the noise subspace

$|\gamma^{(n)}| = |x - U^{(s)}\gamma^{(s)}|$ is the power of x projected into the noise subspace

Once the eigenvectors of the core $(r+1) \times (r+1)$ problem are found, the signal subspace eigenvectors can be updated as

$$\tilde{U} = \left[\tilde{U}^{(s)}, \tilde{U}^{(n)}\right] \tag{7.32}$$

$$= \left[\underbrace{U^{(s)}G^{(s)}}_{n\times r}, \underbrace{U^{(n)}}_{n\times 1}\right] Q_{r+1} \tag{7.33}$$

where updating the new noise eigenvector is not necessary (if the noise subspace is respshericalized). The complexity of the core eigenproblem is $O(r^3)$ and updating the signal eigenvectors is $O(nr^2)$. Thus, the SE update is $O(nr^2)$.

After an update is accomplished, one of the noise eigencomponents is altered by the embedded eigenproblem. To maintain noise subspace sphericity, the noise eigenvalues must be reaveraged before the next SE update can be accomplished. On the other hand, if the noise eigenvalues are not reaveraged, the SE update eventually reverts to a full eigen update.

A whole family of related SS updates is possible by simple modification of the above described process. For example, to obtain SA2, the $\boldsymbol{H}$ transformation in Equation 7.20 would be modified by replacing the I_r with an $r \times r$ Householder matrix that deflates the signal subspace. This would make the core eigenproblem 2×2 and the $\boldsymbol{Q}$ matrix an identity with an embedded 2×2 orthogonal matrix.

7.3.7 Initialization of Subspace and Eigen Tracking Algorithms

It is impossible to give generic initialization requirements that would apply to all subspace tracking algorithms, but one feature that is common to many updating methods is a fading factor. For cold start initialization (e.g., starting from nothing) at $k = 0$, initial convergence can often be sped up by ramping up the fading factor, e.g.,

$$\alpha_k = \left(1 - \frac{1}{k+1}\right)\alpha, \quad k = 0, 1, 2, \ldots \tag{7.34}$$

where α is the final steady state value for the fading factor.

7.3.8 Detection Schemes for Subspace Tracking

Several subspace tracking methods have detection schemes that were specifically designed for them. Xu and Kailath developed a strongly consistent detection scheme for their Lanczos-based method [42]. DeGroat and Dowling adapted information theoretic criteria for use with SA4 [9] and an asymptotic proof of consistency is given in [11]. Stewart proposed the URV update as a rank revealing method [39]. Bin Yang proposed that the eigenvalue estimates from PASTd be used for information theoretic-based rank estimation [43].

7.4 Summary of Subspace Tracking Methods Developed Since 1990

7.4.1 Modified Eigenproblems

An $O(n^2r)$ fast subspace decomposition (FSD) method based on the Lanczos algorithm and a strongly consistent source detection scheme was introduced by Xu et al. [42] (Table 7.1).

TABLE 7.1 Efficient Subspace Tracking Methods Developed Since 1990

Complexity	Subspace or Eigen Tracking Method	Orthog. Span
$O(n^2r)$	FSD [42]	Yes[a]
$O(n^2)$	URV update [39]	Yes
	RRQR [possibly $O(n^3)$] [11]	Yes
	Approximate SVD updates [17,30]	Yes[a]
	Neural network-based updates [35]	No[a]
$O(n^2r)$	Stabilized SE update[b] [8,10]	Yes[a]
	Sphericalized TQR SVD update[b] [12]	Yes[a]
	Sphericalized conjugate gradient SVD update[b] [18]	Yes[a]
	SWEDE [16]	No
	Gradient-based EVD updates with Gram Schmidt orthog. [44,45]	Yes[a]
$O(nr)$	Signal averaged 2-level (SA2) update[b] [8]	Yes
	Signal averaged 4-level (SA4) update[b] [9,11]	Yes[a]
	PAST [43]	No
	PAST with deflation (PASTd) [43]	No[a]
	Sphericalized perturbation-based eigen update (PC method) [5]	Yes[a]
	Sphericalized URV update[b] [34]	Yes

Key: n, no of sensors; r, rank of subspace.
[a] Tracks individual eigencomponents.
[b] Uses sphericalized subspaces.

A TQR iteration-based SVD update was introduced in [12]. To reduce computation to $O(nr^2)$, the noise subspace is sphericalized and deflated. Based on various performance tests, one or two TQR iterations per update yield results that are comparable to the fully converged SVD. Moreover, because the diagonalization process is taking place on a triangular factor, the partially converged, deflated TQR-SVD update is very similar to a deflated URV update [34].

DeGroat and Roberts [10] simplified Bunch's rank-one eigen update [4] and proposed a partial orthogonalization scheme, called pair-wise Gram Schmidt (PGS), to stabilize the eigenvectors. Together with exponential fading to stabilize the eigenvalues, the buildup of roundoff error is essentially controlled and machine precision orthogonality is maintained. For a more complete discussion, see Section 7.3.2. Recently, Gu and Eisenstat [20] presented an improved version of Bunch's rank-one EVD update. The new algorithm contains a more stable way to compute the eigenvectors. DeGroat and Dowling have also developed a family of sphericalized EVD and SVD updates (see Section 7.3.6).

7.4.2 Gradient-Based Eigen Tracking

Jar-Ferr Yang and Hui-Ju Lin [45] proposed a generalized inflation method which extends the gradient-based work of Yang and Kaveh [44]. An $O(nr^2)$ noise sphericalized and deflated conjugate gradient-based eigen tracking method is presented by Fu and Dowling in [18]. This method can be described as an SS update with a conjugate gradient-based eigen tracker at the core.

Bin Yang [43] introduced a projection approximation approach that uses RLS techniques to update the signal subspace. The PAST algorithm computes an arbitrary basis for the signal subspace in $3nr + O(r^2)$ flops per update. The PASTd algorithm (which uses deflation to track the individual eigenvalues and vectors of the signal subspace) requires $4nr + O(n)$ flops per update. Both methods produce eigenvector estimates that are only approximately orthogonal.

Regalia and Loubaton [36] use an adaptive lossless transfer matrix (multivariable lattice filter) excited by sensor output to achieve a condition of maximum "power splitting" between two groups of output bins. The update equations resemble standard gradient descent algorithms, but they do not properly follow the gradient of the error surface. Nonetheless, the convergence speed may be a strong function of the source spectral and spatial characteristics.

Recently, Marcos and Benidir [26] introduced an adaptive subspace-based method that relies on a linear operator, referred to as the Propagator, which exploits the linear independency of the source steering vectors, and which allows the determination of the signal and noise subspaces without any eigendecomposition of the correlation matrix. Two gradient-based adaptive algorithms are proposed for the estimation of the Propagator, and then the basis of the signal subspace. The overall computational complexity of the adaptive Propagator subspace update is $O(nr^2)$.

A family of three perturbation-based EVD tracking methods (denoted PA, PB, and PC) are presented by Champagne [5]. Each method uses perturbation-based approximations to track the eigencomponents. Progressively more simplifications are used to reduce the complexity from $\frac{1}{2}n^3 + O(n^2)$ for PA to $\frac{1}{2}nr^2 + O(nr)$ for PB to $5nr + O(n)$ for PC. Both the PB and PC methods use a sphericalized noise subspace to reduce computation. Thus, PB and PC can be viewed as SS updates that use perturbation-based approximations to track the deflated core eigenproblem. The PC method achieves greater computational simplifications by assuming well-separated eigenvalues. Some special decompositions are also used to reduce the computation of the PC algorithm. Surprisingly, simulations seem to indicate that the PC method achieves good overall performance even when the eigenvalues are not well separated. Convergence rates are also very good for the PC method. However, we have noticed that occasionally spurious frequency estimates may be obtained with PC-based MUSIC. Ironically, the PC estimated subspaces tend to be closer to the true subspaces than other subspace tracking methods that do not exhibit occasionally spurious frequency estimates. Because PC only tracks approximations of the eigencomponents, the orthogonality error is typically much greater than machine precision orthogonality. Nevertheless, partial orthogonality schemes can be used to improve orthogonality and other measures of performance (see Section 7.3.2).

ANN have been developed to find eigen information, e.g., see [27] and [35] as well as the references contained therein. An ANN consists of many richly interconnected simple and similar processing elements (called artificial neurons) operating in parallel. High computational rates (due to massive parallelism) and robustness (due to local neural connectivity) are two important features of ANNs. Most of the eigenvector estimating ANNs appear under the topic of principal component analysis (PCA). The principal eigenvectors are defined as the eigenvectors associated with the larger eigenvalues.

7.4.3 The URV and Rank Revealing QR Updates

The URV update [39] is based on the URVD developed by G.W. Stewart as a two sided generalization of the RRQR methods. The URVD can also be viewed as a generalization of the SVD because U and V are orthogonal matrices and R is an upper triangular matrix. Clearly, the SVD is a special case of the URVD. If $X = URV^{\mathrm{H}}$ is the URVD of X, then the R factor can be rank revealing in the sense that the Euclidean norm of the $n - r$ rightmost columns of R is approximately equal to the Euclidean norm of the $n - r$ smallest singular values of X. Also, the smallest singular value of the first r columns of R is approximately equal to the rth singular value of X. These two conditions effectively partition the corresponding columns of U and V into an r-dimensional dominant subspace and an $(n - r)$-dimensional subdominant subspace that can be used as estimates for the signal and noise subspace spans. The URV update is $O(n^2)$ per update. An RRQR update [that is usually $O(n^2)$ per update] is developed by Bischof and Schroff in [1]. RRQR methods that use the traditional pivoting strategy to maintain a rank revealing structure involve $O(n^3)$ flops per update. An analysis of problems associated with RRQR methods along with a fairly extensive reference list on RRQR methods can be found in [6]. An $O(nr)$ deflated URV update is presented by Rabideau and Steinhardt in [34] (and the references contained therein).

7.4.4 Miscellaneous Methods

Strobach [40] recently introduced a family of low rank or eigensubspace adaptive filters based on orthogonal iteration. The computational complexity ranges from $O(nr^2)$ to $O(nr)$.

A family of subspace methods without eigendecomposition (SWEDE) has been proposed by Eriksson et al. [16]. With SWEDE, portions of the correlation matrix must be updated for each new snapshot vector at a cost of approximately $12nr$ flops. However, the subspace basis (which is computed from the correlation matrix partitions) need only be computed every time a DOA estimate is needed. Computing the subspace estimate is $O(nr^2)$, so if the subspace is computed every kth update, the overall complexity is $O(nr^2/k) + 12nr$ per update. At high SNR, SWEDE performs almost as well as eigen-based MUSIC.

Key References

As previously mentioned, Comon and Golub did a nice survey of SVD tracking methods in 1990 [7]. In 1995, Reddy et al. published a selected overview of eigensubspace estimation methods, including ANN approaches [35]. For a study of URV and RRQR methods, see [6]. Partial orthogonalization schemes are studied in [28]. Finally, of interest is the special issue of *Signal Processing* [41] devoted to some of the issues considered in this chapter.

References

1. Bischof, C.H. and Shroff, G.M., On updating signal subspaces, *IEEE Trans. Signal Process.*, 40(1), 96–105, Jan. 1992.
2. Brockett, R.W., Dynamical systems that sort list, diagonalize matrices and solve linear programming problems, in *Proceedings of the 27th Conference on Decision Control*, Austin, TX, pp. 799–803, 1988.
3. Bunch, J.R. and Nielsen, C.P., Updating the singular value decomposition, *Numer. Math.*, 31, 111–129, 1978.
4. Bunch, J.R., Nielsen, C.P., and Sorensen, D.C., Rank-one modification of the symmetric eigenproblem, *Numer. Math.*, 31, 31–48, 1978.
5. Champagne, B., Adaptive eigendecomposition of data covariance matrices based on first-order perturbations, *IEEE Trans. Signal Process.*, SP-42(10), 2758–2770, Oct. 1994.
6. Chandrasekaran, S. and Ipsen, I.C.F., On rank-revealing factorisations, *SIAM J. Matrix Anal. Appl.*, 15(2), 592–622, April 1994.
7. Comon, P. and Golub, G.H., Tracking a few extreme singular values and vectors in signal processing, *Proc. IEEE*, 78(8), 1327–1343, Aug. 1990.
8. DeGroat, R.D., Non-iterative subspace tracking, *IEEE Trans. Signal Process.*, SP-40(3), 571–577, March 1992.
9. DeGroat, R.D. and Dowling, E.M., Spherical subspace tracking: analysis, convergence and detection schemes, in *Proceedings of the 26th Annual Asilomar Conference on Signals, Systems, and Computers*, Pacific Grove, CA (invited paper), pp. 561–565, Oct. 1992.
10. DeGroat, R.D. and Roberts, R.A., Efficient, numerically stabilized rank-one eigenstructure updating, *IEEE Trans. Acoust. Speech Signal Process.*, ASSP-38(2), 301–316, Feb. 1990.
11. Dowling, E.M., DeGroat, R.D., Linebarger, D.A., and Ye, H., Sphericalized SVD updating for subspace tracking, in Moonen, M. and De Moor, B. (Eds.), *SVD and Signal Processing III: Algorithms, Applications and Architectures*, Elsevier, Amsterdom, the Netherlands, 1995, pp. 227–234.
12. Dowling, E.M., Ammann, L.P., and DeGroat, R.D., A TQR-iteration based SVD for real time angle and frequency tracking, *IEEE Trans. Signal Process.*, 42(4), 914–925, Apr. 1994.
13. Dowling, E.M. and DeGroat, R.D., Adaptation dynamics of the spherical subspace tracker, *IEEE Trans. Signal Process.*, 40(10), 2599–2602, Oct. 1992.

14. Dowling, E.M., DeGroat, R.D., and Linebarger, D.A., Efficient, high performance subspace based tracking problems, in *Adv. Sig. Proc. Algs., Archs. Appl. VI*, SPIE 2563, 253–264, 1995.
15. Edelman, A. and Stewart, G.W., Scaling for orthogonality, *IEEE Trans. Signal Process.*, SP-41(4), 1676–1677, April 1993.
16. Eriksson, A., Stoica, P., and Soderstrom, T., On-line subspace algorithms for tracking moving sources, *IEEE Trans. Signal Process.*, 42(9), 2319–2330, Sept. 1994.
17. Ferzali, W. and Proakis, J.G., Adaptive SVD algorithm and applications, in *SVD and Signal Processing II*, Elsevier, New York, pp. 14–21, 1992.
18. Fu, Z. and Dowling, E.M., Conjugate gradient eigenstructure tracking for adaptive spectral estimation, *IEEE Trans. Signal Process.*, 43(5), 1151–1160, May 1995.
19. Golub, G.H. and VanLoan, C.F., Some modified matrix eigenvalue problems, *SIAM Rev.*, 15, 318–334, 1973.
20. Gu, M. and Eisenstat, S.C., A stable and efficient algorithm for the rank-one modification of the symmetric eigenproblem, *SIAM J. Matrix Anal. Appl.*, 15(4), 1266–1276, Oct. 1994.
21. Karasalo, I., Estimating the covariance matrix by signal subspace averaging, *IEEE Trans. Acoust. Speech Signal Process.*, ASSP-34(1), 8–12, Feb. 1986.
22. Karhunen, J., Adaptive algorithms for estimating eigenvectors of correlation type matrices, in *IEEE International Conference on Acoustics, Speech and Signal Processing*, San Diego, CA, pp. 14.6.1–14.6.4, 1984.
23. Linebarger, D.A., DeGroat, R.D., Dowling, E.M., Stoica, P., and Fudge, G., Incorporating a priori information into MUSIC–algorithms and analysis, *Signal Process.*, 46(1), 85–104, 1995.
24. Linebarger, D.A., DeGroat, R.D., and Dowling, E.M., Efficient direction finding methods employing forward/backward averaging, *IEEE Trans. Signal Process.*, 42(8), 2136–2145, Aug. 1994.
25. Ljung, L., Analysis of recursive stochastic algorithms, *IEEE Trans. Automat. Control*, AC-22(4), 551–575, Aug. 1977.
26. Marcos, S. and Benidir, M., An adaptive subspace algorithm for direction finding and tracking, in *Adv. Sig. Proc. Algs., Archs. Appl. VI*, SPIE 2563, 230–241, 1995.
27. Mathew, G. and Reddy, V.U., Orthogonal eigensubspace estimation using neural networks, *IEEE Trans. Signal Process.*, 42, 1803–1811, July 1994.
28. Mathias, R., Analysis of algorithms for orthogonalizing products of unitary matrices, *J. Numer. Linear Algebr. Appl.*, 3(2), 125–145, 1996.
29. Moonen, M., VanDooren, P., and Vanderwalle, J., A note on efficient, numerically stabilized rank-one eigenstructure updating, *IEEE Trans. Signal Process.*, SP-39(8), 1913–1914, Aug. 1991.
30. Moonen, M., VanDooren, P., and Vanderwalle, J., A singular value decomposition updating algorithm for subspace tracking, *SIAM J. Matrix Anal. Appl.*, 13(4), 1015–1038, Oct. 1992.
31. Oja, E., A simplified neuron model as a principal component analyzer, *J. Math. Biol.*, 15, 267–273, 1982.
32. Owsley, N.L., Adaptive data orthogonalization, in *IEEE International Conference on Acoustics, Speech and Signal Processing*, Tulsa, OK, pp. 109–112, 1978.
33. Parlett, B.N., *The Symmetric Eigenvalue Problem*, Prentice-Hall, Englewood Cliffs, NJ, 1980.
34. Rabideau, D.J., Subspace invariance: The RO-FST and TQR-SVD adaptive subspace tracking algorithms, *IEEE Trans. Signal Process.*, SP-43, 2016–2018, Aug. 1995.
35. Reddy, V.U., Mathew, G., and Paulraj, A., Some algorithms for eigensubspace estimation, *Digit. Signal Process.*, 5, 97–115, 1995.
36. Regalia, P.A. and Loubaton, P., Rational subspace estimation using adaptive lossless filters, *IEEE Trans. Signal Process.*, 40, 2392–2405, Oct. 1992.
37. Schreiber, R., Implementation of adaptive array algorithms, *IEEE Trans. Acoust. Speech Signal Process.*, ASSP-34, 1038–1045, Oct. 1986.
38. Stewart, G.W., *Introduction to Matrix Computations*, Academic Press, New York, 1973.

39. Stewart, G.W., An updating algorithm for subspace tracking, *IEEE Trans. Signal Process.*, SP-40(6), 1535–1541, June 1992.
40. Strobach, P., Fast recursive eigensubspace adaptive filters, in *International Conference on Acoustics, Speech and Signal Processing*, Detroit, MI, pp. 1416–1419, 1995.
41. Viberg, M. and Stoica, P., Eds., *Signal Process.*, 50(1–2) of *Special Issue on Subspace Methods for Detection and Estimation*, April 1996.
42. Xu, G., Zha, H., Golub, G., and Kailath, T., Fast and robust algorithms for updating signal subspaces, *IEEE Trans. Circ. Syst.*, 41(6), 537–549, June 1994.
43. Yang, B., Projection approximation subspace tracking, *IEEE Trans. Signal Process.*, SP-43(1), 95–107, Jan. 1995.
44. Yang, J.F. and Kaveh, M., Adaptive eigensubspace algorithms for direction or frequency estimation and tracking, *IEEE Trans. Acoust. Speech Signal Process.*, ASSP-36(2), 241–251, Feb. 1988.
45. Yang, J.-F. and Lin, H.-J., Adaptive high-resolution algorithms for tracking nonstationary sources without the estimation of source number, *IEEE Trans. Signal Process.*, 42(3), 563–571, March 1994.
46. Yu, K.B., Recursive updating the eigenvalue decomposition of a covariance matrix, *IEEE Trans. Signal Process.*, SP-39(5), 1136–1145, May 1991.

8

Detection: Determining the Number of Sources

8.1 Formulation of the Problem 8-1
8.2 Information Theoretic Approaches 8-3
AIC and MDL • EDC
8.3 Decision Theoretic Approaches 8-6
The Sphericity Test • Multiple Hypothesis Testing
8.4 For More Information 8-10
References 8-10

Douglas B. Williams
Georgia Institute of Technology

The processing of signals received by sensor arrays generally can be separated into two problems: (1) detecting the number of sources and (2) isolating and analyzing the signal produced by each source. We make this distinction because many of the algorithms for separating and processing array signals make the assumption that the number of sources is known *a priori* and may give misleading results if the wrong number of sources is used [3]. A good example are the errors produced by many high resolution bearing estimation algorithms (e.g., MUSIC) when the wrong number of sources is assumed. Because, in general, it is easier to determine how many signals are present than to estimate the bearings of those signals, signal detection algorithms typically can correctly determine the number of signals present even when bearing estimation algorithms cannot resolve them. In fact, the capability of an array to resolve two closely spaced sources could be said to be limited by its ability to detect that there are actually two sources present. If we have a reliable method of determining the number of sources, not only can we correctly use high resolution bearing estimation algorithms, but we can also use this knowledge to utilize more effectively the information obtained from the bearing estimation algorithms. If the bearing estimation algorithm gives fewer source directions than we know there are sources, then we know that there is more than one source in at least one of those directions and have thus essentially increased the resolution of the algorithm. If analysis of the information provided by the bearing estimation algorithm indicates more source directions than we know there are sources, then we can safely assume that some of the directions are the results of false alarms and may be ignored, thus decreasing the probability of false alarm for the bearing estimation algorithms. In this section we will present and discuss the more common approaches to determining the number of sources.

8.1 Formulation of the Problem

The basic problem is that of determining how many signal producing sources are being observed by an array of sensors. Although this problem addresses issues in several areas including sonar, radar, communications, and geophysics, one basic formulation can be applied to all these applications. We will give only a basic, brief description of the assumed signal structure, but more detail can be found in references such as the book by Johnson and Dudgeon [3]. We will assume that an array of M sensors observes signals produced by N_s sources. The array is allowed to have an arbitrary geometry. For our discussion here, we will assume that the sensors are omnidirectional. However, this assumption is only for notational convenience as the algorithms to be discussed will work for more general sensor responses.

The output of the m th sensor can be expressed as a linear combination of signals and noise

$$y_m(t) = \sum_{i=1}^{N_s} s_i(t - \Delta_i(m)) + n_m(t).$$

The noise observed at the mth sensor is denoted by $n_m(t)$. The propagation delays, $\Delta_i(m)$, are measured with respect to an origin chosen to be at the geometric center of the array. Thus, $s_i(t)$ indicates the ith propagating signal observed at the origin, and $s_i(t - \Delta_i(m))$ is the same signal measured by the mth sensor. For a plane wave in a homogeneous medium, these delays can be found from the dot product between a unit vector in the signal's direction of propagation, $\vec{\zeta}_i^{\text{o}}$, and the sensor's location, $\vec{x}_m$,

$$\Delta_i(m) = \frac{\vec{\zeta}_i^{\text{o}} \cdot \vec{x}_m}{c},$$

where c is the plane wave's speed of propagation.

Most algorithms used to detect the number of sources incident on the array are frequency domain techniques that assume the propagating signals are narrowband about a common center frequency, ω^{o}. Consequently, after Fourier transforming the measured signals, only one frequency is of interest and the propagation delays become phase shifts

$$Y_m(\omega^{\text{o}}) = \sum_{i=1}^{N_s} S_i(\omega^{\text{o}}) e^{-j\omega^{\text{o}}\Delta_i(m)} + N_m(\omega^{\text{o}}).$$

The detection algorithms then exploit the form of the spatial correlation matrix, $\mathbf{R}$, for the array. The spatial correlation matrix is the $M \times M$ matrix formed by correlating the vector of the Fourier transforms of the sensor outputs at the particular frequency of interest

$$\mathbf{Y} = [Y_0(\omega^{\text{o}}) \; Y_1(\omega^{\text{o}}) \; \ldots \; Y_{M-1}(\omega^{\text{o}})]^{\text{T}}.$$

If the sources are assumed to be uncorrelated with the noise, then the form of $\mathbf{R}$ is

$$\mathbf{R} = E\{\mathbf{Y}\mathbf{Y}'\} = \mathbf{K}_{\text{n}} + \mathbf{S}\mathbf{C}\mathbf{S}',$$

where

$\mathbf{K}_{\text{n}}$ is the correlation matrix of the noise
$\mathbf{S}$ is the matrix whose columns correspond to the vector representations of the signals
$\mathbf{S}'$ is the conjugate transpose of $\mathbf{S}$
$\mathbf{C}$ is the matrix of the correlations between the signals

Thus, the matrix $\mathbf{S}$ has the form

$$\mathbf{S} = \begin{bmatrix} e^{-j\omega^{\text{o}}\Delta_1(0)} & \ldots & e^{-j\omega^{\text{o}}\Delta_{N_s}(0)} \\ \vdots & & \vdots \\ e^{-j\omega^{\text{o}}\Delta_1(M-1)} & \ldots & e^{-j\omega^{\text{o}}\Delta_{N_s}(M-1)} \end{bmatrix}.$$

If we assume that the noise is additive, white Gaussian noise with power σ_n^2 and that none of the signals are perfectly coherent with any of the other signals, then $\mathbf{K}_{\text{n}} = \sigma_n^2 \mathbf{I}_M$, $\mathbf{C}$ has full rank, and the form of $\mathbf{R}$ is

$$\mathbf{R} = \sigma_n^2 \mathbf{I}_M + \mathbf{S}\mathbf{C}\mathbf{S}'. \tag{8.1}$$

We will assume that the columns of **S** are linearly independent when there are fewer sources then sensors, which is the case for most common array geometries and expected Source locations. As **C** is of full rank, if there are fewer sources than censors then the rank of **SCS**′ is equal to the number of signals incident on the array or, equivalently, the number of sources. If there are N_s sources, then **SCS**′ is of rank N_s and its N_s eigenvalues in descending order are $\delta_1, \delta_2, \ldots, \delta_{N_s}$. The M eigenvalues of $\sigma_n^2 \mathbf{I}_M$ are all equal to σ_n^2, and the eigenvectors are any orthonormal set of length M vectors. So the eigenvectors of **R** are the N_s eigenvectors of **SCS** plus any $M - N_s$ eigenvectors which complete the orthonormal set, and the eigenvalues in descending order are $\sigma_n^2 + \delta_1, \ldots, \sigma_n^2 + \delta_{N_s}, \sigma_n^2, \ldots, \sigma_n^2$. The correlation matrix is generally divided into two parts: the signal-plus-noise subspace formed by the largest eigenvalues $(\sigma_n^2 + \delta_1, \ldots, \sigma_n^2 + \delta_{N_s})$ and their eigenvectors, and the noise subspace formed by the smallest, equal eigenvalues and their eigenvectors. The reason for these labels is obvious as the space spanned by the signal-plus-noise subspace eigenvectors contains the signals and a portion of the noise while the noise subspace contains only that part of the noise that is orthogonal to the signals [3]. If there are fewer sources than sensors, the smallest $M - N_s$ eigenvalues of **R** are all equal and to determine exactly how many sources there are, we must simply determine how many of the smallest eigenvalues are equal. If there are not fewer sources than sensors ($N_s \geq M$), then none of the smallest eigenvalues are equal. The detection algorithms then assume that only the smallest eigenvalue is in the noise subspace as it is not equal to any of the other eigenvalues. Thus, these algorithms can detect up to $M - 1$ sources and for $N_s \geq M$ will say that there are $M - 1$ sources as this is the greatest detectable number. Unfortunately, all that is usually known is $\hat{\mathbf{R}}$, the sample correlation matrix, which is formed by averaging N samples of the correlation matrix taken from the outputs of the array sensors. As $\hat{\mathbf{R}}$ is formed from only a finite number of samples of **R**, the smallest $M - N_s$ eigenvalues of $\hat{\mathbf{R}}$ are subject to statistical variations and are unequal with probability one [4]. Thus, solutions to the detection problem have concentrated on statistical tests to determine how many of the eigenvalues of **R** are equal when only the sample eigenvalues of $\hat{\mathbf{R}}$ are available.

When performing statistical tests on the eigenvalues of the sample correlation matrix to determine the number of sources, certain assumptions must be made about the nature of the signals. In array processing, both deterministic and stochastic signal models are used depending on the application. However, for the purpose of testing the sample eigenvalues, the Fourier transforms of the signals at frequency ω^o; $S_i(\omega^o)$, $i = 1, \ldots, N_s$; are assumed to be zero mean Gaussian random processes that are statistically independent of the noise and have a positive definite correlation matrix **C**. We also assume that the N samples taken when forming $\hat{\mathbf{R}}$ are statistically independent of each other. With these assumptions, the spatial correlation matrix is still of the same form as in Equation 8.1, except that now we can more easily derive statistical tests on the eigenvalues of $\hat{\mathbf{R}}$.

8.2 Information Theoretic Approaches

We will see that the source detection methods to be described all share common characteristics. However, we will classify them into two groups—information theoretic and decision theoretic approaches—determined by the statistical theories used to derive them.

Although the decision theoretic techniques are quite a bit older, we will first present the information theoretic algorithms as they are currently much more commonly used.

8.2.1 AIC and MDL

Akaike information criterion (AIC) and minimum description length (MDL) are both information theoretic model order determination techniques that can be used to test the eigenvalues of a sample correlation matrix to determine how many of the smallest eigenvalues of the correlation matrix are equal. The AIC and MDL algorithms both consist of minimizing a criterion over the number of signals that are detectable, i.e., $N_s = 0, \ldots, M - 1$. To construct these criteria, a family of probability densities,

$f(\mathbf{Y} \mid \theta(N_s))$, $N_s = 0, \ldots, M-1$, is needed, where θ, which is a function of the number of sources, N_s, is the vector of parameters needed for the model that generated the data $\mathbf{Y}$. The criteria are composed of the negative of the log-likelihood function of the density $f(\mathbf{Y}\hat{\theta}(N_s))$, where $\hat{\theta}(N_s)$ is the maximum likelihood estimate of θ for N_s signals, plus an adjusting term for the model dimension. The adjusting term is needed because the negative log-likelihood function always achieves a minimum for the highest dimension model possible, which in this case is the largest possible number of sources. Therefore, the adjusting term will be a monotonically increasing function of N_s and should be chosen so that the algorithm is able to determine the correct model order.

AIC was introduced by Akaike [1]. Originally, the "IC" stood for information criterion and the "A" designated it as the first such test, but it is now more commonly considered an acronym for the "Akaike Information Criterion." If we have N independent observations of a random variable with probability density $g(\mathbf{Y})$ and a family of models in the form of probability densities $f(\mathbf{Y}|\theta)$, where θ is the vector of parameters for the models, then Akaike chose his criterion to minimize

$$I(g; f(\cdot|\theta)) = \int g(\mathbf{Y}) \ln g(\mathbf{Y}) d\mathbf{Y} - \int g(\mathbf{Y}) \ln f(\mathbf{Y}|\theta) d\mathbf{Y} \tag{8.2}$$

which is known as the Kullback–Leibler mean information distance $\frac{1}{N}\text{AIC}(\theta)$ is an estimate of $-E\{\int g(\mathbf{Y}) \ln f(\mathbf{Y}|\theta) d\mathbf{Y}\}$ and minimizing $\text{AIC}(\theta)$ over the allowable values of θ should minimize Equation 8.2. The expression for $\text{AIC}(\theta)$ is

$$\text{AIC}(\theta) = -2 \ln [f(\mathbf{Y}|\hat{\theta}(N_s))] + 2\eta,$$

where η is the number of independent parameters in θ.

Following AIC, MDL was developed by Schwarz [6] using Bayesian techniques. He assumed that the *a priori* density of the observations comes from a suitable family of densities that possess efficient estimates [7]; they are of the form

$$f(\mathbf{Y}|\theta) = \exp(\theta \cdot p(\mathbf{Y}) - b(\theta)).$$

The MDL criterion was then found by choosing the model that is most probable *a posteriori*. This choice is equivalent to selecting the model for which

$$\text{MDL}(\theta) = -\ln [f(\mathbf{Y}|\hat{\theta}(N_s))] + \frac{1}{2}\eta \ln N$$

is minimized. This criterion was independently derived by Rissanen [5] using information theoretic techniques. Rissanen noted that each model can be perceived as encoding the observed data and that the optimum model is the one that yields the minimum code length. Hence, the name MDL comes from "minimum description length."

For the purpose of using AIC and MDL to determine the number of sources, the forms of the log-likelihood function and the adjusting terms have been given by Wax [8]. For N_s signals the parameters that completely parameterize the correlation matrix $\mathbf{R}$ are $\{\sigma_n^2, \lambda_1, \ldots, \lambda_{N_s}, \mathbf{v}_1, \ldots, \mathbf{v}_{N_s}\}$, where λ_i and $\mathbf{v}_i$, $i = 1, \ldots, N_s$, are the eigenvalues and their respective eigenvectors of the signal-plus-noise subspace of the correlation matrix. As the vector of sensor outputs is a Gaussian random vector with correlation matrix $\mathbf{R}$ and all the samples of the sensor outputs are independent, the log-likelihood function of $f(\mathbf{Y}|\theta)$ is

$$\ln f\left(\mathbf{Y}|\sigma_n^2, \lambda_1, \ldots, \lambda_{N_s}, \mathbf{v}_1, \ldots, \mathbf{v}_{N_s}\right) = \pi^{-pN} (\det \mathbf{R})^{-N} \exp\left(-N \text{tr}(\mathbf{R}^{-1}\hat{\mathbf{R}})\right),$$

where

tr(·) denotes the trace of the matrix

$\hat{\mathbf{R}}$ is the sample correlation matrix

$\mathbf{R}$ is the unique correlation matrix formed from the given parameters

The maximum likelihood estimate of the parameters are [2,4]

$$\begin{aligned} \hat{\mathbf{v}}_i &= \mathbf{u}_i; \quad i = 1, \ldots, N_s, \\ \hat{\lambda}_i &= l_i; \quad i = 1, \ldots, N_s, \\ \hat{\sigma}_n^2 &= \frac{1}{M - N_s} \sum_{i=N_s+1}^{M} l_i = \bar{l}, \end{aligned} \tag{8.3}$$

where $l_1, \ldots, l_M$ are the eigenvalues in descending order of $\hat{\mathbf{R}}$ and $\mathbf{u}_i$ are the corresponding eigenvectors. Therefore, the log-likelihood function of $f(\mathbf{Y} \mid \hat{\theta}(N_s))$ is

$$\ln f(\mathbf{Y}|\bar{l}, l_1, \ldots, l_{N_s}, \mathbf{u}_1, \ldots, \mathbf{u}_{N_s}) = \ln \left[\frac{\prod_{i=N_s+1}^{M} l_i^{1/(M-N_s)}}{\frac{1}{M-N_s} \sum_{i=N_s+1}^{M} l_i} \right]^{(M-N_s)N}.$$

Remembering that the eigenvalues of a complex correlation matrix are real and that the eigenvectors are complex and orthonormal, the number of degrees of freedom in the parameters of the model is classically chosen to be $\eta = N_s(2M - N_s)+1$. Noting that any constant term in the criteria which is common to the entire family of models for either AIC or MDL may be ignored, we have the criterion for AIC as

$$\text{AIC}(\hat{N}_s) = -2N \ln \left[\frac{\prod_{i=\hat{N}_s+1}^{M} l_i}{\left[\frac{1}{M-\hat{N}_s} \sum_{i=\hat{N}_s+1}^{M} l_i\right]^{M-\hat{N}_s}} \right] + 2\hat{N}_s(2M - \hat{N}_s); \quad \hat{N}_s = 0, \ldots, M-1$$

and the criterion for MDL as

$$\text{MDL}(\hat{N}_s) = -N \ln \left[\frac{\prod_{i=\hat{N}_s+1}^{M} l_i}{\left[\frac{1}{M-\hat{N}_s} \sum_{i=\hat{N}_s+1}^{M} l_i\right]^{M-\hat{N}_s}} \right] + \frac{1}{2}\hat{N}_s(2M - \hat{N}_s) \ln N; \quad \hat{N}_s = 0, \ldots, M-1.$$

For both of these methods, the estimate of the number of sources is that value of $\hat{N}_s$ which minimizes the criterion. In [9] there is a more thorough discussion concerning determining the number of degrees of freedom and the advantages of choosing instead $\eta = N_s(2M - N_s - 1)$.

In general, MDL is considered to perform better than AIC. Schwarz [6], through his derivation of the MDL criterion, showed that if his assumptions are accepted, then AIC cannot be asymptotically optimal. He also mentioned that MDL tends toward lower-dimensional models than AIC as the model dimension term is multiplied by $\frac{1}{2} \ln N$ in the MDL criterion. Zhao et al. [14] showed that MDL is consistent (the probability of detecting the correct number of sources, i.e., $\Pr(\hat{N}_s = N_s)$, goes to 1 as N goes to infinity), but AIC is not consistent and will tend to overestimate the number of sources as N goes to infinity. Thus, most people in array processing prefer to use MDL over AIC. Interestingly, many statisticians prefer AIC because many of their modeling problems have a very large penalty for underestimating the model order but a relatively mild penalty for overestimating it. Xu and Kaveh [12] have provided a thorough discussion of the asymptotic properties of AIC and MDL, including an examination of their sensitivities to modeling errors and bounds on the probability that AIC will overestimate the number of sources.

8.2.2 EDC

Clearly, the only difference between the implementations of AIC and MDL is the choice of the adjusting term that penalizes for choosing larger model orders. Several people have examined using other adjusting terms to arrive at other criteria. In particular, statisticians at the University of Pittsburgh [13,14] have developed the efficient detection criterion (EDC) procedure which is actually a family of criteria chosen such that they are all consistent. The general form of these criteria is

$$\mathrm{EDC}(\theta) = -\ln\left[f(\mathbf{Y}|\hat{\theta}(N_s))\right] + \eta C_N,$$

where C_N can be any function of N such that

$$(1)\ \lim_{N\to\infty} C_N/N = 0,$$
$$(2)\ \lim_{N\to\infty} C_N/\ln(\ln(N)) = \infty.$$

Thus, for the array processing source detection problem the EDC procedure chooses the value of $\hat{N}_s$ that minimizes

$$\mathrm{EDC}(\hat{N}_s) = -N\ln\left[\frac{\prod_{i=\hat{N}_s+1}^{M} l_i}{\left[\frac{1}{M-\hat{N}_s}\sum_{i=\hat{N}_s+1}^{M} l_i\right]^{M-\hat{N}_s}}\right] + \hat{N}_s(2M-\hat{N}_s)C_N;\quad \hat{N}_s = 0,\ldots,M-1.$$

In their analysis of the EDC procedure, Zhao et al. [14] showed that not only are all the EDC criteria consistent for the data assumptions we have made, but under certain conditions they remain consistent even when the data sample vectors used to form the estimate $\hat{\mathbf{R}}$ are not independent or Gaussian.

The choice of $C_N = \frac{1}{2}\ln(N)$ satisfies the restrictions on C_N and, thus, produces one of the EDC procedures. This particular criterion is identical to MDL and shows that the MDL criterion is included as one of the EDC procedures. Another relatively common choice for C_N is $C_N = \sqrt{N\ln(N)}$.

8.3 Decision Theoretic Approaches

The methods that we term decision theoretic approaches all rely on the statistical theory of hypothesis testing to determine the number of sources. The first of these that we will discuss, the sphericity test, is by far the oldest algorithm for source detection.

8.3.1 The Sphericity Test

Originally, the sphericity test was a hypothesis testing method designed to determine if the correlation (or covariance) matrix, $\mathbf{R}$, of a length M Gaussian random vector is proportional to the identity matrix, $\mathbf{I}_M$, when only $\hat{\mathbf{R}}$, the sample correlation matrix, is known. If $\mathbf{R} \propto \mathbf{I}_M$, then the contours of equal density for the Gaussian distribution form concentric spheres in M-dimensional space. The sphericity test derives its name from being a test of the sphericity of these contours.

The original sphericity test had two possible hypotheses

$$\begin{aligned} H_0: &\quad \mathbf{R} = \sigma_n^2 \mathbf{I}_M \\ H_1: &\quad \mathbf{R} \neq \sigma_n^2 \mathbf{I}_M \end{aligned}$$

for some unknown σ_n^2. If we denote the eigenvalues of $\mathbf{R}$ in descending order by $\lambda_1, \lambda_2, \ldots, \lambda_M$, then equivalent hypotheses are

$$\begin{aligned} H_0: &\quad \lambda_1 = \lambda_2 = \cdots = \lambda_M \\ H_1: &\quad \lambda_1 > \lambda_M. \end{aligned}$$

For the appropriate statistic, $T(\hat{\mathbf{R}})$, the test is of the form

$$T(\hat{\mathbf{R}}) \underset{H_0}{\overset{H_1}{\gtrless}} \gamma,$$

where the threshold, γ, can be set according to the Neyman–Pearson criterion [7]. That is, if the distribution of $T(\hat{\mathbf{R}})$ is known under the null hypothesis, H_0, then for a given probability of false alarm, P_F, we can choose γ such that

$$\Pr(T(\hat{\mathbf{R}}) > \gamma | H_0) = P_F.$$

Using the alternate form of the hypotheses, $T(\hat{\mathbf{R}})$ is actually $T(l_1, l_2, \ldots, l_M)$, and the eigenvalues of the sample correlation matrix are a sufficient statistic for the hypothesis test. The correct form of the sphericity test statistic is the generalized likelihood ratio [4]:

$$T(l_1, l_2, \ldots, l_M) = \ln \left[\frac{\left(\frac{1}{M} \sum_{i=1}^{M} l_i \right)^M}{\prod_{i=1}^{M} l_i} \right]$$

which was also a major component of the information theoretic tests.

For the source detection problem we are interested in testing a subset of the smaller eigenvalues for equality. In order to use the sphericity test, the hypotheses are generally broken down into pairs of hypotheses that can be tested in a series of hypothesis tests. For testing $M - \hat{N}_s$ eigenvalues for equality, the hypotheses are

$$\begin{aligned} H_0: &\quad \lambda_1 \geq \cdots \geq \lambda_{\hat{N}_s} \geq \lambda_{\hat{N}_s+1} = \cdots = \lambda_M \\ H_1: &\quad \lambda_1 \geq \cdots \geq \lambda_{\hat{N}_s} \geq \lambda_{\hat{N}_s+1} > \lambda_M. \end{aligned}$$

We are interested in finding the smallest values of $\hat{N}_s$ for which H_0 is true, which is done by testing $\hat{N}_s = 0$, $\hat{N}_s = 1, \ldots$, until $\hat{N}_s = M - 2$ or the test does not fail. If the test fails for $\hat{N}_s = M - 2$, then we consider none of the smallest eigenvalues to be equal and say that there are $M - 1$ sources. If $\hat{N}_s$ is the smallest value for which H_0 is true, then we say that there are $\hat{N}_s$ sources. There is also a problem involved in setting the desired P_F. The Neyman–Pearson criterion is not able to determine a threshold for given P_F

for the overall detection problem. The best that can be done is to set a P_F for each individual test in the nested series of hypothesis tests using Neyman–Pearson methods. Unfortunately, as the hypothesis tests are obviously not statistically independent and their statistical relationship is not very clear, how this P_F for each test relates to the P_F for the entire series of tests is not known.

To use the sphericity test to detect sources, we need to be able to set accurately the threshold γ according to the desired P_F, which requires knowledge of the distribution of the sphericity test statistic $T(l_{\hat{N}_s+1}, \ldots, l_M)$ under the null hypothesis. The exact form of this distribution is not available in a form that is very useful as it is generally written as an infinite series of Gaussian, chi-squared, or beta distributions [2,4]. However, if the test statistic is multiplied by a suitable function of the eigenvalues of $\hat{\mathbf{R}}$, then its distribution can be accurately approximated as being chi-squared [10]. Thus, the statistic

$$2\left((N-1) - \hat{N}_s - \frac{2(M-\hat{N}_s)^2+1}{6(M-\hat{N}_s)} + \sum_{i=1}^{\hat{N}_s}\left(\frac{l_i}{\bar{l}} - 1\right)^{-2}\right)\ln\left[\frac{\left(\frac{1}{M-\hat{N}_s}\sum_{i=\hat{N}_s+1}^{M} l_i\right)^{M-\hat{N}_s}}{\prod_{i=\hat{N}_s+1}^{M} l_i}\right]$$

is approximately chi-squared distributed with degrees of freedom given by

$$d = (M - \hat{N}_s)^2 - 1,$$

where $\bar{l} = \frac{1}{M-\hat{N}_s}\sum_{i=\hat{N}_s+1}^{M} l_i$.

Although the performance of the sphericity test is comparable to that of the information theoretic tests, it is not as popular because it requires selection of the P_F and calculation of the test thresholds for each value of $\hat{N}_s$. However, if the received data does not match the assumed model, the ability to change the test thresholds gives the sphericity test a robustness lacking in the information theoretic methods.

8.3.2 Multiple Hypothesis Testing

The sphericity test relies on a sequence of binary hypothesis tests to determine the number of sources. However, the optimum test for this situation would be to test all hypotheses simultaneously:

$$\begin{aligned}
\mathbf{H}_0&: \lambda_1 = \lambda_2 = \cdots = \lambda_M \\
\mathbf{H}_1&: \lambda_1 > \lambda_2 = \cdots = \lambda_M \\
\mathbf{H}_2&: \lambda_1 \geq \lambda_2 > \lambda_3 = \cdots = \lambda_M \\
&\vdots \\
\mathbf{H}_{M-1}&: \lambda_1 \geq \lambda_2 \geq \cdots \geq \lambda_{M-1} > \lambda_M
\end{aligned}$$

to determine how many of the smaller eigenvalues are equal. While it is not possible to generalize the sphericity test directly, it is possible to use an approximation to the probability density function (pdf) of the eigenvalues to arrive at a suitable test. Using the theory of multiple hypothesis tests, we can derive a test that is similar to AIC and MDL and is implemented in exactly the same manner, but is designed to minimize the probability of choosing the wrong number of sources.

To arrive at our statistic, we start with the joint pdf of the eigenvalues of the $\boldsymbol{M} \times \boldsymbol{M}$ sample covariance when the $\boldsymbol{M} - \hat{\boldsymbol{N}}_s$ smallest eigenvalues are known to be equal. We will denote this pdf by $f_{\hat{N}_s}(l_1, \ldots, l_M | \lambda_1 \geq \cdots \geq \lambda_{\hat{N}_s+1} = \cdots = \lambda_M)$, where the l_i denote the eigenvalues of the sample matrix and the λ_i are the eigenvalues of the true covariance matrix. The asymptotic expression for $f_{\hat{N}_s}(\cdot)$ is given by Wong et al. [11] for the complex-valued data case as

$$f_{\hat{N}_s}(l_1,\ldots,l_M|\lambda_1 \geq \cdots \geq \lambda_{\hat{N}_s+1} = \cdots = \lambda_M) \approx \frac{n^{mn-\frac{\hat{N}_s}{2}(2M-\hat{N}_s-1)}\pi^{M(M-1)-\frac{\hat{N}_s}{2}(2M-\hat{N}_s-1)}}{\tilde{\Gamma}_M(n)\tilde{\Gamma}_{M-\hat{N}_s}(M-\hat{N}_s)}$$
$$\prod_{i=1}^{M}\lambda_i^{-n}\prod_{i=1}^{M} l_i^{n-M}\exp\left\{-n\sum_{i=1}^{M}\frac{l_i}{\lambda_i}\right\}\prod_{i=\hat{N}_s+1}^{M}\prod_{i<j}^{M}(l_i-l_j)^2$$
$$\prod_{i=1}^{\hat{N}_s}\prod_{i<j}^{\hat{N}_s}\left(\frac{(l_i-l_j)\lambda_i\lambda_j}{\lambda_i-\lambda_j}\right)\prod_{i=1}^{\hat{N}_s}\prod_{j=\hat{N}_s+1}^{M}\left(\frac{(l_i-l_j)\lambda_i\lambda_j}{\lambda_i-\lambda_j}\right)$$

where

$n = N - 1$ is one less than the number of samples

$\tilde{\Gamma}_N(\cdot)$ is the multivariate gamma function for complex-valued data [11]

We then form M likelihood ratios by dividing each joint pdf by $f_{M-1}(\cdot)$ to form

$$\Lambda(\hat{N}_s) = \frac{f_{\hat{N}_s}\left(l_1,\ldots,l_M|\lambda_1 \geq \cdots \geq \lambda_{\hat{N}_s+1} = \cdots = \lambda_M\right)}{f_{M-1}(l_1,\ldots,l_M|\lambda_1 \geq \cdots \geq \lambda_M)}, \quad \hat{N}_s = 0,\ldots,M-1.$$

Assuming that each value of $\hat{N}_s$ is equally likely, then multiple hypothesis testing theory tells us that the value of $\hat{N}_s$ that maximizes $\Lambda(\hat{N}_s)$ is the optimum choice in that it minimizes the probability of choosing the incorrect $\hat{N}_s$ [7]. Because $\Lambda(\hat{N}_s)$ in this form requires knowledge of the unknown parameters λ_i, we must use a generalized likelihood ratio test and independently substitute the maximum likelihood estimates of the λ_i (see Equation 8.3 for these expressions) into both $f_{\hat{N}_s}(\cdot)$, for which we assume $M-\hat{N}_s$ equal λ_i s, and $f_{M-1}(\cdot)$, for which we assume no equal λ_i s, to get our new statistics $\Lambda(\hat{N}_s)$. After much simplification including dropping terms that are common to $\Lambda(\hat{N}_s)$ for every allowable value of $\hat{N}_s$ and then taking the natural logarithm of each $\Lambda(\hat{N}_s)$, we get the statistic

$$\Lambda(\hat{N}_s) = (n-\hat{N}_s)\ln\left[\frac{\prod_{i=\hat{N}_s+1}^{M} l_i}{\left(\frac{1}{M-\hat{N}_s}\sum_{i=\hat{N}_s+1}^{M} l_i\right)^{M-\hat{N}_s}}\right] - \frac{1}{2}\hat{N}_s(2M-\hat{N}_s-1)\ln[n]$$
$$+\ln\left[\frac{\pi^{-\hat{N}_s(2M-\hat{N}_s-1)/2}}{\tilde{\Gamma}_{M-\hat{N}_s}(M-\hat{N}_s)}\right] + \sum_{i=1}^{\hat{N}_s}\sum_{j=\hat{N}_s+1}^{M}\ln\left[\frac{l_i-l_j}{l_i-\bar{l}}\right] - \sum_{i=\hat{N}_s+1}^{M}\sum_{j=i+1}^{M} 2\ln\left[\frac{(l_il_j)^{1/2}}{l_i-l_j}\right]$$

where $\bar{l} = \frac{1}{M-\hat{N}_s}\sum_{i=\hat{N}_s+1}^{M} l_i$.

The terms in the first line of this equation are almost identical to the negative of the MDL criterion, especially when the degrees of freedom recommended in [9] are used. Note that the change in sign is necessary because we are finding the maximum of this criterion, not the minimum. The extra terms on the following line include both the eigenvalues being tested for equality and those not being tested. These extra terms allow this test to outperform the information theoretic techniques, since the use of all the eigenvalues for each value of $\hat{N}_s$ being tested allows this criterion to be more adaptive.

8.4 For More Information

Most of the original papers on model order determination appeared in the statistical literature in journals such as *The Annals of Statistics* and the *Journal of Multivariate Analysis*. However, almost all of the more recent developments that apply these techniques to the source detection problem have appeared in signal processing journals such as the *IEEE Transactions on Signal Processing*. More advanced topics that have been addressed in the signal processing literature but not discussed here include: detecting coherent (i.e., completely correlated) signals, detecting sources in unknown colored noise, and developing more robust source detection methods.

References

1. Akaike, H., A new look at the statistical model identification, *IEEE Trans. Automat. Contr.*, AC-19, 716–723, December 1974.
2. Anderson, T.W., *An Introduction to Multivariate Statistical Analysis*, 2nd edn., John Wiley & Sons, New York, 1984.
3. Johnson, D.H. and Dudgeon, D.E., *Array Signal Processing: Concepts and Techniques*, Prentice-Hall, Englewood Cliffs, NJ, 1993.
4. Muirhead, R.J., *Aspects of Multivariate Statistical Theory*, John Wiley & Sons, New York, 1982.
5. Rissanen, J., Modeling by shortest data description, *Automatica*, 14, 465–471, September 1978.
6. Schwarz, G., Estimating the dimension of a model, *Ann. Stat.*, 6, 461–464, March 1978.
7. Van Trees, H.L., *Detection, Estimation, and Modulation Theory, Part I*, John Wiley & Sons, New York, 1968.
8. Wax, M. and Kailath, T., Detection of signals by information theoretic criteria, *IEEE Trans. Acoust. Speech Signal Process.*, ASSP-33, 387–392, April 1985.
9. Williams, D.B., Counting the degrees of freedom when using AIC and MDL to detect signals, *IEEE Trans. Signal Process.*, 42, 3282–3284, November 1994.
10. Williams, D.B. and Johnson, D.H., Using the sphericity test for source detection with narrowband passive arrays, *IEEE Trans. Acoust. Speech Signal Process.*, 38, 2008–2014, November 1990.
11. Wong, K.M., Zhang, Q.-T., Reilly, J.P., and Yip, P.C., On information theoretic criteria for determining the number of signals in high resolution array processing, *IEEE Trans. Acoust. Speech Signal Process.*, 38, 1959–1971, November 1990.
12. Xu, W. and Kaveh, M., Analysis of the performance and sensitivity of eigendecomposition-based detectors, *IEEE Trans. Signal Process.*, 43, 1413–1426, June 1995.
13. Yin, Y.Q. and Krishnaiah, P.R., On some nonparametric methods for detection of the number of signals, *IEEE Trans. Acoust. Speech Signal Process.*, ASSP–35, 1533–1538, November 1987.
14. Zhao, L.C., Krishnaiah, P.R., and Bai, Z.D., On detection of the number of signals in presence of white noise, *J. Multivar. Anal.*, 20, 1–25, October 1986.

9

Array Processing for Mobile Communications

9.1 Introduction and Motivation 9-1
9.2 Vector Channel Model 9-2
Propagation Loss and Fading • Multipath Effects • Typical Channels • Signal Model • Cochannel Interference • Signal-Plus-Interference Model • Block Signal Model • Spatial and Temporal Structure
9.3 Algorithms for STP 9-9
Single-User ST-ML and ST-MMSE • Multiuser Algorithms • Simulation Example
9.4 Applications of Spatial Processing 9-16
Switched Beam Systems • Space-Time Filtering • Channel Reuse within a Cell
9.5 Summary 9-18
References 9-18

A. Paulraj
Stanford University

C. B. Papadias
Broadband Wireless

9.1 Introduction and Motivation

This chapter reviews the applications of antenna array signal processing to mobile networks. Cellular networks are rapidly growing around the world and a number of emerging technologies are seen to be critical to their improved economics and performance. Among these is the use of multiple antennas and spatial signal processing at the base station. This technology is referred to as Smart Antennas or, more accurately, as space-time processing (STP). STP refers to processing the antenna outputs in both space and time to maximize signal quality.

A cellular architecture is used in a number of mobile/portable communications applications. Cell sizes may range from large macrocells, which serve high speed mobiles, to smaller microcells or very small picocells, which are designed for outdoor and indoor applications. Each of these offers different channel characteristics and, therefore, poses different challenges for STP. Likewise, different service delivery goals such as grade of service and type of service: voice, data, or video, also need specific STP solutions. STP provides three processing leverages. The first is array gain. Multiple antennas capture more signal energy, which can be combined to improve the signal-to-noise ratio (SNR). Next is spatial diversity to combat space-selective fading. Finally, STP can reduce cochannel, adjacent channel, and intersymbol interference.

The organization of this chapter is as follows. In Section 9.2, we describe the vector channel model for a base station antenna array. In Section 9.3 we discuss the algorithms for STP. Section 9.4 outlines the applications of STP in cellular networks. Finally, we conclude with a summary in Section 9.5.

9.2 Vector Channel Model

Channel effects in a cellular radio link arise from multipath propagation and user motion. These create special challenges for STP. A thorough understanding of channel characteristics is the key to developing successful STP algorithms. The main features of a mobile wireless channel are described below.

9.2.1 Propagation Loss and Fading

The signal radiated by the mobile loses strength as it travels to the base station. These losses arise from the mean propagation loss and from slow and fast fading. The mean propagation loss comes from square law spreading, absorption by foliage, and the effect of vertical multipath. A number of models exist for characterizing the mean propagation loss [22,30], which is usually around 40 dB per decade. Slow fading results from shadowing by buildings and natural features and is usually characterized by a log-normal distribution with standard deviation agreed to 8 dB. Fast fading results from multipath scattering in the vicinity of the moving mobile. It is usually Rayleigh distributed. However, if there is a direct path component present, the fading will be Rician distributed.

9.2.2 Multipath Effects

Multipath propagation plays a central role in determining the nature of the channel. By channel we mean the impulse, or frequency response, of the radio channel from the mobile to the output of the antenna array. We refer to it as a vector channel, because we have multiple antennas and, therefore, we have a collection of channels. The mobile radiates omnidirectionally in azimuth using a vertical E-field antenna. The transmitted signal then undergoes scattering, reflection, or diffraction before reaching the base station, where it arrives from different paths, each with its own fading, propagation delay, and angle-of-arrival. This multipath propagation, in conjunction with user motion, determines the behavior of the wireless channel. Multipath scattering arises from three sources (see Figure 9.1). There are scatterers local to the mobile, remote dominant scatterers, and scatterers local to the base. We will now describe these three scattering mechanisms and their effect on the channel.

9.2.2.1 Scatterers Local to Mobile

Scattering local to the mobile is caused by buildings in the vicinity of the mobile (a few tens of meters). Mobile motion and local scattering give rise to Doppler spread, which causes time-selective fading. For a vertical, polarized E-field antenna, it has been shown [22] that the fading signal has a characteristic classical spectrum. For a mobile traveling at 55 MPH, the Doppler spread is about ± 200 Hz in the

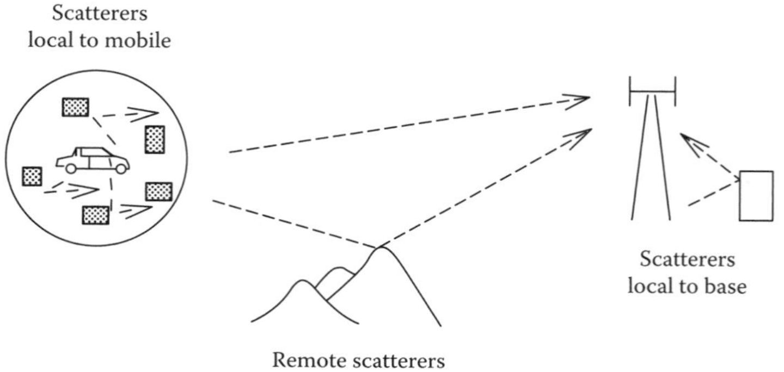

FIGURE 9.1 Multipath propagation has three distinct classes, each of which gives rise to different channel effects.

1900 MHz band. This effect results in rapid signal fluctuations also called time-selective fading. While local scattering contributes to Doppler spread, the delay spread will usually be insignificant because of the small scattering radius. Likewise, the angle spread will also be small.

9.2.2.2 Remote Scatterers

The emerging wave front from the local scatterers may then travel directly to the base and also be scattered toward the base by remote dominant scatterers, giving rise to specular multipath. These remote scatterers can be terrain features or high rise buildings. Remote scattering can cause significant delay and angle spreads. Delay spread causes frequency-selective fading, and the angle spread results in space-selective fading.

9.2.2.3 Scatterers Local to Base

Once these multiple wave fronts reach the base station, they may be scattered further by local structures such as buildings or other structures that are in the vicinity of the base. Such scattering will be more pronounced for low elevation below-roof-top antennas. The scattering local to the base can cause severe angle spread.

9.2.3 Typical Channels

Measurements in macrocells indicate that up to 6–12 paths may be present. Typical channel delay, angle, and one-sided Doppler (1800 MHz) spreads are given in Table 9.1.

A multipath channel structure is illustrated in Figure 9.2. Typical path power and delay statistics can be obtained from the GSM* standard. Angle-of-arrival statistics have been less well studied but several results have been reported (see [1–3]). The resulting channel is shown in Figure 9.3. We show a frequency response at each antenna for a GSM system. Since the channel bandwidth is 200 kHz, it is highly frequency selective in a hilly terrain environment. Also, the large angle spread causes variations of the channel from antenna to antenna. The channel variation in time depends on the Doppler spread. Notice that since GSM uses a short time slot, the channel variation during the time slot is negligible.

9.2.4 Signal Model

We study the case when a single-user transmits and is received at a base station with multiple antennas. The noiseless baseband signal $x_i(t)$ received by the base station at the ith element of an m element antenna array is given by

$$x_i(t) = \sum_{l=1}^{L} a_i(\theta_l)\alpha_l^{\mathrm{R}}(t)u(t - \tau_l) \tag{9.1}$$

TABLE 9.1 Typical Delay, Angle and Doppler Spreads in Cellular Applications

Environment	Delay Spread (μs)	Angle Spread (°)	Doppler Spread (Hz)
Flat rural (macro)	0.5	1	190
Urban (macro)	5	20	120
Hilly (macro)	20	30	190
Microcell (mall)	0.3	120	10
Picocell (indoors)	0.1	360	5

* Global system for Mobile Communications.

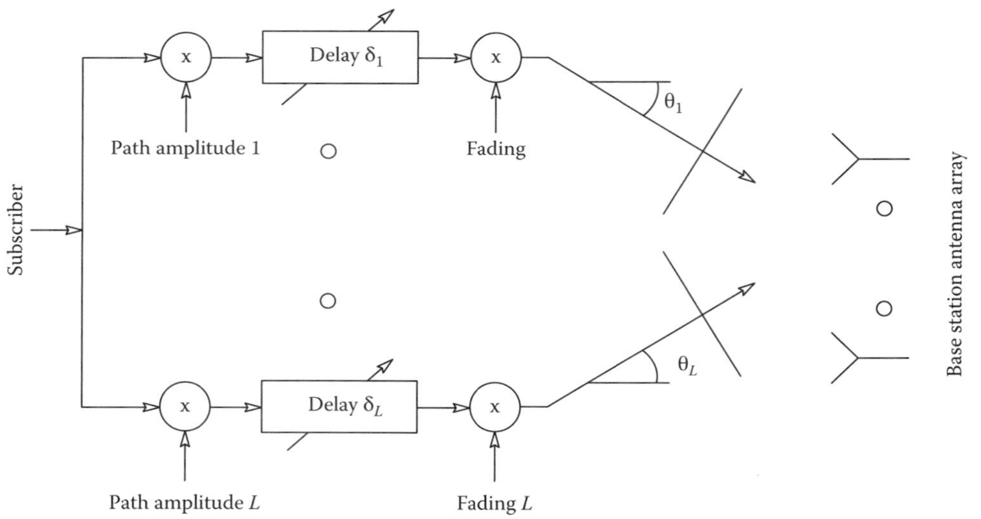

FIGURE 9.2 Multipath model.

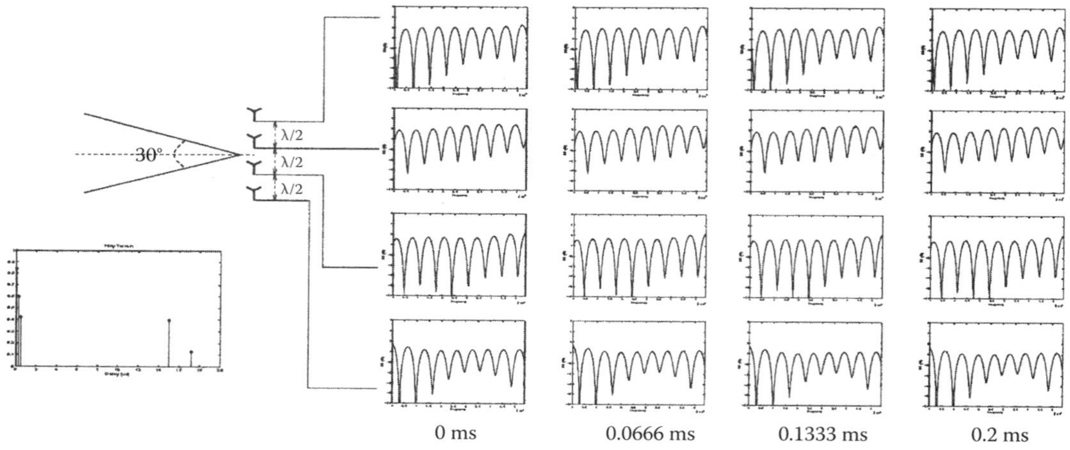

FIGURE 9.3 Channel frequency response at four different antennas for GSM in a typical hilly terrain channel at 1800 MHz. Mobile speed is 100 KPH. The response is plotted at four time instances spaced 66 μs apart.

where

- L is the number of multipaths
- $a_i(\theta_l)$ is the response of the ith element for an lth path from direction θ_l
- $\alpha_l^{\mathrm{R}}(t)$ is the complex path fading
- τ_l is the path delay
- $u(\cdot)$ is the transmitted signal that depends on the modulation waveform and the information data stream

In the IS-54 TDMA standard, $u(\cdot)$ is a $\pi/4$ shifted DQPSK, gray-coded signal that is modulated using a pulse with square-root raised cosine spectrum with excess bandwidth of 0.35. In GSM, a Gaussian minimum shift keying (GMSK) modulation is used. See [12,30,55] for more details. For a linear modulation (e.g., BPSK), we can write

$$u(t) = \sum_{k} g(t - kT)s(k) \tag{9.2}$$

where
$g(\cdot)$ is the pulse-shaping waveform
$s(k)$ represents the information bits

In the above model, we have assumed that the inverse signal bandwidth is large compared to the travel time across the array. For example, in GSM the inverse signal bandwidth is 5 μs, whereas the travel time across the array is, at most, a few *ns*. This is the narrowband assumption in array processing. The signal bandwidth is a sum of the modulation bandwidth and the Doppler spread, with the latter being comparatively negligible. Therefore, the complex envelope of the signal received by different antennas from a given path are identical except for phase and amplitude differences that depend on the path angle-of-arrival, array geometry, and the element pattern. This angle-of-arrival dependent phase and amplitude response at the ith element is $a_i(\theta_l)$ [37].

We collect all the element responses to a path arriving from angle θ_l into an m-dimensional vector, called the array response vector defined as

$$\mathbf{a}(\theta_l) = [a_1(\theta_l)a_2(\theta_l) \ \cdots \ a_m(\theta_l)]^{\mathrm{T}}$$

where $(\cdot)^{\mathrm{T}}$ denotes matrix transpose.

In array processing literature the array vector $\mathbf{a}(\theta)$ is also known as the steering vector. We can rewrite the array output at the base station as

$$\mathbf{x}(t) = \sum_{l=1}^{L} \mathbf{a}(\theta_l)\alpha_l^{\mathrm{R}}(t)u(t - \tau_l) \tag{9.3}$$

where

$$\mathbf{x}(t) = [x_1(t) \ x_2(t) \ \ldots \ x_m(t)]^{\mathrm{T}}$$

$\mathbf{x}(t)$ and $\mathbf{a}(\theta_l)$ are m-dimensional complex vectors. The fade amplitude $|\alpha^{\mathrm{R}}(t)|$ is Rayleigh or Rician distributed depending on the propagation model.

The channel model described above uses physical path parameters such as path gain, delay, and angle of arrival. When the received signal is sampled at the receiver at symbol (or higher) rate, such a model may be inconvenient to use. For linear modulation schemes, it is more convenient to use a "symbol response" channel model.

Such a discrete-time signal model can be obtained easily as follows. Let the continuous-time output from the receive antenna array $\mathbf{x}(t)$ be sampled at the symbol rate at instants $t = t_0 + kT$. The output may be written as

$$\mathbf{x}(k) = \mathbf{H}\mathrm{s}(k) + \mathbf{n}(k) \tag{9.4}$$

where $\mathbf{H}$ is the symbol response channel (a $m \times N$ matrix) that captures the effects of the array response, symbol waveform, and path fading. m is the number of antennas, N is the channel length in symbol periods, and $\mathbf{n}(k)$ is the sampled vector of additive noise. Note that $\mathbf{n}(k)$ may be colored in space and time, as will be shown later. $\mathbf{H}$ is assumed to be time invariant, i.e., α^R is constant. $\mathrm{s}(k)$ is a vector of N consecutive elements of the data sequence and is defined as

$$\mathrm{s}(k) = \begin{bmatrix} s(k) \\ \vdots \\ s(k-N+1) \end{bmatrix} \tag{9.5}$$

It can be shown [49] that the ijth element of the $\mathbf{H}$ is given by

$$[\mathbf{H}]_{ij} = \sum_{l=1}^{L} a_i(\theta_l)\alpha_l^R g((M_d + \Delta - j)T - \tau_l), \quad i = 1\ldots, m; \quad j = 1, \ldots, N \tag{9.6}$$

where

M_d is the maximum path delay

$2\Delta T$ is the duration of the pulse-shaping waveform $g(t)$

9.2.5 Cochannel Interference

In wireless networks a cellular layout with frequency reuse is exploited to support a large number of geographically dispersed users. In TDMA and FDMA networks, when a cochannel mobile operates in a neighboring cell, cochannel interference (CCI) will be present. The average signal-to-interference power ratio (SIR), also called the protection ratio [24], depends on the reuse factor (K). It is 18.7 dB for reuse $K=7$ (IS-54), and 13.8 dB for reuse $K=4$ (GSM). In sectored cells, CCI is significant mainly from cells that lie within the sector beam. The received signal at a base station will therefore be a sum of the desired signal and CCI.

9.2.6 Signal-Plus-Interference Model

The overall signal-plus-interference-and-noise model at the base station antenna array can now be rewritten as

$$\mathbf{x}(k) = \mathbf{H}_s\mathbf{s}_s(k) + \sum_{q=1}^{Q-1} \mathbf{H}_q\mathbf{s}_q(k) + \mathbf{n}(k) \tag{9.7}$$

where $\mathbf{H}_s$ and $\mathbf{H}_q$ are channels for signal and CCI, respectively, while $\mathbf{s}_s$ and $\mathbf{s}_q$ are the corresponding data sequences. Note that Equation 9.7 appears to suggest that the signal and interference are band synchronous. However, this can be relaxed and the time offsets can be absorbed into the channel $\mathbf{H}_q$. In multiuser (MU) cases, all the signals are desired and Equation 9.7 can be rewritten to reflect this situation.

9.2.7 Block Signal Model

It is often convenient to handle signals in blocks. Therefore, we may collect M consecutive snapshots of $\mathbf{x}(\cdot)$ corresponding to time instants $k, \ldots, k+M-1$, (and dropping subscripts for a moment), we get

$$\mathbf{X}(k) = \mathbf{HS}(k) + \mathbf{N}(k) \tag{9.8}$$

where $\mathbf{X}(k)$, $\mathbf{S}(k)$, and $\mathbf{N}(k)$ are defined as

$$\mathbf{X}(k) = [\mathbf{x}(k) \cdots \mathbf{x}(k+M-1)] \quad (m \times M)$$
$$\mathbf{S}(k) = [\mathbf{s}(k) \cdots \mathbf{s}(k+M-1)] \quad (N \times M)$$
$$\mathbf{N}(k) = [\mathbf{n}(k) \cdots \mathbf{n}(k+M-1)] \quad (m \times M)$$

Note that $\mathbf{S}(k)$ by definition is constant along the diagonals and is therefore Toeplitz.

9.2.8 Spatial and Temporal Structure

Given the signal model at Equation 9.8, an important question is whether the unknown channel, $\mathbf{H}$, and data, $\mathbf{s}$, can be determined from the observations $\mathbf{X}$. This leads us to examine the underlying constraints on $\mathbf{H}$ and $\mathbf{S}(\cdot)$ which we call structure.

9.2.8.1 Spatial Structure

From Equation 9.6, the jth column of H is given by

$$\mathbf{H}_{1:m,j} = \sum_{l=1}^{L} a(\theta_l)\alpha_l^{R} g((M_d + \Delta - j)T - \tau_l) \tag{9.9}$$

Spatial structure can help determine $a(\theta_l)$ if the angles of arrival θ_l are known or can be estimated. $a(\theta_l)$ lies on a array manifold A, which is the set of all possible array response vectors indexed by θ.

$$\mathcal{A} = \{a(\theta) | \theta \in \Theta\} \tag{9.10}$$

where Θ is the set of all possible values of θ. $\mathcal{A}$ includes the effect of array geometry, element patterns, inter-element coupling, scattering from support structures, and objects near the base station.

9.2.8.2 Temporal Structure

The temporal structure relates to the properties of the signal $u(t)$ and includes modulation format, pulse-shaping function, and symbol constellation. Some typical temporal structures are

- *Constant modulus*: In many wireless applications, the transmitted waveform has a constant envelope (e.g., in FM modulation). A typical example of a constant envelope waveform is the GMSK modulation used in the GSM cellular system which has the following general form

$$u(t) = e^{j(\omega t + \phi(t))}$$

  where $\phi(t)$ is a Gaussian-filtered phase output of a minimum shift keyed (MSK) signal [40].
- *Finite alphabet*: Another important temporal structure in mobile communication signals is the finite alphabet (FA). This structure underlies all digitally modulated schemes. The modulated signal is a linear or nonlinear map of an underlying finite alphabet. For example, the IS-54 signal is a $\pi/4$ shifted DQPSK signal given by

$$\begin{gathered} u(t) = \sum_p A_p g(t - pT) + j \sum_p B_p g(t - pT) \\ A_p = \cos(\phi_p), \quad B_p = \sin(\phi_p), \quad \phi_p = \phi_{p-1} + \Delta\phi_p \end{gathered} \tag{9.11}$$

where $g(\cdot)$ is the pulse-shaping function (which is a square-root raised cosine function in the case of IS-54), and $\Delta\phi_p$ is chosen from a set of finite phase shifts $\{\frac{5\pi}{4}, \frac{3\pi}{4}, \frac{\pi}{4}, \frac{7\pi}{4}\}$ depending on the data $s(\cdot)$. These finite set of phase shifts represent the FA structure.

- *Distance from Gaussianity*: The distribution of digitally modulated signals is not Gaussian,* and this property can be exploited to estimate the channel from the higher-order moments such as cumulants, see e.g., [15,33]. Clearly constant modulus (CM) signals are non-Gaussian. These higher order statistics (HOS) based methods are usually slower converging than those based on second order statistics.
- *Cyclostationarity*: Recent theoretical results [14,28,39,44] suggest that exploiting the cyclostationary characteristic of the communication signal can lead to second-order statistics based algorithms to identify the channel, **H**, and therefore a more attractive approach than HOS techniques.

  It can be shown [10] that the continuous-time stochastic process $x(t)$ defined in Equation 9.1 (assuming the fade amplitude α^{R} is constant) is cyclostationary. Moreover, the discrete sequence $\{x_i\}$ obtained by sampling $x(t)$ at the symbol rate $1/T$ is wide-sense stationary, whereas the sequence obtained by temporal oversampling (i.e., at a rate higher than $1/T$) or spatial oversampling (multiple antenna elements) is cyclostationary. The cyclostationary signal consists of a number of sampling phases each of which is stationary. A phase corresponds to a shift in the sampling point in temporal oversampling and different antenna element in spatial oversampling.

  The cyclostationary property of sampled communication signals carries important information about the channel phase, which can be exploited in several ways to identify the channel. The cyclostationarity property can also be interpreted as a finite duration property. Put simply, this says that the oversampling increases the number of samples in the signal $\mathbf{x}(t)$ and phases in the channel, **H**, but does not change the value of the data for the duration of the symbol period. This allows **H** to become tall (more rows than columns) and full column rank. Also, the stationarity of the channel makes **H** Toeplitz (or rather block Toeplitz). Tallness and Toeplitz properties are key to the blind estimation of **H**.
- *The temporal manifold*: Just as the array manifold captures spatial wave front information, the temporal manifold captures the temporal pulse-shaping function information [48,49]. We define the temporal manifold $k(\tau)$ as the sampled response of a receiver to an incoming pulse with delay τ. Unlike the array manifold, the temporal manifold can be estimated with good accuracy because it depends only on our knowledge of the pulse-shaping function. Table 9.2 summarizes the duality between the array and the temporal manifold.

The different structures and properties inherent in the signal model are depicted in Figure 9.4

TABLE 9.2 The Duality between the Array and the Time Manifold

Manifold	Indexed by	Characterizes
Array	Angle θ	Antenna array response
Time	Delay τ	Transmitted pulse shape

* The distribution may however, approach Gaussian when constellation shaping is used for spectral efficiency [56].

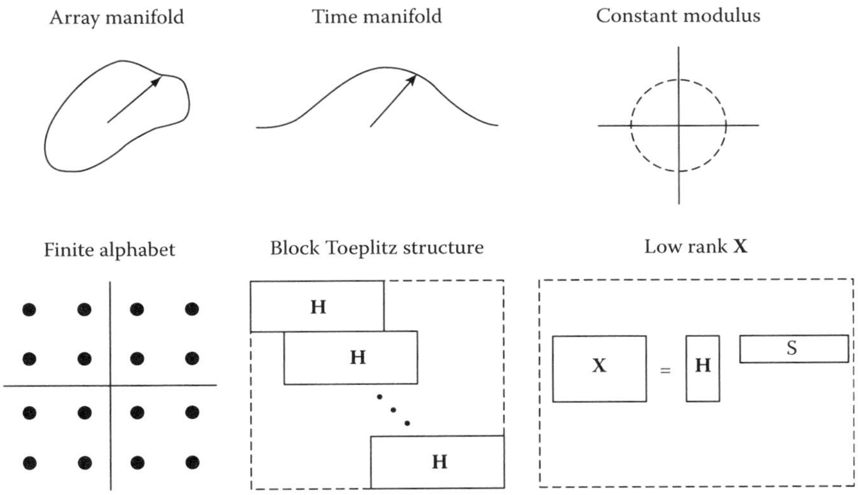

FIGURE 9.4 Space-time structures.

9.3 Algorithms for STP

The history of array signal processing goes back nearly four decades to adaptive antenna combining techniques using phase-lock loops for antenna tracking. An important beginning was made by Howells [21], when he proposed the sidelobe canceller for adaptive nulling, and later Applebaum developed a feedback control algorithm for maximizing SINR. Another significant advance was the least mean square (LMS) algorithm proposed by Widrow [54]. Yet another important milestone was the work of Capon who proposed an adaptive antenna system [8] using a look direction constraint that resulted in the minimum variance distortionless beamformer. Further advances were made by Frost [13] and Griffiths and Jim [17] among several others. See [50] for a review on spatial filtering.

Because of significant delay spread in the channel, array processing in mobile communications can be greatly leveraged by processing the signals in space and in time (STP) to minimize both CCI and inter symbol interference while maximizing SNR. See [35] for a review of channel equalization.

We begin with the single-user case where we are only interested in demodulating the signal of interest. We therefore treat interference from other users as unknown additive noise. This is an interference-suppression approach [53]. Later in the section, we will discuss MU detection which jointly detects all impinging signals.

9.3.1 Single-User ST-ML and ST-MMSE

The first criterion for optimality in STP is maximum likelihood (ML) or is usually referred to as maximum likelihood sequence estimation (MLSE). ST-MLSE seeks to estimate the data sequence that is most likely to have been sent given the received vector signal. Another frequently used criterion is minimum mean square error (MMSE). In ST-MMSE we obtain an estimate of the transmitted signal as a space-time weighted sum of the received signal and seek to minimize the mean square error between the estimate and the true signal at every time instant.

We present ST-MLSE and ST-MMSE in a form that is a space-time extension of the well-known ML and MMSE algorithms.

9.3.1.1 ST-MLSE

With the channel model described by Equation 9.8, we assume that the noise **N** is spatially and temporally white and Gaussian, and that there is no interference. The MLSE problem can be shown to reduce to finding **S** so as to satisfy the following criterion:

$$\min_{\mathbf{S}} \|\mathbf{X} - \mathbf{HS}\|_F^2 \tag{9.12}$$

where the channel **H** is assumed to be known and $\|\cdot\|_F$ denotes Frobenius norm. This is a generalization of the standard MLSE problem where the channel is now defined in space and time. We can, therefore, use a space-time generalization of the well-known Viterbi algorithm (VA) to carry out the search in Equation 9.12 efficiently as is shown in Figure 9.5. See [34] for a discussion on the VA methods.

In the presence of CCI, which is likely to be both spatially and temporally correlated (due to delay spread), the MLSE criterion can be reformulated with a new metric to address the problem. However, the temporal correlation due to delay spread in CCI complicates the implementation of the Viterbi equalizer.

9.3.1.2 ST-MMSE

In the presence of CCI with delay spread, a ST-MMSE receiver is more attractive. This receiver combines the input in space and time to generate an output that minimizes the error between itself and the desired signal (see Figure 9.5). Before proceeding further, we need to introduce some preliminaries.

In a space-time filter (equalizer cum beamformer), **W** has the following form:

$$\mathbf{W}(k) = \begin{bmatrix} w_{11}(k) & \cdots & w_{1M}(k) \\ \vdots & \cdots & \vdots \\ w_{m1}(k) & \cdots & w_{mM}(k) \end{bmatrix} \tag{9.13}$$

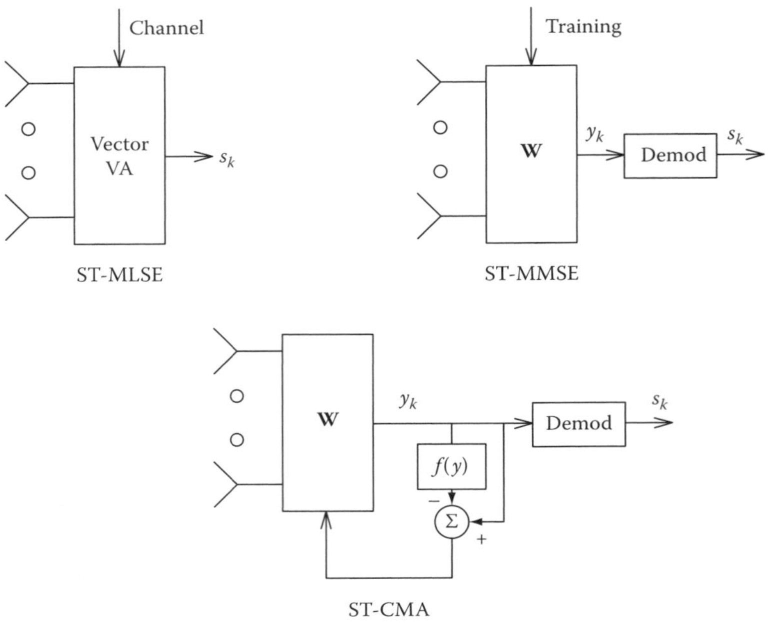

FIGURE 9.5 Different structures for STP.

In order to obtain a convenient formulation for the space-time filter output, we introduce the quantities $W(k)$ and $X(k)$ as follows:

$$\begin{aligned} X(k) &= vec(\mathbf{X}(k))\ (mM \times 1) \\ W(k) &= vec(\mathbf{W}(k))\ (mM \times 1) \end{aligned} \tag{9.14}$$

where the operator vec(·) is defined as

$$\text{vec}([\mathbf{v}_1 \cdots \mathbf{v}_M]) = \begin{bmatrix} \mathbf{v}_1 \\ \vdots \\ \mathbf{v}_M \end{bmatrix}$$

The scalar equalizer output $y(k)$ can then be written as

$$y(k) = W^{\mathrm{H}}(k)X(k) = Tr(\mathbf{W}^{\mathrm{H}}(k)\mathbf{X}(k)) \tag{9.15}$$

where $(\cdot)^{\mathrm{H}}$ denotes complex conjugate transpose.

The ST-MMSE filter chooses the space-time filter weights to approach the minimum mean square error, i.e.,

$$\min_W E\|W^{\mathrm{H}}X(k) - s(k-\zeta)\|_2^2 \tag{9.16}$$

where ζ is a delay chosen to center the space-time filter (the choice of this parameter strongly affects performance). The solution to this least-squares (LS) problem follows from the well-known projection theorem

$$E(X(k)(X^{\mathrm{H}}(k)W - s^*(k-\zeta))) = 0 \tag{9.17}$$

This leads to

$$W = \{E(X(k)X^{\mathrm{H}}(k))\}^{-1}E(X(k)s^*(k-\zeta)) \tag{9.18}$$

where superscript * denotes complex conjugate. If the interference and noise are independent of the signal, the transmitted bit sequence is white, and $M > N$

$$E(X(k)s^*(k-\zeta)) = [0 \cdots 0\ vec^{\mathrm{T}}(\mathbf{H})\, 0 \cdots 0]^{\mathrm{T}} = \bar{H} \tag{9.19}$$

where the number of zeros preceding and succeeding $vec^{\mathrm{T}}(\mathbf{H})$ depends on the choice of ζ. Defining the space-time $mN \times mN$ covariance matrix $\mathbf{R}_{XX} = E(XX^{\mathrm{H}})$, Equation 9.18 takes the familiar form

$$W = \mathbf{R}_{XX}^{-1}\bar{H} \tag{9.20}$$

Note that when $M = N, \bar{H} = vec(\mathbf{H})$. A number of techniques are available in order to solve Equation 9.20, such as the LMS [54] or the recursive least square (RLS) [29]. These have different tradeoffs of computational complexity, tracking capability and steady-state error. See [29] for a discussion. Alternatively, if a block method is used, we can explicitly calculate $\mathbf{R}_{XX}^{-1}$ and then use $\bar{H}$ to find W. This is known as sample matrix inversion (SMI).

The relative performance of ST-MLSE and ST-MMSE schemes is influenced by the dominance of CCI and ISI, and the nature of the channel. When the channel is CCI dominated and contains multipath, a Viterbi equalizer is complicated to implement. In this case, the MMSE approach appears desirable. On the other hand, in an ISI-dominated large delay spread scenario, an MLSE has natural advantages.

Having reviewed the basic approaches to STP, we will now study two important issues: blind vs. nonblind and single vs. MU approaches.

9.3.1.3 Training Signal Methods

In many mobile communications standards such as GSM and IS-54, explicit training signals are inserted inside the TDMA data bursts. These training signals can be used to estimate the channel needed for the MLSE or MMSE receivers.

Let $\mathbf{T}$ be the training sequence arranged in a matrix form (again $\mathbf{T}$ is Toeplitz). Then, during the training burst, the received data is given by

$$\mathbf{X} = \mathbf{HT} + \mathbf{N} \tag{9.21}$$

Clearly $\mathbf{H}$ can be estimated using LS

$$\mathbf{H} = \mathbf{XT}^{\dagger} \tag{9.22}$$

where $\mathbf{T}^{\dagger} = \mathbf{T}^{\mathrm{H}}(\mathbf{TT}^{\mathrm{H}})^{-1}$.

In a ST-MMSE receiver, we need W and this can be computed readily from $\mathbf{H}$ using Equation 9.20.

9.3.1.4 Blind Methods

The term "blind" methods (other names are "self-recovering" or "unsupervised"), do not need training signals and rather exploit the temporal structure such as non-Gaussianity; constant modulus; FA; cyclostationarity, or the spatial structure, such as the array manifold. The performance of blind methods will, of course, be sensitive to the validity of structural properties assumed.

9.3.1.4.1 Spatial Structure or DOA-Based Methods

These techniques use DOA estimates as a basis for determining the optimum beamformer. These methods were developed vigorously in the 1980s in military applications for reception of unknown or noise-like signals. The modern era in DOA estimation began with the MUSIC algorithm first proposed in 1979 by Schmidt and independently by Bienvenu and Kopp [6,36], thus launching the "subspace era" in signal processing. See [52] for a survey. Another class of DOA estimation techniques was launched when Paulraj et al. proposed the ESPRIT algorithm which has striking advantages when compared to MUSIC but needs a special array geometry. See [23] for a survey.

DOA-based methods suffer from serious drawbacks in cellular applications. First, DOA estimation requires an accurate knowledge of the array manifold. This needs expensive calibration support. Next, the number of antennas at cellular base stations vary from four to eight per sector, an insufficient number for the multipath and interference rich cellular environments. Finally, these methods do not exploit the knowledge of the modulation format of the communication signal and the time delay relationship between multipath signals.

A subspace approach can be used to estimate the directions-of-arrival of the impinging wave fronts. The signal model is given by

$$\mathbf{x}(t) = \mathbf{Au}(t) + \mathbf{n}(t) \tag{9.23}$$

where **A** is an $m \times Q$ matrix whose columns are the array response vectors for each wave front (assuming no multipath)

$$\mathbf{A} = [\mathbf{a}(\theta_1) \quad \cdots \quad \mathbf{a}(\theta_Q)],$$

$\mathbf{u}(t)$ contains the fading signals from the Q users

$$\mathbf{u}(t) = [\alpha_1(t)u_1(t-\tau_1)\cdots\alpha_Q(t)u_Q(t-\tau_Q)]^{\mathrm{T}}$$

and

$$u_q(t) = \sum_k s_q(k)g(t-kT)$$

The sampled block signal model then takes the following form

$$\mathbf{X} = \mathbf{AS} + \mathbf{N} \tag{9.24}$$

In the subspace approach, we seek to estimate **A** from the array data by exploiting the underlying array manifold structure. When the number of antennas, m, is greater than the number of signals, Q, the signal $\mathbf{x}(t)$ in the absence of noise is confined to a subspace, referred to as the signal subspace.

We first estimate this signal subspace from the received data **X**. We then search for an $m \times Q$ matrix **A** whose columns lie on the array manifold and whose (column) subspace matches the estimated signal subspace. A good estimate of the signal subspace is given by the first Q dominant eigenvectors of the space-only $m \times m$ covariance matrix $\mathbf{R}_{xx} = E(\mathbf{xx}^{\mathrm{H}})$. If $\mathbf{E}_s$ is a matrix of these eigenvectors, then the subspace fitting approach estimates **A** to minimize the following criterion

$$\min_{\mathbf{A}} \|\mathbf{E}_s - \mathbf{AZ}\|_F^2$$

where **Z** is an arbitrary $Q \times Q$ square matrix.

Once **A** is estimated, we have the array vector for the desired signal. The MMSE and ML estimators (assuming no multipath) of $u_q(t)$ are identical [7] and are given by

$$\mathbf{w}_q = \mathbf{R}_{xx}^{-1}\mathbf{a}(\theta_q) \tag{9.25}$$

$\mathbf{w}_q$ is a (space-only) beamformer that has been studied extensively.

When multipath and delay spread is present, the solution in Equation 9.25 will have a poor performance and improved techniques are needed. If we use a ST-MMSE structure, we can extend the above subspace methods to compute the optimum beamformer given in Equation 9.20.

9.3.1.4.2 Temporal Structure Methods

These techniques include a vast range that spans from the well-studied CM and HOS methods to the more recent second order methods that exploit the cyclostationarity of the received signal.

The fading and dynamics of the mobile propagation channel create special problems for blind techniques, and their performance in mobile channels is only recently gaining attention. A widely known class of simple blind algorithms is the so-called Bussgang class that contains, among others, the CM 1-2, CM 2-2, Sato, and decision-directed (DD) algorithms. See [11,19,20] for a survey of blind algorithms.

Contrary to nonblind techniques, where a training signal drives the recursive algorithms, in the CM approach, we replace the training signals by a modulus corrected version of the output signal. The CM 2-2 minimizes the following cost function

$$\min_{W} J(W) = E\|y(k)|^2 - 1|^2 \tag{9.26}$$

where $y(k)$ is the output of the ST filter (see Equation 9.15).

The resulting LMS-type algorithm is given by

$$W(k+1) = W(k) - \mu X^*(k)\, y(k)(|y(k)|^2 - 1) \tag{9.27}$$

$W(k+1)$, under the right conditions, approaches the optimum ST-MMSE solution in Equation 9.20.

Important performance issues for blind algorithms are speed of convergence, ability to reach the global optimum solution, and capacity to track time varying mobile channels.

9.3.1.4.3 *Polyphase Methods*

Following the path-breaking paper by Tong et al. [44] that presented a blind channel identification method using oversampling and relying only on second order statistics, a number of techniques that exploit cyclostationarity have since dominated the blind-deconvolution literature.

Polyphase methods provide a blind solution by starting with the data

$$\mathbf{X}(k) = \mathbf{HS}(k) + \mathbf{N}(k) \tag{9.28}$$

or its second order statistics. They then extract **H** and **S** by exploiting the tallness structure (obtained via oversampling) of **H** [28,29]. See [25] for a tutorial presentation of polyphase techniques.

9.3.2 Multiuser Algorithms

In MU algorithms, we address the problem of extracting multiple cochannel user signals arriving at an antenna array. Such problems occur in channel reuse within cell (RWC) applications or in situations where we attempt to demodulate the interference signal in order to improve interference suppression. The data model is once again

$$\mathbf{X} = \mathbf{HS} + \mathbf{N} \tag{9.29}$$

where **H** and **S** are suitably defined to include multiple users and are of dimensions $m \times NQ$ and $NQ \times M$, respectively.

We have several approaches that parallel the single-user case. We begin with the ML and MMSE prototypes and then explain in more detail some recently developed blind techniques.

9.3.2.1 Multiuser MLSE and MMSE

If the channels for all arriving signals are known, then we can extend the earlier MLSE to jointly demodulate all the user data sequences. Starting with the data model in Equation 9.29, we can then search for multiple user data sequences that minimize the ML cost function in Equation 9.12. The MU MLSE will have a large number of states in the trellis. Efficient techniques for implementing a Viterbi equalizer need to be developed.

In MU MMSE, we usually estimate each user signal separately using the single-user MMSE processor given in Equation 9.20. In this case, the MMSE treats other user signals as interference with unknown structure. MU techniques either need training signals for all the users or adopt blind methods. The multiple training signals should be designed to have low cross correlation properties so as to minimize cross coupling in the channel estimates.

9.3.2.2 Multiuser Blind Methods

Once again, the techniques are parallel to the single-user spatial and temporal blind methods. The spatial structure MU algorithms are again applicable under the conditions discussed in Section 9.3.1. The approach is identical; we first estimate $\mathbf{A}$ using subspace methods, and the beamformer $\mathbf{w}_q$ for each user follows from Equation 9.25.

We briefly describe some illustrative algorithms.

9.3.2.3 Finite Alphabet FA Method

This approach exploits the FA property of the digitally modulated signals. Assuming no delay spread and perfect MU symbol synchronization, the channel model is given by Equation 9.24, which we repeat here for convenience:

$$\mathbf{X} = \mathbf{AS} + \mathbf{N} \tag{9.30}$$

where both $\mathbf{A}$ and $\mathbf{S}$ are unknown and the additive noise is assumed to be white and Gaussian. The joint ML criterion for this reduces to the familiar minimization problem

$$\min_{\mathbf{A},\mathbf{S}} \|\mathbf{X} - \mathbf{AS}\|_F^2 \tag{9.31}$$

This is a joint ML problem where both the channel and data are unknown. The FA property allows us to solve Equation 9.31 and estimate both $\mathbf{A}$ and $\mathbf{S}$. Since the ML criterion is separable with respect to the unknowns, one approach to minimize the cost function in Equation 9.31 is alternating projections. Starting with an initial estimate of $\mathbf{A}$, we minimize Equation 9.31 with respect to $\mathbf{S}$, keeping $\mathbf{A}$ fixed. This is a data detection problem. With an estimate of $\mathbf{S}$, an improved estimate of $\mathbf{A}$ can be obtained by minimizing Equation 9.31 with respect to $\mathbf{A}$, keeping $\mathbf{S}$ fixed. This is a standard LS problem. We continue this iterative process until a fixed point is reached. The global solution is a fixed point of the iteration. In order to avoid a computationally expensive search, two suboptimal iterative techniques, ILSP and ILSE [42,43], can be used to make this minimization tractable.

Note that the joint ML problem can also be formulated for the single-user case where we estimate the channel and the data jointly using the FA or other signal properties. Joint ML methods are also known as adaptive ML.

9.3.2.4 Finite Alphabet–Oversampling Method

In the presence of delay spread (and unsynchronized symbols), the FA algorithm has to be modified to estimate the space-time channel $\mathbf{H}$ as against the spatial channel $\mathbf{A}$ described earlier. An attractive technique to estimate the temporal channel using polyphase or oversampling method was proposed recently in [47].

We therefore first need to extend the MU data model in Equation 9.30 to incorporate oversampling. Assuming oversampling at P samples per symbol, we define a new $mP \times M$ data matrix $\mathbf{X}$ where each entry is a vector of P data samples per symbol period.

This results once again in the familiar model

$$\mathbf{X} = \mathbf{HS} + \mathbf{N} \tag{9.32}$$

Note that now the dimensions of $\mathbf{X}$, $\mathbf{H}$, and $\mathbf{S}$ are $mP \times M$, $mP \times NQ$, and $NQ \times M$, respectively. As noted earlier, $\mathbf{H}$ is tall, full column rank and has a block Toeplitz structure.

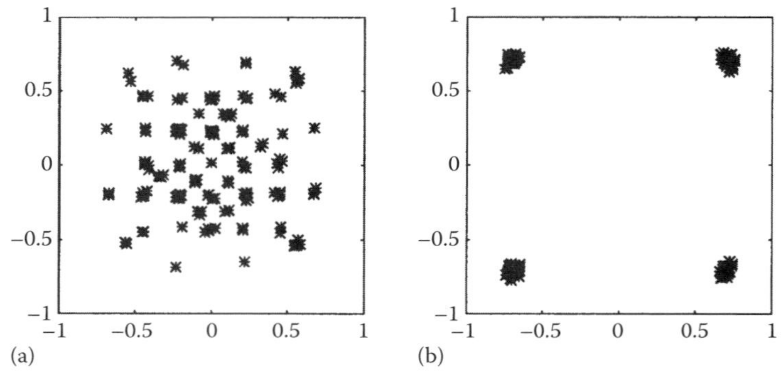

FIGURE 9.6 Interference cancellation using STP. (a) Received signal and (b) after STP processing.

Once again we can estimate **H** and **S** using a joint-ML approach. This reduces to minimizing

$$\min_{\mathbf{H},\mathbf{S}} \|\mathbf{X} - \mathbf{HS}\|_F^2 \tag{9.33}$$

A direct approach to Equation 9.33 is computationally prohibitive. The approach in [47] breaks up the joint problem into two smaller subproblems. First, the channels are equalized by enforcing the low rank and block-Toeplitz structure of **H**. This yields the row subspace of **S**. The FA property can now be enforced to determine the symbols in **S**.

9.3.2.5 Multiuser CM

While the FA approach exploits the FA property, CM is another structure for MU signal separation. When the channel has no delay spread, we have a standard source separation problem, which can be dealt with the so-called *CM array* [4,16,26]. The resulting algorithms are space-only MU counterparts of the temporal CM algorithms: instead of combating the channel ISI in order to retrieve the single-user signal, they try to combat the channel CCI and demodulate the different user signals.

Recent advances in MU CM include an analytical CM (ACM) [32,45], a multistage CM algorithm [38], and related approaches [5] and [9]. The extension to a delay spread environment was proposed in [31].

9.3.3 Simulation Example

Figure 9.6 illustrates the effect of STP in a mobile environment. The channel model chosen was a typical urban channel. A four element linear array with $\frac{\lambda}{2}$ spacing was employed. The desired and interference signals arrived from mean directions of 0° and 45°, respectively. An IS-54 channel interface was used. Figure 9.6a shows the received signal constellation for a simple antenna. Note that this implies that the eye is completely closed. Figure 9.6b shows the constellation after STP, using a ST-MMSE equalizer employing training signals. Note the dramatic improvement in the received constellation.

9.4 Applications of Spatial Processing

In this section, we briefly describe three applications of antennas and STP in cellular base stations.

9.4.1 Switched Beam Systems

Switched beam systems (SBS) consist of a beamformer in the RF stage that forms multiple (nonadaptive) beams, a "sniffer" that determines the beam that has the best SINR and a switch that is used to select the best or best two beams for the receiver. These systems are used as an appliqué unit, where the existing diversity antennas are replaced by a switch-beam antenna system. The SBS operates by sniffer scanning the beamformer outputs to detect the best two beams which are then switched through to the receiver. In order to reduce the probability of incorrect beam selection, the beam outputs are validated by checking the color code (for example CDVCC in IS-54 or SAT tone in AMPS) prior to determining the best beam.

The main advantage of using SBS is the improvement in cell coverage on the reverse link due to the array gain and improved voice quality due to reduced interference. Since the preformed beams are narrower than the sector beamwidth, reduction in interference power is obtained when the desired signal and the interference are separated in angle and fall into different beams. This SINR improvement offers better voice quality and may also allow use of a smaller reuse factor and therefore improve capacity.

The performance of SBS depends on a number of factors, including angle spread of multipath, relative angles-of-arrival of the signals and interference, and the array topology. Performance gains in SBS come from array gain, diversity gain, reduced interference, and trunking efficiency.

9.4.2 Space-Time Filtering

Space-time filtering (STF) applies STP to maximize signal power and minimize intersymbol and CCI. As is evident from earlier sections, STP will be very different for each air interface standard. In GSM, the slot duration is 0.577 ms with 26 training bits and the symbol period is 3.7 μs. Channel equalization is, of course, necessary. The presence of controlled ISI further complicates the equalization problem. It is reasonable to assume that the signal and interference channels are invariant across a slot. In IS-54, the slot duration is 6.66 ms, resulting in substantial variation of the channel over the slot. However, the symbol period is 41.6 μs, and therefore channel equalization is usually not needed. Thus, the STF architectures must find some means to track the signal and interference channels across the burst, using blind techniques, as those described earlier in this chapter.

9.4.3 Channel Reuse within a Cell

RWC refers to the reuse of a channel or radio resource within a cell by exploiting differences in the channels. This is akin to spectrum reuse in cellular systems, where a channel or a spectrum resource used in one cell is reused in another cell separated by sufficient distance such that the CCI is sufficiently small.

When RWC is used in TDMA or FDMA, a cell supports two or more users in a given channel, as against a single user in conventional cells. Antenna arrays and STP is used for joint demodulation of multiple users, assuming such users are sufficiently separated in channels (directions). When two or more users become closely aligned in their channels, they will no longer be separable, and one of the users should be handed off to another frequency or time slot. RWC needs to work on both forward and reverse links, therefore, signal separability must be achieved on both links.

The principal challenge in RWC when used with TDMA or FDMA is to estimate and track the reverse and forward channels to a high degree of accuracy. The problem is further complicated by the near-far problem resulting in power imbalance between users. The ability to estimate and track the reverse link channel depends on angle, delay, and Doppler spreads. The higher these spreads, the higher the sources of channel estimation errors. Therefore, flat rural environments with low angle and delay spreads score over urban and microcells which use antennas below roof top. Also, fixed wireless applications score over mobile applications. In the forward link, we need to once again predict the channel accurately. We can do this by an open loop method, i.e., use the reverse link channel to predict forward channel. Alternatively, we can use feedback from the mobile to estimate the forward channel. For the open loop method, in FDD, angle spread is a source of error and in TDD the Doppler spread is a source of error. Due to these complications, RWC is not a promising technology in most TDMA and FDMA applications.

9.5 Summary

Use of array signal processing or STP is emerging as a powerful tool for improving cellular wireless networks. STP can improve cell coverage, enhance link quality, and increase system capacity. The rapidly varying mobile channel with large multipath delay and angle spreads offer a significant challenge to STP. Effective solutions have to be specific to each air interface and the propagation environment. More work is needed to develop robust STP techniques and to characterize their performance. See [1–3] for a review of the current state of the art in smart antennas technology.

References

1. *First Workshop on Smart Antennas in Wireless Mobile Communications*, Center for Telecommunications and Information Systems Laboratory, Stanford University, Stanford, CA, June 1994.
2. *Second Workshop on Smart Antennas in Wireless Mobile Communications*, Center for Telecommunications and Information Systems Laboratory, Stanford University, Stanford, CA, July 20–21, 1995.
3. *Third Workshop on Smart Antennas in Wireless Mobile Communications*, Center for Telecommunications and Information Systems Laboratory, Stanford University, Stanford, CA, July 25–26, 1996.
4. Agee, B.G., Blind separation and capture of communication signals using multitarget constant modulus beamformer, *Proceedings of the MILCOM'89*, Boston, MA 1989.
5. Batra, A. and Barry, J.R., Blind cancellation of co-channel interference, *Proceedings of the IEEE Globecom Conference*, Singapore, 1995.
6. Bienvenu, G. and Kopp, L., Principe de la goniometri passive adaptative, *Proceedings of the 7 è me Colloque GRETSI*, 106/1–106/10, Nice, France, 1979.
7. Capon, J., High resolution frequency wave number spectrum analysis, *Proc. IEEE*, 57, 1408–1418, 1969.
8. Capon, J., Greenfield, R.J., and Kolker, R.J., Multidimensional maximum likelihood processing of a large aperture seismic array, *Proc. IEEE*, 55, 192–211, February 1967.
9. Castedo, L., Escudero, C.J., and Dapena, A., A blind signal separation method for multiuser communications, *IEEE Trans. Signal Proc., Special Issue on Signal Processing for Advanced Communications*, 45(1), 1343–1348, January 1997.
10. Ding, Z., Blind channel identification and equalization using spectral correlation measurements, Part I: frequency-domain analysis, in *Cyclostationarity in Communications and Signal Processing*, Gardner, W.A. (Ed.), IEEE Press, New York, 1994, pp. 417–436.
11. Duhamel, P., Blind equalization, Tutorial presentation, *International Conference on Acoustics, Speech, and Signal Processing*, Detroit, MI, May 1995.
12. Feher, K., *Wireless Digital Communications*, Feher/Prentice-Hall, Upper Saddle River, NJ, 1995.
13. Frost, O.L., An algorithm for linearly constrained adaptive array processing, *Proc. IEEE*, 60, 926–935, 1972.
14. Gardner, W.A., Ed., *Cyclostationarity in Communications and Signal Processing*, IEEE Press, New York, 1994.
15. Giannakis, G.B. and Mendel, J.M., Identification of nonminimum phase systems using higher order statistics, *IEEE Trans. Acoust. Speech Signal Process.*, 37(3), 360–377, March 1989.
16. Gooch, R.P. and Lundell, J., The CM array: An adaptive beamformer for constant modulus signals, *Proceedings of the ICASSP'86*, pp. 2523–2526, Tokyo, Japan, 1986.
17. Griffiths, L.J. and Jim, C.W., An alternative approach to linearly constrained adaptive beamforming, *IEEE Trans. Antenn. Propag.*, AP-30, 27–34, May 1982.
18. Hansen, L.K. and Xu, G., Geometric properties of the blind digital co-channel communications problem, *Proceedings of the ICASSP'96*, Atlanta, GA, May 1996.

19. Haykin, S., *Blind Deconvolution*, Prentice-Hall, Englewood Cliffs, NJ, 1994.
20. Haykin, S., *Adaptive Filter Theory*, 3rd edn., Prentice-Hall, Englewood Cliffs, NJ, 1995.
21. Howells, P., Intermediate frequency side-lobe canceller, U.S. Patent 3,202,990, August 1965.
22. Jakes, W.C., *Microwave Mobile Communications*, John Wiley & Sons, New York, 1974.
23. Krim, H. and Viberg, M., Two decades of array signal processing research: The parametric approach, *IEEE Signal Process. Mag.*, 13(4), 67–94, July 1996.
24. Lee, W.C., *Mobile Communications—Design Fundamentals*, Howard Sams, Indianapolis, IN, 1986.
25. Liu, H., Xu, G., Tong, L., and Kailath, T., Recent developments in blind channel equalization: From cyclostationarity to subspaces, *Signal Process.*, 50, 83–99, 1996.
26. Lundell, J.D. and Widrow, B., Application of the constant modulus adaptive beamformer to constant and nonconstant modulus signals, in *Proceedings of the Asilomar 21st Conference on Signals, Systems, and Computers*, pp. 432–436, Pacific Grove, CA, November 1991.
27. Matsumoto, T., Nishioka, S., and Hodder, D., Beam selection performance analysis of a switched multi-beam antenna system in mobile communications environments, *Second Workshop on Smart Antennas in Wireless Mobile Communications*, Stanford, CA, July 1995.
28. Moulines, E., Duhamel, P., Cardoso, J.F., and Mayrargue, S., Subspace methods for the blind identification of multichannel FIR filters, *IEEE Trans. Signal Process.*, 43(2), 516–525, 1995.
29. Orfanidis, S.J., *Optimal Signal Processing—An Introduction*, Macmillan Publishing Co., New York, 1985.
30. Pahlavan, K. and Levesque, A.H., *Wireless Information Networks*, John Wiley & Sons, New York, 1995.
31. Papadias, C.B. and Paulraj, A., A constant modulus algorithm for multi-user signal separation in presence of delay spread using antenna arrays, *IEEE Signal Process. Lett.*, 4(6), 178–181, June 1997.
32. Papadias, C.B. and Slock, D.T.M., Towards globally convergent blind equalization of constant modulus signals: a bilinear approach, in *Proceedings of the VII European Signal Processing Conference*, Edinburgh, Scotland, U.K., September 13–16, 1994.
33. Porat, B. and Friedlander, B., Blind equalization of digital communication channels using higher order moments, *IEEE Trans. Acoust. Speech, Signal Process.*, SP-39(2), 522–526, February 1991.
34. Proakis, J.G., *Digital Communications*, McGraw-Hill, New York, 1983.
35. Qureshi, S.U.H., Adaptive equalization, *Proc. IEEE*, 53(12), 1349–1387, September 1985.
36. Schmidt, R.O., Multiple emitter location and signal parameter estimation, in *Proceedings of the RADC Spectrum Estimation Workshop*, pp. 243–258, Griffiss AFB, NY, 1979.
37. Schmidt, R.O., A signal subspace approach to multiple emitter location and spectral estimation, PhD thesis, Stanford University, Stanford, CA, November 1981.
38. Shynk, J.J. and Gooch, R.P., The constant modulus array for co-channel signal copy and direction finding, *IEEE Trans. Signal Process.*, 44(3), 652–660, March 1996.
39. Slock, D.T.M. and Papadias, C.B., Blind fractionally-spaced equalization based on cyclostationarity, in *Proceedings of the Vehicular Technology Conference*, Stockholm, Sweden, June 1994.
40. Steele, R., *Mobile Radio Communications*, Pentech Press, London, U.K., 1992.
41. Swindlehurst, A. and Yang, J., Using least squares to improve blind signal copy performance, *IEEE Signal Process. Lett.*, 1(5), 80–82, May 1994.
42. Talwar, S., Paulraj, A., and Viberg, M., Reception of multiple co-channel digital signals using antenna arrays with applications to PCS, in *Proceedings of the ICC'94*, pp. 700–794, New Orleans, LA, 1994.
43. Talwar, S., Viberg, M., and Paulraj, A., Blind separation of synchronous co-channel digital signals using an antenna array. Part I. Algorithms, *IEEE Trans. Signal Process.*, 44(5), 1184–1197, May 1996.
44. Tong, L., Xu, G., and Kailath, T., Blind identification and equalization of multipath channels: A time domain approach, *IEEE Trans. Inform. Theory*, 40(2), 340–349, March 1994.

45. van der Veen, A., and Paulraj, A., An analytical constant modulus algorithm, *IEEE Trans. Signal Process.*, 44(5), 1136–1195, May 1996.
46. van der Veen, A.J., Talwar, S., and Paulraj, A., Blind identification of FIR channels carrying multiple finite alphabet signals, in *Proceedings of the IEEE ICASSP*, 2, pp. 1213–1216, Detroit, MI, 1995.
47. van der Veen, A.J., Talwar, S., and Paulraj, A., Blind estimation of multiple digital signals transmitted over FIR channels, *IEEE Signal Process. Lett.*, 5(2), 99–102, May 1995.
48. Vanderveen, M.C., Ng, B., Papadias, C.B., and Paulraj, A.J., Joint angle and delay estimation (JADE) for signals in multipath environments, in *Proceedings of the 30th Asilomar Conference on Signals, Systems, and Computers*, Pacific Grove, CA, 1996.
49. Vanderveen, M.C., Papadias, C.B., and Paulraj, A.J., Joint angle and delay estimation (JADE) for multipath signals arriving at an antenna array, *IEEE Commun. Lett.*, 1(1), 12–14, January 1997.
50. Van Veen, B.D. and Buckley, K.M., Beamforming: a versatile approach to spatial filtering, *IEEE ASSP Mag.*, 4–24, April 1988.
51. Verdu, S., Minimum probability of error for asynchronous Gaussian multiple-access channels, *IEEE Trans. Inform. Theory*, 32(1), 85–96, January 1986.
52. Viberg, M. and Stoica, P., Editorial note, in special issue on subspace methods, Part I: Array signal processing and subspace computations, *Signal Process.*, 50(1,2), 1–3, April 1996.
53. Wales, S., Technique for co-channel interference suppression in TDMA mobile radio systems, *IEEE Proc. Commun.*, 142, 106–114, April 1995.
54. Widrow, B. and Stearns, S., *Adaptive Signal Processing*, Prentice-Hall, Englewood Cliffs, NJ, 1985.
55. Yacoub, M.D., *Foundations of Mobile Radio Engineering*, CRC Press, Boca Raton, FL, 1993.
56. Zervas, E., Proakis, J., and Eyuboglu, V., Effects of constellation shaping on blind equalization, *Proc. SPIE*, 1991, 1565, 1991.

10

Beamforming with Correlated Arrivals in Mobile Communications

Victor A. N. Barroso
Instituo Superior Tecnico

José M. F. Moura
Carnegie Mellon University

10.1 Introduction **10**-1
10.2 Beamforming **10**-2
Minimum Output Noise Power Beamforming
10.3 MMSE Beamformer: Correlated Arrivals **10**-8
10.4 MMSE Beamformer for Mobile Communications **10**-11
Model of the Array Output • Maximum Likelihood Estimation of **H**
10.5 Experiments **10**-17
10.6 Conclusions **10**-19
Acknowledgments **10**-20
References **10**-20

10.1 Introduction

The classical definition of a beamformer basically specifies its goal: to estimate the signal waveform arriving at the array from a given direction. Beamformers are spatial processors that combine the signals impinging on an array of captors. Combining the outputs of the captors forms a narrow beam pointing toward the direction of the source (look direction). This narrow beam can discriminate between sources spatially located at distinct sites. This important property of beamformers is used to design techniques that localize active or passive sources particularly in RADAR/SONAR systems.

In the last two decades, beamforming methods have had significant theoretical and practical advances. This, together with other technological advances, has broadened the application of sophisticated beamforming techniques to a diversity of areas, including imaging, geophysical and oceanographic exploration, astrophysical exploration, and biomedical. See [19,20] for an excellent overview of modern beamforming techniques and applications.

Communications is another attractive application area for beamforming. In fact, beamforming has been widely used for directional transmission and reception as well as for sector broadcasting in satellite communications systems. More recently, due to the drastic increase of users in cellular radio systems [10,15,18], including indoors and outdoors mobile systems, it is increasingly being recognized that the design of base station and mobile antennas based on beamforming methods improves significantly the system's spectrum efficiency [1,11]. In turn, this enables accommodating larger numbers of users [3,16]. The most striking argument in favor of using advanced beamforming techniques such as adaptive or blind beamforming for mobile communications is based on the idea of space division multiple access (SDMA) schemes. With SDMA, several mobiles share simultaneously the same frequency channel by creating virtual channels in the spatial domain. Another important argument in favor of using beamforming in cellular radio is that beamforming yields flexible signal processing schemes that properly handle multipath effects which are typical in radio communications. Multipath is the term given when the same signal arrives at the destination through different paths. This may arise when signals bounce off obstacles in their path of propagation. At the receiver, these arrivals are correlated. Their recombination causes severe signal distortions and fading. In limiting cases, the power of the received signal can become so small that the reliability of the data communications link is completely lost.

In this section we design a multichannel beamformer to combat multipath effects. The receiver uses a base station antenna array which handles several radio links operating simultaneously at the same carrier frequency, while preserving the reliability of the communications. The approach relies on statistical signal processing methods, yielding a solution that operates in a blind mode with respect to the parameters that specify the propagation channel. This means that, except for a few quantities related to system specifications, e.g., link budget and array geometry, the receiver that we describe here does not assume any prior knowledge about the locations of the sources and of the structures of the ray arrivals, including directions of arrival and correlations. The simulation results show the excellent performance of this multichannel beamformer in SDMA schemes.

The chapter is organized as follows. In Section 10.2 we introduce the beamforming problem (see also [20]), and classical beamformers such as the delay-and-sum (DS) beamformer, the minimum output noise power beamformer (MNP), and the minimum variance (MV) beamformer. We show that these beamformers present severe drawbacks when operating in multipath environments. Section 10.3 presents a solution to the beamforming problem for the case of correlated arrivals. This solution is based on a minimum mean square error (MMSE) approach. We compare the performance of this beamformer with the performance of the beamformers introduced in Section 10.3. We emphasize, in particular, the case of multipath propagation. In this section, we also discuss issues regarding the implementation of the MMSE beamformer. In Section 10.4, we describe a method to implement the MMSE beamformer in the context of a digital mobile communications system. The method operates in a blind mode and strongly exploits the structure of the received multipath data. Since the propagation channel parameters, e.g., angles of arrivals of the multiple paths, are not known, we estimate them with a maximum likelihood approach supported on a finite mixture distribution model of the array data. We maximize the likelihood function with an iterative scheme. We describe an efficient procedure to initialize the iterative algorithm. In general, this procedure converges rapidly to the global maximum of the likelihood function. Section 10.5 presents simulation results obtained with data synthesized by a simple mobile communications simulator. These results confirm the excellent performance of the MMSE beamformer described in the paper.

10.2 Beamforming

Beamforming is an array processing technique for estimating a desired signal waveform impinging on an array of sensors from a given direction. This technique applies to both narrowband and wideband signals. Here, we will consider only the narrowband case.

Let $s(t)$ be the complex envelope of the source radiated signature. Under the farfield assumption the signal at the receiving array is a planar wave front, see Figure 10.1. In this case, and according to the model derived in [20], the complex envelope of the signal received at each sensor of a uniform and linear array of N omnidirectional sensors is

$$s_n(t) = s(t - \tau_n)e^{-j\omega_0\tau_n}, \tag{10.1}$$

where

ω_0 is the carrier frequency
τ_n is the intersensor propagation delay

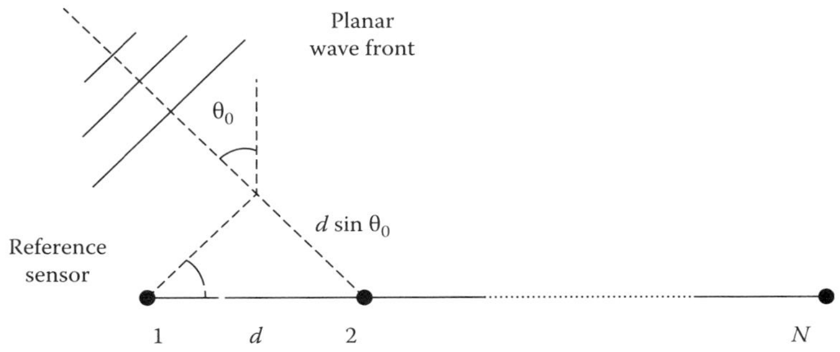

FIGURE 10.1 Source/receiver geometry.

Let d, c, and θ_0 be, respectively, the distance between sensors, the propagation velocity, and the direction of arrival (DOA). The intersensor delays are then

$$\tau_n = \frac{(n-1)d}{c} \sin \theta_0, \quad n = 1, 2, \ldots, N. \tag{10.2}$$

Because of the narrowband assumption, we can make the simplification

$$s(t - \tau_n) \simeq s(t),$$

in Equation 10.1. This means that, for the values of τ_n of interest, the source complex envelope $s(t)$ is slowly varying when compared with the carrier $e^{j\omega_0 t}$.

We model each array sensor by a quadrature receiver, its output being given by

$$z_n(t) = s_n(t) + n_n(t), \tag{10.3}$$

where

$s_n(t)$ is the complex envelope of the signal component

$n_n(t)$ is a complex additive disturbance, such as sensor noise, ambient noise, or another signal interfering with the desired one

Collecting in a vector $\mathbf{z}(t)$ all the responses of the N sensors of the array to a narrowband source coming at the array from the DOA $= \theta_0$, we get the N-dimensional complex vector

$$\mathbf{z}(t) = \mathbf{a}(\theta_0)s(t) + \mathbf{n}(t). \tag{10.4}$$

The vector $\mathbf{a}(\theta_0)$ is referred to as the steering vector for the DOA θ_0.

The elements of the steering vector $\mathbf{a}(\theta_0)$ are given by $a_n(\theta_0) = e^{-j\omega_0\tau_n}$, $n = 1, 2, \ldots, N$. The noise vector $\mathbf{n}(t)$ is an N-dimensional complex vector collecting the N sensor noises $n_n(t)$. In general, it includes components correlated with the desired signal as in multipath propagation environments. With multipath, several replicas of the same signal, each one propagating along a different path, arrive at the array with distinct DOAs.

In beamforming, the goal is to estimate the source signal $s(t)$ given $\mathbf{a}(\theta_0)$. The narrowband beamformer is illustrated in Figure 10.2. The output of the beamformer is

$$y(t) = \mathbf{w}^{\mathrm{H}}\mathbf{z}(t), \tag{10.5}$$

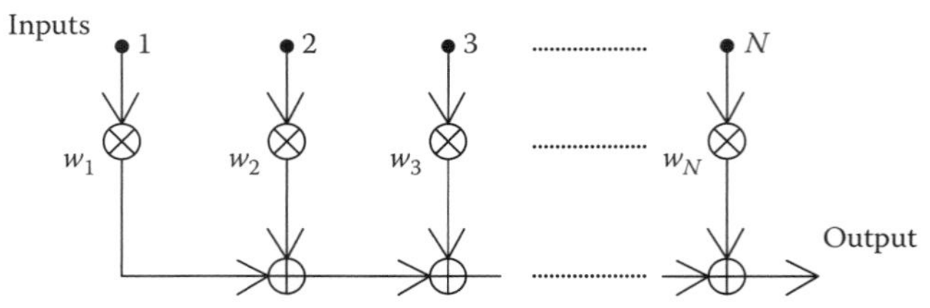

FIGURE 10.2 Narrowband beamformer.

where $\mathbf{w} = [w_1, w_2, w_3, \ldots, w_N]^T$ is a vector of complex weights. We use the notation $\{\cdot\}^T$ to denote vector and matrix transposition, and $\{\cdot\}^H$ for transposition followed by complex conjugation. The beamformer is completely specified by the vector of weights $\mathbf{w}$.

In the absence of the noise term $\mathbf{n}(t)$ in Equation 10.4, it is readily seen that choosing

$$\mathbf{w} = (1/N)\mathbf{a}(\theta_0),$$

the beamformer output is $y(t) = s(t)$. This corresponds to the simplest implementation of the narrowband beamformer, known as the DS beamformer: it combines coherently the signal replicas received at each sensor after compensating for their corresponding relative delays.

The interpretation of the DS beamformer operation is rather intuitive. However, we may ask ourselves the following question: is DS the best we can do to estimate the desired signal when the disturbance $\mathbf{n}(t)$ is present? To answer the question satisfactorily, we begin by noting that, in the presence of noise, the output of the DS beamformer is

$$y(t) = s(t) + (1/N)\mathbf{a}^H(\theta_0)\mathbf{n}(t).$$

The influence of the error term on the estimate $y(t)$ of $s(t)$ depends basically on the structure of $\mathbf{n}(t)$. The optimal design of beamformers depends now on the choice of an adequate optimization criterion that takes into account the disturbance vector, with the goal of improving in some sense the quality of the desired estimate. In the sequel, we will consider several cases of practical interest.

10.2.1 Minimum Output Noise Power Beamforming

To reduce the effect of the error term at the beamformer output, we formulate the beamforming problem as follows:

find the weight vector $\mathbf{w}$ such that the noise output power
$E\{|\mathbf{w}^H\mathbf{n}(t))|^2\}$ is minimized subject to the constraint $\mathbf{w}^H\mathbf{a}(\theta_0) = 1$,

where $E\{\cdot\}$ denotes the statistical average. The cost function is

$$E\{|\mathbf{w}^H\mathbf{n}(t))|^2\} = \mathbf{w}^H\mathbf{R}_n\mathbf{w}, \qquad (10.6)$$

with $\mathbf{R}_n$ the covariance matrix of the disturbance vector $\mathbf{n}(t)$, i.e., $\mathbf{R}_n = E\{\mathbf{n}(t)\mathbf{n}^H(t)\}$. The constraint guarantees that the signal along the look direction θ_0 is not distorted.

The solution to this constrained optimization problem is obtained by Lagrange multipliers techniques. It is given by

$$\mathbf{w} = \left(\mathbf{a}^{\mathrm{H}}(\theta_0)\mathbf{R}_{\mathrm{n}}^{-1}\mathbf{a}(\theta_0)\right)^{-1}\mathbf{R}_{\mathrm{n}}^{-1}\mathbf{a}(\theta_0). \tag{10.7}$$

The vector $\mathbf{w}$ in Equation 10.7 is the gain of the MNP beamformer[20].

When the source signal is uncorrelated with the disturbance,

$$\mathrm{E}\{s(t)\mathbf{n}^{\mathrm{H}}(t)\} = 0,$$

it can be shown that the weight vector (Equation 10.7) of the MNP beamformer takes the form

$$\mathbf{w} = \left(\mathbf{a}^{\mathrm{H}}(\theta_0)\mathbf{R}^{-1}\mathbf{a}(\theta_0)\right)^{-1}\mathbf{R}^{-1}\mathbf{a}(\theta_0), \tag{10.8}$$

where

$$\mathbf{R} = \mathrm{E}\{\mathbf{z}(t)\mathbf{z}^{\mathrm{H}}(t)\}, \tag{10.9}$$

is the covariance matrix of the array data vector $\mathbf{z}$. The vector $\mathbf{w}$ in Equation 10.8 is the gain of the MV beamformer [20]. The MV beamformer minimizes the total output power

$$\mathrm{E}\{|\mathbf{w}^{\mathrm{H}}\mathbf{z}(t))|^2\} = \mathbf{w}^{\mathrm{H}}\mathbf{R}\mathbf{w}$$

subject to $\mathbf{w}^{\mathrm{H}}\mathbf{a}^{\mathrm{H}}(\theta_0) = 1$. The MV beamformer presents an important advantage over the MNP beamformer. While to implement the MNP beamformer we need to know the covariance matrix $\mathbf{R_n}$ of the disturbance vector $\mathbf{n}$, in general, to implement the MV beamformer it is sufficient to estimate the array covariance matrix $\mathbf{R}$ using the available data $\mathbf{z}$.

We discuss how to estimate $\mathbf{R}$. Let T time samples (snapshots) of the array response vector $\mathbf{z}(t)$ be available. An estimate of $\mathbf{R}$ is the data sample covariance matrix $\mathbf{R}_s$:

$$\mathbf{R}_s = \frac{1}{T}\sum_{t=1}^{T}\mathbf{z}(t)\mathbf{z}^{\mathrm{H}}(t). \tag{10.10}$$

Under technical conditions that we will not discuss here, the sample covariance matrix, $\mathbf{R}_s$, converges (in the appropriate sense) to the array covariance matrix $\mathbf{R}$ when T approaches infinity. This means that, for a large enough number T of snapshots, we can replace $\mathbf{R}$ in Equation 10.8 by $\mathbf{R}_s$, without a significant performance degradation.

We provide an alternative interpretation to the MNP beamformer. Using Equation 10.7 in Equation 10.5, and taking into account Equation 10.4, we see that the output of the MNP beamformer has a signal component $s(t)$ and an error term $\mathbf{w}^{\mathrm{H}}\mathbf{n}(t)$ with average power

$$P_{\mathrm{o}} = \mathbf{w}^{\mathrm{H}}\mathrm{E}\{\mathbf{n}(t)\mathbf{n}^{\mathrm{H}}(t)\}\mathbf{w} = \left(\mathbf{a}^{\mathrm{H}}(\theta_0)\mathbf{R}_{\mathrm{n}}^{-1}\mathbf{a}(\theta_0)\right)^{-1}. \tag{10.11}$$

Since the power of the signal is preserved and the MNP beamformer minimizes the power of the noise at its output, the MNP beamformer maximizes the output signal-to-noise ratio (SNR).

We will not discuss in detail the behavior of the MNP and MV beamformers. The reader is referred to the work in [2]. We list some of the properties of the MNP and MV beamformers in two scenarios of practical interest.

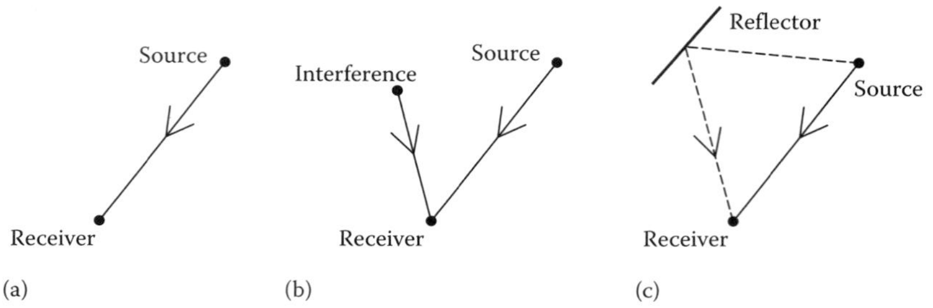

FIGURE 10.3 (a) Single source in white noise; (b) uncorrelated interference; (c) correlated interference.

Case 1: Single Source in White Noise

Here (see Figure 10.3a), we assume that the noise $\mathbf{n}(t)$ is sensor noise. We model it as $\mathbf{n}(t) = \mathbf{u}(t)$ where the components of $\mathbf{u}(t)$ are jointly independent and identically distributed samples of zero mean white noise sequences with variance σ^2, i.e.,

$$\mathbf{R}_n = \mathbf{R}_u = \sigma^2 \mathbf{I},$$

where $\mathbf{I}$ is the identity matrix. The sensor noise models the thermal noise generated at each receiver and is assumed independent of (thus, uncorrelated with) the source signal. If S is the power of the desired signal, the SNR at each sensor is $\text{SNR}_i = S/\sigma^2$. Also, from Equation 10.7, we conclude that when the additive noise is white, as in this case, the MNP beamformer reduces to the DS beamformer.

Moreover, computing the power at the output of the beamformer with Equation 10.11 for this particular situation yields $P_o = \sigma^2/N$. This means that, at the output of the beamformer, the signal-to-noise ratio is $\text{SNR}_o = N\text{SNR}_i$.

We conclude, for the case of a single source in white noise, that the DS beamformer is optimum in the sense of maximizing the output signal-to-noise ratio SNR_o. Further, SNR_o increases linearly with the number N of array sensors.

Case 2: Directional Interferences and White Noise

Now, we assume that the disturbance $\mathbf{n}(t)$ is the superposition of possibly several directional interferences and white noise. Without loss of generality, we consider the case of a single interferer:

$$\mathbf{n}(t) = \mathbf{a}(\theta_i)i(t) + \mathbf{u}(t), \tag{10.12}$$

where

- $i(t)$ is the signal radiated by the interferer
- θ_i is the DOA of the interference signal
- $\mathbf{u}(t)$ is the white noise vector

In general, we assume that $\mathbf{u}(t)$ is uncorrelated with $i(t)$.

Case 2.1: Uncorrelated Arrivals

This is the case where the desired signal and the interference are generated by distinct sources, see Figure 10.3b. It is clear that under this assumption, $s(t)$ and $\mathbf{n}(t)$ are uncorrelated. As we emphasized before, this is the situation where the MNP beamformer (Equation 10.7) is equivalent to the MV beamformer (Equation 10.8).

The covariance of the noise $\mathbf{n}(t)$ is now

$$\mathbf{R}_n = \mathbf{a}(\theta_i)S_i\mathbf{a}^H(\theta_i) + \sigma^2\mathbf{I},$$

where S_i is the average power of the interference $i(t)$. At the DOA $= \theta_i$, the beamformer has an amplitude response

$$\left|\mathbf{w}^H\mathbf{a}(\theta_i)\right| = \frac{|\beta|}{1 + (1 - |\beta|^2)\mathrm{INR}}, \tag{10.13}$$

where

$$\mathrm{INR} = S_i/(\sigma^2/N)$$

is the interference-to-noise ratio (INR), and $\beta = (1/N)\mathbf{a}^H(\theta_i)\mathbf{a}(\theta_0)$ measures the spatial coherence between the desired source and the interference.

Well separated arrivals. When the signal and interference are well separated, their spatial coherence is small, i.e., $|\beta| \ll 1$. In Equation 10.13, the denominator is approximately given by $1 + \mathrm{INR}$. The net effect is that the beamformer output along the interference direction decreases when INR increases. In other words, the MNP and the MV beamformers direct a beam with gain 1 toward the DOA of the desired signal and "null" the interference. The interference canceling property is reflected on the average power of the beamformer output error which is evaluated to

$$P_o = \frac{\sigma^2}{N}\frac{1}{1 - |\beta^2|\frac{\mathrm{INR}}{1+\mathrm{INR}}}. \tag{10.14}$$

For large INR and well-separated DOAs, $P_o \simeq (\sigma^2/N)$. This means that the interference contributes little to the estimation error at the output of the beamformer.

Close arrivals. When the source and the interferer are spatially close, their spatial coherence is large, $|\beta| \simeq 1$, and the output at the interference DOA is $\simeq 1$. This means that the MNP and MV beamformers no longer have the ability to discriminate the two sources.

We conclude from the simple analysis of these two cases that the DOA discrimination capability is strongly related to the "spatial resolution" of the array geometry through the parameter β. In practice, to improve upon the resolution of a linear and uniform array, we increase, when feasible, the number N of sensors. This results in narrower beamwidths which can resolve closer arrivals.

Case 2.2: Correlated Arrivals

This is the case where the interference $i(t)$ is correlated with the desired signal $s(t)$, i.e.,

$$E\{s(t)i^*(t)\} = \rho = |\rho|e^{j\phi_\rho} \neq 0.$$

We denote complex conjugate by $(\cdot)^*$.

With reference to Figure 10.3c, we discuss a simple example where the interference results from a secondary path generated by a reflector (multipath propagation)

$$i(t) = \gamma s(t). \tag{10.15}$$

The complex parameter, γ, accounts for the relative attenuation and delay of the reflected path. The correlation factor, ρ, between $i(t)$ and $s(t)$ is, in this case, given by

$$\rho = \gamma/|\gamma|.$$

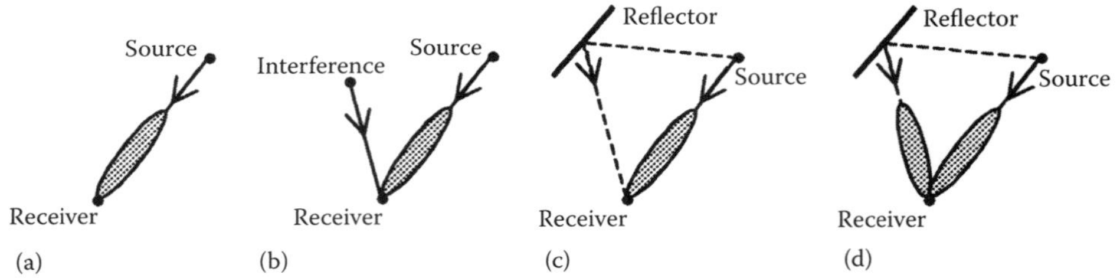

FIGURE 10.4 Artistic representation of alternative solutions to the beamforming problem.

The desired signal and the disturbance vector $\mathbf{n}(t)$ are now correlated and the MV beamformer is no longer equivalent to the MNP beamformer. Recall that the MV beamformer attempts to minimize the total output power under the constraint of a unitary gain at the DOA of the desired source. As the array output vector has a correlated signal component at a different DOA, to minimize the output power may cause the desired signal itself to be strongly attenuated. This is the "signal cancellation" effect, typical of MV beamforming when operating in multipath environments like the one just considered. On the contrary, the behavior of the MNP beamformer is independent of the correlation degree between the desired signal and the disturbance: the MNP beamformer filters out correlated arrivals just as if they were uncorrelated interferences.

To implement the MNP beamformer, besides the DOA of the desired signal, we also need to know the covariance matrix $\mathbf{R}_n$ of the disturbance vector. In general, this covariance is not known *a priori*. It has to be estimated using the available data, and this can be a rather complicated task, not discussed here.

In this section, we discussed the MNP solution to the beamforming problem. The MNP beamformer is optimum in the sense of maximizing the output SNR. When the noise is white, the DS beamformer is recovered as the optimum solution for the single source case. It points a beam toward the source DOA and reduces the sensor noise power by a factor of N, see Figure 10.4a. We also saw that, in the more general situation where the disturbance vector includes directional interferences, the MNP beamformer acts like an interference canceller: it points a beam towards the DOA of the desired signal, while nulling the remaining arrivals regardless of their correlation degrees, see Figure 10.4b and c.

We comment on the adequacy of the MNP solution to the correlated arrivals scenario. By treating the correlated arrivals as an interference, the MNP beamformer neglects the information about the desired signal that may be provided by the reflected path. It is clear that any solution that can combine coherently the information contents of all the correlated arrivals will be more effective in recovering the desired signal from the background noise. This type of solution should behave as a combiner of the outputs of different beams steered toward the DOAs of the correlated replicas of the desired signal, see Figure 10.4d. In the following section, we will see how this solution can be designed using a different optimization criterion when solving the beamforming problem.

10.3 MMSE Beamformer: Correlated Arrivals

In this section, we study a different beamforming technique that, as we will see, is specially suited to multipath propagation environments. The approach is similar to that used in the previous section. The beamformer is still given by Equation 10.5, but we choose a different optimization criterion to solve the beamforming problem. We formulate the problem in the following way:

Find the weight vector $\mathbf{w}$ such that the output error power

$$\mathrm{E}\{|\mathbf{w}^{\mathrm{H}}\mathbf{z}(t) - s(t)|^2\} \text{ is minimized,}$$

i.e., we want to find the weight vector $\mathbf{w}$ that minimizes the mean square error between the beamformer output $\mathbf{y}(t) = \mathbf{w}^H\mathbf{z}(t)$ and the desired signal $s(t)$. The general solution, which we call the Wiener solution, is [2]

$$\mathbf{w} = \mathbf{R}^{-1}\mathbf{r}_{zs}, \tag{10.16}$$

where $\mathbf{r}_{zs} = \mathrm{E}\{\mathbf{z}(t)s^*(t)\}$ is the correlation between the array vector response $\mathbf{z}(t)$ and the desired signal $s(t)$. To understand the behavior of the MMSE beamformer (Equation 10.16), we address the same alternative configurations considered in Case 2 of the previous section.

Directional interference and white noise. In this scenario, the disturbance $\mathbf{n}(t)$ is like in Equation 10.12, so the received array signal is

$$\mathbf{z}(t) = \mathbf{a}(\theta_0)s(t) + \mathbf{a}(\theta_i)i(t) + \mathbf{u}(t),$$

where, as before, $i(t)$ and $\mathbf{u}(t)$ are the interference and the array sensor noise vector, respectively.

Uncorrelated arrivals. The signal $s(t)$ and the interference $i(t)$ are uncorrelated. The correlation between the array vector and the desired signal is

$$\mathbf{r}_{zs} = \mathbf{a}(\theta_0)S,$$

where S is the average power of $s(t)$. The MMSE beamformer weight vector in Equation 10.16 takes the particular form

$$\mathbf{w} = S\mathbf{R}^{-1}\mathbf{a}(\theta_0). \tag{10.17}$$

Comparing Equation 10.17 with Equation 10.8, we conclude that in the present situation, except for a scale factor, the MMSE and the MNP (or MV) beamformers are equivalent. Thus, the MMSE beamformer cancels uncorrelated interferences and directs a beam toward the DOA of the desired signal. However, contrary to what happens with the MNP (or MV) beamformer, the gain of the MMSE beamformer at the look direction is not unity. On the other hand, the MMSE beamformer provides a stronger noise rejection and a smaller output error power than the MNP beamformer. In fact, it can be shown [2] that

$$P_o(\mathrm{MMSE}) \leq \frac{\mathrm{SNR}_o}{1 + \mathrm{SNR}_o} P_o(\mathrm{MNP}), \tag{10.18}$$

where $\mathrm{SNR}_o = S/(\sigma^2/N)$, and $P_o(\mathrm{MNP})$ is the average power of the MNP beamformer output error given by Equation 10.14. Equation 10.18 is particularly significant in low SNR environments.

Correlated arrivals. We take again the interference to be like in Equation 10.15, i.e., $i(t) = \gamma s(t)$. It leads to the correlation

$$\mathbf{r}_{zs} = (\mathbf{a}(\theta_0) + \gamma\mathbf{a}(\theta_i))S,$$

and to the weight vector

$$\mathbf{w} = S\mathbf{R}^{-1}\mathbf{a}(\theta_0) + S\gamma\mathbf{R}^{-1}\mathbf{a}(\theta_i). \tag{10.19}$$

It is clear from Equation 10.19 that the MMSE beamformer directs distinct beams toward the DOAs θ_0 and θ_i of the correlated arrivals in order to combine coherently their respective outputs. If $\mathbf{R}$ has

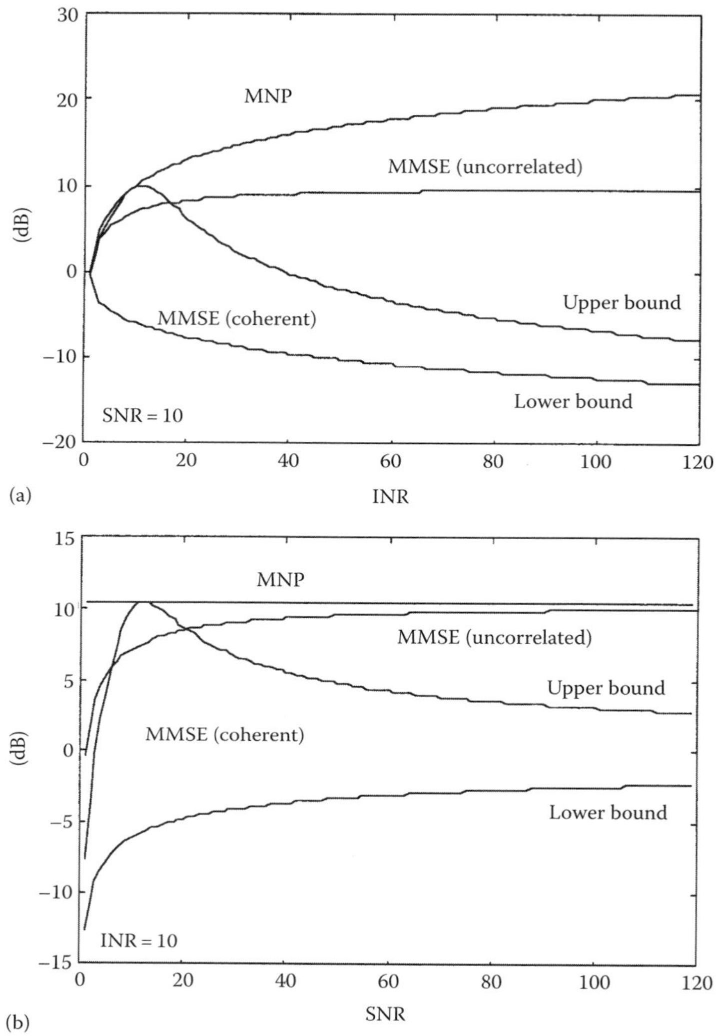

FIGURE 10.5 Output error power. (a) as a function of INR; (b) as a function of SNR. (From Barroso, V.A.N., Moura, J.M.F., and Xavier, J., *IEEE Trans. Signal Process.*, SP-46(3), 737, 1998.)

contributions of other sources uncorrelated with $s(t)$, then these will be filtered out by both beams. This simple example shows how the MMSE beamformer uses the correlated arrivals to improve the output error power.

To have an idea of how the behavior of the MMSE beamformer compares with that of the MNP beamformer, we represent in Figure 10.5 the output error power as a function of the INR and the SNR, for the case of spatially close arrivals. We see from Figure 10.5 that the MMSE beamformer outperforms the MNP beamformer in all scenarios considered. This is particularly apparent in the limiting case of coherent arrivals, when the correlation between the signal and the interference is such that $|\rho| = 1$. The upper and lower bounds of the output error power shown in the figure are determined by $-1 \leq \cos(\phi_\rho) \leq 1$.

To implement the MMSE beamformer in a multipath propagation environment, we need to know the correlation vector $\mathbf{r}_{zs}$ or, alternatively, the DOAs of all correlated replicas of the desired signal as well as their relative attenuations and propagation delays. In practice, $\mathbf{r}_{zs}$ can only be estimated from the data if we have available a reference signal whose correlation with the array vector is similar to that of the desired signal. This is the basic idea underlying adaptive implementations (least mean squares approach); see, e.g., [8,9]. In these implementations, a time sequence, which is known by the receiver, is transmitted by the source and used to adapt the beamforming weights until some error threshold is achieved. In many applications, a reference signal is not available or, as in communications, the transmission of a reference signal may represent a significant waste of the channel capacity. In the next section, we explain how to implement the MMSE beamformer without the need of a reference signal. We do this in the context of a wireless digital communication system.

10.4 MMSE Beamformer for Mobile Communications

In this section, we discuss the implementation of the MMSE beamformer in the context of a digital wireless communication system. The users are mobiles transmitting simultaneously in the same frequency channel. This precludes the use of time/frequency methods to discriminate between the source signals. Here, we assume that the multipath propagation delays are smaller than the baud period (symbol time interval). We develop a multiple output or multichannel MMSE beamformer for this application.

Each mobile is assigned to a specific beamforming processing channel that has the following capabilities:

- Combine the multipath arrivals generated by its assigned mobile.
- Cancel all the arrivals generated by other mobiles.

We illustrate the development of the array receiver in the simple context of Figure 10.6. We consider two mobiles, M_1 and M_2, each one generating a direct path (dm, $m = 1, 2$) and a reflected path (rm, $m = 1, 2$). The propagation channel used by each mobile is characterized by four parameters: two complex numbers, γ_{dm} and γ_{rm}, accounting for the attenuation and the propagation delay in each path, and two DOAs θ_{dm} and θ_{rm}. We model all the parameters that characterize the propagation channels as deterministic unknown parameters. Furthermore, and in contrast with what we assumed in the previous sections, the DOAs of the desired signals are not known a priori. The receiver itself has to track the sources while they are moving. In a mobile communications system, we can safely assume that we know the number of mobiles, their transmitted power, and the sensor noise variance. The number of mobiles is determined by some higher layer of the receiver architecture (e.g., handover procedures to

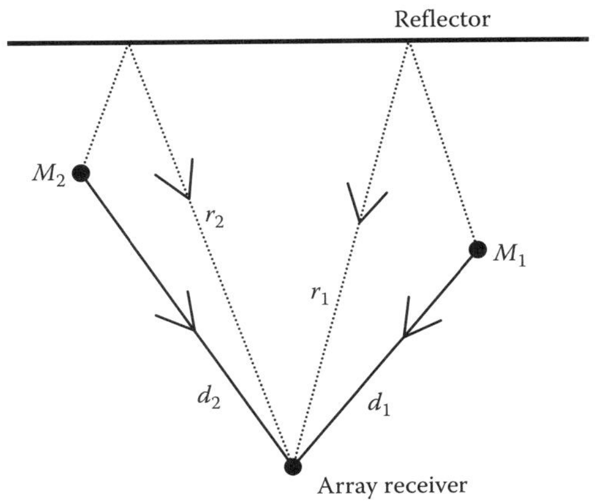

FIGURE 10.6 Example of a mobile communications scenario.

establish the connection between the receiver and each mobile), while the power and the noise variance are system parameters specified by the "link budget."

10.4.1 Model of the Array Output

Let $\mathbf{h}_m$, $m=1,2$, be the single input/multiple output complex transfer function describing, for each source, the channel/receiving array system. Each $\mathbf{h}_m$ is $N \times 1$, where N is the number of array sensors. We write

$$\mathbf{h}_1 = \gamma_{d1}\mathbf{a}(\theta_{d1}) + \gamma_{r1}\mathbf{a}(\theta_{r1}),$$
$$\mathbf{h}_2 = \gamma_{d2}\mathbf{a}(\theta_{d2}) + \gamma_{r2}\mathbf{a}(\theta_{r2}),$$

where $\mathbf{a}(\cdot)$ represents the array steering vector. Let $s_1(t)$ and $s_2(t)$ be the baseband signals transmitted by the mobiles. They are assumed to be independent binary sequences taking the values $+1$ and -1 with equal probability. Thus, the signal vector $\mathbf{s}(t) = [s_1(t), s_2(t)]^{\mathrm{T}}$ has zero mean and covariance matrix $\mathbf{S} = \mathbf{I}$. Defining the $(N \times 2)$ matrix $\mathbf{H} = [\mathbf{h}_1, \mathbf{h}_2]$, the array output vector is

$$\mathbf{z}(t) = \mathbf{H}\mathbf{s}(t) + \mathbf{u}(t), \tag{10.20}$$

where $\mathbf{u}(t)$ represents the complex sensor noise vector. The noise $\mathbf{u}(t)$ will be assumed to be zero mean Gaussian with known covariance matrix $\sigma^2\mathbf{I}$. The goal is to estimate the signal vector $\mathbf{s}(t)$ using a MMSE beamformer which, as we saw in Section 10.3, is the appropriate approach for multipath propagation environments. However, to implement the MMSE beamformer (Equation 10.16), we need to know the correlation between the array output $\mathbf{z}(t)$ and the signal vector $\mathbf{s}(t)$ which in this case is

$$\mathbf{r}_{zs} = \mathrm{E}\{\mathbf{z}(t)\mathbf{s}^{\mathrm{H}}(t)\} = \mathbf{H}.$$

The matrix $\mathbf{H}$ is parameterized by all the unknown θ and γ parameters of our model. Thus, it has to be estimated from the available data. Once we have the estimate of $\mathbf{H}$, we can use it to compute a structured estimate of the array covariance matrix. In this case, where $\mathbf{S} = \mathbf{I}$, this estimate is

$$\mathbf{R} = \mathbf{H}\mathbf{H}^{\mathrm{H}} + \sigma^2\mathbf{I}. \tag{10.21}$$

This estimate is better than the sample covariance matrix computed by Equation 10.10 because it avoids eventual mismatches between the estimate of $\mathbf{H}$ and its actual value embedded in $\mathbf{R}_s$. In the following section, we will describe a method that provides accurate estimates of $\mathbf{H}$.

10.4.2 Maximum Likelihood Estimation of H

Suppose that T is the number of snapshots, i.e., samples, of the array output that are available for estimating the matrix $\mathbf{H}$, while tracking the mobiles. The choice of T should strike a balance between the two following conflicting requirements:

- T should be large enough to guarantee in the appropriate sense a good estimate of $\mathbf{H}$.
- T should be small enough so that we can safely model $\mathbf{H}$ as time invariant.

Fortunately, these two conditions are easily met in the application under study. High data rates usually will be involved in mobile communications. For example, data rates of 2 Mbps or higher are envisaged for the universal mobile telecommunications system (UMTS) and the mobile broadband system (MBS), systems presently under development in Europe. These high data rates, the typical velocities of the mobiles (less than 100 km/h), and the typical mobile/receiver distances ensure that a large amount of data can be collected over a time period during which **H** can be considered to be approximately constant. In other words, the geometry is very slowly varying when compared to the transmission data rates.

Under the assumptions used to establish the model of the array output vector, the "likelihood function" is

$$L(\mathbf{H}) = \prod_{t=1}^{T} \sum_{i=1}^{4} \text{Gauss}_i(\mathbf{H}, \mathbf{z}(t)), \tag{10.22}$$

where the unnormalized Gauss function is

$$\text{Gauss}_i(\mathbf{H}, \mathbf{z}(t)) = \exp\left(- \frac{|\mathbf{z}(t) - \mathbf{H}\mathbf{s}_i|^2}{\sigma^2} \right),$$

and $\{\mathbf{s}_i\}_{i=1}^4$ spans the finite alphabet of the sources, i.e., the four possible realizations of $\mathbf{s}(t)$. In Equation 10.22, we assume that the sampling of the received signals is synchronized with the symbol clock. This is a very strong requirement as time synchronization may be a very difficult problem, specially when the sampling rate equals the data rate. However, the sensitivity of the receiver with respect to timing errors is, in general, efficiently reduced when the sampling rate is made larger than the symbol rate. Although we do not discuss this issue here, the algorithm that we present is easily extended to this case. The value of **H** that maximizes the likelihood function $L(\mathbf{H})$ makes the collected data more likely from a probabilistic point of view. The maximization of Equation 10.22 with respect to **H** is not an easy task. Equation 10.22 is strongly nonlinear in **H** which precludes the analytical solution of the optimization problem. On the other hand, the parameterization of **H** in terms of the channel parameters does not help much because optimization over these parameters requires a multidimensional search approach, which may be excessively time consuming.

10.4.2.1 The EM Algorithm

To maximize Equation 10.22 with respect to **H**, we use an algorithm based on the expectation-maximization (EM) approach [17]. This approach yields an iterative algorithm that, under some technical conditions that are beyond the scope of this paper, is known to converge to the true maximum likelihood estimate. The EM iteration is

$$\mathbf{H}_{l+1} = \left(\sum_{t=1}^{T} \mathbf{z}(t) \frac{\sum_{i=1}^4 \text{Gauss}_i(\mathbf{H}_l, \mathbf{z}(t))\mathbf{s}_i^{\text{H}}}{\sum_{i=1}^4 \text{Gauss}_i(\mathbf{H}_l, \mathbf{z}(t))} \right) \left(\sum_{t=1}^{T} \frac{\sum_{i=1}^4 \text{Gauss}_i(\mathbf{H}_l, \mathbf{z}(t))\mathbf{s}_i\mathbf{s}_i^{\text{H}}}{\sum_{i=1}^4 \text{Gauss}_i(\mathbf{H}_l, \mathbf{z}(t))} \right)^{-1}. \tag{10.23}$$

To run this iterative algorithm, we must be aware of two important aspects:

- In general, this algorithm has a very slow convergence rate.
- Due to strong nonlinearities, it can easily get trapped in a local maximum.

These two aspects reflect a high sensitivity of the EM algorithm with respect to the initial estimate. Special care has to be taken to specify the initial estimate $\mathbf{H}_0$. Next, we describe how to obtain this initial condition, such that the EM iteration converges in just a few iterations to the global maximum of the likelihood function.

10.4.2.1.1 Initialization of the EM Algorithm

In this paragraph, we do not go into the detailed derivation of the initialization procedure. The interested reader is referred to [4]. Here, we describe the basic idea underlying the initialization procedure and present the steps necessary to implement it.

Recall the structure of **H** defined in Section 10.4.1. For the simple example that we have been considering, we write

$$\mathbf{H} = \mathbf{A}(\theta)\Gamma, \tag{10.24}$$

where

$$\mathbf{A}(\theta) = [\mathbf{a}(\theta_{d1}), \mathbf{a}(\theta_{r1}), \mathbf{a}(\theta_{d2}), \mathbf{a}(\theta_{r2})], \tag{10.25}$$

is the $(N \times 4)$ matrix of the steering vectors associated with each incoming path, θ being the vector of the unknown DOAs. The remaining unknown parameters of our model are collected in

$$\Gamma = \begin{bmatrix} \gamma_{d1} & 0 \\ \gamma_{r1} & 0 \\ 0 & \gamma_{d2} \\ 0 & \gamma_{r2} \end{bmatrix}. \tag{10.26}$$

The problem of estimating the initial condition $\mathbf{H}_0$ is decomposed in two steps: (1) estimation of θ and (2) estimation of Γ.

Step (1): Estimation of θ

DOA estimation is a well-studied problem in array processing. The classical approach is based on the MV beamformer [6,12]. The main idea consists in detecting the maxima of the output power, when a set of quantized values of the possible angles of arrival (in the range $[-\pi/2, \pi/2]$ in the case of a linear array) is scanned by varying the look direction specified by the complex weights of the MV beamformer. One of the problems with this approach is that angular resolution is mainly determined by the beamwidth. A narrow beam requires a large number of array sensors. To overcome this limitation, we might use the MUSIC algorithm, a well-known high resolution technique [5,12,13] based on the eigenvalue decomposition of the sample covariance matrix. The angular resolution of the MUSIC algorithm improves as the sample covariance matrix approaches the array covariance matrix. However, these two techniques (MV beamformer and MUSIC) fail when the arrivals are correlated as is the case in mobile communications. The spatial smoothing method [12,14] extends MUSIC to correlated arrivals. The goal of spatial smoothing is to decorrelate the arrivals. It breaks the original array into subarrays. The problem is that the performance of spatial smoothing depends strongly on the correlation degree of the arrivals. To improve the efficiency of spatial smoothing, we have to use a large number of spatial samples, i.e., a larger array.

To circumvent all these difficulties, keeping the array size manageable, and achieving acceptable performance, we use a different approach which is insensitive to the correlation degree of the arrivals. This technique is similar to MUSIC but, rather than relying on the sample covariance matrix (second order statistics), it is based on first order statistics (statistical average) of the array data, see [12] for a detailed derivation of the algorithm. We show how the method is used in the mobile communications problem.

Looking at Equation 10.20, and recalling the assumptions on $\mathbf{s}(t)$ and $\mathbf{u}(t)$, we conclude that $\mathrm{E}\{\mathbf{z}(t)\} = 0$. This seems to contradict our goal of using first order statistics of the array data to estimate the DOAs. However, a more careful analysis of the structure of our model shows that the array samples can be partitioned into a number of sets equal to the cardinality of the source's alphabet (4, in the example that is being used). Each of the sets, $\mathbf{Z}_i$, in the partition has samples of the form

$$\mathbf{z}_i(t) = \mathbf{H}\mathbf{s}_i + \mathbf{u}(t), \quad i = 1, \ldots, 4. \tag{10.27}$$

The statistical average of these samples is

$$\mathbf{x}_i = \mathbf{H}\mathbf{s}_i, \quad i = 1, \ldots, 4. \tag{10.28}$$

This means that the array data is organized in clusters, $\mathbf{Z}_i$, of points centered about each $\mathbf{x}_i$.

To estimate the DOAs, we need to use only one of the $\mathbf{x}_i$'s. Therefore, we are left with finding a scheme to determine one of the sets in the partition of the array data. In a statistical sense, the distance between any two samples of the array data is minimum if the two samples belong to the same partition. In other words, for any pair of time instants, t_l and t_k, $\mathbf{z}(t_l)$ and $\mathbf{z}(t_k)$ belong to the same partition $\mathbf{Z}_i$ if the source signal vectors $\mathbf{s}(t_l)$ and $\mathbf{s}(t_k)$ have the same realization $\mathbf{s}_i$. In this case,

$$\Delta = \mathrm{E}\{|\mathbf{z}(t_l) - \mathbf{z}(t_k)|^2\} = 2N\sigma^2. \tag{10.29}$$

To find one set $\mathbf{Z}_i$ of the partition, after choosing randomly one array sample $\mathbf{z}(t_l)$, we look for all other samples whose distance to $\mathbf{z}(t_l)$ is below some threshold appropriately related with Δ. Once we obtain the set $\mathbf{Z}_i$ in the partition, we estimate the statistical average $\mathbf{x}_i$ of the points in the set by its sample mean.

At the end of this step, we have calculated the first order statistics and are ready to apply the first order statistics algorithm to estimate the DOAs θ [4]. Having the estimate $\hat{\theta}$ of the DOAs, we estimate the matrix of the steering vectors $\mathbf{A}$ in Equation 10.25 by $\hat{\mathbf{A}} = \mathbf{A}(\hat{\theta})$.

Step (2): Estimation of Γ

Recall from Equation 10.24 that $\mathbf{H} = \mathbf{A}(\theta)\Gamma$. Step () estimated $\mathbf{A}$ as $\hat{\mathbf{A}} = \mathbf{A}(\hat{\theta})$. We now proceed with the estimation of Γ. To estimate Γ, we use the estimate $\hat{\mathbf{A}} = \mathbf{A}(\hat{\theta})$ just obtained, as well as the cluster center $\mathbf{x}_i$ on which that estimate was based.

Denote the left pseudoinverse of a matrix $\mathbf{A}$ by $\mathbf{A}^{\dagger}$. It is defined by

$$\mathbf{A}^{\dagger} = (\mathbf{A}^{\mathrm{H}}\mathbf{A})^{-1}\mathbf{A}^{\mathrm{H}}.$$

Recall that the matrix of steering vectors $\mathbf{A}$ defined in Equation 10.25 is a rank 4 $(N \times 4)$ matrix.

We now motivate how to estimate Γ. Define the matrix

$$\mathbf{Q} = \hat{\mathbf{A}}^{\dagger}\,(\mathbf{R} - \sigma^2\mathbf{I})\,\hat{\mathbf{A}}^{\mathrm{H}^{\dagger}}. \tag{10.30}$$

From Equation 10.21, we get

$$\mathbf{R} - \sigma^2\mathbf{I} = \mathbf{H}\mathbf{H}^{\mathrm{H}}.$$

Substituting this equation in 10.30, we get

$$\mathbf{Q} = \hat{\mathbf{A}}^{\dagger}\,\mathbf{H}\mathbf{H}^{\mathrm{H}}\,\hat{\mathbf{A}}^{\mathrm{H}^{\dagger}}.$$

But by Equation 10.24

$$\mathbf{Q} = \hat{\mathbf{A}}^{\dagger}\,\mathbf{A}\Gamma\Gamma^{\mathrm{H}}\mathbf{A}^{\mathrm{H}}\,\hat{\mathbf{A}}^{\mathrm{H}^{\dagger}}. \tag{10.31}$$

Clearly, if $\hat{\mathbf{A}} = \mathbf{A}$, then

$$\mathbf{Q} = \Gamma\Gamma^{\mathrm{H}}. \tag{10.32}$$

Apparently, to estimate Γ, we should compute $\mathbf{Q}$ and then find a factorization of the form (Equation 10.32). This is not so easy because, in general, a unique factorization of a Hermitian matrix like $\mathbf{Q}$ does not exist. Nevertheless, $\mathbf{Q}$ plays a key role in the procedure for estimating Γ. This is essentially based on the singular value decomposition of $\mathbf{Q}$, as we see now. Notice that $\mathbf{Q}$ can be computed from the data using the data sample covariance matrix $\mathbf{R}_s$ given by Equation 10.10 instead of the array data covariance matrix $\mathbf{R}$.

Recalling from Equation 10.26 the structure of Γ, we conclude that $\mathbf{Q}$ should be a (4×4) Hermitian matrix of rank 2. This means that $\mathbf{Q}$ should only have two nonzero eigenvalues (in fact, because $\mathbf{Q}$ is Hermitian, these eigenvalues of $\mathbf{Q}$ are real and positive). The singular value factorization of Q is

$$\mathbf{Q} = \mathbf{V}\Lambda\mathbf{V}^{\mathrm{H}}, \tag{10.33}$$

where Λ is a diagonal matrix whose elements are the nonzero singular values, in this case the eigenvalues of $\mathbf{Q}$, and $\mathbf{V}$ is a matrix whose columns are the orthonormal eigenvectors associated to these eigenvalues.

From Equation 10.26 defining Γ, we see that Γ has itself orthogonal columns. In general, the columns of Γ have different norms. Using the eigenvalue decomposition in Equation 10.33, we conclude that, except for a phase difference, the columns of $\mathbf{V}\Lambda^{1/2}$ equal the columns of Γ. Thus, we can conclude that

$$\mathbf{Q}^{1/2} = \mathbf{V}\Lambda^{1/2} \tag{10.34}$$

$$= \Gamma\mathbf{Diag}(\mathrm{e}^{j\phi_m}), \tag{10.35}$$

where $\mathbf{Diag}(\cdot)$ is a diagonal matrix and the ϕ_m, $m = 1, 2$, measure the phase uncertainty of the columns of $\mathbf{Q}^{1/2}$ with respect to the columns of Γ. The last equality in Equation 10.35 is true only when $\hat{\mathbf{A}} = \mathbf{A}$.

We now consider how to resolve this phase uncertainty. We show that this uncertainty can be resolved up to a sign difference. This means that, at least theoretically, we can achieve an estimate $\hat{\Gamma}$ whose columns are collinear with those of Γ. Using Equation 10.34 and Equation 10.28, it can be shown after some algebra that

$$\mathbf{f} = \mathbf{Q}^{1/2^{\dagger}}\hat{\mathbf{A}}^{\dagger}\mathbf{x}_i\big(= \mathbf{Diag}(\mathrm{e}^{-j\phi_m})\mathbf{s}_i\big). \tag{10.36}$$

The elements of the vector $\mathbf{f}$ have the form $\mathrm{e}^{-j\phi_m}s_{im}$, $m = 1, 2$. Defining $\mathbf{F}$ as the diagonal matrix formed with the elements of $\mathbf{f}$, we define the estimate $\hat{\Gamma}$ of Γ as

$$\hat{\Gamma} = \mathbf{Q}^{1/2}\mathbf{F}(= \Gamma\mathbf{Diag}(\mathbf{s}_i)). \tag{10.37}$$

Recalling that the elements of $\mathbf{s}_i$ are $+1$ or -1, we see that, ideally, the columns of Γ are estimated just with a sign uncertainty. In our application, this sign uncertainty is not relevant. To see this, suppose that one of the columns of $\hat{\Gamma}$ has its sign changed. This sign change is equivalent to an inversion of the data stream generated by one of the mobiles. This inversion of the data stream can be solved using a differential encoding scheme of the transmitted data.

In summary, the algorithm to estimate Γ is as follows:

- Use $\mathbf{R}_s$ to compute $\mathbf{Q}$ as defined by Equation 10.30
- Compute the singular value decomposition of $\mathbf{Q}$ and then form the matrix $\mathbf{Q}^{1/2}$ as given by Equation 10.34
- Compute $\mathbf{f}$ as given by Equation 10.36 and use its elements to form the diagonal matrix $\mathbf{F}$
- Compute $\hat{\Gamma}$ defined in Equation 10.37

Using the results of Steps **(1)** and **(2)**, we compute the estimate used to initialize the EM iteration as:

$$\mathbf{H}_0 = \mathbf{A}(\hat{\theta})\hat{\Gamma}.$$

In practice, it will be shown in the following section by computer simulations that this initial condition enables the EM algorithm to converge to the global maximum of the likelihood function with a small number of iterations. This behavior is maintained even with critical scenarios arising under very specific geometries, see Section 10.5.

The apparent drawback of the proposed initialization procedure is its computational complexity. However, this is not really a problem because we are processing blocks of data with time length T during which the geometry remains, for all practical purposes, unchanged. Also, the changes in geometry between adjacent blocks are small. This means that, except in specific situations, the EM iteration can be initialized using the estimate of $\mathbf{H}$ obtained with the previous block of data. Hence, the algorithm is seldom reinitialized, e.g., when the number of sources present in the cell is changed.

10.5 Experiments

In this section, we present the results obtained with the array receiver designed in the previous section. The context is a cell in a cellular radio mobile communications system. The system is operating at a carrier frequency $f_0 = 1$ GHz and accommodates transmission rates of 1 Mbps with differential encoding binary phase shift keying modulation [7]. All the mobiles present in the cell generate signals that are transmitted simultaneously in the same frequency band. The power emitted by each mobile is assumed to be unity.

The receiving array is a horizontal linear, uniform array, with $N = 19$ identical omnidirectional captors. These captors are separated by half wavelength, i.e., $d = \lambda_0/2$, where $\lambda_0 = c/f_0 = 30$ cm is the center frequency wavelength, yielding an array length of 2.7 m. The geometry of the scenario is similar to that depicted in Figure 10.6, where the reflector is parallel to the array. The reflector is located at a distance of 50 m from the array. The reflector absorbs half of the impinging power. The two mobiles M_1 and M_2 describe circular trajectories around the receiver, from right to left (counterclockwise). Their velocities are 18 km/h for M_1 and 19 km/h for M_2. These are typical speeds for city traffic with slow automobiles circling a rotary. The radii of these trajectories are 30 m and 45 m, respectively. Comparing these distances with the array length, we conclude that we can safely make the planar wave front assumption; in other words, the received signals across the array are planar wave fronts. Finally, the receiver SNR equals 10 dB for the case of the direct path generated by M_1. This means that all the remaining paths have smaller SNRs.

In Figure 10.7, we illustrate examples of the geometries for which we ran the multichannel MMSE beamformer. The data synthesized by the simulator was collected during each second. The data blocks have length 1 Mbit per mobile. The matrix $\mathbf{H}$ is estimated using 3 Kbit (per mobile). In practically all scenarios that we tested, we observed that the EM algorithm converges in just two iterations. The multichannel MMSE beamforming achieved 100% of correct symbol detections for the outputs of both channels. The initialization procedure described in Section 10.4 was necessary only for the first data block (beginning at time $t = 0$). Except for the first data block, the EM algorithm was initialized with

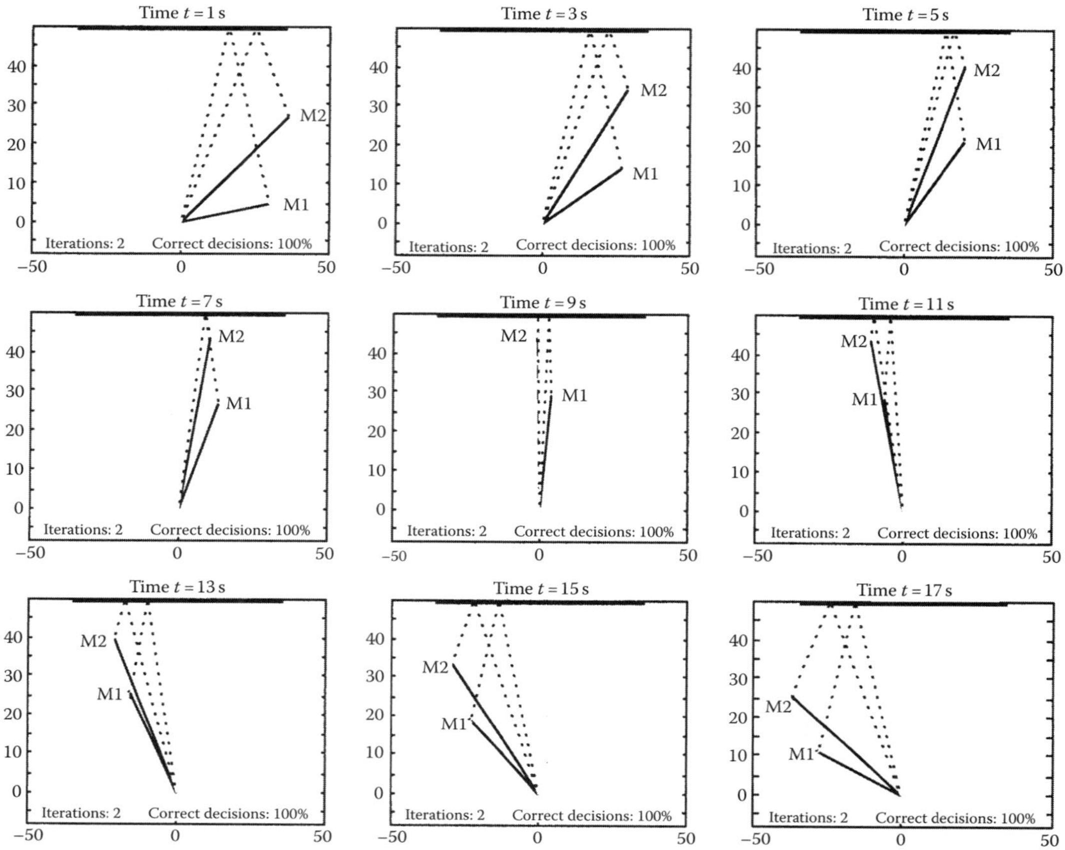

FIGURE 10.7 Simulation scenarios.

the estimate of **H** obtained with the last adjacent data block. Notice that even for scenarios $t = 11$ and $t = 13$, where the DOAs of the direct paths are very close to each other, the EM algorithm was able to keep tracking of the columns of **H**. In fact, each column of **H**, representing the channel/receiving array transfer function, captures the global structure of the multipath propagation channel used by each mobile, which does not depend explicitly on the DOAs. For illustrative purposes, we show in Figure 10.8 the amplitude response of the MMSE and the MNP beamformers as a function of the angle of arrival (beampatterns). The MMSE beampatterns were computed using the weights obtained from the synthesized data, while in the case of the MNP we used the ideal weights. We see that the behavior of both beamformers confirm what was predicted by the simple analysis carried out in Sections 10.2 and 10.3. The MMSE beamformer directs distinct beams toward the arrivals generated by the mobile being tracked and nulls all the remaining arrivals. On the contrary, the MNP beamformer cancels all the incoming wave fronts except the one corresponding to the direct path. The simulations also confirmed the improvement in the output SNR achieved by the MMSE beamformer when compared with the MNP beamformer. In our experiments, a gain close to 3 dB was obtained in almost all scenarios considered.

In this section, we showed the adequacy of the multichannel MMSE beamformer for combating multipath propagation effects as occurring in mobile communications systems. The simulations illustrated the efficiency of the implementation proposed in Section 10.4.

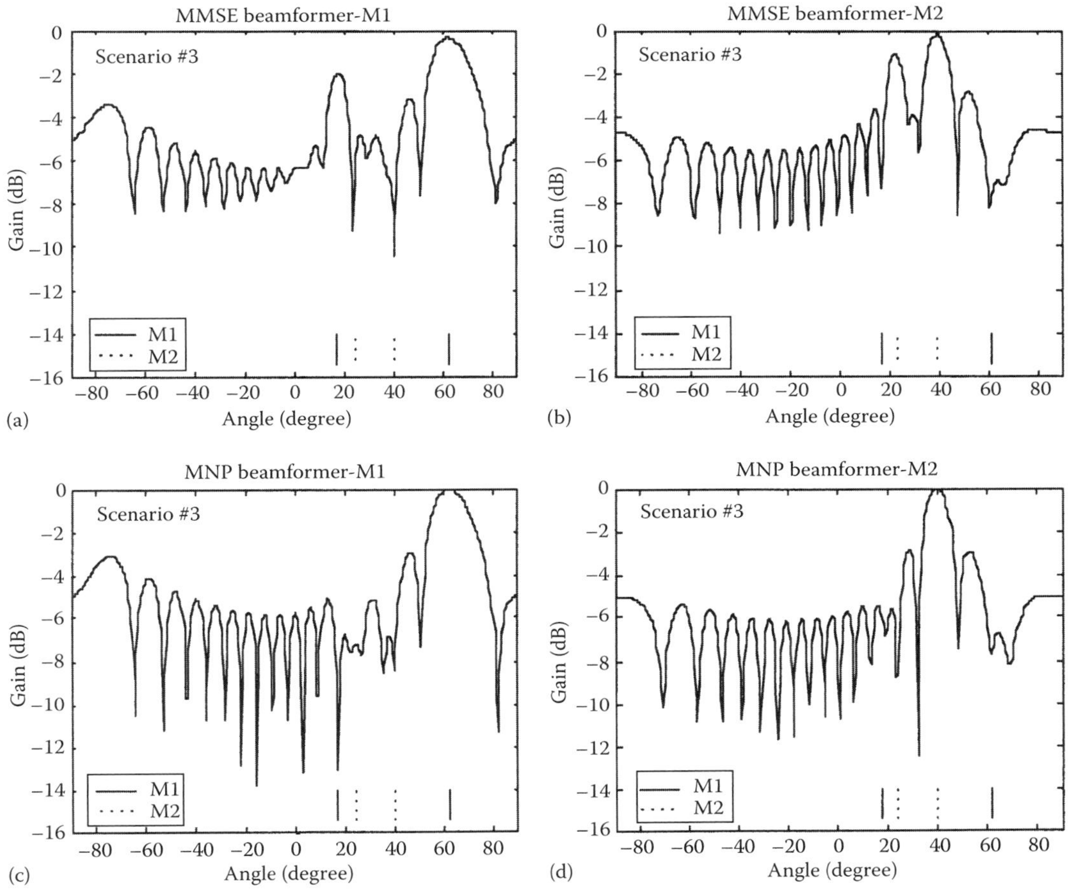

FIGURE 10.8 MMSE and MNP beampatterns: (a, c) channel M_1; (b, d) channel M_2.

10.6 Conclusions

In this article, we addressed the problem of designing beamforming arrays specially suited to operate in multipath environments. The problem is motivated in the context of cellular mobile communication systems. This is a field where array processing is being recognized as a powerful technology that has the potential to improve significantly the traffic capacity of such systems. As we saw, beamforming is an efficient technique for handling several mobiles in the same cell transmitting simultaneously in the same frequency channel. Moreover, specific beamformers can be designed to optimize the receiver with respect to distortions introduced by channels, such as multipath propagation and interferences.

We proposed a solution based on the MMSE beamformer. The study presented in Section 10.3 emphasized the relevant properties of this method. In particular, it was shown that the MMSE beamformer has an important advantage over the MNP introduced in Section 10.2: it combines coherently the correlated signal replicas generated by one source, while canceling the remaining interferences. Recall that the MNP beamformer nulls the secondary multipath arrivals as if they were uncorrelated interferences. Thus, the output SNR of the MMSE beamformer is always better than that of the MNP beamformer, particularly in multipath environments.

We identified the difficulties of implementing the MMSE beamformer. In particular, we pointed out that, except for the case of uncorrelated arrivals, knowledge of the DOA of the desired signal is not sufficient to implement the MMSE beamformer. We showed that, in most cases of interest, we need to know the cross-correlation between the desired signal and the array observations. In general, this cross-correlation is not previously known. It is necessary to estimate it from the available data. In other words, it is necessary to perform a blind estimation of the propagation channel. In Section 10.4 we presented a new method for estimating that correlation. It can be applied in situations where the array data have nonzero first order statistics. The method is insensitive to the correlation degree of the incoming wave fronts. This is the case that is relevant in mobile communications. Basically, the method that we described exploits the finite mixture distribution of the array data, which is organized in clusters with nonzero averages. We used a maximum likelihood approach based on the finite mixture model, the global maximum of the likelihood function being obtained by an EM iterative algorithm. As discussed, initialization of the EM algorithm is a critical issue. For a poor initial guess, the algorithm can diverge or, at most, converge to a local maximum. We provided a technique to initialize the EM algorithm. This initialization method enables the EM algorithm to converge fast to the global maximum of the likelihood function.

We tested the multichannel MMSE beamforming receiver using a simulator for a mobile communications system. We generated a simple scenario with moving sources and multiple replicas. Our results illustrated the efficiency of the receiver. In particular, the results confirm that, under the conditions studied which are typical in cellular communications, the reliability of the system is high. This is a consequence of the way we designed the MMSE beamformer which minimizes the power of the error at the output of the receiver. The MMSE beamformer relies on the accuracy of the blind estimation of the radio propagation channel. The algorithm we described can do a good job of estimating the propagating channel as long as the receiver has the ability to properly track the mobiles. In normal operation, the EM algorithm is initialized using the channel estimates obtained with the previous data block. The receiver is globally reinitialized only in very specific situations, for instance when the number of mobiles is changed. This fact simplifies the computational effort associated with the receiver. Our studies showed that the receiver converges so fast that it operates practically in real time.

Acknowledgments

The authors acknowledge discussions with João Xavier and his help with the development of the simulations software.

References

1. Barroso, V.A.N., Moura, J.M.F., and Xavier, J., Blind array channel division multiple access (AChDMA) for mobile communications, *IEEE Trans. Signal Process.*, see also *IEEE Trans. Acoust., Speech Signal Process.*, 46(3), 737–752, March 1998.
2. Barroso, V.A.N. and Moura, J.M.F., l_2 and l_1 Beamformers: Recursive implementation and performance analysis, *IEEE Trans. Signal Process.*, SP-42(6), 1323–1334, June 1994.
3. Barroso, V.A.N., Rendas, M.J., and Gomes, J., Impact of array processing techniques on the design of mobile communication systems, *Proc. MELECON'94*, Turkey, April 1994.
4. Barroso, V.A.N. and Xavier, J., Blind estimation of multipath channels for the design of array receivers in mobile communications, *Instituto de Sistemas e Robótica*, Internal Report, 1995.
5. Bienvenu, G. and Kopp, L., Adaptivity to background noise spatial coherence for high resolution passive methods, *Proc. IEEE ICASSP'80*, Denver, CO, pp. 307–310, 1980.
6. Capon, J., High resolution frequency-wavenumber spectrum analysis, *Proc. IEEE*, 57, 1408–1418, August 1969.

7. Carlson, A.B., *Communication Systems: An Introduction to Signals and Noise in Electrical Communication*, 3rd edn., McGraw-Hill, New York, 1986.
8. Compton, R.T., Jr., *Adaptive Antennas: Concepts and Performance*, Prentice-Hall, Englewood Cliffs, NJ, 1988.
9. Haykin, S., *Adaptive Filter Theory*, 2nd edn., Prentice-Hall, Englewood Cliffs, NJ, 1991.
10. Lee, W.C.Y., *Mobile Cellular Telecommunications Systems*, McGraw-Hill, New York, 1989.
11. Paulraj, A. and Papadias, C.B., Array processing for mobile communications, *Wireless, Networking, Radar, Sensor Array Processing, and Nonlinear Signal Processing*, CRC Press, Boca Raton, FL, 1997.
12. Pillai, S.U., *Array Signal Processing*, Springer-Verlag, New York, 1989.
13. Schmidt, R.O., Multiple emitter location and signal parameter estimation, *Proc. RADC Spectral Estimation Workshop*, Sunnyvale, CA, pp. 243–258, October 1979.
14. Shan, T.J., Wax, M., and Kailath, T., On spatial smoothing for estimation of coherent signals, *IEEE Trans. Acoust. Speech Signal Process.*, ASSP-33, 806–811, August 1985.
15. Steele, R., *Mobile Radio Communications*, IEEE Press-Pentech Press, London, 1992.
16. Swales, S., Beach, M., Edwards, D., and McGreehan, J., The performance of enhancement of multibeam adaptive base station antennas for cellular mobile radio, *IEEE Trans. Veh. Tech.*, 39(1), 56–67, February 31, 1990.
17. Titterington, D.M., Smith, A.F.M., and Makov, U.E., *Statistical Analysis of Finite Mixture Distributions*, John Wiley & Sons, New York, 1985.
18. Walker, J., *Mobile Information Systems*, Artech House, Norwood, MA, 1990.
19. Van Veen, B.D. and Buckley, K.M., Beamforming: A versatile approach to spatial filtering, *IEEE ASSP Mag.*, 5, 4–24, April 1988.
20. Van Veen, B. and Buckley, K.M., Beamforming techniques for spatial filtering, *IEEE Acoust., Speech, Signal Process. Mag.*, 5(2), 4–24, April 1988.

11

Peak-to-Average Power Ratio Reduction

11.1 Introduction 11-1
11.2 PAR 11-3
Baseband versus Passband PAR • Discrete-Time • IAR
11.3 Nonlinear Peak-Limited Channels 11-6
11.4 Digital Predistortion 11-8
11.5 Backoff 11-10
Time-Invariant Scaling • Time-Varying Scaling
11.6 PAR Reduction 11-14
Transparent PAR Reduction • Receiver-Side Distortion Mitigation • Receiver-Dependent PAR Reduction
11.7 Summary 11-25
References 11-25

Robert J. Baxley
Georgia Tech Research Institute

G. Tong Zhou
Georgia Institute of Technology

11.1 Introduction

Peak-to-average power ratio (PAR) is the most commonly used metric to assess the dynamic range of symbol (also called block)-based signals. The objectives of PAR reduction are

1. To decrease the amount of power required to transmit a signal
2. To reduce the signal's sensitivity to the system nonlinearities
3. To reduce the physical size requirements of the analog components thanks to the smaller dynamic range

A similar metric known as instantaneous-to-average power ratio (IAR) is also frequently used and is preferred for signals that do not have distinct separable symbols. PAR and IAR are important proxy metrics for power efficiency because they dictate the ratio between the peak power, which the transmission power amplifier (PA) must accommodate, and the average transmit power. This peak PA power has a one-to-one relationship with the amount of power consumed by the PA. Thus, it is very important to limit the PAR or IAR of a signal by implementing clever methods, which have the effect of increasing the power efficiency.

As a quick example, denote $P_{\mathrm{m}} = \max_t |x(t)|^2$ the maximum input power, and $P_{\mathrm{i}} = E[|x(t)|^2]$ the average power of the input signal $x(t)$. In the linear scale, the PAR is defined as $P_{\mathrm{m}}/P_{\mathrm{i}}$. In the dB scale, $\mathrm{PAR[dB]} = P_{\mathrm{m}}[\mathrm{dB}] - P_{\mathrm{i}}[\mathrm{dB}]$.

In order to discuss, in particular, the power efficiency, let us assume that the PA is an ideal linear device, i.e., the input signal can be amplified undistorted until the maximum output power of the PA, P_{sat}, is reached. If nonlinear distortion is to be avoided, the power efficiency is the highest when the input signal peak is positioned at the maximum range of the PA. In Figure 11.1, an ideal linear PA output power versus input power characteristic is shown, along with strategies for amplifying two different input signals

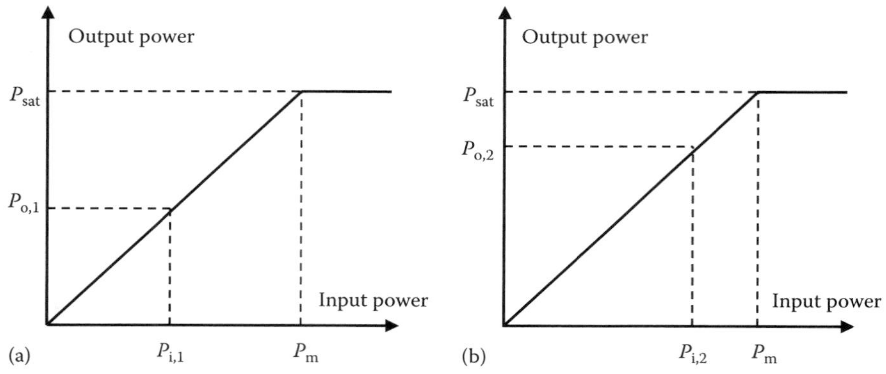

FIGURE 11.1 An ideal linear PA for two input signals with different PAR values. (a) Input-to-output characteristic for high-PAR signal, (b) input-to-output characteristic for same saturation power when the PAR is reduced.

with different PAR values. The power efficiency is proportional to the average output power, which in turn is proportional to the average input power P_i. From Figure 11.1, it is clear that with a low-PAR signal (shown on the right), a higher average power P_i can be achieved, thus yielding a higher power efficiency.

To illustrate the benefits of a power-efficient operation, consider that the average 3G cellular base station consumes 3 kW of power and costs approximately $2600/year at $0.10/kWH. A typical operator may have as many as 20,000 base stations in their network for a total of 60 MW power consumption per year at a cost of $52 million [1]. Just a 10% improvement in power efficiency can lead to millions of dollars in monetary savings per operator. When the power efficiency is improved, less heat is generated as waste, and thus there is a secondary benefit of less air conditioning cost which in turn saves energy. The cost of low power efficiency is not only monetary, as the environmental impact of the worldwide base station power consumption has been estimated at 1.6 million tons of CO_2 per year [1]. Effective PAR reduction can more than double the power efficiency and is an important aspect of worldwide green push to lower energy consumption.

Depending on the signal fidelity figure of merit, several PAR reduction methods are possible. For scenarios where throughput is not important, it is preferable to create a waveform or a set of waveforms that have limited PAR. If the set size is small enough, as it is for radar applications among others, this can be a very effective method for reducing PAR.

For communications scenarios, customized waveforms are not as feasible. As a quick example, consider a signal that is generated from a length-N vector drawn from a quadrature phase shift keying (QPSK) alphabet. For such a signal, 4^N distinct waveforms are possible; with the specific waveform dictated by which bits are used to modulate the signal vector. For even moderately large values of N, it becomes problematic to find a set of 4^N distinct waveforms with low PAR and store them in memory. Accordingly, PAR reduction for communications scenarios tends to be algorithmic and involve online processing.

Another dimension of processing possible for communications involves algorithms that depend on receiver cooperation. For these PAR reduction algorithms, a reversible or partially reversible signal transformation is used by the transmitter to increase the transmitter power efficiency. The receiver, which has prior knowledge of the transformation or set of transformations, uses this information and possibly side information sent by the transmitter, to recover the information bits.

The objective of this chapter is to introduce the fundamentals of PAR reduction. Before the PAR reduction algorithms are discussed, it is important to set the context for PAR reduction in communications system. In particular, how the signal is scaled before transmission has an important effect on the ultimate power efficiency. Without a properly optimized scaling system, the PAR reduction realized may not be harnessed to achieve power-efficiency improvement.

PAR reduction algorithms can be broken into two distinct groups: those that require an informed receiver and those that do not require any receiver modification (i.e., are transparent to the receiver).

In discussing these two groups, we outline the overarching optimization framework that can describe many existing reduction methods. With this framework, we hope to give the reader the underlying idea that characterizes the state of the art in PAR reduction.

11.2 PAR

Ultimately, when characterizing the dynamic range of a signal, the passband signal transmitted through the PA, $x^{(\text{pb})}(t) = \Re\{x^{(\text{bb})}(t)e^{j2\pi f_c t}\}$, is of most interest, where $x^{(\text{bb})}(t)$ denotes the baseband equivalent signal and f_c is the center frequency. However, it is often convenient to work with $x^{(\text{bb})}(t)$, denoted by $x(t)$ from now on for simplicity, which has total two-sided bandwidth, B. At times it is convenient to work with the baseband discrete-time signal, x_n. With this in mind, the PAR of a continuous-time signal is defined as

$$\text{PAR}\{x(t)\} \triangleq \frac{\max_{t\in\mathcal{S}} |x(t)|^2}{\bar{E}[|x(t)|^2]}, \tag{11.1}$$

where

$\mathcal{S}$ is the time period of interest

$\bar{E}[x(t)]$ of a (possibly nonstationary) random process is defined as the time-averaged expectation

$$\bar{E}[x(t)] \triangleq \lim_{T\to\infty} E\left[\frac{1}{T}\int_0^T x(t)\text{d}t\right]. \tag{11.2}$$

11.2.1 Baseband versus Passband PAR

Fortunately, the passband PAR, $\text{PAR}\{x^{(\text{pb})}(t)\}$, is basically twice the baseband PAR, $\text{PAR}\{x(t)\}$, when the passband carrier frequency is sufficiently higher than the signal bandwidth, $f_c \gg B$. This condition holds in practical situations. To show this, note that the peak of the passband signal is upper-bounded by the baseband peaks,

$$\max_{t\in\mathcal{S}} |x(t)|^2 \geq \max_{t\in\mathcal{S}} |x^{(\text{pb})}(t)|^2, \tag{11.3}$$

and that the bound is tight as $f_c \to \infty$, so that

$$\max_{t\in\mathcal{S}} |x(t)|^2 \approx \max_{t\in\mathcal{S}} |x^{(\text{pb})}(t)|^2. \tag{11.4}$$

Therefore, the numerator in the PAR expression in Equation 11.1 for both passband and baseband signals is approximately the same. By expanding the denominator of the passband signal and utilizing Lemma 3.1 of Hasan [2]

$$\bar{E}[|x^{(\text{pb})}(t)|^2] = \bar{E}[|\Re\{x(t)e^{j2\pi f_c t}\}|^2] \tag{11.5}$$

$$= \bar{E}\big[|\Re\{x(t)\}|^2 \cos^2(2\pi f_c t) + |\Im\{x(t)\}|^2 \sin^2(2\pi f_c t) - 2\Re\{x(t)\}\Im\{x(t)\}\cos(2\pi f_c t)\sin(2\pi f_c t)\big] \tag{11.6}$$

$$= \bar{E}\big[\tfrac{1}{2}|\Re\{x(t)\}|^2 + \tfrac{1}{2}|\Im\{x(t)\}|^2\big] \tag{11.7}$$

$$= \tfrac{1}{2}\bar{E}[|x(t)|^2]. \tag{11.8}$$

Therefore, the relationship between the baseband and passband PAR obeys

$$\mathrm{PAR}\{x^{(\mathrm{pb})}(t)\} \approx 2\mathrm{PAR}\{x(t)\}. \tag{11.9}$$

meaning that any baseband PAR changes will result in a proportional change in the passband PAR with a proportionality constant of 2. This is very convenient because it means that PAR reduction algorithms can be analyzed and simulated on baseband signals.

11.2.2 Discrete-Time

From Nyquist theory, it is possible to completely characterize a band-limited signal with a finite number of samples from the continuous-time signal, $x_n \triangleq x(t)|_{t=n/B}$ (Nyquist rate sampling). Discrete-time signals over a period $\mathcal{T}$ have a finite number of samples and are frequently expressed as a vector

$$\mathbf{x} \triangleq [x_1, x_2, \ldots, x_N]^{\mathrm{T}}. \tag{11.10}$$

With this notation, the PAR of a discrete-time symbol can be compactly expressed as

$$\mathrm{PAR}\{\mathbf{x}\} \triangleq \frac{N\|\mathbf{x}\|_\infty^2}{E\left[\|\mathbf{x}\|_2^2\right]}, \tag{11.11}$$

where

$\|\mathbf{x}\|_2^2 = \sum_n |x_n|^2$ is the L_2-norm

$\|\mathbf{x}\|_\infty = \max_n |x_n|$ is the L_∞-norm of the vector $\mathbf{x}$

Because the discrete-time vector $\mathbf{x}$ includes only a subset of the values contained in $x(t)$, it is clear that $\mathrm{PAR}\{x(t)\} \geq \mathrm{PAR}\{\mathbf{x}\}$. In fact, the disparity between the two can be significant. To better estimate the continuous-time PAR, oversampled discrete-time signals, $x_{n/L}$, can be used to collect more points from the continuous-time signal.

The complementary cumulative distribution function (CCDF) of PAR is plotted in Figure 11.2 for various signal types. It is clear from the plot that the PAR is extremely dependent on the signal type, with complex Gaussian signaling having a higher PAR with a higher probability than quadrature amplitude modulation (QAM), for instance.

11.2.3 IAR

One disadvantage of using PAR as a metric is that it is only sensitive to the peak-most value in each symbol. Thus, it only conveys the amount of backoff necessary to avoid clipping at all. In some systems, a certain amount of clipping may be necessary and the PAR may not be a good metric to determine the impact of clipping a particular symbol. For instance, in Figure 11.3, two symbols are plotted that have the same PAR. Despite having the same PAR, symbol 2 has more high-peak samples. In fact, for a clipping level of 5, symbol 2 will experience distortion on four samples while symbol 1 will only experience distortion on one sample. For this reason, some have proposed different metrics other than PAR to assess the peaky nature of signal.

Like PAR, the IAR is also a random variable whose probabilistic distribution reflects the dynamic range of a signal. In fact, the similarities have led some authors to refer to IAR as PAR. However, unlike PAR, IAR does not explicitly depend on the symbol length or any specific time period $\mathcal{S}$. Instead, the IAR is defined as

$$\mathrm{IAR}\{x\} \triangleq \frac{|x|^2}{E\left[|x|^2\right]},$$

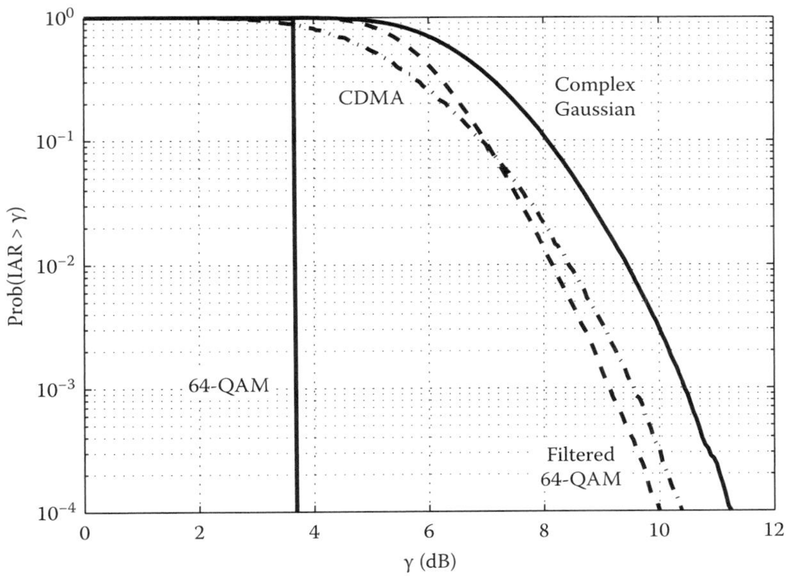

FIGURE 11.2 PAR CCDF of various communications signals at the Nyquist-sampling rate.

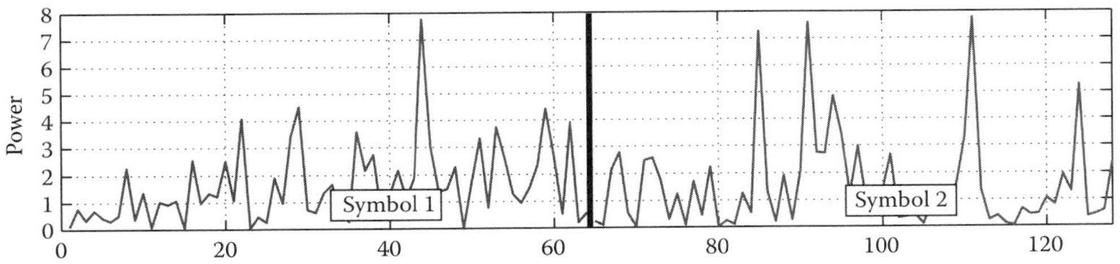

FIGURE 11.3 Two symbols with identical PAR, but different IAR.

where x represents either the discrete or continuous-time signal. The IAR can be described as an average-power-normalized random variable that represents the distribution of the envelope power. Contrastingly, the PAR is an extreme value statistic that quantifies the distribution of only the peak value of a signal for a given time period.

The IAR CCDF is plotted in Figure 11.4, which looks similar to the PAR CCDF in Figure 11.2. Two main distinctions make it easier to tell the two plots apart: (1) for IAR, there is some probability that the IAR will be less than 0 dB and that will be the probability that a sample has less power than the mean power—this is not the case for PAR because the maximum has to always be larger than the mean; and (2) the IAR will always be less than the PAR at every probability level. IAR is a popular metric because, unlike PAR, it is sensitive to all high-power excursions, not just the highest power one.

For discrete-time signals with independent identically distributed (i.i.d.) samples x_n, knowledge of the IAR distribution allows for the quick calculation of the PAR distribution. In practice, the IAR and PAR distributions are characterized through the CCDF. Thus, given the CCDF of the IAR of x_n,

$$\Pr(x_n > \gamma) = f_x(\gamma), \tag{11.12}$$

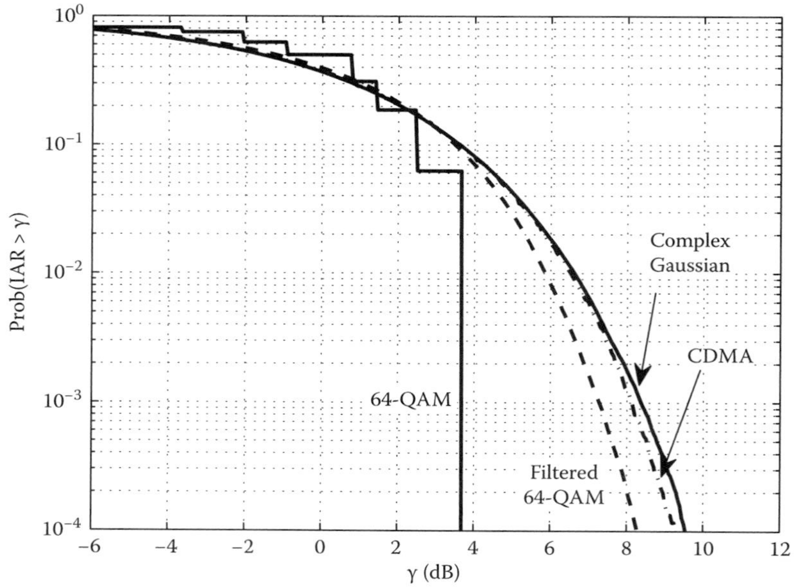

FIGURE 11.4 IAR CCDF of various communications signals at the Nyquist-sampling rate.

the PAR CCDF, when x_n is i.i.d. follows via

$$\Pr(\text{PAR}\{\mathbf{x}\} > \gamma) = 1 - (1 - f_x(\gamma))^N \triangleq xf_{\mathbf{x}}(\gamma). \tag{11.13}$$

When there is interdependence between samples of x_n, Equation 11.13 does not hold. To obtain the PAR CCDF from the IAR CCDF when the samples are not i.i.d. requires integrating over the N-dimensional pdf of $\mathbf{x}$ which, depending on the pdf, may not lead to a closed-form expression.

In some situations, it is of interest to derive the IAR from a known PAR distribution. For the case where the samples that contribute to the PAR are i.i.d., it is clear that Equation 11.36 can simply be inverted to yield

$$\Pr(\text{IAR}\{x_n\} > \gamma) = 1 - (1 - f_{\mathbf{x}}(\gamma))^{1/N}. \tag{11.14}$$

For the non-i.i.d. case, no simple inversion is possible. Because the majority of signal manipulations used in PAR reduction involves creating some amount of interdependence between samples, Equations 11.13 and 11.14 are not useful for analyzing PAR-reduced signals. However, they can be quite useful in analyzing original signals that have not been subject to PAR reduction.

11.3 Nonlinear Peak-Limited Channels

Denote by $x(t)$ and $y(t)$ the baseband equivalent PA input and PA output, respectively. The relationship between the output amplitude $|y(t)|$ and the input amplitude $|x(t)|$ is referred to as the amplitude modulation (AM)/AM characteristic, and the relationship between the output phase deviation $\angle y(t) - \angle x(t)$ and the input amplitude $|x(t)|$ is referred to as the AM/phase modulation (PM) characteristic.

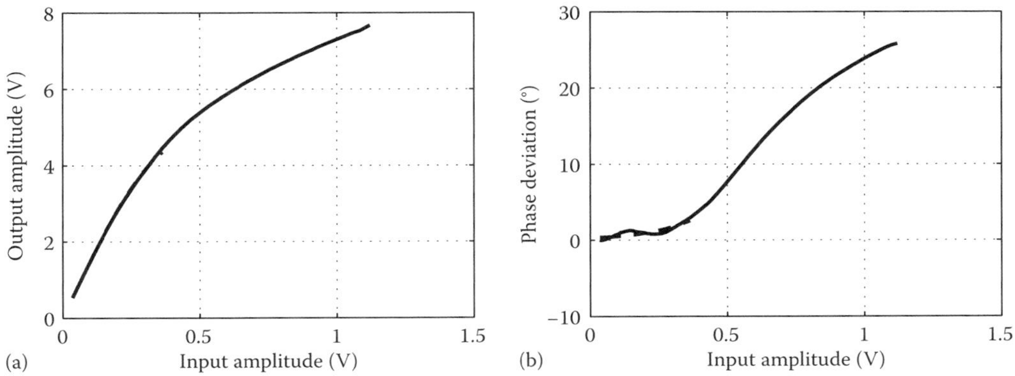

FIGURE 11.5 A third-order polynomial fit to the measured PA characteristics over the interval $|x(t)| \in [0.0354, 0.3701]$. (a) AM/AM. (b) AM/PM.

Figure 11.5a and b shows measured AM/AM and AM/PM characteristics, respectively, of an actual Class AB PA as solid lines. If the input signal is known to have a range of amplitude variations $|x(t)| \in [0.0354, 0.3701]$, then a third-order baseband polynomial PA model,

$$y(t) = (15.0008 + 0.0908j)x(t) + (-23.0826 + 3.3133j)x(t)|x(t)|^2,$$

can produce a fairly good fit to the measured PA characteristics; the model-approximated AM/AM and AM/PM characteristics are shown as dashed lines in Figure 11.5a and b, respectively.

If a large PAR signal is input to the PA so that $|x(t)|$ occupies a larger range, for example, $|x(t)| \in [0.0354, 0.7401]$, then the best third-order baseband polynomial model that we can extract,

$$y(t) = (14.7232 + 0.0468j)x(t) + (-14.4831 + 5.1413j)x(t)|x(t)|^2,$$

still cannot provide a close match to the measured PA characteristics; see the model-approximated AM/AM and AM/PM characteristics as shown in dashed lines in Figure 11.6a and b, respectively. On the other hand, by increasing the polynomial model order to 7,

$$\begin{aligned} y(t) = {} & (15.0372 + 0.0994j)x(t) + (-27.1466 + 1.9232j)x(t)|x(t)|^2 \\ & + (47.5859 + 20.1164j)x(t)|x(t)|^4 + (-38.5259 - 28.9559j)x(t)|x(t)|^6, \end{aligned}$$

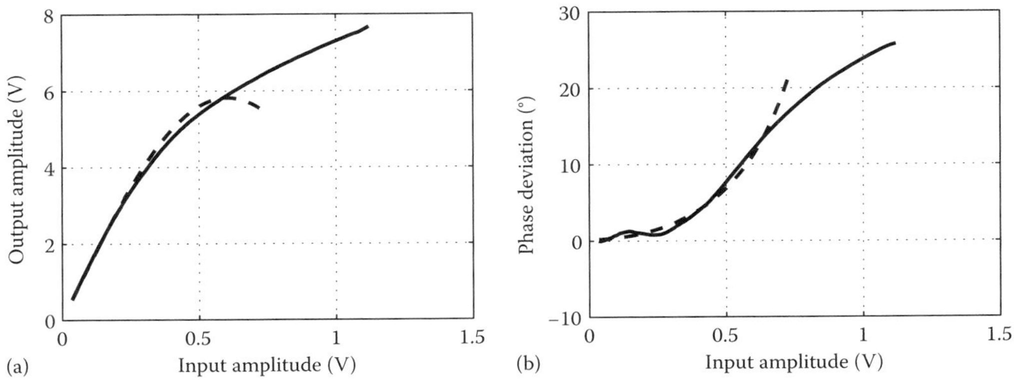

FIGURE 11.6 A third-order polynomial fit to the measured PA characteristics over the interval $|x(t)| \in [0.0354, 0.7401]$. (a) AM/AM. (b) AM/PM.

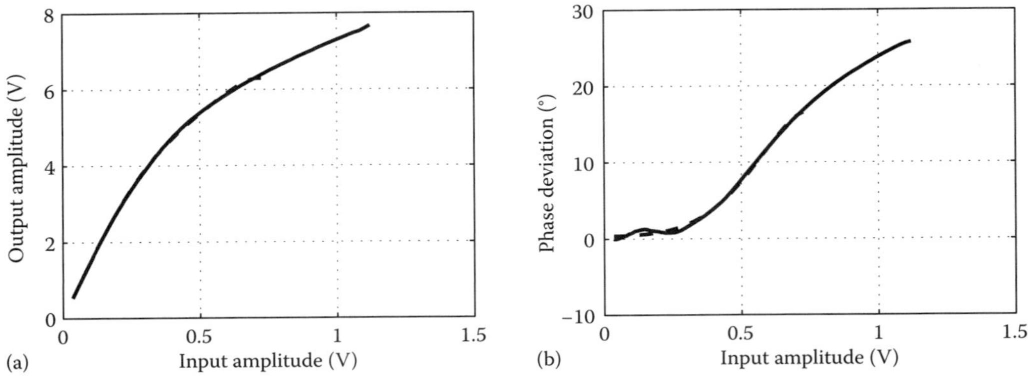

FIGURE 11.7 A seventh-order polynomial fit to the measured PA characteristics over the interval $|x(t)| \in [0.0354, 0.7401]$. (a) AM/AM. (b) AM/PM.

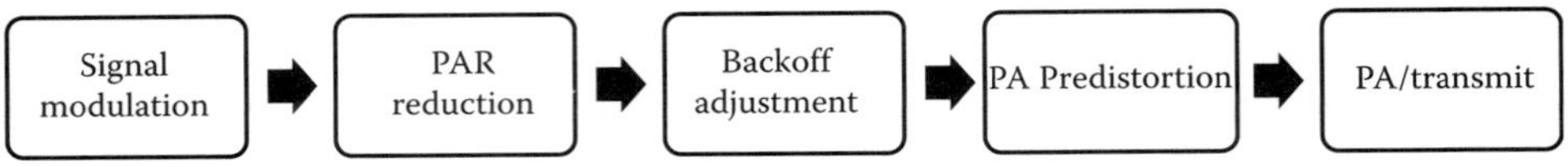

FIGURE 11.8 Basic block diagram for accommodating high-PAR signals.

a satisfactory fit can be achieved; see the model-approximated AM/AM and AM/PM characteristics as shown in dashed lines in Figure 11.7a and b, respectively. The above example shows that when a physical device is inherently nonlinear, a larger dynamic range signal will experience a higher degree of nonlinearity distortions. PAR reduction can reduce the signal's sensitivity toward the system nonlinearity and lead to reduced in-band (IB) distortion (and thus reduced error rates) and out-of-band (OOB) spectral broadening (and thus reduced interference to other users).

As we can see from the IAR and PAR CCDF plots, some signal types suffer from occasional extremely large power excursions. Because PAs are peak-limited devices, if left unmitigated, a large power spike will not be amplified linearly by the PA and may result in damage to the PA. This situation is called saturation which is an even harsher form of nonlinearity.

Thus, it is desirable to mitigate the effects of nonlinear distortion before the signal reaches the PA. An overview block diagram of the system elements used to mitigate PA distortion is provided in Figure 11.8. The process can be broken into three disjoint stages: (1) PAR reduction, (2) signal scaling, and (3) predistortion.

The main focus of this chapter is PAR reduction, but it is important to understand the system blocks downstream from PAR reduction. As we will see, how a signal is scaled is crucial to harnessing PAR reductions.

11.4 Digital Predistortion

All PAs have a peak power limit that is directly related to the direct current (DC) power supplied to the PA. For Class A PAs, this limit is $0.5P_{DC}$ and for Class B and Class B PAs, the limit is $(\pi/4)P_{DC}$ [3]. It is desirable to have linear amplification until saturation so that the overall PA input-to-output function is a soft limiter characteristic as seen in Figure 11.9c. However, there is often a compressing nonlinearity in the PA response prior to saturation. This can be seen in the input–output relationship in Figure 11.9b or the corresponding gain plot in Figure 11.10b. An ideally linear response will have constant gain (constant first derivative) as seen in the digital predistortion (DPD)/PA gain plot in Figure 11.10c.

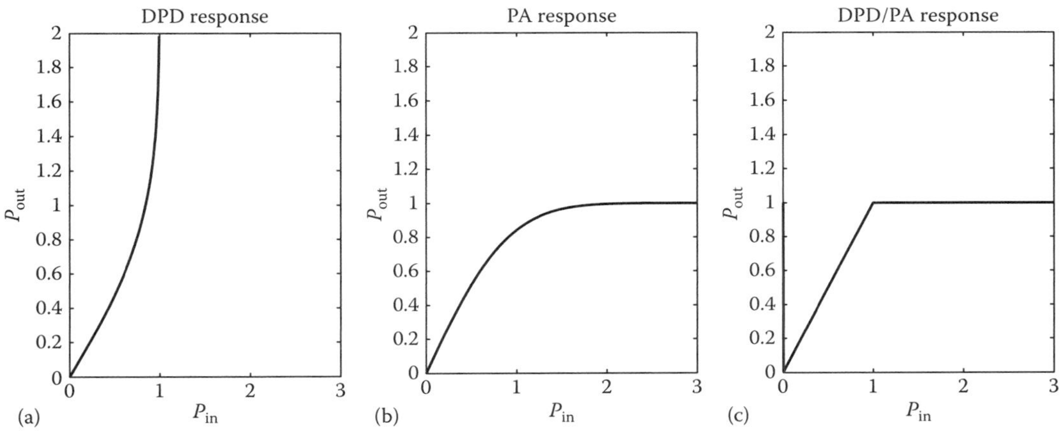

FIGURE 11.9 Input to output power response for (a) DPD, (b) PA response and (c) concatenation of PA and DPD.

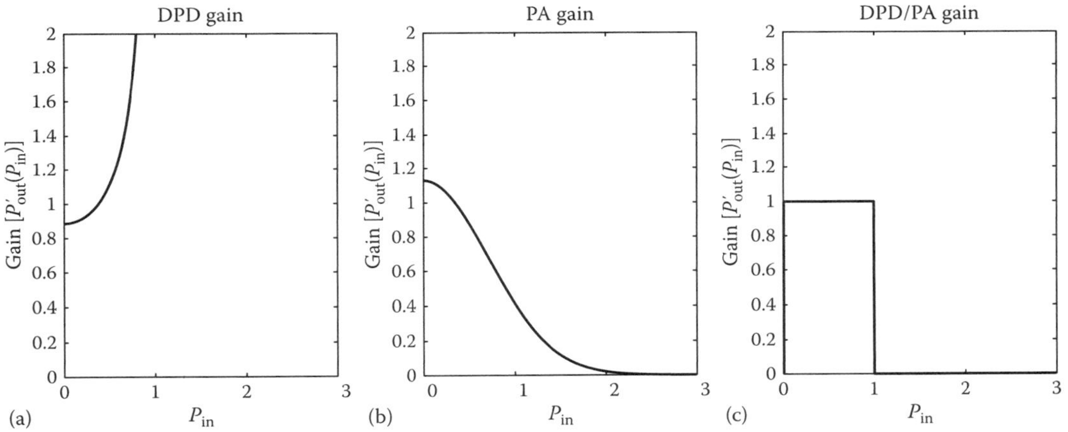

FIGURE 11.10 Input to output power gain for (a) DPD, (b) PA response and (c) concatenation of PA and DPD.

To extend the linear range of the PA, DPD techniques are used where the inverse of the PA function up to the peak limit is applied to a signal prior to the PA. For the PA response in Figure 11.9b, the ideal DPD function is Figure 11.9a. The concatenation of the DPD and PA functions results in the ideal soft limiter characteristic in Figure 11.9c as desired. In the remainder, it is assumed that the linearized PA response has a soft limiter characteristic according to

$$g_{sl}(x) = \begin{cases} x, & |x| \leq 1 \\ e^{j\angle x}, & |x| > 1 \end{cases} \tag{11.15}$$

with a normalized gain of 1.

To accomplish DPD, some mechanism to estimate the PA response is necessary so that the DPD function can be calculated. For extremely stable PAs that vary little from device to device or through changes in temperature of operation, it is possible to design the DPD function off-line and hardwire it into the system. For PAs with time-varying characteristics, one must estimate the PA response adaptively. This can be done with feedback circuitry.

Once the PA response is estimated, the DPD function can be defined through either (1) a look-up table [4,5], (2) a piecewise (non)linear function [6,7], or (3) a continuous function [8,9]. DPD can be further complicated when a PA has prominent memory effects such that previous signal values effect the current PA response and model-based methods have been used [10]. Details for deriving the DPD function from the PA response is beyond the scope of this chapter, but a good overview of the process can be found in [11,12].

11.5 Backoff

For signals that pass through the transmit chain of a communications device, the concern is how the signal should be fed to the PA and how the PA should be biased, so that maximum power efficiency can be realized. There can be two different mechanisms to increase power efficiency: the first involves multiplying the signal by a scaling factor in the digital domain, and the second mechanism involves adjusting the DC power supplied to the PA. Note that because PAR is a scale-invariant metric, a scale factor multiplication will not change the PAR. However, it will be clear that the choice of scaling factor is very important to the system performance and power efficiency. These concepts are illustrated in Figure 11.11.

Figure 11.11a shows the input–output characteristic of an ideal linear (or linearized) PA which saturates when the input power reaches P_{sat}. Assume that the input signal has peak power of $P_m < P_{sat}$ and average power P_i. Let $c = P_{sat}/P_m$. Linear scaling involves digitally multiplying the input signal by $\sqrt{c}$, so that the resulting signal $\sqrt{c}x(t)$ has input–output power relationship shown in Figure 11.11b. Notice that the average power is increased by a factor of c, but P_{sat} and, thus, the DC power consumed by the PA remains unchanged. This will ensure that the PA is operating at maximum efficiency without causing nonlinear distortions. If the scaling factor is greater than $\sqrt{c}$, the PA will be even more efficient at the cost of nonlinear distortion. Figure 11.11b shows the second strategy for transmitting at maximum efficiency, which involves

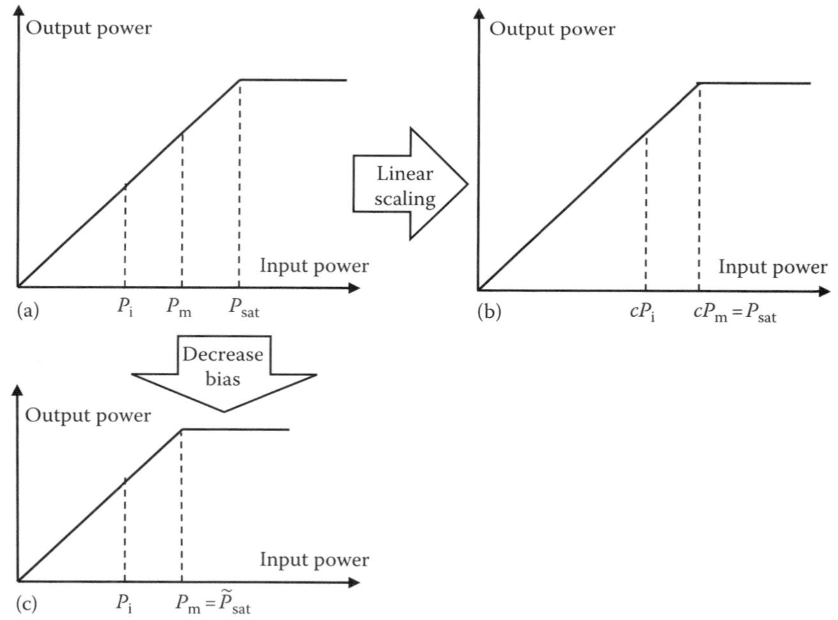

FIGURE 11.11 Illustration of how the input and output power change for linear scaling and PA rebiasing. (a) Original PA characteristic, (b) PA characteristic with linear scaling and (c) PA characteristic after the DC bias point is decreased.

adjusting the bias of the PA so that the new saturation level is $\tilde{P}_{sat} = P_m$. The PA itself will consume less DC power this way for the same amount of transmit power, thus improving the power efficiency.

These two mechanisms for power control have distinct applications and implications for the system. Adjusting the DC power supplied to the PA will change the peak output power and change the amount of power the system consumes. If the DC power could be adjusted at the sample rate of the signal, then high dynamic range signaling would not present a power-efficiency problem. However, in reality, the maximum DC power switching rate may be several orders of magnitude slower than the sample rate of the signal [13]. Because of this, DC power control is not useful for accommodating large short-time signal peaks. Instead, DC power control is more commonly used in multiuser protocols to increase and decrease signal power as dictated by the base station.

In contrast to DC power control, digital scaling can be done very quickly down to the symbol or even sample rate of the signal. Thus, digital scaling is most appropriate for maximizing the power efficiency, or some other metric of interest, for a given DC power. Thus, in this section, it is assumed that the DC power is fixed. Given this constraint, the natural objective is to choose a digital scaling factor that will maximize system performance.

Here, the options for digital scaling* are divided into two groups: (1) time-varying scaling and (2) time-invariant scaling. While in a very general sense both scaling factors may depend on time as the DC power will change over time, here, time-invariant scaling means that for a fixed DC power, the scaling factor will not change with time. Time-varying scaling means that the signal will be scaled as time varies even for a fixed DC power.

For simplicity, assume that the average power of the signal entering the scaling block x_n is $\sigma_x^2 = 1$. Also, assume that the PA has a fixed DC power with response which is exactly the soft limiter response in Equation 11.15. When no scaling is applied to x_n, the system is said to have 0 dB input backoff (IBO), where

$$\text{IBO} = \frac{x_{\max}^2}{E[|\tilde{x}|^2]}, \tag{11.16}$$

and $\tilde{x}$ is the scaled PAR-reduced signal and $x_{\max}$ is the maximum input value before reaching saturation. $\text{IBO} = 1 = 0$ dB follows with no scaling because $x_{\max}^2 = 1$ from Equation 11.15 and $E[|\tilde{x}|^2] = \sigma_x^2 = 1$ by assumption. The output backoff (OBO) is

$$\text{OBO} = \frac{g_{\max}^2}{E[|g(\tilde{x})|^2]}, \tag{11.17}$$

where $g_{\max} = |\max_x g(x)|$. OBO is a useful quantity because it is directly related to the power efficiency

$$\eta = \frac{E[|g(\tilde{x})|^2]}{P(g_{\max}^2)}, \tag{11.18}$$

where $P(\cdot)$ is a PA-dependent function that relates the maximum output power of the PA to the DC power consumed by the PA. When $P(g_{\max}^2)$ is a linear function of $g_{\max}^2$—as it is for Class A PAs, $P_A(g_{\max}^2) = 2g_{\max}^2$ and Class B PAs, $P_B(g_{\max}^2) = g_{\max}^2\pi/4$—it is true that

$$\eta = \frac{\eta_{\max}}{\text{OBO}}. \tag{11.19}$$

* Scaling is a term that is interchangeable with backoff, while backoff more specifically refers to a scaling operation by a factor less than one.

Substituting the maximum efficiency for a Class A PA, $\eta_{max} = 1/2$, we have

$$\eta_A = \frac{1/2}{OBO}, \tag{11.20}$$

and

$$\eta_B = \frac{\pi/4}{OBO}, \tag{11.21}$$

for Class B PA [13]. Thus, OBO for these cases expresses how much efficiency is lost relative to the maximum possible efficiency. With all else equal, a system designer should seek to minimize OBO. However, there are performance trade-offs that are not captured by this simple power-efficiency metric that must be considered.

The relationship between IBO and OBO is more difficult to summarize and depends on the distribution of $\tilde{x}$ and the shape of $g(\cdot)$. For most physical systems, $g(\cdot)$ will be a concave function. It then follows that $OBO \leq IBO$ and that

$$\eta \geq \frac{\eta_{max}}{IBO}. \tag{11.22}$$

In other words, IBO provides a worst-case measure of relative power efficiency.

11.5.1 Time-Invariant Scaling

In a time-invariant or "static" scaling system, the signal to be transmitted is scaled by a constant before going through the PA. That is,

$$\tilde{x}_n = \beta x_n, \quad \forall n. \tag{11.23}$$

Again, because both PAR and IAR are scale-invariant metrics, $PAR\{\tilde{x}\} = PAR\{x\}$ and $IAR\{\tilde{x}\} = IAR\{x\}$. Here, static scaling of β results in an IBO adjustment of $20 \log_{10}(\beta)$ dB. Accordingly, the power efficiency of the system can be quickly bounded based on the choice of β. By only considering η, it would be natural to make β very large in order to maximize efficiency. However, η does not capture how much distortion or spectral spreading a certain choice of β will cause. For signals with finite support like QAM, for certain values of β, there will be no distortion or spectral spreading. On the other hand, signals like orthogonal frequency division multiplexing (OFDM) follow an approximately complex Gaussian distribution and finite but very large support.* For such (near) infinite-support signals, some distortion and spectral spreading will occur and must be taken into account when choosing β.

In practice, when distortion, measured by $\mathcal{D}(x)$, and spectral spreading, measured by $\mathcal{S}(x)$, are subject to known constraints, β should be chosen so that the efficiency is maximized. That is, the goal is to

$$\underset{\beta}{\text{maximize}} \quad E[|g(\beta x)|^2] \tag{11.24}$$

$$\text{subject to} \quad \mathcal{D}(\beta x) \leq th_D \quad \mathcal{S}(\beta x) \leq th_S \tag{11.25}$$

* An OFDM signal is the sum of N finite random variables, so that the output signal approaches complex Gaussian in distribution [14]. But for finite N, the support of an OFDM sample is bounded.

where th_D and th_S are threshold constraints for the distortion and spectral spreading. Obviously, this is a generalization of the problem. For specific constraints, it may be possible to calculate $E[|g(\beta x)|^2]$ and the constraint functions in closed form. With this, the maximizing β can be analytically derived. However, when PAR reduction is involved or more complicated constraints are involved, Monte Carlo methods are usually required to find the optimal static scaling factor β.

As an analytical example, assume x is complex Gaussian distributed with zero mean and unit variance, i.e., $\mathcal{CN}(0,1)$, and the distortion metric $\mathcal{D}(\beta x) = E[|g(\beta x) - \beta x|^2]$ is required to be below th_D, the optimal β follows as the value of β that solves

$$E[|g(\beta x) - \beta x|^2] = \beta^2 e^{-1/\beta^2} - \beta\sqrt{\pi}\,\mathrm{erfc}\left(\frac{1}{\beta}\right) = th_D,$$

which does not have a closed-form solution for β, but such a solution can be quickly found numerically.

Several PAR reduction methods that are designed for static backoff systems implicitly optimize β as part of the algorithm. Other methods reduce the PAR as much as possible and cite the resulting PAR at a certain CCDF level. The implication in this case is that β is set to be equal to that PAR level, and the probability of clipping a "symbol" will equal the probability level of the PAR.

While this is common practice, based on the discussion surrounding Figure 11.4, it is more helpful to gauge the value of β using the IAR CCDF. This allows for the determination of the probability that a certain "sample" will be clipped, which is a better proxy for the amount of distortion incurred by the clipping.

11.5.2 Time-Varying Scaling

The main advantage to static scaling is that it is a receiver ambivalent function. If, instead, it is known that the receiver is capable of estimating a flat fading channel over a collection of samples known as a symbol, with length of at least N, then a time-varying scaling operation is possible.

For time-varying scaling, the idea is to scale each symbol so that no clipping occurs. At the receiver, the scaling is seen transparently as part of the channel and decoded on a symbol-wise basis as part of the channel. For semantic purposes a symbol in this sense may be alternatively referred to as a "packet" or "block" in the parlance of the communication protocol. But here a symbol denotes the period of time for which the channel is reestimated.

Time-varying or "dynamic" scaling is more complicated than static backoff and involves a time-varying scaling function, α_n, so that the scaled signal is

$$\tilde{x}_n = \alpha_n x_n. \tag{11.26}$$

The dynamic scaling factor, α_n, can be chosen in a number of ways, but generally it is chosen adaptively in a way that depends on the peak value of the input symbol over a certain period of samples. The goal is to scale each set of samples so that the peak of the set of samples coincides with the input limit of the PA, x_{max}. A straightforward way to accomplish this feat is to scale each symbol by its peak, so that for each symbol

$$\tilde{x}_n = \begin{cases} \dfrac{x_{max}}{\max_{n\in\mathcal{S}_1}|x_n|}x_n, & n \in \mathcal{S}_1 = \{1,2,\ldots,N\} \\ \dfrac{x_{max}}{\max_{n\in\mathcal{S}_2}|x_n|}x_n, & n \in \mathcal{S}_2 = \{N+1,N+2,\ldots,2N\}. \\ \vdots & \vdots \end{cases} \tag{11.27}$$

The advantage of dynamic scaling is that clipping is completely avoided because the input signal never reaches saturation. The scaling factor can be viewed as part of the multipath channel. However, unlike most multipath channels, the variations caused by linear scaling change from symbol to symbol, which may complicate the receiver-side channel estimation by requiring that the channel be reestimated for every symbol [15,16].

Isolating just the first symbol for simplicity, the average output power of a linear scaling system is expressed as

$$E\left[|\tilde{x}_n|^2\right] = x_{\max}^2 E\left[\left|\frac{x_n}{\max_{n\in\mathcal{S}_1}|x_n|}\right|^2\right] \tag{11.28}$$

$$\approx x_{\max}^2 E[|x_n|^2] E\left[\left|\frac{1}{\max_{n\in\mathcal{S}_1}|x_n|}\right|^2\right] \tag{11.29}$$

$$= x_{\max}^2 E\left[\frac{1}{\text{PAR}}\right]. \tag{11.30}$$

The approximation is required in Equation 11.29 because $\max_{n\in\mathcal{S}_1}|x_n|$ is not independent of x_n. However, for large N, the above approximation becomes extremely tight.

Assuming that an ideal linear or linearized PA is used, the output power for large N follows directly as

$$E\left[|g(\tilde{x}_n)|^2\right] \approx g_{\max}^2 E\left[\frac{1}{\text{PAR}}\right]. \tag{11.31}$$

Therefore, the power efficiency of a dynamic scaling system is

$$\eta \approx \eta_{\max} E\left[\frac{1}{\text{PAR}}\right]. \tag{11.32}$$

Unlike the static scaling case, there is no distortion or spectral spreading to consider, so maximizing efficiency is only a matter of maximizing the harmonic mean of the PAR of the signal, $E\left[\frac{1}{\text{PAR}}\right]$. So, whereas the IAR is the most pertinent metric for a static backoff system, the PAR, and specifically the harmonic mean of the PAR, is the paramount metric for a dynamic scaling architecture. With this in mind, Section 11.6 outlines various algorithms for peak reduction.

11.6 PAR Reduction

PAR reduction was first of interest for radar and speech synthesis applications. In radar, PAR reduction is important because radar systems are peak-power-limited just like communications systems. Accordingly, as has been shown in the signal backoff discussion, low-PAR signaling leads to increased power efficiency. In speech synthesis applications, peaky signals lead to a hard-sounding "computer" voice [17]. For simulated human speech, this characteristic is not desirable. However, in both radar and speech synthesis, maintaining a certain spectral shape is of interest. Therefore, PAR reductions in these fields can be done per spectral shape and are specified by frequency-domain phase sequences. Some examples of low-PAR sequences are the Newmann phase sequence, the Rudin–Shapiro phase sequences, and Galios phase sequences [18]. These phase sequences all produce low-PAR time-domain sequences for a wide variety of spectral masks.

But this is not sufficient for communications systems where random data is modulated, which means that additional degrees of signal freedom beyond spectral shape must be accounted for. Thus, instead of creating one or several low-PAR signals per spectral mask, a communications engineer must create an exponential number of low-PAR sequence, so that there is one for each possible bit combination.

While the preceding sections have been general and apply to any signal type, the following PAR reduction methods are mostly restricted to block signaling like OFDM, code division multiple access (CDMA), and single-carrier (SC) block transmission (see Wang et al. [19] for a comparison of block transmission methods). OFDM is of particular interest because OFDM signals exhibit a Complex Gaussian signal distribution with high probability of large PAR values. Therefore, the following PAR reduction discussion is devoted to OFDM signaling but can be easily extended to any block transmission system. See [20,21] for a thorough introduction to OFDM.

For our purposes, in OFDM, a frequency-domain vector $\mathbf{y}$ of information symbols that have values drawn from a finite constellation, $\mathbf{y} \in \mathcal{A}^N$ (e.g., for QPSK, $\mathcal{A} = \{-1, 1, j, -j\}$), is used to modulate N subcarriers. For transmission, the time-domain symbol is created with an inverse discrete Fourier transform (IDFT)

$$x_n = \frac{1}{\sqrt{N}} \sum_{k=0}^{N-1} y_k e^{j2\pi kn/N}, \quad n \in \{0, 1, \ldots, N-1\}, \tag{11.33}$$

or expressed as in vector notation,

$$\mathbf{x} = \mathbf{Q}\mathbf{y}, \tag{11.34}$$

where $\mathbf{Q}$ is the IDFT matrix with entries $[\mathbf{Q}]_{k,n} = N^{-1/2} \exp\left(j2\pi(n-1)(k-1)/N\right)$, $1 \leq k, n \leq N$.

In OFDM, a cyclic prefix (CP), which is a copy of the last several samples of the payload symbol, is appended to the start of each symbol. The CP is used to diagonalize the channel and to avoid intersymbol interference (ISI) (see Wang et al. [19] for details). Importantly, because the CP is a copy of samples from the payload symbol, it can be ignored for the purposes of PAR reduction.

The frequency-domain received vector after CP removal and fast Fourier transform (FFT) is

$$\mathbf{r} = \mathbf{D_h}\mathbf{y} + \mathbf{n}, \tag{11.35}$$

where

$\mathbf{D_h}$ is a diagonal matrix with diagonal entries that correspond to the channel response, $\mathbf{h}$

$\mathbf{h}$ and $\mathbf{n}$ are the independent channel noises

The distribution of the Nyquist sampled-discrete OFDM symbol, $\mathbf{x}$, can be quickly established using the central limit theorem, which states that the sum of N independent (complex) random variables converges to a (complex) Gaussian random variable as $N \to \infty$. From Brillinger [14], it follows that the IDFT of N i.i.d. random variables, $\mathbf{y}$, results in samples, $\mathbf{x}$, that are also i.i.d. For practical OFDM systems $N > 64$ is sufficient to realize an approximately Gaussian distribution for the elements of $\mathbf{x}$. Thus, the PAR follows as

$$\Pr(\mathrm{PAR}\{\mathbf{x}\} > \gamma) = 1 - (1 - e^{-\gamma})^N. \tag{11.36}$$

In practice, the OFDM symbol created in the digital domain will be oversampled by a factor of L so that

$$x_{n/L} = \frac{1}{\sqrt{NL}} \sum_{k=\mathcal{I}} y_k e^{j2\pi kn/NL}, \quad n \in \{0, 1, \ldots, LN-1\}, \tag{11.37}$$

where $\mathcal{I}$ is the set of nonzero "in-band" subcarriers. Or, in vector notation the oversampled symbol is

$$\mathbf{x} = \mathbf{Q}_L \mathbf{y}_{\mathcal{I}}. \tag{11.38}$$

After the oversampled symbol is created and digitally PAR-reduced, it is passed through the digital-to-analog converter (DAC) and ultimately the continuous-time analog symbol, $x(t)$, will be converted to passband and sent through the PA. An ideal PAR reduction method would reduce the PAR of $x(t)$. But, because PAR reduction is done in the digital domain, the hope is that the PAR is correspondingly reduced in $x(t)$.

Unlike the full band, i.e., the Nyquist-sampled case where all subcarriers are mutually independent, the oversampled symbol does not have independent samples. Significant research effort has been devoted to calculating the CCDF of a continuous-time OFDM symbol. Some researchers employ the joint distribution of derivatives of $x(t)$ to derive the maximum value of a band-limited complex Gaussian process [18,22,23]. The resulting approximation is

$$\Pr(\mathrm{PAR}\{x(t)\} > \gamma) \approx 1 - \exp\left(-N\sqrt{\frac{\pi}{3}}\gamma e^{-\gamma}\right). \tag{11.39}$$

The continuous-time PAR can also be well approximated by using an oversampling factor of $L \geq 4$, so that $x(t) \approx x_{n/L}$, see Figure 11.12.

PAR reduction methods can be classified into two main groups. The first group contains all PAR reduction methods that require a specific receiver algorithm, independent of standard OFDM operation, to decode the PAR-reduced signal. The second group includes all methods that are transparent to the receiver and require nothing beyond the standard OFDM receiver architecture. Despite the fact that the transparent algorithms are designed to not require receiver cooperation, there are receiver-side algorithms that will boost performance, if enough knowledge of the transmitter algorithm is available.

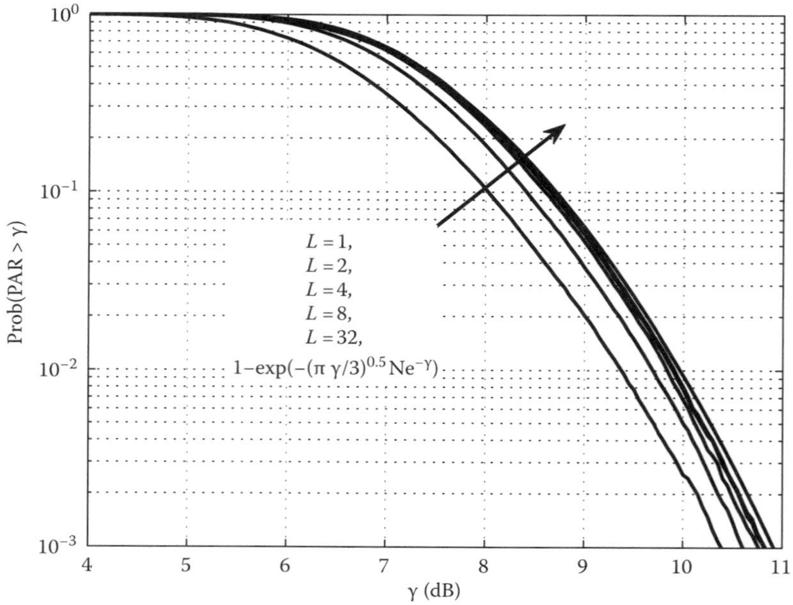

FIGURE 11.12 PAR CCDF with oversampling and the continuous-time approximation.

11.6.1 Transparent PAR Reduction

The transparent methods work by adding a carefully chosen "distortion" signal to the original signal to reduce the PAR so that

$$\mathbf{x} = \mathbf{Q}(\mathbf{y} + \mathbf{d}) \tag{11.40}$$

is transmitted and

$$\mathbf{r} = \mathbf{D_h}(\mathbf{y} + \mathbf{d}) + \mathbf{n} \tag{11.41}$$

is received. Existing PAR reduction methods falling into this category include clipping, clipping and filtering (C&F) [24,25], constrained clipping [26,27], and global distortion optimization [28–30] among others.

For these methods, no receiver modification is necessary because the distortion signal is seen as noise to the receiver, i.e., $\tilde{\mathbf{n}} = \mathbf{d} + \mathbf{n}$. Accordingly, the noise can be shaped to suit system constraints. These constraints can be expressed in a number of ways depending on the system. However, as with the scaling factor optimization, the constraints are usually broken into two groups: IB distortion that affects the user's performance and OOB distortion that affects other users' performance. With this framework, the objective of the transparent methods is to

$$\begin{aligned} \underset{\mathbf{d}}{\text{minimize}} \quad & \text{PAR}\{\mathbf{Q}(\mathbf{y} + \mathbf{d})\} \\ \text{subject to} \quad & \mathcal{D}(\mathbf{d}) \le \text{th}_\text{D} \\ & \mathcal{S}(\mathbf{d}) \le \text{th}_\text{S} \end{aligned} \tag{11.42}$$

which is very similar to the scaling optimization problem. The result is the PAR-reduced vector $\bar{\mathbf{x}} = \mathbf{Q}(\mathbf{y} + \mathbf{d}^{\star})$. Many PAR reduction schemes that have been proposed in the literature can be framed in this way. The various methods are distinguished by either the constraints used or the algorithm used to find the (near) optimal distortion vector $\mathbf{d}^{\star}$. The most common and intuitive pair of constraints involves

$$\mathcal{D}(\mathbf{d}) = \text{EVM} = \|\mathbf{d}_{\mathcal{I}}^{\star}\|_2^2 \tag{11.43}$$

to measure the distortion in band, where $\mathbf{x}_{\mathcal{K}}$ is defined as a vector of values from the set of ascending indices $\mathcal{K}$, i.e.,

$$\mathbf{x}_{\mathcal{K}} = [x_{n_1}, x_{n_2}, \ldots, x_{n_{|\mathcal{K}|}}]^{\text{T}}, \quad \mathcal{K} = \{n_1, n_2, \ldots, n_{|\mathcal{K}|}\}, \tag{11.44}$$

and

$$\mathcal{S}(\mathbf{d}) = I\left(|\mathbf{m}|^2 - |\mathbf{d}_{\mathcal{O}}^{\star}|^2\right), \tag{11.45}$$

where

$I(\cdot)$ is the indicator function which is 0 for negative inputs and 1 otherwise

$\mathbf{m}$ is a vector that specifics the maximum allowed distortion in each band [26]

Any constraint can either be applied on a per-symbol basis as in Equations 11.43 and 11.45 or stochastically to the ensemble symbol:

$$\mathcal{D}_s(\mathbf{d}) = \text{EVM} = E\left[\|\mathbf{d}_{\mathcal{I}}^{\star}\|_2^2\right], \tag{11.46}$$

and

$$\mathcal{S}_s(\mathbf{d}) = I\left(E\left[|\mathbf{m}|^2 - |\mathbf{d}_{\mathcal{O}}^{\star}|^2\right]\right). \tag{11.47}$$

While the performance objectives of a given communications system are usually clear, it may not be straightforward to find an IB and OOB constraint set to optimally meet the performance requirements. Instead, proxy requirements like error vector magnitude (EVM) and spectral mask are often used. Though, other constraints have been proposed to better meet system performance objectives [26,28,30,31]. As an example, one well-researched PAR reduction method named active constellation extension (ACE), explicitly constraints IB distortion so that the constellation minimum distance does not decrease [31].

Once the constraints are agreed upon, the choice of optimization algorithm depends on where in the PAR reduction/complexity trade-space the system can operate. With unlimited complexity, global optimization can be solved for each symbol. However, with the high data rates used in modern communications, it may be that the system is constrained to very modest PAR reduction complexity.

At the highest level of complexity, the global optimal solution to Equation 11.42 can be found using optimization techniques like genetic annealing. For constraints that ensure the problem is convex, the global solution can be found in polynomial time using interior point optimization methods [32].

When only limited complexity is practical, iterative algorithms are frequently used. In an iterative algorithm, for each iteration, the signal of interest is passed through a clipping function or other peak-limited functions. From the resulting symbol $\hat{\mathbf{x}}^{(i)} = g(\hat{\mathbf{x}}^{(i-1)})$, the distortion is calculated to be $\mathbf{Q}^{-1}(\hat{\mathbf{x}}^{(i)} - \hat{\mathbf{x}}^{(0)}) = \mathbf{d}^{(i)}$, where $\hat{\mathbf{x}}^{(0)} = \mathbf{x}$. Next, the distortion is modified by a function $\xi(\cdot)$ so that it meets the IB and OOB constraints, $\hat{\mathbf{d}}^{(i)} = \xi(\mathbf{d}^{(i)})$. Finally, the symbol for the next iteration is generated as $\hat{\mathbf{x}}^{(i+1)} = \hat{\mathbf{x}}^{(0)} + \mathbf{Q}\hat{\mathbf{d}}^{(i)}$. If the maximum allowable number of iterations has been reached, then $\hat{\mathbf{x}}^{(i+1)}$ is transmitted; otherwise the process is repeated.

Many choices are possible for $\xi(\cdot)$ even for a fixed constraint set. For instance, assume it is required that $\|\mathbf{d}\|_2 \leq \text{th}_{\text{evm}}$. To achieve this we could scale all values of $\mathbf{d}$ so that

$$\xi(\mathbf{d}) = \frac{\text{th}_{\text{evm}}}{\|\mathbf{d}\|_2}\mathbf{d}, \tag{11.48}$$

or we could clip the magnitude and maintain the phase of all entries of the distortion vector and set

$$\xi(\mathbf{d}) = \frac{\text{th}_{\text{evm}}}{N} e^{j\angle \mathbf{d}}. \tag{11.49}$$

Still other modifications are possible.

Likewise, the OOB modification function can be defined in a number of ways depending on the constraint. For the popular C&F scheme, the OOB power is set to zero. That is,

$$\xi(\mathbf{d}_{\mathcal{O}}) = 0, \tag{11.50}$$

another possibility is to clip the OOB distortions to exactly the spectral constraint $\mathbf{m}$ according to

$$\xi(\mathbf{d}_{\mathcal{O}}) = e^{j\angle \mathbf{d}_{\mathcal{O}}}|\mathbf{m}|. \tag{11.51}$$

Many variations to this method have been proposed for a variety of constraint functions. The choice of constraint and modification function is very much system-dependent. Figure 11.13 is a plot of various methods: C&F [25], constrained clipping [33], PAR optimization [28], and EVM optimization [30]. The plot shows that EVM optimization and C&F are best suited to a static backoff, while PAR optimization and constrained clipping have lower harmonic mean PAR values and were designed for symbol-wise scaling.

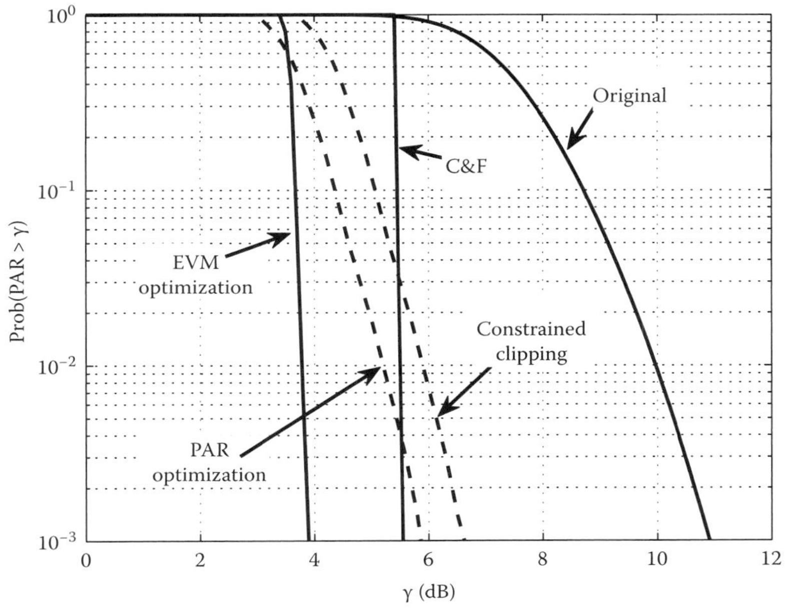

FIGURE 11.13 PAR CCDF of transparent methods. For the plot, $N = 64$ and EVM ≤ 0.1.

11.6.2 Receiver-Side Distortion Mitigation

Note that transparent PAR reduction assumes that the distortion $\mathbf{d}$ will be treated as noise. However, all of these algorithms are deterministic so that there exists a deterministic but not necessarily bijective mapping from $\mathbf{y} \rightarrow \mathbf{d}$. With this, it is reasonable to assume that some information about $\mathbf{d}$ can be gleaned from the received vector $\mathbf{r}$, which is also a function of $\mathbf{y}$. If the receiver can estimate $\mathbf{d}$ from $\mathbf{r}$, then the distortion can be subtracted from the receiver-estimated $\hat{\mathbf{d}}$ to find a closer approximation to the transmitted $\mathbf{y}$.

Receiver distortion mitigation is feasible, if the receiver has perfect knowledge of the transmitter PAR reduction algorithm. Therefore, if we write the PAR reduction algorithm as $p(\mathbf{y}) = \mathbf{y} + \mathbf{d}^{\star}$, the maximum likelihood distortion detection follows as

$$\begin{aligned} \hat{\mathbf{y}}_{\text{ml}} &= \arg \min_{\bar{\mathbf{y}} \in \mathcal{A}^N} \|\mathbf{r} - \mathbf{D_h} p(\bar{\mathbf{y}})\|_2 \\ &= \arg \min_{\bar{\mathbf{y}} \in \mathcal{A}^N} \left\|\mathbf{D_h}(\mathbf{y} + \mathbf{d}^{\star}) + \mathbf{n} - \mathbf{D_h}(\bar{\mathbf{y}} + \bar{\mathbf{d}}^{\star})\right\|_2. \end{aligned} \tag{11.52}$$

Notice that $\hat{\mathbf{y}}_{\text{ml}}$ will not necessarily be equal to $\mathbf{y}$ because the noise may cause another value of $\hat{\mathbf{y}}_{\text{ml}} \neq \mathbf{y}$ to satisfy Equation 11.52. In the general case where $p(\mathbf{y}) \neq [p_1(y_1), p_2(y_2), \ldots, p_N(y_N)]^{\text{T}}$, solving this problem requires searching over the nonconvex set $\mathcal{A}^N$, which is a problem that grows in complexity exponentially with N.

The problem can be simplified [22, p. 129], if we ignore the fact that $\bar{\mathbf{d}}^{\star} \neq \mathbf{d}^{\star}$ and instead assume that they are equal, then

$$\hat{\mathbf{y}} = \arg \min_{\bar{\mathbf{y}} \in \mathcal{A}^N} \|\mathbf{D_h}\mathbf{y} - \mathbf{D_h}\bar{\mathbf{y}} + \mathbf{n}\|_2. \tag{11.53}$$

Because the receiver only has access to $\mathbf{r} = \mathbf{D_h}(\mathbf{y} + \mathbf{d}^*) + \mathbf{n}$ and not $\mathbf{D_h y}$, we can rewrite Equation 11.53 as

$$\hat{\mathbf{y}} = \arg\min_{\bar{\mathbf{y}} \in \mathcal{A}^N} \|\mathbf{r} - \mathbf{D_h d}^* - \mathbf{D_h}\bar{\mathbf{y}}\|_2. \tag{11.54}$$

From this expression, it is clear that individual elements of $\bar{\mathbf{y}}$ do not affect other elements in the vector. Thus, the problem can be transformed into N scalar problems

$$\hat{y}_n = \arg\min_{\bar{y}_n \in \mathcal{A}} |r_n - h_n d_n^* - h_n \bar{y}_n| \tag{11.55}$$

$$= \arg\min_{\bar{y}_n \in \mathcal{A}} \left| \frac{r_n}{h_n} - d_n^* - \bar{y}_n \right|. \tag{11.56}$$

Solving this requires that d_n^* be estimated for each value of n. This can be done iteratively. On each iteration, the transmitted symbol is estimated by a hard decision,

$$\hat{\mathbf{y}}^{(i)} = \arg\min_{\bar{\mathbf{y}} \in \mathcal{A}^N} \|\mathbf{D_h}^{-1}\mathbf{y} + \mathbf{Q}\mathbf{d}^{*,(i-1)} - \bar{\mathbf{y}}\|_2.$$

where the distortion term on the ith iteration is calculated as

$$\mathbf{d}^{*,(i)} = \mathbf{Q}^H \hat{\mathbf{y}}^{(i)} - p(\mathbf{Q}^H \hat{\mathbf{y}}^{(i)})$$

with $\mathbf{d}^{*,(0)}$ initialized to an all-zero vector, i.e.

$$\mathbf{d}^{*,(0)} = \mathbf{0}.$$

Transparent methods are attractive because their implementation does not require an informed receiver. These methods are even more appealing considering a receiver with knowledge of the PAR reduction and physical nonlinearity used at the transmitter can significantly mitigate the distortion.

Figure 11.14 shows the symbol error rate (SER) performance of a clipped 64-QAM OFDM symbol with $N = 64$ subcarriers. The plot shows that a rather amazing SER decrease can be found using iterative distortion cancellation. After only three low-complexity iterations, the algorithm converges to the unclipped SER for the SER range plotted.

11.6.3 Receiver-Dependent PAR Reduction

This section outlines several PAR reduction methods that do not have any backward compatibility to uninformed OFDM systems and require that the receiver have knowledge of the PAR reduction system. Unlike the transparent methods with distortion mitigation, the receiver-dependent distortion methods require very little decoding complexity. Another difference is that the receiver in these methods does not need precise information about the transmitter nonlinearity caused by the physical device characteristics. For a system that is not perfectly predistorted, conveying the transmitter nonlinearity to the receiver may require significant throughput overhead.

11.6.3.1 Multiple Signal Representations

Random search methods operate by creating multiple, equivalent realizations of the signal waveform and achieve a PAR reduction by selecting the lowest-PAR realization for transmission. That is, a set of reversible signal realizations

$$\mathcal{T}_m(\mathbf{y}) = \mathbf{y}^{(m)}, \quad m \in \{1, 2, \ldots, M\}$$

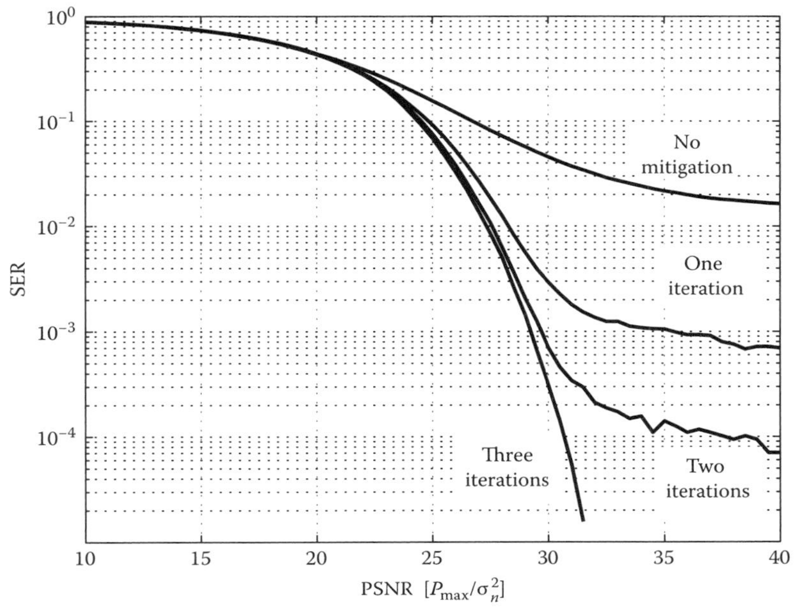

FIGURE 11.14 SER improvement after each iteration of receiver-side distortion mitigation algorithm.

are created. For transmission, the lowest-PAR symbol index is selected

$$\tilde{m} = \arg\min_{m} \text{PAR}\{\mathbf{Q}\mathbf{y}^{(m)}\},$$

so that $\mathbf{x}^{(\tilde{m})} = \mathbf{Q}\mathbf{y}^{(\tilde{m})}$ is transmitted. For simplicity, assume that the channel is flat with gain one, $\mathbf{h} = \mathbf{1}$, where $\mathbf{1}$ is a vector of ones, so that $\mathbf{r} = \mathbf{y}^{(\tilde{m})} + \mathbf{n}$.

At the receiver, the inverse mapping

$$\hat{\mathbf{y}} = \mathcal{T}_{\tilde{m}}^{-1}(\mathbf{r})$$

is used to generate an estimate of the original data. For some methods, this means that the index $\tilde{m}$ must be either transmitted as side information at the cost of $\log_2 M$ bits or $\tilde{m}$ can sometimes be detected blindly. Other methods involve a clever set of mapping functions so that the transmitted index $\tilde{m}$ is not required for decoding and no side information needs to be transmitted.

Many methods are possible for generating equivalent signal realizations $\mathcal{T}_m(\cdot)$. The classic example of a multiple realization PAR reduction method is selected mapping (SLM) [34–38]. In SLM, the constellation points that make up the frequency-domain vector are rotated by a set of phase vectors such that the mth signal realization is

$$\mathcal{T}_m(\mathbf{y}) = \mathbf{D}_{e^{j\theta(m)}}\mathbf{y}.$$

Here, the elements of $\boldsymbol{\theta}^{(m)}$ should be chosen so that each phase vector $e^{j\theta^{(m)}}$ is independent of the other $M-1$ phase vectors [39]. Because this phase sequence selection creates statistically independent mappings, the PAR CCDF for the Nyquist-sampled case can be written as

$$\Pr(\text{PAR}\{\mathbf{x}^{(m)}\} > \gamma) = (1 - (1 - e^{-\gamma})^N)^M.$$

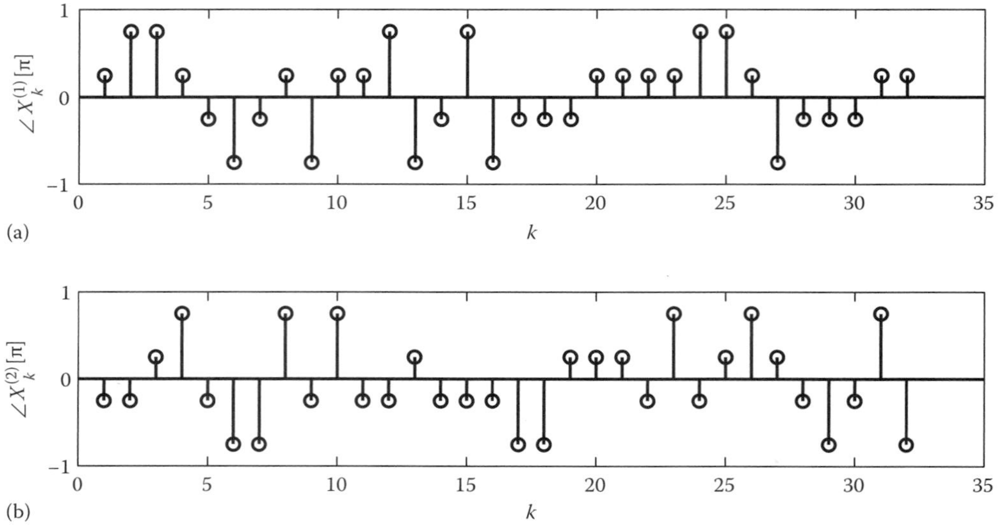

FIGURE 11.15 Phase sequence of the OFDM symbol, before and after phase rotations. (a) Original OFDM symbol, phase sequence. (b) Transformed symbol, phase sequence.

In SLM, the process used to phase the OFDM subcarriers directly impacts how peaky the corresponding time-domain sequence will be. For example, Figure 11.15a shows the phases of a frequency-domain OFDM symbol with $N = 32$ subcarriers. The subcarriers are modulated with QPSK constellation points, $y_k \in \{-1, 1, j, -j\}$, thus the phases take on discrete values $\{\pm\pi/4, \pm 3\pi/4\}$. The corresponding sequence in the discrete-time domain is shown in Figure 11.17a with a PAR value of 10.15 dB. The lone peak at discrete-time index 30 is responsible for the large PAR value.

Suppose that we apply a phase rotation sequence $\theta_k^{(m)}$ as shown in Figure 11.16 to the original frequency-domain OFDM symbol. The resulting phase sequence after the phase rotations is shown in Figure 11.15b. The phase rotation sequence takes on discrete values $\theta_k^{(m)} \in \{-\pi, 0, \pm\pi/4\}$ with roughly equal probabilities. The resulting discrete-time-domain sequence after the frequency-domain phase rotations is shown in Figure 11.17b. The large peak disappeared and the PAR value is reduced to 3.05 dB. Thus, a 7.1 dB PAR reduction was achieved simply through phase changes in the frequency domain.

The sequence in Figure 11.17b is an alternative low-PAR representation of the sequence in Figure 11.17a. If the receiver has knowledge of the phase rotation sequence $\theta^{(\tilde{m})}$, then recovery of the original OFDM symbol is feasible.

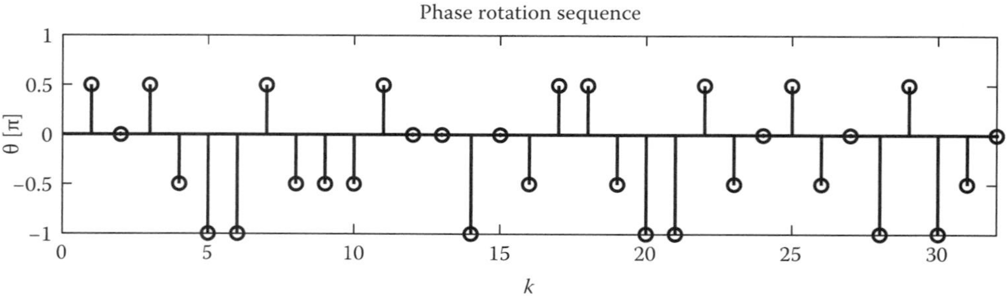

FIGURE 11.16 Phase rotation sequence.

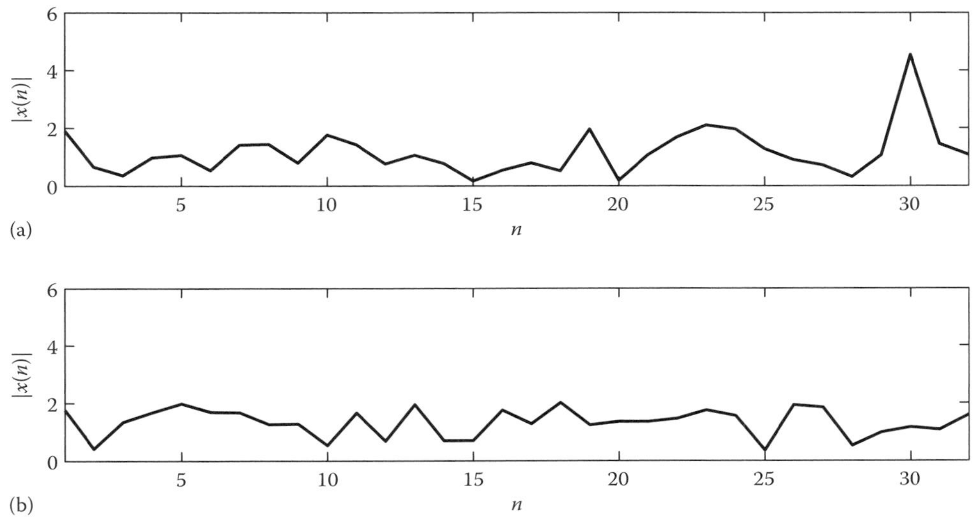

FIGURE 11.17 Discrete-time domain sequence, before and after frequency-domain phase rotations. (a) Original discrete-time sequence. (b) Transformed discrete-time sequence.

In the SLM scheme, the $\theta^{(\tilde{m})}$ sequence itself is not transmitted to the receiver. Typically, a table of M phase rotation sequences is stored at both the transmitter and the receiver. It is thus sufficient that the receiver only knows the index $\tilde{m}$ of the transmitted phase rotation sequence.

One cannot always expect such dramatic PAR reduction as seen in this particular example, since the optimal $\boldsymbol{\theta}^{(m)}$ sequence (for the given OFDM symbol) may not belong to the predefined phase rotation table. However, by trying a number of different phase rotation sequences, impressive PAR reduction results can generally be achieved. Of course, the larger the PAR value to begin with, the more room there is to improve.

For SLM the inverse mapping is simply a phase reversal of the transmit mapping,

$$\mathcal{T}_m^{-1}(\mathbf{r}) = \mathbf{D}_{e^{-j\theta(m)}}\mathbf{r}.$$

From this receiver structure, it is clear that SLM depends on the receiver having knowledge of all possible mappings and being able to detect $\tilde{m}$. Various methods have been proposed to recover $\tilde{m}$ through side information and with blind receiver estimation [40–42].

Another clever method that falls into the multiple mapping framework is tone injection (TI) [22,43]. In TI, the mapping is

$$\mathcal{T}_m(\mathbf{y}) = \mathbf{y} + z\mathbf{t}_m,$$

where z is a constant real number larger than the minimum distance of the constellation $\mathcal{A}$, where $\mathbf{y} \in \mathcal{A}^{\mathbf{N}}$ and $\mathbf{t}_m \in \mathbb{Z}^N$. Unlike SLM, TI does not necessarily preserve the average power of the symbol, i.e., $\|\mathcal{T}_m(\mathbf{y})\|_2^2 \neq \|\mathbf{y}\|_2^2$, which is a disadvantage because it means that more power is required to transmit the same information. The hope is that any average power increase will be outweighed by a power savings achieved through PAR reduction. The drawback with TI is that finding the PAR minimizing vector $\mathbf{t}^{\star}$ is an exponentially complex problem. Accordingly, candidate $\mathbf{t}^{(m)}$ values can be found by randomly searching a predefined subset of the integers.

On the other hand, the inverse mapping in TI has the advantage that no knowledge of $\tilde{m}$ is required to perform the inverse. This is because a simple element-wise modulo operation yields an exact inverse. Thus, for TI

$$\mathcal{T}_m^{-1}(\mathbf{r}) = \mathbf{r} \bmod z,$$

where $\mathbf{r} \bmod z$ is the element-wise modulo-z operation on $\mathbf{r}$, which has no dependence on m. Other mappings not based on the modulo operation and that do not require knowledge of $\tilde{m}$ are possible, but may require that both the transmitter and receiver use a look-up table. TI has the advantage that no look-up table is necessary.

11.6.3.2 Tone Reservation

Tone reservation (TR) is a PAR reduction method, where certain tones that would otherwise carry information symbols are allocated solely to PAR reduction [22,44]. That is, certain portions of the transmit spectrum are "reserved" solely for PAR reduction energy. TR is a receiver-dependent method because in a typical communications system the receiver is expecting to see information on data subcarriers. To an uninformed receiver, the PAR reduction energy would be seen as severe distortion in a subcarrier, where a constellation point is expected. Conversely, an informed receiver will know to ignore reserved tones.

TR can be framed similar to the receiver transparent functions

$$\begin{aligned} &\underset{\mathbf{d}}{\text{minimize}} \quad \text{PAR}\{\mathbf{Q}(\mathbf{y} + \mathbf{d})\} \\ &\text{subject to} \quad \|\mathbf{d}_{\mathcal{W}}\|_2 = 0 \end{aligned} \tag{11.57}$$

where $\mathcal{W}$ is the set of subcarriers that are not reserved for PAR reduction. So, Equation 11.57 will require that the distortion in all nonreserved subcarriers be zero while the distortion in the reserved subcarriers is

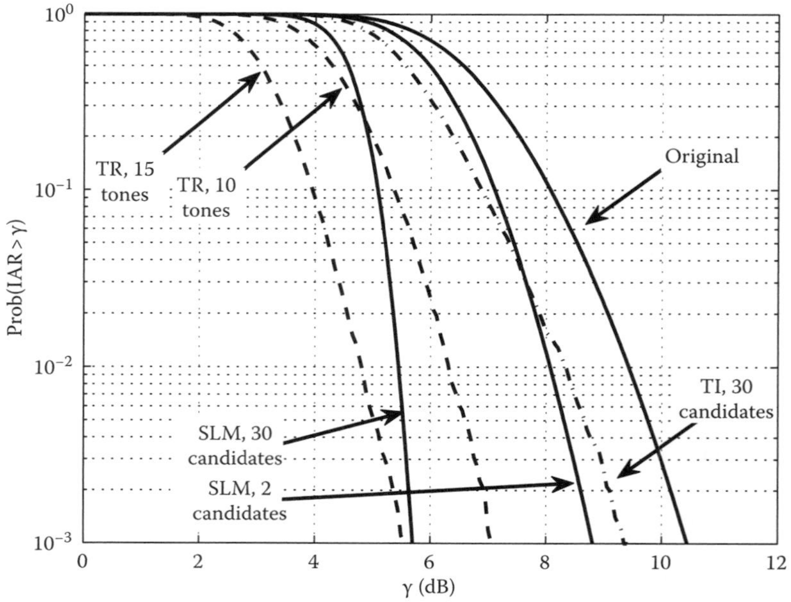

FIGURE 11.18 PAR CCDF of receiver-dependent methods. For the plot, $N = 64$.

optimized to minimize PAR. Obviously for TR to work, the receiver has to have knowledge of the set of subcarriers $\mathcal{W}$. And, it has been shown that this set should be randomly disbursed among the entire bandwidth to achieve the best PAR reduction results [22].

Figure 11.18 is a plot of the performance for several of the receiver-dependent methods. The plot shows the performance for TR with 10 and 15 reserved tones out of a total of $N = 64$ available tones as well as the performance of SLM and TI. TI appears to have the worst PAR performance, but keep in mind that no rate loss is needed to achieve this PAR. On the other hand, TR will lose $10/64 = 15.6\%$ or $15/64 = 23.4\%$ of the possible information rate to achieve these results. Because the loss of rate leads to an effective increase in transmit power, there is an interesting optimization of the number of tones to reserve to maximize the achievable information throughput (mutual information) which is explored in [45].

Several methods have been proposed to avoid side information exchange for SLM, making it possible to utilize SLM without rate loss [40–42]. These methods require more computational resources at the receiver than standard decoding, but if these resources are available, SLM is a very useful method, as can been seen in Figure 11.18.

11.7 Summary

As one can imagine, the methods discussed here are not necessarily mutually exclusive. By utilizing more than one reduction technique, even better PAR reduction is possible compared to each individual scheme. In fact, literature has shown that by synergistically combining methods, it is possible to see PAR reduction performance commensurate with a combined method while avoiding some of the overhead that would be required to utilize each method independently [46,47].

The present chapter has provided a brief overview of the PAR reduction field. This and related topics are still the subject of active research. The results on distortion optimization via interior point optimization are relatively new and such optimization techniques remain a fruitful tool in answering fundamental questions about achievable throughput of PAR-reduced systems. Also unaddressed is a comprehensive comparative analysis of existing constraint functions. With any constraint function possible, there is no evidence that the commonly used Euclidean distance norm distortion constraint is in some sense optimal. Promising alternatives like ACE [44] exist, but it is not clear what the best choice is.

Additionally, interesting new variations on this topic include PAR reduction for multiuser signaling, multiple antenna scenarios, and heterogeneous waveform systems. While some methods can be quickly adjusted to suit one or all of these cases, there is still room for improvement. Also, in these systems, the degrees of freedom is increased and new distortion constraints are present. Presumably, as-yet unproposed methods exist to take advantage of this new versatility.

References

1. J. Walko, Green issues challenge base station power, *EE Times Europe*, September 2007.
2. T. Hasan, Nonlinear time series regression for a class of amplitude modulated consinusoids, *Journal of Time Series Analysis*, 3(2), 109–122, 1982.
3. S. C. Cripps, *RF Power Amplifiers for Wireless Communications*. Norwood, MA: Artech House, 1999.
4. K. J. Muhonen, M. Kavehrad, and R. Krishnamoorthy, Look-up table techniques for adaptive digital predistortion: A development and comparison, *IEEE Transactions on Vehicular Technology*, 49, 1995–2002, September 2000.
5. H.-H. Chen, C.-H. Lin, P.-C. Huang, and J.-T. Chen, Joint polynomial and look-up-table predistortion power amplifier linearization, *IEEE Transactions on Circuits and Systems II: Express Briefs*, 53, 612–616, August 2006.

6. W.-J. Kim, K.-J. Cho, S. P. Stapleton, and J.-H. Kim, Piecewise pre-equalized linearization of the wireless transmitter with a Doherty amplifier, *IEEE Transactions on Microwave Theory and Techniques*, 54, 3469–3478, September 2006.
7. P. Julian, A. Desages, and O. Agamennoni, High-level canonical piecewise linear representation using asimplicial partition, *IEEE Transactions on Circuits and Systems I: Fundamental Theory and Applications*, 46, 463–480, April 1999.
8. C. Eun and E. J. Powers, A new volterra predistorter based on the indirect learning architecture, *IEEE Transactions on Signal Processing*, 45, 223–227, January 1997.
9. R. Raich, H. Qian, and G. T. Zhou, Orthogonal polynomials for power amplifier modeling and predistorter design, *IEEE Transactions on Vehicular Technology*, 53, 1468–1479, September 2004.
10. L. Ding, G. T. Zhou, D. R. Morgan, Z. Ma, J. S. Kenney, J. Kim, and C. R. Giardina, A robust digital baseband predistorter constructed using memory polynomials, *IEEE Transactions on Communications*, 52, 159–165, January 2004.
11. G. T. Zhou, H. Qian, and N. Chen, Communication system nonlinearities: Challenges and some solutions. In *Advances in Nonlinear Signal and Image Processing*, vol. 6 of *EURASIP Book Series on Signal Processing and Communications*. New York: Hindawi, pp. 141–167, 2006.
12. R. Raich, Nonlinear system identification and analysis with applications to power amplifier modeling and power amplifier predistortion. PhD thesis, Georgia Institute of Technology, Atlanta, GA, May 2004.
13. S. A. Maas, *Nonlinear Microwave Circuits*. New York: IEEE Press, 1997.
14. D. Brillinger, *Time Series Data Analysis and Theory*. Philadelphia, PA: SIAM, 2001.
15. H. Ochiai, Performance analysis of peak power and band-limited OFDM system with linear scaling, *IEEE Transactions on Wireless Communications*, 2, 1055–1065, September 2003.
16. C. Zhao, R. J. Baxley, and G. T. Zhou, Peak-to-average power ratio and power efficiency considerations in MIMO-OFDM systems, *IEEE Communications Letters*, 12, 268–270, April 2008.
17. M. R. Schroeder, *Number Theory in Science and Communication*. Berlin, Germany: Springer, 1997.
18. S. Litsyn, *Peak Power Control in Multicarrier Communications*. Cambridge, U.K.: Cambridge University Press, January 2007.
19. Z. Wang, X. Ma, and G. B. Giannakis, OFDM or single-carrier block transmissions? *IEEE Transactions on Communications*, 52, 380–394, March 2004.
20. R. D. J. van Nee, *OFDM for Wireless Multimedia Communications*. Norwood, MA: Artech House Publishers, 1999.
21. Z. Wang and G. B. Giannakis, Wireless multicarrier communications, *IEEE Signal Processing Magazine*, 17, 29–48, May 2000.
22. J. Tellado, *Multicarrier Modulation With Low PAR: Applications to DSL and Wireless*. Norwell, MA: Kluwer Academic Publishers, 2000.
23. H. Ochiai and H. Imai, On the distribution of the peak-to-average power ratio in OFDM signals, *IEEE Transactions on Communications*, 49, 282–289, February 2001.
24. X. Li and L. J. Cimini, Effects of clipping and filtering on the performance of OFDM, *IEEE Communications Letters*, 2, 131–133, May 1998.
25. J. Armstrong, Peak-to-average power reduction for OFDM by repeated clipping and frequency domain filtering, *Electronics Letters*, 38, 246–247, February 2002.
26. R. J. Baxley, C. Zhao, and G. T. Zhou, Constrained clipping for crest factor reduction in OFDM, *IEEE Transactions on Broadcasting*, 52, 570–575, December 2006.
27. C. Zhao, R. J. Baxley, G. T. Zhou, D. Boppana, and J. S. Kenney, Constrained clipping for crest factor reduction in multiple-user ofdm. In *Proceedings IEEE Radio and Wireless Symposium*, Boston, MA, pp. 341–344, January 2007.
28. A. Aggarwal and T. H. Meng, Minimizing the peak-to-average power ratio of OFDM signals using convex optimization, *IEEE Transactions on Signal Processing*, 54, 3099–3110, August 2006.

29. Q. Liu, R. J. Baxley, X. Ma, and G. T. Zhou, Error vector magnitude optimization for OFDM systems with a deterministic peak-to-average power ratio constraint. In *Proceedings IEEE Conference on Information Sciences and Systems*, Princeton, NJ, pp. 101–104, March 2008.
30. Q. Liu, R. J. Baxley, X. Ma, and G. T. Zhou, Error vector magnitude optimization for OFDM systems with a deterministic peak-to-average power ratio constraint, *IEEE Journal on Selected Topics in Signal Processing*, 3(3), 418–429, June 2009.
31. B. S. Krongold and D. L. Jones, PAR reduction in OFDM via active constellation extension, *IEEE Transactions on Broadcasting*, 49, 258–268, September 2003.
32. S. Boyd and L. Vandenberghe, *Convex Optimization*. Cambridge, U.K.: Cambridge University Press, 2004.
33. R. J. Baxley and J. E. Kleider, Embedded synchronization/pilot sequence creation using POCS, In *Proceedings IEEE International Conference on Acoustics, Speech and Signal Processing*, Atlanta, GA, pp. 321–324, May 2006.
34. R. W. Bauml, R. F. H. Fischer, and J. B. Huber, Reducing the peak-to-average power ratio of multicarrier modulation by selected mapping, *Electronics Letters*, 32, 2056–2057, October 1996.
35. R. J. Baxley and G. T. Zhou, Comparing selected mapping and partial transmit sequence for PAR reduction, *IEEE Transactions on Broadcasting*, 53, 797–803, December 2007.
36. P. W. J. Van Eetvelt, G. Wade, and M. Tomlinson, Peak to average power reduction for OFDM schemes by selective scrambling, *Electronics Letters*, 32, 1963–1964, October 1996.
37. D. Mesdagh and P. Spruyt, A method to reduce the probability of clipping in DMT-based transceivers, *IEEE Transactions on Communications*, 44, 1234–1238, October 1996.
38. L. J. Cimini and N. R. Sollenberger, Peak-to-average power ratio reduction of an OFDM signal using partial transmit sequences. In *Proceedings IEEE International Conference on Communications*, vol. 1, Vancouver, BC, pp. 511–515, June 1999.
39. G. T. Zhou and L. Peng, Optimality condition for selected mapping in OFDM, *IEEE Transactions on Signal Processing*, 54, 3159–3165, August 2006.
40. R. J. Baxley and G. T. Zhou, MAP metric for blind phase sequence detection in selected mapping, *IEEE Transactions on Broadcasting*, 51, 565–570, December 2005.
41. N. Chen and G. T. Zhou, Peak-to-average power ratio reduction in OFDM with blind selected pilot tone modulation, *IEEE Transactions on Wireless Communications*, 5, 2210–2216, August 2006.
42. A. D. S. Jayalath and C. Tellambura, SLM and PTS peak-power reduction of OFDM signals without side information, *IEEE Transactions on Wireless Communications*, 4, 2006–2013, September 2005.
43. S. H. Han, J. M. Cioffi, and J. H. Lee, Tone injection with hexagonal constellation for peak-to-average power ratio reduction in OFDM, *IEEE Communications Letters*, 10, 646–648, September 2006.
44. B. S. Krongold and D. L. Jones, An active-set approach for OFDM PAR reduction via tone reservation, *IEEE Transactions on Signal Processing*, 52, 495–509, February 2004.
45. Q. Liu, R. J. Baxley, and G. T. Zhou, Free subcarrier optimization for peak-to-average power ratio minimization in OFDM systems. In *Proceedings IEEE International Conference on Acoustics, Speech and Signal Processing*, Las Vegas, NV, pp. 3073–3076, March 2008.
46. N. Chen and G. T. Zhou, Superimposed training for OFDM: A peak-to-average power ratio analysis, *IEEE Transactions on Signal Processing*, 54, 2277–2287, June 2006.
47. R. J. Baxley, J. E. Kleider, and G. T. Zhou, A method for joint peak-to-average power ratio reduction and synchronization in OFDM. In *Proceedings IEEE Military Communications Conference*, Orlando, FL, pp. 1–6, October 2007.

12

Space-Time Adaptive Processing for Airborne Surveillance Radar

Hong Wang
Syracuse University

12.1 Main Receive Aperture and Analog Beamforming 12-2
12.2 Data to Be Processed 12-3
12.3 Processing Needs and Major Issues 12-4
12.4 Temporal DOF Reduction 12-7
12.5 Adaptive Filtering with Needed and Sample-Supportable DOF and Embedded CFAR Processing 12-8
12.6 Scan-to-Scan Track-before-Detect Processing 12-10
12.7 Real-Time Nonhomogeneity Detection and Sample Conditioning and Selection 12-10
12.8 Space or Space-Range Adaptive Presuppression of Jammers 12-10
12.9 A STAP Example with a Revisit to Analog Beamforming 12-11
12.10 Summary 12-13
References 12-13

Space-time adaptive processing (STAP) is a multidimensional filtering technique developed for minimizing the effects of various kinds of interference on target detection with a pulsed airborne surveillance radar. The most common dimensions, or filtering domains, generally include the azimuth angle, elevation angle, polarization angle, Doppler frequency, etc. in which the relatively weak target signal to be detected and the interference have certain differences. In the following, the STAP principle will be illustrated for filtering in the joint azimuth angle (space) and Doppler frequency (time) domain only.

STAP has been a very active research and development area since the publication of Reed et al.'s seminal paper [1]. With the recently completed Multichannel Airborne Radar Measurement project (MCARM) [2–5], STAP has been established as a valuable alternative to the traditional approaches, such as ultra-low sidelobe beamforming and displaced phase center antenna (DPCA) [6]. Much of STAP research and development efforts have been driven by the needs to make the system affordable, to simplify its front-hardware calibration, and to minimize the system's performance loss in severely nonhomogeneous environments. Figure 12.1 is a general configuration of STAP functional blocks [5,7] whose principles will be discussed in the following sections.

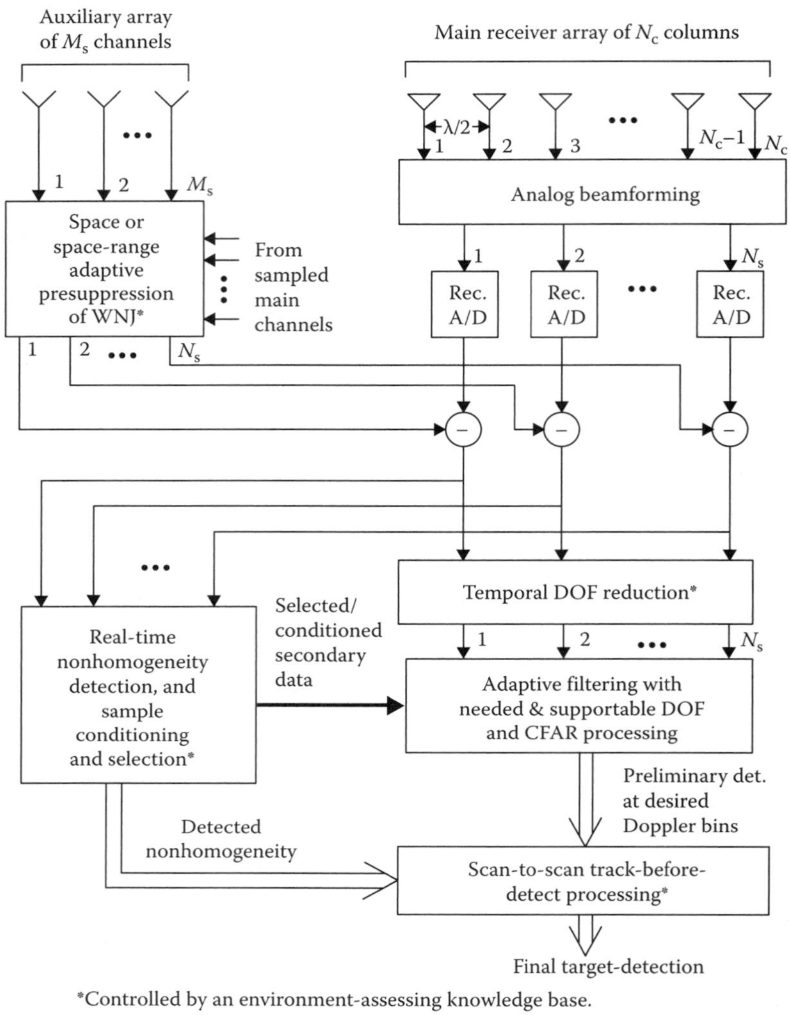

FIGURE 12.1 A general STAP configuration with auxiliary and main arrays.

12.1 Main Receive Aperture and Analog Beamforming

For conceptual clarity, the STAP configuration of Figure 12.1 separates a possibly integrated aperture into two parts: the main aperture which is most likely shared by the radar transmitter, and an auxiliary array of spatially distributed channels for suppression of wideband noise jammers (WNJ). For convenience of discussion, the main aperture is assumed to have N_c columns of elements, with the column spacing equal to a half wavelength and elements in each column being combined to produce a predesigned, non-adaptive elevation beam-pattern.

The size of the main aperture in terms of the system's chosen wavelength is an important system parameter, usually determined by the system specifications of the required transmitter power-aperture product as well as azimuth resolution. Typical aperture size spans from a few wavelengths for some short-range radars to over 60 wavelengths for some airborne early warning systems. The analog beamforming network combines the N_c columns of the main aperture to produce N_s receiver channels whose outputs are digitized for further processing. One should note that the earliest STAP approach presented in [1], i.e., the so-called element-space approach, is a special case of Figure 12.1 when $N_s = N_c$ is chosen.

The design of the analog beamformer affects

1. The system's overall performance (especially in nonhomogeneous environments)
2. Implementation cost
3. Channel calibration burden
4. System reliability
5. Controllability of the system's response pattern

The design principle will be briefly discussed in Section 12.9; and because of the array's element error, column-combiner error, and column mutual-coupling effects, it is quite different from what is available in the adaptive array literature such as [8], where already digitized, perfectly matched channels are generally assumed.

Finally, it should be pointed out that the main aperture and analog beamforming network in Figure 12.1 may also include nonphased-array hardware, such as the common reflector-feed as well as the hybrid reflector and phased-array feed [9]. Also, subarraying such as [10] is considered as a form of analog beamforming of Figure 12.1.

12.2 Data to Be Processed

Assume that the radar transmits, at each look angle, a sequence of N_t uniformly spaced, phase-coherent RF pulses as shown in Figure 12.2 for its envelope only. Each of N_s receivers typically consists of a front-end amplifier, down-converter, waveform-matched filter, and A/D converter with a sampling frequency at least equal to the signal bandwidth. Consider the kth sample of radar return over the N_t pulse repetition intervals (PRI) from a single receiver, where the index "k" is commonly called the range index or cell. The total number of range cells, K_0 is approximately equal to the product of the PRI and signal bandwidth. The coherent processing interval (CPI) is the product of the PRI and N_t; and since a fixed PRI can usually be assumed at a given look angle, CPI and N_t are often used interchangeably.

With N_s receiver channels, the data at the kth range cell can be expressed by a matrix X_k, $N_s \times N_t$, for $k = 1, 2, \ldots, K_0$. The total amount of data visually forms a "cube" shown in Figure 12.3, which is the raw data cube to be processed at a given look angle. It is important to note from Figure 12.3 that the term "time" is associated with the CPI for any given range cell, i.e., across the multiple PRIs, while the term "range" is used within a PRI. Therefore, the meaning of the frequency corresponding to the time is the so-called Doppler frequency, describing the rate of the phase-shift progression of a return component with respect to the initial phase of the phase-coherent pulse train. The Doppler frequency of a return, e.g., from a moving target, depends on the target velocity and direction as well as the airborne radar's platform velocity and direction, etc.

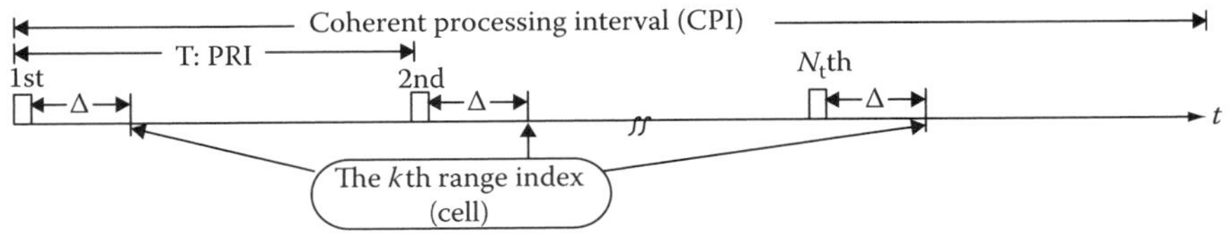

FIGURE 12.2 A sequence of N_t phase-coherent RF pulses (only envelope shown) transmitted at a given angle. The pulse repetition frequency (PRF) is $1/T$.

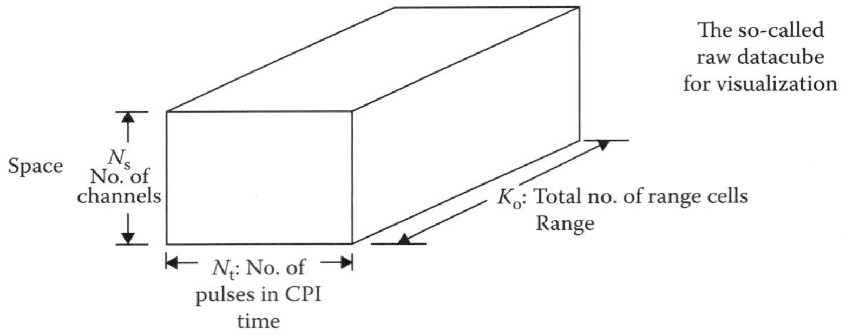

FIGURE 12.3 Raw data at a given look angle and the space, time, and range axes.

12.3 Processing Needs and Major Issues

At a given look angle, the radar is to detect the existence of targets of unknown range and unknown Doppler frequency in the presence of various interference. In other words, one can view the processing as a mapping from the data cube to a range-Doppler plane with sufficient suppression of unwanted components in the data. Like any other filtering, the interference suppression relies on the differences between wanted target components and unwanted interference components in the angle-Doppler domain. Figure 12.4 illustrates the spectral distribution of potential interference in the spatial and temporal (Doppler) frequency domain before the analog beamforming network, while Figure 12.5 shows a typical range distribution of interference power. As targets of interest usually have unknown Doppler frequencies and unknown distances, detection needs to be carried out at sufficiently dense Doppler frequencies along the look angle for each range cell within the system's surveillance volume. For each cell at which target detection is being carried out, some of surrounding cells can be used to produce an estimate of interference statistics (usually up to the second order), i.e., providing "sample support," under the assumption that all cells involved have an identical statistical distribution. Figure 12.4 also shows that,

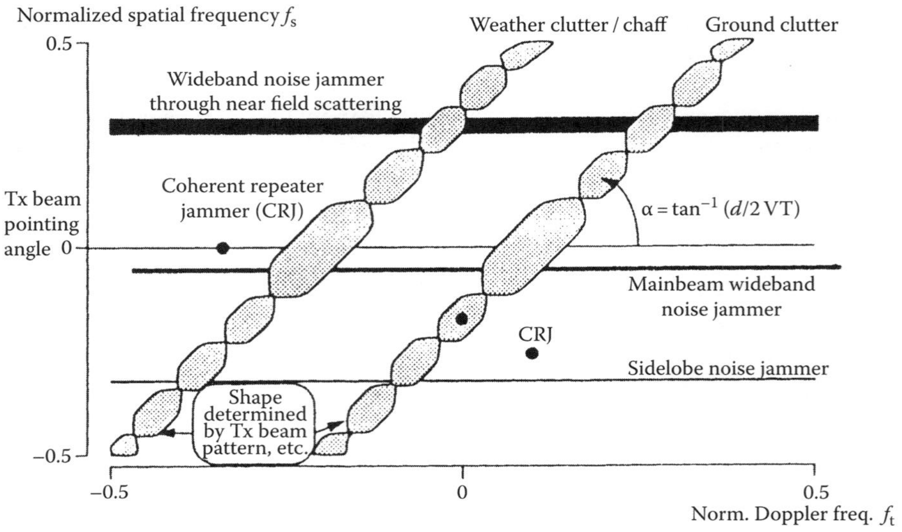

FIGURE 12.4 Illustration of interference spectral distribution for a side-mounted aperture.

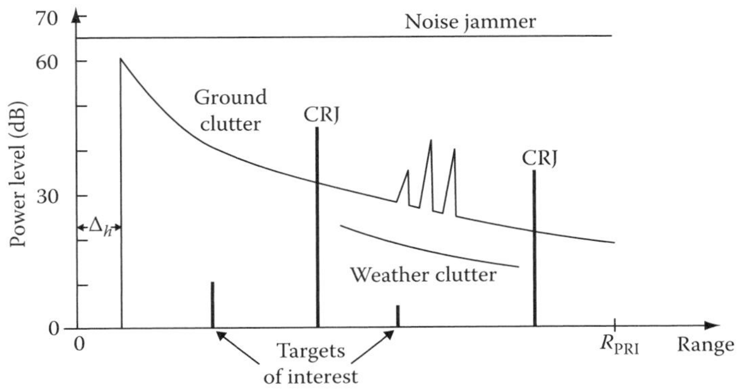

FIGURE 12.5 Illustration of interference-power range distribution, where Δ_h indicates the radar platform height.

in terms of their spectral differences, traditional WNJ, whether entering the system through direct path or multipath (terrain scattering/near-field scattering), require spatial nulling only; while clutter and chaff require angle-Doppler coupled nulling. Coherent repeater jammers (CRJ) represent a nontraditional threat of a target-like spectral feature with randomized ranges and Doppler frequencies, making them more harmful to adaptive systems than to conventional nonadaptive systems [11].

Although Figure 12.5 has already served to indicate that the interference is nonhomogeneous in range, i.e., its statistics vary along the range axis, recent airborne experiments have revealed that its severeness may have long been underestimated, especially over land [3]. Figure 12.6 [5,7] summarizes the sources of various nonhomogeneity together with their main features. As pointed out in [12], a serious problem associated with any STAP approach is its basic assumption that there is a sufficient amount of sample support for its adaptive learning, which is most often void in real environments even in the absence of any nonhomogeneity type of jammers such as CRJ. Therefore, a crucial issue for the

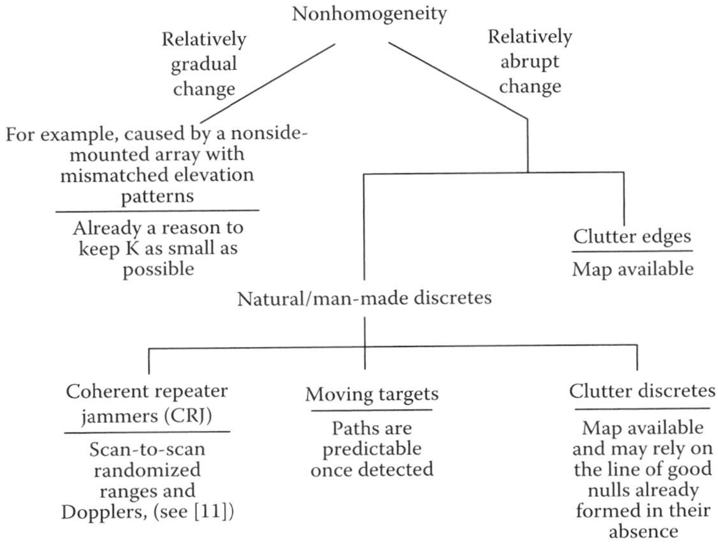

FIGURE 12.6 Typical nonhomogeneities.

success of STAP in real environments is the development of data-efficient STAP approaches, in conjunction with the selection of reasonably identically distributed samples before estimating interference statistics. To achieve a sufficient level of the data efficiency in nonhomogeneous environments, the three most performance- and cost-effective methods are temporal degrees-of-freedom (DOF) reduction, analog beamforming to control the spatial DOF creation, and presuppression of WNJ as shown in Figure 12.1.

Another crucial issue is the affordability of STAP-based systems. As pointed out in [9], phased arrays, especially those active ones (i.e., with the so-called T/R modules), remain very expensive despite the 30 year research and development. For multichannel systems, the cost of adding more receivers and A/D converters with a sufficient quality makes the affordability even worse.

Of course, more receiver channels mean more system's available spatial DOF. However, it is often the case in practice that the excessive amount of the DOF, e.g., obtained via one receiver channel for each column of a not-so-small aperture, is not necessary to the system. Ironically, excessive DOF can make the control of the response pattern more difficult, even requiring significant algorithm constraints [8]; and after all, it has to be reduced to a level supportable by the available amount of reasonably identically distributed samples in real environments. An effective solution, as demonstrated in a recent STAP experiment [13], is via the design of the analog beamformer that does not create unnecessary spatial DOF from the beginning—a sharp contrast to the DOF reduction/constraint applied in the spatial domain.

Channel calibration is a problem issue for many STAP approaches. In order to minimize performance degradation, the channels with some STAP approaches must be matched across the signal band, and steering vectors must be known to match the array. Considering the fact that channels generally differ in both elevation and azimuth patterns (magnitude as well as phase) even at a fixed frequency, the calibration difficulty has been underestimated as experienced in recent STAP experiments [5]. It is still commonly wished that the so-called element-space approaches, i.e., the special case of $N_s = N_c$ in Figure 12.1, with an adaptive weight for each error-bearing "element" which hopefully can be modeled by a complex scalar, could solve the calibration problem at a significantly increased system-implementation cost as each element needs a digitized receiver channel. Unfortunately, such a wish can rarely materialize for a system with a practical aperture size operated in nonhomogeneous environments. With a spatial DOF reduction required by these approaches to bring down the number of adaptive weights to a sample-supportable level, the element errors are no longer directly accessible by the adaptive weights, and thus the wishful "embedded robustness" of these element-space STAP approaches is almost gone. In contrast, the MCARM experiment has demonstrated that, by making best use of what has already been excelled in antenna engineering [13], the channel calibration problem associated with STAP can be largely solved at the analog beamforming stage, which will be discussed in Section 12.9.

The above three issues all relate to the question: "What is the minimal spatial and temporal DOF required?" To simplify the answer, it can be assumed first that clutter has no Doppler spectral spread caused by its internal motion during the CPI, i.e., its spectral width cut along the Doppler frequency axis of Figure 12.4 equals to zero. For WNJ components of Figure 12.4, the required minimal spatial DOF is well established in array processing, and the required minimal temporal DOF is zero as no temporal processing can help suppress these components. The CRJ components appear only in isolated range cells as shown in Figure 12.5, and thus they should be dealt with by sample conditioning and selection so that the system response does not suffer from their random disturbance. With the STAP configuration of Figure 12.1, i.e., presuppression of WNJ and sample conditioning and selection for CRJ, the only interference components left are those angle-Doppler coupled clutter/chaff spectra of Figure 12.4. It is readily available from the two-dimensional filtering theory [14] that suppression of each of these angle-Doppler coupled components only requires one spatial DOF and one temporal DOF of the joint domain processor! In other words, a line of infinitely many nulls can be formed with one spatial DOF and one temporal DOF on top of one angle-Doppler coupled interference component under the assumption that there is no clutter internal motion over the CPI. It is also understandable that, when such an assumption is not valid, one only needs to increase the temporal DOF of the processor so that the null width along the Doppler axis can be correspondingly increased.

For conceptual clarity, $N_s - 1$ will be called the system's available spatial DOF and $N_t - 1$ the system's available temporal DOF. While the former has a direct impact on the implementation cost, calibration burden, and system reliability, the latter is determined by the CPI length and PRI with little cost impact, etc. Mainly due to the nonhomogeneity-caused sample support problem discussed earlier, the adaptive joint domain processor may have its spatial DOF and temporal DOF, denoted by N_{ps} and N_{pt} respectively, different from the system's available by what is so-called DOF reduction. However, the spatial DOF reduction should be avoided by establishing the system's available spatial DOF as close to what is needed as possible from the beginning.

12.4 Temporal DOF Reduction

Typically an airborne surveillance radar has N_t anywhere between 8 and 128, depending on the CPI and PRI. With the processor's temporal DOF, N_{pt}, needed for the adjustment of the null width, normally being no more than $2 \sim 4$, huge DOF reduction is usually performed for the reasons of the sample support and better response-pattern control explained in Section 12.3.

An optimized reduction could be found, given N_t, N_{pt}, and the interference statistics which are still unknown at this stage of processing in practice [7]. There are several nonoptimized temporal DOF reduction methods available, such as the Doppler-domain (joint domain) localized processing (DDL/JDL) [12,15,16] and the PRI-staggered Doppler-decomposed processing (PRI-SDD) [17], which are well behaved and easy to implement. The DDL/JDL principle will be discussed below.

The DDL/JDL consists of unwindowed/untapered DFT of (at least) N_t-point long, operated on each of the N_s receiver outputs. The same $N_{pt} + 1$ most adjacent frequency bins of the DFTs of the N_s receiver outputs form the new data matrix at a given range cell, for detection of a target whose Doppler frequency is equal to the center bin. Figure 12.7 shows an example for $N_s = 3$, $N_t = 8$, and $N_{pt} = 2$. In other words, the DDL/JDL transforms the raw data cube of $N_s \times N_t \times K_0$ into (at least) N_t smaller data cubes, each of $N_s \times (N_{pt} + 1) \times K_0$ for target detection at the center Doppler bin.

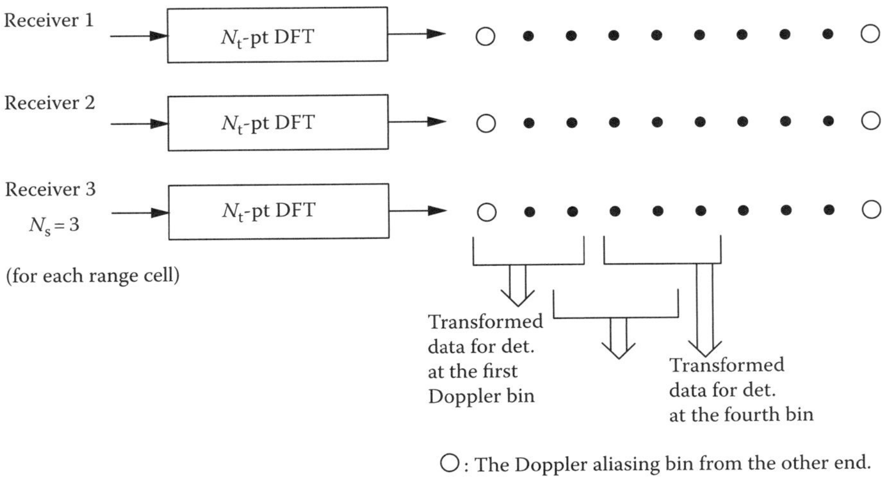

FIGURE 12.7 The DDL/JDL principle for temporal DOF reduction illustrated with $N_s = 3$, $N_t = 8$, and $N_{pt} = 2$.

The DDL/JDL is noticeable for the following features.

1. There is no so-called signal cancellation, as the unwindowed/untapered DFT provides no desired signal components in the adjacent bins (i.e., reference "channel") for the assumed target Doppler frequency.
2. The grouping of $N_{pt}+1$ most adjacent bins gives a high degree of correlation between the interference component at the center bin and those at the surrounding bins—a feature important to cancellation of any spectrum-distributed interference such as clutter. The cross-spectral algorithm [18] also has this feature.
3. The response pattern can be well controlled as N_{pt} can be kept small—just enough for the needed null-width adjustment; and N_{pt} itself easily can be adjusted to fit different clutter spectral spread due to its internal motion.
4. Obviously the DDL/JDL is suitable for parallel processing.

While the DDL/JDL is a typical transformation-based temporal DOF reduction method, other methods involving the use of DFTs are not necessarily transformation-based. An example is the PRI-SDD [17] which applies time-domain temporal DOF reduction on each Doppler component. This explains why the PRI-SDD requires N_{pt} times more DFTs that should be tapered. It also serves as an example that an algorithm classification by the existence of the DFT use may cause a conceptual confusion.

12.5 Adaptive Filtering with Needed and Sample-Supportable DOF and Embedded CFAR Processing

After the above temporal DOF reduction, the dimension of the new data cube to be processed at a given look angle for each Doppler bin is $N_s \times (N_{pt}+1) \times K_0$. Consider a particular range cell at which target detection is being performed. Let $\mathbf{x}$, $N_s(N_{pt}+1) \times 1$, be the stacked data vector of this range cell, which is usually called the primary data vector. Let $\mathbf{y}_1, \mathbf{y}_2, \ldots, \mathbf{y}_k$, all $N_s(N_{pt}+1) \times 1$ and usually called the secondary data, be the same-stacked data vectors of the K surrounding range cells, which have been selected and/or conditioned to eliminate any significant nonhomogeneities with respect to the interference contents of the primary data vector. Let $\mathbf{s}$, $N_s(N_{pt}+1) \times 1$, be the target-signal component of $\mathbf{x}$ with the assumed angle of arrival equal to the look angle and the assumed Doppler frequency corresponding to the center Doppler bin. In practice, a lookup table of the "steering vector" $\mathbf{s}$ for all look-angles and all Doppler bins usually has to be stored in the processor, based on updated system calibration. A class of STAP systems with the steering-vector calibration-free feature has been developed, and an example from [13] will be presented in Section 12.9.

There are two classes of adaptive filtering algorithms: one with a separately designed constant false alarm rate (CFAR) processor, and the other with embedded CFAR processing. The original sample matrix inversion (SMI) algorithm [1] belongs to the former, which is given by

$$\eta_{\mathrm{SMI}} = \left|\hat{\mathbf{w}}_{\mathrm{SMI}}^{H}\mathbf{x}\right|^2 \underset{\mathrm{H}_0}{\overset{\mathrm{H}_1}{\gtrless}} \eta_0 \tag{12.1}$$

where

$$\mathbf{w}_{\mathrm{SMI}} = \hat{\mathbf{R}}^{-1}\mathbf{s} \tag{12.2}$$

and

$$\hat{\mathbf{R}} = \frac{1}{K}\sum_{k=1}^{K}\mathbf{y}_k\mathbf{y}_k^H \tag{12.3}$$

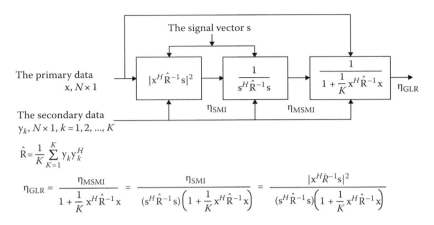

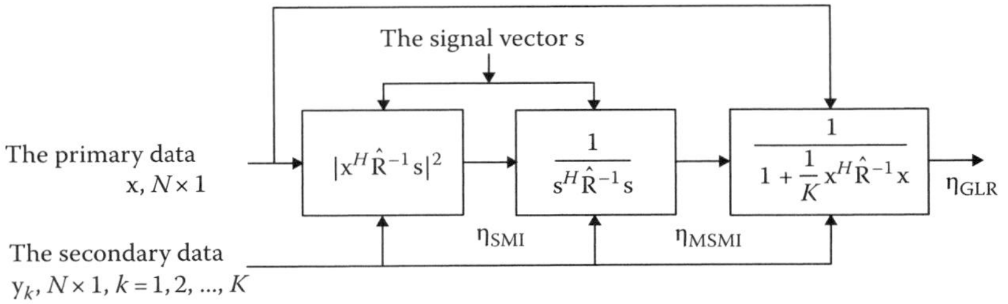

$$\hat{R} = \frac{1}{K}\sum_{K=1}^{K} y_k y_k^H$$

$$\eta_{GLR} = \frac{\eta_{MSMI}}{1+\frac{1}{K}x^H\hat{R}^{-1}x} = \frac{\eta_{SMI}}{(s^H\hat{R}^{-1}s)\left(1+\frac{1}{K}x^H\hat{R}^{-1}x\right)} = \frac{|x^H\hat{R}^{-1}s|^2}{(s^H\hat{R}^{-1}s)\left(1+\frac{1}{K}x^H\hat{R}^{-1}x\right)}$$

FIGURE 12.8 The link among the SMI, modified SMI (MSMI), and GLR where $N = (N_{ps} + 1)(N_{pt} + 1) \times 1$.

The SMI performance under the Gaussian noise/interference assumption has been analyzed in detail [1], and in general it is believed that acceptable performance can be expected if the data vectors are independent and identically distributed (iid) with K, the number of the secondary, being at least two times $N_s(N_{pt} + 1)$. Detection performance evaluation using a SINR-like measure deserves some care when K is finite, even under the iid assumption [19,20].

If the output of an adaptive filter, when directly used for threshold detection, produces a probability of false alarm independent of the unknown interference correlation matrix under a set of given conditions, the adaptive filter is said to have an embedded CFAR. Under the iid Gaussian condition, two well-known algorithms with embedded CFAR are the Modified SMI [21] and Kelly's generalized likelihood ratio detector (GLR) [22], both of which are linked to the SMI as shown in Figure 12.8. The GLR has the following interesting features:

1. $0 < \frac{1}{K}\eta_{GLR} < 1$, which is a necessary condition for robustness in non-Gaussian interference [23]
2. Invariance with respect to scaling all data or scaling **s**
3. One cannot express η_{GLR} as $\hat{\mathbf{w}}^H\mathbf{x}$; and with a finite K, an objective definition of its output SINR becomes questionable

Table 12.1 summarizes the modified SMI and GLR performance, based on [21,24].

It should be noted that the use of the scan-to-scan track-before-detect processor (SSTBD to be discussed in Section 12.6) does not make the CFAR control any less important because the SSTBD itself is not error-free even with the assumption that almost infinite computing power would be available. Moreover, the initial CFAR thresholding can actually optimize the overall performance, in addition to a dramatic reduction of the computation load of the SSTBD processor. Traditionally, filter and CFAR designs have been carried out separately, which is valid as long as the filter is not data-dependent.

TABLE 12.1 Performance Summary of Modified SMI and SLR

Performance Compared	GLR	Modified SMI
Gaussian interference suppression	Similar performance	
Non-Gaussian interference suppression	More robust	Less robust
Rejection of signals mismatched to the steering vector	Better	Worse

Therefore, such a traditional practice becomes questionable for STAP, especially when K is not very large with respect to $N_s(N_{pt} + 1)$, or when some of the secondary data depart from the iid Gaussian assumption that will affect both filtering and CFAR portions. The GLR and Modified SMI start to change the notion that "CFAR is the other guy's job," and their performance has been evaluated in some non-Gaussian interference [21] as well as in some nonhomogeneities [25]. Finally, it should be pointed out that performance evaluation of STAP algorithms with embedded CFAR by an output SINR-like measure may result in underestimating the effects of some nonhomogeneity such as the CRJ [11].

12.6 Scan-to-Scan Track-before-Detect Processing

The surveillance volume is usually visited by the radar many times, and the output data collected over multiple scans (i.e., revisits) are correlated and should be further processed together for the updated and improved final target-detection report. For example, random threshold-crossings over multiple scans due to the noise/interference suppression residue can rarely form a meaningful target trajectory and therefore their effect can be deleted from the final report with a certain level of confidence (but not error-free).

For a conventional ground-based radar, SSTBD processing has been well studied and a performance demonstration can be found in [26]. With a STAP-based airborne system, however, much remains to be researched. One crucial issue, coupled with the initial CFAR control, is to answer what is the optimal or near optimal setting of the first CFAR threshold, given an estimate of the current environment including the detected nonhomogeneity. Further discussion of this subject seems out of the scope of this book and still premature.

12.7 Real-Time Nonhomogeneity Detection and Sample Conditioning and Selection

Recent experience with MCARM Flight 5 data has further demonstrated that successful STAP system operation over land heavily relies on the handling of the nonhomogeneity contamination of samples [3,5], even without intentional nonhomogeneity producing jammers such as CRJ. It is estimated that the total number of reasonably good samples over land may be as few as 10–20. Although some system approaches to obtaining more good samples are available, such as multiband signaling [27,28], it is still essential that a system has the capability of real-time detection of nonhomogeneities, selection of sufficiently good samples to be used as the secondary, and conditioning not-so-good samples in the case of a severe shortage of the good samples. The development of a nonhomogeneity detector can be found in [3], and its integration into the system remains to be a research issue.

Finally, it should be pointed out that the utilization of a sophisticated sample selection scheme makes it nearly unnecessary to look into the so-called training strategy such as sliding window, sliding hole, etc. Also, desensitizing a STAP algorithm via constraints and/or diagonal loading has been found to be less effective than the sample selection [28].

12.8 Space or Space-Range Adaptive Presuppression of Jammers

WNJ have a flat or almost flat Doppler spectrum which means that without multipath/terrain-scattering (TS), only spatial nulling is necessary. Although STAP could handle, at least theoretically, the simultaneous suppression of WNJ and clutter simply with an increase of the processor's spatial DOF (N_{ps}), doing so would unnecessarily raise the size of the correlation matrix which, in turn, requires more samples for its estimation. Therefore, spatial adaptive presuppression (SAPS) of WNJ, followed by STAP-based clutter suppression, is preferred for systems to be operated in severely nonhomogenous environments. Space-range adaptive processing (SRAP) may become necessary in the presence of multipath/TS to exploit the correlation between the direct path and indirect paths for better suppression of the total WNJ effects on the system performance.

The idea of cascading SAPS and STAP itself is not new, and the original work can be found in [29], with other names such as "two step nulling (TSN)" used in [30]. A key issue in applying this idea is the acquisition of the necessary jammer-only statistics for adaptive suppression, free from strong clutter contamination. Available acquisition methods include the use of clutter-free range-cells for low PRF systems, clutter-free Doppler bins for high PRF systems, or receive-only mode between two CPIs. All of these techniques require jammer data to be collected within a restricted region of the available space-time domain, and may not always be able to generate sufficient jammer-only data. Moreover, fast-changing jamming environments and large-scale PRF hopping can also make these techniques unsuitable. Reference [31] presents a new technique that makes use of frequency sidebands close to, but disjointed from, the radar's mainband, to estimate the jammer-only covariance matrix. Such an idea can be applied to a system with any PRF, and the entire or any appropriate portion of the range processing interval (RPI) could be used to collect jammer data. It should be noted that wideband jammers are designed to sufficiently cover the radar's mainband, making sidebands, of more or less bandwidth, containing their energy always available to the new SAPS technique.

The discussion of the sideband-based STAP can be carried out with different system configurations, which determine the details on the sideband-to-mainband jammer information conversion, as well as the mainband jammer-cancellation signal generation. Reference [31] chooses a single array-based system, while a discussion involving an auxiliary-main array configuration can be found in [7].

12.9 A STAP Example with a Revisit to Analog Beamforming

In the early stage of STAP research, it is always assumed that $N_s = N_c$, i.e., each column consumes a digitized receiver channel, regardless of the size of the aperture. More recent research and experiments have revealed that such an "element-space" set up is only suitable for sufficiently small apertures, and the analog beamforming network has become an important integrated part of STAP-based systems with more practical aperture sizes.

The theoretically optimized analog beamformer design could be carried out for any given N_s, which yields a set of N_s nonrealizable beams once the element error, column-combiner error, and column mutual-coupling effects are factored in. A more practical approach is to select, from what antenna design technology has excelled, those beams that also meet the basic requirements for successful adaptive processing, such as the "signal blocking" requirement developed under the generalized sidelobe canceller [32]. Two examples of proposed analog beamforming methods for STAP applications are (1) multiple shape-identical Fourier beams via the Butler matrix [12], and (2) the sum and difference beams [13]. Both selections have been shown to enable the STAP system to achieve near optimal performance with N_s very close to the theoretical minimum of two for clutter suppression.

In the following, the clutter suppression performance of a STAP with the sum (Σ)-difference (Δ) beams is presented using the MCARM Flight 2 data. The clutter in this case was collected from a rural area in the eastern shore region south of Baltimore, Maryland. A known target signal was injected at a Doppler frequency slightly offset from mainlobe clutter and the results compared for the factored approach (FA-STAP) [16] and $\Sigma\Delta$-STAP. A modified SMI processor was used in each case to provide a known threshold level based on a false alarm probability of 10^{-6}. As seen in Figures 12.9 and 12.10, the injected target lies below the detection threshold for FA-STAP, but exceeds the threshold in the case of $\Sigma\Delta$-STAP. This performance was obtained using far fewer samples for covariance estimation in the case of $\Sigma\Delta$-STAP. Also, the $\Sigma\Delta$-STAP uses only two receiver channels, while the FA-STAP consumes all 16 channels.

In terms of calibration burden, the $\Sigma\Delta$-STAP uses two different channels to begin with and its corresponding signal (steering) vector easily remains the simplest form as long as the null of the Δ beam is correctly placed (a job in which antenna engineers have excelled already). In that sense, the $\Sigma\Delta$-STAP is both channel calibration-free and steering-vector calibration-free. On the other hand, keeping the 16 channels of FA-STAP calibrated and updating its steering vector lookup table have been a considerable burden during the MCARM experiment [4].

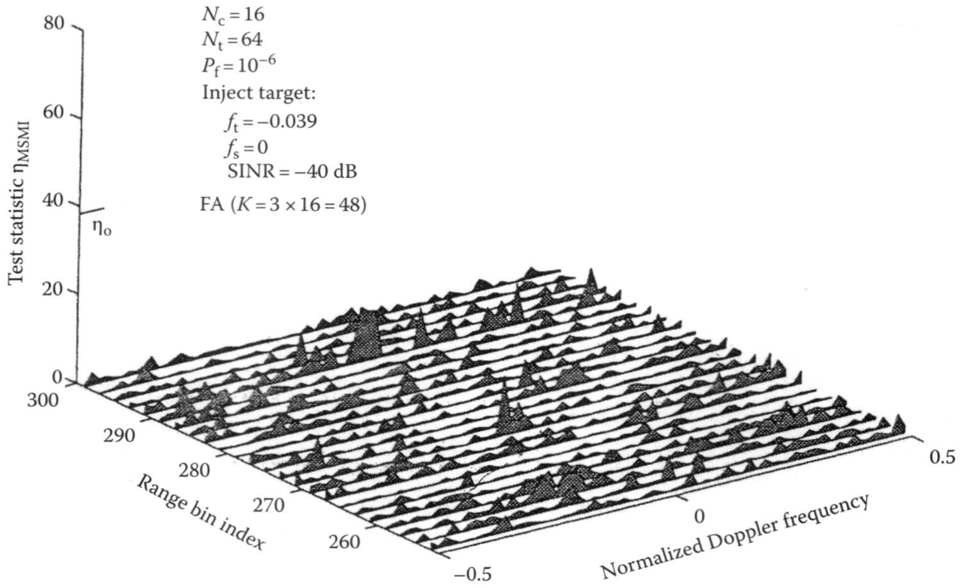

FIGURE 12.9 Range-Doppler plot of MCARM data, factored approach.

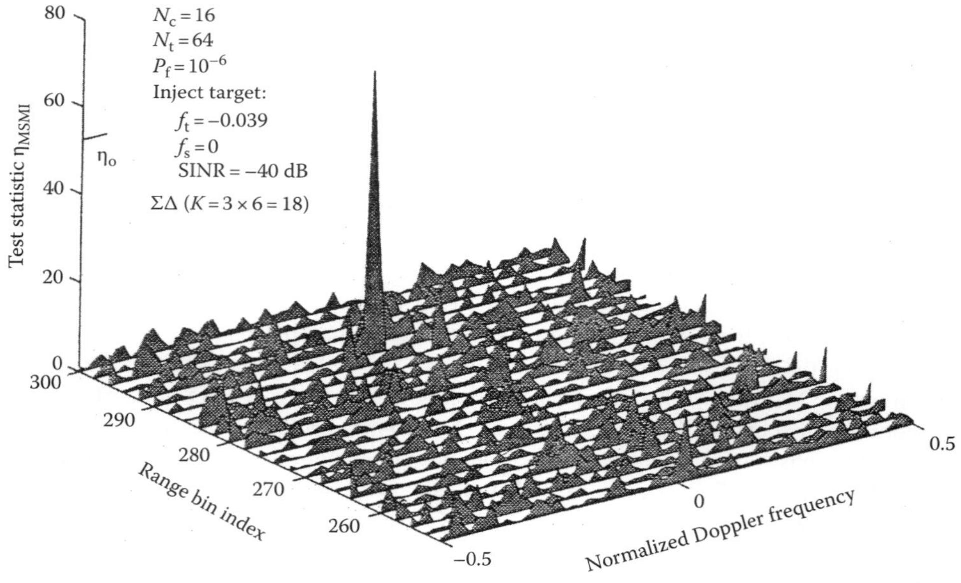

FIGURE 12.10 Range-Doppler plot of MCARM data, ΣΔ-STAP.

Another significant affordability issue is the applicability of ΣΔ-STAP to existing radar systems, both phased array and continuous aperture. Adaptive clutter rejection in the joint angle-Doppler domain can be incorporated into existing radar systems by digitizing the difference channel, or making relatively minor antenna modifications to add such a channel. Such a relatively low cost add-on can significantly improve the clutter suppression performance of an existing airborne radar system, whether its original design is based on low sidelobe beamforming or ΣΔ-DPCA.

While the trend is toward more affordable computing hardware, STAP processing still imposes a considerable burden which increases sharply with the order of the adaptive processor and radar bandwidth. In this respect, ΣΔ-STAP reduces computational requirements in matrix order N^3 adaptive problems. Moreover, the signal vector characteristic (mostly zero) can be exploited to further reduce test statistic numerical computations.

Finally, it should be pointed out that more than one Δ-beam can be incorporated if needed for clutter suppression [33].

12.10 Summary

Over the 22 years from a theoretical paper [1] to the MCARM experimental system, STAP has been established as a valuable alternative to the traditional airborne surveillance radar design approaches. Initially, STAP was viewed as an expensive technique only for newly designed phased arrays with many receiver channels; and now it has become much more affordable for both new and some existing systems. Future challenges lie in the area of real system design and integration, to which the MCARM experience is invaluable.

References

1. Reed, I.S., Mallet, J.D., and Brennan, L.E., Rapid convergence rate in adaptive arrays, *IEEE Trans. Aerosp. Elect. Syst.*, AES-10, 853–863, Nov. 1974.
2. Little, M.O. and Berry, W.P., Real-time multichannel airborne radar measurements, *Proceedings of the IEEE National Radar Conference*, pp. 138–142, Syracuse, NY, May 13–15, 1997.
3. Melvin, W.L., Wicks, M.C., and Brown, R.D., Assessment of multichannel airborne radar measurements for analysis and design of space-time processing architectures and algorithms, *Proceedings of the IEEE 1996 National Radar Conference*, pp. 130–135, Ann Arbor, MI, May 13–16, 1996.
4. Fenner, D.K. and Hoover, Jr., W.F., Test results of a space-time adaptive processing system for airborne early warning radar, *Proceedings of the IEEE 1996 National Radar Conference*, pp. 88–93, Ann Arbor, MI, May 13–16, 1996.
5. Wang, H., Zhang, Y., and Zhang, Q., Lessons learned from recent STAP experiments, *Proceedings of the CIE International Radar Conference*, Beijing, China, Oct. 8–10, 1996.
6. Staudaher, F.M., Airborne MTI, *Radar Handbook*, Skolnik, M.I. (Ed.), McGraw-Hill, New York, 1990, Chapter 16.
7. Wang, H., *Space-Time Processing and Its Radar Applications*, Lecture Notes for ELE891, Syracuse University, Syracuse, NY, Summer 1995.
8. Tseng, C.Y. and Griffiths, L.J., A unified approach to the design of linear constraints in minimum variance adaptive beamformers, *IEEE Trans. Antenn. Propag.*, AP-40(12), 1533–1542, Dec. 1992.
9. Skolnik, M., The radar antenna-circa 1995, *J. Franklin Inst.*, Elsevier Science Ltd., 332B(5), 503–519, 1995.
10. Klemm, R., Antenna design for adaptive airborne MTI, *Proceedings of the 1992 IEE International Conference Radar*, pp. 296–299, Brighton, U.K., Oct. 12–13, 1992.
11. Wang, H., Zhang, Y., and Wicks, M.C., Performance evaluation of space-time processing adaptive array radar in coherent repeater jamming environments, *Proceedings of the IEEE Long Island Section Adaptive Antenna Systems Symposium*, pp. 65–69, Melville, NY, Nov. 7–8, 1994.
12. Wang, H. and Cai, L., On adaptive spatial-temporal processing for airborne surveillance radar systems, *IEEE Trans. Aerosp. Elect. Syst.*, AES-30(3), 660–670, July 1994. Part of this paper is also

in *Proceedings of the 25th Annual Conference on Information Sciences and Systems*, pp. 968–975, Baltimore, MD, March 20–22, 1991, and *Proceedings of the CIE 1991 International Conference on Radar*, pp. 365–368, Beijing, China, Oct. 22–24, 1991.

13. Brown, R.D., Wicks, M.C., Zhang, Y., Zhang, Q., and Wang, H., A space-time adaptive processing approach for improved performance and affordability, *Proceedings of the IEEE 1996 National Radar Conference*, pp. 321–326, Ann Arbor, MI, May 13–16, 1996.
14. Pendergrass, N.A., Mitra, S.K., and Jury, E.I., Spectral transformations for two-dimensional digital filters, *IEEE Trans. Circ. Syst.*, CAS-23(1), 26–35, Jan. 1976.
15. Wang, H. and Cai, L., A localized adaptive MTD processor, *IEEE Trans. Aerosp. Elect. Syst.*, AES-27 (3), 532–539, May 1991.
16. DiPietro, R.C., Extended factored space-time processing for airborne radar systems, *Proceedings of the 26th Asilomar Conference on Signals, Systems, and Computers*, pp. 425–430, Pacific Grove, CA, Nov. 1992.
17. Brennan, L.E., Piwinski, D.J., and Staudaher, F.M., Comparison of space-time adaptive processing approaches using experimental airborne radar data, *IEEE 1993 National Radar Conference*, pp. 176–181, Lynnfield, MA, April 20–22, 1993.
18. Goldstein, J.S., Williams, D.B., and Holder, E.J., Cross-spectral subspace selection for rank reduction in partially adaptive sensor array processing, *Proceedings of the IEEE 1994 National Radar Conference*, Atlanta, GA, May 29–31, 1994.
19. Nitzberg, R., Detection loss of the sample matrix inversion technique, *IEEE Trans. Aerosp. Elect. Syst.*, AES-20, 824–827, Nov. 1984.
20. Khatri, C.G. and Rao, C.R., Effects of estimated noise covariance matrix in optimal signal detection, *IEEE Trans. Acoust. Speech Signal Process.*, ASSP-35(5), 671–679, May 1987.
21. Cai, L. and Wang, H., On adaptive filtering with the CFAR feature and its performance sensitivity to non-Gaussian interference, *Proceedings of the 24th Annual Conference on Information Sciences and Systems*, pp. 558–563, Princeton, NJ, March 21–23, 1990. Also published in *IEEE Trans. Aerosp. Elect. Syst.*, AES-27(3), 487–491, May 1991.
22. Kelly, E.J., An adaptive detection algorithm, *IEEE Trans. Aerosp. Elect. Syst.*, AES-22(1), 115–127, March 1986.
23. Kazakos, D. and Papantoni-Kazakos, P., *Detection and Estimation*, Computer Science Press, New York, 1990.
24. Robey, F.C. et. al., A CFAR adaptive matched filter detector, *IEEE Trans. Aerosp. Elect. Syst.*, AES-28(1), 208–216, Feb. 1992.
25. Cai, L. and Wang, H., Further results on adaptive filtering with embedded CFAR, *IEEE Trans. Aerosp. Elect. Syst.*, AES-30(4), 1009–1020, Oct. 1994.
26. Corbeil, A., Hawkins, L., and Gilgallon, P., Knowledge-based tracking algorithm, *Proceedings of the Signal and Data Processing of Small Targets, SPIE Proceedings Series*, Vol. 1305, Paper 16, pp. 180–192, Orlando, FL, April 16–18, 1990.
27. Wang, H. and Cai, L., On adaptive multiband signal detection with SMI algorithm, *IEEE Trans. Aerosp. Elect. Syst.*, AES-26, 768–773, Sept. 1990.
28. Wang, H., Zhang, Y., and Zhang, Q., A view of current status of space-time processing algorithm research, *Proceedings of the IEEE 1995 International Radar Conference*, pp. 635–640, Alexandria, VA, May 8–11, 1995.
29. Klemm, R., Adaptive air and spaceborne MTI under jamming conditions, *Proceedings of the 1993 IEEE National Radar Conference*, pp. 167–172, Boston, MA, April 1993.
30. Marshall, D.F., A two step adaptive interference nulling algorithm for use with airborne sensor arrays, *Proceedings of the Seventh SP Workshop on SSAP*, Quebec City, Canada, June 26–29, 1994.

31. Rivkin, P., Zhang, Y., and Wang, H., Spatial adaptive pre-suppression of wideband jammers in conjunction with STAP: A sideband approach, *Proceedings of the CIE International Radar Conference*, pp. 439–443, Beijing, China, Oct. 8–10, 1996.
32. Griffiths, L.J. and Jim, C.W., An alternative approach to linearly constrained adaptive beamforming, *IEEE Trans. Antenn. Propag.*, AP-30(1), 27–34, Jan. 1982.
33. Zhang, Y. and Wang, H., Further results of ΣΔ-STAP approach to airborne surveillance radars, *Proceedings of the IEEE National Radar Conference*, pp. 337–342, Syracuse, NY, May 13–15, 1997.

II

Nonlinear and Fractal Signal Processing

Alan V. Oppenheim
Massachusetts Institute of Technology

Gregory W. Wornell
Massachusetts Institute of Technology

13 **Chaotic Signals and Signal Processing** *Alan V. Oppenheim and Kevin M. Cuomo* **13**-1
Introduction • Modeling and Representation of Chaotic Signals • Estimation and Detection • Use of Chaotic Signals in Communications • Synthesizing Self-Synchronizing Chaotic Systems • References

14 **Nonlinear Maps** *Steven H. Isabelle and Gregory W. Wornell* **14**-1
Introduction • Eventually Expanding Maps and Markov Maps • Signals from Eventually Expanding Maps • Estimating Chaotic Signals in Noise • Probabilistic Properties of Chaotic Maps • Statistics of Markov Maps • Power Spectra of Markov Maps • Modeling Eventually Expanding Maps with Markov Maps • References

15 **Fractal Signals** *Gregory W. Wornell* **15**-1
Introduction • Fractal Random Processes • Deterministic Fractal Signals • Fractal Point Processes • References

16 **Morphological Signal and Image Processing** *Petros Maragos* **16**-1
Introduction • Morphological Operators for Sets and Signals • Median, Rank, and Stack Operators • Universality of Morphological Operators • Morphological Operators and Lattice Theory • Slope Transforms • Multiscale Morphological Image Analysis • Differential Equations for Continuous-Scale Morphology • Applications to Image Processing and Vision • Conclusions • Acknowledgment • References

17 **Signal Processing and Communication with Solitons** *Andrew C. Singer* **17**-1
Introduction • Soliton Systems: The Toda Lattice • New Electrical Analogs for Soliton Systems • Communication with Soliton Signals • Noise Dynamics in Soliton Systems • Estimation of Soliton Signals • Detection of Soliton Signals • References

18 Higher-Order Spectral Analysis *Athina P. Petropulu* **18-1**
Introduction • Definitions of HOS • HOS Computation from Real Data • Blind System Identification • HOS for Blind MIMO System Identification • Nonlinear Processes • Conclusions • Acknowledgments • References

TRADITIONALLY, SIGNAL PROCESSING AS A DISCIPLINE has relied heavily on a theoretical foundation of linear time-invariant system theory in the development of algorithms for a broad range of applications. In recent years, a considerable broadening of this theoretical base has begun to take place. In particular, there has been substantial growth in interest in the use of a variety of nonlinear systems with special properties for diverse applications. Promising new techniques for the synthesis and analysis of such systems continue to emerge. At the same time, there has also been rapid growth in interest in systems that are not constrained to be time-invariant. These may be systems that exhibit temporal fluctuations in their characteristics, or, equally importantly, systems characterized by other invariance properties, such as invariance to scale changes. In the latter case, this gives rise to systems with fractal characteristics.

In some cases, these systems are directly applicable for implementing various kinds of signal processing operations such as signal restoration, enhancement, or encoding, or for modeling certain kinds of distortion encountered in physical environments. In other cases, they serve as mechanisms for generating new classes of signal models for existing and emerging applications. In particular, when autonomous or driven by simpler classes of input signals, they generate rich classes of signals at their outputs. In turn, these new classes of signals give rise to new families of algorithms for efficiently exploiting them in the context of applications.

The spectrum of techniques for nonlinear signal processing is extremely broad, and in this chapter we make no attempt to cover the entire array of exciting new directions being pursued within the community. Rather, we present a very small sampling of several highly promising and interesting ones to suggest the richness of the topic.

A brief overview of the specific chapters comprising this part is as follows.

Chapters 13 and 14 discuss the chaotic behavior of certain nonlinear dynamical systems and suggest ways in which this behavior can be exploited. In particular, Chapter 13 focuses on continuous-time chaotic systems characterized by a special self-synchronization property that makes them potentially attractive for a range of secure communications applications. Chapter 14 describes a family of discrete-time nonlinear dynamical and chaotic systems that are particularly attractive for use in a variety of signal processing applications ranging from signal modeling in power converters to pseudorandom number generation and error-correction coding in signal transmission applications.

Chapter 15 discusses fractal signals that arise out of self-similar system models characterized by scale invariance. These represent increasingly important models for a range of natural and man-made phenomena in applications involving both signal synthesis and analysis. Multidimensional fractals also arise in the state-space representation of chaotic signals, and the fractal properties in this representation are important in the identification, classification, and characterization of such signals.

Chapter 16 focuses on morphological signal processing, which encompasses an important class of nonlinear filtering techniques together with some powerful, associated signal representations. Morphological signal processing is closely related to a number of classes of algorithms including order-statistics filtering, cellular automata methods for signal processing, and others. Morphological algorithms are currently among the most successful and widely used nonlinear signal processing techniques in image processing and vision for such tasks as noise suppression, feature extraction, segmentation, and others.

Chapter 17 discusses the analysis and synthesis of soliton signals and their potential use in communication applications. These signals arise in systems satisfying certain classes of nonlinear wave equations. Because they propagate through those equations without dispersion, there has been longstanding interest in their use as carrier waveforms over fiber-optic channels having the appropriate nonlinear characteristics. As they propagate through these systems, they also exhibit a special type of reduced-energy superposition property that suggests an interesting multiplexing strategy for communications over linear channels.

Finally, Chapter 18 discusses nonlinear representations for stochastic signals in terms of their higher-order statistics. Such representations are particularly important in the processing of non-Gaussian signals for which more traditional second-moment characterizations are often inadequate. The associated tools of higher-order spectral analysis find increasing application in many signal detection, identification, modeling, and equalization contexts, where they have led to new classes of powerful signal processing algorithms.

Again, these chapters are only representative examples of the many emerging directions in this active area of research within the signal processing community, and developments in many other important and exciting directions can be found in the community's journal and conference publications.

13

Chaotic Signals and Signal Processing

Alan V. Oppenheim
Massachusetts Institute of Technology

Kevin M. Cuomo
Massachusetts Institute of Technology

13.1 Introduction **13**-1
13.2 Modeling and Representation of Chaotic Signals **13**-1
13.3 Estimation and Detection **13**-3
13.4 Use of Chaotic Signals in Communications **13**-3
Self-Synchronization and Asymptotic Stability • Robustness and Signal Recovery in the Lorenz System • Circuit Implementation and Experiments
13.5 Synthesizing Self-Synchronizing Chaotic Systems **13**-10
References **13**-12

13.1 Introduction

Signals generated by chaotic systems represent a potentially rich class of signals both for detecting and characterizing physical phenomena and in synthesizing new classes of signals for communications, remote sensing, and a variety of other signal processing applications.

In classical signal processing a rich set of tools has evolved for processing signals that are deterministic and predictable such as transient and periodic signals, and for processing signals that are stochastic. Chaotic signals associated with the homogeneous response of certain nonlinear dynamical systems do not fall in either of these classes. While they are deterministic, they are not predictable in any practical sense in that even with the generating dynamics known, estimation of prior or future values from a segment of the signal or from the state at a given time is highly ill-conditioned. In many ways these signals appear to be noise-like and can, of course, be analyzed and processed using classical techniques for stochastic signals. However, they clearly have considerably more structure than can be inferred from and exploited by traditional stochastic modeling techniques.

The basic structure of chaotic signals and the mechanisms through which they are generated are described in a variety of introductory books, e.g., [1,2] and summarized in [3].

Chaotic signals are of particular interest and importance in experimental physics because of the wide range of physical processes that apparently give rise to chaotic behavior. From the point of view of signal processing, the detection, analysis, and characterization of signals of this type present a significant challenge. In addition, chaotic systems provide a potentially rich mechanism for signal design and generation for a variety of communications and remote sensing applications.

13.2 Modeling and Representation of Chaotic Signals

The state evolution of chaotic dynamical systems is typically described in terms of the nonlinear state equation $\dot{x}(t) = F[x(t)]$ in continuous time or $x[n] = F(x[n-1])$ in discrete time. In a signal processing context, we assume that the observed chaotic signal is a nonlinear function of the state and would typically be a scalar time function. In discrete-time, for example, the observation equation would be $y[n] = G(x[n])$. Frequently the observation $y[n]$ is also distorted by additive noise, multipath effects, fading, etc.

Modeling a chaotic signal can be phrased in terms of determining from clean or distorted observations, a suitable state space and mappings $F(\cdot)$ and $G(\cdot)$ that capture the aspects of interest in the observed signal y. The problem of determining from the observed signal a suitable state space in which to model the dynamics is referred to as the embedding problem. While there is, of course, no unique set of state variables for a system, some choices may be better suited than others. The most commonly used method for constructing a suitable state space for the chaotic signal is the method of delay coordinates in which a state vector is constructed from a vector of successive observations.

It is frequently convenient to view the problem of identifying the map associated with a given chaotic signal in terms of an interpolation problem. Specifically, from a suitably embedded chaotic signal it is possible to extract a codebook consisting of state vectors and the states to which they subsequently evolve after one iteration. This codebook then consists of samples of the function F spaced, in general, nonuniformly throughout state space. A variety of both parametric and nonparametric methods for interpolating the map between the sample points in state space have emerged in the literature, and the topic continues to be of significant research interest. In this section we briefly comment on several of the approaches currently used. These and others are discussed and compared in more detail in [4].

One approach is based on the use of locally linear approximations to F throughout the state space [5,6]. This approach constitutes a generalization of autoregressive modeling and linear prediction and is easily extended to locally polynomial approximations of higher order. Another approach is based on fitting a global nonlinear function to the samples in state space [7].

A fundamentally rather different approach to the problem of modeling the dynamics of an embedded signal involves the use of hidden Markov models [8–10]. With this method, the state space is discretized into a large number of states, and a probabilistic mapping is used to characterize transitions between states with each iteration of the map. Furthermore, each state transition spawns a state-dependent random variable as the observation $y[n]$. This framework can be used to simultaneously model both the detailed characteristics of state evolution in the system and the noise inherent in the observed data. While algorithms based on this framework have proved useful in modeling chaotic signals, they can be expensive both in terms of computation and storage requirements due to the large number of discrete states required to adequately capture the dynamics.

While many of the above modeling methods exploit the existence of underlying nonlinear dynamics, they do not explicitly take into account some of the properties peculiar to chaotic nonlinear dynamical systems. For this reason, in principle, the algorithms may be useful in modeling a broader class of signals. On the other hand, when the signals of interest are truly chaotic, the special properties of chaotic nonlinear dynamical systems ought to be taken into account, and, in fact, may often be exploited to achieve improved performance. For instance, because the evolution of chaotic systems is acutely sensitive to initial conditions, it is often important that this numerical instability be reflected in the model for the system. One approach to capturing this sensitivity is to require that the reconstructed dynamics exhibit Lyapunov exponents consistent with what might be known about the true dynamics. The sensitivity of state evolution can also be captured using the hidden Markov model framework since the structural uncertainty in the dynamics can be represented in terms of the probabilistic state transactions. In any case, unless sensitivity of the dynamics is taken into account during modeling, detection and estimation algorithms involving chaotic signals often lack robustness.

Another aspect of chaotic systems that can be exploited is that the long-term evolution of such systems lies on an attractor whose dimension is not only typically nonintegral, but occupies a small fraction of the entire state space. This has a number of important implications both in the modeling of chaotic signals and ultimately in addressing problems of estimation and detection involving these signals. For example, it implies that the nonlinear dynamics can be recovered in the vicinity of the attractor using comparatively less data than would be necessary if the dynamics were required everywhere in state space.

Identifying the attractor, its fractal dimension, and related invariant measures governing, for example, the probability of being in the neighborhood of a particular state on the attractor, are also important aspects of the modeling problem. Furthermore, we can often exploit various ergodicity and mixing properties of chaotic systems. These properties allow us to recover information about the attractor using a single realization of a chaotic signal, and assure us that different time intervals of the signal provide qualitatively similar information about the attractor.

13.3 Estimation and Detection

A variety of problems involving the estimation and detection of chaotic signals arise in potential application contexts. In some scenarios, the chaotic signal is a form of noise or other unwanted interference signal. In this case, we are often interested in detecting, characterizing, discriminating, and extracting known or partially known signals in backgrounds of chaotic noise. In other scenarios, it is the chaotic signal that is of direct interest and which is corrupted by other signals. In these cases we are interested in detecting, discriminating, and extracting known or partially known chaotic signals in backgrounds of other noises or in the presence of other kinds of distortion.

The channel through which either natural or synthesized signals are received can typically be expected to introduce a variety of distortions including additive noise, scattering, multipath effects, etc. There are, of course, classical approaches to signal recovery and characterization in the presence of such distortions for both transient and stochastic signals. When the desired signal in the channel is a chaotic signal, or when the distortion is caused by a chaotic signal, many of the classical techniques will not be effective and do not exploit the particular structure of chaotic signals.

The specific properties of chaotic signals exploited in detection and estimation algorithms depend heavily on the degree of *a priori* knowledge of the signals involved. For example, in distinguishing chaotic signals from other signals, the algorithms may exploit the functional form of the map, the Lyapunov exponents of the dynamics, and/or characteristics of the chaotic attractor such as its structure, shape, fractal dimension and/or invariant measures.

To recover chaotic signals in the presence of additive noise, some of the most effective noise reduction techniques proposed to date take advantage of the nonlinear dependence of the chaotic signal by constructing accurate models for the dynamics. Multipath and other types of convolutional distortion can best be described in terms of an augmented state space system. Convolution or filtering of chaotic signals can change many of the essential characteristics and parameters of chaotic signals. Effects of convolutional distortion and approaches to compensating for it are discussed in [11].

13.4 Use of Chaotic Signals in Communications

Chaotic systems provide a rich mechanism for signal design and generation, with potential applications to communications and signal processing. Because chaotic signals are typically broadband, noise-like, and difficult to predict, they can be used in various contexts in communications. A particularly useful class of chaotic systems are those that possess a self-synchronization property [12–14]. This property allows two identical chaotic systems to synchronize when the second system (receiver) is driven by the first (transmitter). The well-known Lorenz system is used below to further describe and illustrate the chaotic self-synchronization property.

The Lorenz equations, first introduced by E.N. Lorenz as a simplified model of fluid convection [15], are given by

$$\begin{aligned} \dot{x} &= \sigma(y - x) \\ \dot{y} &= rx - y - xz \\ \dot{z} &= xy - bz, \end{aligned} \tag{13.1}$$

where σ, r, and b are positive parameters. In signal processing applications, it is typically of interest to adjust the time scale of the chaotic signals. This is accomplished in a straightforward way by establishing the convention that $\dot{x}, \dot{y}$, and $\dot{z}$ denote $dx/d\tau, dy/d\tau$, and $dz/d\tau$, respectively, where $\tau = t/T$ is normalized time and T is a time scale factor. It is also convenient to define the normalized frequency $\omega = \Omega T$, where Ω denotes the angular frequency in units of rad/s. The parameter values $T = 400\ \mu s$, $\sigma = 16, r = 45.6$, and $b = 4$ are used for the illustrations in this chapter.

Viewing the Lorenz system (Equation 13.1) as a set of transmitter equations, a dynamical receiver system that will synchronize to the transmitter is given by

$$\begin{aligned} \dot{x}_r &= \sigma(y_r - x_r) \\ \dot{y}_r &= rx(t) - y_r - x(t)z_r \\ \dot{z}_r &= x(t)y_r - bz_r. \end{aligned} \tag{13.2}$$

In this case, the chaotic signal $x(t)$ from the transmitter is used as the driving input to the receiver system. In Section 13.4.1, an identified equivalence between self-synchronization and asymptotic stability is exploited to show that the synchronization of the transmitter and receiver is global, i.e., the receiver can be initialized in any state and the synchronization still occurs.

13.4.1 Self-Synchronization and Asymptotic Stability

A close relationship exists between the concepts of self-synchronization and asymptotic stability. Specifically, self-synchronization in the Lorenz system is a consequence of globally stable error dynamics. Assuming that the Lorenz transmitter and receiver parameters are identical, a set of equations that govern their error dynamics is given by

$$\begin{aligned} \dot{e}_x &= \sigma(e_y - e_x) \\ \dot{e}_y &= -e_y - x(t)e_z \\ \dot{e}_z &= x(t)e_y - be_z, \end{aligned} \tag{13.3}$$

where

$$\begin{aligned} e_x(t) &= x(t) - x_r(t) \\ e_y(t) &= y(t) - y_r(t) \\ e_z(t) &= z(t) - z_r(t). \end{aligned}$$

A sufficient condition for the error equations to be globally asymptotically stable at the origin can be determined by considering a Lyapunov function of the form

$$E(\mathbf{e}) = \frac{1}{2}\left(\frac{1}{\sigma}e_x^2 + e_y^2 + e_z^2\right).$$

Since σ and b in the Lorenz equations are both assumed to be positive, E is positive definite and $\dot{E}$ is negative definite. It then follows from Lyapunov's theorem that $\mathbf{e}(t) \to 0$ as $t \to \infty$. Therefore, synchronization occurs as $t \to \infty$ regardless of the initial conditions imposed on the transmitter and receiver systems.

For practical applications, it is also important to investigate the sensitivity of the synchronization to perturbations of the chaotic drive signal. Numerical experiments are summarized in Section 13.4.2, which demonstrates the robustness and signal recovery properties of the Lorenz system.

13.4.2 Robustness and Signal Recovery in the Lorenz System

When a message or other perturbation is added to the chaotic drive signal, the receiver does not regenerate a perfect replica of the drive; there is always some synchronization error. By subtracting the regenerated drive signal from the received signal, successful message recovery would result if the synchronization error was small relative to the perturbation itself. An interesting property of the Lorenz system is that the synchronization error is not small compared to a narrowband perturbation; nevertheless, the message can be recovered because the synchronization error is nearly coherent with the message. This section summarizes experimental evidence for this effect; a more detailed explanation has been given in terms of an approximate analytical model [16].

The series of experiments that demonstrate the robustness of synchronization to white noise perturbations and the ability to recover speech perturbations focus on the synchronizing properties of the transmitter equations (Equation 13.1) and the corresponding receiver equations,

$$\begin{aligned}\dot{x}_r &= \sigma(y_r - x_r)\\ \dot{y}_r &= rs(t) - y_r - s(t)z_r\\ \dot{z}_r &= s(t)y_r - bz_r.\end{aligned} \tag{13.4}$$

Previously, it was stated that with $s(t)$ equal to the transmitter signal $x(t)$, the signals x_r, y_r, and z_r will asymptotically synchronize to x, y, and z, respectively. Below, we examine the synchronization error when a perturbation $p(t)$ is added to $x(t)$, i.e., when $s(t) = x(t) + p(t)$.

First, we consider the case where the perturbation $p(t)$ is Gaussian white noise. In Figure 13.1, we show the perturbation and error spectra for each of the three state variables vs. normalized frequency ω. Note that at relatively low frequencies, the error in reconstructing $x(t)$ slightly exceeds the perturbation of the drive but that for normalized frequencies above 20 the situation quickly reverses. An analytical model closely predicts and explains this behavior [16]. These figures suggest that the sensitivity of synchronization depends on the spectral characteristics of the perturbation signal. For signals that are bandlimited to the frequency range $0 < \omega < 10$, we would expect that the synchronization errors will be larger than the perturbation itself. This turns out to be the case, although the next experiment suggests there are additional interesting characteristics as well.

In a second experiment, $p(t)$ is a low-level speech signal (e.g., a message to be transmitted and recovered). The normalizing time parameter is 400 μs and the speech signal is bandlimited to 4 kHz or equivalently to a normalized frequency ω of 10. Figure 13.2 shows the power spectrum of a representative speech signal and the chaotic signal $x(t)$. The overall chaos-to-perturbation ratio in this experiment is approximately 20 dB.

To recover the speech signal, the regenerated drive signal is subtracted at the receiver from the received signal. In this case, the recovered message is $\hat{p}(t) = p(t) + e_x(t)$. It would be expected that successful message recovery would result if $e_x(t)$ was small relative to the perturbation signal. For the Lorenz system, however, although the synchronization error is not small compared to the perturbation, the message can be recovered because $e_x(t)$ is nearly coherent with the message. This coherence has been confirmed experimentally and an explanation has been developed in terms of an approximate analytical model [16].

13.4.3 Circuit Implementation and Experiments

In Section 13.4.2, we showed that, theoretically, a low-level speech signal could be added to the synchronizing drive signal and approximately recovered at the receiver. These results were based on an analysis of the exact Lorenz transmitter and receiver equations. When implementing synchronized chaotic systems in hardware, the limitations of available circuit components result in approximations of the defining equations. The Lorenz transmitter and receiver equations can be implemented relatively easily with standard analog circuits [17,20,21]. The resulting system performance is in excellent

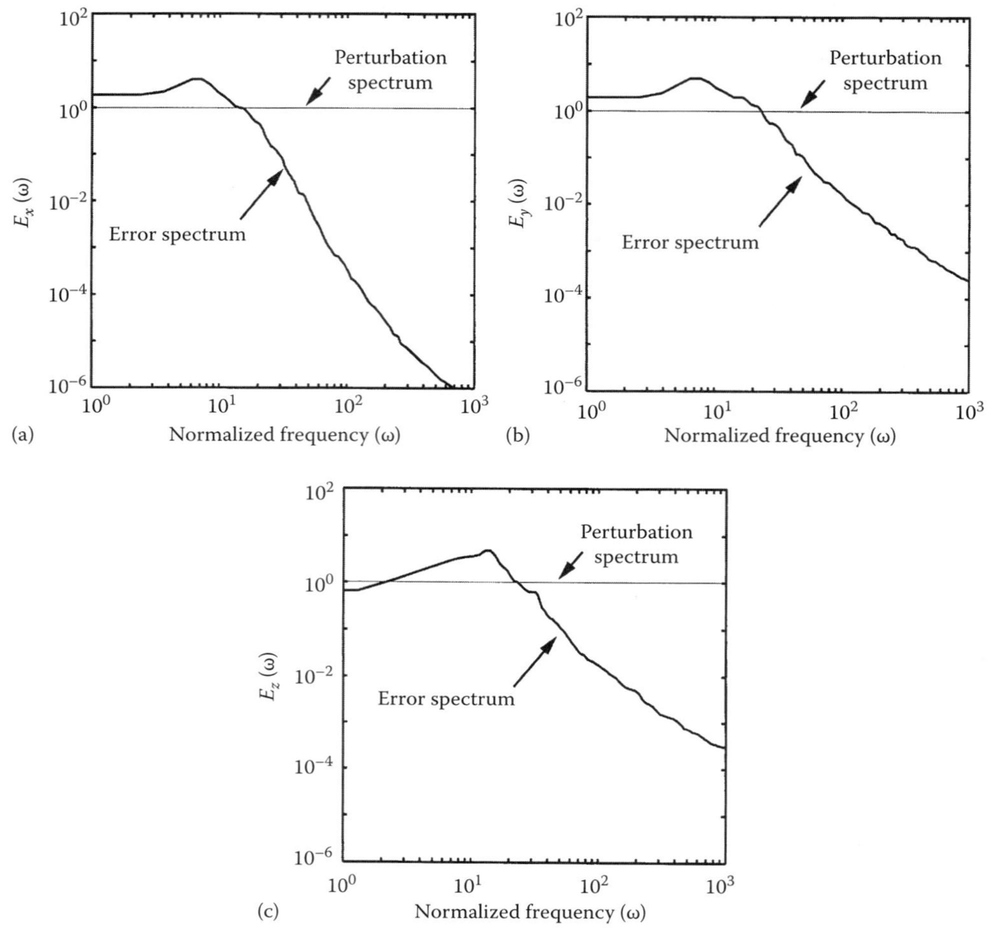

FIGURE 13.1 Power spectra of the error signals: (a) $E_z(\omega)$, (b) $E_y(\omega)$, and (c) $E_z(\omega)$.

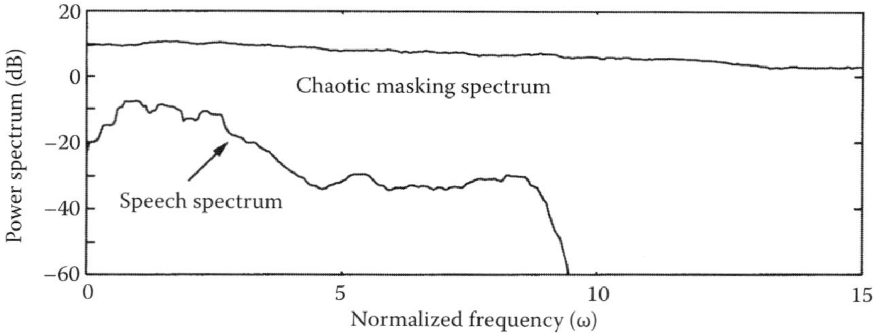

FIGURE 13.2 Power spectra of $x(t)$ and $p(t)$ when the perturbation is a speech signal.

agreement with numerical and theoretical predictions. Some potential implementation difficulties are avoided by scaling the Lorenz state variables according to $u = x/10, v = y/10$, and $w = z/20$. With this scaling, the Lorenz equations are transformed to

$$\begin{aligned} \dot{u} &= \sigma(v - u) \\ \dot{v} &= ru - v - 20uw \\ \dot{w} &= 5uv - bw. \end{aligned} \tag{13.5}$$

For this system, which we refer to as the circuit equations, the state variables all have similar dynamic range and circuit voltages remain well within the range of typical power supply limits. In the following, we discuss and demonstrate some applied aspects of the Lorenz circuits.

In Figure 13.3, we illustrate a communication scenario that is based on chaotic signal masking and recovery [18–21]. In this figure, a chaotic masking signal $u(t)$ is added to the information-bearing signal $p(t)$ at the transmitter, and at the receiver the masking is removed. By subtracting the regenerated drive signal $u_r(t)$ from the received signal $s(t)$ at the receiver, the recovered message is

$$\hat{p}(t) = s(t) - u_r(t) = p(t) + [u(t) - u_r(t)].$$

In this context, $e_u(t)$, the error between $u(t)$ and $u_r(t)$, corresponds directly to the error in the recovered message.

For this experiment, $p(t)$ is a low-level speech signal (the message to be transmitted and recovered). The normalizing time parameter is 400 μs and the speech signal is bandlimited to 4 kHz or, equivalently, to a normalized frequency ω of 10. In Figure 13.4, we show the power spectrum of $p(t)$

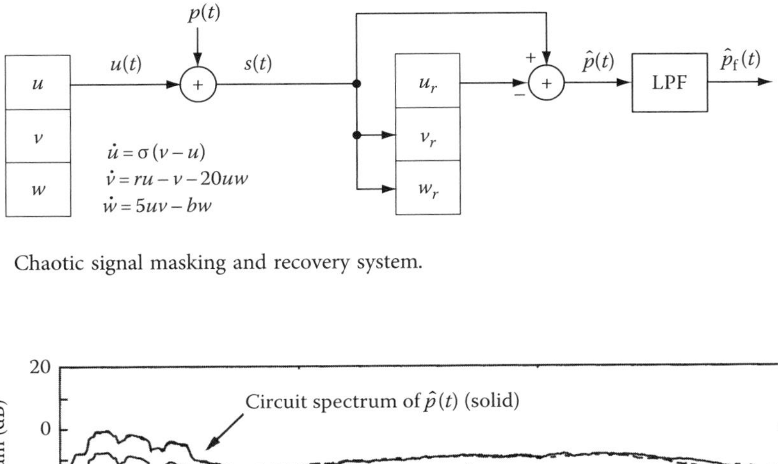

FIGURE 13.3 Chaotic signal masking and recovery system.

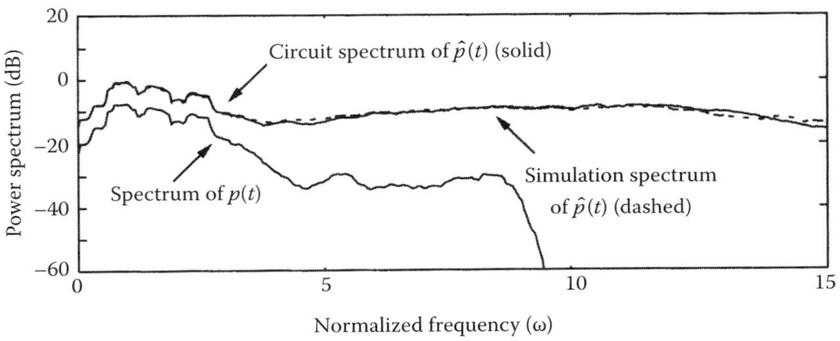

FIGURE 13.4 Power spectra of $p(t)$ and $\hat{p}(t)$ when the perturbation is a speech signal.

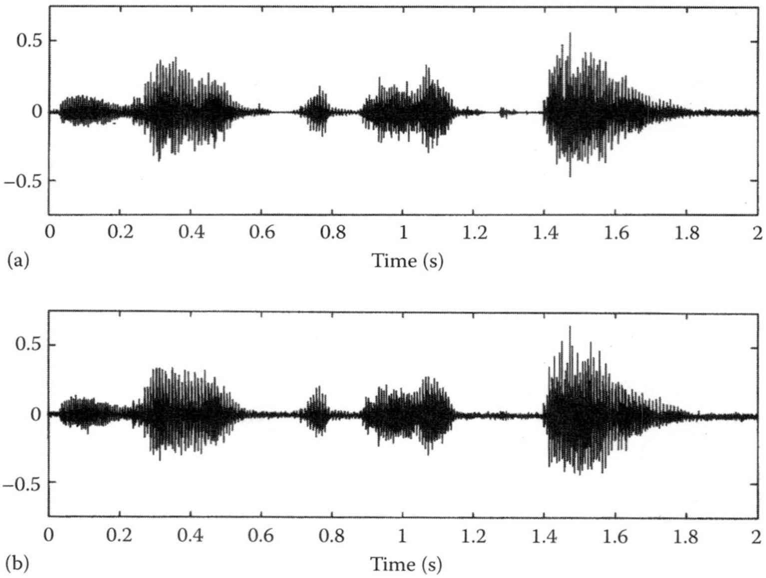

FIGURE 13.5 (a) Recovered speech (simulation) and (b) recovered speech (circuit).

and $\hat{p}(t)$, where $\hat{p}(t)$ is obtained from both a simulation and from the circuit. The two spectra for $\hat{p}(t)$ are in excellent agreement, indicating that the circuit performs very well. Because $\hat{p}(t)$ includes considerable energy beyond the bandwidth of the speech, the speech recovery can be improved by lowpass filtering $\hat{p}(t)$. We denote the lowpass filtered version of $\hat{p}(t)$ by $\hat{p}_{\mathrm{f}}(t)$. In Figure 13.5a and b, we show a comparison of $\hat{p}_{\mathrm{f}}(t)$ from both a simulation and from the circuit, respectively. Clearly, the circuit performs well and, in informal listening tests, the recovered message is of reasonable quality.

Although $\hat{p}_{\mathrm{f}}(t)$ is of reasonable quality in this experiment, the presence of additive channel noise will produce message recovery errors that cannot be completely removed by lowpass filtering; there will always be some error in the recovered message. Because the message and noise are directly added to the synchronizing drive signal, the message-to-noise ratio should be large enough to allow a faithful recovery of the original message. This requires a communication channel that is nearly noise free.

An alternative approach to private communications allows the information-bearing waveform to be exactly recovered at the self-synchronizing receiver(s), even when moderate-level channel noise is present. This approach is referred to as chaotic binary communications [20,21]. The basic idea behind this technique is to modulate a transmitter parameter with the information-bearing waveform and to transmit the chaotic drive signal. At the receiver, the parameter modulation will produce a synchronization error between the received drive signal and the receiver's regenerated drive signal with an error signal amplitude that depends on the modulation. Using the synchronization error, the modulation can be detected.

This modulation/detection process is illustrated in Figure 13.6. To illustrate the approach, we use a periodic square-wave for $p(t)$ as shown in Figure 13.7a. The square-wave has a repetition frequency of approximately 110 Hz with zero volts representing the zero-bit and one volt representing the one-bit. The square-wave modulates the transmitter parameter b with the zero-bit and one-bit parameters given by $b(0) = 4$ and $b(1) = 4.4$, respectively. The resulting drive signal $u(t)$ is transmitted and the noisy received signal $s(t)$ is used as the driving input to the synchronizing receiver circuit. In Figure 13.7b, we show the synchronization error power $e^2(t)$. The parameter modulation produces significant synchronization error during a "1" transmission and very little error during a "0" transmission. It is plausible that a detector based on the average synchronization error power, followed by a threshold device, could yield reliable

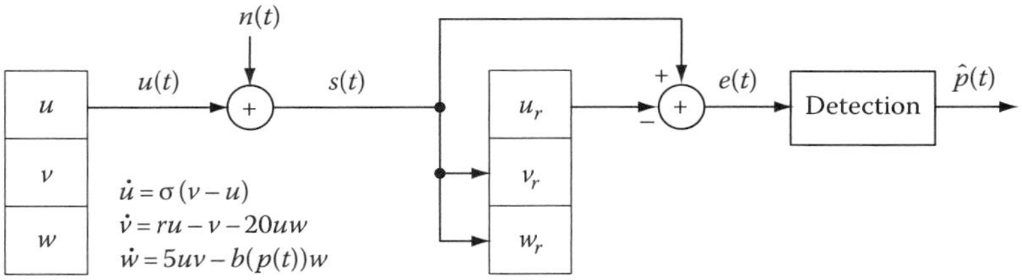

FIGURE 13.6 Communicating binary-valued bit streams with synchronized chaotic systems.

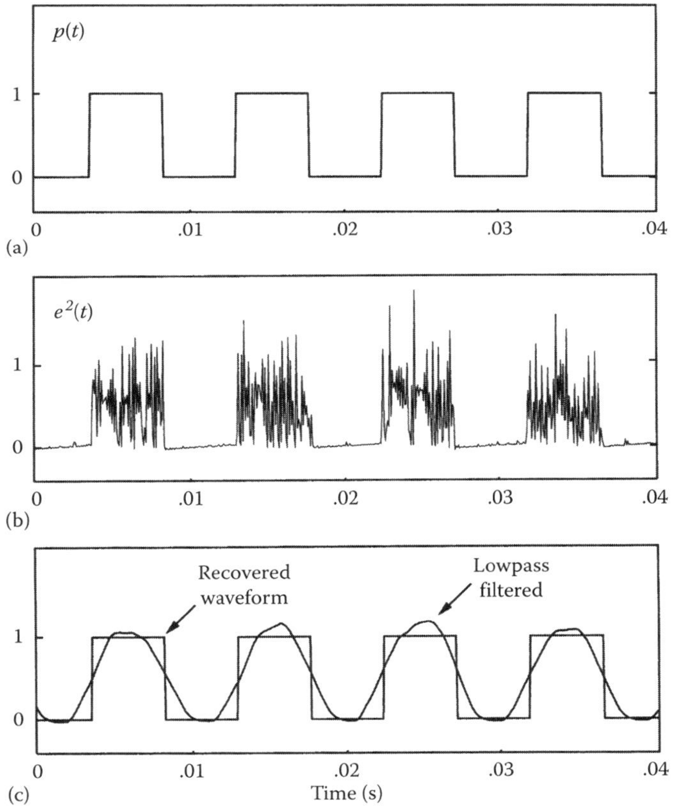

FIGURE 13.7 (a) Binary modulation waveform. (b) Synchronization error power. (c) Recovered binary waveform.

performance. We illustrate in Figure 13.7c that the square-wave modulation can be reliably recovered by lowpass filtering the synchronization error power waveform and applying a threshold test. The threshold device used in this experiment consisted of a simple analog comparator circuit.

The allowable data rate of this communication technique is, of course, dependent on the synchronization response time of the receiver system. Although we have used a low bit rate to demonstrate the technique, the circuit time scale can be easily adjusted to allow much faster bit rates.

While the results presented above appear encouraging, there are many communication scenarios where it is undesirable to be restricted to the Lorenz system, or for that matter, any other low-dimensional chaotic system. In private communications, for example, the ability to choose from a wide variety of synchronized chaotic systems would be highly advantageous. In the next section, we briefly describe an approach for synthesizing an unlimited number of high-dimensional chaotic systems. The significance of this work lies in the fact that the ability to synthesize high-dimensional chaotic systems further enhances their applicability for practical applications.

13.5 Synthesizing Self-Synchronizing Chaotic Systems

An effective approach to synthesis is based on a systematic four step process. First, an algebraic model is specified for the transmitter and receiver systems. As shown in [22,23], the chaotic system models can be very general; in [22] the model represents a large class of quadratically nonlinear systems, while in [23] the model allows for an unlimited number of Lorenz oscillators to be mutually coupled via an N-dimensional linear system.

The second step in the synthesis process involves subtracting the receiver equations from the transmitter equations and imposing a global asymptotic stability constraint on the resulting error equations. Using Lyapunov's direct method, sufficient conditions for the error system's global stability are usually straightforward to obtain. The sufficient conditions determine constraints on the free parameters of the transmitter and receiver which guarantee that they possess the global self-synchronization property.

The third step in the synthesis process focuses on the global stability of the transmitter equations. First, a family of ellipsoids in state space is defined and then sufficient conditions are determined which guarantee the existence of a "trapping region." The trapping region imposes additional constraints on the free parameters of the transmitter and receiver equations.

The final step involves determining sufficient conditions that render all of the transmitter's fixed points unstable. In most cases, this involves numerically integrating the transmitter equations and computing the system's Lyapunov exponents and/or attractor dimension. If stable fixed points exist, the system's bifurcation parameter is adjusted until they all become unstable. Below, we demonstrate the synthesis approach for linear feedback chaotic systems (LFBCSs).

LFBCSs are composed of a low-dimensional chaotic system and a linear feedback system as illustrated in Figure 13.8. Because the linear system is N-dimensional, considerable design flexibility is possible with LFBCSs. Another practical property of LFBCSs is that they synchronize via a single drive signal while exhibiting complex dynamics.

While many types of LFBCSs are possible, two specific cases have been considered in detail: (1) the chaotic Lorenz signal $x(t)$ drives an N-dimensional linear system and the output of the linear system is added to the equation for $\dot{x}$ in the Lorenz system; and (2) the Lorenz signal $z(t)$ drives an N-dimensional linear system and the output of the linear system is added to the equation for $\dot{z}$ in the Lorenz system. In both cases, a complete synthesis procedure was developed.

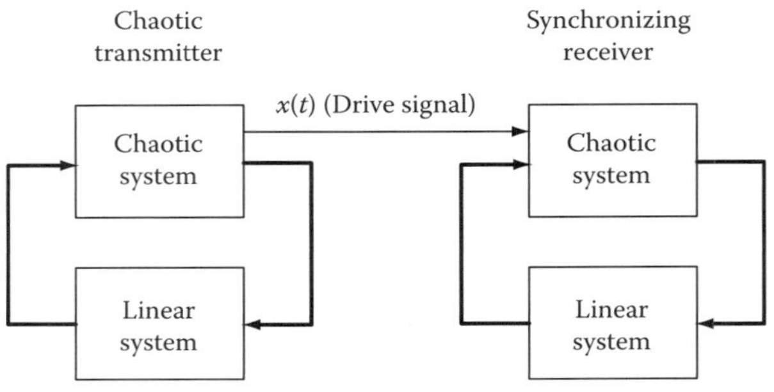

FIGURE 13.8 Linear feedback chaotic systems.

Below, we summarize the procedure; a complete development is given elsewhere [24].

Synthesis procedure

1. Choose any stable A matrix and any $N \times N$ symmetric positive definite matrix Q.
2. Solve $PA + A^{\mathrm{T}}P + Q = 0$ for the positive definite solution P.
3. Choose any vector B and set $C = -B^{\mathrm{T}}P/r$.
4. Choose any D such that $\sigma - D > 0$.

The first step of the procedure is simply the self-synchronization condition; it requires the linear system to be stable. Clearly, many choices for A are possible. The second and third steps are akin to a negative feedback constraint, i.e., the linear feedback tends to stabilize the chaotic system. The last step in the procedure restricts $\sigma - D > 0$ so that the $\dot{x}$ equation of the Lorenz system remains dissipative after feedback is applied.

For the purpose of demonstration, consider the following five-dimensional x-input/x-output LFBCS.

$$\begin{aligned}
\dot{x} &= \sigma(y - x) + v \\
\dot{y} &= rx - y - xz \\
\dot{z} &= xy - bz \\
\begin{bmatrix} \dot{l}_1 \\ \dot{l}_2 \end{bmatrix} &= \begin{bmatrix} -\frac{1}{2} & 10 \\ -10 & -\frac{1}{2} \end{bmatrix}\begin{bmatrix} l_1 \\ l_2 \end{bmatrix} + \begin{bmatrix} 1 \\ 1 \end{bmatrix} x \\
v &= -\begin{bmatrix} 1 & 1 \end{bmatrix}\begin{bmatrix} l_1 \\ l_2 \end{bmatrix}
\end{aligned} \tag{13.6}$$

It can be shown in a straightforward way that the linear system satisfies the synthesis procedure for suitable choices of P, Q, and R. For the numerical demonstrations presented below, the Lorenz parameters chosen are $\sigma = 16$ and $b = 4$; the bifurcation parameter r will be varied.

In Figure 13.9, we show the computed Lyapunov dimension as r is varied over the range, $20 < r < 100$. This figure demonstrates that the LFBCS achieves a greater Lyapunov dimension than

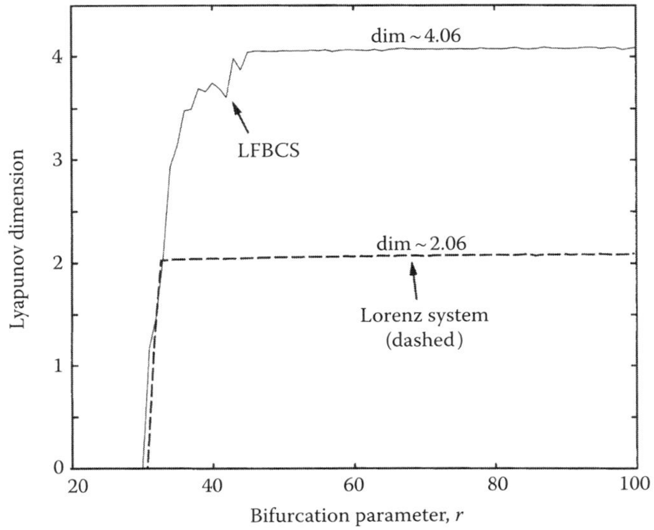

FIGURE 13.9 Lyapunov dimension of a 5-D LFBCS.

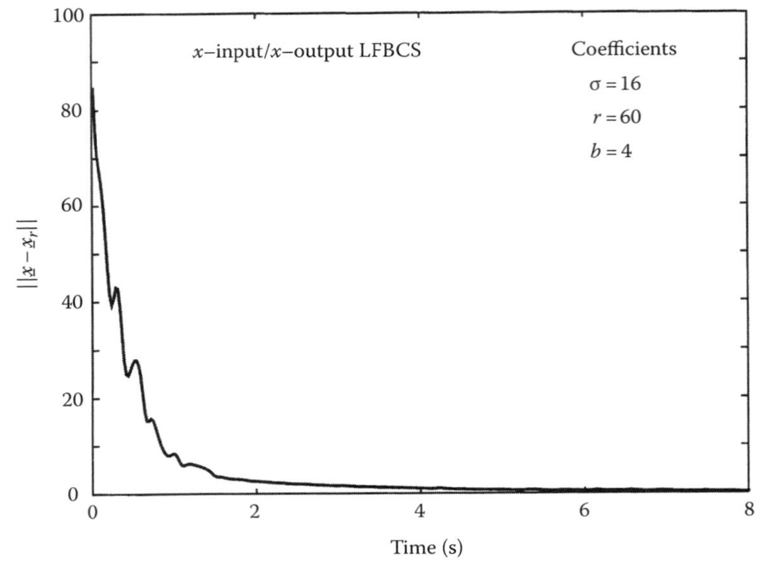

FIGURE 13.10 Self-synchronization in a 5-D LFBCS.

the Lorenz system without feedback. The Lyapunov dimension could be increased by using more states in the linear system. However, numerical experiments suggest that stable linear feedback creates only negative Lyapunov exponents, limiting the dynamical complexity of LFBCSs. Nevertheless, their relative ease of implementation is an attractive practical feature.

In Figure 13.10, we demonstrate the rapid synchronization between the transmitter and receiver systems. The curve measures the distance in state space between the transmitter and receiver trajectories when the receiver is initialized from the zero state. Synchronization is maintained indefinitely.

References

1. Moon, F., *Chaotic Vibrations*, John Wiley & Sons, New York, 1987.
2. Strogatz, S. H., *Nonlinear Dynamics and Chaos: With Applications to Physics, Biology, Chemistry, and Engineering*, Addison-Wesley, Reading, MA, 1994.
3. Abarbanel, H. D. I., Chaotic signals and physical systems, *Proceedings of the 1992 IEEE ICASSP*, San Francisco, CA, IV, 113–116, Mar. 1992.
4. Sidorowich, J. J., Modeling of chaotic time series for prediction, interpolation and smoothing, *Proceedings of the 1992 IEEE ICASSP*, San Francisco, CA, IV, pp. 121–124, Mar. 1992.
5. Singer, A., Oppenheim, A. V., and Wornell, G., Codebook prediction: A nonlinear signal modeling paradigm, *Proceedings of the 1992 IEEE ICASSP*, San Francisco, CA, V, pp. 325–328, Mar. 1992.
6. Farmer, J. D and Sidorowich, J. J., Predicting chaotic time series, *Phys. Rev. Lett.*, 59, 845–848, Aug. 1987.
7. Haykin, S. and Leung, H., Chaotic signal processing: First experimental radar results, *Proceedings of the 1992 IEEE ICASSP*, San Francisco, CA, IV, pp. 125–128, 1992.
8. Meyers, C., Kay, S., and Richard, M., Signal separation for nonlinear dynamical systems, *Proceedings of the 1992 IEEE ICASSP*, San Francisco, CA, IV, pp. 129–132, Mar. 1992.
9. Hsu, C. S., *Cell-to-Cell Mapping*, Springer-Verlag, New York, 1987.
10. Meyers, C., Singer, A., Shin, B., and Church, E., Modeling chaotic systems with hidden Markov models, *Proceedings of the 1992 IEEE ICASSP*, San Francisco, CA, IV, pp. 565–568, Mar. 1992.

11. Isabelle, S. H., Oppenheim, A. V., and Wornell, G. W., Effects of convolution on chaotic signals, *Proceedings of the 1992 IEEE ICASSP*, San Francisco, CA, IV, pp. 133–136, Mar. 1992.
12. Pecora, L. M. and Carroll, T. L., Synchronization in chaotic systems, *Phys. Rev. Lett.*, 64(8), 821–824, Feb. 1990.
13. Pecora, L. M. and Carroll, T. L., Driving systems with chaotic signals, *Phys. Rev. A*, 44(4), 2374–2383, Aug. 1991.
14. Carroll, T. L. and Pecora, L. M., Synchronizing chaotic circuits, *IEEE Trans. Circuits Syst.*, 38(4), 453–456, Apr. 1991.
15. Lorenz, E. N., Deterministic nonperiodic flow, *J. Atmospheric Sci.*, 20(2), 130–141, Mar. 1963.
16. Cuomo, K. M., Oppenheim, A. V., and Strogatz, S. H., Robustness and signal recovery in a synchronized chaotic system, *Int. J. Bifurcation Chaos*, 3(6), 1629–1638, Dec. 1993.
17. Cuomo, K. M. and Oppenheim, A. V., Synchronized chaotic circuits and systems for communications, Technical Report 575, MIT Research Laboratory of Electronics, 1992.
18. Cuomo, K. M., Oppenheim, A. V., and Isabelle, S. H., Spread spectrum modulation and signal masking using synchronized chaotic systems, Technical Report 570, MIT Research Laboratory of Electronics, 1992.
19. Oppenheim, A. V., Wornell, G. W., Isabelle, S. H., and Cuomo, K. M., Signal processing in the context of chaotic signals, in *Proceedings of the 1992 IEEE ICASSP*, San Francisco, CA, IV, pp. 117–120, Mar. 1992.
20. Cuomo, K. M. and Oppenheim, A. V., Circuit implementation of synchronized chaos with applications to communications, *Phys. Rev. Lett.*, 71(1), 65–68, July 1993.
21. Cuomo, K. M., Oppenheim, A. V., and Strogatz, S. H., Synchronization of Lorenz-based chaotic circuits with applications to communications, *IEEE Trans. Circuits Syst*, 40(10), 626–633, Oct. 1993.
22. Cuomo, K. M., Synthesizing self-synchronizing chaotic systems, *Int. J. Bifurcation Chaos*, 3(5), 1327–1337, Oct. 1993.
23. Cuomo, K. M., Synthesizing self-synchronizing chaotic arrays, *Int. J. Bifurcation Chaos*, 4(3), 727–736, June 1994.
24. Cuomo, K. M., Analysis and synthesis of self-synchronizing chaotic systems, PhD thesis, Massachusetts Institute of Technology, Cambridge, MA, Feb. 1994.

14

Nonlinear Maps

14.1 Introduction 14-1
14.2 Eventually Expanding Maps and Markov Maps 14-2
Eventually Expanding Maps
14.3 Signals from Eventually Expanding Maps 14-4
14.4 Estimating Chaotic Signals in Noise 14-4
14.5 Probabilistic Properties of Chaotic Maps 14-5
14.6 Statistics of Markov Maps 14-8
14.7 Power Spectra of Markov Maps 14-10
14.8 Modeling Eventually Expanding Maps with Markov Maps 14-12
References 14-12

Steven H. Isabelle
Massachusetts Institute of Technology

Gregory W. Wornell
Massachusetts Institute of Technology

14.1 Introduction

One-dimensional nonlinear systems, although simple in form, are applicable in a surprisingly wide variety of engineering contexts. As models for engineering systems, their richly complex behavior has provided insight into the operation of, for example, analog-to-digital converters [1], nonlinear oscillators [2], and power converters [3]. As realizable systems, they have been proposed as random number generators [4] and as signal generators for communication systems [5,6]. As analytic tools, they have served as mirrors for the behavior of more complex, higher dimensional systems [7–9].

Although one-dimensional nonlinear systems are, in general, hard to analyze, certain useful classes of them are relatively well understood. These systems are described by the recursion:

$$x[n] = f(x[n-1]), \tag{14.1a}$$

$$y[n] = g(x[n]), \tag{14.1b}$$

initialized by a scalar initial condition $x[0]$, where $f(\cdot)$ and $g(\cdot)$ are real-valued functions that describe the evolution of a nonlinear system and the observation of its state, respectively. The dependence of the sequence $x[n]$ on its initial condition is emphasized by writing $x[n] = f^n(x[0])$, where $f^n(\cdot)$ represents the n-fold composition of $f(\cdot)$ with itself.

Without further restrictions of the form of $f(\cdot)$ and $g(\cdot)$, this class of systems is too large to easily explore. However, systems and signals corresponding to certain "well-behaved" maps $f(\cdot)$ and observation functions $g(\cdot)$ can be rigorously analyzed. Maps of this type often generate chaotic signals—loosely speaking, bounded signals that are neither periodic nor transient—under easily verifiable conditions. These chaotic signals, although completely deterministic, are in many ways analogous to stochastic processes. In fact, one-dimensional chaotic maps illustrate in a relatively simple setting that the distinction between deterministic and stochastic signals is sometimes artificial and can be profitably emphasized or de-emphasized according to the needs of an application. For instance, problems of signal recovery from noisy observations are often best approached with a deterministic emphasis, while certain signal generation problems [10] benefit most from a stochastic treatment.

14.2 Eventually Expanding Maps and Markov Maps

Although signal models of the form [1] have simple, one-dimensional state spaces, they can behave in a variety of complex ways that model a wide range of phenomena. This flexibility comes at a cost, however; without some restrictions on its form, this class of models is too large to be analytically tractable. Two tractable classes of models that appear quite often in applications are eventually expanding maps and Markov maps.

14.2.1 Eventually Expanding Maps

Eventually expanding maps—which have been used to model sigma–delta modulators [11], switching power converters [3], other switched flow systems [12], and signal generators [6,13]—have three defining features: they are piecewise smooth, they map the unit interval to itself, and they have some iterate with slope that is everywhere greater than unity. Maps with these features generate time series that are chaotic, but on average well behaved. For reference, the formal definition is as follows, where the restriction to the unit interval is convenient but not necessary.

Definition 14.1: A nonsingular map f: [0,1] → [0,1] is called "eventually expanding" if

1. There is a set of partition points $0 = a_0 < a_1 < \cdots\ a_N = 1$ such that restricted to each of the intervals $\mathcal{V}_i = [a_{i-1}, a_i)$, called partition elements, the map $f(\cdot)$ is monotonic, continuous, and differentiable.
2. The function $1/|f'(x)|$ is of bounded variation [14]. (In some definitions, this smoothness condition on the reciprocal of the derivative is replaced with a more restrictive bounded slope condition, i.e., there exists a constant B such that $|f'(x)| < B$ for all x.)
3. There exists a real $\lambda > 1$ and a integer m such that

$$\left|\frac{\mathrm{d}}{\mathrm{d}x} f^m(x)\right| \geq \lambda,$$

wherever the derivative exists. This is the eventually expanding condition.

Every eventually expanding map can be expressed in the form

$$f(x) = \sum_{i=1}^{N} f_i(x)\chi_i(x), \tag{14.2}$$

where each $f_i(\cdot)$ is continuous, monotonic, and differentiable on the interior of the ith partition element and the indicator function $\chi_i(x)$ is defined by

$$\chi_i(x) = \begin{cases} 1, & x \in \mathcal{V}_i, \\ 0, & x \notin \mathcal{V}_i. \end{cases} \tag{14.3}$$

This class is broad enough to include, for example, discontinuous maps and maps with discontinuous or unbounded slope. Eventually expanding maps also include a class that is particularly amenable to analysis—the Markov maps.

Markov maps are analytically tractable and broadly applicable to problems of signal estimation, signal generation, and signal approximation. They are defined as eventually expanding maps that are piecewise-linear and have some extra structure.

Definition 14.2: A map f: [0,1] → [0,1] is an "eventually expanding, piecewise-linear, Markov map" if f is an eventually expanding map with the following additional properties:

1. The map is piecewise-linear, i.e., there is a set of partition points $0 = a_0 < a_1 < \cdots < a_N = 1$ such that restricted to each of the intervals $\mathcal{V}_i = [a_{i-1}, a_i)$, called partition elements, the map $f(\cdot)$ is affine, i.e., the functions $f_i(\cdot)$ on the right side of Equation 14.2 are of the form

$$f_i(x) = s_i x + b_i.$$

2. The map has the Markov property that partition points map to partition points, i.e., for each i, $f(a_i) = a_j$ for some j.

Every Markov map can be expressed in the form

$$f(x) = \sum_{i=1}^{N} (s_i x + b_i)\chi_i(x), \tag{14.4}$$

where $s_i \neq 0$ for all i. Figure 14.1 shows the Markov map

$$f(x) = \begin{cases} (1-a)x/a + a, & 0 \leq x \leq a, \\ (1-x)/(1-a), & a < x \leq 1, \end{cases} \tag{14.5}$$

which has partition points $\{0, a, 1\}$, and partition elements $\mathcal{V}_1 = [0, a)$ and $\mathcal{V}_2 = [a, 1)$.

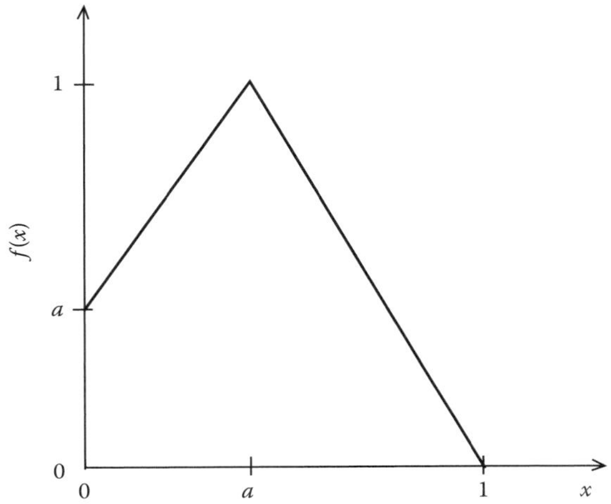

FIGURE 14.1 An example of a piecewise-linear Markov map with two partition elements.

Markov maps generate signals with two useful properties: they are, when suitably quantized, indistinguishable from signals generated by Markov chains; they are close, in a sense, to signals generated by more general eventually expanding maps [15]. These two properties lead to applications of Markov maps for generating random numbers and approximating other signals. The analysis underlying these types of applications depends on signal representations that provide insight into the structure of chaotic signals.

14.3 Signals from Eventually Expanding Maps

There are several general representations for signals generated by eventually expanding maps. Each provides different insights into the structure of these signals and proves useful in different applications. First, and most obviously, a sequence generated by a particular map is completely determined by (and is thus represented by) its initial condition $x[0]$. This representation allows certain signal estimation problems to be recast as problems of estimating the scalar initial condition. Second, and less obviously, the quantized signal $y[n] = g(x[n])$, for $n \geq 0$ generated by Equation 14.1 with $g(\cdot)$ defined by

$$g(x) = i, \quad x \in \mathcal{V}_i, \tag{14.6}$$

uniquely specifies the initial condition $x[0]$ and hence the entire state sequence $x[n]$. Such quantized sequences $y[n]$ are called the symbolic dynamics associated with $f(\cdot)$ [7]. Certain properties of a map, such as the collection of initial conditions leading to periodic points, are most easily described in terms of its symbolic dynamics. Finally, a hybrid representation of $x[n]$ combining the initial condition and symbolic representations

$$\mathbf{H}[N] = \{g(x[0]), \ldots, g(x[N]), x[N]\},$$

is often useful.

14.4 Estimating Chaotic Signals in Noise

The hybrid signal representation described in Section 14.3 can be applied to a classical signal processing problem—estimating a signal in white Gaussian noise. For example, suppose the problem is to estimate a chaotic sequence $x[n]$, $n = 0, \ldots, N-1$ from the noisy observations

$$r[n] = x[n] + w[n], \quad n = 0, \ldots, N-1, \tag{14.7}$$

where

$w[n]$ is a stationary, zero-mean white Gaussian noise sequence with variance σ_w^2
$x[n]$ is generated by iterating (Equation 14.1) from an unknown initial condition

Because $w[n]$ is white and Gaussian, the maximum likelihood estimation problem is equivalent to the constrained minimum distance problem

$$\underset{x[n]:\, x[i] = f(x[i-1])}{\text{minimize}}, \quad \varepsilon[N] = \sum_{k=0}^{N} (r[k] - x[k])^2, \tag{14.8}$$

and to the scalar problem

$$\underset{x[0] \in [0,1]}{\text{minimize}}, \quad \varepsilon[N] = \sum_{k=0}^{N} \left(r[k] - f^k(x[0])\right)^2. \tag{14.9}$$

Thus, the maximum likelihood problem can, in principle, be solved by first estimating the initial condition, then iterating (Equation 14.1) to generate the remaining estimates. However, the initial condition is often difficult to estimate directly because the likelihood function (Equation 14.9), which is highly irregular with fractal characteristics, is unsuitable for gradient-descent type optimization [16]. Another solution divides the domain of $f(\cdot)$ into subintervals and then solves a dynamic programming problem [17]; however, this solution is, in general, suboptimal, and computationally expensive.

Although the maximum likelihood problem described above need not, in general, have a computationally efficient recursive solution, it does have one when, for example, the map $f(\cdot)$ is a symmetric tent map of the form

$$f(x) = \beta - 1 - \beta|x|, \quad x \in [-1, 1], \tag{14.10}$$

with parameter $1 < \beta \leq 2$ [5]. This algorithm solves for the hybrid representation of the initial condition from which an estimate of the entire signal can be determined. The hybrid representation is of the form

$$\mathbf{H}[N] = \{y[0], \ldots, y[N], x[N]\},$$

where each $y[i]$ takes one of two values which, for convenience, we define as $y[i] = \mathrm{sgn}(x[i])$. Since each $y[n]$ can independently takes one of two values, there are 2^N feasible solutions to this problem and a direct search for the optimal solution is thus impractical even for moderate values of N.

The resulting algorithm has computational complexity that is linear in the length of the observation, N. This efficiency is the result of a special "separation property," possessed by the map [10]: given $y[0], \ldots, y[i-1]$ and $y[i+1], \ldots, y[N]$ the estimate of the parameter $y[i]$ is independent of $y[i+1], \ldots, y[N]$. The algorithm is as follows. Denoting by $\hat{\phi}[n|m]$ the ML estimates of any sequence $\phi[n]$ given $r[k]$ for $0 \leq k \leq m$, the ML solution is of the form,

$$\hat{x}[n|n] = \frac{(\beta^2 - 1)\beta^{2n} r[n] + (\beta^{2n} - 1)\hat{x}[n|n-1]}{\beta^{2(n+1)} - 1}, \tag{14.11}$$

$$\hat{y}[n|N] = \mathrm{sgn}\hat{x}[n|n], \tag{14.12}$$

$$\hat{x}_{\mathrm{ML}}[n|n] = \mathcal{L}_\beta(\hat{x}[n|n]), \tag{14.13}$$

where $\hat{x}[n|n-1] = f(\hat{x}[n-1|n-1])$, the initialization is $\hat{x}[0|0] = r[0]$, and the function $\mathcal{L}_\beta(\hat{x}[n|n])$, defined by

$$\mathcal{L}_\beta(x) = \begin{cases} x, & x \in (-1, \beta - 1), \\ -1, & x \leq -1, \\ \beta - 1, & x \geq \beta - 1, \end{cases} \tag{14.14}$$

serves to restrict the ML estimates to the interval $x \in (-1, \beta - 1)$. The smoothed estimates $\hat{x}_{\mathrm{ML}}[n|N]$ are obtained by converting the hybrid representation to the initial condition and then iterating the estimated initial condition forward.

14.5 Probabilistic Properties of Chaotic Maps

Almost all waveforms generated by a particular eventually expanding map have the same average behavior [18], in the sense that the time average

$$\bar{h}(x[0]) = \lim_{n\to\infty} \frac{1}{n} \sum_{k=0}^{n-1} h(x[k]) = \lim_{n\to\infty} \frac{1}{n} \sum_{k=0}^{n-1} h\big(f^k(x[0])\big) \tag{14.15}$$

exists and is essentially independent of the initial condition $x[0]$ for sufficiently well-behaved functions $h(\cdot)$. This result, which is reminiscent of results from the theory of stationary stochastic processes [19], forms the basis for a probabilistic interpretation of chaotic signals, which in turn leads to analytic methods for characterizing their time-average behavior.

To explore the link between chaotic and stochastic signals, first consider the stochastic process generated by iterating (Equation 14.1) from a random initial condition $x[0]$, with probability density function $p_0(\cdot)$. Denote by $p_n(\cdot)$ the density of the nth iterate $x[n]$. Although, in general, the members of the sequence $p_n(\cdot)$ will differ, there can exist densities, called "invariant densities," that are time-invariant, i.e.,

$$p_0(\cdot) = p_1(\cdot) = \cdots = p_n(\cdot) = p(\cdot). \tag{14.16}$$

When the initial condition $x[0]$ is chosen randomly according to an invariant density, the resulting stochastic process is stationary [19] and its ensemble averages depend on the invariant density. Even when the initial condition is not random, invariant densities play an important role in describing the time-average behavior of chaotic signals. This role depends on, among other things, the number of invariant densities that a map possesses.

A general one-dimensional nonlinear map may possess many invariant densities. For example, eventually expanding maps with N partition elements have at least one and at most N invariant densities [20]. However, maps can often be decomposed into collections of maps, each with only one invariant density [19], and little generality is lost by concentrating on maps with only one invariant density. In this special case, the results that relate the invariant density to the average behavior of chaotic signals are more intuitive.

The invariant density, although introduced through the device of a random initial condition, can also be used to study the behavior of individual signals. Individual signals are connected to ensembles of signals, which correspond to random initial conditions, through a classical result due to Birkhoff, which asserts that the time average $\bar{h}(x[0])$ defined by Equation 14.15 exists whenever $f(\cdot)$ has an invariant density. When the $f(\cdot)$ has only one invariant density, the time-average is independent of the initial condition for almost all (with respect to the invariant density $p(\cdot)$) initial conditions and equals

$$\lim_{n\to\infty} \frac{1}{n} \sum_{k=0}^{n-1} h(x[k]) = \lim_{n\to\infty} \frac{1}{n} \sum_{k=0}^{n-1} h\left(f^k(x[0])\right) = \int h(x)p(x)\mathrm{d}x. \tag{14.17}$$

where the integral is performed over the domain of $f(\cdot)$ and where $h(\cdot)$ is measurable.

Birkhoff's theorem leads to a relative frequency interpretation of time-averages of chaotic signals. To see this, consider the time-average of the indicator function $\tilde{\chi}_{[s-\varepsilon,s+\varepsilon]}(x)$, which is zero everywhere but in the interval $[s-\varepsilon, s+\varepsilon]$ where it is equal to unity. Using Birkhoff's theorem with Equation 14.17 yields

$$\lim_{n\to\infty} \frac{1}{n} \sum_{k=0}^{n-1} \tilde{\chi}_{[s-\varepsilon,s+\varepsilon]}(x[k]) = \int \tilde{\chi}_{[s-\varepsilon,s+\varepsilon]}(x)p(x)\mathrm{d}x \tag{14.18}$$

$$= \int_{[s-\varepsilon,s+\varepsilon]} p(x)\mathrm{d}x \tag{14.19}$$

$$\approx 2\varepsilon p(s), \tag{14.20}$$

where Equation 14.20 follows from Equation 14.19 when ε is small and $p(\cdot)$ is sufficiently smooth. The time-average (Equation 14.18) is exactly the fraction of time that the sequence $x[n]$ takes values in the interval $[s-\varepsilon, s+\varepsilon]$. Thus, from Equation 14.20, the value of the invariant density at any point s is approximately proportional to the relative frequency with which $x[n]$ takes values in a small neighborhood of the point. Motivated by this relative frequency interpretation, the probability that an arbitrary function $h(x[n])$ falls into an arbitrary set A can be defined by

$$\Pr\{h(x) \in A\} = \lim_{n\to\infty} \frac{1}{n} \sum_{k=0}^{n-1} \tilde{\chi}_A(h(x[k])). \tag{14.21}$$

Using this definition of probability, it can be shown that for any Markov map, the symbol sequence $y[n]$ defined in Section 14.3 is indistinguishable from a Markov chain in the sense that

$$\Pr\{y[n]|y[n-1],\ldots,y[0]\} = \Pr\{y[n]|y[n-1]\}, \tag{14.22}$$

holds for all n [21]. The first-order transition probabilities can be shown to be of the form

$$\Pr(y[n]|y[n-1]) = \frac{|\mathcal{V}_{y[n]}|}{|s_{y[n]}||\mathcal{V}_{y[n-1]}|},$$

where the s_i are the slopes of the map $f(\cdot)$ as in Equation 14.4 and $|\mathcal{V}_{y[n]}|$ denotes the length of the interval $\mathcal{V}_{y[n]}$. As an example, consider the asymmetric tent map

$$f(x) = \begin{cases} x/a & 0 \le x \le a, \\ (1-x)/(1-a), & a < x \le 1, \end{cases}$$

with parameter in the range $0 < a < 1$ and a quantizer $g(\cdot)$ of the form (Equation 14.6). The previous results establish that $y[n] = g(x[n])$ is equivalent to a sample sequence from the Markov chain with transition probability matrix

$$[P]_{ij} = \begin{bmatrix} a & 1-a \\ a & 1-a \end{bmatrix},$$

where $[P]_{ij} = \Pr\{y[n] = i|y[n-1] = j\}$. Thus, the symbolic sequence appears to have been generated by independent flips of a biased coin with the probability of heads, say, equal to a. When the parameter takes the value $a = 1/2$, this corresponds to a sequence of independent equally likely bits. Thus, a sequence of Bernoulli random variables can been constructed from a deterministic sequence $x[n]$. Based on this remarkable result, a circuit that generates statistically independent bits for cryptographic applications has been designed [4].

Some of the deeper probabilistic properties of chaotic signals depend on the integral Equation 14.17, which in turn depends on the invariant density. For some maps, invariant densities can be determined explicitly. For example, the tent map (Equation 14.10) with $\beta = 2$ has invariant density

$$p(x) = \begin{cases} 1/2, & -1 \le x \le 1, \\ 0, & \text{otherwise}, \end{cases}$$

as can be readily verified using elementary results from the theory of derived distributions of functions of random variables [22]. More generally, all Markov maps have invariant densities that are piecewise-constant function of the form

$$\sum_{i=1}^{n} c_i \chi_i(x), \tag{14.23}$$

where c_i are real constants that can be determined from the map's parameters [23]. This makes Markov maps especially amenable to analysis.

14.6 Statistics of Markov Maps

The transition probabilities computed above may be viewed as statistics of the sequence $x[n]$. These statistics, which are important in a variety of applications, have the attractive property that they are defined by integrals having, for Markov maps, readily computable, closed-form solutions. This property holds more generally—Markov maps generate sequences for which a large class of statistics can be determined in closed form. These analytic solutions have two primary advantages over empirical solutions computed by time averaging: they circumvent some of the numerical problems that arise when simulating the long sequences of chaotic data that are necessary to generate reliable averages; and they often provide insight into aspects of chaotic signals, such as dependence on a parameter, that could not be easily determined by empirical averaging.

Statistics that can be readily computed include correlations of the form

$$R_{f;h_0,h_1,\ldots,h_r}[k_1,\ldots,k_r] = \lim_{L\to\infty} \frac{1}{L} \sum_{n=0}^{L-1} h_0(x[n])h_1(x[n+k_1])\cdots h_r(x[n+k_r]) \tag{14.24}$$

$$= \int h_0(x[n])h_1(x[n+k_1])\cdots h_r(x[n+k_r])p(x)\mathrm{d}x, \tag{14.25}$$

where the $h_i(\cdot)'$s are suitably well-behaved but otherwise arbitrary functions, the k_i's are nonnegative integers, the sequence $x[n]$ is generated by Equation 14.1, and $p(\cdot)$ is the invariant density. This class of statistics includes as important special cases the autocorrelation function and all higher order moments of the time series. Of primary importance in determining these statistics is a linear transformation called the Frobenius–Perron (FP) operator, which enters into the computation of these correlations in two ways. First, it suggests a method for determining an invariant density. Second, it provides a "change of variables" within the integral that leads to simple expressions for correlation statistics.

The definition of the FP operator can be motivated by using the device of a random initial condition $x[0]$ with density $p_0(x)$ as in Section 14.5. The FP operator describes the time evolution of this initial probability density. More precisely, it relates the initial density to the densities $p_n(\cdot)$ of the random variables $x[n] = f^n(x[0])$ through the equation

$$p_n(x) = P_f^n p_0(x), \tag{14.26}$$

where P_f^n denotes the n-fold self-composition of P_f. This definition of the FP operator, although phrased in terms of its action on probability densities, can be extended to all integrable functions. This extended operator, which is also called the FP operator, is linear and continuous. Its properties are closely related to the statistical structure of signals generated by chaotic maps (see [9] for a thorough discussion of these issues). For example, the evolution Equation 14.26 implies that an invariant density of a map is a fixed point of its FP operator, that is, it satisfies

$$p(x) = P_f p(x). \tag{14.27}$$

This relation can be used to determine explicitly the invariant densities of Markov maps [23], which may in turn be used to compute more general statistics.

Using the change of variables property of the FP operator, the correlation statistic (Equation 14.25) can be expressed as the ensemble average

$$R_{f;h_0,h_1,\ldots,h_r}[k_1,\ldots,k_r] = \int h_r(x)P_f^{k_r-k_{r-1}}\Big\{h_{r-1}(x)\cdots P_f^{k_2-k_1}\Big\{h_1(x)P_f^{k_1}\{h_0(x)p(x)\}\Big\}\cdots\Big\}\mathrm{d}x. \tag{14.28}$$

Although such integrals are, for general one-dimensional nonlinear maps, difficult to evaluate, closed-form solutions exist when $f(\cdot)$ is a Markov map—a development that depends on an explicit expression for FP operator.

The FP operator of a Markov map has a simple, finite-dimensional matrix representation when it operates on certain piecewise polynomial functions. Any function of the form

$$h(\cdot) = \sum_{i=0}^{K} \sum_{j=1}^{N} a_{ij} x^i \chi_j(x) \tag{14.29}$$

can be represented by an $N(K+1)$-dimensional coordinate vector with respect to the basis

$$\{\theta_1(x), \theta_2(x), \ldots, \theta_{N(K+1)}\} = \{\chi_1(x), \ldots, \chi_N(x), x\chi_1(x), \ldots, x\chi_N(x), \ldots, x^K\chi_1(x), \ldots, x^K\chi_N(x)\}. \tag{14.30}$$

The action of the FP operator on any such function can be expressed as a matrix-vector product: when the coordinate vector of $h(x)$ is $\mathbf{h}$, the coordinate vector of $q(x) = P_f h(x)$ is

$$\mathbf{q} = \mathbf{P}_K \mathbf{h},$$

where $\mathbf{P}_K$ is the square $N(K+1)$-dimensional, block upper-triangular matrix

$$\mathbf{P}_K = \begin{bmatrix} \mathbf{P}_{00} & \mathbf{P}_{01} & \cdots & \cdots & \mathbf{P}_{0K} \\ 0 & \mathbf{P}_{11} & \mathbf{P}_{12} & \cdots & \mathbf{P}_{1K} \\ \vdots & \vdots & \vdots & \vdots & \vdots \\ 0 & 0 & \cdots & \cdots & \mathbf{P}_{KK} \end{bmatrix}, \tag{14.31}$$

and where each nonzero $N \times N$ block is of the form

$$\mathbf{P}_{ij} = \binom{j}{i} \mathbf{P}_0 \mathbf{B}^{j-i} \mathbf{S}^j \quad \text{for } j \geq i. \tag{14.32}$$

The $N \times N$ matrices $\mathbf{B}$ and $\mathbf{S}$ are diagonal with elements $B_{ii} = -b_i$ and $S_{ii} = 1/s_i$, respectively, while $\mathbf{P}_0 = \mathbf{P}_{00}$ is the $N \times N$ matrix with elements

$$[\mathbf{P}_0]_{ij} = \begin{cases} 1/|s_j|, & i \in \mathcal{I}_j, \\ 0, & \text{otherwise.} \end{cases} \tag{14.33}$$

The invariant density of a Markov map, which is needed to compute the correlation statistic Equation 14.25, can be determined as the solution of an eigenvector problem. It can be shown that such invariant densities are piecewise constant functions so that the fixed point equation (Equation 14.27) reduces to the matrix expression

$$\mathbf{P}_0 \mathbf{p} = \mathbf{p}.$$

Due to the properties of the matrix $\mathbf{P}_0$, this equation always has a solution that can be chosen to have nonnegative components. It follows that the correlation statistic equation (Equation 14.29) can always be expressed as

$$R_{f;h_0,h_1,\ldots,h_r}[k_1, \ldots, k_r] = \mathbf{g}_1^{\mathrm{T}} \mathbf{M} \mathbf{g}_2, \tag{14.34}$$

where **M** is a basis correlation matrix with elements

$$[\mathbf{M}]_{ij} = \int \theta_i(x)\theta_j(x)\mathrm{d}x. \tag{14.35}$$

and $\mathbf{g}_i$ are the coordinate vectors of the functions

$$\mathbf{g}_1(x) = h_r(x), \tag{14.36}$$

$$\mathbf{g}_2(x) = P_f^{k_r-k_{r-1}}\left\{h_{r-1}(x)\cdots P_f^{k_2-k_1}\left\{h_1(x)P_f^{k_1}\{h_0(x)p(x)\}\right\}\cdots\right\}. \tag{14.37}$$

By the previous discussion, the coordinate vectors $\mathbf{g}_1$ and $\mathbf{g}_2$ can be determined using straightforward matrix-vector operations. Thus, expression (Equation 14.34) provides a practical way of exactly computing the integral Equation 14.29, and reveals some important statistical structure of signals generated by Markov maps.

14.7 Power Spectra of Markov Maps

An important statistic in the context of many engineering applications is the power spectrum. The power spectrum associated with a Markov map is defined as the Fourier transform of its autocorrelation sequence

$$R_{xx}[k] = \int x[n]x[n+k]p(x)\mathrm{d}x, \tag{14.38}$$

which, using Equation 14.34 can be rewritten in the form

$$R_{xx}[k] = \mathbf{g}_1^{\mathrm{T}}\mathbf{M}_1\mathbf{P}_1^k\tilde{\mathbf{g}}_2, \tag{14.39}$$

where $\mathbf{P}_1$ is the matrix representation of the FP operator restricted to the space of piecewise linear functions, and where $\mathbf{g}_1$ is the coordinate vector associated with the function x, and where $\tilde{\mathbf{g}}_2$ is the coordinate vector associated with $\tilde{g}_2(x) = xp(x)$.

The power spectrum is obtained from the Fourier transform of Equation 14.39, yielding,

$$S_{xx}(e^{j\omega}) = \mathbf{g}_1^{\mathrm{T}}\mathbf{M}_1\left(\sum_{k=-\infty}^{+\infty}\mathbf{P}_1^{|k|}e^{-j\omega k}\right)\tilde{\mathbf{g}}_2. \tag{14.40}$$

This sum can be simplified by examining the eigenvalues of the FP matrix $\mathbf{P}_1$. In general, $\mathbf{P}_1$ has eigenvalues whose magnitude is strictly less than unity, and others with unit-magnitude [9]. Using this fact, Equation 14.40 can be expressed in the form

$$S_{xx}(e^{j\omega}) = \mathbf{h}_1^{\mathrm{T}}\mathbf{M}\left(\mathbf{I}-\Gamma_2e^{-j\omega}\right)^{-1}\left(\mathbf{I}-\Gamma_2^2\right)\left(\mathbf{I}-\Gamma_2e^{j\omega}\right)^{-1}\tilde{\mathbf{g}}_2 + \sum_{i=1}^{m}C_i\delta(\omega-\omega_i), \tag{14.41}$$

where Γ_2 has eigenvalues that are strictly less than one in magnitude, and C_i and ω_i depend on the unit magnitude eigenvalues of $\mathbf{P}_1$.

As Equation 14.41 reflects, the spectrum of a Markov map is a linear combination of an impulsive component and a rational function. This implies that there are classes of rational spectra that can be generated not only by the usual method of driving white noise through a linear time-invariant filter with a rational system function, but also by iterating deterministic nonlinear dynamics. For this reason it is natural to view chaotic signals corresponding to Markov maps as "chaotic autoregressive moving average (ARMA) processes." Special cases correspond to the "chaotic white noise" described in [5] and the first-order autoregressive processes described in [24].

Consider now a simple example involving the Markov map defined in Equation 14.5 and shown in Figure 14.1. Using the techniques described above, the invariant density is determined to be the piecewise-constant function

$$p(x) = \begin{cases} 1/(1+a), & 0 \le x \le a, \\ 1/(1-a^2), & a \le x \le 1. \end{cases}$$

Using Equation 14.41 and a parameter value $a = 8/9$, the rational part of the autocorrelation sequence associated with $f(\cdot)$ is determined to be

$$S_{xx}(z) = -\frac{42{,}632}{459} \frac{36z^{-1} - 145 + 36z}{(9+8z)(9+8z^{-1})(64z^2 + z + 81)(64z^{-2} + z^{-1} + 81)}. \tag{14.42}$$

The power spectrum corresponding to evaluating Equation 14.42 on the unit circle $z = e^{j\omega}$ is plotted in Figure 14.2, along with an empirical spectrum computed by periodogram averaging with a window length of 128 on a time series of length 50,000. The solid line corresponds to the analytically obtained expression (Equation 14.42), while the circles represent the spectral samples estimated by periodogram averaging.

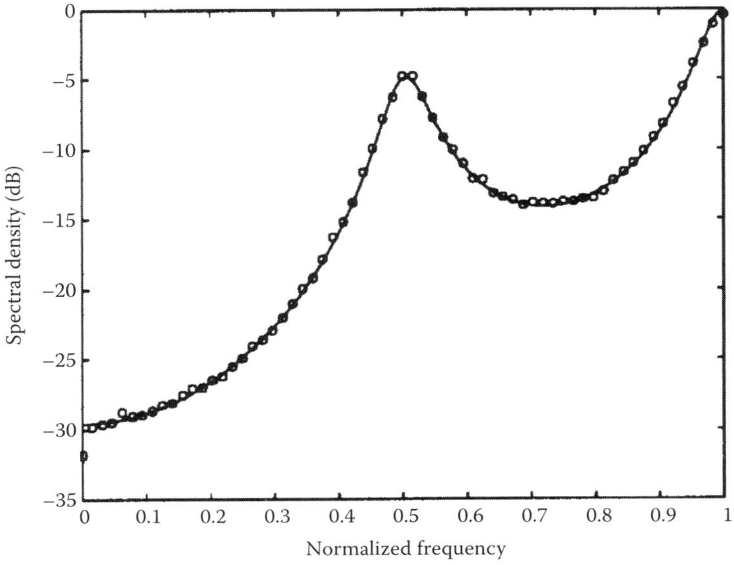

FIGURE 14.2 Comparison of analytically computed power spectrum to empirical power spectrum for the map of Figure 14.1. The solid line indicates the analytically computed spectrum, while the circles indicate the samples of the spectrum estimated by applying periodogram averaging to a time series of length 50,000.

14.8 Modeling Eventually Expanding Maps with Markov Maps

One approach to studying the statistics of more general eventually expanding maps involves approximation by Markov maps—the statistics of any eventually expanding map can be approximated to arbitrary accuracy by those of some Markov map. This approximation strategy provides a powerful method for analyzing chaotic time series from eventually expanding maps: first approximate the map by a Markov map, then use the previously described techniques to determine its statistics. In order for this approach to be useful, an appropriate notion, the approximation quality, and a constructive procedure for generating an approximate map are required.

A sequence of piecewise-linear Markov maps $\hat{f}_i(\cdot)$ with statistics that converge to those of a given eventually expanding map $f(\cdot)$ is said to "statistically converge" to $f(\cdot)$. More formally:

Definition 14.3: Let $f(\cdot)$ be an eventually expanding map with a unique invariant density $p(\cdot)$. A sequence of maps $\{\hat{f}_i(\cdot)\}$ statistically converges to $f(\cdot)$ if each $\hat{f}_i(\cdot)$ has a unique invariant density $p_i(\cdot)$ and

$$R_{\hat{f}_i,h_0,h_1,\ldots,h_r}[k_1,\ldots,k_r] \to R_{f,h_0,h_1,\ldots,h_r}[k_1,\ldots,k_r] \quad \text{as } i \to \infty,$$

for any continuous $h_j(\cdot)$ and all finite k_j and finite r.

Any eventually expanding map $f(\cdot)$ is the limit of a sequence of Markov maps that statistically converges and can be constructed in a straightforward manner. The idea is to define a Markov map on an increasingly fine set of partition points that includes the original partition points of $f(\cdot)$. Denote by $\mathcal{Q}$ the set of partition points of $f(\cdot)$, and by $\mathcal{Q}_i$ the set of partition points of the ith map in the sequence of Markov map approximations. The sets of partition points for the increasingly fine approximations are defined recursively via

$$\mathcal{Q}_i = \mathcal{Q}_{i-1} \cup f^{-1}(\mathcal{Q}_{i-1}). \tag{14.43}$$

In turn, each approximating map $\hat{f}_i(\cdot)$ is defined by specifying its value at the partition points $\mathcal{Q}_i$ by a procedure that ensures that the Markov property holds [15]. At all other points, the map $\hat{f}_i(\cdot)$ is defined by linear interpolation.

Conveniently, if $f(\cdot)$ is an eventually expanding map in the sense of Definition 14.1, then the sequence of piecewise-linear Markov approximations $\hat{f}_i(\cdot)$ obtained by the above procedure statistically converges to $f(\cdot)$, i.e., converges in the sense of Definition 14.3. This means that, for sufficiently large i, the statistics of $\hat{f}_i(\cdot)$ are close to those of $f(\cdot)$. As a practical consequence, the correlation statistics of the eventually expanding map $f(\cdot)$ can be approximated by first determining a Markov map $\hat{f}_k(\cdot)$ that is a good approximation to $f(\cdot)$, and then finding the statistics of Markov map using the techniques described in Section 14.6.

References

1. Feely, O. and Chua, L.O., Nonlinear dynamics of a class of analog-to-digital converters, *Intl. J. Bifurcat. Chaos Appl. Sci. Eng.*, 2(2), 325–340, June 1992.
2. Tang, Y.S., Mees, A.I., and Chua, L.O., Synchronization and chaos, *IEEE Trans. Circ. Syst.*, CAS-30(9), 620–626, 1983.
3. Deane, J.H.B. and Hamill, D.C., Chaotic behavior in a current-mode controlled DC-DC converter, *Electron. Lett.*, 27, 1172–1173, 1991.

4. Espejo, S., Martin, J.D., and Rodriguez-Vazquez, A., Design of an analog/digital truly random number generator, in *1990 IEEE International Symposium Circular Systems*, Murray Hill, NJ, pp. 1368–1371, 1990.
5. Papadopoulos, H.C. and Wornell, G.W., Maximum likelihood estimation of a class of chaotic signals, *IEEE Trans. Inform. Theory*, 41, 312–317, January 1995.
6. Chen, B. and Wornell, G.W., Efficient channel coding for analog sources using chaotic systems, In *Proceedings of the IEEE GLOBECOM*, London, November 1996.
7. Devaney, R., *An Introduction to Chaotic Dynamical Systems*, Addison-Wesley, Reading, MA, 1989.
8. Collet, P. and Eckmann, J.P., *Iterated Maps on the Interval as Dynamical Systems*, Birkhauser, Boston, MA, 1980.
9. Lasota, A. and Mackey, M., *Probabilistic Properties of Deterministic Systems*, Cambridge University Press, Cambridge, U.K., 1985.
10. Richard, M.D., Estimation and detection with chaotic systems, PhD thesis, MIT, Cambridge, MA. Also RLE Tech. Rep. No. 581, February 1994.
11. Risbo, L., On the design of tone-free sigma-delta modulators, *IEEE Trans. Circ. Syst. II*, 42(1), 52–55, 1995.
12. Chase, C., Serrano, J., and Ramadge, P.J., Periodicity and chaos from switched flow systems: Contrasting examples of discretely controlled continuous systems, *IEEE Trans. Automat. Contr.*, 38, 71–83, 1993.
13. Chua, L.O., Yao, Y., and Yang, Q., Generating randomness from chaos and constructing chaos with desired randomness, *Intl. J. Circ. Theory App.*, 18, 215–240, 1990.
14. Natanson, I.P., *Theory of Functions of a Real Variable*, Frederick Ungar Publishing, New York, 1961.
15. Isabelle, S.H., A signal processing framework for the analysis and application of chaos, PhD thesis, MIT, Cambridge, MA. Also RLE Tech. Rep. No. 593, February 1995.
16. Myers, C., Kay S., and Richard, M., Signal separation for nonlinear dynamical systems, in *Proceedings of the International Conference Acoustic Speech, Signal Processing*, San Francisco, CA, 1992.
17. Kay, S. and Nagesha, V., Methods for chaotic signal estimation, *IEEE Trans. Signal Process.*, 43(8), 2013, 1995.
18. Hofbauer, F. and Keller, G., Ergodic properties of invariant measures for piecewise monotonic transformations, *Math. Z.*, 180, 119–140, 1982.
19. Peterson, K., *Ergodic Theory*, Cambridge University Press, Cambridge, U.K., 1983.
20. Lasota, A. and Yorke, J.A., On the existence of invariant measures for piecewise monotonic transformations, *Trans. Am. Math. Soc.*, 186, 481–488, December 1973.
21. Kalman, R., Nonlinear aspects of sampled-data control systems, in *Proceedings of the Symposium Nonlinear Circuit Analysis*, New York, pp. 273–313, April 1956.
22. Drake, A.W., *Fundamentals of Applied Probability Theory*, McGraw-Hill, New York, 1967.
23. Boyarsky, A. and Scarowsky, M., On a class of transformations which have unique absolutely continuous invariant measures, *Trans. Am. Math. Soc.*, 255, 243–262, 1979.
24. Sakai, H. and Tokumaru, H., Autocorrelations of a certain chaos, *IEEE Trans. Acoust., Speech, Signal Process.*, 28(5), 588–590, 1990.

15

Fractal Signals

Gregory W. Wornell
Massachusetts Institute of Technology

15.1 Introduction 15-1
15.2 Fractal Random Processes 15-1
Models and Representations for $1/f$ Processes
15.3 Deterministic Fractal Signals 15-8
15.4 Fractal Point Processes 15-10
Multiscale Models • Extended Markov Models
References 15-13

15.1 Introduction

Fractal signal models are important in a wide range of signal-processing applications. For example, they are often well-suited to analyzing and processing various forms of natural and man-made phenomena. Likewise, the synthesis of such signals plays an important role in a variety of electronic systems for simulating physical environments. In addition, the generation, detection, and manipulation of signals with fractal characteristics has become of increasing interest in communication and remote-sensing applications.

A defining characteristic of a fractal signal is its invariance to time- or space dilation. In general, such signals may be one-dimensional (e.g., fractal time series) or multidimensional (e.g., fractal natural terrain models). Moreover, they may be continuous- or discrete time in nature, and may be continuous or discrete in amplitude.

15.2 Fractal Random Processes

Most generally, fractal signals are signals having detail or structure on all temporal or spatial scales. The fractal signals of most interest in applications are those in which the structure at different scales is similar. Formally, a zero-mean random process $x(t)$ defined on $-\infty < t < \infty$ is statistically self-similar if its statistics are invariant to dilations and compressions of the waveform in time. More specifically, a random process $x(t)$ is statistically self-similar with parameter H if for any real $a > 0$ it obeys the scaling relation $x(t) \stackrel{\Delta}{=} a^{-H}x(at)$, where $\stackrel{\Delta}{=}$ denotes equality in a statistical sense. For strict-sense self-similar processes, this equality is in the sense of all finite-dimensional joint probability distributions. For wide-sense self-similar processes, the equality is interpreted in the sense of second-order statistics, i.e., the

$$R_x(t,s) \stackrel{\Delta}{=} E[x(t)x(s)] = a^{-2H}R_x(at,as)$$

A sample path of a self-similar process is depicted in Figure 15.1.

While regular self-similar random processes cannot be stationary, many physical processes exhibiting self-similarity possess some stationary attributes. An important class of models for such phenomena are

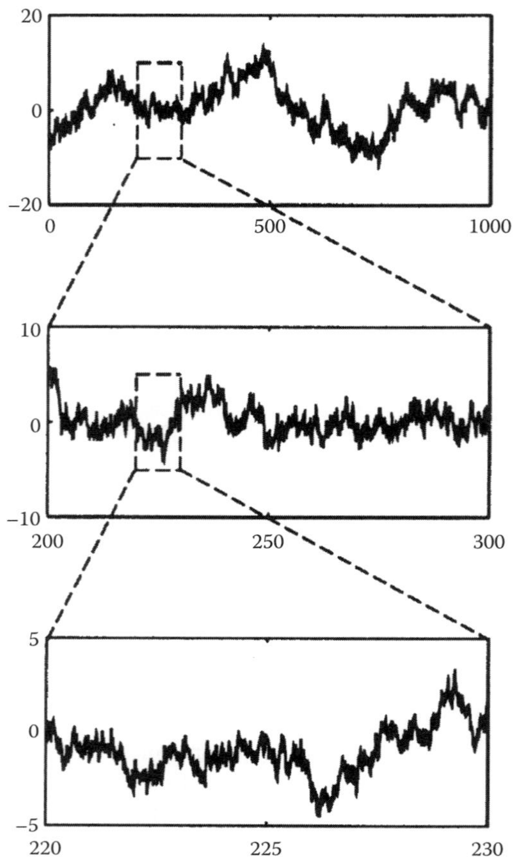

FIGURE 15.1 A sample waveform from a statistically scale-invariant random process, depicted on three different scales.

referred to as "$1/f$ processes." The $1/f$ family of statistically self-similar random processes are empirically defined as processes having measured power spectra obeying a power–law relationship of the form

$$S_x(\omega) \sim \frac{\sigma_x^2}{|\omega|^\gamma} \tag{15.1}$$

for some spectral parameter γ related to H according to $\gamma = 2H + 1$.

Generally, the power–law relationship (Equation 15.1) extends over several decades of frequency. While data length typically limits access to spectral information at lower frequencies, and data resolution typically limits access to spectral content at higher frequencies, there are many examples of phenomena for which arbitrarily large data records justify a $1/f$ spectrum of the form shown in Equation 15.1 over all accessible frequencies. However, Equation 15.1 is not integrable and hence, strictly speaking, does not constitute a valid power spectrum in the theory of stationary random processes. Nevertheless, a variety of interpretations of such spectra have been developed based on notions of generalized spectra [1–3].

As a consequence of their inherent self-similarity, the sample paths of $1/f$ processes are typically fractals [4]. The graphs of sample paths of random processes are one-dimensional curves in the plane; this is their "topological dimension." However, fractal random processes have sample paths that are so irregular that their graphs have an "effective" dimension that exceeds their topological dimension of unity. It is this effective dimension that is usually referred to as the "fractal" dimension of the graph. However, it is important to note that the notion of fractal dimension is not uniquely defined. There are several different definitions of fractal dimension from which to choose for a given application—each with subtle but significant differences [5]. Nevertheless, regardless of the particular definition, the fractal dimension D of the graph of a fractal function typically ranges between $D = 1$ and $D = 2$. Larger values of D correspond to functions whose graphs are increasingly rough in appearance and, in an appropriate sense, fill the plane in which the graph resides to a greater extent. For $1/f$ processes, there is an inverse relationship between the fractal dimension D and the self-similarity parameter H of the process: An increase in the parameter H yields a decrease in the dimension D, and vice versa. This is intuitively reasonable, since an increase in H corresponds to an increase in γ, which, in turn, reflects a redistribution of power from high to low frequencies and leads to sample functions that are increasingly smooth in appearance.

A truly enormous and tremendously varied collection of natural phenomena exhibit $1/f$-type spectral behavior over many decades of frequency. A partial list includes (see, e.g., [4,6–9] and the references therein) geophysical, economic, physiological, and biological time series; electromagnetic and resistance fluctuations in media; electronic device noises; frequency variation in clocks and oscillators; variations in music and vehicular traffic; spatial variation in terrestrial features and clouds; error behavior and traffic patterns in communication networks.

While $\gamma \approx 1$ in many of these examples, more generally $0 \leq \gamma \leq 2$. However, there are many examples of phenomena in which γ lies well outside this range. For $\gamma \geq 1$, the lack of integrability of Equation 15.1 in a neighborhood of the spectral origin reflects the preponderance of low-frequency energy in the corresponding processes. This phenomenon is termed the infrared catastrophe. For many physical phenomena, measurements corresponding to very small frequencies show no low-frequency roll-off, which is usually understood to reveal an inherent nonstationarity in the underlying process. Such is the case for the Wiener process (regular Brownian motion), for which $\gamma = 2$. For $\gamma \leq 1$, the lack of integrability in the tails of the spectrum reflects a preponderance of high-frequency energy and is termed the ultraviolet catastrophe. Such behavior is familiar for generalized Gaussian processes such as stationary white Gaussian noise ($\gamma = 0$) and its usual derivatives. When $\gamma = 1$, both catastrophes are experienced. This process is referred to as "pink" noise, particularly in the audio applications where such noises are often synthesized for use in room equalization.

An important property of $1/f$ processes is their persistent statistical dependence. Indeed, the generalized Fourier pair [10]

$$\frac{|\tau|^{\gamma-1}}{2\Gamma(\gamma)\cos(\gamma\pi/2)} \overset{F}{\longleftrightarrow} \frac{1}{|\omega|^{\gamma}} \tag{15.2}$$

valid for $\gamma > 0$ but $\gamma \neq 1, 2, 3, \ldots$, reflects that the autocorrelation $R_x(\tau)$ associated with the spectrum (Equation 15.1) for $0 < \gamma < 1$ is characterized by slow decay of the form $R_x(\tau) \sim |\tau|^{\gamma-1}$.

This power–law decay in correlation structure distinguishes $1/f$ processes from many traditional models for time series analysis. For example, the well-studied family of autoregressive moving-average (ARMA) models have a correlation structure invariably characterized by exponential decay. As a consequence, ARMA models are generally inadequate for capturing long-term dependence in data.

One conceptually important characterization for $1/f$ processes is that based on the effects of band-pass filtering on such processes [11]. This characterization is strongly tied to empirical characterizations of $1/f$ processes, and is particularly useful for engineering applications. With this characterization, a $1/f$ process is formally defined as a wide-sense statistically self-similar random process having the property that when filtered by some arbitrary ideal band-pass filter (where $\omega = 0$ and $\omega = \pm\infty$ are strictly not in the passband), the resulting process is wide-sense stationary and has finite variance.

Among a variety of implications of this definition, it follows that such a process also has the property that when filtered by any ideal band-pass filter (again such that $\omega = 0$ and $\omega = \pm\infty$ are strictly not in the passband), the result is a wide-sense stationary process with a spectrum that is $\sigma_x^2/|\omega|^\gamma$ within the passband of the filter.

15.2.1 Models and Representations for $1/f$ Processes

A variety of exact and approximate mathematical models for $1/f$ processes are useful in signal-processing applications. These include fractional Brownian motion, generalized ARMA, and wavelet-based models.

15.2.1.1 Fractional Brownian Motion and Fractional Gaussian Noise

Fractional Brownian motion and fractional Gaussian noise have proven to be useful mathematical models for Gaussian $1/f$ behavior. In particular, the fractional Brownian motion framework provides a useful construction for models of $1/f$-type spectral behavior corresponding to spectral exponents in the range $-1 < \gamma < 1$ and $1 < \gamma < 3$; see, for example [4,7]. In addition, it has proven useful for addressing certain classes of signal-processing problems; see, for example [12–15].

Fractional Brownian motion is a nonstationary Gaussian self-similar process $x(t)$ with the property that its corresponding self-similar increment process

$$\Delta x(t;\varepsilon) \triangleq \frac{x(t+\varepsilon) - x(t)}{\varepsilon}$$

is stationary for every $\varepsilon > 0$.

A convenient though specialized definition of fractional Brownian motion is given by Barton and Poor [12]:

$$x(t) \triangleq \frac{1}{\Gamma(H+1/2)}\left[\int_{-\infty}^{0}\left(|t-\tau|^{H-1/2} - |\tau|^{H-1/2}\right)w(\tau)d\tau + \int_{0}^{t}|t-\tau|^{H-1/2}w(\tau)d\tau\right] \qquad (15.3)$$

where

$0 < H < 1$ is the self-similarity parameter
$w(t)$ is a zero-mean, stationary white Gaussian noise process with unit spectral density

When $H = 1/2$, Equation 15.3 specializes to the Wiener process, i.e., classical Brownian motion. Sample functions of fractional Brownian motion have a fractal dimension (in the Hausdorff–Besicovitch sense) given by [4,5]

$$D = 2 - H.$$

Moreover, the correlation function for fractional Brownian motion is given by

$$R_x(t,s) = E[x(t)x(s)] = \frac{\sigma_H^2}{2}\left(|s|^{2H} + |t|^{2H} - |t-s|^{2H}\right),$$

where

$$\sigma_H^2 = \text{var}\, x(1) = \Gamma(1-2H)\frac{\cos(\pi H)}{\pi H}.$$

The increment process leads to a conceptually useful interpretation of the derivative of fractional Brownian motion: As $\varepsilon \to 0$, fractional Brownian motion has, with $H' = H - 1$, the generalized derivative [12]

$$x'(t) = \frac{d}{dt}x(t) = \lim_{\varepsilon \to 0} \Delta x(t;\varepsilon) = \frac{1}{\Gamma(H'+1/2)} \int_{-\infty}^{t} |t-\tau|^{H'-1/2} w(\tau)d\tau, \tag{15.4}$$

which is termed fractional Gaussian noise. This process is stationary and statistically self-similar with parameter H'. Moreover, since Equation 15.4 is equivalent to a convolution, $x'(t)$ can be interpreted as the output of an unstable linear time-invariant (LTI) system with impulse response

$$\upsilon(t) = \frac{1}{\Gamma(H-1/2)} t^{H-3/2} u(t)$$

driven by $w(t)$. Fractional Brownian motion $x(t)$ is recovered via

$$x(t) = \int_{0}^{t} x'(t)dt.$$

The character of the fractional Gaussian noise $x'(t)$ depends strongly on the value of H. This follows from the autocorrelation function for the increments of fractional Brownian motion, viz.,

$$\begin{aligned} R_{\Delta x}(\tau;\varepsilon) &\triangleq E[\Delta x(t;\varepsilon)\Delta x(t-\tau;\varepsilon)] \\ &= \frac{\sigma_H^2 \varepsilon^{2H-2}}{2}\left[\left(\frac{|\tau|}{\varepsilon}+1\right)^{2H} - 2\left(\frac{|\tau|}{\varepsilon}\right)^{2H} + \left(\frac{|\tau|}{\varepsilon}-1\right)^{2H}\right], \end{aligned}$$

which at large lags ($|\tau| \gg \varepsilon$) takes the form

$$R_{\Delta x}(\tau) \approx \sigma_H^2 H(2H-1)|\tau|^{2H-2}. \tag{15.5}$$

Since the right-hand side of Equation 15.5 has the same algebraic sign as $H - 1/2$, for $1/2 < H < 1$ the process $x'(t)$ exhibits long-term dependence, i.e., persistent correlation structure; in this regime, fractional Gaussian noise is stationary with autocorrelation

$$R_{x'}(\tau) = E[x'(t)x'(t-\tau)] = \sigma_H^2 (H'+1)(2H'+1)|\tau|^{2H'},$$

and the generalized Fourier pair (Equation 15.2) suggests that the corresponding power spectral density can be expressed as $S_{x'}(\omega) = 1/|\omega|^{\gamma'}$, where $\gamma' = 2H' + 1$. In other regimes, for $H = 1/2$ the derivative $x'(t)$ is the usual stationary white Gaussian noise, which has no correlation, while for $0 < H < 1/2$, fractional Gaussian noise exhibits persistent anticorrelation.

A closely related discrete-time fractional Brownian motion framework for modeling $1/f$ behavior has also been extensively developed based on the notion of fractional differencing [16,17].

15.2.1.2 ARMA Models for $1/f$ Behavior

Another class of models that has been used for addressing signal-processing problems involving $1/f$ processes is based on a generalized ARMA framework. These models have been used both in signal-modeling and -processing applications, as well as in synthesis applications as $1/f$ noise generators and simulators [18–20].

One such framework is based on a "distribution of time constants" formulation [21,22]. With this approach, a $1/f$ process is modeled as the weighted superposition of an infinite number of independent random processes, each governed by a distinct characteristic time-constant $1/\alpha > 0$. Each of these random processes has correlation function $R_\alpha(\tau) = e^{-\alpha|\tau|}$ corresponding to a Lorentzian spectra of the form $S_\alpha(\omega) = 2\alpha/(\alpha^2 + \omega^2)$, and can be modeled as the output of a causal LTI filter with system function $Y_\alpha(s) = \sqrt{2\alpha}/(s+\alpha)$ driven by an independent stationary white noise source. The weighted superposition of a continuum of such processes has an effective spectrum

$$S_x(\omega) = \int_0^\infty S_\alpha(\omega) f(\alpha) d\alpha, \tag{15.6}$$

where the weights $f(\alpha)$ correspond to the density of poles or, equivalently, relaxation times. If an unnormalizable, scale-invariant density of the form $f(\alpha) = \alpha^{-\gamma}$ is chosen for $0 < \gamma < 2$, the resulting spectrum (Equation 15.6) is $1/f$, i.e., of the form in Equation 15.1.

More practically, useful approximate $1/f$ models result from using a countable collection of single time-constant processes in the superposition. With this strategy, poles are uniformly distributed along a logarithmic scale along the negative part of the real axis in the s-plane. The process $x(t)$ synthesized in this manner has a nearly $1/f$ spectrum in the sense that it has a $1/f$ characteristic with superimposed ripple that is uniform-spaced and of uniform amplitude on a log–log frequency plot. More specifically, when the poles are exponentially spaced according to

$$\alpha_m = \Delta^m, \quad -\infty < m < \infty, \tag{15.7}$$

for some $1 < \Delta < \infty$, the limiting spectrum

$$S_x(\omega) = \sum_m \frac{\Delta^{(2-\gamma)m}}{\omega^2 + \Delta^{2m}} \tag{15.8}$$

satisfies

$$\frac{\sigma_L^2}{|\omega|^\gamma} \leq S_x(\omega) \leq \frac{\sigma_U^2}{|\omega|^\gamma} \tag{15.9}$$

for some $0 < \sigma_L^2 \leq \sigma_U^2 < \infty$, and has exponentially spaced ripple such that for all integers k

$$|\omega|^\gamma S_x(\omega) = |\Delta^k \omega|^\gamma S_x(\Delta^k \omega). \tag{15.10}$$

As Δ is chosen closer to unity, the pole spacing decreases, which results in a decrease in both the amplitude and spacing of the spectral ripple on a log–log plot.

The $1/f$ model that results from this discretization may be interpreted as an infinite-order ARMA process, i.e., $x(t)$ may be viewed as the output of a rational LTI system with a countably infinite number of both poles and zeros driven by a stationary white noise source. This implies, among other properties, that the corresponding space descriptions of these models for long-term dependence require infinite numbers of state variables. These processes have been useful in modeling physical $1/f$ phenomena, see, for example [23–25]; and practical signal-processing algorithms for them can often be obtained by extending classical tools for processing regular ARMA processes.

The above method focuses on selecting appropriate pole locations for the extended ARMA model. The zero locations, by contrast, are controlled indirectly, and bear a rather complicated relationship to the pole locations. With other extended ARMA models for $1/f$ behavior, both pole and zero locations are explicitly controlled, often with improved approximation characteristics [20]. As an example, [6,26] describe a construction as filtered white noise where the filter structure consists of a cascade of first-order sections each with a single pole and zero. With a continuum of such sections, exact $1/f$ behavior is obtained. When a countable collection of such sections is used, nearly $1/f$ behavior is obtained as before. In particular, when stationary white noise is driven through an LTI system with a rational system function

$$Y(s) = \prod_{m=-\infty}^{\infty} \left[\frac{s + \Delta^{m+\gamma/2}}{s + \Delta^m} \right], \qquad (15.11)$$

the output has power spectrum

$$S_x(\omega) \propto \prod_{m=-\infty}^{\infty} \left[\frac{\omega^2 + \Delta^{2m+\gamma}}{\omega^2 + \Delta^{2m}} \right]. \qquad (15.12)$$

This nearly $1/f$ spectrum also satisfies both Equations 15.9 and 15.10. Comparing the spectra (Equations 15.12 and 15.8) reveals that the pole placement strategy for both is identical, while the zero placement strategy is distinctly different.

The system function (Equation 15.11) associated with this alternative extended ARMA model lends useful insight into the relationship between $1/f$ behavior and the limiting processes corresponding to $\gamma \to 0$ and $\gamma \to 2$. On a logarithmic scale, the poles and zeros of Equation 15.11 are each spaced uniformly along the negative real axis in the s-plane, and to the left of each pole lies a matching zero, so that poles and zeros are alternating along the half-line. However, for certain values of γ, pole-zero cancellation takes place. In particular, as $\gamma \to 2$, the zero pattern shifts left canceling all poles except the limiting pole at $s = 0$. The resulting system is therefore an integrator, characterized by a single state variable, and generates a Wiener process as anticipated. By contrast, as $\gamma \to 0$, the zero pattern shifts right canceling all poles. The resulting system is therefore a multiple of the identity system, requires no state variables, and generates stationary white noise as anticipated.

An additional interpretation is possible in terms of a Bode plot. Stable, rational system functions composed of real poles and zeros are generally only capable of generating transfer functions whose Bode plots have slopes that are integer multiples of $20 \log_{10} 2 \approx 6$ dB/octave. However, a $1/f$ synthesis filter must fall off at $10\gamma \log_{10} 2 \approx 3\gamma$ dB/octave, where $0 < \gamma < 2$ is generally not an integer. With the extended ARMA models, a rational system function with an alternating sequence of poles and zeros is used to generate a stepped approximation to a -3γ dB/octave slope from segments that alternate between slopes of -6 and 0 dB/octave.

15.2.1.3 Wavelet-Based Models for $1/f$ Behavior

Another approach to $1/f$ process modeling is based on the use of wavelet basis expansions. These lead to representations for processes exhibiting $1/f$-type behavior that are useful in a wide range of signal-processing applications.

Orthonormal wavelet basis expansions play the role of Karhunen–Loève type expansions for $1/f$-type processes [11,27]. More specifically, wavelet basis expansions in terms of uncorrelated random variables constitute very good models for $1/f$-type behavior. For example, when a sufficiently regular orthonormal wavelet basis $\{\psi_n^m(t) = 2^{m/2}\psi(2^m t - n)\}$ is used, expansions of the form

$$x(t) = \sum_m \sum_n x_n^m \psi_n^m(t),$$

where the x_n^m are a collection of mutually uncorrelated, zero-mean random variables with the geometric scale-to-scale variance progression

$$\text{var } x_n^m = \sigma^2 2^{-\gamma m}, \tag{15.13}$$

lead to a nearly $1/f$ power spectrum of the type obtained via the extended ARMA models. This behavior holds regardless of the choice of wavelet within this class, although the detailed structure of the ripple in the nearly $1/f$ spectrum can be controlled by judicious choice of the particular wavelet.

More generally, wavelet decompositions of $1/f$-type processes have a decorrelating property. For example, if $x(t)$ is a $1/f$ process, then the coefficients of the expansion of the process in terms of a sufficiently regular wavelet basis, i.e., the

$$x_n^m = \int_{-\infty}^{+\infty} x(t)\psi_n^m(t)dt$$

are very weakly correlated and obey the scale-to-scale variance progression (Equation 15.13). Again, the detailed correlation structure depends on the particular choice of wavelet [3,11,28,29].

This decorrelating property is exploited in many wavelet-based algorithms for processing $1/f$ signals, where the residual correlation among the wavelet coefficients can usually be ignored. In addition, the resulting algorithms typically have very efficient implementations based on the discrete wavelet transform. Examples of robust wavelet-based detection and estimation algorithms for use with $1/f$-type signals are described in [11,27,30].

15.3 Deterministic Fractal Signals

While stochastic signals with fractal characteristics are important models in a wide range of engineering applications, deterministic signals with such characteristics have also emerged as potentially important in engineering applications involving signal generation ranging from communications to remote sensing.

Signals $x(t)$ of this type satisfying the deterministic scale-invariance property

$$x(t) = a^{-H}x(at) \tag{15.14}$$

for all $a > 0$, are generally referred to in mathematics as homogeneous functions of degree H. Strictly homogeneous functions can be parameterized with only a few constants [31], and constitute a rather limited class of models for signal generation applications. A richer class of homogeneous signal models is obtained by considering waveforms that are required to satisfy Equation 15.14 only for values of a that are integer powers of two, i.e., signals that satisfy the dyadic self-similarity property $x(t) = 2^{-kH}x(2^k t)$ for all integers k.

Homogeneous signals have spectral characteristics analogous to those of $1/f$ processes, and have fractal properties as well. Specifically, although all nontrivial homogeneous signals have infinite energy and many have infinite power, there are classes of such signals with which one can associate a generalized $1/f$-like Fourier transform, and others with which one can associate a generalized $1/f$-like power spectrum. These two classes of homogeneous signals are referred to as energy- and power-dominated, respectively [11,32]. An example of such a signal is depicted in Figure 15.2.

Orthonormal wavelet basis expansions provide convenient and efficient representations for these classes of signals. In particular, the wavelet coefficients of such signals are related according to

$$x_n^m = \int_{-\infty}^{+\infty} x(t)\psi_n^m(t) = \beta^{-m/2}q[n],$$

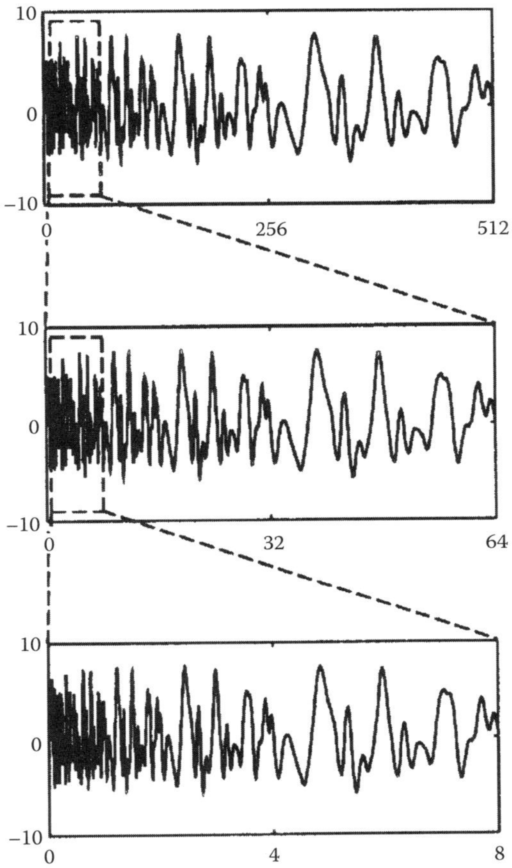

FIGURE 15.2 Dilated homogeneous signal.

where $q[n]$ is termed a generating sequence and $\beta = 2^{2H+1} = 2^{\gamma}$. This relationship is depicted in Figure 15.3, where the self-similarity inherent in these signals is immediately captured in the time–frequency portrait of such signals as represented by their wavelet coefficients. More generally, wavelet expansion naturally lead to "orthonormal self-similar bases" for homogeneous signals [11,32]. Fast synthesis and analysis algorithms for these signals are based on the discrete wavelet transform.

For some communications applications, the objective is to embed an information sequence into a fractal waveform for transmission over an unreliable communication channel. In this context, it is often natural for $q[n]$ to be the information bearing sequence such as a symbol stream to be transmitted, and the corresponding modulation

$$x(t) = \sum_m \sum_n x_n^m \psi_n^m(t)$$

to be the fractal waveform to be transmitted. This encoding, referred to as "fractal modulation" [32] corresponds to an efficient diversity transmission strategy for certain classes of communication channels. Moreover, it can be viewed as a multirate modulation strategy in which data are transmitted simultaneously at multiple rates, and is particularly well-suited to channels having the characteristic that they are "open" for some unknown time interval T, during which they have some unknown bandwidth W and a

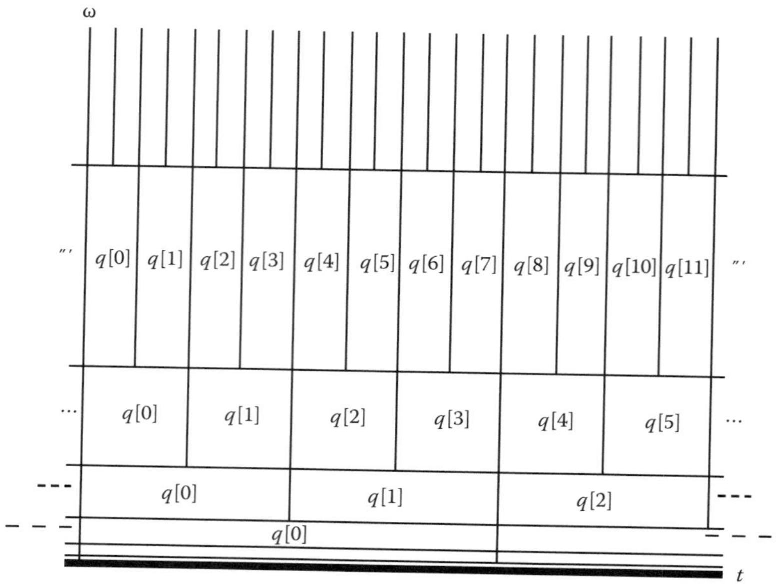

FIGURE 15.3 The time–frequency portrait of a homogeneous signal.

particular signal-to-noise ratio (SNR). Such a channel model can be used, for example, to capture both characteristics of the transmission medium, such as in the case of meteor-burst channels, the constraints inherent in disparate receivers in broadcast applications, and/or the effects of jamming in military applications.

15.4 Fractal Point Processes

Fractal point processes correspond to event distributions in one or more dimensions having self-similar statistics, and are well-suited to modeling, among other examples, the distribution of stars and galaxies, demographic distributions, the sequence of spikes generated by auditory neural firing in animals, vehicular traffic, and data traffic on packet-switched data communication networks [4,33–36].

A point process is said to be self-similar if the associated counting process $N_X(t)$, whose value at time t is the total number of arrivals up to time t, is statistically invariant to temporal dilations and compressions, i.e., $N_X(t) \stackrel{\Delta}{=} N_X(at)$ for all $a > 0$, where the notation $\stackrel{\Delta}{=}$ again denotes statistical equality in the sense of all finite-dimensional distributions. An example of a sample path for such a counting process is depicted in Figure 15.4.

Physical fractal point process phenomena generally also possess certain quasi-stationary attributes. For example, empirical measurements of the statistics of the interarrival times $X[n]$, i.e., the time interval between the $(n-1)$th and nth arrivals, are consistent with a renewal process. Moreover, the associated interarrival density is a power–law, i.e.,

$$f_X(x) \sim \frac{\sigma_x^2}{x^\gamma} u(x), \tag{15.15}$$

where $u(x)$ is the unit-step function. However, Equation 15.15 is an unnormalizable density, which is a reflection of the fact that a point process cannot, in general, be both self-similar and renewing. This is

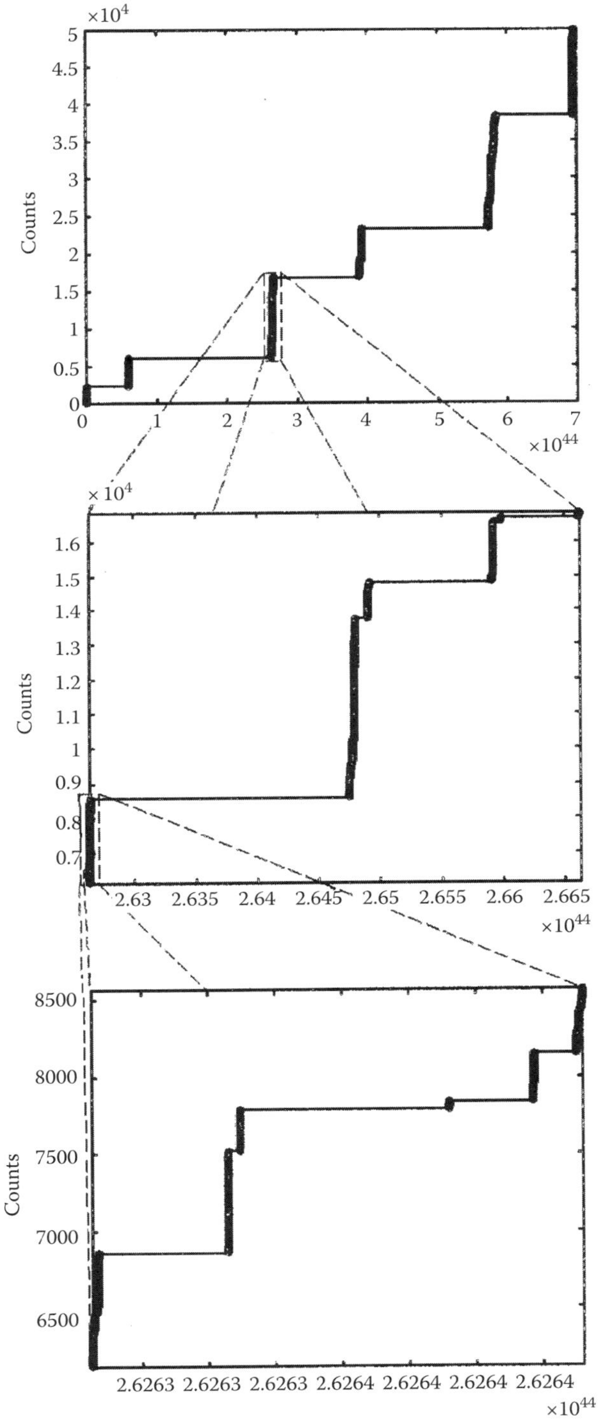

FIGURE 15.4 Dilated fractal renewal process sample path.

analogous to the result that a continuous process cannot, in general, be simultaneously self-similar and stationary.

However, self-similar processes can possess a milder "conditionally renewing" property [37,38]. Such processes are referred to as fractal renewal processes and have an effectively stationary character. The shape parameter γ in the unnormalizable interarrival density (Equation 15.15) is related to the fractal dimension D of the process via [4] $D=\gamma-1$, and is a measure of the extent to which arrivals cover the line.

15.4.1 Multiscale Models

As in the case of continuous fractal processes, multiscale models are both conceptually and practically important representations for discrete fractal processes. As an example, one useful class of multiscale models corresponds to a mixture of simple Poisson processes on different timescales [37]. The construction of such processes involves a collection $\{N_{W_A}(t)\}$ of mutually independent Poisson counting processes such that $N_{W_A}(t) \triangleq N_{W_0}(e^{-A}t)$. The process $N_{W_0}(t)$ is a prototype whose mean arrival rate we denote by λ, so that the mean arrival rates of the constituent processes are related according to $\lambda_A = e^{-A}\,\lambda$. A random mixture of this continuum of Poisson processes yields a fractal renewal process when the index choice $A[n]$ for the nth arrival is distributed according to the extended exponential density $f_A(a) \sim \sigma_A^2 e^{-(\gamma-1)a}$. In particular, the first interarrival of the composite process is chosen to be the first arrival of the Poisson process indexed by $A[1]$, the second arrival of the composite process is chosen to be the next arrival in the Poisson process indexed by $A[2]$, and so on.

Useful alternative but equivalent constructions result from exploiting the memoryless property of Poisson processes. For example, interarrival times can be generated according to $X[n] = W_{A[n]}[n]$ or

$$X[n] = e^{A[n]} W_0[n], \tag{15.16}$$

where $W_A[n]$ is the nth interarrival time for the Poisson process indexed by A. The synthesis (Equation 15.16) is particularly appealing in that it requires access to only exponential random variables that can be obtained in practice from a single prototype Poisson process. The construction (Equation 15.16) also leads to the interpretation of a fractal point process as a Poisson process in which the arrival rate is selected randomly and independently after each arrival (and held constant between consecutive arrivals). Related doubly stochastic process models are described by Johnson and Kumar [39].

In addition to their use in applications requiring the synthesis of fractal point processes, these multiscale models have also proven useful in signal estimation problems. For these kinds of signal analysis applications, it is frequently convenient to replace the continuum Poisson mixture with a discrete Poisson mixture. Typically, a collection of constituent Poisson counting processes $N_{W_M}(t)$ is used, where M is an integer-valued scale index, and where the mean arrival rates are related according to $\lambda_M = \rho^{-M}\lambda$ for some λ. In this case, the scale selection is governed by an extended geometric probability mass function of the form $p_M(m) \sim \sigma_M^2 \rho^{-(\gamma-1)m}$. This discrete synthesis leads to processes that are approximate fractal renewal processes, in the sense that the interarrival densities follow a power law with a typically small amount of superimposed ripple. A number of efficient algorithms for exploiting such models in the development of robust signal estimation algorithms for use with fractal renewal processes are described in [37].

From a broader perspective, the Poisson mixtures can be viewed as a nonlinear multiresolution signal analysis framework that can be generalized to accommodate a broad class of point process phenomena. As such, this framework is the point process counterpart to the linear multiresolution signal analysis framework based on wavelets that is used for a broad class of continuous-valued signals.

15.4.2 Extended Markov Models

An equivalent description of the discrete Poisson mixture model is in terms of an extended Markov model. The associated multiscale pure-birth process, depicted in Figure 15.5, involves a state space

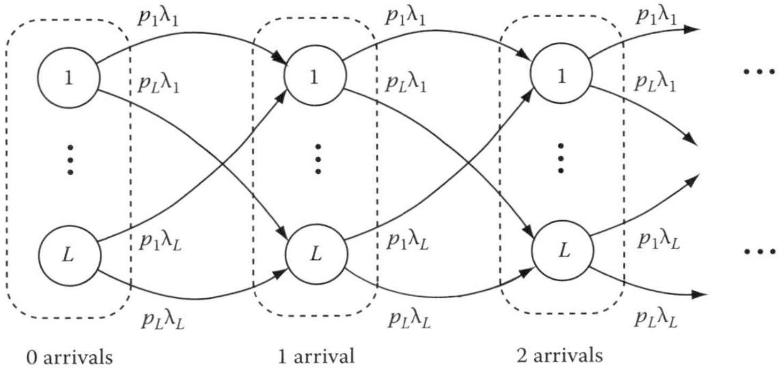

FIGURE 15.5 Multiscale pure-birth process corresponding to Poisson mixture.

consisting of a set of "superstates," each of which corresponds to fixed number of arrivals (births). Included in a superstate is a set of states corresponding to the scales in the Poisson mixture. Hence, each state is indexed by an ordered pair (i, j), where i is the superstate index and j is the scale index within each superstate.

The extended Markov model description has proven useful in analyzing the properties of fractal point processes under some fundamental transformations, including superposition and random erasure. These properties, in turn, provide key insight into the behavior of merging and branching traffic at nodes in data communication, vehicular, and other networks. See, for example [40].

Other important classes of fractal point process transformations that arise in applications involving queuing. And the extended Markov model also plays an important role in analyzing fractal queues. To address these problems, a multiscale birth–death process model is generally used [40].

References

1. Mandelbrot, B.B. and Van Ness, H.W., Fractional Brownian motions, fractional noises and applications, *SIAM Rev.*, 10, 422–436, October 1968.
2. Mandelbrot, B., Some noises with $1/f$ spectrum: A bridge between direct current and white noise, *IEEE Trans. Inform. Theory*, IT-13, 289–298, April 1967.
3. Flandrin, P., On the spectrum of fractional Brownian motions, *IEEE Trans. Inform. Theory*, IT-35, 197–199, January 1989.
4. Mandelbrot, B.B., *The Fractal Geometry of Nature*, Freeman, San Francisco, CA, 1982.
5. Falconer, K., *Fractal Geometry: Mathematical Foundations and Applications*, John Wiley & Sons, New York, 1990.
6. Keshner, M.S., $1/f$ noise, *Proc. IEEE*, 70, 212–218, March 1982.
7. Pentland, A.P., Fractal-based description of natural scenes, *IEEE Trans. Pattern Anal. Machine Intell.*, PAMI-6, 661–674, November 1984.
8. Voss, R.F., $1/f$ (flicker) noise: A brief review, *Proc. Ann. Symp. Freq. Contr.*, 40–46, 1979.
9. van der Ziel, A., Unified presentation of $1/f$ noise in electronic devices: Fundamental $1/f$ noise sources, *Proc. IEEE*, 76(3), 233–258, March 1988.
10. Champeney, D.C., *A Handbook of Fourier Theorems*, Cambridge University Press, Cambridge, England, 1987.
11. Wornell, G.W., *Signal Processing with Fractals: A Wavelet-Based Approach*, Prentice-Hall, Upper Saddle River, NJ, 1996.
12. Barton, R.J. and Poor, V.H., Signal detection in fractional Gaussian noise, *IEEE Trans. Inform. Theory*, IT-34, 943–959, September 1988.

13. Lundahl, T., Ohley, W.J., Kay, S.M., and Siffert, R., Fractional Brownian motion: A maximum likelihood estimator and its application to image texture, *IEEE Trans. Medical Imaging,* MI-5, 152–161, September 1986.
14. Deriche, M. and Tewfik, A.H., Maximum likelihood estimation of the parameters of discrete fractionally differenced Gaussian noise process, *IEEE Trans. Signal Process.,* 41, 2977–2989, October 1993.
15. Deriche, M. and Tewfik, A.H., Signal modeling with filtered discrete fractional noise processes, *IEEE Trans. Signal Process.,* 41, 2839–2849, September 1993.
16. Granger, C.W. and Joyeux, R., An introduction to long memory time series models and fractional differencing, *J. Time Ser. Anal.,* 1 (1), 15–29, 1980.
17. Hosking, J.R.M., Fractional differencing, *Biometrika,* 68 (1), 165–176, 1981.
18. Pellegrini, B., Saletti, R., Neri, B., and Terreni, P., $1/f^{\nu}$ Noise generators, in *Noise in Physical Systems and 1/f Noise,* D'Amico A. and Mazzetti, P. (Eds.), North-Holland, Amsterdam, 1986, pp. 425–428.
19. Corsini, G. and Saletti, R., Design of a digital $1/f^{\nu}$ noise simulator, in *Noise in Physical Systems and 1/f Noise,* Van Vliet, C.M. (Ed.), World Scientific, Singapore, 1987, pp. 82–86.
20. Saletti, R., A comparison between two methods to generate $1/f^{\gamma}$ noise, *Proc. IEEE,* 74, 1595–1596, November 1986.
21. Bernamont, J., Fluctuations in the resistance of thin films, *Proc. Phys. Soc.,* 49, 138–139, 1937.
22. van der Ziel, A., On the noise spectra of semi-conductor noise and of flicker effect, *Physica,* 16 (4), 359–372, 1950.
23. Machlup, S., Earthquakes, thunderstorms and other $1/f$ noises, in *Noise in Physical Systems,* Meijer, P.H.E., Mountain, R.D., and Soulen, Jr. R.J. (Eds.), National Bureau of Standards, Washington, D.C., Special Publ. No. 614, 1981, pp. 157–160.
24. West, B.J. and Shlesinger, M.F., On the ubiquity of $1/f$ noise, *Int. J. Mod. Phys.,* 3(6), 795–819, 1989.
25. Montroll, E.W. and Shlesinger, M.F., On $1/f$ noise and other distributions with long tails, *Proc. Natl. Acad. Sci.,* 79, 3380–3383, May 1982.
26. Oldham, K.B. and Spanier, J., *The Fractional Calculus,* Academic Press, New York, 1974.
27. Wornell, G.W., Wavelet-based representations for the $1/f$ family of fractal processes, *Proc. IEEE,* 81, 1428–1450, October 1993.
28. Flandrin, P., Wavelet analysis and synthesis of fractional Brownian motion, *IEEE Trans. Info. Theory,* IT-38, 910–917, March 1992.
29. Tewfik, A.H. and Kim, M., Correlation structure of the discrete wavelet coefficients of fractional Brownian motion, *IEEE Trans. Inform. Theory,* IT-38, 904–909, March 1992.
30. Wornell, G.W. and Oppenheim, A.V., Estimation of fractal signals from noisy measurements using wavelets, *IEEE Trans. Signal Process.,* 40, 611–623, March 1992.
31. Gel'fand, I.M., Shilov, G.E., Vilenkin, N.Y., and Graev, M.I., *Generalized Functions,* Academic Press, New York, 1964.
32. Wornell, G.W. and Oppenheim, A.V., Wavelet-based representations for a class of self-similar signals with application to fractal modulation, *IEEE Trans. Inform. Theory,* 38, 785–800, March 1992.
33. Schroeder, M., *Fractals, Chaos, Power Laws,* W.H. Freeman, New York, 1991.
34. Teich, M.C., Johnson, D.H., Kumar, A.R., and Turcott, R.G., Rate fluctuations and fractional power-law noise recorded from cells in the lower auditory pathway of the cat, *Hearing Res.,* 46, 41–52, June 1990.
35. Leland, W.E., Taqqu, M.S., Willinger, W., and Wilson, D.V., On the self-similar nature of ethernet traffic, *IEEE/ACM Trans. Network.,* 2, 1–15, February 1994.
36. Paxson, V. and Floyd, S., Wide area traffic: The failure of Poisson modeling, *IEEE/ACM Trans. Network.,* 3(3), 226–244, 1995.
37. Lam, W.M. and Wornell, G.W., Multiscale representation and estimation of fractal point processes, *IEEE Trans. Signal Process.,* 43, 2606–2617, November 1995.

38. Mandelbrot, B.B., Self-similar error clusters in communication systems and the concept of conditional stationarity, *IEEE Trans. Commun. Technol.,* COM-13, 71–90, March 1965.
39. Johnson, D.H. and Kumar, A.R., Modeling and analyzing fractal point processes, in *Proceedings of the International Conference on Acoustics, Speech, Signal Processing,* Albuquerque, NM, 1990.
40. Lam, W.M. and Wornell, G.W., Multiscale analysis of fractal point processes and queues, in *Proceedings of the International Conference on Acoustics, Speech, Signal Processing,* Cambridge, MA, 1996.

16

Morphological Signal and Image Processing

16.1 Introduction 16-1

16.2 Morphological Operators for Sets and Signals 16-2
Boolean Operators and Threshold Logic • Morphological Set Operators • Morphological Signal Operators and Nonlinear Convolutions

16.3 Median, Rank, and Stack Operators 16-7

16.4 Universality of Morphological Operators 16-8

16.5 Morphological Operators and Lattice Theory 16-11

16.6 Slope Transforms 16-14

16.7 Multiscale Morphological Image Analysis 16-16
Binary Multiscale Morphology via Distance Transforms • Multiresolution Morphology

16.8 Differential Equations for Continuous-Scale Morphology 16-19

16.9 Applications to Image Processing and Vision 16-21
Noise Suppression • Feature Extraction • Shape Representation via Skeleton Transforms • Shape Thinning • Size Distributions • Fractals • Image Segmentation

16.10 Conclusions 16-28

Acknowledgment 16-28

References 16-28

Petros Maragos
National Technical University of Athens

16.1 Introduction

This chapter provides a brief introduction to the theory of morphological signal processing and its applications to image analysis and nonlinear filtering. By "morphological signal processing" we mean a broad and coherent collection of theoretical concepts, mathematical tools for signal analysis, nonlinear signal operators, design methodologies, and applications systems that are based on or related to mathematical morphology (MM), a set- and lattice-theoretic methodology for image analysis. MM aims at quantitatively describing the geometrical structure of image objects. Its mathematical origins stem from set theory, lattice algebra, convex analysis, and integral and stochastic geometry. It was initiated mainly by Matheron [42] and Serra [58] in the 1960s. Some of its early signal operations are also found in the work of other researchers who used cellular automata and Boolean/threshold logic to analyze binary image data in the 1950s and 1960s, as surveyed in [49,54]. MM has formalized these earlier operations and has also added numerous new concepts and image operations. In the 1970s it was extended to gray-level images [22,45,58,62]. Originally MM was applied to analyzing images from geological or biological specimens. However, its rich theoretical framework, algorithmic efficiency, easy implementability on special hardware, and suitability for many shape-oriented problems have propelled its widespread diffusion and adoption by many academic and industry groups in many countries as one among the dominant image analysis methodologies. Many of these research groups have also extended the theory and applications of MM. As a result, MM nowadays offers many theoretical and algorithmic tools to and inspires new directions in many research areas from the fields of signal processing, image processing and machine vision, and pattern recognition.

As the name "morphology" implies (study/analysis of shape/form), morphological signal processing can quantify the shape, size, and other aspects of the geometrical structure of signals viewed as image objects, in a rigorous way that also agrees with human intuition and perception. In contrast, the traditional tools of linear systems and Fourier analysis are of limited or no use for solving geometry-based problems in image processing because they do not directly address the fundamental issues of how to quantify shape, size, or other geometrical structures in signals and may distort important geometrical features in images. Thus, morphological systems are more suitable than linear systems for shape analysis. Further, they offer simple and efficient solutions to other nonlinear problems, such as non-Gaussian noise suppression or envelope estimation. They are also closely related to another class of nonlinear systems, the median, rank, and stack operators, which also outperform linear systems in non-Gaussian noise suppression and in signal enhancement with geometric constraints. Actually, rank and stack operators can be represented in terms of elementary morphological operators. All of the above, coupled with the rich mathematical background of MM, make morphological signal processing a rigorous and efficient framework to study and solve many problems in image analysis and nonlinear filtering.

16.2 Morphological Operators for Sets and Signals

16.2.1 Boolean Operators and Threshold Logic

Early works in the fields of visual pattern recognition and cellular automata dealt with analysis of binary digital images using local neighborhood operations of the Boolean type. For example, given a sampled* binary image signal $f[x]$ with values 1 for the image foreground and 0 for the background, typical signal transformations involving a neighborhood of n samples whose indices are arranged in a window set $W = \{y_1, y_2, \ldots, y_n\}$ would be

$$\psi_b(f)[x] = b(f[x - y_1], \ldots, f[x - y_n])$$

where $b(v_1, \ldots, v_n)$ is a Boolean function of n variables. The mapping $f \mapsto \psi_b(f)$ is a nonlinear system, called a Boolean operator. By varying the Boolean function b, a large variety of Boolean operators can be obtained; see Table 16.1 where $W = \{-1, 0, 1\}$. For example, choosing a Boolean AND for b would shrink the input image foreground, whereas a Boolean OR would expand it.

TABLE 16.1 Discrete Set Operators and Their Generating Boolean Function

Set Operator $\Psi(X), X \subseteq \mathbb{Z}$	Boolean Function $b(v_1, v_2, v_3)$
Erosion $X \Theta \{-1, 0, 1\}$	$v_1 v_2 v_3$
Dilation: $X \oplus \{-1, 0, 1\}$	$v_1 + v_2 + v_3$
Median: $X \Box_2 \{-1, 0, 1\}$	$v_1 v_2 + v_1 v_3 + v_2 v_3$
Hit-Miss: $X \otimes (\{-1, 1\}, \{0\})$	$v_1 \bar{v}_2 v_3$
Opening: $X \circ \{0, 1\}$	$v_1 v_2 + v_2 v_3$
Closing: $X \bullet \{0, 1\}$	$v_2 + v_1 v_3$

* $x \in \mathbb{R}^d$ $f(x)$ $x \in \mathbb{Z}^d$ $f[x]$.

Two alternative implementations and views of these Boolean operations are (1) thresholded convolutions, where a binary input is linearly convolved with an n-point mask of ones and then the output is thresholded at 1 or n to produce the Boolean OR or AND, respectively, and (2) min/max operations, where the moving local minima and maxima of the binary input signal produce the same output as Boolean AND/OR, respectively. In the thresholded convolution interpretation, thresholding at an intermediate level r between 1 and n produces a binary rank operation of the binary input data (inside the moving window). For example, if $r = (n + 1)/2$, we obtain the binary median filter whose Boolean function expresses the majority voting logic; see the third example of Table 16.1. Of course, numerous other Boolean operators are possible, since there are 2^{2^n} possible Boolean functions of n variables. The main applications of such Boolean signal operations have been in biomedical image processing, character recognition, object detection, and general 2D shape analysis. Detailed accounts and more references of these approaches and applications can be found in [49,54].

16.2.2 Morphological Set Operators

Among the new important conceptual leaps offered by MM was to use sets to represent binary image signals and set operations to represent binary image transformations. Specifically, given a binary image, let its foreground be represented by the set X and its background by the set complement X^c. The Boolean OR transformation of X by a (window) set B (local neighborhood of pixels) is mathematically equivalent to the Minkowski set addition $\oplus$, also called dilation, of X by B:

$$X \oplus B \equiv \{x + y : x \in X, y \in B\} = \bigcup_{y \in B} X_{+y} \tag{16.1}$$

where $X_{+y} \equiv \{x + y : x \in X\}$ is the translation of X along the vector y. Likewise, if $B^r \equiv \{x : -x \in B\}$ denotes the reflection of B with respect to the axes' origin, the Boolean AND transformation of X by the reflected B is equivalent to the Minkowski set subtraction [24] $\ominus$, also called erosion, of X or B:

$$X \ominus B \equiv \{x : B_{+x} \subseteq X\} = \bigcap_{y \in B} X_{-y} \tag{16.2}$$

In applications, B is usually called a structuring element and has a simple geometrical shape and a size smaller than the image set X. As shown in Figure 16.1, erosion shrinks the original set, whereas dilation expands it.

The erosion (Equation 16.2) can also be viewed as Boolean template matching since it gives the center points at which the shifted structuring elements fits inside the image foreground. If we now consider a set A probing the image foreground set X and another set B probing the background X^c, the set of points at which the shifted pair (A, B) fits inside the images is the hit-miss transformation of X by (A, B):

$$X \otimes (A, B) \equiv \{x : A_{+x} \subseteq X, B_{+x} \subseteq X^c\} \tag{16.3}$$

In the discrete case, this can be represented by a Boolean product function whose uncomplemented (complemented) variables correspond to points of $A(B)$; see Table 16.1. It has been used extensively for binary feature detection [58] and especially in document image processing [8,9].

Dilating an eroded set by the same structuring element in general does not recover the original set but only a part of it, its opening. Performing the same series of operations to the set complement yields a set containing the original, its closing. Thus, cascading erosion and dilation gives rise to two new operations, the opening $X \circ B \equiv (X \ominus B) \oplus B$ and the closing $X \bullet B \equiv (X \oplus B) \ominus B$ of X by B. As shown in Figure 16.1, the opening suppresses the sharp capes and cuts the narrow isthmuses of X, whereas the

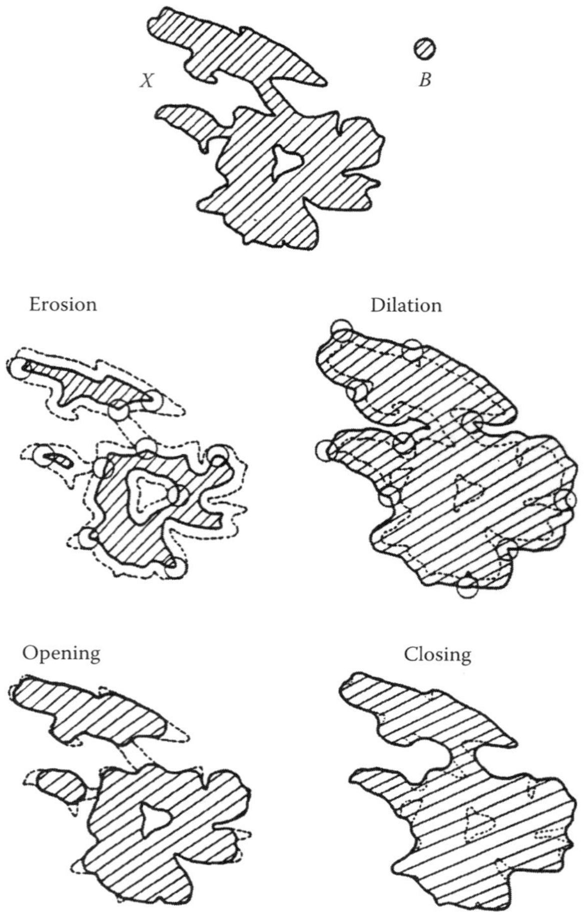

FIGURE 16.1 Erosion, dilation, opening, and closing of X (binary image of an island) by a disk B centered at the origin. The shaded curve to the boundary of the original set X.

closing fills in the thin gulfs and small holes. Thus, if the structuring element B has a regular shape, both opening and closing can be thought of as nonlinear filters which smooth the contours of the input signal.

These set operations make MM more general than previous approaches because it unifies and systematizes all previous digital and analog binary image operations, mathematically rigorous and notationally elegant since it is based on set theory, and intuitive since the set formalism is easily connected to mathematical logic. Further, the basic morphological set operators directly relate to the shape and size of binary images in a way that has many common points with human perception about geometry and spatial reasoning.

16.2.3 Morphological Signal Operators and Nonlinear Convolutions

In the 1970s, morphological operators were extended from binary to gray-level images and real-valued signals. Going from sets to functions was made possible by using set representations of signals and transforming these input sets via morphological set operations. Thus, consider a signal $f(x)$ defined on the d-dimensional continuous or discrete domain $\mathbb{D} = \mathbb{R}^d$ or $\mathbb{Z}^d$ and assuming values in $\bar{\mathbb{R}} = \mathbb{R} \cup \{-\infty, \infty\}$. Thresholding the signal at all amplitude values v produces an ensemble of threshold binary signals

$$\theta_v(f)(x) \equiv 1 \quad \text{if } f(x) \geq v, \quad \text{and } 0 \text{ else} \tag{16.4}$$

represented by the threshold sets [58]

$$\Theta_v(f) \equiv \{x \in \mathbb{D} : f(x) \geq v\}, \quad -\infty < v < +\infty \tag{16.5}$$

The signal can be exactly reconstructed from all its thresholded versions since

$$f(x) = \sup\{v \in \mathbb{R} : x \in \Theta_v(f)\} = \sup\{v \in \mathbb{R} : \theta_v(f)(x) = 1\} \tag{16.6}$$

Transforming each threshold set by a set operator Ψ and viewing the transformed sets as threshold sets of a new signal creates a flat signal operator ψ whose output is

$$\psi(f)(x) = \sup\{v \in \mathbb{R} : x \in \Psi[\Theta_v(f)]\} \tag{16.7}$$

Using set dilation and erosion in place of Ψ, the above procedure creates the two most elementary morphological signal operators: the dilation and erosion of a signal $f(x)$ by a set B:

$$(f \oplus B)(x) \equiv \bigvee_{y \in B} f(x - y) \tag{16.8}$$

$$(f \ominus B)(x) \equiv \bigwedge_{y \in B} f(x + y) \tag{16.9}$$

where $\vee$ denotes supremum (or maximum for finite B) and $\wedge$ denotes infimum (or minimum for finite B). These gray-level morphological operations can also be created from their binary counterparts using concepts from fuzzy sets where set union and intersection becomes maximum and minimum on gray-level images [22,45]. As Figure 16.2 shows, flat erosion (dilation) of a function f by a small convex set B reduces (increases) the peaks (valleys) and enlarges the minima (maxima) of the function. The flat opening $f \circ B = (f \ominus B) \oplus B$ of f by B smooths the graph of f from below by cutting down its peaks, whereas the closing $f \bullet B = (f \oplus B) \ominus B$ smoothes it from above by filling up its valleys.

More general morphological operators for gray-level 2D image signals $f(x)$ can be created [62] by representing the surface of f and all the points underneath by a 3D set $U(f) = \{(x,v) : v \leq f(x)\}$, called its umbra; then dilating or eroding $U(f)$ by the umbra of another signal g yields the umbras of two new signals, the dilation or erosion of f by g, which can be computed directly by the formulae:

$$(f \oplus g)(x) \equiv \bigvee_{y \in \mathbb{D}} f(x - y) + g(y) \tag{16.10}$$

$$(f \ominus g)(x) \equiv \bigwedge_{y \in \mathbb{D}} f(x + y) - g(y) \tag{16.11}$$

and two supplemental rules for adding and subtracting with infinities: $r \pm s = -\infty$ if $r = -\infty$ or $s = -\infty$, and $+\infty - r = +\infty$ if $r \in \mathbb{R} \cup \{+\infty\}$. These two signal transformations are nonlinear and translation-invariant. Their computational structure closely resembles that of a linear convolution $(f \star g)[x] = \sum_y f[x - y]g[y]$ if we correspond the sum of products to the supremum of sums in the dilation. Actually, in the areas of convex analysis [50] and optimization [6], the operation (Equation 16.10) has been known as the supremal convolution. Similarly, replacing $-g(-x)$ with $g(x)$ in the erosion (Equation 16.11) yields the infimal convolution

$$(f \cdot g)(x) \equiv \bigwedge_{y \in D} f(x - y) + g(y) \tag{16.12}$$

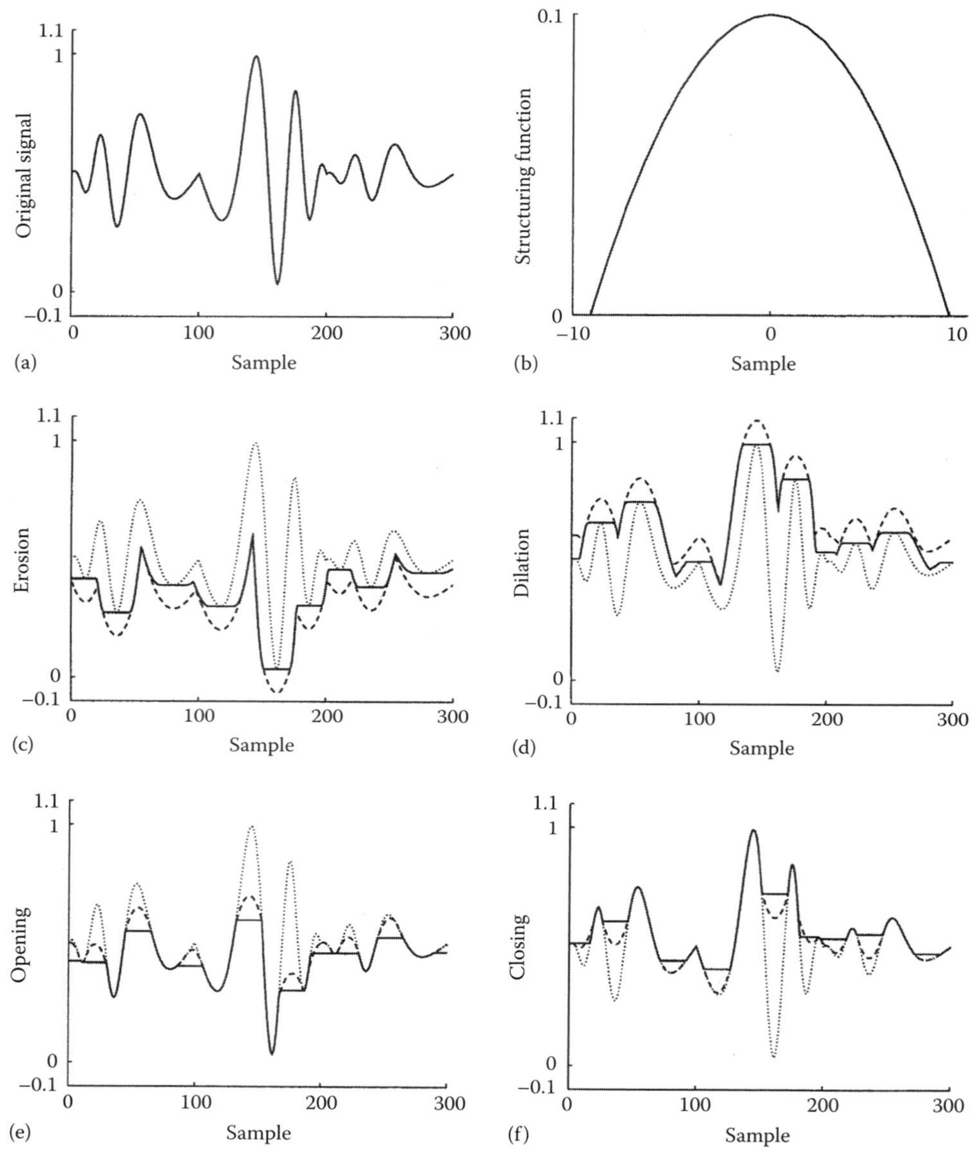

FIGURE 16.2 (a) Original signal f. (b) Structuring function g (a parabolic pulse). (c) Erosion $f \ominus g$ with dashed line and flat erosion $f \ominus B$ with solid line, where the set $B = \{x \in \mathbb{Z} : |x| \leq 10\}$ is the support of g. Dotted line shows original signal f. (d) Dilation $f \oplus g$ (dashed line) and flat dilation $f \oplus B$ (solid line). (e) Opening $f \circ g$ (dashed line) and flat opening $f \circ B$ (solid line). (f) Closing $f \bullet g$ (dashed line) and flat closing $f \bullet B$ (solid line).

The nonlinearity of $\oplus$ and $\ominus$ causes some differences between these signal operations and the linear convolutions. A major difference is that serial or parallel interconnections of systems represented by linear convolutions are equivalent to an overall linear convolution, whereas interconnections of dilations and erosions lead to entirely different nonlinear systems. Thus, there is an infinite variety of nonlinear operators created by cascading dilations and erosions or by interconnecting them in parallel via max/min or addition. Two such useful examples are the opening $\circ$ and closing $\bullet$:

TABLE 16.2 Definitions of Operator Properties

Property	Set Operator Ψ	Signal Operator ψ
Translation-Invar.	$\Psi(X_{+y}) = \Psi(X)_{+y}$	$\psi[f(x-y)+c] = c + \psi(f)(x-y)$
Shift-Invariant	$\Psi(X_{+y}) = \Psi(X)_{+y}$	$\psi[f(x-y)] = \psi(f)(x-y)$
Increasing	$X \subseteq Y \Rightarrow \Psi(X) \subseteq \Psi(Y)$	$f \leq g \Rightarrow \psi(f) \leq \psi(g)$
Extensive	$X \subseteq \psi(X)$	$f \leq \psi(f)$
Antiextensive	$\Psi(X) \subseteq X$	$\psi(f) \leq f$
Idempotent	$\Psi(\Psi(X)) = \Psi(X)$	$\psi(\psi(f)) = \psi(f)$

TABLE 16.3 Properties of Basic Morphological Signal Operators

Property	Dilation	Erosion	Opening	Closing
Duality	$f \oplus g = -[(-f) \ominus g^r]$		$f \circ g = -[(-f) \bullet g^r]$	
Distributivity	$(\vee_i f_i) \oplus g = \vee_i f_i \oplus g$	$(\wedge_i f_i) \ominus g = \wedge_i f_i \ominus (g \ominus h)$	No	no
Composition	$(f \oplus g) \oplus h = f \oplus (g \oplus h)$	$(f \ominus g) \ominus h = f \ominus (g \oplus h)$		
Extensive	Yes if $g(0) \geq 0$	No	No	No
Antiextensive	No	Yes if $g(0) \geq 0$	Yes	No
Commutative	$f \oplus g = g \oplus f$	No	No	No
Increasing	Yes	Yes	Yes	Yes
Translation-Invar.	Yes	Yes	Yes	Yes
Idempotent	No	No	Yes	Yes

$$f \circ g \equiv (f \ominus g) \oplus g \tag{16.13}$$

$$f \bullet g \equiv (f \oplus g) \ominus g \tag{16.14}$$

which act as nonlinear smoothers.

Figure 16.2 shows that the four basic morphological transformations of a 1D signal f by a concave even function g with a compact support B have similar effects as the corresponding flat transformations by the set B. Among the few differences, the erosion (dilation) of f by g subtracts from (adds to) f the values of the moving template g during the decrease (increase) of signal peaks (valleys) and the broadening of the local signal minima (maxima) that would incur during erosion (dilation) by B. Similarly, the opening (closing) of f by g cuts the peaks (fills up the valleys) inside which no translated version of $g(-g)$ can fit and replaces these eliminated peaks (valleys) by replicas of $g(-g)$. In contrast, the flat opening or closing by B only cuts the peaks or fills valleys and creates flat plateaus in the output.

The four above morphological operators of dilation, erosion, opening, and closing have a rich collection of algebraic properties, some of which are listed in Tables 16.2 and 16.3, which endow them with a broad range of applications, make them rigorous, and lead to a variety of efficient serial or parallel implementations.

16.3 Median, Rank, and Stack Operators

Flat erosion and dilation of a discrete-domain signal $f[x]$ by a finite window $W = \{y_1, \ldots, y_n\} \subseteq Z^d$ is a moving local minimum or maximum. Replacing min/max with a more general rank leads to rank operators. At each location $x \in Z^d$, sorting the signal values within the reflected and shifted n-point window $(W^r)_{+x}$ in decreasing order and picking the pth largest value, $p = 1, 2, \ldots, n = \text{card}(W)$, yields the output signal from the pth rank operator:

$$(f_{\bullet p} W)[x] \equiv p\text{th rank of } (f[x - y_1], \ldots, f[x - y_n]) \tag{16.15}$$

For odd n and $p = (n + 1)/2$ we obtain the median operator. If the input signal is binary, the output is also binary since sorting preserves a signal's range. Representing the input binary signal with a set $S \subseteq Z^d$, the output set produced by the pth rank set operators is

$$S_{\bullet p} W \equiv \{x : \text{card}((W^r)_{+x} \cap S) \geq p\} \tag{16.16}$$

Thus, computing the output from a set rank operator involves only counting of points and no sorting.

All rank operators commute with thresholding [21,27,41,45,58,65]; i.e.,

$$\Theta_v [f_{\bullet p} W] = [\Theta_v(f)]_{\bullet p} W, \forall v, \forall p. \tag{16.17}$$

This property is also shared by all morphological operators that are finite compositions or maxima/minima of flat dilations and erosions, e.g., openings and closings, by finite structuring elements. All such signal operators ψ that have a corresponding set operator Ψ and commute with thresholding can be alternatively implemented via threshold superposition [41,58] as in Equation 16.7. Namely, to transform a multilevel signal f by ψ is equivalent to decomposing f into all its threshold sets, transforming each set by the corresponding set operator Ψ, and reconstructing the output signal $\psi(f)$ via its thresholded versions. This allows us to study all rank operators and their cascade or parallel (using $\vee, \wedge$) combinations by focusing on their corresponding binary operators. Such representations are much simpler to analyze and they suggest alternative implementations that do not involve numeric comparisons or sorting.

Binary rank operators and all other binary discrete translation-invariant finite-window operators can be described by their generating Boolean function; see Table 16.1. Thus, in synthesizing discrete multilevel signal operators from their binary countparts via threshold superposition all that is needed is knowledge of this Boolean function. Specifically, transforming all the threshold binary signals $\theta_v(f)[x]$ of an input signal $f[x]$ with an increasing Boolean function $b(u_1, \ldots, u_n)$ (i.e., containing no complemented variables) in place of the set operator Ψ in Equation 16.7 creates a large variety of nonlinear signal operators via threshold superposition, called stack filters [41,70]

$$\phi_b(f)[x] \equiv \sup\{v : b(\theta_v(f)[x - y_1], \ldots, \theta_v(f)[x - y_n]) = 1\} \tag{16.18}$$

For example, ϕ_b becomes the pth rank operator if b is equal to the sum $\binom{n}{p}$ product terms where each contains one distinct p-point subset from the n variables. In general, the use of Boolean functions facilitates the design of such discrete flat operators with determinable structural properties. Since each increasing Boolean function can be uniquely represented by an irreducible sum (product) of product (sum) terms, and each product (sum) term corresponds to an erosion (dilation), each stack filter can be represented as a finite maximum (minimum) of flat erosions (dilations) [41].

16.4 Universality of Morphological Operators

Dilations or erosions, the basic nonlinear convolutions of morphological signal processing, can be combined in many ways to create more complex morphological operators that can solve a broad variety of problems in image analysis and nonlinear filtering. In addition, they can be implemented using simple and fast software or hardware; examples include various digital [58,61] and analog, i.e., optical or hybrid optical-electronic implementations [46,63]. Their wide applicability and ease of implementation poses the question which signal processing systems can be represented by using dilations and erosions as the basic building blocks. Toward this goal, a theory was introduced in [33,34] that represents a broad class of nonlinear and linear operators as a minimal combination of erosions or dilations. Here we summarize the main results of this theory, in a simplified way, restricting our discussion only to signals with discrete domain $\mathbb{D}=\mathbb{Z}^d$.

Consider a translation-invariant set operator Ψ on the class $P(\mathrm{D})$ of all subsets of $\mathbb{D}$. Any such Ψ is uniquely characterized by its kernel that is defined [42] as the subclass $\mathrm{Ker}(\Psi) \equiv \{X \in \mathcal{P}(\mathbb{D}) : 0 \in \Psi(X)\}$ of input sets, where 0 is the origin of $\mathbb{D}$. If Ψ is also increasing, then it can be represented [42] as the union of erosions by its kernel sets and as the intersection of dilations by the reflected kernel sets of its dual operator $\Psi^d(X) \equiv [\Psi(X^c)]^c$. This kernel representation can be extended to signal operators ψ on the class Fun $(\mathbb{D}, \bar{\mathbb{R}})$ of signals with domain $\mathbb{D}$ and range $\bar{\mathbb{R}}$. The kernel of ψ is defined as the subclass $\mathrm{Ker}(\psi) = \{f \in \mathrm{Fun}\ (\mathbb{D}, \bar{\mathbb{R}}) : [\psi(f)](0) \geq 0\}$ of input signals. If ψ is translation-invariant and increasing, then it can be represented [33,34] as the pointwise supremum of erosions by its kernel functions, and as the infimum of dilations by the reflected kernel functions of its dual operator $\psi^d(f) \equiv -\psi(-f)$.

The two previous kernel representations require an infinite number of erosions or dilations to represent a given operator because the kernel contains an infinite number of elements. However, we can find more efficient (requiring less erosions) representations by using only a substructure of the kernel, its basis. The basis Bas $(\cdot)$ of a set (signal) operator is defined [33,34] as the collection of kernel elements that are minimal with respect to the ordering $\subseteq$ $(\leq)$.

If a translation-invariant increasing set operator Ψ is also upper semicontinuous, i.e., obeys a monotonic continuity where $\Psi\left(\bigcap_n X_n\right) = \bigcap_n \Psi(X_n)$ for any decreasing set sequence X_n, then Ψ has a nonempty basis and can be represented via erosions only by its basis sets. If the dual Ψ^d is also upper semicontinuous, then its basis sets provide an alternative representation of Ψ via dilations:

$$\Psi(X) = \bigcup_{A \in \mathrm{Bas}(\Psi)} X \ominus A = \bigcap_{B \in \mathrm{Bas}(\Psi^d)} X \oplus B^r \tag{16.19}$$

Similarly, any signal operator ψ that is translation-invariant, increasing, and upper semicontinuous (i.e., $\psi(\wedge_n f_n) = \wedge_n \psi(f_n)$ for any decreasing function sequence f_n) can be represented as the supremum of erosions by its basis functions, and (if ψ^d is upper semicontinuous) as the infimum of dilations by the reflected basis functions of its dual operators:

$$\psi(f) = \bigvee_{g \in \mathrm{Bas}(\psi)} f \ominus g = \bigwedge_{h \in \mathrm{Bas}(\psi^d)} f \oplus h^r \tag{16.20}$$

where $h^r(x) \equiv h(-x)$. Finally, if ϕ is a flat signal operator as in Equation 16.7 that is translation-invariant and commutes with thresholding, then ϕ can be represented as a supremum of erosions by the basis sets of its corresponding set operator Φ:

$$\phi(f) = \bigvee_{A \in \mathrm{Bas}(\Phi)} f \ominus A = \bigwedge_{B \in \mathrm{Bas}(\Phi^d)} f \oplus B^r \tag{16.21}$$

While all the above representations express translation-invariant increasing operators via erosions or dilations, operators that are not necessarily increasing can be represented [4] via operations closely related to hit-miss transformations.

Representing operators that satisfy a few general properties in terms of elementary morphological operations can be applied to more complex morphological systems and various other filters such as linear rank, hybrid linear/rank, and stack filters, as the following examples illustrate.

Example 16.1: Morphological Filters

All systems made up of serial or sup/inf combinations of erosions, dilations, opening, and closings admit a basis, which is finite if the system's local definition depends on a finite window. For example, the set opening $\Phi(X) = X \circ A$ has as a basis the set collection Bas $(\Phi) = \{A_{-a} : a \in A\}$. Consider now 1D discrete-domain signals and let $A = \{-1, 0, 1\}$. Then, the basis of Φ has three sets: $G_1 = A_{-1}$, $G_2 = A$, $G_3 = A_{+1}$. The basis of the dual operator $\Phi^d(X) = X \bullet A$ has four sets: $H_1 = \{0\}$, $H_2 = \{-2, 1\}$, $H_3 = \{-1, 2\}$, $H_4 = \{-1, 1\}$. The flat signal operator corresponding to Φ is the opening $\phi(f) = f \circ A$. Thus, from Equation 16.21, the signal opening can also be realized as a max (min) of local minima (maxima):

$$(f \circ A)[x] = \bigvee_{i=1}^{3} \left\{ \bigwedge_{y \in G_i} f[x+y] \right\} = \bigwedge_{k=1}^{4} \left\{ \bigvee_{y \in H_k} f[x+y] \right\} \tag{16.22}$$

Example 16.2: Linear Filters

A linear shift-invariant filter is translation-invariant and increasing (see Table 16.2 for definitions) if its impulse response is everywhere nonnegative and has area equal to one. Consider the 2-point FIR filter $\psi(f)[x] = af[x] + (1 - a)f[x - 1]$, where $0 < a < 1$. The basis of ψ consists of all functions $g[x]$ with $g[0] = r \in \mathbb{R}, g[-1] = -ar/(1 - a)$, and $g[x] = -\infty$ for $x \neq 0, -1$. Then Equation 16.20 yields

$$af[x] + (1 - a)f[x - 1] = \bigvee_{r \in R} \left[\min\left\{ f[x] - r, f[x - 1] + \frac{ar}{1 - a} \right\} \right] \tag{16.23}$$

which expresses a linear convolution as a supremum of erosions. FIR linear filters have an infinite basis, which is a finite-dimensional vector space.

Example 16.3: Median Filters

All rank operators have a finite basis; hence, they can be expressed as a finite max-of-erosions or min-of-dilations. Further, they commute with thresholding, which allows us to focus only on their binary versions. For example, the set median by the window $W = \{-1, 0, 1\}$ has three basis sets: $\{-1, 0\}, \{-1, 1\}$, and $\{0, 1\}$. Hence, Equation 16.21 yields

$$\text{median}(f[x - 1], f[x], f[x + 1]) = \max \left\{ \begin{array}{c} \min\,(f[x - 1], f[x]) \\ \min\,[f(x - 1), f(x + 1)] \\ \min\,[f(x), f(x + 1)] \end{array} \right\} \tag{16.24}$$

Example 16.4: Stack Filters

Stack filters (Equation 16.18) are discrete translation-invariant flat operators ϕ_b, locally defined on a finite window W, and are generated by a increasing Boolean function $b(v_1, \ldots, v_n)$, where $n = \text{card}(W)$. This function corresponds to a translation-invariant increasing set operator Φ. For example, consider 1D signals, let $W = \{-2, -1, 0, 1, 2\}$ and

$$b(v_1, \ldots, v_5) = v_1v_2v_3 + v_2v_3v_4 + v_3v_4v_5 = v_3(v_1 + v_4)(v_2 + v_4)(v_2 + v_5) \tag{16.25}$$

This function generates via threshold superposition the flat opening $\phi_b(f) = f \circ A, A = \{-1, 0, 1\}$, of Equation 16.22. There is one-to-one correspondence between the three prime implicants of b and the erosions (local min) by the three basis sets of Φ, as well as between the four prime implicates of β and the dilations (local max) by the four basis sets of the dual Φ^d. In general, given b, Φ or ϕ_b is found by replacing Boolean AND/OR with set $\cap/\cup$ or with min/max, respectively. Conversely, given ϕ_b, we can find its generating Boolean function from the basis of its set operator (or directly from its max/min representation if available) [41].

The above examples show the power of the general representation theorems. An interesting applications of these results is the design of morphological systems via their basis [5,20,31]. Given the wide applicability of erosions/dilations, their parallelism, and their simple implementations, the previous theorems theoretically support a general purpose vision (software or hardware) module that can perform erosions/dilations, based on which numerous other complex image operations can be built.

16.5 Morphological Operators and Lattice Theory

In the late 1980s and 1990s a new and more general formalization of morphological operators was introduced [26,51,52,59], which views them as operators on complete lattices. A complete lattice is a set $\mathcal{L}$ equipped with a partial ordering $\leq$ such that $(\mathcal{L}, \leq)$ has the algebraic structure of a partially ordered set (poset) where the supremum and infimum of any of its subsets exist in $\mathcal{L}$. For any subset $\mathcal{K} \subseteq \mathcal{L}$, its supremum $\vee\mathcal{K}$ and infimum $\wedge\mathcal{K}$ are defined as the lowest (with respect to $\leq$) upper bound and greatest lower bound of $\mathcal{K}$, respectively. The two main examples of complete lattices used in morphological processing are (1) the set space $\mathcal{P}(\mathbb{D})$ where the $\vee/\wedge$ lattice operations are the set union/intersection, and (2) the signal space Fun $(\mathbb{D}, \bar{\mathbb{R}})$ where the $\vee/\wedge$ lattice operations are the supremum/infimum of sets of real numbers. Increasing operators on $\mathcal{L}$ are of great importance because they preserve the partial ordering, and among them four fundamental examples are

$$\delta \text{ is dilation} \Longleftrightarrow \delta\left(\bigvee_{i\in I} f_i\right) = \bigvee_{i\in I} \delta(f_i) \tag{16.26}$$

$$\varepsilon \text{ is erosion} \Longleftrightarrow \varepsilon\left(\bigwedge_{i\in I} f_i\right) = \bigwedge_{i\in I} \varepsilon(f_i) \tag{16.27}$$

$$\alpha \text{ is opening} \Longleftrightarrow \alpha \text{ is increasing, idempotent, and antiextensive} \tag{16.28}$$

$$\beta \text{ is closing} \Longleftrightarrow \beta \text{ is increasing, idempotent, and extensive} \tag{16.29}$$

where I is an arbitrary index set.

The above definitions allow broad classes of signal operators to be grouped as lattice dilations, erosions, openings, or closing and their common properties to be studied under the unifying lattice framework. Thus, the translation-invariant morphological dilations, erosions, openings, and closings we saw before are simply special cases of their lattice counterparts. Next, we see some examples and applications of the above general definitions.

Example 16.5: Dilation and Translation-Invariant Systems

Consider a signal operator that is shift-invariant and obeys a supremum-of-sums superposition:

$$\mathcal{D}\left[\bigvee_i c_i + f_i(x)\right] = \bigvee_i c_i + \mathcal{D}[f_i(x)] \tag{16.30}$$

Then $\mathcal{D}$ is both a lattice dilation and translation-invariant (DTI). We call it a DTI system in analogy to linear time-invariant (LTI) systems that are shift-invariant and obey a linear (sum-of-products) superposition. As an LTI system corresponds in the time domain to a linear convolution with its impulse response, a DTI system can be represented as a supremal convolution with its upper "impulse response" $g_\vee(x)$ defined as its output when the input is the upper zero impulse $\imath(x)$, defined in Table 16.4. Specifically,

$$\mathcal{D} \text{ is DTI} \Longleftrightarrow \mathcal{D}(f) = f \oplus g_\vee, \quad g_\vee \equiv \mathcal{D}(\imath) \tag{16.31}$$

TABLE 16.4 Examples of Upper Slope Transform

Signal: $f(s)$	Transform: $F_v(a)$
$t(x - x_0) \equiv 0$ if $x \equiv x_0$ and $-\infty$ else	$-ax_0$
$a_0 x$	$-t(a - a_0)$
$\lambda(x) \equiv 0$ if $x \geq 0$ and $-\infty$ else	$-\lambda(a)$
$a_0 x + \lambda(x)$	$-\lambda(a - a_0)$
$\begin{cases} 0, & \lvert x\rvert \leq r \\ -\infty, & \lvert x\rvert > r \end{cases}$	$r\lvert a\rvert$
$-a_0\lvert x\rvert, a_0 > 0$	$\begin{cases} 0, & \lvert a\rvert \leq a_0 \\ +\infty, & \lvert a\rvert > a_0 \end{cases}$
$\sqrt{1 - x^2}, \lvert x\rvert \leq 1$	$\sqrt{1 + a^2}$
$-(\lvert x\rvert^p)/p, p > 1$	$(\lvert a\rvert^q)/q, 1/p + 1/q = 1$
$\exp(x)$	$a(1 - \log a)$

A similar class is the erosion and translation-invariant (ETI) systems ε which are shift-invariant and obey an infimum-of-sums superposition as in Equation 16.30 but with $\vee$ replaced by $\wedge$. Such systems are equivalent to infimal convolutions with their lower impulse response $g_\wedge = \varepsilon(-\imath)$, defined as the system's output due to the lower impulse $-\imath(x)$. Thus, DTI and ETI systems are uniquely determined in the time/spatial domain by their impulse responses, which also control their causality and stability [37].

Example 16.6: Shift-Varying Dilation

Let $\delta_B(f) = f \oplus B$ be the shift-invariant flat dilation of Equation 16.8. In applying it to nonstationary signals, the need may arise to vary the moving window B by actually having a family of windows $B(x)$, possibly varying at each location x. This creates the new operator

$$\delta_B(f)(x) = \bigvee_{y \in B(x)} f(x - y) \tag{16.32}$$

which is still a lattice dilation, i.e., it distributes over suprema, but it is shift-varying.

Example 16.7: Adjunctions

An operator pair (ε, δ) is called an adjunction if $\delta(f) \leq g \leq \Longleftrightarrow f \leq \varepsilon(g)$ for all $f, g \in \mathcal{L}$. Given a dilation δ, there is a unique erosion ε such that (ε, δ) is adjunction, and vice versa. Further, if (ε, δ) is an adjunction, then δ is a dilation, ε is an erosion, $\delta\varepsilon$ is an opening, and $\varepsilon\delta$ is a closing. Thus, from any adjunction we can generate an opening via the composition of its erosion and dilation. If ε and δ are the translation-invariant morphological erosion and dilation in Equations 16.11 and 16.10, then $\delta\varepsilon$ coincides with the translation-invariant morphological opening of Equation 16.13. But there are also numerous other possibilities.

Example 16.8: Radial Opening

If a 2D image f contains 1D objects, e.g., lines, and B is a 2D convex structuring element, then the opening or closing of f by B will eliminate these 1D objects. Another problem arises when f contains large-scale objects with sharp corners that need to be preserved; in such cases opening or closing f by a disk B will round these corners. These two problems could be avoided in some cases if we replace the conventional opening with

$$\alpha(f) = \bigvee_{\theta} f \circ L_\theta \tag{16.33}$$

where the sets L_θ are rotated versions of a line segment L at various angles $\theta \in [0, 2\pi)$. The operator α, called radial opening, is a lattice opening in the sense of Equation 16.28. It has the effect of preserving an object in f if this object is left unchanged after the opening by L_θ in at least one of the possible orientations θ.

Example 16.9: Opening by Reconstruction

Consider a set $X = \bigcup_i X_i$ as a union of disjoint connected components X_i and let $M \subseteq X_j$ be a marker in the jth component; i.e., M could be a single point or some feature set in X that lies only in X_j. Then, define the conditional dilation of M by B within X as

$$\delta_{B|X}(M) \equiv (M \oplus B) \cap X \qquad (16.34)$$

If B is a disk with a radius smaller than the distance between X_j and any of the other components, then by iterating this conditional dilation we can obtain in the limit

$$MR_{B|X}(M) = \lim_{n \to \infty} \underbrace{\left(\delta_{B|X} \cdots (\delta_{B|X}(\delta_{B|X}(M))\right)}_{n \text{ times}} \qquad (16.35)$$

the whole component X_j. The operator MR is a lattice opening, called opening by reconstruction, and its output is called the morphological reconstruction of the component from the marker. An example is shown in Figure 16.3. It can extract large-scale components of the image from knowledge only of a smaller marker inside them.

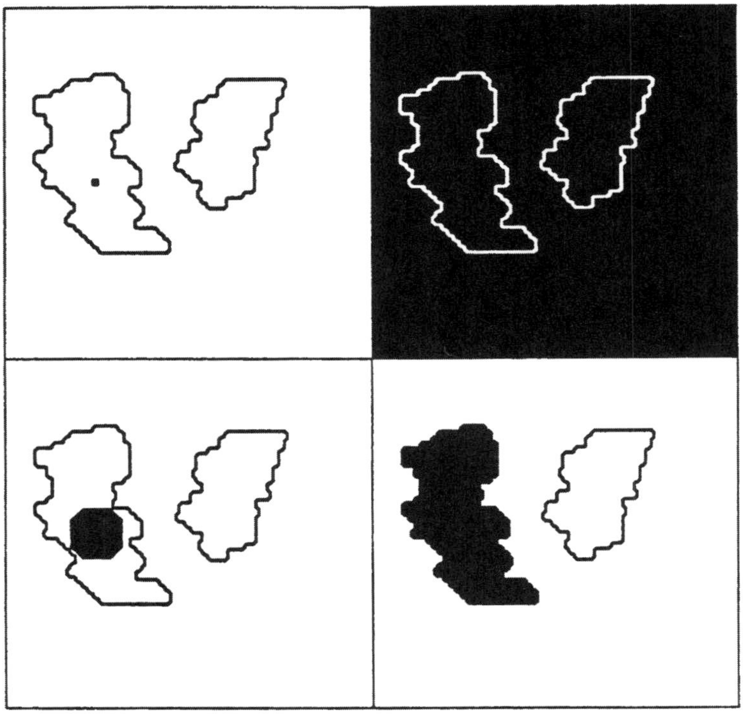

FIGURE 16.3 Let X be the union of the two region boundaries in the top left image, and let M be the single-point marker inside the left region. Top right shows the complement X^c. If $Y_0 = M$ and B is a disk-like set whose radius does not exceed the width of the region boundary, iterating the conditional dilation $Y_i = (Y_{i-1} \oplus B) \cap X^c$, for $i = 1, 2, 3, \ldots$, yields in the limit (reached at $i = 18$ in this case) the interior Y_∞ of the left region via morphological reconstruction, shown in bottom right. (Bottom left shows an intermediate result for $i = 9$.)

16.6 Slope Transforms

Fourier transforms are among the most useful linear signal transformations because they enable us to analyze the processing of signals by LTI systems in the frequency domain, which could be more intuitive or easier to implement. Similarly, there exist some nonlinear signal transformations, called slope transforms, which allow the analysis of the dilation and erosion translation-invariant (DTI and ETI) systems in a transform domain, the slope domain. First, we note that the lines $f(x) = ax + b$ are eigenfunctions of any DTI system $\mathcal{D}$ or ETI system E because

$$\begin{aligned} D[ax+b] &= ax + b + G_{\vee}(a), \quad G_{\vee}(a) \equiv \bigvee_x g_{\vee}(x) - ax \\ E[ax+b] &= ax + b + G_{\wedge}(a), \quad G_{\wedge}(a) \equiv \bigwedge_x g_{\wedge}(x) - ax \end{aligned} \tag{16.36}$$

with corresponding eigenvalues $G_{\vee}(a)$ and $G_{\wedge}(a)$, which are called, respectively, the upper and lower slope response of the DTI and ETI system. They measure the amount of shift in the intercept of the input lines with slope a and are conceptually similar to the frequency response of LTI systems.

Then, by viewing the slope response as a signal transform with variable the slope $a \in \mathrm{R}$, we define [37] for a 1D signal $f : \mathbb{D} \to \bar{\mathbb{R}}$ its upper slope transform $F_{\vee}$ and its lower slope transform* $F_{\wedge}$ as the functions

$$F_{\vee}(a) \equiv \bigvee_{x \in \mathrm{D}} f(x) - ax \tag{16.37}$$

$$F_{\wedge}(a) \equiv \bigwedge_{x \in \mathrm{D}} f(x) - ax \tag{16.38}$$

Since $f(x) - ax$ is the intercept of a line with slope a passing from the point $(x, f(x))$ on the signal's graph, for each a the upper (lower) slope transform of f is the maximum (minimum) value of this intercept, which occurs when the above line becomes a tangent. Examples of slope transforms are shown in Figure 16.4. For differentiable signals, f, the maximization or minimization of the intercept $f(x) - ax$

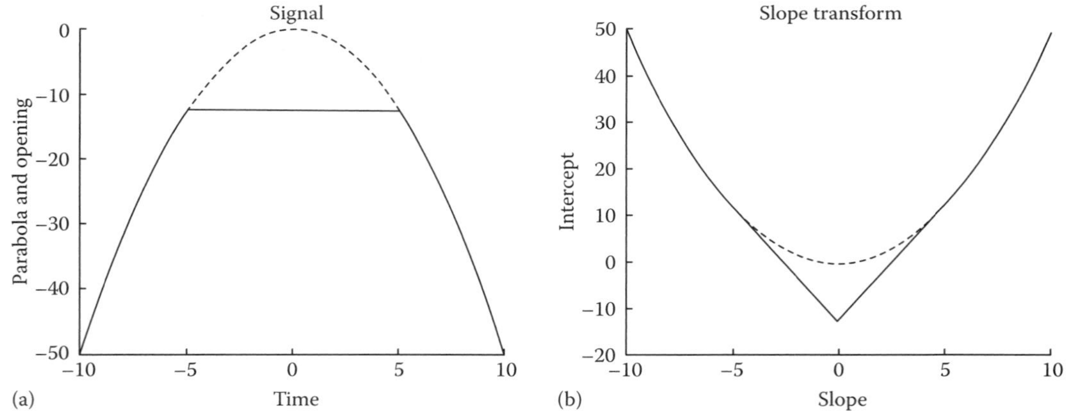

FIGURE 16.4 (a) Original parabola signal $f(x) = -x^2/2$ (in dashed line) and its morphological opening (in solid line) by a flat structuring element $[-5, 5]$. (b) Upper slope transform $F_{\vee}(a)$ of the parabola (in dashed line) and of its opening (in solid line).

* In convex analysis [50], to convex function h there uniquely corresponds its Fenchel conjugate $h^*(a) = \bigvee_x ax - h(x)$, which is the negative of the lower slope transform of h.

can also be done by finding the stationary point(s) x^* such that $\mathrm{d}f(x^*)/\mathrm{d}x = a$. This extreme value of the intercept is the Legendre transform of f:

$$F_L(a) \equiv f\left((\mathrm{d}f/\mathrm{d}x)^{-1}(a)\right) - a[(\mathrm{d}f/\mathrm{d}x)^{-1}(a)] \tag{16.39}$$

It is extensively used in mathematical physics. If the signal $f(x)$ is concave or convex and has an invertible derivative, its Legendre transform is single-valued and equal (over the slope regions it is defined) to the upper or lower transform; e.g., see the last three examples in Table 16.4. If f is neither convex nor concave or if it does not have an invertible derivative, its Legendre transform becomes a set $F_L(a) = \{f(x^*) - ax^*: \mathrm{d}f(x^*)/\mathrm{d}x = a\}$ of real numbers for each a. This multivalued Legendre transform, defined and studied in [19] as a "slope transform," has properties similar to those of the upper/lower slope transform, but there are also some important differences [37].

The upper and lower slope transform have a limitation in that they do not admit an inverse for arbitrary signals. The closest to an "inverse" upper slope transform is

$$\hat{f}(x) \equiv \bigwedge_{a\in \mathrm{R}} F_\vee(a) + ax \tag{16.40}$$

which is equal to f only if f is concave; otherwise, $\hat{f}$ covers f from above by being its smallest concave upper envelope. Similarly, the supremum over a of all lines $F_\wedge(a) + ax$ creates the greatest convex lower envelope $f(x)$ of f, which plays the role of an "inverse" lower slope transform and is equal to f only if f is convex. Thus, for arbitrary signals we have $f \leq f \leq \hat{f}$.

Tables 16.4 and 16.5 list several examples and properties of the upper slope transform. The most striking is that (dilation) supremal convolution in the time/space domain corresponds to addition in the slope domain. Note the analogy with LTI systems where linearly convolving two signals corresponds to multiplying their Fourier transforms. Very similar properties also hold for the lower slope transform, the only differences being the interchange of suprema with infima, concave with convex, and the supremal $\oplus$ with the infimal convolution $\square$.

The upper/lower slope transforms for discrete-domain and/or multidimensional signals are defined as in the 1D continuous case by replacing the real variable x with an integer and/or multidimensional variable, and their properties are very similar or identical to the ones for signals defined on $\mathbb{R}$. See [37,38] for details.

One of the most useful applications of LTI systems and Fourier transform is the design of frequency-selective filters. Similarly, it is also possible to design morphological systems that have a slope selectivity. Imagine a DTI system that rejects all line components with slopes in the band $[-a_0, a_0]$ and passes all the rest unchanged. Then its slope response would be

$$G(a) = 0 \quad \text{if } |a| \leq a_0 \quad \text{and} \quad +\infty \text{ else} \tag{16.41}$$

TABLE 16.5 Properties of Upper Slope Transform

Signal: $f(x)$	Transform: $F_{\vee(a)}$
$\bigvee_i c_i + f_i(x)$	$\bigvee_i c_i + F_i(a)$
$f(x - x_0)$	$F(a) - ax_0$
$f(x) + a_0 x$	$F(a - a_0)$
$f(rx)$	$F(a/r)$
$f(x) \oplus g(x)$	$F(a) + G(a)$
$\bigvee_y f(x) + g(x + y)$	$F(-a) + G(a)$
$f(x) \leq g(x) \forall x$	$F(a) \leq G(a) \forall a$
$g(x) = \begin{cases} f(x), & \lvert x\rvert \leq r \\ -\infty, & \lvert x\rvert > r \end{cases}$	$G(a) = F(a) \square r\lvert a\rvert$

This is an ideal-cutoff slope bandpass filter. In the time domain it acts as a supremal convolution with its impulse response

$$g(x) = -a_0|x| \tag{16.42}$$

However, $f \oplus g$ is a noncausal infinite-extent dilation, and hence not realizable. Instead, we could implement it as a cascade of a causal dilation by the half-line $g_1(x) = -a_0x + \lambda(x)$ followed by an anticausal dilation by another half-line $g_2(x) = a_0x + \lambda(-x)$, where $\lambda(x)$ is the zero step defined in Table 16.4. This works because $g = g_1 \oplus g_2$. For a discrete-time signal $f[x]$, this slope-bandpass filtering could be implemented via the recursive max-sum difference equation $f_1[x] = \max(f_1[x] - a_0, f[x])$ run forward in time, followed by another difference equation $f_2[x] = \max(f_2[x+1] + a_0, f_1[x])$ run backward in time. The final result would be $f_2 = f \oplus g$. Such slope filters are useful for envelope estimation [37].

16.7 Multiscale Morphological Image Analysis

Multiscale signal analysis has recently emerged as a useful framework for many computer vision and signal processing tasks. Examples include: (1) detecting geometrical features or other events at large scales and then refining their location or value at smaller scales, (2) video and audio data compression using multiband frequency analysis, and (3) measurements and modeling of fractal signals. Most of the work in this area has obtained multiscale signal versions via linear multiscale smoothing, i.e., convolutions with a Gaussian with a variance proportional to scale [15,53,72]. There is, however, a variety of nonlinear smoothing filters, including the morphological openings and closings [35,42,58] that can provide a multiscale image ensemble and have the advantage over the linear Gaussian smoothers that they do not blur or shift edges, as shown in Figure 16.5. There we see that the gray-level close-openings by reconstruction are especially useful because they can extract the exact outline of a certain object by locking on it while smoothing out all its surroundings; these nonlinear smoothers have been applied extensively in multiscale image segmentation [56]. The use of morphological operators for multiscale signal analysis is not limited to operations of a smoothing type; e.g., in fractal image analysis, erosion and dilation can provide multiscale distributions of the shrink-expand type from which the fractal dimension can be computed [36].

Overall, many applications of morphological signal processing such as nonlinear smoothing, geometrical feature extraction, skeletonization, size distributions, and segmentation, inherently require or can benefit from performing morphological operations at multiples scales. The required building blocks for a morphological scale-space are the multiscale dilations and erosions. Consider a planar compact convex set $B = \{(x,y) : \|(x,y)\|_p \leq 1\}$ that is the unit ball generated by the L_p norm, $p = 1, 2, \ldots, \infty$. Then the simplest multiscale dilation and erosion of a signal $f(x,y)$ at scales $t > 0$ are the multiscale flat sup/inf convolutions by $tB = \{tz : z \in B\}$

$$\delta(x,y,t) \equiv (f \oplus tB)(x,y) \tag{16.43}$$

$$\varepsilon(x,y,t) \equiv (f \ominus tB)(x,y) \tag{16.44}$$

which apply both to gray-level and binary images.

16.7.1 Binary Multiscale Morphology via Distance Transforms

Viewing the boundaries of multiscale erosions/dilations of a binary image by disks as wave fronts propagating from the original image boundary at uniform unit normal velocity and assigning to each pixel the time t of wave front arrival creates a distance function, called the distance transform [10]. This

FIGURE 16.5 (a) Original image and its multiscale smoothings via: (b–d) Gaussian convolution at scales 2, 4, 16; (e–g) close-opening by a square at scales 2, 4, 16; (h–j) close-opening by reconstruction at scales 2, 4, 16.

transform is a compact way to represent their multiscale dilations and erosions by disks and other polygonal structuring elements whose shape depends on the norm $\|\cdot\|_p$ used to measure distances. Formally, the distance transform of the foreground set F of a binary image is defined as

$$D_p(F)(x,y) \equiv \bigwedge_{(v,u)\in F^c} \{\|(x-v, y-u)\|_p\} \tag{16.45}$$

Thresholding the distance transform at various levels $t > 0$ yields the erosions of the foreground F (or the dilation of the background F^c) by the norm-induced ball B at scale t:

$$F \ominus tB = \Theta_t[D_p(F)] \tag{16.46}$$

Another view of the distance transform results from seeing it as the infimal convolution of the $(0, +\infty)$ indicator function of F^c,

$$I_{F^c}(x) \equiv 0 \quad \text{if } x \in F^c \quad \text{and} \quad +\infty \text{ else} \tag{16.47}$$

with the norm-induced conical structuring function:

$$D_p(F)(x) = I_{F^c}(x) \square \|x\|_p \tag{16.48}$$

Recognizing $g_\wedge(x) = \|x\|_p$ as the lower impulse response of an ETI system with slope response

$$G_\wedge(a) = 0 \quad \text{if } \|a\|_q \leq 1 \quad \text{and} \quad -\infty \text{ else} \tag{16.49}$$

where $1/p + 1/q = 1$, leads to seeing the distance transform as the output of an ideal-cutoff slope-selective filter that rejects all input planes whose slope vector falls outside the unit ball with respect to the $\|\cdot\|_q$ norm, and passes all the rest unchanged.

To obtain isotropic distance propagation, the Euclidean distance transform is desirable because it gives multiscale morphology with the disk as the structuring element. However, since this has a significant computational complexity, various techniques are used to obtain approximations to the Euclidean distance transform of discrete images at a lower complexity. A general such approach is the use of discrete distances [54] and their generalization via chamfer metrics [11]. Given a discrete binary image $f[i,j] \in \{0, +\infty\}$ with 0 marking background/source pixels and $+\infty$ marking foreground/object pixels, its global chamfer distance transform is obtained by propagating local distances within a small neighborhood mask. An efficient method to implement it is a two-pass sequential algorithm [11,54] where for a 3×3 neighborhood the min-sum difference equation

$$u_n[i,j] = \min(u_{n-1}[i,j], u_n[i-1,j]+a, u_n[i,j-1]+a, u_n[i-1,j-1]+b, u_n[i+1,j-1]+b) \tag{16.50}$$

is run recursively over the image domain: first ($n = 1$), in a forward scan starting from $u_0 = f$ to obtain u_1, and second ($n = 2$) in a backward scan on u_1 using a reflected mask to obtain u_2, which is the final distance transform. The coefficients a and b are the local distances within the neighborhood mask. The unit ball associated with chamfer metrics is a polygon whose approximation of the disk improves by increasing the size of the mask and optimizing the local distances so as to minimize the error in approximating the true Euclidean distances. In practice, integer-valued local distances are used for faster implementation of the distance transform. If (a,b) is $(1,1)$ or $(1,\infty)$, the chamfer ball becomes a square or rhombus, respectively, and the chamfer distance transform gives poor approximations to multiscale morphology with disks. The commonly used $(a = 3,\ b = 4)$ chamfer metric gives a maximum absolute error of about 6%, but even better approximations can be found by optimizing a, b.

16.7.2 Multiresolution Morphology

In certain multiscale image analysis tasks, the need also arises to subsample the multiscale image versions and thus create a multiresolution pyramid [15,53]. Such concepts are very similar to the ones encountered in classical signal decimation. Most research in image pyramids has been based on linear smoothers. However, since morphological filters preserve essential shape features, they may be superior in many applications. A theory of morphological decimation and interpolation has been developed in [25] to address these issues which also provides algorithms on reconstructing a signal after morphological smoothing and decimation with quantifiable error. For example, consider a binary discrete image represented by a set X that is smoothed first to $Y = X \circ B$ via opening and then down-sampled to $Y \cap S$ by intersecting it with a periodic sampling set S (satisfying certain conditions). Then the Hausdorff distance between the smoothed signal Y and the interpolation (via dilation) $(Y \cap S) \oplus B$ of its down-sampled version does not exceed the radius of B. These ideas also extend to multilevel signals.

16.8 Differential Equations for Continuous-Scale Morphology

Thus far, most of the multiscale image filtering implementations have been discrete. However, due to the current interest in analog VLSI and neural networks, there is renewed interest in analog computation. Thus, continuous models have been proposed for several computer vision tasks based on partial differential equations (PDEs). In multiscale linear analysis [72] a continuous (in scale t and spatial argument x, y) multiscale signal ensemble

$$\gamma(x, y, t) = f(x, y) * G_t(x, y), \quad G_t(x, y) = \frac{\exp\left[-(x^2 + y^2)/4t\right]}{\sqrt{4\pi t}} \tag{16.51}$$

is created by linearly convolving an original signal f with a multiscale Gaussian function G_t whose variance $(2t)$ is proportional to the scale parameter t. The Gaussian multiscale function γ can be generated [28] from the linear diffusion equation

$$\frac{\partial \gamma}{\partial t} = \frac{\partial^2 \gamma}{\partial x^2} + \frac{\partial^2 \gamma}{\partial y^2} \tag{16.52}$$

starting from the initial condition $\gamma(x, y, 0) = f(x, y)$.

Motivated by the limitations or inability of linear systems to successfully model several image processing problems, several nonlinear PDE-based approaches have been developed. Among them, some PDEs have been recently developed to model multiscale morphological operators as dynamical systems evolving in scale-space [1,14,66].

Consider the multiscale morphological flat dilation and erosion of a 2D image signal $f(x, y)$ by the unit-radius disk at scales $t \geq 0$ as the space-scale functions $\delta(x, y, t)$ and $\varepsilon(x, y, t)$ of Equations 16.43 and 16.44. Then [14] the PDE generating these multiscale flat dilations is

$$\frac{\partial \delta}{\partial t} = \|\nabla \delta\| = \sqrt{\left(\frac{\partial \delta}{\partial x}\right)^2 + \left(\frac{\partial \delta}{\partial y}\right)^2} \tag{16.53}$$

and for the erosions is $\partial \varepsilon / \partial t = -\|\nabla \varepsilon\|$. These morphological PDEs directly apply to binary images because flat dilations/erosions commute with thresholding and hence, when the gray-level image is dilated/eroded, each one of its thresholded versions representing a binary image is simultaneously dilated/eroded by the same element and at the same scale.

In equivalent formulations [10,57,66], the boundary of the original binary image is considered as a closed curve and this curve is expanded perpendicularly at constant unit speed. The dilation of the original image with a disk of radius t is the expanded curve at time t. This propagation of the image boundary is a special case of more general curvature-dependent propagation schemes for curve evolution studied in [47]. This general curve evolution methodology was applied in [57] to obtain multiscale morphological dilations/erosions of binary images, using an algorithm [47] where the original curve is first embedded in the surface of a 2D continuous function $\Phi_0(x, y)$ as its zero level set and then the evolving 2D curve is obtained as the zero level set of a 2D function $\Phi(x, y, t)$ that evolves from the initial condition $\Phi(x, y, 0) = \Phi_0(x, y)$ according to the PDE $\partial\Phi/\partial t = \|\nabla\Phi\|$. This function evolution PDE makes zero level sets expand at unit normal speed and is identical to the PDE Equation 16.53 for flat dilation by disk. The main steps in its numerical implementations [47] are

$$
\begin{aligned}
\Phi_{i,j}^n &= \text{estimate of } \Phi(i\Delta x, j\Delta_y, n\Delta t) \text{ on a grid}\\
D_x^+ &= \left(\Phi_{i+1,j}^n - \Phi_{i,j}^n\right)\Big/\Delta x, \quad D_x^- = \left(\Phi_{i,j}^n - \Phi_{i-1,j}^n\right)\Big/\Delta x\\
D_y^+ &= \left(\Phi_{i,j+1}^n - \Phi_{i,j}^n\right)\Big/\Delta y, \quad D_y^- = \left(\Phi_{i,j}^n - \Phi_{i,j-1}^n\right)\Big/\Delta y\\
G^2 &= \min{}^2(0, D_x^-) + \max{}^2(0, D_x^+) + \min{}^2(0, D_y^-) + \max{}^2(0, D_y^+)\\
\Phi_{i,j}^n &= \Phi_{i,j}^{n-1} + G\Delta t, \quad n = 1, 2, \ldots, (R/\Delta t)
\end{aligned}
$$

where

R is the maximum scale (radius) of interest
$\Delta x, \Delta y$ are the spatial grid spacings
Δt is the time (scale) step

Continuous multiscale morphology using the above curve evolution algorithm for numerically implementing the dilation PDE yields better approximations to disks and avoids the abrupt shape discretization inherent in modeling digital multiscale using discrete polygons [16,57]. Comparing it to discrete multiscale morphology using chamfer distance transforms, we note that for binary images: (1) the chamfer distance transform is easier to implement and yields similar errors for small scale dilations/erosions; (2) implementing the distance transform via curve evolution is more complex, but at medium and large scales gives a better and very close approximation to Euclidean geometry, i.e., to morphological operations with the disk structuring element (see Figure 16.6).

FIGURE 16.6 Distance transforms of a binary image, shown as intensity images modulo 20, obtained using (a) Metric $\|\cdot\|_\infty$ (chamfer metric with local distances (1,1)), (b) chamfer metric with 3×3 neighborhood and local distances (24,34)/25, and (c) curve evolution.

16.9 Applications to Image Processing and Vision

There are numerous applications of morphological image operators to image processing and computer vision. Examples of broad application areas include biomedical image processing, automated visual inspection, character and document image processing, remote sensing, nonlinear filtering, multiscale image analysis, feature extraction, motion analysis, segmentation, and shape recognition. Next we shall review a few of these applications to specific problems of image processing and low/mid-level vision.

16.9.1 Noise Suppression

Rank filters and especially medians have been applied mainly to suppress impulse noise or noise whose probability density has heavier tails than the Gaussian for enhancement of image and other signals [2,12,27,64,65], since they can remove this type of noise without blurring edges, as would be the case for linear filtering. The rank filters have also been used for envelope detection. In their behavior as nonlinear smoothers, as shown in Figure 16.7, the medians act similarly to an "open-closing" $(f \circ B) \bullet B$ by a convex set B of diameter about half the diameter of the median window. The open–closing has the advantages over the median that it requires less computation and decomposes the noise suppression task into two independent steps, i.e., suppressing positive spikes via the opening and negative spikes via the closing. Further, cascading open–closings $\beta_t\alpha_t$ at multiple scales $t = 1, \ldots, r$, where $\alpha_t(f) = f \circ tB$ and

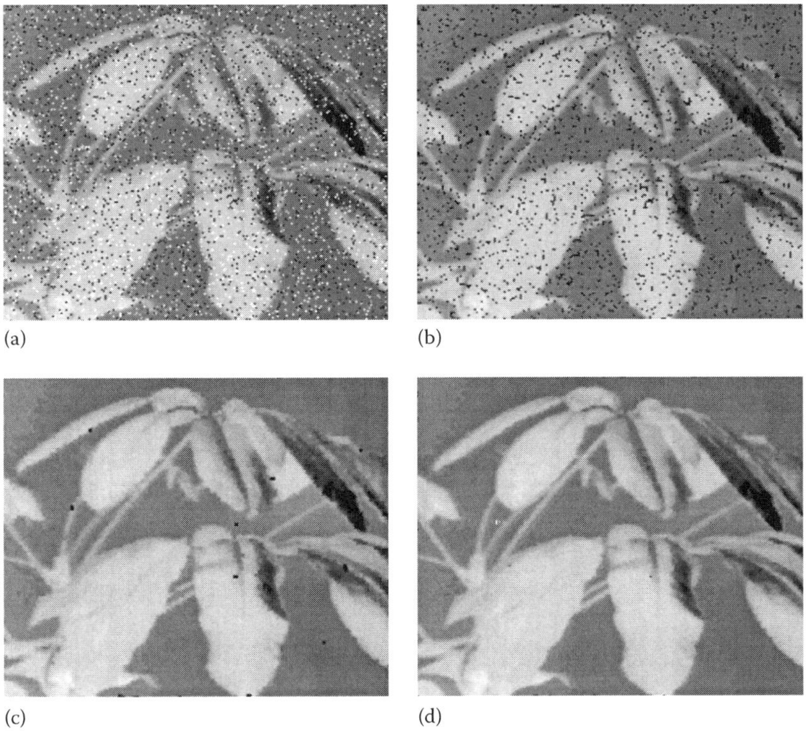

FIGURE 16.7 (a) Noisy image f, corrupted with salt-and-pepper noise of probability 10%. (b) Opening $f \circ B$ of f by a 2×2-pixel square B. (c) Open-closing $(f \circ B) \bullet B$. (d) Median of f by a 3×3-pixel square window.

$\beta_t(f) = f \bullet tB$, generates a class of efficient nonlinear smoothing filters $\beta_r\alpha_r \cdots \beta_2\alpha_2\beta_1\alpha_1$, called alternating sequential filters, which smooth progressively from the smallest scale possible up to a maximum scale r and have a broad range of applications [59,60,62].

16.9.2 Feature Extraction

Residuals between a signal and some morphologically transformed versions of it can extract line- or blob-type features or enhance their contrast. An example is the difference between the flat dilation and erosion of an image f by a symmetric disk-like set B whose diameter, diam (B), is very small;

$$\text{edge}(f) = \frac{(f \oplus B) - (f \ominus B)}{\text{diam}(B)} \qquad (16.54)$$

If f is binary, edge (f) extracts its boundary. If f is gray-level, the above residual enhances its edges [7,58] by yielding an approximation to $\| \nabla f \|$, which is obtained in the limit of Equation 16.54 as diam $(B) \rightarrow 0$ (see Figure 16.8). This morphological edge operator can be made more robust for edge detection by first smoothing the input image signal and compares favorably with other gradient approaches based on linear filtering.

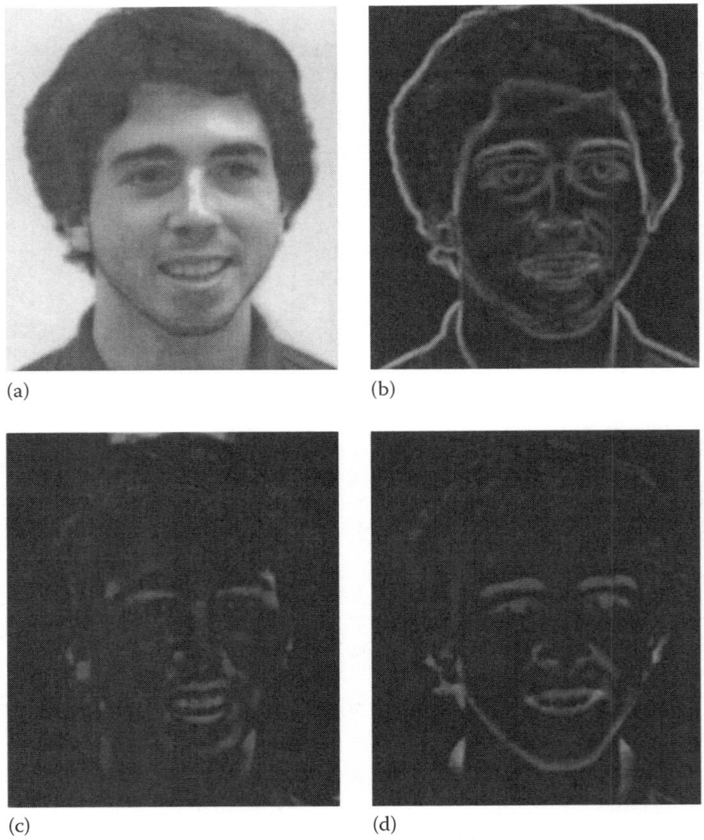

FIGURE 16.8 (a) Image f. (b) Edge enhancement: dilation-erosion residual $f \oplus B - f \ominus B$, where B is a 21-pixel octagon. (c) Peak detection: opening residual $f - f \circ B^{\oplus 3}$. (d) Valley detection: closing residual $f \bullet B^{\oplus 3} - f$.

Another example involves subtracting the opening of a signal f by a compact convex set B from the input signal yields an output consisting of the signal peaks whose support cannot contain B. This is the top-hat transformation [43,58]

$$\text{peak}(f) = f - (f \circ B) \tag{16.55}$$

and can detect bright blobs, i.e., regions with significantly brighter intensities relative to the surroundings. Similarly, to detect dark blobs, modeled as intensity valleys, we can use the closing residual operator $f \mapsto (f \bullet B) - f$ (see Figure 16.8). The morphological peak/valley extractors, in addition to their being simple and efficient, have some advantages over curvature-based approaches.

16.9.3 Shape Representation via Skeleton Transforms

There are applications in image processing and vision where a binary shape needs to be summarized down to its thin medial axis and then reconstructed exactly from this axial information. This process, known as medial axis (or skeleton) transform has been studied extensively for shape representation and description [10,54]. Among many approaches, it can also be obtained via multiscale morphological operators, which offer as a by-product a multiscale representation of the original shape via its skeleton components [39,58]. Let $X \subseteq \mathbb{Z}^2$ represent the foreground of a finite discrete binary image and let $B \subseteq \mathbb{Z}^2$ be a convex disk-like set at scale 1 and $B^{\oplus n}$ be its multiscale version at scale $n = 1, 2, \ldots$ The nth skeleton component of X is the set

$$S_n = (X \ominus B^{\oplus n}) \backslash [(X \ominus B^{\oplus n}) \circ B], \quad n = 0, 1, \ldots, N \tag{16.56}$$

where

$\backslash$ denotes the difference
n is a discrete scale parameter
$N = \max\{n : X \ominus B^{\oplus n} \neq \emptyset\}$ is the maximum scale

The S_n are disjoint subsets of X, whose union is the morphological skeleton of X.

The morphological skeleton transform of X is the finite sequence $(S_0, S_1, \ldots, S_N)$. The union of all the S_n s dilated by a n-scale disk reconstructs exactly the original shape; omitting the first k components leads to a smooth partial reconstruction, the opening of X at scale k:

$$X \circ B^{\oplus k} = \bigcup_{k \leq n \leq N} S_n \oplus B^{\oplus n}, \quad 0 \leq k \leq N \tag{16.57}$$

Thus, we can view the S_n as "shape components," where the small-scale components are associated with the lack of smoothness of the boundary of X, whereas skeleton components of large-scale indices n are related to the bulky interior parts of X that are shaped similarly to $B^{\oplus n}$. Figure 16.9 shows a detailed description of the skeletal decomposition and reconstruction of an image.

Several generalizations or modifications of the morphological skeletonization include: using structuring elements different than disks that might result in fewer skeletal points, or removing redundant points from the skeleton [29,33,39]; using different structuring elements for each skeletonization step [23,33]; using lattice generalizations of the erosions and openings involved in skeletonization [30]; image representation based on skeleton-like multiscale residuals [23]; and shape decomposition based on residuals between image parts and maximal openings [48]. In addition to its general use for shape analysis, a major application of skeletonization has been binary image coding [13,30,39].

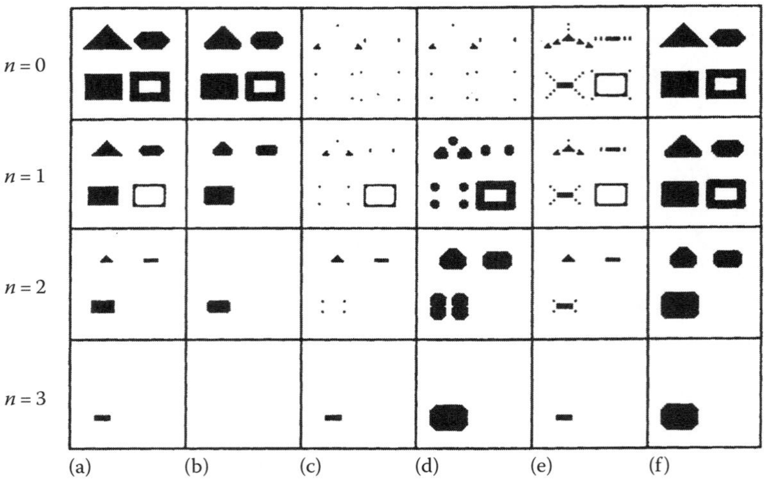

FIGURE 16.9 Morphological skeletonization of a binary image X (top left image) with respect to a 3×3-pixel square structuring element B. (a) Erosions $X \ominus B^{\oplus n}$, $n = 0, 1, 2, 3$. (b) Openings of erosions $(X \ominus B^{\oplus n}) \circ B$. (c) Skeleton subsets S_n. (d) Dilated skeleton subsets $S_n \oplus B^{\oplus n}$. (e) Partial unions of skeleton subsets $\cup_{N \geq k \geq n} S_k$. (f) Partial unions of dilated skeleton subsets $\cup_{N \geq k \geq n} S_k \oplus B^{\oplus k}$.

16.9.4 Shape Thinning

The skeleton is not necessarily connected; for connected skeletons see [3]. Another approach for summarizing a binary shape down to a thin medial axis that is connected but does not necessarily guarantee reconstruction is via thinning. Morphological thinning is defined [58] as the difference between the original set X (representing the foreground of a binary image) and a set of feature locations extracted via hit-miss transformations by pairs of foreground-background probing sets (A_i, B_i) designed to detect features that thicken the shape's axis:

$$X\text{ú}\ \{(A_i, B_i)\}_{i=1}^{n} \equiv X \Big\backslash \bigcup_{i=1}^{n} X \otimes (A_i, B_i) \tag{16.58}$$

Usually each hit-miss by a pair (A_i, B_i) detects a feature at some orientation, and then the difference from the original peels off this feature from X. Since this feature might occur at several orientations, the above thinning operator is applied iteratively by rotating its set of probing elements until there is no further change in the image. Thinning has been applied extensively to character images. Examples are shown in Figure 16.10, where each thinning iteration used $n = 3$ template pairs (A_i, B_i) for the hit-miss transformations of Equation 16.58 designed in [8].

16.9.5 Size Distributions

Multiscale openings $X \mapsto X \circ rB$ and closings $X \mapsto X \bullet rB$ of compact sets X in R^d by convex compact structuring elements rB, parameterized by a scale parameter $r \geq 0$, are called granulometries and can unify all sizing (sieving) operations [42]. Because they satisfy a monotonic ordering

$$\cdots X \circ sB \subseteq X \circ rB \subseteq \cdots \subseteq X \subseteq \cdots X \bullet rB \subseteq X \bullet sB \subseteq \cdots, \quad r < s \tag{16.59}$$

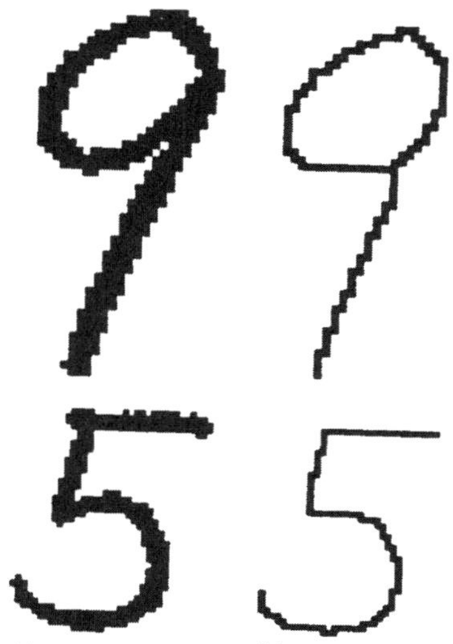

FIGURE 16.10 (a) Shows binary images of handwritten characters. (b) Shows their thinned version.

if we measure the volume (or area) of these sets as a function of scale, this function will also satisfy the same ordering and hence create size distributions. Further, taking its derivative leads to a size density function (or size histogram in the discrete case)

$$h(r) \equiv \begin{cases} -\dfrac{d\ \mathrm{vol}(X \circ rB)}{dr}, & r \geq 0 \\ \dfrac{d\ \mathrm{vol}(X \bullet |r|B)}{d|r|}, & r < 0 \end{cases} \tag{16.60}$$

This conveys several types of information useful for shape description and multiscale image analysis. For example, the boundary roughness of X relative to B manifests itself as contributions in the lower-size part of the size histogram. Long capes or bulky protruding parts in X that consist of patterns sB show up as isolated impulses in the histogram around positive $r = s$. Finally, the size density can be defined for "negative" sizes by using closings instead of openings; in this case impulses at negative sizes indicate the existence of prominent intruding gulfs or holes in X.

If X is a random set [42], then probabilistic measures of its size distribution have been used extensively in image analysis applications to petrography and biology [58]. All of the above ideas can be extended to gray-level images [35]. One application of gray-level size distributions is texture classification [17].

16.9.6 Fractals

A large variety of natural image objects (e.g., clouds, coastlines, mountains, islands, trees, leaves, etc.) can be modeled with fractals [32]. Fractals are mathematical sets with a very high level of geometrical complexity; formally, their Hausdorff dimension is larger than their topological dimension. An important characteristic of fractals to measure for purposes of shape description or classification is their fractal dimension. Among the various methods [32] to estimate the fractal dimension D of the surface of a set $F \subseteq \mathrm{R}^3$, the covering method is based conceptually on Minkowski's idea of finding the area of irregular sets; dilate them with spheres of radius r, find the volume $V(r)$ of the dilated set, and set its area equal to $\lim_{r \downarrow 0} A(r)$, where $A(r) = V(r)/2r$. Further, the fractal dimension of F can be found by

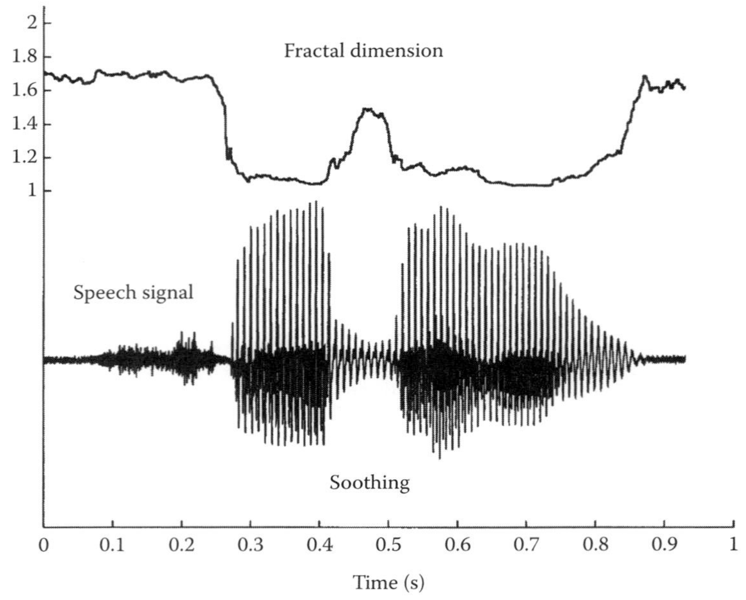

FIGURE 16.11 Speech waveform of the word "soothing" sampled at 10 kHz and its short-time fractal dimension over 10-ms speech segments, computed every 1 ms and post-smoothed by a 3-point median filter. The short-time fractal dimension increases with the amount of turbulence existing during production of the corresponding sound, having a small value for vowels, medium for weak voiced fricatives, and high for unvoiced fricatives.

$$D = \lim_{r \downarrow 0} \frac{\log [V(r)/r^3]}{\log [(1/r)]} \tag{16.61}$$

The intuitive meaning of D is that $V(r) \approx (\text{constant}) \cdot r^{3-D}$ as $r \downarrow 0$, from which D can be estimated by least-squares fitting a straight line to a log-log plot of $V(r)$.

The theory of morphological operators allows us to find more efficient implementations of the above idea when F is the graph of a 2D function $f(x, y)$. Then, instead of multiscale 3D set dilations of F by spheres, it is computationally more efficient to perform 2D multiscale signal dilations and erosions of f by disks rB and measure the multiscale volumes by

$$V(r) = \iint [(f \oplus rB)(x, y) - (f \ominus rB)(x, y)] \mathrm{d}x\mathrm{d}y \tag{16.62}$$

Thus, morphological flat dilations and erosions are used to create a volume-blanket as a layer either covering or being peeled off from the surface of f at various scales. This morphological covering method can also be applied to 1D signals $f(x)$ by replacing volumes with areas and disks rB with horizontal linear segments $[-r, r]$; such a 1D application is shown in Figure 16.11.

16.9.7 Image Segmentation

One of the most powerful and advanced tools of MM is the watershed transformation [7] as applied to image segmentation. Let us regard the gray-level image to be segmented as a topographic relief and assume a drop of water falling at a point on it and flowing down along a steep slope path until it is trapped in a local minimum M of the relief. The set of points such that a drop falling on them eventually reaches M is the catchment basin associated with the minimum M. The union of the boundaries of the different catchment basins of the image constitute its watershed. Thus, the watershed consists of contours located on crest lines separating adjacent minima.

To ease the segmentation of the original image f, the watershed transformation is usually applied to its gradient magnitude $g = \|\nabla f\|$, which has higher contrast. However, direct computation of the watershed of g usually leads to poor results, i.e., oversegmentation of f, because, even after smoothing f or g, the latter often exhibits far too many minima. One of the best solutions to this problem is to use markers for the regions to be extracted. A marker is a small connected component of pixels, a feature, located inside a region. Once the markers have been extracted, the gradient image g is modified via morphological reconstruction so that these markers are imposed as the only minima of the modified function while preserving the highest crest lines of g located between two markers. Then, computing the watershed of the modified g usually provides a good segmentation whose quality depends mainly on the markers and somewhat on g and the initial smoothing of f. An example is shown in Figure 16.12. The power of this approach as well as its difficulty lies in the choice of the markers. Efficient ways to choose markers as well as fast algorithms for the watershed computation are detailed in [44,69]. This watershed methodology has already proved to be very useful in various fields of image analysis, ranging from medical imaging to material sciences, remote sensing, and digital elevation models.

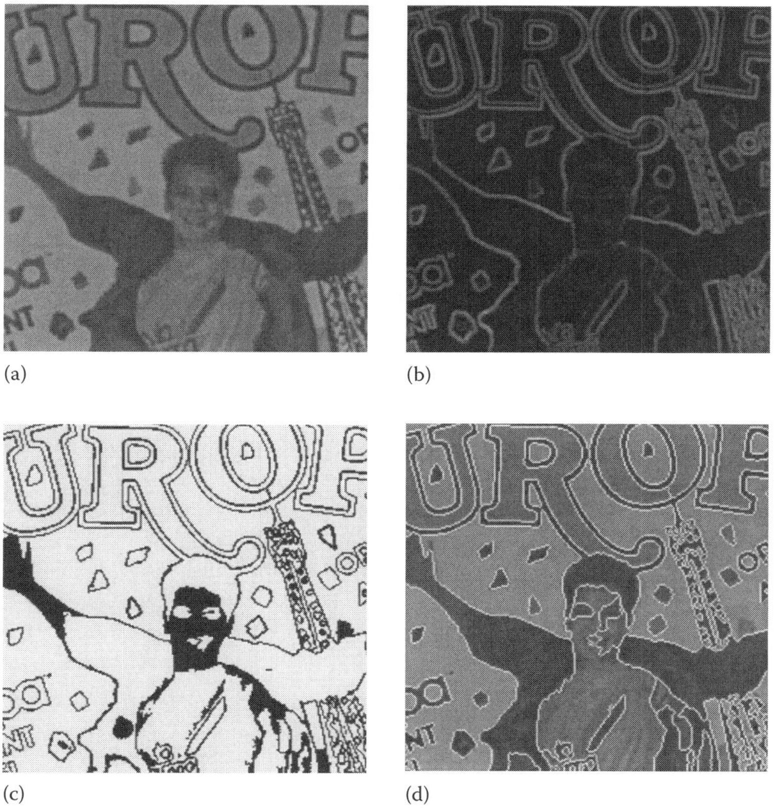

FIGURE 16.12 (a) Image f. (b) Edge enhancement (magnitude of gradient) of f. (c) Markers. (d) Watershed.

16.10 Conclusions

This chapter has provided a brief introduction of the theory of morphological signal processing and its applications to image analysis and nonlinear filtering. This methodology nowadays offers a large diversity of theoretical and algorithmic ideas that provide useful tools and inspire new directions in the following research areas from the fields of signal processing, image processing and machine vision, and pattern recognition: nonlinear filtering, nonlinear signal and system representation, image feature extraction, multiscale analysis and geometry-driven diffusion, image segmentation, region-based image coding, motion analysis, automated visual inspection, and detection/estimation in random sets.

Some attractive aspects of morphological signal operators for efficiently solving problems in the above areas include: (1) suitability for geometry-related signal/image analysis problems; (2) unification power because they can be defined both for numerical signals as well as for more abstract data using their lattice generalizations; (3) simplicity of software or hardware implementations of the basic operators; and (4) existence of efficient algorithms for implementing complex morphological systems [68].

Three current research areas where successful future developments may significantly broaden and improve the applicability of morphological signal processing are: (A) Optimal design of nonlinear systems based on morphological and related signal operators, where, despite their numerous applications, very few ideas exist for their optimal design. (The current three main approaches are: (1) designing binary systems as a finite union of erosions [20,31] or hit-miss operations [5] using the morphological representation theory of [33,34] or [4]; (2) designing stack filters via threshold decomposition and linear programming [18]; (3) gradient-based optimization of morphological/rank filters either via simulated annealing [71] or via a least-mean-square algorithm and adaptive filtering [55].) (B) The continuous (differential) approach to MM via PDEs and exploitation of its exciting relationships to the physics of wave propagation and eikonal optics [38,57,67]. (C) Development of morphological systems for image pattern recognition by exploiting the efficiency of morphological operators for shape analysis and their logic-related structure.

Acknowledgment

This chapter was written while the author's research work was supported by the U.S. National Science Foundation under Grant MIP-94-21677.

References

1. Alvarez, L. and Morel, J.M., Formalization and computational aspects of image analysis, *Acta Numer.*, 1–59, 1994.
2. Arce, G.R., Gallagher, N.C., and Nodes, T.A., Median filters: Theory for one- and two-dimensional filters, in *Advances in Computer Vision and Image Processing,* Vol. 2, Huang, T.S. (Ed.), JAI Press, Greenwich, CT, 1986.
3. Arcelli, C., Cordella, L., and Levialdi, S., From local maxima to connected skeletons, *IEEE Trans. Pattern Anal. Mach. Intell.,* PAMI-3, 134–143, March 1981.
4. Banon, G.J.F. and Barrera, A.J., Minimal representations for translation-invariant set mappings by mathematical morphology, *SIAM J. Appl. Math.,* 51, 1782–1798, Dec. 1991.
5. Barrera, A.J., Salas, B.G.P., and Hashimoto, C.R.F., Set operations on closed intervals and their applications to the programming of Mach's, in *Mathematical Morphology and Its Application to Image and Signal Processing,* Maragos, P., Schafer, R.W., and Butt, M.A. (Eds.), Kluwer Academic Publishing, Atlanta, GA, 1996.
6. Bellman, R. and Karush, W., On the maximum transform, *J. Math. Anal. Appl.,* 6, 67–74, 1963.
7. Beucher, S. and Lantuejoul, C., Use of watersheds in contour detection, *Proceedings of the International Workshop on Image Processing: Real-time Edge & Motion Detection/Estimation,* Rennes, France, 1979.

8. Bloomberg, D.S., Connectivity-preserving morphological image transformations, in *Visual Communications and Image Processing'91,* Tzou, K.-H. and Koga, T. (Eds.), Boston, MA, *Proc. SPIE,* 1606, 1991.
9. Bloomberg, D.S., Multiresolution morphological analysis of document images, in *Visual Communications and Image Processing'92,* Maragos, P. (Ed.), Boston, MA, *Proc. SPIE,* 1818, 648–662, 1992.
10. Blum, H., Biological shape and visual science (part I), *J. Theory Biol.,* 38, 205–287, 1973.
11. Borgefors, G., Distance transformations in digital images, *Comp. Vis. Graph. Image Process.,* 34, 344–371, 1986.
12. Bovik, A.C., Huang, T.S., and Munson, Jr., D.C., A generalization of median filtering using linear combinations of order statistics, *IEEE Trans. Acoust. Speech Signal Process.,* 31, 1342–1349, Dec. 1983.
13. Brandt, J.W., Jain, A.K., and Algazi, V.R., Medial axis representation and encoding of scanned documents, *J. Vis. Commun. Image Represent.,* 2, 151–165, June 1991.
14. Brockett, R.W. and Maragos, P., Evolution equations for continuous-scale morphological filtering, *IEEE Trans. Signal Process.,* 42, 3377–3386, Dec. 1994.
15. Burt, P.J. and Adelson, E.H., The Laplacian pyramid as a compact image code, *IEEE Trans. Commun.,* 31, 532–540, April 1983.
16. Butt, M.A. and Maragos, P., Comparison of multiscale morphology approaches: PDE implemented via curve evolution versus chamfer distance transform, in *Mathematical Morphology and Its Application to Image and Signal Processing,* Maragos, P., Schafer, R.W. and Butt, M.A. (Eds.), Kluwer Academic Publishing, Atlanta, GA, 1996.
17. Chen, Y. and Dougherty, E.R., Gray-scale morphological granulometric texture classification, *Opt. Eng.,* 33, 2713–2722, Aug. 1994.
18. Coyle, E.J. and Lin, J.H., Stack filters and the mean absolute error criterion, *IEEE Trans. Acoust. Speech Signal Process.,* 36, 1244–1254, Aug. 1988.
19. Dorst, L. and van der Boomgaard, R., Morphological signal processing and the slope transform, *Signal Process.,* 38, 79–98, July 1994.
20. Dougherty, E.R., Optimal mean-square *N*-observation digital morphological filters: I. Optimal binary filters, *CVGIP: Image Underst.,* 55, 36–54, Jan. 1992.
21. Fitch, J.P., Coyle, E.J., and Gallagher, Jr., N.C., Median filtering by threshold decomposition, *IEEE Trans. Acoust. Speech, Signal Process.,* 32, 1183–1188, Dec. 1984.
22. Goetcherian, V., From binary to grey tone image processing using fuzzy logic concepts, *Pattern Recogn.,* 12, 7–15, 1980.
23. Goutsias, J. and Shonfeld, D., Morphological representation of discrete and binary images, *IEEE Trans. Signal Process.,* 39, 1369–1379, June 1991.
24. Hadwiger, H., *Vorlesungen über Inhalt, Oberfläche, und Isoperimetrie,* Springer-Verlag, Berlin, 1957.
25. Haralick, R.M., Zhuang, X., Lin, C., and Lee, J.S.J., The digital morphological sampling theorem, *IEEE Trans. Acoust. Speech, Signal Process,* 37, 2067–2090, Dec. 1989.
26. Heijmans, H.J.A.M. and Ronse, C., The algebraic basis of mathematical morphology—part I: Dilations and erosions, *Comput. Vis. Graph. Image Process.,* 50, 245–295, 1990.
27. Justusson, B.I., Median filtering: Statistical properties, in *Two-Dimensional Digital Signal Processing II: Transforms and Median Filters,* Huang, T.S. (Ed.), Springer-Verlag, New York, 1981.
28. Koenderink, J.J., The structure of images, *Biol. Cybern.,* 50, 363–370, 1984.
29. Kresh, R. and Malah, D., Morphological reduction of skeleton redundancy, *Signal Process.,* 38, 143–151, 1994.
30. Kresh, R., Morphological image representation for coding applications, PhD thesis, Technion, Israel, June 1995.
31. Loce, R.P. and Dougherty, E.R., Facilitation of optimal binary morphological filter design via structuring element libraries and design constraints, *Opt. Eng.,* 31, 1008–1025, May 1992.

32. Mandelbrot, B.B., *The Fractal Geometry of Nature,* Freeman, San Francisco, CA, 1982.
33. Maragos, P., A unified theory of translation-invariant systems with applications to morphological analysis and coding of images, PhD thesis, Georgia Institute of Technology, Atlanta, GA, July 1985.
34. Maragos, P., A representation theory for morphological image and signal processing, *IEEE Trans. Pattern Anal. Mach. Intell.,* 11, 586–599, June 1989.
35. Maragos, P., Pattern spectrum and multiscale shape representation, *IEEE Trans. Pattern Anal. Mach. Intell.,* 11, 701–716, July 1989.
36. Maragos, P., Fractal signal analysis using mathematical morphology, in *Advances in Electronics and Electron Physics,* vol. 88, Hawkes, P. and Kazan, B. (Eds.), Academic Press, New York, 1994, pp. 199–246.
37. Maragos, P., Morphological systems: Slope transforms and max–min difference and differential equations, *Signal Process.,* 38, 57–77, July 1994.
38. Maragos, P., Differential morphology and image processing, *IEEE Trans. Image Process.,* 5, 922–937, June 1996.
39. Maragos, P. and Schafer, R.W., Morphological skeleton representation and coding of binary images, *IEEE Trans. Acoust. Speech, Signal Process.,* 34, 1228–1244, Oct. 1986.
40. Maragos, P. and Schafer, R.W., Morphological filters—Part I: Their set-theoretic analysis and relations to linear shift-invariant filters, *IEEE Trans. Acoust. Speech, Signal Process.,* 35, 1153–1169, Aug. 1987.
41. Maragos, P. and Schafer, R.W., Morphological filters—Part II: Their relations to median, order-statistic, and stack filters, *IEEE Trans. Acoust. Speech, Signal Process.,* 35, 1170–1184, Aug. 1987; *ibid,* 37, 597, April 1989.
42. Matheron, G., *Random Sets and Integral Geometry,* John Wiley & Sons, New York, 1975.
43. Meyer, F., Contrast feature extraction, in *Quantitative Analysis of Microstructures in Materials Science, Biology and Medicine*, Chermant, J.L. (Ed.), Special Issues of Practical Metallography, Riederer-Verlag, Stuttgart, Germany, 1978, pp. 374–380.
44. Meyer, F. and Becheur, S., Morphological segmentation, *J. Vis. Commun. Image Represent.,* 1, 21–46, Sept. 1990.
45. Nakagawa, Y. and Rosenfeld, A., A note on the use of local min and max operations in digital picture processing, *IEEE Trans. Syst., Man, Cybern.,* 8, 632–635, 1978.
46. O'Neil, K.S. and Rhodes, W.T., Morphological transformations by hybrid optical-electronic methods, in *Hybrid Image Processing,* Casasent, D. and Tescher, A. (Eds.), Orlando, FL, *Proc. SPIE,* 638, 41–44, 1986.
47. Osher, S. and Sethian, J.A., Fronts propagating with curvature-dependent speed: Algorithms based on Hamilton-Jacobi formulations, *J. Comput. Phys.,* 79, 12–49, 1988.
48. Pitas, I. and Venetsanopoulos, A., Morphological shape decomposition, *IEEE Trans. Pattern Anal. Mach. Intell.,* 12, 38–45, 1990.
49. Preston, Jr., K. and Duff, M.J.B., *Modern Cellular Automata,* Plenum Press, New York, 1984.
50. Rockafellar, R.T., *Convex Analysis,* Princeton University Press, Princeton, NJ, 1972.
51. Roerdink, J.B.T.M., Mathematical morphology with non-commutative symmetry groups, in *Mathematical Morphology in Image Processing,* Dougherty, E.R. (Ed.), Marcel Dekker, New York, 1993.
52. Ronse, C. and Heijmans, H.J.A.M., The algebraic basis of mathematical morphology—Part II: Openings and closings, *CVGIP: Image Underst.,* 54, 74–97, 1991.
53. Rosenfeld, A., Ed., *Multiresolution Image Processing and Analysis,* Springer-Verlag, New York, 1984.
54. Rosenfeld, A. and Kak, A.C., *Digital Picture Processing,* vols. 1 and 2, Academic Press, New York, 1982.
55. Salembier, P., Structuring element adaptation for morphological filters, *J. Vis. Commun. Image Represent.,* 3, 115–136, June 1992.
56. Salembier, P. and Serra, J., Morphological multiscale image segmentation, in *Visual Communications and Image Processing'92,* Maragos, P. (Ed.), *Proc. SPIE,* 1818, 620–631, 1992.

57. Sapiro, G., Kimmel, R., Shaked, D., Kimia, B., and Bruckstein, A., Implementing continuous-scale morphology via curve evolution, *Pattern Recogn.*, 26(9), 1363–1372, 1993.
58. Serra, J., *Image Analysis and Mathematical Morphology,* Academic Press, New York, 1982.
59. Serra, J., Ed., *Image Analysis and Mathematical Morphology, Vol. 2: Theoretical Advances,* Academic Press, New York, 1988.
60. Schonfeld, D. and Goutsias, J., Optimal morphological pattern restoration from noisy binary images, *IEEE Trans. Pattern Anal. Machine Intell.,* 13, 14–29, Jan. 1991.
61. Sternberg, S.R., Cellular computers and biomedical image processing, in *Biomedical Images and Computers,* Sklansky, J. and Bisconte, J.C. (Eds.), Springer-Verlag, Berlin, 1982.
62. Sternberg, S.R., Grayscale morphology, *Comput. Vis. Graph. Image Process.,* 35, 333–355, 1986.
63. Szoplik, T., Ed., *Selected Papers on Morphological Image Processing: Principles and Optoelectronic Implementations*, SPIE Press, Bellingham, WA, 1996.
64. Tukey, J.W., *Exploratory Data Analysis,* Addison-Wesley, Reading, MA, 1977.
65. Tyan, S.G., Median filtering: Deterministic properties, in *Two-Dimensional Digital Signal Processing II: Transforms and Median Filters,* Huang, T.S. (Ed.), Springer-Verlag, New York, 1981.
66. van der Boomgaard, R. and Smeulders, A., The morphological structure of images: The differential equations of morphological scale-space, *IEEE Trans. Pattern Anal. Mach. Intell.,* 16, 1101–1113, Nov. 1994.
67. Verbeek, P.W. and Verwer, B.J.H., Shading from shape, the eikonal equation solved by grey-weighted distance transform, *Pattern Recogn. Lett.,* 11, 618–690, 1990.
68. Vincent, L., Morphological algorithms, in *Mathematical Morphology in Image Processing,* Dougherty, E.R. (Ed.), Marcel Dekker, New York, 1993.
69. Vincent, L. and Soille, P., Watersheds in digital spaces: An efficient algorithm based on immersion simulations, *IEEE Trans. Pattern Anal. Mach. Intell.,* 13, 583–598, June 1991.
70. Wendt, P.D., Coyle, E.J., and Gallagher, N.C., Stack filters, *IEEE Trans. Acoust. Speech Signal Process.,* 34, 898–911, Aug. 1986.
71. Wilson, S.S., Training structuring elements in morphological networks, in *Mathematical Morphology in Image Processing,* Dougherty, E.R. (Ed.), Marcel Dekker, New York, 1993.
72. Witkin, A.P., Scale-space filtering, in *Proceedings of the International Joint Conference on Artificial Intelligence,* Karlsruhe, Germany, 1983.

17

Signal Processing and Communication with Solitons

17.1 Introduction 17-1
17.2 Soliton Systems: The Toda Lattice 17-2
The Inverse Scattering Transform
17.3 New Electrical Analogs for Soliton Systems 17-6
Toda Circuit Model of Hirota and Suzuki • Diode Ladder Circuit Model for Toda Lattice • Circuit Model for Discrete-KdV
17.4 Communication with Soliton Signals 17-9
Low Energy Signaling
17.5 Noise Dynamics in Soliton Systems 17-11
Toda Lattice Small Signal Model • Noise Correlation • Inverse Scattering-Based Noise Modeling
17.6 Estimation of Soliton Signals 17-15
Single-Soliton Parameter Estimation: Bounds • Multisoliton Parameter Estimation: Bounds • Estimation Algorithms • Position Estimation • Estimation Based on Inverse Scattering
17.7 Detection of Soliton Signals 17-20
Simulations
References 17-22

Andrew C. Singer
Sanders (A Lockhead Martin Company)

17.1 Introduction

As we increasingly turn to nonlinear models to capture some of the more salient behavior of physical or natural systems that cannot be expressed by linear means, systems that support solitons may be a natural class to explore because they share many of the properties that make linear time-invariant (LTI) systems attractive from an engineering standpoint. Although nonlinear, these systems are solvable through inverse scattering, a technique analogous to the Fourier transform for linear systems [1]. Solitons are eigenfunctions of these systems which satisfy a nonlinear form of superposition. We can therefore decompose complex solutions in terms of a class of signals with simple dynamical structure. Solitons have been observed in a variety of natural phenomena from water and plasma waves [7,12] to crystal lattice vibrations [2] and energy transport in proteins [7]. Solitons can also be found in a number of man-made media including superconducting transmission lines [11] and nonlinear circuits [6,13]. Recently, solitons have become of significant interest for optical telecommunications, where optical pulses have been shown to propagate as solitons for tremendous distances without significant dispersion [4].

We view solitons from a different perspective. Rather than focusing on the propagation of solitons over nonlinear channels, we consider using these nonlinear systems to both generate and process signals for transmission over traditional linear channels. By using solitons for signal synthesis, the corresponding nonlinear systems become specialized signal processors which are naturally suited to a number of complex signal processing tasks. This section can be viewed as an exploration of the properties of solitons as signals. In the process, we explore the potential application of these signals in a multiuser wireless communication context. One possible benefit of such a strategy is that the soliton signal dynamics provide a mechanism for simultaneously decreasing transmitted signal energy and enhancing communication performance.

17.2 Soliton Systems: The Toda Lattice

The Toda lattice is a conceptually simple mechanical example of a nonlinear system with soliton solutions.* It consists of an infinite chain of masses connected with springs satisfying the nonlinear force law $f_n = a(e^{-b(y_n - y_{n-1})} - 1)$, where f_n is the force on the spring between masses with displacements y_n and y_{n-1} from their rest positions. The equations of motion for the lattice are given by

$$m\ddot{y}_n = a(e^{-b(y_n - y_{n-1})} - e^{-b(y_{n+1} - y_n)}), \tag{17.1}$$

where
- m is the mass
- a and b are constants

This equation admits pulse-like solutions of the form

$$f_n(t) = \left(\frac{m}{ab}\right)\beta^2 \text{sech}^2(\sinh^{-1}(\sqrt{m/ab}\beta)n - \beta t), \tag{17.2}$$

which propagate as compressional waves stored as forces in the nonlinear springs. A single right-traveling wave $f_n(t)$ is shown in Figure 17.1a.

This compressional wave is localized in time, and propagates along the chain maintaining constant shape and velocity. The parameter β appears in both the amplitude and the temporal- and spatial-scales of this one parameter family of solutions giving rise to tall, narrow pulses which propagate faster than small, wide pulses. This type of localized pulse-like solution is often referred to as a "solitary wave."

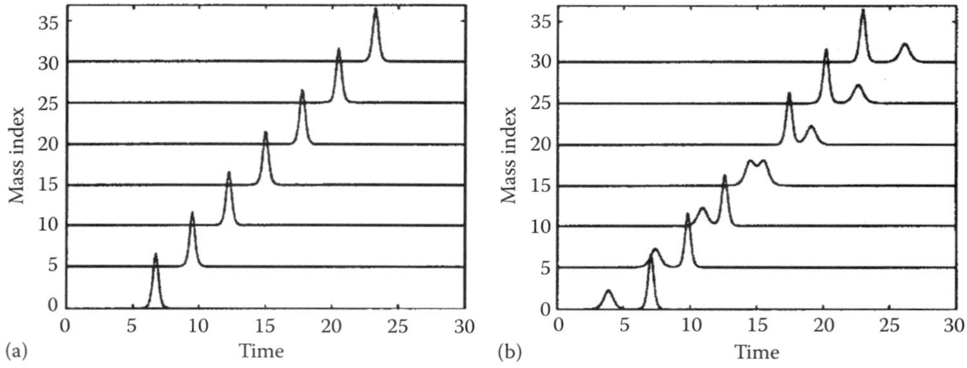

FIGURE 17.1 Propagating wave solutions to the Toda lattice equations. Each trace corresponds to the force $f_n(t)$ stored in the spring between mass n and $n-1$. (a) Solitary wave. (b) Two solitons.

* A comprehensive treatment of the lattice and its associated soliton theory can be found in the monograph by Toda [18].

The study of solitary wave solutions to nonlinear equations dates back to the work of John Scott Russell in 1834 and perhaps the first recorded sighting of a solitary wave. Scott Russell's observations of an unusual water wave in the Union Canal near Edinburgh, Scotland, are interpreted as a solitary wave solution to the Korteweg deVries (KdV) equation [12].* In a 1965 paper, Zabusky and Kruskal performed numerical experiments with the KdV equation and noticed that these solitary wave solutions retained their identity upon collision with other solitary waves, which prompted them to coin the term "soliton" implying a particle-like nature. The ability to form solutions to an equation from a superposition of simpler solutions is the type of behavior we would expect for linear wave equations. However, that nonlinear equations such as the KdV or Toda lattice equations permit such a form of superposition is an indication that they belong to a rather remarkable class of nonlinear systems.

An example of this form of soliton superposition is illustrated in Figure 17.1b for two solutions of the form of Equation 17.2. Note that as a function of time, a smaller, wider soliton appears before a taller, narrower one. However, as viewed by, e.g., the 30th mass in the lattice, the larger soliton appears first as a function of time. Since the larger soliton has arrived at this node before the smaller soliton, it has therefore traveled faster. Note that when the larger soliton catches up to the smaller soliton as viewed on the fifteenth node, the combined amplitude of the two solitons is actually less than would be expected for a linear system, which would display a linear superposition of the two amplitudes. Also, the signal shape changes significantly during this nonlinear interaction.

An analytic expression for the two-soliton solution for $\beta_1 > \beta_2 > 0$ is given by [6]

$$f_n(t) = \frac{m}{ab} \frac{\beta_1^2 \operatorname{sech}^2(\eta_1) + \beta_2^2 \operatorname{sech}^2(\eta_2) + A \operatorname{sech}^2(\eta_1)\operatorname{sech}^2(\eta_2)}{(\cosh(\phi/2) + \sinh(\phi/2)\tanh(\eta_1)\tanh(\eta_2))^2}, \tag{17.3}$$

where

$$\begin{aligned}
A &= \sinh(\phi/2)\left((\beta_1^2 + \beta_2^2)\sinh(\phi/2) + 2\beta_1\beta_2\cosh(\phi/2)\right) \\
\phi &= \ln\left(\frac{\sinh((p_1 - p_2)/2)}{\sinh((p_1 + p_2)/2)}\right) \\
\beta_i &= \sqrt{ab/m}\,\sinh(p_i) \\
\eta_i &= p_i n - \beta_i(t - \delta_i)
\end{aligned} \tag{17.4}$$

Although Equation 17.3 appears rather complex, Figure 17.1b illustrates that for large separations, $|\delta_1 - \delta_2|$, $f_n(t)$ essentially reduces to the linear superposition of two solitons with parameters β_1 and β_2. As the relative separation decreases, the multiplicative cross term becomes significant, and the solitons interact nonlinearly. This asymptotic behavior can also be evidenced analytically

$$\begin{aligned}
f_n(t) &= \frac{m}{ab}\beta_1^2 \operatorname{sech}^2(p_1 n - \beta_1(t - \delta_1) \pm \phi/2) \\
&\quad + \frac{m}{ab}\beta_2^2 \operatorname{sech}^2(p_2 n - \beta_2(t - \delta_2) \mp \phi/2), \quad t \to \pm\infty,
\end{aligned} \tag{17.5}$$

where each component soliton experiences a net displacement ϕ from the nonlinear interaction. The Toda lattice also admits periodic solutions which can be written in terms of Jacobian elliptic functions [18].

* A detailed discussion of linear and nonlinear wave theory including KdV can be found in [21].

An interesting observation can be made when the Toda lattice equations are written in terms of the forces,

$$\frac{d^2}{dt^2}\ln\left(1+\frac{f_n}{a}\right)=\frac{b}{m}(f_{n+1}-2f_n+f_{n-1}). \tag{17.6}$$

If the substitution $f_n(t)=\frac{d^2}{dt^2}\ln\phi_n(t)$ is made into Equation 17.6, then the lattice equations become

$$\frac{m}{ab}\left(\dot{\phi}_n^2-\phi_n\ddot{\phi}_n\right)=\phi_n^2-\phi_{n-1}\phi_{n+1}. \tag{17.7}$$

In view of the Teager energy operator introduced by Kaiser in [8], the left-hand side of Equation 17.7 is the Teager instantaneous-time energy at the node n, and the right-hand side is the Teager instantaneous-space energy at time t. In this form, we may view solutions to Equation 17.7 as propagating waveforms that have equal Teager energy as calculated in time and space, a relationship also observed by Kaiser [9].

17.2.1 The Inverse Scattering Transform

Perhaps the most significant discovery in soliton theory was that under a rather general set of conditions, certain nonlinear evolution equations such as KdV or the Toda lattice could be solved analytically. That is, given an initial condition of the system, the solution can be explicitly determined for all time using a technique called inverse scattering. Since much of inverse scattering theory is beyond the scope of this section, we will only present some of the basic elements of the theory and refer the interested reader to [1].

The nonlinear systems that have been solved by inverse scattering belong to a class of systems called conservative Hamiltonian systems. For the nonlinear systems that we discuss in this section, an integral component of their solution via inverse scattering lies in the ability to write the dynamics of the system implicitly in terms of an operator differential equation of the form

$$\frac{dL(t)}{dt}=B(t)L(t)-L(t)B(t), \tag{17.8}$$

where

$L(t)$ is a symmetric linear operator
$B(t)$ is an antisymmetric linear operator

Both $L(t)$ and $B(t)$ depend explicitly on the state of the system.

Using the Toda lattice as an example, the operators L and B would be the symmetric and antisymmetric tridiagonal matrices

$$L=\begin{bmatrix}\ddots & a_{n-1} & \\ a_{n-1} & b_n & a_n \\ & a_n & \ddots\end{bmatrix},\quad B=\begin{bmatrix}\ddots & -a_{n-1} & \\ a_{n-1} & 0 & -a_n \\ & a_n & \ddots\end{bmatrix}, \tag{17.9}$$

where $a_n=e^{(y_n-y_{n+1})/2}/2$, and $b_n=\dot{y}_n/2$, for mass positions y_n in a solution to Equation 17.1. Written in this form, the entries of the matrices in Equation 17.8 yield the following equations

$$\begin{aligned}\dot{a}_n&=a_n(b_n-b_{n+1}),\\ \dot{b}_n&=2\left(a_{n-1}^2-a_n^2\right).\end{aligned} \tag{17.10}$$

These are equivalent to the Toda lattice equations, Equation 17.1, in the coordinates a_n and b_n. Lax has shown [10] that when the dynamics of such a system can be written in the form of Equation 17.8, then the eigenvalues of the operator $L(t)$ are time-invariant, i.e., $\dot{\lambda}=0$. Although each of the entries of $L(t)$, $a_n(t)$, and $b_n(t)$ evolve with the state of a solution to the Toda lattice, the eigenvalues of $L(t)$ remain constant.

If we assume that the motion on the lattice is confined to lie within a finite region of the lattice, i.e., the lattice is at rest for $|n| \to \infty$, then the spectrum of eigenvalues for the matrix $L(t)$ can be separated into two sets. There is a continuum of eigenvalues $\lambda \in [-1, 1]$ and a discrete set of eigenvalues for which $|\lambda_k| > 1$. When the lattice is at rest, the eigenvalues consist only of the continuum. When there are solitons in the lattice, one discrete eigenvalue will be present for each soliton excited. This separation of eigenvalues of $L(t)$ into discrete and continuous components is common to all of the nonlinear systems solved with inverse scattering.

The inverse scattering method of solution for soliton systems is analogous to methods used to solve linear evolution equations. For example, consider a linear evolution equation for the state $y(x, t)$. Given an initial condition of the system, $y(x, 0)$, a standard technique for solving for $y(x, t)$ employs Fourier methods. By decomposing the initial condition into a superposition of simple harmonic waves, each of the component harmonic waves can be independently propagated. Given the Fourier decomposition of the state at time t, the harmonic waves can then be recombined to produce the state of the system $y(x, t)$. This process is depicted schematically in Figure 17.2a.

An outline of the inverse scattering method for soliton systems is similar. Given an initial condition for the nonlinear system, $y(x, 0)$, the eigenvalues λ and eigenfunctions $\psi(x, 0)$ of the linear operator $L(0)$ can be obtained. This step is often called "forward scattering" by analogy to quantum mechanical scattering, and the collection of eigenvalues and eigenfunctions is called the nonlinear spectrum of the system in analogy to the Fourier spectrum of linear systems. To obtain the nonlinear spectrum at a point in time t, all that is needed is the time evolution of the eigenfunctions, since the eigenvalues do not change with time. For these soliton systems, the eigenfunctions evolve simply in time, according to linear differential equations. Given the eigenvalue-eigenfunction decomposition of $L(t)$, through a process called "inverse scattering," the state of the system $y(x, t)$ can be completely reconstructed. This process is depicted in Figure 17.2b in a similar fashion to the linear solution process.

For a large class of soliton systems, the inverse scattering method generally involves solving either a linear integral equation or a linear discrete-integral equation. Although the equation is linear, finding its solution is often very difficult in practice. However, when the solution is made up of pure solitons, then the integral equation reduces a set of simultaneous linear equations.

Since the discovery of the inverse scattering method for the solution to KdV, there has been a large class of nonlinear wave equations, both continuous and discrete, for which similar solution methods have been obtained. In most cases, solutions to these equations can be constructed from a nonlinear superposition of soliton solutions. For a comprehensive study of inverse scattering and equations solvable by this method, the reader is referred to the text by Ablowitz and Clarkson [1].

$$y(x, 0) \xrightarrow{\text{F.T.}} k, Y(k, 0) \qquad\qquad y(x, 0) \xrightarrow{\text{F.S.}} \lambda, \psi(x, 0)$$
$$\downarrow e^{-j\omega(k)t} \qquad\qquad\qquad\qquad\qquad \downarrow e^{j\beta(k)t}$$
$$y(x, t) \xleftarrow{\text{I.F.T.}} k, Y(k, t) \qquad\qquad y(x, t) \xleftarrow{\text{I.S.}} \lambda, \psi(x, t)$$

(a) \qquad\qquad\qquad\qquad\qquad (b)

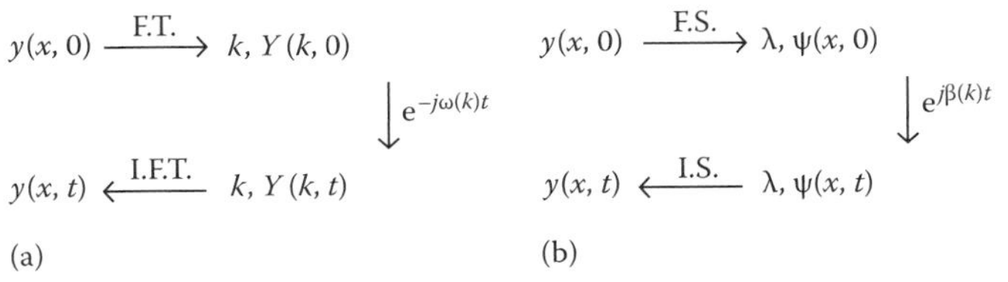

FIGURE 17.2 Schematic solution to evolution equations: (a) linear and (b) soliton.

17.3 New Electrical Analogs for Soliton Systems

Since soliton theory has its roots in mathematical physics, most of the systems studied in the literature have at least some foundation in physical systems in nature. For example, KdV has been attributed to studies ranging from ion-acoustic waves in plasma [22] to pressure waves in liquid gas bubble mixtures [12]. As a result, the predominant purpose of soliton research has been to explain physical properties of natural systems. In addition, there are several examples of man-made media that have been designed to support soliton solutions and thus exploit their robust propagation. The use of optical fiber solitons for telecommunications and of Josephson junctions for volatile memory cells are two practical examples [11,12].

Whether its goal has been to explain natural phenomena or to support propagating solitons, this research has largely focused on the properties of propagating solitons through these nonlinear systems. In this section, we will view solitons as signals and consider exploiting some of their rich signal properties in a signal processing or communication context. This perspective is illustrated graphically in Figure 17.3, where a signal containing two solitons is shown as an input to a soliton system which can either combine or separate the component solitons according to the evolution equations. From the "solitons-as-signals" perspective, the corresponding nonlinear evolution equations can be viewed as special-purpose signal processors that are naturally suited to such signal processing tasks as signal separation or sorting. As we shall see, these systems also form an effective means of generating soliton signals.

17.3.1 Toda Circuit Model of Hirota and Suzuki

Motivated by the work of Toda on the exponential lattice, the nonlinear LC ladder network implementation shown in Figure 17.4 was given by Hirota and Suzuki in [6]. Rather than a direct analogy to the Toda lattice, the authors derived the functional form of the capacitance required for the LC line to be equivalent. The resulting network equations are given by

$$\frac{d^2}{dt^2}\ln\left(1+\frac{V_n(t)}{V_0}\right)=\frac{1}{LC_0V_0}(V_{n-1}(t)-2V_n(t)+V_{n+1}(t)), \tag{17.11}$$

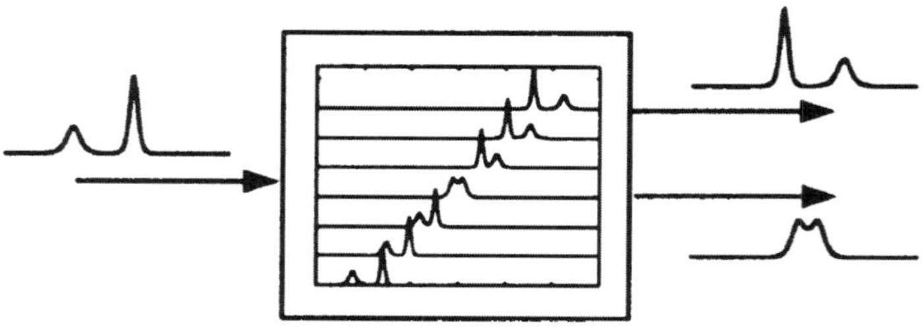

FIGURE 17.3 Two-soliton signal processing by a soliton system.

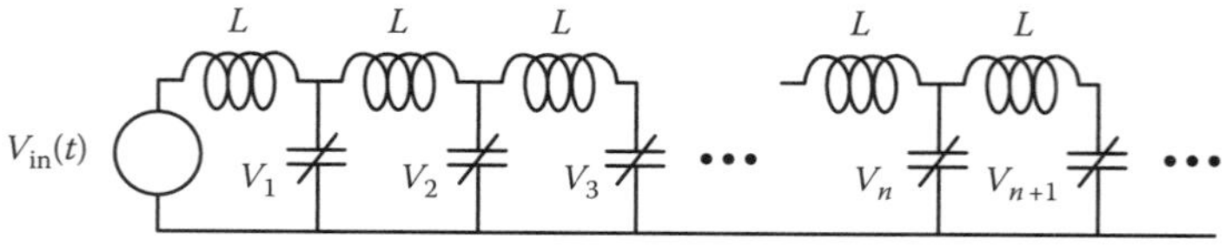

FIGURE 17.4 Nonlinear LC ladder circuit of Hirota and Suzuki.

which is equivalent to the Toda lattice equation for the forces on the nonlinear springs given in Equation 17.6. The capacitance required in the nonlinear LC ladder is of the form

$$C(V) = \frac{C_0 V_0}{V_0 + V}, \tag{17.12}$$

where V_0 and C_0 are constants representing the bias voltage and the nominal capacitance, respectively. Unfortunately, such a capacitance is rather difficult to construct from standard components.

17.3.2 Diode Ladder Circuit Model for Toda Lattice

In [14], the circuit model shown in Figure 17.5a is presented which accurately matches the Toda lattice and is a direct electrical analog of the nonlinear spring mass system. When the shunt impedance Z_n has the voltage–current relation $\ddot{v}_n(t) = \alpha(i_n(t) - i_{n+1}(t))$, then the governing equations become

$$\frac{d^2 v_n(t)}{dt^2} = \alpha I_s (e^{(v_{n-1}(t) - v_n(t))/v_t} - e^{(v_n(t) - v_{n+1}(t))/v_t}), \tag{17.13}$$

or,

$$\frac{d^2}{dt^2} \ln\left(1 + \frac{i_n(t)}{I_s}\right) = \frac{\alpha}{v_t}(i_{n-1}(t) - 2i_n(t) + i_{n+1}(t)), \tag{17.14}$$

where $i_1(t) = i_{\text{in}}(t)$. These are equivalent to the Toda lattice equations with $a/m = \alpha I_s$ and $b = 1/v_t$. The required shunt impedance is often referred to as a double capacitor, which can be realized using ideal operational amplifiers in the gyrator circuit shown in Figure 17.5b, yielding the required impedance of $Z_n = \alpha/s^2 = R_3/R_1R_2C^2s^2$ [13].

This circuit supports a single-soliton solution of the form

$$i_n(t) = \beta^2 \text{sech}^2(pn - \beta\tau), \tag{17.15}$$

where

$$\beta = \sqrt{I_s} \sinh(p)$$

$$\tau = t\sqrt{\alpha/v_t}$$

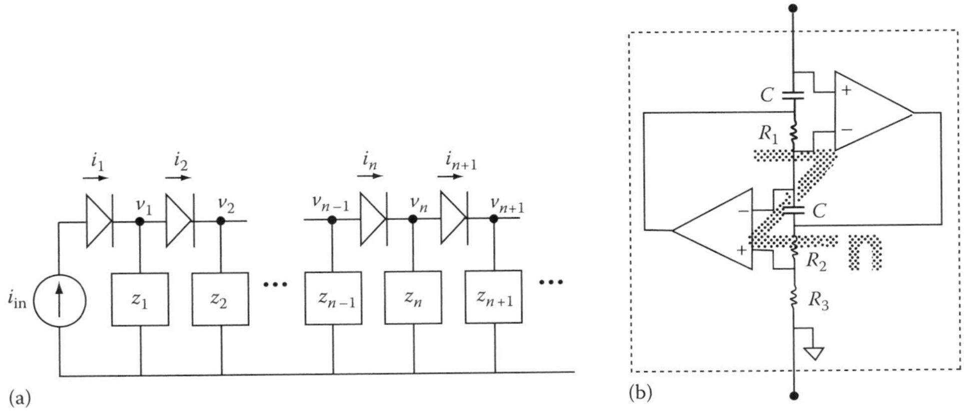

FIGURE 17.5 Diode ladder network in (a), with z_n realized with a double capacitor as shown in (b).

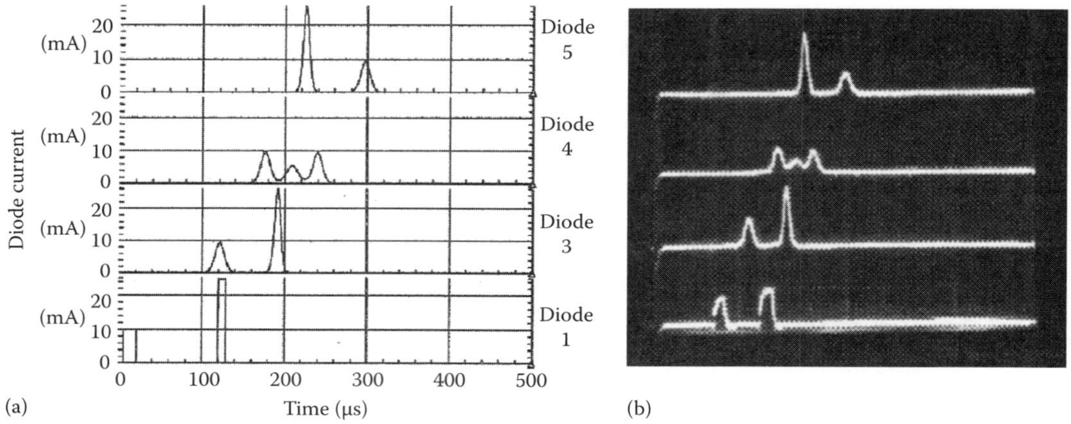

FIGURE 17.6 Evolution of a two-soliton signal through the diode lattice. Each horizontal trace shows the current through one of the diodes 1, 3, 4, and 5. (a) Hspice simulation. (b) Oscilloscope traces.

The diode ladder circuit model is very accurate over a large range of soliton wavenumbers, and is significantly more accurate than the LC circuit of Hirota and Suzuki. Shown in Figure 17.6a is an HSPICE simulation with two solitons propagating in the diode ladder circuit.

As illustrated in the bottom trace of Figure 17.6a, a soliton can be generated by driving the circuit with a square pulse of approximately the same area as the desired soliton. As seen on the third node in the lattice, once the soliton is excited, the nonsoliton components rapidly become insignificant.

A two-soliton signal generated by a hardware implementation of this circuit is shown on the oscilloscope traces in Figure 17.6b. The bottom trace in the figure corresponds to the input current to the circuit, and the remaining traces, from bottom to top, show the current through the third, fourth, and fifth diodes in the lattice.

17.3.3 Circuit Model for Discrete-KdV

The discrete-KdV equation (dKdV), sometimes referred to as the nonlinear ladder equations [1], or the Kac and van Moerbeke system (KM) [17] is governed by the equation

$$\dot{u}_n(t) = e^{u_{n-1}(t)} - e^{u_{n+1}(t)}. \tag{17.16}$$

In [14], the circuit shown in Figure 17.7, is shown to be governed by the dKdV equation

$$\dot{v}_n(t) = \frac{I_s}{C}\left(e^{v_{n-1}(t)/v_t} - e^{v_{n+1}(t)/v_t}\right), \tag{17.17}$$

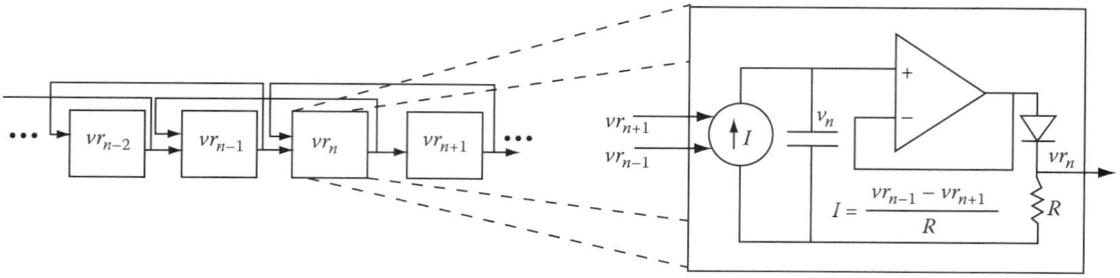

FIGURE 17.7 Circuit model for dKdV.

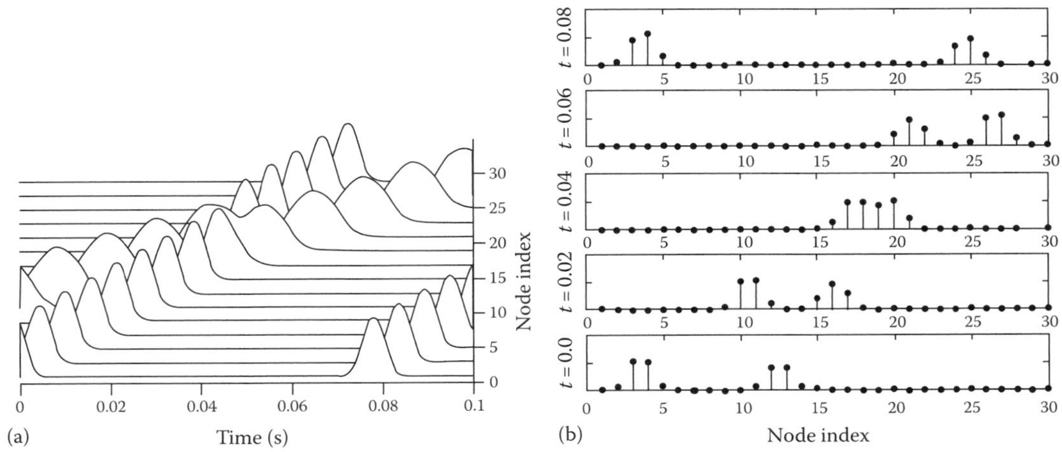

FIGURE 17.8 (a) The normalized node capacitor voltages, $v_n(t)/v_t$ for each node is shown as a function of time. (b) The state of the circuit is shown as a function of node index for five different sample times. The bottom trace in the figure corresponds to the initial condition.

where

I_s is the saturation current of the diode

C is the capacitance

v_t is the thermal voltage

Since this circuit is first-order, the state of the system is completely specified by the capacitor voltages.

Rather than processing continuous-time signals as with the Toda lattice system, we can use this system to process discrete-time solitons as specified by v_n. For the purposes of simulation, we consider the periodic dKdV equation by setting $v_{n+1}(t) = v_0(t)$ and initializing the system with the discrete-time signal corresponding to a listing of node capacitor voltages. We can place a multisoliton solution in the circuit using inverse scattering techniques to construct the initial voltage profile. The single-soliton solution to the dKdV system is given by

$$v_n(t) = \ln\left(\frac{\cosh(\gamma(n-2)-\beta t)\cosh(\gamma(n+1)-\beta t)}{\cosh(\gamma(n-1)-\beta t)\cosh(\gamma n-\beta t)}\right), \quad (17.18)$$

where $\beta = \sinh(2\gamma)$. Shown in Figure 17.8, is the result of an HSPICE simulation of the circuit with 30 nodes in a loop configuration.

17.4 Communication with Soliton Signals

Many traditional communication systems use a form of sinusoidal carrier modulation, such as amplitude modulation (AM) or frequency/phase modulation (FM/PM) to transmit a message-bearing signal over a physical channel. The reliance upon sinusoidal signals is due in part to the simplicity with which such signals can be generated and processed using linear systems. More importantly, information contained in sinusoidal signals with different frequencies can easily be separated using linear systems or Fourier techniques. The complex dynamic structure of soliton signals and the ease with which these signals can be both generated and processed with analog circuitry renders them potentially applicable in the broad context of communication in an analogous manner to sinusoidal signals.

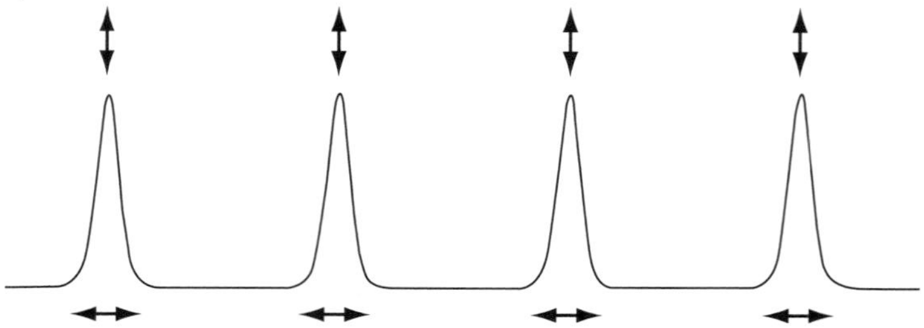

FIGURE 17.9 Modulating the relative amplitude or position of soliton carrier signal for the Toda lattice.

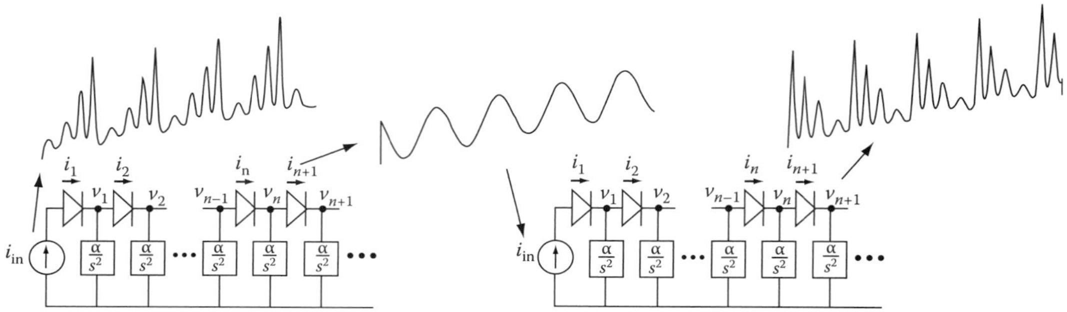

FIGURE 17.10 Multiplexing of a four-soliton solution to the Toda lattice.

We define a soliton carrier as a signal that is composed of a periodically repeated single-soliton solution to a particular nonlinear system. For example, a soliton carrier signal for the Toda lattice is shown in Figure 17.9. As a Toda lattice soliton carrier is generated, a simple AM scheme could be devised by slightly modulating the soliton parameter β, since the amplitude of these solitons is proportional to β^2. Similarly, an analog of FM or pulse-position modulation could be achieved by modulating the relative position of each soliton in a given period, as shown in Figure 17.9.

As a simple extension, these soliton modulation techniques can be generalized to include multiple solitons in each period and accommodate multiple information-bearing signals, as shown in Figure 17.10 for a four-soliton example using the Toda lattice circuits presented in [14]. In the figure, a signal is generated as a periodically repeated train of four solitons of increasing amplitude. The relative amplitudes or positions of each of the component solitons could be independently modulated about their nominal values to accommodate multiple information signals in a single-soliton carrier.

The nominal soliton amplitudes can be appropriately chosen so that as this signal is processed by the diode ladder circuit, the larger amplitude solitons propagate faster than the smaller solitons, and each of the solitons can become nonlinearly superimposed as viewed at a given node in the circuit. From an input–output perspective, the diode ladder circuit can be used to make each of the solitons coincidental in time. As indicated in the figure, this packetized soliton carrier could then be transmitted over a wireless communication channel. At the receiver, the multisoliton signal can be processed with an identical diode ladder circuit which is naturally suited to perform the nonlinear signal separation required to demultiplex the multiple soliton carriers. As the larger amplitude solitons emerge before the smaller, after a given number of nodes, the original multisoliton carrier re-emerges from the receiver in amplitude-reversed order. At this point, each of the component soliton carriers could be demodulated to recover the individual message signals it contains. Aside from a packetization of the component solitons, we will see that multiplexing the soliton carriers in this fashion can lead to an increased energy efficiency for such carrier modulation schemes, making such techniques particularly attractive for a broad range of portable wireless and power-limited communication applications.

Since the Toda lattice equations are symmetric with respect to time and node index, solitons can propagate in either direction. As a result, a single diode ladder implementation could be used as both a modulator and demodulator simultaneously. Since the forward propagating solitons correspond to positive eigenvalues in the inverse scattering transform and the reverse propagating solitons have negative eigenvalues, the dynamics of the two signals will be completely decoupled.

A technique for modulation of information on soliton carriers was also proposed by Hirota et al. [15,16]. In their work, an amplitude and phase modulation of a two-soliton solution to the Toda lattice were presented as a technique for private communication. Although their signal generation and processing methods relied on an inexact phenomenon known as recurrence, the modulation paradigm they presented is essentially a two-soliton version of the carrier modulation paradigm presented in [14].

17.4.1 Low Energy Signaling

A consequence of some of the conservation laws satisfied by the Toda lattice is a reduction of energy in the transmitted signal for the modulation techniques of this section. In fact, as a function of the relative separation of two solitons, the minimum energy of the transmitted signal is obtained precisely at the point of overlap. This can be shown [14] for the two-soliton case by analysis of the form of the equation for the energy in the waveform, $v(t) = f_n(t)$,

$$E = \int_{-\infty}^{\infty} v(t; \delta_1, \delta_2)^2 dt, \qquad (17.19)$$

where $v(t; \delta_1, \delta_2)$ is given in Equation 17.3. In [14] it is proven that E is exactly minimized when $\delta_1 = \delta_2$, i.e., the two solitons are mutually colocated. Significant energy reduction can be achieved for a fairly wide range of separations and amplitudes, indicating that the modulation techniques described here could take advantage of this reduction.

17.5 Noise Dynamics in Soliton Systems

In order to analyze the modulation techniques presented here, accurate models are needed for the effects of random fluctuations on the dynamics of soliton systems. Such disturbances could take the form of additive or convolutional corruption incurred during terrestrial or wired transmission, circuit thermal noise, or modeling errors due to system deviation from the idealized soliton dynamics. A fundamental property of solitons is that they are stable in the presence of a variety of disturbances.

With the development of the inverse scattering framework and the discovery that many soliton systems were conservative Hamiltonian systems, many of the questions regarding the stability of soliton solutions are readily answered. For example, since the eigenvalues of the associated linear operator remain unchanged under the evolution of the dynamics, then any solitons that are initially present in a system must remain present for all time, regardless of their interactions. Similarly, the dynamics of any nonsoliton components that are present in the system are uncoupled from the dynamics of the solitons. However, in the communication scenario discussed in [14], soliton waveforms are generated and then propagated over a noisy channel. During transmission, these waveforms are susceptible to additive corruption from the channel. When the waveform is received and processed, the inverse scattering framework can provide useful information about the soliton and noise content of the received waveform.

In this section, we will assume that soliton signals generated in a communication context have been transmitted over an additive white Gaussian noise (AWGN) channel. We can then consider the effects of additive corruption on the processing of soliton signals with their nonlinear evolution equations.

Two general approaches are taken to this problem. The first approach primarily deals with linearized models and investigates the dynamic behavior of the noise component of signals composed of an information bearing soliton signal and additive noise. The second approach is taken in the framework of inverse scattering and is based on some results from random matrix theory. Although the analysis techniques developed here are applicable to a large class of soliton systems, we focus our attention on the Toda lattice as an example.

17.5.1 Toda Lattice Small Signal Model

If a signal that is processed in a Toda lattice receiver contains only a small amplitude noise component, then the dynamics of the receiver can be approximated by a small signal model,

$$\frac{d^2 V_n(t)}{dt^2} = \frac{1}{LC}(V_{n-1}(t) - 2V_n(t) + V_{n+1}(t)), \tag{17.20}$$

when the amplitude of $V_n(t)$ is appropriately small.

If we consider processing signals with an infinite linear lattice and obtain an input–output relationship where a signal is input at the zeroth node and the output is taken as the voltage on the Nth node, it can be shown that the input–output frequency response of the system can be given by

$$H_N(j\omega) = \begin{cases} e^{-2j\sin^{-1}(\omega\sqrt{LC}/2)N}, & |\omega| < 2/\sqrt{LC}, \\ e^{[j\pi - 2\cosh^{-1}(\omega\sqrt{LC}/2)]N}, & \text{else}, \end{cases} \tag{17.21}$$

which behaves as a low pass filter, and for $N \gg 1$, approaches

$$|H_N(j\omega)|^2 = \begin{cases} 1, & |\omega| < \omega_c = 2/\sqrt{LC}, \\ 0, & \text{else}. \end{cases} \tag{17.22}$$

Our small signal model indicates that in the absence of solitons in the received signal, small amplitude noise will be processed by a low pass filter. If the received signal also contains solitons, then the small signal model of Equation 17.20 will no longer hold. A linear small signal model can still be used if we linearize Equation 17.11 about the known soliton signal. Assuming that the solution contains a single soliton in small amplitude noise, $V_n(t) = S_n(t) + v_n(t)$, we can write Equation 17.11 as an exact equation that is satisfied by the nonsoliton component

$$\frac{d^2}{dt^2}\ln\left(1 + \frac{v_n(t)}{1 + S_n(t)}\right) = \frac{1}{LC}(v_{n-1}(t) - 2v_n(t) + v_{n+1}(t)), \tag{17.23}$$

which can be viewed as the fully nonlinear model with a time-varying parameter, $(1 + S_n(t))$. As a result, over short time scales relative to $S_n(t)$, we would expect this model to behave in a similar manner to the small signal model of Equation 17.20. With $v_n(t) \ll (1 + S_n(t))$, we obtain

$$\frac{d^2}{dt^2}\frac{v_n(t)}{1 + S_n(t)} \approx \frac{1}{LC}(v_{n-1}(t) - 2v_n(t) + v_{n+1}(t)). \tag{17.24}$$

When the contribution from the soliton is small, Equation 17.24 reduces to the linear system of Equation 17.20. We would therefore expect that both before and after a soliton has passed through the lattice, the system essentially low pass filters the noise. However, as the soliton is processed, there will be a time-varying component to the filter.

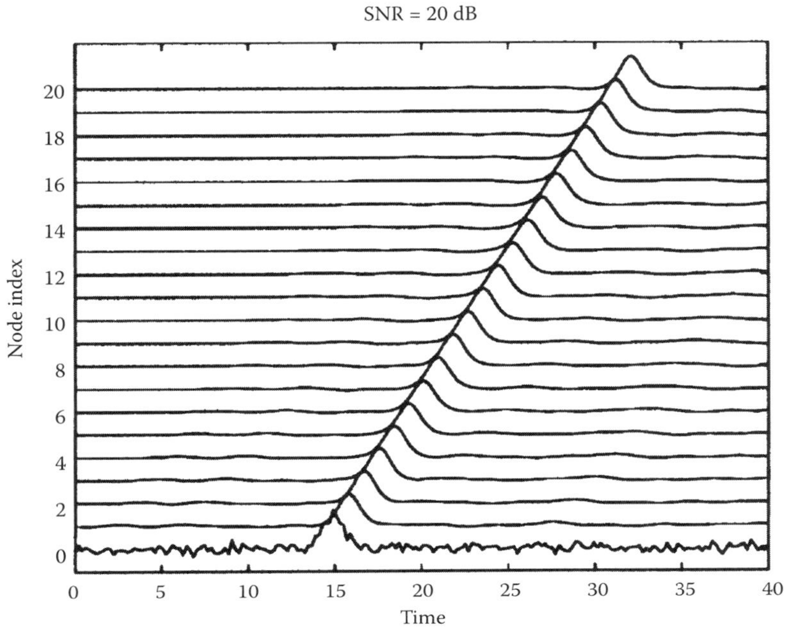

FIGURE 17.11 Response to a single soliton with $\beta = \sinh(1)$ in 20 dB Gaussian noise.

To confirm the intuition developed through small signal analyses, the fully nonlinear dynamics are shown in Figure 17.11 in response to a single soliton at 20 dB signal-to-noise ratio (SNR). As expected, the response to the lattice is essentially the unperturbed soliton with an additional low pass perturbation. The spectrum of the noise remains essentially flat over the bandwidth of the soliton and is attenuated out of band.

17.5.2 Noise Correlation

The statistical correlation of the system response to the noise component can also be estimated from our linear analyses. Given that the lattice behaves as a low pass filter, the small amplitude noise $v_n(t)$ is zero mean and has an autocorrelation function given by

$$R_{n,n}(\tau) = E\{v_n(t)v_n(t+\tau)\} \approx N_0 \frac{\sin(\omega_c\tau)}{\pi\tau}, \tag{17.25}$$

and a variance $\sigma_{v_n}^2 \approx N_0\omega_c/\pi$, for $n \gg 1$.

Although the autocorrelation of the noise at each node is only affected by the magnitude response of Equation 17.21, the cross-correlation between nodes is also affected by the phase response. The cross-correlation between nodes m and n is given by

$$R_{m,n}(\tau) = R_{m,m}(\tau) * h_{n-m}(-\tau), \tag{17.26}$$

where $h_m(\tau)$ is the inverse Fourier transform of $H_m(j\omega)$ in Equation 17.21. Since $h_m(\tau) * h_m(-\tau)$ approaches the impulse response of an ideal low pass filter for $m \gg 1$, we have

$$R_{m,n}(\tau) \approx N_0 \frac{\sin(\omega_c\tau)}{\pi\tau} * h_{n-m}(\tau). \tag{17.27}$$

For small amplitude noise, the correlation structure can be examined through the linear lattice, which acts as a dispersive low pass filter. A corresponding analysis of the nonlinear system in the presence of solitons is prohibitively complex. However, we can explore the analyses numerically by linearizing the dynamics of the system about the known soliton trajectory.

From our earlier linearized analyses, the linear time-varying small signal model can be viewed over short time scales as a LTI chain, with a slowly varying parameter. The resulting input–output transfer function can be viewed as a low pass filter with time varying cutoff frequency equal to ω_c when a soliton is far from the node, and to $\omega_0\sqrt{1+V_n^0}$ as a soliton passes through. Thus, we would expect the variance of the node voltage to rise from a nominal value as a soliton passes through. This intuition can be verified experimentally by numerically integrating the corresponding Riccati equation for the node covariance and computing the resulting variance of the noise component on each node. Since the lattice was assumed initially at rest, there will be a startup transient, as well as an initial spatial transient at the beginning of the lattice, after which the variance of the noise is amplified from the nominal variance as each soliton passes through, confirming our earlier intuition.

17.5.3 Inverse Scattering-Based Noise Modeling

The inverse scattering transform provides a particularly useful mechanism for exploring the long term behavior of soliton systems. In a similar manner to the use of the Fourier transform for describing the ability of linear processors to extract a signal from a stationary random background, the nonlinear spectrum of a received soliton signal in noise can effectively characterize the ability of the nonlinear system to extract or process the component solitons. In this section, we focus on the effects of random perturbations on the dynamics of solitons in the Toda lattice from the viewpoint of inverse scattering.

As seen in Section 17.2.1, the dynamics of the Toda lattice may be described by the evolution of the matrix

$$L(t) = \begin{bmatrix} \ddots & a_{n-1}(t) & \\ a_{n-1}(t) & b_n(t) & a_n(t) \\ & a_n(t) & \ddots \end{bmatrix}, \tag{17.28}$$

whose eigenvalues outside the range $|\lambda| \leq 1$ give rise to soliton behavior. By considering the effects of small amplitude perturbations to the sequences $a_n(t)$ and $b_n(t)$ on the eigenvalues of $L(t)$, we can observe the effects on the soliton dynamics through the eigenvalues corresponding to solitons.

Following [20], we write the $N \times N$ matrix L as $L = L_0 + D$, where L_0 is the unperturbed symmetric matrix, and D is the symmetric random perturbation. To second-order, the eigenvalues are given by

$$\lambda_g = \mu_g + \hat{d}_{gg} - \sum_{i=1,\, i\neq g}^{N} \frac{\hat{d}_{gi}\hat{d}_{ig}}{\mu_{ig}}, \tag{17.29}$$

where μ_g is the gth eigenvalue of L_0, $\mu_{ig} = \mu_i - \mu_g$, and $\hat{d}_{ij}$ are the elements of the matrix $\hat{D}$ defined by $\hat{D} = C^\top DC$, and C is a matrix that diagonalizes L, $C^\top L_0 C = \mathrm{diag}(\mu_1, \ldots, \mu_N)$.

To second-order, the means of the eigenvalues are given by

$$E\{\lambda_g\} = \mu_g - \sum_{i=1, i\neq g}^{N} \frac{\hat{d}_{gi}\hat{d}_{ig}}{\mu_{ig}} \tag{17.30}$$

indicating that the eigenvalues of L are asymptotically (SNR $\to \infty$) unbiased estimates of the eigenvalues of L_0. To first-order, $\lambda_g \approx \mu_g - \hat{d}_{gg}$, and $\hat{d}_{gg}$ is a linear combination of the elements of D,

$$\hat{d}_{gg} = \sum_{r=1,s=1}^{N} c_{gr}c_{gs}d_{rs}. \tag{17.31}$$

Therefore, if the elements of D are jointly Gaussian, then to first-order, the eigenvalues of L will be jointly Gaussian, distributed about the eigenvalues of L_0.

The variance of the eigenvalues can be shown to be approximately given by

$$\mathrm{Var}(\lambda_g) \approx \frac{\sigma_\beta^2 + 2\sigma_\alpha^2(1 + \cos(4\pi g/N))}{N}, \tag{17.32}$$

to second-order, where σ_β^2 and σ_α^2 are the variances of the iid perturbations to b_n and a_n, respectively. This indicates that the eigenvalues of L are consistent estimates of the eigenvalues of L_0.

To first-order, when processing small amplitude noise alone, the noise only excites eigenvalues distributed about the continuum, corresponding to nonsoliton components. When solitons are processed in small amplitude noise, to first-order, there is a small Gaussian perturbation to the soliton eigenvalues as well.

17.6 Estimation of Soliton Signals

In the communication techniques suggested in Section 17.4, the parameters of a multisoliton carrier are modulated with message-bearing signals and the carrier is then processed with the corresponding nonlinear evolution equation. A potential advantage to transmission of this packetized soliton carrier is a net reduction in the transmitted signal energy. However, during transmission, the multisoliton carrier signal can be subjected to distortions due to propagation, which we have assumed can be modeled as AWGN. In this section, we investigate the ability of a receiver to estimate the parameters of a noisy multisoliton carrier. In particular, we consider the problems of estimating the scaling parameters and the relative positions of component solitons of multisoliton solutions, once again focusing on the Toda lattice as an example. For each of these problems, we derive Cramér–Rao lower bounds (CRBs) for the estimation error variance through which several properties of multisoliton signals can be observed. Using these bounds, we will see that although the net transmitted energy in a multisoliton signal can be reduced through nonlinear interaction, the estimation performance for the parameters of the component solitons can also be enhanced. However, at the receiver there are inherent difficulties in parameter estimation imposed by this nonlinear coupling. We will see that the Toda lattice can act as a tuned receiver for the component solitons, naturally decoupling them so that the parameters of each soliton can be independently estimated. Based on this strategy, we develop robust algorithms for maximum likelihood (ML) parameter estimation. We also extend the analogy of the inverse scattering transform as an analog of the Fourier transform for linear techniques, by developing a ML estimation algorithm based on the nonlinear spectrum of the received signal.

17.6.1 Single-Soliton Parameter Estimation: Bounds

In our simplified channel model, the received signal $r(t)$ contains a soliton signal $s(t)$ in an AWGN background $n(t)$ with noise power N_0. A bound on the variance of an estimate of the parameter β may be useful in determining the demodulation performance of an AM-like modulation or pulse amplitude modulation (PAM), where the component soliton wavenumbers are slightly amplitude modulated by a message-bearing waveform. When $s(t)$ contains a single soliton for the Toda lattice, $s(t) = \beta^2 \operatorname{sech}^2(\beta\, t)$, the variance of any unbiased estimator $\hat{\beta}$ of β must satisfy the CRB [19],

$$\mathrm{Var}(\hat{\beta}) \geq \frac{N_0}{\int_{t_i}^{t_f} \left(\frac{\partial s(t;\beta)}{\partial \beta}\right)^2 dt}, \tag{17.33}$$

where the observation interval is assumed to be $t_i < t < t_f$. For the infinite observation interval, $-\infty < t < \infty$, the CRB (Equation 17.33) is given by

$$\mathrm{Var}(\hat{\beta}) \geq \frac{N_0}{\left(\frac{8}{3} + \frac{4\pi^2}{45}\right)\beta} \approx \frac{N_0}{3.544\beta}. \tag{17.34}$$

A slightly different bound may be useful in determining the demodulation performance of an FM-like modulation or PPM, where the soliton position, or time-delay, is slightly modulated by a message-bearing waveform. The fidelity of the recovered message waveform will be directly affected by the ability of a receiver to estimate the soliton position. When the signal $s(t)$ contains a single soliton $s(t) = \beta^2 \operatorname{sech}^2(\beta(t - \delta))$, where δ is the relative position of the soliton in a period of the carrier, the CRB for $\hat{d}$ is given by

$$\mathrm{Var}(\hat{\delta}) \geq \frac{N_0}{\int_{t_i}^{t_f} 4\beta^6 \operatorname{sech}^4(\beta(t-\delta)) \tanh^2(\beta(t-\delta))dt} = \frac{N_0}{\left(\frac{16}{15}\right)\beta^5}. \tag{17.35}$$

As a comparison, for estimating the time of arrival of the raised cosine pulse, $\beta^2(1 + \cos(2\pi\beta(t - \delta)))$, the CRB for this more traditional pulse position modulation would be

$$\mathrm{Var}(\hat{\delta}) \geq \frac{N_0}{\pi^2 \beta^5}, \tag{17.36}$$

which has the same dependence on signal amplitude as Equation 17.35. These bounds can be used for multiple-soliton signals if the component solitons are well separated in time.

17.6.2 Multisoliton Parameter Estimation: Bounds

When the received signal is a multisoliton waveform where the component solitons overlap in time, the estimation problem becomes more difficult. It follows that the bounds for estimating the parameters of such signals must also be sensitive to the relative positions of the component solitons.

We will focus our attention on the two-soliton solution to the Toda lattice, given by Equation 17.3. We are generally interested in estimating the parameters of the multisoliton carrier for an unknown relative spacing among the solitons present in the carrier signal. Either the relative spacing of the solitons has been modulated and is therefore unknown, or the parameters β_1 and β_2 are slightly modulated and the induced phase shift in the received solitons, ϕ, is unknown. For large separations, $\delta = \delta_1 - \delta_2$, the CRB for estimating the parameters of either of the component solitons will be unaffected by the parameters of the other soliton. As shown in Figure 17.12, when the component solitons are well separated, the CRB for either β_1 or β_2 approaches the CRB for estimation of a single soliton with that parameter value in the same level of noise. The bounds for estimating β_1 and β_2 are shown in Figure 17.12 as a function of the relative separation, δ.

Note that both of the bounds are reduced by the nonlinear superposition, indicating that the potential performance of the receiver is enhanced by the nonlinear superposition. However, if we let the parameter difference $\beta_2 - \beta_1$ increase, we notice a different character to the bounds. Specifically, we maintain $\beta_1 = \sinh(2)$, and let $\beta_1 = \sinh(1.25)$. The performance of the larger soliton is inhibited by the nonlinear superposition, while the smaller soliton is still enhanced. In fact, the CRB for the smaller soliton becomes

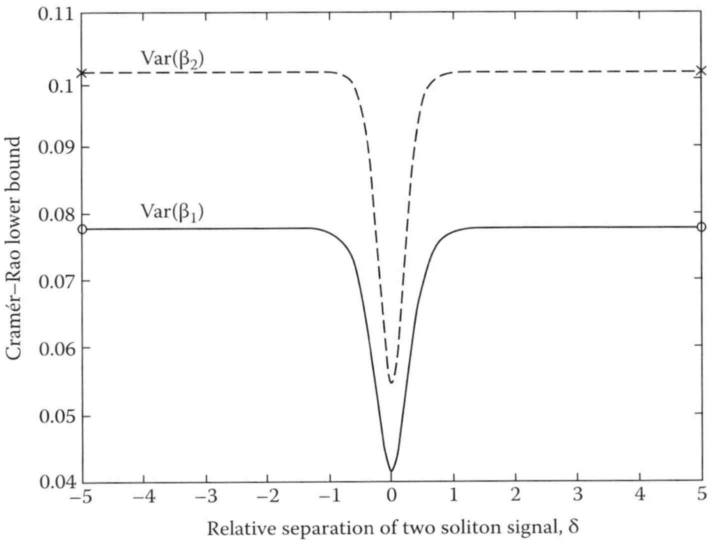

FIGURE 17.12 The CRB for estimating $\beta_1 = \sinh(2)$ and $\beta_2 = \sinh(1.75)$ with all parameters unknown in AWGN with $N_0 = 1$. The bounds are shown as a function of the relative separation, $\delta = \delta_1 - \delta_2$. The CRB for estimating β_1 and β_2 of a single soliton with the same parameter value is indicated with "o" and "×" marks, respectively.

lower than that for the larger soliton near the range $\delta = 0$. This phenomenon results from the relative sensitivity of the signal $s(t)$ to each of the parameters β_1 and β_2. The ability to simultaneously enhance estimation performance while decreasing signal energy is an inherently nonlinear phenomena.

Combining these results with the results of Section 17.4.1, we see that the nonlinear interaction of the component solitons can simultaneously enhance the parameter estimation performance and reduce the net energy of the signal. This property may make superimposed solitons attractive for use in a variety of communication systems.

17.6.3 Estimation Algorithms

In this section we will present and analyze several parameter estimation algorithms for soliton signals. Again, we will focus on the diode ladder circuit implementation of the Toda lattice equations, Equation 17.14. As motivation, consider the problem of estimating the position, δ, of a single-soliton solution $s(t;\delta) = \beta^2 \text{sech}^2(\beta(t-\delta))$, with the parameter β known. This is a classical time-of-arrival estimation problem. For observations $r(t) = s(t) + n(t)$, where $n(t)$ is a stationary white Gaussian process, the ML estimate is given by the value of the parameter δ which minimizes the expression

$$\hat{\delta} = \arg\min_{\tau} \int_{t_i}^{t_f} (r(t) - s(t-\tau))^2 dt. \tag{17.37}$$

Since the replica signals all have the same energy, we can represent the minimization in Equation 17.37 as a maximization of the correlation

$$\hat{\delta} = \arg\min_{\tau} \int_{t_i}^{t_f} r(t)s(t-\tau) dt. \tag{17.38}$$

It is well known that an efficient way to perform the correlation (Equation 17.38) with all of the replica signals $s(t-\tau)$ over the range $\delta_{min} < \tau < \delta_{max}$, is through convolution with a matched filter followed by a peak-detector [19].

When the signal $r(t)$ contains a multisoliton signal, $s(t; \underline{\beta}, \underline{\delta})$, where we wish to estimate the parameter vector $\underline{\delta}$, the estimation problem becomes more involved. If the component solitons are well separated in time, then the ML estimator for the positions of each of the component solitons would again involve a matched filter processor followed by a peak-detector for each soliton. If the component solitons are not well separated and are therefore nonlinearly combined, the estimation problems are tightly coupled and should not be performed independently. The estimation problems can be decoupled by preprocessing the signal $r(t)$ with the Toda lattice. By setting $i_{in}(t) = r(t)$, that is the current through the first diode in the diode ladder circuit, then as the signal propagates through the lattice, the component solitons will naturally separate due to their different propagation speeds.

Defining the signal and noise components as viewed on the kth node in the lattice as $s_k(t)$ and $n_k(t)$, respectively, i.e., $i_k(t) = s_k(t) + n_k(t)$, where $n_0(t)$ is the stationary white Gaussian noise process $n(t)$, in Section 17.5, we saw that in the high SNR limit, $n_k(t)$ will be low pass and Gaussian. In this limit, the ML estimator for the positions, δ_i, can again be formulated using matched filters for each of the component solitons. Since the lattice equations are invertible, at least in principle through inverse scattering, then the ML estimate of the parameter $\underline{\delta}$ based on $r(t)$ must be the same as the estimate based on $i_n(t) = T(r(t))$, for any invertible transformation $T(\cdot)$. If the component solitons are well separated as viewed on the Nth node of the lattice, $i_N(t)$, then an ML estimate based on observations of $i_N(t)$ will reduce to the aggregate of ML estimates for each of the separated component solitons on low pass Gaussian noise. For soliton position estimation, this amounts to a bank of matched filters. We can view this estimation procedure as a form of nonlinear matched filtering, whereby first, dynamics matched to the soliton signals are used to perform the necessary signal separation, and then filters matched to the separated signals are used to estimate their arrival time.

17.6.4 Position Estimation

We will focus our attention on the two-soliton signal (Equation 17.3). If the component solitons are well-separated as viewed on the Nth node of the Toda lattice, the signal appears to be a linear superposition of two solitons,

$$\begin{aligned} i_N(t) \approx\ & \beta_1^2 \text{sech}^2(\beta_1(t-\delta_1) - p_1 N - \phi/2) \\ & + \beta_2^2 \text{sech}^2(\beta_2(t-\delta_2) - p_2 N + \phi/2), \end{aligned} \tag{17.39}$$

where $\phi/2$ is the time-shift incurred due to the nonlinear interaction. Matched filters can now be used to estimate the time of the arrival of each soliton at the Nth node. We formulate the estimate

$$\hat{\delta}_1 = \left(t_{N,1}^a - \frac{p_1 N + \phi/2}{\beta_1} \right), \quad \hat{\delta}_2 = \left(t_{N,2}^a - \frac{p_2 N - \phi/2}{\beta_2} \right), \tag{17.40}$$

where $t_{N,i}^a$ is the time of arrival of the ith soliton on node N. The performance of this algorithm for a two-soliton signal with $\underline{\beta} = [\sinh(2), \sinh(1.5)]$ is shown in Figure 17.13. Note that although the error variance of each estimate appears to be a constant multiple of the CRB, the estimation error variance approaches the CRB in an absolute sense as $N_0 \to 0$.

17.6.5 Estimation Based on Inverse Scattering

The transformation $L(t) = T\{r(t)\}$, where $L(t)$ is the symmetric matrix from the inverse scattering transform, is also invertible in principle. Therefore, an ML estimate based on the matrix $L(t)$ must be

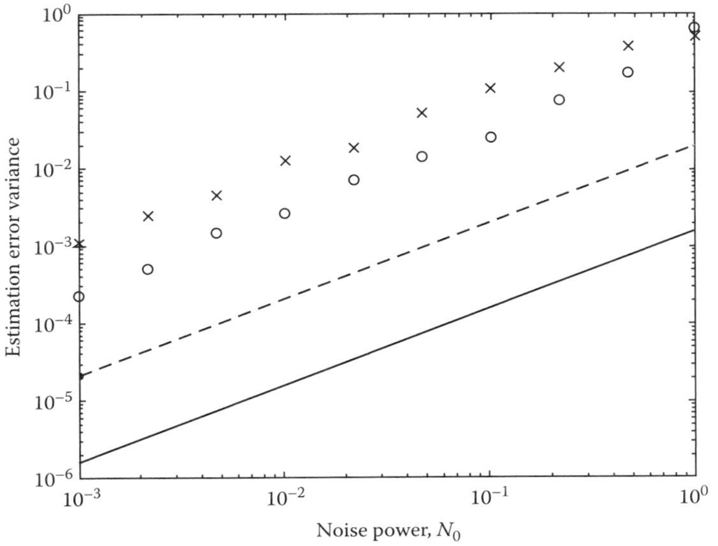

FIGURE 17.13 The CRBs for δ_1 and δ_2 are shown with solid and dashed lines, while the estimation error results of 100 Monte Carlo trials are indicated with "o" and "×" marks, respectively.

the same as an ML estimate based on $r(t)$. We therefore seek to form an estimate of the parameters of the signal $r(t)$ by performing the estimation in the nonlinear spectral domain. This can be accomplished by viewing the Toda lattice as a nonlinear filterbank which projects the signal $r(t)$ onto the spectral components of $L(t)$. This use of the inverse scattering transform is analogous to performing frequency estimation with the Fourier transform.

If $v_n(t)$ evolves according to the Toda lattice equations, then the eigenvalues of the matrix, $L(t)$ are time-invariant, where $a_n(t) = \frac{1}{2}e^{(v_n(t)-v_{n+1}(t))/2}$, and $b_n = \dot{v}_n(t)/2$. Further, the eigenvalues of $L(t)$ for which $|\lambda_i| > 1$ correspond to soliton solutions, with $\beta_i = \sinh(\cosh^{-1}(\lambda_i)) = \sqrt{\lambda_i^2 - 1}$. The eigenvalues of $L(t)$ are, to first-order, jointly Gaussian and distributed about the true eigenvalues corresponding to the original multisoliton signal, $s(t)$. Therefore, estimation of the parameters β_i from the eigenvalues of $L(t)$ as described above constitutes a ML approach in the high SNR limit.

The parameter estimation algorithm now amounts to an estimation of the eigenvalues of $L(t)$. Note that since $L(t)$ is tridiagonal, very efficient techniques for eigenvalue estimation may be used [3]. The estimate of the parameter β is then found by the relation $\hat{\beta}_i = \sqrt{\lambda_i^2 - 1}$, where $|\lambda_i| > 1$, and the sign of β_i can be recovered from the sign of λ_i. Clearly if there is a prespecified number of solitons, k, present in the signal, then the k largest eigenvalues would be used for the estimation. If the number k were unknown, then a simultaneous detection and estimation algorithm would be required.

An example of the joint estimation of the parameters of a two-soliton signal is shown in Figure 17.14a. The estimation error variance decreases with the noise power at the same exponential rate as the CRB.

To verify that the performance of the estimation algorithm has the same dependence on the relative separation of solitons as indicated in Section 17.6.2, the estimation error variance is also indicated in Figure 17.14b vs. the relative separation, δ. In the figure, the mean-squared parameter estimation error for each of the parameters β_i are shown along with their corresponding CRB. At least empirically, we see that the fidelity of the parameter estimates are indeed enhanced by their nonlinear interaction, even though this corresponds to a signal with lower energy, and therefore lower observational SNR.

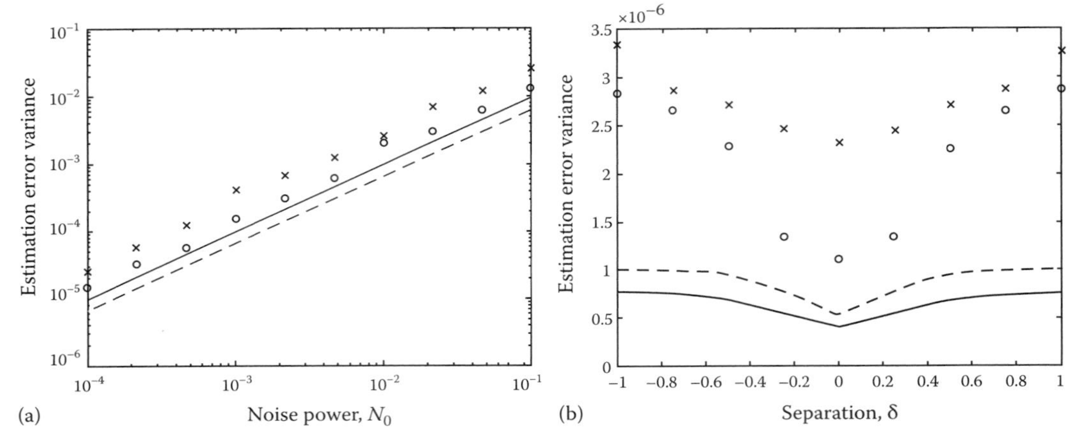

FIGURE 17.14 The estimation error variance for the inverse scattering-based estimates of $\beta_1 = \sinh(2)$, $\beta_2 = \sinh(1.5)$. The bounds for β_1 and β_2 are indicated with solid and dashed lines, respectively. The estimation results for 100 Monte Carlo trials with a diode lattice of $N = 10$ nodes for β_1 and β_2 are indicated by the points labeled "o" and "×," respectively. (a) Error variance vs. N_0. (b) Error variance vs. δ.

17.7 Detection of Soliton Signals

The problem of detecting a single soliton or multiple nonoverlapping solitons in AWGN falls within the theory of classical detection. The Bayes optimal detection of a known or multiple known signals in AWGN can be accomplished with matched filter processing. When the signal $r(t)$ contains a multisoliton signal where the component solitons are not resolved, the detection problem becomes more involved. Specifically, consider a signal comprising a two-soliton solution to the Toda lattice, where we wish to decide which, if any, solitons are present. If the relative positions of the component solitons are known *a priori*, then the detection problem reduces to deciding which among four possible known signals is present,

$$\begin{aligned} H_0&: r(t) = n(t), \\ H_1&: r(t) = s_1(t) + n(t), \\ H_2&: r(t) = s_2(t) + n(t), \\ H_{12}&: r(t) = s_{12}(t) + n(t), \end{aligned}$$

where $s_1(t), s_2(t)$, and $s_{12}(t)$ are soliton one, soliton two, and the multisoliton signals, respectively. Once again, this problem can be solved with standard Gaussian detection theory.

If the relative positions of the solitons are unknown, as would be the case for a modulated soliton carrier, then the signal $s_{12}(t)$ will vary significantly as a function of the relative separation. Similarly, if the signals are to be transmitted over a soliton channel where different users occupy adjacent soliton wavenumbers, any detection at the receiver would have to be performed with the possibility of another soliton component present at an unknown position. We therefore obtain a composite hypothesis testing problem, whereby under each hypothesis, we have

$$\begin{aligned} H_0&: r(t) = n(t), \\ H_1&: r(t) = s_1(t;\delta_1) + n(t), \\ H_2&: r(t) = s_2(t;\delta_2) + n(t), \\ H_{12}&: r(t) = s_{12}(t;\underline{\delta}) + n(t), \end{aligned}$$

where $\underline{\delta} = [\delta_1, \delta_2]^\top$. The general problem of detection with an unknown parameter, $\underline{\delta}$, can be handled in a number of ways. For example, if the parameter can be modeled as random and the distribution for the parameter were known, $p_{\underline{\delta}}(\underline{\delta})$, along with the distributions $p_{r|\underline{\delta},H}(R|\underline{\delta}, H_i)$ for each hypothesis, then the Bayes or Neyman–Pearson criteria can be used to formulate a likelihood ratio test. Unfortunately, even when the distribution for the parameter $\underline{\delta}$ is known, the likelihood ratios cannot be found in closed form for even the single-soliton detection problem.

Another approach that is commonly used in radar processing [5,19] applies when the distribution of δ does not vary rapidly over a range of possible values while the likelihood function has a sharp peak as a function of δ. In this case, the major contribution to the integral in the averaged likelihood function is due to the region around the value of δ for which the likelihood function is maximum, and therefore this value of the likelihood function is used as if the maximizing value, $\hat{\delta}_{ML}$, were the actual value. Since $\hat{\delta}_{ML}$ is the ML estimate of δ based on the observation, $r(t)$, such techniques are called "ML detection." Also, the term "generalized likelihood ratio test" (GLRT) is used since the hypothesis test amounts to a generalization of the standard likelihood ratio test.

If we plan to employ a GLRT for the multisoliton detection problem, we are again faced with the need for an ML estimate of the position, $\underline{\delta}_{ML}$. A standard approach to such problems would involve turning the current problem into one with hypotheses H_0, H_1, and H_2 as before, and an additional M hypotheses—one for each value of the parameter $\underline{\delta}$ sampled over a range of possible values. The complexity of this type of detection problem increases exponentially with the number of component solitons, N_s, resulting in a hypothesis testing problem with $O((M+1)^{N_s})$ hypotheses.

However, as with the estimation problems in Section 17.6, the detection problems can be decoupled by preprocessing the signal $r(t)$ with the Toda lattice. If the component solitons separate as viewed on the Nth node in the lattice, then the detection problem can be more simply formulated using $i_N(t)$. The invertibility of the lattice equations implies that a Bayes optimal decision based on $r(t)$ must be the same as that based on $i_N(t)$. Since the Bayes optimal decision can be performed based on the likelihood function $\Lambda(r(t))$, and $\Lambda(i_N(t)) = \Lambda(T\{r(t)\}) = \Lambda(r(t))$, the optimal decisions based on $r(t)$ and $i_N(t)$ must be the same for any invertible transformation $T\{\cdot\}$. Although we will be using a GLRT, where the value of $\underline{\delta}_{ML}$ is used for the unknown positions of the multisoliton signal, since the ML estimates based on $r(t)$ and $i_N(t)$ must also be the same, the detection performance of a GLRT using those estimates must also be the same. Since at high SNR, the noise component of the signal $i_N(t)$ can be assumed low pass and Gaussian, the GLRT can be performed by preprocessing $r(t)$ with the Toda lattice equations followed by matched filter processing.

17.7.1 Simulations

To illustrate the algorithm, we consider the hypothesis test between H_0 and H_{12}, where the separation of the two solitons, $\delta_1 - \delta_2$, varies randomly in the interval $[-1/\beta_2, 1/\beta_2]$. The detection processor comprises a Toda lattice of $N = 20$ nodes, with the detection performed based on the signal $i_{10}(t)$. To implement the GLRT, we search over a fixed time interval about the expected arrival time for each soliton. In this manner we obtain a sequence of 1000 Monte Carlo values of the processor output for each hypothesis. A set of Monte Carlo runs has been completed for each of three different levels of the noise power, N_0.

The receiver operating characteristic (ROC) for the soliton with $\beta_2 = \sinh(1.5)$ is shown in Figure 17.15, where the probability of detection, P_D, for this hypothesis test is shown as a function of the false alarm probability, P_F. For comparison, we also show the ROC that would result from a detection of the soliton alone, at the same noise level and with the time-of-arrival known. The detection index, $d = \sqrt{E/N_0}$, is indicated for each case, where E is the energy in the component soliton. The corresponding results for the larger soliton are qualitatively similar, although the detection indices for that soliton alone, with $\beta_1 = \sinh(2)$, are 5.6, 4, and 3.3, respectively. Therefore, the detection probabilities are considerably higher for a fixed probability of false alarm. Note that the detection performance for the smaller soliton is

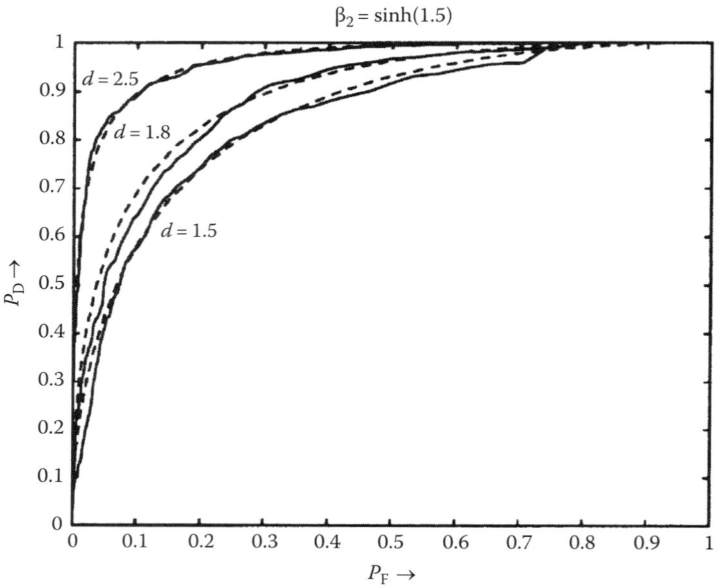

FIGURE 17.15 A set of empirically generated ROCs are shown for the detection of the smaller soliton from a two-soliton signal. For each of the three noise levels, the ROC for detection of the smaller soliton alone is also indicated along with the corresponding detection index, d.

well modeled by the theoretical performance for detection of the smaller soliton alone. This implies, at least empirically, that the ability to detect the component solitons in a multisoliton signal appears to be unaffected by the nonlinear coupling with other solitons. Further, although the unknown relative separation results in significant waveform uncertainty and would require a prohibitively complex receiver for standard detection techniques, Bayes optimal performance can still be achieved with a minimal increase in complexity.

References

1. Ablowitz, M.J. and Clarkson, A.P., *Solitons, Nonlinear Evolution Equations and Inverse Scattering*, London Mathematical Society Lecture Note Series, Vol. 149, Cambridge University Press, Cambridge, U.K., 1991.
2. Fermi, E., Pasta, J.R., and Ulan, S.M., Studies of nonlinear problems, in *Collected Papers of E. Fermi*, vol. II, pp. 977–988, University of Chicago Press, Chicago, IL, 1965.
3. Golub, G.H. and Van Loan, C.F., *Matrix Computations*, The Johns Hopkins University Press, Baltimore, MD, 1989.
4. Haus, H.A., Molding light into solitons, *IEEE Spectrum*, 30(3), 48–53, March 1993.
5. Helstrom, C.W., *Statistical Theory of Signal Detection*, 2nd edn., Pergamon Press, New York, 1968.
6. Hirota, R. and Suzuki, K., Theoretical and experimental studies of lattice solitons in nonlinear lumped networks, *Proc. IEEE*, 61(10), 1483–1491, October 1973.
7. Infeld, E. and Rowlands, R., *Nonlinear Waves, Solitons and Chaos*, Cambridge University Press, New York, 1990.
8. Kaiser, J.F., On a simple algorithm to calculate the "energy" of a signal, in *Proceedings of the International Conference Acoustics Speech, Signal Processing*, pp. 381–384, Albuquerque, NM, 1990.
9. Kaiser, J.F., Personal communication, June 1994.

10. Lax, P.D., Integrals of nonlinear equations of evolution and solitary waves, *Comm. Pure Appl. Math.*, XXI, 467–490, 1968.
11. Scott, A.C., *Active and Nonlinear Wave Propagation in Electronics*, Wiley-Interscience, New York, 1970.
12. Scott, A.C., Chu, F.Y.F., and McLaughlin, D., The soliton: A new concept in applied science, *Proc. IEEE*, 61(10), 1443–1483, October 1973.
13. Singer, A.C., A new circuit for communication using solitons, in *Proceedings of the IEEE Workshop on Nonlinear Signal and Image Processing*, vol. I, pp. 150–153, Halkidiki, Greece, 1995.
14. Singer, A.C., Signal processing and communication with solitons, PhD thesis, Massachusetts Institute of Technology, Cambridge, MA, February 1996.
15. Suzuki, K., Hirota, R., and Yoshikawa, K., Amplitude modulated soliton trains and coding-decoding applications, *Int. J. Electron.*, 34(6), 777–784, 1973.
16. Suzuki, K., Hirota, R., and Yoshikawa, K., The properties of phase modulated soliton trains, *Jpn. J. Appl. Phys.*, 12(3), 361–365, March 1973.
17. Toda, M., *Theory of Nonlinear Lattices*, Springer Series in *Solid-State Science*, Vol. 20, Springer-Verlag, New York, 1981.
18. Toda, M., *Nonlinear Waves and Solitons, Mathematics and Its Applications*, Kluwer Academic Publishers, Boston, MA, 1989.
19. Van Trees, H.L., *Detection, Estimation, and Modulation Theory: Part I Detection, Estimation and Linear Modulation Theory*, John Wiley & Sons, New York, 1968.
20. vom Scheidt, J. and Purkert, W., *Random Eigenvalue Problems, Probability and Applied Mathematics*, North-Holland, New York, 1983.
21. Whitham, G.B., *Linear and Nonlinear Waves*, Wiley, New York, 1974.
22. Zabusky, N.J. and Kruskal, M.D., Interaction of solitons in a collisionless plasma and the recurrence of initial states, *Phys. Rev. Lett.*, 15(6), 240–243, August 1965.

18

Higher-Order Spectral Analysis

18.1 Introduction ... 18-1
18.2 Definitions of HOS ... 18-3
Moments and Cumulants of Random Variables • Moments and Cumulants of Stationary Random Processes • Linear Processes
18.3 HOS Computation from Real Data ... 18-6
Indirect Method • Direct Method
18.4 Blind System Identification ... 18-8
Bicepstrum-Based System Identification • A Frequency-Domain Approach • Parametric Methods
18.5 HOS for Blind MIMO System Identification ... 18-11
Resolving the Phase Ambiguity
18.6 Nonlinear Processes ... 18-16
18.7 Conclusions ... 18-19
Acknowledgments ... 18-19
References ... 18-19

Athina P. Petropulu
Drexel University

18.1 Introduction

Power spectrum estimation techniques have proved essential in many applications, such as communications, sonar, radar, speech/image processing, geophysics, and biomedical signal processing [22,25,26,35]. In power spectrum estimation, the process under consideration is treated as a superposition of statistically uncorrelated harmonic components. The distribution of power among these frequency components is the power spectrum. As such, phase relations between frequency components are suppressed. The information in the power spectrum is essentially present in the autocorrelation sequence, which would suffice for the complete statistical description of a Gaussian process of known mean. However, there are applications where there is a wealth of information in higher-order spectra (HOS) (of order greater than 2) [29]. The third-order spectrum is commonly referred to as bispectrum, the fourth-order one as trispectrum, and in fact, the power spectrum is also a member of the HOS class; it is the second-order spectrum. HOS consist of higher-order moment spectra, which are defined for deterministic signals, and cumulant spectra, which are defined for random processes.

HOS possess many attractive properties. For example, HOS suppress additive Gaussian noise, thus providing high signal-to-noise ratio domains, in which one can perform detection, parameter estimation, or even signal reconstruction. The same property of HOS can provide a means of detecting and characterizing deviations of the data from the Gaussian model. Cumulant spectra of order greater than two preserve phase information. In modeling of time series, second-order statistics (SOS) (autocorrelation) have been heavily used because they are the result of least-squares optimization criteria. However, an accurate phase reconstruction in the autocorrelation domain can be achieved only if the signal is minimum phase. Nonminimum phase signal reconstruction can be achieved only in the HOS domain. Due to this property, HOS have been used extensively in system identification problems. Figure 18.1 shows two signals, a nonminimum phase and a minimum phase one, with identical magnitude spectra but different phase spectra. Although power spectrum cannot distinguish between the two signals, the

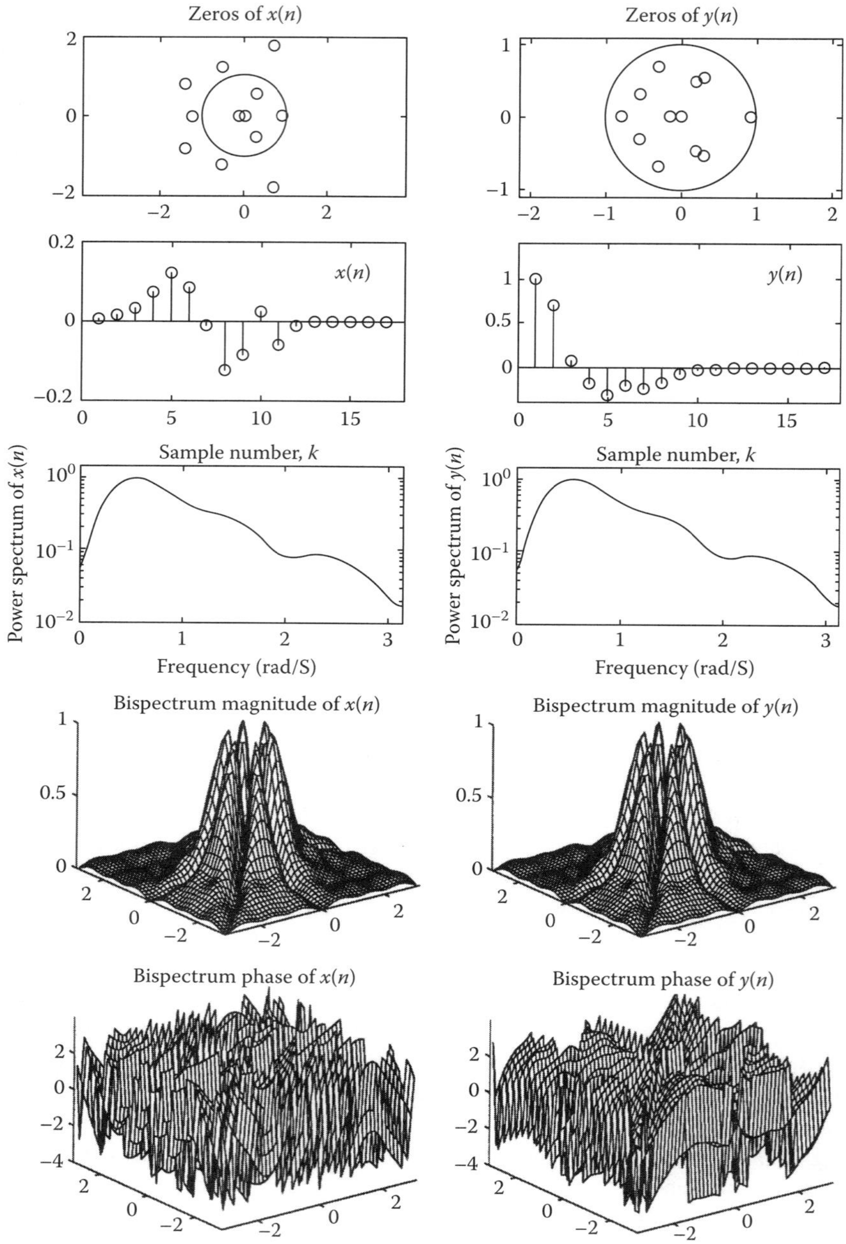

FIGURE 18.1 $x(n)$ is a nonminimum phase signal and $y(n)$ is a minimum phase one. Although their power spectra are identical, their bispectra are different since they contain phase information.

bispectrum that uses phase information can. Being nonlinear functions of the data, HOS are quite natural tools in the analysis of nonlinear systems operating under a random input. General relations for arbitrary stationary random data passing through an arbitrary linear system exist and have been studied extensively. Such expression, however, is not available for nonlinear systems, where each type of nonlinearity must be studied separately. Higher-order correlations between input and output can detect and characterize certain nonlinearities [43], and for this purpose several HOS-based methods have been developed.

The organization of this chapter is as follows. First, the definitions and properties of cumulants and HOS are introduced. Then, two methods for the estimation of HOS from finite length data are outlined and the asymptotic statistics of the obtained estimates are presented. Following that we present some methods for blind system identification, first for a single-input single-output system and then for a multiple-input multiple-output (MIMO) system. Then the use of HOS in the identification of some particular nonlinear systems is briefly discussed. The chapter closes with some concluding remarks, and pointers to HOS software.

18.2 Definitions of HOS

18.2.1 Moments and Cumulants of Random Variables

The "joint moments" of order r of the real random variables $x_1, \ldots, x_n$ are given by [5,34]

$$\begin{aligned}\mathrm{Mom}\left[x_1^{k_1}, \ldots, x_n^{k_n}\right] &= E\left\{x_1^{k_1}, \ldots, x_n^{k_n}\right\} \\ &= (-j)^r \frac{\partial^r \Phi(\omega_1, \ldots, \omega_n)}{\partial \omega_1^{k_1} \ldots \partial \omega_n^{k_n}}\Bigg|_{\omega_1=\cdots=\omega_n=0}\end{aligned} \tag{18.1}$$

where

$k_1 + \cdots + k_n = r$

$\Phi()$ is their joint characteristic function

The "joint cumulants" are defined as

$$\mathrm{Cum}\left[x_1^{k_1}, \ldots, x_n^{k_n}\right] = (-j)^r \frac{\partial^r \ln \Phi(\omega_1, \ldots, \omega_n)}{\partial \omega_1^{k_1} \ldots \partial \omega_n^{k_n}}\Bigg\}\Bigg|_{\omega_1=\cdots=\omega_n=0} \tag{18.2}$$

All the above definitions involve real random variables. For "complex random variables," the nth-order cumulant sequence has 2^n representations; in each representation, each of the n terms may appear as conjugated or unconjugated.

Next, we present some important properties of moments and cumulants.

(P1) For $a_1, \ldots, a_n$ constants, it holds:

$$\mathrm{Mom}[a_1x_1, a_2x_2, \ldots, a_nx_n] = \left(\prod_{i=1}^{n} a_i\right)\mathrm{Mom}[x_1, \ldots, x_n]$$

$$\mathrm{Cum}[a_1x_1, a_2x_2, \ldots, a_nx_n] = \left(\prod_{i=1}^{n} a_i\right)\mathrm{Cum}[x_1, \ldots, x_n]$$

(P2) Moments and cumulants are symmetric in their arguments. For example, Cum $[x_1, x_2] =$ Cum $[x_2, x_1]$.

(P3) If among the random variables $x_1, \ldots, x_n$, there is a subset that is independent of the rest, then $\mathrm{Cum}[x_1, x_2, \ldots, x_n] = 0$. In general, the corresponding moment will not be zero.

(P4) Cumulants and moments are additive in their arguments, i.e.,

$$\mathrm{Cum}[x_1 + y_1, x_2, \ldots, x_n] = \mathrm{Cum}[x_1, x_2, \ldots, x_n] + \mathrm{Cum}[y_1, x_2, \ldots, x_n]$$
$$\mathrm{Mom}[x_1 + y_1, x_2, \ldots, x_n] = \mathrm{Mom}[x_1, x_2, \ldots, x_n] + \mathrm{Mom}[y_1, x_2, \ldots, x_n]$$

(P5) If the random variable sets $\{x_1, \ldots, x_n\}$ and $\{y_1, \ldots, y_n\}$ are independent, then it holds:

$$\mathrm{Cum}[x_1 + y_1, \ldots, x_n + y_n] = \mathrm{Cum}[x_1, x_2, \ldots, x_n] + \mathrm{Cum}[y_1, y_2, \ldots, y_n]$$

In general,

$$\mathrm{Mom}[x_1 + y_1, \ldots, x_n + y_n] \neq \mathrm{Mom}[x_1, x_2, \ldots, x_n] + \mathrm{Mom}[y_1, y_2, \ldots, y_n]$$

(P6) If the random variable $x_1, \ldots, x_n$ is jointly Gaussian, then all cumulants of order greater than 2 are zero.

18.2.2 Moments and Cumulants of Stationary Random Processes

For a stationary discrete-time random process $x(k)$, (k denotes discrete time), the "moments" of order n are equal to

$$m_n^x(\tau_1, \tau_2, \ldots, \tau_{n-1}) = E\{x(k)x(k+\tau_1)\cdots x(k+\tau_{n-1})\} \tag{18.3}$$

where $E\{.\}$ denotes expectation. The nth-order cumulants are defined as

$$c_n^x(\tau_1, \tau_2, \ldots, \tau_{n-1}) = \mathrm{Cum}[x(k), x(k+\tau_1), \ldots, x(k+\tau_{n-1})] \tag{18.4}$$

The nth-order "cumulants" are functions of the moments of order up to n:

First-order cumulants:

$$c_1^x = m_1^x = E\{x(k)\} \text{ (mean)} \tag{18.5}$$

Second-order cumulants:

$$c_2^x(\tau_1) = m_2^x(\tau_1) - \left(m_1^x\right)^2 \text{ (covariance)} \tag{18.6}$$

Third-order cumulants:

$$c_3^x(\tau_1, \tau_2) = m_3^x(\tau_1, \tau_2) - \left(m_1^x\right)\left[m_2^x(\tau_1) + m_2^x(\tau_2) + m_2^x(\tau_2 - \tau_1)\right] + 2\left(m_1^x\right)^3 \tag{18.7}$$

Fourth-order cumulants:

$$\begin{aligned} c_4^x(\tau_1, \tau_2, \tau_3) = {} & m_4^x(\tau_1, \tau_2, \tau_3) - m_2^x(\tau_1)m_2^x(\tau_3 - \tau_2) - m_2^x(\tau_2)m_2^x(\tau_3 - \tau_1) - m_2^x(\tau_3)m_2^x(\tau_2 - \tau_1) \\ & - m_1^x\left[m_3^x(\tau_2 - \tau_1, \tau_3 - \tau_1) + m_3^x(\tau_2, \tau_3) + m_3^x(\tau_2, \tau_4) + m_3^x(\tau_1, \tau_2)\right] \\ & + \left(m_1^x\right)^2\left[m_2^x(\tau_1) + m_2^x(\tau_2) + m_2^x(\tau_3) + m_2^x(\tau_3 - \tau_1) + m_2^x(\tau_3 - \tau_2) \right. \\ & \left. + m_1^x(\tau_2 - \tau_1)\right] - 6\left(m_1^x\right)^4 \end{aligned} \tag{18.8}$$

where

$m_3^x(\tau_1, \tau_2)$ is the third-order moment sequence
m_1^x is the mean

The general relationship between cumulants and moments can be found in Nikias and Petropulu [29].

By substituting $\tau_1 = \tau_2 = \tau_3 = 0$ in Equations 18.6 through 18.8 we get the variance, the skewness, and the kurtosis of the process, respectively.

Given the real stationary processes $x_1(k), x_2(k), \ldots, x_n(k)$, the nth-order "cross-cumulant" sequence is

$$c_{x_1,x_2,\ldots,x_n}(\tau_1, \tau_2, \ldots, \tau_{n-1}) \stackrel{\Delta}{=} \text{Cum}[x_1(k), x_2(k+\tau_1), \ldots, x_n(k+\tau_{n-1})] \quad (18.9)$$

HOS are defined in terms of either cumulants (e.g., "cumulant spectra") or moments (e.g., "moment spectra").

Assuming that the nth-order cumulant sequence is absolutely summable, the nth-order cumulant spectrum of $x(k)$, $C_n^x(\omega_1, \omega_2, \ldots, \omega_{n-1})$, exists, and is defined to be the $(n-1)$-dimensional Fourier transform of the nth-order cumulant sequence. The cumulant spectrum of order $n=2$ is referred to as power spectrum, the cumulant spectrum of order $n=3$ is referred to as "bispectrum," while the cumulant spectrum of order $n=4$ is referred to as "trispectrum."

In an analogous manner, "moment spectrum" is the multidimensional Fourier transform of the moment sequence.

The nth-order "cross-spectrum" of stationary random processes $x_1(k), \ldots, x_n(k)$ is the n–1th-order Fourier transform of the cross-cumulant sequence.

18.2.3 Linear Processes

Consider a real process $x(n)$ that is generated by exciting a linear time-invariant (LTI) system with impulse response $h(k)$ with a stationary, zero-mean, non-Gaussian process $v(k)$ that is nth-order white, i.e.,

$$c_n^v(\tau_1, \tau_2, \ldots, \tau_{n-1}) = \gamma_n^v \delta(\tau_1, \tau_2, \ldots, \tau_{n-1}) \quad (18.10)$$

It holds $x(n) = \sum_k h(k)v(n-k)$, and the nth-order cumulants of the process are

$$\begin{aligned}
c_n^x(\tau_1, \ldots, \tau_{n-1}) &= \text{Cum}[x(k), x(k+\tau_1), \ldots, x(k+\tau_{n-1}) \\
&= \text{Cum}\left[\sum_{k_1} h(k_1)v(k-k_1), \ldots, \sum_{k_n} h(k_n)v(k+\tau_{n-1}-k_n)\right] \\
&= \sum_{k_1} \text{Cum}\left[h(k_1)v(k-k_1), \ldots, \sum_{k_n} h(k_n)v(k+\tau_{n-1}-k_n)\right] \quad \text{(via P4)} \\
&= \sum_{k_1}\sum_{k_2}\cdots\sum_{k_n} \text{Cum}[h(k_1)v(k-k_1), \ldots, h(k_n)v(k+\tau_{n-1}-k_n)] \quad \text{(via P4)} \\
&= \sum_{k_1}\sum_{k_2}\cdots\sum_{k_n} h(k_1)h(k_2)\cdots h(k_n)\text{Cum}[v(k-k_1), \ldots, v(k+\tau_{n-1}-k_n)] \quad \text{(via P1)} \\
&= \gamma_n^v \sum_k h(k)h(k+\tau_1)\cdots h(k+\tau_{n-1}) \quad \text{via Equation 18.10}
\end{aligned} \quad (18.11)$$

The corresponding nth-order cumulant spectrum equals

$$C_n^x(\omega_1, \omega_2, \ldots, \omega_{n-1}) = \gamma_n^v H(\omega_1)\cdots H(\omega_{n-1})H(-\omega_1 - \cdots - \omega_{n-1}) \quad (18.12)$$

Cumulant spectra are more useful in processing random signals than moment spectra since they possess properties that the moment spectra do not share: (1) the cumulants of the sum of two independent random processes equal the sum of the cumulants of the process; (2) the cumulant spectra of order >2 are zero, if the underlying process is Gaussian; (3) the cumulants quantify the degree of statistical dependence of time series; (4) the cumulants of higher-order white noise are multidimensional impulses, and the corresponding cumulant spectra are flat.

18.3 HOS Computation from Real Data

The definitions of cumulants presented in Section 18.2.3 are based on expectation operations, and they assume infinite length data. In practice, we always deal with data of finite length, therefore, the cumulants can only be approximated. Two methods for cumulants and spectra estimation are presented next for the third-order case.

18.3.1 Indirect Method

Let $x(k)$, $k = 1, \ldots, N$ be the available data.

1. Segment the data into K records of M samples each. Let $x^i(k)$, $k = 1, \ldots, M$, represent the ith record.
2. Subtract the mean of each record.
3. Estimate the moments of each segments $x^i(k)$ as follows:

$$m_3^{x_i}(\tau_1, \tau_2) = \frac{1}{M} \sum_{l=l_1}^{l_2} x^i(l) x^i(l + \tau_1) x^i(l + \tau_2), \quad l_1 = \max(0, -\tau_1, -\tau_2)$$
$$l_2 = \min(M - 1, M - 2), \quad |\tau_1| < L, |\tau_2| < L, \quad i = 1, 2, \ldots, K \tag{18.13}$$

   Since each segment has zero-mean, its third-order moments and cumulants are identical, i.e., $c_3^{x_i}(\tau_1, \tau_2) = m_3^{x_i}(\tau_1, \tau_2)$.

4. Compute the average cumulants as

$$\hat{c}_3^x(\tau_1, \tau_2) = \frac{1}{K} \sum_{i=1}^{K} m_3^{x_i}(\tau_1, \tau_2) \tag{18.14}$$

5. Obtain the third-order spectrum (bispectrum) estimate as

$$\hat{C}_3^x(\omega_1, \omega_2) = \sum_{\tau_1=-L}^{L} \sum_{\tau_2=-L}^{L} \hat{C}_3^x(\tau_1, \tau_2) e^{-j(\omega_1 \tau_1 + \omega_2 \tau_2)} w(\tau_1, \tau_2) \tag{18.15}$$

   where
   $L < M - 1$
   $w(\tau_1, \tau_2)$ is a two-dimensional window of bounded support, introduced to smooth out edge effects

The bandwidth of the final bispectrum estimate is $\Delta = 1/L$.

A complete description of appropriate windows that can be used in Equation 18.15 and their properties can be found in [29]. A good choice of cumulant window is

$$w(\tau_1, \tau_2) = d(\tau_1) d(\tau_2) d(\tau_1 - \tau_2) \tag{18.16}$$

where

$$d(\tau) = \begin{cases} \frac{1}{\pi} \left| \sin \frac{\pi \tau}{L} \right| + \left(1 - \frac{|\tau|}{L}\right) \cos \frac{\pi \tau}{L} & |\tau| \le L \\ 0 & |\tau| > L \end{cases} \tag{18.17}$$

which is known as the minimum bispectrum bias supremum [30].

18.3.2 Direct Method

Let $x(k)$, $k = 1, \ldots, N$ be the available data.

1. Segment the data into K records of M samples each. Let $x^i(k)$, $k = 1, \ldots, M$, represent the ith record.
2. Subtract the mean of each record.
3. Compute the discrete Fourier transform $X^i(k)$ of each segment, based on M points, i.e.,

$$X^i(k) = \sum_{n=0}^{M-1} x^i(n)e^{-j\frac{2\pi}{M}nk}, \quad k = 0, 1, \ldots, M-1, \quad i = 1, 2, \ldots, K \tag{18.18}$$

4. The discrete third-order spectrum of each segment is obtained as

$$C_3^{x_i}(k_1, k_2) = \frac{1}{M} X^i(k_1)X^i(k_2)X^i(-k_1 - k_2), \quad i = 1, \ldots, K \tag{18.19}$$

   Due to the bispectrum symmetry properties, $C_3^{x_i}(k_1, k_2)$ need to be computed only in the triangular region $0 \leq k_2 \leq k_1, k_1 + k_2 < M/2$.
5. In order to reduce the variance of the estimate note that additional smoothing over a rectangular window of size $(M_3 \times M_3)$ can be performed around each frequency, assuming that the third-order spectrum is smooth enough, i.e.,

$$\tilde{C}_3^{x_i}(k_1, k_2) = \frac{1}{M_3^2} \sum_{n_1=-M_3/2}^{M_3/2-1} \sum_{n_2=-M_3/2}^{M_3/2-1} C_3^{x_i}(k_1 + n_1, k_2 + n_2) \tag{18.20}$$

6. Finally, the discrete third-order spectrum is given as the average overall third-order spectra, i.e.,

$$\hat{C}_3^{x}(\omega_1, \omega_2) = \frac{1}{K} \sum_{i=1}^{K} \tilde{C}_3^{x_i}(\omega_1, \omega_2), \quad \omega_i = \frac{2\pi}{M} k_i, \quad i = 1, 2 \tag{18.21}$$

The final bandwidth of this bispectrum estimate is $\Delta = M_3/M$, which is the spacing between frequency samples in the bispectrum domain.

For large N, and as long as

$$\Delta \to 0, \quad \text{and} \quad \Delta^2 N \to \infty \tag{18.22}$$

both the direct and the indirect methods produce asymptotically unbiased and consistent bispectrum estimates, with real and imaginary part variances [41]:

$$\begin{aligned} \operatorname{var}\left(\operatorname{Re}\left[\hat{C}_3^x(\omega_1, \omega_2)\right]\right) &= \operatorname{var}\left(\operatorname{Im}\left[\hat{C}_3^x(\omega_1, \omega_2)\right]\right) = \frac{1}{\Delta^2 N} C_2^x(\omega_1)C_2^x(\omega_2)C_2^x(\omega_1 + \omega_2) \\ &= \begin{cases} \frac{VL^2}{MK} C_2^x(\omega_1)C_2^x(\omega_2)C_2^x(\omega_1 + \omega_2) & \text{indirect} \\ \frac{M}{KM_3^2} C_2^x(\omega_1)C_2^x(\omega_2)C_2^x(\omega_1 + \omega_2) & \text{direct} \end{cases} \end{aligned} \tag{18.23}$$

where V is the energy of the bispectrum window.

From the above expressions, it becomes apparent that the bispectrum estimate variance can be reduced by increasing the number of records, or reducing the size of the region of support of the window in the cumulant domain (L), or increasing the size of the frequency smoothing window (M_3), etc. The relation between the parameters M, K, L, M_3 should be such that Equation 18.22 is satisfied.

18.4 Blind System Identification

Consider $x(k)$ generated as shown in Figure 18.2. Estimation of the system impulse response based solely on the system output is referred to as blind system identification. If the system is minimum-phase, estimation can be carried out using SOS only. In most applications of interest, however, the system is a nonminimum phase. For example, blind channel estimation/equalization in communications, or estimation of reverberation from speech recordings, can be formulated as a blind system estimation problem where the channel is a nonminimum phase.

In the following, we focus on nonminimum phase systems. If the input has some known structure, for example, if it is cyclostationary, then estimation can be carried out using SOS of the system output. However, if the input is stationary independent identically distributed (i.i.d.) non-Gaussian, then system estimation is only possible using HOS. There is extensive literature on HOS-based blind system estimation [3,4,7–9,11–21,28,31,44–48,50–58]. In this section, we present three different approaches, i.e., based on the bicepstrum, based on the selected slices of the bispectrum, and a parametric approach, in which a parametric model is fitted to the data and the model parameters are estimated based on third-order statistics.

All methods refer to the system of Figure 18.2, where the input $v(k)$ is stationary, zero-mean non-Gaussian, i.i.d., nth-order white, and $h(k)$ is a LTI nonminimum phase system, with system function $H(z)$. The noise $w(k)$ is zero-mean Gaussian, independent of $v(k)$.

Via properties (P5) and (P6) and using Equation 18.10, the nth-order cumulant of the $x(k)$, with $n > 2$, equals

$$\begin{aligned} c_n^x(\tau_1,\ldots,\tau_{n-1}) &= c_n^y(\tau_1,\ldots,\tau_{n-1}) + c_n^w(\tau_1,\ldots,\tau_{n-1}) \\ &= c_n^y(\tau_1,\ldots,\tau_{n-1}) \qquad (18.24) \\ &= \gamma_n^v \sum_{k=0}^{\infty} h(k)h(k+\tau_1)\cdots h(k+\tau_{n-1}) \qquad (18.25) \end{aligned}$$

The bispectrum of $x(k)$ is

$$C_3^x(\omega_1,\omega_2) = \gamma_3^v H(\omega_1)H(\omega_2)H(-\omega_1-\omega_2) \qquad (18.26)$$

18.4.1 Bicepstrum-Based System Identification

Let us assume that $H(z)$ has no zeros on the unit circle. Taking the logarithm of $C_3^x(\omega_1,\omega_2)$ followed by an inverse 2-D Fourier transform, we obtain the bicepstrum $b_x(m,n)$ of $x(k)$. The resulting bicepstrum is zero everywhere except along the axes, $m=0$, $n=0$, and the diagonal $m=n$ [33], where it is equal to the complex cepstrum of $h(k)$ [32], i.e.,

$$b_x(m,n) = \begin{cases} \hat{h}(m) & m \neq 0, n = 0 \\ \hat{h}(n) & n \neq 0, m = 0 \\ \hat{h}(-n) & m = n, m \neq 0 \\ \ln\left(c\gamma_n^v\right) & m = n = 0, \\ 0 & \text{elsewhere} \end{cases} \qquad (18.27)$$

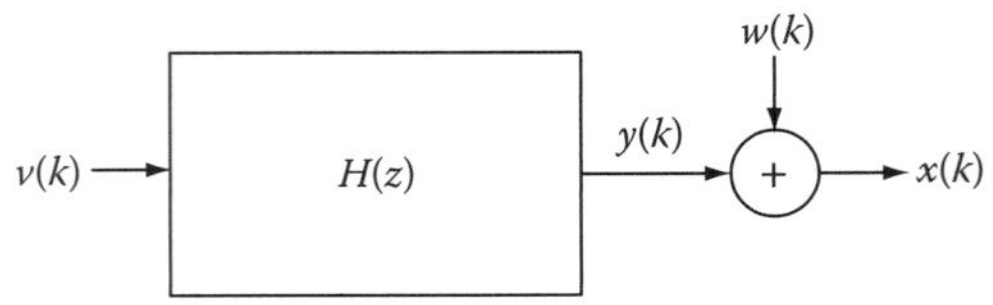

FIGURE 18.2 Single channel model.

where $\hat{h}(n)$ denotes complex cepstrum, i.e., $\hat{h}(n) = F^{-1}[\ln(H(\omega))]$, where $F^{-1}(.)$ denotes inverse Fourier transform. From Equation 18.27, the system impulse response $h(k)$ can be reconstructed from $b_x(m, 0)$ (or $b_x(0, m)$, or $b_x(m, m)$), within a constant and a time delay, via inverse cepstrum operations.

The main difficulty with cepstrum operations is taking the logarithm of a complex number, i.e., $\ln(z) = \ln(|z|) + j\arg(z)$. The term $\arg(z)$ is defined up to an additive multiple of 2π. When applying log operation to $C_3^x(\omega_1, \omega_2)$, an integer multiple of 2π needs to be added to the phase at each (ω_1, ω_2), in order to maintain a continuous phase. This is called phase unwrapping and is a process that involves complexity and is sensitive to noise. Just using the principal argument of the phase will not result in a correct system estimate. To avoid phase unwrapping, the bicepstrum can be estimated using the group delay approach:

$$b_x(m, n) = \frac{1}{m} F_2^{-1}\left\{\frac{F_2\left[\tau_1 c_3^x(\tau_1, \tau_2)\right]}{C_3^x(\omega_1, \omega_2)}\right\}, \quad m \neq 0 \tag{18.28}$$

with $b_x(0, n) = b_x(n, 0)$, and $F_2\{\cdot\}$ and $F_2^{-1}\{\cdot\}$ denoting 2-D Fourier transform operator and its inverse, respectively. The cepstrum of the system impulse response can also be computed directly from the cumulants of the system output based on the equation [33]:

$$\sum_{k=1}^{\infty} k\hat{h}(k)\left[c_3^x(m-k, n) - c_3^x(m+k, n+k)\right] + k\hat{h}(-k)\left[c_3^x(m-k, n-k) - c_3^x(m+k, n)\right] = mc_3^x(m, n) \tag{18.29}$$

If $H(z)$ has no zeros on the unit circle its cepstrum decays exponentially; thus, Equation 18.29 can be truncated to yield an approximate equation. An overdetermined system of truncated equations can be formed for different values of m and n, which can be solved for $\hat{h}(k)$, $k = \ldots, -1, 1, \ldots$. The system response $h(k)$ can then be recovered from its cepstrum via inverse cepstrum operations.

The above described bicepstrum approach results in estimates with small bias and variance as compared to many other approaches [33,36].

A similar methodology can be applied for system estimation using fourth-order statistics. The inverse Fourier transform of the logarithm of the trispectrum, or otherwise tricepstrum, $t_x(m, n, l)$, of $x(k)$ is also zero everywhere expect along the axes and the diagonal $m = n = l$ [21]. Along these lines it equals the complex cepstrum; thus, $h(k)$ can be recovered from slices of the tricepstrum based on inverse cepstrum operations.

For the case of nonlinear processes, the bicepstrum is nonzero everywhere [17]. The distinctly different structure of the bicepstrum corresponding to linear and nonlinear processes can be used to test for deviations from linearity [17].

18.4.2 A Frequency-Domain Approach

Let us assume that $H(z)$ has no zeros on the unit circle. Let us take ω to be discrete frequency taking values in $\{\omega = (2\pi/N)k, k = 0, \ldots, N-1\}$, and for notational simplicity, let us denote the bispectrum at $\omega_1 = (2\pi/N)k_1$, $\omega_2 = (2\pi/N)k_2$ as $C_3^x(k_1, k_2)$.

Based on Equation 18.26, it can be easily shown [38] that, for some $r, \ell \in [0, \ldots, N-1]$, it holds

$$\log H(i+r) - \log H(i) = \log H(\ell + r) - \log H(\ell) + \log C_3^y(-i-r-\ell, \ell) - \log C_3^x(-i-r-\ell, \ell + r) \tag{18.30}$$

for $i = 0, \ldots, N-1$. One can also show that

$$c_{\ell,r} \stackrel{\Delta}{=} \log H(\ell + r) - \log H(\ell) = \frac{1}{N}\sum_{k=0}^{N-1}\left[\log C_3^y(k, \ell + r) - \log C_3^y(k, \ell)\right] \tag{18.31}$$

Thus, Equation 18.30 becomes

$$\log H(i+r) - \log H(i) = \log C_3^y(-i-r-\ell,\ell) - \log C_3^x(-i-r-\ell,\ell+r) + c_{\ell,r} \tag{18.32}$$

Let us assume that $H(0)=1$. This introduces a scalar ambiguity to the solution; however, there is an inherent scalar ambiguity to the problem anyway. Based on Equation 18.32 for $i=0,\ldots,N-1$, let us form the matrix equation

$$\mathbf{A}\tilde{\mathbf{h}} = c_\ell \tag{18.33}$$

where

$\mathbf{c}_\ell$ is a vector whose ith element equals $\log C_3^y(-i-r-\ell,\ell) - \log C_3^x(-i-r-\ell,\ell+r) + c_{\ell,r}$
$\tilde{\mathbf{h}} = [\log H(1),\ldots,\log H(N-1)]^{\mathrm{T}}$
$\mathbf{A}$ is an $(N-1)\times(N-1)$ matrix defined as follows

Let $\tilde{\mathbf{A}}$ be an $N\times N$ circulant matrix whose first row contains all zeros expect at the first and the $r+1$ entries, where it contains -1 and 1, respectively. Then $\mathbf{A}$ equals what is left of $\tilde{\mathbf{A}}$ after discarding its first column and its last row.

It can be shown that as long as r and N are coprime integers, $\det(\mathbf{A})=1$, thus $\mathbf{A}$ is invertible, and Equation 18.33 can be solved for $\mathbf{h}$. Finally, the estimate of $[H(1),\ldots,H(N-1)]^{\mathrm{T}}$ can be obtained by taking exp $(\tilde{\mathbf{h}})$. The value $H(0)$ is set to 1. We should note that the log operations used in this approach can be based on the principal argument of the phase, in other words, no phase unwrapping is required. When using principal arguments of phases, the recovered frequency response $H(k)$ differs from the true one by a scalar and a linear phase term of the form $\exp\left(j\frac{2\pi}{N}km_\ell\right)$. Equivalently, the impulse response is recovered within a scalar and a circular shift by m_ℓ, which depends on the particular choice of ℓ.

The ability to select slices allows one to avoid regions in which the bispectrum estimate exhibits high variance, or regions where the bispectrum values are low. A methodology to select good slices was proposed in [38]. Estimates corresponding to different pairs of slices can be combined, after taking care of the corresponding circular shift, in order to improve the system estimate. More details and extensions to HOS can be found in [38].

18.4.3 Parametric Methods

Let us model $H(z)$ (see Figure 18.2) as an autoregressive moving average (ARMA) system, with autoregressive (AR) and moving average (MA) orders p and q, respectively. Input and output signals are related as

$$\sum_{i=0}^{p} a(i)y(k-i) = \sum_{j=0}^{q} b(j)v(k-j) \tag{18.34}$$

where $a(i)$, $b(j)$ represent the AR and MA parameters of the system.

Equations analogous to the Yule–Walker equations [26] can be derived based on third-order cumulants of $x(k)$, i.e.,

$$\sum_{i=0}^{p} a(i)c_3^x(\tau-i,j) = 0, \quad \tau > q \tag{18.35}$$

or

$$\sum_{i=1}^{p} a(i)c_3^x(\tau - i, j) = -c_3^x(\tau, j), \quad \tau > q \tag{18.36}$$

where it was assumed $a(0) = 1$. Concatenating Equation 18.36 for $\tau = q+1, \ldots, q+M$, $M \geq 0$ and $j = q-p, \ldots, q$, the matrix equation

$$\mathbf{Ca} = \mathbf{c} \tag{18.37}$$

can be formed, where **C** and **c** are a matrix and a vector, respectively, formed by third-order cumulants of the process according to Equation 18.36, and the vector **a** contains the AR parameters. The vector **a** can be obtained by solving Equation 18.37. If the AR-order p is unknown and Equation 18.37 is formed based on an overestimate of p, the resulting matrix **C** has always rank p. In this case, the AR parameters can be obtained using a low-rank approximation of **C** [19].

Using the estimated AR parameters, a pth-order filter with transfer function $\hat{A}(z) = 1 + \sum_{i=1}^{p} a(i)z^{-1}$ can be constructed. Based on the filtered through $A(z)$ process $x(k)$, i.e., $\tilde{x}(k)$, or otherwise known as the residual time series [19], the MA parameters can be estimated via any MA method [27], for example [20]:

$$b(k) = \frac{c_3^{\tilde{x}}(q, k)}{c_3^{\tilde{x}}(q, 0)}, \quad k = 0, 1, \ldots, q \tag{18.38}$$

Practical problem associated with the parametric approach is sensitivity to model-order mismatch. A significant amount of research has been devoted to the ARMA parameter estimation problem. A good review of the literature on this topic can be found in [27,29].

18.5 HOS for Blind MIMO System Identification

A more general system identification problem than the one discussed in Section 18.4, is the identification of a MIMO system (Figure 18.3). The goal here is, given the n system outputs and some statistical knowledge about the r inputs, to estimate $H(z)$, and subsequently recover the input signals (sources). This problem is also referred to as blind source separation and is in the heart of many important applications. For example, in speech enhancement in the presence of competing speakers, an array of microphones is used to obtain multiple recordings, based on which the signals of interest can be estimated. The microphone measurements can be viewed as the outputs of a MIMO system representing the acoustic environment. MIMO models arise frequently in digital multiuser/multiaccess communications systems, digital radio with diversity, multisensor sonar/radar systems [50,53]. They also arise in biomedical measurements, when recordings of a distributed array of sensors, placed on the skin, are used to pick up signals originating from inside the body.

A special case of MIMO systems are the memoryless systems, where the cross-channels are just scalars. HOS have played a key role in blind MIMO system estimation for both memoryless and convolutive systems. Identification of memoryless systems excited by white independent non-Gaussian inputs has been studied under the name independent component analysis (ICA) (see [12] and references therein), or

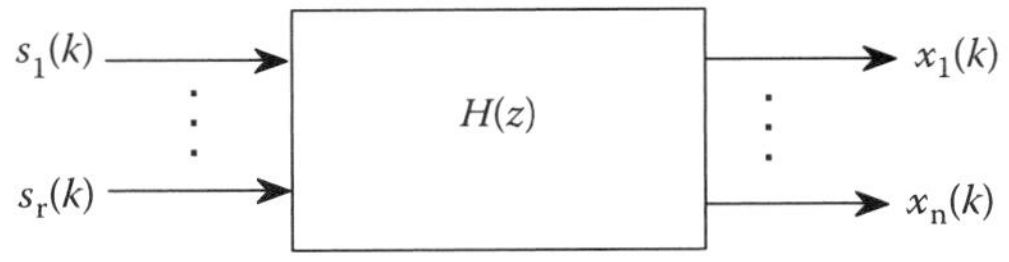

FIGURE 18.3 A MIMO system.

separation of instantaneous mixtures. ICA-based methods search for a linear transformation of the system output that minimizes the statistical dependence between its components. Due to the "central limit theorem," a mixing of non-Gaussian signals tends to produce almost Gaussian outputs. The linear transformation sought here should maximize the non-Gaussianity of the resulting signals. Non-Gaussianity can be measured by the kurtosis (fourth-order statistics), among other criteria. Solutions to the same problem have been proposed based on minimization of contrast functions [13], or multi-linear singular-value decomposition [14,15].

In this section, we focus on the identification of convolutive MIMO systems. Most of the blind convolutive MIMO system identification methods in the literature exploit either SOS or HOS. SOS-based methods, as opposed to HOS-based ones, do not require long data in order to obtain good estimates and involve low complexity. Examples of SOS-based methods can be found in [1,3,4,16,52]. All these methods require channel diversity and they apply to nonwhite inputs only. On the other hand, HOS-based methods provide system information without requiring channel diversity, and also can deal with white inputs as long as they are non-Gaussian. Examples of HOS MIMO methods can be found in [7,9,11,18,24,53].

Among the possible approaches, frequency-domain methods offer certain advantages over time domain ones: they do not require system length information, and also, their formulation can take advantage of existing results for the memoryless MIMO problem. Indeed, in the frequency domain, at each frequency, the convolutive problem is transformed into a scalar one. However, an additional step is required to resolve the frequency-dependent permutation, scaling, and phase ambiguities.

Next we outline the frequency-domain approach of [11], that was proposed for the convolutive MIMO case with white independent inputs.

Let us consider the $(n \times n)$ MIMO estimation problem. The more general case of $n \times r$ is studied in [11,58].

Let $\mathbf{s}(k) = [s_1(k) \cdots s_n(k)]^T$ be a vector of n statistically independent zero-mean stationary sources, $\mathbf{h}(l)$ the impulse response matrix whose (i,j) element is denoted by $\{h_{ij}(l)\}$; $\mathbf{x}(k) = [x_1(k) \cdots x_n(k)]^T$ the vector of observations, and $\mathbf{n}(k) = [n_1(k) \cdots n_n(k)]^T$ represent observation noise. Note all the variables can be real or complex-valued.

The MIMO system output equals

$$\mathbf{x}(k) = \sum_{l=0}^{L-1} \mathbf{h}(l)\mathbf{s}(k-l) + \mathbf{n}(k) \tag{18.39}$$

where L is the length of the longest $h_{ij}(k)$.

Again, let ω denote discrete frequency, i.e., $\omega = \frac{2\pi}{N}k$, $k = 0, \ldots, N-1$ with $N > L$. Let $\mathbf{H}(\omega)$ be a $n \times n$ matrix whose (i,j) element equals N-point $(N > L)$ discrete Fourier transform of $h_{ij}(k)$ at frequency ω.

Let us also assume that the inputs are zero-mean, non-Gaussian, independent i.i.d., stationary process, each with nonzero skewness, and unit variance, i.e., $\gamma_{s_i}^2 = 1$ for $i = 1, \ldots, n$. The noise processes $n_i(.)$, $i = 1, \ldots, n$ are zero-mean Gaussian stationary random processes, mutually independent, independent of the inputs, with variance σ_n^2.

The mixing channels are generally complex. The matrix $\mathbf{H}(\omega)$ is invertible for all ω's. In addition, there exist a nonempty subset of ω's, denoted by ω_*, and a nonempty subset of the indices $1, \ldots, n$, denoted l_*, so that for $l \in l_*$ and $\omega \in \omega_*$, the lth row of matrix $\mathbf{H}(\omega)$ has elements with magnitudes that are mutually different.

The power spectrum of the received signal vector equals

$$\mathbf{P}_X(\omega) = \mathbf{H}^*(\omega)\mathbf{H}^T(\omega) + E\{\mathbf{n}^*(\omega)\mathbf{n}^T(\omega)\} \tag{18.40}$$

$$= \mathbf{H}^*(\omega)\mathbf{H}^T(\omega) + \sigma_n^2\mathbf{I} \tag{18.41}$$

Let us define the whitening matrix $\mathbf{V}(\omega)$ $(n \times n)$, as follows:

$$\mathbf{V}(\omega)\left(\mathbf{P}_X(\omega) - \sigma_n^2\mathbf{I}\right)\mathbf{V}(\omega)^H = \mathbf{I} \tag{18.42}$$

The existence of $\mathbf{V}(\omega)$ is guaranteed by the assumptions stated above.

Based on the assumption of i.i.d. input signals, the cross-cumulant of the received signals $x_l(k), x_i^\star(k), x_j(k)$ equals

$$\begin{aligned} c_{lij}^3(\tau,\rho) &\stackrel{\Delta}{=} \mathrm{Cum}\left[x_l(m), x_i^\star(m+\tau), x_j(m+\rho)\right] \\ &= \sum_{p=1}^{n} \gamma_{s_p}^3 \sum_{m=0}^{L-1} h_{lp}(m)h_{ip}^\star(m+\tau)h_{jp}(m+\rho) \end{aligned} \tag{18.43}$$

where $\gamma_{s_p}^3 = \mathrm{Cum}\left[s_p(k), s_p^\star(k), s_p(k)\right]$. The cross-bispectrum of $x_l(k), x_i^\star(k), x_j(k)$, defined as the two-dimensional Discrete Fourier Transform of $c_{lij}^3(\tau,\rho)$, equals

$$C_{lij}^3(\omega_1,\omega_2) = \sum_{p=1}^{n} \gamma_{s_p}^3 H_{lp}(-\omega_1-\omega_2)H_{ip}^\star(-\omega_1)H_{jp}(\omega_2) \tag{18.44}$$

Let us now define a matrix $\mathbf{C}_l^3(\omega,\beta-\omega)$, whose (i,j)th element is equal to $C_{lij}^3(\omega,\beta-\omega)$. Then, from Equation 18.44 we get

$$\mathbf{C}_l^3(\omega,\beta-\omega) = \mathbf{H}^\star(-\omega)\begin{bmatrix} \gamma_{s_1}^3 H_{l1}(-\beta) & 0 & \cdots & 0 \\ 0 & \gamma_{s_2}^3 H_{l2}(-\beta) & \cdots & 0 \\ \vdots & \vdots & \ddots & \vdots \\ 0 & 0 & \cdots & \gamma_{s_n}^3 H_{ln}(-\beta) \end{bmatrix}\mathbf{H}^T(\beta-\omega) \tag{18.45}$$

Let us also define $\mathbf{Y}_l^3(\omega,\beta-\omega)$ as follows:

$$\mathbf{Y}_l^3(\omega,\beta-\omega) \stackrel{\Delta}{=} \mathbf{V}(-\omega)\mathbf{C}_l^3(\omega,\beta-\omega)\mathbf{V}(\beta-\omega)^H \tag{18.46}$$

Combining Equations 18.42 and 18.46 we get

$$\mathbf{Y}_l^3(\omega,\beta-\omega) = \mathbf{W}(-\omega)\begin{bmatrix} \gamma_{s_1}^3 H_{l1}(-\beta) & 0 & \cdots & 0 \\ 0 & \gamma_{s_2}^3 H_{l2}(-\beta) & \cdots & 0 \\ \vdots & \vdots & \ddots & \vdots \\ 0 & 0 & \cdots & \gamma_{s_n}^3 H_{ln}(-\beta) \end{bmatrix}\mathbf{W}(\beta-\omega)^H \tag{18.47}$$

where

$$\mathbf{W}(\omega) = \mathbf{V}(\omega)\mathbf{H}(\omega)^\star \tag{18.48}$$

It can be easily verified using Equation 18.42 that $\mathbf{W}(\omega)$ is an orthonormal matrix, i.e., $\mathbf{W}(\omega)\,\mathbf{W}(\omega)^H = \mathbf{I}$.

Equation 18.47 can be viewed as a singular-value decomposition (SVD) of $\mathbf{Y}_l^3(\omega,\beta-\omega)$.

For each ω let us take β so that $-\beta \in \omega_\star$, and also take $l \in l_\star$. Based on our assumptions, $\mathbf{Y}_l^3(\omega,\beta-\omega)$ has full rank, thus $\mathbf{W}(-\omega)$ is unique up to column permutation and phase ambiguities. Since all SVDs use the same ordering of singular values, the column permutation is the same for all ωs.

Although l and $-\beta$ could take any value in $l_\star$ and $\omega_\star$, respectively, the corresponding $\mathbf{W}(\omega)$s cannot be easily combined to obtain a more robust solution, as each set of ls and $-\beta$s corresponds to a different permutation matrix. An alternative approach to utilize the redundant information can be based on the ideas of "joint diagonalization," proposed in [8] for ICA. Joint diagonalization of several matrices, i.e., $\{\mathbf{M}_l\}_{l=1}^{K}$, is a way to define an average eigenstructure shared by the matrices. If the matrices are given in the form: $\mathbf{M}_l = \mathbf{U}\mathbf{D}_l\mathbf{U}^{\mathrm{H}}$, where $\mathbf{U}$ is a unitary matrix and each $\mathbf{D}_l$ is a diagonal matrix, then the criterion defined as

$$T(\mathbf{V}) \triangleq \sum_{l=1}^{K} \mathrm{off}\left(\mathbf{V}^{\mathrm{H}}\mathbf{M}_l\mathbf{V}\right) \tag{18.49}$$

where off($\mathbf{B}$) denotes the sum of the square absolute values of the nondiagonal elements of $\mathbf{B}$, reaches its minimum value at $\mathbf{U} = \mathbf{V}$, and at the minimum, $T(\mathbf{U}) = 0$. In the case where the matrices are arbitrary, an approximate joint diagonalizer is still defined as a unitary minimizer of Equation 18.49.

These ideas can be applied for the joint diagonalization of the matrices $\{\mathbf{M}_{l,\beta}\}$ defined as

$$\mathbf{M}_{l\beta} \triangleq \mathbf{Y}_l^3(\omega, \beta - \omega)^{\mathrm{H}}\mathbf{Y}_l^3(\omega, \beta - \omega) \tag{18.50}$$

$$= \mathbf{W}(-\omega)\mathbf{D}_{l,\beta}\mathbf{W}(-\omega)^{\mathrm{H}}, \quad l = 1, \ldots, n, \quad \beta \in \omega_\star \tag{18.51}$$

where $\mathbf{D}_{l\beta} = \mathrm{Diag}\left(\ldots, \left|\gamma_{s_i}^3 H_{li}(-\beta)\right|^2, \ldots\right)$. Thus, $\mathbf{W}(-\omega)$ can be found as the matrix that minimizes Equation 18.49 over all possible ls and all ωs, and satisfies $T(\mathbf{W}(\omega)) = 0$. However, a phase ambiguity still exists, since if $\mathbf{W}(\omega)$ is a joint diagonalizer so is matrix $\mathbf{W}(\omega)e^{j\mathbf{\Phi}(\omega)}$, where $\Phi(\omega)$ is a real diagonal matrix.

18.5.1 Resolving the Phase Ambiguity

Let $\hat{\mathbf{W}}(\omega)$ be the eigenvector matrix obtained via joint diagonalization or SVD. Due to the phase ambiguity mentioned above, it holds

$$\hat{\mathbf{W}}(\omega) = \mathbf{W}(\omega)e^{j\mathbf{\Phi}(\omega)} \tag{18.52}$$

where $\mathbf{W}(\omega)$ as defined in Equation 18.48.

Let us define

$$\begin{aligned}\hat{\mathbf{H}}_{\Phi}(\omega) &\triangleq \left[\mathbf{V}(\omega)\right]^{-1}\hat{\mathbf{W}}(\omega)\right]^{\star} \\ &= \mathbf{H}(\omega)\mathbf{P}e^{-j\mathbf{\Phi}(\omega)}\end{aligned} \tag{18.53}$$

The solution $\hat{\mathbf{H}}_{\Phi}(\omega)$ could be used in an inverse filtering scheme to decouple the inputs, but would leave a shaping ambiguity in each input. To obtain a solution, we need to eliminate the phase ambiguity.

Let us define

$$\begin{aligned}\mathbf{Q}(\omega) &\triangleq \left(\hat{\mathbf{H}}_{\Phi}(\omega)^{\star}\right)^{-1}\mathbf{C}_l^3(-\omega, \omega + \alpha)\left(\hat{\mathbf{H}}_{\Phi}(\omega + \alpha)^{\mathrm{T}}\right)^{-1} \\ &= e^{-j\mathbf{\Phi}(\omega)}\mathbf{P}^{\mathrm{T}}\begin{bmatrix} \gamma_{s_1}^3 H_{l1}(-\alpha) & 0 & \cdots & 0 \\ 0 & \gamma_{s_2}^3 H_{l2}(-\alpha) & \cdots & 0 \\ \vdots & \vdots & \ddots & 0 \\ 0 & 0 & \cdots & \gamma_{s_n}^3 H_{ln}(-\alpha) \end{bmatrix}\mathbf{P}e^{j\mathbf{\Phi}(\omega+\alpha)}\end{aligned} \tag{18.54}$$

for α being any of the N discrete frequencies $\left\{\frac{2\pi}{N}l, l = 0, \ldots, N-1\right\}$.

Based on the properties of permutation matrices, it can be seen that $\mathbf{Q}(\omega)$ is a diagonal matrix. The following relation holds for the phases of the quantities involved in Equation 18.54:

$$\mathbf{\Phi}(\omega+\alpha)-\mathbf{\Phi}(\omega)=\mathbf{\Psi}(\omega)-\mathbf{\Theta}(\alpha) \tag{18.55}$$

where

$$\mathbf{\Psi}(\omega)=\arg\{\mathbf{Q}(\omega)\} \tag{18.56}$$

and

$$\mathbf{\Theta}(\alpha)=\arg\left\{\mathbf{P}^{\mathrm{T}}\mathrm{Diag}\left(\ldots,\gamma_{s_i}^3 H_{li}(-\alpha),\ldots\right)\mathbf{P}\right\} \tag{18.57}$$

with both $\mathbf{\Psi}(\omega)$ and $\mathbf{\Theta}(\alpha)$ being diagonal.

At this point "arg{·}" denotes phase taking any value in $(-\infty,\infty)$, as opposed to modulo 2π phase.

Summing up Equation 18.55 over all discrete frequencies, i.e., $\left\{\omega=\frac{2\pi}{N}k,\quad k=0,\ldots,N-1\right\}$, yields a zero left-hand side, thus we get

$$\mathbf{\Theta}(\alpha)=\frac{1}{N}\sum_{k=0}^{N-1}\mathbf{\Psi}\left(\frac{2\pi}{N}k\right) \tag{18.58}$$

which implies that $\mathbf{\Theta}(\alpha)$ can actually be computed from $\mathbf{Q}(\omega)$.

Let $\Phi_{ii}(\omega)$ and $\Psi_{ii}(\omega)$ denote the ith diagonal elements of matrices $\mathbf{\Phi}(\omega)$ and $\mathbf{\Psi}(\omega)$, respectively. Let us also define the vectors:

$$\mathbf{\Phi}_i\triangleq\left[\Phi_{ii}\left(\frac{2\pi}{N}1\right),\ldots,\Phi_{ii}\left(\frac{2\pi}{N}(N-1)\right)\right]^{\mathrm{T}},\quad i=1,\ldots,n \tag{18.59}$$

$$\mathbf{\Psi}_i\triangleq\left[\Psi_{ii}\left(\frac{2\pi}{N}1\right),\ldots,\Psi_{ii}\left(\frac{2\pi}{N}(N-1)\right)\right]^{\mathrm{T}},\quad i=1,\ldots,n \tag{18.60}$$

Let $\alpha=\frac{2\pi}{N}k_\alpha$, where k_α is an integer in $[0,\ldots,N-1]$. For k_α and N coprime it holds:

$$\mathbf{\Phi}_i=\mathbf{A}_{N,k_\alpha}^{-1}\mathbf{\Psi}_i+\Theta_{ii}(\alpha)\mathbf{A}_{N,k_\alpha}^{-1}\mathbf{1}_{(N-1)\times 1}-\Phi_{ii}(0)\mathbf{1}_{(N-1)\times 1},\quad i=1,\ldots,n \tag{18.61}$$

where

$\Theta_{ii}(\alpha)$ is the (i, i) element of the diagonal matrix $\mathbf{\Theta}(\alpha)$

$\mathbf{1}_{(N-1)\times 1}$ denotes a $(N-1)\times 1$ vector whose elements are equal to 1

$\mathbf{A}_{N,k_\alpha}$ is an $(N-1)\times(N-1)$ matrix

Matrix $\mathbf{A}_{N,k_\alpha}$ is constructed as follows. Let $\tilde{\mathbf{A}}$ be an $N\times N$ circulant matrix whose first row contains all zeros expect at the first and the $k_\alpha+1$ entries, where it contains -1 and 1, respectively. Then $\mathbf{A}$ equals what is left of $\tilde{\mathbf{A}}$ after discarding its rightmost column and its last row.

As long as N and k_α are coprime, $\mathbf{A}$ has full rank and $\det(\mathbf{A})=-1$.

Next we provide an example of matrix $\mathbf{A}$ corresponding to $N=8$ and $k_\alpha=5$.

$$\mathbf{A}_{8,5}=\begin{bmatrix} -1 & 0 & 0 & 0 & 0 & 1 & 0 \\ 0 & -1 & 0 & 0 & 0 & 0 & 1 \\ 0 & 0 & -1 & 0 & 0 & 0 & 0 \\ 1 & 0 & 0 & -1 & 0 & 0 & 0 \\ 0 & 1 & 0 & 0 & -1 & 0 & 0 \\ 0 & 0 & 1 & 0 & 0 & -1 & 0 \\ 0 & 0 & 0 & 1 & 0 & 0 & -1 \end{bmatrix} \tag{18.62}$$

Although Equation 18.61 provides a closed-form solution for $\mathbf{\Phi}_i$, and thus for $\mathbf{\Phi}(\omega)$, it is not a convenient formula for phase computation. Ideally we would like the phase expression to involve the modulo 2π phases only, so that no phase unwrapping is necessary. It turns out that $\mathbf{\Phi}_i$ can actually be based on modulo 2π phases, and this is due to the special properties of matrix $\mathbf{A}_{N,k_\alpha}$.

It was shown in [11], that if principal arguments of phases are used, the resulting phase contains a constant phase ambiguity and a linear phase ambiguity. A system estimate defined on principal arguments-based phase estimate equals

$$\begin{aligned}\hat{\mathbf{H}}(\omega) &\triangleq \hat{\mathbf{H}}_\Phi(\omega)e^{j\hat{\mathbf{\Phi}}(\omega)} \\ &= \mathbf{H}(\omega)\mathbf{P}e^{-j\mathbf{\Phi}(0)-j\omega\mathbf{M}}, \quad \omega \neq 0\end{aligned} \tag{18.63}$$

where $\mathbf{M}$ is a constant integer diagonal matrix. Thus, $\mathbf{H}(\omega)$ is estimated within a constant permutation matrix, a constant diagonal matrix, and a linear phase term.

To see the effect of the estimate of Equation 18.63 on the inputs, let us consider an inverse filter operation:

$$\hat{\mathbf{H}}^{-1}(\omega)\mathbf{x}(\omega) = \left(\mathbf{P}e^{-j\mathbf{\Phi}(0)-j\omega\mathbf{M}}\right)^{-1}\mathbf{s}(\omega) \tag{18.64}$$

Thus, the result of the inverse filtering is a vector whose elements are the input signals multiplied by the diagonal elements of the permuted matrix $e^{j\mathbf{\Phi}(0)+j\omega\mathbf{M}}$.

The prewhitening matrix, $\mathbf{V}(\omega)$, was necessary in order to apply joint diagonalization. Prewhitening is employed in the majority of HOS-based blind MIMO estimation methods [3,8,9,28]. However, this is a sensitive process as it tends to lengthen the global system response, and as a result increases complexity and estimation errors.

The need for whitening can be obviated by a decomposition that does not require unitary matrices. One such approach is the parallel factorization (PARAFAC) decomposition, which is a low-rank decomposition of three- or higher-way arrays [10]. The PARAFAC decomposition can be thought as an extension of SVD to multiway arrays, where uniqueness is guaranteed even if the nondiagonal matrices involved are nonunitary. Blind MIMO system estimation based on PARAFAC decomposition of HOS-based tensors of the system output can be found in [2]. In [58], it was shown that PARAFAC decomposition of a K–way tensor, constructed based on Kth-order statistics of the system outputs, allows for identification of a general class of convolutive and systems that can have more inputs than outputs.

18.6 Nonlinear Processes

Despite the fact that progress has been established in developing the theoretical properties of nonlinear models, only a few statistical methods exist for detection and characterization of nonlinearities from a finite set of observations. In this section, we will consider nonlinear Volterra systems excited by Gaussian

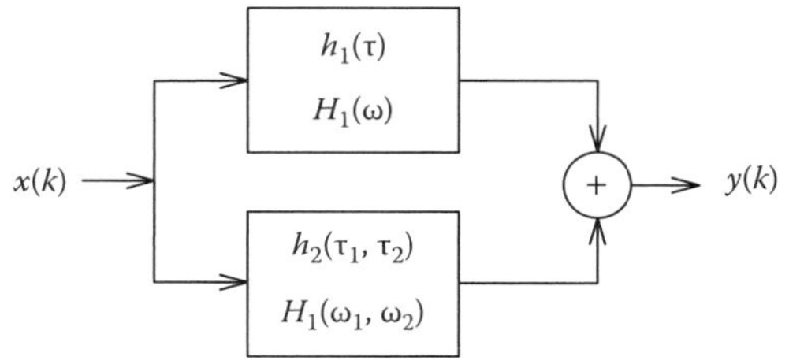

FIGURE 18.4 Second-order Volterra system. Linear and quadratic parts are connected in parallel.

stationary inputs. Let $y(k)$ be the response of a discrete-time invariant pth-order Volterra filter whose input is $x(k)$. Then,

$$y(k) = h_0 + \sum_i \sum_{\tau_1,\ldots,\tau_i} h_i(\tau_1,\ldots,\tau_i)x(k-\tau_1)\ldots x(k-\tau_i) \tag{18.65}$$

where $h_i(\tau_1,\ldots,\tau_i)$ are the Volterra kernels of the system, which are symmetric functions of their arguments; for causal systems $h_i(\tau_1,\ldots,\tau_i) = 0$ for any $\tau_i < 0$.

The output of a second-order Volterra system when the input is zero-mean stationary is

$$y(k) = h_0 + \sum_{\tau_1} h_1(\tau_1)x(k-\tau_1) + \sum_{\tau_1}\sum_{\tau_2} h_2(\tau_1,\tau_2)x(k-\tau_1)x(k-\tau_2) \tag{18.66}$$

Equation 18.66 can be viewed as a parallel connection of a linear system $h_1(\tau_1)$ and a quadratic system $h_2(\tau_1, \tau_2)$ as illustrated in Figure 18.4. Let

$$c_2^{xy}(\tau) = E\{x(k+\tau)[y(k) - m_1^y]\} \tag{18.67}$$

be the cross-covariance of input and output, and

$$c_3^{xxy}(\tau_1,\tau_2) = E\{x(k+\tau_1)x(k+\tau_2)[y(k) - m_1^y]\} \tag{18.68}$$

be the third-order cross-cumulant sequence of input and output.

It can be shown that the system's linear part can be identified by

$$H_1(-\omega) = \frac{C_2^{xy}(\omega)}{C_2^x(\omega)} \tag{18.69}$$

and the quadratic part by

$$H_2(-\omega_1,-\omega_2) = \frac{C_3^{xxy}(\omega_1,\omega_2)}{2C_2^x(\omega_1)C_2^x(\omega_2)} \tag{18.70}$$

where $C_2^{xy}(\omega)$ and $C_3^{xxy}(\omega_1,\omega_2)$ are the Fourier transforms of $c_2^{xy}(\tau)$ and $c_3^{xxy}(\tau_1,\tau_2)$, respectively. It should be noted that the above equations are valid only for Gaussian input signals. More general results assuming non-Gaussian input have been obtained in [23,37]. Additional results on particular nonlinear systems have been reported in [6,42].

An interesting phenomenon caused by a second-order nonlinearity is the quadratic phase coupling. There are situations where nonlinear interaction between two harmonic components of a process contribute to the power of the sum and/or difference frequencies. The signal

$$x(k) = A\cos(\lambda_1 k + \theta_1) + B\cos(\lambda_2 k + \theta_2) \tag{18.71}$$

after passing through the quadratic system:

$$z(k) = x(k) + \varepsilon x^2(k), \quad \varepsilon \neq 0 \tag{18.72}$$

contains cosinusoidal terms in $(\lambda_1, \theta_1), (\lambda_2, \theta_2), (2\lambda_1, 2\theta_1), (2\lambda_2, 2\theta_2), (\lambda_1 + \lambda_2, \theta_1 + \theta_2), (\lambda_1 - \lambda_2, \theta_1 - \theta_2)$. Such a phenomenon that results in phase relations that are the same as the frequency relations is called quadratic phase coupling. Quadratic phase coupling can arise only among harmonically related components. Three frequencies are harmonically related when one of them is the sum or difference of the other two. Sometimes it is important to find out if peaks at harmonically related positions in the power spectrum are in fact phase-coupled. Due to phase suppression, the power spectrum is unable to provide an answer to this problem.

As an example, consider the process [39]:

$$x(k) = \sum_{i=1}^{6} \cos(\lambda_i k + \phi_i) \tag{18.73}$$

where $\lambda_1 > \lambda_2 > 0, \lambda_4 + \lambda_5 > 0, \lambda_3 = \lambda_1 + \lambda_2, \lambda_6 = \lambda_4 + \lambda_5, \phi_1, \ldots, \phi_5$ are all independent, uniformly distributed random variables over $(0, 2\pi)$, and $\phi_6 = \phi_4 + \phi_5$. Among the six frequencies, $(\lambda_1, \lambda_2, \lambda_3)$ and $(\lambda_4, \lambda_5, \lambda_6)$ are harmonically related, however, only λ_6 is the result of phase coupling between λ_4 and λ_5. The power spectrum of this process consists of six impulses at λ_i, $i = 1, \ldots, 6$ (see Figure 18.5), offering no indication whether each frequency component is independent or a result of frequency coupling. On the other hand, the bispectrum of $x(k)$, $C_3^x(\omega_1, \omega_2)$ (evaluate in its principal region) is zero everywhere, except at point (λ_4, λ_5) of the (ω_1, ω_2) plane, where it exhibits an impulse (Figure 18.5b). The peak indicates that only λ_4, λ_5 are phase-coupled.

The bicoherence index, defined as

$$P_3^x(\omega_1, \omega_2) = \frac{C_3^x(\omega_1, \omega_2)}{\sqrt{C_2^x(\omega_1) C_2^x(\omega_2) C_2^x(\omega_1 + \omega_2)}} \tag{18.74}$$

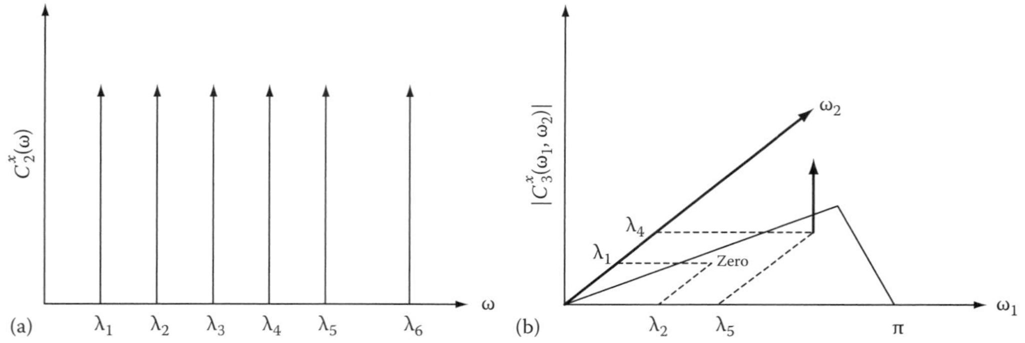

FIGURE 18.5 Quadratic phase coupling. (a) The power spectrum of the process described in Equation 18.73 cannot determine what frequencies are coupled. (b) The corresponding magnitude bispectrum is zero everywhere in the principle region, except at points corresponding to phase-coupled frequencies.

has been extensively used in practical situations for the detection and quantification of quadratic phase coupling. The value of the bicoherence index at each frequency pair indicates the degree of coupling among the frequencies of that pair. Almost all bispectral estimators can be used in Equation 18.74. However, estimates obtained based on parametric modeling of the bispectrum have been shown to yield superior resolution [39,40] than the ones obtained with conventional methods.

18.7 Conclusions

In the mid-1980s, when power spectrum-based techniques were dominating the signal processing literature, there was an explosion of interest in HOS as they appeared to provide solutions to pressing problems. Initially there was a lot of interest in the noise suppression ability of HOS. However, soon came the realization that perhaps estimation issues such as complexity and need for long data lengths stood in the way of achieving the desirable noise suppression. Then, the focus shifted to more difficult problems, where HOS could really make a difference despite their complexity, simply because of the signal-related information that they contain. Today, HOS are viewed as an indispensable tool for modeling of time series, blind source separation and blind system estimation problems.

Software for signal processing with HOS can be found at http://www.mathworks.com/matlabcentral developed by A. Swami, or at www.ece.drexel.edu/CSPL developed by A. Petropulu's research group.

Acknowledgments

Parts of this chapter were based on [29]. The author would like to thank Prof. C.L. Nikias for helpful discussions and Dr. U. Abeyratne for providing the figures.

References

1. K. Abed-Meraim, P. Loubaton, and E. Moulines, A subspace algorithm for certain blind identificaion problems, *IEEE Trans. Inform. Theory*, 43, 499–511, March 1997.
2. T. Acar, A.P. Petropulu, and Y. Yu, Blind MIMO system estimation based on PARAFAC decomposition of tensors formed based HOS of the system output, *IEEE Trans. Signal Process.*, 54(11), 4156–4168, November 2006.
3. A. Belouchrani, K.A. Meraim, J.F. Cardoso, and E. Moulines, A blind source separation technique using second-order statistics, *IEEE Trans. Signal Process.*, 45(2), 434–444, February 1997.
4. I. Bradaric, A.P. Petropulu, and K.I. Diamantaras, Blind MIMO FIR channel identification based on second-order spectra correlations, *IEEE Trans. Signal Process.*, 51(6), 1668–1674, June 2003.
5. D.R. Brillinger and M. Rosenblatt, Computation and interpretation of kth-order spectra, in *Spectral Analysis of Time Series*, B. Harris, ed., John Wiley & Sons, New York, pp. 189–232, 1967.
6. D.R. Brillinger, The identification of a particular nonlinear time series system, *Biometrika*, 64(3), 509–515, 1977.
7. V. Capdevielle, Ch. Serviere, and J.L. Lacoume, Blind separation of wide-band sources in the frequency domain, In *Proc. ICASSP-95*, vol. 3, pp. 2080–2083, Detroit, MI, 1995.
8. J.F. Cardoso and A. Souloumiac, Blind beamforming for non-Gaussian signals, *IEE Proceedings-F*, 140(6), 362–370, December 1993.
9. M. Castella, J.C. Pesquet, and A.P. Petropulu, Family of frequency and time-domain contrasts for blind separation of convolutive mixtures of temporally dependent signals, *IEEE Trans. Signal Process.*, 53(1), 107–120, January 2005.
10. R.B. Cattell, "Parallel proportional profiles" and other principles for determining the choice of factors by rotation, *Psychometrica*, 9, 267–283, 1944.
11. B. Chen and A.P. Petropulu, Frequency domain MIMO system identification based on second and higher-order statistics, *IEEE Trans. Signal Process.*, 49(8), 1677–1688, August 2001.

12. P. Comon, Independent component analysis, a new concept? *Signal Process., Elsevier*, 36, 287–314, April 1994.
13. P. Comon, Contrasts for multichannel blind deconvolution, *IEEE Signal Process. Lett.*, 3, 209–211, July 1996.
14. L.D. De Lathauwer, B. De Moor, and J. Vandewalle, Dimensionality reduction in higher-order-only ICA, *Proceedings of the IEEE Signal Processing Conference on Higher-Order Statistics*, Banff, Albetra, Canada, pp. 316–320, July 1997.
15. L.D. De Lathauwer, Signal processing based on multilinear algebra, PhD thesis, Departement Elektrotechniek (ESAT), Katholieke Universiteit Leuven, Leuven, Belgium, September 1997.
16. K.I. Diamantaras, A.P. Petropulu, and B. Chen, Blind two-input-two-output FIR channel identification based on second-order statistics, *IEEE Trans. Signal Process.*, 48(2), 534–542, February 2000.
17. A.T. Erdem and A.M. Tekalp, Linear bispectrum of signals and identification of nonminimum phase FIR systems driven by colored input, *IEEE Trans. Signal Process.*, 40, 1469–1479, June 1992.
18. G.B. Giannakis and J.M. Mendel, Identification of nonminimum phase system using higher order statistics, *IEEE Trans. Acoust., Speech, Signal Process.*, 37, 360–377, March 1989.
19. G.B. Giannakis and J.M. Mendel, Cumulant-based order determination of non-Gaussian ARMA models, *IEEE Trans. Acoust., Speech Signal Process.*, 38, 1411–1423, 1990.
20. G.B. Giannakis, Cumulants: A powerful tool in signal processing, *Proc. IEEE*, 75, 1987.
21. D. Hatzinakos and C.L. Nikias, Blind equalization using a tricepstrum-based algorithm, *IEEE Trans. Commun.*, 39(5), 669–682, May 1991.
22. S. Haykin, *Nonlinear Methods of Spectral Analysis*, 2nd edn., Springer-Verlag, Berlin, Germany, 1983.
23. M.J. Hinich, Identification of the coefficients in a nonlinear time series of the quadratic type, *J. Econ.*, 30, 269–288, 1985.
24. Y. Inouye and K. Hirano, Cumulant-based blind identification of linear multi-input-multi-output systems driven by colored inputs, *IEEE Trans. Signal Process.*, 45(6), 1543–1552, June 1997.
25. S.M. Kay, *Modern Spectral Estimation*, Prentice-Hall, Inc., Englewood Cliffs, NJ, 1988.
26. S.L. Marple, Jr., *Digital Spectral Analysis with Applications*, Prentice-Hall, Inc., Englewood Cliffs, NJ, 1987.
27. J.M. Mendel, Tutorial on higher-order statistics (spectra) in signal processing and system theory: Theoretical results and some applications, *IEEE Proc.*, 79, 278–305, March 1991.
28. E. Moreau and J.-C. Pesquet, Generalized contrasts for multichannel blind deconvolution of linear systems, *IEEE Signal Process. Lett.*, 4, 182–183, June 1997.
29. C.L. Nikias and A.P. Petropulu, *Higher-Order Spectra Analysis: A Nonlinear Signal Processing Framework*, Prentice Hall, Inc., Englewood Cliffs, NJ, 1993.
30. C.L. Nikias and M.R. Raghuveer, Bispectrum estimation: A digital signal processing framework, *Proc. IEEE*, 75(7), 869–891, July 1987.
31. C.L. Nikias and H.-H. Chiang, Higher-order spectrum estimation via noncausal autoregressive modeling and deconvolution, *IEEE Trans. Acoust., Speech Signal Process.*, 36(12), 1911–1913, December 1988.
32. A.V. Oppenheim and R.W. Schafer, *Discrete-Time Signal Processing*, Prentice-Hall, Englewood Cliffs, NJ, 1989.
33. R. Pan and C.L. Nikias, The complex cepstrum of higher order cumulants and nonminimum phase system identification, *IEEE Trans. Acoust., Speech Signal Process.*, 36(2), 186–205, February 1988.
34. A. Papoulis, *Probability Random Variables and Stochastic Processes*, McGraw-Hill, New York, 1984.
35. A.P. Petropulu, Higher-order spectra in biomedical signal processing, *CRC Press Biomedical Engineering Handbook*, CRC Press, Boca Raton, FL, 1995.
36. A.P. Petropulu and C.L. Nikias, The complex cepstrum and bicepstrum: Analytic performance evaluation in the presence of Gaussian noise, *IEEE Trans. Acous. Speech Signal Process., Special Mini-Section on Higher-Order Spectral Analysis*, ASSP-38(7), July 1990.

37. E.L. Powers, C.K. Ritz, et al., Applications of digital polyspectral analysis to nonlinear systems modeling and nonlinear wave phenomena, *Workshop on Higher-Order Spectral Analysis*, pp. 73–77, Vail, CO, June 1989.
38. H. Pozidis and A.P. Petropulu, System reconstruction from selected regions of the discretized higher-order spectrum, *IEEE Trans. Signal Process.*, 46(12), 3360–3377, December 1998.
39. M.R. Raghuveer and C.L. Nikias, Bispectrum estimation: A parametric approach, *IEEE Trans. Acoust. Speech Signal Process.*, ASSP 33(5), 1213–1230, October 1985.
40. M.R. Raghuveer and C.L. Nikias, Bispectrum estimation via AR modeling, *Signal Process.*, 10, 35–48, 1986.
41. T. Subba Rao and M.M. Gabr, An introduction to bispectral analysis and bilinear time series models, *Lecture Notes in Statistics*, Vol. 24, Springer-Verlag, New York, 1984.
42. N. Rozario and A. Papoulis, The identification of certain nonlinear systems by only observing the output, *Workshop on Higher-Order Spectral Analysis*, pp. 73–77, Vail, CO, June 1989.
43. M. Schetzen, *The Volterra and Wiener Theories on Nonlinear System*, updated edition, Krieger Publishing Company, Malabar, FL, 1989.
44. O. Shalvi and E. Weinstein, New criteria for blind deconvolution of nonminimum phase systems (channels), *IEEE Trans. Inform. Theory*, 36, 312–321, March 1990.
45. S. Shamsunder and G.B. Giannakis, Multichannel blind signal separation and reconstruction, *IEEE Trans. Speech Audio Process.*, 5(6), 515–527, November 1997.
46. A. Swami, G.B. Giannakis, and S. Shamsunder, Multichannel ARMA processes, *IEEE Trans. Signal Process.*, 42(4), 898–913, April, 1994.
47. L. Tong and S. Perreau, Multichannel blind identification: From subspace to maximum likelihood methods, *Proc. IEEE*, 86(10), 1951–1968, October 1998.
48. A. Swami and J.M. Mendel, ARMA parameter estimation using only output cumulants, *IEEE Trans. Acoust. Speech Signal Process.*, 38, 1257–1265, July 1990.
49. L.J. Tick, The estimation of transfer functions of quadratic systems, *Technometrics*, 3(4), 562–567, November 1961.
50. M. Torlak and G. Xu, Blind multiuser channel estimation in asynchronous CDMA systems, *IEEE Trans. Signal Process.*, 45(1), 137–147, January 1997.
51. L. Tong, Y. Inouye, and R. Liu, A finite-step global convergence algorithm for parameter estimation of multichannel MA processes, *IEEE Trans. Signal Process.*, 40, 2547–2558, October 1992.
52. J.K. Tugnait and B. Huang, Multistep linear predictors-based blind identificaion and equalization by the subspace method: Identifibiality results, *IEEE Trans. Signal Process.*, 48, 26–38, January 2000.
53. J.K. Tugnait, Identification and deconvolution of multichannel linear non-Gaussian processes using higher order statistics and inverse filter criteria, *IEEE Trans. Signal Process.*, 45(3), 658–672, March 1997.
54. A.-J. van der Veen, S. Talwar, and A. Paulraj, Blind estimation of multiple digital signals transmitted over FIR channels, *IEEE Signal Process. Lett.*, 2(5), 99–102, May 1995.
55. A.-J. van der Veen, S. Talwar, and A. Paulraj, A subspace approach to blind space-time signal processing for wireless communication systems, *IEEE Trans. Signal Process.*, 45(1), 173–190, January 1997.
56. E. Weinstein, M. Feder, and A. Oppenheim, Multi-channel signal separation by decorrelation, *IEEE Trans. Speech Audio Process.*, 1(4), 405–413, October 1993.
57. D. Yellin and E. Weinstein, Multi-channel signal separation: Methods and analysis, *IEEE Trans. Signal Process.*, 44, 106–118, January 1996.
58. Y. Yu and A.P. Petropulu, PARAFAC-based blind estimation of possibly underdetermined convolutive MIMO systems, *IEEE Trans. Signal Process.*, 56(1), 111–124, January 2008.

III

DSP Software and Hardware

Vijay K. Madisetti
Georgia Institute of Technology

19 Introduction to the TMS320 Family of Digital Signal Processors
Panos Papamichalis **19**-1
Introduction • Fixed-Point Devices: TMS320C25 Architecture and Fundamental Features • TMS320C25 Memory Organization and Access • TMS320C25 Multiplier and ALU • Other Architectural Features of the TMS320C25 • TMS320C25 Instruction Set • Input/Output Operations of the TMS320C25 • Subroutines, Interrupts, and Stack on the TMS320C25 • Introduction to the TMS320C30 Digital Signal Processor • TMS320C30 Memory Organization and Access • Multiplier and ALU of the TMS320C30 • Other Architectural Features of the TMS320C30 • TMS320C30 Instruction Set • Other Generations and Devices in the TMS320 Family • References

20 Rapid Design and Prototyping of DSP Systems *T. Egolf, M. Pettigrew, J. Debardelaben, R. Hezar, S. Famorzadeh, A. Kavipurapu, M. Khan, Lan-Rong Dung, K. Balemarthy, N. Desai, Yong-kyu Jung, and Vijay K. Madisetti* **20**-1
Introduction • Survey of Previous Research • Infrastructure Criteria for the Design Flow • The Executable Requirement • The Executable Specification • Data and Control Flow Modeling • Architectural Design • Performance Modeling and Architecture Verification • Fully Functional and Interface Modeling and Hardware Virtual Prototypes • Support for Legacy Systems • Conclusions • Acknowledgments • References

21 Baseband Processing Architectures for SDR *Yuan Lin, Mark Woh, Sangwon Seo, Chaitali Chakrabarti, Scott Mahlke, and Trevor Mudge* **21**-1
Introduction • SDR Overview • Workload Profiling and Characterization • Architecture Design Trade-Offs • Baseband Processor Architectures • Cognitive Radio • Conclusion • References

22 Software-Defined Radio for Advanced Gigabit Cellular Systems *Brian Kelley* **22**-1
Introduction • Waveform Signal Processing • Communication Link Capacity Extensions • RF, IF, and A/D Systems Trends • Software Architectures • Reconfigurable SDR Architectures for Gigabit Cellular • Conclusion • References

THE PRIMARY TRAITS OF EMBEDDED SIGNAL processing systems that distinguish them from general purpose computer systems are their predictable reactions to real-time* stimuli from the environment, their form- and cost-optimized design, and their compliance with required or specified modes of response behavior and functionality [1].

Other traits that they share with other forms of digital products include the need for reliability, fault-tolerance, and maintainability, to name just a few. An embedded system usually consists of hardware components such as memories, application-specific ICs (ASICs), processors, DSPs, buses, analog–digital interfaces, and also software components that provide control, diagnostic, and application-specific capabilities required of it. In addition, they often contain electromechanical (EM) components such as sensors and transducers and operate in harsh environmental conditions. Unlike general purpose computers, they may not allow much flexibility in support of a diverse range of programming applications, and it is not unusual to dedicate such systems to specific application. Embedded systems, thus, range from simple, low-cost sensor/actuator systems consisting of a few tens of lines of code and 8/16-bit processors (CPU) (e.g., bank ATM machines) to sophisticated high-performance signal processing systems consisting of runtime operating system support, tens of x86-class processors, digital signal processing (DSP) chips, interconnection networks, complex sensors, and other interfaces (e.g., radar-based tracking and navigational systems). Their lack of flexibility may be apparent when one considers that an ATM machine cannot be easily programmed to support additional image processing tasks, unless upgraded in terms of resources. Finally, embedded systems typically do not support direct user interaction in terms of higher-order programming languages (HOLs) such as Fortran or C, but allow users to provide inputs that are sensor- or menu-driven. The debug and diagnostic interfaces, however, support HOLs and other lower level software and hardware programmability.

Embedded systems, in general, may be classified into one of the following four general categories of products. The prices are indicative of the multibillion dollar marketplace in 1996, and their relative magnitudes are more significant than their actual values. The relationship of the categories to dollar cost is intentional and is an early harbinger of the fact that underlying cost and performance tradeoffs motivate and drive most of the system design and prototyping methodologies.

Commodity DSP products: High-volume market and valued at less than $300 a piece. These include CD players, recorders, VCRs, facsimile and answering machines, telemetry applications, simple signal processing filtering packages, etc., primarily aimed at the highly competitive mass–volume consumer market.

Portable DSP products: High-volume market and valued at less than $800. These include portable and hand-held low-power electronic products for man–machine communications such as DSP boards, digital audio, security systems, modems, camcorders, industrial controllers, scanners, communications equipment, and others.

Cost-performance DSP products: High-volume market, and valued at less than $3000. These products trade off cost for performance, and include DSP products such as video teleconferencing equipment, laptops, audio, telecommunications switches, high-performance DSP boards and coprocessors, and DSP CAD packages for hardware and software design.

High-performance products: Low-to-moderate volume market, and valued at over $8000. These products include high-end workstations with DSP coprocessors, real-time signal processors, real-time database processing systems, digital HDTV, radar signal processor systems, avionics and military systems, and sensor and data processing hardware and software systems. This class of products contains a significant amount of software compared to the earlier classes, which often focus on large volume, low-cost, hardware-only solutions.

* Real-time indicates behavior related to wall-clock time and does not necessarily imply a quick response.

It may be useful to classify high-performance products further into three categories:

- *Real-time embedded control systems*: These systems are characterized by the following features: interrupt driven, large numerical processing requirements, small databases, tight real-time constraints, well-defined user interface, and requirements and design driven by performance requirements. Examples include an aircraft control system, or a control system for a steel plant.
- *Embedded information systems*: These systems are characterized by the following features: transaction-based, moderate numerical/DSP processing, flexible time constraints, complex user interfaces, and requirements and design driven by user interface. Examples include accounting and inventory management systems.
- *Command, control, communication, and intelligence (C4I) systems*: These systems are characterized by large numerical processing, large databases, moderate to tight real-time constraints, flexible and complex user interfaces, and requirements and design driven by performance and user interface. Examples include missile guidance systems, radar-tracking systems, and inventory and manufacturing control systems.

These four categories of embedded systems can be further distinguished in terms of other metrics such as computing speed (integer or floating point performance), input/output transfer rates, memory capacities, market volume, environmental issues, typical design and development budgets, lifetimes, reliability issues, upgrades, and other lifecycle support costs. Another interesting fact is that the higher the software value in a product, the greater its profitability margin. Recent studies by Andersen Consulting have shown that profit margin pressures are increasing due to increasing semiconductor content in systems' sales' values. In 1985, silicon represented 9.5% of a system's value. By 1995, that had shot up to 19.1%. The higher the silicon content, the greater the pressure on margins resulting in lower profits. In PCs, integrated circuit components represent 30%–35% of the sales value and the ratio is steadily increasing. More than 50% of value of the new network computers (NCs) is expected to be in integrated circuits. In the area of DSPs, we estimate that this ratio is about 20%.

In this part, Chapter 19 by Panos Papamichalis outlines the programmable DSP families developed by Texas Instruments, the leading organization in this area. Chapter 20 by Egolf et al. discusses how signal processing systems are designed and integrated using a novel top down design approach developed as part of DARPA's RASSP program.

Reference

1. Madisetti, V.K., *VLSI Digital Signal Processors*, IEEE Press, Piscataway, NJ, 1995.

19

Introduction to the TMS320 Family of Digital Signal Processors

Panos Papamichalis
Texas Instruments

19.1 Introduction 19-1
19.2 Fixed-Point Devices: TMS320C25 Architecture and Fundamental Features 19-2
19.3 TMS320C25 Memory Organization and Access 19-8
19.4 TMS320C25 Multiplier and ALU 19-10
19.5 Other Architectural Features of the TMS320C25 19-13
19.6 TMS320C25 Instruction Set 19-14
19.7 Input/Output Operations of the TMS320C25 19-16
19.8 Subroutines, Interrupts, and Stack on the TMS320C25 19-16
19.9 Introduction to the TMS320C30 Digital Signal Processor 19-17
19.10 TMS320C30 Memory Organization and Access 19-23
19.11 Multiplier and ALU of the TMS320C30 19-25
19.12 Other Architectural Features of the TMS320C30 19-25
19.13 TMS320C30 Instruction Set 19-27
19.14 Other Generations and Devices in the TMS320 Family 19-30
References 19-34

This chapter discusses the architecture and the hardware characteristics of the TMS320 family of digital signal processors (DSPs). The TMS320 family includes several generations of programmable processors with several devices in each generation. Since the programmable processors are split between fixed- and floating-point devices, both categories are examined in some detail. The TMS320C25 serves here as a simple example for the fixed-point processor family, while the TMS320C30 is used for the floating-point family.

19.1 Introduction

Since its introduction in 1982 with the TMS32010 processor, the TMS320 family of DSPs has been exceedingly popular. Different members of this family were introduced to address the existing needs for real-time processing, but then, designers capitalized on the features of the devices to create solutions and products in ways never imagined before. In turn, these innovations fed the architectural and hardware configurations of newer generations of devices.

Digital signal processing encompasses a variety of applications, such as digital filtering, speech and audio processing, image and video processing, and control. All DSP applications share some common characteristics:

- The algorithms used are mathematically intensive. A typical example is the computation of an FIR filter, implemented as sum of products. This operation involves a lot of multiplications combined with additions.
- DSP algorithms must typically run in real time: that is, the processing of a segment of the arriving signal must be completed before the next segment arrives, or else data will be lost.
- DSP techniques are under constant development. This implies that DSP systems should be flexible to support changes and improvements in the state of the art. As a result, programmable processors have been the preferred way of implementation. In recent times, though, fixed-function devices have also been introduced to address high-volume consumer applications with low-cost requirements.

These needs are addressed in the TMS320 family of DSPs by using appropriate architecture, instruction sets, I/O capabilities, as well as the raw speed of the devices. However, it should be kept in mind that these features do not cover all the aspects describing a DSP device, and especially a programmable one. Availability and quality of software and hardware development tools (such as compilers, assemblers, linker, simulators, hardware emulators, and development systems), application notes, third-party products and support, hot-line support, etc., play an important role on how easy it will be to develop an application on the DSP. The TMS320 family has very extensive such support, but its description goes beyond the scope of this chapter. The interested reader should contact the TI DSP hotline (Tel. 713-274-2320).

For the purposes of this chapter, two devices have been selected to be highlighted from the Texas Instruments TMS320 family of DSPs. One is the TMS320C25, a 16 bit, fixed-point DSP, and the other is the TMS320C30, a 32 bit, floating-point DSP. As a short-hand notation, they will be called 'C25 and 'C30, respectively. The choice was made so that both fixed-point issues are considered.

There have been newer (and more sophisticated) generations added to the TMS320 family but, since the objective of this chapter is to be more tutorial, they will be discussed as extensions of the 'C25 and the 'C30. Such examples are other members of the 'C2x and the 'C3x generations, as well as the TMS320C5x generation ('C5x for short) of fixed-point devices, and the TMS320C4x ('C4x) of floating-point devices. Customizable and fixed-function extensions of this family of processors will be also discussed.

Texas Instruments, like all vendors of DSP devices, publishes detailed User's Guides that explain at great length the features and the operation of the devices. Each of these User's Guides is a pretty thick book, so it is not possible (or desirable) to repeat all this information here. Instead, the objective of this chapter is to give an overview of the basic features for each device. If more detail is necessary for an application, the reader is expected to refer to the User's Guides. If the User's Guides are needed, it is very easy to obtain them from Texas Instruments.

19.2 Fixed-Point Devices: TMS320C25 Architecture and Fundamental Features

The Texas Instruments TMS320C25 is a fast, 16 bit, fixed-point DSP. The speed of the device is 10 MHz, which corresponds to a cycle time of 100 ns. Since the majority of the instructions execute in a single cycle, the figure of 100 ns also indicates how long it takes to execute one instruction. Alternatively, we can say that the device can execute 10 million instructions per second (MIPS). The actual signal from the external oscillator or crystal has a frequency four times higher, at 40 MHz. This frequency is then divided on-chip to generate the internal clock with a period of 100 ns. Figure 19.1 shows the relationship between the input clock CLKIN from the external oscillator, and the output clock CLKOUT. CLKOUT is the same

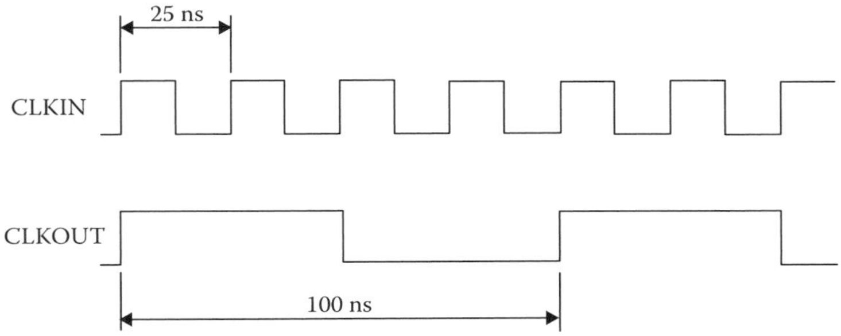

FIGURE 19.1 Clock timing of the TMS320C25. CLKIN, external oscillator; CLKOUT, clock of the device.

as the clock of the device, and it is related to CLKIN by the equation CLKOUT = CLKIN/4. Note that in Figure 19.1 the shape of the signal is idealized ignoring rise and fall times.

Newer versions of the TMS320C25 operate in higher frequencies. For instance, there is a spinoff that has a cycle time of 80 ns, resulting in a 12.5 MIPS operation. There are also slower (and cheaper) versions for applications that do not need this computational power.

Figure 19.2 shows in a simplified form the key features of the TMS320C25. The major parts of the DSP are the memory, the central processing unit (CPU), the ports, and the peripherals. Each of these parts will be examined in more detail later. The on-chip memory consists of 544 words of RAM (read/write memory) and 4K words of ROM (read-only memory). In the notation used here, 1K = 1024 words, and 4K = 4 × 1024 = 4096 words. Each word is 16 bit wide and, when some memory size is given, it is

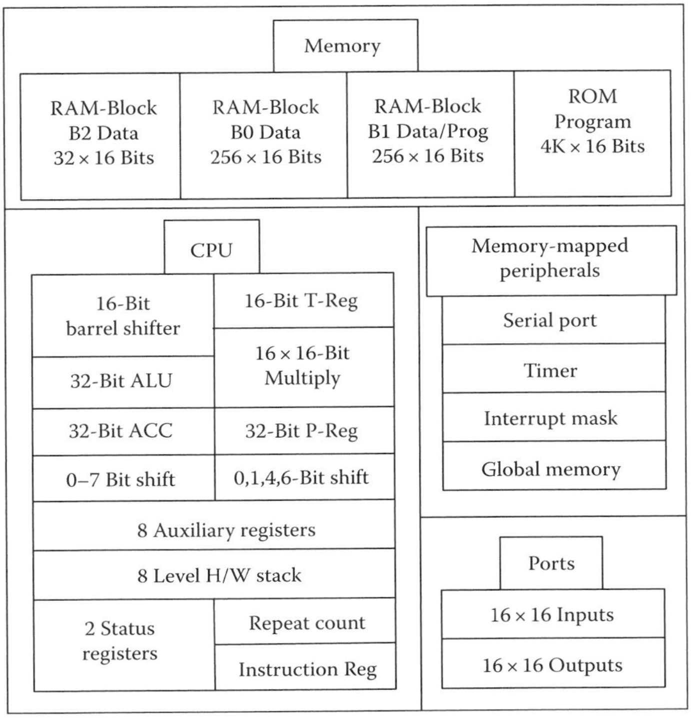

FIGURE 19.2 Key architectural features of the TMS320C25.

measured in 16 bit words, and not in bytes (as is the custom in microprocessors). Of the 544 words of RAM, 256 words can be used as either program or data memory, while the rest are only data memory. All 4K of on-chip ROM is program memory. Overall, the device can address 64K words of data memory and 64K words of program memory. Except for what resides on-chip, the rest of the memory is external, supplied by the designer.

The CPU is the heart of the processor. Its most important feature, distinguishing it from the traditional microprocessors, is a hardware multiplier that is capable of performing a 16×16 bit multiplication in a single cycle. To preserve higher intermediate accuracy of results, the full 32 bit product is saved in a product register. The other important part of the CPU is the arithmetic logic unit (ALU) that performs additions, subtractions, and logical operations. Again, for increased intermediate accuracy, there is a 32 bit accumulator to handle all the ALU operations.

All the arithmetic and logical functions are accumulator based. In other words, these operations have two operands, one of which is always the accumulator. The result of the operation is stored in the accumulator.

Because of this approach the form of the instructions is very simple indicating only what the other operand is. This architectural philosophy is very popular but it is not universal. For instance, as is discussed later, the TMS320C30 takes a different approach, where there are several "accumulators" in what is called a register file.

Other components of the TMS320C25 CPU are several shifters to facilitate manipulation of the data and increase the throughput of the device by performing shifting operations in parallel with other functions. As part of the CPU, there are also eight auxiliary registers (ARs) that can be used as memory pointers or loop counters. There are two status registers, and an 8-deep hardware stack. The stack is used to store the memory address where the program will continue execution after a temporary diversion to a subroutine.

To communicate with external devices, the TMS320C25 has 16 input and 16 output parallel ports. It also has a serial port that can serve the same purpose. The serial port is one of the peripherals that have been implemented on chip. Other peripherals include the interrupt mask, the global memory capability, and a timer. The above components of the TMS320C25 are examined in more detail below.

The device has 68 pins that are designated to perform certain functions, and to communicate with other devices on the same board. The names of the signals and the corresponding definitions appear in Table 19.1. The first column of the table gives the pin names. Note that a bar over the name indicates that the pin is in the active position when it is electrically low. For instance, if the pins take the voltage levels of 0 and 5 V, a pin indicated with an overbar is asserted when it is set at 0 V. Otherwise, assertion occurs at 5 V. The second column indicates if the pin is used for input to the device or output from the device or both. The third column gives a description of the pin functionality.

Understanding the functionality of the device pins is as important as understanding the internal architecture because it provides the designer with the tools available to communicate with the external world. The DSP device needs to receive data and, often, instructions from the external sources, and send the results back to the external world. Depending on the paths available for such transactions, the design of a program can take very different forms. Within this framework, it is up to the designer to generate implementations that are ingenious and elegant.

The TMS320C25 has its own assembly language to be programmed. This assembly language consists of 133 instructions that perform general-purpose and DSP-specific functions. Familiarity with the instruction set and the device architecture are the two components of efficient program implementation. High-level-language compilers have also been developed that make the writing of programs an easier task. For the TMS320C25, there is a C compiler available. However, there is always a loss of efficiency when programming in high-level languages, and this may not be acceptable in computation-bound real-time systems. Besides, for complete understanding of the device it is necessary to consider the assembly language.

A very important characteristic of the device is its Harvard architecture. In Harvard architecture (see Figure 19.3), the program and data memory spaces are separated and they are accessed by

TABLE 19.1 Names and Functionality of the 68 Pins of the TMS320C25

Signals	I/O/Z[a]	Definition
V_{CC}	I	5 V supply pins
V_{SS}	I	Ground pins
X1	O	Output from internal oscillator for crystal
X2/CLKIN	I	Input to internal oscillator from crystal or external clock
CLKOUT1	O	Master clock output (crystal or CLKIN frequency/4)
CLKOUT2	O	A second clock output signal
D15-D0	I/O/Z	16 bit data bus D15 (MSB) through DO (LSB). Multiplexed between program, data, and I/O spaces.
A15-A0	O/Z	16 bit address bus A15 (MSB) through AO (LSB)
$\overline{PS}$, $\overline{DS}$, $\overline{IS}$	O/Z	Program, data, and I/O space select signals
R/$\overline{W}$	O/Z	Read/write signal
$\overline{STRB}$	O/Z	Strobe signal
$\overline{RS}$	I	Reset input
$\overline{INT}$ 2-$\overline{INT}$ 0	I	External user interrupt inputs
MP/$\overline{MC}$	I	Microprocessor/microcomputer mode select pin
$\overline{MSC}$	O	Microstate complete signal
$\overline{IACK}$	O	Interrupt acknowledge signal
READY	I	Data ready input. Asserted by external logic when using slower devices to indicate that the current bus transaction is complete.
$\overline{BR}$	O	Bus request signal. Asserted when the TMS320C25 requires access to an external global data memory space.
XF	O	External flag output (latched software-programmable signal)
$\overline{HOLD}$	I	Hold input. When asserted. TMS320C25 goes into an idle mode and places the data, address, and control lines in the high impedance state.
$\overline{HOLDA}$	O	Hold acknowledge signal
$\overline{SY}\,\overline{NC}$	I	Synchronization input
$\overline{BIO}$	I	Branch control input. Polled by BIOZ instruction.
DR	I	Serial data receive input
CLKR	I	Clock for receive input for serial port
FSR	I	Frame synchronization pulse for receive input
DX	O/Z	Serial data transmit output
CLKX	I	Clock for transmit output for serial port
FSX	I/O/Z	Frame synchronization pulse for transmit. Configurable as either an input or an output.

Note: The first column is the pin name; the second column indicates if it is an input or an output pin; the third column gives a description of the pin functionality.
[a] I/O/Z denotes input/output/high-impedance state.

different buses. One bus accesses the program memory space to fetch the instructions, while another bus is used to bring operands from the data memory space and store the results back to memory. The objective of this approach is to increase the throughput by bringing instructions and data in parallel. An alternate philosophy is the von Neuman architecture. The von Neuman architecture (see Figure 19.4) uses a single bus and a unified memory space. Unification of the memory space is convenient for partitioning it between program and data, but it presents a bottleneck since both data and program instructions must use the same path and, hence, they must be multiplexed. The Harvard architecture of multiple buses is used in DSPs because the increased throughput is of paramount importance in real-time systems.

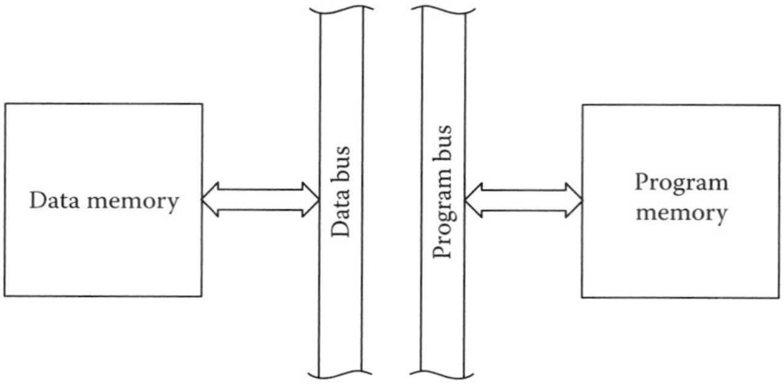

FIGURE 19.3 Simplified block diagram of the Harvard architecture.

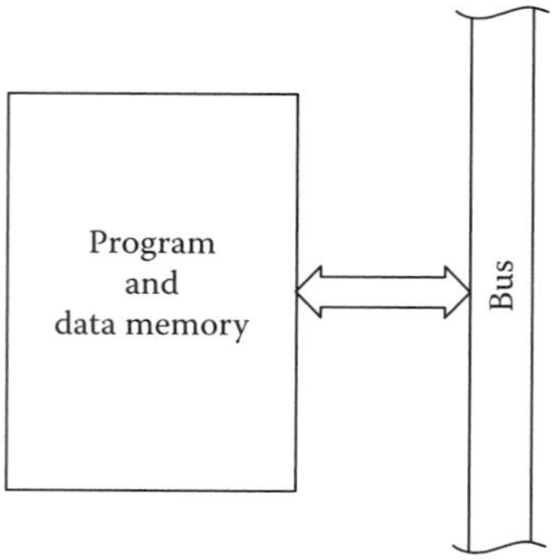

FIGURE 19.4 Simplified block diagram of the von Neuman architecture.

The difference of the architectures is important because it influences the programming style. In Harvard architecture, two memory locations can have the same address, as long as one of them is in the data space and the other is in the program space. Hence, when the programmer uses an address label, he has to be alert as to what space he is referring. Another restriction of the Harvard architecture is that the data memory cannot be initialized during loading because loading refers only to placing the program on the memory (and the program memory is separate from the data memory). Data memory can be initialized during execution only. The programmer must incorporate such initialization in his program code. As it will be seen later, such restrictions have been removed from the TMS320C30 while retaining the convenient feature of multiple buses.

Figure 19.5 shows a functional block diagram of the TMS320C25 architecture. The Harvard architecture of the device is immediately apparent from the separate program and data buses. What is not apparent is that the architecture has been modified to permit communication between the two buses. Through such communication, it is possible to transfer data between the program and memory spaces. Then, the program memory space also can be used to store tables. The transfer takes place by using special instructions such as TBLR (Table Read), TBLW (Table Write), and BLKP (Block transfer from Program memory).

As shown in the block diagram, the program ROM is linked to the program bus, while data RAM blocks B1 and B2 are linked to the data bus. The RAM block B0 can be configured either as program or data memory (using the instructions CNFP and CNFD), and it is multiplexed with both buses. The different segments, such as the multiplier, the ALU, the memories, etc., are examined in more detail below.

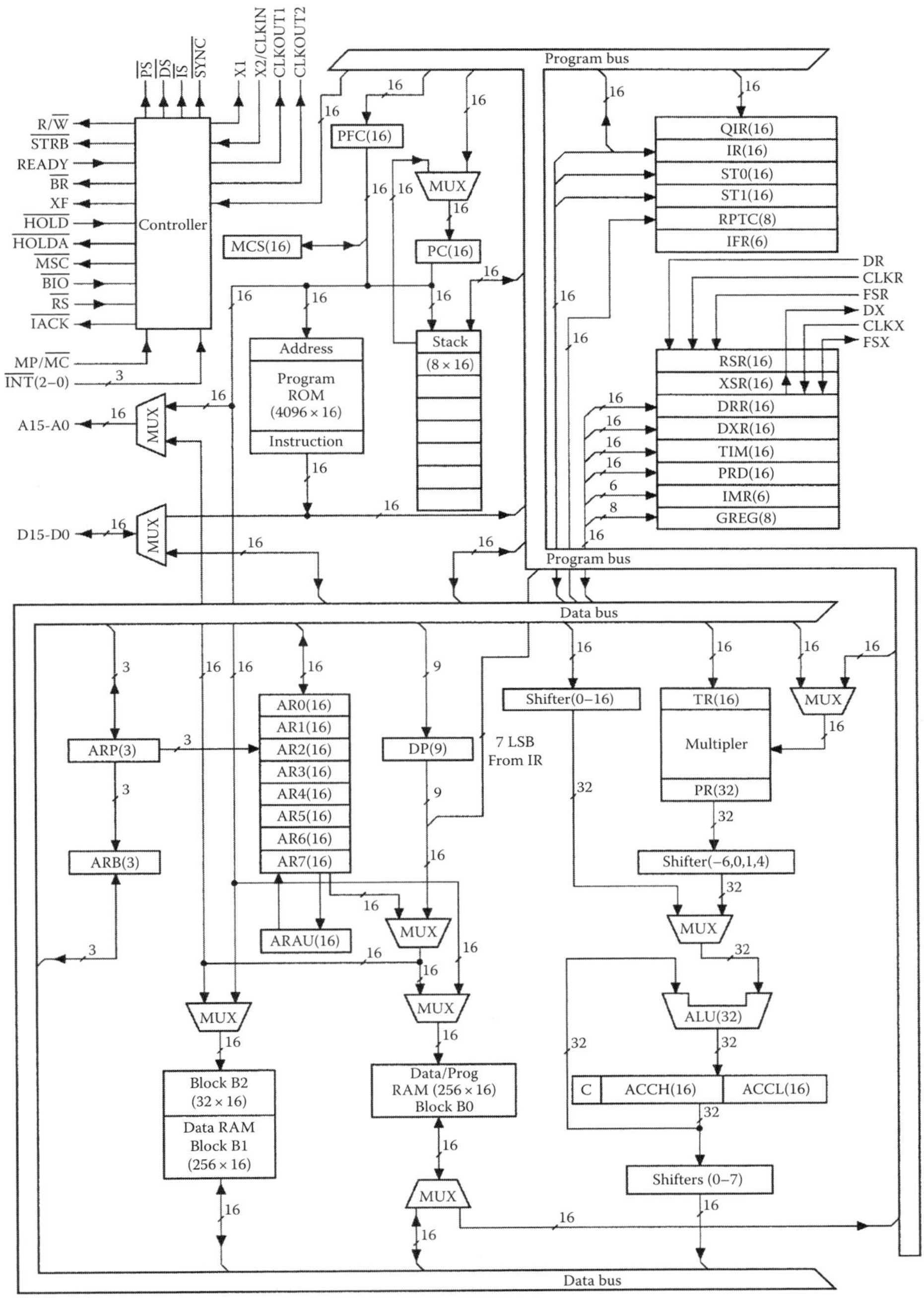

Legend:

ACCH = Accumulator high
ACCL = Accumulator low
ALU = Arithmetic logic unit
ARAU = Auxiliary register arithmetic unit
ARS = Auxiliary register pointer buffer
ARP = Auxiliary register pointer
DP = Data memory page pointer
DRR = Serial port data receive register
DXR = Serial port data transmit register

IFR = Interrupt flag register
IMR = Interrupt mask register
IR = Instruction register
MCS = Microcall stack
QIR = Queue instruction register
PR = Product register
PRD = Period register for timer
TIM = Timer
TR = Temporary register

PC = Program counter
PFC = Prefetch counter
RPTC = Repeat instruction counter
GREG = Global memory allocation register
RSR = Serial port receive shift register
XSR = Serial port transmit shift register
AR0-AR7 = Auxiliary registers
ST0,ST1 = Status registers

FIGURE 19.5 Functional block diagram of the TMS320C25 architecture.

19.3 TMS320C25 Memory Organization and Access

Besides the on-chip memory (RAM and ROM), the TMS320C25 can access external memory through the external bus. This bus consists of the 16 address pins A0–A15, and the 16 data pins D0–D15. The address pins carry the address to be accessed, while the data pins carry the instruction word or the operand, depending on whether program or data memory is accessed. The bus can access either program or data memory, the difference indicated by which of the pins PS and DS (with overbars) becomes active. The activation is done automatically when, during the execution, an instruction or a piece of data needs to be fetched. Since the address is 16 bit wide, the maximum memory space is 64K words for program and 64K words for data.

The device starts execution after a reset signal, i.e., after the RS pin is pulled low for a short period. The execution always begins at program memory location 0, where there should be an instruction to direct the program execution to the appropriate location. This direction is accomplished by a branch instruction

B PROG

which loads the program counter with the program memory address that has the label PROG (or any other label you choose). Then, execution continues from the address PROG, where, presumably, a useful program has been placed.

It is clear that the program memory location 0 is very important, and you need to know where it is physically located. The TMS320C25 gives you the flexibility to use as location 0 either the first location of the on-chip ROM, or the first location of the external memory. In the first case, we say that the device operates in the microcomputer mode, while in the second one it is in the microprocessor mode. In the microprocessor mode, the on-chip ROM is ignored altogether. You can choose between the two modes by pulling the device MP/MC high or low. The microcomputer mode is useful for production purposes, while for laboratory and development work the microprocessor mode is used exclusively.

Figure 19.6 shows the memory configuration of the TMS320C25, where the microprocessor and microcomputer configurations of the program memory are depicted separately. The data memory is partitioned in 512 sections, called pages, of 128 words each. The reason of the partitioning is for

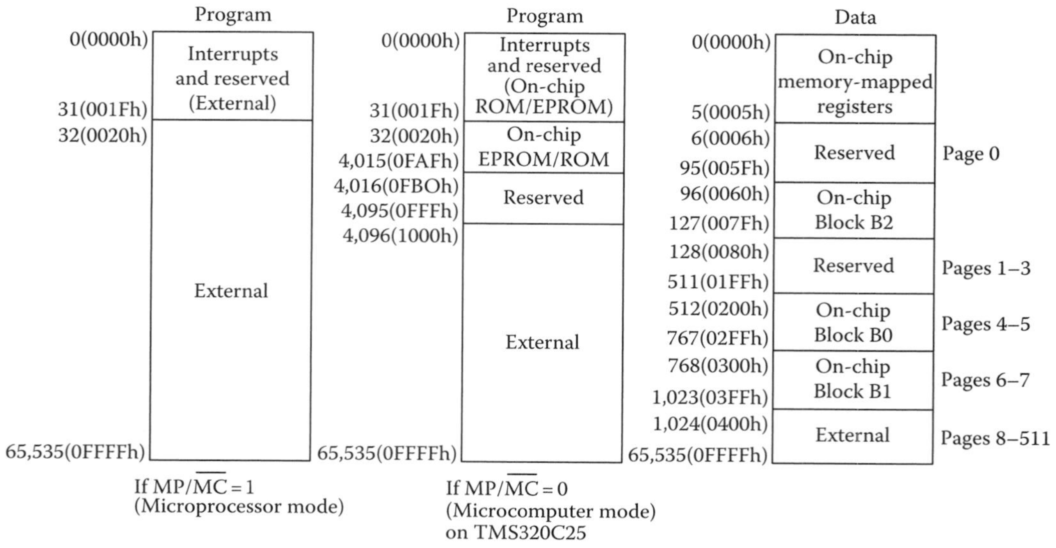

FIGURE 19.6 Memory maps for program and data memory of the TMS320C25.

addressing purposes, as discussed below. Memory boundaries of the 64K memory space are shown in both decimal and hexadecimal notation (hexadecimal notation indicated by an "h" or "H" at the end.) Compare this map with the block diagram in Figure 19.5.

As mentioned earlier, in two-operand operations, one of the operands resides in the accumulator, and the result is also placed in the accumulator. (The only exceptions is the multiplication operation examined later.) The other operand can either reside in memory or be part of the instruction. In the latter case, the value to be combined with the accumulator is explicitly specified in the instruction, and this addressing mode is called immediate addressing mode. In the TMS320C25 assembly language, the immediate addressing mode instructions are indicated by a "K" at the end of the instruction.

For example, the instruction

ADDK 5

increments the contents of the accumulator by 5.

If the value to be operated upon resides in memory, there are two ways to access it: either by specifying the memory address directly (direct addressing) or by using a register that holds the address of that number (indirect addressing).

As a general rule, it is desirable to describe an instruction as briefly as possible so that the whole description can be held in one 16 bit word. Then, when the program is executed, only one word needs to be fetched before all the information from the instruction is available for execution. This is not always possible and there are two-word instructions as well, but the chip architects always strive to achieve one-word instructions. In the direct addressing mode, full description of a memory address would require a 16 bit word by itself because the memory space is 64K words. To reduce that requirement, the memory space is divided in 512 pages of 128 words each. An instruction using direct addressing contains the 7 bit indicating what word you want to access within a page. The page number (9 bit) is stored in a separate register (actually, part of a register), called the Data Page (DP) pointer. You store the page number in the DP pointer by using the instructions LDP (Load Data Page pointer) or LDPK (Load Data Page pointer immediate).

In the indirect addressing mode, the data memory address is held in a register that acts as a memory pointer. There are eight such registers available, called ARs, AR0–AR7. The ARs can also be used for other functions, such as loop counters, etc. To save bits in the instruction, the AR used as memory pointer is not indicated explicitly, but it is stored in a separate register (actually, part of a register), the auxiliary register pointer (ARP). In other words, there is the concept of the "current register." In an operation using indirect addressing, the contents of the current AR point to the desired memory location. The current AR is specified by the contents of the ARP as shown in Figure 19.7. In an instruction, indirect addressing is indicated by an asterisk.

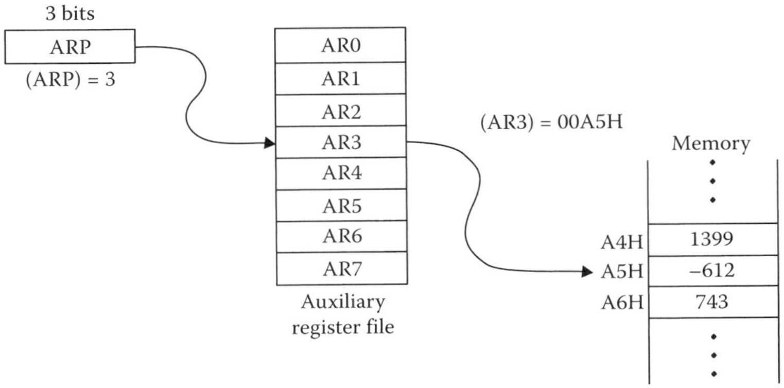

FIGURE 19.7 Example of indirect addressing mode.

TABLE 19.2 Operations That Can Be Performed in Parallel with Indirect Addressing

Notation	Operation
ADD *	No manipulation of AR or ARP
ADD *, Y	Y → ARP
ADD * +	AR(ARP)+1 → AR(ARP)
ADD *+,Y	AR(ARP)+1 → AR(ARP)
	Y → ARP
ADD *−	AR(ARP) − 1 → AR(ARP)
ADD *−, Y	AR(ARP) − 1 → AR(ARP)
	Y → ARP
ADD *0+	AR(ARP) + AR0 → AR(ARP)
ADD *0+, Y	AR(ARP) + AR0 → AR(ARP)
	Y → ARP
ADD *0−	AR(ARP)-AR0 → AR(ARP)
ADD *0−, Y	AR(ARP)-AR0 → AR(ARP)
	Y → ARP
ADD *BR0+	AR(ARP) + rcAR0 → AR(ARP)
ADD *BR0+, Y	AR(ARP) + rcAR0 → AR(ARP)
	Y → ARP
ADD *BR0−	AR(ARP)-rcAR0 → AR(ARP)
ADD *BR0−, Y	AR(ARP)-rcAR0 → AR(ARP)
	Y → ARP

Note: $Y = 0, \ldots, 7$ is the new "current" AR. AR(ARP) is the AR pointed to by the ARP. BR, bit reversed; rc, reverse carry.

A "+" sign at the end of an instruction using indirect addressing means "after the present memory access, increment the contents of the current AR by 1." This is done in parallel with the load-accumulator operation. The above autoincrementing of the AR is an optional operation that offers additional flexibility to the programmer. And it is not the only one available. The TMS320C25 has an auxiliary register arithmetic unit (ARAU, see Figure 19.5) that can execute such operations in parallel with the CPU, and increase the throughput of the device in this way. Table 19.2 summarizes the different operations that can be done while using indirect addressing. As seen from this table, the contents of an AR can be incremented or decremented by 1, incremented or decremented by the contents of AR0, and incremented or decremented by AR0 in a bit-reversed fashion. The last operation is useful when doing fast Fourier transforms. The bit-reversed addressing is implemented by adding AR0 with reverse carry propagation, an operation explained in the *TMS320C25 User's Guide*. Additionally, it is possible to load at the same time the ARP with a new value, thus saving an extra instruction.

19.4 TMS320C25 Multiplier and ALU

The heart of the TMS320C25 is the CPU consisting, primarily, of the multiplier and the ALU. The hardware multiplier can perform a 16 × 16 bit multiplication in a single machine cycle. This capability is probably the major distinguishing feature of DSPs because it permits high throughput in numerically intensive algorithms.

Associated with the multiplier, there are two registers that hold operands and results. The T-register (for temporary register) holds one of the two factors. The other factor comes from a memory location. Again, this construct, with one implied operand residing in the T-register, permits more compact instruction words. When multiplier and multiplicand (two 16 bit words) are multiplied together,

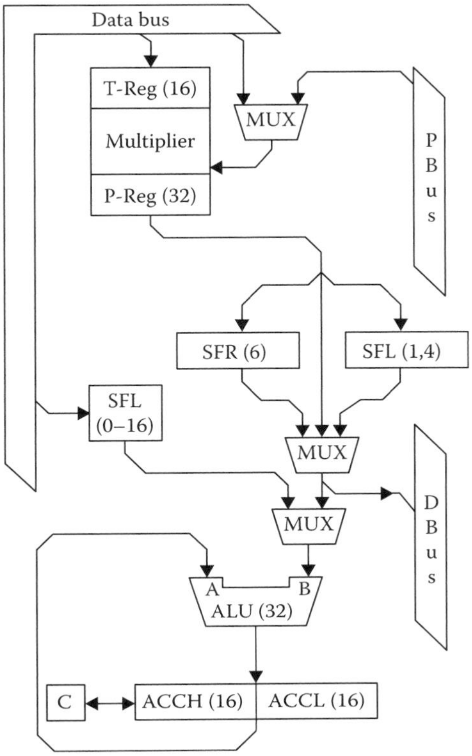

FIGURE 19.8 Diagram of the TMS320C25 multiplier and ALU.

the result is 32 bit long. In traditional microprocessors, this product would have been truncated to 16 bit, and presented as the final result. In DSP applications, though, this product is only an intermediate result in a long stream of multiply-adds, and if truncated at this point, too much computational noise would be introduced to the final result. To preserve higher final accuracy, the full 32 bit result is held in the P-register (for product register). This configuration is shown in Figure 19.8 which depicts the multiplier and the ALU of the TMS320C25.

Actually, the P-register is viewed as two 16 bit registers concatenated. This viewpoint is convenient if you need to save the product using the instructions SPH (store product high) and SPL (store product low). Otherwise, the product can operate on the accumulator, which is also 32 bit wide. The contents of the product register can be loaded on the accumulator, overwriting whatever was there, using the PAC (product to accumulator) instruction. It can also be added to or subtracted from the accumulator using the instructions APAC or SPAC.

When moving the contents of the T-register to the accumulator, you can shift this number using the built-in shifters. For instance you can shift the result left by 1 or 4 locations (essentially multiplying it by 2 or 16), or you can shift it right by 6 (essentially dividing it by 64). These operations are done automatically, without spending any extra machine cycles, simply by setting the appropriate product mode with SPM instruction. Why would you want to do such shifting? The left shifts have as a main purpose to eliminate any extra sign bits that would appear in computations. The right shift scales down the result and permits accumulation of several products before you start worrying about overflowing the accumulator.

At this point, it is appropriate to discuss the data formats supported on the TMS320C25. This device, as most fixed-point processors, uses two's-complement notation to represent the negative numbers.

In two's-complement notation, to form the negative of a given number, you take the complement of that number and you add 1. In two's-complement notation, the most significant bit (MSB, the left-most bit) of a positive number is zero, while the MSB of a negative number is one. In the 'C25, the two's-complement numbers are sign-extended, which means that, if the absolute value of the number is not large enough to fill all the bits of the word, there will be more than one sign bits.

As seen from Figure 19.8, the multiplier path is not the only way to access the accumulator. Actually, the ALU and the accumulator support a wealth of arithmetic (ADD, SUB, etc.) and logical (OR, AND, XOR, etc.) instructions, in addition to load and store instructions for the accumulator (LAC, ZALH, SACL, SACH, etc.).

An interesting characteristic of the TMS320C25 architecture is the existence of several shifters that can perform such shifts in parallel with other operations. Except for the right shifter at the multiplier, all the other shifters are left shifters. An input shifter to the ALU and the accumulator can shift the input value to the left by up to 16 locations, while output shifters from the accumulator can shift either the high or the low part of the accumulator by up to 7 locations to the left.

A construct that appears very often in mathematical computations is the sum of products. Sums of products appear in the computation of dot products, in matrix multiplication, and in convolution sums for filtering, among other applications. Since it is important to carry out this computation as fast as possible for real-time operation, all DSPs have special instructions to speed up this particular function.

The TMS320C25 has the instruction LTA which loads the T-register and, in parallel with that, adds the previous product (which already resides in the P-register) to the accumulator. LTS subtracts the product from the accumulator. Another instruction, LTD, does the same thing as LTA, but it also moves the value that was just loaded on the T-register to the next higher location in memory. This move realizes the delay line that is needed in filtering applications. LTA, when combined with the MPY instruction, can implement very efficiently the sum of products.

For even higher efficiency, there is a MAC instruction that combines LTA and MPY. An additional MACD instruction combines LTD and MPY. The increased efficiency is achieved by using both the data

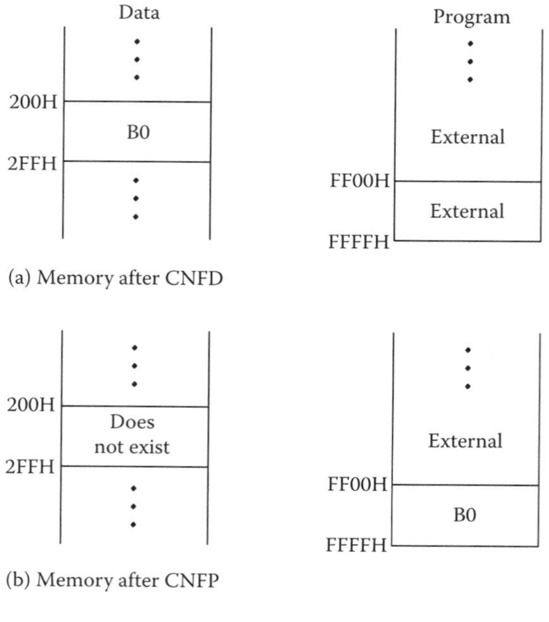

FIGURE 19.9 Partial memory configuration of the TMS320C25 after the CNFD (a) and the CNFP (b) instructions.

and the program buses to bring in the operands of the multiplication. The data coming from the data bus can be traced in memory by an AR, using indirect addressing. The data coming from the program bus are traced by the program counter (actually, the prefetch counter, PFC) and, hence, they must reside in consecutive locations of program memory. To be able to modify the data and then use it in such multiply–add operations, the TMS320C25 permits reconfiguration of block B0 in the on-chip memory. B0 can be configured either as program or as data memory, as shown in Figure 19.9, using the CNFD and CNFP instructions.

19.5 Other Architectural Features of the TMS320C25

The TMS320C25 has many interesting features and capabilities that can be found in the User's Guide [1]. Here, we present briefly only the most important of them.

The program counter is a 16 bit register, hidden from the user, which contains the address of the next instruction word to be fetched and executed. Occasionally, the program execution may be redirected, for instance, through a subroutine call. In this case, it is necessary to save the contents of the program counter so that the program flow continues from the correct instruction after the completion of the subroutine call. For this purpose, a hardware stack is provided to save and recover the contents of the program counter.

The hardware stack is a set of eight registers, of which only the top one is accessible to the user. Upon a subroutine call, the address after the subroutine call is pushed on the stack, and it is reinstated in the program counter when the execution returns from the subroutine call. The programmer has control over the stack by using the PUSH, PSHD, POP, and POPD instructions. The PUSH and POP operations push the accumulator on the stack or pop the top of the stack to the accumulator respectively. PSHD and POPD do the same functions but with memory locations instead of the accumulator.

Occasionally the program execution in a processor must be interrupted in order to take care of urgent functions, such as receiving data from external sources. In these cases, a special signal goes to the processor, and an interrupt occurs. The interrupts can be internal or external. During an interrupt, the processor stops execution, wherever it may be, pushes the address of the next instruction on the stack, and starts executing from a predetermined location in memory. The interrupt approach is appropriate when there are functions or devices that need immediate attention. On the TMS320C25, there are several internal and external interrupts, which are prioritized, i.e., when several of the interrupts occur at the same time, the one with the highest priority is executed first. Typically, the memory location where the execution is directed to during an interrupt contains a branch instruction. This branch instruction directs the program execution to an area in the program memory where an interrupt service routine exists. The interrupt service routine will perform the tasks that the interrupt has been designed for, and then return to the execution of the original program.

Besides the external hardware interrupts (for which there are dedicated pins on the device), there are internal interrupts generated by the serial port and the timer. The serial port provides direct communication with serial devices, such as codecs, serial analog-to-digital converters, etc. In these devices, the data are transmitted serially, one bit at a time, and not in parallel, which would require several parallel lines. When 16 bit has been input, the 16 bit word can be retrieved from the DRR (data receive register). Conversely, to transmit a word, you put it in the DXR (data transmit register). These two registers occupy data memory locations 0 and 1, respectively, and they can be treated like any other memory location.

The timer consists of a period register and a timer register. At the beginning of the operation, the contents of the period register are loaded on the timer register, which is then decremented at every machine cycle. When the value of the timer register reaches zero, it generates a timer interrupt, the period register is loaded again on the timer register, and the whole operation is repeated.

19.6 TMS320C25 Instruction Set

The TMS320C25 has an instruction set consisting of 133 instructions. Some of these assembly language instructions perform general purpose operations, while others are more specific to DSP applications. This section discusses examples of instructions selected from different groups. For a detailed description of each instruction, the reader is referred to the *TMS320C25 User's Guide* [1].

Each instruction is represented by one or two 16 bit words. Part of the instruction is a unique code identifying the operation to be performed, while the rest of the instruction contains information on the operation. For instance, this additional information determines if direct or indirect addressing is used, if there is a shift of the operand, what is the address of the operand, etc. In the case of two-word instructions, the second word is typically a 16 bit constant or program memory address. As it should be obvious, a two-word instruction takes longer to execute because it has to fetch two words, and it should be avoided if the same operation could be accomplished with a single-word instruction.

For example, if you want to load the accumulator with the contents of the memory location 3FH, shifting it to the left by 8 locations at the same time, you can write the instruction

```
LAC    3FH,8
```

The above instruction, when encoded, is represented by the word 283FH. The left-most 4 bit in this example, i.e., 0010, represents the "opcode" of the instruction. The opcode is the unique identifier of the instruction. The next 4 bit, 1000, is the shift of the operand. Then there is one bit (zero in this case) to signal that the direct addressing mode is used, and the last 7 bit is the operand address 3Fh (in hexadecimal).

Below, some of the more typical instructions are listed, and the ones that have an important interpretation are discussed. It is a good idea to review carefully the full set of instructions so that you know what tools you have available to implement any particular construct. The instructions are grouped here by functionality.

The accumulator and memory reference instructions involve primarily the ALU and the accumulator. Note that there is a symmetry in the instruction set. The addition instructions have counterparts for subtraction, the direct and indirect-addressing instructions have complementary immediate instructions, and so on.

ABS	Absolute value of accumulator
ADD	Add to accumulator with shift
ADDH	Add to high accumulator
ADDK	Add to accumulator short immediate
AND	Logical AND with accumulator
LAC	Load accumulator with shift
SACH	Store high accumulator with shift
SACL	Store low accumulator with shift
SUB	Subtract from accumulator with shift
SUBC	Subtract conditionally
ZAC	Zero accumulator
ZALH	Zero low accumulator and load high accumulator

Operations involving the accumulator have versions affecting both the high part and the low part of the accumulator. This capability gives additional flexibility in scaling, logical operations, and double-precision arithmetic.

For example, let location A contain a 16 bit word that you want to scale down dividing by 16, and store the result in B. The following instructions perform this operation:

```
LAC     A,12   Load ACC with A shifted by 12 locations
SACH    B      Store ACCH to B:B = A/16
```

The ARs and DP pointer instructions deal with loading, storing, and modifying the ARs and the DP pointer. Note that the ARs and the ARP can also be modified during operations using indirect addressing. Since this last approach has the advantage of making the modifications in parallel with other operations, it is the most common method of AR modification.

LAR	Load auxiliary register
LARP	Load auxiliary register pointer
LDP	Load data memory page pointer
MAR	Modify auxiliary register
SAR	Store auxiliary register

The multiplier instructions are more specific to signal-processing applications.

APAC	Add P-register to accumulator
LT	Load T-register
LTD	Load T-register, accumulate previous product, and move data
MAC	Multiply and accumulate
MACD	Multiply and accumulate with data move
MPY	Multiply
MPYK	Multiply immediate
PAC	Load accumulator with P-register
SQRA	Square and accumulate

Note that the instructions that perform multiplication and accumulation at the same time do not accumulate the present product but the result of an earlier multiplication. This result is found in the P-register. The square and accumulate function, SQRA, is a special case of the multiplication that appears often enough to prompt the inclusion of this specific instruction.

The branch instructions correspond to the GOTO instruction of high-level languages. They redirect the flow of the execution either unconditionally or depending on some previous result.

B	Branch unconditionally
BANZ	Branch on auxiliary register nonzero
BGEZ	Branch if accumulator $>=0$
CALA	Call with subroutine address in the accumulator
CALL	Call subroutine
RET	Return from subroutine

The CALL and RET instructions go together because the first one pushes the return address on the stack, while the second one pops the address from the stack into the program counter. The BANZ instruction is very helpful in loops where an AR is used as a loop counter. BANZ tests the AR, modifies it, and branches to the indicated address.

The I/O operations are, probably, among the most important in terms of final system configuration, because they help the device interact with the rest of the world. Two instructions that perform that function are the IN and OUT instructions.

BLKD	Block move from data memory to data memory
IN	Input data from port
OUT	Output data to port
TBLR	Table read
TBLW	Table write

The IN and OUT instructions read from or write to the 16 input and the 16 output ports of the TMS320C25. Any transfer of data goes to a specified memory location. The BLKD instruction permits movement of data from one memory location to another without going through the accumulator. To make such a movement effective, though, it is recommended to use BLKD with a repeat instruction, in which case every data move takes only one cycle.

The TBLR and TBLW instructions represent a modification to the Harvard architecture of the device. Using them, data can be moved between the program and the data spaces. In particular, if any tables have been stored in the program memory space they can be moved to data memory before they can be used. That is how the terminology of the instructions originated.

Some other instructions include

DINT	Disable interrupts
EINT	Enable interrupts
IDLE	Idle until interrupt
RPT	Repeat instruction as specified by data memory value
RPTK	Repeat instruction as specified by immediate value

19.7 Input/Output Operations of the TMS320C25

During program execution on a DSP, the data are moved between the different memory locations, on-chip and off-chip, as well as between the accumulator and the memory locations. This movement is necessary for the execution of the algorithm that is implemented on the processor. However, there is a need to communicate with the external world in order to receive data that will be processed, and return the processed results.

Devices communicate with the external world through their external memory or through the serial and parallel ports. Such a communication can be achieved, for instance, by sharing the external memory. Most often, the communication with the external world takes place through the external parallel or serial ports that the device has. Some devices may have ports of only one kind, serial or parallel, but most modern processors have both types. The two kinds of ports differ in the way in which the bits are read. In a parallel port, there is a physical line (and a processor pin) dedicated to every bit of a word. For example, if the processor reads in words that are 16 bit wide, as is the case with the TMS320C25, it has 16 lines available to read a whole word in a single operation. Typically, the same pins that are used for accessing external memory are also used for I/O.

The TMS320C25 has 16 input and 16 output ports that are accessed with the IN and OUT instructions. These instructions transfer data between memory locations and the I/O port specified.

19.8 Subroutines, Interrupts, and Stack on the TMS320C25

When writing a large program, it is advisable to structure it in a modular fashion. Such modularity is achieved by segmenting the program in small, self-contained tasks that are encoded as separate routines. Then, the overall program can be simply a sequence of calls to these subroutines, possibly with some "glue" code. Constructing the program as a sequence of subroutines has the advantage that it produces a much more readable algorithm that can greatly help in debugging and maintaining it. Furthermore, each subroutine can be debugged separately, which is far easier than trying to uncover programming errors in a "spaghetti-code" program.

Typically, the subroutine is called during the program execution with an instruction such as

```
CALL SUBRTN
```

where SUBRTN is the address where the subroutine begins. In this example, SUBRTN would be the label of the first instruction of the subroutine. The assembler and the linker resolve what the actual value is. Calling a subroutine has the following effects:

- Increments the program counter (PC) by one and pushes its contents on the top of the stack (TOS). The TOS now contains the address of the instruction to be executed after returning from the subroutine.
- Loads the address SUBRTN on the PC.
- Starts execution from where the PC is pointing at (i.e., from location SUBRTN).

At the end of the subroutine execution, a return instruction (RET) will pop the contents of the top of the stack on the program counter, and the program will continue execution from that location.

The stack is a set of memory locations where you can store data, such as the contents of the PC. The difference from regular memory is that the stack keeps track of the location where the most recent data were stored. This location is the TOS. The stack is implemented either in hardware or software.

The TMS320C25 has a hardware stack that is eight locations deep. When a piece of data is put ("pushed") on the stack, everything already there is moved down by one location. Notice that the contents of the last location (bottom of the stack) are lost. Conversely, when a piece of data are retrieved from the stack (it is "popped"), all the other locations are moved up by one location. Pushing and popping always occur at the top of the stack.

The interrupt is a special case of subroutine. The TMS320C25 supports interrupts generated either internally or from external hardware. An interrupt causes a redirection of the program execution in order to accomplish a task. For instance, data may be present at an input port, and the interrupt forces the processor to go and "service" this port (inputting the data). As another example, an external D/A converter may need a sample from the processor, and it uses an interrupt to indicate to the DSP device that it is ready to receive the data. As a result, when the processor is interrupted, it knows by the nature of the interrupt that it has to go and do a specific task, and it does just that.

The performance of the designated task is done by the interrupt service routine (ISR). An ISR is like a subroutine with the only difference on the way it is accessed, and in the functions performed upon return. When an interrupt occurs, the program execution is automatically redirected to specific memory locations, associated with each interrupt. As explained earlier, the TMS320C25 continues execution from a specified memory location which, typically, contains a branch instruction to the actual location of the interrupt service routine.

The return from the interrupt service routine, like in a subroutine, pops the top of the stack to the program counter. However, it has the additional effect of re-enabling the interrupts. This is necessary because when an interrupt is serviced, the first thing that happens is that all interrupts are disabled to avoid confusion from additional interrupts. Re-enabling is done explicitly in the TMS320C25 (by using the EINT command).

19.9 Introduction to the TMS320C30 Digital Signal Processor

The Texas Instruments TMS320C30 is a floating-point processor that has some commonalities with the TMS320C25, but that also has a lot of differences. The differences are due more to the fact that the TMS320C30 is a newer processor than that it is a floating-point processor. The TMS320C30 is a fast, 32 bit, DSP that can handle both fixed-point and floating-point operations. The speed of the device is 16.7 MHz, which corresponds to a cycle time of 60 ns. Since the majority of the instructions execute in a single cycle (after the pipeline is filled), the figure of 60 ns also indicates how long it takes to execute one instruction. Alternatively, we can say that the device can execute 16.7 MIPS. Another figure of merit is based on the fact that the device can perform a floating-point multiplication and addition in a single cycle. Then, it is said that the device has a (maximum) throughput of 33 million floating-point operations per second (MFLOPS).

The actual signal from the external oscillator or crystal has a frequency twice that of the internal device speed, at 33.3 MHz (and period of 30 ns). This frequency is then divided on-chip to generate the internal clock with a period of 60 ns. Newer versions of the TMS320C30 and other members of the 'C3x generation operate in higher frequencies.

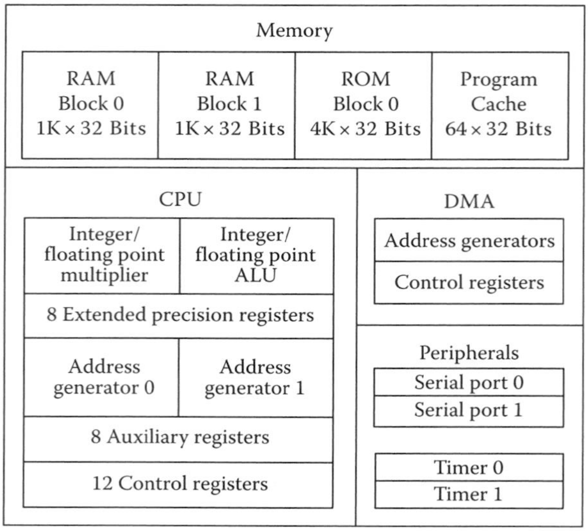

FIGURE 19.10 Key architectural features to the TMS320C30.

Figure 19.10 shows in a simplified form the key features of the TMS320C30. The major parts of the DSP processor are the memory, the CPU, the peripherals, and the direct memory access (DMA) unit. Each of these parts are examined in more detail later in this chapter. The on-chip memory consists of 2K words of RAM and 4K words of ROM. There is also a 64-word long program cache. Each word is 32 bit wide and the memory sizes for the TMS320C30 are measured in 32 bit words, and not in bytes. The memory (RAM or ROM) can be used to store either program instructions or data. This presents a departure from the practice of separating the two spaces that the TMS320C25 uses, combining features of a von Neuman architecture with a Harvard architecture. Overall, the device can address 16 M words of memory through two external buses. Except for what resides on-chip, the rest of the memory is external, supplied by the designer.

The CPU is the heart of the processor. It has a hardware multiplier that is capable of performing a multiplication in a single cycle. The multiplication can be between two 32 bit floating point numbers, or between two integers. To achieve a higher intermediate accuracy of results, the product of two floating-point numbers is saved as a 40 bit result. In integer multiplication, two 24 bit numbers are multiplied together to give a 32 bit result. The other important part of the CPU is ALU that performs additions, subtractions, and logical operations. Again, for increased intermediate accuracy, the ALU can operate on 40 bit long floating-point numbers and generates results that are also 40 bit long.

The 'C30 can handle both integers and floating-point numbers using corresponding instructions. There are three kinds of floating-point numbers, as shown in Figure 19.11: short, single-precision, and extended-precision. In all three kinds, the number consists of an exponent e, a sign *s*, and a mantissa *f*. Both the mantissa (part of which is the sign) and the exponent are expressed in two's-complement notation.

In the short floating-point format, the mantissa consists of 12 bit and the exponent of 4 bit. The short format is used only in immediate operands, where the actual number to operate upon becomes part of the instruction. The single-precision format is the regular format representing the numbers in the TMS320C30, which is a 32 bit device. It has 24 bit for mantissa and 8 bit for exponent. Finally, the extended-precision format is encountered only in the extended-precision registers, to be discussed below. In this case, the exponents is also 8 bit long, but the mantissa is 32 bit, giving extra precision. The mantissa is normalized so that it has a magnitude $|f|$ such that $1.0 =< |f| < 2.0$.

The integer formats supported in the TMS320C30 are shown in Figure 19.12. Both the short and the single-precision integer formats represent the numbers in two's-complement notation. The short

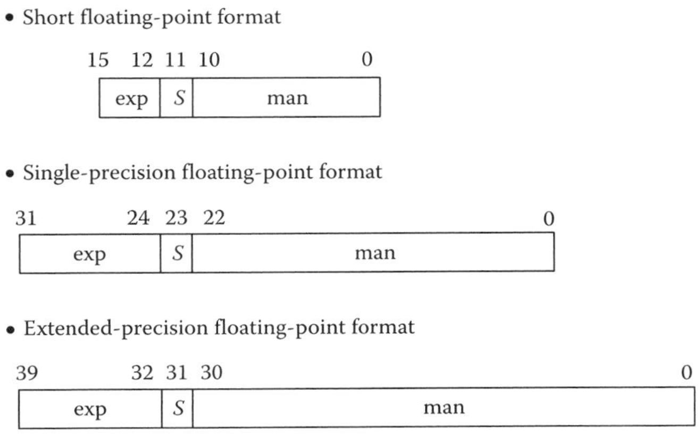

FIGURE 19.11 TMS320C30 floating point formats.

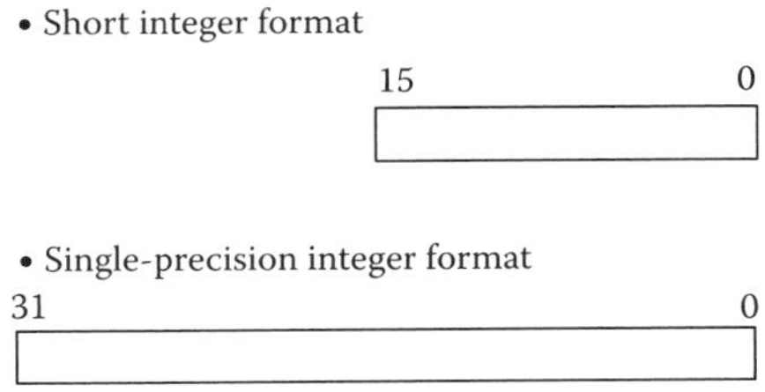

FIGURE 19.12 TMS320C30 integer (fixed-point) formats.

format is used in immediate operands, where the actual number to be operated upon is part of the instruction itself.

All the arithmetic and logical functions are register based. In other words, the destination and at least one source operand in every instruction are register file associated with the TMS320C30 CPU. Figure 19.13 shows the components of the register file. There are eight extended-precision registers, R0–R7, that can be used as general purpose accumulators for both integer and floating-point arithmetic. These registers are 40 bit wide. When they are used in floating-point operations, the top 8 bit is the exponent and the bottom 32 bit is the mantissa of the number. When they are used as integers, the bottom 32 bit is the integer, while the top 8 bit is ignored and are left intact.

The eight ARs, AR0–AR7, are designated to be used as memory pointers or loop counters. When treated as memory pointers, they are used during the indirect addressing mode, to be examined below. AR0–AR7 can also be used as general-purpose registers but only for integer arithmetic.

Additionally, there are 12 control registers designated for specific purposes. These registers too can be treated as general purpose registers for integer arithmetic if they are not used for their designated purpose. Examples of such control registers are the status register, the stack pointer, the block repeat registers, and the index registers.

To communicate with the external world, the TMS320C30 has two parallel buses, the primary bus and the expansion bus. It also has two serial ports that can serve the same purpose. The serial ports are part of the peripherals that have been implemented on chip. Other peripherals include the DMA unit, and two timers. These components of the TMS320C30 are examined in more detail in the following.

The device has 181 pins that are designated to perform certain functions, and to communicate with other devices on the same board. The names of the signals and the corresponding definitions appear in Table 19.3. The first column of the table gives the pin names; the second one indicates if the pin is used

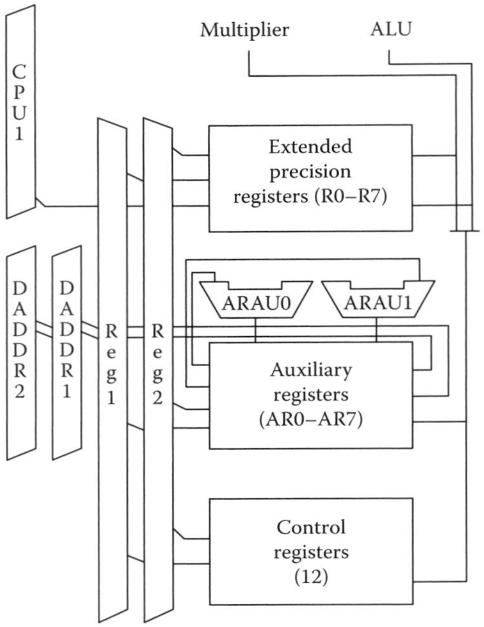

FIGURE 19.13 TMS320C30 register file.

for input or output; the third column gives a description of the pin functionality. Note that a bar over the name indicates that the pin is in the active position when it is electrically low. The second column indicates if the pin is used for input to the device, output from the device, or both.

The TMS320C30 has its own assembly language consisting of 114 instructions that perform general-purpose and DSP-specific functions. High-level language compilers have also been developed that make the writing of programs an easier task. The TMS320C30 was designed with a high-level language compiler in mind, and its architecture incorporates some appropriate features. For instance, the presence

TABLE 19.3 Names and Functionality of the 181 Pins of the TMS320C30

Signal	I/O	Description
D(31-0)	I/O	32 bit data port of the primary bus
A(23-0)	O	24 bit address port of the primary bus
$R/\overline{W}$	O	Read/write signal for primary bus interface
$\overline{STRB}$	O	External access strobe for the primary bus
$\overline{RDY}$	I	Ready signal
$\overline{HOLD}$	I	Hold signal for primary bus
$\overline{HOLDA}$	O	Hold acknowledge signal for primary bus
XD(31-0)	I/O	32 bit data port of the expansion bus
XA(12-0)	O	13 bit address port of the expansion bus
$XR/\overline{W}$	O	Read/write signal for expansion bus interface
MSTRB	O	External access strobe for the expansion bus
IOSTRB	O	External access strobe for the expansion bus
$\overline{XRDY}$	I	Ready signal
$\overline{RESET}$	I	Reset
$\overline{INT}$(3-0)	I	External interrupts
$\overline{IACK}$	O	Interrupt acknowledge signal

TABLE 19.3 (continued) Names and Functionality of the 181 Pins of the TMS320C30

Signal	I/O	Description
MC/$\overline{\text{MP}}$	I	Microcomputer/microprocessor mode pin
XF(1-0)	I/O	External flag pins
CLKX(1-0)	I/O	Serial port (1-0) transmit clock
DX(1-0)	O	Data transmit output for port (1-0)
FSX(1-0)	I/O	Frame synchronization pulse for transmit
CLKR(1-0)	I/O	Serial port (1-0) receive clock
DR(1-0)	I	Data receive for serial port (1-0)
FSR(1-0)	I	Frame synchronization pulse for receive
TCLK(1-O)	I/O	Timer (1-0) clock
V_{DD}, etc.	I	12 + 5 V supply pins
V_{SS}, etc.	I	11 ground pins
X1	O	Output pin from internal oscillator for the crystal
X2/CLKIN	I	Input pin to the internal oscillator from the crystal
H1, H3	O	External H1, H3 clock. H1 = H3 = 2 CLKIN
EMU, etc.	I/O	20 Reserved and miscellaneous pins

of the software stack, the register file, and the large memory space were to a large extent motivated by compiler considerations.

The TMS320C30 combines the features of the Harvard and the von Neuman architectures to offer more flexibility. The memory is a unified space where the designer can select the places for loading program instructions or data. This von Neuman feature maximizes the efficient use of the memory. On the other hand, there are multiple buses to access the memory in a Harvard style, as shown in Figure 19.14.

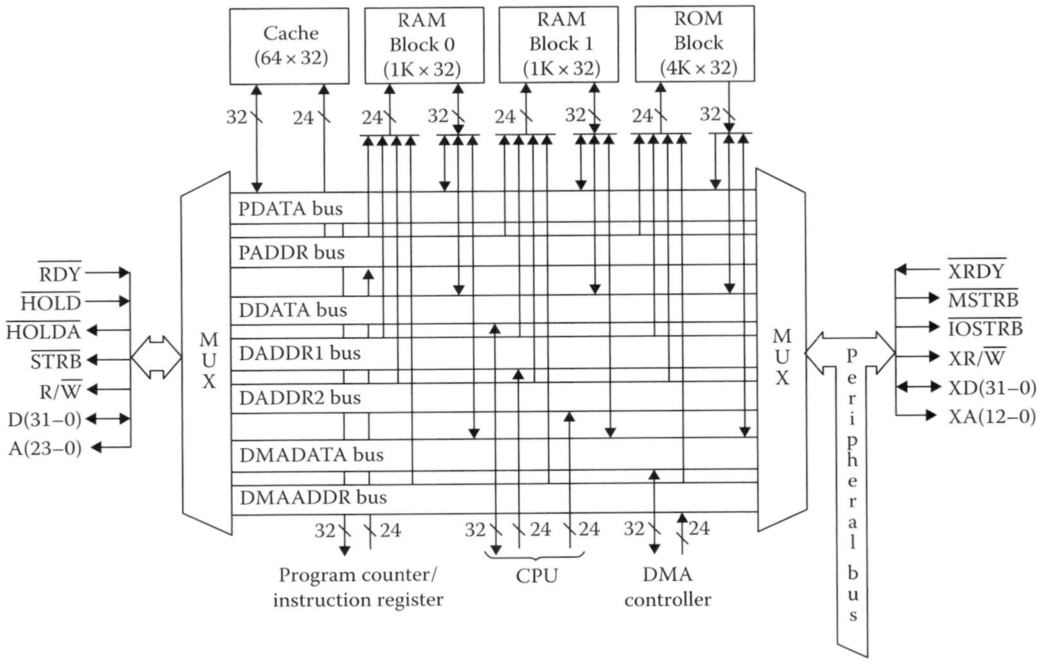

FIGURE 19.14 Internal bus structure of the TMS320C30.

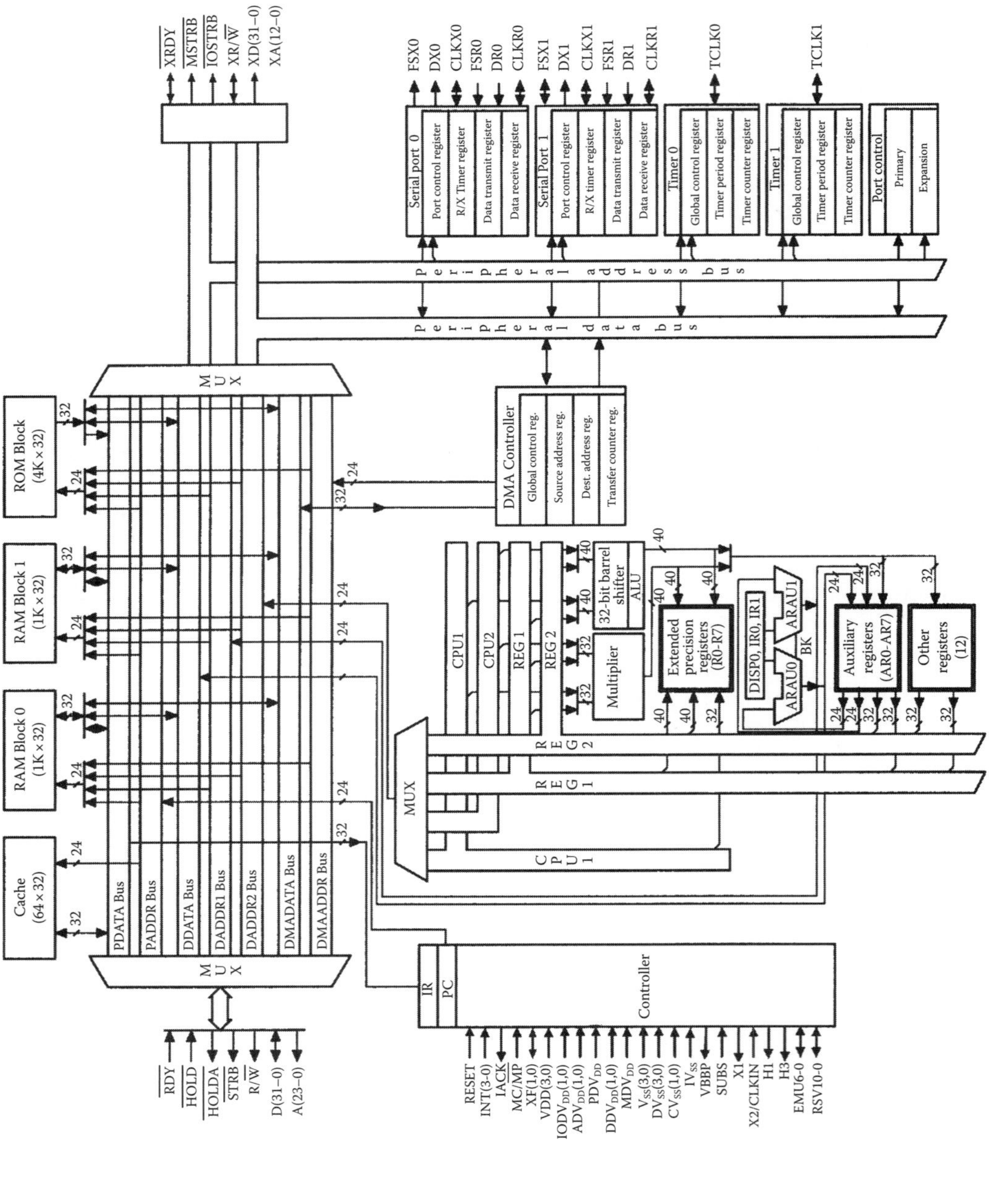

FIGURE 19.15 Functional block diagram of the TMS320C30 architecture.

Two of the buses are used for the program, to carry the instruction address and fetch the instruction. Three buses are associated with data: two of those carry data addresses, so that two memory accesses can be done in the same machine cycle. The third bus carries the data. The reason that one bus is sufficient to carry the data is that the device needs only one-half of a machine cycle to fetch an operand from the internal memory. As a result, two data fetches can be accomplished in one cycle over the same bus.

The last two buses are associated with the DMA unit, which transfers data in parallel with and transparently to the CPU. Because of the multiple buses, program instructions and data operands can be moved simultaneously increasing the throughput of the device. Of course, it is conceivable that too many accesses can be attempted to the same memory area, causing access conflicts. However, the TMS320C30 has been designed to resolve such conflicts automatically by inserting the appropriate delays in instruction execution. Hence, the operations always give the correct results.

Figure 19.15 shows a functional block diagram of the TMS320C30 architecture with the buses, the CPU, and the register file. It also points out the peripheral bus with the associated peripherals. Because of the peripheral bus, all the peripherals are memory-mapped, and any operations with them are seen by the programmer as accesses (reads/writes) to the memory.

19.10 TMS320C30 Memory Organization and Access

The TMS320C30 has on-chip 2K words (32 bit wide) of RAM and 4K of ROM. This memory can be accessed twice in a single cycle, a fact that is reflected in the instruction set, which includes three-operand instructions: two of the operands reside in memory, while the third operand is the register where the result is placed.

Besides the on-chip memory, the TMS320C30 can access external memory through two external buses, the primary and the expansion. The primary bus consists of 24 address pins A0–A23, and 32 data pins D0–D31. As the number of address pins suggests, the maximum memory space available is 16M words. Not all of that, though, resides on the primary bus. The primary bus has 16M words minus the on-chip memory, and minus the memory available on the expansion bus.

The expansion bus has 13 address pins, XA0–XA12, and 32 data pins, XD0–XD31. The 13 address pins can address 8K words of memory. However, there are two strobes, MSTRB and IOSTRB, that select two different segments of 8K of memory. In other words, the total memory available on the expansion bus is 16K. The differences between the two strobes is in timing. The timing differences can make one of the memory spaces more preferable to the other in certain applications, such as peripheral devices.

As mentioned earlier, the destination operand is always a register in the register file (except for storing a result, where, of course, the destination is a memory location.) The register can also be one of the source operands. It is possible to specify a source operand explicitly and include it in the instruction. This addressing mode is called immediate addressing mode. The immediate constant should be accommodated by a 16 bit wide word, as discussed earlier in the data formats.

For example, if it is desired to increment the (integer) contents of the register R0 by 5, the following instruction can be used:

```
ADDI   5,R0
```

To increment the (floating-point) contents of the register R3 by −2.75, you can use the instruction

```
ADDF   −2.75,R3
```

If the value to be operated upon resides in memory, there are two ways to access it: either by specifying the memory address directly (direct addressing) or by using an AR holding that address and, hence, pointing to that number indirectly (indirect addressing). In the direct addressing mode, full description of a memory address would require a 24 bit word because the memory space is 16M words. To reduce that requirement, the memory space is divided in 256 pages of 64K words each. An instruction using direct addressing contains the 16 bit indicating what word you want to access within a page. The page number (8 bit) is stored in one of the control registers, the DP pointer. The DP pointer can be modified by using either a load instruction or the pseudo-instruction LDP. During assembly time, LDP picks the top 8 bit of a memory address and places them in the DP register.

Of course, if several locations need to be accessed in the same page, you can set the DP pointer only once. Since the majority of the routines written are expected to be less than 64K words long, setting the DP register at the beginning of the program suffices. The exception to that would be placing the code over the boundary of two consecutive pages.

In the indirect addressing mode, the data memory address is held in a register that acts as a memory pointer. There are eight such registers available, AR0–AR7. These registers can also be used for other functions, such as loop counters or general purpose registers. If they are used as memory pointers, they are explicitly specified in the instruction. In an instruction, indirect addressing is indicated by an asterisk preceding the AR.

For example, the instruction

```
LDF    * AR3++,R0    Load R0 with −612
```

loads R0 with the contents of the memory location pointed at by AR3.

The "++" sign in the above instruction means "after the present memory access, increment the contents of the current AR by 1." This is done in parallel with the load-register operation.

The above autoincrementing of the AR is an optional operation that offers additional flexibility to the programmer, and it is not the only one available. The TMS320C30 has two ARAUs (ARAU0 and ARAU1) that can execute such operations in parallel with the CPU, and increase the throughput of the device in this way. The primary function of ARAU0 and ARAU1 is to generate the addresses for accessing operands.

Table 19.4 summarizes the different operations that can be done while using indirect addressing. As seen from this table, the contents of an AR can be incremented or decremented before or after accessing the memory location. In the case of premodification, this modification can be permanent or temporary. When an auxiliary register ARn, $n = 0$–7, is modified, the displacement disp is either a constant (0–255) or the contents of one of the two index registers IR0, IR1 in the register file. If the displacement is missing, a 1 is implied. The AR contents can be incremented or decremented in a circular fashion, or incremented by the contents of IR0 in a bit-reversed fashion.

The last two kinds of operation have special purposes. Circular addressing is used to create a circular buffer, and it is helpful in filtering applications. Bit-reversed addressing is useful when doing fast Fourier transforms. The bit-reversed addressing is implemented by adding IR0 with reverse carry propagation, an operation explained in the *TMS320C30 User's Guide*.

The TMS320C30 has a software stack that is part of its memory. The software stack is implemented by having one of the control registers, the SP, point to the next available memory location. Whenever a subroutine call occurs, the address to return to after the subroutine completion is pushed on the stack (i.e., it is written on the memory location that SP is pointing at), and SP is incremented by one. Upon return from a subroutine, the SP is decremented by one and the value in that memory location is copied on the program counter.

Since the SP is a regular register, it can be read or written to. As a result, you can specify what part of the memory is used for the stack by initializing SP to the appropriate address. There are specific instructions to push on or pop from the stack any of the registers in the register file: PUSH, POP for integer values, PUSHF, POPF for floating-point numbers. Such instructions can use the stack to pass arguments to subroutines or to save information during an interrupt. In other words, the stack is a convenient scratch-pad that you designate at the beginning, so that you do not have to worry where to store some temporary values.

TABLE 19.4 Operations That Can Be Performed in Parallel with Indirect Addressing in the TMS320C30

Notation	Operation	Description
* ARn	addr = ARn	Indirect without modification
*+ ARn(disp)	addr = ARn + disp	With predisplacement add
*− ARn(disp)	addr = ARn − disp	With predisplacement subtract
*++ ARn(disp)	addr = ARn − disp	With predisplacement add and modify
	ARn = ARn + disp	
*−− ARn(disp)	addr = ARn − disp	With predisplacement subtract and modify
	ARn = ARn − disp	
* ARn++(disp)	addr = ARn	With postdisplacement add and modify
	ARn = ARn + disp	
* ARn−−(disp)	addr = ARn	With postdisplacement subtract and modify
	ARn = ARn − disp	
* ARn++(disp)%	addr = ARn	With postdisplacement add and circular modify
	ARn = circ(ARn + disp)	
* ARn−−(disp)%	addr = ARn	With postdisplacement subtract and circular modify
	ARn = circ(ARn − disp)	
* ARn++(IR0)B	addr = ARn	With postdisplacement add and bit-reversed modify
	ARn = rc(ARn + IR0)	

Note: B, bit reversed; circ, circular modification; rc, reverse carry.

19.11 Multiplier and ALU of the TMS320C30

The heart of the TMS320C30 is the CPU consisting, primarily, of the multiplier and the ALU. The CPU configuration is shown in Figure 19.16 which depicts the multiplier and the ALU of the TMS320C30. The hardware multiplier can perform both integer and floating-point multiplications in a single machine cycle.

The inputs to the multiplier come from either the memory or the registers of the register file. The outputs are placed in the register file. When multiplying floating-point numbers, the inputs are 32 bit long (8 bit exponent and 24 bit mantissa), and the result is 40 bit wide directed to one of the extended precision registers. If the input is longer than 32 bit (extended precision) or shorter than 32 bit (short format) it is truncated or extended, respectively, by the device to become a 32 bit number before the operation. Multiplication of integers consists of multiplying two 24 bit numbers to generate a 32 bit result. In this case, the registers used can be any of the registers in the register file.

The other major part of the CPU is the ALU. The ALU can also take inputs from either the memory or the register file and perform arithmetic or logical operations. Operations on floating-point numbers can be done on 40 bit wide inputs (8 bit exponent and 32 bit mantissa) to give also 40 bit results. Integer operations are done on 32 bit numbers. Associated with the ALU, there is a barrel shifter that can perform either a right-shift or a left-shift of a register's contents for any number of locations in a single cycle. The instructions for shifting are ASH (Arithmetic SHift) and LSH (Logical SHift).

19.12 Other Architectural Features of the TMS320C30

The TMS320C30 has many interesting features and capabilities. For a full account, the reader is urged to look them up in the User's Guide [2]. Here, we briefly present only the most important of them so that you have a global view of the device and its salient characteristics.

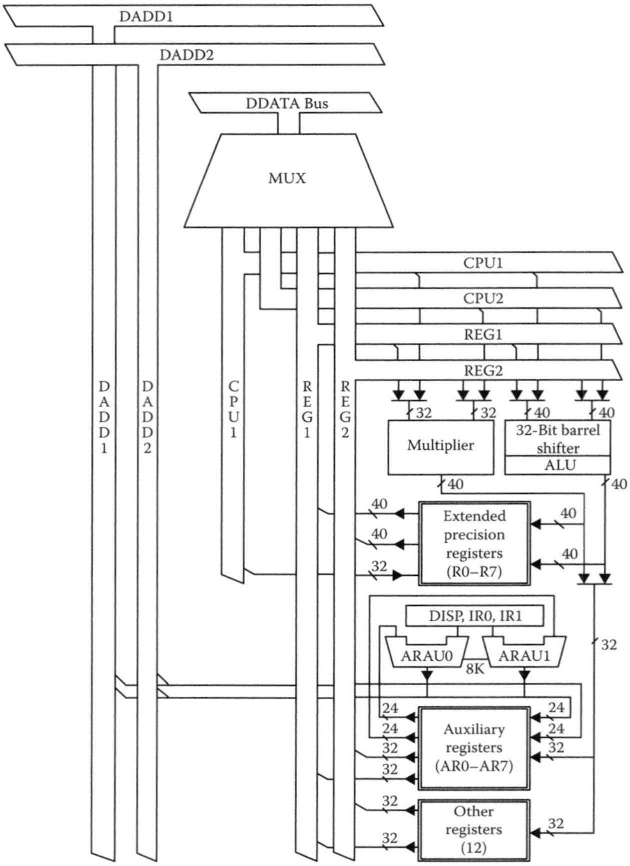

FIGURE 19.16 Central processing unit (CPU) of the TMS320C30.

The TMS320C30 is a very fast device, and it can execute very efficiently instructions from the on-chip memory. Often, though, it is necessary to use external memory for program storage. The existing memory devices either are not as fast as needed, or are quite expensive. To ameliorate this problem, the TMS320C30 has 64 words of program cache on-chip. When executing a program from external memory, every instruction is stored on the cache as it is brought in. Then, if the same instruction needs to be executed again (as is the case for instructions in a loop), it is not fetched from the external memory but from the cache. This approach speeds up the execution, but it also frees the external bus to fetch, for instance, operands. Obviously, the cache is most effective for loops that are shorter than 64 words long, something usual in DSP applications. On the other hand, it does not offer any advantages in the case of straight-line code. However, the structure of DSP problems suggests that the cache is a feature that can be put to good use.

In the instruction set of the 'C30 there is the RPTS (RePeaT Single) instruction

RPTS N

that repeats the following instruction N + 1 times. A more generalized repeated mode is implemented by the RPTB (RePeaT Block) instruction that repeats a number of times all the instructions between RPTB and a label that is specified in the block-repeat instruction. The number of repetitions is one more than the number stored in the repeat count register, RC, one of the control registers in the register file.

For example the following instructions are repeated one time more than the number included in the RC.

```
LDI    63,RC              ; The loop is to be repeated 64 times
       RPTB   LOOP        ; Repeat up to the label LOOP
       LDI    * AR0,R0    ; Load the number on R0
       ADDI   1,R0        ; Increment it by 1
LOOP   STI    R0,* AR0++  ; Store the result; point to the next
                          ; number; and loop back
```

Besides RC, there are two more control registers used with the block-repeat instruction. The repeat-start (RS) contains the beginning of the loop, and the repeat-end (RE) the end of the loop. These registers are initialized automatically by the processor, but they are available to the user in case he needs to save them.

On the TMS320C30, there are several internal and external interrupts, which are prioritized, i.e., when several of the interrupts occur at the same time, the one with the highest priority is executed first.

Besides the reset signal, there are four external interrupts, INT0–INT3. Internally, there are the receive and transmit interrupts of the serial ports, and the timer interrupts. There is also an interrupt associated with the DMA. Typically, the memory location where the execution is directed to during an interrupt contains the address where an interrupt service routine starts. The interrupt service routine will perform the tasks for which the interrupt has been designed, and then return to the execution of the original program. All the interrupts (except the reset) are maskable, i.e., they can be ignored by setting the interrupt enable (IE) register to appropriate values. Masking of interrupts, as well as the memory locations where the interrupt addresses are stored, are discussed in the *TMS320C30 User's Guide* [2].

Each of the two serial ports provides direct communication with serial devices, such as codes, serial analog-to-digital converters, etc. In these devices, the data are transmitted serially, one bit at a time, and not in parallel, which would require several parallel lines. The serial ports have the flexibility to consider the incoming stream as 8, 16, 24, or 32 bit words. Since they are memory-mapped, the programmer goes to certain memory locations to read in or write out the data.

Each of the two timers consists of a period register and a timer register. At the beginning of the operation, the contents of the timer register are incremented at every machine cycle. When the value of the timer register becomes equal to the one in the period register, it generates a timer interrupt, the period register is zeroed out, and the whole operation is repeated.

A very interesting addition to the TMS320C30 architecture is the DMA unit. The DMA can transfer data between memory locations in parallel with the CPU execution. In this way, blocks of data can be transferred transparently, leaving the CPU free to perform computational tasks, and thus increasing the device throughput.

The DMA is controlled by a set of registers, all of which are memory mapped: You can modify these registers by writing to certain memory locations. One register is the source address from where the data are coming. The destination address is where the data are going. The transfer count register specifies how many transfers will take place. A control register determines if the source and the destination addresses are to be incremented, decremented, or left intact after every access. The programmer has several options of synchronizing the DMA data transfers with interrupts or leaving them asynchronous.

19.13 TMS320C30 Instruction Set

The TMS320C30 has an instruction set consisting of 114 instructions. Some of these instructions perform general purpose operations, while others are more specific to DSP applications. The instruction set of the TMS320C30 presents an interesting symmetry that makes programming very easy. Instructions that can be used with integer operands are distinguished from the same instructions for floating-point numbers with the suffix I versus F. Instructions that take three operands are distinguished from the ones with two operands by using the suffix 3. However, since the assembler permits elimination of the symbol 3, the notation becomes even simpler.

A whole new class of TMS320C30 instructions (as compared to the TMS320C25) are the parallel instructions. Any multiplier or ALU operation can be performed in parallel with a store instruction. Additionally, two stores, two loads, or a multiply and an add/subtract can be performed in parallel. Parallel instructions are indicated by placing two vertical lines in front of the second instruction.

For example, the following instruction adds the contents of * AR3 to R2 and puts the result in R5. At the same time, it stores the previous contents of the R5 into the location * AR0.

```
   ADDF  * AR3++,R2,R5
|| STF     R5,*AR0—
```

Note that the parallel instructions are not really two instructions but one, which is also different from its two components. However, the syntax used helps remembering the instruction mnemonics. One of the most important parallel instructions for DSP applications is the parallel execution of a multiplication with an addition or subtraction. This single-cycle multiply-accumulate is very important in the computation of dot products appearing in vector arithmetic, matrix multiplication, digital filtering, etc.

For example, assume that we want to take the dot product of two vectors having 15 points each. Assume that AR0 points to one vector and AR1 to the other. The dot product can be computed with the following code:

```
LDF  0.0,R2                      ; Initialize R2 = 0.0
     LDF 0.0,R0                  ; Initialize R0 = 0.0
     RPTS 14                     ; Repeat loop (single instruction)
     MPYF *AR0++, *AR1++, R0     ; Multiply two points, and
||   ADDF R0,R2                  ; Accumulate previous product
     ADDF R0,R2                  ; Accumulate last product
```

After the operation is completed, R2 holds the dot product.

Before proceeding with the instructions, it is important to understand the working of the device pipeline. At every instant in time, there are four execution units operating in parallel in the TMS320C30: the fetch, decode, read, and execute unit, in order of increasing priority. The fetch unit fetches the instruction; the decode unit decodes the instruction and generates the addresses; the read unit reads the operands from the memory or the registers; the execute unit performs the operation specified in the instruction. Each one of these units takes one cycle to complete. So, an instruction in isolation takes, actually, four cycles to complete. Of course, you never run a single instruction alone.

In the pipeline configuration, as shown in Figure 19.17, when an instruction is fetched, the previous instruction is decoded. At the same time, the operands of the instruction before that are read, while the third instruction before the present one is executed. So, after the pipeline is full, each instruction takes a single cycle to execute.

Is it true that all the instructions take a single cycle to execute? No. There are some instructions, like the subroutine calls and the repeat instructions, that need to flush the pipeline before proceeding. The regular branch instructions also need to flush the pipeline. All the other instructions, though, should take one cycle to execute, if there are no pipeline conflicts.

There are a few reasons that can cause pipeline conflicts, and if the programmer is aware of where the conflicts occur, he can take steps to reorganize his code and eliminate them. In this way, the device throughput is maximized. The pipeline conflicts are examined in detail in the User's Guide [2].

The load and store instructions can load a word into a register, store the contents of a register to memory, or manipulate data on the system stack. Note that the instructions with the same functionality

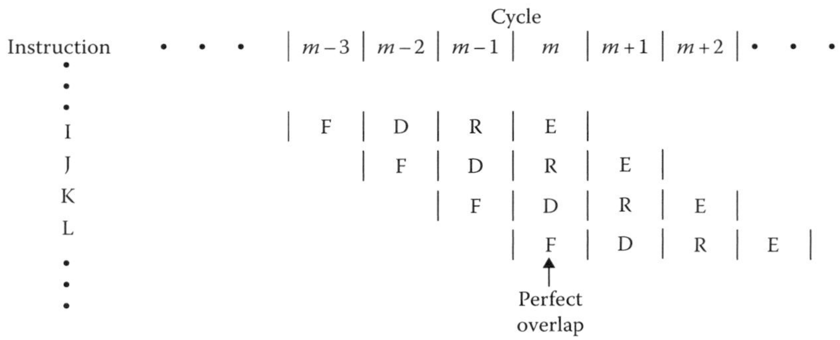

FIGURE 19.17 Pipeline structure of the TMS320C30.

that operate on integers or floating-point numbers are presented together in the following selective listing.

LDF, LDI	Load a floating-point or integer value
LDFcond, LDIcond	Load conditionally
POPF, POP	Pop value from stack
PUSHF, PUSH	Push value on stack
STF, STI	Store value to memory

The conditional loads perform the indicated load only if the condition tested is true. The condition tested is, typically, the sign of the last performed operation.

The arithmetic instructions include both multiplier and ALU operations.

ABSF, ABSI	Absolute value
ADDF, ADDI	Add
CMPF, CMPI	Compare values
FIX, FLOAT	Convert between fixed- and floating point
MPYF, MPYI	Multiply
NEGF, NEGI	Negate
SUBF, SUBI	Subtract
SUBRF, SUBRI	Reverse subtract

The difference between the subtract and the reverse subtract instructions is that the first one subtracts the first operand from the second, while the second one subtracts the second operand from the first.

The logical instructions always operate on integer (or unsigned) operands.

AND	Bitwise logical AND
ANDN	Bitwise logical AND with complement
LSH	Logical shift
NOT	Bitwise logical complement
OR	Bitwise logical OR
XOR	Bitwise exclusive OR

The logical shift differs from an arithmetic shift (which is part of the arithmetic instructions) in that, on a right shift, the logical shift fills the bits to the left with zeros. The arithmetic shift sign extends the (integer) number.

The program control instructions include the branch instructions (corresponding to GOTO of a high-level languages), and the subroutine call and return instructions.

Bcond[D]	Branch conditionally [with delay]
CALL, CALLcond	Call or call conditionally a subroutine
RETIcond, RETScond	Return from interrupt or subroutine conditionally
RPTB, RPTS	Repeat block or repeat a single instruction.

The branch instructions can have an optional "D" at the end to convert them into delayed branches. The delayed branch does the same operation as a regular branch but it takes fewer cycles. A regular branch needs to flush the pipeline before proceeding with the next instruction because it is not known in advance if the branch will be taken or not. As a result, a regular branch costs four machine cycles. If, however, there are three instructions that can be executed no matter if the branch is taken or not, a delayed branch can be used. In a delayed branch, the three instructions following the branch instruction are executed before the branch takes effect. This reduces the effective cost of the delayed branch to one cycle.

19.14 Other Generations and Devices in the TMS320 Family

So far, the discussion in this chapter has focused on two specific devices of the TMS320 family in order to examine in detail their features. However, the TMS320 family consists of five generations (three fixed-point and two floating-point) of DSPs (as well as the latest addition, the TMS320C8x generation, also known as MVP, multimedia video processors). The fixed-point devices are members of the TMS320C1x, TMS320C2x, or TMS320C5x generation, and the floating-point devices belong to the TMS320C3x or TMS320C4x generation.

The TMS320C5x generation is the highest-performance generation of the TI 16 bit fixed-point DSPs. The 'C5x performance level is achieved through a faster cycle time, larger on-chip memory space, and systematic integration of more signal-processing functions. As an example, the TMS320C50 (Figure 19.18) features large on-chip RAM blocks. It is source-code upward-compatible with the first- and second-generation TMS320 devices.

Some of the key features of the TMS320C5x generation are listed below. Specific devices that have a particular feature are enclosed in parentheses.

- CPU
  - 25, 35, 50 ns single-cycle instruction execution time
  - Single-cycle multiply/accumulate for program code
  - Single-cycle/single-word repeats and block repeats for program code
  - Block memory moves
  - Four-deep pipeline
  - Indexed-addressing mode
  - Bit-reversed/indexed-addressing mode to facilitate FFTs
  - Power-down modes
  - 32 bit ALU, 32 bit accumulator, and 32 bit accumulator buffer
  - Eight ARs with a dedicated arithmetic unit for indirect addressing
  - 16 bit parallel logic unit (PLU)
  - 16×16 bit parallel multiplier with a 32 bit product capacity
  - To 16 bit right and left barrel-shifters
  - 64 bit incremental data shifter
  - Two indirectly addressed circular data buffers for circular addressing

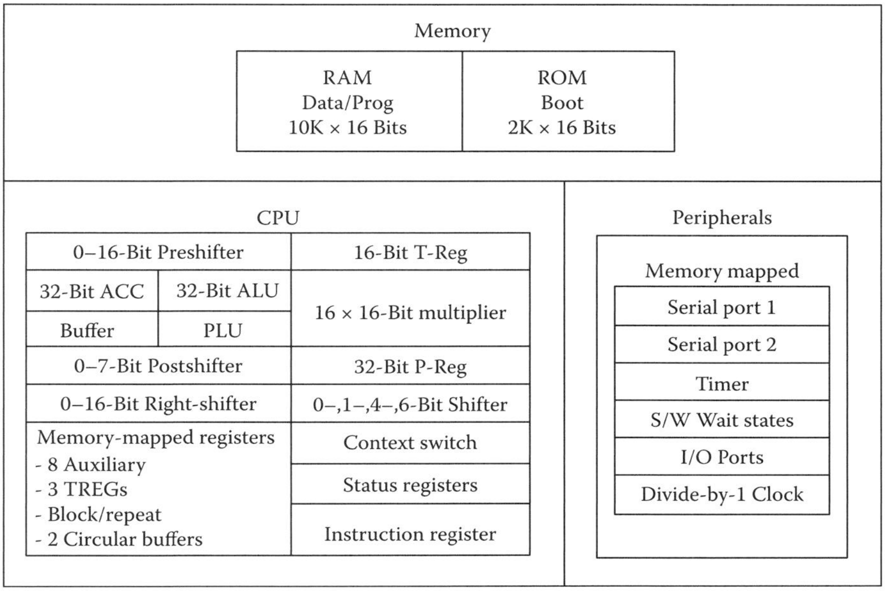

FIGURE 19.18 TMS320C50 block diagram.

- Peripherals
  - Eight-level hardware stack
  - Eleven context-switch registers to shadow the contents of strategic CPU-controlled registers during interrupts
  - Full-duplex, synchronous serial port, which directly interfaces to codec
  - Time-division multiplexed (TDM) serial port (TMS320C50/C51/C53)
  - Interval timer with period and control registers for software stops, starts, and resets
  - Concurrent external DMA performance, using extended holds
  - On-chip clock generator
  - Divide-by-one clock generator (TMS320C50/C51/C53)
  - Multiply-by-two clock generator (TMS320C52)
- Memory
  - 10K × 16 bit single cycle on-chip program/data RAM (TMS320C50)
  - 2K × 16 bit single cycle on-chip program/data RAM (TMS320C51)
  - 1K × 16 RAM (TMS320C52)
  - 4K × 16 RAM (TMS320C53)
  - 2K × 16 bit single cycle on-chip boot ROM (TMS320C50)
  - 8K × 16 bit single cycle on-chip boot ROM (TMS320C51)
  - 4K × 16 ROM (TMS320C52)
  - 16K × 16 ROM (TMS320C53)
  - 1056 × 16 bit dual-access on-chip data/program RAM
- Memory interfaces
  - Sixteen programmable software wait-state generators for program, data, and I/O memories
  - 224K-Word × 16 bit maximum addressable external memory space

Table 19.5 shows the overall TMS320 family. It provides a tabulated overview of each member's memory capacity, number of I/O ports (by type), cycle time, package type, technology, and availability.

TABLE 19.5 TMS320 Family Overview

		Memory (Words)				I/O[a]						
		On-Chip		Off-Chip								
Data Type	Device	RAM	ROM	EPROM	Dat/Pro	Ser	Par	DMA	Com	On-Chip Timer	Cycle Time (ns)	Package
Fixed-point (16 bit word size)	TMS320C10[b]	144	1.5K	—	—/4K	—	8 × 16	—	—	—	200	DIP/PLCC
	TMS320C10–14	144	1.5K	—	—/4K	—	8 × 16	—	—	—	280	DIP/PLCC
	TMS320C10–25[b]	144	1.5K	—	—/4K	—	8 × 16	—	—	—	160	DIP/PLCC
	TMS320C14	256	4K	—	—/4K	1	7 × 16	—	—	4	160	PLCC
	TMS320E14[b]	256	—	4K	—/4K	1	7 × 16	—	—	4	160	CERQUAD
	TMS320E14–25[b]	256	—	4K	—/4K	1	7 × 16	—	—	4	167	CERQUAD
	TMS320P14	256	—	4K	—/4K	1	7 × 16	—	—	4	160	PLCC
	TMS320C15[b]	256	4K	—	—/4K	—	8 × 16	—	—	—	200	DIP/PLCC/PQFP
	TMS320C15–25[b]	256	4K	—	—/4K	—	8 × 16	—	—	—	160	DIP/PLCC
	TMS320E15[b]	256	—	4K	—/4K	—	8 × 16	—	—	—	200	DIP/CER-QUAD
	TMS320E15–25	256	—	4K	—/4K	—	8 × 16	—	—	—	160	DIP/CER-QUAD
	TMS320LC15	256	4K	—	—/4K	—	8 × 16	—	—	—	200	DIP/PLCC
	TMS320P15	256	—	4K	—/4K	—	8 × 16	—	—	—	200	DIP/PLCC
	TMS320P15–25	256	—	4K	—/4K	—	8 × 16	—	—	—	160	DIP/PLCC
	TMS320C16	256	8K	—	—/64K	—	8 × 16	—	—	—	114	PQFP
	TMS320LC16	256	8K	—	—/64K	—	8 × 16	—	—	—	250	PQFP
	TMS320C17	256	4K	—	—/—	2	6 × 16	—	—	1	200/160	DIP/PLCC
	TMS320E17	256	—	4K	—/—	2	6 × 16	—	—	1	200/160	DIP
						I/O[a,c]						
						Ser	Par	DMA	Com			
Fixed-point (16 bit word size)	TMS320LC17	256	4K	—	—/—	2	6 × 16	—	—	1	200	DIP/PLCC
	TMS320P17	256	—	4K	—/—	2	6 × 16	—	—	1	160	DIP
	TMS320C25[b]	544	4K	—	64K/64K	1	16 × 16	Ext	—	1	100	PGA/PLCC/PQFP
	TMS320C25–33	544	4K	—	64K/64K	1	16 × 16	Ext	—	1	120	PLCC
	TMS320C25–50[b]	544	4K	—	64k/64K	1	16 × 16	Ext	—	1	80	PGA/PLCC

	TMS320E25	544	—	4K	64K/64K	1	16 × 16	Ext	—	1	100	PQFP/PLCC
	TMS320C26[b]	1.5K	—	—	64K/64K	1	16 × 16	Ext	—	1	100	PLCC
	TMS320C28	544	8K	—	64k/64K	1	16 × 16	Ext	—	1	100	PQFP/PLCC
	TMS320C28–50	544	8K	—	64K/64K	1	16 × 16	Ext	—	1	80	PQFP/PLCC
	TMS320C50[b]	10K	BL	—	64K/64K	2	64K × 16[d]	Ext	—	1	50/35/25/20[e]	PQFP
	TMS320C51	2K	8K	—	64K/64K	2	64K × 16[d]	Ext	—	1	50/35/25/20[e]	PQFP/TQFP
	TMS210BC51	2K	BL	—	64K/64K	2	64K × 16[d]	Ext	—	1	50/35/25/20[e]	PQFP/TQFP
	TMS320C52	1K	4K	—	64K/64K	1	64K × 16[d]	Ext	—	1	50/35/25/20[e]	PQFP/TQFP
	TMS320BC52	1K	BL	—	64K/64K	1	64K × 16[d]	Ext	—	1	50/35/25/20[e]	PQFP/TQFP
	TMS320C53	4K	16K	—	64K/64K	2	64K × 16[d]	Ext	—	1	50/35/25/20[e]	PQFP/TQFP
	TMS320BC53	4K	BL	—	64K/64K	2	64K × 16[d]	Ext	—	1	50/35/25/20[e]	PQFP/TQFP
Floating-point (32 bit word size)	TMS320C30	2K	4K	—	16M[f]	2	16M × 32[g]	Int/Ext	—	2(6)[d]	60	PGA and PQFP
	TMS320C30–50	2K	4K	—	16M[f]	2	16M × 32[g]	Int/Ext	—	2(6)[d]	40	PGA and PQFP
	TMS320C30–27	2K	4K	—	16M[f]	2	16M × 32[g]	Int/Ext	—	2(6)[d]	74	PGA and PQFP
	TMS320C30–40	2K	4K	—	16M[f]	2	16M × 32[g]	Int/Ext	—	2(6)[d]	50	PGA and PQFP
	TMS320C30–50	2K	4K	—	16M[f]	2	16M × 32[g]	Int/Ext	—	2(6)[d]	40	PGA and PQFP
	TMS320C31[b]	2K	[e]	—	16M[f]	1	16M × 32	Int/Ext	—	2(4)[d]	60	PQFP
	TMS320LC31	2K	[e]	—	16M[f]	1	16M × 32	Int/Ext	—	2(4)[d]	60	PQFP
	TMS320C31–27	2K	[e]	—	16M[f]	1	16M × 32	Int/Ext	—	2(4)[d]	74	PQFP
	TMS320C31–40	2K	[e]	—	16M[f]	1	16M × 32	Int/Ext	—	2(4)[d]	50	PQFP
	TMS320C31–50	2K	[e]	—	16M[f]	1	16M × 32	Int/Ext	—	2(4)[d]	40	PQFP
	TMS320C40	2K	4K[e]	—	4G[f]	—	4G × 32[g]	Int/Ext	6	2	40	PGA
	TMS320C40–40	2K	4K[e]	—	4G[f]	—	4G × 32[g]	Int/Ext	6	2	50	PGA

[a] Ser, serial; Par, parallel; DMA, direct memory access (Int, internal; Ext, external); Com, parallel communication ports.
[b] A military version is available/planned; contact the nearest TI field sales office for availability.
[c] Programmed transcoders (TMS320SS16 and TMS320SA32) are also available.
[d] Includes the use of serial port timers.
[e] Preprogrammed ROM bootloader.
[f] Single logical memory space for program, data, and I/O; not including on-chip RAM, peripherals, and reserved spaces.
[g] Dual buses.

Many features are common among these TMS320 processors. When the term TMS320 is used, it refers to all five generations of DSP devices. When referring to a specific member of the TMS320 family (e.g., TMS320C15), the name also implies enhanced-speed in megahertz (−14, −25, etc.), erasable/programmable (TMS320E15), low-power (TMS320LC15), and one-time programmable (TMS320P15) versions. Specific features are added to each processor to provide different cost/performance alternatives.

References

1. *TMS320C2x User's Guide*, Texas Instruments, Dallas, TX.
2. *TMS320C3x User's Guide*, Texas Instruments, Dallas, TX.

20

Rapid Design and Prototyping of DSP Systems

T. Egolf,
M. Pettigrew,
J. Debardelaben,
R. Hezar,
S. Famorzadeh,
A. Kavipurapu,
M. Khan,
Lan-Rong Dung,
K. Balemarthy,
N. Desai,
Yong-kyu Jung, and
Vijay K. Madisetti
Georgia Institute of Technology

20.1 Introduction 20-2
20.2 Survey of Previous Research 20-4
20.3 Infrastructure Criteria for the Design Flow 20-5
20.4 The Executable Requirement 20-7
An Executable Requirements Example: MPEG-1 Decoder
20.5 The Executable Specification 20-10
An Executable Specification Example: MPEG-1 Decoder
20.6 Data and Control Flow Modeling 20-15
Data and Control Flow Example
20.7 Architectural Design 20-16
Cost Models • Architectural Design Model
20.8 Performance Modeling and Architecture Verification 20-25
A Performance Modeling Example: SCI Networks • Deterministic Performance Analysis for SCI • DSP Design Case: Single Sensor Multiple Processor
20.9 Fully Functional and Interface Modeling and Hardware Virtual Prototypes 20-33
Design Example: I/O Processor for Handling MPEG Data Stream
20.10 Support for Legacy Systems 20-37
20.11 Conclusions 20-38
Acknowledgments 20-38
References 20-39

The rapid prototyping of application-specific signal processors (RASSP) [1–3] program of the United States Department of Defense (ARPA and Tri-Services) targets a 4X improvement in the design, prototyping, manufacturing, and support processes (relative to current practice). Based on a current practice study (1993) [4], the prototyping time from system requirements definition to production and deployment, of multiboard signal processors, is between 37 and 73 months. Out of this time, 25–49 months are devoted to detailed hardware/software (HW/SW) design and integration (with 10–24 months devoted to the latter task of integration). With the utilization of a promising top-down hardware-less codesign methodology based on VHSIC hardware description language (VHDL) models of HW/SW components at multiple abstractions, reduction in design time has been shown especially in the area of HW/SW integration [5]. The authors describe a top-down design approach in VHDL starting with the capture of system requirements in an executable form and through successive stages of design refinement, ending with a detailed hardware design. This HW/SW codesign process is based on the

RASSP program design methodology called virtual prototyping, wherein VHDL models are used throughout the design process to capture the necessary information to describe the design as it develops through successive refinement and review. Examples are presented to illustrate the information captured at each stage in the process. Links between stages are described to clarify the flow of information from requirements to hardware.

20.1 Introduction

We describe a RASSP-based design methodology for application specific signal processing systems which supports reengineering and upgrading of legacy systems using a virtual prototyping design process. The VHDL [6] is used throughout the process for the following reasons. (1) It is an IEEE standard with continual updates and improvements; (2) it has the ability to describe systems and circuits at multiple abstraction levels; (3) it is suitable for synthesis as well as simulation; and (4) it is capable of documenting systems in an executable form throughout the design process.

A "virtual prototype" (VP) is defined as an executable requirement or specification of an embedded system and its stimuli describing it in operation at multiple levels of abstraction. "Virtual prototyping" is defined as the top-down design process of creating a VP for hardware and software cospecification, codesign, cosimulation, and coverification of the embedded system. The proposed top-down design process stages and corresponding VHDL model abstractions are shown in Figure 20.1. Each stage in the process serves as a starting point for subsequent stages. The testbench developed for requirements

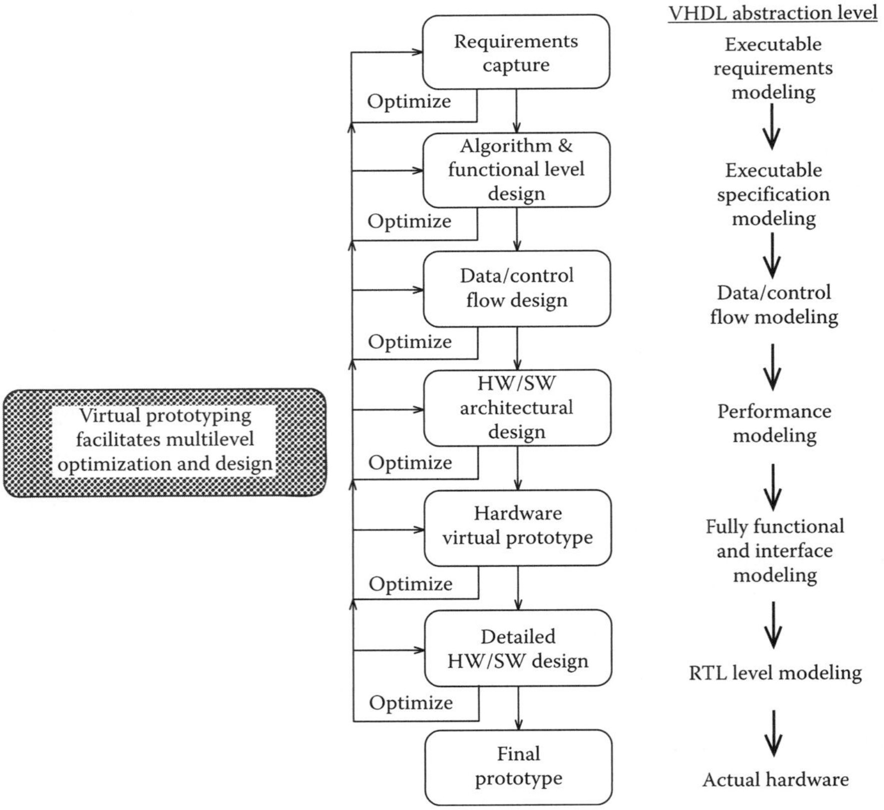

FIGURE 20.1 The VHDL top-down design process.

capture is used for design verification throughout the process. More refined subsystem, board, and component level testbenches are also developed in-cycle for verification of these elements of the system.

The process begins with requirements definition which includes a description of the general algorithms to be implemented by the system. An algorithm is here defined as a system's signal processing transformations required to meet the requirements of the high level paper specification. The model abstraction created at this stage, the "executable requirement," is developed as a joint effort between contractor and customer in order to derive a top-level design guideline which captures the customer intent. The executable requirement removes the ambiguity associated with the written specification. It also provides information on the types of signal transformations, data formats, operational modes, interface timing data and control, and implementation constraints. A description of the executable requirement for an MPEG decoder is presented later. Section 20.4 addresses this subject in more detail.

Following the executable requirement, a top-level "executable specification" is developed. This is sometimes referred to as functional level VHDL design. This executable specification contains three general categories of information: (1) the system timing and performance, (2) the refined internal function, and (3) the physical constraints such as size, weight, and power. System timing and performance information include I/O timing constraints, I/O protocols, and system computational latency. Refined internal function information includes algorithm analysis in fixed/floating point, control strategies, functional breakdown, and task execution order. A functional breakdown is developed in terms of primitive signal processing elements (PEs) which map to processing hardware cells or processor specific software libraries later in the design process. A description of the executable specification of the MPEG decoder is presented later. Section 20.5 investigates this subject in more detail.

The objective of data and control flow modeling is to refine the functional descriptions in the executable specification and capture concurrency information and data dependencies inherent in the algorithm. The intent of the refinement process is to generate multiple implementation independent representations of the algorithm. The implementations capture potential parallelism in the algorithm at a primitive level. The primitives are defined as the set of functions contained in a design library consisting of signal processing functions such as Fourier transforms or digital filters at course levels and of adders and multipliers at more fine-grained levels. The control flow can be represented in a number of ways ranging from finite state machines for low level hardware to run-time system controllers with multiple application data flow graphs. Section 20.6 investigates this abstraction model.

After defining the functional blocks, data flow between the blocks, and control flow schedules, hardware–software design trade-offs are explored. This requires architectural design and verification. In support of architecture verification, "performance level modeling" is used. The performance level model captures the time aspects of proposed design architectures such as system throughput, latency, and utilization. The proposed architectures are compared using cost function analysis with system performance and physical design parameter metrics as input. The output of this stage is one or few optimal or nearly optimal system architectural choice(s). In this stage, the interaction between hardware and software is modeled and analyzed. In general, models at this abstraction level are not concerned with the actual data in the system but rather the flow of data through the system. An abstract VHDL data type known as a token captures this flow of data. Examples of performance level models are shown later. Sections 20.7 and 20.8 address architecture selection and architecture verification, respectively.

Following architecture verification using performance level modeling, the structure of the system in terms of PEs, communications protocols, and input/output requirements is established. Various elements of the defined architecture are refined to create hardware virtual prototypes. "Hardware virtual prototypes" are defined as "software simulatable" models of hardware components, boards, or systems containing sufficient accuracy to guarantee their successful realization in actual hardware. At this abstraction level, fully functional models (FFMs) are utilized. FFMs capture both internal and external (interface) functionality completely. Interface models capturing only the external pin behavior are also used for hardware virtual prototyping. Section 20.9 describes this modeling paradigm.

Application specific component designs are typically done in-cycle and use register transfer level (RTL) model descriptions as input to synthesis tools. The tool then creates gate level descriptions and final layout information. The RTL description is the lowest level contained in the virtual prototyping process and will not be discussed in this chapter because existing RTL methodologies are prevalent in the industry.

At least six different HW/SW codesign methodologies have been proposed for rapid prototyping in the past few years. Some of these describe the various process steps without providing specifics for implementation. Others focus more on implementation issues without explicitly considering methodology and process flow. In the next section, we illustrate the features and limitations of these approaches and show how they compare to the proposed approach.

Following the survey, Section 20.3 lays the groundwork necessary to define the elements of the design process. At the end of the chapter, Section 20.10 describes the usefulness of this approach for life cycle support and maintenance.

20.2 Survey of Previous Research

The codesign problem has been addressed in recent studies by Thomas et al. [7], Kumar et al. [8], Gupta and De Micheli [9], Kalavade and Lee [10,11], and Ismail and Jerraya [12]. A detailed taxonomy of HW/SW codesign was presented by Gajski and Vahid [13]. In the taxonomy, the authors describe the desired features of a codesign methodology and show how existing tools and methods try to implement them. However, the authors do not propose a method for implementing their process steps. The features and limitations of the latter approaches are illustrated in Figure 20.2 [14]. In the table, we show how these approaches compare to the approach presented in this chapter with respect to some desired attributes of a codesign methodology. Previous approaches lack automated architecture selection tools, economic cost models, and the integrated development of testbenches throughout the design cycle. Very few approaches allow for true HW/SW cosimulation where application code executes on a simulated version of the target hardware platform.

DSP Codesign Features	TA93	KA93	GD93	KL93 KL94	IJ95	Proposed Method
Executable functional specification	✓	✓	✓	✓	✓	✓
Executable timing specification		✓	✓			✓
Automated architecture selection						✓
Automated partitioning			✓	✓		✓
Model-based performance estimation		✓	✓			✓
Economic cost/profit estimation models						✓
HW/SW cosimulation				✓		✓
Uses IEEE standard languages		✓				✓
Integrated testbench generation						✓

FIGURE 20.2 Features and limitations of existing codesign methodologies.

20.3 Infrastructure Criteria for the Design Flow

Four enabling factors must be addressed in the development of a VHDL model infrastructure to support the design flow mentioned in the introduction. These include model verification/validation, interoperability, fidelity, and efficiency.

Verification, as defined by IEEE/ANSI, is the process of evaluating a system or component to determine whether the products of a given development phase satisfy the conditions imposed at the start of that phase. Validation, as defined by IEEE/ANSI, is the process of evaluating a system or component during or at the end of the development process to determine whether it satisfies the specified requirements. The proposed methodology is broken into the design phases represented in Figure 20.1 and uses black- and white-box software testing techniques to verify, via a structured simulation plan, the elements of each stage. In this methodology, the concept of a reference model, defined as the next higher model in the design hierarchy, is used to verify the subsequently more detailed designs. For example, to verify the gate level model after synthesis, the test suite applied to the RTL model is used. To verify the RTL level model, the reference model is the FFM. Moving test creation, test application, and test analysis to higher levels of design abstraction, the test description developed by the test engineer is more easily created and understood. The higher functional models are less complex than their gate level equivalents. For system and subsystem verification, which include the integration of multiple component models, higher level models improve the overall simulation time. It has been shown that a processor model at the fully functional level can operate over 1000 times faster than its gate level equivalent while maintaining clock cycle accuracy [5]. Verification also requires efficient techniques for test creation via automation and reuse and requirements compliance capture and test application via structured testbench development.

Interoperability addresses the ability of two models to communicate in the same simulation environment. Interoperability requirements are necessary because models usually developed by multiple design teams and from external vendors must be integrated to verify system functionality. Guidelines and potential standards for all abstraction levels within the design process must be defined when current descriptions do not exist. In the area of fully functional and RTL modeling, current practice is to use IEEE Std 1164-1993 nine-valued logic packages [15]. Performance modeling standards are an ongoing effort of the RASSP program.

Fidelity addresses the problem of defining the information captured by each level of abstraction within the top-down design process. The importance of defining the correct fidelity lies in the fact that information not relevant within a model at a particular stage in the hierarchy requires unnecessary simulation time. Relevant information must be captured efficiently so simulation times improve as one moves toward the top of the design hierarchy. Figure 20.3 describes the RASSP taxonomy [16] for accomplishing this objective. The diagram illustrates how a VHDL model can be described using five resolution axes; temporal, data value, functional, structural, and programming level. Each line is continuous and discrete labels are positioned to illustrate various levels ranging from high to low resolution. A full specification of a model's fidelity requires two charts, one to describe the internal attributes of the model and the second for the external attributes. An "*X*" through a particular axis implies the model contains no information on the specific resolution. A compressed textual representation of this figure will be used throughout the remainder of the chapter. The information is captured in a 5-tuple as follows

$$\{(\text{Temporal Level}), (\text{Data Value}), (\text{Function}), (\text{Structure}), (\text{Programming Level})\}$$

The temporal axis specifies the time scale of events in the model and is analogous to precision as distinguished from accuracy. At one extreme, for the case of purely functional models, no time is modeled. Examples include fast Fourier transform (FFT) and finite impulse response (FIR) filtering procedural calls. At the other extreme, time resolutions are specified in gate propagation delays. Between the two extremes, models may be time accurate at the clock level for the case of fully functional processor models, at the instruction cycle level for the case of performance level processor models, or at the system

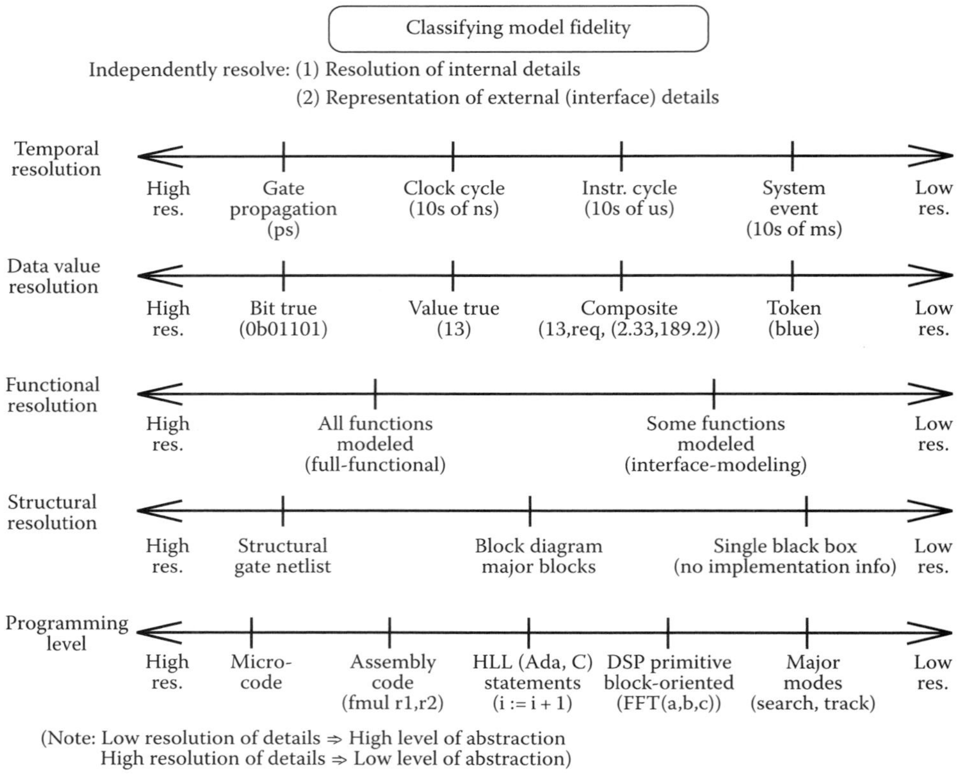

FIGURE 20.3 A model fidelity classification scheme.

level for the case of application graph switching. In general, higher resolution models require longer simulation times due to the increased number of event transactions.

The data value axis specifies the data resolution used by the model. For high resolution models, data is represented with bit true accuracy and is commonly found in gate level models. At the low end of the spectrum, data is represented by abstract token types where data is represented by enumerated values, for example, "blue." Performance level modeling uses tokens as its data type. The token only captures the control information of the system and no actual data. For the case of no data, the axis would be represented with an "X." At intermediate levels, data is represented with its correct value but at a higher abstraction (i.e., integer or composite types, instead of the actual bits). In general, higher resolutions require more simulation time.

Functional resolution specifies the detail of device functionality captured by the model. At one extreme, no functions are modeled and the model represents the processing functionality as a simple time delay (i.e., no actual calculations are performed). At the high end, all the functions are implemented within the model. As an example, for a processor model, a time delay is used to represent the execution of a specific software task at low resolutions while the actual code is executed on the model for high resolution simulations. As a rule of thumb, the more functions represented, the slower the model executes during simulation.

The structural axis specifies how the model is constructed from its constituent elements. At the low end, the model looks like a black box with inputs and outputs but no detail as to the internal contents. At the high end the internal structure is modeled with very fine detail, typically as a structural net list of lower level components. In the middle, the major blocks are grouped according to related functionality.

The final level of detail needed to specify a model is its programmability. This describes the granularity at which the model interprets software elements of a system. At one extreme, pure hardware is specified and the model does not interpret software, for example, a special purpose FFT processor hard wired for 1024 samples. At the other extreme, the internal microcode is modeled at the detail of its datapath control. At this resolution, the model captures precisely how the microcode manipulates the datapath elements. At decreasing resolutions the model has the ability to process assembly code and high-level languages as input. At even lower levels, only digital signal processing (DSP) primitive blocks are modeled. In this case, programming consists of combining functional blocks to define the necessary application. Tools such as MATLAB®/Simulink® provide examples for this type of model granularity. Finally, models can be programmed at the level of the major modes. In this case, a run-time system is switched between major operating modes of a system by executing alternative application graphs.

Finally, efficiency issues are addressed at each level of abstraction in the design flow. Efficiency will be discussed in coordination with the issues of fidelity where both the model details and information content are related to improving simulation speed.

20.4 The Executable Requirement

The methodology for developing signal processing systems begins with the definition of the system requirement. In the past, common practice was to develop a textual specification of the system. This approach is flawed due to the inherent ambiguity of the written description of a complex system. The new methodology places the requirements in an executable format enforcing a more rigorous description of the system. Thus, VHDL's first application in the development of a signal processing system is an "executable requirement" which may include signal transformations, data format, modes of operation, timing at data and control ports, test capabilities, and implementation constraints [17]. The executable requirement can also define the minimum required unit of development in terms of performance (e.g., SNR, throughput, latency, etc.). By capturing the requirements in an executable form, inconsistencies and missing information in the written specification can also be uncovered during development of the requirements model.

An executable requirement creates an "environment" wherein the surroundings of the signal processing system are simulated. Figure 20.4 illustrates a system model with an accompanying testbench. The testbench generates control and data signals as stimulus to the system model. In addition, the testbench receives output data from the system model. This data is used to verify the correct operation of the system model. The advantages of an executable requirement are varied. First, it serves as a mechanism to define and refine the requirements placed on a system. Also, the VHDL source code along with supporting textual description becomes a critical part of the requirements documentation and life cycle support of the system. In addition, the testbench allows easy examination of different command sequences and data sets. The testbench can also serve as the stimulus for any number of designs. The development of different system models can be tested within a single simulation environment using the same testbench. The requirement is easily adaptable to changes that can occur in lower levels of the design process. Finally, executable requirements are formed at all levels of abstraction and create a documented history of the design process. For example, at the system level, the environment may consist of image data from a camera while at the application-specific integrated circuit (ASIC) level it may be an interface model of another component.

The RASSP program, through the efforts of MIT Lincoln Laboratory, created an executable requirement [18] for a synthetic aperture radar (SAR) algorithm and documented many of the lessons learned in implementing this stage in the top-down design process. Their high level requirements model served as the baseline for the design of two SAR systems developed by separate contractors, Lockheed Sanders and Martin Marietta Advanced Technology Labs. A testbench generation system for capturing high level requirements and automating the creation of VHDL is presented in [19]. In the following sections, we present the details of work done at Georgia Tech in creating an executable requirement and specification for an MPEG-1 decoder.

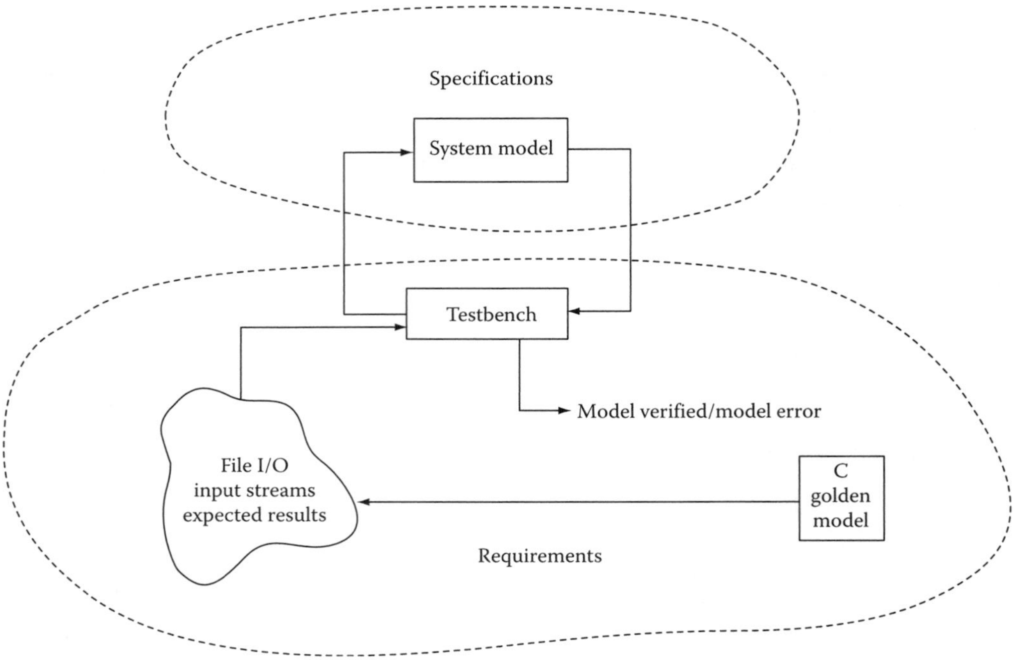

FIGURE 20.4 Illustration of the relation between executable requirements and specifications.

20.4.1 An Executable Requirements Example: MPEG-1 Decoder

MPEG-1 is a video compression–decompression standard developed under the International Standard Organization originally targeted at CD-ROMs with a data rate of 1.5 Mb/s [20]. MPEG-1 is broken into three layers: system, video, and audio. Table 20.1 depicts the system clock frequency requirement taken from layer 1 of the MPEG-1 document.* The system time is used to control when video frames are decoded and presented via decoder and presentation time stamps contained in the ISO 11172 MPEG-1 bitstream. A VHDL executable rendition of this requirement is illustrated in Figure 20.5.

The testbench of this system uses an MPEG-1 bitstream created from a "golden C model" to ensure correct input. A public-domain C version of an MPEG encoder created at UC Berkeley [21] was used as the golden C model to generate the input for the executable requirement. From the testbench, an MPEG bitstream file is read as a series of integers and transmitted to the MPEG decoder model at a constant rate of 174,300 Bytes/s along with a system clock and a control line named mpeg_go which activates the decoder. Only 50 lines of VHDL code are required to characterize the top level testbench. This is due to

TABLE 20.1 MPEG-1 System Clock Frequency Requirement Example

Layer 1—System Requirement Example from ISO 11172 Standard	
System clock frequency	The value of the system clock frequency is measured in Hz and shall meet the following constraints: 90,000−4.5$ Hz ≤ 90,000 + 4.5$ Hz Rate of change of system_clock_frequency ≤ 250 $*$ 10^{-6} Hz/s

* Our efforts at Georgia Tech have only focused on layers 1 and 2 of this standard.

```
- system_time_clk process is a clock process that counts at a rate
- of 90 kHz as per MPEG-I requirement. In addition, it is updated by
- the value of the incoming SCR fields read from the ISO11172 stream.
-
system_time_clock : PROCESS(stc_strobe,sys_clk)
   VARIABLE clock_count : INTEGER := 0;
   VARIABLE SCR, system_time_var : bit33;
   CONSTANT clock_divider : INTEGER := 2;
BEGIN
   IF mpeg_go = '1' THEN
   - if stc_strobe is high then update system_time value to latest SCR
   IF (stc_strobe = '1') AND (stc_strobe'EVENT) THEN
      system_time <= system_clock_ref;
      clock_count := 0; - reset counter used for clock downsample
   ELSIF (sys_clk = '1') AND (sys_clk'EVENT) THEN
      clock_count := clock_count + 1;
      IF clock_count MOD clock_divider = 0 THEN
        system_time_var := system_time + one;
        system_time <= system_time_var;
      END IF;
   END IF;
   END IF;
END PROCESS system_time_clock;
```

FIGURE 20.5 System clock frequency requirement example translated to VHDL.

the availability of the golden C MPEG encoder and a shell script which wraps around the output of the golden C MPEG encoder bitstream with system layer information. This script is necessary because there are no "complete" MPEG software codecs in the public domain, i.e., they do not include the system information in the bitstream. Figure 20.6 depicts the process of verification using golden C models. The golden model generates the bitstream sent to the testbench. The testbench reads the bitstream as a series of integers. These are in turn sent as data into the VHDL MPEG decoder model driven with appropriate clock and control lines. The output of the VHDL model is compared with the output of the golden model (also available from Berkeley) to verify the correct operation of the VHDL decoder. A warning message alerts the user to the status of the model's integrity.

The advantage of the configuration illustrated in Figure 20.6 is its reusability. An obvious example is MPEG-2 [22], another video compression–decompression standard targeted for the all-digital transmission of broadcast TV quality video at coded bit rates between 4 and 9 Mb/s. The same testbench structure could be used by replacing the golden C models with their MPEG-2 counterparts. While the system layer information encapsulation script would have to be changed, the testbench itself remains the same because the interface between an MPEG-1 decoder and its surrounding environment is identical to the interface for an MPEG-2 decoder. In general, this testbench configuration could be used for a wide class of video decoders. The only modifications would be the golden C models and the interface between the VHDL decoder model and the testbench. This would involve making only minor alterations to the testbench itself.

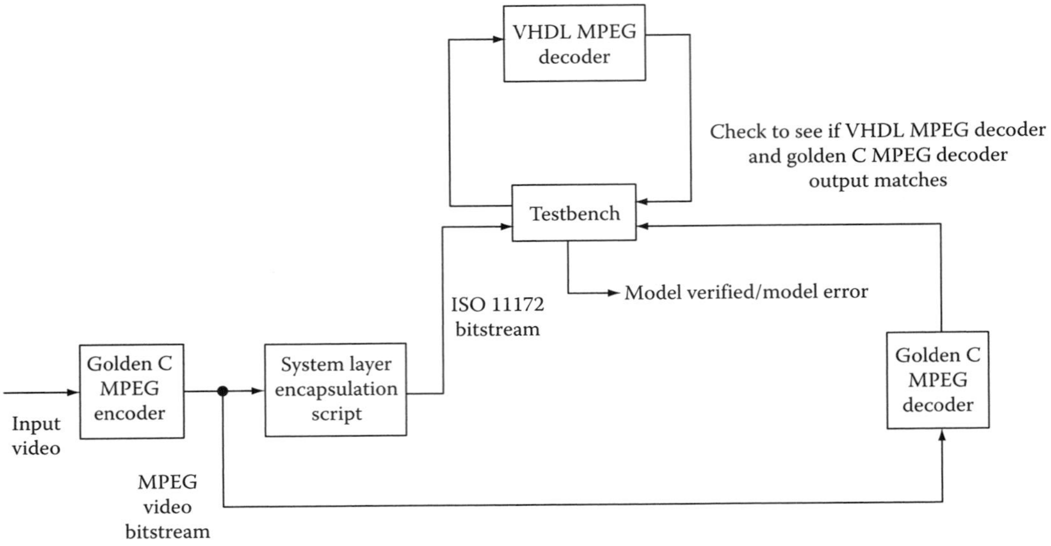

FIGURE 20.6 MPEG-1 decoder executable requirement.

20.5 The Executable Specification

The executable specification depicted in Figure 20.4 processes and responds to the outside stimulus, provided by the executable requirement, through its interface. It reflects the particular function and timing of the intended design. Thus, the executable specification describes the behavior of the design and is timing accurate without consideration of the eventual implementation. This allows the user to evaluate the completeness, logical correctness, and algorithmic performance of the system through the testbench. The creation of this formal specification helps identify and correct functional errors at an early stage in the design and reduce total design time [13,16,23,24].

The development of an executable specification is a complex task. Very often, the required functionality of the system is not well-understood. It is through a process of learning, understanding, and defining that a specification is crystallized. To specify system functionality, we decompose it into elements. The relationship between these elements is in terms of their execution order and the data passing between them. The executable specification captures

- The refined internal functionality of the unit under development (some algorithm parallelism, fixed/floating point bit level accuracies required, control strategies, functional breakdown, and task execution order)
- Physical constraints of the unit such as size, weight, area, and power
- Unit timing and performance information (I/O timing constraints, I/O protocols, and computational complexity)

The purpose of VHDL at the executable specification stage is to create a formalization of the elements in a system and their relationships. It can be thought of as the high level design of the unit under development. And although we have restricted our discussion to the system level, the executable specification may describe any level of abstraction (algorithm, system, subsystem, board, device, etc.).

The allure of this approach is based on the user's ability to see what the performance "looks" like. In addition, a stable test mechanism is developed early in the design process (note the complementary relation between the executable requirement and specification). With the specification precisely defined, it becomes easier to integrate the system with other concurrently designed systems. Finally, this executable approach facilitates the reuse of system specifications for the possible redesign of the system.

In general, when considering the entire design process, executable requirements and specifications can potentially cover any of the possible resolutions in the fidelity classification chart. However, for any particular specification or requirement, only a small portion of the chart will be covered. For example, the MPEG decoder presented in this and the previous section has the fidelity information represented by the 5-tuple below,

Internal: {(Clock cycle), (Bit true $\rightarrow$ Value true), (All), (Major blocks), (X)}

External: {(Clock cycle), (Value true), (Some), (Black box), (X)}

where (Bit true $\rightarrow$ Value true) means all resolutions between bit true and value true inclusive.

From an internal viewpoint, the timing is at the system clock level, data is represented by bits in some cases and integers in others, the structure is at the major block level, and all the functions are modeled. From an external perspective, the timing is also at the system clock level, the data is represented by a stream of integers, the structure is seen as a single black box fed by the executable requirement and from an external perspective the function is only modeled partially because this does not represent an actual chip interface.

20.5.1 An Executable Specification Example: MPEG-1 Decoder

As an example, an MPEG-1 decoder executable specification developed at Georgia Tech will be examined in detail. Figure 20.7 illustrates how the system functionality was broken into a discrete number of elements. In this diagram each block represents a process and the lines connecting them are signals. Three major areas of functionality were identified from the written specification: memory, control, and the video decoder itself. Two memory blocks, video_decode_memory and system_level_memory are clearly labeled. The present_frame_to_decode_file process contains a frame reorder buffer which holds a frame until its presentation time. All other VHDL processes with the exception of decode_video_frame_process are control processes and pertain to the systems layer of the MPEG-1 standard. These processes take the incoming MPEG-1 bitstream and extract system layer information. This information is stored in the system_level_memory process where other control processes and the video decoder can access pertinent data. After removing the system layer information from the MPEG-1 bitstream, the remainder is placed in the video_decode_memory. This is the input buffer to the video decoder. It should be noted that although MPEG-1 is capable of up to 16 simultaneous video streams multiplexed into the MPEG-1 bitstream only one video stream was selected for simplicity.

The last process, decode_video_frame_process, contains all the subroutines necessary to decode the video bitstream from the video buffer (video_decode_memory). MPEG video frames are broken into three types: (I)ntra, (P)redictive, and (B)idirectional. I frames are coded using block discrete cosine transform (DCT) compression. Thus, the entire frame is broken into 8×8 blocks, transformed with a DCT and the resulting coefficients transmitted. P frames use the previous frame as a prediction of the current frame. The current frame is broken into 16×16 blocks. Each block is compared with a corresponding search window (e.g., 32×32, 48×48) in the previous frame. The 16×16 block within the search window which best matches the current frame block is determined. The motion vector identifies the matching block within the search window and is transmitted to the decoder. B frames are similar to P frames except a previous frame and a future frame are used to estimate the best matching block from either of these frames or an average of the two. It should be noted that this requires the encoder and decoder to store these two reference frames.

The functions contained in the decode_video_frame_process are shown in Figure 20.8. In the diagram, there are three main paths representing the procedures or functions in the executable specification which

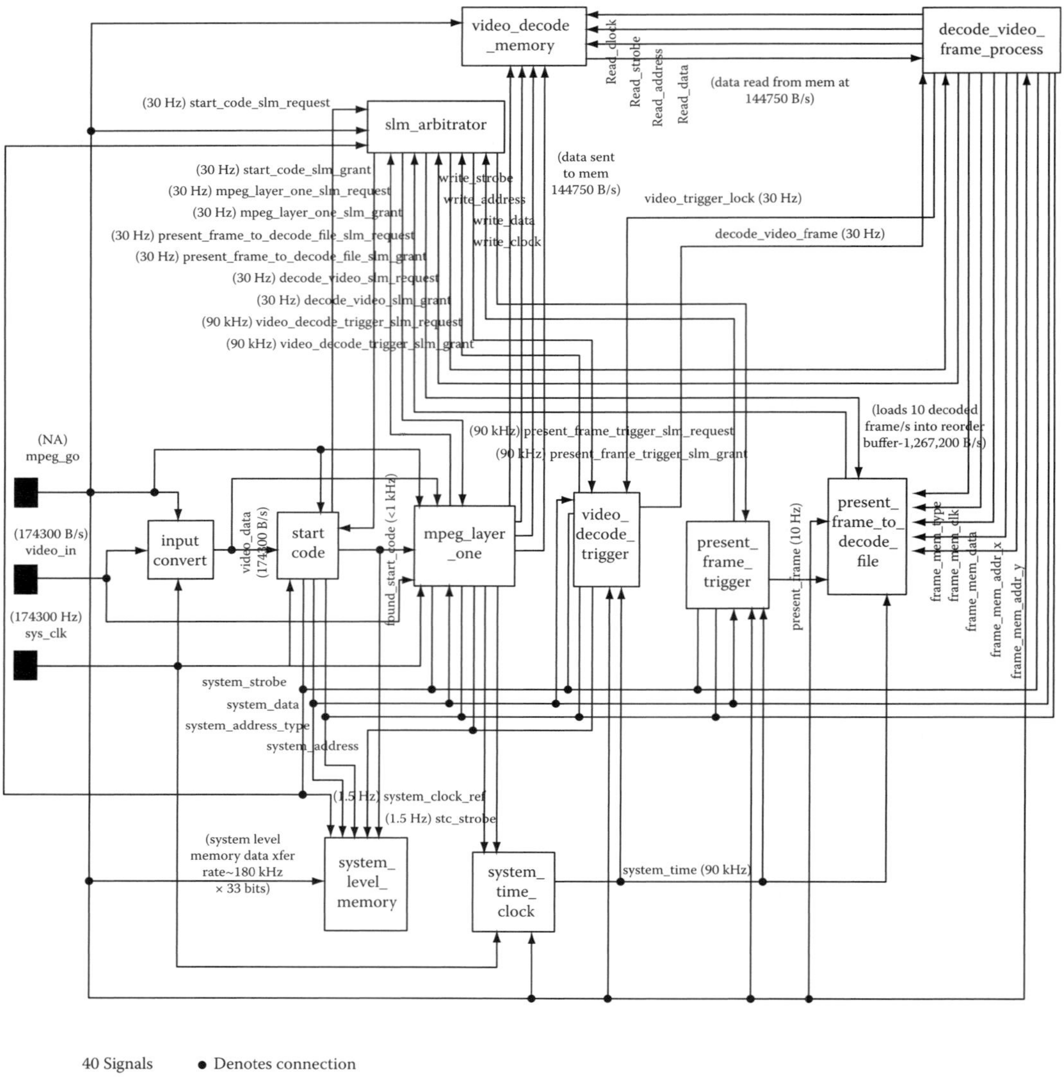

FIGURE 20.7 System functionality breakdown for MPEG-1 decoder.

process the I, P, or B frame, respectively. Each box below a path encloses all the procedures executed from within that function. Beneath each path is an estimate of the number of computations required to process each frame type. Comparing the three executable paths in this diagram, one observes the large similarity between each path. Overall, only 25 unique routines are called to process the video frame. By identifying key functions within the video decoding algorithm itself, efficient and reusable code can be created. For instance, the data transmitted from the encoder to the decoder is compressed using a Huffman scheme. The procedures vlc, advance_bit, and extract_n_bits perform the Huffman decode function and miscellaneous parsing of the MPEG-1 video bitstream. Thus, this set of procedures can be used in each frame type execution path. Reuse of these procedures can be applied in the development of an MPEG-2 decoder executable specification. Since MPEG-2 is structured as a super set of the syntax defined in MPEG-1, there are many procedures that can be utilized with only minor modifications. Other procedures such as motion_compensate_forward and idct can be reused in a variety of DCT-based video compression algorithms.

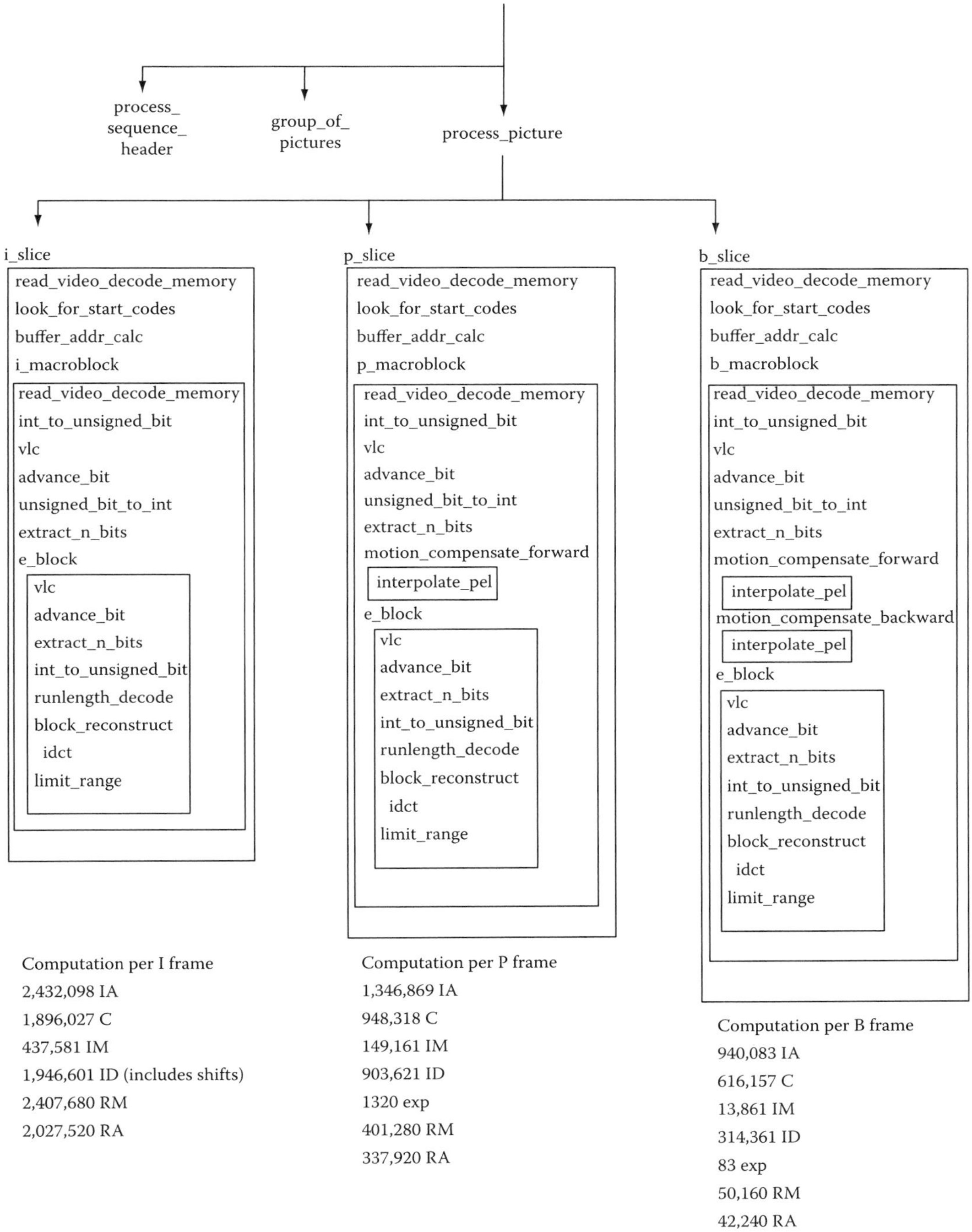

FIGURE 20.8 Description of procedural flow within MPEG-1 decoder executable specification.

TABLE 20.2 Computational Complexity of Some Specification Procedures

Procedure	Int Adds	Int Div	Comp	Int Mult	Exp	Real Add	Real Mult
vlc	—	—	2	—	—	—	—
Advance_bit	10	16	9	—	—	—	—
int_to_unsigned_bit	8	16	8	—	—	—	—
extract_n_bits	24	16	20	—	—	—	—
look_for_start_codes	9	16	10	—	—	—	—
Runlength_decode	2	—	1	1	—	—	—
block_reconstruct	66	64	258	193	—	—	—
idct	—	—	—	—	—	1024	1216
Qmotion_compensate_forward	1422	646	1549	16	—	—	—

The executable specification also allows detailed analysis of the computational complexity on a procedural level. Table 20.2 lists the computational complexity of some of the procedures identified in Figure 20.8. This breakdown identifies what areas of the algorithm are the most computationally intensive and the numbers were arrived at through a data flow analysis of the VHDL code. Within the MPEG-1 video decoder algorithm, the most intense computational loads occur in the inverse DCT and motion compensation procedures. Thus, such an analysis can alert the user early in the design process to potential design issues. While parallelism is a logical topic for the data and control flow modeling section, preliminary investigations can be made from the executable specification itself. With the specifications captured in a language, execution order and data passing between procedures are known precisely. This knowledge facilitates the user in extracting potential parallelism from the specification. From the MPEG-1 decoder executable specification, potential parallelism can be seen in several areas. In an I frame, no data dependencies are present between each 8×8 block. Therefore, an inverse DCT could potentially be performed on each 8×8 block in parallel. In P and B frames, data dependencies occur between consecutive 16×16 blocks (called macroblocks) but no data dependencies occur between slices (a grouping of consecutive macroblocks). Thus, parallelism is potentially exploitable at the slice and macroblock level. This information is passed to the data/control flow modeling phase where more detailed analysis of parallelism is done. It is also possible to delve into implementation requirement issues at the executable specification level. Fixed vs. floating point trade-offs can be examined in detail. The necessary accuracy and resolution required to meet system requirements can be determined through the use of floating and fixed point packages written in VHDL. At Georgia Tech, fixed point packages have been developed. These packages allow the user to experiment with the executable specification and see the effect finite bit accuracy has on the system model. In addition, packages have been developed which implement specific arithmetic architectures such as the ADSP 2100 [25]. This analysis results in additional design requirements being passed to hardware and software developers in later design phases.

Finally, the executable specification allows the explicit capture of internal timing and control flow requirements of the MPEG-1 decoding algorithm itself. The written document is imprecise about the details of how timing considerations for presentation and decoder time stamps will be handled. The control necessary to trigger present and decode video frame events is difficult to articulate in a written form. The most difficult aspects of coding the executable specification for a MPEG-1 decoder were these considerations. The decoder itself hinges on developing a mechanism for robustly determining when to decode or present a frame in the buffer. Events must be triggered using a system time clock which is updated from the input bitstream itself. This task is handled by five processes (start_code, mpeg_layer_one, video_decode_trigger, present_frame_trigger, present_frame_to_decode_file) grouped around a common memory (system_level_memory). This memory was necessary to allow each concurrent process to access timing information extracted from the system layer of the input bitstream. These timing and control considerations had to fit into a larger system timing requirement. For a MPEG-1 decoder, the most critical timing constraints are initial latency and the fixed presentation rate (e.g., 30 frames/s). All other timing considerations were driven by this requirement.

20.6 Data and Control Flow Modeling

This modeling level captures data and control flow information in the system algorithms. The objective of data flow modeling is to refine the functional descriptions in the executable specification and capture concurrency information and data dependencies inherent in the algorithm. The output of the refinement process is one or a few manually generated implementation independent representations of the algorithm. These multiple implementations capture potential algorithmic parallelism at a primitive level where primitives are defined as that set of functions contained in a design library. The primitives are signal processing functions such as FFTs or filter routines at coarse-grained levels to adders and multipliers at more fine-grained levels. The breakdown of primitive elements depend on the granularity exploited by the algorithm as well as potential architectural design paradigms to which the algorithm is mapped. For example, if the design paradigm demands architectures using multiple commercial-off-the-shelf (COTS) reduced instruction set computer (RISC) processors, the primitives consist of signal processing functional block level elements such as FFTs or FIR filters which exist as performance optimized library elements available for the specific processor. For custom computationally intense designs, the data flow of the algorithm may be dissected into lower primitive components such as adders and multipliers using bit-slice architectures. In our design flow, the fidelity captured by data/control flow models is shown below:

$$\text{Internal: } \{(\text{X}), (\text{Value true} \rightarrow \text{Composite}), (\text{All}), (\text{X}), (\text{Major modes})\}$$
$$\text{External: } \{(\text{X}), (\text{Value true} \rightarrow \text{Composite}), (\text{X}), (\text{X}), (\text{X})\}.$$

Because the models are purely functional and their major objective is to refine the internal representation of the algorithm, there is no time information captured by its internal or external representation as illustrated by the "X." The internal data processed by the model and external data loaded into the model are typically represented by standard data types such as "float" and/or "integer" and in some cases by composite data types such as records or arrays. All internal functionality is represented and is verified using the same data presented to the executable specification. No function is captured via external interfaces since data is input to the model through file input/output. The data processed by the executable specification is also processed by the data/control flow model. No internal or external structural information is captured since the model is implementation independent. Its level of programmability is represented at the application graph level. The applications are major modes of the system under investigation and hence at a low resolution. In general, because the primitive elements can represent adders and/or multipliers, programmability for data/control flow models can resolve to higher resolutions including the microcode level.

The implementation independent representations are compared with the executable specification using the test data supplied by the requirements development phase to verify compliance with the original algorithm design. The representations are then input to the architecture selection phase and, with additional metrics, determine the final architecture of the system.

Signal processing applications inherently follow the data flow execution model. Processing graph methodology (PGM) [26] from Naval Research Laboratory was developed specifically to capture signal processing applications. PGM supports specification of full system data flow and its associated control. An application is first captured as a graph, where nodes of the graph represent processing and edges represent queues that hold intermediate data between nodes. The scheduling criteria for each node is based on the state of its corresponding input/output queues. Each queue in the graph can be linked to one node at a time. Associated with each queue is a control block structure containing information such as size, current amount of data, and threshold. A run-time system provides a set of procedures used by each node to check the availability of data from the upstream queue or available space in the downstream queue. Applications consist of one or more graphs, one or more I/O procedures, and a run-time system interfaced with one or more command programs. The PGM graphs serve as the implementation independent representation of the algorithm discussed earlier. An example of a 2-D FFT PGM graph is presented in the next section.

Under the support of the RASSP program, a set of tools is being developed by Management Communications and Control, Inc. (MCCI) and Lockheed Martin Advance Technology Laboratories [27,28]. The toolset automates the translation of software architecture specifications to design implementations of application and control software for a signal processing system. HW/SW architectures are presented to the autocoding toolset as PGM application data flow graphs along with a candidate architectures file and graph partition lists. The lists are generated by HW/SW partitioning tools. The proposed partitions are then simulated for performance and verified against the top level specification for correct functionality. The verified partition graphs are then used as inputs to detailed design level autocode tools that generate actual source code. The source code implements the partitions processing specifications using the target processor's math library. It also produces a memory map converting all queues and variables to static buffers. Finally the application graph, with its set of source files, are translated to run-time data structures that are used by the run-time system to create an executable image of the application as distributed tasks on the target processors.

Other tools provide paths from specification to hardware and are briefly mentioned. The Ptolemy [29,30] design system from the University of California at Berkeley provides a synchronous data flow domain which can be used to perform system level simulations. Silage, another product of UC Berkeley is a data flow modeling language. Data flow language (DFL), a commercial version of Silage is used in Mentor Graphics' DSP Station to perform algorithm/architecture trade-offs. It also provides a path to synthesis as a high-level design entry tool.

20.6.1 Data and Control Flow Example

An example of a small PGM application is presented in Figure 20.9. The graph represents a two-dimensional FFT program implemented in PGM. The graph captures both the functionality and the data flow aspects of the application. The source data is read from a file and represents the I/O processor that would normally provide the input data stream. The data are then distributed to a number of queues serving as inputs to the FFT primitives that perform the operations on the rows of the input stream. The output of the FFT primitives flow to another set of queues that are input to the corner turn graph. Once the data are sorted correctly, they are sent to the input queues of the column FFT primitives. The graph is then executed by the simulator where the functionality, queue sizes, and communication between nodes are examined. This same graph is input to the HW/SW partitioning tools that generate the partition list. Given the partition list and the hardware configuration file, the autocode toolset generates the load image for the target platform.

20.7 Architectural Design

Signal processing systems are characterized as having high throughput requirements as well as stringent physical constraints. However, due to economic objectives, signal processing systems must also be developed and produced at minimal cost, while meeting time-to-market constraints in order to maximize product profits. Such cost-effective systems can only be produced by applying a high degree of cost emphasis during the early stages of design. Although the conceptual design process typically involves less than 4% of the total prototyping time and cost, it accounts for more than 70% of a system's life cycle cost. Consequently, the goal of the architecture designer is to optimize preliminary architectural design

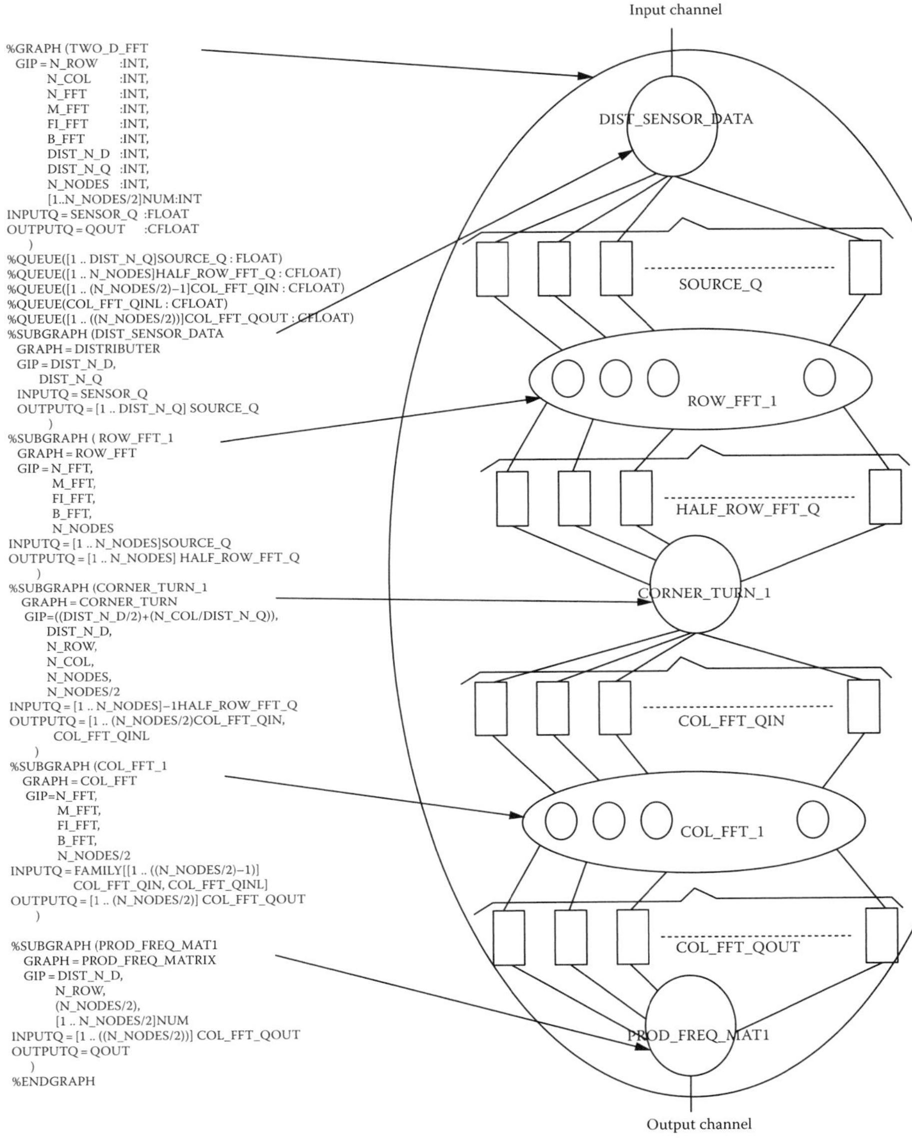

FIGURE 20.9 Example PGM application graph.

decisions with respect to the dominant system-level cost elements such as acquisition costs, maintenance costs, and time-to-market costs, while satisfying performance and physical constraints.

20.7.1 Cost Models

Current rapid prototyping design methodologies have overlooked an important characteristic of software prototyping. Various parametric studies based on historical project data show that software is difficult to design and test if "slack" margins for hardware CPU and memory resources are overly restrictive [31]. Severe resource constraints may require software developers to interact directly with the operating system and/or hardware in order to optimize the code to meet system requirements. Such

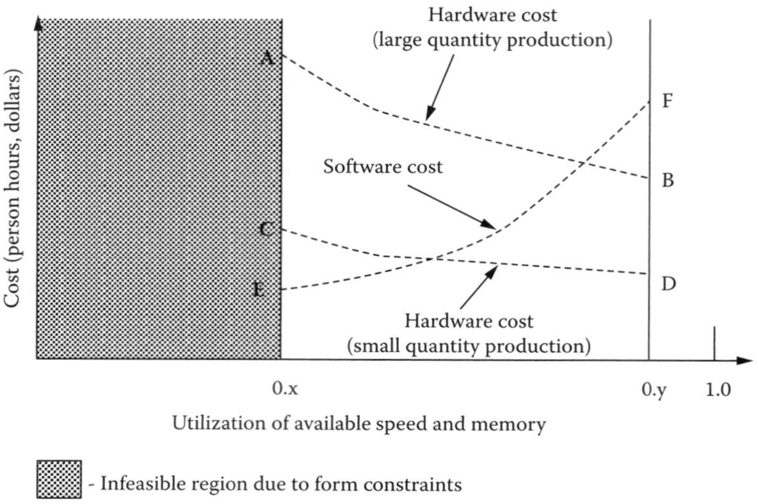

FIGURE 20.10 Hardware/software prototyping costs.

constrained architectures particularly increase the integration and test phase because resource constraints usually are not pushed until all software pieces come together. In systems in which most hardware is simply COTS parts, the time and cost of software prototyping and design can dominate the schedule and budget. If physical constraints permit, the hardware platform can be relaxed to achieve significant reductions in overall development cost and time. This principle of software prototyping is illustrated in Figure 20.10 [14,32]. The figure shows how system costs are dominated by software costs, especially at low production volumes, when CPU and memory utilization is high. However, as the utilization of these hardware resources is reduced, the software costs decrease drastically. Most parametric software cost estimation models quantitatively represent this principle. For example, the embedded mode revised intermediate COCOMO (REVIC) [33] software development cost and time models can be written as follows:

$$S_C = C_s\left[3.312 \times L^{1.2} \times F_E \times F_M \times \prod_{i=1}^{17} F_i\right] \quad (20.1)$$

$$S_T = 4.376\left[3.312 \times L^{1.2} \times F_E \times F_M \times \prod_{i=1}^{17} F_i\right]^{0.32} \quad (20.2)$$

where

S_C refers to the software development cost in dollars
S_T depicts development time in months
C_s is the software labor cost per person-month of effort
L denotes the number of delivered source instructions (thousands) including application code, OS kernel services, control and diagnostics, and support software

The F_i's represent additional cost drivers which model the effect of personnel, computer, product, and project attributes on software cost. F_E and F_M are effort adjustment factors which denote the effect of the execution time margin and storage margin on development cost. The relation between these effort adjustment factors and CPU and memory utilization is shown in Table 20.3. Linear interpolation is used to determine the effort multiplier values for utilizations between the given data points displayed in the table.

Despite the fact that many signal processing systems are being implemented with purely software solutions due to flexibility and scalability requirements, the combination of high throughput

TABLE 20.3 Execution Time and Main Storage Constraint Effort Multipliers

Rating	Utilization	F_E	F_M
Nominal	Up to 50%	1.00	1.00
High	70%	1.11	1.06
Very high	85%	1.30	1.21
Extra high	95%	1.66	1.56

requirements and stringent form factor constraints sometimes necessitate the need for implementing part of the system with dedicated hardware elements such as ASICs or field programmable gate arrays (FPGAs). Even though ASICs can provide sizable increases in performance and size efficiency, they come with a heavy development cost penalty which can usually only be justified by high volume production. In order to quantify this effect for trade-off analysis, parametric hardware cost models can be used. For example, the parametric cost model presented in [34,35] for ASIC design provides the following hardware development time and cost relations for ASIC development:

$$H_C = C_h\{(1 + D)^{YR}[A + B(S_h)^H]\} \tag{20.3}$$

$$H_T = 3.5\{(1 + D)^{YR}[A + B(S_h)^H]\}^{0.34} \tag{20.4}$$

where
- YR is 1984 minus the year of the bulk of the design effort
- C_h is the hardware labor cost per person-month
- A, B, D, and H are parameters of the model

D is the average annual improvement factor; A is the startup manpower; B is a measure of the productivity; and H is a measure of economies/diseconomies of scale. FPGAs provide more flexibility and lower cost penalties than ASICs at the expense of performance and size efficiency. From the FPGA cost model presented in [36], we will assume that FPGA development time and cost can be modeled as roughly one-third of that obtained for a comparable size ASIC. Although rising HW/SW development costs have very detrimental effects on life cycle cost, the effect of an increase in development time can be much more devastating.

Time-to-market costs can often outweigh design, prototyping, and production costs. A recent survey showed that being 6 months late to market resulted in an average of 33% profit loss. Engineering managers stated that they would rather have a 100% overrun in design and prototyping costs than be three months late to market with a product. Early market entry allows for increased brand name recognition, market share, and product yields. Market research performed by Logic Automation has shown that the demand and potential profits for a new HW/SW product can be modeled by a triangular window of opportunity as shown in Figure 20.11 [36]. Figure 20.11 illustrates the effect of delivering a product to market late. The shaded region of the triangle signifies the loss of revenue due to late entry in the market. This loss in revenue can be quantitatively stated as follows:

$$R_L = R_0 \times D(3W - D)/(2W^2) \tag{20.5}$$

where
- R_0 refers to the expected product revenue
- D is the delay (months) in delivering a product to market
- W is half the product lifetime (months)

Therefore, in order to maximize profits, the product must be on the market by the start of the demand window.

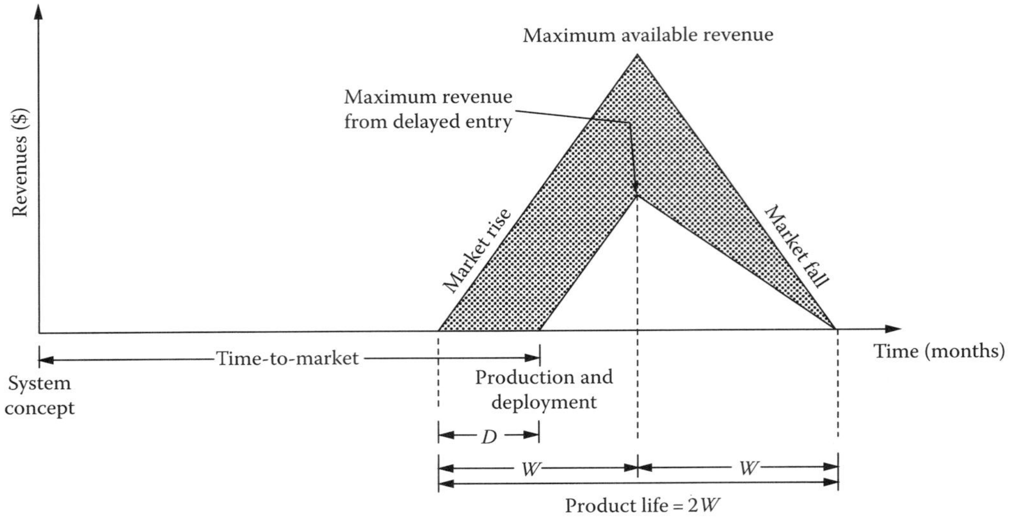

FIGURE 20.11 Time-to-market cost model.

20.7.2 Architectural Design Model

In this section, we present a cost-driven approach to conceptual architecture design. The conceptual architecture design process consists of HW/SW partitioning, architecture selection, and software partitioning. As input, this design stage accepts the application data/control flow graph, system-level performance requirements, form factor constraints, schedule constraints, and HW/SW reuse library parameters. As output, this stage produces an architecture candidate which serves as input to the architectural verification stage. VHDL performance models, described in Section 20.8, are used to verify the architecture candidate. This model is known as the "conceptual prototype" of the system. If the conceptual prototype does not meet performance specifications, updated performance parameters are back annotated to the architecture design process, and the process is repeated.

The architectural design problem can be modeled as a constrained optimization problem. The objective of cost-effective architecture design is to choose a HW/SW architecture which minimizes total life cycle cost, while maximizing potential product profits subject to performance and form factor constraints. Our approach quantitatively models the architecture design process by formulating it as a nonlinear mixed-integer programming problem. In order to provide support for high performance signal processing applications, we assume a distributed memory architecture composed of multiple programmable processors, ASICs, I/O devices, and/or FPGAs connected over a crossbar network. The goal of the architectural design process in this context is to determine the number and type of programmable processors (i860, SHARC, etc.), the memory capacity, the number of dedicated hardware elements (ASIC, FPGA), and to map the data/control flow graph nodes to the architectural elements in a manner that optimally meets design and economic objectives.

An example mathematical programming formulation that can be used to model the architecture design process is shown in Figure 20.12.

Objective: Minimize software development costs, hardware dev./prod. costs, NRE costs, and revenue losses

$$\sum_{i=1}^{2} C_h h_i + C_{\text{NRE}} \sum_{j=1}^{N} \sum_{i=1}^{N} a_i^{\text{asic}} \delta_{ij} + V(C^{\text{mem}} \mu) + V\left(\sum_{i=1}^{M} C_i^{\text{proc}} \rho_i\right) + V\left(\sum_{j=1}^{N} \sum_{i=1}^{N} C^{\text{fpga}} a_i^{\text{fpga}} \beta_{ij}\right)$$
$$+ V\left(\sum_{j=1}^{N} \sum_{i=1}^{N} C^{asic} a_i^{asic} \delta_{ij}\right) + R_0 d(3W - d)/(2W^2) + C_s$$

subject to:

Software development cost definition constraint

$$-Zw_{ij} \le \left(8 - 3.312L^{1.2} f_i^E f_j^M \prod_{i=1}^{17} F_i\right) \le Zw_{ij}$$

where

$$f_1^E = 1, \quad f_1^M = 1$$
$$f_2^E = 0.55u_p + 0.725, \quad f_2^M = 0.3u_m + 0.85$$
$$f_3^E = 1.27u_p + 0.223, \quad f_3^M = u_m + 0.36$$
$$f_4^E k = 3.6u_p - 1.76, \quad f_4^M = 3.5u_m + 1.765$$

Software development time definition constraint

$$4.376_8^{0.32} - (d_s^+ - d_s^-) = T_S$$

ASIC development cost definition constraint

$$h_1 - (1 + D)^{YR}\left[A + B\left(\sum_{j=1}^{N} \sum_{i=1}^{N} a_i^{\text{asic}} \delta_{ij}\right)^H\right] = 0$$

ASIC development time definition constraint

$$h_c \ge (1 + D)^{YR}\left[A + B\left(\sum_{i=1}^{N} a_i^{\text{asic}} \delta_{ij}\right)^H\right] \quad \text{for } j = 1, \ldots, N \;\; 3.5h_c^{0.34} - (d_{\text{asic}}^+ - d_{\text{asic}}^-) = T_S$$

FPGA development cost definition constraint

$$h_2 - \left\{(1 + D)^{YR}\left[A + B\left(\sum_{j=1}^{N} \sum_{i=1}^{N} a_i^{\text{fpga}} \beta_{ij}\right)^H\right]\right\} \Big/ 3 = 0$$

FPGA development time definition constraint

$$h_c' \ge (1 + D)^{YR}\left[A + B\left(\sum_{i=1}^{N} a_i^{\text{fpga}} \beta_{ij}\right)^H\right] \quad \text{for } j = 1, \ldots, N \;\; \left(3.5\left(h_c'\right)^{0.34}\right) \Big/ 3 - \left(d_{\text{fpga}}^+ - d_{\text{fpga}}^-\right) = T_S$$

FIGURE 20.12 Architecture design model.

(continued)

System-level performance constraint

$$\sum_{k=1}^{N}\sum_{j=1}^{M}\sum_{i=1}^{N} t_{ij}^{\text{proc}}\alpha_{ijk} + \sum_{j=1}^{N}\sum_{i=1}^{N} t_{ij}^{\text{asic}}\delta_{ij} + \sum_{j=1}^{N}\sum_{i=1}^{N} t_{ij}^{\text{fpga}}\beta_{ij} - \sum_{i=1}^{M}\rho_i((O_i^P + (O^M(\rho_i - 1)) - C_{\text{comm}}$$
$$+ \sum_{k=1}^{N}\sum_{j=1}^{M}\sum_{i=1}^{N-1}\sum_{l=i+1}^{N} c_{il}\alpha_{ijk}\alpha_{ljk} + \sum_{j=1}^{N}\sum_{i=1}^{N-1}\sum_{k=i+1}^{N} c_{ik}\beta_{ij}\beta_{kj} + \sum_{j=1}^{N}\sum_{i=1}^{N-1}\sum_{k=i+1}^{N} c_{ik}\delta_{ij}\delta_{kj} \geq T_{\text{SYS}}$$

Hardware/Software partitioning constraint

$$\sum_{k=1}^{N}\sum_{j=1}^{M}\alpha_{ijk} + \sum_{j=1}^{N}(\beta_{ij} + \delta_{ij}) = 1 \quad \text{for } i = 1, 2, \ldots, N$$

Time-to-market constraints

$$d \geq d_s^+, d \geq d_{\text{asic}}^+, d \geq d_{\text{fpga}}^+ d \leq W$$

Memory utilization definition constraint

$$u_m\mu = \sum_{k=1}^{N}\sum_{j=1}^{M}\sum_{i=1}^{N} m_i\alpha_{ijk} + O^{\text{MEM}}$$

Local memory size definition

$$u_m l_{jk}^{ni} = \sum_{i=1}^{N} m_i\alpha_{ijk} + O^{lm} \quad \text{for all } j, k$$

Total number of programmable processors definition

$$Zn_{jk} \geq \sum_{i=1}^{N}\alpha_{ijk} \quad \text{for all } j, k$$
$$\rho_i = \sum_{j=1}^{N} n_{ij} \quad \text{for } i = 1, 2, \ldots, M$$

ASIC Yield constraint

$$\sum_{i=1}^{N} a_i^{\text{asic}}\delta_{ij} \leq (1/K)\ln(1/Y), \quad \text{where } j = 1, 2, \ldots, N$$

FPGA size constraint

$$\sum_{i=1}^{N} a_i^{\text{fpga}}\beta_{ij} \leq U^{\text{ipga}} \quad \text{for all } j$$

FIGURE 20.12 (continued)

System-level area and power constraints

$$\sum_{i=1}^{M} a_i^{\text{proc}} \rho_i + a_{\text{mcm}}\mu + \sum_{j=1}^{N}\sum_{i=1}^{N} a_i^{\text{asic}} \delta_{ij} + \sum_{j=1}^{N}\sum_{i=1}^{N} a_i^{\text{fpga}} \beta_{ij} \leq U_A$$

$$\sum_{i=1}^{M} p_i^{\text{proc}} \rho_i + p_{\text{mcm}}\mu + \sum_{j=1}^{N}\sum_{i=1}^{N} p_i^{\text{asic}} \delta_{ij} + \sum_{j=1}^{N}\sum_{i=1}^{N} p_i^{\text{fpga}} \beta_{ij} \leq U_P$$

Load balancing constraints

$$u_p - \Delta_p - Z(1 - n_{jk}) \leq \left(\sum_{i=1}^{N} t_{ij}^{\text{proc}} \alpha_{ijk} + O_j^P\right) \Big/ S_j \leq u_p + \Delta_p + Z(1 - n_{jk}) \quad \text{for all } j, k$$

Processor utilization definition constraint

$$u_p\left(\sum_{i=1}^{N} S_i \rho_i\right) = \sum_{k=1}^{N}\sum_{j=1}^{M}\sum_{i=1}^{N} t_{ij}^{\text{proc}} \alpha_{ijk} + \sum_{i=1}^{M} \rho_i\left(O_i^P + O^M(\rho_i - 1)\right)$$

Interval relation

$$w_{ij} = 1 - y_i g_j \quad \text{for } i, j = 1, 2, 3, 4$$

Memory utilization interval constraints

$$u_m \leq 0.5_{g1} + 0.7_{g2} + 0.85_{g3} + 0.95_{g4}$$

$$u_m \geq 0.5_{g2} + 0.7_{g3} + 0.85_{g4}, \quad \sum_{i=1}^{4} g_i = 1$$

Processor utilization interval constraints

$$u_p \leq 0.5_{y1} + 0.7_{y2} + 0.85_{y3} + 0.95_{y4}$$

$$u_p \geq 0.5_{y2} + 0.7_{y3} + 0.85_{y4}$$

where

$$\sum_{i=1}^{4} y_i = 1$$

FIGURE 20.12 (continued)

The major decision variables of the model are defined as follows:

$\alpha_{ijk} = 1$	if DFG task i is implemented on the kth programmable processor of type j, otherwise 0
$\beta_{ij} = 1$	if DFG task i is implemented on the jth FPGA, otherwise 0
$\delta_{ij} = 1$	if DFG task i is implemented on the jth ASIC, otherwise 0
ρ_i	the number of programmable processors of type i in the architecture
μ	the number of DRAM chips in the architecture
d	the delay in delivering the product to market(months)
u_{p}	the overall processor utilization
u_{m}	the overall memory utilization

y_i is a binary variable which signifies the processor utilization interval
g_i is a binary variable which signifies the memory utilization interval
f_i^E the execution time constraint effort multiplier for processor utilization interval i
f_i^M the memory constraint effort multiplier for memory utilization interval i
s the software development effort (person-months)
h_i the hardware development effort for dedicated hardware type i (person-months)
$n_{ij} = 1$ if processor j of type i is included in the architecture, otherwise 0
l_{jk}^m the local memory allocated to processor k of type j

Additional model parameters include

C_{NRE} the overall nonrecurring engineering cost per unit ASIC die area
N the number of tasks (nodes) in the data/control flow graph; also, the maximum number of computational elements in the architecture
V the production volume
C_i^{proc} the procurement cost for a programmable processor of type i
C^{mem} the procurement cost for per DRAM chip
M the number of different types of programmable processors
C^{fpga} the production cost per unit size for an FPGA
C^{asic} the production cost per unit size for an ASIC
T_S the amount of time after system concept when the product should be delivered to market
Z a very large number
Y the minimum allowable ASIC yield
K the process defect density
U^{fpga} the size limit on a single FPGA
U_A the area limit on the system
U_P the power limit on the system
S_i the peak throughput of processor i
t_{ij}^{proc} the throughput of task i implemented on processor j
t_i^{asic} the throughput of task i implemented on an asic
t_i^{fpga} the throughput of task i implemented on an fpga
O_i^P the processor overhead (scheduling, resource management, dispatching, context switching) (ops/s)
O^M multiprocessor overhead factor
T_{sys} the minimum required system level throughput
m_i the memory requirement for task i
O^{MEM} the total memory required for OS services, control and diagnostics, and support software
O^{lm} the local memory required for OS services, control and diagnostics, and support software
Δ_p is a constant that defines the desired utilization range on a processor
c_{kl} the throughput penalty for transferring data between tasks k and l off-chip
C_{comm} the worst case total throughput penalty due to transferring data between tasks off-chip

We are currently using the GAMS [37] optimization system to solve examples of this form. More specifically, the GAMS/DICOPT nonlinear mixed-integer programming package is being employed. DICOPT utilizes nonlinear optimization programs such as MINOS or CONOPT and mixed-integer programming packages such as ontrack systems limited (OSL) to rigorously solve these problems. Linearization techniques such as those described in [38] are also being applied to the models to improve computational efficiency.

Interestingly, while the rapid prototyping community has largely ignored rigorous integer programming methods for "quick" simplified heuristics, the communications industry (e.g., AT&T, Airlines reservation systems, etc.) routinely uses optimization algorithms with variables numbering in a few tens of thousands and more. The authors feel that complex nonlinear and multiobjective functions cannot be optimized via the "human-in-the-optimization-loop" methods, and any extra effort spent in the conceptual phase of the design process is time well spent.

20.8 Performance Modeling and Architecture Verification

The selection process for possible system architectures was discussed in the previous section. Performance models [39–42] are used to verify that the architectures adhere to specific time-critical system constraints. The advantages of performance models include

- They capture the time aspects of the system under development (i.e., throughput, latency, and resource utilization) and present this information for rapid evaluation.
- They verify the performance of proposed architectures found in the architectural design stage.
- They allow for true HW/SW codesign by simulating the behavior of software on performance models of processor hardware. The model of software can take many forms, one being a simple delay which models the performance of a library software primitive executing on a specific processor. For example, an Analog Devices SHARC 2106X chip executes a FFT in shorter time than a Texas Instruments C30 processor. The performance model captures this information through the use of "generic" parameters and when simulated, uses this parameter to determine how long the processor will be utilized while performing the function [40,41].
- They provide the capability for modeling operating system effects on multiprocessor network architectures [41].
- Performance model development time is shorter when compared with that of FFMs and hence library population can be done in-cycle.

The fidelity attributes of token-based performance models used for system architecture verification through simulation are

Internal: {(Clock cycle $\rightarrow$ System event), (X), (X), (X), (Assembly $\rightarrow$ Primitive)}
External: {(Clock cycle $\rightarrow$ System event), (X), (X), (Full Structure $\rightarrow$ Black box), (X)}.

Temporal information for both the internal and external attributes are captured at multiple levels depending on the application modeled. For example, the system event level can capture large blocks of data passing over an interconnect network or the simulating of a large time slice of processing on a single processor. System events occur in the 10s of microseconds to 10s of milliseconds time span and potentially could contain millions of actual clock cycles. The clock cycles, however, are not simulated, only the time events where information is interchanged or processed. At higher resolutions, details may be required to capture how an interconnect network handles data streams from multiple processors at the clock cycle level. The performance models we use for architecture verification fall into these categories and an example is presented later. Performance level models do not capture the function or data values of the system but focus on its time aspects. Internally, a performance model has no structure, but externally, the structure can be represented by any level in the resolution hierarchy. For example, a network architecture model consisting of multiple processors and interconnect ASICs could be described by first instantiating each of the components in a network model, then by connecting them using a performance model of a particular bus or interconnect protocol. At the other extreme, the model may be represented by a black box that outputs tokens based on a specified control input with the internal details represented by abstract behavior. On the RASSP program, the programmability generally was captured at the DSP function primitive level where signal processing procedures (FFT, etc.) are scheduled on performance models of processors. The processor models can, however, be defined for much higher resolutions where the primitives represent assembly level instructions.

The efficiency of these models is very high because the code is written at the behavioral level of abstraction in VHDL. Signals are used to pass abstract data types known as tokens between component elements. The tokens are record types in VHDL and take the form as shown in Figure 20.13. All protocol

```
-- ucue : the basic unresolved virtual packet type.
TYPE ucue IS
RECORD
  -- basic information field
  c_id      : integer;
  dest_id   : integer;    -- destination
  src_id    : integer;    -- source
  c_type    : cue_type;  -- packet type
    -- *** TYPE cue_type IS
    -- *** (DATA, READ, WRITE, CONTROL, USER1, USER2);
    -- *** USER1 and USER2 are used to declare some special
    -- *** packet type.
    -- *** e.g. ''echo packet'' and ''idle symbol'' in SCI
c_size   :size_type;
    -- *** TYPE size_type IS RANGE 0 to INTEGER'high
    -- *** UNITS
    -- *** bit_unit;
    -- *** byte    = 8 bit_unit;
    -- *** kbyte   = 1024 byte;
    -- ***  ... ... ... ... ... ...
    -- ***  ... ... ... ... ... ...
    -- ***  ... ... ... ... ... ...
    -- *** END UNITS;
block_size   : size_type;
  -- *** the block size of response requested by the packet.
init_time   : TIME;  -- initial time
  -- protocol field
priority   : INTEGER;
c_state    : cue_state_type; -- packet state
  -- *** TYPE cue_state_type IS (IDLE,REQ,ACK,BUSY,
  -- *** USERSTATE1, USERSTATE2);
protocol   : STRING(1 TO 8);
  -- *** the protocol name, e.g. ''pt_to_pt''
collisions   : INTEGER;
retries      : INTEGER;
route       : INTEGER;
-- user-defined field
int_user1    : INTEGER;
int_user2    : INTEGER;
real_user1   : INTEGER;
bits_user1   : BIT_VECTOR(15 downto 0);
  -- *** used to set some special flags or handshaking signals . . .
  -- *** e.g. ''burst'' in HIPPI . . .
END RECORD;
```

FIGURE 20.13 The declaration of "cue." The information of each communication transaction is contained in three fields—basic information field, protocol field, and user-defined field.

handling information is carried within it. This data type is referred to as "cue" throughout the remainder of the chapter. The "cue" data type is used as a virtual packet to pass information between elements in a system architecture. It contains fields for capturing statistical information about how the data is passed through a processor network (priority, collisions, retries, routes), information on the source and destination of packet (src_id, dest_id), packet size (c_size, resp_size), packet identification number (c_id), and transmission status (c_state). There are also user defined fields that allow the performance model designer to implement model specific details (int_user1, int_user2, real_user1, real_user2).

Interoperability is determined by the ability of models with a similar external fidelity to communicate. For performance models to achieve this goal, a token format must be defined and standardized. Currently, there are no standards to meet this demand. Protocol converters can be developed to link performance models with alternate token structures but a standard is encouraged. RASSP is pursuing a standard.

20.8.1 A Performance Modeling Example: SCI Networks

A scalable coherent interface (SCI) performance model has been developed as an example. The executable SCI model can serve as an executable specification for the communication protocol. Figure 20.14 illustrates the SCI node interface structure. ***Linc*** transmits or bypasses packets, and performs the primary SCI protocols. ***REC_QUEUE*** and ***TR_QUEUE*** are first-in-first-out (FIFO) buffers which store receive packets and transmit packets. ***Processor*** contains a ***responder*** and a ***res_handler***; ***responder*** generates the response packets for the request packets who ask for responses, and ***res_handler*** serves as a response packet consumer. The "packet generator" is used to create cues according to the required communication patterns. Connecting the SCI node interface, designers can easily construct an SCI network and create their inputs to evaluate the performance results.

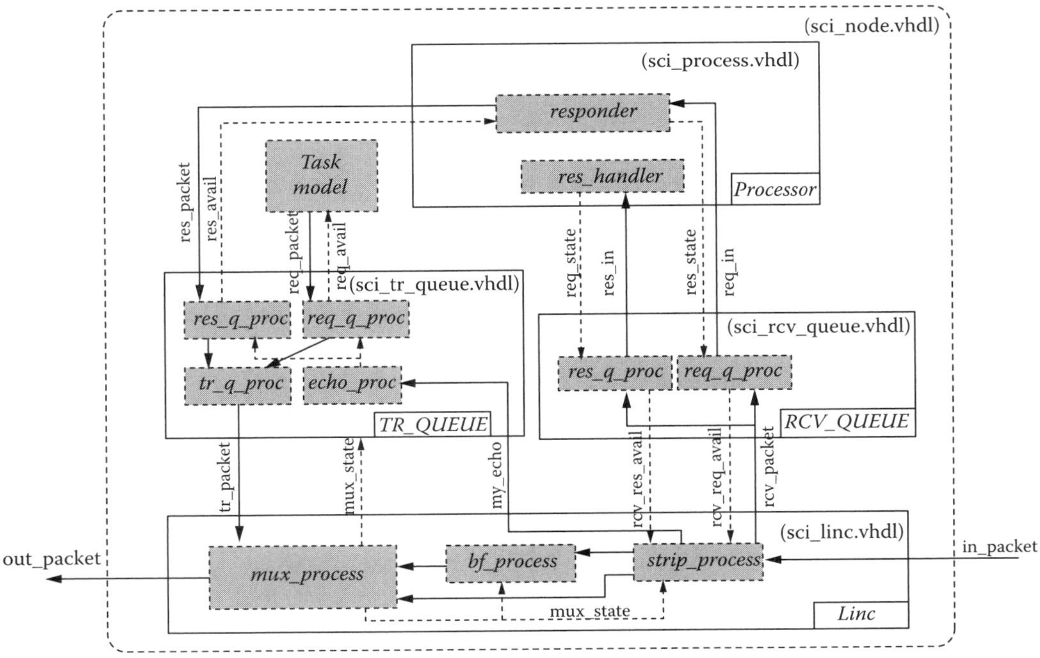

FIGURE 20.14 The SCI node interface performance model.

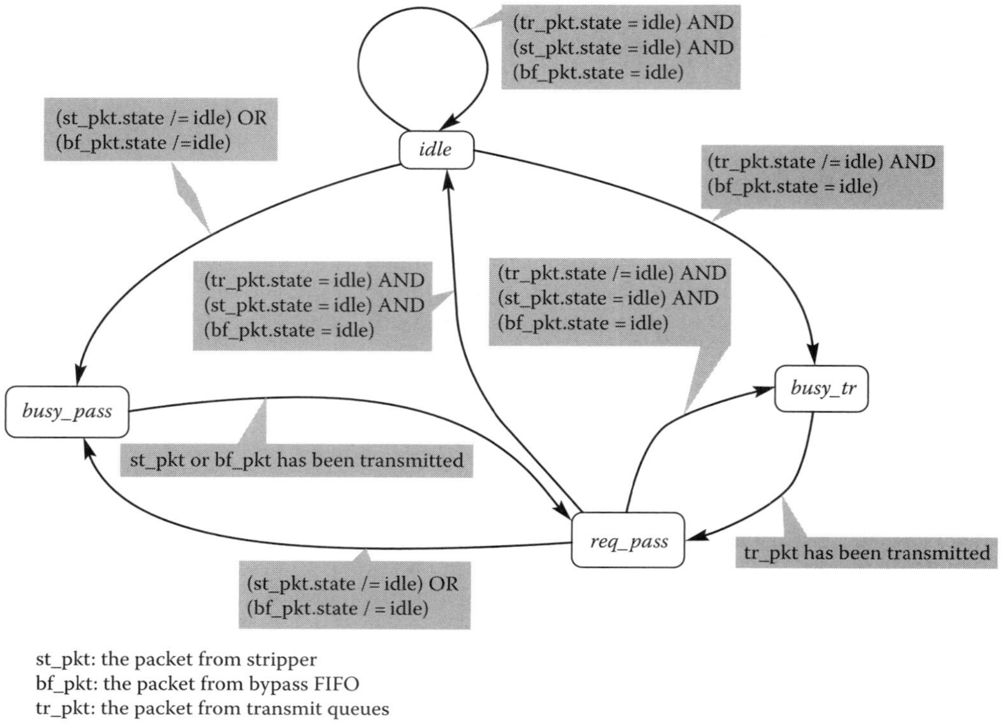

FIGURE 20.15 The state diagram of MUX.

Among the processes of the SCI node, ***mux_process*** dominates the primary communication protocol. The SCI protocol can be converted to the state diagram shown in Figure 20.15. Based on the state diagram, the VHDL representation of ***mux_process*** at the performance level is written as shown in Figure 20.16. MUX_Process is activated by st_pkt, bf_pkt, tr_pkt, and MUX_State and changes MUX_State according to the state diagram.

20.8.2 Deterministic Performance Analysis for SCI

A key requirement of real-time DSP architectures is determinism. Most designers would like to know the guaranteed worst-case performance rather than the average or peak performance. In order to make the performance determinable, an SCI network must satisfy the following constraints:

Step 1. The size of each packet is deterministic.
Step 2. The interprocessor communications are deterministic.
Step 3. All arrival packets are accepted.

That is to say, packets should not be retransmitted in an SCI ring. The retry packets might prevent sending the fresh packets and make the throughput and latency unpredictable.

Definition 20.1: If a packet is retransmitted from the transmit queues, it is called a retry packet. If a packet is transmitted for the first time, it is called a fresh packet.

```
MUX_Process : PROCESS(st_pkt, bf_pkt, tr_pkt, MUX_State)
  VARIABLE temp_pkt        : ucue := init_pkt;
  VARIABLE time_busy       : TIME;

BEGIN
  IF (now = 0 ns) THEN
    MUX_State <= idle;
  END IF;

  CASE MUX_State IS
    WHEN idle =>
      IF (tr_pkt.state = idle) THEN
        IF (st_pkt'Event AND st_pkt.state /= idle) THEN
          time_busy := TransferTime(st_pkt.size);
            --*** calculate the transfer time of the packet
          Out_Token <= st_pkt after WIRE_DELAY+OUT_DELAY;
          MUX_State <= busy_pass;
          MUX_State <= req_pass after time_busy;
        ELSE
          IF (bf_pkt.state /= idle) THEN
            time_busy := TransferTime(bf_pkt.size);
            Out_Token <= bf_pkt after WIRE_DELAY+OUT_DELAY;
            MUX_State <= busy_pass;
            MUX_State <= req_pass after time_busy;
          END IF;
        END IF;
      ELSE
        time_busy := TransferTime(tr_pkt.size);
        Out_Token <= tr_pkt after WIRE_DELAY+OUT_DELAY;
        --insert idle symbol
        Out_Token <= idle_pkt
                      after time_busy+WIRE_DELAY+OUT_DELAY;
        MUX_State <= busy_tr;
        MUX_State <= req_pass after time_busy + SCI_BASE_TIME;
          --*** SCI_BASE_TIME := 2 ns;
      END IF;

    WHEN busy_tr = >

    WHEN busy_pass =>

    WHEN req_pass =>
      IF (st_pkt'Event AND st_pkt.state /= idle) THEN
        time_busy := TransferTime(st_pkt.size);
```

FIGURE 20.16 The VHDL process of MUX.

(continued)

```
        Out_Token <= st_pkt after WIRE_DELAY+OUT_DELAY;
        MUX_State <= busy_pass;
        MUX_State <= req_pass after time_busy;
      ELSIF (bf_pkt.state /= idle) THEN
        time_busy := TransferTime(bf_pkt.size);
        Out_Token <= bf_pkt after WIRE_DELAY+OUT_DELAY;
        MUX_State <= busy_pass;
        MUX_State <= req_pass after time_busy;
      ELSIF (tr_pkt.state /= idle) THEN
        time_busy := TransferTime(tr_pkt.size);
        Out_Token <= tr_pkt after WIRE_DELAY+OUT_DELAY;
        Out_Token <= idle_pkt
                     after time_busy+WIRE_DELAY+OUT_DELAY;
        MUX_State <= busy_tr;
        time_busy := time_busy + SCI_BASE_TIME;
        MUX_State <= req_pass after time_busy;
      ELSE
        MUX_State <= idle;
      END IF;
  END CASE;
END PROCESS MUX_Process;
```

FIGURE 20.16 (continued)

A basic SCI network is a unidirectional SCI ring. The maximum number of nodes traversed by a packet is equal to N in an SCI ring, and the worst-case path contains a MUX, a stripper, and $(N-2)$ links. So, we find the worst-case latency, $L_{\text{worst-case}}$, is

$$L_{\text{worst-case}} = T_{\text{MUX}} + T_{\text{wire}} + T_{\text{stripper}} + (N-2) \cdot T_{\text{linc}} \quad (20.6)$$

$$= (N-1) \cdot T_{\text{linc}} - T_{\text{FIFO}} \quad (20.7)$$

where

T_{linc}, the link delay, is equal to $T_{\text{MUX}} + T_{\text{FIFO}} + T_{\text{stripper}} + T_{\text{wire}}$
T_{MUX} is the MUX delay
T_{wire} is the wire delay between nodes
T_{stripper} is the stripper delay
T_{FIFO} is the full bypass FIFO delay

The SCI link bandwidth, BW_{link}, is equal to 1 byte per second per link; the maximum bandwidth of an SCI ring is proportional to the number of nodes:

$$\text{BW}_{\text{ring}} = N \cdot \text{BW}_{\text{link}} \text{ (bytes/second)} \quad (20.8)$$

where N is the number of nodes. Now let us consider the bandwidth of an SCI node. Since each link transmits the packets issued by all nodes in the ring, BW_{link} is shared by not only transmitting packets but passing packets, echo packets, and idle symbols.

$$\mathrm{BW}_{\mathrm{link}} = \mathrm{bw}_{\mathrm{transmitting}} + \mathrm{bw}_{\mathrm{passing}} + \mathrm{bw}_{\mathrm{echo}} + \mathrm{bw}_{\mathrm{idle}} \tag{20.9}$$

where

$\mathrm{bw}_{\mathrm{transmitting}}$ is the consumed bandwidth of transmitting packets
$\mathrm{bw}_{\mathrm{passing}}$ is the consumed bandwidth of passing packets
$\mathrm{bw}_{\mathrm{echo}}$ is the consumed bandwidth of echo packets
$\mathrm{BW}_{\mathrm{idle}}$ is the consumed bandwidth of idle symbols

Assuming that the size of the send packets is fixed, we find $\mathrm{bw}_{\mathrm{transmitting}}$ is

$$\mathrm{bw}_{\mathrm{transmitting}} = \mathrm{BW}_{\mathrm{link}} \cdot \frac{N_{\mathrm{transmitting}} \cdot D_{\mathrm{packet}}}{D_{\mathrm{link}}} \tag{20.10}$$

$$= \mathrm{BW}_{\mathrm{link}} \cdot \frac{N_{\mathrm{transmitting}} \cdot D_{\mathrm{packet}}}{(N_{\mathrm{passing}} + N_{\mathrm{transmitting}}) \cdot (D_{\mathrm{packet}} + 16) + N_{\mathrm{echo}} \cdot 8 + N_{\mathrm{idle}} \cdot 2} \tag{20.11}$$

where

D_{packet} is the data size of a transmitting packet
D_{link} is the number of bytes passed through the link
$N_{\mathrm{transmitting}}$ is the number of transmitting packets
N_{passing} is the number of passing packets
N_{echo} is the number of echo packets
N_{idle} is the number of idle symbols

A transmitting packet consists of an unbroken sequence of data symbols with a 16-byte header that contains address, command, transaction identifier, and status information. The echo packet uses an 8-byte subset of the header while idle symbols require only 2 bytes of overhead. Because each packet is followed by at least an idle symbol, the maximum $\mathrm{bw}_{\mathrm{transmitting}}$ is

$$\mathrm{BW}_{\mathrm{transmitting}} = \mathrm{BW}_{\mathrm{link}} \cdot \frac{N_{\mathrm{transmitting}} \cdot D_{\mathrm{packet}}}{(N_{\mathrm{passing}} + N_{\mathrm{transmitting}}) \cdot (D_{\mathrm{packet}} + 18) N_{\mathrm{echo}} \cdot 10} \tag{20.12}$$

However, $\mathrm{BW}_{\mathrm{transmitting}}$ might be consumed by retry packets; the excessive retry packets will stop sending fresh packets. In general, when the processing rate of arrival packets, $R_{\mathrm{processing}}$, is less than the arrival rate of arrival packets, R_{arrival}, the excessive arrival packets will not be accepted and their retry packets will be transmitted by the sources. This cause for rejecting an arrival packet is the so-called queue contention. The number of retry packets will increase with time because retry packets increase the arrival rate. Once $\mathrm{bw}_{\mathrm{transmitting}}$ is saturated with fresh packets and retry packets, the transmission of fresh packets is stopped resulting in an increase in the number of retry packets transmitted.

Besides queue contention, incorrect packets cause the rejection of an arrival packet. This indicates a possible component malfunction. No matter what the cause, the retry packets should not exist in a real-time system in that two primary requirements of real-time DSP are data correctness and guaranteed timing behavior.

20.8.3 DSP Design Case: Single Sensor Multiple Processor

Figure 20.17 shows a DSP system with a sensor and N PEs. This system is called the single sensor multiple processor (SSMP). In this system, the sensor uniformly transmits packets to each PE and the sampling rate of the sensor is R_{input}.

For the node i, if the arrival rate, $R_{\mathrm{arrival},i}$, is greater than the processing rate, $R_{\mathrm{processing},i}$, receive queue contention will occur and unacceptable arrival packets will be sent again from the sensor node. Retry

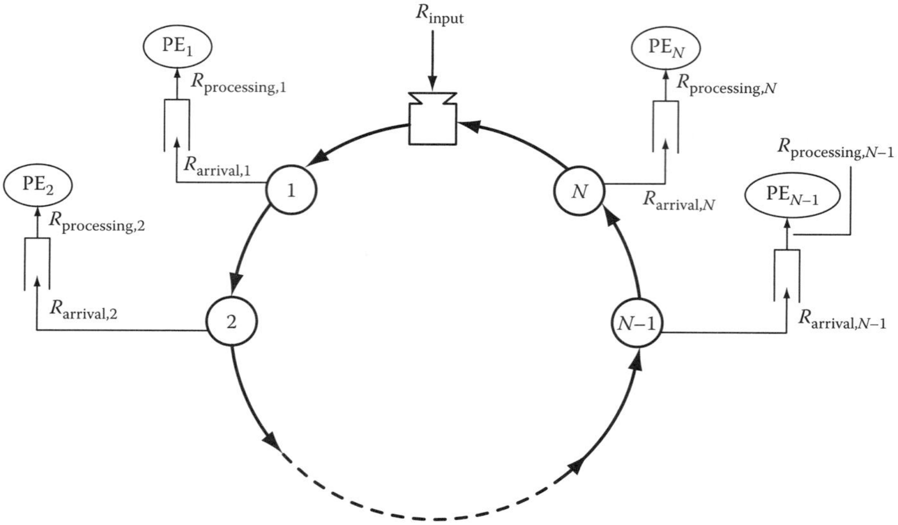

FIGURE 20.17 The SSMP architecture. The sensor uniformly transmits packets to each PE and the sampling rate of sensor is R_{input}; so, the arrival rate of each node is $\frac{R_{\text{input}}}{N}$.

packets increase the arrival rate and result in more retry packets transmitted by the sensor node. Since the bandwidth of the retry packets and fresh packets is limited by $\text{BW}_{\text{transmitting}}$, the sensor will stop reading input data when the bandwidth is saturated. For a real-time DSP, the input data should not be suspended, thus, the following inequality has to be satisfied to avoid the retry packets:

$$\frac{R_{\text{input}}}{N} \leq R_{\text{processing}} \tag{20.13}$$

Because the output link of the sensor node will only transmit the transmitting packets, the maximum transmitting bandwidth is

$$\text{BW}_{\text{transmitting}} = \text{BW}_{\text{link}} \cdot \frac{D_{\text{packet}}}{D_{\text{packet}} + 18} \tag{20.14}$$

and the limitation of R_{input} is

$$R_{\text{input}} \leq \text{BW}_{\text{link}} \cdot \frac{D_{\text{packet}}}{D_{\text{packet}} + 18} \tag{20.15}$$

We now assume a SSMP system design with a 10 MB/s sampling rate and five PEs where the computing task of each PE is a 64-point FFT. Since each packet contains 64 32-bit floating-point data, D_{packet} is equal to 256 bytes.

From Equation 20.13 the processing rate must be greater than 2 MB/s, so the maximum processing time for each packet is equal to 128 μs. Because an n-point FFT needs $\frac{n}{2}\log_2 n$ butterfly operations and each butterfly needs 10 FLOPs [44], the computing power of each PE should be greater than 15 MFLOPS. From a design library we pick *i*860 s to be the PEs and a single SCI ring to be the communication element in that *i*860 provides 59.63 MFLOPS for 64-point FFT and $BW_{\text{transmitting}}$ of a single SCI ring is 934.3 MB/s which satisfies Equation 20.15. Using 5 *i*860s, the total computing power is equal to 298.15 MFLOPS. The simulation result is shown in Figure 20.18a. The result shows that retry packets for $R_{\text{input}} = 10$ MB/s do not exist.

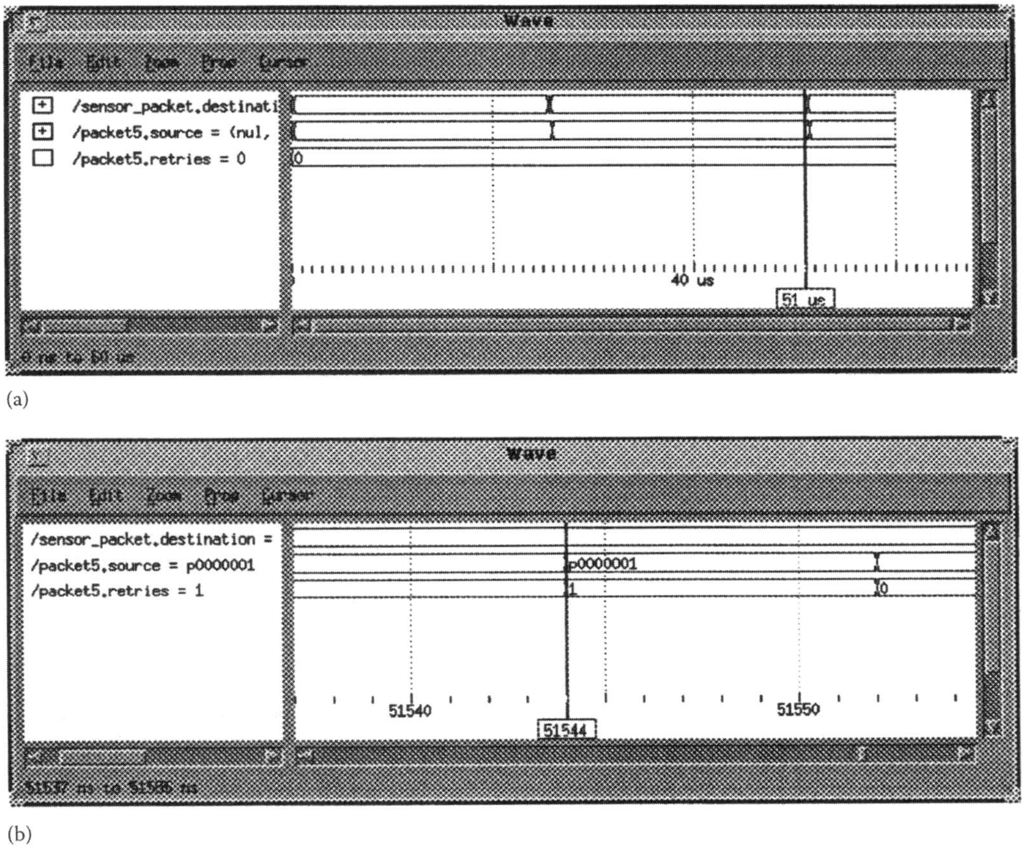

FIGURE 20.18 The simulation of SSMP using *i*860 and SCI ring. (a) The result of SSMP with $R_{\text{input}} = 10$ MB/s. The value of "packet5.retries" shows that there does not exist any request of retry packets. (b) The result of SSMP with $R_{\text{input}} = 100$ MB/s. A retry packet is requested at 25,944 ns.

As stated earlier, if the processing rate is less than the arrival rate, the retry packets will be generated and the input stream will be stopped. Hence, we changed R_{input} from 10–100 MB/s to test whether the *i*860 can process a sampling rate as high as 100 MB/s. Upon simulating the performance model, we found that the sensor node received an echo packet which asked for retransmitting a packet at 25,944 ns in Figure 20.18b; thus, we have to substitute another processor with higher MFLOPS for *i*860 to avoid the occurrence of the retry packets.

Under the RASSP program, an additional example where performance level models were used to help define the system architecture can be found in Paulson and Kindling [45].

20.9 Fully Functional and Interface Modeling and Hardware Virtual Prototypes

Fully functional and interface models support the concept of a hardware virtual prototype (HVP). The HVP is defined as a software representation of a hardware component, board, or system containing sufficient accuracy to guarantee its successful hardware system-level realization [5,47]. The HVP adopts as its main goals (1) verification of the design correctness by eliminating hardware design errors from the in-cycle design loop, (2) decreasing the design process time through first time correctness, (3) allowing concurrent codevelopment of hardware and software, (4) facilitating rapid HW/SW integration, and (5) generation of models to support future system upgrades and maintenance. This model abstraction captures all the documented functionality and interface timing of the unit under development.

Following architectural trade studies, a high level design of the system is determined. This high level design consists of COTS parts, in-house design library elements, and/or new application specific designs to be done in-cycle. At this level, it is assumed the COTS parts and in-house designs are represented by previously verified FFMs of the devices. FFMs of in-cycle application specific designs serve as high level models useful for system level simulation. They also serve as "golden" models for verification of the synthesizable RTL level representation and define its testbench. For system level simulations, this high level model can improve simulation speed by an order of magnitude while maintaining component interface timing fidelity.

The support infrastructure required for FFMs is the existence of a library of component elements and appropriate hardware description language (HDL) simulation tools. The types of components contained in the library should include models of processors, buses/interconnects, memories, programmable logic, controllers, and medium and large scale integrated circuits. Without sufficient libraries, the development of complex models within the in-cycle design loop can diminish the usefulness of this design philosophy by increasing the design time. The model fidelity used for hardware virtual prototyping can be classified as

$$\text{Internal: } \{(\text{Gate} \rightarrow \text{Clock cycle}), (\text{Bit true} \rightarrow \text{Token}), (\text{All}), (\text{Major blocks}), (\text{Micro code} \rightarrow \text{Assembly})\}$$

$$\text{External: } \{(\text{Gate} \rightarrow \text{Clock cycle}), (\text{Bit true}), (\text{All}), (\text{Full Structure}), (\text{X})\}.$$

Internally and externally, the temporal information of the device should be at least clock cycle accurate. Therefore, internal and external signal events should occur as expected relative to clock edges. For example, if an address line is set to a value after a time of 3 ns from the falling edge of a clock based on the specification for the device, then the model shall capture it. The model shall also contain hooks, via generic parameters, to set the time related parameters. The user selectable generic parameters are placed in a VHDL package and represent the minimum, typical, and maximum setup times for the component being modeled.

Internal data can be represented by any value on the axis, while the interface must be bit true. For example, in the case of an internal 32-bit register, the value could be represented by an integer or a 32-bit vector. Depending on efficiency issues, one or the other choice is selected. The external data resolution must capture the actual hardware pinout footprint and the data on these lines must be bit true. For example, an internally generated address may be in integer format but when it attempts to access external hardware, binary values must be placed on the output pins of the device. The internal and external functionality is represented fully by definition.

Structurally, because the external pins must match those of the actual device, the external resolution is as high as possible and therefore the device can be inserted as a component into a larger system if it satisfies the interoperability constraints. Internally, the structure is composed of high level blocks rather than detailed gates. This improves efficiency because we minimize the signal communication between processes and/or component elements.

Programmability is concerned with the level of software instructions interpreted by the component model. When developing HVPs, the programmable devices are typically general purpose, digital, or video signal processors. In these devices, the internal model executes either microcode or the binary form of assembly instructions and the fidelity of the model captures all the functionality enabling this. This facilitates HW/SW codevelopment and cosimulation. For example, in [5], a processor model of the Intel *i*860 was used to develop and test over 700 lines of Ada code prior to actual hardware prototyping.

An important requirement for FFMs to support reuse across designs and rapid systems development is the ability to operate in a seamless fashion with models created by other design teams or external vendors. In order to ensure interoperability, the IEEE standard nine value logic package* is used for all models. This improves technology insertion for future design system upgrades by allowing segments of the design to be replaced with new designs which follow the same interoperability criteria.

Under the RASSP program, various design efforts utilized this stage in the design process to help achieve first pass success in the design of complex signal processing systems. The Lockheed Sanders team developed an infrared search and track (IRST) system [46,47] consisting of 192 Intel *i*860 processors using a Mercury RACEWAY network along with custom hardware for data input buffering and distribution and video output handling. The HVP served to find a number of errors in the original design both in hardware and software. Control code was developed in Ada and executed on the HVP prior to actual hardware development. Another example where HVPs were used can be found in [48].

20.9.1 Design Example: I/O Processor for Handling MPEG Data Stream

In this example, we present the design of an I/O processor for the movement of MPEG-1 encoder data from its origin at the output of the encoder to the memory of the decoder. The encoded data obtained from the source is transferred to the VME bus through a slave interface module which performs the proper handshaking. Upon receiving a request for data (AS low,WRITE high) and a valid address, the data is presented on the bus in the specified format (the mode of transfer is dictated by the VME signals LWORD,DS0,DS1 and AM[0.5]). The VME DTACK signal is then driven low by the slave indicating that the data is ready on the bus after which the master accepts the data. It repeats this cycle if more data transfer is required, otherwise it releases the bus. In the simulation of the I/O architecture in Figure 20.19 a quad-byte-block transfer (QBBT) was done.

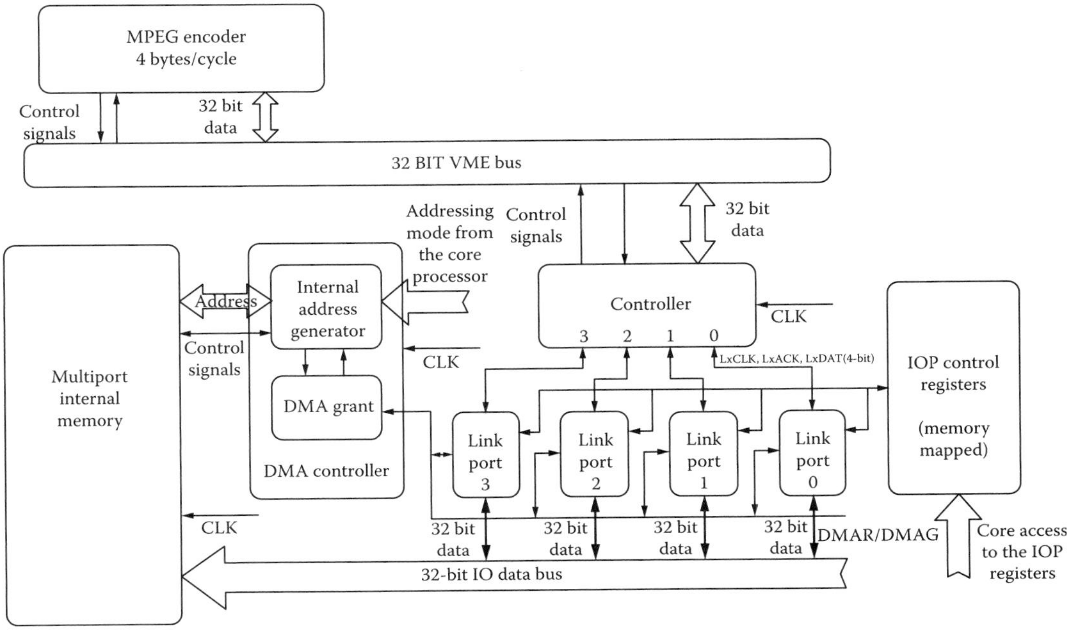

FIGURE 20.19 The system I/O architecture.

* IEEE 1164-1993 Standard Multi-value Logic System for VHDL Model Interoperability.

The architecture of the I/O processor is described below. The link ports were chosen for the design since they were an existing element in our design library and contain the same functionality as the link ports on the Analog Devices 21060 digital signal processor. The circuit's ASIC controller is designed to interface to the VME bus, buffer data, and distribute it to the link ports. To achieve a fully pipelined design, it contains a 32-bit register buffer both at the input and outputs. The 32-bit data from the VME is read into the input buffer and transferred to the next empty output register. The output registers send the data by unpacking. The unpacking is described as follows: at every rising edge of the clock (LxCLK) a 4-bit nibble of the output register, starting from the LSB, is sent to the link port data line (LxDAT) if the link port acknowledge (LxACK) signal is high. Link ports that are clocked by LxCLK, running at the twice the core processor's clock rate, read the data from the controller ports with the rising edge of the LxCLK signal. When their internal buffers are full they deassert LxACK to stop the data transfer.

Since we have the option of transferring data to the link ports at twice the processor's clock rate, four link ports were devoted to this data transfer to achieve a fully pipelined architecture and maximize utilization of memory bandwidth. With every rising edge of the processor clock (CLK) a new data can be read into the memory. Figure 20.20 shows the pipelined data transfer to the link ports where DATx represents a 4-bit data nibble. As seen from the table, Port0 can start sending the new 32-bit data immediately after it is done with the previous one. Time multiplexing among the ports is done by the use of a token. The token is transferred to the next port circularly with the rising edge of the processor clock. When the data transfer is complete (buffer is empty), each port of the controller deasserts the corresponding LxCLK which disables the data transfer to the link ports. LxCLKs are again clocked when the transfer of a new frame starts. The slave address, the addressing mode, and the data transfer mode require setups for each transfer. The link ports, IOP registers, DMA control units, and multiport memory models were available in our existing library of elements and they were integrated with the VME bus model library element. However, the ASIC controller was designed in-cycle to perform the interface handshaking. In the design of the ASIC, we made use of the existing library elements, i.e., I/O processor link ports,

CLK	1		2		3		4		5		6		7		8	
LxCLK	1	2	3	4	5	6	7	8	9	10	11	12	13	14	15	16
PORT0	DAT0	DAT1	DAT2	DAT3	DAT4	DAT5	DAT6	DAT7								
PORT1		DAT0	DAT1	DAT2	DAT3	DAT4	DAT5	DAT6	DAT7							
PORT2			DAT0	DAT1	DAT2	DAT3	DAT4	DAT5	DAT6	DAT7						
PORT3				DAT0	DAT1	DAT2	DAT3	DAT4	DAT5	DAT6	DAT7					

(a)

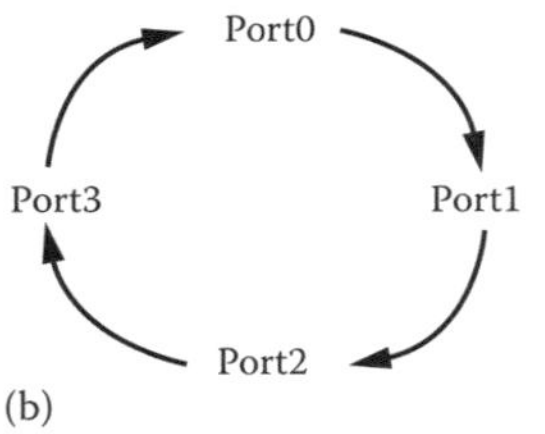

(b)

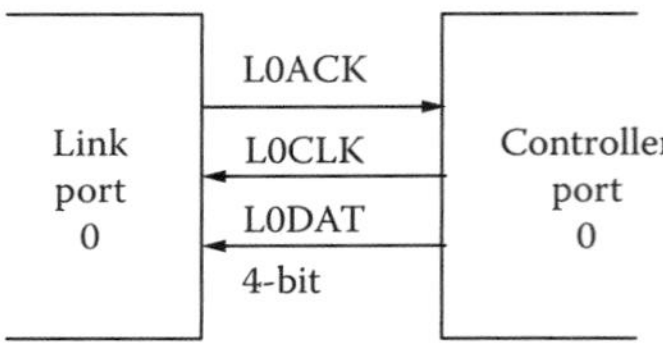

(c)

FIGURE 20.20 (a) Table showing the full pipelining, (b) token transfer, and (c) signals between a link port and the associated controller port.

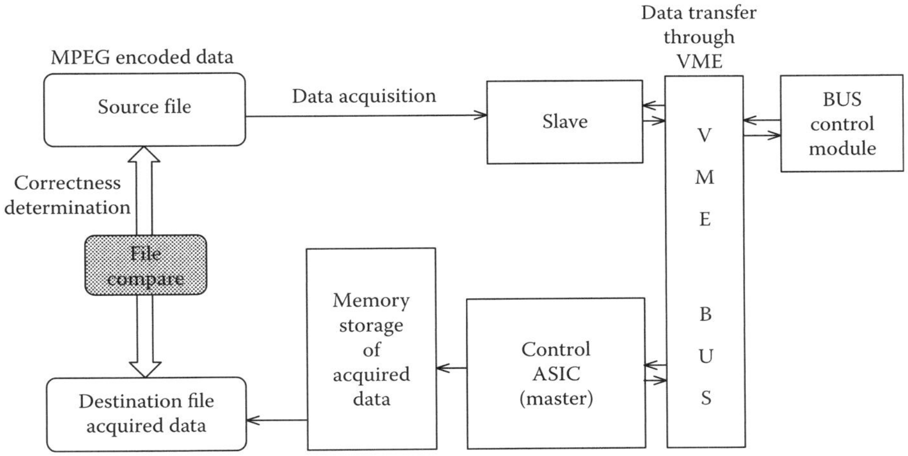

FIGURE 20.21 Data comparison mechanism.

to improve the design time. To verify the performance and correctness of the design, the comparison mechanism we used is shown in Figure 20.21. The MPEG-1 encoder data is stored in a file prior to being sent over the VME bus via master-slave handshaking. It passes through the controller design and link ports to local memory. The memory then dumps its contents to a file which is compared to the original data. The comparisons are made by reading the files in VHDL and doing a bit by bit evaluation. Any discrepancies are reported to the designer.

The total simulation time required for the transfer of a complete frame of data (28 kB) to the memory was approximately 19 min of CPU time and 1 h of wall clock time. These numbers indicate the usefulness of this abstraction level in the design hierarchy. The goal is to prove correctness of design and not simulate algorithm performance. Algorithm simulations at this level would be time prohibitive and must be moved to the performance level of abstraction.

20.10 Support for Legacy Systems

A well-defined design process capturing systems requirements through iterative design refinement improves system life cycle and supports the reengineering of existing legacy systems. With the system captured using the top-down evolving design methodology, components, boards, and/or subsystems can be replaced and redesigned from the appropriate location within the design flow. Figure 20.22 shows examples of possible scenarios. For example, if a system upgrade requires a change in a major system operating mode (e.g., search/track), then the design process can be reentered at the executable requirements or specification stage with the development of an improved algorithm. The remaining system functionality can serve as the environment testbench for the upgrade. If the system upgrade consists of a new processor design to reduce board count or packaging size, then the design flow can be reentered at the hardware virtual prototyping phase using FFMs. The improved hardware is tested using the previous models of its surrounding environment. If an architectural change using an improved interconnect technology is required, the performance modeling stage is entered. In most cases, only a portion of the entire system is affected, therefore, the remainder serves as a testbench for the upgrade. The test vectors developed in the initial design can be reused to verify the current upgrade.

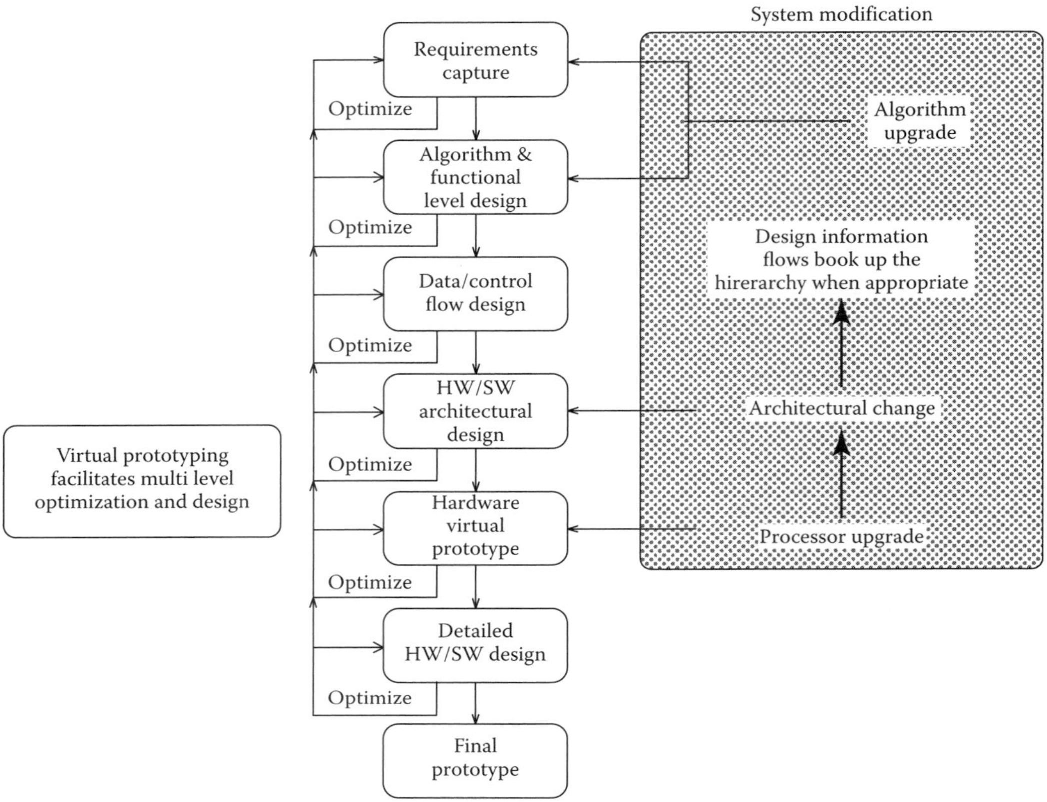

FIGURE 20.22 Reengineering of legacy systems in VHDL.

20.11 Conclusions

In this chapter, we have presented a top-down design process based on the RASSP virtual prototyping methodology. The process starts by capturing the system requirements in an executable form and through successive stages of design refinement, and ends with a detailed hardware design. VHDL models are used throughout the design process to both document the design stages and provide a common language environment for which to perform requirements simulation, architecture verification, and hardware virtual prototyping. The fidelity of the models contain the necessary information to describe the design as it develops through successive refinement and review. Examples were presented to illustrate the information captured at each stage in the process. Links between stages were described to clarify the flow of information from requirements to hardware. Case studies were referenced to point the reader to more detail on how the methodology performs in practice. Tools are being developed by RASSP participants to automate the process at each of the design stages and references are provided for more information.

Acknowledgments

This research was supported in part by DARPA ETO (F33615-94C-1493) as part of the RASSP Program 1994–1997. The authors would like to thank all the RASSP program participants for their effort in creating and demonstrating the usefulness of the methodology and its effectiveness in achieving improvements in the overall design process.

References

1. Richards, M.A., The rapid prototyping of application specific signal processors (RASSP) program: Overview and accomplishments, in *Proceedings of the First Annual RASSP Conference*, pp. 1–8, Arlington, VA, August 1994. URL: http://rassp.scra.org/public/confs/1st/papers.html#RASSP P
2. Hood, W., Hoffman M., Malley, J. et al., RASSP program overview, in *Proceedings of the Second Annual RASSP Conference*, pp. 1–18, Arlington, VA, July 24–27, 1995. URL: http://rassp.scra.org/public/confs/2nd/papers.html
3. Saultz, J.E., Lockheed Martin advanced technology laboratories RASSP second year overview, in *Proceedings of the Second Annual RASSP Conference*, pp. 19–31, Arlington, VA, July 24–27, 1995. URL: http://rassp.scra.org/public/confs/2nd/papers.html#saultz
4. Madisetti, V., Corley, J., and Shaw, G., Rapid prototyping of application-specific signal processors: Educator/facilitator current practice (1993) model and challenges, in *Proceedings of the Second Annual RASSP Conference*, Arlington, VA, July 1995. URL: http://rassp.scra.org/public/confs/2nd/papers.html#current
5. Madisetti, V.K. and Egolf, T.W., Virtual prototyping of embedded microcontroller-based DSP systems, *IEEE Micro*, 15(5), 3188–3208, October 1995.
6. ANSI/IEEE Std 1076-1993 IEEE Standard VHDL Language Reference Manual (1-55937-376-8), Order Number [SH16840].
7. Thomas, D., Adams, J., and Schmit, H., A model and methodology for hardware-software codesign, *IEEE Des. Test Comput.*, 10(3), 6–15, September 1993.
8. Kumar, S., Aylor, J., Johnson, B., and Wulf, W., A framework for hardware/software codesign, *Computer*, 26(12), 39–45, December 1993.
9. Gupta, R. and De Micheli, G., Hardware-software cosynthesis for digital systems, *IEEE Des. Test Comput.*, 10(3), 42–45, September 1993.
10. Kalavade, A. and Lee, E., A hardware-software codesign methodology for DSP applications, *IEEE Des. Test Comput.*, 10(3), 16–28, September 1993.
11. Kalavade, A. and Lee, E., A global criticality/local phase driven algorithm for the constrained hardware/software partitioning problem, in *Proceedings of the Third International Workshop on Hardware/Software Codesign*, Grenoble, France, September 1994.
12. Ismail, T. and Jerraya, A., Synthesis steps and design models for codesign, *Computer*, 28(2), 44–52, February 1995.
13. Gajski, D. and Vahid, F., Specification and design of embedded hardware-software systems, *IEEE Des. Test Comput.*, 12(1), 53–67, Spring 1995.
14. DeBardelaben, J. and Madisetti, V., Hardware/software codesign for signal processing systems—A survey and new results, in *Proceedings of the 29th Annual Asilomar Conference on Signals, Systems, and Computers*, Arlington, VA, November 1995.
15. IEEE Std 1164-1993 IEEE Standard Multivalue Logic System for VHDL Model Interoperability (Std_logic_1164) (1-55937-299-0), Order Number [SH16097].
16. Hein, C., Carpenter, T., Kalutkiewicz, P., and Madisetti, V., RASSP VHDL modeling terminology and taxonomy—Revision 1.0, in *Proceedings of the Second Annual RASSP Conference*, pp. 273–281, Arlington, VA, July 24–27, 1995. URL: http://rassp.scra.org/public/confs/2nd/papers.html#taxonomy
17. Anderson, A.H. et al., VHDL executable requirements, in *Proceedings of the First Annual RASSP Conference*, pp. 87–90, Arlington, VA, August 1994. URL: http://rassp.scra.org/public/confs/1st/papers.html#VER
18. Shaw, G.A. and Anderson A.H., Executable requirements: Opportunities and impediments, in *IEEE Proceedings of the International Conference on Acoustics, Speech, and Signal Processing*, pp. 1232–1235, Atlanta, GA, May 7–10, 1996.

19. Frank, G.A., Armstrong, J.R., and Gray, F.G., Support for model-year upgrades in VHDL test benches, in *Proceedings of the Second Annual RASSP Conference*, pp. 211–215, Arlington, VA, July 24–27, 1995. URL: http://rassp.scra.org/public/confs/2nd/papers.html
20. ISO/IEC 11172, Information technology—coding of moving picture and associated audio for digital storage media at up to about 1.5 Mbit/s, 1993.
21. Rowe, L.A., Patel, K. et al., mpeg_encode/mpeg_play, Version 1.0, available via anonymous ftp at ftp://mm-ftp.cs.berkeley.edu/pub/multimedia/mpeg/bmt1r1.tar.gz, Computer Science Department, EECS University of California at Berkeley, Berkeley, CA, May 1995.
22. ISO/IEC 13818, Coding of moving pictures and associated audio, November 1993.
23. Tanir, O. et al., A specification-driven architectural design environment, *Computer*, 26(6), 26–35, June 1995.
24. Vahid, F. et al., SpecCharts: A VHDL front-end for embedded systems, *IEEE Trans. Comput.-Aided Des. Integr. Circ. Syst.*, 14(6), 694–706, June 1995.
25. Egolf, T.W., Famorzadeh, S., and Madisetti, V.K., Fixed-point codesign in DSP, *VLSI Signal Processing Workshop*, Vol. 8, La Jolla, CA, Fall 1994.
26. Naval Research Laboratory, Processing graph method tutorial, January 8, 1990.
27. Robbins, C.R., Autocoding in Lockheed Martin ATL-camden RASSP hardware/software codesign, in *Proceedings of the Second Annual RASSP Conference*, pp. 129–133, Arlington, VA, July 24–27, 1995. URL: http://rassp.scra.org/public/confs/2nd/papers.html
28. Robbins, C.R., Autocoding: An enabling technology for rapid prototyping, in *IEEE Proceedings of the International Conference on Acoustics, Speech, and Signal Processing*, pp. 1260–1263, Atlanta, GA, May 7–10, 1996. URL: http://rassp.scra.org/public/confs/2nd/papers.html
29. System-Level Design Methodology for Embedded Signal Processors, URL: http://ptolemy.eecs.berkeley. edu/ptolemyrassp.html
30. Publications of the DSP Design Group and the Ptolemy Project, URL: http://ptolemy.eecs.berkeley.edu/papers/publications.html/index.html
31. Boehm, B., *Software Engineering Economics*, Prentice-Hall, Englewood Cliffs, NJ, 1981.
32. Madisetti, V. and Egolf, T., Virtual prototyping of embedded microcontroller-based DSP systems, *IEEE Micro*, 15(5), October 1995.
33. U.S. Air Force Analysis Agency, *REVIC Software Cost Estimating Model User's Manual Version 9.2*, December 1994.
34. Fey, C., Custom LSI/VLSI chip design productivity, *IEEE J. Solid State Circ.*, sc-20(2), 216–222, April 1985.
35. Paraskevopoulos, D. and Fey, C., Studies in LSI technology economics III: Design schedules for application-specific integrated circuits, *IEEE J. Solid-State Circ.*, sc-22(2), 223–229, April 1987.
36. Liu, J., Detailed model shows FPGAs' true costs, *EDN*, 153–158, May 11, 1995.
37. Brooke, A., Kendrick, D., and Meeraus, A., *Release 2.25 GAMS: A User's Guide*, Boyd & Fraser, Danvers, MA, 1992.
38. Oral, M. and Kettani, O., A linearization procedure for quadratic and cubic mixed-integer problems, *Oper. Res.*, 40(1), 109–116, 1992.
39. Rose, F., Steeves, T., and Carpenter, T., VHDL performance modeling, in *Proceedings of the First Annual RASSP Conference*, pp. 60–70, Arlington, VA, August 1994. URL: http://rassp.scra.org/public/confs/1st /papers.html#VHDL P
40. Hein, C. and Nasoff, D., VHDL-based performance modeling and virtual prototyping, in *Proceedings of the Second Annual RASSP Conference*, pp. 87–94, Arlington, VA, July 24–27, 1995. URL: http://rassp.scra.org/public/confs/2nd/papers.html
41. Steeves, T., Rose, F., Carpenter, T., Shackleton, J., and von der Hoff, O., Evaluating distributed multiprocessor designs, in *Proceedings of the Second Annual RASSP Conference*, pp. 95–101, Arlington, VA, July 24–27, 1995. URL: http://rassp.scra.org/public/confs/2nd/papers.html

42. Commissariat, H., Gray, F., Armstrong, J., and Frank, G., Developing re-usable performance models for rapid evaluation of computer architectures running DSP algorithms, in *Proceedings of the Second Annual RASSP Conference*, pp. 103–108, Arlington, VA, July 24–27, 1995. URL: http://rassp.scra.org/public/confs/2nd/papers.html
43. Athanas, P.M. and Abbott, A.L., Real-time image processing on a custom computing platform, *Computer*, 28(2), 16–24, February 1995.
44. Madisetti, V.K., *VLSI Digital Signal Processors: An Introduction to Rapid Prototyping and Design Synthesis*, IEEE Press, Piscataway, NJ, 1995.
45. Paulson, R.H., Kindling: A RASSP application case study, in *Proceedings of the Second Annual RASSP Conference*, pp. 79–85, Arlington, VA, July 24–27, 1995. URL: http://rassp.scra.org/public/confs/2nd/papers.html
46. Vahey, M. et al., Real time IRST development using RASSP methodology and Process, in *Proceedings of the Second Annual RASSP Conference*, pp. 45–51, Arlington, VA, July 24–27, 1995. URL: http://rassp.scra.org/public/confs/2nd/papers.html
47. Egolf, T., Madisetti, V., Famorzadeh, S., and Kalutkiewicz, P., Experiences with VHDL models of COTS RISC processors in virtual prototyping for complex systems synthesis, in *Proceedings of the VHDL International Users' Forum (VIUF)*, San Diego, CA, Spring 1995.
48. Rundquist, E.A., RASSP benchmark 1: Virtual prototyping of a synthetic aperture radar processor, in *Proceedings of the Second Annual RASSP Conference*, pp. 169–175, Arlington, VA, July 24–27, 1995. URL: http://rassp.scra.org/public/ confs/2nd/papers.html

21

Baseband Processing Architectures for SDR

Yuan Lin
University of Michigan at Ann Arbor

Mark Woh
University of Michigan at Ann Arbor

Sangwon Seo
University of Michigan at Ann Arbor

Chaitali Chakrabarti
Arizona State University

Scott Mahlke
University of Michigan at Ann Arbor

Trevor Mudge
University of Michigan at Ann Arbor

21.1 Introduction 21-1
21.2 SDR Overview 21-3
21.3 Workload Profiling and Characterization 21-4
W-CDMA Physical Layer Processing • W-CDMA Workload Profiling
21.4 Architecture Design Trade-Offs 21-7
8 and 16 Bit Fixed-Point Operations • Vector-Based Arithmetic Computations • Control Plane versus Data Plane • Scratchpad Memory versus Cache • Algorithm-Specific ASIC Accelerators
21.5 Baseband Processor Architectures 21-9
SODA Processor • ARM Ardbeg Processor • Other SIMD-Based Architectures • Reconfigurable Architectures
21.6 Cognitive Radio 21-16
21.7 Conclusion 21-16
References 21-17

21.1 Introduction

Wireless communication has become one of the dominating applications in today's world. Mobile communication devices are the largest consumer electronic group in terms of volume. In 2007, there was an estimated 3.3 billion mobile telephone subscriptions. This number is roughly half of the world's population. Applications like web browsing, video streaming, e-mail, and video conferencing have all become key applications for mobile devices. As technology becomes more advanced, users will require more functionality from their mobile devices and more bandwidth to support them. Furthermore, in recent years, we have seen the emergence of an increasing number of wireless protocols that are applicable to different types of networks. Figure 21.1 lists some of these wireless protocols and their application domains, ranging from the home and office WiFi network to the citywide cellular networks. The next-generation mobile devices are expected to enable users to connect to information ubiquitously from every corner of the world.

One of the key challenges in realizing ubiquitous communication is the seamless integration and utilization of multiple existing and future wireless communication networks. In many current wireless communication solutions, the physical layer of the protocols is implemented with nonprogrammable

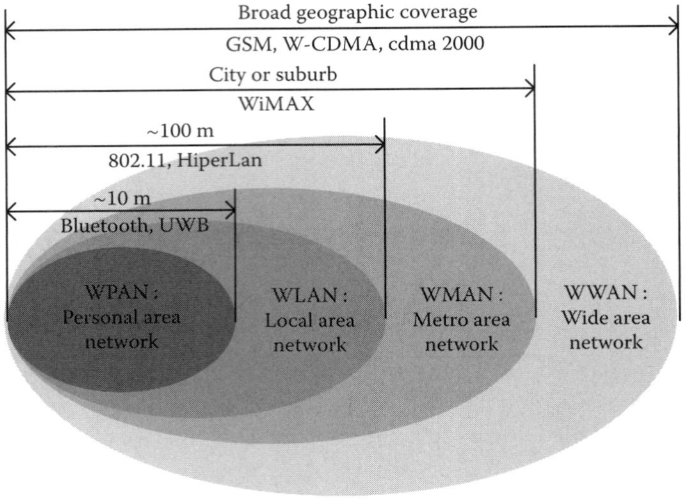

FIGURE 21.1 Categories of wireless networks.

application-specific integrated circuit (ASIC) processors. The communication device consists of multiple processors, one for each wireless protocol. Such a solution is not scalable and is infeasible in the long run. Software-defined radio (SDR) promises to deliver a cost-effective and flexible solution by implementing a wide variety of wireless protocols in software, and running them on the same hardware platform. A software solution offers many potential advantages, including but not limited to the following:

- A programmable SDR processor would allow multimode operation, running different protocols depending on the available wireless network, global system for mobile communications (GSM) in Europe, code division multiple access (CDMA) in the United States, and some parts of Asia, and 802.11 in coffee shops. This is possible with less hardware than custom implementations.
- A protocol implementation's time to market would be shorter because it would reuse the hardware. The hardware integration and software development tasks would progress in parallel.
- Prototyping and bug fixes would be possible for next-generation protocols on existing silicon through software changes. The use of a programmable solution would support the continuing evolution of specifications; after the chipset's manufacture, developers could deploy algorithmic improvements by changing the software without redesign.
- Chip volumes would be higher because the same chip would support multiple protocols without requiring hardware changes.

Designing a SDR processor for mobile communication devices must address two key challenges—meeting the computational requirements of wireless protocols while operating under the power budget of a mobile device. The operation throughput requirements of current third-generation (3G) wireless protocols are already an order of magnitude higher than the capabilities of modern digital signal processing (DSP) processors. This gap is likely to grow in the future. Figure 21.2 shows the computation and power demands of a typical 3G wireless protocol. Although most DSP processors operate at an efficiency of approximately 10 million operations per second (Mops) per milliwatt (mW), the typical wireless protocol requires 100 Mops/mW.

This chapter presents the challenges and trade-offs in designing architectures for baseband processing in mobile communication devices. It gives an overview of baseband processing in SDR, followed by workload and performance analysis of a representative protocol. Next it describes the architectural features of two low-power baseband architectures, signal processing on demand architecture (SODA), and Ardbeg, followed by brief descriptions of other representative processor prototypes. It briefly introduces cognitive radio (CR) as the next challenge and concludes the chapter.

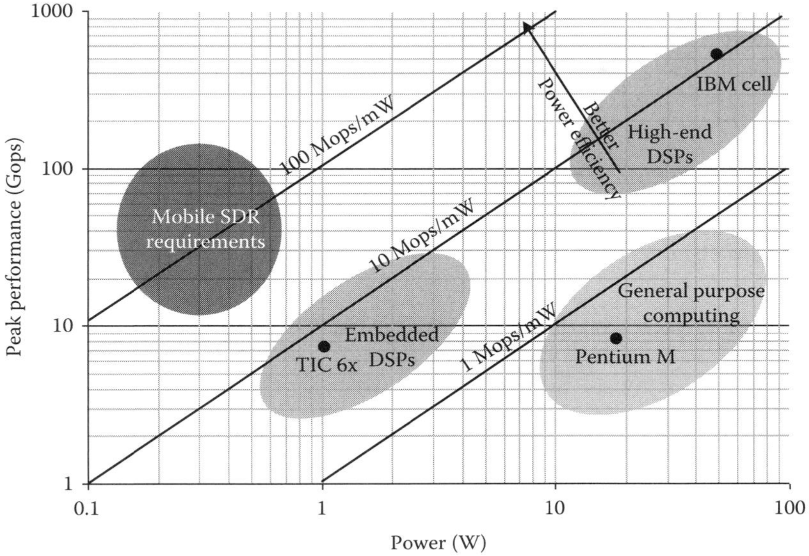

FIGURE 21.2 Throughput and power requirements of typical 3G wireless protocols. The results are calculated for 16 bit fixed-point operations.

21.2 SDR Overview

SDR promises to solve the problems of supporting multiple wireless protocols and addresses future challenges. The SDR forum, which is a consortium of service operators, designers, and system integrators, defines SDR as

> A collection of hardware and software technologies that enable reconfigurable system architectures for wireless networks and user terminals. SDR provides an efficient and comparatively inexpensive solution to the problem of building multimode, multiband, multifunctional wireless devices that can be enhanced using software upgrades. As such, SDR can really be considered an enabling technology that is applicable across a wide range of areas within the wireless industry.

Figure 21.3 shows the architecture for a typical 3G cellular phone. The architecture includes four major blocks: analog front-end, digital baseband, protocol processor, and application processor. The physical

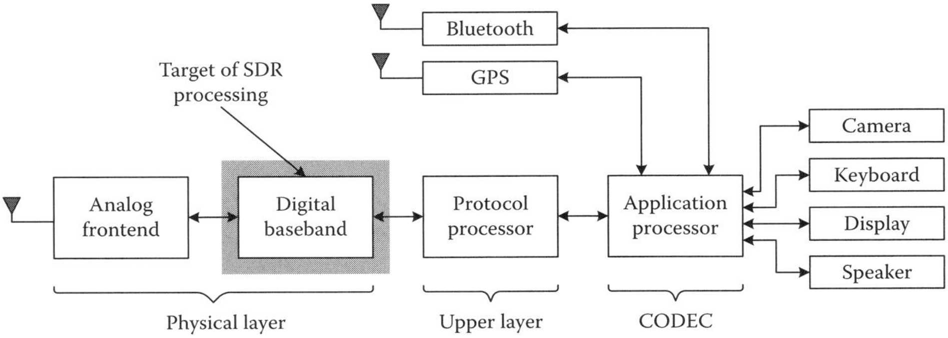

FIGURE 21.3 Architecture of a 3G cellular phone.

layer of wireless protocols includes both the analog front-end and the digital baseband. The analog front-end is usually implemented with analog ASICs. The digital baseband block performs the remaining physical layer operations and is also typically implemented with ASICs. The upper layers are implemented by the protocol processor and application processor, which are usually system on chips (SoCs) and consist of general purpose embedded DSP processors.

The objective of SDR is to replace the baseband ASICs with a programmable hardware platform, and implement the baseband processing in software. Designing programmable analog front-ends is quite a challenge and is beyond the scope of this chapter. Here, we focus on design of programmable digital baseband processing engines for SDR.

21.3 Workload Profiling and Characterization

21.3.1 W-CDMA Physical Layer Processing

We select the wide-band code division multiple access (W-CDMA) protocol as a representative wireless workload case study for designing the SDR processor. This section provides a brief summary of its algorithms and characteristics. A more detailed analysis can be found in [14].

The W-CDMA system is one of the dominant 3G wireless communication networks where the goal is multimedia service including video telephony on a wireless link [12]. It improves over prior cellular protocols by increasing the data rate from 64 Kbps to 2 Mbps. The protocol stack of the W-CDMA system consists of several layers. At the bottom of the stack is the physical layer which is responsible for overcoming errors induced by an unreliable wireless link. The next layer is the medium access control (MAC) layer which resolves contention in shared radio resources. The upper-layer protocols including MAC are implemented on a general purpose processor due to their relatively low computation requirements. In this section, we focus on the computation model of the W-CDMA physical layer.

Figure 21.4 shows a high-level block diagram of the digital processing in the W-CDMA physical layer. It shows that the physical layer contains a set of disparate DSP kernels that work together as one system. There are four major components: filtering, modulation, channel estimation, and error correction.

21.3.1.1 Filtering

Filtering algorithms are used to suppress signals transmitted outside of the allowed frequency band so that interference with other frequency bands are minimized. The finite impulse response (FIR) filter operations in W-CDMA can be very easily parallelized.

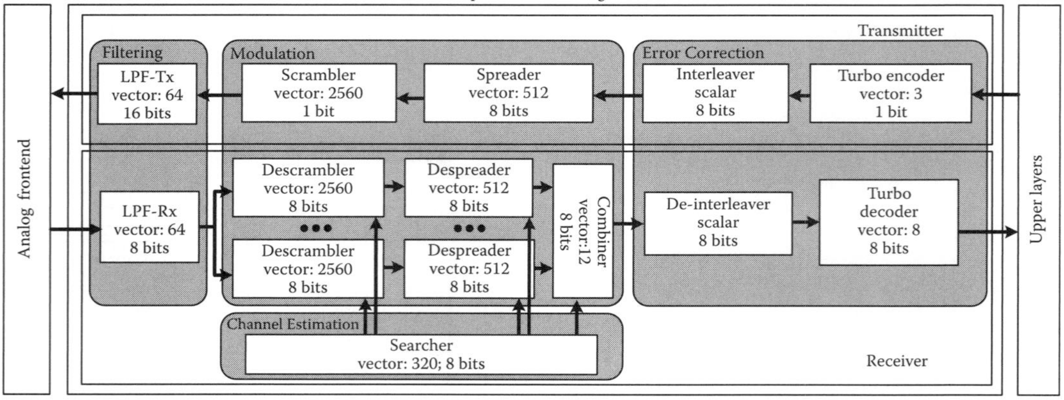

FIGURE 21.4 Physical layer operation of W-CDMA wireless protocol. Each block includes the algorithm's name, vector or scalar computation, vector width, and the data precision. The algorithms are grouped into four categories, shown in shaded boxes: filtering, modulation, channel estimation, and error correction.

21.3.1.2 Modulation

Modulation algorithms map source information onto the signal waveforms of the transmitter, and the receiver demodulates the signal waveforms back into source information. Two sets of codes are used for modulation: channelization codes and scrambling codes. Channelization codes are used so that the same radio medium can be used to transmit multiple different signal streams. Scrambling codes are used to extract the signal of a specific terminal among many transmitting terminals. On the receiver side, despreader is used to decode the channelization codes and descrambler is used to decode the scrambling codes. Demodulation requires the transmitter and the receiver to be perfectly synchronized. However, radio transmission suffers from multipath fading effect, where multiple delayed versions of the same signal stream are received due to environment interference. A searcher is used to find the synchronization point of each of the delayed signal streams and each of these delayed signals is decoded with its own despreader and descrambler. The decoded output of the despreader/descrambler pairs are then combined together as the demodulated output.

21.3.1.3 Channel Estimation

Channel estimation algorithms calculate the channel conditions to synchronize the two communicating terminals to ensure lockstep communication between the sender and the receiver. W-CDMA uses a searcher as its channel estimation algorithm. Searcher is called once per W-CDMA frame. There are two types of searchers—full searcher and quick searcher. In a group of eight frames, the full searcher is used for the first frame, and the quick searcher is used for the remaining seven frames. Both types of searchers consist of four steps: correlation, filtering out high-frequency noise, detecting peaks, and global peak detection.

21.3.1.4 Error Control Coding Algorithms

Error control coding (ECC) algorithms are used to combat noisy channel conditions. The sender encodes the original data sequence with a coding scheme that inserts systematic redundancies into the output, which is decoded by the receiver to find the most likely original data sequence. Two types of ECC algorithms are used—convolutional coding and Turbo coding. Turbo coding is used for the 2 Mbps data channel, while convolutional coding is used for all the other channels. For decoding, the Viterbi decoder is used for convolutional codes and the Turbo decoder is used for Turbo codes. Turbo decoding is usually the most computationally intensive algorithm in baseband processing. The corresponding decoder consists of two component decoders that are typically of type software output Viterbi algorithm (SOVA) or based on maximum a posteriori (MAP) and connected together by interleavers. The interleaving pattern is specified by the W-CDMA standard, but the choice of component decoder is left to the designers.

21.3.2 W-CDMA Workload Profiling

Table 21.1 shows the result of workload profiling for W-CDMA in active mode with a throughput of 2 Mbps [14]. The first column lists the W-CDMA algorithms. The second column lists the corresponding configurations for each of the algorithms. The third and fourth column lists the vector computation information for the algorithms. The fifth column lists the data precision width. The last column shows the peak workload of the algorithms. It is the minimum performance needed to sustain 2 Mbps throughput, under the worst wireless channel conditions. For this analysis the W-CDMA model was compiled with an Alpha gcc compiler, and executed on M5 architectural simulator [3]. The peak workload of each algorithm was calculated by dividing the maximum number of processing cycles by its execution time.

The analysis shows that there are a set of key DSP algorithms that are responsible for the majority of the computation. These algorithms include the FIR filter, searcher, Turbo decoder, descrambler, and despreader. The workloads of Viterbi and Turbo decoder require further verification because their

TABLE 21.1 Workload Analysis Result of W-CDMA Physical Layer Processing

Algorithms	Configurations	Vector Comp.	Vector Length	Bit Width	Comp. Mcycles/s
		W-CDMA (2 Mbps)			
Scrambler	Defined in W-CDMA standard	Yes	2,560	1, 1	240
Descrambler[a]	12 fingers, 3 base stations	Yes	2,560	1, 8	2,600
Spreader	Spreading factor = 4	Yes	512	8	300
Despreader[a]	12 fingers, 3 base stations	Yes	512	8	3,600
PN code (Rx)	3 base stations	No	1	8	30
PN code (Tx)	Defined in W-CDMA standard	No	1	8	10
Combiner[a]	2 Mbps data rate	Partial	12	8	100
FIR (Tx)	4 filters × 65 coeff × 3.84 Msps	Yes	64	1, 16	7,900
FIR (Rx)	2 filters × 65 coeff × 7.68 Msps	Yes	64	8, 8	3,900
Searcher[a]	3 base stations, 320 windows	No	320	1, 8	26,500
Interleaver	1 frame	No	1	8	10
Deinterleaver	1 frame	Partial	1	8	10
Turbo Enc.	$K=4$	Yes	3	1, 1	100
Turbo Dec.[a]	$K=4$, 5 iterations	Yes	8	8, 8	17,500

Note: "Vector Comp." indicates whether the algorithm contains vector-based arithmetic operations. "Vector Width" lists the native computation vector width. "Bit Width" lists the data precision width. "Comp. Mcycle/s" lists the peak workload of running the algorithm on a general purpose processor.

[a] These algorithms have dynamically changing workloads that are dependent on channel conditions.

processing times are not fixed. They are based on estimates that are derived from the protocol standard specifications for the different transmission channels that use these decoders.

21.3.2.1 Instruction Type Breakdown

Figure 21.5 shows the instruction type breakdown for the W-CDMA physical layer and their weighted average for the case when the terminal is in the active state and has peak workload [14]. Instructions are

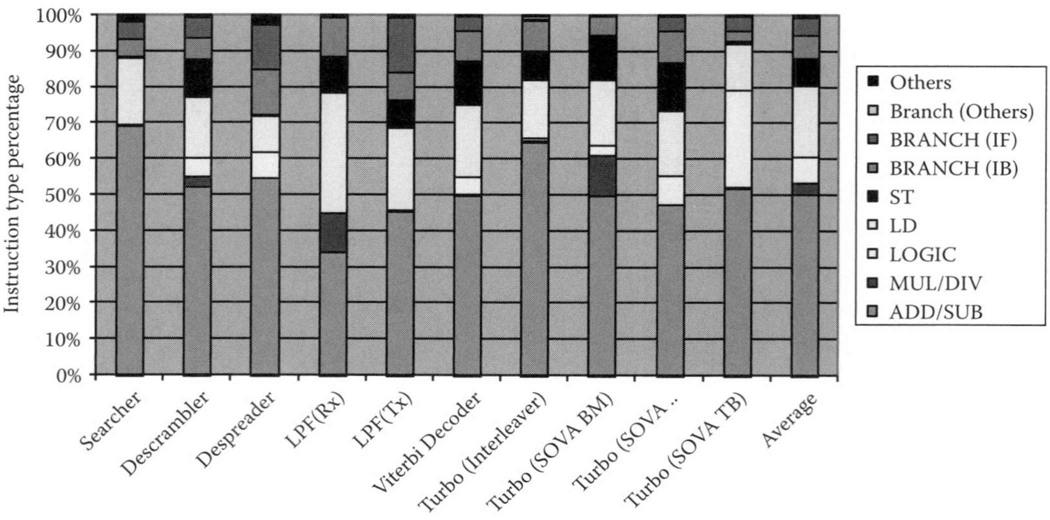

FIGURE 21.5 W-CDMA instruction type breakdown.

grouped into seven categories: add/sub; multiply/divide; logic; load; store; branches; and miscellaneous instructions. All computations are done using fixed-point arithmetic. The first thing to notice is the high percentage of add/subtract instructions. They account for almost 50% of all instructions, with the searcher at about 70%. This is because the searcher's core operation is the inner product of two vectors, and the multiplication here can be simplified into a conditional complement which can again be implemented by a subtraction. Other hot spots, like the Turbo decoder, also have a large number of addition operations. The second thing to notice is the lack of multiplications/divisions (<3% on average). This is because the multiplications of major algorithms are simplified into logical or arithmetic operations as discussed earlier. The multiplication of the combiner and Turbo encoder is not as significant because their workload is very small. One exception is multiplication in the low pass filter for the receiver (LPF-Rx). Figure 21.5 also shows that the number of load/store operations is significant. This is because most algorithms consist of loading two operands and storing the operation result. Results also show that the portion of branch operations is about 10% on average. Most frequent branch patterns are loops with a fixed number of iterations corresponding to a vector size, and a conditional operation on vector variables. There are a few while or do-while loops, and most loops are 1–2-levels deep.

21.3.2.2 Parallelism in the Protocol

To meet the real-time W-CDMA performance requirement in software, the inherent algorithmic parallelism must be exploited. Columns 3 and 4 of Table 21.1 show the potential parallelism that can be exploited either through data-level parallelism (DLP) or thread-level parallelism (TLP). Most of the algorithms operate on one-dimensional vectors. Therefore, we define parallelism as the maximum vector width for these algorithms' arithmetic operations, which is shown in column 4.

From this result, we can see that the searcher, filter, scrambler, and descrambler all contain very wide vector operations. They also have a fair amount of TLP. For instance, the searcher operation can be executed as 5120 concurrent threads. For the case of scrambler and descrambler, TLP can be obtained by bisecting a wide vector into smaller ones. Although sliced vectors are not perfectly uncorrelated, we can execute the smaller vector operations with negligible dependency. Another example is the Turbo decoder, which when implemented in a straightforward manner, contains limited vector and TLP. This can be mitigated by use of sliding window techniques where multiple blocks of data can be processed at the same time.

21.3.2.3 Stream Computation

Another important characteristic of the W-CDMA protocol is that it is a streaming computation system. Here, multiple DSP kernel algorithms are connected together in feed-forward pipelines and data streams through these pipelines sequentially. There is very little data temporal locality, which means that data do not get reused. Thus, cache structures provide little additional benefits, in terms of power and performance, over software-controlled scratchpad memories.

21.4 Architecture Design Trade-Offs

This section describes some of the key challenges and trade-offs in designing processor architectures for supporting mobile SDR.

21.4.1 8 and 16 Bit Fixed-Point Operations

In W-CDMA wireless protocol, the majority of the algorithms operate on 1 to 8 bit data, with some algorithms operating on 16 bit data. In 802.11a wireless protocol, the majority of the DSP algorithms operate on 16 bit fixed-point data, with a few algorithms operating on 8 bit data. A few DSP algorithms, such as fast Fourier transform (FFT) for the next-generation 4G wireless protocols, may require 24 bit or 32 bit fixed-point precision. There are almost no wireless protocols that require floating-point support.

21.4.2 Vector-Based Arithmetic Computations

Computations in a typical DSP application consists of vector and matrix arithmetic operations. This is true for wireless protocols as well. However, it is important to note that wireless protocols' computations are mostly on one-dimensional vectors.

DSP computations on vectors and matrices can be broken into two key components: data computation and data rearrangement. Data computation includes addition, multiplication, and other arithmetic and logic operations. Data rearrangement reorders the elements in vectors and matrices. In most DSP computations, it is not the data computation, but the data rearrangement that is the real performance bottleneck. Data computation limitations can always be addressed by adding more hardware computation units (such as adders and multipliers). However, there is no single common hardware solution that can accelerate all rearrangement operations for any arbitrary-sized vectors and matrices. For example, FFT requires a butterfly operation, where pairs of data values in a vector are swapped. In comparison, the ECC decoder may require an interleaver that transposes a matrix by blocks. The optimal hardware accelerators for these two patterns are quite different. An efficient design must exploit the specific data rearrangement patterns of the kernel algorithms. Since most of the vector arrangement patterns do not change during run-time, the interconnects between the arithmetic and logic units (ALUs) and memories can be tailored to these algorithms.

21.4.3 Control Plane versus Data Plane

Complete software implementations of wireless protocols usually require two separate steps: (1) implementation of the DSP algorithms; and (2) implementation of the protocol system. The DSP algorithms are computational intensive kernels that have relatively simple data-independent control structures. The protocol system has relatively light computation load, but complicated data-dependent control behavior. Therefore, most commercial SDR solutions include a two-tiered architecture: a set of data processors that is responsible for heavy duty data processing; and a control processor that handles the system operations, and manages the data processors through remote-procedure-calls and DMA operations.

21.4.4 Scratchpad Memory versus Cache

In addition to data computation, wireless protocols must also handle the data communication between algorithms. These interalgorithm communications are usually implemented by data streaming buffers. While the processors operate on the current data, the next batch of data can be streamed between the memories and register files (RFs). There is very little data locality between the algorithms. This suggests that a scratchpad memory is better suited for SDR than a cache-based memory. Scratchpad memories are easier to implement and consume less power. However, they do not have the hardware support for managing data consistencies, which means that memory management must be performed explicitly by software. This adds extra burden and complexity for the programmers and compilers. In addition, executing memory management instructions prevents processors from doing DSP computations, which may significantly degrade the overall performance. Therefore, many scratchpad memory systems provide direct memory access (DMA) units. DMAs can move large blocks of data between the memories without interrupting the processors. However, these devices must also be explicitly managed by software.

21.4.5 Algorithm-Specific ASIC Accelerators

SDR is a solution where the entire wireless protocol is implemented in software. However, for a SDR solution to be competitive with existing ASIC solutions, both in terms of performance and power, ASIC accelerators are sometimes used for ECC and filtering. This is because these two groups of algorithms are often the most computationally intensive algorithms in a wireless protocol, and different wireless protocols often use very similar algorithms for filtering and ECC. Previous studies have shown that there is an approximately 4–5× energy efficiency between an ASIC and a programmable solution for these two groups of algorithms. Consequently, there have been multiple efforts both in the academia and the industry on developing combined Viterbi and Turbo ASIC accelerators. Texas Instrument already sells a Viterbi accelerator, as a part of a SoC package, which can support configurations needed for multiple wireless protocols.

21.5 Baseband Processor Architectures

Many baseband processing architectures have been proposed in the last few years. They can be broadly categorized into single-instruction, multiple data (SIMD)-based architectures and reconfigurable architectures [20]. SIMD-based architectures usually consist of one or few high-performance DSP processors that are typically connected together through a shared bus, and managed through a general purpose control processor. Some SIMD-based architectures also have a shared global memory connected to the bus. Both SODA [15] and Ardbeg [23], that we will describe in details, fall under the SIMD-based architecture category. Reconfigurable architectures, on the other hand, are usually made up of many simpler processing elements (PEs). Depending on the particular design, these PEs range from fine-grain ALU units to coarse-grain ASICs. The PEs are usually connected through a reconfigurable fabric. Compared with SIMD-based design, reconfigurable architectures are more flexible at the cost of higher power. In this section, we first describe SODA and Ardbeg processors followed by descriptions of other SIMD and reconfigurable SDR architectures.

21.5.1 SODA Processor

SODA processor architecture is an academic research prototype for mobile SDR [15]. It is a SIMD-based DSP architecture designed to meet both the performance and power requirements for two representative protocols, W-CDMA and IEEE 802.11a. The SODA multiprocessor architecture is shown in Figure 21.6. It consists of multiple PEs, a scalar control processor, and global scratchpad memory, all connected through a shared bus. Each SODA PE consists of five major components: (1) an SIMD pipeline for supporting vector operations; (2) a scalar pipeline for sequential operations; (3) two local scratchpad memories for the SIMD pipeline and the scalar pipeline; (4) an address generation unit (AGU) pipeline for providing the addresses for local memory access; and (5) a programmable DMA unit to transfer data between memories and interface with the outside system. The SIMD pipeline, scalar pipeline, and the AGU pipeline execute in very long instruction word (VLIW)-styled lockstep, controlled with one program counter (PC). The DMA unit has its own PC, its main purpose is to perform memory transfers and data rearrangement. It is also the only unit that can initiate memory access with the global scratchpad memory.

The SIMD pipeline consists of a 32-lane 16 bit datapath, with 32 arithmetic units working in lockstep. It is designed to handle computationally intensive DSP algorithms. Each datapath includes a 2 read-port, 1 write-port 16 entry register file, and one 16 bit ALU with multiplier. The multiplier takes two execution cycles when running at the targeted 400 MHz. Intraprocessor data movements are supported through the SIMD shuffle network (SSN), as shown in Figure 21.7. The SSN consists of a shuffle exchange (SE) network, an inverse shuffle exchange (ISE) network, and a feedback path. By including both the SE and ISE networks, the number of iterations can be reduced. In addition to the SSN network, a straight-through connection is also provided for data that does not need to be permutated.

The SIMD pipeline can take one of its source operands from the scalar pipeline. This is done through the scalar-to-vector (STV) registers, shown in the SIMD pipeline portion of Figure 21.6. The STV contains four 16 bit registers, which only the scalar pipeline can write, and only the SIMD pipeline can read. The SIMD pipeline can read 1, 2, or all 4 STV register values and replicate them into

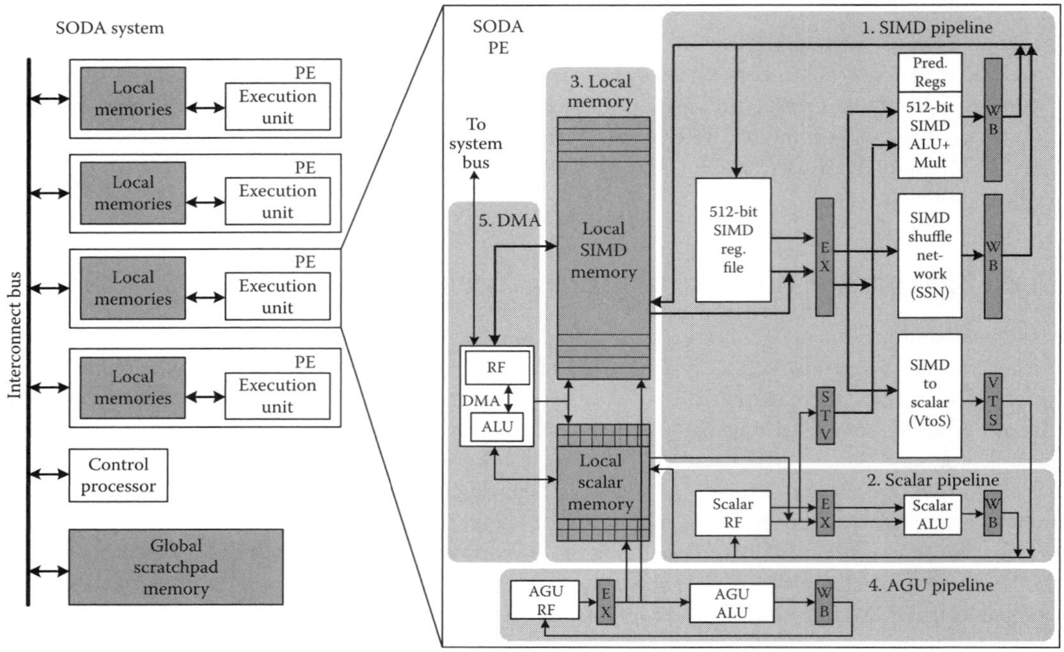

FIGURE 21.6 SODA architecture. The system consists of four data PEs, one control processor, and global scratchpad memory, all connected through a shared bus. Each PE consists of a 32-lane 16 bit SIMD pipeline, a 16 bit scalar pipeline, two local scratchpad memories, an AGU for calculating memory addresses, and a DMA unit for interprocessor data transfer.

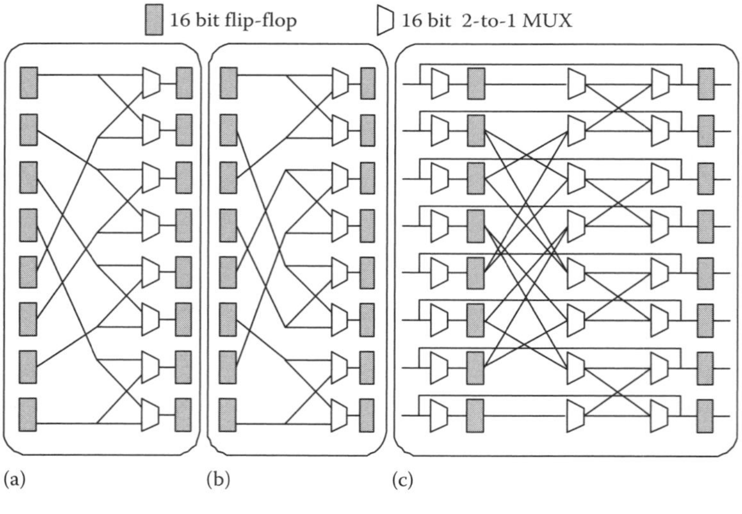

FIGURE 21.7 8-Wide SSN. (a) 8-Wide shuffle exchange network; (b) 8-wide inverse shuffle exchange network; (c) 8-wide SSN with shuffle exchange, inverse shuffle exchange and feedback path.

32-element SIMD vectors. SIMD-to-scalar operations transfer values from the SIMD pipeline into the scalar pipeline. This is done through the vector-to-scalar (VTS) registers, shown in Figure 21.6. There are several SIMD reduction operations that are supported in SODA, including vector summation, finding the minimum and the maximum.

The DMA controller is responsible for transferring data between memories. It is the only component in the processor that can access the SIMD, scalar and global memories. The DMA is also implemented as a slave device controlled by the scalar pipeline. However, it can also execute its own instructions on its internal register file and ALU, similar to the scalar pipeline. This gives the DMA the ability to access the memory in a wide variety of application-specific patterns without assistance from the master processor.

21.5.1.1 Arithmetic Data Precision

SODA PE only provides support for 8 and 16 bit fixed-point operations. This is because for both W-CDMA and 802.11a wireless protocols, the majority of the DSP algorithms, operate on 8 bit or 16 bit fixed-point data. Each lane in the SODA PE supports 16 bit fixed-point arithmetic operations. The AGU registers are 12 bit, but only support 8 bit addition and subtraction. This is because AGU is used for software management of data buffers, in which 8 bit are sufficient. The higher 4 bit are used to address different PEs, as well as different memory buffers within PEs.

21.5.1.2 Vector Permutation Operations

With SODA's SSN network (shown in Figure 21.7), any 32-wide vector permutation can be performed with a maximum of nine cycles. Combining with predicated move operations, the SSN network can support any vector length permutation. However, for every 32-wide vector permutation, one additional instruction must be used to setup the permutation network. This is because each multiplexor (MUX) within the shuffle network requires its own control bit. Each iteration through the SSN requires 64 bit. For a maximum of nine iterations, this requires 576 bit of control information. The SSN is not very efficient if there are many permutation patterns that are used frequently. However, the majority of the algorithms in SDR only use one or a few shuffle patterns, which makes the network setup overhead not significant.

21.5.1.3 Performance

For W-CDMA and 802.11a, the SODA architecture achieves large speedups over a general purpose Alpha processor. For example, W-CDMA's searcher algorithm requires 26.5 giga operations per second (Gops) on the general purpose processor and requires only 200 Mops on SODA. The performance improvements are mainly due to SODA's wide-SIMD execution.

The Register Transfer Language (RTL) Verilog model of SODA was synthesized in TSMC 180 nm technology. The results show that for a clock frequency of 400 MHz; SODA consumes 2.95 W for W-CDMA 2 Mbps system and 3.2 W for 802.11a 24 Mbps system. However, with technology scaling, the power numbers are expected to reduce to acceptable levels such as 450 mW for 90 nm technology and 250 mW for 65 nm technology.

21.5.2 ARM Ardbeg Processor

Ardbeg is a commercial prototype based on the SODA architecture [23]. The Ardbeg system architecture is shown in Figure 21.8. It consists of two PEs, each running at 350 MHz in 90 nm technology, an ARM general purpose controller, and a global scratchpad memory. In addition, it includes an accelerator dedicated to Turbo decoding. Each Ardbeg PE supports 512 bit SIMD data, and can operate with 8, 16, and 32 bit fixed-point and 16 bit block floating-point (BFP) precision. It supports VLIW execution on its SIMD pipeline, allowing a maximum of two different SIMD instructions to be issued in parallel. The instruction set for Ardbeg is derived from NEON [1].

21.5.2.1 SIMD Permutation Support

The Ardbeg PE employs a 128-lane 8 bit 7-stage Banyan shuffle network. This network is designed to support the most common vector permutation operations in a single cycle. In addition, it also provides a few key permutation patterns designed for interleaving. Interleaving is essentially a long vector permutation operation, where the vector width is far greater than the SIMD width. This is a challenge because

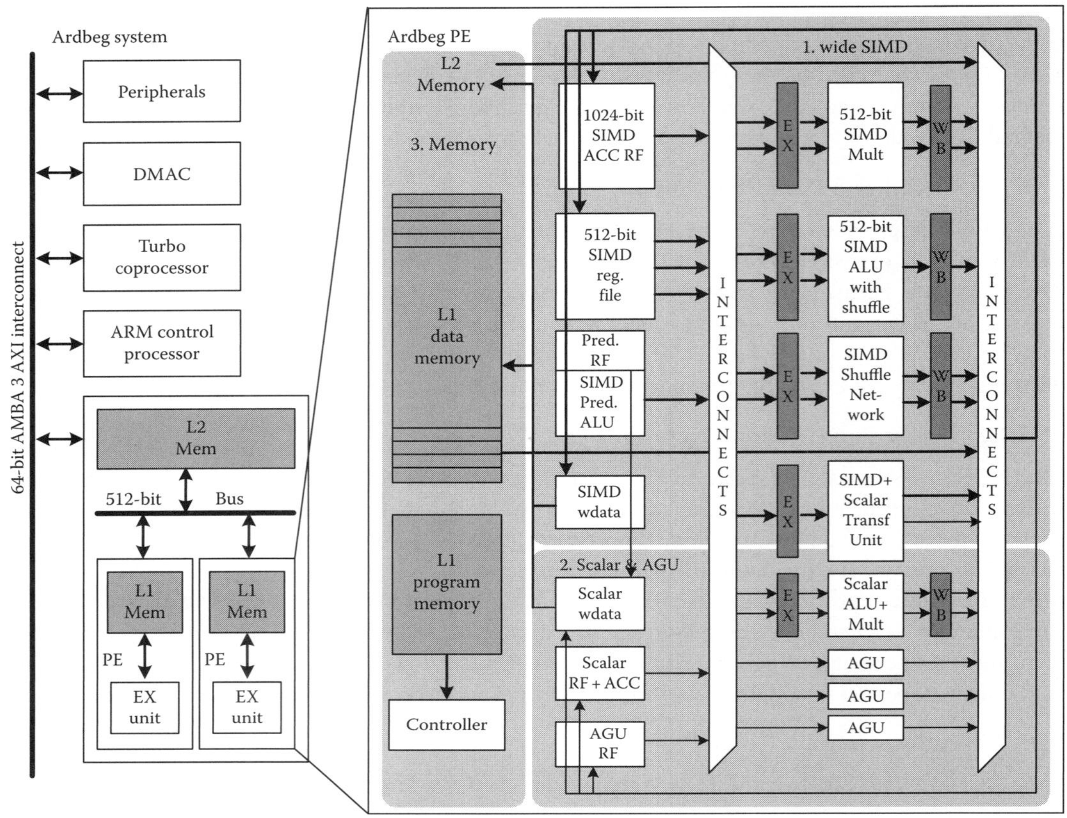

FIGURE 21.8 Ardbeg System Architecture. It consists of two data PEs, a Turbo coprocessor, an ARM control processor and peripherals. Each PE consists of a 512 bit SIMD pipeline and a 16 bit scalar pipeline.

the shuffle network can only permute vector patterns of SIMD width. However, by designing the network to support a small set of permutation patterns, the permutation time can be significantly reduced even in these cases.

21.5.2.2 Partial VLIW SIMD Execution

VLIW execution is supported in Ardbeg, but with restrictions on the combinations of instructions that can be issued in a cycle. This results in slower speedup than a full 2-issue VLIW, but provides better energy efficiency due to lesser number of SIMD register file ports. For instance, combining overlapping memory accesses with SIMD computation is beneficial because most DSP algorithms are streaming. Similarly, combining SIMD arithmetic/multiplication with SIMD-scalar transfer is beneficial for filter-based algorithms and combining SIMD multiply with move operation is beneficial for FFT-based algorithms.

21.5.2.3 Block Floating-Point Support

Contemporary wireless protocols require large-sized FFTs. BFP provides near floating-point precision without its high power and area costs. A key operation in BFP is finding the maximum value among a set of numbers. While DSP processors support this operation in software, in Ardbeg, a special instruction is implemented that finds the maximum value in a 32-lane 16 bit SIMD vector. Though BFP is currently used only for FFT computations, any algorithm which requires higher precision can utilize the BFP instruction extensions.

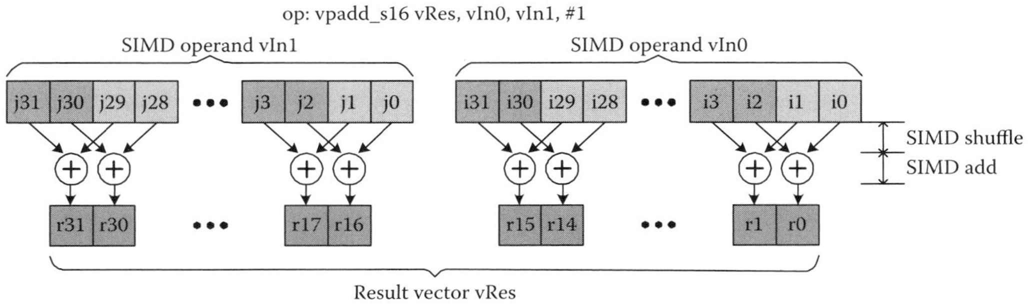

FIGURE 21.9 Ardbeg's pair-wise butterfly SIMD operation implemented using a fused permute and ALU operation. The figure shows pairs of a 2-element butterfly operation. Ardbeg supports pairs of 1-, 2-, 4-, 8-, and 16-element butterfly of 8 and 16 bit. This butterfly operation uses the inverse perfect shuffle pattern.

21.5.2.4 Fused Permute-and-ALU Operations

In many DSP algorithms it is common to first permute the vectors before performing arithmetic operations. An example is the butterfly operation in FFT, where vectors are first shuffled in a butterfly pattern before vector adds and subtracts are performed. The Ardbeg PE is optimized for these operations by including two shuffle networks. The 128-lane SSN is a separate unit that can support many different permutation patterns. In addition, a smaller 1024 bit 1-stage shuffle network is included in front of the SIMD ALU. This 1-stage shuffle network only supports inverse perfect shuffle patterns between different groups of lanes. The different pair-wise butterfly operations are shown in Figure 21.9. In each case, the shuffle and the add operations are performed in the same cycle. A 2048-point FFT is able to gain 25% speedup using fused butterfly operations.

21.5.2.5 Performance

Three different wireless communication protocols, W-CDMA, 802.11a, digital video broadcasting-terrestrial/handheld (DVB-T/H), are implemented on the Ardbeg architecture and the performance compared to SODA. The results show that Ardbeg's three enhancements—optimized wide-SIMD design such as wider SSN and reduced latency functional units (FUs), long instruction word (LIW) support for wide-SIMD, and algorithm-specific hardware accelerator—achieve large speedup ranging from 1.5× to 7× over SODA. In addition, the evaluation shows that the Ardbeg processor synthesized in 90 nm technology running at 350 MHz is able to support the 3G wireless processing within the 500 mW power budget of typical mobile devices.

21.5.3 Other SIMD-Based Architectures

In this section, we describe several other SIMD-based architectures proposed in the industry and the academia. Like SODA and Ardbeg, most of these architectures consist of one or more high-performance SIMD DSP processors which are connected via a shared bus. These architectures usually use software-managed scratchpad data memories to meet the real-time constraints. Many of them support VLIW execution by allowing concurrent memory and SIMD arithmetic operations. Some solutions also choose to incorporate accelerators for ECC algorithms, such as those based on convolutional and Turbo codes.

21.5.3.1 EVP

The embedded vector processor (EVP) [22] from NXP consists of a 16-lane 16 bit SIMD datapath, one scalar datapath, programmable memory, and VLIW controller. It allows five vector operations, four scalar operations, three address operations, and one loop control to be executed at once. The SIMD datapath consists of vector memory, 16 vector registers, load/store unit, ALU, MAC/shift unit, intravector unit, and code generation unit. The 16 bit datapath supports 8 and 32 bit data to allow word-level parallelism. It also supports data types such as complex numbers. In the SIMD datapath, the shuffle unit rearranges the elements of a single vector according to any pattern; the intravector unit supports summation, maximum, and split operations; and the code generation unit supports different CDMA-codes and cyclic redundancy checks (CRC) checks.

21.5.3.2 DXP

The Deep eXecution Processor (DXP) [13] from Icera is a 2-LIW 4-lane SIMD architecture. Its key features are deeply pipelined execution units and a programmable configuration map which holds pseudostatic configurations and constants for the execution units. In the SIMD execution datapath, SIMD ALUs are chained to exploit the characteristics of streaming data. The chained operation saves register file access power at the cost of a less flexible SIMD datapath. Icera's processors do not use any hardware accelerators.

21.5.3.3 TigerSHARC

The TigerSHARC [9] implementation, ADSP-TS001, from Analog Devices adopts several features found in general purpose processors such as a register-based load-store architecture with a static superscalar dispatch mechanism, a highly parallel short-vector-oriented memory architecture, support for multiple data types including 32 bit single-precision floating-point and 8 bit/16 bit fixed-point, parallel arithmetic instructions for two floating-point multiply-accumulate (MAC) operation or eight 16 bit MACs, 128-entry four-way set associative branch target buffer, and 128 architecturally visible, fully interlocked registers in four orthogonal RFs. The TigerSHARC architecture provides concurrent SIMD arithmetic operations by having two 4-lane SIMD computation blocks controlled with two instructions. It fetches four instructions and issues one to four instructions per clock cycle. The 128 32 bit registers are memory mapped and divided into four separate RFs of size 32×32 bit. The multiple data type supports subword parallelism in addition to inherent SIMD data parallelism.

21.5.3.4 MuSIC-1

The MuSIC-1[4] processor from Infineon consists of four SIMD core clusters, a general purpose processor, shared memory and dedicated programmable processors for FIR filter and Turbo/Viterbi decoder. Each SIMD core contains four PEs and supports special instructions and LIW features for arithmetic operations, local memory accesses, and inter-PE communications. The general purpose processor runs control programs to provide the PE controller with instruction addresses. The code and data are stored in an external memory, and the on-chip memory consists of multiple banks to support simultaneous accesses.

21.5.3.5 Sandblaster

The Sandblaster [11] architecture from Sandbridge Technology is an example of a multithreaded SIMD vector architecture. It consists of a reduced instruction set computer (RISC)-based integer execution unit and multiple SIMD vector units. In addition, multiple copies of I-cache and data memory are available for each thread. Each instruction has four fields: load/store, ALU, integer multiplier, and vector multiplier. Therefore, it can issue up to four simultaneous instructions where one may be a data-parallel vector operation. The architecture also uses token-triggered threading, which consumes much less power than simultaneous multithreading because it issues instructions in round-robin fashion. The Sandblaster architecture supports up to eight threads.

21.5.3.6 SIMT

The single-instruction stream multiple task architecture (SIMT) [18] from Linkoping University consists of Complex MAC (CMAC) SIMD units, Complex ALU (CALU) SIMD units, multiple memory banks, on-chip network, accelerators, and a controller unit. The CMAC unit consists of four CMAC lanes, each of which uses 14×14 bit complex multipliers and has eight 2×40 bit accumulator registers. The CALU unit is similar to the CMAC unit except with simplified complex multiplier supporting only $0, +/-1$, $+/-i$ multiplications. The controller efficiently manages the two SIMD units and the memory so that several threads can be run simultaneously. The memory is organized into multiple banks and each bank contains its own AGU to minimize memory access conflicts. The programmable reconfigurable crossbar switch is used as the on-chip network.

21.5.4 Reconfigurable Architectures

Reconfigurable architectures usually consist of many small PEs, which are connected through an interconnection fabric. These architectures can be categorized as either homogeneous or heterogeneous based on the type of PE. In addition, these PEs range from fine-grain lookup tables (LUTs) to coarse-grain ALU units and even ASICs.

21.5.4.1 ADRES

An architecture for dynamically reconfigurable embedded system (ADRES) [16] from IMEC is an example of a coarse-grain reconfigurable tile architecture, which tightly couples a VLIW processor and a coarse-grain reconfigurable matrix. This tightly coupled system has the advantages of shared resources, reduced communication costs, improved performance, and simplified programming model. The VLIW processor and the reconfigurable matrix share FUs and RFs. For the reconfigurable matrix part, there are many reconfigurable cells (RCs) which comprise FUs, RFs, and configuration Random Access Memory (RAM). The RCs are connected to the nearest neighbor RCs and RCs within the same row or column in the tile. Therefore, kernels with a high level of DLP are assigned to the ADRES tiles, whereas sequential codes are run on the VLIW processor. In ADRES architecture, the data communication is performed through the shared RF and memory, which is more compiler-friendly than the message-passing method. In addition, the ADRES relies on modulo scheduling and traditional VLIW compiler support to exploit both DLP and instruction level parallelism (ILP) to maximize PE utilization.

21.5.4.2 Montium

The Montium [21] architecture from Delft University is a coarse-grained reconfigurable processor targeting 16 bit algorithms. Montium consists of two parts: (1) communication and configuration unit (CCU) and (2) Montium tile processor (TP). The CCU configures the Montium TP and parts of the CCU for either "block-based communication" mode or "streaming communication" mode. The TP consists of five Montium ALUs and ten local memories, which are vertically segmented into five processing part arrays (PPAs). A relatively simple sequencer controls the entire PPA and selects configurable PPA instructions that are stored in the decoders. Montium ALU consists of four FUs in level 1 followed by a multiplier and an adder in level 2. Neighboring ALUs can also communicate directly on level 2 in the TP. Finally, each local Static Random Access Memory (SRAM) is 16 bit wide and accompanies each AGU, and the memory can be used for both integer and fixed-point LUTs.

21.5.4.3 Adapt2400 ACM

Adapt2400 adaptive computing machine (ACM) [19] from QuickSilver consists of two basic components: nodes and matrix interconnection network (MIN). Nodes are the computing resources in the ACM architecture that perform actual work. Each node has its own controller, memory, and computational resources, so that it can independently executes algorithms that are downloaded in the form of binary files, called SilverWare. A node is capable of implementing a first come, first serve, non pre-emptive multitasking with the support of a hardware task manager. The MIN ties the heterogeneous nodes together and provides the backbone for carrying data, SilverWare, and control information. It is hierarchically structured, and data within the MIN is transported in single 32 bit word packets to other nodes or external interface. This heterogeneous coarse-grain reconfigurable architecture cooperates with InSpire SDK Tool Set to provide integrated scalable hardware/software platform.

21.5.4.4 XiRisc

XiRisc [8] from XiSystem is an example of a fine-grain reconfigurable architecture. It is a VLIW RISC processor with two concurrent execution datapaths and a set of DSP-like FUs that are shared between two datapaths. The concurrent execution path represented by a pipelined configurable gate array (PiCoGA) provides a run-time extension of the processor instruction set architecture (ISA) for application-specific functions. The PiCoGA is a programmable pipelined datapath composed of an array or rows, which can function as customized pipeline stages. Each row is composed of 16 reconfigurable logic cells (RLCs) containing two 4-input 2-output LUTs, four registers, and dedicated logic for a fast carry chain. Each RLC is connected to the others through a programmable interconnection matrix with 2 bit granularity switches. XiRisc exploits the synergy of the different execution units, ranging from a 32 bit dedicated MAC unit to bit-level processing blocks on PiGoGA, thereby increasing the execution efficiency and saving energy.

21.6 Cognitive Radio

Wireless networks typically have to operate on a fixed spectrum assignment, as a result of which, the spectrum is severely underutilized. CR built on SDR has the ability to change the transmitter parameters and operating characteristics based on interaction with its environment. It reduces spectrum underutilization through dynamic and opportunistic access to licensed and unlicensed bands. In order to support dynamic spectrum management, CR has to provide (1) spectrum sensing, the ability to monitor available spectrum bands, (2) spectrum decision, the capability to select the optimum spectrum band, (3) spectrum sharing, the coordination of users' transmission, and (4) spectrum mobility, the ability to ensure smooth transmission during spectrum changes [2]. Standards such as IEEE 802.11 (WiFi), IEEE 802.15.4 (Zigbee), and IEEE 802.16 (Wimax) implement some level of cognitive functionality.

There are a handful of hardware architectures for CR. Most of them are heterogeneous computing platforms consisting of field programmable gate arrays (FPGAs), DSP processors along with the analog front-end. The FPGA-based BEE2 platform has been used for implementing matched filter, energy detection, and cyclostationary detection for spectrum sensing [6]. Cooperative sensing using ultra-wide-band transmission has also been mapped onto this platform [7]. MAC layer-based optimization has been shown to be effective for CRs, and a SDR-based implementation is described in [5]. Cog-Net, a FPGA-based platform for CR, has been developed for the SDR modem, MAC, and network engines [17]. Finally, a comprehensive design that includes FPGA-based universal software radio peripheral, GNU radio, and Cell Broadband Engine has been developed in [10]. All these designs are flexible but have high power and are not suitable for mobile devices. The challenge is to design a CR architecture that achieves a balance between programmability and low power.

21.7 Conclusion

Wireless communications will soon be ubiquitous. Users will be able to access high-speed Internet with their mobile devices in every corner of the globe. To support this, mobile communication devices will have to be able to connect to a multitude of different wireless networks. SDR promises to deliver a cost-effective communication solution by supporting different wireless protocols in software. In this chapter, we presented low-power digital baseband processors for SDR. We analyzed the characteristics of the signal processing kernels, and described the design rationales behind SODA and its commercial prototype, Ardbeg. Both are multiprocessor wide-SIMD architectures that exploit the data-level parallelism, the thread-level parallelism and the streaming nature of the inter-kernel computations. We also described the salient features of other existing baseband architectures that are either SIMD-based or reconfigurable tile-based. We hope that the lessons learned from designing SDR architectures will accelerate the development of low-power architectures for mobile devices that support CR and other emerging wireless technologies.

References

1. ARM Advanced SIMD Extension (NEON Technology). [Online]: http://www.arm.com/products/CPUs/NEON.html
2. I.F. Akyildiz, W.-Y. Lee, M.C. Vuran, and S. Mohanty. Next generation/dynamic spectrum access/cognitive radio wireless networks: A survey. *Computer Networks: The International Journal of Computer and Telecommunications Networking*, 50(13): 2127–2159, September 2006.
3. N.L. Binkert, R.G. Dreslinski, L.R. Hsu, K.T. Lim, A.G. Saidi, and S.K. Reinhardt. The M5 simulator: Modeling networked systems. *IEEE Micro*, 26(1): 52–60, July/August 2006.
4. H. Bluethgen, C. Grassmann, W. Raab, U. Ramacher, and J. Hausner. A programmable baseband platform for software defined radio. In *Proceedings of the 2004 SDR Technical Conference and Product Exposition*, Phoenix, AZ, November 2004.
5. B. Bougard, S. Pollin, J. Craninckx, A. Bourdoux, L. Ven der Perre, and F. Catthoor. Green reconfigurable radio system. *IEEE Signal Processing Magazine*, 24(3): 90–101, May 2007.
6. D. Cabric. Addressing the feasibility of cognitive radio. *IEEE Signal Processing Magazine*, 25(6): 85–93, November 2008.
7. D. Cabric, I.D. O'Donnell, M.S.-W Chen, and R.W. Brodersen. Spectrum sharing radios. *IEEE Circuits and Systems Magazine*, 6(2): 30–45, 2006.
8. A. Cappelli, A. Lodi, M. Bocchi, C. Mucci, M. Innocenti, C. De Bartolomeis, L. Ciccarelli, R. Giansante, A. Deledda, F. Campi, M. Toma, and R. Guerrieri. XiSystem: A XiRisc-based SoC with a reconfigurable IO module. In *IEEE International Solid-State Circuits Conference (ISSCC)*, San Francisco, CA, pp. 196–593, February 2005.
9. J. Fridman and Z. Greenfield. The TigerSharc DSP architecture. *IEEE Micro*, 20(1): 66–76, January 2000.
10. F. Ge, Q. Chen, Y. Wang, C.W. Bostian, T.W. Rondeau, and B. Le. Cognitive radio: From spectrum sharing to adaptive learning and reconfiguration. In *IEEE Aerospace Conference*, Big Sky, MT, pp. 1–10, March 2008.
11. J. Glossner, D. Iancu, M. Moudgill, G. Nacer, S. Jinturkar, S. Stanley, and M. Schulte. The sandbridge SB3011 platform. *EURASIP Journal on Embedded Systems*, 2007(1): 16, 2007.
12. H. Holma and A. Toskala. *WCDMA for UMTS: Radio Access For Third Generation Mobile Communications*. John Wiley & Sons, New York, 2001.
13. S. Knowles. Ths SoC Future is Soft. In *IEE Cambridge Processor Seminar*, December 2005. [Online]: http://www.iee-cambridge.org.uk/arc/seminar05/slides/SimonKnowles.pdf
14. H. Lee, Y. Lin, Y. Harel, M. Woh, S. Mahlke, T. Mudge, and K. Flautner. Software defined radio—A high performance embedded challenge. In *Proceedings of the 2005 International Conference on High Performance Embedded Architectures and Compilers (HiPEAC)*, Barcelona, Spain, pp. 6–26, November 2005.
15. Y. Lin, H. Lee, M. Woh, Y. Harel, S. Mahlke, T. Mudge, C. Chakrabarti, and K. Flautner. SODA: A low-power architecture for software radio. In *Proceedings of the 33rd Annual International Symposium on Computer Architecture*, Boston, MA, pp. 89–101, July 2006.

16. B. Mei, S. Vernalde, D. Verkest, H. De Man, and R. Lauwereins. ADRES: An architecture with tightly coupled VLIW processor and coarse-grained reconfigurable Matrix. In *13th International Conference on Field-Programmable Logic and Applications*, Berlin, Germany, pp. 61–70, January 2003.
17. Z. Miljanic, I. Seskar, K. Le, and D. Raychaudhuri. The WINLAB network centric cognitive radio hardware platform WiNC2R. In *Cognitive Radio Oriented Wireless Networks and Communications*, Orlando, FL, pp. 155–160, August 2007.
18. A. Nilsson and D. Liu. Area efficient fully programmable baseband processors. In *International Symposium on Systems, Architectures, Modeling and Simulation (SAMOS)*, Greece, pp. 333–342, July 2007.
19. B. Plunkett and J. Watson. *Adapt2400 ACM Architecture Overview*. QuickSilver Technology Inc., San Jose, CA, January 2004.
20. U. Ramacher. Software-defined radio prospects for multistandard mobile phones. *Computer*, 40(10): 62–69, 2007.
21. G.K. Rauwerda, P.M. Heysters, and G.J.M. Smit. Towards software defined radios using coarse-grained reconfigurable hardware. *IEEE Transactions Very Large Scale Integration (VLSI) Systems*, 16(1): 3–13, January 2008.
22. K. van Berkel, F. Heinle, P.P.E. Meuwissen, K. Moerman, and M. Weiss. Vector processing as an enabler for Software-Defined Radio in handsets from 3G+WLAN onwards. *EURASIP Journal on Applied Signal Processing*, 2005(1): 2613–2625, January 2005.
23. M. Woh, Y. Lin, S. Seo, S. Mahlke, T. Mudge, C. Chakrabarti, R. Bruce, D. Kershaw, A. Reid, M. Wilder, and K. Flautner. From soda to scotch: The evolution of a wireless baseband processor. *Microarchitecture, 2008. MICRO-41. 2008 41st IEEE/ACM International Symposium on Microarichetures*, Lake Como, Italy, pp. 152–163, November 2008.

22

Software-Defined Radio for Advanced Gigabit Cellular Systems

22.1 Introduction 22-1
22.2 Waveform Signal Processing 22-2
22.3 Communication Link Capacity Extensions 22-4
22.4 RF, IF, and A/D Systems Trends 22-5
22.5 Software Architectures 22-6
22.6 Reconfigurable SDR Architectures for Gigabit Cellular 22-7
22.7 Conclusion 22-11
References 22-12

Brian Kelley
The University of Texas at San Antonio

The evolution of software defined radio (SDR) systems has placed us at a crossroads. SDR systems increasingly find themselves vital to the ecosystem of defense communications, civilian cellular and noncellular wireless systems. Continued CMOS digital device shrinkages and ever increasing clock and A/D rates enable more and more opportunities for direct RF sampling of the antenna port waveforms. The enabled digital samples can be processed in flexible software rather than fixed hardware. This leads to a desire for multiprotocol support to enable multiple services, a desire to incorporate the most sophisticated signal processing algorithms to achieve maximum network capacity, and license band sensing and avoidance regimes in support of cognitive radio. In this paper, we present a survey of the important five trends in SDR system infrastructure: waveform signal processing, capacity extending communication radio link processing, RF, IF or A/D band-pass modulation and demodulation, software architectures, and reconfigurable hardware architectures. Finally, we discuss SDR systems that can support the broadband, gigabit cellular protocol known as 3GPP long term evolution-advanced (LTE-advanced).

22.1 Introduction

Many of the modern concepts of SDR were first outlined in the early 1990s by Mitola [23]. Mitola defined software radio (see [3,23,37,38]) as "a set of digital signal processing primitives (and) a meta-level system for combining the primitives into a communication system function and a set of target processors on which the software radio is hosted for real-time communications." Alternatively, the software radio forum defines SDR (see [12]) as "concepts, technologies and standards for wireless communications systems and devices, to support the needs of all user domains including consumer, commercial, public safety, military markets, and stakeholders such as regulatory authorities and a collection of interfaces and standards."

In general, we note a consensus definition among various authors considers SDR as a software programmable baseband and RF communication protocol system with flexible RF modulation and demodulation tunable to any RF frequency band and capable of transmitting and receiving any signaling waveform and bandwidth [13]. Research in this area has, to a large extent, matured so that modern wireless system frameworks exist in which radio communication algorithms can be embodied almost entirely in reprogrammable software, as opposed to hardware.

22.2 Waveform Signal Processing

One significant challenge for SDRs relates to the waveform signal processing algorithms and commensurate digital signal processing to optimally estimate the transmitted waveform. The SDR waveform dilemma can then be stated as follows: irrespective of the maximum computing capability of an SDR, we can define set of radio algorithms which can yield superior network capacity, but outstrip the SDRs native computing limits. We note that MLSE demodulation of communication systems yields optimum demodulation and detection results, but results in exponential increases in computational requirements. Communication algorithms therefore utilize suboptimal algorithms for decoding, demodulation, and equalization. However, even suboptimal algorithms that approach the performance of optimal Shannon limits require high computational efficiency.

From the standpoint of the waveform signal processing, particularly as we migrate to gigabit cellular, approaching Shannon's limit is the ultimate goal that determines the selection of signal processing algorithms. Network capacity goals are counterbalanced by SDR complexity considerations. We note that the SDR receiver complexity, specifically for multicarrier modulation systems, (see [26]) can be defined as

$$C_{\text{omp}} = O\left(N_{\text{iter}} N_{\text{re}} 2^{\log(M) \times (N_{\text{TxA}} + N_{\text{RxA}})}\right) \tag{22.1}$$

where

N_{iter} number of iterations of the message passing equalization and decoding algorithms
N_{re} number of subcarriers or resource elements of the multicarrier system
M number of points in the constellation
N_{TxA} number of transmit antennas
N_{RxA} number of receive antennas

Thus, if we define $N_{\text{iter}} \approx 8$ iteration message passing Turbo decoding, $N_{\text{re}} = 900$, $N_{\text{TxA}} = 4$ number of transmit antennas, and $N_{\text{RxA}} = 4$ receive antennas, complexity is $C_{\text{omp}} = O(7200 \times 2^{8 \log(M)})$!

In Figure 22.1, we illustrate the fact that traditional GSM cellular systems had surplus computing capability which insured the ability for instruction reprogramming. We also note that margin requirements and design modifications are modest. However, the trend line in Figure 22.1 toward gigabit SDR indicate computational requirements that exceed the capability of the traditional DSP architectures based upon Moore's law improvement trends. We discuss this further in Section 22.6.

The significant breadth of topics associated with waveform signal processing algorithms requires lengthy exposition. We summarize the most prescient waveform signal processing and radio link management algorithms in Tables 22.1 and 22.2, respectively. Note that the signal processing algorithms in Table 22.2 are defined according to quality of service (QOS), throughput, or cost. Of all the algorithms, the parameter estimation algorithms involving estimation of frequency offsets, timing estimation, and channel estimation can have a dramatic impact on SDR architectures and system performance. Closely related to this are the estimation algorithms associated with smart antenna processing and MIMO. We point out that in [18], the SDR forum formed an entire group known as the smart antenna workshop group (SAWG) just to define the software radio interfaces and definitions to cope with MIMO signal processing (heretofore, references to MIMO lump beamforming, diversity, and spatial multiplexing).

One of the most important class of algorithms currently under investigation involves those related to cognitive radio [7,4,35,36]. After studies indicated that the efficient use of license spectrum, rather than

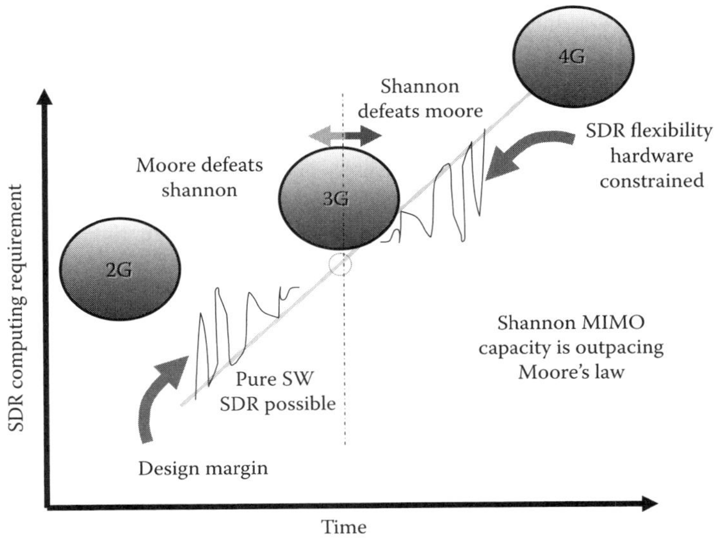

FIGURE 22.1 Traditional 2G wireless (e.g., GSM) can be run entirely on software programmable DSPs. As capacity utilization has increased, due to improved Shannon communication, algorithm computational requirements increasingly constrain flexibility of the SDR system.

TABLE 22.1 Algorithms Important to Advanced Wireless Systems

Waveform Algorithm	Key Enabler	Comment or Description
Space time coding	QOS	TX/RX diversity; 2×2 Alamouti diversity yields 20 dB improvement
Macro Rx diversity	QOS	Information Tx coordinated from differrent cell sites
Beamforming	QOS	Requires reverse link channel knowledge
Channel, doppler timing synchronization, freq. estimation	QOS	High velocity mobiles require doppler frequency estimation. Chanel estimation is normally expensive.
QPP interleaver	QOS	Lowers latency; enable parallel operation
OFDM-A	Throughput	Orthogonal frequency division multiple access
High order modulation	Throughput	(e.g., 256-ary QAM)
MIMO spatial multiplexing	Throughput	Requires precoding
Adaptive modulation	Throughput	Adaptive modulation usually combined with coding (AMC)
Bandwidth scalability	Low cost	Enables network deployment in more bandwidth configurations
Spectrally efficient	Low cost	Supports more users
Lower peak to average	Lower cost	Power amplifier operation

TABLE 22.2 Radio Link Management Signal Processing Algorithms

Radio Link Algorithm	Key Enabler	Comment or Description
Turbo codes LDPCH codes	Radio link management	Adaptive coding rate via puncturing
Hybrid-ARQ	Radio link management	Packet retransmitted and combined with incremental redundancy and chase combining
ARQ	Radio link management	Service data unit segmentation is link dependent
Network scheduling	Radio link management	Greedy scheduling based on channel profile maximizes capacity, but is not "fair" to each user

lack of frequency spectrum, is a bottleneck in U.S. spectrum allocation policy, the FCC has encouraged the development of cognitive radio systems. Such systems can sense the presence of licensed signal transmission power, thereby altering their transmission and reception parameter to avoid interference. In Globecom [7] Motorola labs has shown a cognitive radio prototype that is capable of performing channel allocation within 20 MHz bandwidth (fine sensing) using spectrum correlation density technique. Another spectrum sensing technique is based on wavelets suggested by Georgia Tech and implemented in [8,9]. Researchers at Berkeley's Wireless Research Center have studied the spectrum sensing implementation consideration for cognitive radios in [10,11]. In [6], Lyrtech presented their cognitive radio architecture in support of the IEEE 802.22 standard (www.lyrtech.com) adopting Welch's periodogram [4] using a cyclo stationary feature detector [14].

22.3 Communication Link Capacity Extensions

For cellular SDR systems, wireless network capacity is defined as the throughput per user per Hertz per cell (b/s/Hz/cell) of the profiled user. The profiled user could either be the average user or the bottom 5% of users (i.e., can be chosen using many different criteria). This capacity is ultimately bounded by Shannon's capacity, which is in turn depends on the MIMO channel statistics and number of antennas, $M = M_{\mathrm{T}} = M_{\mathrm{R}}$:

$$C = \log\left[\det\left(I_M + \frac{P}{\sigma^2}HH^H\right)\right] \approx \sum_{i=1}^{M} \log_2\left(1 + \lambda_i \frac{P_i}{\sigma_i^2}\right) \qquad (22.2)$$

where

$\lambda_i = \text{eigenvalue}(HH^H)$ is non-zero for $i = 1, 2, \ldots r, \sigma_i^2$ is the noise variance of the ith received antenna

P_i represents the optimum power allocation across the transmit array defined by water filling (see [27])

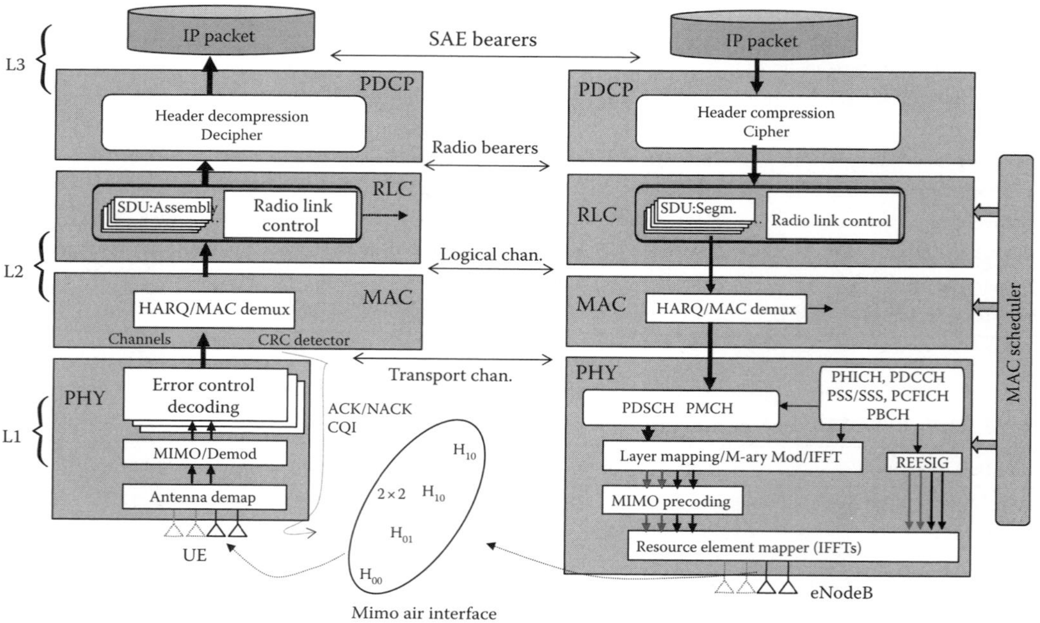

FIGURE 22.2 3GPP LTE downlink SDR architecture.

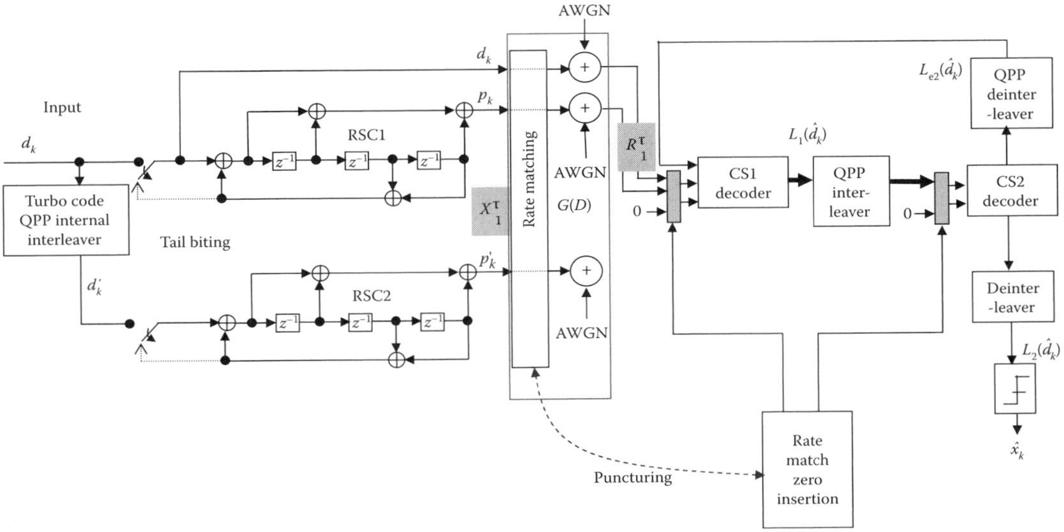

FIGURE 22.3 Iterative Turbo codes and other iterative estimation algorithm consume significant computational resources. Parallel memory access contention using contention free interleaving can enable parallel Turbo decoding structures.

$$\sum_{i=1}^{r} p_i = P$$

For the above scenario, note that the capacity, C, approaches infinity at a rate of $\log_2(M_r)$ presuming a sufficiently high SNR. Therefore, when SNR and bandwidth are fixed, increasing the capacity of the wireless network implies increasing the number of MIMO transmit and receive antennas assuming we are fortunate enough to experience a MIMO channel with sufficiently rich scatters in the environment.

Figure 22.2 illustrates a diagram of an advanced downlink wireless SDR oriented for 3GPP cellular communications. From the standpoint of error correction, message passing decoding, potentially combined with iterative equalization, represents another means for extending the link. Many advanced wireless systems rely either upon low density parity check codes [10,26], or turbo codes [16,29,32] for error correction. We note that the 3GPP LTE, in particular, utilizes a Turbo code system along with a unique quadratic permutation polynomial (QPP) interleaver design to enable parallel memory access [23,29]. This in turn enables fast parallel Turbo decoding (i.e., for reduced latency). The turbo coder and decoder for 3GPP-LTE are illustrated in Figure 22.3.

Other techniques used to improve the link are hybrid-ARQ [29,34], multiuser network scheduling in which the network schedules the kth user's transmission according to the channel presented between the kth user and the system base station, labeled eNodeB in 3GPP-LTE [28] (Figure 22.4).

22.4 RF, IF, and A/D Systems Trends

SDRs are enabled, in part, by ever increasing A/D converters rates. Walden [25,26] projects 1 GHz, 12 bit A/Ds by circa 2015. Direct sampling of the antenna ports enables large swaths of RF circuitry to be replaced with flexible, programmable digital signal processing. Hence, embedded SDR transceivers with full programmability of 500 MHz RF should be commonplace in the next decade. ADCs used in high performance broadband circuits include successive approximation multistage flash, time interleaved filter-banks, and band-pass delta-sigma modulators [3,26].

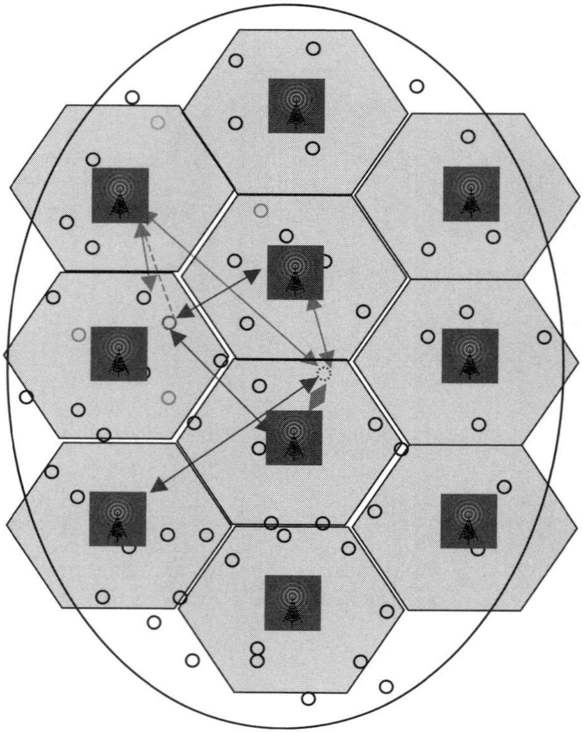

FIGURE 22.4 Cellular network diagram. System capacity is optimized through network scheduling.

As part of the core physical layer in most next generation wireless systems, the DC carrier is unused. One advantage of this is that direct conversion receivers have gained increasing popularity among RF radio system engineers. Direct conversion receivers perform one level of frequency conversion from RF to baseband as opposed to utilizing traditional super heterodyne transceiver techniques. Due to the fact that direct conversion receivers have difficulty accurately converting the DC resource element (i.e., subcarrier) component of the received signal, many multicarrier modulation systems do not utilize the DC component for communication of signaling information.

Other common signal processing algorithms in advanced multicarrier modulations systems involve accurate estimation and correction associated with relative gain amplitude imbalance on the I and Q baseband path, digital frequency offset correction, and integrated digital-analog gain control. In addition, OFDM systems have high, naturally occurring peak-to-average-ratios (PAR). Unfortunately, power amplifier efficiencies are low at high PARs. This is problematic in regards to power amplifier cost and average power usage. It has become increasingly common for PAR minimization protocols and signal processing algorithms to be directly integrated into OFDM uplink signaling [28].

22.5 Software Architectures

For SDR, the software architectures are arguably one of the most identifiable system components. In [5], the authors discuss methods for software efficiency using a VLIW machine with system capture described in MATLAB®. The authors then refine the code in multistage process using an electronic system level (ESL) design methodology flow.

Much research [8,13,15,20], attempts to integrate data flow description, statically schedule dataflow (SSDF) (data flow with compile time static scheduling and fixed consumer producer tokens), or cyclo stationary dataflow (CSDF) procedures. One drawback of these methods is the challenge presented by incremental modification after the design algorithms are mapped to hardware.

In [10–12] Seo, Mudge, Zhu, and Chakrabarti, discuss a very flexible SIMD-based SDR architecture designated as SODA. In [14] the authors conclude that SODA based architecture could meet 4G wireless computing requirements, but would necessitate operation at 20 GHz.

The Department of Defence's (DoD) Joint Tactical Radio Systems (JTRS) Joint Program Executive Office (JPEO) has defined a standard known as the software communication architecture (SCA). The SCA enables integration of hardware and software components from different vendors into seamless products. Open standards improve portability of waveforms to multiple platforms and simultaneously increase the complexity of individual waveforms processing. The object management group (OMG), has developed the unified modeling language (UML) based on SCA [18]. Both the OMG and software defined radio forum (SDR-Forum) have also developed software standards for smart antennas under the SAWG.

22.6 Reconfigurable SDR Architectures for Gigabit Cellular

Figure 22.5 indicates the accelerating trends toward gigabit wireless cellular systems. Unfortunately, traditional SDR design styles can prove problematic (see [14]) for emerging generations of gigabit cellular. In our SDR system, we begin by defining SDR as "an RF and baseband flexible radio architecture which, under software instruction control is RF, software, and hardware reprogrammable." Our definition of SDR differs from many prior authors in regards to our further delineation of the term flexibility:

$$\text{Flexibility} \propto \min\{k_0 \text{ permutations}, k_1 \text{ design margin}, k_2 \text{ modification time}\} \tag{22.3}$$

Often SDR flexibility is implicitly defined from the standpoint of combinatorial permutations. An N instruction processor with a program memory of size P can perform P^N unique permutations of program

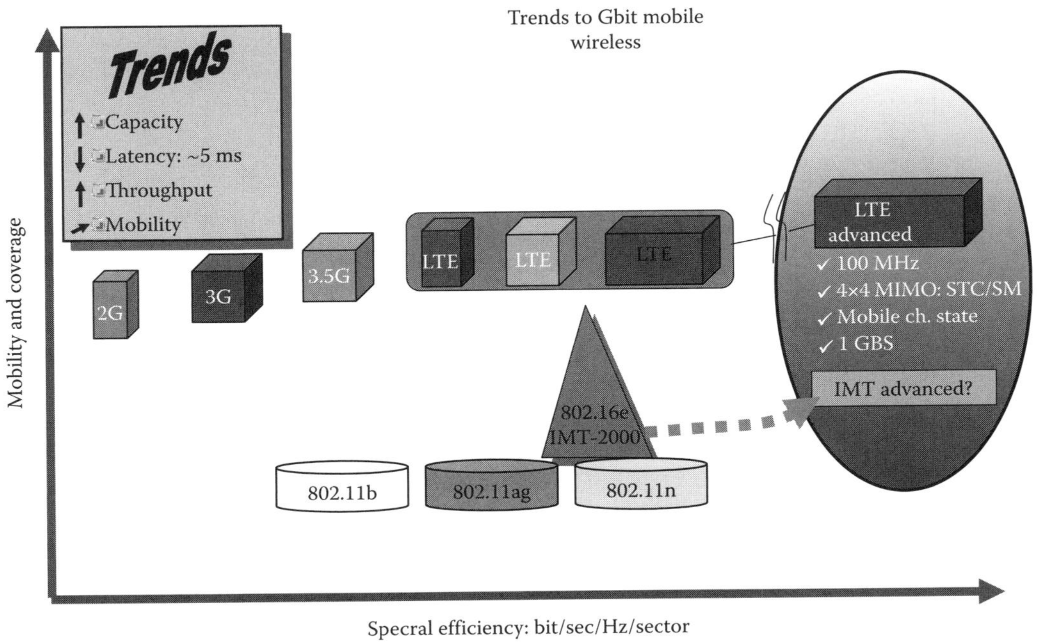

FIGURE 22.5 Trends in Gigabit-cellular mobile wireless.

operations. The ability to permute is only one aspect of flexibility. Design-margin flexibility and modification-time flexibility are two often overlooked aspects. The constants in Equation 22.3, k_0, k_1, and k_2, normalize for dimensional aspects of each component.

In most radio systems, design-margin targets are indicated. Typically this includes Eb/No, packet error rates, power consumption, and latencies. For instance, due to finite battery capacities in mobile systems, mode dependent power consumption tends to rate as highly important as a design parameter. We rate systems barely exceeding the battery capacity targets as inflexible In addition to power consumption, other items typically considered are RF requirements, co-channel interference, adjacent channel performance, cost, memory, size, and packet error rate constraints as a function of Eb/No and the air interface channel model [31,33]. Systems with high degrees of permutation margin, but close to the edge of the design margin space can easily lack extensibility for the most modest of requirement specification changes. Under this premise, an infinitely permutable system on the precipice of design margin constraints lacks a key flexibility component.

Design modification time is another key, often overlooked, component associated with flexibility. Gate level modifiable systems offer more flexibility from a permutation standpoint, but lack flexibility in regards to the amount of time it takes to modify the gate design. We need to question presumptive notions regarding the ease of software changes, particularly, when legacy protocol and algorithms are commonplace. When large bodies of software rely on underlying SDR components, the capability to modify may be much faster than the capability to verify. Thus, modification flexibility may require both fast reconfiguration of their system state, information hiding, and isolation of system substates. Design modification need not be real-time, however. The design time associated with off-line system reprogramming favors systems embedded with fast reprogramming characteristics. In this instance, reusability is also favored relative to systems with slow, complex reconfiguration regimes.

Under design modification flexibility components with high level, flexible, object oriented reprogramming with its commensurate reusability, inheritance, information hiding, and class extensibility is favored over unstructured gate, register level programming, or procedural programs.

These considerations are illustrated in Figures 22.6 and 22.7. Figure 22.6 illustrates the effect of incorporating a revised notion of flexibility in the definitions of SDR. Figure 22.8 demonstrates the architectural method that can be utilized to solve the design time modification issue. With flexible advance cellular protocols containing thousands of pages of requirements (www.3gpp.org), design time

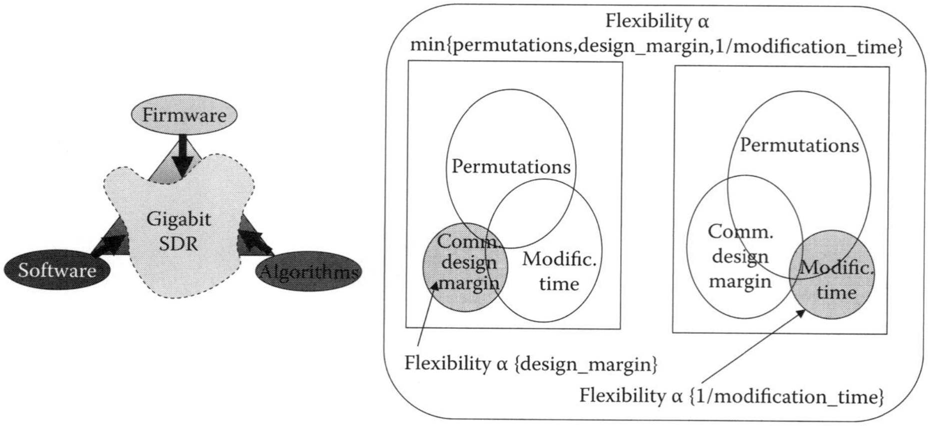

FIGURE 22.6 (a) Choice of communication algorithms affects firmware tradeoffs; (b) In gigabit SDR design, we implicitly constrain flexibility though permutation flexibility, design margin flexibility, and design modification time flexibility.

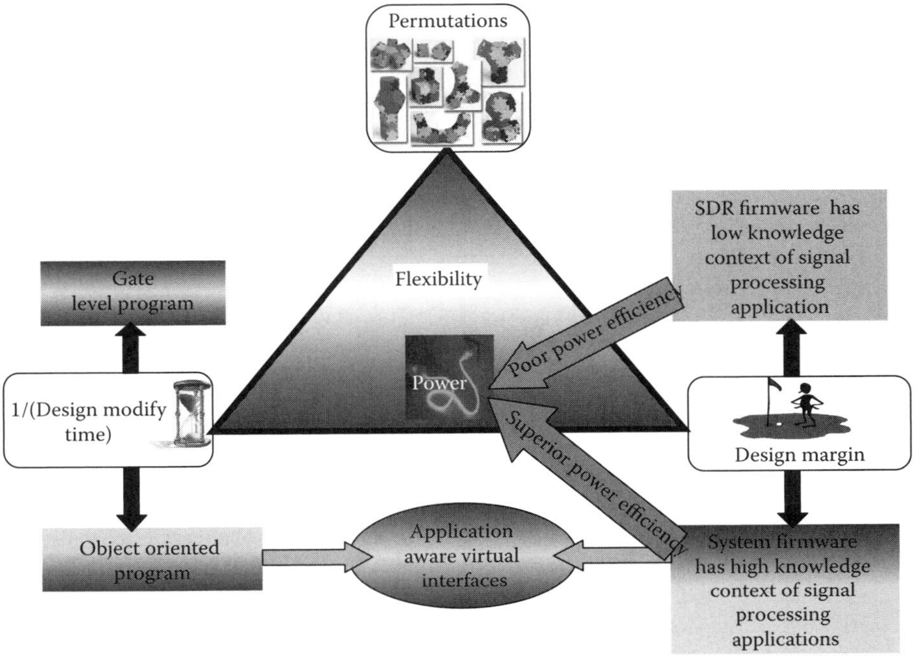

FIGURE 22.7 Flexibility is the key to SDR extensibility. Flexibility is a function of permutations, design modification time, and design margin. Systems with awareness of the algorithms built in consume less power.

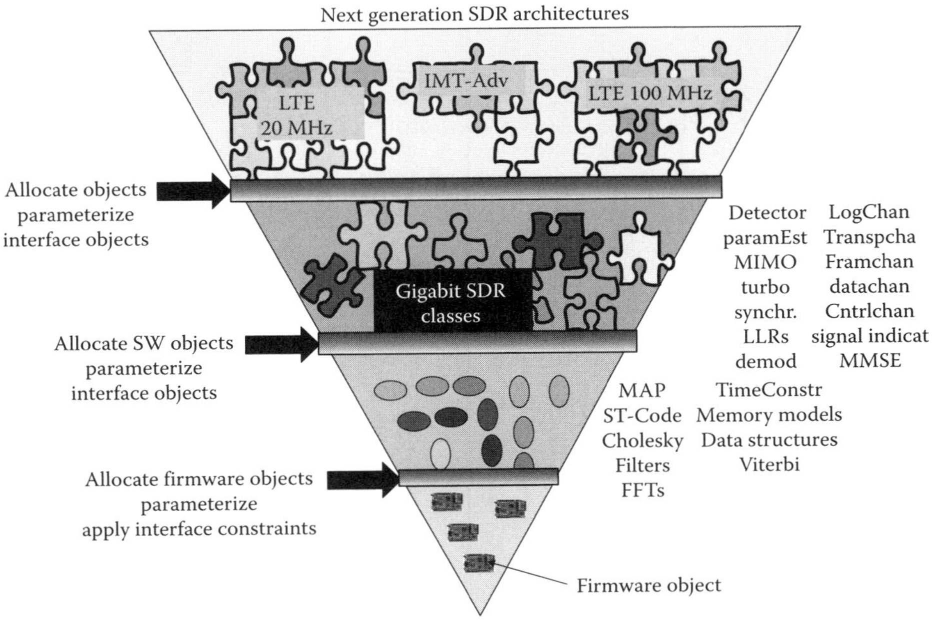

FIGURE 22.8 Hierarchical model of design. MCO objects are allocated via dynamic constructor and destructor calls to enable task parallel signal processing computation. Hierarchically, firmware objects are assembled into radio system primitive objects.

modification is a key flexibility limitation in many systems, particularly those with low level programming approaches. Three key methods to improve design time flexibility are (1) virtual interfaces, (2) object oriented descriptions of hardware and software extensible radio systems, and (3) hierarchical design. This is illustrated in Figure 22.8. Such virtual interfaces are important to isolate new software changes from hardware changes and to allow new micro compute object (MCO) hardware to be introduced in future designs without impacting software. This increases flexibility by lowering modification time. We also note that there is a key tradeoff in the choice of algorithm and the degree to which the algorithms are suitably migrated to either software or firmware. This tradeoff is illustrated in Figure 22.9b.

Our solution is to define ALU firmware systems with hardware acceleration complexes defined as programmable, extensible, and invokeable micro-level compute objects. The specific design of the MCO is described in Figure 22.9. In Figure 22.9b, the MCOs, are allocated on an as needed bases to meet the computational requirements. If the computing margin is low, excess MCOs are added to the hardware complex to insure an extensibility of computational requirements. Allocation of MCOs is distributed according to the dataflow and MIP allocation to enable optimal demodulation of successive packets arrivals and to avoid bus congestion. In turn, each MCO contains collections of software primitive classes with objects instantiated and constructed prior to run time and dynamically parameterizable during run time. Virtual interfaces are created so as to emulate algorithmic parameters. This first level interface exists between the MCO level and the software primitive level so as to abstract the hardware configuration from the software algorithms. This is illustrated in the second level in Figure 22.8. The second level of Figure 22.8 contains algorithmic class primitives such as butterfly, correlate, sort, convolution, ComplexMAC, and ComplexDIV instructions.

The third level of Figure 22.8 communicates with the second level through virtual interfaces which abstracts level-3 firmware constraints from algorithmic parameters. Virtual interfaces are important to

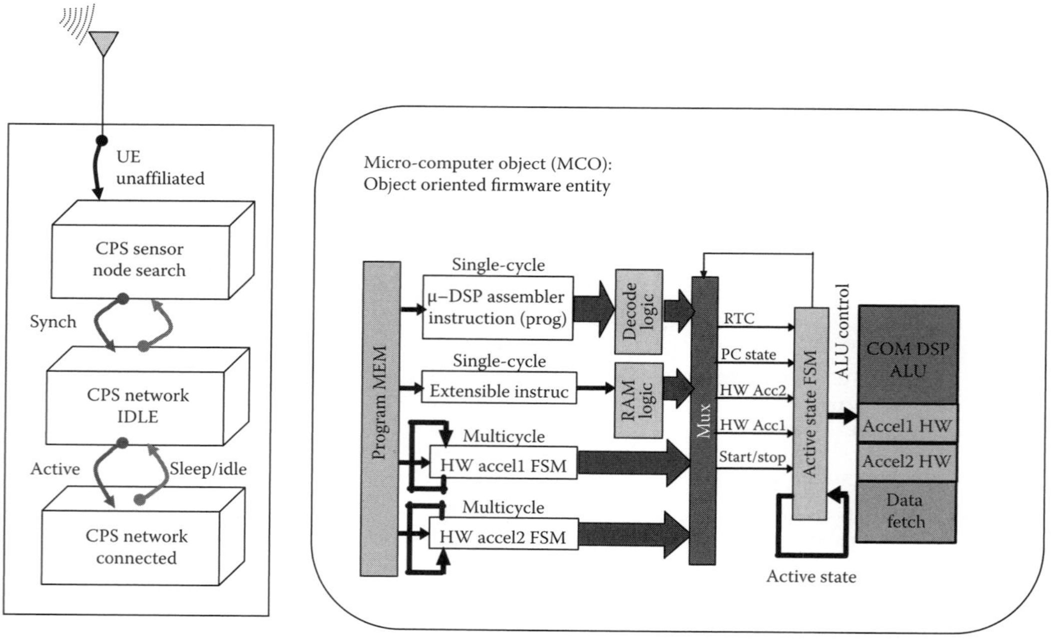

FIGURE 22.9 (a) MCO firmware has facilities for synchronizes to network and for discontinuous low power operation; (b) MCO architecture. The MCO firmware enables normal instructions, extensible (postdesign or late binding) instructions and HW accelerated instructions.

allow new hardware and firmware complexes to be installed without affecting system level software (i.e., design time modification flexibility). Examples of some of the algorithms integrated at the third level of hierarchy as instantiated objects are detectors, precoders, MIMO space time decoders, FFTs, multicarrier equalizers, and MAP decoders.

Finally, at fourth level in Figure 22.8, we incorporate communication layer virtual interfaces and communication object models representing the communication standard integrated into the SDR. Note that the SDR at the top level of Figure 22.8 could represent a 20 MHz version of 3GPP-LTE using spaced time coding and 16 QAM or 100 MHz version of 3GPP-LTE advanced using spatial multiplexing and 64 QAM.

One challenge with gigabit SDRs mentioned above is the fact that we are often margin constrained due to operation closer to the Shannon limit. The near-ergodic capacity algorithms (e.g., Turbo decoding, iterative equalization, Sphere-decoding, MLSE) [34] can virtually always be defined for superior sensitivity performance in a way that outstrips the computational limits of the SDR. One solution to mitigate this is to optimize the MCO design for improved computational efficiency relative to the standard programmable DSP methods of computation. Higher levels of efficiency imply hardware accelerated instructions, extensible instruction sets, and specialized multicycle instructions. These capabilities are indicated in the design architecture of the MCO indicated in Figures 22.9a and b.

In order to increase flexibility for such systems we must add algorithm optimized extensible instruction sets and hardware accelerated instructions. We make special note of the fact that the instruction set control-words of the extensible instruct sets are designed to be much larger than typical program words. Their word-size is on the same order as the logic decoder generated control. This is also true of the hardware accelerated control logic. The rational underlying the long control words of these units is that they are predesigned to have maximum control over the entire set of internal registers, MUXs, memories, and ALU operators. For instance, extensible instructions are loaded into a RAM registers file prior to run time based upon the specifics of the new instructions that we desire to incorporate. We note that the addition of extensible instructions occurs after design of the MCO.

All instructions, whether multicycle or extensible, are integrated into the normal program instruction flow associated with program memory. Thus, a program instruction referring to an extensible instruction initiates a write of new logical control information from the extensible instruction memory that has been pre-stored in the extensible instruction register file. Multicycle, extensible programs could be formed by storing multiple single cycle extensible instructions references in particular subsections of program memory. Hardware accelerated multicycle instructions are integrated as a single instruction in main program memory.

However, a reference to a multicycle instruction initiates a state change from the system level finite state machine (FSM) to the multicycle FSM machine. Note that multicycle instructions are all run to completion, but must be designed into the MCO at design time. In part, acceleration of the hardware occurs for extensible and multicycle instructions due to the finer control granularity capable over the entire set of register transfer level (RTL) resources available in the MCO. In addition, HW accelerated instructions typically have adjunct hardware associated with the ALU that is optimized for specific mathematical computations (e.g., $1/\mathrm{sqrt}(x), \exp(jx), \exp(x)$).

22.7 Conclusion

We have presented a survey of SDR and proposed a new way forward for gigabit cellular SDR systems. We note that both design margin constraints and design modification characteristics are as important to account for as the ability of the system to permute or reconfigure. A new architecture approach has been proposed based concepts gleamed from a renewed focus on the concept of flexibility.

References

1. Mitola, J., Software radios: Survey, critical evaluation, and future directions technologies, *Software Radio Techn. Selected Read*, IEEE Press, p. 3, 1999.
2. SDR Forum, www.sdrforum.org
3. Kelley, B., Jointly optimized software radio for low power 4G systems, *Invited Talk, Proceedings of the IEEE Asilomar Conference on Systems, Signals, and Computers*, Pacific Grove, CA, pp. 113–117, Nov. 2007.
4. Ghosh, C. and Agravwal, D.P., Channel assignment with route discovery (CARD) using cognitive radio in multi-channel multi-radio wireless mesh networks, *IEEE Conference on Software Defined Radio*, Reston, VA, pp. 36–41, Sept. 2006.
5. Van der Perre, L. et al., Efficient SW design and SW design efficiency: Fuel for software defined radios, *10th Annual Symposium on Spread Spectrum Techniques and Applications, ISSSTA*, pp. 391–394, Bologna, Italy, Aug. 2008.
6. Tachwali, Y., Chmeiseh, M., Basma, F., and Refai, H.H., A frequency agile implementation for IEEE 802.22 using software defined radio platform, *Global Telecommunications Conference, 2008. GLOBECOM '08, IEEE*, New Orleans, LO, pp. 1–6, 2008.
7. Taubenheim, D., Chiou, W., Correal, N., Gorday, P., Kyperountas, S., Machan, S., Pham, M., Shi, Q., Callaway, E., and Rachwalski, R., Implementing an experimental cognitive radio system for DySPAN, *Global Telecommunications Conference, 2007. GLOBECOM '07, IEEE*, Washington, D.C., pp. 4040–4044, Nov. 26–30, 2007.
8. Tribble, A.C., The software defined radio: Fact and fiction, *IEEE Radio and Wireless Symposium*, Orlando, FL, pp. 5–8, Jan. 2008.
9. Boyd, J. and Schlenzig, J., Directional networks for above 2 GHz software defined radios, *Military and Communications Conference, IEEE MILCOM*, 5, 3176–3180, 2005.
10. Seo, S., Mudge, T., Zhu, Y., and Chakrabarti, C., Design and analysis of LDPC decoders for software defined radio, *IEEE Workshop on Signal Processing*, Shanghai, China, pp. 210–215, 2007.
11. Lin, Y. Lee, H., Woh, M., Harel, Y., Mahke, S., Chakrabarti, C., and Flautne, K., SODA: A low-power architecture for software radio, *Proceedings of the 33rd Annual International Symposium on Computer Architecture (ISCA)*, Boston, MA, pp. 89–101, 2006.
12. Lin, Y., Lee, H., Woh, M. Harel, Y., Mahke, S., Chakrabarti, C., and Flautner, K., SODA: A high performance DSP architecture of software defined radio, *IEEE Micro*, 27(1), 114–123, Feb. 2007.
13. Lin, Y., Kudlur, M., Mahlke, S., and Mudge, T., Hierarchical coarse-grained stream compilation for software defined radio, *ACM*, Salzburg, Austria, pp. 115–124, Oct., 2007.
14. Woh, M., Seo, S., Lee, H., Lin, Y., Mahlke, S., Mudge, T., Chakrabarti, C., and Flautner, K., The next generation challenge for software defined radio, *Proceedings Lecture Notes in Computer Science, SAMOS*, Greece, Springer-Verlag, pp. 343–354, 2007.
15. Lin, Y., Choi, Y., Mahlke, S., Mudge, T., and Chakrabarti, C., A parameterized dataflow language extension for embedded streaming systems, *International Conference on Embedded Computer Systems Architectures, Modeling, and Simulation*, Samos, Greece, July 2008.
16. Lin, Y., Mahlke, S., Mudge, T., Chakrabarti, C., Reic, A., and Flautner, K., Design and implementation of turbo decoders for software defined radio, *IEEE Workshop on Signal Processing Systems Design and Implementation*, Banff, AB, Canada, pp. 22–27, Oct. 2006.
17. Lee, H. and Mudge, T., A dual-processor solution for the MAC layer of a software radio terminal, *ACM*, San Francisco, CA, Oct. 2005.
18. Hyeon, S., Kim, J., and Choi, S., Topics in radio communications: Evolution and standardization of the smart antenna system for software defined radio, *IEEE Communication Magazine*, 68–74, Sep. 2008.
19. Silverman, S.J., Game theory and software defined radio, *Military Communications Conference: MILCOM*, Washington, D.C., pp. 1–7, Oct. 2006.

20. Berg, H., Brunelli, C., and Lucking, U., Analyzing models of computation for software defined radio applications, *International Symposium of Systems on Chip*, Tampere, Finland, pp. 1–4, Nov. 2008.
21. Kimmell, E., Komp, G., Minden, J., Evans, P., and Alexander, Synthesizing software defined radio components from rosetta, *Forum On Specification Verification and Design Languages*, Stuttgart, Germany, pp. 148–153, Sep. 2008.
22. Kempf, S., Wallentowitz, G., Ascheid, R., Leupers, H., and Meyr, A workbench for analytical and simulation based design space exploration of software defined radio, *22nd International Conference on VLSI Design*, New Delhi, India, pp. 281–286, 2009.
23. Kelley, B., 3GPP long term evolution aspects and migration to 4G cellular systems, *IEEE Radio and Wireless Workshop on 3GPP Long Term Evolution*, San Diego, CA, Jan. 2009.
24. Walden, R.H., Analog-to-digital converter survey and analysis, *IEEE Journal on Selected Areas in Communications*, 17(4), 539–550, 1999.
25. Walden, R.H., Analog-to-digital converters and associated IC technologies, *IEEE Compound Semiconductor Integrated Circuits Symposium: CSICS-08*, Monterey, CA, pp. 1–2, Oct. 2008.
26. Goldsmith, A., Next-generation wireless systems challenges and solution, *3GPP LTE Workshop*: 2009, *IEEE Radio and Wireless Conference*, San Diego, CA, 2009.
27. Biglieri, E., Calderbank, R., Constantinides, A., Goldsmith, A., Paulraj, A., and Vincent Poor, H., *MIMO Wireless Communications*, Cambridge University Press, New York, pp. 29–45, 2007.
28. 3GPP TS 36.211: 3GPP TS 36.211 v1.2.0, 3GPP TSPG RAN Physical Channels and Modulation (Release 8).
29. 3GPP TS 36.212: 3GPP TS 36.211 v1.0.0, 3GPP TSPG RAN: Mutiplexing and Channel Coding (Release 8).
30. Norsworth, S.R., Schreier, R., and Temes, G.C., *Delta-Sigma Data Converters*, IEEE Press, Piscataway, NJ, 1996.
31. 3GPP TS 36.201: 3GPP TSG-RAN EUTRA LTE Physical Layer General Description, R8 WG1, September 2007.
32. Vucetic, B. and Yuan, J., *Turbo Codes, Principles and Applications*, Kluwer Academic Publishers, Norwell, MA, 2002.
33. Holma, H. and Toskala, A., *HSDPA/HSUPA for UMTS: High Speed Radio Access for Mobile Communications*. John Wiley & Sons, New York, 2006.
34. Goldsmith, A., *Wireless Communications*, Cambridge University Press, Cambridge, U.K., 2005.
35. Haykin, S., Cognitive radio: Brain-empowered wireless communications, *IEEE Jourmal Selected Areas in Communications*, 23(2), 201–220, Feb. 2005.
36. Nandagopal, S., Cordeiro, C., and Challapali, K., Spectrum agile radios: Utilization and sensing architectures, *Proceedings of IEEE DySPAN 2005*, Baltimore, MD, Nov. 11, 2005.
37. SDR Forum, www.sdrforum.org.
38. Dillinger, M., Madani, K., and Alonistioti, N., *Software Defined Radios: Architectures, Sysetms, and Functions*, John Wiley & Sons, New York, 2003.

IV

Advanced Topics in DSP for Mobile Systems

Vijay K. Madisetti
Georgia Institute of Technology

23 **OFDM: Performance Analysis and Simulation Results under Mobile Environments** *Mishal Al-Gharabally and Pankaj Das* **23**-1
Introduction • System Model • Performance Analysis • Numerical and Simulation Results • Conclusion • References

24 **Space–Time Coding and Application in WiMAX** *Naofal Al-Dhahir, Robert Calderbank, Jimmy Chui, Sushanta Das, and Suhas Diggavi* **24**-1
Introduction • Space–Time Codes: A Primer • Space–Time Block Codes • Application of Space–Time Coding in WiMAX • A Novel Quaternionic Space–Time Block Code • Simulation Results • Appendix A: Quaternions • References

25 **Exploiting Diversity in MIMO-OFDM Systems for Broadband Wireless Communications** *Weifeng Su, Zoltan Safar, and K. J. Ray Liu* **25**-1
Introduction • MIMO-OFDM System Model and Code Design Criteria • Full-Diversity SF Codes Design • Full-Diversity STF Code Design • Simulation Results • Conclusion • References

26 **OFDM Technology: Fundamental Principles, Transceiver Design, and Mobile Applications** *Xianbin Wang, Yiyan Wu, and Jean-Yves Chouinard* **26**-1
Overview of OFDM Principles • OFDM-Based Mobile Broadcasting Standards • Frequency and Timing Synchronization for OFDM-Based Wireless Systems • Summary • References

27 **Space–Time Block Coding** *Mohanned O. Sinnokrot and Vijay K. Madisetti* **27**-1
Introduction • System and Channel Model • Design Criteria of Space–Time Codes • Survey of Space–Time Block Codes • Summary and Comparison of Space–Time Block Codes • Future Research Topics • References

28 **A Multiplexing Approach to the Construction of High-Rate Space–Time Block Codes** *Mohanned O. Sinnokrot and Vijay K. Madisetti* **28**-1
Introduction • System and Channel Model • Key Properties for Reduced Complexity Decoding • The R Matrix for Space–Time Block Codes • Rate-Two Space–Time Block Codes for Two Transmit Antennas • High-Rate Space–Time Block Codes for Four Transmit Antennas • Conclusion • References

29 Soft-Output Detection of Multiple-Input Multiple-Output Channels
David L. Milliner and John R. Barry **29**-1
Introduction • System Model and Notation • Problem Statement • List Tree Search • Discussion • Further Information • References

30 Lattice Reduction–Aided Equalization for Wireless Applications *Wei Zhang and Xiaoli Ma* **30**-1
Introduction • Lattice Reduction • Lattice Reduction-Aided Equalizers • Applications • Conclusions • References

31 Overview of Transmit Diversity Techniques for Multiple Antenna Systems
D. A. Zarbouti, D. A. Kateros, D. I. Kaklamani, and G. N. Prezerakos **31**-1
Introduction • System Model • Antenna and Temporal Diversity Techniques for OFDM • Antenna and Frequency Diversity Techniques for OFDM • Antenna, Time, and Frequency Diversity Techniques for OFDM • References

23

OFDM: Performance Analysis and Simulation Results for Mobile Environments

Mishal Al-Gharabally
College of Engineering and Petroleum

Pankaj Das
University of California

23.1 Introduction 23-1
23.2 System Model 23-2
23.3 Performance Analysis 23-4
Single-Input-Single-Output Systems • Space-Time-Block-Coded System
23.4 Numerical and Simulation Results 23-16
23.5 Conclusion 23-20
References 23-20

23.1 Introduction

Orthogonal frequency division multiplexing (OFDM) is recently becoming very popular, and has already been chosen as the transmission method for many wireless communications standards, such as the IEEE 802.11 wireless local area networks (WLAN) and IEEE 802.16 wireless metropolitan area networks (WMAN) also known as WiMAX. The basic principle of OFDM is to divide the available broadband wireless channel into a large number of narrowband orthogonal subcarriers. This enables OFDM systems to combat the effects of the channel frequency selectivity with low receiver complexity. Unfortunately, OFDM is very sensitive to time variations of the wireless channel. Channel time variations destroy the orthogonality of the OFDM subcarriers and cause energy to leak from one subcarrier to the adjacent subcarriers. The intercarrier interference (ICI) caused by this energy leakage severely degrades the performance of OFDM, and introduces an irreducible error floor.

The WMAN-OFDM PHY (IEEE 802.16-2004) is based on OFDM modulation and designed for non-line-of-sight (NLOS) operation in the frequency bands below 11 GHz [1], and it is very important to understand the abilities and limitations of OFDM systems by investigating their performance under different transmission conditions.

This chapter begins with a detailed description of the system model. The various parts of the uncoded OFDM system are explained in Section 23.2. In Section 23.3, the performance analysis of the OFDM system discussed in Section 23.2 is analyzed. In Section 23.3.1, the performance of the system is analyzed assuming a single antenna at both the transmitter and the receiver. It will be shown that the time variation of the wireless channel will destroy the orthogonality between OFDM subcarriers. This loss of orthogonality will cause energy to leak from one subcarrier to the adjacent subcarriers, and hence, ICI will be present at the receiver. A general case where the receiver does not have a perfect estimate of the wireless channel is considered. The effect of both ICI and channel estimation errors on the quality of the transmission is quantified by deriving a compact expression for the bit-error probability (BEP). In Section 23.3.2, the system model is extended to include space time coding as described in Section 8.3.8 of the IEEE 802.16-2004 standard documentation [1]. It will be shown that in addition to ICI, inter-antenna interference (IAI) caused by the loss of orthogonality of the space time code will also be present, hence, the conventional detection scheme will fail to detect the transmitted OFDM signal, and an alternative detection method that will eliminate IAI and significantly reduce ICI is proposed. Finally, Section 23.4 contains various performance plots obtained by both numerical and simulation analysis of the systems discussed in Sections 23.3.1 and 23.3.2.

23.2 System Model

Figure 23.1 shows the block diagram of a conventional uncoded OFDM system. The input bit stream is first grouped into blocks of size $\log_2(M)$, and then sent to the subcarrier modulator to produce $X(k)$. $X(k)$ is the complex valued data symbol in frequency domain that is modulated onto the kth subcarrier. $X(k)$ may take any value from an M-dimensional constellation depending on the type of mapping used, such as BPSK, QPSK, or QAM for coherent demodulation, or DPSK for noncoherent demodulation. In the IEEE 802.16-2004 standard, the data modulation of choice is M-QAM, where M stands for the number of symbols in the constellation map. Typical values for M are 16, 64, and 256, depending on the required data rate and channel conditions. A block of N complex valued data symbols $\{X(k)\}_{k=0}^{N-1}$ are grouped and converted into parallel (column vector) to form the input to the OFDM modulator. The OFDM modulator consists of an inverse discrete Fourier transform (IDFT) block, the output of the IDFT block is the modulated OFDM symbol in time domain and is represented by

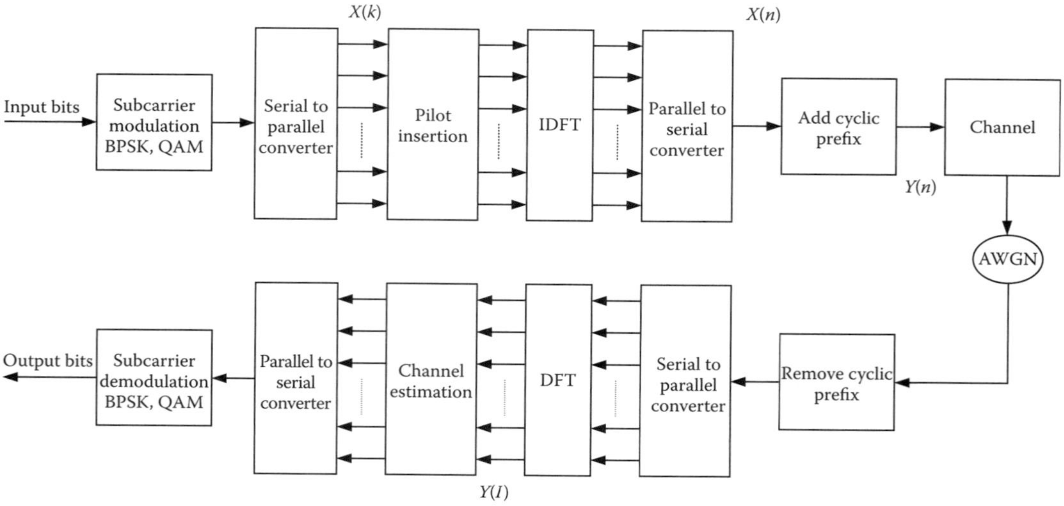

FIGURE 23.1 Uncoded OFDM system.

$$x(t) = \frac{1}{N}\sum_{k=0}^{N-1} X(k)e^{j2\pi f_k t}$$

$$x(n) = x(t)|_{t=\frac{nT}{N}} = \frac{1}{N}\sum_{k=0}^{N-1} X(k)e^{\frac{j2\pi f_k nT}{N}}$$

$$= \frac{1}{N}\sum_{k=0}^{N-1} X(k)e^{\frac{j2\pi nk}{N}}. \quad 0 \le n \le N-1 \tag{23.1}$$

where

f_k is the frequency of the kth subcarrier
T is the symbol duration
$j = \sqrt{-1}$

After the OFDM modulator, the modulated OFDM symbol $x(n)$ is parallel to serial converted, and the last G samples of $x(n)$ are copied to the beginning to form the guard interval, which protects the received OFDM symbol from intersymbol interference (ISI) caused by the channel frequency selectivity (delay spread). To ensure that the guard interval will eliminate ISI, the duration of G is chosen to be no less than the channel delay spread. The modulated OFDM symbol after adding the guard interval is

$$x^g(n) = \frac{1}{N}\sum_{k=0}^{N-1} X(k)e^{\frac{j2\pi nk}{N}} \quad G \le n \le N-1 \tag{23.2}$$

$x^g(n)$ is then upconverted before being sent through the frequency selective and time-variant channel. As a practical example from the IEEE 802.16-2004, the following parameters are used:

- Number of subcarriers $N = 256$, 1024
- Guard length $G = 1/4$, 1/8, 1/16, and 1/32 of the total number of subcarriers

The Stanford University Interim (SUI) channel model consists of a set of six empirical channels that cover most of the terrain types that are typical of the continental United States. The definition of each of the six channels is given in Ref. [2]. Table 23.1 shows the parameters for SUI-6 channel.

Each channel can be modeled as a tapped delay line with time-variant tap gains $h(l, n)$, where l and n are the path and time indices, respectively. The tap gains $\{h(l, n)\}_{l=0}^{L-1}$ are complex Gaussian random variables ($h(l, n) = |h(l, n)|e^{j\phi(n)}$) that are independent and uncorrelated for different values of l; this can be achieved by assuming the wide-sense stationary uncorrelated scattering channels (WSSUS).

If the length of the guard interval G is chosen such that the ISI is negligible, then the received baseband OFDM symbol after removing the first G samples is given by

$$y(n) = \sum_{l=0}^{L-1} h(l, n)x(n-l) + w(n) \tag{23.3}$$

where $w(n)$ is a complex additive white Gaussian noise (AWGN) with one-sided power spectral density N_0.

TABLE 23.1 SUI-6 Channel Model

	Tap 1	Tap 2	Tap 3	Units
Delay	0	14	20	Ms
Power	0	−10	−14	dB
Doppler	0.4	0.3	0.5	Hz

23.3 Performance Analysis

23.3.1 Single-Input-Single-Output Systems

At the receiver, the received symbol is first demodulated by the N-point discrete Fourier transform (DFT), and the demodulated OFDM symbol at the ith subcarrier is given by

$$\begin{aligned}
Y(i) &= \sum_{n=0}^{N-1} y(n)e^{\frac{-j2\pi ni}{N}} \\
&= \sum_{n=0}^{N-1}\sum_{l=0}^{L-1} h(l,n)x(n-l)e^{\frac{-j2\pi ni}{N}} + \sum_{n=0}^{N-1} w(n)e^{\frac{-j3\pi ni}{N}} \\
&= \sum_{n=0}^{N-1}\sum_{l=0}^{L-1} h(l,n)\frac{1}{N}\sum_{k=0}^{N-1} X(k)e^{\frac{-2\pi(n-l)k}{N}}e^{\frac{-j2\pi ni}{N}} + W(i) \\
&= \frac{1}{N}\sum_{k=0}^{N-1} X(k)\sum_{n=0}^{N-1}\sum_{l=0}^{L-1} h(l,n)e^{\frac{j2\pi(k-i)n}{N}}e^{\frac{-j2\pi lk}{N}} + W(i)
\end{aligned} \tag{23.4}$$

Splitting Equation 23.4 into two parts, namely, for $i=k$ and $i \neq k$, respectively, the received OFDM symbol at the kth subcarrier can be represented in the following compact form:

$$\begin{aligned}
Y(k) &= \frac{1}{N}\sum_{n=0}^{N-1} H(k,n)X(k) + \frac{1}{N}\sum_{n=0}^{N-1}\sum_{i=0,i\neq k}^{N-1} H(i,n)X(i)e^{\frac{j2\pi(i-k)n}{N}} + W(k) \\
&= \alpha(k)X(k) + \beta(k) + W(k)
\end{aligned} \tag{23.5}$$

where $H(k, n)$ is the time-varying frequency response of the channel. The first term above represents the desired signal, which consists of the transmitted symbol $X(k)$ multiplied by the multiplicative distortion $\alpha(k)$. $\beta(k)$ is the ICI owing to the loss of subcarrier orthogonality and $W(k)$ is the Fourier transform of $w(n)$. To show how the expression of the received signal will differ if the channel is time invariant, that is, how does the channel time variation cause ICI, let us start from the orthogonality principle,

$$\frac{1}{N}\sum_{n=0}^{N-1} e^{\frac{j2\pi(u-v)n}{N}} = \begin{cases} 1 & u = v \\ 0 & u \neq v \end{cases} \tag{23.6}$$

Now recall the expression of the ICI

$$\beta(k) = \frac{1}{N}\sum_{n=0}^{N-1}\sum_{i=0,i\neq k}^{N-1} H(i,n)X(i)e^{\frac{j2\pi(i-k)n}{N}}$$

if the channel is time invariant, the above expression reduces to

$$\beta(k) = \sum_{i=0,i\neq k}^{N-1} H(i)X(i)\frac{1}{N}\sum_{n=0}^{N-1} e^{\frac{j2\pi(i-k)n}{N}}$$

and from Equation 23.6, it can be clearly seen that the ICI term vanishes, and the channel will only induce a multiplicative distortion. To gain more insight into the system, it is sometimes desirable to express the system in matrix form. The received signal (Equation 23.3) can be expressed in matrix form as such

$$\mathbf{y} = \mathbf{h}\mathbf{x} + \mathbf{w} \tag{23.7}$$

where boldface letters represent vector or matrix quantities. The variables in Equation 23.7 are defined as follows:

$$\begin{aligned}\mathbf{y} &= [y(0)y(1)\cdots y(N-1)]^{\mathrm{T}}\\ \mathbf{x} &= [x(0)x(1)\cdots x(N-1)]^{\mathrm{T}}\\ \mathbf{w} &= [w(0)w(1)\cdots w(N-1)]^{\mathrm{T}}\end{aligned}$$

and the channel matrix $\mathbf{h}$ is given by

$$\mathbf{h} = \begin{bmatrix} h(0,0) & 0 & \cdots & 0 & h(0,L-1) & \cdots & h(0,1) \\ h(1,1) & h(1,0) & 0 & \cdots & 0 & \ddots & \vdots \\ \vdots & \ddots & \ddots & \ddots & \ddots & \ddots & h(L-2,L-1) \\ h(L-1,L-1) & \ddots & \ddots & h(L-1,0) & 0 & \ddots & 0 \\ 0 & h(L,L-1) & \ddots & \ddots & h(L,0) & \ddots & \vdots \\ \vdots & \ddots & \ddots & \cdots & \ddots & \ddots & 0 \\ 0 & \cdots & 0 & h(N-1,L-1) & \cdots & h(N-1,1) & h(N-1,0) \end{bmatrix} \tag{23.8}$$

A very important observation here is that the channel matrix $\mathbf{h}$ is not circulant, which is a direct consequence of the channel time variation. A desirable property of circulant matrices is that they can be diagonalized by pre- and postmultiplication with a unitary matrix. Let $\mathbf{F}$ be the DFT matrix defined as

$$\mathbf{F} = [F_{p,q}]_{N\times N} \quad F_{p,q} = \frac{1}{\sqrt{N}}\mathrm{e}^{\frac{-j2\pi pq}{N}} \tag{23.9}$$

Applying the DFT to the received OFDM vector in Equation 23.7 we get

$$\begin{aligned}\mathbf{Y} &= \mathbf{F}\mathbf{y} = \mathbf{F}(\mathbf{h}\mathbf{x} + \mathbf{w})\\ &= \mathbf{F}\mathbf{h}\mathbf{x} + \mathbf{F}\mathbf{w}\end{aligned} \tag{23.10}$$

From Equation 23.1, we have $\mathbf{x} = \mathbf{F}^{\mathrm{H}}\mathbf{X}$, where $\mathbf{F}^{\mathrm{H}}$ is the IDFT matrix. The received OFDM vector then becomes

$$\begin{aligned}&= \mathbf{F}\mathbf{h}\mathbf{F}^{\mathbf{H}}\mathbf{X} + \mathbf{F}\mathbf{w}\\ &= \mathbf{H}\mathbf{X} + \mathbf{W}\end{aligned} \tag{23.11}$$

where $(.)^{\mathbf{H}}$ is the conjugate transpose (Hermitian) operation and $\mathbf{H} = \mathbf{F}\mathbf{h}\mathbf{F}^{\mathrm{H}}$ is given by

$$\mathbf{H} = [H_{p,q}]_{N\times N} \quad H_{p,q} = \frac{1}{N}\sum_{n=0}^{N-1} H(q,n)\mathrm{e}^{\frac{j2\pi(q-p)n}{N}} \tag{23.12}$$

If the channel is time invariant, then **h** has a circulant structure, and since the DFT matrix **F** is a unitary matrix, this implies that $\mathbf{H} = \mathbf{FhF}^{\mathrm{H}}$ is a diagonal matrix. It can now be stated that mathematically, the ICI is caused by the nondiagonal elements of **H**. The received OFDM vector is now given as

$$\begin{bmatrix} Y(0) \\ \vdots \\ Y(N-1) \end{bmatrix} = \begin{bmatrix} H_{0,0} & \cdots & H_{0,N-1} \\ \vdots & \ddots & \vdots \\ H_{N-1,0} & \cdots & H_{N-1,N-1} \end{bmatrix} \begin{bmatrix} X(0) \\ \vdots \\ X(N-1) \end{bmatrix} + \begin{bmatrix} W(0) \\ \vdots \\ W(N-1) \end{bmatrix} \tag{23.13}$$

The performance of the above system will be analyzed by deriving the BEP. The receiver is assumed to have an estimate of the channel at each subcarrier, namely, $\alpha(k)$, but this estimate is not necessarily accurate. Obtaining good channel estimates for a time-varying channel is a complicated task, and the quality of the channel estimates will determine the overall performance of the OFDM system. The process of obtaining the channel estimates is out of the scope of this chapter. For the interested reader, Refs. [3–6] provide a good sample of the work done on channel estimation for OFDM system using pilot symbol-aided modulation (PSAM). Since this chapter is mainly concerned with the overall performance analysis, the quality of the channel estimates will be measured by the complex correlation coefficient ρ between the channel $\alpha(k)$ and its estimate $g(k)$.

$$\rho = \frac{E[\alpha g^*]}{\sqrt{E[|g|^2]E[|\alpha|^2]}} = \frac{Re\{E[\alpha g^*]\} + jIm\{E[\alpha g^*]\}}{\sigma_\alpha \sigma_g} \tag{23.14}$$

where we have dropped the subcarrier index for simplicity of notation. $Re\{.\}$ and $Im\{.\}$ denote the real and imaginary parts, respectively, $2\sigma_g^2 = E[|g|^2]$ is the variance of the channel estimate and $E[.]$ the expectation operation. To show how ρ can be obtained, two common channel estimation schemes are considered. First, if the channel estimation scheme results in an additive error ($g = \alpha + \eta$), where η is a complex Gaussian random variable with zero mean and variance $2\sigma_\eta^2$, then ρ is given by Ref. [9]

$$\rho = \frac{\sigma_\alpha}{\sqrt{\sigma_\alpha^2 + \sigma_\eta^2}}$$

Second, if g is obtained from minimum mean square error channel estimation, then ρ is given by

$$\rho = \frac{\sigma_g}{\sigma_\alpha}$$

Starting from the received signal at the kth subcarrier (Equation 23.5), after dropping the subcarrier index for notational simplicity

$$Y = \alpha X + \beta + W$$

The ICI is modeled as a zero-mean Gaussian random variable with variance equal to σ_β^2. A proof of the Gaussianity of the ICI is available in Ref. [7]. Assuming M-QAM modulation, the equalized OFDM symbol is given by

$$Z = \frac{Yg^*}{|g|^2} = \frac{\alpha g^* X}{|g|^2} + \frac{(\beta + W)g^*}{|g|^2} \tag{23.15}$$

The direct approach to find the average BEP is to solve the following triple integration:

$$P_e = \int_0^{\infty}\int_0^{\infty}\int_0^{2\pi} P(e|\alpha|, |g|, \phi) f(|\alpha|, |g|, \phi) d\alpha \; dg \; d\phi$$

where

$P(e|\alpha|, |g|, \phi)$ is the BEP conditioned on $|\alpha|$, $|g|$ and the phase difference between α and g, namely, ϕ
$f(|\alpha|, |g|, \phi)$ is the joint probability density function (pdf) of $|\alpha|$, $|g|$, and ϕ

To avoid having to solve a triple integration, a different approach is proposed in Ref. [7] to simplify the problem.

Since α and g are jointly Gaussian [8], conditioned on g, the true channel α can be expressed in terms of the channel estimate g as follows [9]:

$$\alpha = \rho \frac{\sigma_\alpha}{\sigma_g} g + U \tag{23.16}$$

where U is a complex Gaussian random variable with independent inphase and quadrature Gaussian components. The random variable U has zero mean and variance equal to $\sigma_U^2 = 2\sigma_\alpha^2(1 - |\rho|^2)$ The variance of U represents the variance of the channel estimation error, while ρ represents the quality of the estimation. Substituting Equation 23.16 into Equation 23.15, we get

$$Z = \rho \frac{\sigma_\alpha}{\sigma_g} X + \frac{Ug^* X}{|g|^2} + \frac{\beta g^*}{|g|^2} + \frac{W g^*}{|g|^2} \tag{23.17}$$

Now expanding U and W into inphase and quadrature component, the equalized OFDM symbol in Equation 23.17 becomes

$$\begin{aligned} Z &= \rho \frac{\sigma_\alpha}{\sigma_g} X + \frac{\beta g^*}{|g|^2} + \frac{g^*(XU_I + W_I)}{|g|^2} + j\frac{g^*(XU_Q + W_Q)}{|g|^2} \\ &= \rho \frac{\sigma_\alpha}{\sigma_g} X + \xi_1 + \xi_2 + j\xi_3 \end{aligned} \tag{23.18}$$

Conditioned on g, ξ_1, ξ_2, and ξ_3 are Gaussian random variables with zero mean and variances given by

$$\begin{aligned} \sigma_{\xi_1}^2 &= \frac{\sigma_\beta^2}{|g|^2} \\ \sigma_{\xi_{2,3}}^2 &= \frac{\sigma_\alpha^2(1 - |\rho|^2 + \sigma_W^2}{|g|^2} \end{aligned}$$

where it is assumed that $E[|X|^2] = 1$. From the expression of the equalized OFDM symbol in Equation 23.18, it can been seen that the problem of finding the average-error probability is greatly simplified, and reduced to solving a single integral to average over $|g|^2$ as opposed to triple integrations. To show the simplicity of the approach, the BEP for 16-QAM is derived. Figure 1 of Ref. [10] shows the 16-QAM constellation with gray coding. Assuming that each symbol contains four bits ($b_1b_2b_3b_4$), the BEP for bit b_1, conditioned on g, is given by

$$P_e(b_1|g) = \frac{1}{2}[\, Pr(Z_I < 0|X_I = d, g) + Pr(Z_I < 0|X_I = 3d, g)] \tag{23.19}$$

where

Z_I and X_I are the real parts of Z and X, respectively
d is the minimum distance between two QAM symbols in the constellation

Similarly, the conditional error probability for bit b_3 is

$$P_e(b_3|g) = \frac{1}{2}\left[Pr(|Z_I| > 2d|X_I = d, g) + Pr(|Z_I| < 2d|X_I = 3d, g)\right] \tag{23.20}$$

Define $\nu = \xi_1 + \xi_2$, ν is then a zero-mean Gaussian random variable with variance equal to $\sigma_\nu^2 = \sigma_{\xi_1}^2 + \sigma_{\xi_1}^2$, where it is assumed that the thermal noise is independent of the ICI. Then

$$P_e(b_1|g) = \frac{1}{2}\left[Q\left(\sqrt{\frac{\rho^2\left(\sigma_\alpha^2/\sigma_g^2\right)d^2}{\sigma_\nu^2}}\right) + Q\left(\sqrt{\frac{9\rho^2\left(\sigma_\alpha^2/\sigma_g^2\right)d^2}{\sigma_\nu^2}}\right)\right] \tag{23.21}$$

Substituting $\sigma_{\xi_1}^2$ and $\sigma_{\xi_2}^2$ into σ_ν^2, the conditional BEP for bit b_1 becomes

$$P_e(b_1|g) = \frac{1}{2}\left[Q\left(\sqrt{\frac{\rho^2\left(\sigma_\alpha^2/\sigma_g^2\right)d^2|g|^2}{\sigma_\beta^2 + \sigma_\alpha^2(1-\rho^2) + \sigma_W^2}}\right) + Q\left(\sqrt{\frac{9\rho^2\left(\sigma_\alpha^2/\sigma_g^2\right)d^2|g|^2}{\sigma_\beta^2 + \sigma_\alpha^2(1-\rho^2) + \sigma_W^2}}\right)\right] \tag{23.22}$$

Assuming a Rayleigh fading channel, the pdf of $|g|^2$ is $f_{|g|^2}(\lambda) = 1/2\sigma_g^2 e^{-\lambda/2\sigma_g^2}$, and the average BER for bit b_1 is

$$P_e(b_1) = \frac{1}{2}(I_1 + I_2) \tag{23.23}$$

where

$$I_1 = \int_0^\infty Q\left(a_1\sqrt{|g|^2}\right)\frac{1}{2\sigma_g^2}e^{-\frac{\lambda}{2\sigma_g^2}}d\lambda$$

$$I_2 = \int_0^\infty Q\left(a_2\sqrt{|g|^2}\right)\frac{1}{2\sigma_g^2}e^{-\frac{\lambda}{2\sigma_g^2}}d\lambda$$

and

$$a_1^2 = \frac{\rho^2\left(\sigma_\alpha^2/\sigma_g^2\right)d^2}{\sigma_\beta^2 + \sigma_\alpha^2(1-\rho^2) + \sigma_W^2}, \quad a_2^2 = 9a_1^2$$

By using Equation 5.6 in Ref. [11] we get

$$I_1 = \frac{1}{2}\left[1 - \sqrt{\frac{\rho^2\sigma_\alpha^2 d^2}{\sigma_\beta^2 + \sigma_\alpha^2(1-\rho^2) + \sigma_W^2 + 9\rho^2\sigma_\alpha^2 d^2}}\right]$$

$$I_2 = \frac{1}{2}\left[1 - \sqrt{\frac{9\rho^2\sigma_\alpha^2 d^2}{\sigma_\beta^2 + \sigma_\alpha^2(1-\rho^2) + \sigma_W^2 + 9\rho^2\sigma_\alpha^2 d^2}}\right]$$

then

$$P_e(b_1) = \frac{1}{2}\left[1 - \frac{1}{2}\sqrt{\frac{\rho^2\sigma_\alpha^2 d^2}{\sigma_\beta^2 + \sigma_\alpha^2(1-\rho^2) + \sigma_W^2 + \rho^2\sigma_\alpha^2 d^2}} - \frac{1}{2}\sqrt{\frac{9\rho^2\sigma_\alpha^2 d^2}{\sigma_\beta^2 + \sigma_\alpha^2(1-\rho^2) + \sigma_W^2 + \rho^2\sigma_\alpha^2 d^2}}\right] \quad (23.24)$$

If we assume that the channel is time invariant ($\sigma_\beta^2 = 0$) and that perfect knowledge of the channel ($\rho = 1$) is available, then $P_e(b_1)$ reduces to the well-known result [12], namely,

$$P_e(b_1) = \frac{1}{2}\left[1 - \frac{1}{2}\sqrt{\frac{2\ snr}{5 + 2\ snr}} - \frac{1}{2}\sqrt{\frac{18\ snr}{5 + 18\ snr}}\right]$$

where $d^2 = 2E_b/5$ and $snr = \sigma_\alpha^2 E_b/\sigma_W^2$. Furthermore, if the channel is completely unknown ($\rho = 0$), then the average-error probability approaches 0.5, which is intuitively correct. By using similar steps as above, $P_e(b_3)$ is equal to

$$P_e(b_3) = \frac{1}{2}\left[1 - \frac{1}{2}\sqrt{\frac{d^2\eta_1^2\sigma_g^2}{\sigma_\beta^2 5\sigma_\alpha^2(1-\rho^2) + \sigma_W^2 + d^2\eta_1^2\sigma_g^2}} - \frac{1}{2}\sqrt{\frac{d^2\eta_2^2\sigma_g^2}{\sigma_\beta^2 5\sigma_\alpha^2(1-\rho^2) + \sigma_W^2 + d^2\eta_2^2\sigma_g^2}} - \frac{1}{2}\sqrt{\frac{d^2\eta_3^2\sigma_g^2}{\sigma_\beta^2 5\sigma_\alpha^2(1-\rho^2) + \sigma_W^2 + d^2\eta_3^2\sigma_g^2}} + \frac{1}{2}\sqrt{\frac{d^2\eta_4^2\sigma_g^2}{\sigma_\beta^2 5\sigma_\alpha^2(1-\rho^2) + \sigma_W^2 + d^2\eta_4^2\sigma_g^2}}\right] \quad (23.25)$$

where

$\eta_1 = 2 - \rho\sigma_\alpha/\alpha_g$
$\eta_2 = 2 + \rho\sigma_\alpha/\alpha_g$
$\eta_3 = -2 + 3\rho\sigma_\alpha/\alpha_g$
$\eta_4 = 2 + 3\rho\sigma_\alpha/\sigma_g$

Owing to the symmetry of the 16-QAM constellation, it can be easily seen that $P_e(b_1) = P_e(b_2)$ and $P_e(b_3) = P_e(b_4)$. Hence, the final expression for the average BER is given by

$$P_e = \frac{1}{2}\,[P_e(b_1) + P_e(b_3)]$$

To complete the derivation, the power of the ICI, σ_β^2, needs to be derived. From Ref. [13],

$$\begin{aligned}\sigma_\beta^2 &= E[|\beta(k)|^2] \\ &= \frac{1}{N^2}\sum_{\substack{i=0\\i\neq k}}^{N-1} E[|X(i)|^2] \times \sum_{n_1=0}^{N-1}\sum_{n_2=0}^{N-1} E[H(n_1,i)H^*(n_2,i)]e^{\frac{-j2\pi(n_1-n_2)(k-i)}{N}} \\ &= \frac{E_s}{N^2}\sum_{\substack{i=0\\i\neq k}}^{N-1}\left[N + 2\sum_{n=1}^{N-1}(N-n)\times J_0\left(\frac{2\pi f_d Tn}{N}\right)\cos\left(\frac{2\pi n(k-i)}{N}\right)\right]\end{aligned}$$

The above procedure can be generalized to any constellation size and data modulation scheme. To investigate the performance of a specific channel estimation scheme, one has to only derive the appropriate value of ρ for that scheme, and express the true channel α in terms of its estimate g to simplify the problem.

23.3.2 Space-Time-Block-Coded System

In Section 23.2.1, it is shown that the time variation of the wireless channel severely degrades the performance of OFDM; this degradation in performance is caused by the loss of orthogonality between subcarriers. In space-time-block-coded (STBC)-OFDM, the effect of wireless channel variation is more profound than that of single-input-single-output (SISO)-OFDM, especially when orthogonal STBC are used to code across space and time. It will be shown that along with ICI, channel variations cause IAI between two different transmit antennas.

In the IEEE 802.16-2004 standard, Alamouti STBC [14] may be used on the downlink to provide second order (space) transmit diversity [1]. There are two transmit antennas on the base station (BS) side and one receive antenna on the subscriber station (SS) side. This scheme requires multiple-input-single-output channel estimation, which is out of the scope of this chapter. Figure 23.2 shows the STBC-OFDM as proposed by the standard documentation. Each transmit antenna has its own OFDM chain, and both antennas transmit at the same time two different OFDM symbols $x_1(n)$ and $x_2(n)$. The two symbols are transmitted twice from the two transmit antennas to get second-order diversity.

The transmission scheme can be described as follows: at time $2n$, OFDM symbols $x_1(n)$ and $x_2(n)$ are transmitted from the first and second antennas, respectively; at time $2n+1, -x_2^*(n)$ and $x_1^*(n)$ are transmitted from the first and second antennas, respectively, where $(.)^*$ denotes the conjugate operation. The code matrix of the above scheme is given by

$$\begin{bmatrix} x_1(n) & x_2(n) \\ -x_2^*(n) & x_1^*(n) \end{bmatrix} \tag{23.26}$$

Starting from the matrix representation of the received OFDM vector, and following the same notations in Equations 23.10 and 23.11, it can easily be seen that the received OFDM vectors for the $2n$ and $2n+1$ transmission instances are given by

$$\mathbf{Y}^{2n} = \mathbf{H}_1^{2n}\mathbf{X}_1 + \mathbf{H}_2^{2n}\mathbf{X}_2 + \mathbf{W}^{2n} \tag{23.27}$$

$$\mathbf{Y}^{2n+1} = -\mathbf{H}_1^{2n+1}\mathbf{X}_2^* + \mathbf{H}_2^{2n+1}\mathbf{X}_1^* + \mathbf{W}^{2n+1} \tag{23.28}$$

where

$\mathbf{Y}^i$ and $\mathbf{W}^i$ are the received OFDM vector and AWGN at time $i=2n, 2n+1$
$\mathbf{H}_j^i$ is the channel matrix from transmit antenna $j=1,2$ at time i
$\mathbf{X}_j$ the jth transmitted OFDM vectors

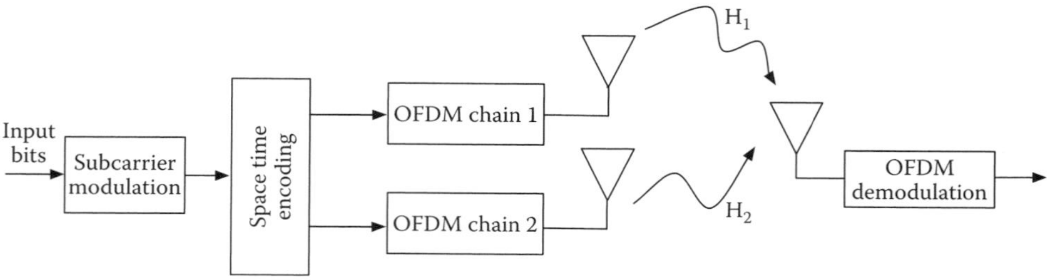

FIGURE 23.2 STBC-OFDM (Alamouti).

The above system can be written in the following compact matrix representation:

$$\begin{bmatrix} \mathbf{Y}^{2n} \\ \mathbf{Y}^{2n+1^*} \end{bmatrix} = \begin{bmatrix} \mathbf{H}_1^{2n} & \mathbf{H}_2^{2n} \\ \mathbf{H}_2^{2n+1^*} & -\mathbf{H}_1^{2n+1^*} \end{bmatrix} \begin{bmatrix} \mathbf{X}_1 \\ \mathbf{X}_2 \end{bmatrix} + \begin{bmatrix} \mathbf{W}^{2n} \\ \mathbf{W}^{2n+1^*} \end{bmatrix} \tag{23.29}$$

$$\mathbf{Y} = \mathbf{HX} + \mathbf{W}$$

where

$\mathbf{Y}, \mathbf{X}, \mathbf{W} \in C^{2N\times 1}$

$\mathbf{H} \in C^{2N\times 2N}$

By carefully analyzing Equation 23.29, the following two problems can be observed if the channel is time varying.

- Similar to the SISO system in Equation 23.13, the channel matrices $\mathbf{H}_j^i$ are not diagonal.
- $\mathbf{H}_1^{2n} \neq \mathbf{H}_1^{2n+1}$ and $\mathbf{H}_2^{2n} \neq \mathbf{H}_2^{2n+1}$.

As a direct consequence of the first problem, the system performance will be degraded from ICI caused by the nondiagonal elements of $\mathbf{H}_j^i$. This ICI is because of channel variations and is independent of the coding scheme used. The second problem causes IAI between the two transmit antennas, which further degrades system performance. Analogous to ICI, the IAI results from the loss of orthogonality of the channel matrix, a major property of the Alamouti coding scheme; hence, IAI depends on the code used, and is present only in orthogonal STBC schemes.

The effect of the time variation of the wireless channel can also be seen by observing the received signal at a specific subcarrier. The received OFDM signal at the kth subcarrier is given by

$$Y^{2n} = \alpha_1^{2n} X_1 + \beta_1^{2n} + \alpha_2^{2n} X_2 + \beta_2^{2n} + W^{2n} \tag{23.30}$$

$$Y^{2n+1} = -\alpha_1^{2n+1} X_2^* - \beta_1^{2n+1} + \alpha_2^{2n+1} X_1^* + \beta_2^{2n+1} + W^{2n+1} \tag{23.31}$$

and in matrix form

$$\begin{bmatrix} Y^{2n} \\ Y^{2n+1^*} \end{bmatrix} = \begin{bmatrix} \alpha_1^{2n} & \alpha_2^{2n} \\ \alpha_2^{2n+1^*} & -\alpha_1^{2n+1^*} \end{bmatrix} \begin{bmatrix} X_1 \\ X_2 \end{bmatrix} + \begin{bmatrix} \beta_1^{2n} + \beta_2^{2n} \\ \beta_2^{2n+1} - \beta_1^{2n+1} \end{bmatrix} + \begin{bmatrix} W^{2n} \\ W^{2n+1^*} \end{bmatrix} \tag{23.32}$$

where the subcarrier index is dropped for notational simplicity. α_j^i and β_j^i are the same as those defined in Equation 23.5. The two problems discussed above can be clearly seen here and restated as such:

- Time variation of the wireless channel causes ICI
- The channel matrix is not orthogonal anymore, because $\alpha_j^{2n} \neq \alpha_j^{2n+1}$; hence, IAI is present

Figure 23.3 shows the block diagram of the proposed detector. The first step in decoding the transmitted symbols, as proposed by Alamouti, is to multiply the received OFDM vector in Equation 23.29 by the transpose conjugate of the channel matrix; this operation is referred to as space time combining.

$$\begin{aligned} \mathbf{Z} &= \mathbf{H}^{\mathrm{H}}\mathbf{Y} = \mathbf{H}^{\mathrm{H}}(\mathbf{HX} + \mathbf{W}) \\ &= \mathbf{H}^{\mathrm{H}}\mathbf{HX} + \mathbf{H}^{\mathrm{H}}\mathbf{W} \\ &= \mathbf{AX} + \hat{\mathbf{W}} \end{aligned} \tag{23.33}$$

$$\begin{bmatrix} \mathbf{Z}^{2n} \\ \mathbf{Z}^{2n+1} \end{bmatrix} = \begin{bmatrix} \mathbf{A}_1 & \mathbf{A}_2 \\ \mathbf{A}_3 & \mathbf{A}_4 \end{bmatrix} \begin{bmatrix} \mathbf{X}_1 \\ \mathbf{X}_2 \end{bmatrix} + \begin{bmatrix} \hat{\mathbf{W}}^{2n} \\ \hat{\mathbf{W}}^{2n+1} \end{bmatrix} \tag{23.34}$$

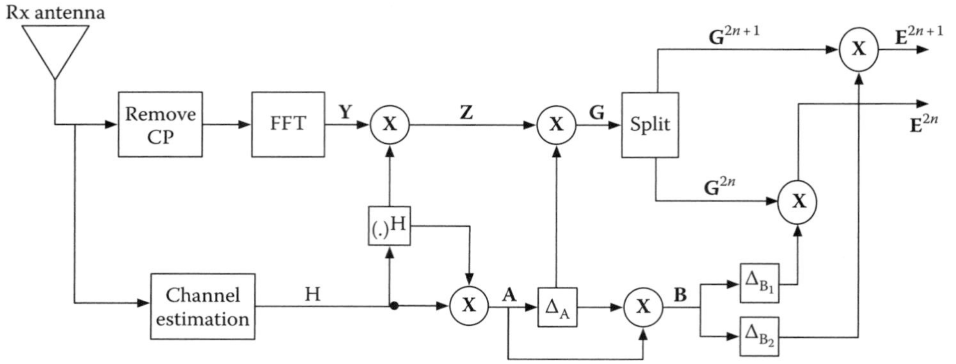

FIGURE 23.3 Block diagram of the proposed detection for STBC-OFDM.

where

$$\mathbf{A} = \mathbf{H}^{\mathrm{H}}\mathbf{H}$$

$$\mathbf{A}_1 = \mathbf{H}_1^{2n^{\mathrm{H}}}\mathbf{H}_1^{2n} + \mathbf{H}_2^{2n+1^{\mathrm{T}}}\mathbf{H}_2^{2n+1^{*}}$$

$$\mathbf{A}_2 = \mathbf{H}_1^{2n^{\mathrm{H}}}\mathbf{H}_2^{2n} - \mathbf{H}_2^{2n+1^{\mathrm{T}}}\mathbf{H}_1^{2n+1^{*}}$$

$$\mathbf{A}_3 = \mathbf{H}_2^{2n^{\mathrm{H}}}\mathbf{H}_1^{2n} - \mathbf{H}_1^{2n+1^{\mathrm{T}}}\mathbf{H}_2^{2n+1^{*}}$$

$$\mathbf{A}_4 = \mathbf{H}_1^{2n^{\mathrm{H}}}\mathbf{H}_1^{2n} + \mathbf{H}_2^{2n+1^{\mathrm{T}}}\mathbf{H}_2^{2n+1^{*}}$$

and $\mathbf{A}_1, \mathbf{A}_2, \mathbf{A}_3$, and $\mathbf{A}_4$ are all $\in C^{N\times N}$. If the channel is time invariant, then $\mathbf{H}$ is an orthogonal matrix and $\mathbf{A}_2 = \mathbf{A}_3 = 0$; therefore, Equation 23.34 reduces to

$$\begin{bmatrix} \mathbf{Z}^{2n} \\ \mathbf{Z}^{2n+1} \end{bmatrix} = \begin{bmatrix} \mathbf{A}_1 & \mathbf{0} \\ \mathbf{0} & \mathbf{A}_4 \end{bmatrix} \begin{bmatrix} \mathbf{X}_1 \\ \mathbf{X}_2 \end{bmatrix} + \begin{bmatrix} \hat{\mathbf{W}}^{2n} \\ \hat{\mathbf{W}}^{2n+1} \end{bmatrix} \tag{23.35}$$

where
- $\mathbf{A}_1$ and $\mathbf{A}_4$ are diagonal matrices
- $\mathbf{0}$ are matrices with zeros in all its elements

It can be seen from Equation 23.35 that $\mathbf{X}_1$ and $\mathbf{X}_2$ are completely decoupled, and can accurately be detected using the maximum likelihood (ML) detector proposed in Ref. 13.

In a mobile channel environment, the transmitted symbols $\mathbf{X}_1$ and $\mathbf{X}_2$ cannot be decoupled using the Alamouti detection scheme, and a different approach should be taken. Here, a new two-stage detection scheme that will reduce the effect of both ICI and IAI and enable us to accurately detect the transmitted symbols is presented.

Starting from the matrix representation of the output of the space time combiner

$$\begin{bmatrix} \mathbf{Z}^{2n} \\ \mathbf{Z}^{2n+1} \end{bmatrix} = \begin{bmatrix} \mathbf{A}_1 & \mathbf{A}_2 \\ \mathbf{A}_3 & \mathbf{A}_4 \end{bmatrix} \begin{bmatrix} \mathbf{X}_1 \\ \mathbf{X}_2 \end{bmatrix} + \begin{bmatrix} \hat{\mathbf{W}}^{2n} \\ \hat{\mathbf{W}}^{2n+1} \end{bmatrix}$$

$$\mathbf{Z} = \mathbf{A}\mathbf{X} + \hat{\mathbf{W}}$$

In the first stage of the detector, multiply the output of the space time combiner by the following decoupling matrix [15]:

$$\boldsymbol{\Delta}_{\mathrm{A}} = \begin{bmatrix} \mathbf{I}_N & -\mathbf{A}_2\mathbf{A}_4^{-1} \\ -\mathbf{A}_3\mathbf{A}_1^{-1} & \mathbf{I}_N \end{bmatrix} \quad (23.36)$$

where

$\boldsymbol{\Delta}_{\mathrm{A}}$ is a decoupling matrix constructed from the elements of $\mathbf{A}$

$\mathbf{I}_N$ is an $N \times N$ identity matrix

The result of the first stage would be

$$\begin{aligned} \mathbf{G} &= \boldsymbol{\Delta}_{\mathrm{A}}\mathbf{Z} = \boldsymbol{\Delta}_{\mathrm{A}}\mathbf{A}\mathbf{X} + \boldsymbol{\Delta}_{\mathrm{A}}\hat{\mathbf{W}} \\ &= \mathbf{B}\mathbf{X} + \tilde{\mathbf{W}} \end{aligned} \quad (23.37)$$

where the matrices in Equation 23.37 are defined as follows:

$$\mathbf{G} = \Delta_{\mathrm{A}}\mathbf{Z} = \begin{bmatrix} \mathbf{G}^{2n} \\ \mathbf{G}^{2n+1} \end{bmatrix} \in C^{2N\times 1} \quad (23.38)$$

$$\mathbf{B} = \Delta_{\mathrm{A}}\mathbf{A} = \begin{bmatrix} \mathbf{B}_1 & 0 \\ 0 & \mathbf{B}_2 \end{bmatrix} \in C^{2N\times 2N} \quad (23.39)$$

$$\mathbf{B}_1 = \mathbf{A}_1 - \mathbf{A}_2\mathbf{A}_4^{-1}\mathbf{A}_3 \in C^{N\times N} \quad (23.40)$$

$$\mathbf{B}_2 = \mathbf{A}_4 - \mathbf{A}_3\mathbf{A}_1^{-1}\mathbf{A}_2 \in C^{N\times N} \quad (23.41)$$

and the output of the first stage is

$$\begin{bmatrix} \mathbf{G}^{2n} \\ \mathbf{G}^{2n+1} \end{bmatrix} = \begin{bmatrix} \mathbf{B}_1 & 0 \\ 0 & \mathbf{B}_2 \end{bmatrix} \begin{bmatrix} \mathbf{X}_1 \\ \mathbf{X}_2 \end{bmatrix} + \begin{bmatrix} \tilde{\mathbf{W}}^{2n} \\ \tilde{\mathbf{W}}^{2n+1} \end{bmatrix} \quad (23.42)$$

It can be seen from Equation 23.42 that the transmitted symbols $\mathbf{X}_1$ and $\mathbf{X}_2$ are completely decoupled, but the effect of ICI is still present because $\mathbf{B}_1$ and $\mathbf{B}_2$ are not diagonal. $\mathbf{B}_1$ and $\mathbf{B}_2$ have the following structure:

$$\mathbf{B}_1 = \begin{bmatrix} b^1_{0,0} & b^1_{0,1} & \cdots & b^1_{0,N-1} \\ b^1_{1,0} & b^1_{1,1} & \cdots & b^1_{1,N-1} \\ \vdots & \vdots & \ddots & \vdots \\ b^1_{N-1,0} & b^1_{N-1,1} & \cdots & b^1_{N-1,N-1} \end{bmatrix}$$

$$\mathbf{B}_2 = \begin{bmatrix} b^2_{0,0} & b^2_{0,1} & \cdots & b^2_{0,N-1} \\ b^2_{1,0} & b^2_{1,1} & \cdots & b^2_{1,N-1} \\ \vdots & \vdots & \ddots & \vdots \\ b^2_{N-1,0} & b^2_{N-1,1} & \cdots & b^2_{N-1,N-1} \end{bmatrix}$$

where $b^j_{p,q}$ is the (p, q) element of $\mathbf{B}_j$. Similar to what was discussed earlier, the nondiagonal elements of $\mathbf{B}_1$ and $\mathbf{B}_2$ are the ICI caused by channel variations. The goal of the second stage of the detector is to reduce the effect of ICI.

In the second-detection stage, split the output of the first stage into even and odd subcarriers such as

$$\mathbf{G}^{2n} = \begin{bmatrix} \mathbf{G}_e^{2n} \\ \mathbf{G}_o^{2n} \end{bmatrix} = \begin{bmatrix} \mathbf{B}_{11} & \mathbf{B}_{12} \\ \mathbf{B}_{13} & \mathbf{B}_{14} \end{bmatrix} \begin{bmatrix} \mathbf{X}_{1e} \\ \mathbf{X}_{1o} \end{bmatrix} + \begin{bmatrix} \tilde{\mathbf{W}}_e^{2n} \\ \tilde{\mathbf{W}}_o^{2n} \end{bmatrix} \quad (23.43)$$

$$\mathbf{G}^{2n+1} = \begin{bmatrix} \mathbf{G}_e^{2n+1} \\ \mathbf{G}_o^{2+1} \end{bmatrix} = \begin{bmatrix} \mathbf{B}_{21} & \mathbf{B}_{22} \\ \mathbf{B}_{23} & \mathbf{B}_{24} \end{bmatrix} \begin{bmatrix} \mathbf{X}_{2e} \\ \mathbf{X}_{2o} \end{bmatrix} + \begin{bmatrix} \tilde{\mathbf{W}}_e^{2n+1} \\ \tilde{\mathbf{W}}_o^{2n+1} \end{bmatrix} \quad (23.44)$$

where

$$\mathbf{G}_e^i = [G^i(0)G^i(2)\cdots G^i(N-2)]^T \in C^{\frac{N}{2}\times 1}, \quad i = 2n, 2n+1 \quad (23.45)$$

$$\mathbf{G}_o^i = [G^i(1)G^i(3)\cdots G^i(N-1)]^T \in C^{\frac{N}{2}\times 1}, \quad i = 2n, 2n+1 \quad (23.46)$$

$$\mathbf{X}_{je} = [X_j(0)X_j(2)\cdots X_j(N-2)]^T \in C^{\frac{N}{2}\times 1}, \quad j = 1, 2 \quad (23.47)$$

$$\mathbf{X}_{jo} = [X_j(1)X_i(3)\cdots X_i(N-1)]^T \in C^{\frac{N}{2}\times 1}, \quad j = 1, 2 \quad (23.48)$$

and

$$\mathbf{B}_{k1} = \begin{bmatrix} b_{0,0}^k & b_{0,2}^k & \cdots & b_{0,N-2}^k \\ b_{2,0}^k & b_{2,2}^k & \cdots & b_{2,N-2}^1 \\ \vdots & \vdots & \ddots & \vdots \\ b_{N-2,0}^k & b_{N-2,2}^k & \cdots & b_{N-2,N-2}^k \end{bmatrix} \in C^{\frac{N}{2}\times\frac{N}{2}}, \quad k = 1, 2 \quad (23.49)$$

$$\mathbf{B}_{k2} = \begin{bmatrix} b_{0,1}^k & b_{0,3}^k & \cdots & b_{0,N-1}^k \\ b_{2,1}^k & b_{2,3}^k & \cdots & b_{2,N-1}^k \\ \vdots & \vdots & \ddots & \vdots \\ b_{N-2,1}^k & b_{N-2,3}^k & \cdots & b_{N-2,N-1}^k \end{bmatrix} \in C^{\frac{N}{2}\times\frac{N}{2}}, \quad k = 1, 2 \quad (23.50)$$

$$\mathbf{B}_{k3} = \begin{bmatrix} b_{1,0}^k & b_{1,2}^k & \cdots & b_{1,N-2}^k \\ b_{3,0}^k & b_{3,2}^k & \cdots & b_{3,N-2}^k \\ \vdots & \vdots & \ddots & \vdots \\ b_{N-1,0}^k & b_{N-1,2}^k & \cdots & b_{N-1,N-2}^k \end{bmatrix} \in C^{\frac{N}{2}\times\frac{N}{2}}, \quad k = 1, 2 \quad (23.51)$$

$$\mathbf{B}_{k4} = \begin{bmatrix} b_{1,1}^k & b_{1,3}^k & \cdots & b_{1,N-1}^k \\ b_{3,1}^k & b_{3,3}^k & \cdots & b_{3,N-1}^k \\ \vdots & \vdots & \ddots & \vdots \\ b_{N-1,1}^k & b_{N-1,3}^k & \cdots & b_{N-1,N-1}^k \end{bmatrix} \in C^{\frac{N}{2}\times\frac{N}{2}}, \quad k = 1, 2 \quad (23.52)$$

An interesting result from the above operation is that the matrices $\mathbf{B}_{k1}$ and $\mathbf{B}_{k4}$ contain partial ICI, in the sense that they have ICI contribution from half of the subcarriers only, and $\mathbf{B}_{k2}$ and $\mathbf{B}_{k3}$ are full ICI matrices. It is desired now to eliminate $\mathbf{B}_{k2}$ and $\mathbf{B}_{k3}$. To achieve this goal, multiply $\mathbf{G}^{2n}$ and $\mathbf{G}^{2n+1}$ in Equations 23.43 and 23.44 with the following matrices respectively.

$$\mathbf{\Delta}_{\mathrm{B}_1} = \begin{bmatrix} \mathbf{I}_{\frac{N}{2}} & -\mathbf{B}_{12}\mathbf{B}_{14}^{-1} \\ -\mathbf{B}_{13}\mathbf{B}_{11}^{-1} & \mathbf{I}_{\frac{N}{2}} \end{bmatrix} \tag{23.53}$$

$$\mathbf{\Delta}_{\mathrm{B}_2} = \begin{bmatrix} \mathbf{I}_{\frac{N}{2}} & -\mathbf{B}_{22}\mathbf{B}_{24}^{-1} \\ -\mathbf{B}_{23}\mathbf{B}_{21}^{-1} & \mathbf{I}_{\frac{N}{2}} \end{bmatrix} \tag{23.54}$$

That is,

$$\mathbf{E}^{2n} = \begin{bmatrix} \mathbf{E}_{\mathrm{e}}^{2n} \\ \mathbf{E}_{o}^{2n} \end{bmatrix} = \mathbf{\Delta}_{\mathrm{B}_1}\mathbf{G}^{2n} = \mathbf{\Delta}_{\mathrm{B}_1}\begin{bmatrix} \mathbf{B}_{11} & \mathbf{B}_{12} \\ \mathbf{B}_{13} & \mathbf{B}_{14} \end{bmatrix}\begin{bmatrix} \mathbf{X}_{1\mathrm{e}} \\ \mathbf{X}_{1\mathrm{o}} \end{bmatrix} + \mathbf{\Delta}_{\mathrm{B}_1}\begin{bmatrix} \tilde{\mathbf{W}}_{\mathrm{e}}^{2n} \\ \tilde{\mathbf{W}}_{\mathrm{o}}^{2n} \end{bmatrix} \tag{23.55}$$

$$\mathbf{E}^{2n+1} = \begin{bmatrix} \mathbf{E}_{\mathrm{e}}^{2n+1} \\ \mathbf{E}_{\mathrm{o}}^{2n+1} \end{bmatrix} = \mathbf{\Delta}_{\mathrm{B}_2}\mathbf{G}^{2n+1} = \mathbf{\Delta}_{\mathrm{B}}2\begin{bmatrix} \mathbf{B}_{21} & \mathbf{B}_{22} \\ \mathbf{B}_{23} & \mathbf{B}_{24} \end{bmatrix}\begin{bmatrix} \mathbf{X}_{2\mathrm{e}} \\ \mathbf{X}_{2\mathrm{o}} \end{bmatrix} + \mathbf{\Delta}_{\mathrm{B}_2}\begin{bmatrix} \tilde{\mathbf{W}}_{\mathrm{e}}^{2n+1} \\ \tilde{\mathbf{W}}_{\mathrm{o}}^{2n+1} \end{bmatrix} \tag{23.56}$$

and the result is

$$\begin{bmatrix} \mathbf{E}_{\mathrm{e}}^{2n} \\ \mathbf{E}_{\mathrm{o}}^{2n} \end{bmatrix} = \begin{bmatrix} \mathbf{\Lambda}_1 & 0 \\ 0 & \mathbf{\Lambda}_2 \end{bmatrix}\begin{bmatrix} \mathbf{X}_{1\mathrm{e}} \\ \mathbf{X}_{1\mathrm{o}} \end{bmatrix} + \begin{bmatrix} \tilde{\mathbf{W}}_{\mathrm{e}}^{2n} \\ \tilde{\mathbf{W}}_{\mathrm{o}}^{2n} \end{bmatrix} \tag{23.57}$$

$$\begin{bmatrix} \mathbf{E}_{\mathrm{e}}^{2n+1} \\ \mathbf{E}_{\mathrm{o}}^{2n+1} \end{bmatrix} = \begin{bmatrix} \mathbf{\Lambda}_3 & 0 \\ 0 & \mathbf{\Lambda}_4 \end{bmatrix}\begin{bmatrix} \mathbf{X}_{2\mathrm{e}} \\ \mathbf{X}_{2\mathrm{o}} \end{bmatrix} + \begin{bmatrix} \tilde{\mathbf{W}}_{\mathrm{e}}^{2n+1} \\ \tilde{\mathbf{W}}_{\mathrm{o}}^{2n+1} \end{bmatrix} \tag{23.58}$$

where

$$\mathbf{\Lambda}_1 = \mathbf{B}_{11} - \mathbf{B}_{12}\mathbf{B}_{14}^{-1}\mathbf{B}_{13} \in C^{\frac{N}{2}\times\frac{N}{2}} \tag{23.59}$$

$$\mathbf{\Lambda}_2 = \mathbf{B}_{14} - \mathbf{B}_{13}\mathbf{B}_{11}^{-1}\mathbf{B}_{12} \in C^{\frac{N}{2}\times\frac{N}{2}} \tag{23.60}$$

$$\mathbf{\Lambda}_3 = \mathbf{B}_{21} - \mathbf{B}_{22}\mathbf{B}_{24}^{-1}\mathbf{B}_{23} \in C^{\frac{N}{2}\times\frac{N}{2}} \tag{23.61}$$

$$\mathbf{\Lambda}_4 = \mathbf{B}_{24} - \mathbf{B}_{23}\mathbf{B}_{21}^{-1}\mathbf{B}_{22} \in C^{\frac{N}{2}\times\frac{N}{2}} \tag{23.62}$$

and the final expression would be

$$\mathbf{E}_{\mathrm{e}}^{2n} = \mathbf{\Lambda}_1\mathbf{X}_{1\mathrm{e}} + \bar{\mathbf{W}}_{\mathrm{e}}^{2n} \tag{23.63}$$

$$\mathbf{E}_{\mathrm{o}}^{2n} = \mathbf{\Lambda}_2\mathbf{X}_{1\mathrm{o}} + \bar{\mathbf{W}}_{\mathrm{o}}^{2n} \tag{23.64}$$

$$\mathbf{E}_{\mathrm{e}}^{2n+1} = \mathbf{\Lambda}_3\mathbf{X}_{2\mathrm{e}} + \bar{\mathbf{W}}_{\mathrm{e}}^{2n+1} \tag{23.65}$$

$$\mathbf{E}_{\mathrm{o}}^{2n+1} = \mathbf{\Lambda}_4\mathbf{X}_{2\mathrm{o}} + \bar{\mathbf{W}}_{\mathrm{o}}^{2n+1} \tag{23.66}$$

After the second stage, the two objectives have been achieved, namely, successful decoupling of the transmitted symbols and reduction of ICI. The transmitted symbols can now be detected by using the ML detector proposed by Alamouti [14]. The average BEP expression for the above system can be derived by first writing Equations 23.63 through 23.66 in a scalar form. Next, the power of the ICI is derived using an approach similar to the one used in deriving σ_{β}^2 in Section 23.3.1. Finally, using Equations 23.15

through 23.25, the expression for the average BEP can be derived. The actual expressions are omitted from this chapter because of space limitations.

23.4 Numerical and Simulation Results

This section presents some simulation results of BEP for the two systems considered earlier in this chapter. First, the BEP of SISO-OFDM in the case of perfect channel estimation is considered. Figures 23.4 and 23.5 show the average BEP of 16-QAM OFDM with 256 subcarriers for different mobile speeds with 20 and 1.75 MHz channel bandwidth, respectively. Clearly, the average BEP suffer from error floor owing to channel variations. The error floor is more severe in Figure 23.5 because the symbol duration is larger and the system is more susceptible to channel variations. Figure 23.6 compares the BEP at 40 dB E_b/N_0 plotted against different mobile speeds for a 16-QAM OFDM system with 256 subcarriers and different symbol duration.

The average BEP performance of SISO-OFDM with channel estimation errors is shown in Figures 23.7 and 23.8. Figure 23.7 shows the average BEP performance of 16-QAM OFDM with 256 subcarriers, 20 MHz channel bandwidth, 3 km/h mobile speed, and different values of the complex correlation coefficient ρ. Figure 23.8 shows the average BEP performance at 100 km/h mobile speed, using the same system parameters used in Figure 23.7. The combined effect of channel variation and channel estimation errors can be clearly seen from the large error floor. Also, with channel estimation errors, the error floor occurs at low values of E_b/N_0.

Next, the performance of STBC-OFDM is considered. Figures 23.9 through 23.11 compare the average BEP performance of STBC-OFDM with and without IAI cancellation and ICI reduction. In Figure 23.9, the average BEP performance of 16-QAM STBC-OFDM with 256 subcarriers and channel bandwidth of 20 MHz is plotted versus E_b/N_0. It can be seen that the standard detection will cause a high-error floor, especially at high mobile speeds. The improvement in BEP when employing the proposed detection can

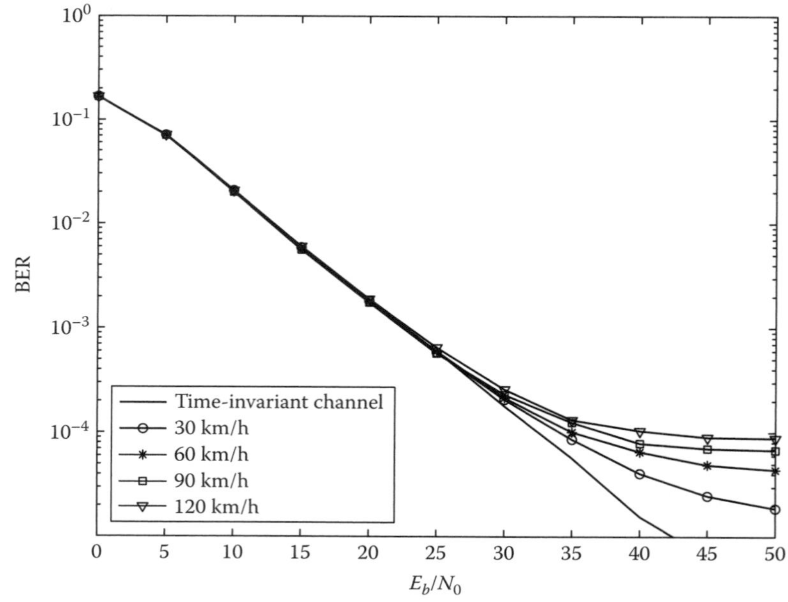

FIGURE 23.4 Average BEP for 16-QAM SISO-OFDM with 256 subcarriers and 11.2 μs symbol duration for different mobile speeds with perfect channel estimation.

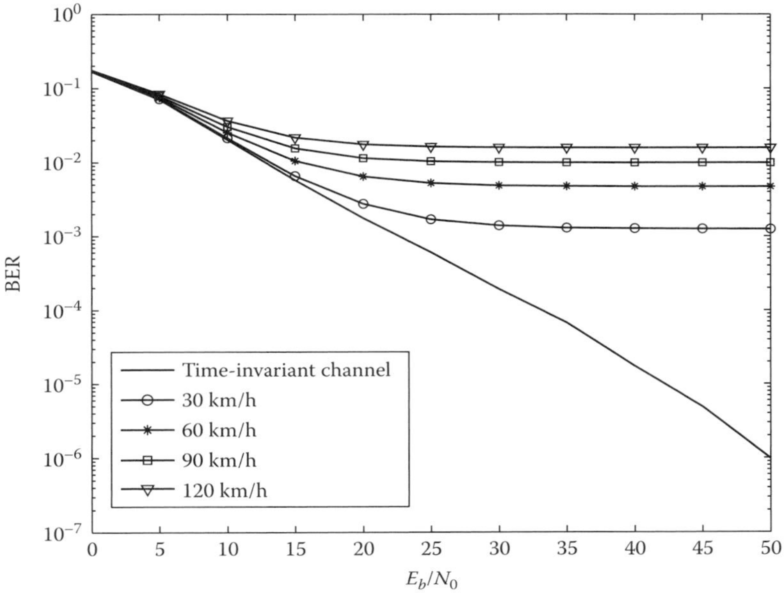

FIGURE 23.5 Average BEP for 16-QAM SISO-OFDM with 256 subcarriers and 148 μs symbol duration for different mobile speeds with perfect channel estimation.

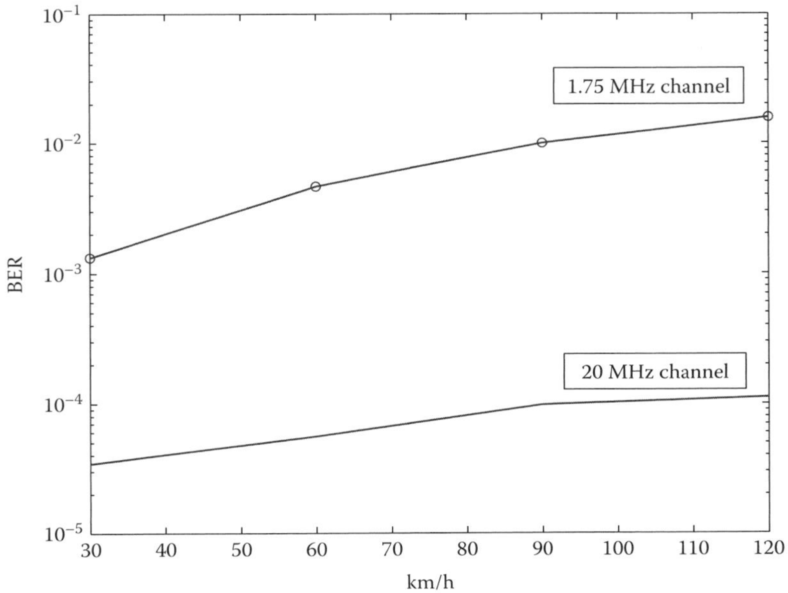

FIGURE 23.6 Average BEP comparison at 40 E_b/N_0 with perfect channel estimation.

also be seen. Figures 23.10 and 23.11 show the average BEP for 1.75 MHz channel bandwidth. Despite the significant improvement in average BEP, the system still suffers from high-error floor compared to the 20 MHz channel bandwidth case. The reason is that, although the number of interfering subcarriers is reduced by half when using the proposed detector, the power of the ICI from the first and second

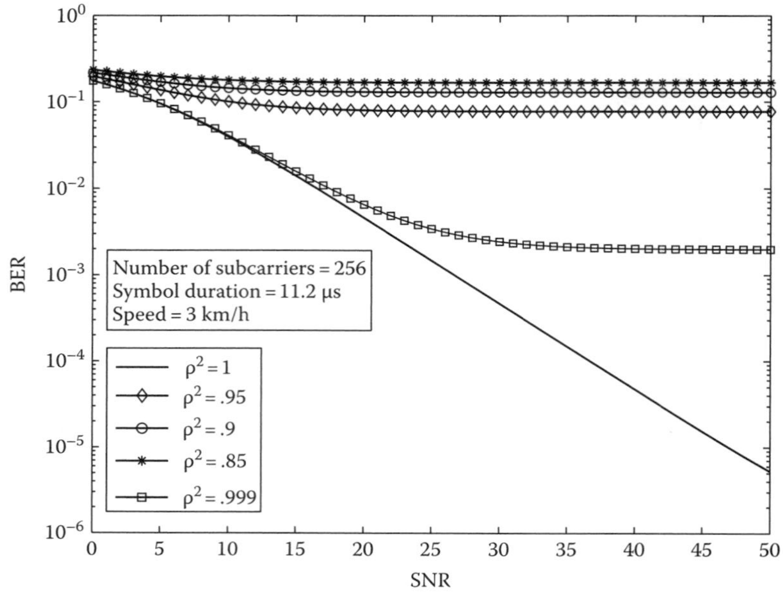

FIGURE 23.7 Average BEP for 16-QAM SISO-OFDM with 256 subcarriers and 11.2 μs symbol duration at 3 km/h with channel estimation errors.

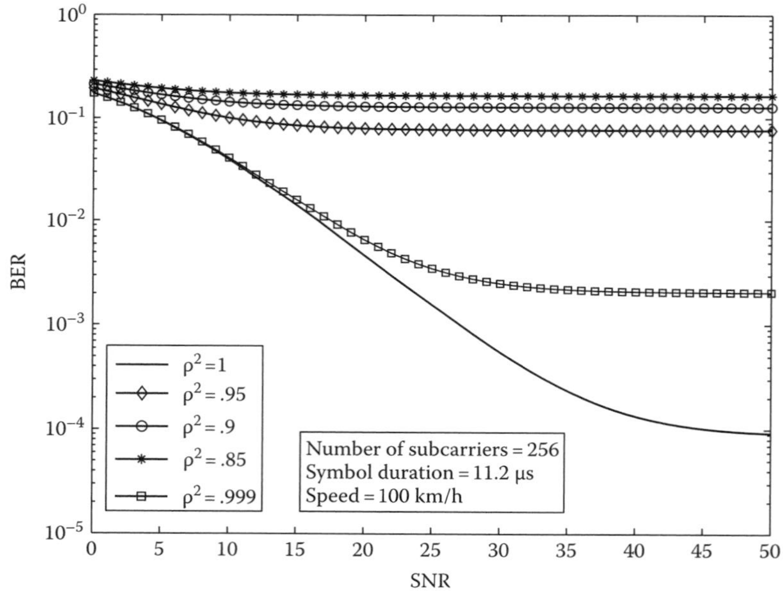

FIGURE 23.8 Average BEP for 16-QAM SISO-OFDM with 256 subcarriers and 11.2 μs symbol duration at 100 km/h with channel estimation errors.

adjacent subcarriers is still high. One way to improve the performance is to repeat the second stage of the proposed detector. This will further eliminate some of the high-power interfering subcarriers, and lower the error floor. A trade-off between performance and complexity should determine the number of times the second stage is repeated.

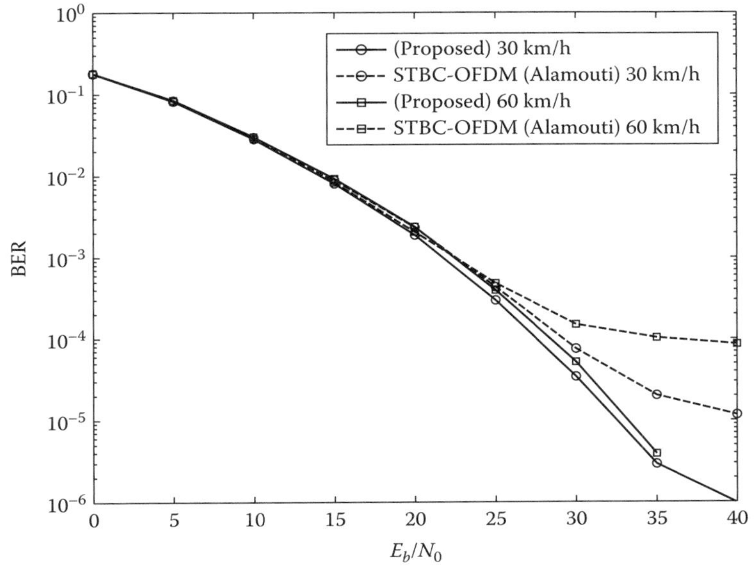

FIGURE 23.9 Average BEP comparison between the standard Alamouti and the proposed detection schemes for 16-QAM MISO-OFDM with 256 subcarriers and 11.2 μs symbol duration at 30 and 60 km/h.

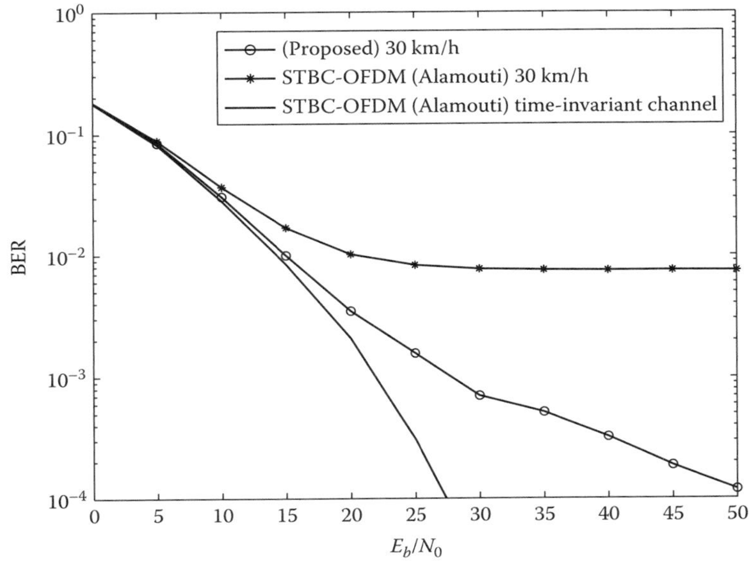

FIGURE 23.10 Average BEP comparison between the standard Alamouti and the proposed detection schemes for 16-QAM MISO-OFDM with 256 subcarriers and 148 μs symbol duration at 30 km/h.

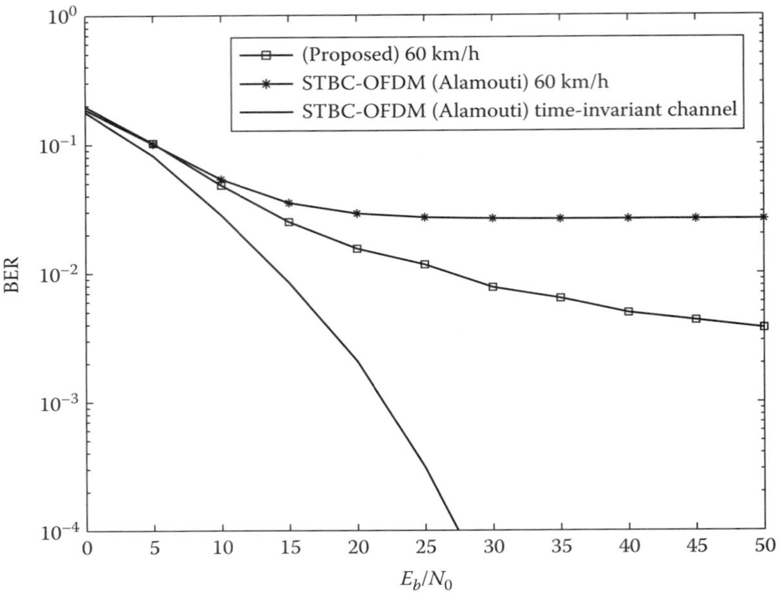

FIGURE 23.11 Average BEP comparison between the standard Alamouti and the proposed detection schemes for 16-QAM MISO-OFDM with 256 subcarriers and 148 μs symbol duration at 60 km/h.

23.5 Conclusion

In this chapter, the performance of both SISO- and STBC-OFDM systems are analyzed in mobile environments. In the SISO-OFDM case, the average BEP is derived assuming the general case when the receiver does not have perfect knowledge of the channel state information. It was shown that by conveniently expressing the actual channel in terms of the channel estimates, the BEP expressions can simply be derived by solving a single integral as opposed to the threefold integral suggested in previous literature. For the STBC-OFDM, it was shown that if the channel is time varying, IAI will also be present in addition to ICI, and the conventional Alamouti-detection scheme will fail to detect the transmitted OFDM signal. To overcome this problem, a two-stage detection scheme was suggested. The first stage of the new detector eliminates IAI, where the second stage reduces ICI. To further improve the BEP performance, the second step can be repeated many times to reach the desired performance level. A trade-off between performance and complexity should determine the number of times the second stage is repeated. Simulation results show the advantages of employing the new detection scheme as opposed to the conventional Alamouti scheme. All simulations are performed using the same parameters of the IEEE 802.16-2004 standard.

References

1. IEEE Standard for Local and Metropolitan Area Networks, Part 16: Air Interface for Fixed Broadband Wireless Access Systems, IEEE 2004.
2. V. Erceg, et al. *Channel Models for Fixed Wireless Applications*, Contribution IEEE 802.16.3c-01, February 29, 2001.
3. S. Coleri, M. Ergen, A. Puri, and A. Bahai, Channel estimation techniques based on pilot arrangement in OFDM systems, *IEEE Trans. Broadcast*, 48(3), 223–229, September 2002.

4. S.G. Kang, Y.M. Ha, and E.K. Joo, A comparative investigation on channel estimation algorithms for OFDM in mobile communications, *IEEE Trans. Broad.*, 49(2), 142–154, June 2003.
5. O. Edford, et al., OFDM channel estimation by singular value decomposition, *IEEE Trans. Commun.*, 46(7), 931–939, July 1998.
6. M. Morelli and U. Mengali, A comparison of pilot-aided channel estimation methods for ofdm systems, *IEEE Trans. Signal Process.*, 49(12), 3065–3073, December 2001.
7. M. Al-Gharabally and P. Das, On the performance of OFDM systems in time varying channels with channel estimation error. In *Proceedings of the IEEE ICC-2006*, Istanbul, Turkey, vol. 11, pp. 5180–5185, June 2006.
8. Z.-Z. Chang and Y.T. Su, Performance analysis of equalized ofdm systems in Rayleigh fading, *IEEE Trans. Wireless Commun.*, 1(4), 721–732, October 2002.
9. R. Annavajjala, P.C. Cosman, and L.B. Milstein, Performance analysis of linear modulation schemes with generalized diversity combining on Rayleigh fading channels with noisy channel estimates, *IEEE Trans. Inform. Theory*, under revision.
10. L. Cao and N.C. Beaulieu, Exact error-rate analysis of diversity 16-qam with channel estimation error, *IEEE Trans. Commun.*, 52(6), 1019–1029, June 2004.
11. M.K. Simon and M.S. Alouini, *Digital Communications Over Generalized Fading Channels: A Unified Approach to Performance Analysis*, Wiley & Sons, New York, 2000.
12. M. Surendra Raju, A. Ramesh, and A. Chockalingam, BER analysis of QAM with transmit diversity in Rayleigh fading channels, *IEEE GLOBECOM 2003*, 2, 641–645, December 2003.
13. Y.-S. Choi, P.J. Voltz, and F.A. Cassara, On channel estimation and detection for multicarrier signals in fast and selective Rayleigh fading channels, *IEEE Trans. Commun.*, 49(8), 1375–1376, August 2001.
14. S.M. Alamouti, A simple transmitter diversity scheme for wireless communications, *IEEE J. Select. Areas Commun.*, 16, 1451–1458, October 1998.
15. W.M. Younis, A.H. Sayed, and N. Al-Dhahir, Efficient adaptive receivers for joint equalization and interface cancellation in multiuser space-time block-coded systems, *IEEE Trans. Signal Process.*, 51(11), 2849–2862, November 2003.

24

Space–Time Coding and Application in WiMAX

Naofal Al-Dhahir
The University of Texas at Dallas

Robert Calderbank
Princeton University

Jimmy Chui
Princeton University

Sushanta Das
Phillips Research N.A.

Suhas Diggavi
Ecole Polytechnique

24.1 Introduction 24-1
24.2 Space–Time Codes: A Primer 24-3
System Model: Quasistatic Rayleigh Fading Channel • Diversity Gain and Coding Gain • Trade-Offs between Diversity and Rate • The ISI Channel
24.3 Space–Time Block Codes 24-7
Spatial Multiplexing • The Alamouti Code • The Golden Code • Other Space–Time Block Codes
24.4 Application of Space–Time Coding in WiMAX 24-10
Space–Time Coding in OFDM • Channel Estimation • A Differential Alamouti Code
24.5 A Novel Quaternionic Space–Time Block Code 24-12
Code Construction • Coherent Maximum-Likelihood Decoding • An Efficient Decoder • A Differential Quaternionic Code
24.6 Simulation Results 24-15
Appendix A: Quaternions 24-17
References 24-19

24.1 Introduction

Next-generation wireless systems aim to support both voice and high capacity flexible data services with limited bandwidth. Multiplicative and additive distortions inherent to the wireless medium make this difficult, and extensive research efforts are focused on developing efficient technologies to support reliable wireless communications. One such technology is multiple-input multiple-output (MIMO) communication systems with multiple antennas at both the transmitter and the receiver. Information-theoretic analysis in Refs. [1,2] shows that multiple antennas at the transmitter and the receiver enable very high data rates. Another key technology enabling high-rate communications over the wireless channels is orthogonal frequency division multiplexing (OFDM) [3]. In this chapter, we address and investigate a special class of MIMO, namely, space–time block codes (STBCs), and its application in WiMAX, the next-generation OFDM system based on IEEE 802.16 standard [4,5].

The main attribute that dictates the performance of wireless communications is uncertainty (randomness). Randomness exists in the users' transmission channels, as well as in the users' geographical locations. The spatial separation of antennas results in additional randomness. Space–time codes, introduced in Ref. [6], improve the reliability of communication over fading channels by correlating the transmit signals in both spatial and temporal dimensions, thus attaining diversity. We broadly define diversity as the method of conveying information through multiple independent instantiations of these random attenuations. The inherent undesirable trait of randomness is used as the foundation to enhance performance!

Space–time coding has received considerable attention in both academic and industrial circles:

- First, it improves the downlink performance without the need for multiple receive antennas at the terminals. For example, space–time coding techniques in wideband CDMA achieve substantial capacity gains owing to the resulting smoother fading which, in turn, makes power control more effective and reduces the transmitted power [7].
- Second, it can be elegantly combined with channel coding, as shown in Ref. [6], which realizes a coding gain in addition to the spatial diversity gain.
- Third, it does not require channel state information (CSI) at the transmitter, i.e., it operates in open-loop mode. Thus, it eliminates the need for an expensive and, in case of rapid channel fading, unreliable reverse link.
- Finally, it has been shown to be robust against nonideal operating conditions, such as antenna correlation, channel estimation errors, and Doppler effects [8,9].

An elegant and simple subclass of space–time codes are STBCs, which are able to provide high information rate to serve a large number of users over a wide coverage area with adequate reliability. This chapter describes the principal codes in this class that appear in the IEEE 802.16-2004 standard [4] and its 802.16e-2005 amendment [5]. We also examine other prospective candidates, including a novel nonlinear design based on quaternions.

The WiMAX Forum [10], the associated industry consortium of IEEE Std. 802.16, promises to deliver broadband connectivity with a data rate up to 40 Mbps and a coverage radius of 3–10 km using multiple-antenna technology. The IEEE 802.16-2004 standard is designed for stationary transmission, and the 802.16e amendment deals with both stationary and mobile transmissions. The standard and the amendment define the physical (PHY) layer specifications used in WiMAX. We identify two important features of WiMAX: (1) it uses OFDM that has an inherent robustness against multipath propagation and frequency-selective channels and (2) it allows the choice of multiple-antenna systems, e.g., MIMO systems and adaptive-antenna systems (AASs). Multiple-antenna systems can be implemented very easily with OFDM. The AAS transmits multiple spatially separated overlapped signals using space-division multiple access and MIMO often uses spatial multiplexing (e.g., BLAST). Even though both of these schemes are modeled as high data rate providers in WiMAX, the simplest example of multiple-antenna systems is the well-known Alamouti code [11]. The use of multiple antennas at the transmitter and the receiver enhances the system spectral efficiency, supports better error rate, and increases coverage area. These benefits come at no extra cost of bandwidth and power.

We demonstrate the value of using multiple antennas and STBCs in WiMAX by examining the performance gain of our nonlinear quaternionic code, which utilizes four transmit antennas and achieves full diversity, and comparing it with a single-input single-output (SISO) implementation. The gain in signal-to-noise ratio (SNR), achieved through the use of this novel 4×4 STBC over SISO in a WiMAX environment, translates to a 50% increase in the cell coverage area assuming only one receive antenna. By adding a second receive antenna, the percentage increase becomes 166%.

In the following sections, we will review the details of multiple-antenna transmission schemes and the design criteria for space–time codes. We also present codes that appear in the standard, as well as other notable codes, and relate them to the underlying theory. We examine how STBCs are used in practice, including implementation in OFDM and nonideal considerations. At the end of this chapter, we present our rate-1 full-diversity code for four transmit antennas, and demonstrate the value of this STBC in the WiMAX environment.

24.2 Space–Time Codes: A Primer

24.2.1 System Model: Quasistatic Rayleigh Fading Channel

WiMAX is a broadband transmission system, and thus inherently has inter-symbol interference (ISI) due to the frequency-selective nature of the wireless channel. The use of OFDM in WiMAX divides the entire bandwidth into many parallel narrowband channels, each with flat-fading characteristics. In this section, we focus our discussion on the theory of space–time coding on flat-fading channels. Our motivation for this approach is to emulate the design and application of space–time coding in WiMAX. At the end of this section, we examine the ISI channel in more detail and discuss its implications on more optimal designs for space–time codes.

The challenge of communication over Rayleigh fading channels is that the error probability decays only inversely with SNR, compared with the exponential decay observed on AWGN channels. Space–time coding [6] is a simple method that enhances reliability by increasing the decay of error probability through diversity.

There has been extensive work on the design of space–time codes since their introduction. In this section, we describe the basic design principles of space–time codes. We define the notions of rate and diversity, and examine the governing trade-off law between the two.

We formulate the system model as follows. The channel model is a MIMO quasistatic Rayleigh flat-fading channel with M_t transmit antennas and M_r receive antennas. The quasistatic assumption indicates that the channel gain coefficients remain constant for the duration of the codeword and change independently for each instantiation. The flat-fading assumption allows each transmitted symbol to be represented by a single tap in the discrete-time model with no ISI. We assume independent Rayleigh coefficients, i.e., each fading coefficient is an i.i.d. circular-complex normal random variable $\mathcal{CN}(0, 1)$. White Gaussian noise is also added at the receiver. The system model also assumes that the receiver has perfect CSI, whereas the transmitter does not have any CSI.

This system model can be described in matrix notation. The transmitted (received) codewords are represented by matrices, where the rows are indexed by the transmit (receive) antennas, the columns are indexed by time slots in a data frame, and the entries are the transmitted (received) symbols in baseband. For each codeword transmission,

$$\mathbf{R} = \mathbf{HX} + \mathbf{Z} \tag{24.1}$$

where

$\mathbf{R} \in \mathbb{C}^{M_r \times T}$ is the received signal matrix

$\mathbf{H} \in \mathbb{C}^{M_r \times M_t}$ the quasistatic channel matrix representing the channel gains of the $M_r M_t$ paths from each transmit antenna to each receive antenna

$\mathbf{X} \in \mathbb{C}^{M_t \times T}$ the transmitted codeword

$\mathbf{Z} \in \mathbb{C}^{M_r \times T}$ the additive white Gaussian noise where each entry has distribution $\mathcal{CN}(0, \sigma^2)$

There is also a power constraint, where the transmitted symbols have an average power of P, where the average is taken across all codewords over both spatial and temporal components. We assume that $P = 1$.

In this context, the length T of a codeword is a design parameter. In practice, T must be less than the coherence interval of the channel, to satisfy the quasistatic assumption.

24.2.2 Diversity Gain and Coding Gain

For a SISO channel, Equation 24.1 simplifies to

$$\mathbf{r}[m] = hx[m] + \mathbf{z}[m]$$

We use lowercase variables to emphasize the vector nature in the SISO environment. The instantaneous received SNR is the product $|h|^2\text{SNR}$. If $|h|^2\text{SNR} \gg 1$ then the separation between signal points is significantly larger than the standard deviation of the Gaussian noise, and error probability is very small since the tail of the Q-function decays rapidly. On the contrary, if $|h|^2\text{SNR} \ll 1$ then the separation between signal points is much less than the standard deviation of the noise, and the error probability is significant. Error events in the high SNR regime most often occur because the channel is in a deep fade ($|h|^2\text{SNR} < 1$), and not as a result of high additive noise. For the Rayleigh fading channel, the probability of error (for each bit and for the codeword) is proportional to 1/SNR at sufficiently high SNR.

The independent Rayleigh fading encountered across different codewords can be exploited to provide diversity. By repeating the same codeword M times, the probability of error will be proportional to $1/\text{SNR}^M$. Reliable communication for a particular codeword is possible when at least one of the M transmissions encounter favorable conditions, that is, does not see a deep fade. A supplementary outer code can help achieve better performance through additional coding gain.

Diversity is also introduced through multiple antennas at the transmitter. During the transmission of a single codeword, there are M_tM_r observed independent fades. Space–time codes correlate the transmitted symbols in a codeword, which protect each (uncoded) symbol up to M_tM_r times for each independent fade it encounters. If at least one path is strong, reliable communication for that symbol is possible. Diversity can also occur at the receiver if it is equipped with multiple antennas, provided that the receive antennas are spaced sufficiently far enough so that the correlation of the fade coefficients is negligible.

So far, we have qualitatively discussed diversity as the notion of sending correlated symbols across multiple paths from the transmitter to the receiver. Diversity is quantified through the notion of diversity gain or diversity order, which represents the decay of error probability in the high SNR regime.

Definition 24.1: A coding scheme with an average error probability $\mathbb{P}_e(\text{SNR})$ as a function of SNR that behaves as

$$\lim_{\text{SNR}\to\infty} \frac{\log[\mathbb{P}_e(\text{SNR})]}{\log(\text{SNR})} = -d \tag{24.2}$$

is said to have a diversity gain of d.

In words, a scheme with diversity order d has an error probability at high SNR behaving as* $\mathbb{P}_e(\text{SNR}) \doteq \text{SNR}^{-d}$.

One can approximate the performance of a space–time code by determining the worst pairwise error probability (PEP) between two candidate codewords. This leads to the rank criterion for determining the diversity order of a space–time code [6,12]. The PEP between two codewords X_i and X_j can be determined by properties of the difference matrix $\Delta(X_i, X_j) = X_i - X_j$. When there is no ambiguity we will denote the difference simply by Δ.

$$\begin{aligned} \mathbb{P}(X_i \to X_j) &= \mathbb{E}_{\text{H}}[\mathbb{P}(X_i \to X_j|\text{H})] \\ &= E_{\text{H}}\left[Q\left(\frac{\|\text{H}\Delta\|}{\sqrt{2N_0}}\right)\right] \end{aligned} \tag{24.3}$$

* We use the notation $\doteq$ to denote exponential equality, i.e., $g(\text{SNR}) \doteq \text{SNR}^a$ means that $\lim_{\text{SNR}\to\infty} \log g(\text{SNR})/\log \text{SNR} = a$. Moreover, if $g(\text{SNR}) \doteq f(\text{SNR})$, it means that $\lim_{\text{SNR}\to\infty} \log g(\text{SNR})/\log \text{SNR} = \lim_{\text{SNR}\to\infty} \log f(\text{SNR})/\log \text{SNR}$.

Under the Rayleigh fading assumption, Equation 24.3 can be bounded above using the Chernoff bound, as in Ref. [6]. If q is the rank of Δ and $\{\lambda_n\}_{n=1}^{q}$ are the nonzero eigenvalues of $\mathbf{\Delta\Delta}^*$, then the upper bound is given by

$$\mathbb{P}(X_i \rightarrow X_j) \leq \left(\prod_{n=1}^{q} \lambda_n\right)^{-M_r} \left(\frac{\text{SNR}}{4}\right)^{-qM_r} \tag{24.4}$$

It can be shown that the exact expression for the asymptotic PEP at high SNR is a multiplicative constant (dependent only on M_r and q) of the upper bound given in Equation 24.3 [13].

For a space–time code with a fixed-rate codebook $\mathcal{C}$, the worst PEP corresponds to the pair of codewords for which the difference matrix has the lowest rank and the lowest product measure (the product of the nonzero eigenvalues). By a simple union-bound argument, it follows that the diversity gain d is given by

$$d = M_\text{r} \min_{X_i \neq X_j \in c} \text{rank}[\Delta(X_i, X_j)] \tag{24.5}$$

The expression in Equation 24.4 also leads one to the following definition of coding gain.

Definition 24.2: The coding gain for a space–time code with a fixed-rate codebook is given by the quantity

$$\left(\prod_{n=1}^{q} \lambda_n\right)^{1/q} \tag{24.6}$$

that corresponds to the worst PEP.

Both the coding gain and the diversity order contribute to the error probability. Hence, two criteria for code design, as indicated in Ref. [6], are to design the codebook X such that the following are satisfied.

Rank criterion: Maximize the minimum rank of the difference $X_i - X_j$ over all distinct pairs of space–time codewords X_i, X_j.

Determinant criterion: For a given diversity d, maximize the minimum product of the nonzero singular values of the difference $X_i - X_j$ over all distinct pairs of space–time codewords X_i, X_j whose difference has rank d.

We note that these criteria optimize the code for the high SNR regime. Optimizing for lower SNR values implies using the Euclidean distance metric [14]. Recent results also suggest that examining the effect of multiple interactions between codewords is necessary for a more accurate comparison between codes [15].

24.2.3 Trade-Offs between Diversity and Rate

A natural question that arises is to determine how many codewords can we have, which allow us to attain a certain diversity order. One point of view is to fix the constellation, and examine the trade-off between the number of codewords and the code's diversity gain. A second is to allow the constellation to grow as SNR increases. This latter viewpoint is motivated because the capacity of the multiple-antenna channel grows with SNR behaving as $\min(M_\text{r}, M_\text{t})\log(\text{SNR})$ [1,16], at high SNR even for finite M_r, M_t.

24.2.3.1 Trade-Off for Fixed Constellations

For the quasistatic Rayleigh flat-fading channel, this has been examined in Ref. [6] where the following result was obtained.

THEOREM 24.1 [6]:

For a static constellation per transmitted symbol, if the diversity order of the system is qM_r, then the rate R that can be achieved (in terms of symbols per symbol-time period) is bounded as

$$R \leq M_t - q + 1 \tag{24.7}$$

24.2.3.2 Diversity-Multiplexing Trade-Off

Zheng and Tse [16] define a multiplexing gain of a transmission scheme as follows.

Definition 24.3: A coding scheme that has a transmission rate of $R(\mathrm{SNR})$ as a function of SNR is said to have a multiplexing gain r if

$$\lim_{\mathrm{SNR}\to\infty} \frac{R(\mathrm{SNR})}{\log(\mathrm{SNR})} = r \tag{24.8}$$

Therefore, the system has a rate of $r \log(\mathrm{SNR})$ at high SNR. The main result in Ref. [16] states that

---

THEOREM 24.2 [16]:

For $T \geq M_t + M_r - 1$, and $K = \min(M_t, M_r)$, the optimal trade-off curve $d^*(\mathrm{r})$ is given by the piece-wise linear function connecting points in $[k, d^*(k)]$, $k = 0, \ldots, K$ where

$$d^*(k) = (M_r - k)(M_t - k) \tag{24.9}$$

Both Theorems 24.1 and 24.2 show the tension between achieving high-rate and high-diversity. If $r = k$ is an integer, the result can be interpreted as using $M_r - k$ receive antennas and $M_t - k$ transmit antennas to provide diversity while using k antennas to provide the multiplexing gain. Clearly, this result means that one can get large rates that grow with SNR if we reduce the diversity order from the maximum achievable. This diversity-multiplexing trade-off implies that a high multiplexing gain comes at the price of decreased diversity gain and is associated with a trade-off between error probability and rate.

This tension between rate and diversity (reliability) demonstrates that different codes will be suitable for different situations. The code choice is by design, and can be influenced by factors such as quality of service and maximum tolerable delay. Section 24.3 demonstrates a selection of various space–time transmission schemes, including those that appear in the IEEE 802.16 standards. These codes range from those that maximize spatial multiplexing (V-BLAST) to codes with full-diversity gain (the Alamouti code, the Golden Code, and our code based on quaternions).

24.2.4 The ISI Channel

We have justified our use of the idealized flat-fading channel due to the narrowband characteristics for each tone in an OFDM system. This allows for simple code design criteria, as each tone is coded independent of the others. The theory of space–time codes is not limited to flat-fading, however, and can be extended to ISI channels such as frequency-selective channels or channels with multipath. The general model for ISI channels is almost identical to Equation 24.1 except the model now includes a new parameter ν, the number of taps required to characterize a given ISI channel.

The rank and determinant criteria for the ISI channel are similar to those given in Section 24.2.2. The main problem in practice is to construct such codes that do not have large decoding complexity. The trade-off between performance and complexity is more prominent for ISI code design. A thorough examination can be found in Ref. [17].

The parameter ν allows an increase in the maximum diversity gain for the system by a factor of ν. Intuitively, this is a result of using frequency as an additional means of diversity for a frequency-selective channel. As such, it affects the tension between rate and diversity. For example, the maximum achievable rate for fixed-rate codes given a diversity order of qM_r is

$$R \leq M_t - (q/\nu) + 1 \tag{24.10}$$

The diversity-multiplexing trade-off for the SISO ISI channel is [18]

$$d^*(r) = \nu(1 - r) \tag{24.11}$$

For a SISO OFDM system with N tones, if there are ν i.i.d. ISI taps, then the frequency-domain taps separated by N/ν are going to be independent. For simplicity, we assume that all tones are in use and that N/ν is an integer. Therefore, we have N/ν sets of ν parallel channels, where within each set the frequency-domain fading are independent. Then by using a diversity-multiplexing optimal code for the parallel fading channel for each of these sets, we can achieve Equation 24.11. Trivially, for one degree of freedom, i.e., $\min(M_t, M_r) = 1$, the SISO result can be extended to $d^*(r) = \nu\max(M_t, M_r) \times (1 - r)$, and the same coding architecture applies.

In WiMAX, frequency diversity is achieved by using an outer code and frequency interleaving (e.g., Ref. [4], Sections 8.3.3 and 8.4.9). The use of this outer code, along with STBCs designed for the flat-fading channel, achieves a very good improvement in performance, which we will demonstrate in Section 24.6.

24.3 Space–Time Block Codes

24.3.1 Spatial Multiplexing

One strategy for constructing codes is to maximize rate and achieve the greatest spectral efficiency. Spatial multiplexing achieves this goal by transmitting uncorrelated data over space and time, and relies on a sophisticated decoding mechanism to separate the data streams at the receiver. Strictly speaking, spatial multiplexing is not a method of space–time coding; it does not provide transmit diversity. However, it can be considered an extreme case in the diversity-multiplexing trade-off curve; it trades off diversity for maximal rate. The performance of these codes is highly dependent on the decoding algorithm at the receiver and the instantaneous channel characteristics.

Codes exploiting spatial multiplexing occur in the standard for two transmit antennas (Section 8.4.8.3.3, Code B), three transmit antennas (Section 8.4.8.3.4, Code C), and four transmit antennas (Section 8.4.8.3.5, Code C). The symbols may support different code rates.

$$\begin{pmatrix} x_1 \\ x_2 \end{pmatrix}; \quad \begin{pmatrix} x_1 \\ x_2 \\ x_3 \end{pmatrix}; \quad \begin{pmatrix} x_1 \\ x_2 \\ x_3 \\ x_4 \end{pmatrix}$$

These constructions fall under the Bell Labs layered space–time architecture (BLAST) framework [1,19], as V-BLAST codes. A major challenge in realizing this significant additional throughput gain in practice is the development of cost-effective low-complexity and highly optimal receivers. The receiver signal-processing functions are similar to a decision feedback equalizer operating in the spatial domain where the nulling operation is performed by the feedforward filter and the interference cancellation operation is performed by the feedback filter [20]. As with all feedback-based detection schemes, V-BLAST suffers from error propagation effects.

24.3.2 The Alamouti Code

The Alamouti code [11] was discovered as a method to provide transmit diversity in the same manner maximum-ratio receive combining provides receive diversity. This is achieved by correlating the transmit symbols spatially across two transmit antennas, and temporally across two consecutive time intervals. In the notation of Equation 24.1, the Alamouti code encodes two symbols x_1 and x_2 as the following 2×2 matrix:

$$(x_1, x_2) \rightarrow \begin{pmatrix} x_1 & x_2 \\ -x_2^* & x_1^* \end{pmatrix} \tag{24.12}$$

Unlike spatial multiplexing codes, the Alamouti code achieves the same rate as SISO but attains maximum diversity gain for two transmit antennas. Achieving diversity is a second strategy for designing space–time codes, and it comes at the expense of reduced rate.

The coding and decoding mechanisms for the Alamouti code are remarkably simple, and equally effective. It is not surprising to see the appearance of the Alamouti code many times in the standard (e.g., Section 8.3.8.2 in Ref. [4], Section 8.4.8.3.3 in Ref. [5], and variants for three and four antennas described later in this section). Next, we demonstrate the simplicity of Alamouti code decoding for the case of one receive antenna.

The receive antenna obtains the signals r_1, r_2 over the two consecutive time slots for the corresponding code word. They are given by

$$\begin{pmatrix} r_1 \\ -r_2^* \end{pmatrix} = \begin{pmatrix} h_1 & h_2 \\ -h_2^* & h_1^* \end{pmatrix} \begin{pmatrix} x_1 \\ -x_2^* \end{pmatrix} + \begin{pmatrix} z_1 \\ -z_2^* \end{pmatrix} \tag{24.13}$$

where h_1, h_2 are the path gains from the two transmit antennas to the mobile, and the noise samples z_1, z_2 are independent samples of a zero-mean complex Gaussian random variable with noise energy N_0 per complex dimension. Thus

$$\mathrm{r} = \mathbf{H}\mathrm{s} + \mathrm{z} \tag{24.14}$$

where the matrix $\mathbf{H}$ is orthogonal. If the receiver knows the path gains h_1 and h_2, then it is able to form

$$\mathbf{H}^*\mathrm{r} = \|\mathrm{h}\|^2\mathrm{s} + \mathrm{z}' \tag{24.15}$$

where the new noise term z' remains white. This allows the linear-complexity maximum-likelihood (ML) decoding of x_1, x_2 to be done independently rather than jointly.

We note that the Alamouti code provides the same rate as an equivalent SISO channel. Using the Alamouti code provides us with higher reliability, owing to the increased diversity.

The Alamouti code is defined for only two transmit antennas. In the standard, variants using three and four transmit antennas are described as well. These codes are constructed by using only two transmit antennas over two consecutive time intervals, leaving the other antenna(s) effectively in an off state.

To increase rate, the off states can be removed and replaced by extra symbols. For three transmit antennas, these extra symbols are transmitted in an uncorrelated fashion; for four transmit antennas, the extra symbols are sent in a correlated manner by means of the Alamouti code. Note that in each scenario, each unique symbol uses the same transmitted power. These may be considered as a combination of spatial multiplexing in conjunction with Alamouti codewords.

Examples of three-antenna Alamouti-based space–time codes are listed in Ref. [5], Section 8.4.8.3.4. Codes A_1 and B_1 have the following form:

$$A_1 = \begin{pmatrix} x_1 & x_2 & 0 & 0 \\ -x_2^* & x_1^* & x_3 & x_4 \\ 0 & 0 & -x_4^* & x_3^* \end{pmatrix} \quad B_1 = \begin{pmatrix} \sqrt{\frac{3}{4}} & 0 & 0 \\ 0 & \sqrt{\frac{3}{4}} & 0 \\ 0 & 0 & \sqrt{\frac{3}{2}} \end{pmatrix} \begin{pmatrix} x_1 & x_2 & x_3 & x_4 \\ -x_2^* & x_1^* & -x_4^* & x_3^* \\ x_5 & x_6 & x_7 & x_8 \end{pmatrix}$$

The four-antenna Alamouti variants are similar and can be found in Ref. [5], Section 8.4.8.3.5. Two of the codes are provided below. We note that the decoding for Code B can be achieved by successive cancellation [21,22].

$$A = \begin{pmatrix} x_1 & x_2 & 0 & 0 \\ -x_2^* & x_1^* & 0 & 0 \\ 0 & 0 & x_3 & x_4 \\ 0 & 0 & -x_4^* & x_3^* \end{pmatrix} \quad B = \begin{pmatrix} x_1 & x_2 & x_5 & x_6 \\ -x_2^* & x_1^* & x_7 & x_8 \\ x_3 & x_4 & -x_6^* & x_5^* \\ -x_4^* & x_3^* & -x_8^* & x_7^* \end{pmatrix}$$

24.3.3 The Golden Code

Of the above codes, one provides diversity (Alamouti) while the other provides rate (spatial multiplexing). Can we obtain a code that achieves the diversity order of Alamouti and the rate of spatial multiplexing? According to the diversity multiplexing trade-off (Theorem 24.2), it appears that such a code may exist. The answer is in the affirmative, and is given by the Golden code.

The Golden code is a space–time code for a system with two transmit antennas. It has the property that four complex symbols can be encoded over two time slots, yet achieves full diversity. This code was independently discovered in Refs. [23,24].

The Golden code encodes four QAM symbols x_1, x_2, x_3, x_4 to the following 2×2 matrix:

$$(x_1, x_2, x_3, x_4) \to \frac{1}{\sqrt{5}} \begin{pmatrix} \alpha(x_1 + \theta x_2) & \alpha(x_3 + \theta x_4) \\ i\bar{\alpha}(x_3 + \bar{\theta} x_4) & \bar{\alpha}(x_1 + \bar{\theta} x_2) \end{pmatrix} \tag{24.16}$$

where

$$\theta = \frac{1 + \sqrt{5}}{2} \qquad \alpha = 1 + i - i\theta$$

$$\bar{\theta} = \frac{1 - \sqrt{5}}{2} = 1 - \theta \qquad \bar{\alpha} = 1 + i - i\bar{\theta}$$

In Ref. [25], it is shown that nonzero matrices of this form (and hence any nonzero difference matrix) have nonvanishing determinant. That is to say, the set of all determinants can never be arbitrarily made close to 0. Indeed, the minimum absolute value for the determinant of nonzero matrices of this code is 0.2. Decoding of the Golden code can be done with sphere decoding [26–28].

It can be shown that the Alamouti code, as well as the spatial multiplexing code, does not achieve the diversity-multiplexing trade-off, whereas the Golden code does [23]. This appealing property is a reason for its appearance in the standard (Section 8.4.8.3.3, Code C in Ref. [5]). The code in the standard is a variation of the Golden code (Equation 24.16) [29].

24.3.4 Other Space–Time Block Codes

In the remainder of this section, we examine other codes that do not make an appearance in the standard.

Two attractive properties of the Alamouti STBC are the ability to separate symbols at the receiver with linear processing and the achievement of maximal diversity gain. These properties define the broader class of STBCs, namely orthogonal STBCs [30]. Orthogonal designs achieve maximal diversity at a linear (in constellation size) decoding complexity. Tarokh et al. prove that the only full-rate complex orthogonal design only exists for the case of two transmit antennas, and is given by the Alamouti code. As the number of transmit antennas increases, the available rate becomes unattractive [31].

For four transmit antennas, the maximum rate is 3/4, and one such example is presented in Ref. [30]:

$$\begin{bmatrix} x_1 & x_2 & x_3 & 0 \\ -\bar{x}_2 & \bar{x}_1 & 0 & x_3 \\ \bar{x}_3 & 0 & -\bar{x}_1 & x_2 \\ 0 & \bar{x}_3 & -\bar{x}_2 & -x_1 \end{bmatrix} \tag{24.17}$$

Given that full-rate orthogonal designs are limited in number, it is natural to relax the requirement that linear processing at the receiver be able to separate all transmitted symbols. The lack of a full-rate complex design for even four transmit antennas motivated Jafarkhani [32] to consider the quasiorthogonal STBC

$$\mathbf{S} = \left[\begin{array}{cc|cc} x_1 & x_2 & x_3 & x_4 \\ -x_2^* & x_1^* & -x_4^* & x_3^* \\ \hline x_3^* & -x_4^* & x_1^* & x_2^* \\ x_4 & -x_3 & -x_2 & x_1 \end{array}\right]$$

where the structure of each 2×2 block is identical to that of the Alamouti code. ML decoding can be achieved by processing pairs of symbols, namely (x_1, x_4) and (x_2, x_3) for this code.

This construction only achieves a diversity gain of two for every receive antenna. Full diversity can be achieved at the expense of signal constellation expansion, for example, by rotating the symbols x_1 and x_3 (see e.g., Ref. [33]). The optimal angle of rotation depends on the base constellation.

There exist many other space–time codes in literature, including trellis-based constructions (including super-orthogonal codes [34]), linear dispersion codes [35], layered space–time coding [36], threaded algebraic space–time codes [37], and more recently, perfect STBCs [38]. The Golden code can be considered as an instance of the last two constructions.

24.4 Application of Space–Time Coding in WiMAX

24.4.1 Space–Time Coding in OFDM

STBCs based on the flat-fading design has a straightforward implementation in OFDM systems. The modulated symbols are encoded into space–time codes before the IFFT operation. Figure 24.1 depicts the implementation. Like the SISO case, multipath along each transmitter–receiver path is mitigated by

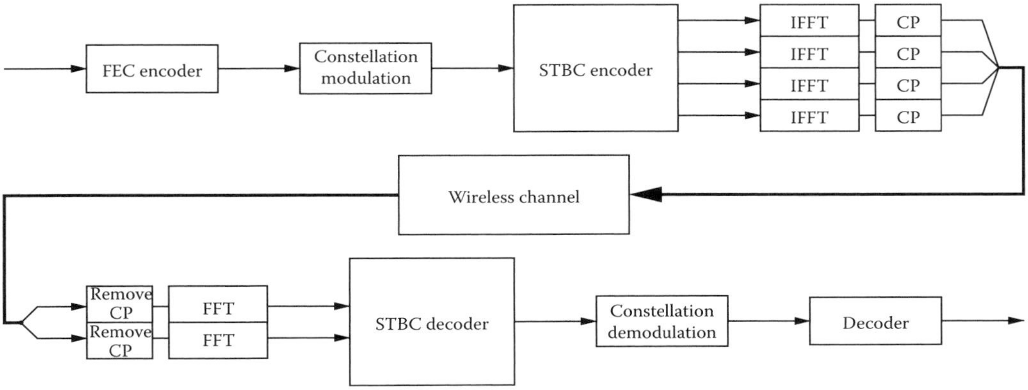

FIGURE 24.1 Implementation of a STBC in WiMAX.

use of the cyclic prefix. The only additional condition is that the coherence time of the physical channel must exceed that of T times the duration of an OFDM symbol, where T is the length of the codeword. During the span of T OFDMA symbols, each subcarrier in use transmits one space–time code.

24.4.2 Channel Estimation

Our assumption of perfect CSI at the receiver will not be achieved in practice. Measurements of the channel path gains must be made at the receiver. Estimating the channel gains is typically accomplished by using preambles or by inserting pilot tones within a data frame, at the expense of a slight loss in rate. We note that for an OFDM system, it is not necessary to make channel measurements for every tone due to the correlation in the frequency domain. However, the quality of these measurements depends heavily on the placement of pilots (e.g., [39–41]).

Estimating the path gains can be accomplished in a very straightforward manner for orthogonal codes using pilot tones. We describe the procedure for the Alamouti code.

We assume that two pilot tones x_1 and x_2 are transmitted, in the form of Equation 24.12, whose values are known to the receiver. The receiver obtains the signals r_1 and r_2 where

$$\mathbf{r} = \begin{pmatrix} r_1 \\ r_2 \end{pmatrix} = \begin{pmatrix} x_1 & -x_2^* \\ x_2 & x_1^* \end{pmatrix} \begin{pmatrix} h_1 \\ h_2 \end{pmatrix} + \begin{pmatrix} z_1 \\ z_2 \end{pmatrix} = \mathbf{Xh} + \mathbf{z}$$

The optimal estimates for h_1, h_2 can be obtained by linear processing at the receiver, and are given by

$$\tilde{\mathbf{h}} = \begin{pmatrix} \tilde{h}_1 \\ \tilde{h}_2 \end{pmatrix} = \frac{1}{\|x\|^2} \mathbf{X}^* \mathbf{r} = \begin{pmatrix} h_1 & + & \tilde{z}_1 \\ h_2 & + & \tilde{z}_2 \end{pmatrix}$$

where

$$\tilde{z}_1 = \frac{x_1 z_1 - x_2^* z_2}{\|x\|^2}; \quad \tilde{z}_2 = \frac{x_2 z_1 + x_1^* z_2}{\|x\|^2}$$

These channel estimates can then be used to detect the next pair of code symbols. After the next code symbols are decoded, the channel estimate can be updated using those decoded symbols in place of the pilot symbols. When the channel variation is slow, the receiver improves stability of the decoding algorithm by averaging old and new channel estimates. This differential detection requires that the transmission begins with a pair of known symbols, and will perform within 3 dB of coherent detection where the CSI is perfectly known at the receiver.

24.4.3 A Differential Alamouti Code

In some circumstances, it is desirable to forgo the channel estimation module to keep the receiver complexity low. Channel estimates may also be unreliable due to motion. Under such circumstances, differential decoding algorithms become attractive despite their SNR loss from coherent decoding. In this section, we describe the differential encoding and decoding algorithm for the Alamouti code [42,43]. We note that differential coding is not in the standard.

In Section 24.3.2, we described the Alamouti code whose channel model was represented as

$$\begin{pmatrix} r_1 \\ -r_2^* \end{pmatrix} = \begin{pmatrix} h_1 & h_2 \\ -h_2^* & h_1^* \end{pmatrix} \begin{pmatrix} x_1 \\ -x_2^* \end{pmatrix} + \begin{pmatrix} z_1 \\ -z_2^* \end{pmatrix}$$

This expression can be manipulated further to

$$\begin{pmatrix} r_1 & r_2 \\ -r_2^* & r_1^* \end{pmatrix} = \begin{pmatrix} h_1 & h_2 \\ -h_2^* & h_1^* \end{pmatrix} \begin{pmatrix} x_1 & x_2 \\ -x_2^* & x_1^* \end{pmatrix} + \text{noise}$$

In the absence of noise, this can be rewritten as $\mathbf{R}(k) = \mathbf{HX}(k)$. Under quasistatic conditions, it follows that $\mathbf{R}^*(k-1)\mathbf{R}(k) = \|\mathbf{h}\|^2\mathbf{X}^*(k-1)\mathbf{X}(k)$.

Thus, we define the differential transmission rule

$$\mathbf{X}(k) = \bar{\mathbf{X}}^{-1}(k-1)\mathbf{U}(k) \tag{24.18}$$

where $\mathbf{U}(k)$ is the information matrix, which has Alamouti form. By setting the transmission rule in this manner, the receiver merely performs matrix multiplication to determine the transmitted matrix.

A method using differential Alamouti for ISI channels is also discussed in Ref. [43].

24.5 A Novel Quaternionic Space–Time Block Code

24.5.1 Code Construction

In the following section, we revisit the problem of designing orthogonal STBC for four transmit antennas. We present a novel full-rate full-diversity orthogonal STBC for four transmit antennas and in its applications in broadband wireless communication (WiMAX) environment. This code is constructed by means of a 2×2 array over the quaternions, thus resulting in a 4×4 array over the complex field $\mathbb{C}$. The code is orthogonal over $\mathbb{C}$ but is not linear. The structure of the code is a generalization of the 2×2 Alamouti code [11], and reduces to it if the 2×2 quaternions in the code are replaced by complex numbers. For QPSK modulation, the code has no constellation expansion and enjoys a simple ML decoding algorithm. We also develop a differential encoding and decoding algorithm for this code. Another reason for our interest in orthogonal designs is that they limit the SNR loss incurred by differential decoding to its minimum of 3 dB from coherent decoding. A brief overview of equations is provided in the appendix.

We consider the STBC

$$\begin{bmatrix} p & q \\ -\bar{q} & \dfrac{\bar{q}\bar{p}q}{\|q\|^2} \end{bmatrix}$$

where the entries are quaternions. We may replace the quaternions p and q by the corresponding Alamouti 2×2 blocks to obtain a 4×4 STBC with complex entries.

$$\begin{bmatrix} \mathbf{P} & \mathbf{Q} \\ -\bar{\mathbf{Q}} & \dfrac{\bar{\mathbf{Q}}\bar{\mathbf{P}}\mathbf{Q}}{\|\mathbf{Q}\|^2} \end{bmatrix} \begin{bmatrix} \bar{\mathbf{P}} & -\mathbf{Q} \\ \bar{\mathbf{Q}} & \dfrac{\bar{\mathbf{Q}}\mathbf{P}\mathbf{Q}}{\|\mathbf{Q}\|^2} \end{bmatrix} = (\|p\|^2 + \|q\|^2)\mathrm{I} \tag{24.19}$$

Observe that the rows of this code are orthogonal with respect to the standard inner product operation. Since QPSK signaling corresponds to choosing the quaternions p and q from the set $(\pm 1 \pm i \pm j \pm k)/2$, there is no constellation expansion because $(\bar{q}\bar{p}q)/\|q\|^2$ is always a quaternion of this same form. However, multiplication of quaternions is not commutative and it is not possible to have a 2×2 linear code over the quaternions with orthogonal rows and orthogonal columns [30].

24.5.2 Coherent Maximum-Likelihood Decoding

We can represent the model in Equation 24.1 by quaternionic algebra. For simplicity, let us consider $M_r = 1$; all arguments can be easily generalized to $M_r > 1$. Consider the 2×2 complex matrices formed as

$$\begin{aligned} \mathbf{R}_1 &= \begin{bmatrix} r(0) & r(1) \\ -\bar{r}(1) & \bar{r}(0) \end{bmatrix}; \quad \mathbf{R}_2 = \begin{bmatrix} r(2) & r(3) \\ -\bar{r}(3) & \bar{r}(2) \end{bmatrix} \\ \mathbf{H}_1 &= \begin{bmatrix} H(1,1) & H(1,2) \\ -\bar{H}(1,2) & \bar{H}(1,1) \end{bmatrix}; \quad \mathbf{H}_2 = \begin{bmatrix} H(1,3) & H(1,4) \\ -\bar{H}(1,4) & \bar{H}(1,3) \end{bmatrix} \end{aligned} \tag{24.20}$$

where $\mathbf{H}(u, v)$ is the (u, v)th component of the channel matrix $\mathbf{H}$. Then we can rewrite Equation 24.1 for our code as

$$[\mathbf{R}_1\mathbf{R}_2] = [\mathbf{H}_1\mathbf{H}_2] \begin{bmatrix} \mathbf{P} & \mathbf{Q} \\ -\bar{\mathbf{Q}} & \dfrac{\bar{\mathbf{Q}}\bar{\mathbf{P}}\mathbf{Q}}{\|\mathbf{Q}\|^2} \end{bmatrix} + [\mathbf{Z}_1\mathbf{Z}_2] \tag{24.21}$$

where the noise vectors are also replaced by corresponding quaternionic matrices of the forms given in Equation 24.20.

From Equation 24.21, the ML decoding rule is given by*

$$\begin{aligned} \{\hat{\mathbf{P}}, \hat{\mathbf{Q}}\} &= \arg\min_{\mathbf{P},\mathbf{Q}} \left\| [\mathbf{R}_1\mathbf{R}_2] - [\mathbf{H}_1\mathbf{H}_2] \begin{bmatrix} \mathbf{P} & \mathbf{Q} \\ -\bar{\mathbf{Q}} & \dfrac{\bar{\mathbf{Q}}\bar{\mathbf{P}}\mathbf{Q}}{\|\mathbf{Q}\|^2} \end{bmatrix} \right\|^2 \\ &= \arg\max_{\mathbf{P},\mathbf{Q}} \Re \left\{ \mathrm{trace} \left([\mathbf{R}_1\mathbf{R}_2] \begin{bmatrix} \bar{\mathbf{P}} & -\mathbf{Q} \\ \bar{\mathbf{Q}} & \dfrac{\bar{\mathbf{Q}}\mathbf{P}\mathbf{Q}}{\|\mathbf{Q}\|^2} \end{bmatrix} \begin{bmatrix} \bar{\mathbf{H}}_1 \\ \bar{\mathbf{H}}_2 \end{bmatrix} \right) \right\} \end{aligned} \tag{24.22}$$

* Assuming that $\|\mathbf{P}\|$ and $\|\mathbf{Q}\|$ are constant.

24.5.3 An Efficient Decoder

We can write Equation 24.21 in quaternionic algebra as follows

$$[r_1 r_2] = [h_1 h_2]\begin{bmatrix} p & q \\ -\bar{q} & \dfrac{\bar{q}\bar{p}q}{\|q\|^2} \end{bmatrix} + [z_1 z_2] \tag{24.23}$$

where we have defined h_1,h_2 as the quaternions corresponding to the matrices $\mathbf{H}_1, \mathbf{H}_2$ given in Equation 24.21. Being inspired by the simplicity of decoding scheme of the standard Alamouti code [11] through linear combinations of the received signals, we generalize the idea and derive the following four expressions by linearly combining the received signals in Equation 24.23. Interested readers can find the detailed derivations of the linear combination process in Ref. [44].

$$\Re(\tilde{r}_1) = \Re(h_1\bar{r}_1 + r_2\bar{h}_2) = \left(\|h_1\|^2 + \|h_2\|^2 p_0 + \Re(\tilde{z}_0)\right) \tag{24.24}$$

$$\Re(\tilde{r}_2) \stackrel{\text{def}}{=} \Re[h_1 i\bar{r}_1 + r_2(i\mathbf{T}_q)\bar{h}_2] = \left(\|h_1\|^2 + \|h_2\|^2\right)p_1 + \Re(\tilde{z}_1) \tag{24.25}$$

$$\Re(\tilde{r}_3) \stackrel{\text{def}}{=} \Re[h_1 j\bar{r}_1 + r_2(j\mathbf{T}_q)\bar{h}_2] = \left(\|h_1\|^2 + \|h_2\|^2\right)p_2 + \Re(\tilde{z}_2) \tag{24.26}$$

and

$$\Re(\tilde{r}_4) \stackrel{\text{def}}{=} \Re[h_1 k\bar{r}_1 + r_2(k\mathbf{T}_q)\bar{h}_2] = \left(\|h_1\|^2 + \|h_2\|^2\right)p_3 + \Re(\tilde{z}_3) \tag{24.27}$$

Decoding proceeds as follows. First, p_0 is calculated by applying a hard slicer to the left-hand side of Equation 24.24. Next, as discussed in the appendix, there are eight choices for the transformation $\mathbf{T}q$ where each can be used to calculate a candidate for the triplet (p_1, p_2, p_3) by applying a hard slicer to the left-hand sides of Equations 24.25 through 24.27. For each choice of $\mathbf{T}q$, there are two choices of q (sign ambiguity). Finally, the 16 candidates for (p, q) are compared using the ML metric in Equation 24.22 to obtain the decoded QPSK information symbols. We have proved that the statistics $R(\tilde{r}_1)$ through $R(\tilde{r}_4)$ are sufficient for ML decoding [44]. In addition, we emphasize that there is no loss of optimality in applying the hard QPSK slicer operation to Equation 24.24 through 24.27 since the noise samples are zero-mean uncorrelated Gaussian.

24.5.4 A Differential Quaternionic Code

Our starting point is the input–output relationship in Equation 24.23 that can be written in compact matrix notation as follows

$$\mathbf{r}^{(k)} = \mathbf{h}\mathbf{C}^{(k)} + \mathbf{z}^{(k)} \tag{24.28}$$

Consider the following differential encoding rule

$$\mathbf{C}^{(k)} = \mathbf{C}^{(k-1)}\mathbf{U}^{(k)} \tag{24.29}$$

where the information matrix

$$\mathbf{U}^{(k)} = \begin{bmatrix} \mathbf{P} & \mathbf{Q} \\ -\bar{\mathbf{Q}} & \dfrac{\bar{\mathbf{Q}}\bar{\mathbf{P}}\mathbf{Q}}{\|\mathbf{Q}\|^2} \end{bmatrix}$$

Therefore, we have

$$\mathbf{r}^{(k)} = \mathbf{h}\mathbf{C}^{(k-1)}\mathbf{U}^{(k)} + \mathbf{z}^{(k)} \tag{24.30}$$

from which we can write

$$\begin{aligned}\mathbf{r}^{(k)} &= (\mathbf{r}^{(k-1)} - \mathbf{z}^{(k-1)})\mathbf{U}^{(k)} + \mathbf{z}^{(k)} \\ &= \mathbf{r}^{(k-1)}\mathbf{U}^{(k)} + \underbrace{\mathbf{z}^{(k)} - \mathbf{z}^{(k-1)}\mathbf{U}^{(k)}}_{\tilde{\mathbf{z}}^{(k)}}\end{aligned} \tag{24.31}$$

This equation has identical form to the received signal equation in the coherent case **except** for the two main differences.

- The previous output vector $\mathbf{r}^{(k-1)}$ in Equation 24.31 plays the role of the channel coefficient vector and is known at the receiver.
- Since $\mathbf{U}^{(k)}$ is a unitary matrix by construction, the equivalent noise vector $\tilde{\mathbf{z}}^{(k)}$ will also be zero-mean white Gaussian (such as $\mathbf{z}^{(k)}$ and $\mathbf{z}^{(k-1)}$) but with twice the variance.

Hence, the same efficient ML coherent decoding algorithm applies in the differential case as well but at an additional 3 dB performance penalty at high SNR.

24.6 Simulation Results

We present simulation results on the performance of our proposed STBC with the efficient ML decoding algorithm. We assume QPSK modulation, a single antenna at the receiver (unless otherwise stated), and no CSI at the transmitter.

We start by investigating the resulting performance degradation when the assumption of perfect CSI at the receiver is not satisfied. We consider two scenarios. In the first scenario, no CSI is available at the receiver and the differential encoding/decoding scheme of Section 24.5.4 is used. Figure 24.2 shows that the SNR penalty from coherent decoding (with perfect CSI) is 3 dB at high SNR. In the second scenario, the coherent ML decoder uses estimated CSI acquired by transmitting a pilot codeword of the same quaternionic structure and using a simple matched filter operation at the receiver to calculate the CSI vector. Figure 24.3 shows that the performance loss due to channel estimation is about 2–3 dB, which is comparable to the differential technique.

Next, we compare the performance of both the schemes in a time-varying channel. The pilot-based channel estimation scheme will suffer performance degradation since the channel estimate will be outdated due to the Doppler effect. To mitigate this effect, we need to increase the frequency of pilot codeword insertion as the Doppler frequency increases, which in turn increases the training overhead. We assume a fixed pilot insertion rate of one every 20 codewords; that is, a training overhead of only 5%. Similarly, the differential scheme will also suffer performance degradation since the assumption of a constant channel over two consecutive codewords (i.e., eight symbol intervals) is no longer valid. Figure 24.4 shows that for high mobile speeds ($\geq$60 mph), an error floor occurs for both schemes.

Both schemes achieve comparable performance for low (pedestrian, $\leq$5 mph) speeds, but the pilot-based scheme performs better at moderate to high speeds at the expense of a more complex receiver (to perform channel estimation) and the pilot transmission overhead.

To investigate the performance of our proposed quaternionic code in a wireless broadband environment, we assume the widely-used Stanford University Interim (SUI) channel models [45] where each of the three-tap SUI channel models is defined for a particular terrain type with varying degree of Ricean fading **K** factors and Doppler frequency. We combine our quaternionic STBC with OFDM transmission where each codeword is now transmitted over four consecutive OFDM symbol durations (for each tone).

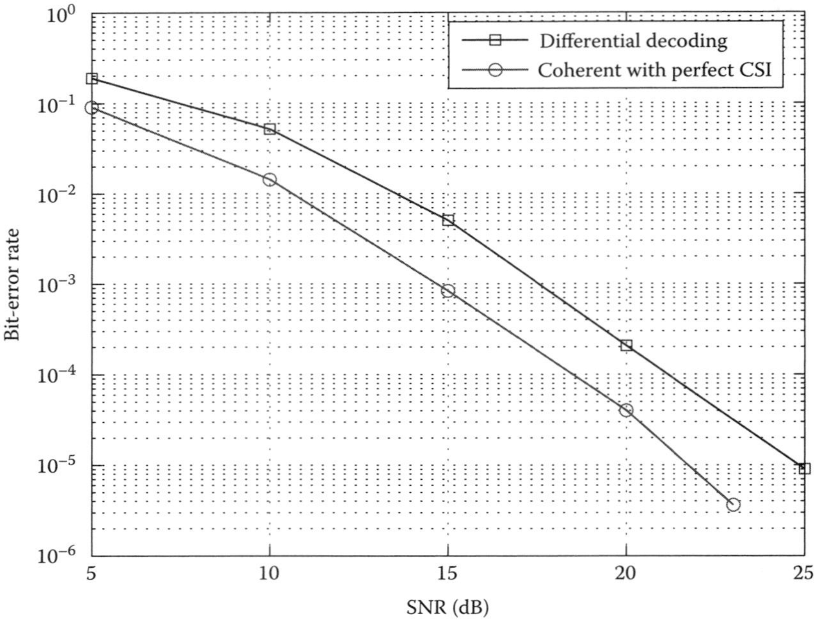

FIGURE 24.2 Performance comparison between coherent and differential decoding in quasistatic fading.

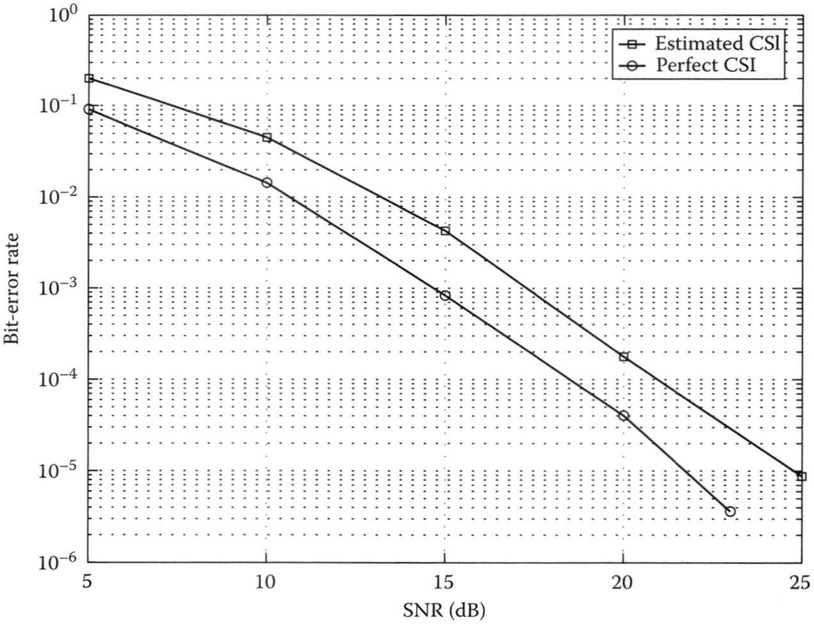

FIGURE 24.3 Performance comparison between perfect CSI and estimated CSI in quasistatic fading.

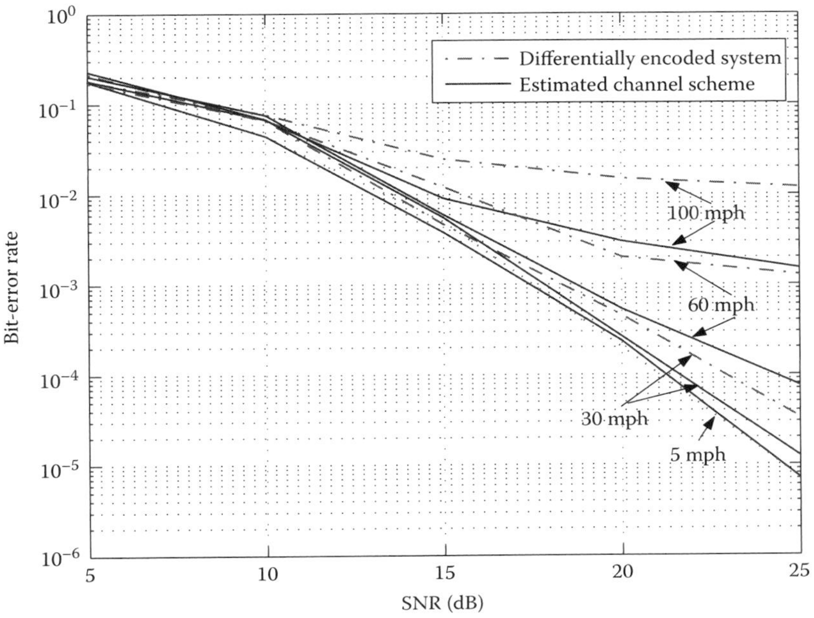

FIGURE 24.4 Performance comparison between differential and pilot-based decoding schemes in time-varying channel.

In our simulations, we use 256 subcarriers and a cyclic prefix length of 64 samples. We simultaneously transmit 256 codewords from four transmit antennas over four OFDM symbols and assume that the channel remains fixed over that period. We also use a Reed–Solomon RS(255, 163) outer code and frequency-interleave of the coded data before transmitting through the channel. This simulation model is not compliant with WiMAX, but it captures the concept of the technologies used in WiMAX.

Figure 24.5 illustrates the significant performance gains achieved by our proposed 4-TX STBC in the 802.16 environment as compared to SISO transmission. To put these SNR gains in perspective, at $\text{BER} = 10^{-3}$, these SNR gains translate to a 50% increase in the cell coverage area assuming one receive antenna. By adding a second receive antenna, the percentage increase becomes 166%.*

Appendix A: Quaternions

Quaternions are a noncommutative extension of the complex numbers. In the mid-nineteenth century, Hamilton discovered quaternions and was so pleased that he immediately carved the following message into the Brougham bridge in Dublin [46].

$$i^2 = j^2 = k^2 = ijk = -1 \tag{24.A.1}$$

This equation is the fundamental equation for the quaternions. A quaternion can be written uniquely as a linear combination of the four basis quaternions $1, i, j, k$, over the reals:

$$q \stackrel{\text{def}}{=} q_0 + q_1 i + q_2 j + q_3 k \tag{24.A.2}$$

* These calculations assume a path loss exponent of 4, which is recommended for the SUI-3 channel model with a Base Station height of 50 m [45].

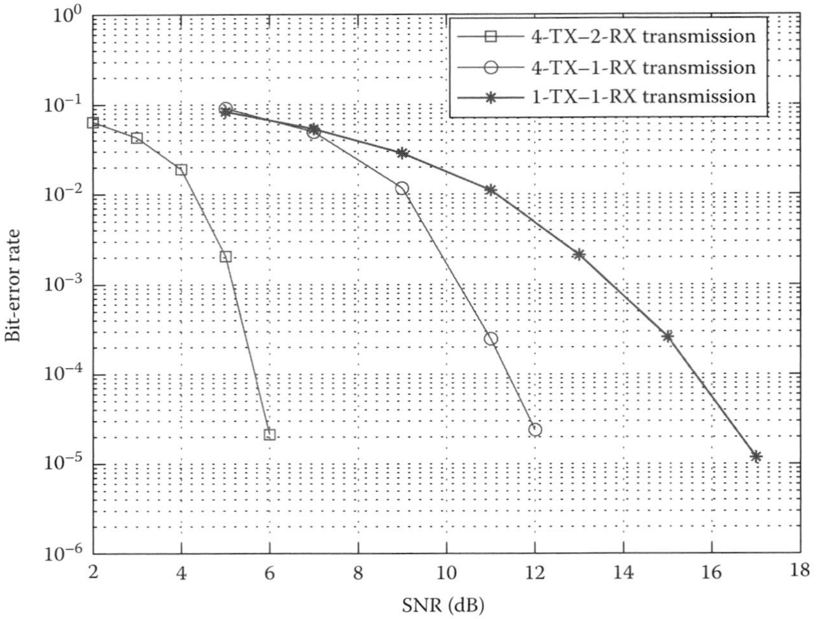

FIGURE 24.5 Performance comparison between our proposed code (with one and two receive antennas) and SISO transmission. Both are combined with OFDM in an 802.16 scenario.

It is a noncommutative group, as can be seen by the relations $ij=-ji=k$, $jk=-kj=i$, and $ki=-ik=j$. These relations follow from Equation 24.A.1 and associativity.

Quaternions can be viewed as a 4×4 matrix algebra over the real numbers **R**, where right multiplication by the quaternion q is described by

$$x_0+x_1i+x_2j+x_3k \equiv [x_0x_1x_2x_3] \to [x_0x_1x_2x_3]\begin{bmatrix} q_0 & q_1 & q_2 & q_3 \\ -q_1 & q_0 & -q_3 & q_2 \\ -q_2 & q_3 & q_0 & -q_1 \\ -q_3 & -q_2 & q_1 & q_0 \end{bmatrix} \tag{24.A.3}$$

The conjugate quaternion $\bar{q}$ is given by $\bar{q} \overset{\text{def}}{=} q_0 - q_1 i - q_2 j - q_3 k$ and we have

$$q\bar{q} = \|q\|^2 = q_0^2 + q_1^2 + q_2^2 + q_3^2 \tag{24.A.4}$$

We may also view quaternions as pairs of complex numbers, where the product of quaternions (v, w) and (v', w') is given by

$$(v,w)(v',w') = (vv' - \bar{w}'w, vw' + \bar{v}'w) \tag{24.A.5}$$

These are Hamilton's biquaternions (see Ref. [47]), and right multiplication by the biquaternion (v, w) is described by

$$x_0 + x_1 i + y_0 j + y_1 k \equiv [xy] \to [xy]\begin{bmatrix} v & w \\ -\bar{w} & \bar{v} \end{bmatrix} \tag{24.A.6}$$

The matrices

$$\begin{bmatrix} 1 & 0 \\ 0 & 1 \end{bmatrix}; \quad \begin{bmatrix} i & 0 \\ 0 & -i \end{bmatrix}; \quad \begin{bmatrix} 0 & 1 \\ -1 & 0 \end{bmatrix}; \quad \begin{bmatrix} 0 & i \\ i & 0 \end{bmatrix} \tag{24.A.7}$$

describe right multiplication by 1, i, j, and k, respectively.

There is an isomorphism between the quaternions q, 4×4 real matrices, and 2×2 complex matrices:

$$q \cong \begin{bmatrix} q_0 & q_1 & q_2 & q_3 \\ -q_1 & q_0 & -q_3 & q_2 \\ -q_2 & q_3 & q_0 & -q_1 \\ -q_3 & -q_2 & q_1 & q_0 \end{bmatrix} \cong \begin{bmatrix} q^c(0) & q^c(1) \\ -q^{-c}(1) & q^{-c}(0) \end{bmatrix} = \mathbf{Q} \tag{24.A.8}$$

where $q^c(0), q^c(1) \in \mathbb{C}$, and $q^c(0) = q0 + iq_1, q^c(1) = q_2 + iq_3$. Therefore, we can interchangeably use the matrix representation for the quaternions to demonstrate their properties. We will represent the 2×2 complex version of q by $\mathbf{Q}$. The norms have the following relationship:

$$\|q\|^2 = q_0^2 + q_1^2 + q_2^2 + q_3^2 = |q^c(0)|^2 + |q^c(1)|^2 = \|\mathbf{Q}\|^2 \tag{24.A.9}$$

The matrix representing right multiplication by the biquaternion (v, w) is the 2×2 STBC introduced by Alamouti [11]. Note that the rows and columns are orthogonal with respect to the standard inner product

$$[x\ y] \cdot [x'\ y'] = x\bar{x}' + y\bar{y}' \tag{24.A.10}$$

There is a classical correspondence between unit quaternions and rotations in $\mathbb{R}^3$ given by

$$q \rightarrow \mathbf{T}_q: p \rightarrow \bar{q}pq \tag{24.A.11}$$

where we have identified vectors in $\mathbb{R}^4$ with quaternions $p = p0 + p_1 i + p_2 j + p_3 k$ [46]. The transformation $\mathbf{T}_q$ fixes the real part $\Re(p)$ of the quaternion p, and if $q = q0 + q_1 i + q_2 j + q_3 k$, then $\mathbf{T}_q$ describes rotation about the axis (q_1, q_2, q_3) through an angle 2θ where $\cos(\theta) = q_0$ and $\sin(\theta) = \sqrt{q_1^2 + q_2^2 + q_3^2}$. For example, if $q = \pm(1 + i + j + k)/2$, the transformation $\mathbf{T}_q$ and its effect on i, j, and k are as follows:

$$\mathbf{T}_q = \begin{bmatrix} 1 & 0 & 0 & 0 \\ 0 & 0 & 0 & 1 \\ 0 & 1 & 0 & 0 \\ 0 & 0 & 1 & 0 \end{bmatrix}; \quad \begin{matrix} i \rightarrow k \\ j \rightarrow i \\ k \rightarrow j \end{matrix} \tag{24.A.12}$$

The eight transformations $\mathbf{T}_q$ together with their effect on i, j, and k can be found in Ref. [44]. A transformation $\mathbf{T}_q$ either maps $i \rightarrow \pm k$; $k \rightarrow \pm j$; $j \rightarrow \pm i$ or maps $i \rightarrow \pm j$; $\rightarrow \pm k$; $k \rightarrow i$. In both cases, the product of the signs is equal to 1.

References

1. G. J. Foschini, Layered space-time architecture for wireless communication in a fading environment when using multi-element antennas, *Bell Labs Tech. J.*, 1(2), 41–59, Fall 1996.
2. E. Telatar, Capacity of multi-antenna Gaussian channels, *European Trans. Telecommun.*, 10(6), 585–595, Nov./Dec. 1999.
3. Y. Li and G. L. Stüber, *Orthogonal Frequency Division Multiplexing for Wireless Communications*, Springer, New York, 2006.

4. IEEE Standard for Local and Metropolitan Area Networks Part 16: Air Interface for Fixed Broadband Wireless Access Systems, IEEE Std. 802.16-2004, 2004.
5. IEEE Std 802.16e-2005 and IEEE Std 802.16-2004/Cor 1-2005. Amendment and Corrigendum to IEEE Std 802.16-2004, IEEE, Std. 802.16e-2005, 2005.
6. V. Tarokh, N. Seshadri, and A. R. Calderbank, Space-time codes for high data rate wireless communication: Performance criterion and code construction, *IEEE Trans. Inform. Theory*, 44(2), 744–765, March 1998.
7. S. Parkvall, M. Karlsson, M. Samuelsson, L. Hedlund, and B. Göransson, Transmit diversity in WCDMA: Link and system level results, in *Proceedings of the Vehicular Technology Conference 2000-Spring*, Tokyo, Japan, May 2000.
8. A. F. Naguib, V. Tarokh, N. Seshadri, and A. R. Calderbank, A space-time coding modem for high-data-rate wireless communications, *IEEE J. Select. Areas Commun.*, 16(8), 1459–1478, Oct. 1998.
9. V. Tarokh, A. Naguib, N. Seshadri, and A. R. Calderbank, Space-time codes for high data rate wireless communication: Performance criteria in the presence of channel estimation errors, mobility, and multiple paths, *IEEE Trans. Commun.*, 47(2), 199–207, Feb. 1999.
10. WiMAX forum [Online]. Available: http//www.wimaxforum.org, 2006.
11. S. M. Alamouti, A simple transmit diversity technique for wireless communications, *IEEE J. Select. Areas Commun.*, 16(8), 1451–1458, Oct. 1998.
12. J. Guey, M. P. Fitz, M. R. Bell, and W. Kuo, Signal design for transmitter diversity wireless communication systems over Rayleigh fading channels, *IEEE Trans. Commun.*, 47(4), 527–537, April 1999.
13. H. Lu, Y. Wang, P. V. Kumar, and K. M. Chugg, Remarks on space-time codes including a new lower bound and an improved code, *IEEE Trans. Inform. Theory*, 49(10), 2752–2757, Oct. 2003.
14. H. Jafarkhani, *Space–time Coding: Theory and Practice*. Cambridge University Press, New York, 2005.
15. J. Chui and A. R. Calderbank, Effective coding gain for space–time codes, in *Proceedings of the IEEE International Symposium on Information Theory* (*ISIT*), Seattle, WA, July 2006.
16. L. Zheng and D. N. Tse, Diversity and multiplexing: A fundamental tradeoff in multiple-antenna channels, *IEEE Trans. Inform. Theory*, 49(5), 1073–1096, May 2003.
17. S. N. Diggavi, N. Al-Dhahir, A. Stamoulis, and A. R. Calderbank, Great expectations: The value of spatial diversity to wireless networks, *Proc. IEEE*, 92(2), 219–270, Feb. 2004, invited paper.
18. L. Grokop and D. Tse, Diversity/multiplexing tradeoff in ISI channels, in *Proceedings of the IEEE International Symposium on Information Theory* (*ISIT*), Chicago, IL, June/July 2004.
19. G. J. Foschini, G. Golden, R. Valenzuela, and P. Wolniansky Simplified processing for high spectral efficiency wireless communication employing multi-element arrays, *IEEE J. Select. Areas Commun.*, 17(11), 1841–1852, Nov. 1999.
20. W. Choi, R. Negi, and J. M. Cioffi, Combined ML and DFE decoding for the V-BLAST system, *Proc. ICC*, 3, 1243–1248, June 2003.
21. A. F. Naguib and N. Seshadri, Combined interference cancellation and ML decoding of space–time block codes, in *Proceedings Communications Theory Mini-Conference, Held in Conjunction with Globecomm '98*, Sydney, Australia, pp. 7–15, 1998.
22. S. Sirianunpiboon, A. R. Calderbank, and S. D. Howard, Space-polarization-time codes for diversity gains across line of sight and rich scattering environments, *Trans. Inform. Theory*, 2006 (submitted).
23. H. Yao and G. W. Wornell, Achieving the full mimo diversity-multiplexing frontier with rotation-based space-time codes, in *Proceedings of the Allerton Conference on Communication, Control, Computing*, Monticello, IL, Oct. 2003.
24. P. Dayal and M. K. Varanasi, An optimal two transmit antenna space-time code and its stacked extensions, *IEEE Trans. Inform. Theory*, 51(12), 4348–4355, Dec. 2005.

25. J.-C. Belfiore, G. Rekaya, and E. Viterbo, The golden code: A 2×2 full-rate space-time code with nonvanishing determinants, *IEEE Trans. Inform. Theory*, 51(4), 1432–1436, April 2005.
26. E. Viterbo and J. Boutros, A universal lattice code decoder for fading channels, *IEEE Trans. Inform. Theory*, 45(5), 1639–1642, July 1999.
27. M. O. Damen, A. Chkeif, and J.-C. Belfiore, Lattice code decoder for space-time codes, *IEEE Commun. Lett.*, 4(5), 161–163, May 2000.
28. U. Fincke and M. Pohst, Improved methods for calculating vectors of short length in a lattice, including a complexity analysis, *Math. Comp.*, 44(170), 463–471, April 1985.
29. S. J. Lee et al., A space–time code with full-diversity and rate 2 for 2 transmit antenna transmission, in *IEEE 802.16 Session #34*, 2004, Contribution IEEE C802.16e-04/434r2, Nov. 2004.
30. V. Tarokh, H. Jafarkhani, and A. R. Calderbank, Space-time block codes from orthogonal designs, *IEEE Trans. Inform. Theory*, 45(5), 1456–1467, July 1999.
31. H. Wang and X.-G. Xia, Upper bounds of rates of complex orthogonal space-time block codes, *IEEE Trans. Inform. Theory*, 49(10), 2788–2796, Oct. 2003.
32. H. Jafarkhani, A quasi-orthogonal space-time block code, *IEEE Trans. Commun.*, 49(1), 1–4, Jan. 2001.
33. N. Sharma and C. B. Papadias, Improved quasi-orthogonal codes through constellation rotation, *IEEE Trans. Commun.*, 51(3), 332–335, March 2003.
34. H. Jafarkhani and N. Seshadri, Super-orthogonal space-time trellis codes, *IEEE Trans. Inform. Theory*, 49(4), 937–950, April 2003.
35. B. Hassibi and B. Hochwald, High-rate codes that are linear in space and time, *IEEE Trans. Inform. Theory*, 48(7), 1804–1824, July 2002.
36. H. E. Gamal and A. R. Hammons, Jr., A new approach to layered space-time coding and signal processing, *IEEE Trans. Inform. Theory*, 47(6), 2321–2334, Sept. 2001.
37. M. O. Damen, H. El Gamal, and N. C. Beaulieu, Linear threaded algebraic space-time constellations, *IEEE Trans. Inform. Theory*, 49(10), 2372–2388, Oct. 2003.
38. F. Oggier, G. Rekaya, J.-C. Belfiore, and E. Viterbo, Perfect space-time block codes, *IEEE Trans. Inform. Theory*, 52(9), 3885–3902, Sept. 2006.
39. J. Cavers, An analysis of pilot symbol assisted modulation for Rayleigh fading channels (mobile radio), *IEEE Trans. Veh. Technol.*, 40(4), 686–693, Nov. 1991.
40. R. Negi and J. Cioffi, Pilot tone selection for channel estimation in a mobile OFDM system, *IEEE Trans. Consum. Electron.*, 44(3), 1122–1128, Aug. 1998.
41. S. Ohno and G. B. Giannakis, Capacity maximizing MMSE-optimal pilots for wireless OFDM over frequency-selective block Rayleigh-fading channels, *IEEE Trans. Inform. Theory*, 50(9), 2138–2145, Sept. 2004.
42. V. Tarokh and H. Jafarkhani, A differential detection scheme for transmit diversity, *IEEE J. Select. Areas Commun.*, 18(7), 1169–1174, July 2000.
43. S. N. Diggavi, N. Al-Dhahir, A. Stamoulis, and A. R. Calderbank, Differential space-time coding for frequency-selective channels, *IEEE Commun. Lett.*, 6(6), 253–255, June 2002.
44. R. Calderbank, S. Das, N. Al-Dhahir, and S. Diggavi, Construction and analysis of a new quaternionic space-time code for 4 transmit antennas, *Commun. Inform. Syst.* (Special issue dedicated to the 70th birthday of Thomas Kailath: Part I), 5(1), 97–122, 2005.
45. V. Erceg, K. V. S. Hari, M. S. Smith, D. S. Baum, et al., Channel models for fixed wireless applications, in IEEE 802.16 Broadband Wireless Access Working Group, IEEE 802.16a-03/01, 2003.
46. J. H. Conway and D. Smith, *On Quaternions and Octonions*. AK Peters, Ltd., Wellesley, MA, 2003.
47. B. L. van der Waerden, *A History of Algebra: From Al-Khwarizmi to Emmy Noether*. Springer, New York, 1985.

25

Exploiting Diversity in MIMO-OFDM Systems for Broadband Wireless Communications

Weifeng Su
State University of New York at Buffalo

Zoltan Safar
IT University of Copenhagen

K. J. Ray Liu
University of Maryland

25.1 Introduction 25-1
25.2 MIMO-OFDM System Model and Code Design Criteria 25-3
System Model • Code Design Criteria
25.3 Full-Diversity SF Codes Design 25-6
Obtaining Full-Diversity SF Codes from ST Codes via Mapping • Full-Rate and Full-Diversity SF Code Design
25.4 Full-Diversity STF Code Design 25-12
Repetition-Coded STF Code Design • Full-Rate Full-Diversity STF Code Design
25.5 Simulation Results 25-15
25.6 Conclusion 25-20
References 25-20

25.1 Introduction

WiMAX is a broadband wireless solution that is likely to play an important role in providing ubiquitous voice and data services to millions of users in the near future, both in rural and urban environment. It is based on the IEEE 802.16 air interface standard suite [1,2], which provides the wireless technology for both fixed, nomadic, and mobile data access. Wireless systems have the capacity to cover large geographic areas without the need for costly cable infrastructure to each service access point, so WiMAX has the potential to prove to be a cost-effective and quickly deployable alternative to cabled networks, such as fiber optic links, cable modems, or digital subscriber lines (DSL).

The driving force behind the development of the WiMAX system has been the desire to satisfy the emerging need for high data rate applications, e.g., voice over IP, video conferencing, interactive gaming, and multimedia streaming. However, recently performed system-level simulation results indicate that the performance of the current WiMAX system may not be able to satisfy such needs. In Ref. [3], the downlink (DL) performance of a 10 MHz WiMAX system was evaluated in a 1/1 frequency reuse scenario with two transit and two receive antennas, and 10–14 Mbit/s average total sector throughput was obtained. The average DL cell throughput of a 5 MHz WiMAX system in Ref. [4] was found to be around 5 Mbit/s, also with two transit and two receive antennas and 1/1 frequency reuse. As a consequence, it seems that further performance improvement is necessary to be able to support interactive multimedia applications that require the user data rates in excess of 0.5–1 Mbit/s. The authors of Ref. [4] provide a list of techniques to achieve this: hybrid automatic repeat-request (ARQ), interference cancellation, adaptive per-subcarrier power allocation, and frequency-domain scheduling.

This list can be appended with one more item: improved multiantenna coding techniques. So far, the only open-loop multiple-input multiple-output (MIMO) coding method adopted by the WiMAX forum has been the Alamouti's 2×2 orthogonal design [7] (in different variants). Since for high-mobility users the closed-loop transmission techniques, such as frequency-domain scheduling or beam forming, are not available due to the feedback loop delay, more powerful open-loop MIMO coding methods could help to achieve even higher data rates than it is possible now with the current WiMAX system. One way to characterize the performance of MIMO systems is by their "diversity" order, which is the asymptotic slope of the bit-error rate (BER) versus signal-to-noise ratio (SNR) curve, i.e., it describes how fast the BER decreases as the SNR increases. The achieved diversity order depends on both the MIMO channel (its spatial, spectral, and temporal structure) and on the applied MIMO coding and decoding method, which can exploit a part or all of the available diversity in the MIMO channel.

The problem of designing MIMO coding and modulation techniques to improve the performance of wireless communication systems has attracted considerable attention in both industry and academia. In case of narrowband wireless communications, where the fading channel is frequency nonselective, abundant coding methods [5–10], termed as space–time (ST) codes, have been proposed to exploit the spatial and temporal diversities. In case of broadband wireless communications, where the fading channel is frequency-selective, orthogonal frequency division multiplexing (OFDM) modulation can be used to convert the frequency-selective channel into a set of parallel flat-fading channels, providing high spectral efficiency and eliminating the need for high-complexity equalization algorithms. To have the "best of both worlds," MIMO systems and OFDM modulation can be combined, resulting in MIMO-OFDM systems. This combination seemed so attractive that several communication systems, including the MIMO option in WiMAX, were based on it.

There are two major coding approaches for MIMO-OFDM systems. One is the space–frequency (SF) coding approach, where coding is applied within each OFDM block to exploit the spatial and frequency diversities. The other one is the space–time–frequency (STF) coding approach, where the coding is applied across multiple OFDM blocks to exploit the spatial, temporal, and frequency diversities. Early works on SF coding [11–16] used ST codes directly as SF codes, i.e., previously existing ST codes were used by replacing the time domain with the frequency domain (OFDM tones). The performance criteria for SF-coded MIMO-OFDM systems were derived in Refs. [16,17], and the maximum achievable diversity was found to be LM_tM_r, where M_t and M_r are the number of transmit and receive antennas, respectively, and L is the number of delay paths in frequency-selective fading channels. It has been shown in Ref. [17] that the way of using ST codes directly as SF codes can achieve only the spatial diversity, but not the full spatial and frequency diversity LM_tM_r. Later, in Refs. [18–20], systematic SF code design methods were proposed that could guarantee to achieve the maximum diversity.

One may also consider STF coding across multiple OFDM blocks to exploit all of the spatial, temporal, and frequency diversities. The STF coding strategy was first proposed in Ref. [21] for two transmit antennas and further developed in Refs. [22–24] for multiple transmit antennas. Both Refs. [21,24] assumed that the MIMO channel stays constant over multiple OFDM blocks; however, STF coding under this assumption cannot provide any additional diversity compared to the SF coding approach [26]. In Ref. [23], an intuitive explanation on the equivalence between antennas and OFDM tones was presented from the viewpoint of channel capacity. In Ref. [22], the performance criteria for STF codes were derived, and an upper bound on the maximum achievable diversity order was established. However, there was no discussion in Ref. [22] whether the upper bound can be achieved or not, and the proposed STF codes were not guaranteed to achieve the full spatial, temporal, and frequency diversities. Later, in Ref. [26], we proposed a systematic method to design full-diversity STF codes for MIMO-OFDM systems.

In this chapter, we review SF/STF code design criteria and summarize our findings on SF/STF coding for MIMO-OFDM systems [19,20,26]. This chapter is organized as follows. First, we describe a general MIMO-OFDM system model and review code design criteria. Second, we introduce two SF code design methods that can guarantee to achieve full spatial and frequency diversity based on Refs. [19] and [20]. Then, we summarize our results on STF coding based on Ref. [26]. Finally, some simulation results are presented and some conclusions are drawn.

25.2 MIMO-OFDM System Model and Code Design Criteria

We first describe a general STF-coded MIMO-OFDM system and discuss its performance criteria. Since SF coding, where coding is applied within each OFDM block, is a special case of STF coding, the performance criteria of SF-coded MIMO-OFDM systems can be easily obtained from that of STF codes, as shown at the end of this section.

25.2.1 System Model

We consider a general STF-coded MIMO-OFDM system with M_t transmit antennas, M_r receive antennas, and N subcarriers. Suppose that the frequency-selective fading channels between each pair of transceiver antennas have L independent delay paths and the same power delay profile. The MIMO channel is assumed to be constant over each OFDM block, but it may vary from one OFDM block to another. At the kth OFDM block, the channel coefficient from transmit antenna i to receive antenna j at time τ can be modeled as

$$h_{i,j}^k(\tau) = \sum_{l=0}^{L-1} \alpha_{i,j}^k(l)\delta(\tau - \tau_l) \tag{25.1}$$

where

τ_l is the delay

$\alpha_{i,j}^k(l)$ the complex amplitude of the lth path between transmit antenna i and receive antenna j

The $\alpha_{i,j}^k(l)$ are modeled as zero-mean, complex Gaussian random variables with variances δ_l^2 and $\sum_{l=0}^{L-1} \delta_l^2 = 1$. From Equation 25.1, the frequency response of the channel is given by $H_{i,j}^k(f) = \sum_{l=0}^{L-1} \alpha_{i,j}^k(l)e^{-j2\pi f \tau_l}$ and $j = \sqrt{-1}$.

We consider STF coding across M_t transmit antennas, N OFDM subcarriers, and K consecutive OFDM blocks (the $K=1$ case corresponds to SF coding).

Each STF codeword can be expressed as a $KN \times M_t$ matrix

$$C = \left[C_1^T C_2^T \ldots C_k^T\right]^T \tag{25.2}$$

where the channel symbol matrix C_k is given by

$$C_k = \begin{bmatrix} c_1^k(0) & c_2^k(0) & \cdots & c_{M_t}^k(0) \\ c_1^k(1) & c_2^k(1) & \cdots & c_{M_t}^k(1) \\ \vdots & \vdots & \ddots & \vdots \\ c_1^k(N-1) & c_2^k(N-1) & \cdots & c_{M_t}^k(N-1) \end{bmatrix} \tag{25.3}$$

in which $c_i^k(n)$ is the channel symbol transmitted over the nth subcarrier by transmit antenna i in the kth OFDM block. The STF code is assumed to satisfy the energy constraint $E\|C\|_F^2 = KNM_t$, where $\|C\|_F$ is the Frobenius norm of C. During the kth OFDM block period, the transmitter applies an N-point inverse fast Fourier transform (IFFT) to each column of the matrix C_k. After appending a cyclic prefix, the OFDM symbol corresponding to the ith $(i = 1, 2, \ldots, M_t)$ column of C_k is transmitted by transmit antenna i.

At the receiver, after removing the cyclic prefix and applying FFT, the received signal at the nth subcarrier at receive antenna j in the kth OFDM block is given by

$$y_j^k(n) = \sqrt{\frac{\rho}{M_t}} \sum_{i=1}^{M_t} c_i^k(n) H_{i,j}^k(n) + z_j^k(n) \tag{25.4}$$

where

$$H_{i,j}^k(n) = \sum_{l=0}^{L-1} \alpha_{i,j}^k(l) e^{-j2\pi n \Delta f \tau_l} \tag{25.5}$$

is the channel frequency response at the nth subcarrier between transmit antenna i and receive antenna j; $\Delta f = 1/T$ the subcarrier separation in the frequency domain; and T the OFDM symbol period. We assume that the channel state information $H_{i,j}^k(n)$ is known at the receiver, but not at the transmitter. In Equation 25.4, $z_j^k(n)$ denotes the additive white complex Gaussian noise with zero-mean and unit variance at the nth subcarrier at receive antenna j in the kth OFDM block. The factor $\sqrt{\rho/M_t}$ in Equation 25.4 ensures that ρ is the average SNR at each receive antenna.

25.2.2 Code Design Criteria

We discuss the STF code design criteria based on the pairwise error probability of the system. The channel frequency response vector between transmit antenna i and receive antenna j over K OFDM blocks will be denoted by

$$H_{i,j} = \left[H_{i,j}(0) H_{i,j}(1) \ldots H_{i,j}(KN-1)\right]^T \tag{25.6}$$

where we use the notation $H_{i,j}((k-1)N + n) \triangleq H_{i,j}^k(n)$ for $1 \le k \le K$. Using the notation $w = e^{-j2\pi\Delta f}$, $H_{i,j}$ can be decomposed as

$$H_{i,j} = (I_K \otimes W) A_{i,j} \tag{25.7}$$

where

$$W = \begin{bmatrix} 1 & 1 & \cdots & 1 \\ w^{\tau_0} & w^{\tau_1} & \cdots & w^{\tau_{L-1}} \\ \vdots & \vdots & \ddots & \vdots \\ w^{(N-1)\tau_0} & w^{(N-1)\tau_1} & \cdots & w^{(N-1)\tau_{L-1}} \end{bmatrix}$$

which is related to the delay distribution, and

$$A_{i,j} = \left[\alpha_{i,j}^1(0)\alpha_{i,j}^1(1) \ldots \alpha_{i,j}^1(L-1) \ldots \alpha_{i,j}^K(0)\alpha_{i,j}^K(1) \ldots \alpha_{i,j}^K(L-1)\right]^T$$

which is related to the power distribution of the channel impulse response. In general, W is not a unitary matrix. If all of the L delay paths fall at the sampling instances of the receiver, W is part of the discrete Fourier transform (DFT)-matrix, which is unitary. From Equation 25.7, the correlation matrix of the channel frequency response vector between transmit antenna i and receive antenna j can be calculated as

$$R_{i,j} = E\left\{H_{i,j}H_{i,j}^{\mathcal{H}}\right\} = (I_K \otimes W)E\left\{A_{i,j}A_{i,j}^{\mathcal{H}}\right\}\left(I_K \otimes W^{\mathcal{H}}\right)$$

We assume that the MIMO channel is spatially uncorrelated, i.e., the channel coefficients $\alpha_{i,j}^{k}(l)$ are independent for different indices (i,j). So we can define the time correlation at lag m as $r_T(m) = E\left\{\alpha_{i,j}^{k}(l)\alpha_{i,j}^{k+m^*}(l)\right\}$. Thus, the correlation matrix $E\left\{A_{i,j}A_{i,j}^{\mathcal{H}}\right\}$ can be expressed as

$$E\left\{A_{i,j}A_{i,j}^{\mathcal{H}}\right\} = R_{\mathrm{T}} \otimes \Lambda \tag{25.8}$$

where
$\Lambda = \mathrm{diag}\{\delta_0^2, \delta_1^2, \ldots, \delta_{L-1}^2\}$,
R_{T} is the temporal correlation matrix of size $K \times K$.

We can also define the frequency correlation matrix, R_{F}, as $R_{\mathrm{F}} = E\left\{H_{i,j}^{k}H_{i,j}^{k\,\mathcal{H}}\right\}$, where $H_{i,j}^{k} = \left[H_{i,j}^{k}(0), \ldots, H_{i,j}^{k}(N-1)\right]^{\mathrm{T}}$. Then, $R_{\mathrm{F}} = W\Lambda W^{\mathcal{H}}$. As a result, we arrive at

$$R_{i,j} = R_{\mathrm{T}} \otimes \left(W\Lambda W^{\mathcal{H}}\right) = R_{\mathrm{T}} \otimes R_{\mathrm{F}} \triangleq R \tag{25.9}$$

where the correlation matrix R is independent of the transceiver antenna indices i and j.

For two distinct STF codewords C and $\tilde{C}$, we denote

$$\Delta \triangleq (C - \tilde{C})(C - \tilde{C})^{\mathcal{H}} \tag{25.10}$$

Then, the pairwise error probability between C and $\tilde{C}$ can be upper bounded as [26]

$$P(C \to \tilde{C}) \leq \binom{2\nu M_{\mathrm{r}} - 1}{\nu M_{\mathrm{r}}}\left(\prod_{i=1}^{\nu}\lambda_i\right)^{-M_{\mathrm{r}}}\left(\frac{\rho}{M_{\mathrm{t}}}\right)^{-\nu M_{\mathrm{r}}} \tag{25.11}$$

where ν is the rank of $\Delta \circ R$; $\lambda_1, \lambda_2, \ldots, \lambda_\nu$ the nonzero eigenvalues of $\Delta \circ R$; and o denotes the Hadamard product.* The minimum value of the product $\prod_{i=1}^{\nu}\lambda_i$ over all pairs of distinct signals C and $\tilde{C}$ is termed as "coding advantage," denoted by

$$\zeta_{\mathrm{STF}} = \min_{C \neq \tilde{C}} \prod_{i=1}^{\nu}\lambda_i \tag{25.12}$$

* Suppose that $A = \{a_{i,j}\}$ and $B = \{b_{i,j}\}$ are two matrices of size $m \times n$. The *Hadamard product* of A and B is defined as $A \circ B = \{a_{i,j}\, b_{i,j}\}_{1 \leq i \leq m, 1 \leq j \leq n}$.

Based on the performance upper bound, two STF code design criteria were proposed in Ref. [26].

- *Diversity (rank) criterion*: The minimum rank of $\Delta \circ R$ over all pairs of distinct codewords C and $\tilde{C}$ should be as large as possible.
- *Product criterion*: The coding advantage or the minimum value of the product $\prod_{i=1}^{\nu} \lambda_i$ over all pairs of distinct signals C and $\tilde{C}$ should also be maximized.

If the minimum rank of $\Delta \circ R$ is ν for any pair of distinct STF codewords C and $\tilde{C}$, we say that the STF code achieves a diversity order of νM_r. For a fixed number of OFDM blocks K, number of transmit antennas M_t, and correlation matrices R_T and R_F, the "maximum achievable diversity" or "full diversity" is defined as the maximum diversity order that can be achieved by STF codes of size $KN \times M_t$.

According to the rank inequalities on Hadamard products [37], we have

$$\text{rank}(\Delta \circ R) \leq \text{rank}(\Delta)\text{rank}(R_T)\text{rank}(R_F)$$

Since the rank of Δ is at most M_t and the rank of R_F is at most L, we obtain

$$\text{rank}(\Delta \circ R) \leq \min\{LM_t\text{rank}(R_T), KN\} \tag{25.13}$$

Thus, the maximum achievable diversity is at most $\min\{LM_tM_r\text{rank}(R_T), KNM_r\}$ in agreement with the results of Ref. [22]. However, there is no discussion in Ref. [22] on whether this upper bound can be achieved or not. As we will see later, this upper bound can indeed be achieved. We also observe that if the channel stays constant over multiple OFDM blocks ($\text{rank}(R_T) = 1$), the maximum achievable diversity is only $\min\{LM_tM_r, KNM_r\}$. In this case, STF coding cannot provide additional diversity advantage compared to the SF coding approach.

Note that the above analytical framework includes ST and SF codes as special cases. If we consider only one subcarrier ($N = 1$) and one delay path ($L = 1$), then the channel becomes a single-carrier, time-correlated, flat-fading MIMO channel. The correlation matrix R simplifies to $R = R_T$, and the code design problem reduces to that of ST code design, as described in Ref. [27]. In the case of coding over a single OFDM block ($K = 1$), the correlation matrix R becomes $R = R_F$, and the code design problem simplifies to that of SF codes, as discussed in Refs. [19,20].

25.3 Full-Diversity SF Codes Design

We introduce in this section two systematic SF code design methods, where coding is applied within each OFDM block. The first method is to obtain full-diversity SF codes from ST codes via mapping [19], which shows that by using a simple repetition mapping, full-diversity SF codes can be constructed from any ST (block or trellis) code designed for quasi-static flat Rayleigh fading channels. The other method is to design spectrally efficient SF codes that can guarantee full rate and full diversity for MIMO-OFDM systems with arbitrary power delay profiles [20]. Note that in case of SF coding ($K = 1$), each SF codeword C in Equation 25.2 has a size of $N \times M_t$ and the correlation matrix $R = R_F$ has a size of $N \times N$.

25.3.1 Obtaining Full-Diversity SF Codes from ST Codes via Mapping

For any given integer l ($1 \leq l \leq L$), assume that $lM_t \leq N$ (the number of OFDM subcarriers N is generally larger than LM_t) and k is the largest integer such that $klM_t \leq N$. Suppose that there is a ST encoder with output matrix G. (For ST block encoder, G is a concatenation of some block codewords. For ST trellis encoder, G corresponds to a path of length kM_t starting and ending at the zero state.) Then, a full-diversity SF code C of size $N \times M_t$ can be obtained by mapping the ST codeword G as follows:

$$C = \begin{bmatrix} M_l(G) \\ 0_{(N-klM_t)\times M_t} \end{bmatrix} \tag{25.14}$$

where

$$\mathcal{M}_l(G) = [I_{kM_t} \otimes \mathbf{1}_{l\times 1}]G \quad (25.15)$$

in which $\mathbf{1}_{l\times 1}$ is an all one matrix of size $l \times 1$. Actually, the resulting SF code C is obtained by repeating each row of G l times and adding some zeros. The zero padding used here ensures that the SF code C has size $N \times M_t$, and typically the size of the zero padding is small. The following theorem states that if the employed ST code G has full diversity for flat-fading channels, the SF code constructed by Equation 25.14 will achieve a diversity of at least lM_tM_r [19].

THEOREM 25.1

Suppose that the frequency-selective channel has L independent paths and the maximum path delay is less than one OFDM block period. If an ST (block or trellis) code designed for M_t transmit antennas achieves full diversity for quasi-static flat-fading channels, then the SF code obtained from this ST code via the mapping $\mathcal{M}_l$ $(1 \leq l \leq L)$ defined in Equation 25.15 will achieve a diversity order of at least min $\{lM_tM_r, NM_r\}$.

Moreover, the SF code obtained from an ST block code of square size via the mapping $\mathcal{M}_l$ $(1 \leq l \leq L)$ achieves a diversity of lM_tM_r exactly. Since the maximum achievable diversity is upper bounded by $\min\{LM_tM_r, NM_r\}$; therefore, according to Theorem 25.1, the SF code obtained from a full-diversity ST code via the mapping $\mathcal{M}_L$ defined in Equation 25.15 achieves the maximum achievable diversity $\min\{LM_tM_r, NM_r\}$. We can see that the coding rate of the resulting full-diversity SF codes obtained via the mapping $\mathcal{M}_l$ (Equation 25.15) is $1/l$ times that of the corresponding ST codes, which, however, is larger than that in Ref. [18]. For example, for a system with two transmit antennas, eight subcarriers, and a two-ray delay profile, the coding rate of the full-diversity SF codes introduced here is 1/2, while the coding rate in Ref. [18] is only 1/4. Note that the simple repetition mapping is independent of particular ST codes, so all the existing ST block and trellis codes achieving full spatial diversity in quasi-static flat-fading environment can be used to design full-diversity SF codes for MIMO-OFDM systems.

To the end of this subsection, we characterize the coding advantage of the resulting SF codes in terms of the coding advantage of the underlying ST codes. We also analyze the effect of the delay distribution and the power distribution on the performance of the proposed SF codes. The "coding advantage" or "diversity product" of a full-diversity ST code for quasi-static flat-fading channels has been defined as [9,27] $\zeta_{ST} = \min_{G \neq \tilde{G}} \left|\prod_{i=1}^{M_t} \beta_i\right|^{1/2M_t}$, where $\beta_1, \beta_2, \ldots, \beta_{M_t}$ are the nonzero eigenvalues of $(G - \tilde{G})(G - \tilde{G})^{\mathcal{H}}$ or any pair of distinct ST codewords G and $\tilde{G}$. We have the following result [19].

THEOREM 25.2

The diversity product of the full-diversity SF code is bounded by that of the corresponding ST code as follows:

$$\sqrt{\eta_L}\Phi\zeta_{ST} \leq \zeta_{SF} \leq \sqrt{\eta_1}\Phi\zeta_{ST} \quad (25.16)$$

where
$\Phi = \left(\prod_{l=0}^{L-1} \delta_1\right)^{1/L}$
η_1 and η_L are the largest and smallest eigenvalues, respectively, of the matrix H defined as

$$H = \begin{bmatrix} H(0) & H(1)^* & \cdots & H(L-1)^* \\ H(1) & H(0) & \cdots & H(L-2)^* \\ \vdots & \vdots & \ddots & \vdots \\ H(L-1) & H(L-2) & \cdots & H(0) \end{bmatrix}_{L\times L} \quad (25.17)$$

and the entries of H are given by $H(n) = \sum_{l=0}^{L-1} e^{-j2\pi n\Delta f\tau_1}$ for $n = 0, 1, \ldots, L-1$.

From Theorem 25.2, we can see that the larger the coding advantage of the ST code, the larger the coding advantage of the resulting SF code, suggesting that to maximize the performance of the SF codes, we should look for the best-known ST codes existing in the literature. Moreover, the coding advantage of the SF code depends on the power delay profile. First, it depends on the power distribution through the square root of the geometric average of path powers, i.e., $\Phi = \left(\prod_{l=0}^{L-1} \delta_1\right)^{1/L}$. Since the sum of the powers of the paths is unity, this implies that the best performance is expected in case of uniform power distribution (i.e., $\delta_l^2 = 1/L$). Second, the entries of the matrix H defined in Equation 25.17 are functions of the path delays, so the coding advantage also depends on the delay distribution of the paths.

25.3.2 Full-Rate and Full-Diversity SF Code Design

In this subsection, we describe a systematic method to obtain full-rate SF codes achieving full diversity [20]. Specifically, we design a class of SF codes that can achieve a diversity order of $\Gamma M_t M_r$ for any fixed integer Γ $(1 \le \Gamma \le L)$.

We consider a coding strategy where each SF codeword C is a concatenation of some matrices G_p:

$$C = \left[G_1^T G_2^T \ldots G_P^T 0_{N-P\Gamma M_t}^T\right]^T \quad (25.18)$$

where $P = \lfloor N/(\Gamma M_t)\rfloor$, and each matrix G_p, $p = 1, 2, \ldots, P$, is of size ΓM_t by M_t. The zero padding in Equation 25.18 is used if the number of subcarriers N is not an integer multiple of ΓM_t. Each matrix $G_p (1 \le p \le P)$ has the same structure given by

$$G = \sqrt{M_t}\,\mathrm{diag}(X_1, X_2, \ldots, X_{M_t}) \quad (25.19)$$

where $\mathrm{diag}(X_1, X_2, \ldots, X_{M_t})$ is a block diagonal matrix, $X_i = [x_{(i-1)\Gamma+1}\, x_{(i-1)\Gamma+2} \cdots x_{i\Gamma}]^T$, $i = 1, 2, \ldots, M_t$, and all x_k, $k = 1, 2, \ldots, \Gamma M_t$, are complex symbols and will be specified later. The energy constraint is $E\left(\sum_{k=1}^{\Gamma M_t} |x_k|^2\right) = \Gamma M_t$. For a fixed p, the symbols in G_p are designed jointly, but the designs of G_{p1} and G_{p2}, $p_1 \ne p_2$, are independent of each other. The symbol rate of the code is $P\Gamma M_t/N$, ignoring the cyclic prefix. If N is a multiple of ΓM_t, the symbol rate is 1. If not, the rate is less than 1, but since usually N is much greater than ΓM_t, the symbol rate is very close to 1. We have the following sufficient conditions for the SF codes described above to achieve a diversity order of $\Gamma M_t M_r$ [20].

THEOREM 25.3

For any SF code constructed by Equations 25.18 and 25.19, if $\prod_{k=1}^{\Gamma M_t} |x_k - \tilde{x}_k| \ne 0$ for any pair of distinct sets of symbols $\mathbf{X} = [x_1\ x_2 \ldots x_{\Gamma M_t}]$ and $\tilde{\mathbf{X}} = [\tilde{x}_1\ \tilde{x}_2 \ldots \tilde{x}_{\Gamma M_t}]$, then the SF code achieves a diversity order of $\Gamma M_t M_r$, and the diversity product is

$$\zeta_{SF} = \zeta_{in}|\det(Q_0)|^{\frac{1}{2\Gamma}} \tag{25.20}$$

where ζ_{in} is the intrinsic diversity product of the SF code defined as

$$\zeta_{in} = \frac{1}{2}\min_{\mathbf{X}\neq\tilde{\mathbf{X}}}\left(\prod_{k=1}^{\Gamma M_t}|x_k - \tilde{x}_k|\right)^{\frac{1}{\Gamma M_t}} \tag{25.21}$$

and $Q_0 = W_0\text{diag}(\delta_0^2, \delta_1^2, \ldots, \delta_{L-1}^2)W_0^{\mathcal{H}}$, in which

$$W_0 = \begin{bmatrix} 1 & 1 & \cdots & 1 \\ w^{\tau_0} & w^{\tau_1} & \cdots & w^{\tau_{L-1}} \\ \vdots & \vdots & \ddots & \vdots \\ w^{(\Gamma-1)\tau_0} & w^{(\Gamma-1)\tau_1} & \cdots & w^{(\Gamma-1)\tau_{L-1}} \end{bmatrix}_{\Gamma\times L}$$

From Theorem 25.3, we observe that $|\det(Q_0)|$ depends only on the power delay profile of the channel, and the intrinsic diversity product ζ_{in} depends only on $\min_{\mathbf{X}\neq\tilde{\mathbf{X}}}\left(\prod_{k=1}^{\Gamma M_t}|x_k - \tilde{x}_k|\right)^{1/(\Gamma M_t)}$, which is called the "minimum product distance" of the set of symbols $\mathbf{X} = [x_1\ x_2 \ldots x_{\Gamma M_t}]$ [28,29]. Therefore, given the code structure Equation 25.38, it is desirable to design the set of symbols $\mathbf{X}$ such that the minimum product distance is as large as possible, a problem that leads to design signal constellations for Rayleigh fading channels [30,31]. A detailed review of the signal design can be found in Ref. [20].

In the sequel, we would like to maximize the coding advantage of the proposed full-rate full-diversity SF codes by permutations. Note that if the transmitter has no *a priori* knowledge about the channel, the performance of the SF codes can be improved by random interleaving, as it can reduce the correlation between adjacent subcarriers. However, if the power delay profile of the channel is available at the transmitter side, further improvement can be achieved by developing a permutation (or interleaving) method that explicitly takes the power delay profile into account. In the following, we assume that the power delay profile of the channel is known at the transmitter. Our objective is to develop an optimum permutation method such that the resulting coding advantage is maximized [20].

THEOREM 25.4

For any subcarrier permutation, the diversity product of the resulting SF code based on Equations 25.18 and 25.19 is

$$\zeta_{SF} = \zeta_{in}\cdot\zeta_{ex} \tag{25.22}$$

where

ζ_{in} is the intrinsic diversity products defined in Equation 25.21

ζ_{ex} is the extrinsic diversity products defined as

$$\zeta_{ex} = \left(\prod_{m=1}^{M_t}\left|\det\left(V_m\Lambda V_m^{\mathcal{H}}\right)\right|\right)^{\frac{1}{2\Gamma M_t}} \tag{25.23}$$

in which $\Lambda = \mathrm{diag}(\delta_0^2, \delta_1^2, \ldots, \delta_{L-1}^2)$ and

$$V_m = \begin{bmatrix} 1 & 1 & \cdots & 1 \\ w^{[n_{(m-1)\Gamma+2}-n_{(m-1)\Gamma+1}]\tau_0} & w^{[n_{(m-1)\Gamma+2}-n_{(m-1)\Gamma+1}]\tau_1} & \cdots & w^{[n_{(m-1)\Gamma+2}-n_{(m-1)\Gamma+1}]\tau_{L-1}} \\ \vdots & \vdots & \ddots & \vdots \\ w^{[n_{m\Gamma}-n_{(m-1)\Gamma+1}]\tau_0} & w^{[n_{m\Gamma}-n_{(m-1)\Gamma+1}]\tau_1} & \cdots & w^{[n_{m\Gamma}-n_{(m-1)\Gamma+1}]\tau_{L-1}} \end{bmatrix} \quad (25.24)$$

Moreover, the extrinsic diversity product ζ_{ex} is upper bounded as

(i) $\zeta_{ex} \leq 1$
and more precisely,
(ii) if we sort the power profile $\delta_0, \delta_1, \ldots, \delta_{L-1}$ in a nonincreasing order as $\delta_{l_1} \geq \delta_{l_2} \geq \cdots \geq \delta_{l_L}$, then

$$\zeta_{ex} = \left(\prod_{i=1}^{\Gamma} \delta_{l_i} \right)^{\frac{1}{\Gamma}} \left| \prod_{m=1}^{M_t} \det\left(V_m V_m^{\mathcal{H}}\right) \right|^{\frac{1}{2\Gamma M_t}} \quad (25.25)$$

where equality holds when $\Gamma = L$. As a consequence, $\zeta_{ex} \leq \sqrt{L}\left(\prod_{i=1}^{\Gamma} \delta_{l_i}\right)^{1/\Gamma}$.

We observe that the extrinsic diversity product ζ_{ex} depends on the power delay profile in two ways. First, it depends on the power distribution through the square root of the geometric average of the largest Γ path powers, i.e., $\left(\prod_{i=1}^{\Gamma} \delta_{l_i}\right)^{1/\Gamma}$. In case of $\Gamma = L$, the best performance is expected if the power distribution is uniform (i.e., $\delta_l^2 = 1/L$) since the sum of the path powers is unity. Second, the extrinsic diversity product ζ_{ex} also depends on the delay distribution and the applied subcarrier permutation. In contrast, the intrinsic diversity product, ζ_{in}, is not affected by the power delay profile or the permutation method, and it depends only on the signal constellation and the SF code design.

By carefully choosing the applied permutation method, the overall performance of the SF code can be improved by increasing the value of the extrinsic diversity product ζ_{ex}. Toward this end, we consider a specific permutation strategy as follows. We decompose any integer n $(0 \leq n \leq N-1)$ as

$$n = e_1\Gamma + e_0 \quad (25.26)$$

where $0 \leq e_0 \leq \Gamma - 1$, $e_1 = \lfloor \frac{n}{\Gamma} \rfloor$, and $\lfloor x \rfloor$ denotes the largest integer not greater than x. For a fixed integer $\mu(\mu \geq 1)$, we further decompose e_1 in Equation 25.26 as

$$e_1 = \upsilon_1\mu + \upsilon_0 \quad (25.27)$$

where $0 \leq \upsilon_0 \leq \mu - 1$ and $\upsilon_1 = \lfloor \frac{e_1}{\mu} \rfloor$. We permute the rows of the $N \times M_t$ SF codeword constructed from Equations 25.37 and 25.38 in such a way that the nth $(0 \leq n \leq N-1)$ row of C is moved to the $\sigma(n)$th row, where

$$\sigma(n) = \upsilon_1\mu\Gamma + e_0\mu + \upsilon_0 \quad (25.28)$$

in which e_0, υ_0, and υ_1 come from Equations 25.26 and 25.27. We call the integer μ as the "separation factor." The separation factor μ should be chosen such that $\sigma(n) \leq N$ for any $0 \leq n \leq N-1$, or equivalently, $\mu \leq \lfloor N/\Gamma \rfloor$. Moreover, to guarantee that the mapping Equation 25.28 is one-to-one over the set $\{0, 1, \ldots, N-1\}$ (i.e., it defines a permutation), μ must be a factor of N. The role of the permutation specified in Equation 25.28 is to separate two neighboring rows of C by μ subcarriers. The following result characterizes the extrinsic diversity product of the SF code that is permuted with the above described method [20].

THEOREM 25.5

For the permutation specified in Equation 25.28 with a separation factor μ, the extrinsic diversity product of the permuted SF code is

$$\zeta_{ex} = \left|\det\left(V_0 \Lambda V_0^{\mathcal{H}}\right)\right|^{\frac{1}{2\Gamma}} \tag{25.29}$$

where

$$V_0 = \begin{bmatrix} 1 & 1 & \cdots & 1 \\ w^{\mu\tau_0} & w^{\mu\tau_1} & \cdots & w^{\mu\tau_{L-1}} \\ w^{2\mu\tau_0} & w^{2\mu\tau_1} & \cdots & w^{2\mu\tau_{L-1}} \\ \vdots & \vdots & \ddots & \vdots \\ w^{(\Gamma-1)\mu\tau_0} & w^{(\Gamma-1)\mu\tau_1} & \cdots & w^{(\Gamma-1)\mu\tau_{L-1}} \end{bmatrix}_{\Gamma\times L} \tag{25.30}$$

Moreover, if $\Gamma = L$, the extrinsic diversity product ζ_{ex} can be calculated as

$$\zeta_{ex} = \left(\prod_{l=0}^{L-1}\delta_l\right)^{\frac{1}{L}} \left\{\prod_{0\le l_1<l_2\le L-1}\left|2\sin\left[\frac{\mu(\tau_{l_2}-\tau_{l_1})\pi}{T}\right]\right|\right\}^{\frac{1}{L}} \tag{25.31}$$

The permutation (Equation 25.28) is determined by the separation factor μ. Our objective is to find a separation factor μ_{op} that maximizes the extrinsic diversity product ζ_{ex}, i.e., $\mu_{op} = \arg,$ $\max_{1\le\mu\le\lfloor N/\Gamma\rfloor}\left|\det\left(V_0\Lambda V_0^{\mathcal{H}}\right)\right|$. If $\Gamma = L$, the optimum separation factor μ_{op} can be expressed as

$$\mu_{op} = \arg \max_{1\le\mu\le\lfloor N/\Gamma\rfloor} \prod_{0\le l_1<l_2\le L-1}\left|\sin\left[\frac{\mu(\tau_{l_2}-\tau_{l_1})\pi}{T}\right]\right| \tag{25.32}$$

which is independent of the path powers. The optimum separation factor can be easily found via low complexity computer search. However, in some cases, closed form solutions can also be obtained:

- If $\Gamma = L = 2$, the extrinsic diversity product ζ_{ex} is

$$\zeta_{ex} = \sqrt{\delta_0\delta_1}\left|2\sin\left[\frac{\mu(\tau_1-\tau_0)\pi}{T}\right]\right|^{\frac{1}{2}} \tag{25.33}$$

Suppose that the system has $N = 128$ subcarriers, and the total bandwidth is BW = 1 MHz. Then, the OFDM block duration is $T = 128$ μs without the cyclic prefix. If $\tau_1 - \tau_0 = 5$ μs, then $\mu_{op} = 64$ and $\zeta_{ex} = \sqrt{2\delta_0\delta_1}$. If $\tau_1 - \tau_0 = 20$ μs, then $\mu_{op} = 16$ and $\zeta_{ex} = \sqrt{2\delta_0\delta_1}$. In general, if $\tau_1 - \tau_0 = 2^a b$ μs, where a is a nonnegative integer and b is an odd integer, $\mu_{op} = 128/2^{a+1}$. In all of these cases, the extrinsic diversity product is $\zeta_{ex} = \sqrt{2\delta_0\delta_1}$, which achieves the upper bound in Theorem 25.4.

- Assume that $\tau_l - \tau_0 = lN_0T/N$, $l = 1, 2, \ldots, L-1$, and N is an integer multiple of LN_0, where N_0 is a constant and not necessarily an integer. If $\Gamma = L$ or $\delta_0^2 = \delta_1^2 = \cdots = \delta_{L-1}^2 = 1/L$, the optimum separation factor is

$$\mu_{op} = \frac{N}{LN_0} \tag{25.34}$$

and the corresponding extrinsic diversity product is $\zeta_{ex} = \sqrt{L}\left(\prod_{l=0}^{L-1} \delta_l\right)^{1/L}$. In particular, in case of $\delta_0^2 = \delta_1^2 = \cdots = \delta_{L-1}^2 = 1/L$, $\zeta_{ex} = 1$. In both cases, the extrinsic diversity products achieve the upper bounds of Theorem 25.4. Note that if $\tau_l = lT/N$ for $l = 0, 1, \ldots, L-1$, $\Gamma = L$ and N is an integer multiple of L, the permutation with the optimum separation factor $\mu_{op} = N/L$ is similar to the optimum subcarrier grouping method proposed in Ref. [25], which is not optimal for arbitrary power delay profiles.

25.4 Full-Diversity STF Code Design

In this section, we review two STF code design methods to achieve the maximum achievable diversity order $\min\{LM_tM_r\text{rank}(R_T), KNM_r\}$ [26]. Assume that the number of subcarriers N is not less than LM_t, so the maximum achievable diversity order is $LM_tM_r\text{rank}(R_T)$.

25.4.1 Repetition-Coded STF Code Design

First, we try to systematically design full-diversity STF codes by taking advantage of the full-diversity SF codes discussed in Section 25.3. Specifically, assume that C_{SF} is a full-diversity SF code of size $N \times M_t$, then we can construct a full-diversity STF code, C_{STF}, by repeating C_{SF} K times (over K OFDM blocks) as follows:

$$C_{STF} = \mathbf{1}_{k\times 1} \otimes C_{SF} \tag{25.35}$$

where $\mathbf{1}_{k\times 1}$ is an all one matrix of size $k \times 1$. Denote

$$\Delta_{STF} = (C_{STF} - \tilde{C}_{STF})(C_{STF} - \tilde{C}_{STF})^{\mathcal{H}}$$

and

$$\Delta_{SF} = (C_{SF} - \tilde{C}_{SF})(C_{SF} - \tilde{C}_{SF})^{\mathcal{H}}$$

Then, we have

$$\Delta_{STF} = \left[\mathbf{1}_{k\times 1} \otimes (C_{SF} - \tilde{C}_{SF})\right]\left[\mathbf{1}_{1\times k} \otimes (C_{SF} - \tilde{C}_{SF})^{\mathcal{H}}\right] = \mathbf{1}_{k\times k} \otimes \Delta_{SF}$$

Thus,

$$\Delta_{STF} \circ R = (\mathbf{1}_{k\times k} \otimes \Delta_{SF}) \circ (R_T \otimes R_F) = R_T \otimes (\Delta_{SF} \circ R_F)$$

Since the SF code C_{SF} achieves full diversity in each OFDM block, the rank of $\Delta_{SF} \circ R_F$ is LM_t. Therefore, the rank of $\circ\Delta_{STF} \circ R$ is $LM_t\text{rank}(R_T)$, so C_{STF} in Equation 25.35 is guaranteed to achieve a diversity order of $LM_tM_r\text{rank}(R_T)$.

We observe that the maximum achievable diversity depends on the rank of the temporal correlation matrix R_T. If the fading channels are constant during K OFDM blocks, i.e., $\text{rank}(R_T) = 1$, then the maximum achievable diversity order for STF codes (coding across several OFDM blocks) is the same as that for SF codes (coding within one OFDM block). Moreover, if the channel changes independently in time, i.e., $R_T = I_K$, the repetition structure of STF code C_{STF} in Equation 25.35 is sufficient, but not necessary, to achieve the full diversity.

In this case,

$$\Delta \circ R = \text{diag}(\Delta_1 \circ R_F, \Delta_2 \circ R_F, \ldots, \Delta_k \circ R_F)$$

where $\Delta_k = (C_k - \tilde{C}_k)(C_k - \tilde{C}_k)^{\mathcal{H}}$ for $1 \le k \le K$. Thus, in this case, the necessary and sufficient condition to achieve full diversity KLM_tM_r is that each matrix Δ_k o R_F be of rank LM_t over all pairs of distinct codewords simultaneously for all $1 \le k \le K$.

Note that the above repetition-coded STF code design ensures full diversity at the price of symbol rate decrease by a factor of $1/K$ (over K OFDM blocks) compared to the symbol rate of the underlying SF code. The advantage of this approach is that any full-diversity SF code (block or trellis) can be used to design full-diversity STF codes.

25.4.2 Full-Rate Full-Diversity STF Code Design

Second, we would like to design a class of STF codes that can guarantee a diversity order of ΓM_tM_rrank (R_T) for any given integer $\Gamma(1 \le \Gamma \le L)$ by extending the full-rate full-diversity SF code construction method (coding over one OFDM block, i.e., the $K = 1$ case) as discussed in Section 25.3.2.

We consider an STF code structure consisting of STF codewords C of size KN by M_t:

$$C = \left[C_1^T C_2^T \ldots C_K^T\right]^T \quad (25.36)$$

where

$$C_k = \left[G_{k,1}^T G_{k,2}^T \ldots G_{k,p}^T 0_{N-P\Gamma M_t}^T\right]^T \quad (25.37)$$

for $k = 1, 2, \ldots, K$. In Equation 25.37, $P = \lfloor N/(\Gamma M_t) \rfloor$, and each matrix $G_{k,p}(1 \le k \le K,\ 1 \le p \le P)$ is of size ΓM_t by M_t. The zero padding in Equation 25.37 is used if the number of subcarriers N is not an integer multiple of ΓM_t. For each p $(1 \le p \le P)$, we design the code matrices $G_{1,p}$, $G_{2,p}, \ldots,$ $G_{K,p}$ jointly, but the designs of G_{k_1,p_1} and G_{k_2,p_2}, $p_1 \ne p_2$, are independent of each other. For a fixed p $(1 \le p \le P)$, let

$$G_{k,p} = \sqrt{M_t}\text{diag}(X_{k,1}, X_{k,2}, \ldots, X_{k,M_t}), \quad k = 1, 2, \ldots, K \quad (25.38)$$

where $\text{diag}(X_{k,1}, X_{k,2}, \ldots, X_{k,M_t})$ is a block diagonal matrix, $X_{k,i} = \left[x_{k,(i-1)\Gamma+1} x_{k,(i-1)\Gamma+2} \ldots x_{k,i\Gamma}\right]^T$, $i = 1, 2, \ldots, M_t$, and all $x_{k,j}$, $j = 1, 2, \ldots, \Gamma M_t$, are complex symbols and will be specified later. The energy normalization condition is $E\left(\sum_{k=1}^{K}\sum_{j=1}^{\Gamma M_t}|x_{k,j}|^2\right) = K\Gamma M_t$. The symbol rate of the proposed scheme is $P\Gamma M_t/N$, ignoring the cyclic prefix. If N is a multiple of ΓM_t, the symbol rate is 1. If not, the rate is less than 1, but since usually N is much greater than ΓM_t, the symbol rate is very close to 1.

The following theorem provides a sufficient condition for the STF codes described above to achieve a diversity order of ΓM_tM_rrank(R_T). For simplicity, we use the notation $\mathbf{X} = [x_{1,1} \ldots x_{1,\Gamma M_t} \ldots x_{K,1} \ldots x_{K,\Gamma M_t}]$ and $\tilde{\mathbf{X}} = [\tilde{x}_{1,1} \ldots \tilde{x}_{1,\Gamma M_t} \ldots \tilde{x}_{K,1} \ldots \tilde{x}_{K,\Gamma M_t}]$. Moreover, for any $n \times n$ nonnegative definite matrix A, we denote its eigenvalues in a nonincreasing order as $\text{eig}_1(A) \ge \text{eig}_2(A) \ge \cdots \ge \text{eig}_n(A)$.

THEOREM 25.6

For any STF code constructed by Equations 25.36 through 25.38, if $\prod_{k=1}^{K} \prod_{j=1}^{\Gamma M_t} \times |x_{k,j} - \tilde{x}_{k,j}| \neq 0$ for any pair of distinct symbols $\mathbf{X}$ and $\tilde{\mathbf{X}}$, then the STF code achieves a diversity order of $\Gamma M_t M_r \text{rank}(R_T)$, and the coding advantage is bounded by

$$(M_t \delta_{\min})^{\Gamma M_t \text{rank}(R_T)} \Phi \leq \zeta_{\text{STF}} \leq (M_t \delta_{\max})^{\Gamma M_t \text{rank}(R_T)} \Phi \tag{25.39}$$

where

$$\begin{aligned} \delta_{\min} &= \min_{\mathbf{X} \neq \tilde{\mathbf{X}}} \min_{1 \leq k \leq K, 1 \leq j \leq \Gamma M_t} |x_{k,j} - \tilde{x}_{k,j}|^2 \\ \delta_{\max} &= \max_{\mathbf{X} \neq \tilde{\mathbf{X}}} \max_{1 \leq k \leq K, 1 \leq j \leq \Gamma M_t} |x_{k,j} - \tilde{x}_{k,j}|^2 \\ \Phi &= |\det(Q_0)|^{M_t \text{rank}(R_T)} \prod_{i=1}^{\text{rank}(R_T)} (\text{eig}_i(R_T))^{\Gamma M_t} \end{aligned} \tag{25.40}$$

and

$$Q_0 = W_0 \text{diag}(\delta_0^2, \delta_1^2, \ldots, \delta_{L-1}^2) W_0^{\mathcal{H}} \tag{25.41}$$

$$W_0 = \begin{bmatrix} 1 & 1 & \cdots & 1 \\ w^{\tau_0} & w^{\tau_1} & \cdots & w^{\tau_{L-1}} \\ \vdots & \vdots & \ddots & \vdots \\ w^{(\Gamma-1)\tau_0} & w^{(\Gamma-1)\tau_1} & \cdots & w^{(\Gamma-1)\tau_{L-1}} \end{bmatrix}_{\Gamma \times L} \tag{25.42}$$

Furthermore, if the temporal correlation matrix R_T is of full rank, i.e., $\text{rank}(R_T) = K$, the coding advantage is

$$\zeta_{\text{STF}} = \delta M_t^{K \Gamma M_t} |\det(R_T)|^{\Gamma M_t} |\det(Q_0)|^{K M_t} \tag{25.43}$$

where

$$\delta = \min_{\mathbf{X} \neq \tilde{\mathbf{X}}} \prod_{k=1}^{K} \prod_{j=1}^{\Gamma M_t} |x_{k,j} - \tilde{x}_{k,j}|^2 \tag{25.44}$$

We observe from Theorem 25.6 that with the code structure specified in Equations 25.36 through 25.38, it is not difficult to achieve the diversity order of $\Gamma M_t M_r \text{rank}(R_T)$ if we have signals satisfying $\prod_{k=1}^{K} \prod_{j=1}^{\Gamma M_t} |x_{k,j} - \tilde{x}_{k,j}| \neq 0$ for distinct symbols $\mathbf{X}$ and $\tilde{\mathbf{X}}$. However, it is challenging to design a set of complex symbol vectors, $\mathbf{X} = [x_{1,1} \ldots x_{1,\Gamma M_t} \ldots x_{K,1} \ldots x_{K,\Gamma M_t}]$, such that the coding advantage ζ_{STF} is as large as possible. One approach is to maximize $\delta_{\min}$ and $\delta_{\max}$ in Equation 25.39 according to the lower and upper bounds of the coding advantage. Another approach is to maximize δ in Equation 25.44. We would like to design signals according to the second criterion for two reasons. First, the coding advantage ζ_{STF} in Equation 25.43 is determined by δ in closed form although this closed form only holds with the assumption that the temporal correlation matrix R_T is of full rank. Second, the problem of designing $\mathbf{X}$ to maximize δ is related to the problem of constructing signal constellations for Rayleigh fading channels which have been well developed [28–31]. In the literature, δ is called the "minimum product distance" of the set of symbols $\mathbf{X}$ [28,29].

We summarize some existing results on designing **X** to maximize the minimum product distance δ as follows. For simplicity, denote $\mathcal{L} = K\Gamma M_t$, and assume that Ω is a constellation such as quadrature amplitude modulation (QAM), pulse amplitude modulation PAM, and so on. The set of complex symbol vectors is obtained by applying a transform over $\mathcal{L}$-dimensional signal set $\Omega^{\mathcal{L}}$ [30,31]. Specifically,

$$\mathbf{X} = S \cdot \frac{1}{\sqrt{\mathcal{L}}} V(\theta_1, \theta_2, \ldots, \theta_{\mathcal{L}}) \tag{25.45}$$

where $S = [s_1\ s_2 \ldots s_{\mathcal{L}}] \in \Omega^K$ is a vector of arbitrary channel symbols to be transmitted, and $V(\theta_1, \theta_2, \ldots, \theta_{\mathcal{L}})$ a Vandermonde matrix with variables $\theta_1, \theta_2, \ldots, \theta_{\mathcal{L}}$ [37]:

$$V(\theta_1, \theta_2, \ldots, \theta_{\mathcal{L}}) = \begin{bmatrix} 1 & 1 & \cdots & 1 \\ \theta_1 & \theta_1 & \cdots & \theta_{\mathcal{L}} \\ \vdots & \vdots & \ddots & \vdots \\ \theta_1^{\mathcal{L}-1} & \theta_2^{\mathcal{L}-1} & \cdots & \theta_{\mathcal{L}}^{\mathcal{L}-1} \end{bmatrix} \tag{25.46}$$

The optimum θ_l, $1 \leq l \leq \mathcal{L}$, has been specified for different $\mathcal{L}$ and Ω. For example, if Ω is a QAM constellation and $\mathcal{L} = 2^s (s \geq 1)$, the optimum θ_l values were given by [30,31]

$$\theta_l = e^{j\frac{4l-3}{2\mathcal{L}}\pi}, \quad l = 1, 2, \ldots, \mathcal{L} \tag{25.47}$$

In case of $\mathcal{L} = 2^s \cdot 3^t (s \geq 1, t \geq 1)$, a class of θ_l values were given in Ref. [31] as

$$\theta_l = e^{j\frac{6l-5}{3\mathcal{L}}\pi}, \quad l = 1, 2, \ldots, \mathcal{L} \tag{25.48}$$

For more details and other cases of Ω and $\mathcal{L}$, we refer the reader to Refs. [30,31].

The STF code design discussed in this subsection achieves full symbol rate, which is much larger than that of the repetition coding approach. However, the maximum-likelihood decoding complexity of this approach is high. Its complexity increases exponentially with the number of OFDM blocks, K, while the decoding complexity of the repetition-coded STF codes increases only linearly with K. Fortunately, sphere decoding methods [32,33] can be used to reduce the complexity.

25.5 Simulation Results

We simulated MIMO-OFDM systems with the proposed SF/STF code designs and present average BER performance curves as functions of the average SNR. The simulated systems have $M_t = 2$ transmit antennas and $M_r = 1$ receive antenna, and the OFDM modulation has $N = 128$ subcarriers. We considered two fading channel models. The first one is a two-ray, equal-power delay profile with a delay τ μs between the two rays and each ray was modeled as a zero-mean, complex Gaussian random variable with variance 0.5. The second channel model is a more realistic six-ray typical urban (TU) power delay profile [36] with delay distribution {0.0, 0.2, 0.5, 1.6, 2.3, 5.0 μs} and power distribution {0.189, 0.379, 0.239, 0.095, 0.061, 0.037}.

First, we simulated a full-diversity SF code obtained from ST code via mapping as proposed in Equations 25.14 and 25.15. We used the orthogonal ST block codes for two transmit antennas as follows [7]:

$$G_2 = \begin{bmatrix} x_1 & x_2 \\ -x_2^* & x_1^* \end{bmatrix} \tag{25.49}$$

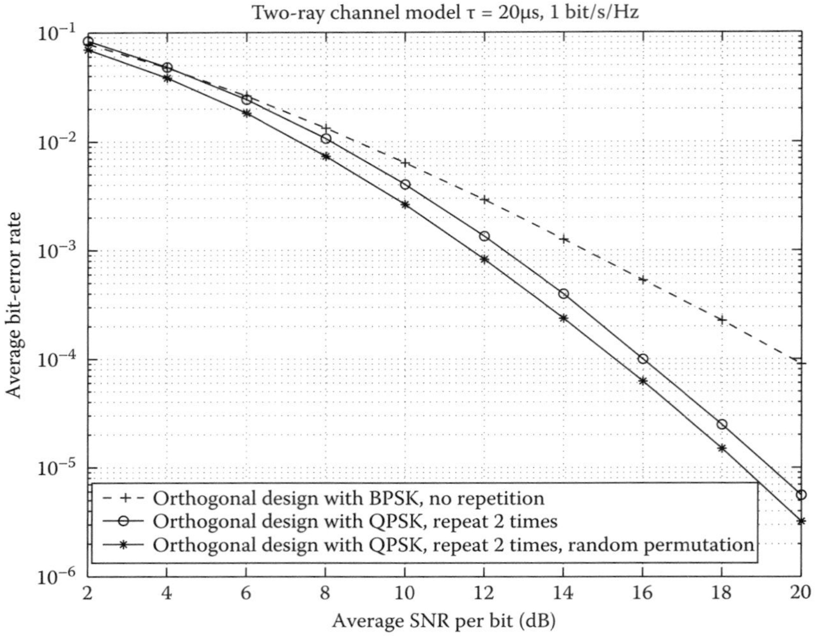

FIGURE 25.1 Performances of the full-diversity SF code obtained from orthogonal ST code via mapping in a two-ray fading model with $\tau = 20\ \mu s$.

where x_1 and x_2 are phase shift keying (PSK) or QAM information symbols. The SF code was obtained by repeating each row of the orthogonal ST code (Equation 25.49) twice. We simulated the SF code over the two-ray fading channel model in which we considered two cases: (a) $\tau = 5\ \mu s$ and (b) $\tau = 20\ \mu s$ with OFDM bandwidth BW = 1 MHz. We compared the performances of the full-diversity SF code (with row repetition) with that of the scheme by applying the orthogonal design directly (without row repetition). We used binary phase shift keying (BPSK) modulation for the nonrepeated case and quadrature phase shift keying (QPSK) for the repeated case. Therefore, both schemes had a same spectral efficiency of 1 bit/s/Hz. From Figure 25.1, we can see that in case of $\tau = 20\ \mu s$, the performance curve of the full-diversity SF code has a steeper slope than that of the code without repetition, i.e., the obtained SF code has higher diversity order. We can observe a performance improvement of about 4 dB at a BER of 10^{-4}. From Figure 25.2, we observe that the performance of the full-diversity SF code degraded significantly from $\tau = 20\ \mu s$ case to $\tau = 5\ \mu s$ case, which is consistent with the theoretical result that the coding advantage depends on the delay distribution of the multiple paths. Note that the correlation of the channel frequency responses at adjacent subcarriers can severely degrade the performance of the SF codes, so we may apply a random permutation to break down the correlation to improve the code performance. We considered a random permutation generated by the Takeshita–Constello method [34], which is given by

$$\sigma(n) = \left\lfloor \frac{n(n+1)}{2} \right\rfloor \bmod N, \quad n = 0, 1, \ldots, N-1 \tag{25.50}$$

From Figures 25.1 and 25.2, we can see that by applying a random permutation over the channel frequency responses, the performance of the full-diversity SF code can be further improved and the performance improvement depends on the channel power delay distribution.

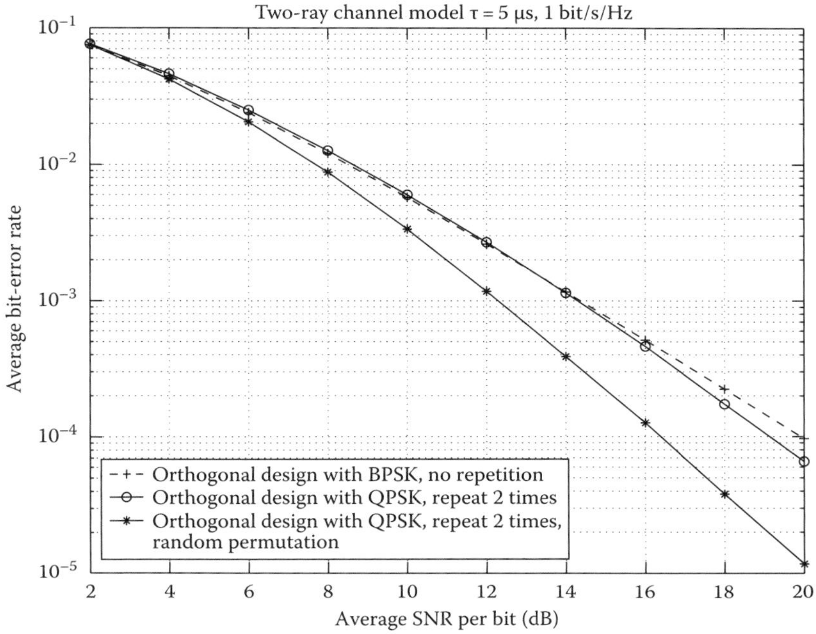

FIGURE 25.2 Performances of the full-diversity SF code obtained from orthogonal ST code via mapping in a two-ray fading model with $\tau = 5$ μs.

Second, we simulated the full-rate full-diversity SF code design proposed in Equations 25.18 and 25.19. We considered an SF code for $M_t = 2$ transmit antennas with the following structure

$$G = \sqrt{2}\begin{bmatrix} x_1 & 0 \\ x_2 & 0 \\ 0 & x_3 \\ 0 & x_4 \end{bmatrix} \tag{25.51}$$

The symbols x_1, x_2, x_3, x_4 were obtained as $[x_1x_2x_3x_4] = [s_1s_2s_3s_4]\cdot\frac{1}{2}V(\theta, -\theta, j\theta, -j\theta)$, where s_1, s_2, s_3, s_4 were chosen from QPSK constellation ($s_i \in \{\pm 1, \pm j\}$), $V(\cdot)$ is the Vandermonde matrix defined in Equation 25.46, and $\theta = e^{j\pi/8}$. This code targets a frequency diversity order of $\Gamma = 2$, thus it achieves full diversity only if the number of delay paths is $L \leq 2$. We simulated the SF code over the more realistic six-ray TU fading model with two scenarios: (a) BW = 1 MHz, and (b) BW = 4 MHz. We first compared the SF code performances by using three different permutation schemes: no permutation, random permutation, and the proposed optimum permutation. The random permutation was generated by the Takeshita–Constello method in Equation 25.50. From Figures 25.3 and 25.4, we can see that the performance of the proposed SF code with the random permutation is much better than that without permutation. If we apply the proposed optimum permutation with a separation factor $\mu = 64$, the code performance can further be improved and there is an additional performance gain of 1.5 and 1 dB at a BER of 10^{-5} in case of BW = 1 MHz and BW = 4 MHz, respectively.

We also compared the full-rate full-diversity SF code design (Equation 25.51) with the full-diversity SF code obtained from the orthogonal ST code (Equation 25.49) with repetition. The full-diversity SF code based on Equation 25.49 used 16-QAM modulation to maintain the same spectral efficiency of 2 bits/s/Hz as that of Equation 25.51. We observe from Figure 25.3 that in case of BW = 1 MHz, the SF code in Equation 25.51 outperforms the SF code based on repetition by about 2 dB at a BER of 10^{-4}.

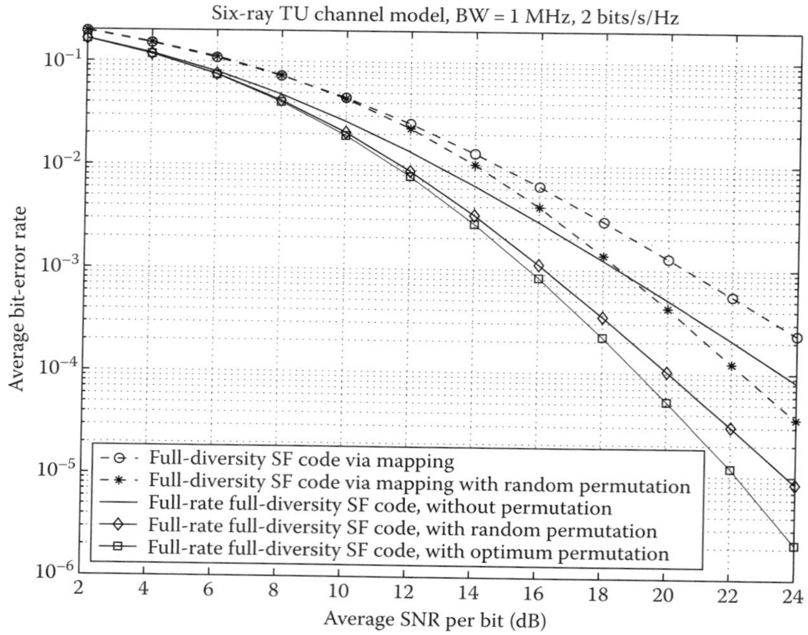

FIGURE 25.3 Performances of the full-diversity SF codes with different permutation strategies in the six-ray TU fading model, BW = 1 MHz.

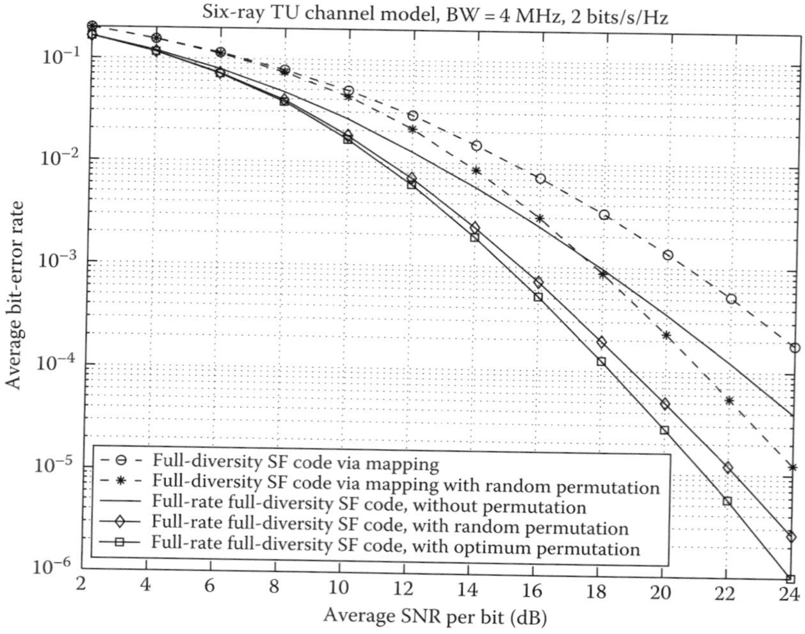

FIGURE 25.4 Performances of the full-diversity SF codes with different permutation strategies in the six-ray TU fading model, BW = 4 MHz.

With random permutation (Equation 25.50), the performance of the full-rate SF code is better than that of the code from orthogonal design by about 2.5 dB at a BER of 10^{-4}. Moreover, we have an additional performance improvement of 1 dB at a BER of 10^{-4} achieved by the full-rate SF code when the optimum permutation strategy is applied. In case of BW = 4 MHz, from Figure 25.4, we can see that without permutation, the performance of the SF code in Equation 25.51 is better than that of the code from orthogonal design by about 3 dB at a BER of 10^{-4}. With random permutation, the SF code (Equation 25.51) outperforms the SF code based on Equation 25.49 by about 2 dB at a BER of 10^{-4}. Compared to the SF code from orthogonal design with the random permutation, the full-rate SF code with the optimum permutation has a total gain of 3 dB at a BER of 10^{-4}.

Finally, we simulated a full-rate full-diversity STF code based on Equations 25.36 through 25.38 for $M_t = 2$ transmit antennas with $\Gamma = 2$ and over $K = 1, 2, 3$ OFDM blocks. The set of complex symbol vectors **X** was obtained via Equation 25.45 by applying Vandermonde transforms over a signal set Ω^{4K} for $K = 1, 2, 3$. The Vandermonde transforms were determined for different K values according to Equations 25.47 and 25.48. The constellation Ω was chosen to be BPSK. Thus, the spectral efficiency the resulting STF codes were 1 bit/s/Hz (omitting the cyclic prefix), which is independent of the number of jointly encoded OFDM blocks, K. We considered the six-ray TU channel model and assumed that the fading channel is constant within each OFDM block period but varies from one OFDM block period to another according to a first-order Makovian model [35]: $\alpha_{i,j}^{k}(l) = \varepsilon\alpha_{i,j}^{k-1}(l) + \eta_{i,j}^{k}(l)$, $0 \le l \le L-1$, where the constant ε $(0 \le \varepsilon \le 1)$ determines the amount of temporal correlation, and $\eta_{i,j}^{k}(l)$ is a zero-mean, complex Gaussian random variable with variance $\delta_l\sqrt{1-\varepsilon^2}$. If $\varepsilon = 0$, there is no temporal correlation (independent fading), while if $\varepsilon = 1$, the channel stays constant over multiple OFDM blocks. We considered three temporal correlation scenarios: $\varepsilon = 0$, $\varepsilon = 0.8$, and $\varepsilon = 0.95$. The performance of the full-rate STF codes are depicted in Figure 25.5 for the three different temporal correlation scenarios. From the figures, we observe that the diversity order of the STF codes increases with the number of

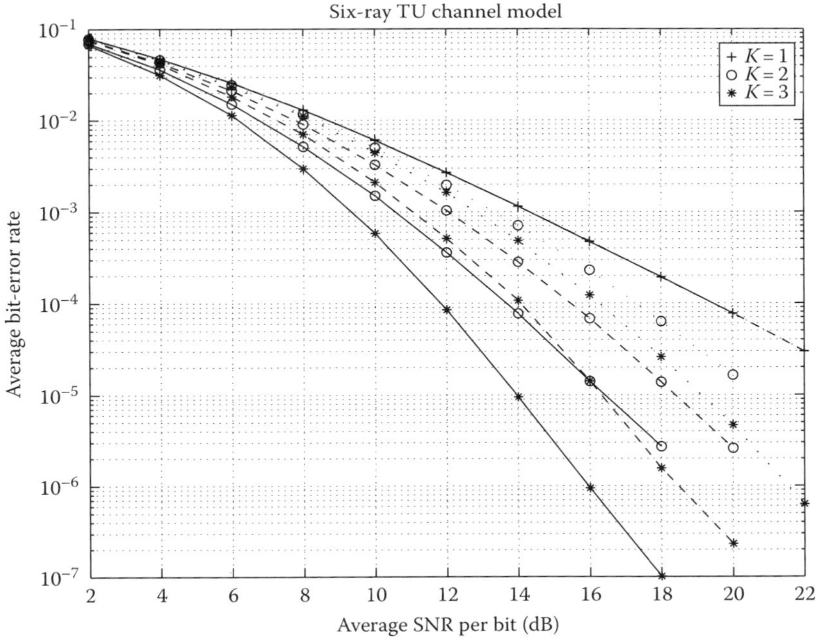

FIGURE 25.5 Performances of the full-rate full-diversity STF codes over $K = 1, 2, 3$ OFDM blocks. Note that the dotted curves show the results for the channel with correlation $\varepsilon = 0.95$, the broken curves for the channel with correlation $\varepsilon = 0.8$, and the solid curve for channel with correlation $\varepsilon = 0$.

jointly encoded OFDM blocks, K. However, the improvement of the diversity order depends on the temporal correlation. The performance gain obtained by coding across multiple OFDM blocks decreases as the correlation factor ε increases.

25.6 Conclusion

In this chapter, we reviewed code design criteria for MIMO-OFDM systems and summarized our results on SF and STF code design [19,20,26]. We explored different coding approaches for MIMO-OFDM systems by taking into account all opportunities for performance improvement in the spatial, the temporal, and the frequency domains in terms of the achievable diversity order. For SF coding, where the coding is applied within each OFDM block, we proposed two systematic SF code design methods that can guarantee to achieve the full-diversity order of LM_tM_r, where factor L comes from the frequency diversity owing to the delay spread of the channel. For STF coding, where coding is applied over multiple OFDM blocks, we developed two STF code design methods by taking advantage of the SF code design methodology. The proposed STF code design methods can guarantee the maximum achievable diversity order of LM_tM_rT, where T is the rank of the temporal correlation matrix of the fading channel. The simulation results demonstrate that using the proposed coding methods, considerable performance improvement can be achieved compared to previously existing SF and STF coding approaches. We also observed that the performance of the obtained SF and TF codes depend heavily on the channel power delay profile, and there is trade-off between the diversity order and the spectral efficiency of the code.

References

1. Air interface for fixed broadband wireless access systems, *IEEE Standard 802.16-2004*, October 2004.
2. Air interface for fixed and mobile broadband wireless access systems, *IEEE Standard Amendment P802.16e/D12*, February 2005.
3. Mobile WiMAX—Part I: A technical overview and performance evaluation, *WiMAX Forum*, June 2006.
4. A. Ghosh and D. Wolter, Broadband wireless access with WiMax/802.16: Current performance benchmarks and future potential, *IEEE Commun. Mag.*, 43(2), 129–136, February 2005.
5. J.-C. Guey, M.P. Fitz, M.R. Bell, and W.-Y. Kuo, Signal design for transmitter diversity wireless communication systems over Rayleigh fading channels, *IEEE Trans. Commun.*, 47, 527–537, April 1999.
6. V. Tarokh, N. Seshadri, and A.R. Calderbank, Space-time codes for high data rate wireless communication: Performance criterion and code construction, *IEEE Trans. Info. Theory*, 44(2), 744–765, 1998.
7. S. Alamouti, A simple transmit diversity technique for wireless communications, *IEEE JSAC*, 16(8), 1451–1458, 1998.
8. V. Tarokh, H. Jafarkhani, and A.R. Calderbank, Space-time block codes from orthogonal designs, *IEEE Trans. Info. Theory*, 45(5), 1456–1467, 1999.
9. B.M. Hochwald and T.L. Marzetta, Unitary space-time modulation for multiple-antenna communication in Rayleigh flat fading, *IEEE Trans. Info. Theory*, 46(2), 543–564, 2000.
10. M.O. Damen, K. Abed-Meraim, and J.C. Belfiore, Diagonal algebraic space-time block codes, *IEEE Trans. Info. Theory*, 48(3), 628–636, March 2002.
11. D. Agrawal, V. Tarokh, A. Naguib, and N. Seshadri, Space-time coded OFDM for high data-rate wireless communication over wideband channels, *Proceedings of the IEEE VTC*, pp. 2232–2236, Ottawa, Ontario, Canada, 1998.
12. K. Lee and D. Williams, A space-frequency transmitter diversity technique for OFDM systems, *Proceedings of the IEEE GLOBECOM*, vol. 3, pp. 1473–1477, San Francisco, CA, 2000.
13. R. Blum, Y. Li, J. Winters, and Q. Yan, Improved space-time coding for MIMO-OFDM wireless communications, *IEEE Trans. Commun.*, 49(11), 1873–1878, 2001.

14. Y. Gong and K.B. Letaief, An efficient space-frequency coded wideband OFDM system for wireless communications, *Proceedings of the IEEE ICC*, vol. 1, pp. 475–479, New York, 2002.
15. Z. Hong and B. Hughes, Robust space-time codes for broadband OFDM systems, *Proceedings of the IEEE WCNC*, vol. 1, pp. 105–108, Orlando, FL, 2002.
16. B. Lu and X. Wang, Space-time code design in OFDM systems, *Proceedings of the IEEE GLOBECOM*, pp. 1000–1004, San Francisco, CA, November 2000.
17. H. Bölcskei and A.J. Paulraj, Space-frequency coded broadband OFDM systems, *Proceedings of the IEEE WCNC*, pp. 1–6, Chicago, IL, September 2000.
18. H. Bölcskei and A.J. Paulraj, Space-frequency codes for broadband fading channels, *Proceedings of the ISIT'2001*, p. 219, Washington, DC, June 2001.
19. W. Su, Z. Safar, M. Olfat, and K.J.R. Liu, Obtaining full-diversity space-frequency codes from space-time codes via mapping, *IEEE Trans. Signal Process.* (Special Issue on MIMO Wireless Communications), 51(11), 2905–2916, November 2003.
20. W. Su, Z. Safar, and K.J.R. Liu, Full-rate full-diversity space-frequency codes with optimum coding advantage, *IEEE Trans. Info. Theory*, 51(1), 229–249, January 2005.
21. Y. Gong and K.B. Letaief, Space-frequency-time coded OFDM for broad-band wireless communications, *Proceedings of the IEEE GLOBECOM*, San Antonio, TX, November 2001.
22. B. Lu, X. Wang, and K.R. Narayanan, LDPC-based space-time coded OFDM systems over correlated fading channels: Performance analysis and receiver design, *IEEE Trans. Commun.*, 50(1), 74–88, January 2002.
23. A.F. Molisch, M.Z. Win, and J.H. Winters, Space-time-frequency (STF) coding for MIMO-OFDM systems, *IEEE Commun. Lett.*, 6(9), 370–372, September 2002.
24. Z. Liu, Y. Xin, and G. Giannakis, Space-time-frequency coded OFDM over frequency selective fading channels, *IEEE Trans. Signal Processing*, 50 2465–2476, October 2002.
25. Z. Liu, Y. Xin, and G.B. Giannakis, Linear constellation precoding for OFDM with maximum multipath diversity and coding gains, *IEEE Trans. Commun.*, 51(3), 416–427, March 2003.
26. W. Su, Z. Safar, and K.J.R. Liu, Towards maximum achievable diversity in space, time and frequency: Performance analysis and code design, *IEEE Trans. Wireless Commun.*, 4(4), 1847–1857, July 2005.
27. W. Su, Z. Safar, and K.J.R. Liu, Space-time signal design for time-correlated Rayleigh fading channels, *Proceedings of the IEEE ICC*, vol. 5, pp. 3175–3179, Anchorage, Alaska, May 2003.
28. C. Schlegel and D.J. Costello, Jr., Bandwidth efficient coding for fading channels: Code construction and performance analysis, *IEEE JSAC*, 7, 1356–1368, December 1989.
29. K. Boullé and J.C. Belfiore, Modulation schemes designed for the Rayleigh channel, *Proceedings of the CISS'92*, pp. 288–293, Princeton University, NJ, March 1992.
30. X. Giraud, E. Boutillon, and J.C. Belfiore, Algebraic tools to build modulation schemes for fading channels, *IEEE Trans. Info. Theory*, 43(2), 938–952, May 1997.
31. J. Boutros and E. Viterbo, Signal space diversity: A power- and bandwidth-efficient diversity technique for the Rayleigh fading channel, *IEEE Trans. Info. Theory*, 44(4), 1453–1467, July 1998.
32. M. Damen, A. Chkeif, and J. Belfiore, Lattice code decoder for space-time codes, *IEEE Commun. Lett.*, 4(5), 161–163, 2000.
33. Z. Safar, W. Su, and K.J.R. Liu, A fast sphere decoding algorithm for space-frequency codes, *EURASIP J. App. Signal Process.*, 2006, 1–14, 2006.
34. O.Y. Takeshita and D.J. Constello, Jr., New classes of algebraic interleavers for turbo-codes, *Proceedings of the ISIT*, p. 419, Cambridge, MA, 1998.
35. H.S. Wang and N. Moayeri, Finite-state Markov channel—a useful model for radio communication channels, *IEEE Trans. Veh. Tech.*, 44(1), 163–171, 1996.
36. G. Stuber, *Principles of Mobile Communication*, Kluwer Academic Publishers, Norwell, MA, 2001.
37. R.A. Horn and C.R. Johnson, *Topics in Matrix Analysis*, Cambridge University Press, Cambridge, U.K., 1991.

26

OFDM Technology: Fundamental Principles, Transceiver Design, and Mobile Applications

Xianbin Wang
University of Western Ontario

Yiyan Wu
Communications Research Centre

Jean-Yves Chouinard
Laval University

26.1 Overview of OFDM Principles **26**-2
Principles of DFT-Based OFDM System • ISI Mitigation through Cyclic Prefix • Bit Rate of OFDM Systems • In-Band Pilots and OFDM Channel Estimation • Modulation Schemes for Subchannels • OFDM Error Probability Performances

26.2 OFDM-Based Mobile Broadcasting Standards **26**-12
DVB-H Digital Mobile Broadcasting System • MediaFLO Digital Mobile Broadcasting System • T-DMB Digital Mobile Broadcasting System

26.3 Frequency and Timing Synchronization for OFDM-Based Wireless Systems **26**-17
Timing Offset Estimation Techniques • Frequency Offset and Estimation Techniques • Joint Estimation of the Frequency and Timing Offsets • Fast Synchronization for DVB-H System

26.4 Summary **26**-31

References **26**-31

Orthogonal frequency division multiplexing (OFDM) is the primary modulation technique for digital broadband communication and broadcasting systems, including the digital video broadcasting (DVB)[1] and digital audio broadcasting (DAB).[2] The fundamental principle of OFDM technology is the concept of multichannel modulation that divides a broadband wireless channel into a number of parallel subchannels, or subcarriers, so that multiple symbols can be sent in parallel. Earlier overviews of the OFDM system and its applications can be found in Bingham[3] and Zou and Wu.[4]

OFDM has received considerable attention during the last two decades due to its robustness against intersymbol interference (ISI) and multipath distortion, low implementation complexity, and high spectral efficiency. With the introduction of the parallel transmission concept, the symbol duration in OFDM becomes significantly longer, compared to single-carrier transmission with the same channel condition and given data rate. Consequently, the impact of the ISI in OFDM system is substantially reduced. This is why OFDM became the primary technology for wireless communications, where multipath distortion is very common. The type of OFDM that we describe in this chapter uses the discrete Fourier transform (DFT)[5] with a cyclic prefix.[6] The DFT (implemented with a fast Fourier transform [FFT]) and the cyclic prefix have made OFDM both practical and attractive to the communication system designer.

OFDM has been applied in a wide variety of digital communications applications, including DVB, DAB, wireless LAN, and several evolving industry standard. A similar multichannel modulation scheme, discrete multitone (DMT) modulation, has been developed for static channels such as the digital subscriber loop.[7] DMT also uses DFTs and cyclic prefixes but has the additional feature of bit loading, which is generally not used in OFDM, although related ideas can be found in Wesel.[8]

One of the principal disadvantages of OFDM is its sensitivity to synchronization errors, characterized mainly by the so-called frequency and timing offsets. Frequency offset causes a reduction of desired signal amplitude in the output decision variable and introduces intercarrier interference (ICI) due to the loss of orthogonality among subcarriers. Timing offset results in the rotation of the OFDM subcarrier signal constellation. As a result, an OFDM system cannot recover the transmitted signal without a near-perfect synchronization, especially when high-order quadrature amplitude modulation (QAM) is used. As such, OFDM-based mobile broadcast systems are very sensitive to synchronization errors. In this chapter, the impact of the synchronization errors, including carrier frequency and timing offsets, are analyzed. Various techniques for the estimation and tracking of the frequency and timing offsets are overviewed and discussed.

The organization of this chapter is as follows. A brief introduction to OFDM systems is first presented. Then the generation, equalization, and demodulation of the OFDM signals for wireless communications are overviewed. OFDM-based mobile broadcasting standards are introduced as the application examples of the technology. The impact of the synchronization errors on the performance of OFDM systems is then considered. Synchronization techniques for OFDM-based mobile broadcast systems are analyzed, with special emphasis on the DVB-Terrestrial (DVB-T) and DVB-Handheld (DVB-H) standards' pilot and frame structures.

26.1 Overview of OFDM Principles

Digital communication involves the transmission of information in digital form from an information source to one or several destinations. In an ideal channel, there is no ISI caused by multipath channel distortion, and error-free transmission can be achieved. However, this condition could not be satisfied with channel distortion as in a wireless communication environment due to the multipath propagation effects, including reflection and scattering. Equalization and channel control coding methods can be applied to achieve robust transmission. A time domain equalizer could be used to shorten the effective channel impulse response duration, or length, of a dispersive channel,[9] and whose coefficients are updated with an adaptive algorithm like that of Kalman filtering or the gradient algorithm, for instance. However, adaptive equalization could considerably increase the system implementation complexity, and the convergence of such an equalizer is not guaranteed.

This critical problem associated with time domain equalization convinced researchers to investigate other modulation schemes. Because the maximum delay spread present in the channel is fixed, a solution to overcome the spreading of the channel impulse response and ISI would consist in using several carriers in parallel instead of one, as shown in Figure 26.1. The main feature of multicarrier modulation (MCM) techniques is to divide a wideband channel into a number of orthogonal narrowband subchannels. This is accomplished by modulating parallel information data at a much lower rate on a number, N, of subcarriers. Because the symbol duration for each subcarrier is multiplied by this factor N, the ratio of the maximum delay to the modulation period can be reduced significantly with a large number N of subcarriers.[4]

In early MCM technologies like conventional frequency division multiplexing, there is a guard band between adjacent subcarriers so as to be able to isolate them at the receiver using conventional band-pass filters. However, the bandwidth can be used much more efficiently in MCM systems if the spectra of subcarriers are permitted to overlap. By using subcarriers separated by a frequency difference that is the

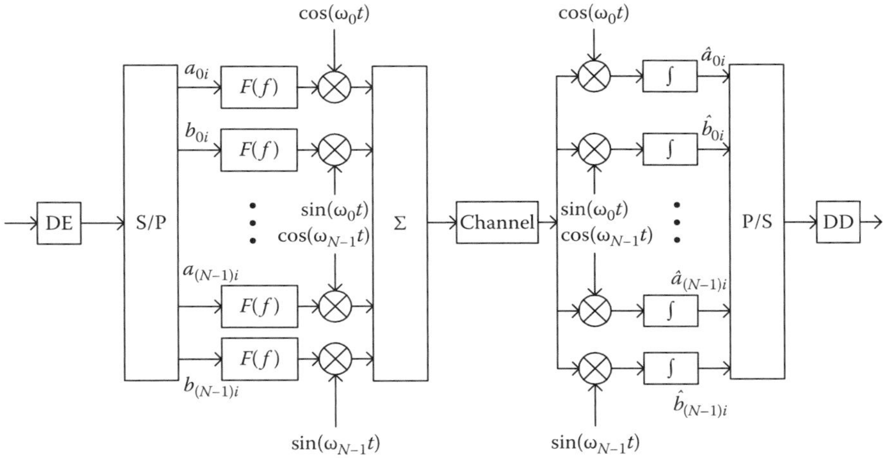

FIGURE 26.1 Overview of the OFDM Principle.

reciprocal of the symbol duration, the orthogonality between the multiplexed tones can be realized. In this context, the MCM is called OFDM.

OFDM is a form of MCM that was first introduced more than three decades ago.[10–12] The first multichannel modulation systems appeared in the 1950s as military radio links, which were systems best characterized as frequency division multiplexed systems. The first OFDM schemes were presented by Chang[11] and Saltzberg.[12] Actual use of OFDM was limited, and the practicability of the concept was questioned. However, OFDM was made more practical through the work of Chang and Gibby,[13] Weinstein and Ebert,[5] Peled and Ruiz,[6] and Hirosaki.[14] OFDM embodies the use of parallel subchannels to transmit information over dispersive channels with impairments. There are two main features of this technique. One is that it can increase the bit rate of the channel because of its high spectral efficiency. The other is that it can mitigate ISI and impulsive noise effectively because the symbol duration in OFDM is much longer than it is in single-carrier modulation with the same data rate. This technique has been known by many names: MCM, orthogonally multiplexed QAM, DMT, parallel data transmission, and so on.

26.1.1 Principles of DFT-Based OFDM System

Implementation of the OFDM system can be achieved through different approaches, including frequency division multiplexing, discrete Fourier transformation, as well as wavelet transformation. However, the majority of the OFDM systems today are based on inverse fast Fourier transform (IFFT) and FFT for modulation and demodulation, respectively. Using this method, both transmitter and receiver can be implemented using efficient FFT techniques, which reduce the number of operations from N^2 to $N\log_2 N$.

Consider the IDFT/DFT-based OFDM system in Figure 26.2.[15] First, the serial binary data stream passes the data encoder (DE), which is used to map $\log_2 M$ binary data onto a two-dimensional M-ary digital modulation signal constellation. The resultant symbol (i.e., M-ary signal) stream is grouped into blocks, each block containing N symbols. Thus, an M-ary signal (data) sequence $(d_0, d_1, d_2, \ldots, d_{N-1})$ is produced, where d_k is a complex number $d_k = a_k + jb_k$. Then, the N serial data symbols are converted to parallel and an IFFT is performed. The output of the inverse discrete Fourier transform (IDFT) is

$$S_n = \sum_{k=0}^{N-1} d_k e^{-j(2\pi nk/N)} = \sum_{k=0}^{N-1} d_k e^{-j2\pi f_k t_n}, \tag{26.1}$$

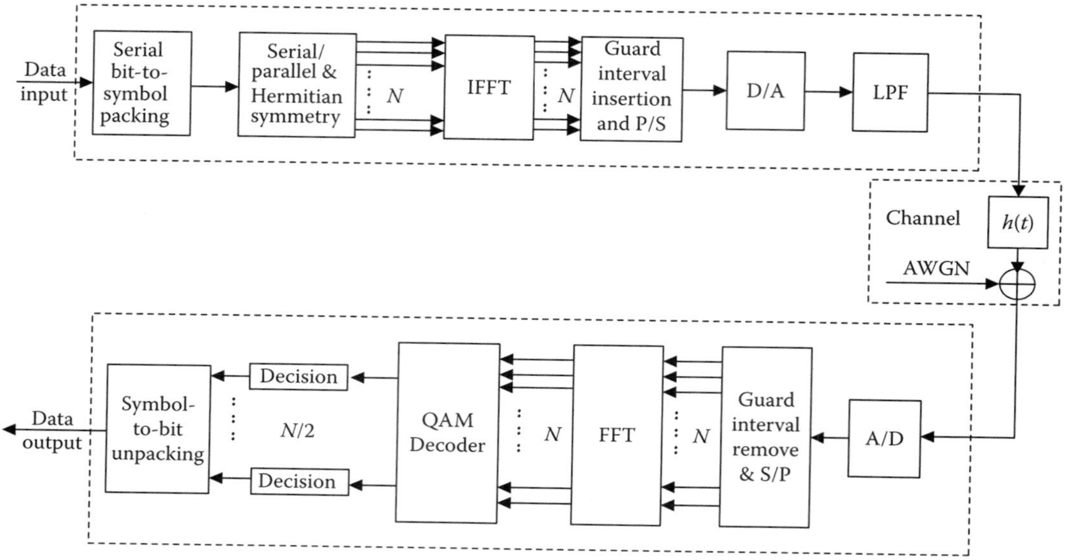

FIGURE 26.2 FFT/IFFT-based OFDM System.

where

$$f_k = k/(NT) \quad \text{and} \quad t_n = nT,$$

where T is an arbitrarily chosen symbol duration of the serial data sequence d_k. The real part of the vector S_n has components

$$S'_n = \sum_{k=0}^{N-1} (a_k \cos 2\pi f_k t_n + b_k \sin 2\pi f_k t_n). \tag{26.2}$$

If these components are sent to a low-pass filter at fixed time intervals T, the desired OFDM signal is obtained as

$$S(t) = \sum_{k=0}^{N-1} (a_k \cos 2\pi f_k t + b_k \sin 2\pi f_k t). \tag{26.3}$$

If we consider an infinite transmission time, the OFDM signal becomes

$$S(t) = \sum_{j=-\infty}^{\infty} \sum_{k=0}^{N-1} (a_{kj} \cos 2\pi f_k t + b_{kj} \sin 2\pi f_k t)\Pi(t - jT), \tag{26.4}$$

where $\Pi(t) = \begin{cases} 1, & 0 \le t \le T \\ 0, & \text{elsewhere} \end{cases}$, is a unit rectangular window function.

Figure 26.3 gives an example of the construction of an OFDM signal in which the emitted symbols are from an alphabet of a quadrature phase shift keying (QPSK) constellation, that is, $\{1 + j, 1 - j, -1 + j, -1 - j\}$. Figure 26.3a and b shows, respectively, the data for the real and imaginary parts of complex data $c_k = a_k + jb_k$. Figure 26.3c depicts their corresponding waveforms for each

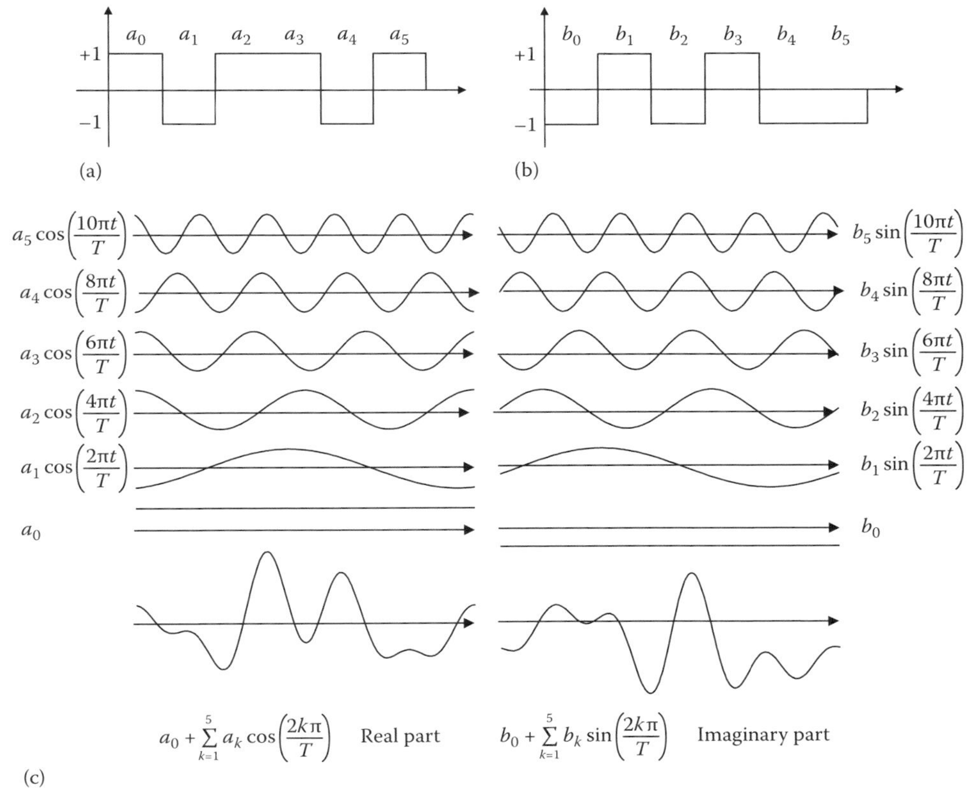

FIGURE 26.3 Construction of an OFDM Signal.

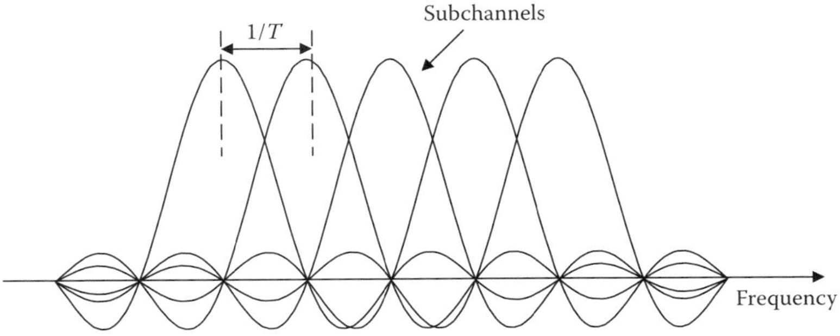

FIGURE 26.4 Signal amplitude spectrum of each subchannel.

subchannel. From Equations 26.3 and 26.4, the signal amplitude spectrum of each subchannel can be shown in Figure 26.4: the subchannels do overlap. When the emitted symbols are independent and have equal probabilities, the corresponding power spectral density of the OFDM signal can be easily calculated. Assume that all the carriers are modulated in an independent way. The power spectral density of the transmitted signal is obtained by the sum of the power spectral density of all the subcarriers.

The Fourier transform of each carrier is the convolution of Fourier transformation of a sine wave with the Fourier transform of rectangular function $\Pi(t)$. An example is given in Figure 26.5 for the case of

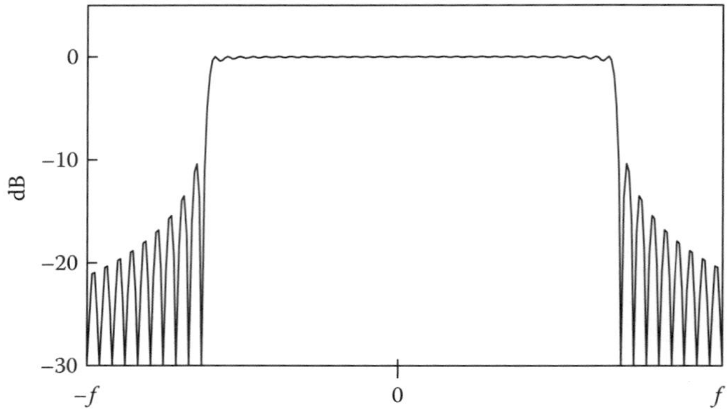

FIGURE 26.5 Fourier transform of the OFDM carrier (N = 32).

$N = 32$. It should be noted that the value of N considered in Figure 26.5 is used to make the diagram clear: In practice, the value of N is considerably larger. It should also be noted that even if the secondary sidelobes have a high amplitude, their width is proportional to $1/NT$, and their relative widths therefore decrease rapidly as N increases. The spectrum of an OFDM signal then tends asymptotically toward an ideal rectangular spectrum.

The demodulation process is based on the following orthogonal conditions:

$$\int_0^T a_k \cos(2\pi f_k t)\cos(2\pi f_{k'} t)dt = \begin{cases} 0 & \text{if } k' \neq k \\ a_k \frac{T}{2} & \text{if } k' = k, \end{cases} \tag{26.5}$$

$$\int_0^T \cos(2\pi f_k t)\sin(2\pi f_{k'} t)dt = 0. \tag{26.6}$$

However if a phase shift, ϕ_k, is introduced to the nonideal channel, the above equations will become

$$\int_0^T a_k \cos(2\pi f_k t + \phi_k)\cos(2\pi f_{k'} t)dt = \begin{cases} 0 & \text{if } k' \neq k \\ \frac{T}{2} a_k \cos(\phi_k) & \text{if } k' = k, \end{cases} \tag{26.7}$$

and

$$\int_0^T a_k \cos(2\pi f_k t + \phi_k)\sin(2\pi f_{k'} t)dt = \begin{cases} 0 & \text{if } k \neq k' \\ a_k \frac{T}{2} \sin(\phi_k) & \text{if } k = k'. \end{cases} \tag{26.8}$$

Obviously, the loss of orthogonality will cause intrachannel interference (ICI) between the in-phase and quadrature components of each subcarrier. However, ICI can be eliminated through channel estimation and equalization.

The implementation of the DFT-based OFDM can be efficiently realized with the FFT. An example of an eight-point radix-2 FFT is illustrated in Figure 26.6, where W is the twiddle coefficient.[16]

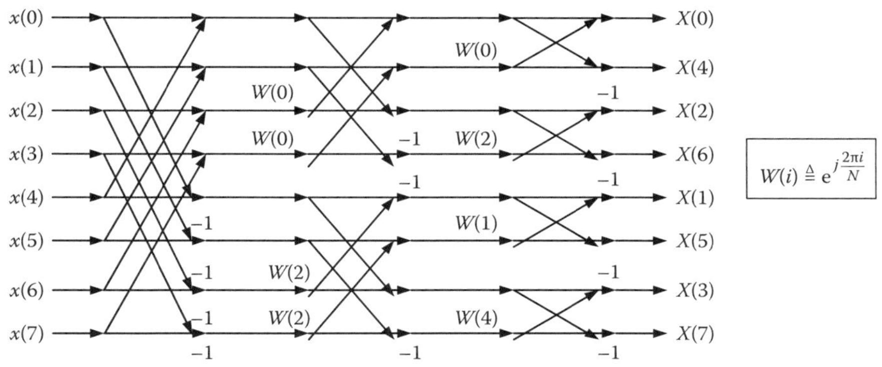

FIGURE 26.6 Structure of the FFT computation.

26.1.2 ISI Mitigation through Cyclic Prefix

In the presence of ISI caused by the transmission channel, the properties of orthogonality between the subcarriers are no longer maintained. In this situation, it is impossible to isolate an interval of T seconds in the received signal containing information from only one symbol. The signal is thus corrupted by ISI.

One can approach asymptotically toward a solution to the ISI problem by increasing indefinitely the number of subcarriers N. This would give rise to an increased symbol duration for a specific channel with a given data rate. However, this method is limited by technological limitations such as phase noise affecting the oscillators at the receiver and impractical implementation complexity. Another solution is to deliberately sacrifice some of the transmission time by preceding each OFDM symbol with a guard interval, to eliminate the ISI problem, as explained in the following.[17] The duration of each symbol is changed to $(T - \tau_p)$ seconds with a guard interval of τ_p seconds. When the guard interval is longer than the duration of the channel impulse response, or the multipath delay, then ISI can be eliminated.

At the receiver, only the "useful" signal is demodulated and the guard interval is discarded. The use of a guard interval results in a loss in transmission capacity of $10\log_{10}(T'/T)$, with $T' = T - \tau_p$. This capacity reduction can in practice be kept below 1 dB and can be largely compensated by the system advantages, such as high bandwidth efficiency and ISI-free transmission.

In practice, the equivalent widely used method to combat ISI consists in adding a cyclic prefix to an OFDM symbol as indicated in Figure 26.7.[18] The reason for its popularity is that it is easy to implement in digital form. In this approach, the guard interval is a cyclic extension of the IFFT output sequence. If N is the original OFDM block length and the channel's impulse response $h(n)$ has length G, and assuming that the length of the added cyclic extension is also G, then the cyclically extended OFDM block has a new length of $(N+G)$. The original symbol sequence S is cyclically extended to form the new symbol sequence S^g with a cyclic prefix as follows:

$$S = \{S_0, S_1, S_2, \ldots, S_{N-1}\}, \quad (26.9)$$

$$S^g = \{S_{N-G}, \ldots, S_{N-1}, S_0, S_1, S_2, \ldots, S_{N-G}, \ldots, S_{N-1}\} \quad (26.10)$$

For instance, for $N=6$ and $G=3$, we have a new cyclically extended symbol sequence of length $(N+G)=9$ described by the above equation. Thus, as seen by the finite-length impulse response of the channel $h(n)$ of length G, each extended symbol sequence <equ_0026.eps> appears as if we had repeated the original symbol sequence S periodically. The cyclically extended sequence S^g convolved with the impulse response of the channel sequence $h(n)$ appears as if it was convolved with a periodic sequence consisting of repeated S's. Therefore, using the cyclic extension, the new OFDM symbol of

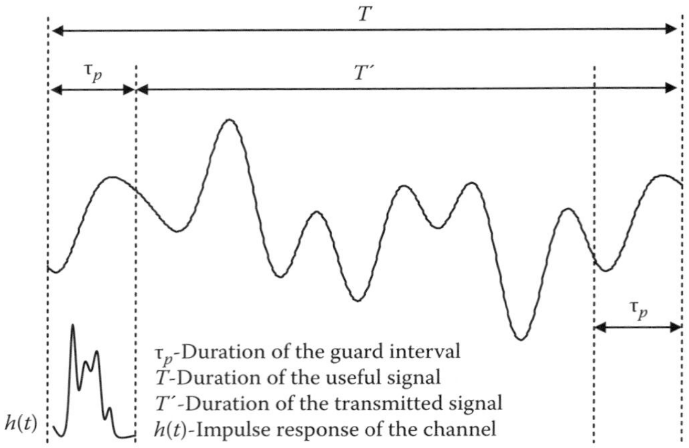

FIGURE 26.7 Addition of cyclic prefix to the OFDM symbol.

length $(N + G)$ sampling periods suffers no longer from ISI. For the previous example with $N = 6$, $G = 3$, and $N + G = 9$, we can obtain a subset of six samples at the receiver as follows:

$$\underline{R} = \underline{S} \cdot \begin{bmatrix} 0 & h_0 & h_1 & h_2 & 0 & 0 \\ 0 & 0 & h_0 & h_1 & h_2 & 0 \\ 0 & 0 & 0 & h_0 & h_1 & h_2 \\ h_2 & 0 & 0 & 0 & h_0 & h_1 \\ h_1 & h_2 & 0 & 0 & 0 & h_0 \\ h_0 & h_1 & h_2 & 0 & 0 & 0 \end{bmatrix}, \tag{26.11}$$

where

$\underline{S}$ and $\underline{R}$ are transmitted and received signal vectors (both of them are length N row vectors)
$h = (h_0, h_1, h_2)$ is the impulse response of the channel

Equivalently, there exists a cyclic convolution between $\underline{S}$ and h, and the following DFT transform pair holds:

$$\underline{S} \otimes h \Leftrightarrow \text{DFT}[\underline{S}] \cdot H(k), \tag{26.12}$$

where

$\otimes$ denotes the cyclic convolution operation
$H(k)$ is the Fourier transform of h

With the knowledge of $H(k)$ at the kth subcarrier, we can mitigate the intrachannel interference coming from $H(k)$ inside each symbol.

26.1.3 Bit Rate of OFDM Systems

If we consider that each carrier conveys a symbol taken in a two-dimensional constellation with 2^a points and is modulated during T seconds, the bit rate can be shown to be[17]

$$D = \frac{Na}{T} \quad [\text{bit/s}]. \tag{26.13}$$

As indicated in Figure 26.4, the frequency spacing between two adjacent subcarriers is $1/T$. Three sidelobes are also counted at each side of the OFDM spectrum border when we determine the bandwidth of the signal. Therefore, the total bandwidth occupied by the N carriers is then given by[4,17]

$$W = \frac{N-1}{T} + 2\frac{3}{T} = \frac{N+5}{T}. \tag{26.14}$$

Thus, the spectral efficiency is

$$\eta = \frac{D}{W} = a\frac{N}{N+5} \quad \text{[bit/s/Hz]}. \tag{26.15}$$

Asymptotically, η tends toward a bits/s/Hz when N increases, and OFDM can be considered an optimum modulation for spectral efficiency. If we take the guard interval τ_p into consideration, W and η will become

$$W' = \frac{N-1}{T'} + \frac{6}{T'} = \frac{N+5}{T-\tau_p}, \tag{26.16}$$

$$\eta' = \frac{D}{W'} = \frac{aN(T-\tau_p)}{(N+5)T} \quad \text{[bit/s/Hz]}. \tag{26.17}$$

26.1.4 In-Band Pilots and OFDM Channel Estimation

In-band pilots, that is, subcarriers modulated with symbols known to the receiver, are normally used for channel estimation and synchronization purposes in conventional OFDM systems. Channel response at pilot frequency is obtained at the receiver side by demodulating the pilot. Assume we have an OFDM symbol $\mathbf{x}$ denoted by 1 from the transmitter; the signal at the kth subcarrier after FFT is

$$Y_k = X_k H_k + W_k, \quad 0 \le k \le N-1 \tag{26.18}$$

The estimate of the channel frequency response, at pilot subcarrier p based on least square estimation, is given by

$$\hat{H}_p = \frac{Y_p}{X_p}. \tag{26.19}$$

The overall frequency response of the channel for a given OFDM symbol is obtained by the interpolation of the channel responses at all pilot frequencies, as shown in Figure 26.8. Due to the varying nature of the channel in both frequency and time domains, different pilot patterns can be used to improve the performance of the corresponding channel estimator. The design of pilot patterns for a given channel relies mainly on the time and frequency selectivity of the channel. Four different pilot patterns normally used in OFDM systems are illustrated in Figure 26.9. The criteria for choosing a specific pilot pattern for an OFDM system rely mainly on the time and frequency selectivity of the channel model. For instance, the top-right pilot pattern shown in Figure 26.9 is suitable for low to medium frequency-selective fast fading channels, whereas the pattern at the bottom left of the figure performs better in severely frequency-selective and slow fading channels. Hybrid pilot patterns can also be used in certain applications; for instance, DVB-T employs pilot arrangements as those shown in the top-left and bottom-right areas of Figure 26.9. As each pilot can be regarded as one sample of the channel response in the frequency domain at a given time, the sampling theorem can then be applied for the design of the pilot patterns for OFDM systems. For an accurate channel estimation, the interval between the two adjacent pilots must be limited by the two-dimensional version of the sampling theorem.

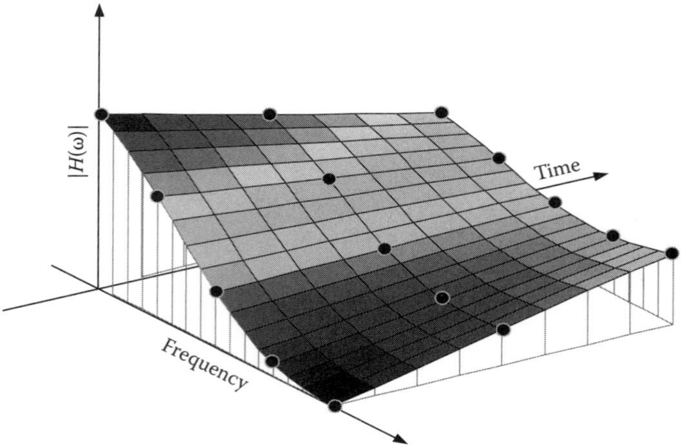

FIGURE 26.8 Overall frequency response of the OFDM channel.

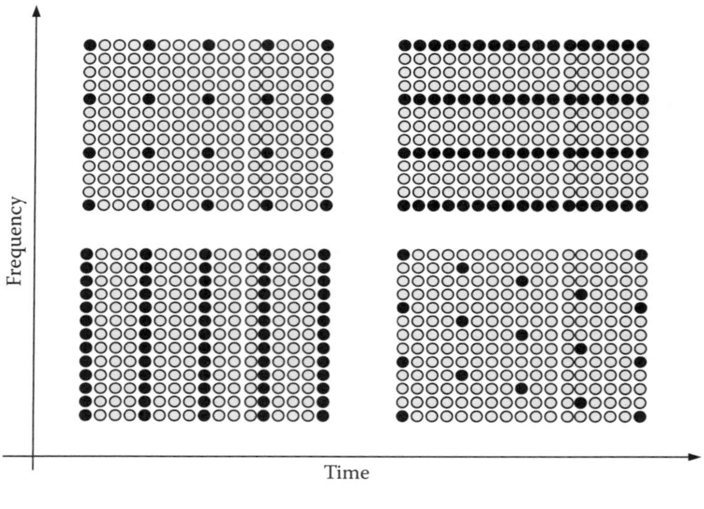

FIGURE 26.9 Four different pilot patterns used in OFDM.

26.1.5 Modulation Schemes for Subchannels

OFDM systems can be classified into two categories according to the modulation schemes for each subchannel: (1) coherent OFDM systems for which coherent modulation such as M-ary phase shift keying (MPSK) and M-ary quadrature amplitude modulation (MQAM) are used as subchannel modulation schemes, and (2) noncoherent OFDM systems for which noncoherent modulation such as differential phase shift keying (DPSK) is used as the subchannel modulation scheme.

Several factors influence the choice of a digital modulation scheme for an OFDM system. A desirable modulation scheme provides low-bit-error rates at low received signal-to-noise ratios (SNR), performs well in dispersive channel conditions, occupies a minimum of transmission bandwidth, and is not complex and cost effective to implement. Existing modulation schemes do not simultaneously satisfy all of these requirements: trade-offs have to be made depending on the requirements of the particular application when selecting a digital modulation scheme. In the following, two digital modulation schemes are briefly introduced, and these schemes will be used as subcarrier modulation schemes in the OFDM systems, whose performance will be studied.

M-ary phase shift keying (MPSK) and M-ary differential phase shift keying (MDPSK): In digital phase modulation, the signal waveforms are represented as[14]

$$s_m(t) = \mathrm{Re}[g(t)\exp(j2\pi f_c t + \theta_m)], \quad \text{for} \quad m = 1, 2, \ldots, M, \quad \text{and} \quad 0 \le t \le T, \qquad (26.20)$$

where
$g(t)$ is the signal pulse shape
T is the symbol duration

$$\theta_m = 2\pi(m-1)/M$$

are the M possible phases of the carrier that convey the transmitted information. Digital phase modulation is usually called phase shift keying (PSK). The mapping, or assignment of k information bits to the $M = 2^k$ possible phases, can be done in such a way that the most likely errors caused by noise will result in a single bit error. This mapping scheme is called Gray bit mapping. The signal space diagrams for $M = 2$, 4, and 8 with Gray bit mapping are illustrated in Figure 26.10. A differentially encoded phase-modulated scheme is called MDPSK.

M-ary quadrature amplitude modulation (MQAM): It is to be noted here that MQAM is employed for each subchannel of the OFDM system. The signal space diagram is rectangular, as shown in Figure 26.11 for 16QAM. The bandwidth efficiency of MQAM modulation is identical to that of MPSK modulation. In terms of power efficiency, MQAM is superior to MPSK because MQAM efficiently uses the signal constellation space to increase the distance between constellation points (and hence reduce the probability of error detection). MQAM is widely used in mobile wireless and fixed line communication systems.

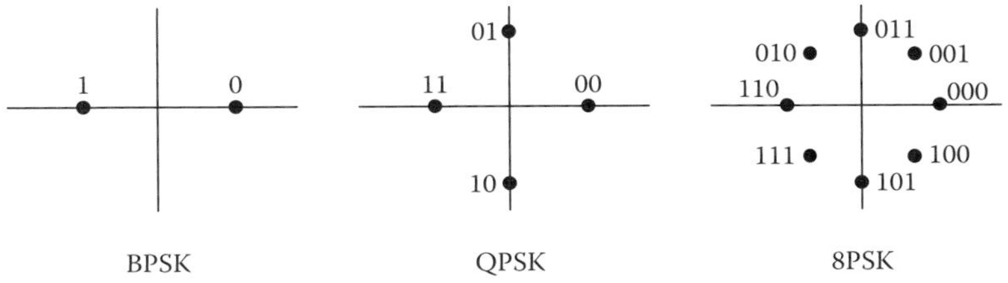

FIGURE 26.10 Signal space patterns for M = 2, 4, and 8.

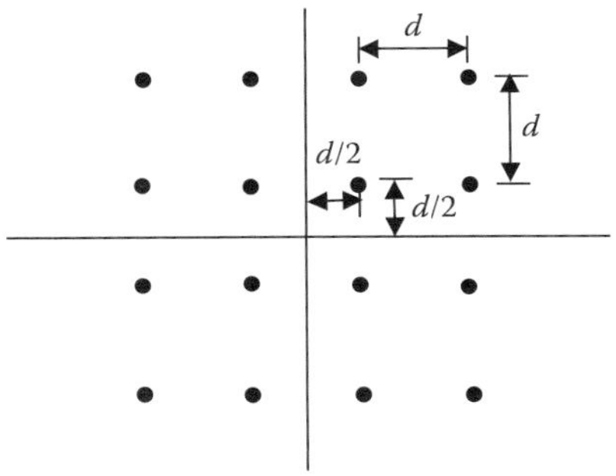

FIGURE 26.11 Signal space pattern for 16 QAM.

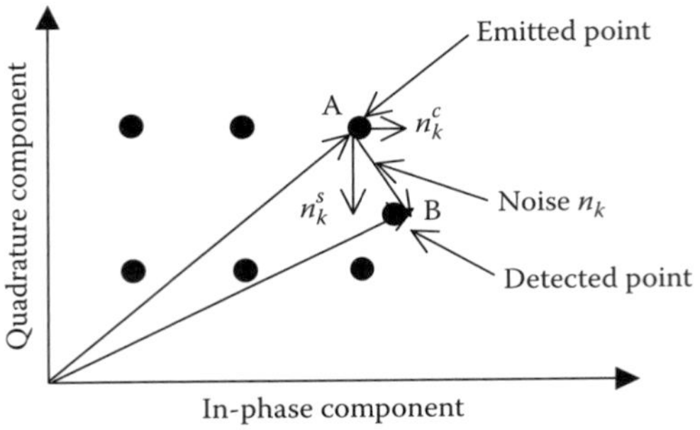

FIGURE 26.12 Euclidean distances and relationship to the error.

26.1.6 OFDM Error Probability Performances

The error probability performance of OFDM systems is closely related to the Euclidean distances between the points in the signal constellation. Each carrier is modulated with signal points taken from a two-dimensional signal constellation. The constellation can be different from one carrier to another, and the Euclidean distance between the points will establish the robustness of each subcarrier. The more spread the constellation is (i.e., with the maximum Euclidean distance for a given energy and given number of points), the better the system performance is.

To improve the power efficiency of OFDM systems, square constellations are often used. An important point to determine is the relation between the distance d separating two adjacent signal points and the mean energy of the constellations, which is defined as[17]

$$\bar{E}_c^2 = \frac{1}{2} \sum_{k=0}^{2^c-1} \frac{a_k^2 + b_k^2}{2^c} = \bar{a}_k^2 = \bar{b}_k^2, \tag{26.21}$$

where c is the number of the constellation points. An error will occur if the noise component is larger than half the distance between two points in the constellation in each subchannel,[17] as shown in Figure 26.12.

Let A be the emitted signal point and B be the detected signal point. n_k is the noise vector and is represented by an in-phase component n_k^c and quadrature component n_k^s. It can be shown that the noise samples n_k^c and n_k^s are always uncorrelated. As a result, the error probability for QAM symbols can be determined from the symbol error probability of the two pulse amplitude modulation (PAM) systems. Detailed analysis can be found in de Couasnon et al.[17] and Proakis.[19]

26.2 OFDM-Based Mobile Broadcasting Standards

The digitization of traditional broadcast systems has made significant progress in recent years. In many countries, the new digital broadcasting systems are based on OFDM technologies. Recently the broadcast industry has turned its eyes to mobile multimedia broadcasting, including mobile TV. Mobile TV still has many obstacles to overcome, but at the moment it looks very likely to be the next killer application in broadcast industry. There are several OFDM-based systems that can provide multimedia broadcasting services, including DVB-H,[20] MediaFLO,[21] and T-DMB (digital multimedia broadcasting).[22] So far, there have only been a few real implementations and many pilot projects. This is mainly due to the fact that there are a couple of competing technologies and the winning technology is yet to be determined. DVB-H technology is the leader at the moment, as it has several major industry players backing it and running pilot projects around the world. This section describes the three most promising mobile multimedia broadcasting technologies at the current time: DVB-H, MediaFLO, and T-DMB.

26.2.1 DVB-H Digital Mobile Broadcasting System

The first mobile broadcast technology discussed here is the DVB-H standard. A comprehensive overview of the DVB-H system can be found in ETSI EN 302 304[20] and Kornfeld and May.[23] DVB-H is a new standard that emphasizes mobile features for DVB-T. It is the latest development from the European DVB standard family, targeted for handheld devices like mobile phones and personal digital assistants (PDAs). The enhancement of DVB-T with DVB-H introduces the timing-slicing technique to save battery power, improved performance with multiprotocol encapsulation–forward error correction (MPE–FEC), and the hybrid networks for mobile handheld reception. As DVB-H is built upon DVB-T, an overview of DVB-T will be given first.

26.2.1.1 DVB-T System

The DVB-T system was developed by an European consortium of public and private sector organizations—the DVB project. The DVB-T specification is part of a family of specifications also covering satellite (DVB-S) and cable (DVB-C) operations. This family allows for digital video and digital audio distribution as well as transport of forthcoming multimedia services. For terrestrial broadcasting, the system was designed to operate within the existing UHF spectrum allocated to analog PAL and SECAM television standard transmissions. Although the system was developed for 8 MHz channels, it can be scaled to different channel bandwidths, that is, 6, 7, or 8 MHz, with corresponding scaling in the data capacity. The net bit rate available in the 8 MHz channel ranges between 4.98 and 31.67 Mbits/s, depending on the choice of channel coding parameters, modulation types, and guard interval duration.

The system was essentially designed with built-in flexibility, to be able to adapt to different types of channels. It is capable of coping not only with Gaussian channels, but also with Ricean and Rayleigh channels. The system is robust to interference from delayed signals, with echoes resulting from either terrain or building reflections. The system uses OFDM modulation with a large number of carriers per channel modulated in parallel via an FFT process. It has two operational modes: a 2k mode, which uses a 2k FFT, and an 8k mode, which requires an 8k FFT. The system makes provisions for selection between different levels of QAM modulation and different inner code rates and also allows two-level hierarchical channel coding and modulation. Moreover, a guard interval with selectable width separates the transmitted symbols, which allows the system to support different network configurations, such as large-area single-frequency networks (SFNs) and single-transmitter operation. The 2k mode is suitable for single-transmitter operation and small-SFNs with limited distance between transmitters, whereas the 8k mode can be used for both single-transmitter operation and small- and large-SFNs.

The DVB-T standard was first published in 1997 and was not targeted for mobile receivers. Nevertheless, following the positive test results, DVB-T mobile services were launched in Singapore and Germany. Despite the success of mobile DVB-T reception, its major downfall has been the battery life. The current and projected power consumption is too high to support mobile devices that are supposed to last a long period with a single battery charge. Another issue for DVB-T, which is improved in DVB-H, is IP Datacasting.* IP datacasting will facilitate the interoperability of telecommunications and broadcasting networks, a complex topic involving detailed work on the interface at different service levels. Although DVB-T is the world's most used digital terrestrial television system, the current situation is most obviously going to be changed by DVB-H or MediaFLO.

* IP Datacasting: Internet Protocol datacasting.

26.2.1.2 DVB-H System Overview

The objective of DVB-H is to provide an efficient way for carrying multimedia data over digital terrestrial broadcasting networks to handheld terminals. It is the latest development within the set of DVB transmission standards. The DVB-H transmission system is built from the capabilities of DVB-T, but it overcomes the two key limitations of DVB-T technology. It extends the battery life of the handheld device and improves the robustness of the mobile reception in fading environments. DVB-H uses a power-saving technique based on the time-multiplexed transmission of different broadcast services. The technique, called time slicing, results in a large battery power-saving effect. Additionally, time slicing allows soft handover if the receiver moves from one broadcast cell to another with only one receiving front end. For reliable transmission in poor signal reception conditions, an enhanced error-protection scheme on the link layer is introduced. This scheme is called MPE–FEC. MPE–FEC employs powerful channel coding on top of the channel coding included in the DVB-T specification and offers a degree of time interleaving. DVB-H also provides an enhanced signaling channel for improving access to the various services. DVB-H, as a transmission standard for mobile broadcasting, also specifies the physical layer as well as the elements of the lowest protocol layers.

Furthermore, the DVB-H standard features an additional transmission mode, the 4k mode, offering additional flexibility in designing SFNs, which are still well suited for mobile reception. DVB-H allows this additional 4k mode to be used, which is created via a 4096-point IDFT in the OFDM modulator. The 4k mode represents a compromise solution between the DVB-T 2k and 8k modes. It allows for a doubling of the transmitter distance in SFNs compared to the 2k mode and, when compared to the 8k mode, is less susceptible to the impairment effects caused by Doppler shifts in the case of mobile reception. The 4k mode also offers a new degree of network planning flexibility. Because DVB-T does not include this mode, it may only be used in dedicated DVB-H networks.

26.2.1.3 Time-Slicing Technique for Power Reduction

A major problem for mobile broadcasting receivers is the limited battery capacity. Therefore, being compatible with DVB-T would place a burden on the DVB-H receiver because demodulating a high-data-rate stream like the DVB-T involves significant power dissipation at the mobile receiver. The major disadvantage of DVB-T is the fact that the whole data stream has to be decoded before any one of the services (i.e., TV programs) of the multiplexed DVB-H data stream can be accessed. Time slicing is a power-saving technique that takes advantage of the fact that the service the user wants to watch or listen to is only transmitted for a fraction of the time since there are multiple services carried in a multiplexed stream. This allows for the RF front end to be turned off when the desired service signal is not being transmitted. This allows a significant amount of power to be saved, since the RF front end's amplifiers are relatively inefficient in terms of power consumption because OFDM reception requires highly linear RF amplifiers, and the higher the required linearity of the amplifier, the lower the power efficiency. With DVB-H, service multiplexing is performed in a pure time division multiplex fashion, as illustrated in Figure 26.13. The data of one particular service is therefore not transmitted continuously but in compact periodical bursts with transmission interruptions in between. Multiplexing of several services leads again to a continuous, uninterrupted transmitted stream with a constant data rate.

Consequently, battery power saving is made possible through receiving the broadcast services in short time burst signals. The terminal synchronizes to the bursts for the wanted service but switches to a power-save mode during the time intervals when other services are being transmitted. The power-save time between bursts, relative to the on-time required for the reception of an individual service, is a direct measure of the power saving provided by DVB-H. Bursts entering the receiver have to be buffered and read out at the service data rate. The position of the bursts is signaled in terms of the relative time difference between two consecutive bursts of the same service. A lead time for powering up the front end of the mobile receiver, for resynchronization and channel estimation, and so on, has to be taken into account. Depending on the ratio of on-time/power-save time, the resulting power saving may be more than 90%.

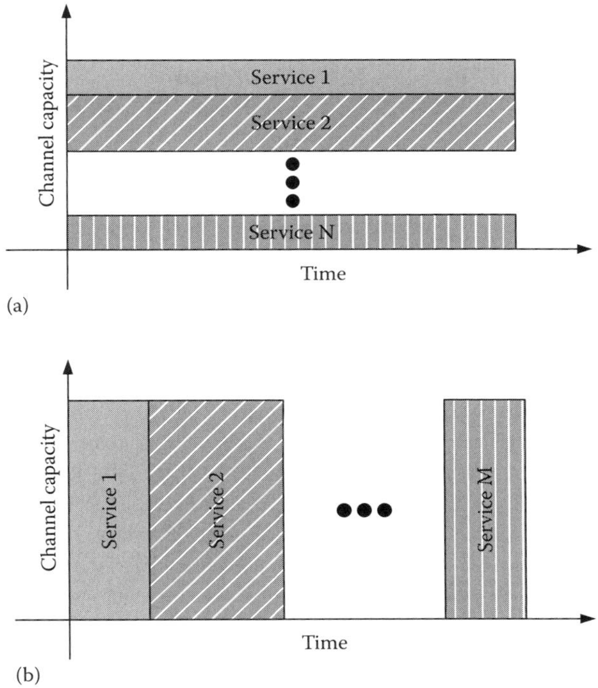

FIGURE 26.13 Service multiplexing in DVB-H with time division multiplexing.

In addition, time slicing offers another benefit for the design of mobile terminals. The long power-save periods may be used to search for other channels in neighboring radio cells offering the selected service. Smooth channel handover can be achieved at the border between two cells providing the mobile broadcast service. Both the monitoring of the services in adjacent cells and the reception of the selected service data can be realized with the one receiving front end.

26.2.1.4 Multiprotocol Encapsulation–Forward Error

26.2.1.4.1 Correction

Different from other DVB transmission systems that are based on the DVB transport stream from the MPEG-2 standard, the DVB-H system is using the IP as the interface to higher protocol layers, which allows the DVB-H system to be combined with other IP-based networks. The introduction of MPE–FEC[23] also improves the system performance under mobile receiving conditions. DVB-H terminals are expected to be used in various situations of reception: indoor/outdoor while the user is static, a pedestrian, or mobile. This time-slicing technique imposes the implementation of a long time interleaver to mitigate the deep-fading impairments experienced in mobile reception. For this purpose, DVB-H defined an additional protection through MPE–FEC. Nevertheless, the MPEG-2 transport stream is still used by the base layer. The IP data is embedded into the transport stream by means of MPE, an adaptation protocol defined in the DVB data broadcast specification. On the level of the MPE, an additional stage of FEC is added. MPE–FEC complements the physical layer FEC of the underlying DVB-T standard, with the purpose of reducing the SNR requirements for reception by a handheld device. Intensive testing of DVB-H, which was carried out by DVB member companies, showed that the use of MPE–FEC results in a coding gain of about 7 dB over DVB-T systems.

MPE–FEC processing is located on the link layer at the level of the IP input streams before they are encapsulated by means of the MPE. MPE–FEC, MPE, and the time-slicing technique were defined jointly and directly aligned with each other. The IP input streams provided by different sources as individual elementary streams are multiplexed according to the time-slicing method.[23] The MPE–FEC error protection is first calculated separately for each individual elementary stream. Then IP packets are encapsulated and embedded into the transport stream. All relevant data processing is carried out before the transport stream interface to guarantee compatibility to a DVB-T transmission network.

MPE–FEC and time slicing are closely related techniques. Both are applied on the elementary stream level, and one time-slicing burst includes the content of exactly one MPE–FEC frame. Separating the IP data and parity check bytes of each burst makes the use of MPE–FEC decoding in the receiver optional because the application data can be recovered while ignoring the parity information when the reception condition is good.

26.2.1.5 Transmission Parameter Signaling for DVB-H

Transmission parameter signaling (TPS) in DVB standards creates a reserved information channel that provides tuning parameters to the receiver. In DVB-T, 23 of 68 TPS bits in a frame are currently used to carry information about the transmission mode and a cell identifier. The signaling of parameters of the DVB-H elementary streams in the multiplex uses an extension of the TPS. The new elements of the TPS channel provide the information that time-sliced DVB-H elementary streams are available in the multiplex and indicate whether MPE–FEC protection is used in at least one of the elementary streams.[23] In addition, broadcasting of the cell identifier known as an optional element of DVB-T is made mandatory for DVB-H. The availability of this cell identifier simplifies the detection of neighboring network cells in which the selected same service is available.

26.2.2 MediaFLO Digital Mobile Broadcasting System

MediaFLO is a technology developed by Qualcomm based on FLO (Forward Link Only) technology. It is an OFDM-based air interface designed specifically for multicasting a significant volume of multimedia content to wireless handsets.[21] The MediaFLO system consists of two components: FLO technology and media distribution system (MDS). FLO technology is designed for markets where regulations permit high-power transmission from a single tower or a small number of towers. FLO can also be deployed across wide-area regions using a network of transmitters, spaced 60 km apart.

The MDS enables the efficient delivery of high-quality network-scheduled video content for viewing by masses of mobile subscribers. The MDS seamlessly handles multiple content streams from multiple sources and plays them on client software in the most popular video and audio formats. Because MDS is air interface independent, it can be deployed on any IP packet data network or current point-to-point third-generation (3G) wireless networks, and will scale easily for tomorrow's multicast networks. The MDS provides the tools to assimilate and aggregate content, bundle channels into subscription packages, and ultimately merchandise and deliver this content securely to wireless operator target subscribers. Wireless operators can also leverage additional MDS features to entice users to engage with other media (video on demand, music on demand, ring tones, games, etc.) over their 3G networks while viewing content delivered over FLO.

26.2.3 T-DMB Digital Mobile Broadcasting System

The DMB system can operate via satellite (S-DMB) or terrestrial (T-DMB) wireless links. DMB is based on the Eureka 147 standard, also known as DAB standard, and shares some similarities with DVB-H.[22] It is operated in band III from 174 to 230 MHz and in L band from 1452 to 1492 MHz. It is a narrowband solution for mobile broadcasting. T-DMB services started in South Korea in December 2005, and some pilots began in Europe in 2006, for instance, in Germany, France, and the United Kingdom. DAB technology has a vast amount of users and about 800 services worldwide. Most of them are directed to mobile radio users and will not affect mobile TV users.

26.3 Frequency and Timing Synchronization for OFDM-Based Wireless Systems

At the front end of the mobile terminal, the received broadcast signals are subject to synchronization errors due to the variation of oscillator frequency and sample clock differences. The demodulation of the received OFDM signal to baseband, possibly via an intermediate frequency, involves oscillators whose frequencies may not be perfectly aligned with the transmitter frequencies. This results in a carrier frequency offset. Also, demodulation of the OFDM signal usually introduces phase noise acting as an unwanted phase modulation of the carrier wave. Both carrier frequency offset and phase noise degrade the performance of an OFDM system. The most important effect of a frequency offset between transmitter and receiver is a loss of orthogonality between the subcarriers, resulting in ICI. The characteristics of this ICI are similar to white Gaussian noise and lead to a degradation of the SNR.[24] For both AWGN and fading channels, this degradation increases with the square of the number of subcarriers. Like frequency offsets, phase noise and sample clock offsets cause ICI, and thus a degradation of the SNR. However, for a DVB-like OFDM system, Muschallik[25] concludes that phase noise is not performance limiting in properly designed consumer receivers for OFDM.

When the baseband signal is sampled at the analog-to-digital (A/D) converter, the sample clock frequency at the receiver may not be the same as that at the transmitter side. This sample clock offset not only causes errors, but also may cause the duration of an OFDM symbol at the receiver to be different from that at the transmitter. If the OFDM symbol synchronization is derived from the sample clock, this generates variations in the symbol clock, that is, timing offsets. Because the receiver needs to determine accurately when the OFDM symbol begins for proper demodulation with the FFT, a symbol synchronization algorithm at the receiver is usually necessary. Symbol synchronization also compensates for propagation delay changes in the channel. The effects of synchronization errors have been investigated by, among others, Moose,[26] Wei and Schlegel,[27] and Garcia Armada and Calvo.[28] The degradation due to symbol timing errors is not graceful. If the length of the cyclic prefix exceeds the length of the channel impulse response, a receiver can capture an OFDM symbol anywhere in the region where the symbol appears to be cyclic, without sacrificing orthogonality. A small timing error only appears as pure phase rotations of the data symbols and may be compensated by a channel equalizer, still preserving the system's orthogonality. A large timing error resulting in capturing a symbol outside this allowable interval, on the other hand, causes ISI and ICI and leads to performance degradation. Pollet et al.[29] showed that the degradation due to a sample clock frequency offset differs from subcarrier to subcarrier—the highest subcarrier experiencing the largest SNR loss.

Summarizing, oscillator phase noise and sample clock variations generate ICI but seldom limit the system performance. Frequency offsets and timing offsets (symbol clock offsets), however, generally need to be tracked and compensated at the receiver. We now give a brief review of some recently proposed frequency and timing estimators for OFDM, and then describe one of these methods, based on the cyclic prefix, in more detail.

26.3.1 Timing Offset Estimation Techniques

Timing offset estimators have been addressed in a number of publications (see references 26, 27, and 30–37). We divide these estimators conceptually into two approaches. For the first approach, the authors of references 26 and 30–32 assume that transmitted data symbols are known at the receiver. This can in practice be accomplished by transmitting known pilot symbols according to some protocol. The unknown symbol timing and carrier frequency offset may then be estimated from the received signal. The insertion of pilot symbols usually implies a reduction of the data rate. An example of such a pilot-based algorithm is found in Warner and Leung.[30] Joint time and frequency offset estimators based on this concept are described in Classen and Meyr[31] and Schmidl and Cox,[32] and in Moose[26] the repetition of an OFDM symbol supports the estimation of a frequency offset.

The second approach, considered by the authors of references 33, 34, 36, and 37, uses statistical redundancy in the received signal. The transmitted OFDM signal is modeled as a Gaussian process. The offset values are then estimated by exploiting the intrinsic redundancy provided by the L samples constituting the cyclic prefix. The basic idea behind these methods is that the cyclic prefix of the transmitted OFDM signal yields information about where an OFDM symbol is likely to begin. Moreover, the transmitted signal's redundancy also contains useful information about the carrier frequency offset. Tourtier et al.[33] observe that the statistic

$$\xi(m) = \sum_{k=m}^{m+L-1} |r(k) - r(k+N)| \tag{26.22}$$

contains information about the time offset. This statistic, implemented with a sliding sum, identifies samples of the cyclic prefix by the sum of L consecutive differences. The statistic is likely to become small when index m is close to the beginning of the OFDM symbol. Sandell et al.,[35] van de Beek et al.,[36] and later Lee and Cheon[37] use the statistic

$$\gamma(m) = \sum_{k=m}^{m+L-1} r(k) \cdot r(k+N)^*, \tag{26.23}$$

where $r(k+N)^*$ is the complex conjugate of $r(k+N)$, to estimate the time offset. This statistic is the sum of L consecutive correlations, and its magnitude is likely to become large when the index m is close to the start of the OFDM symbol.

However, the above time synchronization techniques can only provide a coarse synchronization time at integer signal samples. Therefore, residual timing offset is unavoidable after the synchronization using either Equation 26.22 or 26.23. A residual time offset Δn, which is normalized to the sampling interval, is considered next to evaluate its impact on the system performance of the OFDM-based broadcasting system.

To estimate the impact of the timing offset Δn, consider the following OFDM symbol given by the N-point complex modulation sequence:

$$x_n = \frac{1}{\sqrt{N}} \sum_{k=-K}^{K} X_k e^{j2\pi\frac{nk}{N}}, \quad n = 0, 1, 2, \ldots, N-1 \tag{26.24}$$

It consists of $(2K+1)$ complex sinusoids or subcarriers, which have been modulated with $(2K+1)$ complex data symbols X_k. The subcarriers are mutually orthogonal within the symbol interval, that is,

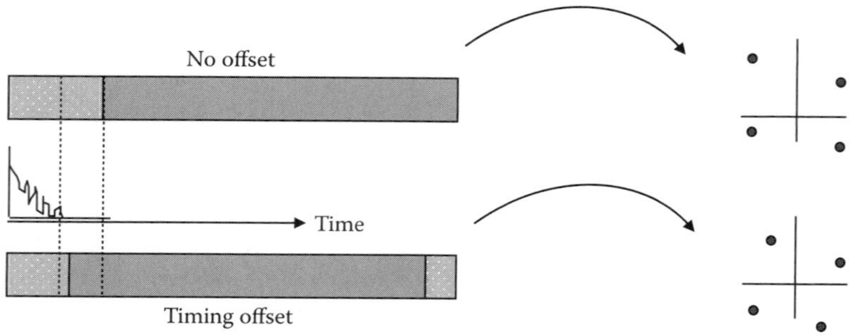

FIGURE 26.14 Impact of a timing offset.

$$\sum_{n=0}^{N-1} x_{nk}x_{nl}^{\star} = \frac{1}{N}|X_k|^2\delta_{kl}, \tag{26.25}$$

where δ_{kl} is the Kronecker Delta function. After passing through a band-pass channel, the complex envelope of the received sequence can be expressed as[26]

$$r_n = \frac{1}{\sqrt{N}}\sum_{k=-K}^{K} X_k H_k \cdot e^{j2\pi\frac{(n+\Delta n)k}{N}} + w_n \quad n = 0, 1, \ldots, N-1; \quad N \geq 2K+1, \tag{26.26}$$

where
- H_k is the channel transfer function at the kth carrier frequency
- Δk is the relative frequency offset (the ratio of the actual frequency offset to the subcarrier spacing)
- Δn is the relative timing offset (the ratio of the timing offset to the sampling interval)
- w_n is the sample of a complex Gaussian random variable with zero mean and variance $\sigma_w^2 = \frac{1}{2}E[|w_n|^2]$

After the DFT demodulation, the kth element of the DFT sequence R_k is[26]

$$\begin{aligned} R_k &= \text{DFT}_N\{r_n\} \\ &= X_k H_k . e^{j2\pi k \Delta n/N} + W_k \end{aligned} \tag{26.27}$$

W_k denotes the Gaussian noise component for the kth subcarrier. The impact of the timing offset can be evaluated by the above equation. A phase shift will be introduced to all the subcarriers. With the help of the in-band pilot, the phase shift corresponding to the residual timing offset for each subcarrier can then be easily estimated and compensated.

26.3.2 Frequency Offset and Estimation Techniques

As discussed earlier, frequency offset is caused by a carrier frequency mismatch between the transmitter and receiver oscillators. Frequency offset is especially problematic in OFDM systems compared to single-carrier systems. To achieve a negligible bit error rate (BER) degradation, the tolerable frequency offset should be within the order of 1% of the subcarrier spacing, which is unlikely achievable in an OFDM system using low-cost commercial crystals without applying any frequency offset compensation techniques. The frequency offset is divided into an integer part and a fractional part, that is,

$$\Delta f_c T = K_{\Delta f} + \Delta k, \tag{26.28}$$

where
- $K_{\Delta f}$ is an integer
- $\Delta k \in (-1/2, 1/2)$

The integer part $K_{\Delta f}$ can be found by a simple frequency domain correlation between the demodulated OFDM symbol and the in-band pilot. The fractional part is estimated by correlating the signal samples at an offset T (OFDM symbol duration) using the cyclic prefix

$$\Delta\hat{k} = \frac{1}{2\pi}\arg\left\{\sum_{n=0}^{L-1} r(n)r^{*}(n+N)\right\} \tag{26.29}$$

because the signal samples of the cyclic prefix and its counterpart at offset T are identical except for a phase factor caused by the frequency offset.

26.3.3 Joint Estimation of the Frequency and Timing Offsets

The estimation techniques discussed in Sections 26.3.1 and 26.3.2 deal with the frequency and timing offsets separately. In practice, joint estimation of the two different offsets is often used to improve the estimation accuracy. Let us consider an OFDM symbol with a normalized frequency offset Δk and a timing offset Δn. After passing through a band-pass channel, the complex envelope of the received sequence for the OFDM symbol under consideration can be expressed as[26,38,39]

$$r_n = \frac{1}{\sqrt{N}} \sum_{k=-K}^{K} X_k H_k \cdot e^{j2\pi\frac{(n+\Delta n)(k+\Delta k)}{N}} + w_n \quad n = 0, 1, \ldots, N-1; \quad N \geq 2K+1, \tag{26.30}$$

where w_n is the sample of a complex Gaussian random variable with zero mean and variance $\sigma_w^2 = \frac{1}{2}E[|w_n|^2]$. After the demodulation, the kth element of the DFT output R_k is[26]

$$\begin{aligned} R_k &= \text{DFT}_N\{r_n\} \\ &= X_k H_k \left\{ \frac{\sin(\pi \Delta k)}{N \sin(\pi \Delta k/N)} \right\} \cdot e^{j\pi\Delta k(N-1)/N} \cdot e^{j2\pi k \Delta n/N} \cdot e^{j2\pi \Delta k \Delta n/N} \\ &\quad + I_k + W_k, \end{aligned} \tag{26.31}$$

where I_k denotes ICI caused by the frequency offset,

$$I_k = \sum_{\substack{l=-K \\ l \neq K}}^{K} X_l H_l \left\{ \frac{\sin[\pi \Delta k]}{N \sin[\pi(l-k+\Delta k)/N]} \right\} \cdot e^{j\pi(N-1)(l-k+\Delta k)/N} \cdot e^{j2\pi k \Delta n/N} \cdot e^{j2\pi \Delta k \Delta n/N}, \tag{26.32}$$

and W_k denotes the Gaussian noise component in the frequency domain for the kth subcarrier after the demodulation. Because each ICI sample is the summation of $(N-1) \times N$ samples, and N is usually sufficiently large, the central limit theorem can be used to approximate its statistics. Consequently, the ICI can be regarded as Gaussian distributed. The demodulation decision variable can be expressed as

$$R_k = S_k + V_k, \tag{26.33}$$

where $V_k = W_k + I_k \cdot I_k$ is a two-dimensional Gaussian distributed variable because it is the summation of two independent Gaussian distributed random variables. Because the in-phase and quadrature components of V_k are mutually independent, the joint probability density function (pdf) of the in-phase and quadrature components of the kth subcarrier can be written as[39]

$$q_{\Re[R_k],\Im[R_k]} = \frac{1}{2\pi\sigma_V^2} \exp\left[-\frac{(\alpha - \Re[S_k])^2 + (\beta - \Im[S_k])^2}{2\pi\sigma_V^2} \right], \tag{26.34}$$

where $\Re[R_k]$, $\Im[R_k]$ and $\Re[S_k]$, $\Im[S_k]$ denote the real and imaginary parts of R_k and S_k, respectively. Now convert the above equation to a polar coordinate system to obtain the magnitude and phase information, Γ_k and Φ_k; we have

$$q_{\Gamma_k,\Phi_k} = \frac{\gamma_k}{2\pi\sigma_V^2} \exp\left[-\frac{\gamma_k^2 + \varepsilon_k - 2\gamma_k\sqrt{\varepsilon_k}\cos\varphi_k}{2\sigma_V^2} \right], \tag{26.35}$$

where

$$\varphi_k = \tan^{-1}\left(\frac{\Im[R_k]}{\Re[R_k]}\right) - \tan^{-1}\left(\frac{\Im[S_k]}{\Re[S_k]}\right), \tag{26.36}$$

$$\gamma_k = |R_k| = \sqrt{(\Re[R_k])^2 + (\Im[R_k])^2}, \tag{26.37}$$

and

$$\varepsilon_k = |S_k|^2. \tag{26.38}$$

The pdf of the phase φ_k can be obtained by integrating the above equation with respect to γ_k[38,39]:

$$q_{\Phi_k} = \frac{e^{-\lambda_k^2}}{2\pi} + \lambda_k \frac{\cos\varphi_k}{2\sqrt{\pi}} e^{-(\lambda \sin\varphi_k)^2}\{1 + \operatorname{erf}(\lambda_k \cos\varphi_k)\}, \tag{26.39}$$

where

$$\lambda_k = \sqrt{\frac{\varepsilon_k}{2\sigma_V^2}}. \tag{26.40}$$

The following pdf $p(\varphi_k)$ well approximates q_{Φ_k},[39] which is symmetric and bell shaped over the range of interest of λ_k. That is,

$$p(\varphi_k) = \frac{1}{\sigma_{\varphi_k}\sqrt{2\pi}} \exp\left[-\frac{\varphi_k^2}{2\sigma_{\varphi_k}^2}\right], \tag{26.41}$$

with

$$\sigma_{\varphi_k} = \sqrt{\frac{\sigma_V^2}{\varepsilon_k}}. \tag{26.42}$$

The frequency and timing offsets after the acquisition are to be estimated by demodulating the synchronization preamble or the pilot tones. The observed phase of the rotated constellation in the demodulated OFDM synchronization symbol for the kth subcarrier with frequency and timing offset is represented as

$$\phi_k = \frac{\pi\Delta k(N-1) + 2\pi k\Delta n + 2\pi\Delta k\Delta n}{N} + \varphi_k, \tag{26.43}$$

where φ_k is the phase shift error due to ICI and Gaussian noise. As the product of the frequency and timing offsets is usually very small in practice, Equation 26.43 can be well approximated as

$$\phi_k \approx \frac{\pi\Delta k(N-1) + 2\pi k\Delta n}{N} + \varphi_k. \tag{26.44}$$

Define a linear observation model for frequency and timing offset estimation, in the form of

$$\bar{\phi} = H\bar{\theta} + \bar{\varphi}, \tag{26.45}$$

where

$$\bar{\theta} = \left[\frac{\pi\Delta k(N-1)}{N} \quad \frac{2\pi\Delta n}{N}\right]^{\mathrm{T}} \tag{26.46}$$

$$H = \begin{bmatrix} 1 & 1 & \cdots & 1 \\ -K & -K+1 & \cdots & K \end{bmatrix}^{\mathrm{T}}, \tag{26.47}$$

with $[\,]^{\mathrm{T}}$ being the transpose of the matrix. Then, the least squares line fitting of q is given by[40]

$$\bar{\theta} = (H^{\mathrm{T}}H)^{-1}H^{\mathrm{T}}\bar{\phi} = \begin{bmatrix} \frac{\sum_{k=-K}^{K}\phi_k}{2K+1} \\ \frac{\sum_{k=-K}^{K}k\phi_k}{\sum_{k=-K}^{K}k^2} \end{bmatrix}. \tag{26.48}$$

This estimator is unbiased, that is, the mean of the estimated frequency and timing offsets is zero. Therefore, the residual frequency and timing offsets after synchronization can be regarded as Gaussian random variables with the same variances as the estimated frequency and timing offsets $\Delta\hat{k}$ and $\Delta\hat{n}$, which can be found through the above equation. Let

$$\Delta k' = \Delta k - \Delta\hat{k} \tag{26.49}$$

and

$$\Delta n' = \Delta n - \Delta\hat{n}. \tag{26.50}$$

The variances of the residual frequency and timing offsets can be easily determined as[40]

$$\sigma_{\Delta k'}^2 = \frac{N^2\sum_{k=-K}^{K}\sigma_{\varphi_k}^2}{\pi^2(N-1)^2(2K+1)^2} \tag{26.51}$$

and

$$\sigma_{\Delta n'}^2 = \frac{N^2\sum_{k=-K}^{K}k^2\sigma_{\varphi_k}^2}{4\pi^2\left(\sum_{k=-K}^{K}k^2\right)^2}. \tag{26.52}$$

26.3.3.1 OFDM System Performance with Residual Frequency and Timing Offsets

After the estimation and compensation of the frequency and timing offsets, residual offsets still exist in both frequency and time domains due to the presence of the interference during the synchronization process. To evaluate the impact of such residual offsets on the OFDM system performance, denote the residual frequency and timing offsets as $\Delta k' = \Delta k - \Delta\hat{k}$ and $\Delta n' = \Delta n - \Delta\hat{n}$, respectively. Consider the demodulated output of the synchronized OFDM receiver,

$$R_k = X_kH_k \cdot \left\{\frac{\sin(\pi\Delta k')}{N\sin(\pi\Delta k'/N)}\right\} \cdot e^{j\pi\Delta k'(N-1)/N} \cdot e^{j2\pi k\Delta n'/N} \cdot e^{j2\pi\Delta k'\Delta n'/N} + I_k + W_k, \tag{26.53}$$

where

$$I_k = \sum_{\substack{l=-K \\ l \neq K}}^{K} (X_l H_l) \cdot \left\{ \frac{\sin(\pi \Delta k')}{N \sin(\pi(l-k+\Delta k')/N)} \right\} \cdot e^{j\pi(N-1)(l-k+\Delta k')/N} \cdot e^{j2\pi k \Delta' n/N} \cdot e^{j2\pi \Delta k' \Delta n'/N}. \quad (26.54)$$

Before a decision is made, the decision variable is usually normalized according to the modulation scheme at the transmitter side as

$$\begin{aligned} R_k &= X_k \cdot e^{j\pi\Delta k'(N-1)/N} \cdot e^{j2\pi k \Delta' n/N} \cdot e^{j2\pi \Delta k' \Delta n'/N} + \frac{I_k + W_k}{H_k\{\sin(\pi\Delta k')/(N \sin(\pi \Delta k'/N))\}} \\ &= X_k + n_{\text{syn}} + I'_k + W'_k, \end{aligned} \quad (26.55)$$

where n_{syn} is defined as

$$n_{\text{syn}} = \sum_{n=1}^{\infty} \frac{[(j\pi\Delta k'(N-1) + j2\pi k \Delta n' + j2\pi \Delta k' \Delta n')/N]^n}{n!} \cdot X_k \quad (26.56)$$

$$I'_k = \sum_{\substack{l=-K \\ l \neq K}}^{K} \frac{X_l H_l}{H_k} \cdot \frac{\sin(\pi\Delta k'/N)}{\sin(\pi(l-k+\Delta k')/N)} \cdot e^{-j\pi(l-k)/N} \cdot e^{j\pi\Delta k'(N-1)/N} \cdot e^{j2\pi k \Delta n'/N} \cdot e^{j2\pi \Delta k' \Delta n'/N} \quad (26.57)$$

and

$$W'_k = \frac{W_k N \sin(\pi \Delta k'/N)}{H_k \sin(\pi \Delta k')}. \quad (26.58)$$

Because $\Delta k'$ and $\Delta n'$ are usually very small, the variance of I'_k and V'_k can be approximated as ($\langle \cdot \rangle$ is the time average operator)

$$\begin{aligned} \sigma^2_{I_{k'}} &= \sum_{\substack{l=-K \\ l \neq k}}^{K} \left\langle \left(\frac{X_l H_l}{H_k} \right)^2 \right\rangle \cdot \left\langle \frac{\sin^2(\pi\Delta k'/N)}{\sin^2(\pi(l-k+\Delta k')/N)} \right\rangle \\ &\approx \sum_{\substack{l=-K \\ l \neq k}}^{K} \left\langle \left(\frac{X_l H_l}{H_k} \right)^2 \right\rangle \cdot \left\langle \frac{\pi^2 \sigma^2_{\Delta k'}}{N^2 \sin^2(\pi(l-k+\Delta k')/N)} \right\rangle \end{aligned} \quad (26.59)$$

$$\sigma^2_{W_{k'}} = \frac{\sigma^2_W}{|H_k|^2} \quad (26.60)$$

$$\sigma^2_{n_{\text{syn}}} = \left| \sum_{n=1}^{\infty} \frac{[(j\pi\Delta k'(N-1) + j2\pi k \Delta n' + j2\pi \Delta k' \Delta n')/N]^n}{n!} \right|^2 E(X_k). \quad (26.61)$$

It is obvious in Equations 26.55 through 26.61 that the decision variable R_k can be reduced as the summation of the desired signal component, and that the variance of Gaussian noise can be determined from Equations 26.59 through 26.61.

To study the impact of $\Delta k'$ and $\Delta n'$ on the symbol error rate (SER) performance, an OFDM system with QAM modulation scheme is considered. For rectangular signal constellations with $L = 2^{B_k}$, the QAM signal is equivalent to two PAM signals modulated on quadrature carriers, each with $\sqrt{L} = 2^{B_k/2}$

signal levels.[19] Because the signals in the in-phase and quadrature components can be perfectly separated at the demodulator, the error probability for QAM is easily determined from the probability of error for the constituent PAM signals. Specifically, the SER of the $\sqrt{L}$-ary PAM for the kth subcarrier can be estimated by[39]

$$P'_k(x) = 2\left(1 - \frac{1}{\sqrt{L}}\right)Q\left[\sqrt{\frac{3}{L-1}\frac{P_s(k)}{\left(\sigma^2_{I'_k} + \sigma^2_{W'_k} + \sigma^2_{n_{\text{syn}}}\right)}}\right], \tag{26.62}$$

where $Q(\alpha)$ is the error function

$$Q(\alpha) = \int_{\alpha}^{\infty} \frac{1}{\sqrt{2\pi}} e^{-\frac{y^2}{2}} dy \tag{26.63}$$

and $P_s(k)$ is the signal power in the decision variable of $\sqrt{L}$-ary PAM:

$$P_s(k) = E\{a_k^2\} = \frac{1}{2}E\{a_k^2 + b_k^2\} \tag{26.64}$$

The SER for the kth subcarrier of the OFDM system is

$$\begin{aligned} P_k(x) &= 1 - (1 - P'_k)^2 \\ &\approx 4\left(1 - \frac{1}{\sqrt{L}}\right)Q\left[\sqrt{\frac{3}{L-1}\frac{P_s(k)}{\left(\sigma^2_{I'_k} + \sigma^2_{W'_k} + \sigma^2_{n_{syn}}\right)}}\right]. \end{aligned} \tag{26.65}$$

The overall SER with the impact of residual synchronization error can then be evaluated from the error probability contribution for each subcarrier, $P_e(k)$, as

$$P_e = \frac{1}{N}\sum_{k=-K}^{K} P_e(k). \tag{26.66}$$

The joint estimation of frequency and timing offsets as well as the impact of the residual synchronization errors has been studied in Wang et al.[39] The root mean square (RMS) values of the residual estimation errors against the received SNR are shown in Figures 26.15 and 26.16, respectively. At high SNRs, it is observed that the estimation error levels off where the ICI is dominant over Gaussian noise due to the loss of orthogonality among the OFDM subcarriers. The SER of different subcarriers in the presence of the residual frequency and timing offsets is plotted in Figures 26.17 and 26.18 for 256 OFDM subcarriers. The initial frequency and timing offsets from the coarse acquisition, Δn and Δk, are set to moderate values.[41] As expected, the impact of the residual offsets on the OFDM system performance is different from one subcarrier to another. The higher the subcarrier index is (with reference to the center of the channel), the poorer the performance that the subcarrier has. Note that the reference of the subcarrier index in Figures 26.17 and 26.18 is at the center of the channel. It can be seen, from Figures 26.17 and 26.18, that the SER increment among all the subcarriers increases with the SNR, that is, the sensitivity of the OFDM system to synchronization offsets increases with the SNR. A minimum of the SER can also be

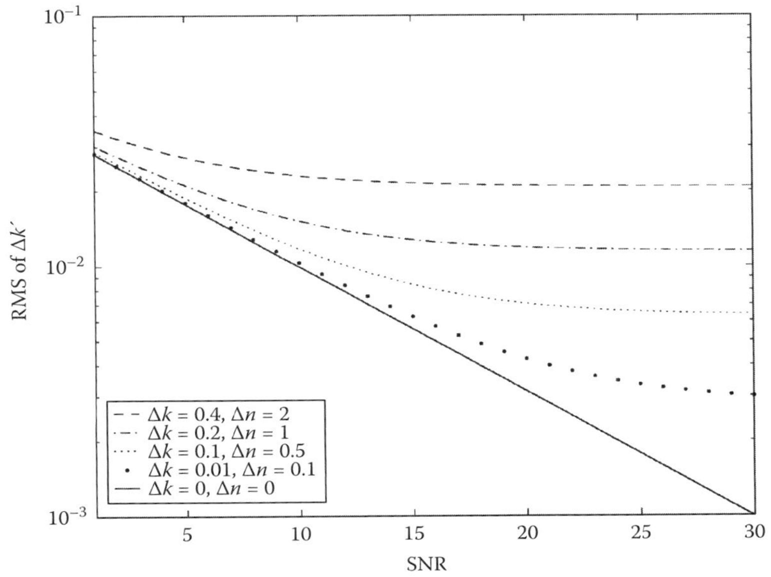

FIGURE 26.15 RMS values of residual estimation errors against SNR.

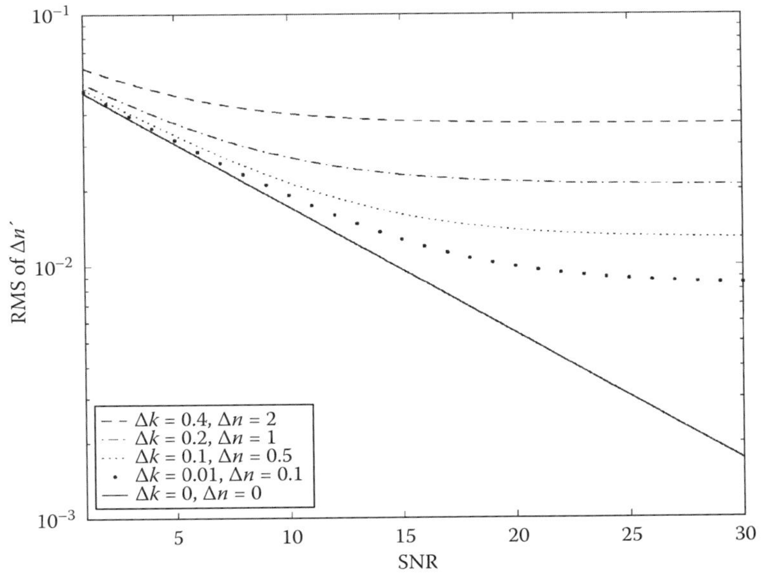

FIGURE 26.16 RMS values of residual estimation errors against SNR.

observed in the middle of the subcarriers. This is based on the assumption that the carrier frequency is in the middle of the OFDM signal spectrum. Variances of residual frequency and timing offsets should be determined using Equations 26.51 and 26.52. It is also observed that the overall SER of the OFDM system is dominated by the subcarriers having larger indexes. The system performance is simulated based on the

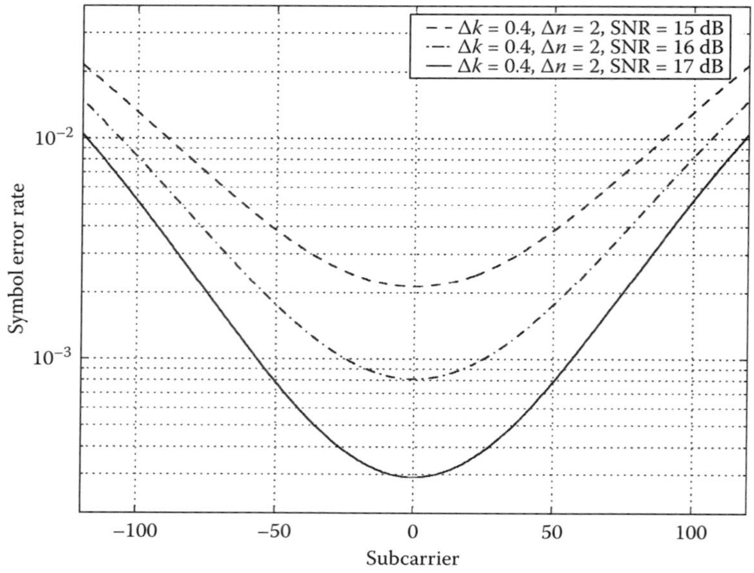

FIGURE 26.17 Sensitivity of OFDM system to synchronization offsets versus SNR.

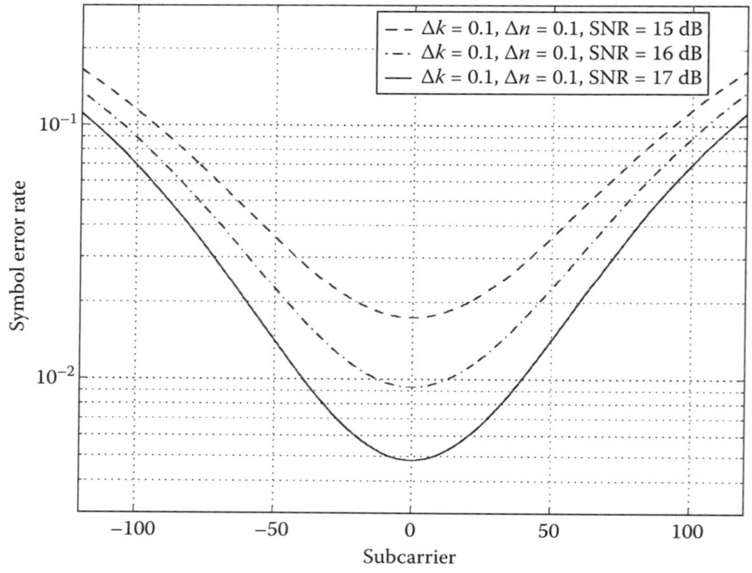

FIGURE 26.18 Sensitivity of OFDM system to synchronization offsets versus SNR.

assumption of an ideal frequency domain equalizer to remove $A(k)$. An imperfect frequency domain equalizer will lead to deterioration of the system performance. Because the sensitivity of the OFDM system to synchronization offsets increases with SNR, higher-order QAM modulation constellations, which require higher SNR, need better and more accurate synchronization systems.

26.3.4 Fast Synchronization for DVB-H System

Various synchronization algorithms have been discussed in the previous sections. As an OFDM-based system, DVB-H can achieve synchronization using these techniques. However, an extra synchronization requirement is needed for the optimal performance of burst transmission of DVB-H. In this section, a fast synchronization technique for DVB-H system is discussed.[42]

One major problem for the design of DVB-H terminals is the limited battery life. Being compatible with DVB-T would place a burden on the DVB-H terminal because demodulating and decoding a broadband, high-data-rate stream like the DVB-T stream involves constant power dissipation for the tuner and demodulator. A considerable drawback for battery-operated DVB-T terminals is that the whole data stream has to be decoded before any one of the multiplexed data streams can be accessed. In DVB-H, service multiplexing is performed in a pure time division multiplex fashion. The data for a particular service is therefore not transmitted continuously, but instead in short periodic bursts with interruptions in between. This bursty signal can be received time selectively: The terminal synchronizes to the bursts of the wanted service but switches to a power-save mode during the intermediate time when other services are transmitted. This technique is called the time-slicing technique, described in Section 26.2.1.3. The power-save time between bursts, relative to the on-time required for the reception of an individual broadcast service, is a direct measure of the power saving provided by the DVB-H receiver. Time slicing offers another benefit for the receiving terminal architecture. The rather long power-save periods may be used to search for channels in neighboring radio cells offering the selected service. This way, a channel handover can be performed at the border between two cells, which remains imperceptible to the user. Both the monitoring of the services in adjacent cells and the reception of the selected service data can then be realized with the same front end.[43]

When the handheld terminal requires a constant lower bit rate, the duration of each burst will be very short. Therefore, the synchronization times of the DVB-H receiver must be rigorously minimized to fully exploit the benefits of time slicing. For the conventional DVB-T receiver, synchronization is usually achieved in two steps: The coarse time acquisition based on cyclic prefix and the removal of the residual timing offset by comparing the demodulated and transmitted pilots. However, identification of pilots is usually based on TPS pilots for the DVB system, which introduces a long delay of up to 68 OFDM symbols. In Schwoerer and Vesma,[43,44] two fast synchronization schemes were proposed for DVB-H based on the scattered pilots in the frequency domain. However, the power-based scattered pilot synchronization may not work effectively in frequency-selective channels. For the correlation-based scattered pilot approach, four OFDM symbols are needed before the synchronization can be achieved.

In Wang et al.,[42] a fast time synchronization technique for DVB-H using only one OFDM symbol is proposed. This new technique is based on the time domain correlation between the received signal and four local reference symbols generated from in-band pilots, including the scattered pilots and continual pilots, as shown in Figure 26.19. To reduce the computational complexity, a coarse time window can be derived from the cyclic prefix. The proposed time synchronization method can be

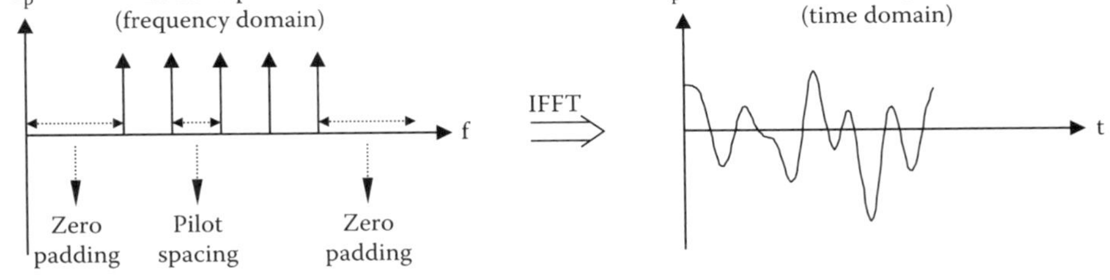

FIGURE 26.19 Fast time synchronization method for DVB-H.

achieved in two steps: A coarse OFDM symbol synchronization using the cyclic prefix and a pilot symbol synchronization based on a time sequence generated from scattered pilots. Consider a DVB-H symbol given by the N-point complex modulation sequence:

$$s(n) = \frac{1}{\sqrt{N}} \sum_{k=0}^{N-1} X_k e^{\frac{j2\pi nk}{N}}, \quad n = 0, 1, 2, \ldots, N-1. \tag{26.67}$$

The OFDM symbol consists of N complex sinusoids or subcarriers modulated with the complex data X_k, which can be divided into two different sets: The data symbol to be transmitted and the pilot symbols for channel synchronization and estimation. A cyclic prefix is inserted to protect the OFDM signal from ISI. The cyclic nature of the OFDM signal provides a straightforward way to achieve the coarse time synchronization using

$$C = \sum_{m=0}^{L-1} r(m) r^*(m+N). \tag{26.68}$$

The correlation function in Equation 26.68 has a triangular shape but can be corrupted by other interferences, as illustrated in Figure 26.20.

Once the coarse timing is achieved, the position of embedded pilots has to be identified so that the channel can be estimated and the residual time offset can be derived more accurately using the in-band pilots. Here, the timing offset is obtained from the time domain correlation of the received signal and four local references generated from the scattered pilots. Denote the subcarrier sets for these two symbol sets as data carrier D and pilots p respectively. Equation 26.67 may be reorganized as[42]

$$s(n) = d(n) + p_i(n), \tag{26.69}$$

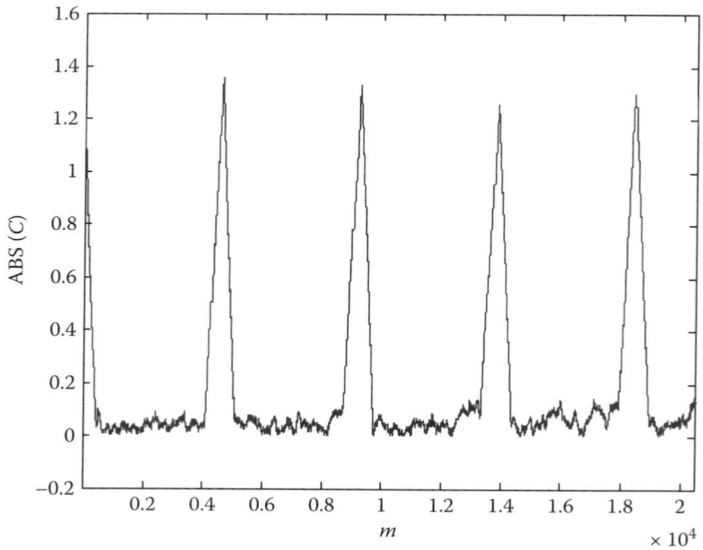

FIGURE 26.20 Coarse time synchronization.

where the data symbols are

$$d(n) = \frac{1}{\sqrt{N}} \sum_{k \in D} X_k e^{j\frac{2\pi nk}{N}} \tag{26.70}$$

and the pilots are

$$p_i(n) = \frac{1}{\sqrt{N}} \sum_{k \in P_i} X_k e^{j\frac{2\pi nk}{N}}, \tag{26.71}$$

where i is the index of the pilot p and ranges from 0 to 3 due to the shifting of the scattered pilots. As a result, p will repeat itself every four OFDM symbols. After passing through a multipath channel characterized by its complex impulse response $h(n)$, the received signal $r(n)$ can be written as

$$r(n) = d(n) \otimes h(n) + p_i(n) \otimes h(n) + w(n). \tag{26.72}$$

Identification of the scattered pilots is based on the correlation of the received signal and the four local pilot sequences p. When the correct local reference is selected, a correlation peak will be observed. Under this condition, the correlation function between the received signal $r(n)$ and the time domain pilot sequence $p_i(n)$ is given by[42]

$$R_{rp} = R_{pp} \otimes h(n) + R_{dp} \otimes h(n) + w(n) \otimes h(n), \tag{26.73}$$

where R_{xy} denotes the correlation between signals x and y. Using the central limit theorem (assuming a sufficiently large value of N), R_{dp} can be approximated as a Gaussian-distributed variable with mean zero and a variance of $\sigma_d^2\sigma_p^2/N$. Similarly, the autocorrelation function of $p(n)$ can be formulated as

$$\begin{aligned} R_{pp}(m) &= \frac{1}{N}\sum_{n=0}^{N-1} p_i(n)p_i^*(n-m), \\ R_{pp}(m) &= \begin{cases} \frac{1}{N}\sum\limits_{n=0}^{N-1} p_i(n)p_i^*(n) \approx \sigma_p^2, & m = 0 \\ w(m), & m \neq 0, \end{cases} \end{aligned} \tag{26.74}$$

where $w(m)$ is a Gaussian noise with a variance of σ_p^4/N. Note that the convolution of a Gaussian noise with $h(n)$ is still Gaussian distributed. For convenience of the analysis, we assume that R'_{rp} has the same duration as $h(n)$

$$R'_{rp}(n) \approx w_1(n) + h(n)\sigma_p^2, \tag{26.75}$$

where $w_1(n)$ is the combined interference from the second term in Equation 26.73 and $w(m)$ in Equation 26.74. An example of the correlation function of the scattered pilot sequence is plotted in Figure 26.21.

The scattered pilot position is therefore identified with Equation 26.74 when it achieves its maximum with one of the four local pilot sequences. The channel estimation normalized to the main path can therefore be obtained through the cross-correlation between the received signal and the time domain pilot sequence as

$$\hat{h}(n) = \frac{R'_{rp}(n)}{R'_{rp,\max}} = \frac{R'_{rp}(n)}{h_{\max}\sigma_p^2} + w_2(n), \tag{26.76}$$

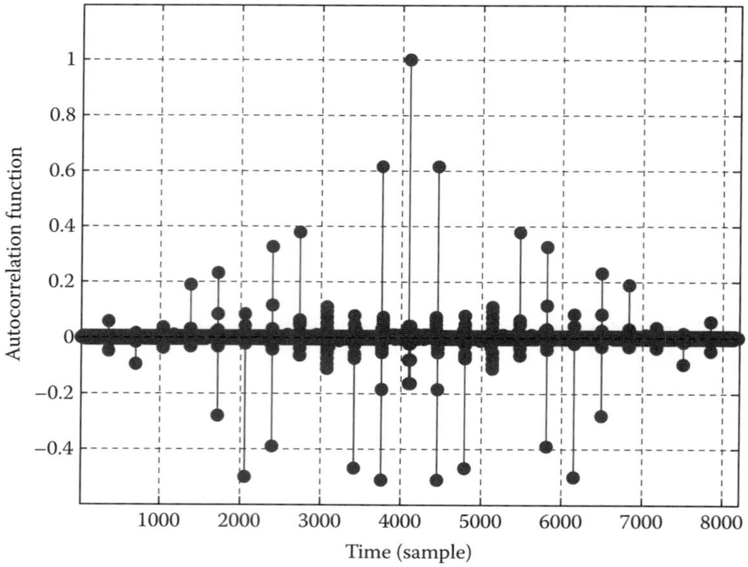

FIGURE 26.21 Correlation function of scattered pilot sequence.

where the subscript "max" denotes the index of the main path, and $w_2(n)$ is the interference term from $w_1(n)$. One example for the proposed fast synchronization technique can be found in Figures 26.22 and 26.23. Once the scattered pilots are identified, the residual timing offset can be easily determined by the phase shift of the FFT output for the scattered pilots.

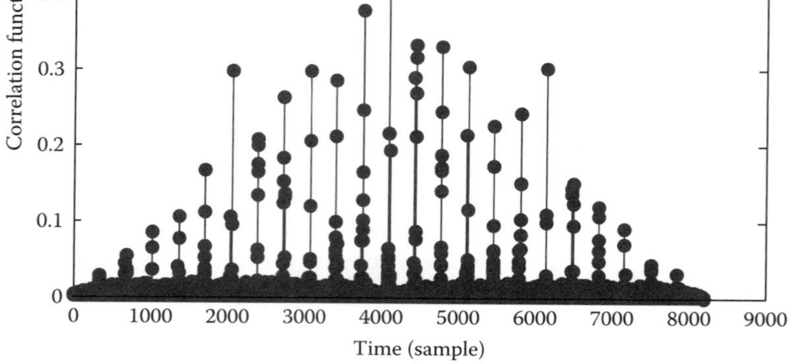

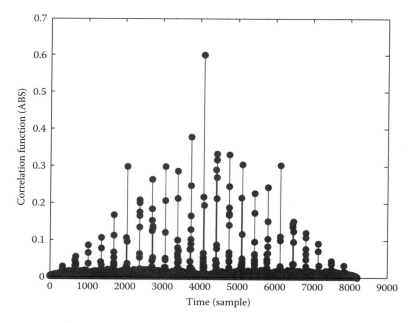

FIGURE 26.22 Fast synchronization technique.

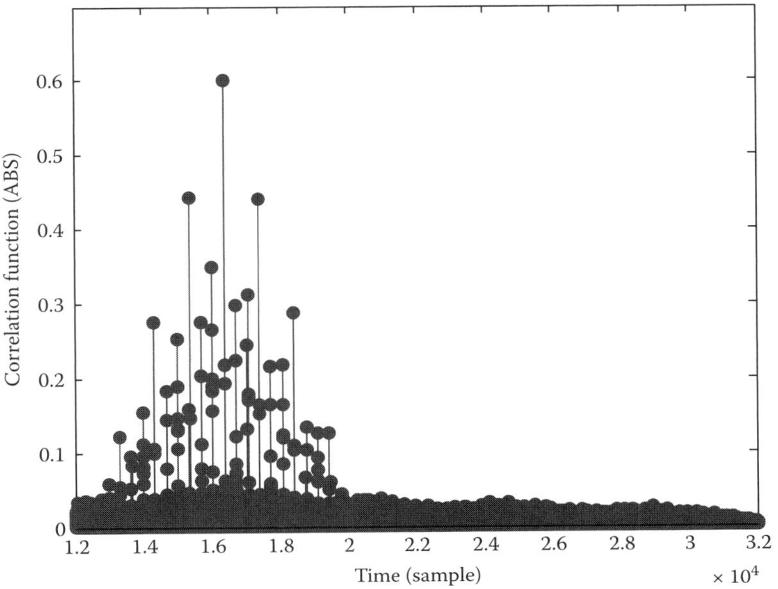

FIGURE 26.23 Fast synchronization technique.

26.4 Summary

OFDM is the primary modulation technique for broadband digital communication systems. The fundamental principles of OFDM system are presented, including the ISI mitigation through cyclic prefix and various channel estimation techniques. One of the principal disadvantages of OFDM systems is their inherent sensitivity to synchronization errors, caused mainly by the so-called frequency and timing offsets. Frequency offset causes a reduction of desired signal amplitude in the output decision variable and introduces ICI due to the loss of orthogonality among subcarriers. Timing offset results in the rotation of the OFDM subcarrier constellation. As a result, an OFDM system cannot recover the transmitted signal without a near-perfect synchronization. In this chapter, the impacts of the synchronization errors, including carrier frequency and timing offsets, were analyzed. Various techniques for the estimation and tracking of the frequency and timing offsets were overviewed and discussed. The effects of the synchronization errors on the system performance of OFDM are also considered. As an application example, the OFDM-based mobile broadcast standards are also overviewed in this chapter. Synchronization techniques for OFDM-based DVB-T and DVB-H broadcast systems are analyzed, with special consideration given to the DVB pilot and frame structures.

References

1. European Telecommunications Standards Institute. 1997. *Radio Broadcasting Systems; Digital Audio Broadcasting (DAB) to Mobile, Portable and Fixed Receivers*. ETS 300 401, 2nd ed. Valbonne, France.
2. European Telecommunications Standards Institute. 1997. *Digital Video Broadcasting (DVB); Framing Structure, Channel Coding and Modulation for Digital Terrestrial Television*. ETS EN 300 744, v.1.1.2.
3. J. A. C. Bingham. 1990. Multicarrier modulation for data transmission: An idea whose time has come. *IEEE Communications Magazine* 28:5–14.

4. W. Y. Zou and Y. Wu. 1995. COFDM: An overview. *IEEE Transactions on Broadcasting* 41:1–8.
5. S. B. Weinstein and P. M. Ebert. 1971. Data transmission by frequency-division multiplexing using the discrete Fourier transform. *IEEE Transactions on Communications* 19:628–634.
6. A. Peled and A. Ruiz. 1980. Frequency domain data transmission using reduced computational complexity algorithms. In *Proceedings of the IEEE International Conference on Acoustics, Speech, and Signal Processing (ICASSP'80)*, Denver, CO, pp. 964–967.
7. ANSI. 1995. *Network and Customer Installation Interfaces—Asymmetric Digital Subscriber Line (ADSL) Metallic Interface*. ANSI standard T1.413.
8. R. Wesel. 1995. Fundamentals of coding for broadcast OFDM. In *Proceedings of the 29th Asilomar Conference on Signals, Systems & Computers*, ACM, Pacific Grove, CA, pp. 2–6.
9. B. R. Saltzberg. 1998. Comparison of single-carrier and multitone digital modulation for ADSL applications. *IEEE Communications Magazine* 36:114–121.
10. H. F. Harmuth. 1960. On the transmission of information by orthogonal time functions. *AIEE Transactions* I:248–255.
11. R. W. Chang. 1966. Synthesis of band-limited orthogonal signals for multichannel data transmission. *Bell System Technical Journal* 45:1775–1796.
12. B. R. Saltzberg. 1967. Performance of an efficient parallel data transmission system. *IEEE Transactions on Communications Technology* COM-15:805–811.
13. R. W. Chang and R. A. Gibby. 1968. Theoretical study of performance of an orthogonal multiplexing data transmission scheme. *IEEE Transactions on Communications* 16:529–540.
14. B. Hirosaki. 1981. An orthogonally multiplexed QAM system using the discrete Fourier transform. *IEEE Transactions on Communications* 29:982–989.
15. Y. Wu and W. Y. Zou. 1995. Orthogonal frequency division multiplexing: A multi-carrier modulation scheme. *IEEE Transactions on Consumer Electronics* 41:392–399.
16. A. V. Oppenheim and R. W. Schafer. 1989. *Discrete Signal Processing*. Englewood Cliffs, NJ: Prentice Hall.
17. T. de Couasnon, R. Monnier, and J. B. Rault. 1994. OFDM for digital TV broadcasting. *Signal Processing* 39:1–32.
18. A. Ruiz, J. M. Cioffi, and S. Kasturia. 1992. Discrete multiple tone modulation with coset coding for the spectrally shaped channel. *IEEE Transactions on Communication* 40:1012–1029.
19. J. G. Proakis. 1995. *Digital Communications*. 3rd edn. New York: McGraw-Hill.
20. *Digital Video Broadcasting (DVB): Transmission System for Handheld Terminals (DVB-H)*. ETSI EN 302 304, v.1.1.1.
21. M. R. Chari, F. Ling, A. Mantravadi, R. Krishnamoorthi, R. Vijayan, G. K. Walker, and R. Chandhok. 2007. FLO physical layer: An overview. *IEEE Transactions on Broadcasting* 53:145–160.
22. S. Cho, G. Lee, B. Bae, K. Yang, C.-H. Ahn, S.-I. Lee, and C. Ahn. 2007. System and services of terrestrial digital multimedia broadcasting (T-DMB). *IEEE Transactions on Broadcasting* 53:171–178.
23. M. Kornfeld and G. May. 2007. DVB-H and IP datacast-broadcast to handheld devices. *IEEE Transactions on Broadcasting* 53:161–170.
24. T. Pollet, M. van Bladel, and M. Moeneclaey. 1995. BER sensitivity of OFDM systems to carrier frequency offset and Wiener phase noise. *IEEE Transactions on Communications* 43:191–193.
25. C. Muschallik. 1995. Influence of RF oscillators on an OFDM signal. *IEEE Transactions on Consumer Electronics* 41:592–603.
26. P. H. Moose. 1994. A technique for orthogonal frequency division multiplexing frequency offset correction. *IEEE Transactions on Communications* 42:2908–2914.
27. L. Wei and C. Schlegel. 1995. Synchronization requirements for multiuser OFDM on satellite mobile and two-path Rayleigh fading channels. *IEEE Transactions on Communications* 43:887–895.

28. A. Garcia Armada and M. Calvo. 1998. Phase noise and sub-carrier spacing effects on the performance of an OFDM communication system. *IEEE Communications Letters* 2:11–13.
29. T. Pollet, P. Spruyt, and M. Moeneclaey. 1994. The BER performance of OFDM systems using non-synchronized sampling. In *Proceedings of the IEEE GLOBECOM'94*, San Francisco, CA, pp. 253–257.
30. W. D. Warner and C. Leung. 1993. OFDM/FM frame synchronization for mobile radio data communication. *IEEE Transactions on Vehicular Technology* 42:302–313.
31. F. Classen and H. Meyr. 1994. Frequency synchronization algorithms for OFDM systems suitable for communication over frequency-selective fading channels. In *Proceedings of the IEEE Vehicular Technology Conference* (*VTC'94*), Stockholm, Sweden, pp. 1655–1659.
32. T. M. Schmidl and C. Cox. 1997. Robust frequency and timing synchronization for OFDM. *IEEE Transactions on Communications* 45:1613–1621.
33. P. J. Tourtier, R. Monnier, and P. Lopez. 1993. Multicarrier modem for digital HDTV terrestrial broadcasting. *Signal Processing: Image Communication* 5:379–403.
34. F. Daffara and O. Adami. 1995. A new frequency detector for orthogonal multicarrier transmission techniques. In *Proceedings of the Vehicular Technology Conference* (*VTC'95*), Chicago, IL, pp. 804–809.
35. M. Sandell, J. J. van de Beek, and P. O. Börjesson. 1995. Timing and frequency synchronization in OFDM systems using the cyclic prefix. In *Proceedings of the IEEE International Symposium on Synchronization*, Essen, Germany, pp. 16–19.
36. J. J. van de Beek, M. Sandell, and P. O. Börjesson. 1997. ML estimation of time and frequency offsets in OFDM systems. *IEEE Transactions on Signal Processing* 45:1800–1805.
37. D. Lee and K. Cheon. 1997. A new symbol timing recovery algorithm for OFDM systems. *IEEE Transactions on Consumer Electronics* 43:767–775.
38. K. W. Kang, J. Ann, and H. S. Lee. 1994. Decision-directed maximum-likelihood estimation of OFDM frame synchronization offset. *Electronics Letters* 30:2153–2154.
39. X. Wang, T. T. Tjhung, Y. Wu, and B. Caron. 2003. SER performance evaluation and optimization of OFDM system with residual frequency and timing offsets from imperfect synchronization. *IEEE Transactions on Broadcasting* 49:170–177.
40. J. L. Melsa and D. L. Cohn. 1978. *Decision and Estimation Theory*. New York: McGraw-Hill.
41. H. Minn, M. Zeng, and V.K. Bhargava. 2000. On timing offset estimation for OFDM systems. *IEEE Communications Letters* 4:242–244.
42. X. Wang, Y. Wu, and J.-Y. Chouinard. 2006. A fast synchronization technique for DVB-H System using in-band pilots and cyclic prefix. In *Proceedings of the IEEE ICCE*, pp. 407–408.
43. L. Schwoerer. 2004. Fast pilot synchronization schemes for DVB-H. In *Proceedings of the 4th International Multi-Conference Wireless and Optical Communications*, Banff, Canada, pp. 420–424.
44. L. Schwoerer and J. Vesma. 2003. Fast scattered pilot synchronization for DVB-T and DVB-H. Paper presented at *Proceedings of the 8th International OFDM Workshop*, Germany.

27

Space–Time Block Coding

Mohanned O. Sinnokrot
Georgia Institute of Technology

Vijay K. Madisetti
Georgia Institute of Technology

27.1 Introduction 27-1
27.2 System and Channel Model 27-2
27.3 Design Criteria of Space–Time Codes 27-4
27.4 Survey of Space–Time Block Codes 27-5
Orthogonal Space–Time Block Codes • Diagonal Algebraic Space–Time Block Codes • Quasiorthogonal Space–Time Block Codes • Single-Symbol Decodable Space–Time Block Codes • Semiorthogonal Algebraic Space–Time Block Codes • Threaded Algebraic Space–Time Block Codes • Perfect Space–Time Block Codes
27.5 Summary and Comparison of Space–Time Block Codes 27-11
27.6 Future Research Topics 27-14
References 27-14

27.1 Introduction

Multiple-input multiple-output (MIMO) technology is one of the most significant and promising advances in digital communications because it offers higher transmission rates and improved bit-error-rate (BER) performance over single antenna systems. The use of multiple antennas to increase the throughput of the communication channel or increase its reliability has been quantified by defining three related terms: rate, multiplexing gain, and diversity gain. The rate of a communication system defines the number of transmitted symbols per channel use. The multiplexing and diversity gains define how fast the rate increases or how fast the error probability decreases, respectively, with increase in signal-to-noise ratio (SNR) in the high SNR region [1]. For any number of antennas, there is a continuous trade-off between the diversity and multiplexing gains. The diversity gain and code rate, however, can be simultaneously maximized.

A key idea in multiple antenna systems is *space–time coding*, in which the time dimension inherent in digital communications is complemented with the spatial dimension inherent in the use of multiple antennas. A key benefit of space–time coding is the ability to turn multipath propagation, an impairment in single-antenna wireless communications, into a benefit for multiple antenna systems by taking advantage of the random fading in increasing the transmission rate of the communication link. The prospect of improving the wireless communication link and data rates at no cost of extra spectrum is largely responsible for incorporating MIMO technology into commercial wireless products and standards such as W-CDMA, CDMA-2000, IEEE 802.11n (WiFi), and IEEE 802.16e (WiMAX) [2].

The key benefits of multiple antenna systems can be understood by studying two concepts: *spatial diversity* and *spatial multiplexing*. The spatial diversity concept states that the error probability decays or vanishes exponentially with the *diversity order*, which is the number of decorrelated spatial branches between the transmitter and the receiver. The spatial diversity benefit can be understood intuitively by considering a communication system with two transmit antennas and one receive antenna that encodes the data jointly across the two antennas. If one of the paths between the two transmitters and the receiver is in deep fade, the receiver can still recover the transmitted data, if the other path is in a good enough state to allow for reliable detection. The spatial multiplexing concept states that under certain conditions, a number of independent data streams can be sent from each of the transmit antennas such that they are decoded reliably at the receiver. The number of independent data streams is known as the *multiplexing order*, and it is the minimum of the number of transmit antennas and the number of receive antennas. As a result of the two different benefits of MIMO channels, early transmission schemes typically fell into two categories: diversity maximization schemes or rate maximization schemes. However, recent work in space–time code design has shifted away from increasing the diversity order or transmission rate alone to increasing both simultaneously.

The delay diversity scheme of Wittneben [3] inspired the first attempt to develop space–time codes in [4]. The key development of the space–time coding concept is due to Tarokh et al. in [5], in which they proposed *space–time trellis codes*. These space–time trellis codes had a high decoding complexity and required a vector Viterbi algorithm at the receiver for decoding. In addressing the high decoding complexity of space–time trellis codes, *space–time block codes* (STBCs) were discovered. The first STBC is due to Alamouti [6], who invented a remarkable scheme for two transmit antennas. The Alamouti STBC allowed the receiver to harness the spatial diversity of the transmit antennas, while maintaining simple maximum-likelihood (ML) decoding at the receiver.

We will restrict our attention to space–time block coding rather than space–time trellis coding, since block-based designs currently dominate the literature. This is mainly due to the high decoding complexity of space–time trellis codes which, for a fixed number of antennas, grows exponentially as a function of the diversity level and transmission rate.

In the remainder of this chapter, we describe the MIMO communication system model and the channel model. Next, we briefly review the design criteria of STBCs in quasistatic-fading channels. Then, we provide a survey of STBCs and compare them in terms of their code rate and decoding complexity. Finally, we discuss some potential future research topics in the area of space–time block coding.

27.2 System and Channel Model

We consider a MIMO system with M transmit antennas and N receive antennas. The transmitted code word of a STBC can be expressed as a $T \times M$ matrix:

$$C = \begin{bmatrix} c_1[1] & c_2[1] & \cdots & c_M[1] \\ c_1[2] & c_2[2] & \cdots & c_M[2] \\ \vdots & \vdots & \ddots & \vdots \\ c_1[T] & c_2[T] & \cdots & c_M[T] \end{bmatrix},$$

where $c_m[t]$ denotes the symbol transmitted from antenna $m \in \{1, \ldots, M\}$ at time $t \in \{1, \ldots, T\}$. The received signal $y_n[t]$ at receive antenna $n \in \{1, \ldots, N\}$ at time t is given by

$$y_n[t] = \sum_{m=1}^{M} h_{m,n}[t]c_m[t] + w_n[t],$$

where

$w_n[t]$ is the complex additive white Gaussian noise at receive antenna n at time t

$h_{m,n}[t]$ is the channel coefficient between the mth transmit antenna and nth receive antenna at time t

For quasistatic fading, the channel coefficient $h_{m,n}[t] = h_{m,n}$ is independent of time t. The transmitted complex information symbols are drawn from q-ary quadrature amplitude modulation (QAM) alphabet,

q-ary hexagonal (HEX) alphabet, or q-ary phase shift keying (PSK) alphabet. We next define what it means to be a linear STBC and define the code rate and other terms used in describing the code rate.

Definition 27.1: A *linear* STBC is a $T \times M$ matrix whose entries are complex linear combinations of P complex information symbols $x_p = x_p^{\mathrm{R}} + \mathrm{i}x_p^{\mathrm{I}}, p \in \{1, \ldots, P\}$, and their complex conjugates, where the superscripts "R" and "I" denote the real and imaginary components of the complex symbol x_p, respectively.

Definition 27.2: The *code rate* of a $T \times M$ STBC transmitting P complex information symbols over T symbol periods is $\kappa = P/T$ symbols per channel use.

We restrict our attention to linear STBCs because they can be decoded using linear processing algorithms such as the sphere decoding algorithm [7,8], which has an expected complexity that is polynomial in the number of antennas over a wide range of SNR values [9,10]. We next briefly discuss ML decoding at the receiver.

Assuming that perfect channel state information (CSI) is known at the receiver, but not at the transmitter, the ML decoder at the receiver chooses P complex information symbols x_p, $p \in \{1, \ldots, P\}$, that minimize the metric:

$$f(x_1, \ldots, x_p) = \sum_{t=1}^{T} \sum_{n=1}^{N} \left| y_n[t] - \sum_{m=1}^{M} h_{m,n}[t] c_m[t] \right|^2.$$

We next define three terms related to ML decoding.

Definition 27.3: The *decoding complexity* of an ML decoder is the number of metric computations required to reach the ML decision. The decoding complexity cannot exceed q^P metric computations, which is the *worst-case decoding complexity* achieved by an exhaustive search ML decoder.

Definition 27.4: A linear STBC is said to Γ-*group decodable*, if the ML decoding metric can be decoupled into a linear sum of Γ-independent metrics such that each metric consists of the symbols from only one group.

Definition 27.5: A linear STBC is said to be *separable* if $\Gamma > 1$.

One of the important design goals of space–time coding is the construction of codes with reduced decoding complexity such that the worst-case decoding complexity is less than q^P.

Definition 27.6: The worst-case ML decoding complexity of a Γ-group decodable STBC with B symbols in each group is Γq^B, where q is the cardinality of the modulation alphabet.

Before we discuss the design criteria of STBCs, we discuss an example to clarify the definitions of worst-case ML decoding complexity, group decodability and separability of STBCs.

Example 27.1

Consider three 4 × 4 STBCs $\mathbf{C}_X$, $\mathbf{C}_Y$, and $\mathbf{C}_Z$ transmitting four complex information symbols drawn from an arbitrary q-ary alphabet. Assume that the metric function can be written as follows for the three codes:

$$\mathbf{C}_X : f(x_1, \ldots, x_4) = f(x_1) + f(x_2) + f(x_3) + f(x_4)$$

$$\mathbf{C}_Y : f(x_1, \ldots, x_4) = f(x_1, x_2) + f(x_3, x_4)$$

$$\mathbf{C}_Z : f(x_1, \ldots, x_4) = f(x_1, x_2, x_3, x_4)$$

According to the definitions, the STBC $\mathbf{C}_X$ is separable or four-group decodable since we can separate the decoding of the transmitted symbols into four groups, each containing one complex symbol ($\Gamma = 4$, $B = 1$). Similarly, STBC $\mathbf{C}_Y$ is separable or two-group decodable since we can separate the decoding of the transmitted symbols into two groups, each containing two complex symbols ($\Gamma = 2$, $B = 2$). Finally, STBC $\mathbf{C}_Z$ is not separable since all the symbols have to be decoded jointly ($\Gamma = 1$, $B = 4$). Therefore, STBC $\mathbf{C}_X$ has the lowest decoding complexity, followed by $\mathbf{C}_Y$ and then $\mathbf{C}_Z$, which does not offer any reduction in complexity. Furthermore, the worst-case ML decoding complexity of $\mathbf{C}_X$, $\mathbf{C}_Y$, and $\mathbf{C}_Z$ is $4q$, $2q^2$, and q^4, respectively.

27.3 Design Criteria of Space–Time Codes

The maximum spatial diversity order for the M-input N-output channel is the product of the number of transmit and the number of receive antennas, namely MN. The following two design criteria for space–time codes over quasistatic-fading channels were derived in [5]:

- *Rank criterion*: To achieve diversity gain of MN, the difference matrix $\Delta\mathbf{C} = \mathbf{C} - \tilde{\mathbf{C}}$ must have full rank (i.e., rank M) for any pair of distinct code words $\mathbf{C}$ and $\tilde{\mathbf{C}}$.
- *Determinant criterion*: To further optimize performance, a code that has full rank should be chosen to maximize the asymptotic coding gain:

$$\delta_{\text{gain}} = \min_{\mathbf{C} \neq \tilde{\mathbf{C}}} \det(\mathbf{C} - \tilde{\mathbf{C}})^*(\mathbf{C} - \tilde{\mathbf{C}}).$$

Both the diversity and coding gain improve the system performance by decreasing the error rate, but they do so in two different ways. While the diversity gain improves system performance by increasing the magnitude of the slope of the error-rate curve, the coding gain shifts the error-rate curve to the left. Furthermore, the SNR advantage due to diversity gain increases with increase in SNR, but remains constant with the coding gain at high-enough SNR. A schematic highlighting the differences between the diversity and coding gain in improving performance is shown in Figure 27.1 [11]. We next give two definitions related to the rank and determinant criteria.

Definition 27.7: A STBC is said to be *fully diverse* or has *full diversity*, if the difference matrix $\Delta\mathbf{C} = \mathbf{C} - \tilde{\mathbf{C}}$ has rank M for any pair of distinct code words $\mathbf{C}$ and $\tilde{\mathbf{C}}$.

Definition 27.8: A STBC is said to have the *nonvanishing determinant* (NVD) property, if the coding gain for a fully diverse code does not decay to zero as the constellation size grows.

Therefore, according to the previous discussion, a STBC that is fully diverse will exhibit improved performance due to the increase in the slope of the error-rate curve. A STBC satisfying the NVD property will maintain its good performance as we increase the spectral efficiency of the code by using higher-order modulations.

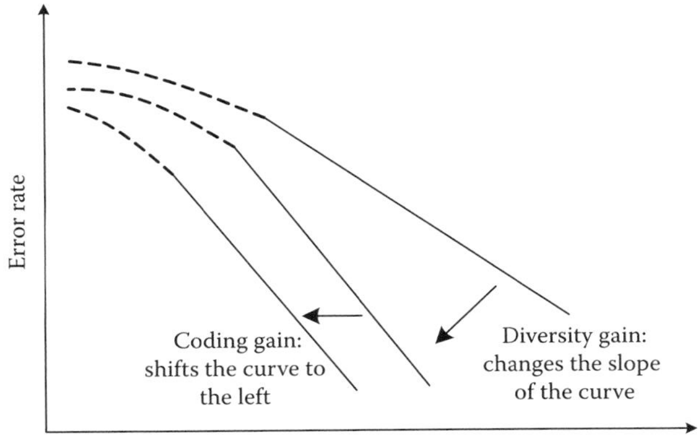

FIGURE 27.1 Schematic showing difference between coding gain and diversity gain.

Design criteria similar to the rank and determinant criteria were proposed for rapid-fading channels in [5] and are known as the distance and product criteria. We remark that space–time codes that are designed for the quasistatic channel achieve at least the same diversity gain in rapid-fading channels. This fact was proven in Theorem 6.1 in [12] and ensures that a well-designed space–time code for quasistatic fading will continue to perform well in rapid fading since it maintains the same diversity gain. However, the space–time codes discussed in this survey do not exploit the temporal diversity of the rapid-fading channel since they are designed explicitly for the quasistatic-fading channel. Examples of space–time codes designed for the rapid-fading channel can be found in [13,14]. These codes, however, suffer from a high decoding complexity and as noted in [13], simple decoding methods are needed for these codes to be practical. As a result, we restrict our discussion to the STBCs designed for the quasistatic-fading channel.

27.4 Survey of Space–Time Block Codes

Having presented the channel model and design criteria of STBCs, we next provide a survey of the different constructions of STBCs in literature and compare them in terms of their code rate, delay, and decoding complexity. We review some of the most important constructions of STBCs including orthogonal designs, quasiorthogonal designs, diagonal algebraic and threaded algebraic STBCs, perfect space–time codes, single-symbol decodable (SSD) codes, and semiorthogonal algebraic space–time (SAST) codes. We begin our survey with a discussion of orthogonal STBCs.

27.4.1 Orthogonal Space–Time Block Codes

Orthogonal STBCs are an important family of linear space–time codes that achieve full diversity, while decoupling the ML detection of the transmitted symbols such that each transmitted symbol is detected separately from the other transmitted symbols. More precisely, an orthogonal STBC is a linear space–time code that satisfies the following orthogonality property:

$$\mathbf{C}^{*}\mathbf{C} = \sum_{p=1}^{P} |x_p|^2 \cdot I$$

The first orthogonal STBC is due to Alamouti [6], who constructed a full-rate STBC for two transmit antennas. Tarokh et al. [15] constructed orthogonal STBCs for real and complex constellations for arbitrary number of antennas. In particular, Tarokh et al. constructed real orthogonal STBCs with full rate for any number of transmit antennas and complex orthogonal STBCs with full rate for two antennas, rate 3/4 for three and four antennas and rate 1/2 for more than four antennas. Orthogonal STBCs for two [6], three, and four transmit antennas [16] are given by:

$$\mathbf{C}_2^{\text{Ortho}} = \begin{bmatrix} x_1 & x_2 \\ -x_2^\star & x_1^\star \end{bmatrix}, \quad \mathbf{C}_3^{\text{Ortho}} = \begin{bmatrix} x_1 & 0 & -x_2^\star \\ 0 & x_1 & -x_3 \\ x_2 & x_3^\star & x_1^\star \\ -x_3 & x_2^\star & 0 \end{bmatrix}, \quad \text{and} \quad \mathbf{C}_4^{\text{Ortho}} = \begin{bmatrix} x_1 & 0 & -x_2^\star & x_3^\star \\ 0 & x_1 & -x_3 & -x_2 \\ x_2 & x_3^\star & x_1^\star & 0 \\ -x_3 & x_2^\star & 0 & x_1^\star \end{bmatrix}.$$

The construction of maximal-rate complex orthogonal STBCs was studied by Liang in [17]. Liang not only determined the maximal rate achieved by complex orthogonal designs for arbitrary number of antennas, but also gave a systematic construction of maximal-rate complex orthogonal STBCs. Furthermore, the constructed codes were also shown to be delay optimal for six transmit antennas or less. A delay optimal orthogonal design refers to a design that achieves the maximal rate while minimizing the code length T. Delay optimality is a desirable property because it minimizes the decoding delay at the receiver since the receiver has to receive the entire block before it starts decoding it. The maximal rate of a complex orthogonal STBC with M transmit antennas is given by [17]:

$$\kappa = \frac{\lceil M/2 \rceil + 1}{2\lceil M/2 \rceil}.$$

The minimal delay for the maximal-rate complex orthogonal designs for M antennas is given by [17,18]:

$$D_{\min} = \begin{cases} \frac{1}{2\kappa}\binom{M}{\lceil M/2 \rceil} & M = 4l \\ \frac{1}{\kappa}\binom{M}{\lceil M/2 \rceil} & M \neq 4l \end{cases}.$$

We give a summary of the maximal rate and the minimum delay for complex orthogonal STBCs for 2–16 transmit antennas in Table 27.1. As can be seen from Table 27.1, not only does the code rate tend to 1/2

TABLE 27.1 Maximal Rate and Decoding Delay of Complex Orthogonal STBCs

N	κ	$D_{\min}$
2	1	2
3	3/4	4
4	3/4	4
5	2/3	15
6	2/3	30
7	5/8	56
8	5/8	56
9	3/5	210
10	3/5	420
11	7/12	792
12	7/12	792
13	4/7	3,003
14	4/7	6,006
15	9/16	11,440
16	9/16	11,440

as the number of antennas increases, but also the code length becomes prohibitively large for practical implementations. These two drawbacks motivated the construction of diagonal algebraic STBCs and quasiorthogonal STBCs, which we discuss next.

27.4.2 Diagonal Algebraic Space–Time Block Codes

Diagonal algebraic space–time block (DAST) codes are a family of linear space–time codes constructed by the use of rotated constellations [19,20]. The word algebraic in the description of DAST codes comes from the fact that rotation matrices used in DAST codes were constructed using algebraic number field theory [21,22]. The word diagonal refers to the structure of the code matrix, wherein the rotated information symbols are spread over the diagonal of the square code matrix. The DAST block codes not only achieve full-diversity order for arbitrary number of transmit antennas, but they also achieve rate-one transmission. For M transmit antennas, the DAST block codes transmit M information symbols in M signal intervals as follows:

$$\mathbf{C}_M^{\text{DAST}} = \begin{bmatrix} u_1 & 0 & \cdots & 0 \\ 0 & u_2 & \cdots & 0 \\ \vdots & \ddots & \ddots & \vdots \\ 0 & \cdots & 0 & u_M \end{bmatrix},$$

where

$(u_1, u_2, \ldots, u_M)^{\text{T}} = \Phi_M\,(x_1, x_2, \ldots, x_M)^{\text{T}}$

Φ_M is the complex or real rotation matrix that ensures a full-diversity DAST code

The DAST block codes were developed for q-QAM modulations and outperform complex orthogonal designs when the number of transmit antennas is greater than 2 (i.e., $M > 2$). The cost of improved performance compared to complex orthogonal designs is increased decoding complexity. In general, DAST block codes with QAM modulation and complex rotation matrices require the joint detection of $2M$ real PAM symbols, or equivalently M complex QAM symbols. Therefore, the DAST codes with complex rotation matrices do not offer any reduction in decoding complexity. With real rotation matrices, however, the DAST codes require the joint detection of M real symbols for M transmit antennas. Although complex rotation matrices result in higher decoding complexity than real rotations, they have a better BER performance. For example, for four transmit antennas, complex rotation matrices offer 0.7 dB gain over real rotations. However, it was concluded in [20] that real rotations are better in terms of the complexity-performance tradeoff since the slight increase in performance might not justify the significant increase in decoding complexity.

27.4.3 Quasiorthogonal Space–Time Block Codes

Quasiorthogonal space–time codes relax the orthogonality constraint to enable rate-one transmission, at the expense of an increase in decoding complexity. For example, quasiorthogonal codes for four antennas were proposed independently by Jafarkhani [23], Tirkkonen et al. [24], and Papadias and Foschini [25]; these rate-one codes have two drawbacks: they are not fully diverse, and they require pair-wise complex symbol decoding.

The first drawback of quasiorthogonal STBCs was eliminated through constellation rotation. For example, rate-one and full-diversity quasiorthogonal codes with rotation were proposed by Tirkkonen [26], Sharma and Papadias [27], Su and Xia [28], and Wang and Xia [29]. While these quasiorthogonal codes outperformed orthogonal codes at all spectral efficiencies for four transmit antennas, they required pair-wise complex symbol decoding. Full-diversity rate-one quasiorthogonal codes for an arbitrary number of antennas were proposed by Sharma and Papadias [30]. The quasiorthogonal STBC for M transmit antennas, where M is a power of 2, is given by

$$\mathbf{C}_M^{\text{Quasi}}(x_1,\ldots,x_M) = \begin{bmatrix} \mathbf{C}_{M/2}^{\text{Quasi}}(y_1,\ldots,y_{M/2}) & \mathbf{C}_{M/2}^{\text{Quasi}}(y_{M/2+1},\ldots,y_M) \\ -\mathbf{C}_{M/2}^{\text{Quasi}}(y_{M/2+1}^*,\ldots,y_M^*) & \mathbf{C}_{M/2}^{\text{Quasi}}(y_1^*,\ldots,y_{M/2}^*) \end{bmatrix},$$

where

$y_m = x_m e^{j\phi_m}$ and $m \in \{1,\ldots,M\}$ are the rotated information symbols

$\mathbf{C}_{M/2}^{\text{Quasi}}(y_1,\ldots,y_{M/2})$ and $\mathbf{C}_{M/2}^{\text{Quasi}}(y_{M/2+1},\ldots,y_M)$ are the STBCs for $M/2$ transmit antennas in the rotated symbols y_1 through $y_{M/2}$ and symbols $y_{M/2+1}$ through y_M, respectively

The rotation angles are chosen to ensure full diversity and maximize the coding gain. By definition, $\mathbf{C}_1^{\text{Quasi}}(x_1) = x_1$, and the quasiorthogonal code for two transmit antennas is the orthogonal Alamouti STBC. For M transmit antennas, where M is not a power of two, the quasiorthogonal space–time code can be obtained by deleting any $L - M$ columns of $\mathbf{C}_L^{\text{Quasi}}(x_1,\ldots,x_L)$, where $L = 2^{\lceil \log_2 M \rceil}$. For example, the quasiorthogonal STBC for three antennas is obtained by deleting any one column from the code matrix for four antennas. The STBCs for three and four transmit antennas are then given by

$$\mathbf{C}_3^{\text{Quasi}} = \begin{bmatrix} x_1 & x_2 & x_3 e^{j\phi} \\ -x_2^* & x_1^* & -x_4^* e^{-j\phi} \\ -x_3 e^{j\phi} & -x_4 e^{j\phi} & x_1 \\ x_4^* e^{-j\phi} & -x_3^* e^{-j\phi} & -x_2^* \end{bmatrix} \quad \text{and} \quad \mathbf{C}_4^{\text{Quasi}} = \begin{bmatrix} x_1 & x_2 & x_3 e^{j\phi} & x_4 e^{j\phi} \\ -x_2^* & x_1^* & -x_4^* e^{-j\phi} & x_3^* e^{-j\phi} \\ -x_3 e^{j\phi} & -x_4 e^{j\phi} & x_1 & x_2 \\ x_4^* e^{-j\phi} & -x_3^* e^{-j\phi} & -x_2^* & x_1^* \end{bmatrix},$$

where $\phi = \pi/4$ is the angle that maximizes the coding gain and ensures full diversity. As can be seen from the equation above, the code for three antennas was obtained by deleting the last column of the code matrix for four antennas.

Because quasiorthogonal STBCs achieve full diversity and rate-one transmission, they outperform orthogonal designs at all spectral efficiencies for complex constellations. The improved performance, however, comes at the expense of increased decoding complexity. In particular, quasiorthogonal STBCs require the joint decoding of $2^{\lceil \log_2 M \rceil - 1}$ complex information symbols for M antennas. For example, for four transmit antennas, quasiorthogonal designs require the joint decoding of two complex information symbols. Similarly, the DAST block codes with real rotations require the joint decoding of four real PAM symbols, or equivalently two complex QAM symbols. For the particular configuration of three or four transmit antennas, a family of linear space–time codes that require the joint decoding of two real symbols, or equivalently one complex symbol, was developed. We next discuss this family of SSD space–time codes.

27.4.4 Single-Symbol Decodable Space–Time Block Codes

SSD STBCs are a family of rate-one, fully diverse nonorthogonal space–time codes for three and four transmit antennas with ML decoding that requires only pair-wise decoding of real symbols [31–34]. Since a pair of real symbols defines a single complex symbol, these codes are said to be SSD.

A framework for the construction of SSD space–time codes was presented in [34]. The encoder decomposes into a concatenation of three steps, as shown in Figure 27.2. The encoder starts with a vector

FIGURE 27.2 Encoding architecture for SSD STBCs.

of information symbols $\mathbf{x} = (x_1, x_2, x_3, x_4)^T$ chosen from a conventional q-ary QAM alphabet. The first step is to distort the alphabet in some way; the codes of [31,32] rotate each alphabet by an angle of ϕ such that $a_p = x_p e^{j\phi}$, while the code of [34] stretches the alphabet by a factor K such that $a_p = Kx_p^R + jx_p^I$. The rotation angle and scaling factor are chosen to maximize the coding gain. The purpose of the rotation or scaling is to ensure full diversity.

The second step is to interleave the coordinates of $\mathbf{a}$, yielding $\mathbf{s} = \Pi(\mathbf{a})$. The interleavers act on the real and imaginary parts separately, so that $[s_1^R, s_1^I, s_2^R, s_2^I, s_3^R, s_3^I, s_4^R, s_4^I] = [a_1^R, a_1^I, a_2^R, a_2^I, a_3^R, a_3^I, a_4^R, a_4^I]\Pi$, where Π is an 8×8 permutation matrix (so that its columns are a permutation of the columns of the identity matrix, with the possibility of sign inversion), and s_p^R and s_p^I denote the real and imaginary parts, respectively, of s_p. The interleaver is used to achieve full diversity while maintaining single-symbol decodability for the codes of [31,32] and also to ensure that the peak-to-average-power ratio is the same as the underlying QAM alphabet for the code of [34]. The final step is to encode $\mathbf{s}$ using a conventional space–time block encoder $\mathbf{G}(\cdot)$, yielding $\mathbf{C} = \mathbf{G}(\mathbf{s})$. We next give an example of the SSD code of [32] for four antennas.

In terms of Figure 27.2, the code of [32] is specified by $\phi = (1/2)\tan^{-1}(2)$, $\Pi = [\mathbf{e}_1, \mathbf{e}_6, \mathbf{e}_3, \mathbf{e}_8, \mathbf{e}_5, \mathbf{e}_2, \mathbf{e}_7, \mathbf{e}_4]$, where $\mathbf{e}_i$ is the ith column of the 8×8 identity matrix, and

$$\mathbf{G}(s) = \sqrt{2}\begin{bmatrix} \mathbf{A}(s_1, s_2) & 0 \\ 0 & \mathbf{A}(s_3, s_4) \end{bmatrix},$$

where $A(s_i, s_{i+1}) = \begin{bmatrix} s_i & s_{i+1} \\ -s_{i+1}^* & s_i^* \end{bmatrix}$ is the Alamouti space–time code [6]. The constant $\sqrt{2}$ ensures that the average transmit power is identical to that of the underlying alphabet. Therefore, in terms of the rotated information symbols $a_p = x_p e^{j\phi}$, the SSD space–time code of [32] is

$$C^{\text{SSD}} = \sqrt{2}\begin{bmatrix} a_1^R + ja_3^I & a_2^R + ja_4^I & 0 & 0 \\ -a_2^R + ja_4^I & a_1^R - ja_3^I & 0 & 0 \\ 0 & 0 & a_3^R + ja_1^I & a_4^R + ja_2^I \\ 0 & 0 & -a_4^R + ja_2^I & a_3^R - ja_1^I \end{bmatrix}.$$

Having discussed SSD codes, we next discuss the semiorthogonal algebraic STBCs, which have the lowest decoding complexity of any rate-one full-diversity STBC for any number of transmit antennas.

27.4.5 Semiorthogonal Algebraic Space–Time Block Codes

The SAST block codes are a family of linear space–time codes that achieve full rate and full diversity for any number of transmit antennas [35]. The word algebraic in the description of SAST codes comes from the fact that they use the same real rotation matrices of the DAST codes [21,22] (see also [36]), which were constructed using algebraic number field theory. In fact, SAST codes are constructed using the DAST codes as we will discuss shortly. The word semiorthogonal refers to a property of the code matrix, where half the columns of the code matrix are orthogonal to the other half.

An equivalent form of the rate-one SAST code for M transmit antennas, where M is even, is [35]

$$C_M^{\text{SAST}}(x_1, \ldots, x_M) = \begin{bmatrix} \mathbf{C}_{M/2}^{\text{DAST}}(x_1, \ldots, x_{M/2}) & \mathbf{C}_{M/2}^{\text{DAST}}(x_{M/2+1}, \ldots, x_M) \\ -\mathbf{C}_{M/2}^{\text{DAST}}(x_{M/2+1}^*, \ldots, x_M^*) & \mathbf{C}_{M/2}^{\text{DAST}}(x_1^*, \ldots, x_{M/2}^*) \end{bmatrix}.$$

The SAST code for $M - 1$ antennas is obtained by deleting the last column of the SAST code for M antennas. This version of SAST codes differs from the original in [35] only by the fact that we used the DAST matrices instead of using circulant matrices. The circulant matrix reduces the peak-to-average-power ratio but has no impact on the diversity, coding gain, and decoding complexity. In fact, any circulant matrix is diagonalizable using the Fourier transform matrix. We simply use DAST matrices to simplify the presentation.

The semiorthogonal property of SAST codes allows for the separate decoding of symbols x_1 through $x_{M/2}$ from symbols $x_{M/2+1}$ through x_M. Furthermore, the use of real rotation matrices allows for separate decoding of the real and imaginary parts of the QAM symbols. As a result, the SAST block codes for M and $M-1$ antennas require the joint decoding of only $M/2$ real PAM symbols, or equivalently $M/4$ complex QAM symbols. For the case of two transmit antennas, the SAST code simplifies to the Alamouti STBC. For the case of three and four transmit antennas, the SAST codes require the joint decoding of only two real symbols, which is the same decoding complexity as the SSD codes. In fact, the SAST code for four transmit antenna is equivalent to the SSD code of [31] and has identical diversity order, coding gain and BER performance. The SAST codes have the lowest decoding complexity of any rate-one STBC and achieve comparable BER performance to the best codes.

All of the space–time codes that we have discussed thus far achieve a maximum rate of one symbol per signaling interval. We next discuss two families of STBCs whose transmission rate is higher than one; the threaded algebraic and perfect STBCs.

27.4.6 Threaded Algebraic Space–Time Block Codes

The threaded algebraic space–time (TAST) codes are a family of linear space–time codes that are fully diverse and achieve arbitrary rate for arbitrary number of transmit antennas. The TAST block codes layer or thread rate-one DAST block codes to achieve maximal rate of M symbols per signaling interval for M transmit antennas. As a result, they can also achieve other rates by simply puncturing or deleting layers. For example, for a four transmit and four receive antenna system, the TAST block codes of rate $\kappa \in \{1, 2, 3, 4\}$ are easily obtained by puncturing or deleting $4-\kappa$ layers. We next discuss the construction of TAST block codes. We begin by presenting the notation that will be used in describing the TAST codes and perfect STBCs.

Let $\mathbf{x}_l = [x_{l,0}, x_{l,1}, \ldots, x_{l,M-2}, x_{l,M-1}]^T$ denote the vector of M information symbols for the lth *thread*, where $x_{l,m}$ is a complex information symbol drawn from a q-QAM or q-HEX alphabet. Furthermore, let Φ_M denote an $M \times M$ unitary *rotation* or *generator* matrix, which can be complex or real, and let $\mathbf{e}_m$ denote the mth column of the $M \times M$ identity matrix. Finally, let $\mathbf{D}_l = \mathrm{diag}(\Phi_M \mathbf{x}_l)$ be the diagonal matrix with diagonal elements consisting of a rotated version of the lth symbol vector.

In terms of the above definitions, the rate-κ TAST code for an M-input N-output MIMO system is [37]

$$C_M^{\mathrm{TAST}} = \sum_{l=1}^{\kappa} \mathbf{D}_l (\gamma \mathbf{U})^{l-1},$$

where

$\mathbf{U} = [\mathbf{e}_M, \mathbf{e}_1, \ldots, \mathbf{e}_{M-2}, \mathbf{e}_{M-1}]$

γ is a unit-magnitude complex number that ensures a full-diversity TAST code

The value γ depends on both the QAM modulation alphabet size and number of transmit antennas M.

The decoding complexity of TAST codes depends on the transmission rate. When complex generator matrices are used, the TAST codes do not offer any reduction in decoding complexity and they require the joint detection of κM symbols for ML decoding. However, when real rotation matrices are used, the worst-case decoding complexity can be reduced by exploiting the fact that the last layer of the TAST code has a worst-case decoding complexity of $2q^{M/2}$. Since the first $\kappa - 1$ layers have a worst-case decoding complexity of $q^{(\kappa-1)M}$; the overall worst-case decoding complexity is then $q^{(\kappa-1)M} \times 2q^{M/2} = 2q^{(\kappa-1/2)M}$.

27.4.7 Perfect Space–Time Block Codes

Perfect STBCs are a family of linear space–time codes that were proposed for two, three, four, and six antennas in [38] and later generalized for any number of antennas in [39]. These codes are termed perfect because they have full diversity, a nonvanishing determinant for increasing spectral efficiency, uniform average transmitted energy per antenna, and achieve rate-M symbols per channel use for M antennas. Similar to TAST codes, perfect space–time codes layer or thread rate-one diagonal codes to achieve maximal rate of M symbols per signaling interval and can also achieve other rates by puncturing layers. We next discuss encoding of the rate-M perfect code.

The rate-M *perfect* space–time code for an M-input M-output MIMO system is [38,39]

$$\mathbf{C}_M^{\text{PERFECT}} = \sum_{l=1}^{M} \mathbf{D}_l \mathbf{J}^{l-1},$$

where $\mathbf{J} = [\gamma \mathbf{e}_M, \mathbf{e}_1, \ldots, \mathbf{e}_{M-2}, \mathbf{e}_{M-1}]$ and γ is a unit-magnitude complex number that ensures a full-diversity perfect space–time code. The value of γ depends on the number of transmit antennas M.

The decoding complexity of the perfect space–time codes also depends on the transmission rate. By using real rotation matrices and the same decoding strategy as the TAST codes, the worst-case ML decoding complexity is also $2q^{(\kappa-1/2)M}$, the same as the TAST code.

27.5 Summary and Comparison of Space–Time Block Codes

We compare orthogonal, quasiorthogonal, DAST, SAST, TAST, SSD, and perfect STBCs in terms of their code rate, delay, and worst-case ML decoding complexity in Table 27.2.

The following assumptions were made in comparing these STBCs:

- The comparison assumes quasistatic-fading channel.
- The comparison is for an arbitrary number of antennas M, except for the SSD codes and quasiorthogonal codes where $M \in \{3, 4\}$ and $M > 2$, respectively.
- We assumed real rotation matrices for DAST, TAST, and perfect space–time codes.
- The complex information symbols are assumed to be drawn from q-ary QAM alphabet with the exception of the perfect space–time block, where the modulation alphabet can also be q-ary HEX (e.g., three and six transmit antenna).

TABLE 27.2 Comparison of Different STBCs in Terms of Rate, Delay, and Decoding Complexity

Space–Time Code	Rate (κ)	Delay (T)	Decoding Complexity
Orthogonal [15–18]	$\frac{\lceil M/2 \rceil + 1}{2\lceil M/2 \rceil}$	$\begin{cases} \frac{1}{2\kappa}\binom{M}{\lceil M/2 \rceil} & M = 4l \\ \frac{1}{\kappa}\binom{M}{\lceil M/2 \rceil} & M \neq 4l \end{cases}$	$\begin{cases} \binom{M}{\lceil M/2 \rceil}\sqrt{q} & M = 4l \\ 2\binom{M}{\lceil M/2 \rceil}\sqrt{q} & M \neq 4l \end{cases}$
Quasiorthogonal [30]	1	$4\lceil M/4 \rceil$	$2q^{2\lceil M/4 \rceil}$
DAST [19]	1	M	$2q^{M/2}$
SSD [31–34]	1	4	$4q$
SAST [35]	1	$2\lceil M/2 \rceil$	$4q^{\lceil M/2 \rceil/2}$
TAST [37]	$\{1, \ldots, M\}$	M	$2q^{(\kappa-1/2)M}$
Perfect [38]	$\{1, \ldots, M\}$	M	$2q^{(\kappa-1/2)M}$

We also compare the BER performance of STBCs for three configurations; two transmit and one receive antenna, two transmit and two receive antennas, and four transmit and one receive antenna. For the TAST and perfect STBCs, the transmission rate will be set to $\kappa = \min(M, N)$ in order to facilitate low-complexity decoding using the sphere decoding algorithm.

For the two transmit antennas and one receive antenna, we compare the orthogonal STBC (Alamouti code) with the DAST code. We note that for two transmit antennas, SAST codes and quasiorthogonal codes yield the Alamouti space–time code as a special case. Furthermore, the TAST and perfect codes yield an equivalent form of the DAST code for two antennas when the rate $\kappa = 1$. Finally, SSD codes either yield the Alamouti space–time code or DAST code as a special case. Hence, for two transmit antennas and one receive antenna, we only compare the Alamouti and DAST block code as shown in Figure 27.3. The comparison is for 4-QAM and 16-QAM modulation. The Alamouti block code outperforms the DAST code by 2 dB and 3 dB for 4-QAM and 16-QAM, respectively. Furthermore, the Alamouti block code has lower decoding complexity. The Alamouti STBC remains unmatched in terms of performance and decoding complexity for the two transmit antennas and one receive antenna.

For the two transmit antennas and two receive antennas, we compare the rate-one Alamouti STBC and rate-two perfect space–time code as shown in Figure 27.4. To avoid cluttering the plot, we omit simulation results for TAST codes and simply note that they perform within 1 dB of the perfect code. Similarly, we omit simulation results for DAST codes and note that they perform within 2–4 dB of the Alamouti block code.

For the two codes to transmit at the same spectral efficiency, the Alamouti code transmits M^2-QAM symbols while the rate-two perfect code transmits M-QAM symbols. For spectral efficiency of 4 b/s/Hz, the Alamouti space–time code transmits 16-QAM symbols, while the perfect STBC transmits 4-QAM symbols. The perfect code outperforms the Alamouti code by 1 dB in the high SNR region. For spectral efficiency of 8 b/s/Hz, the Alamouti code transmits 256-QAM symbols, while the perfect code transmits 16-QAM symbols. The perfect code outperforms the Alamouti code by 3.5 dB. This result is expected

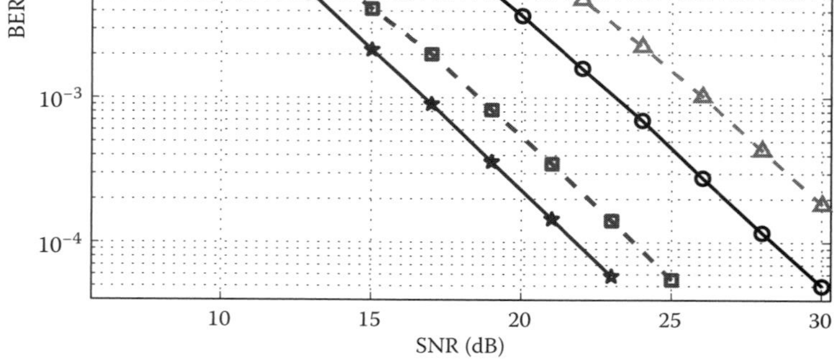

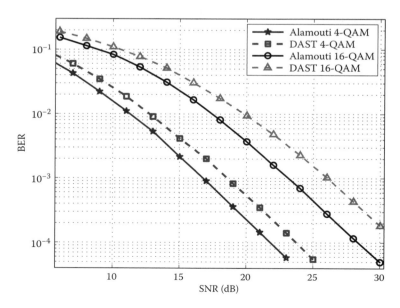

FIGURE 27.3 Performance of the Alamouti and DAST code for two transmit and one receive antenna.

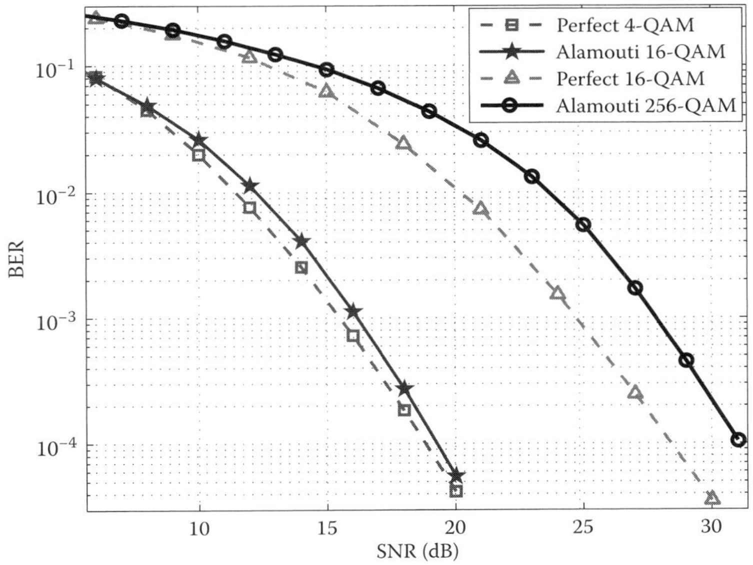

FIGURE 27.4 Performance of the perfect code and Alamouti code for two transmit and two receive antennas.

since the Alamouti code has multiplexing gain of 1 b/s/Hz per 3 dB, while the perfect code has multiplexing gain of 2 b/s/Hz per 3 dB. Both the Alamouti and perfect STBC have diversity order of four, which is confirmed by the fact that they both have the same slope.

We end the discussion of the two transmit antenna cases by noting that the Alamouti space–time code has a worst-case decoding complexity of $4\sqrt{q}$, while the rate-two perfect code has been shown recently to have a worst-case decoding complexity of $2q^3$ [40]. The complexity-performance trade-off of the Alamouti and perfect code helps explain why both codes have been incorporated in the WiMAX standard [41]. The Alamouti STBC is optimal in terms of the diversity-multiplexing trade-off for two transmit antennas and one receive antenna, while maintaining a low ML decoding complexity. Because the rate of the Alamouti code is one symbol per channel use, it does not achieve the optimal diversity-multiplexing trade-off for two transmit and two receive antennas. The rate-one and rate-two perfect codes, on the other hand, achieve the optimal diversity-multiplexing trade-off for one and two receive antennas, respectively. For one receive antenna, the Alamouti code not only has better performance than the rate-one perfect code, but it also has a lower decoding complexity. For two receive antennas, the rate-two perfect code has superior performance compared to the Alamouti code, albeit at a much higher decoding complexity. Both the Alamouti and perfect code are part of the WiMAX standard, and are referred to as "matrix A" and "matrix C," respectively.

For the four transmit antennas and one receive antenna configuration, we compare the rate-1/2 orthogonal code, rate-one quasiorthogonal code, rate-one SAST code, and rate-one TAST code, which is identical to the DAST code. To avoid cluttering the plot, we omit simulation results for rate-one perfect code and simply note that they perform almost identically to the rate-one TAST code. For rate-one transmission, quasiorthogonal codes are the best performing space–time codes for more than two transmit antennas as shown in Figure 27.5. The improved performance compared to SAST and orthogonal codes, however, comes at the cost of higher decoding complexity. In particular, quasiorthogonal codes have a worst-case decoding complexity of $2q^2$, the same as the TAST code, while the orthogonal and SAST codes have worst-case decoding complexity of $8\sqrt{q}$ and $4q$, respectively.

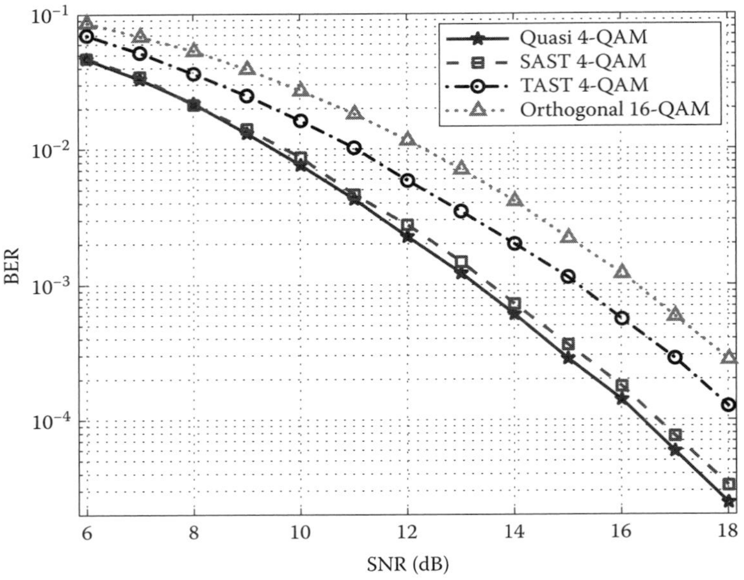

FIGURE 27.5 Performance of quasiorthogonal, SAST, rate-one TAST, and rate-half orthogonal block codes for four transmit antennas and one receive antenna.

27.6 Future Research Topics

An interesting topic for future research is the development of *separable* high-rate STBCs. As discussed in Section 27.4, a rate-one full-diversity STBC was proposed in [35] for an arbitrary number of antennas. The SAST codes in [35] are four-group decodable and required the joint detection of M real symbols in each group, for $2M$ and $2M - 1$ transmit antennas. To the authors' knowledge, however, a separable rate-two full-diversity STBC has not been reported in the literature and the design of high-rate group-decodable (i.e., separable) STBCs is an open research problem.

A few attempts have been reported in the literature on the construction of separable STBCs, with rates higher than one. Most notably, two constructions of rate-5/4 symbols per channel use for four transmit antennas were reported in [42,43]. The proposed code in [42] has only a second-order transmit diversity and as a result, orthogonal and quasiorthogonal STBCs outperform the codes in [42] at high SNR values. The proposed code in [43] is two-group decodable, requiring the joint detection of five real symbols per group.

An interesting open research problem is the existence of separable rate-two STBCs. One approach for solving this problem would be to establish necessary conditions for the existence of full diversity and high-rate separable space–time codes. The insight offered by the results obtained from the necessary conditions for the existence of such codes can guide future research efforts. In particular, if it is proven that separable high-rate space–time codes can exist, then a systematic design of such codes can be undertaken. If it is proven that separable high-rate space–time codes cannot exist, then relaxing the full diversity or rate-two requirement and reexamining the design problem can also be undertaken.

References

1. L. Zheng and D. Tse, Diversity and multiplexing: A fundamental tradeoff in multiple antenna channels, *IEEE Trans. Inf. Theory*, 49(4), 1073–1096, May 2003.
2. D. Gesbert, M. Shafi, D.S. Shiu, P. Smith, and A. Naguib, From theory to practice: An overview of MIMO space–time coded wireless systems, *IEEE J. Sel. Areas Commun.*, 21, 281–302, April 2003.

3. A. Wittneben, A new bandwidth efficient transmit antenna modulation diversity scheme for linear digital modulation, in *IEEE International Conference on Communications* (*ICC 93*), vol. 3, pp. 1630–1634, Geneva, Switzerland, May 1993.
4. N. Seshadri and J.H. Winters, Two schemes for improving the performance of frequency-division duplex (FDD) transmission systems using transmitter antenna diversity, *Int. J. Wireless Inf. Netw.*, 1, 49–60, January 1994.
5. V. Tarokh, N. Seshadri, and A.R. Calderbank, Space–time codes for high data rate wireless communication: Performance criterion and code construction, *IEEE Trans. Inf. Theory*, 44, 744–765, March 1998.
6. S.M. Alamouti, A simple transmit diversity technique for wireless communications, *IEEE J. Sel. Areas Commun.*, 16, 1451–1458, October 1998.
7. U. Fincke and M. Pohst, Improved methods for calculating vectors of short length in a lattice, including a complexity analysis, *Math. Comput.*, 44, 463–471, April 1985.
8. C.P. Schnorr and M. Euchner, Lattice basis reduction: Improved practical algorithms and solving subset sum problems, *Math. Program.*, 66, 181–191, 1994.
9. B. Hassibi and H. Vikalo, On the sphere-decoding algorithm. I. Expected complexity, *IEEE Trans. Signal Process.*, 53, 2806–2818, August 2005.
10. H. Vikalo and B. Hassibi, On the sphere-decoding algorithm. II. Generalization, second-order statistics, and applications to communications, *IEEE Trans. Signal Process.*, 53, 2819–2834, August 2005.
11. A. Paulraj, R. Nabar, and D. Gore, *Introduction to Space–Time Wireless Communications*, Cambridge University Press, Cambridge, U.K., 2003.
12. V. Tarokh, A. Naguib, N. Seshadri, and A.R. Calderbank, Space–time codes for high data rate wireless communications: Performance criteria in the presence of channel estimation errors, mobility, and multiple paths, *IEEE Trans. Commun.*, 47, 199–207, February 1999.
13. Q. Yan and R. Blum, Robust space–time block coding for rapid fading channels, in *Proceedings of the IEEE GLOBECOM*, vol. 1, pp. 460–464, San Antonio, TX, 2001.
14. S. Zummo and S. Al-Semari, Space–time coded QPSK for rapid fading channels, *Proceedings of the IEEE International symposium on Personal, Indoor and Mobile Radio Communications* (PIMRC), vol. 1, pp. 504–508, London, U.K., 2000.
15. V. Tarokh, H. Jafarkhani, and A. Calderbank, Space–time block codes from orthogonal designs, *IEEE Trans. Inf. Theory*, 45(5), 1456–1467, July 1999.
16. G. Ganesan and P. Stoica, Space–time block codes: A maximum SNR approach, *IEEE Trans. Inf. Theory*, 47, 1650–1656, May 2001.
17. X.-B. Liang, Orthogonal designs with maximal rates, *IEEE Trans. Inf. Theory*, 49(10), 2468–2503, October 2003.
18. Z. Li-gang, M. Jian-song, L. Xin, and D. Wei, Complete design for complex orthogonal designs with maximal rates and minimal decoding delays, *Sixth International Conference on ITS Telecommunications Proceedings*, pp. 390–393, Chengdu, China, June 2006.
19. M.O. Damen, K. Abed-Meraim, and J.-C. Belfiore, Diagonal algebraic space–time block codes, *IEEE Trans. Inf. Theory*, 48(3), 628–636, March 2002.
20. M.O. Damen and N.C. Beaulieu, On diagonal algebraic space–time block codes, *IEEE Trans. Commun.*, 51(6), 911–919, June 2003.
21. J. Boutros and E. Viterbo, Signal space diversity: A power and bandwidth efficient diversity technique for the Rayleigh fading channel, *IEEE Trans. Inf. Theory*, 44, 1453–1467, July 1998.
22. J.-C. Belfiore, X. Giraud, and J. Rodriguez, Linear labeling for joint source channel coding, in *Proceedings of the IEEE International symposium on Information Theory* (*ISIT*), Sorrento, Italy, June 2000.
23. H. Jafarkhani, A quasi-orthogonal space–time block code, *IEEE Trans. Commun.*, 49(1), 1–4, January 2001.

24. O. Tirkkonen, A. Boariu, and A. Hottinen, Minimal nonorthogonality rate 1 space–time block code for 3+ Tx antennas, in *Proceedings of the IEEE Sixth International Symposium on Spread-Spectrum Techniques and Applications (ISSSTA)*, pp. 429–432, Parsippany, NJ, September 2000.
25. C.B. Papadias and G.J. Foschini, Capacity-approaching space–time codes for systems employing four transmitter antennas, *IEEE Trans. Inf. Theory*, 49(3), 726–732, March 2003.
26. O. Tirkkonen, Optimizing space–time block codes by constellation rotations, in *Proceedings of the Finnish Wireless Communications Workshop (FWCW)*, Finland, pp. 59–60, October 2001.
27. N. Sharma and C. Papadias, Improved quasi-orthogonal codes through constellation rotation, *IEEE Trans. Commun.*, 51(3), 332–335, March 2003.
28. W. Su and X.-G. Xia, Signal constellations for quasi-orthogonal space–time block codes with full diversity, *IEEE Trans. Inf. Theory*, 50(10), 2331–2347, October 2004.
29. D. Wang and X.-G. Xia, Optimal diversity product rotations for quasi-orthogonal STBC with MPSK symbols, *IEEE Commun. Lett.*, 9, 420–422, May 2005.
30. N. Sharma and C.B. Papadias, Full rate full diversity linear quasi-orthogonal space–time codes for any transmit antennas, *EURASIP J. Appl. Signal Process.*, 9, 1246–1256, August 2004.
31. C. Yuen, Y.L. Guan, and T.T. Tjhung, Quasi-orthogonal STBC with minimum decoding complexity, *IEEE Trans. Wireless Commun.*, 4(5), 2089–2094, September 2005.
32. M.Z.A. Khan and B.S. Rajan, Single-symbol maximum likelihood decodable linear STBCs, *IEEE Trans. Inf. Theory*, 52(5), 2062–2091, May 2006.
33. P. Marsch, W. Rave, and G. Fettweis, Quasi-orthogonal STBC using stretched constellations for low detection complexity, in *Proceedings of the Wireless Communications and Networking Conference (WCNC)*, Hong Kong, China, pp. 757–761, March 2007.
34. M.O. Sinnokrot and J.R. Barry, A single-symbol decodable space–time block code with full rate and low peak-to-average-power ratio to appear, *IEEE International Symposium on Personal, Indoor and Mobile Radio Communications (PIMRC)*, Cannes, France, September 15–18, 2008.
35. D.N. Dao, C. Yuen, C. Tellambura, Y.L. Guan, and T.T. Tjhung, Four-group decodable space–time block codes, *IEEE Trans. Signal Process.*, 56, 424–430, January 2008.
36. F. Oggier and E. Viterbo, Full diversity rotations. [Online] Available: www1.tlc.polito.it/~viterbo/rotations/rotations.html
37. M.O. Damen, H.E. Gamal, and N.C. Beaulieu, Linear threaded algebraic space–time constellations, *IEEE Trans. Inf. Theory*, 49, 2372–2388, October 2003.
38. F. Oggier, G. Rekaya, J.-C. Belfiore, and E. Viterbo, Perfect space–time block codes, *IEEE Trans. Inf. Theory*, 52(9), 3885–3902, September 2006.
39. P. Elia, B.A. Sethuraman, and P. Vijay Kumar, Perfect space–time codes with minimum and non-minimum delay for any number of transmit antennas, *IEEE Trans. Inf. Theory*, 53(11), 3853–3868, November 2007.
40. M.O. Sinnokrot and J.R. Barry, The golden code is fast decodable, to appear, *IEEE Global Telecommunications Conference* (Globecom), New Orleans, LA, November 30–December 4, 2008.
41. IEEE 802.16e-2005: IEEE Standard for Local and Metropolitan Area Networks—Part 16: Air Interface for Fixed and Mobile Broadband Wireless Access Systems—Amendment 2: Physical Layer and Medium Access Control Layers for Combined Fixed and Mobile Operation in Licensed Bands, February 2006.
42. J. Chung, S.H. Nam, and C.-S. Hwang, High rate space–time block codes, *IEICE Trans. Commun.*, E89-B(4), 1420–1422, April 2006.
43. C. Yuen, Y.L. Guan, and T.T. Tjhung, On the search for high-rate quasi-orthogonal space–time block code, *Int. J. Wireless Inf. Networks*, 13(4), October 2006.

28

A Multiplexing Approach to the Construction of High-Rate Space–Time Block Codes

Mohanned O. Sinnokrot
Georgia Institute of Technology

Vijay K. Madisetti
Georgia Institute of Technology

28.1 Introduction 28-1
28.2 System and Channel Model 28-2
28.3 Key Properties for Reduced Complexity Decoding 28-4
28.4 The R Matrix for Space–Time Block Codes 28-8
Orthogonal Space–Time Block Codes • Quasiorthogonal Space–Time Block Codes • Diagonal Algebraic Space–Time Block Codes • Semiorthogonal Algebraic Space–Time Block Codes
28.5 Rate-Two Space–Time Block Codes for Two Transmit Antennas 28-12
28.6 High-Rate Space–Time Block Codes for Four Transmit Antennas 28-13
28.7 Conclusion 28-18
References 28-18

28.1 Introduction

The design of space–time block codes that harness the diversity and rate benefits of the multiple-input multiple-output (MIMO) channel remains elusive. The goal of this chapter is to provide an insight into the design of certain families of space–time block codes that are being considered or have already been incorporated into the next generation wireless technology. We restrict our attention to the design and decoding of the high-rate space–time block codes for the two-input two-output and the four-input two-output channels in which the transmission rate is higher than one symbol per signaling interval.

The first central theme of this chapter is multiplexing. In particular, we view the high-rate space–time block codes for the 2×2 and 4×2 channels as a multiplexing (combination) of two rate-one space–time block codes, with the exception of the high-rate space–time block code proposed in [1], which can be viewed as a combination of rate-3/4 orthogonal space–time block codes. We will refer to the rate-one and rate-3/4 codes used to generate the high-rate space–time codes as parent codes.

The second central theme of the chapter is decoding complexity. None of the high-rate space–time block codes are *separable* and as a result, the decoding cannot be done on two or more groups of independent symbols (in fact, the design of separable high-rate space–time block codes remains an open research problem). Consequently, the worst-case decoding complexity will depend on the worst-case decoding complexity of the multiplexed parent codes. The choice of the parent code and multiplexing strategy affects not only the worst-case decoding complexity, but also the performance. An interesting result that we will obtain later in the chapter is that the best performing parent code does not generate the best performing high-rate code. However, the lowest-complexity parent code generates the lowest-complexity high-rate code.

In the remainder of this chapter, we describe the MIMO system and the channel model (Section 28.2). In Section 28.3, we discuss the key properties for reduced complexity decoding of space–time codes. In Section 28.4, we provide a brief survey of the decoding complexity of the rate-one and rate-3/4 space–time block codes (parent codes) used to generate the high-rate codes. In Section 28.5, we discuss the construction and the worst-case decoding complexity of high-rate space–time codes for the 2×2 channel. We discuss the construction and the worst-case decoding complexity of high-rate space–time codes for the 4×2 channel in Section 28.6. Finally, we conclude the chapter in Section 28.7.

28.2 System and Channel Model

We consider a MIMO system with M transmit antennas and N receive antennas. The transmitted code word of a space–time block code can be expressed as a $T \times M$ matrix:

$$\mathbf{C} = \begin{bmatrix} c_1[1] & c_2[1] & \cdots & c_M[1] \\ c_1[2] & c_2[2] & \cdots & c_M[2] \\ \vdots & \vdots & \ddots & \vdots \\ c_1[T] & c_2[T] & \cdots & c_M[T] \end{bmatrix},$$

where $c_m[t]$ denotes the symbol transmitted from antenna $m \in \{1, \ldots, M\}$ at time $t \in \{1, \ldots, T\}$.

Definition 28.1: The *code rate* of a $T \times M$ space–time block code transmitting P complex information symbols over T symbol periods is $\kappa = P/T$ symbols per channel use.

Assuming a quasistatic channel with coherence time T, the received signal $y_n[t]$ at receive antenna $n \in \{1, \ldots, N\}$ at time t is given by

$$y_n[t] = \sum_{m=1}^{M} h_{m,n} c_m[t] + w_n[t],$$

where

$w_n[t]$ is the complex additive white Gaussian noise at receive antenna n at time t
$h_{m,n}$ is the channel coefficient between the mth transmit antenna and nth receive antenna

Let $\tilde{y}_n[t]$ (resp. $\tilde{w}_n[t]$) designate the received signal $y_n[t]$ (resp. $w_n[t]$) or its conjugate $y_n^*[t]$ (resp. $w_n^*[t]$), then the received vector in matrix form is given by

$$\mathbf{y} = \mathbf{H}\mathbf{x} + \mathbf{w},$$

where

$\mathbf{y} = [\tilde{y}_1[1] \ \cdots \ \tilde{y}_1[T] \ \cdots \ \tilde{y}_N[1] \ \cdots \ \tilde{y}_N[T]]^{\mathrm{T}}$ is the vector of the received signals from all the antennas at all time intervals
$\mathbf{H}$ is the $NT \times P$ effective channel matrix
$\mathbf{x} = [x_1 \ \cdots \ x_P]^{\mathrm{T}}$ is the vector of transmitted symbols
$\mathbf{w} = [\tilde{w}_1[1] \ \cdots \ \tilde{w}_1[T] \ \cdots \ \tilde{w}_N[1] \ \cdots \ \tilde{w}_N[T]]^{\mathrm{T}}$ is the noise from all the antennas at all time intervals

Example 28.1

Consider the 2 × 2 Alamouti space–time block code [2] transmitting two complex information symbols in two time intervals. The code matrix $\mathbf{C}$ is given by

$$\mathbf{C} = \begin{bmatrix} x_1 & x_2 \\ -x_2^* & x_1^* \end{bmatrix}.$$

Assuming a single receive antenna, the received signal at the two time instances is given by

$$\begin{aligned} y_1[1] &= h_{1,1}x_1 + h_{2,1}x_2 + w_1[1], \\ y_1[2] &= -h_{1,1}x_2^* + h_{2,1}x_1^* + w_1[2]. \end{aligned}$$

In matrix form, the received signal is given by

$$\mathbf{y} = \begin{bmatrix} y_1[1] \\ y_1^*[2] \end{bmatrix} = \begin{bmatrix} h_{1,1} & h_{2,1} \\ h_{2,1}^* & -h_{1,1}^* \end{bmatrix} \begin{bmatrix} x_1 \\ x_2 \end{bmatrix} + \begin{bmatrix} w_1[1] \\ w_1^*[2] \end{bmatrix} = \mathbf{Hx} + \mathbf{w}.$$

Example 28.2

Consider the 2 × 2 diagonal algebraic space–time block (DAST) code [3] transmitting two complex information symbols in two time intervals. The code matrix $\mathbf{C}$ is given by

$$\mathbf{C} = \begin{bmatrix} u_1 & 0 \\ 0 & u_2 \end{bmatrix},$$

where $\begin{bmatrix} u_1 \\ u_2 \end{bmatrix} = \begin{bmatrix} \alpha & \beta \\ -\beta & \alpha \end{bmatrix} \begin{bmatrix} x_1 \\ x_2 \end{bmatrix}$, $\alpha = \cos(\theta)$, $\beta = \sin(\theta)$, and $\theta = \frac{1}{2}\tan^{-1}(2)$. Assuming a single receive antenna, the received signal at the two time instances is given by

$$\begin{aligned} y_1[1] &= h_{1,1}u_1 + w_1[1] = \alpha h_{1,1}x_1 + \beta h_{1,1}x_2 + w_1[1], \\ y_1[2] &= h_{2,1}u_2 + w_1[2] = -\beta h_{2,1}x_1 + \alpha h_{2,1}x_2 + w_1[2]. \end{aligned}$$

In matrix form, the received signal is given by

$$\mathbf{y} = \begin{bmatrix} y_1[1] \\ y_1[2] \end{bmatrix} = \begin{bmatrix} \alpha h_{1,1} & \beta h_{1,1} \\ -\beta h_{2,1} & \alpha h_{2,1} \end{bmatrix} \begin{bmatrix} x_1 \\ x_2 \end{bmatrix} + \begin{bmatrix} w_1[1] \\ w_1[2] \end{bmatrix} = \mathbf{Hx} + \mathbf{w}$$

For the remainder of the chapter, we will use the matrix representation for the received signal. Assuming that the perfect channel state information (CSI) is known at the receiver, but not at the transmitter, the maximum-likelihood (ML) decoder at the receiver chooses the information symbol vector $\hat{\mathbf{x}}$ such that:

$$\hat{\mathbf{x}} = \arg\min_{\mathbf{x}} \|\mathbf{y} - \mathbf{Hx}\|^2.$$

We next define two important terms related to ML decoding: *decoding complexity* and *worst-case decoding complexity*.

Definition 28.2: The *decoding complexity* of an ML decoder is the number of metric computations required to reach the ML decision. The *worst-case decoding complexity* achieved by an exhaustive search ML decoder is q^P metric computations, where q is the alphabet size.

A central theme in this chapter is the worst-case decoding complexity of space–time block codes. Because the structure of the space–time block code induces certain properties in the effective channel matrix, we will examine the worst-case decoding complexity in terms of the properties of the effective channel matrix. In particular, we examine the properties of the **R** matrix in the orthogonal-triangular (QR) decomposition of the matrix **H**. A QR decomposition of the matrix **H** can be obtained by applying the Gram-Schmidt procedure to the columns of $\mathbf{H}=[\mathbf{h}_1,\ldots,\mathbf{h}_P]$ to obtain $\mathbf{H}=\mathbf{QR}$, where $\mathbf{Q}=[\mathbf{q}_1,\ldots,\mathbf{q}_P]$ is unitary and **R** is upper triangular with nonnegative real diagonal elements, so that the entry of **R** in row i and column j is $r_{i,j}=\mathbf{q}_i^\star\mathbf{h}_j$. Because **Q** is unitary such that $\mathbf{QQ}^\star=\mathbf{Q}^\star\mathbf{Q}=\mathbf{I}$, where **I** is the identity matrix, the ML decoder can choose $\hat{\mathbf{x}}$ that minimizes:

$$\hat{\mathbf{x}}=\arg\min_{\mathbf{x}}\|\mathbf{y}-\mathbf{QRx}\|^2=\arg\min_{\mathbf{x}}\|\mathbf{Q}^\star\mathbf{y}-\mathbf{Rx}\|^2=\arg\min_{\mathbf{x}}\|\mathbf{v}-\mathbf{Rx}\|^2,$$

where $\mathbf{v}=\mathbf{Q}^\star\mathbf{y}$. Because the ML cost function depends on **R**, the worst-case decoding complexity also depends of the properties of the **R** matrix.

In the next section, we discuss the properties of the **R** matrix in the QR decomposition of the effective channel matrix **H** that lead to a reduction in the worst-case decoding complexity.

28.3 Key Properties for Reduced Complexity Decoding

In this section, we discuss the key properties of reduced complexity decoding of space–time block codes in terms of the resulting **R** matrix of the QR decomposition of the effective channel matrix **H**. We start the discussion with a simple example.

Example 28.3

Consider the QR decomposition of the effective channel matrix for the Alamouti space–time block code discussed in Example 28.1.

$$\mathbf{H}=\begin{bmatrix} h_{1,1} & h_{2,1}\\ h_{2,1}^\star & -h_{1,1}^\star\end{bmatrix}=\begin{bmatrix}\frac{h_{1,1}}{|h_{1,1}|^2+|h_{2,1}|^2} & \frac{h_{2,1}}{|h_{1,1}|^2+|h_{2,1}|^2}\\ \frac{h_{2,1}^\star}{|h_{1,1}|^2+|h_{2,1}|^2} & \frac{-h_{1,1}^\star}{|h_{1,1}|^2+|h_{2,1}|^2}\end{bmatrix}\begin{bmatrix}|h_{1,1}|^2+|h_{2,1}|^2 & 0\\ 0 & |h_{1,1}|^2+|h_{2,1}|^2\end{bmatrix}=\mathbf{QR}.$$

For the Alamouti space–time block code, it is well known that the symbols x_1 and x_2 are separately decodable. Furthermore, if the information symbols are drawn from a square or a rectangular q-ary quadrature amplitude modulation (QAM) alphabet, then their real and imaginary components are also separately decodable. Hence, the Alamouti space–time block code is single real symbol decodable. Furthermore, the worst-case decoding complexity of the Alamouti code is $\mathcal{O}(1)$. This is because for certain alphabets, like QAM, we can decide on the transmitted symbol with a slicer whose complexity does not grow with the size of the alphabet; rather, the worst-case complexity of a QAM slicer is $\mathcal{O}(1)$. Specifically, a QAM slicer can be implemented as a pair of pulse-amplitude modulation (PAM) slicers, with each requiring a single multiply, a single rounding operation, a single addition, and a single hard-limiting operation, none of which depends on q.

The fact that the Alamouti code is single real symbol decodable is easily seen from the **R** matrix. Because the element $r_{1,2}$ is zero, there is no interference between the symbols x_1 and x_2, and hence, we can decode them separately. Furthermore, because $r_{1,1}$ and $r_{2,2}$ are real, the real and imaginary components of x_1 and x_2 are also separately decodable, assuming a QAM alphabet.

As discussed in Example 28.3, the reduction in decoding complexity of the Alamouti space–time block code follows directly from the presence of zero and real elements in the **R** matrix.

More formally, the number and the location of the zero and the real elements of the **R** matrix can potentially lead to a reduction in decoding complexity by exploiting one or more of the following three properties:

1. Some groups of symbols can be decoded separately or independently from other groups.
2. The real and the imaginary parts of the elements of a group can be decoded separately.
3. In the absence of some symbols, some symbols are separately decodable from others.

We note that the early designs of space–time block codes focused on exploiting the first and the second properties. In other words, early designs focused on *separable* space–time block codes such that the decoding can be done on separate groups of symbols. The third property has only been recently utilized and exploited in decoding and designing space–time block codes [4–8]. We next discuss an example highlighting different forms of the **R** matrix and the resulting worst-case decoding complexity.

Example 28.4

Consider the **R** matrix in the QR decomposition of seven different 4×4 effective channel matrices as shown in Table 28.1. Assume that the transmitted symbols are drawn from q-ary QAM alphabet. We will use the following representation for the matrix elements: white, gray, and black designate the zero, real, and complex elements, respectively.

We start with matrix A. Because the only nonzero elements are the diagonal elements, the symbols x_1 through x_4 are separately decodable. Furthermore, since the diagonal elements are always real by definition, the real and the imaginary parts of the symbols are also separable. As a result, the worst-case decoding complexity is $\mathcal{O}(1)$.

For matrix B, the 2×2 subblock consisting of the elements $r_{1,3}, r_{1,4}, r_{2,3}$, and $r_{2,4}$ is zero. As a result, the symbols $\{x_1, x_2\}$ are separable from the symbols $\{x_3, x_4\}$. Furthermore, the elements $r_{1,2}$ and $r_{3,4}$ are real and hence, the real and the imaginary components of the symbols $\{x_1, x_2\}$ and $\{x_3, x_4\}$ are separable. The decoding can be done over four independent or separate groups; $\Re\{x_1, x_2\}$, $\Im\{x_1, x_2\}$, $\Re\{x_3, x_4\}$, and $\Im\{x_3, x_4\}$. To determine the worst-case decoding complexity, consider the decoding of $\Re\{x_1, x_2\}$. Since there are $\sqrt{q}$ ways to pick the PAM symbol $\Re(x_1)$, and the symbol $\Re(x_2)$ can be decoded with a simple slicer after canceling the interference from $\Re(x_1)$, the worst-case decoding complexity is $\mathcal{O}(\sqrt{q})$.

For matrix C, the symbols $\{x_1, x_2\}$ are also separable from the symbols $\{x_3, x_4\}$, since the 2×2 subblock consisting of the elements $r_{1,3}, r_{1,4}, r_{2,3}$, and $r_{2,4}$ is zero. However, the elements $r_{1,2}$ and $r_{3,4}$ are complex, and hence, there is no further reduction in complexity. To determine the worst-case decoding complexity, consider the decoding of $\{x_1, x_2\}$. Since there are q ways to pick the symbol x_1, and the symbol x_2 can be decoded with a simple slicer after canceling the interference from x_1, the worst-case decoding complexity is $\mathcal{O}(q)$.

For matrix D, all the elements are real, and hence the real part of the transmitted symbols is separable from the imaginary parts. Therefore, the decoding can be done over two independent groups; $\Re\{x_1, x_2, x_3, x_4\}$ and $\Im\{x_1, x_2, x_3, x_4\}$. To determine the worst-case decoding complexity, consider the decoding of $\Re\{x_1, x_2, x_3, x_4\}$. There are $\sqrt{q} \times \sqrt{q} \times \sqrt{q}$ ways to pick the three PAM symbols $\Re\{x_1, x_2, x_3\}$, and the symbol x_4 can be decoded with a simple slicer after canceling the interference from $\Re\{x_1, x_2, x_3\}$. Therefore, the worst-case decoding complexity is

$$\mathcal{O}(\sqrt{q} \times \sqrt{q} \times \sqrt{q}) = \mathcal{O}(q^{1.5}).$$

For matrices A through D, the decoding of the transmitted symbols was done over groups of two or more. Specifically, the decoding was done over 8, 4, 2, and 2 groups for matrices A, B, C, and D, respectively. For matrices E through G, the decoding cannot be done over independent groups of symbols. A reduction in decoding complexity, however, might still be possible. We next discuss the worst-case decoding complexity for matrices E, F, and G.

TABLE 28.1 The **R** Matrix and Its Relationship to Worst-Case Decoding Complexity

R Matrix	Worst-Case Decoding Complexity
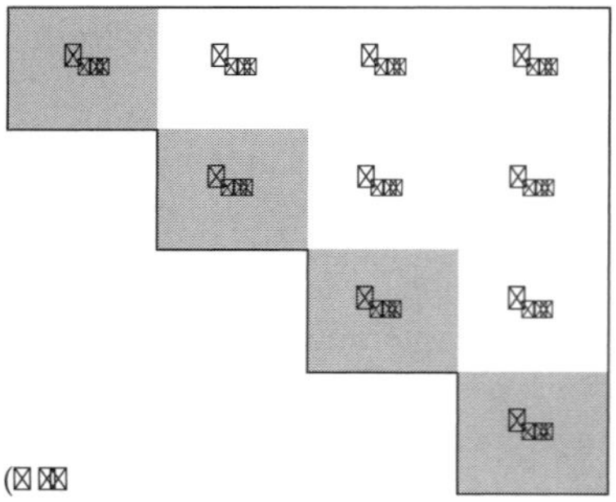	$1 \Rightarrow$ All symbol separately decodable $2 \Rightarrow$ Real and imaginary parts separately decodable $\Rightarrow \mathcal{O}(1)$
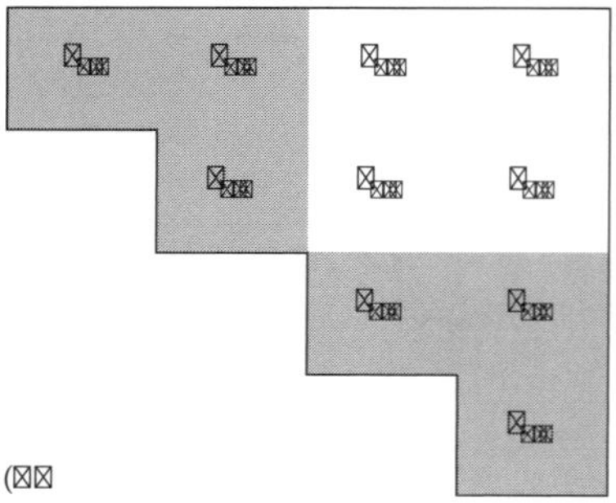	$1 \Rightarrow$ Symbols (x_1, x_2) are separately decodable from (x_3, x_4) $2 \Rightarrow$ (a) $\Re(x_1, x_2)$ separately decodable from $\Im(x_1, x_2)$ (b) $\Re(x_3, x_4)$ separately decodable from $\Im(x_3, x_4)$ $\Rightarrow \mathcal{O}(\sqrt{q})$
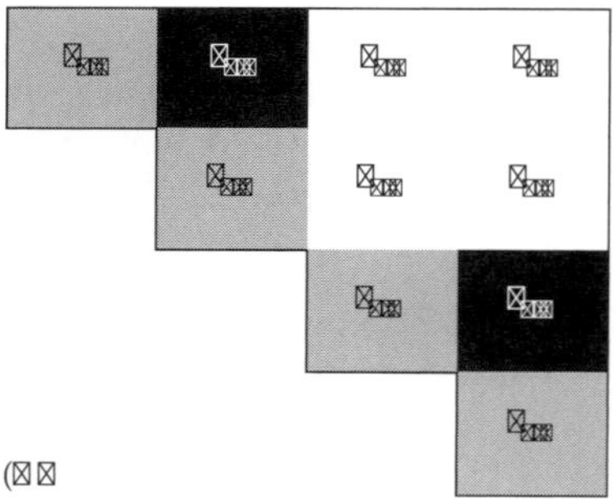	$1 \Rightarrow$ symbols (x_1, x_2) are separately decodable from (x_3, x_4) $\Rightarrow \mathcal{O}(q)$
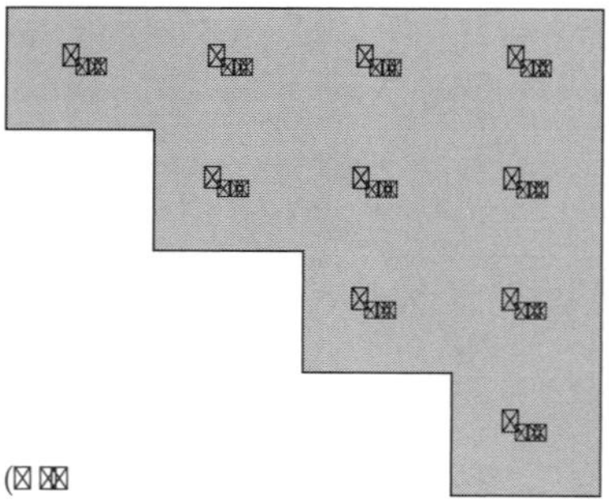	$2 \Rightarrow \Re(x_1, x_2, x_3, x_4)$ separable from $\Im(x_1, x_2, x_3, x_4)$ $\Rightarrow \mathcal{O}(q^{1.5})$

TABLE 28.1 (continued) The **R** Matrix and Its Relationship to Worst-Case Decoding Complexity

R Matrix	Worst-Case Decoding Complexity
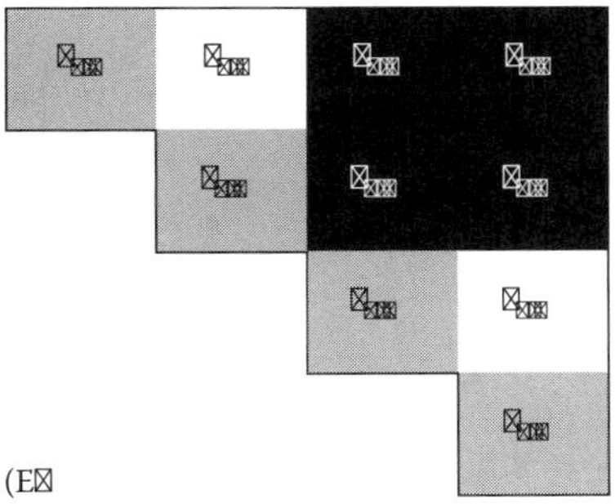 (E⊠	$3 \Rightarrow$ In the absence of (x_3, x_4): 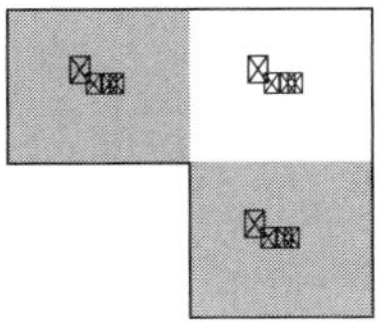 $1 \Rightarrow x_1$ separable from x_2 $2 \Rightarrow \Re(x_1, x_2)$ separable from $\Im(x_1, x_2)$ $\Rightarrow \mathcal{O}(q^2)$
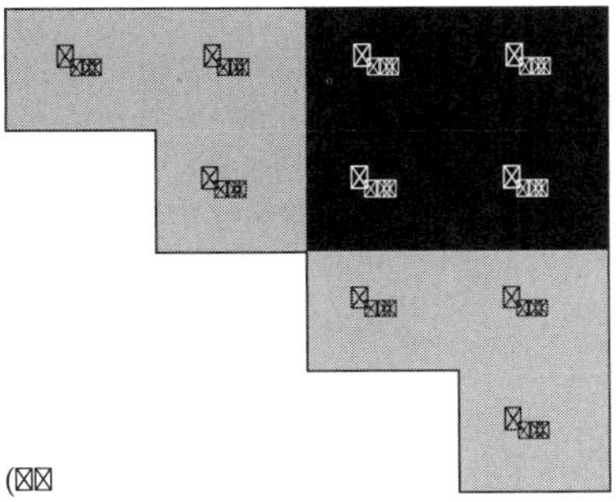 (⊠⊠	$3 \Rightarrow$ In the absence of (x_3, x_4): 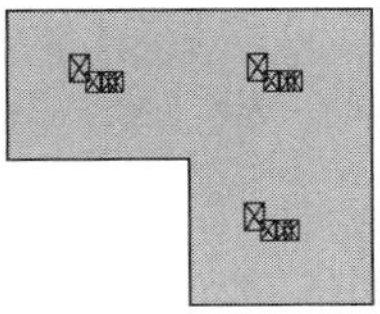 $2 \Rightarrow \Re(x_1, x_2)$ separable from $\Im(x_1, x_2)$ $\Rightarrow \mathcal{O}(q^{2.5})$
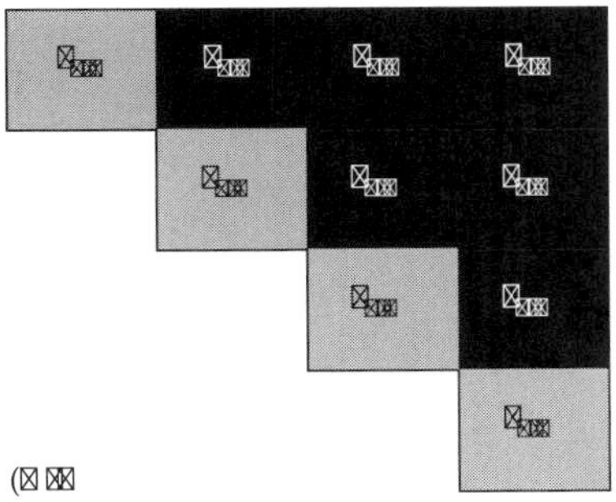 (⊠⊠⊠	$\Rightarrow \mathcal{O}(q^3)$

For matrix E, we note that the element $r_{1,2}$ is zero. This is significant, since in the absence of interference from the symbols x_3 and x_4, we can decode the symbols x_1 and x_2 independently. In fact, in the absence of x_3 and x_4, the decoding complexity of x_1 and x_2 is $\mathcal{O}(1)$, since the decoding can be done with a slicer over four independent groups; $\Re(x_1)$, $\Im(x_1)$, $\Re(x_2)$, and $\Im(x_2)$. Therefore, the worst-case decoding complexity is $\mathcal{O}(q^2)$, where the factor q^2 comes from the number of ways to pick the pair $\{x_3, x_4\}$.

For matrix F, we note that the element $r_{1,2}$ is real. This is also significant, since in the absence of interference from the symbols x_3 and x_4, we can decode $\Re\{x_1, x_2\}$ and $\Im\{x_1, x_2\}$ independently. In the absence of x_3 and x_4, the decoding complexity of $\{x_1, x_2\}$ is $\mathcal{O}(\sqrt{q})$. This is because the decoding can be done over two independent groups; $\Re\{x_1, x_2\}$ and $\Im\{x_1, x_2\}$, and the decoding complexity of each group is $\mathcal{O}(\sqrt{q})$. Therefore, the worst-case decoding complexity is $\mathcal{O}(q^2 \times q^{0.5}) = \mathcal{O}(q^{2.5})$, where the factor q^2 comes from the number of ways to pick the pair $\{x_3, x_4\}$ and the factor $q^{0.5}$ comes from the decoding complexity of $\{x_1, x_2\}$, in the absence of x_3 and x_4.

Finally, for matrix G, all the matrix elements are complex, except for the diagonal elements. As a result, there is no reduction in decoding complexity beyond the reduction offered by a practical implementation of a slicer. Specifically, in the absence of the symbols x_2, x_3, and x_4, the decoder decides on the symbol x_1 with a slicer. As a result, the worst-case decoding complexity is $\mathcal{O}(q^3)$, where the factor q^3 comes from the number of ways to choose $\{x_2, x_3, x_4\}$.

In the next section, we discuss the general form of the **R** matrix for the parent space–time codes that are used to generate high-rate space–time codes. In particular, we discuss a general representation of the **R** matrix for orthogonal, quasiorthogonal, diagonal algebraic, and semiorthogonal algebraic space–time (SAST) block codes.

28.4 The R Matrix for Space–Time Block Codes

We next discuss the general form of the **R** matrix for the orthogonal codes, the quasiorthogonal codes, the DAST block codes, and the SAST block codes. These families of space–time block codes have a transmission rate of one symbol per signaling interval; with the exception of orthogonal codes for more than two transmit antennas whose rate is strictly less than one. For clarity of exposition, we will consider the case of a single receive antenna in this section. We first discuss the general form of the **R** matrix for orthogonal space–time block codes.

28.4.1 Orthogonal Space–Time Block Codes

Orthogonal space–time block codes are characterized by low ML decoding complexity, such that each transmitted symbol is detected separately from the other transmitted symbols [2,9,10]. Because orthogonal designs transmit a symbol and its conjugate, it is not always possible to represent the received vector in matrix form in the complex domain. However, a matrix form of the received vector can be expressed in the real domain as follows:

$$\begin{bmatrix} \Re(y_1[1]) \\ \Im(y_1[1]) \\ \vdots \\ \Re(y_1[T]) \\ \Im(y_1[T]) \end{bmatrix} = \mathbf{H} \begin{bmatrix} \Re(x_1) \\ \Im(x_1) \\ \vdots \\ \Re(x_P) \\ \Im(x_P) \end{bmatrix} + \begin{bmatrix} \Re(w_1[1]) \\ \Im(w_1[1]) \\ \vdots \\ \Re(w_1[T]) \\ \Im(w_1[T]) \end{bmatrix},$$

where $\mathbf{H} = [h_1 \quad h_2 \quad \cdots \quad h_{2P}]$ is the $2T \times 2P$ real effective channel matrix. All the columns in **H** are orthogonal to each other and have the same norm and hence, the resulting **R** matrix is diagonal and given by the general form:

$$\mathbf{R} = \begin{bmatrix} r_{1,1} & 0 & \cdots & 0 \\ 0 & r_{2,2} & \cdots & 0 \\ \vdots & \ddots & \ddots & \vdots \\ 0 & \cdots & 0 & r_{2P,2P} \end{bmatrix},$$

where $r_{1,1} = r_{2,2} = \cdots = r_{2P,2P} = \|h_1\|$. Although orthogonal space–time block codes are trivially decodable without resorting to QR decomposition, we will see later in the discussion that multiplexed orthogonal designs for two and four transmit antennas are not separable. As a result, a significant reduction in decoding complexity is possible by exploiting the special properties of the **R** matrix of the effective channel matrix.

28.4.2 Quasiorthogonal Space–Time Block Codes

Quasiorthogonal space–time codes enable rate-one transmission, at the expense of an increase in decoding complexity compared to orthogonal codes [11–17]. In particular, full diversity quasiorthogonal space–time block codes require the joint decoding of $M/2$ complex information symbols for M antennas, where M is a power of 2. The general form of the **R** matrix is then given by

$$\mathbf{R} = \begin{bmatrix} r_{1,1} & \cdots & r_{1,M/2} & 0 & \cdots & 0 \\ & \ddots & \vdots & \vdots & \ddots & \vdots \\ & & r_{M/2,M/2} & 0 & \cdots & 0 \\ & & & r_{M/2+1,M/2+1} & \cdots & r_{M/2+1,M} \\ & & & & \ddots & \vdots \\ & & & & & r_{M,M} \end{bmatrix}.$$

Except for the diagonal elements, the remaining elements in the **R** matrix are complex, in general.

28.4.3 Diagonal Algebraic Space–Time Block Codes

The DAST block codes are a family of linear space–time codes constructed by the use of rotated constellations [3,18]. For M transmit antennas, the DAST block codes transmit M information symbols in M signal intervals as follows:

$$\mathbf{C}_M^{\text{DAST}} = \begin{bmatrix} u_1 & 0 & \cdots & 0 \\ 0 & u_2 & \cdots & 0 \\ \vdots & \ddots & \ddots & \vdots \\ 0 & \cdots & 0 & u_M \end{bmatrix},$$

where $(u_1, u_2, \ldots, u_M)^{\mathrm{T}} = \Phi_M(x_1, x_2, \ldots, x_M)^{\mathrm{T}}$, and Φ_M is the complex or real rotation matrix that ensures a full diversity DAST code. The general form of the **R** matrix is then given by

$$\mathbf{R} = \begin{bmatrix} r_{1,1} & r_{1,2} & \cdots & r_{1,M} \\ 0 & r_{2,2} & \cdots & r_{2,M} \\ \vdots & \ddots & \ddots & \vdots \\ 0 & \cdots & 0 & r_{M,M} \end{bmatrix}.$$

If the generator matrix Φ_M is real, then all the elements of the **R** matrix are real. However, if Φ_M is complex, then the only real elements are the diagonal elements. Therefore, complex generator matrices have higher decoding complexity, but they outperform real generator matrices in terms of the bit-error-rate (BER) performance. We will assume the use of real generator matrices in constructing DAST codes for the remainder of the chapter because they offer a better complexity-performance trade-off than complex generator matrices since the slight increase in performance might not justify the significant increase in decoding complexity [18].

28.4.4 Semiorthogonal Algebraic Space–Time Block Codes

In SAST block codes, half the columns of the code matrix are orthogonal to the other half [19]. For M transmit antennas, where M is even, the semiorthogonal property of SAST codes allows for the separate decoding of symbols x_1 through $x_{M/2}$ from symbols $x_{M/2+1}$ through x_{2M}. Furthermore, the real and the imaginary parts of the QAM symbols are separately decodable due to the use of the real generator matrices. As a result, the SAST block codes for M antennas require the joint decoding of only $M/2$ real PAM symbols and the $\mathbf{R}$ matrix has the general form:

$$\mathbf{R} = \begin{bmatrix} r_{1,1} & \cdots & r_{1,M/2} & 0 & \cdots & 0 \\ & \ddots & \vdots & \vdots & \ddots & \vdots \\ & & r_{M/2,M/2} & 0 & \cdots & 0 \\ & & & r_{M/2+1,M/2+1} & \cdots & r_{M/2+1,M} \\ & & & & \ddots & \vdots \\ & & & & & r_{M,M} \end{bmatrix}.$$

The $\mathbf{R}$ matrix for SAST codes has the same general form as the quasiorthogonal codes, but all the elements of the $\mathbf{R}$ matrix are real. We end this section by giving examples of the code matrices for two and four transmit antennas along with the general form of the $\mathbf{R}$ matrix. We express the $\mathbf{R}$ matrix in the real domain for orthogonal designs (Table 28.2).

TABLE 28.2 The **R** Matrix for Orthogonal, Quasiorthogonal, DAST, and SAST Block Codes

M	$\mathbf{C}$	$\mathbf{R}$	Decoding Complexity
Orthogonal space–time block codes			
2	$\begin{bmatrix} x_1 & x_2 \\ -x_2^* & x_1^* \end{bmatrix}$	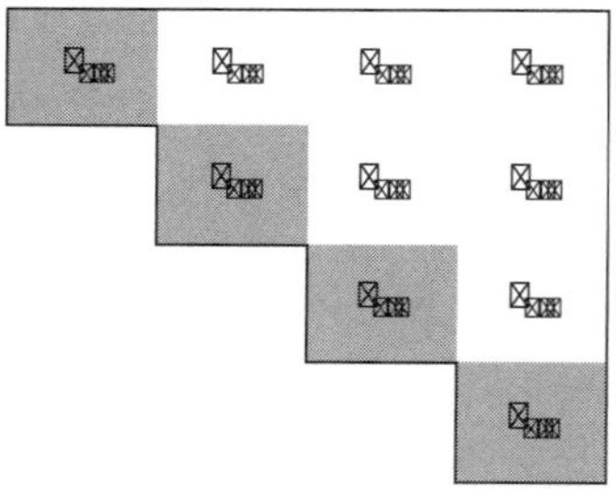	$\mathcal{O}(1)$
4	$\begin{bmatrix} x_1 & x_2 & x_3 & 0 \\ -x_2^* & x_1^* & 0 & x_3 \\ -x_3^* & 0 & x_1^* & -x_2 \\ 0 & -x_3^* & x_2^* & x_1 \end{bmatrix}$	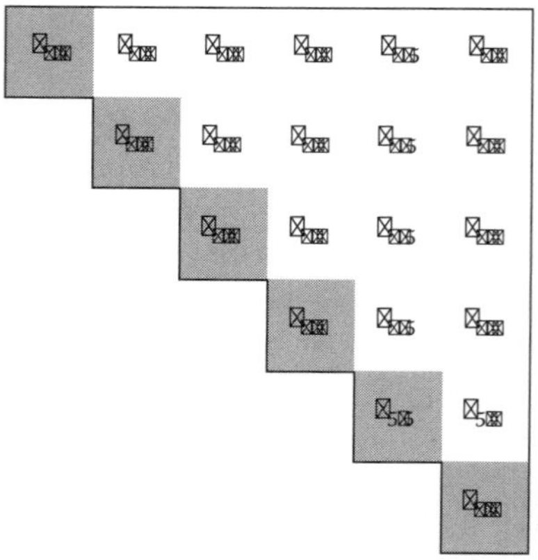	$\mathcal{O}(1)$

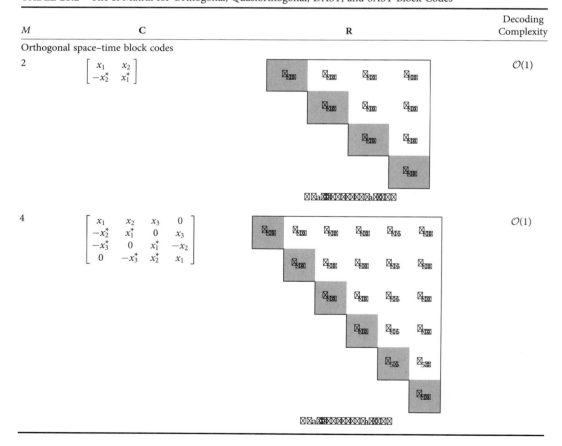

TABLE 28.2 (continued) The **R** Matrix for Orthogonal, Quasiorthogonal, DAST, and SAST Block Codes

M	**C**	**R**	Decoding Complexity
Quasiorthogonal space–time block codes			
4	$\begin{bmatrix} x_1 & x_2 & x_3 & x_4 \\ -x_2^* & x_1^* & -x_4^* & x_3^* \\ -x_3^* & -x_4^* & x_1^* & x_2^* \\ x_4 & -x_3 & -x_2 & x_1 \end{bmatrix}$ Not full diversity	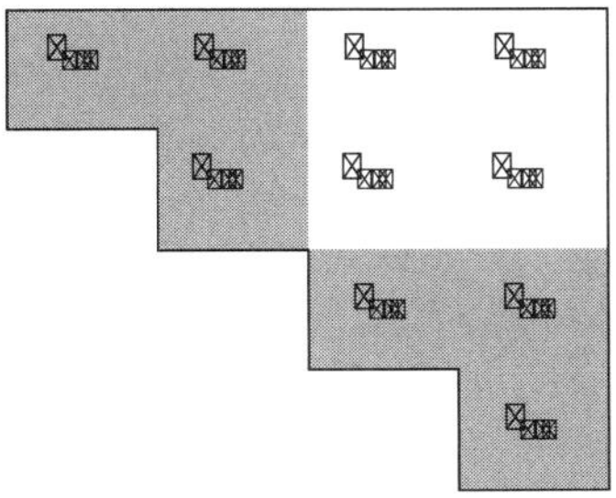	$\mathcal{O}(\sqrt{q})$
4	$\begin{bmatrix} x_1 & x_2 & x_3 e^{j\phi} & x_4 e^{j\phi} \\ -x_2^* & x_1^* & -x_4^* e^{-j\phi} & x_3^* e^{-j\phi} \\ -x_3 e^{j\phi} & -x_4 e^{j\phi} & x_1 & x_2 \\ x_4^* e^{-j\phi} & -x_3^* e^{-j\phi} & -x_2^* & x_1^* \end{bmatrix}$ Full diversity with $\phi = \pi/4$	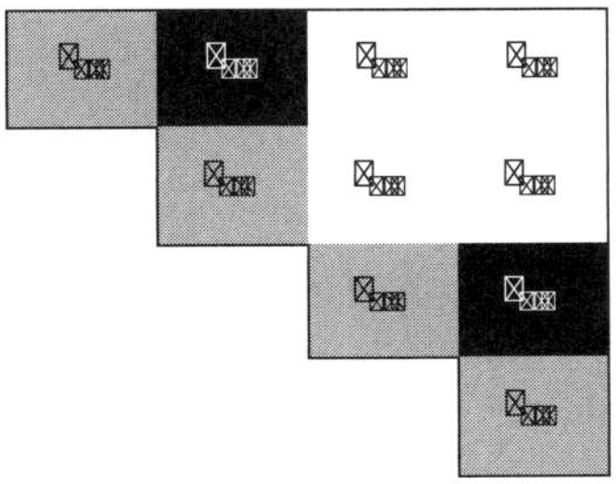	$\mathcal{O}(q)$
DAST block codes			
2	$\begin{bmatrix} \alpha x_1 + \beta x_2 & 0 \\ 0 & -\beta x_1 + \alpha x_2 \end{bmatrix}$ $\alpha = \cos(\theta)$, $\beta = \sin(\theta)$ and $\theta = \frac{1}{2}\tan^{-1}(2)$	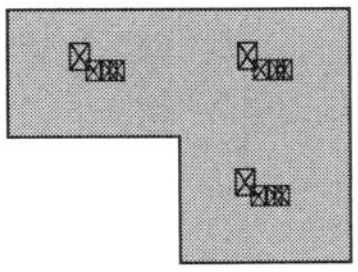	$\mathcal{O}(\sqrt{q})$
4	$\begin{bmatrix} u_1 & 0 & \cdots & 0 \\ 0 & u_2 & \cdots & 0 \\ \vdots & \ddots & u_3 & \vdots \\ 0 & \cdots & 0 & u_4 \end{bmatrix}$ $\begin{bmatrix} u_1 \\ u_2 \\ u_3 \\ u_4 \end{bmatrix} = \begin{bmatrix} 0.201 & 0.326 & 0.486 & 0.786 \\ -0.326 & 0.201 & -0.786 & 0.486 \\ -0.486 & -0.786 & 0.201 & 0.326 \\ 0.786 & -0.486 & -0.326 & 0.201 \end{bmatrix} \begin{bmatrix} x_1 \\ x_2 \\ x_3 \\ x_4 \end{bmatrix}$	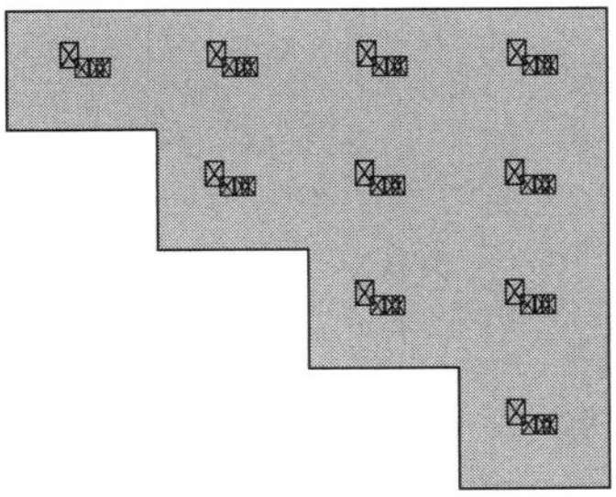	$\mathcal{O}(q^{1.5})$
SAST block codes			
4	$\begin{bmatrix} u_1 & 0 & u_3 & 0 \\ 0 & u_2 & 0 & u_4 \\ -u_3^* & 0 & u_1^* & 0 \\ 0 & -u_4^* & 0 & u_2^* \end{bmatrix}$ $\begin{bmatrix} u_1 \\ u_2 \end{bmatrix} = \begin{bmatrix} \alpha & \beta \\ -\beta & \alpha \end{bmatrix} \begin{bmatrix} x_1 \\ x_2 \end{bmatrix}$, $\begin{bmatrix} u_3 \\ u_4 \end{bmatrix} = \begin{bmatrix} \alpha & \beta \\ -\beta & \alpha \end{bmatrix} \begin{bmatrix} x_3 \\ x_4 \end{bmatrix} \alpha = \cos(\theta)$, $\beta = \sin(\theta)$, and $\theta = \frac{1}{2}\tan^{-1}(2)$	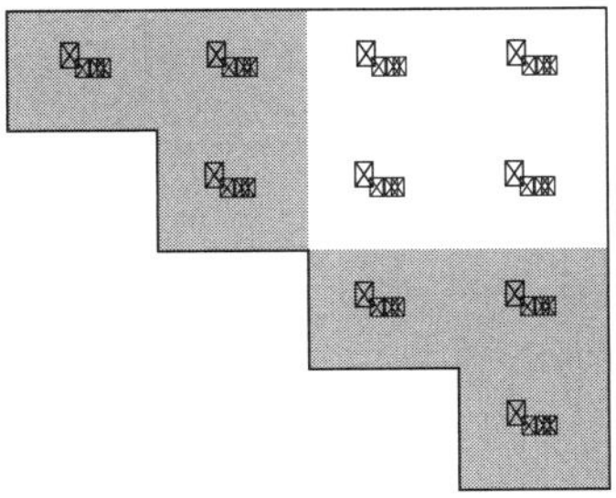	$\mathcal{O}(\sqrt{q})$

28.5 Rate-Two Space–Time Block Codes for Two Transmit Antennas

In this section, we discuss the construction of two families of rate-two space–time block codes for the 2×2 channel. The starting point is the rate-one space–time block code for the 2×1 channel. We begin the discussion by introducing a common framework for the construction of the rate-two space–time block codes as the sum of rate-one codes. In particular, the two families of codes may be described by the following 2×2 space–time code:

$$\mathbf{C}(x_1, x_2, x_3, x_4) = \mathbf{C}_p(x_1, x_2) + \mathbf{\Phi}_L \mathbf{C}_p(\bar{x}_3, \bar{x}_4)\mathbf{\Phi}_R,$$

where

$\mathbf{C}_p(x_1, x_2)$ is the rate-one encoder (parent code) for symbols x_1 and x_2

$\begin{bmatrix} \bar{x}_3 \\ \bar{x}_4 \end{bmatrix} = \mathbf{\Phi}_P \begin{bmatrix} x_3 \\ x_4 \end{bmatrix}$

$\mathbf{\Phi}_P$ is a unitary precoding matrix

$\mathbf{\Phi}_L$ and $\mathbf{\Phi}_R$ are 2×2 matrices that multiply $\mathbf{C}_p(x_1, x_2)$ on the left and right, respectively

The four parameters $\mathbf{C}_p$, $\mathbf{\Phi}_P$, $\mathbf{\Phi}_L$, and $\mathbf{\Phi}_R$ are chosen to ensure full diversity, maximize the coding gain and reduce decoding complexity.

One possibility for $\mathbf{C}_p$ is the Alamouti block code; the other is the DAST block code for two transmit antennas. We note that for two transmit antennas, the quasiorthogonal and the SAST codes reduce to the orthogonal Alamouti space–time block code. By multiplexing two DAST code words, we can obtain the golden code [20,21], and by multiplexing two Alamouti code words, we can obtain the overlaid Alamouti codes [5–8]. The four parameters for the construction of the golden and overlaid Alamouti codes are given in Table 28.3.

The constants φ_1, φ_2, and γ are as follows:

$$\varphi_1 = \frac{1+j}{\sqrt{7}}, \quad \varphi_2 = \frac{1+2j}{\sqrt{7}}, \quad \text{and} \quad \gamma = (1-\sqrt{7}) + j(1+\sqrt{7}).$$

For two transmit antennas, the Alamouti space–time code is not only lower in complexity than the DAST code, but it also performs better in terms of the required signal-to-noise ratio (SNR) required to achieve a target BER. However, the multiplexed DAST code (i.e., golden code) performs better than the multiplexed Alamouti codes (i.e., overlaid Alamouti codes) as shown in Figure 28.1 for 4-QAM and 16-QAM.

In terms of decoding complexity, however, the overlaid Alamouti codes have lower worst-case decoding complexity than the golden code. The golden and overlaid Alamouti codes are not separable. However, a reduction in decoding complexity is possible by examining the **R** matrix, given in Table 28.4.

TABLE 28.3 Parameters for the Construction of the Golden and Overlaid Alamouti Codes

Code	C_p	Φ_P	Φ_L	Φ_R
Golden code [20,21]	DAST	I	$e^{j\pi/4}\mathrm{I}$	$\begin{bmatrix} 0 & 1 \\ 1 & 0 \end{bmatrix}$
Overlaid Alamouti 1 [Tirkonnen et al.] [5,8]	Alamouti	$\begin{bmatrix} \varphi_1 & \varphi_2 \\ -\varphi_2^* & \varphi_1^* \end{bmatrix}$	$\begin{bmatrix} 1 & 0 \\ 0 & -1 \end{bmatrix}$	I
Overlaid Alamouti 2 [Sezginer and Sari] [6]	Alamouti	I	$\frac{\gamma}{\lvert\gamma\rvert}\begin{bmatrix} 1 & 0 \\ 0 & -j \end{bmatrix}$	I

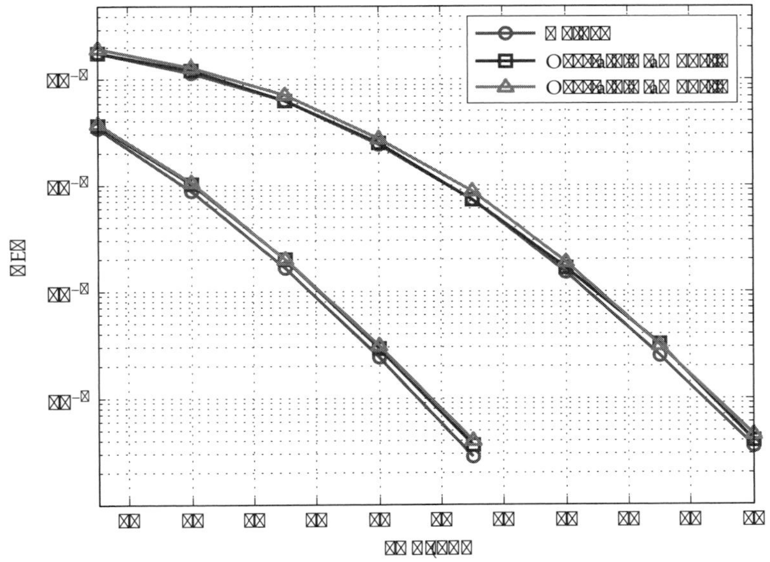

FIGURE 28.1 Performance of the golden and overlaid Alamouti codes for the two-input two-output channel.

For the golden code, the **R** matrix is similar in form to matrix F of Example 28.4. As a result, the worst-case decoding complexity is $\mathcal{O}(q^{2.5})$. The **R** matrix for the overlaid Alamouti codes is similar in form to matrix E of Example 28.4, and hence, the worst-case decoding complexity is $\mathcal{O}(q^2)$.

We end this section by noting that the rate-two space–time block codes for the 2×2 channel illustrate the difficulty of constructing high-rate codes. The difficulty arises from situations where the best rate-one or parent code does not always generate the best performing rate-two space–time block code. However, designing high-rate codes by multiplexing lower-rate parent codes is a reasonable strategy since it simplifies the design problem into two simpler problems. The first is to design a lower-rate parent code and the second is to multiplex or combine these parent codes into a high-rate space–time code.

28.6 High-Rate Space–Time Block Codes for Four Transmit Antennas

In this section, we discuss the construction of four families of high-rate space–time block codes for the 4×2 channel. Similar to the rate-two codes for the 2×2 channel, the starting point is the rate-one or rate-3/4 space–time block code for the 4×1 channel. Three of these families have a rate-two since they are based on multiplexing two rate-one codes. The fourth family has a rate of 1.5 symbols per signaling interval, since it is based on multiplexing two rate-3/4 orthogonal codes. Let κ be the rate of the code, then the four families of codes may be described by the following 4×4 space–time code:

$$\mathbf{C}(x_1, \ldots, x_{4\kappa}) = \mathbf{C}_p(x_1, \ldots, x_{2\kappa}) + \mathbf{\Phi}_L \mathbf{C}_p(\bar{x}_{2\kappa+1}, \ldots, \bar{x}_{4\kappa}) \mathbf{\Phi}_R,$$

where

$\mathbf{C}_p(x_1, \ldots, x_{2\kappa})$ is the rate $\frac{\kappa}{2}$ parent code for symbols x_1 through $x_{2\kappa}$

$[\bar{x}_{2\kappa+1}, \ldots, \bar{x}_{4\kappa}]^{\mathrm{T}} = \mathbf{\Phi}_P [x_{2\kappa+1}, \ldots, x_{4\kappa}]^{\mathrm{T}}$

$\mathbf{\Phi}_P$ is a unitary precoding matrix

$\mathbf{\Phi}_L$ and $\mathbf{\Phi}_R$ are 4×4 matrices that multiply $\mathbf{C}_p(x_1, \ldots, x_{2\kappa})$ on the left and right, respectively

TABLE 28.4 Decoding Complexity and **R** Matrix of the Golden and Overlaid Alamouti Codes

Space–Time Code	**R**	Decoding Complexity
Golden code [20]	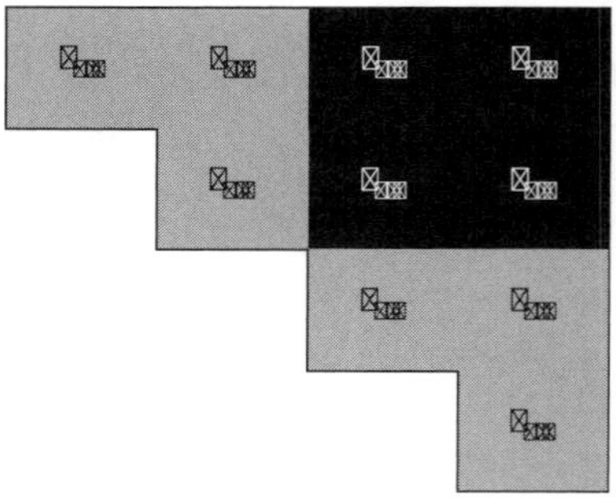	$\mathcal{O}(q^{2.5})$
Overlaid Alamouti 1 [Tirkonnen et al.] [5,8]	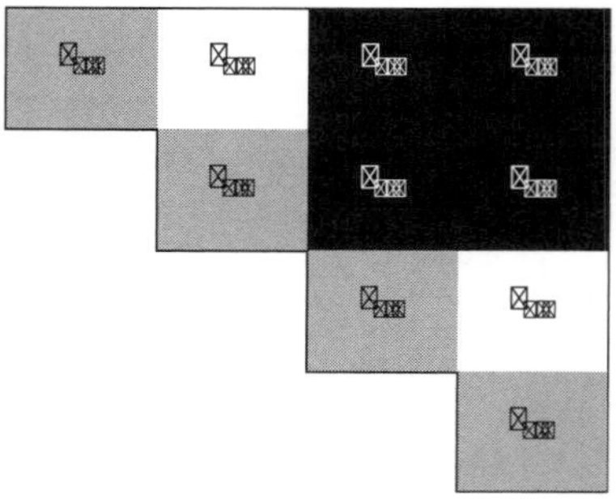	$\mathcal{O}(q^2)$
Overlaid Alamouti 2 [Sezginer and Sari] [6]	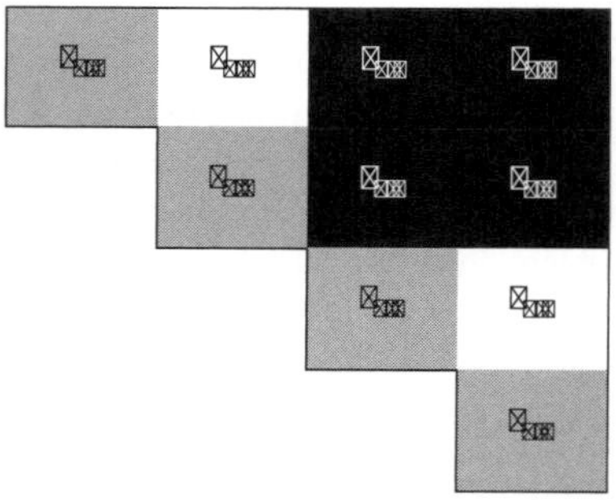	$\mathcal{O}(q^2)$

Based on our brief survey, there are four possibilities for the parent code; rate-3/4 orthogonal code, rate-one quasiorthogonal, rate-one DAST code, and rate-one SAST code. These four choices of the parent code give rise to four families of codes as shown in Table 28.5.

We compare the performance of the fast-decodable, embedded-Alamouti space–time (EAST) and threaded algebraic space–time (TAST) codes for 4-QAM alphabet in Figure 28.2. We see that the EAST and fast-decodable code not only outperform the TAST code, but they are also lower in complexity. The performance of the multiplexed orthogonal code is shown in Figure 28.3 for 4-QAM alphabet along with the EAST that has been punctured so that it also achieves a rate of 1.5 symbols per signaling interval. We omit the details for brevity, but it can be shown that the EAST code has a worst-case decoding complexity of $\mathcal{O}(q^{2.5})$. Therefore, the EAST code also outperforms the multiplexed orthogonal code and is lower in complexity.

TABLE 28.5 Parameters for the Construction of High-Rate Codes for the 4 × 2 Channel

Code	$\mathbf{C}_p$	$\mathbf{\Phi}_P$	$\mathbf{\Phi}_L$	$\mathbf{\Phi}_R$
Multiplexed orthogonal [1]	Orthogonal	$\frac{1}{\sqrt{5}}\begin{bmatrix} 1 & 1+i & 1+i & 1+i \\ -1+i & 1 & 1-i & 1+i \\ -1-i & 1+i & 1 & 1+i \\ 1-i & -1-i & -1+i & 1 \end{bmatrix}$	$\begin{bmatrix} 1 & 0 & 0 & 0 \\ 0 & 1 & 0 & 0 \\ 0 & 0 & -1 & 0 \\ 0 & 0 & 0 & -1 \end{bmatrix}$	I
Fast-decodable[a] [7]	Quasiorthogonal	$\frac{1}{2}\begin{bmatrix} e^{j\varphi_1} & e^{j\varphi_1} & e^{j\varphi_1} & e^{j\varphi_1} \\ e^{j\varphi_2} & -je^{j\varphi_2} & -e^{j\varphi_2} & je^{j\varphi_2} \\ e^{j\varphi_3} & -e^{j\varphi_3} & e^{j\varphi_3} & -e^{j\varphi_3} \\ e^{j\varphi_4} & je^{j\varphi_4} & -e^{j\varphi_4} & -je^{j\varphi_4} \end{bmatrix}$	$\begin{bmatrix} 1 & 0 & 0 & 0 \\ 0 & 1 & 0 & 0 \\ 0 & 0 & -1 & 0 \\ 0 & 0 & 0 & -1 \end{bmatrix}$	I
EAST [22]	SAST	I	$e^{j\pi/4}\mathbf{I}$	$\begin{bmatrix} 0 & 1 & 0 & 0 \\ 1 & 0 & 0 & 0 \\ 0 & 0 & 0 & 1 \\ 0 & 0 & 1 & 0 \end{bmatrix}$
TAST [23]	DAST	I	$e^{j\pi/24}\mathbf{I}$	$\begin{bmatrix} 0 & 1 & 0 & 0 \\ 0 & 0 & 1 & 0 \\ 0 & 0 & 0 & 1 \\ 1 & 0 & 0 & 0 \end{bmatrix}$

[a] $[\varphi_1\ \varphi_2\ \varphi_3\ \varphi_4]^T = [e^{j2\pi/7}\ e^{j4\pi/7}\ e^{j10\pi/7}\ e^{j12\pi/7}]$ for 4-QAM modulation alphabet and $[\varphi_1\ \varphi_2\ \varphi_3\ \varphi_4]^T = [e^{j6\pi/17}\ e^{j8\pi/17}\ e^{j10\pi/17}\ e^{j26\pi/17}]$ for 16-QAM.

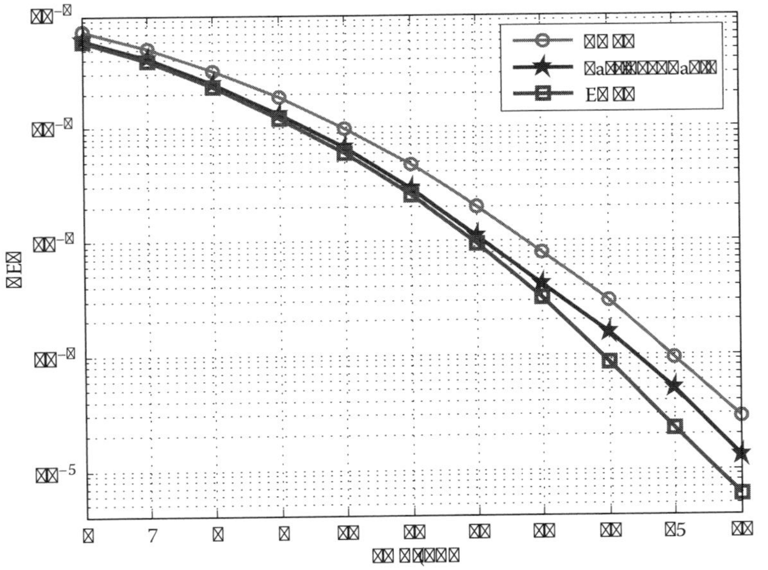

FIGURE 28.2 Performance of the fast-decodable, EAST, and TAST codes for the four-input two-output channel.

Similar to the high-rate codes for the 2 × 2 channel, the high-rate codes for the 4 × 2 channel are not separable. However, a reduction in decoding complexity is possible by exploiting the special structure of the **R** matrix, given in Table 28.6.

For the multiplexed orthogonal code, we note that in the absence of interference from the symbols x_4, x_5, and x_6, we can decode the symbols x_1, x_2, and x_3 with a slicer over six independent groups; $\Re(x_1)$, $\Im(x_1)$, $\Re(x_2)$, $\Im(x_2)$, $\Re(x_3)$, and $\Im(x_3)$. Therefore, the worst-case decoding complexity is $\mathcal{O}(q^3)$, where the factor q^3 comes from the number of ways to pick $\{x_4, x_5, x_6\}$.

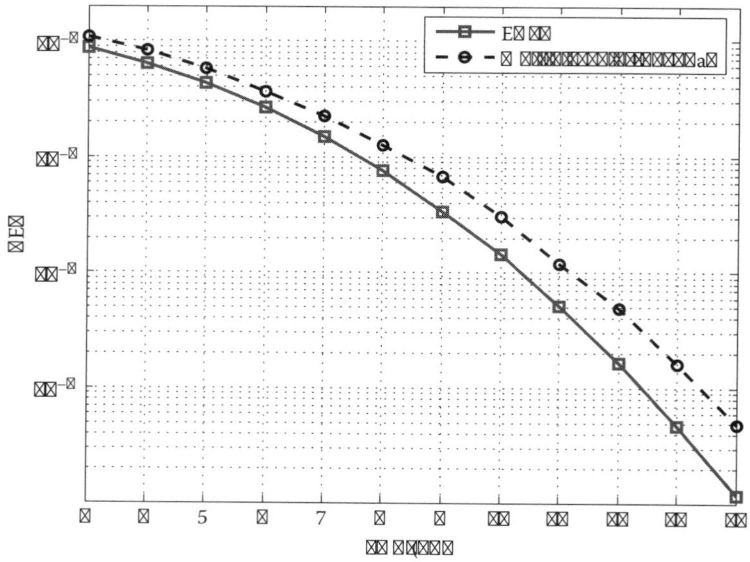

FIGURE 28.3 Performance of the EAST and multiplexed orthogonal code for the four-input two-output channel.

For the fast-decodable codes proposed in [7], the worst-case decoding complexity is $\mathcal{O}(q^{4.5})$. This is because in the absence of interference from symbols x_5, x_6, x_7, and x_8, the remaining symbols $\{x_1, x_2, x_3, x_4\}$ have a worst-case decoding complexity of $\mathcal{O}(q^{0.5})$ (see matrix **R** in Example 28.4). As a result, the worst-case decoding complexity is $\mathcal{O}(q^4 \times q^{0.5}) = \mathcal{O}(q^{4.5})$, where the factor q^4 comes from the number of ways to pick $\{x_5, x_6, x_7, x_8\}$ and the factor $q^{0.5}$ comes from the decoding complexity of $\{x_1, x_2, x_3, x_4\}$ in the absence of $\{x_5, x_6, x_7, x_8\}$.

For the EAST block codes proposed in [22], the **R** matrix is identical in form to the **R** matrix of the fast-decodable codes. Hence, the worst-case decoding complexity for the EAST codes is also $\mathcal{O}(q^{4.5})$.

TABLE 28.6 Decoding Complexity and **R** Matrix of the Fast-Decodable, Multiplexed Orthogonal, EAST, and TAST Codes

Code	**R**	Decoding Complexity
Multiplexed orthogonal [1]	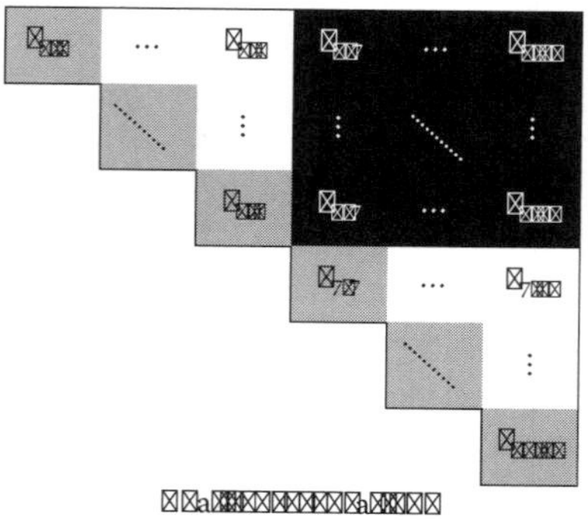	$\mathcal{O}(q^{3.5})$

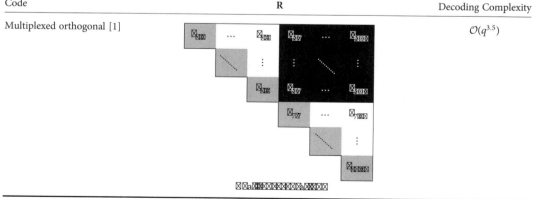

TABLE 28.6 (continued) Decoding Complexity and **R** Matrix of the Fast-Decodable, Multiplexed Orthogonal, EAST, and TAST Codes

Code	**R**	Decoding Complexity
Fast-decodable [7]	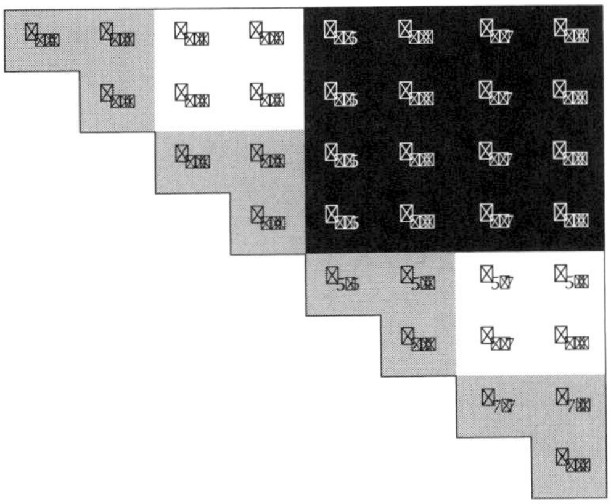	$\mathcal{O}(q^{4.5})$
EAST [22]		$\mathcal{O}(q^{4.5})$
TAST [23]	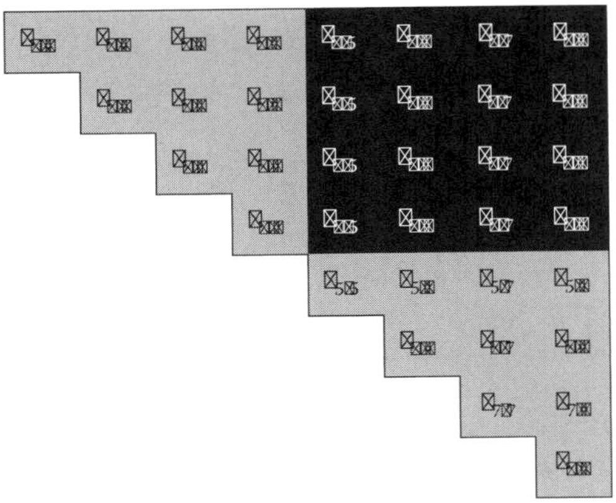	$\mathcal{O}(q^{5.5})$

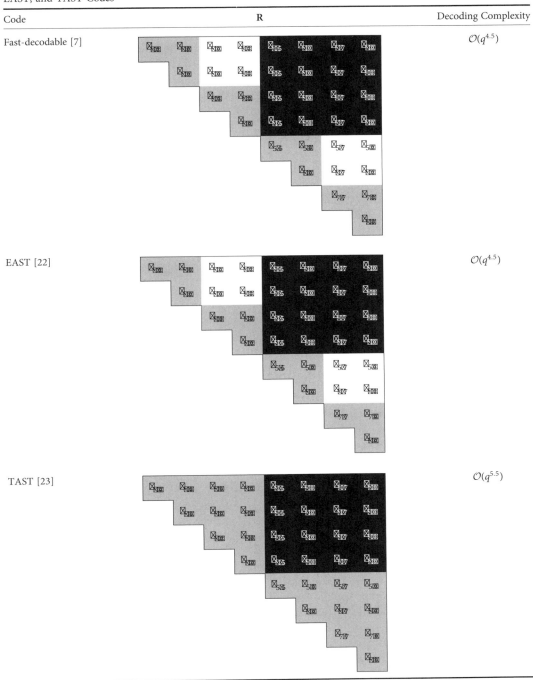

Finally, for the TAST codes proposed in [23], the worst-case decoding complexity is $\mathcal{O}(q^{5.5})$. This is because in the absence of interference from symbols x_5, x_6, x_7, and x_8, the remaining symbols $\{x_1, x_2, x_3, x_4\}$ have a worst-case decoding complexity of $\mathcal{O}(q^{1.5})$ (see matrix D in Example 28.4). As a result, the worst-case decoding complexity is $\mathcal{O}(q^4 \times q^{1.5}) = \mathcal{O}(q^{5.5})$, where the factor q^4 comes from the number of ways to pick $\{x_5, x_6, x_7, x_8\}$ and the factor $q^{1.5}$ comes from the decoding complexity of $\{x_1, x_2, x_3, x_4\}$ in the absence of $\{x_5, x_6, x_7, x_8\}$.

28.7 Conclusion

In this chapter, we introduced a general framework for the construction of high-rate space–time block codes for the 2×2 and 4×2 channels. These high-rate codes were constructed by multiplexing two lower-rate parent codes. Interestingly, starting with the best performing parent code does not always generate the best performing high-rate code. However, starting with the lowest-complexity parent code generates the lowest-complexity high-rate code. As a result, the design of high-rate codes with low ML decoding complexity does not necessarily entail a significant performance loss. In fact, for the 4×2 channel, the best performing space–time code is also the lowest in decoding complexity. We also discussed the worst-case ML decoding complexity for these high-rate codes. In particular, we showed that in the absence of interference from some symbols, the remaining symbols could be decoded efficiently resulting in reduction in the worst-case decoding complexity.

References

1. S. Sirianunpiboon, Y. Wu, R. A. Calderbank, and S. D. Howard, Fast optimal decoding of multiplexed orthogonal designs by conditional optimization, Submitted to *IEEE Trans. Inf. Theory*, 2008.
2. S. M. Alamouti, A simple transmit diversity technique for wireless communications, *IEEE J. Sel. Areas Commun.*, 16, 1451–1458, Oct. 1998.
3. M. O. Damen, K. Abed-Meraim, and J.-C. Belfiore, Diagonal algebraic space-time block codes, *IEEE Trans. Inf. Theory*, 48(3), 628–636, Mar. 2002.
4. M. O. Sinnokrot and J. R. Barry, The golden code is fast decodable, in *IEEE Global Telecommunications Conference* (Globecom), New Orleans, LA, pp. 1–5, Nov. 2008.
5. J. Paredes, A. B. Gershman, and M. G. Alkhanari, A 2×2 space-time code with non-vanishing determinants and fast maximum likelihood decoding, in *Proceedings of the IEEE International Conference on Acoustics, Speech, and Signal Processing*, ICASSP 2007, Honolulu, HI, pp. 877–880, Apr. 2007.
6. S. Sezginer and H. Sari, Full-rate full-diversity 2×2 space-time codes for reduced decoding complexity, *IEEE Commun. Lett.*, 11(12), 1–3, Dec. 2007.
7. E. Biglieri, Y. Hong, and E. Viterbo, On fast decodable space-time block codes, *IEEE Trans. Inf. Theory*, 55(2), 524–530, 2009.
8. O. Tirkkonen and R. Kashaev, Combined information and performance optimization of linear MIMO modulations, in *Proceedings of the IEEE International symposium on Information Theory (ISIT 2002)*, Lausanne, Switzerland, p. 76, June 2002.
9. V. Tarokh, H. Jafarkhani, and A. Calderbank, Space-time block codes from orthogonal designs, *IEEE Trans. Inf. Theory*, 45(5), 1456–1467, July 1999.
10. G. Ganesan and P. Stoica, Space–time block codes: A maximum SNR approach, *IEEE Trans. Inf. Theory*, 47, 1650–1656, May 2001.
11. H. Jafarkhani, A quasi-orthogonal space-time block code, *IEEE Trans. Commun.*, 49(1), 1–4, Jan. 2001.
12. O. Tirkkonen, A. Boariu, and A. Hottinen, Minimal nonorthogonality rate 1 space-time block code for 3+ Tx antennas, in *Proceedings of the IEEE 6th International symposium on Spread-Spectrum Techniques and Applications (ISSSTA)*, Parsippany, NJ, pp. 429–432, Sept. 2000.

13. C. B. Papadias and G. J. Foschini, Capacity-approaching space-time codes for systems employing four transmitter antennas, *IEEE Trans. Inf. Theory*, 49(3), 726–732, Mar. 2003.
14. O. Tirkkonen, Optimizing space-time block codes by constellation rotations, in *Proceedings of the Finnish Wireless Communication Workshop (FWCW)*, Finland, pp. 59–60, Oct. 2001.
15. N. Sharma and C. Papadias, Improved quasi-orthogonal codes through constellation rotation, *IEEE Trans. Commun.*, 51(3), 332–335, Mar. 2003.
16. W. Su and X.-G. Xia, Signal constellations for quasi-orthogonal space-time block codes with full diversity, *IEEE Trans. Inf. Theory*, 50(10), 2331–2347, Oct. 2004.
17. N. Sharma and C. B. Papadias, Full rate full diversity linear quasi-orthogonal space-time codes for any transmit antennas, *EURASIP J. Appl. Signal Process.*, 2004(9), 1246–1256, Aug. 2004.
18. M. O. Damen and N. C. Beaulieu, On diagonal algebraic space-time block codes, *IEEE Trans. Commun.*, 51(6), 911–919, June 2003.
19. D. N. Dao, C. Yuen, C. Tellambura, Y. L. Guan, and T. T. Tjhung, Four-group decodable space-time block codes, *IEEE Trans. Signal Process.*, 56, 424–430, Jan. 2008.
20. J. -C. Belfiore, G. Rekaya, and E. Viterbo, The golden code: A 2×2 full rate space-time code with non vanishing determinants, *IEEE Trans. Inf. Theory*, 51(4), 1432–1436, Apr. 2005.
21. P. Dayal and M. K. Varanasi, An optimal two transmit antenna space-time code and its stacked extensions, *IEEE Trans. Inf. Theory*, 51(12), 4348–4355, Dec. 2005.
22. M. O. Sinnokrot, J. R. Barry, and V. Madisetti, Embedded Alamouti space-time codes for high rate and low decoding complexity, *Asilomar Conference on Signals, Systems, and Computers (Asilomar 2008)*, Pacific Grove, CA, Oct. 26–29, 2008.
23. M. O. Damen, H. E. Gamal, and N. C. Beaulieu, Linear threaded algebraic space-time constellations, *IEEE Trans. Inf. Theory*, 49, 2372–2388, Oct. 2003.

29

Soft-Output Detection of Multiple-Input Multiple-Output Channels

David L. Milliner
Georgia Institute of Technology

John R. Barry
Georgia Institute of Technology

29.1 Introduction 29-1
29.2 System Model and Notation 29-2
29.3 Problem Statement 29-3
29.4 List Tree Search 29-7
Optimal Search • Suboptimal Search • Candidate Adding
29.5 Discussion 29-17
Performance Comparison
29.6 Further Information 29-21
References 29-21

29.1 Introduction

The use of multiple antennas at the transmitter and the receiver leads to a multiple-input multiple-output (MIMO) channel that can significantly increase the data rate and the reliability of a wireless communication link, without the need for increased transmit power or additional bandwidth. MIMO technology is thus a key enabling technology for many next-generation wireless services, such as real-time streaming of video conferences on portable devices, and seamless nationwide high-speed wireless Internet connectivity.

Like any communication system, a MIMO system relies on error-control coding to ensure reliable communication in the face of noise. And while error-control coding is relatively easy to implement at the transmitter, even for a MIMO system, the problem of error-control decoding at the receiver is greatly complicated by the presence of a MIMO channel. In fact, an optimal receiver that jointly accounts for the mutual interference of a MIMO channel and error-control coding is completely intractable for any realistic code and channel. For this reason, a practical MIMO receiver will decompose the detection problem into two steps: (1) MIMO detection, which is the topic of this chapter, and (2) error-control decoding.

The role of the MIMO detector is to account for the MIMO channel, largely ignoring the presence of the coding, while the role of the decoder is to account for the presence of the error-control code, largely ignoring the effects of the MIMO channel.

Ignoring the presence of coding is tantamount to assuming that the coded bits are independent and equally likely to be 0 or 1. With this assumption, the best the MIMO detector can do is to compute the *a posteriori* probability (APP) for each of the coded bits, which is the conditional probability that each coded bit is 1 given the observation of the channel output.

A suboptimal MIMO detector would make a hard decision about each coded bit by comparing each APP to a threshold of one-half. A MIMO detector that produces binary decisions about the coded bits is called a "hard-output" detector; anything else is called a "soft-output" detector. A MIMO detector that quantizes each APP to two or more bits of precision is an example of soft-output detector, while the full APP detector is the ultimate soft-output detector.

The problem of soft-output detection is critically important for two fundamental reasons: first, the performance of the error-control decoder depends critically on how well its inputs approximate the true APPs, and second, the high complexity of soft-output detection can easily dominate the other receiver tasks such as error-control decoding. Specifically, the complexity of exact APP computation grows exponentially with both the number of transmit antennas and the number of bits per transmitted symbol, and is prohibitively complex even for MIMO systems with moderately small antenna arrays and alphabets. In summary, the soft-output detector is often the critical determining factor in both the performance and the complexity of the overall system. Fortunately, there are approximate APP detectors with significantly reduced complexity that perform close to exact APP detectors. The significant impact of soft detection on the performance and complexity of MIMO systems makes an efficient and accurate soft-output detector essential for modern communication systems.

This chapter begins by reviewing the MIMO channel model. Next, we describe the ultimate soft-output detector that computes the APPs exactly. We then present a summary of the suboptimal soft-output detectors that offer a better performance-complexity trade-off. We then discuss the performance and computational complexity of the different algorithms. We conclude the chapter with suggestions for further reading.

29.2 System Model and Notation

Throughout this chapter we adopt the N_t-input N_r-output MIMO channel of Figure 29.1, so that the equivalent complex baseband model is

$$\mathbf{r} = \mathbf{Ha} + \mathbf{w}, \tag{29.1}$$

where

$\mathbf{a} \in \mathbb{C}^{[N_t \times 1]}$ is the vector of transmitted symbols, one for each transmit antenna
$\mathbf{r} \in \mathbb{C}^{[N_r \times 1]}$ is the vector of received samples, one for each receive antenna
$\mathbf{w} \in \mathbb{C}^{[N_r \times 1]}$ is additive noise
$\mathbf{H} \in \mathbb{C}^{[N_r \times N_t]}$ is the channel matrix

We are thus assuming a single-carrier narrowband flat-fading channel. We assume additive white Gaussian noise (AWGN), so that the components of $\mathbf{w}$ are zero-mean, circularly symmetric, i.i.d. complex Gaussian random variables with variance N_0, so that $\varepsilon[\mathbf{ww}^*] = N_0\mathbf{I}$, where $\mathbf{w}^*$ denotes the complex conjugate of $\mathbf{w}$.

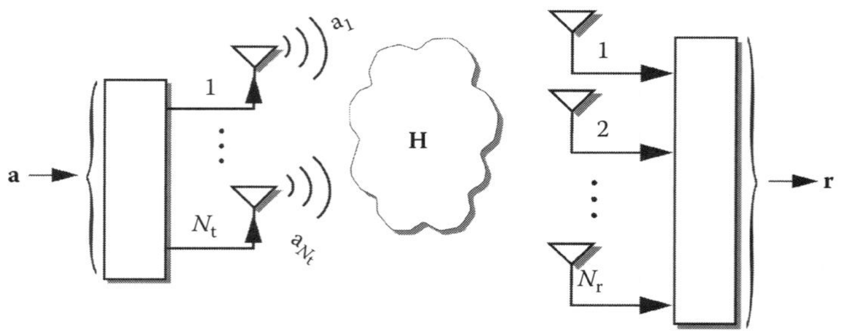

FIGURE 29.1 A MIMO channel.

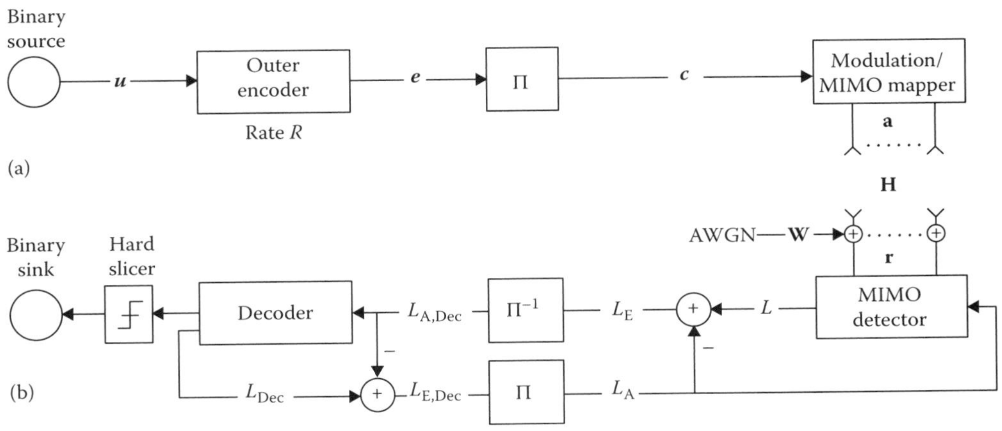

FIGURE 29.2 Coded system model depicting (a) MIMO transmitter and (b) MIMO receiver.

The entry in the *i*th row and the *j*th column of **H** represents the complex channel gain between transmit antenna *j* and receive antenna *i*. We assume Rayleigh fading, typical of non-line-of-sight communication systems, so that the entries of **H** are independant and identically-distributed (i.i.d.) complex Gaussian random variables. We assume that the entries of **a** are chosen uniformly and independently from the same complex q-ary quadrature amplitude modulation (QAM) alphabet $\mathcal{A}$ with energy E/N_t, where $q = |\mathcal{A}|$. The signal-to-noise ratio (SNR) at any receive antenna is $\text{SNR} = E/N_0$. To avoid the complexity challenge inherent in underdetermined systems [1], we assume $N_r \geq N_t$. Additionally, we assume that the receiver knows the channel perfectly.

We consider the simple model for the transmitter shown in Figure 29.2a [2]. The input is a vector **u** of i.i.d. uniform information bits that is encoded and interleaved, perhaps using a turbo or low-density parity check (LDPC) code. We then partition the coded bit stream into blocks **c** of bN_t bits and map each block onto a vector **a** whose N_t component symbols are taken from the complex alphabet $\mathcal{A}$ of size $q = |\mathcal{A}| = 2^b$, where b is the number of bits per symbol. The vector of transmitted symbols **a** is sent through the MIMO channel model of Figure 29.1 to produce the vector of received samples **r** at the receiver.

In Figure 29.2b we show a MIMO receiver based on the turbo principle [3]. The first block in the receiver is the MIMO detector, which is the focus of this chapter. As described in Section 29.1, the detector's job is to produce hard or soft estimates L for the bit stream **c**. After the detector, the receiver subtracts off any *a priori* information from the decoder, i.e. L_A, and then deinterleaves this signal producing $L_{A,\text{Dec}}$, the *a priori* information for the decoder.

The decoder improves the estimate for **u** by exploiting the presence of the error correcting code. In an iterative detection-decoding system, the receiver sends the output of the decoder block back to the MIMO detector as *a priori* information. This iterative approach yields improved error rate performance at the cost of increased receiver complexity. Iterative detection-decoding is, however, unnecessary for us to enjoy the benefits of soft-output detection. The decoder in a noniterative receiver will exhibit improved performance when their inputs are soft.

29.3 Problem Statement

The aim of a soft-output detector is to calculate or approximate the APP for each of the code bits c_j in a given signaling interval, where $j \in \{1, \ldots, bN_t\}$ is the bit index. This probability is conveniently represented by the so-called a posteriori log-likelihood ratio (LLR):

$$L(c_j|\mathbf{r}) := \ln \frac{\Pr[c_j = +1|\mathbf{r}]}{\Pr[c_j = -1|\mathbf{r}]}. \tag{29.2}$$

The sign of $L(c_j|\mathbf{r})$ is the maximum a posteriori (MAP) estimate for c_j, and the magnitude represents the reliability of the estimate. Larger magnitudes correspond to higher reliability, and smaller magnitudes indicate low reliability. In particular, the extreme case of $L=0$ indicates that c_j is equally likely to be $+1$ and -1.

Applying Bayes' rule to Equation 29.2 yields

$$\begin{aligned} L(c_j|\mathbf{r}) &= \ln\frac{f(\mathbf{r}|c_j=+1)(\Pr[c_j=+1]/f(\mathbf{r}))}{f(\mathbf{r}|c_j=-1)(\Pr[c_j=-1]/f(\mathbf{r}))} \\ &= \ln\frac{\Pr[c_j=+1]}{\Pr[c_j=-1]} + \ln\frac{f(\mathbf{r}|c_j=+1)}{f(\mathbf{r}|c_j=-1)} \\ &= L_{\mathrm{A}}(c_j) + L_{\mathrm{E}}(c_j|\mathbf{r}), \end{aligned} \tag{29.3}$$

where $\Pr[c_j=+1]$ and $\Pr[c_j=-1]$ are the *a priori* probabilities that bit c_j is 1 or -1, respectively, and where

$$L_{\mathrm{A}}(c_j) = \ln\frac{\Pr[c_j=+1]}{\Pr[c_j=-1]} \tag{29.4}$$

is the *a priori* LLR for the *j*th bit. The second term L_{E} in Equation 29.3 represents the *extrinsic* contribution to the a posteriori LLR. Using the law of total probability, it can be written as

$$L_{\mathrm{E}}(c_j|\mathbf{r}) = \ln\frac{f(\mathbf{r}|c_j=+1)}{f(\mathbf{r}|c_j=-1)} = \ln\frac{\sum_{\hat{\mathbf{c}}\in\mathcal{X}_j^{+1}} f(\mathbf{r}|\hat{\mathbf{c}})}{\sum_{\hat{\mathbf{c}}\in\mathcal{X}_j^{-1}} f(\mathbf{r}|\hat{\mathbf{c}})}, \tag{29.5}$$

where we rely on a partitioning of the vector alphabet into two, depending on whether the bit of interest is 1 or -1. Specifically, the set of possible $\mathbf{c}$ vectors $\mathcal{X}=\{\pm1\}^{bN_t}$ is partitioned into $\mathcal{X}_j^{+1}$ and $\mathcal{X}_j^{-1}$, where $\mathcal{X}_j^{+1}$ denotes the set of 2^{bN_t-1} vectors $\mathbf{c}\in\mathcal{X}$ for which $c_j=+1$, and $\mathcal{X}_j^{+1}$ denotes the set of 2^{bN_t-1} vectors $\mathbf{c}\in\mathcal{X}$ for which $c_j=-1$. For use later in the chapter, let us similarly define $\mathcal{Z}=\mathcal{A}^{N_t}$ as the set of all possible symbol vectors $\mathbf{a}$, one for each binary vector $\mathbf{c}\in\mathcal{X}$, as determined by the mapping from coded bits to transmitted symbols. Similar to $\mathcal{X}_j^{\pm1}$, let $\mathcal{Z}_j^{\pm1}$ denote the partitioning of $\mathcal{Z}$ depending on whether the *j*th bit used to form $\hat{\mathbf{a}}\in\mathcal{Z}$ is 1 or -1.

Figure 29.3 shows an example of partitioning for a vector alphabet after linear transformation by $\mathbf{H}$. The vector alphabet size is 32, which might arise from a binary scalar alphabet and $N_t=5$. The partitions in Figure 29.3a and b correspond to bits one and two from the binary transmission vector.

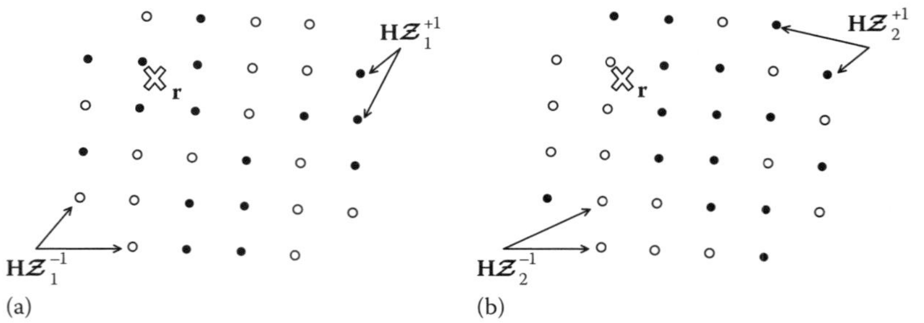

FIGURE 29.3 Partitioning of the set of transmission vectors $\mathcal{Z}$ after linear transformation by a channel $\mathbf{H}$ for the bit mapping corresponding to the (a) first bit transmitted from the first antenna C_1 and the (b) first bit transmitted from the second antenna C_2. Open and closed circles correspond to valid transmission vectors where the bit of interest is -1 and $+1$, respectively.

Since the noise is AWGN, the condition probability function $f(\mathbf{r}|\hat{\mathbf{c}})$ reduces to [4]

$$f(\mathbf{r}|\hat{\mathbf{c}}) = \frac{1}{(\pi N_0)^{N_r}} \exp\left(\frac{-\|\mathbf{r} - \mathbf{Ha}(\hat{\mathbf{c}})\|^2}{N_0}\right), \tag{29.6}$$

where $\mathbf{a}(\hat{\mathbf{c}}) \in \mathcal{Z}$ is the unique vector of transmitted symbols associated with the bit vector $\hat{\mathbf{c}}$. Substituting Equation 29.6 into Equation 29.5 and Equation 29.5 into Equation 29.3 yields

$$L(c_j|\mathbf{r}) = L_A(c_j) + \underbrace{\ln \frac{\sum_{\hat{\mathbf{c}} \in \mathcal{X}_j^{+1}} \exp\{-\|\mathbf{r} - \mathbf{Ha}(\hat{\mathbf{c}})\|^2/N_0\}}{\sum_{\hat{\mathbf{c}} \in \mathcal{X}_j^{-1}} \exp\{-\|\mathbf{r} - \mathbf{Ha}(\hat{\mathbf{c}})\|^2/N_0\}}}_{L_E(c_j|\mathbf{r})}, \tag{29.7}$$

where L is broken into an *a priori* component L_A and an extrinsic component L_E.

In the context of the iterative receiver shown in Figure 29.2, the *a priori* information L_A is provided to the detector via feedback from the error-control decoder, and the extrinsic information L_E is the extra contribution to the a posteriori LLR that was gleaned from detector, above and beyond what was provided by the decoder. Additionally, $\Pr[c_k = \hat{c}_k]$ is the *a priori* probability that $c_k = \hat{c}_k$. Computing Equation 29.7 for a given bit c_j requires knowledge of the received vector $\mathbf{r}$, the channel $\mathbf{H}$, the mapping $\mathbf{a}(\cdot)$ from bits to symbols, and any *a priori* information, if available.

Exact evaluation of Equation 29.7 requires that a computation of the form $\|\mathbf{r} - \mathbf{Ha}\|^2$ be computed q^{N_t} times, and this would have to be done a total of bN_t times each signaling interval, once for each bit. As an example, for a four-input MIMO system with each input coming from a 64-QAM alphabet, the total number of possible transmission vectors which are necessary to exactly compute Equation 29.7 for just one bit is over 16 million! Clearly a lower-complexity solution is needed. Additionally, the exponential operation must be applied to each of these 16 million squared norms, resulting in an extremely high computational complexity. Methods for reducing the computational complexity of Equation 29.7 comprise the remainder of this chapter.

A common approximation for Equation 29.7 is to use the *max-log* approximation:

$$L(c_j|\mathbf{r}) \approx L_A(c_j) + \max_{\hat{\mathbf{c}} \in \mathcal{X}_j^{+1}} \left\{\frac{-\|\mathbf{r} - \mathbf{Ha}(\hat{\mathbf{c}})\|^2}{N_0}\right\} - \max_{\hat{\mathbf{c}} \in \mathcal{X}_j^{-1}} \left\{\frac{-\|\mathbf{r} - \mathbf{Ha}(\hat{\mathbf{c}})\|^2}{N_0}\right\}. \tag{29.8}$$

The max-log approximation is based on the assumption that the exponential term with maximum argument in the sum of exponentials will dominate the summation. By avoiding the sum of exponentials only one exponential term remains in the numerator and one in the denominator of Equation 29.7. After some simplifications, the result is Equation 29.8. This approximation is widely accepted because of its relatively small performance loss [5,6]. The max-log approximation does not, however, reduce the problem size of soft-output MIMO detection. Specifically, a brute-force search for the $\mathbf{a}(\hat{\mathbf{c}}) \in \mathcal{Z}$ minimizing $\|\mathbf{r} - \mathbf{Ha}(\hat{\mathbf{c}})\|^2$ in Equation 29.8 would still need to consider q^{N_t} possibilities.

Although at a glance it might appear from Equation 29.8 that the receiver will need to perform a pair of optimizations for each of the bN_t bits of interest, for a total of $2bN_t$ optimizations per signaling interval, in fact the MAP solution will always be one of each pair. So to compute Equation 29.8 for each of the bN_t bits, it would be sufficient to find the MAP solution once, and then, for each of the bN_t bits, to find a counterhypothesis.

Suppose we form a *list* containing the MAP candidate along with the bN_t counterhypothesis candidates. This would allow us to constrain the optimizations in Equation 29.8 to this list without degrading performance. In doing so, we can reduce the size of each optimization search space from 2^{bN_t-1} to only bN_t. We can further constrain our optimization search from an exact counterhypothesis list to a list of candidates $\mathcal{L} \subseteq \mathcal{Z}$ with minimum squared Euclidean distance to the received vector. This further reduces the complexity of forming the list and arguably performs just as well as the exact counterhypothesis list because a far-away counterhypothesis does not contribute significantly anyway. Consequently, the "minimum-distance list" is of great importance in soft-output MIMO detection. The search to find this minimum-distance list will dominate the remainder of this chapter.

List detection is the process of finding a list of candidates $\mathcal{L} \subseteq \mathcal{Z}$. From this list, the max-log formula 29.8 is approximated using:

$$L(c_j|\mathbf{r}) \approx L_A(c_j) + \max_{\hat{\mathbf{a}} \in \mathcal{L} \cap \mathcal{Z}_j^{+1}} \left\{ \frac{-\|\mathbf{r} - \mathbf{H}\hat{\mathbf{a}}\|^2}{N_0} \right\} - \max_{\hat{\mathbf{a}} \in \mathcal{L} \cap \mathcal{Z}_j^{-1}} \left\{ \frac{-\|\mathbf{r} - \mathbf{H}\hat{\mathbf{a}}\|^2}{N_0} \right\}. \tag{29.9}$$

Despite the notational changes to operate on the vector alphabet $\mathcal{Z}$ instead of the binary vector $\mathcal{X}$, the key difference between Equations 29.8 and 29.9 is the insertion of the list $\mathcal{L}$, where $|\mathcal{L}|$ is typically much less than $|\mathcal{Z}|$, i.e., $|\mathcal{L}| \ll |\mathcal{Z}|$. The list length $\ell = |\mathcal{L}|$ plays a critical roll in the overall complexity and performance. The value for ℓ should be large enough to closely approximate exact performance, but small enough to maintain low detection complexity [7].

Two example lists, for the received vector $\mathbf{r}$ from Figure 29.3, are depicted in Figure 29.4. The elements within the boundaries of the circular region depicted in Figure 29.4a corresponds to the minimum-distance list. The elements within the boundaries of the region in Figure 29.4b correspond to a suboptimal list. The performance of a system using the list in Figure 29.4b would suffer relative to one using the list in Figure 29.4a because the list does not include the elements to optimize Equation 29.8 for each of the five transmission bits.

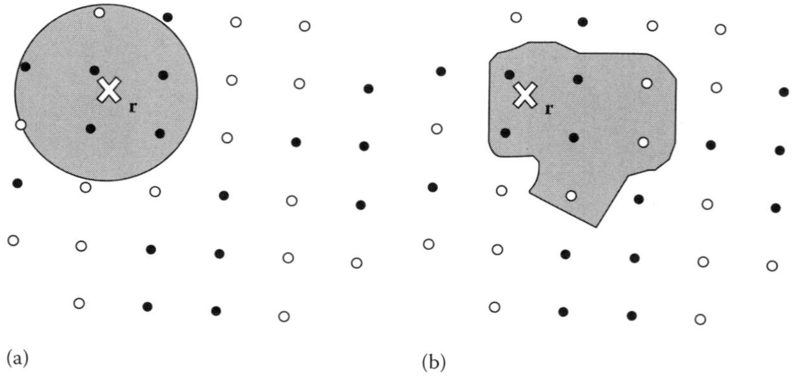

FIGURE 29.4 A minimum distance list (a) contains the ℓ candidates closest to $\mathbf{r}$, while a suboptimal list (b) might contain different candidates.

In the remainder of this chapter, we will assume that there is no *a priori* information; this will simplify our presentation while still retaining the essential features of the problem. With this assumption Equation 29.9 reduces to

$$L(c_j|\mathbf{r}) \approx \frac{1}{N_0}\left(\min_{\hat{\mathbf{a}}\in\mathcal{L}\cap\mathcal{Z}_j^{+1}} \|\mathbf{r}-\mathbf{H}\hat{\mathbf{a}}\|^2 - \min_{\hat{\mathbf{a}}\in\mathcal{L}\cap\mathcal{Z}_j^{-1}} \|\mathbf{r}-\mathbf{H}\hat{\mathbf{a}}\|^2\right). \tag{29.10}$$

In terms of the partition into white and black points, as shown in Figure 29.4, one of the minimizations of Equation 29.10 will produce the squared distance to the nearest black point, while the other minimization will produce the squared distance to the nearest white point. Specifically we see that, because the suboptimal list of Figure 29.4b excludes the closest white point, it will overestimate the reliability of the bit in question.

In the next section, we demonstrate how to efficiently find the minimum-distance list both exactly and approximately.

29.4 List Tree Search

A useful construct for efficiently finding a list of candidates is that of a tree [8–10]. The detection tree can be derived and interpreted in two ways: either geometrically or algebraically.

The geometric view is based on the fact that the candidate vectors after the channel fall on a lattice whenever the alphabets are QAM. And any lattice can be decomposed into the union of multiple sublattices that are translated relative to each other. Therefore, the squared distance from a received vector to any point on the lattice is easily expressed in terms of the projection of the received vector onto the hyperplane spanned by the lattice point's sublattice; namely, the squared distance decomposes into the sum of the squared distance from the received vector to the projection vector, plus the squared distance from the projection vector to the sublattice point. And this latter term can be computed recursively based on the same principle.

As an example of this geometric construction [11], consider the example shown in Figure 29.5 for a 2-transmitter system with an antipodal alphabet $\mathcal{A} = \{+1, -1\}$. Figure 29.5a depicts a received vector $\mathbf{r}$ and the valid transmission vectors after being transformed by the channel. These detection vectors are denoted with a gray circle, a black circle, a gray-black, and a black-gray circle corresponding to the valid vectors $\{-1, -1\}$, $\{+1, +1\}$, $\{-1, +1\}$, and $\{+1, -1\}$, respectively.

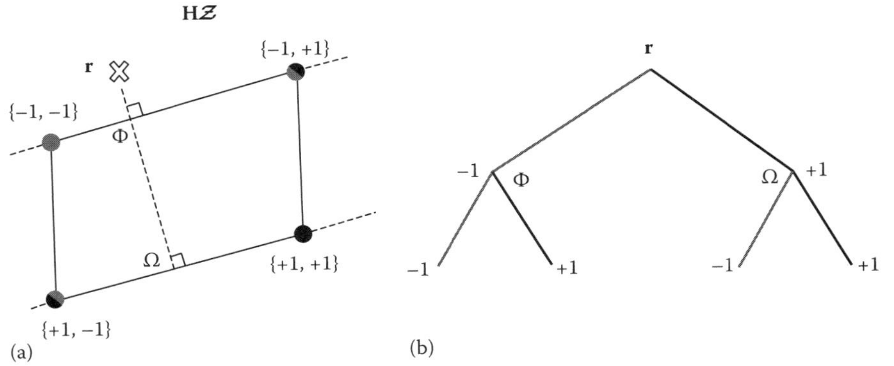

FIGURE 29.5 A q-ary tree for $N_t = 2$ and $\mathcal{Z} = \{-1, -1\}, \{-1, +1\}, \{+1, +1\}$, and $\{+1, -1\}$ ($q = 2$).

Ultimately we are interested in computing the squared Euclidean distance from the received vector to a valid transmission vector in the detection space, i.e., $\|\mathbf{r} - \mathbf{H}\hat{\mathbf{a}}\|^2$. For the purposes of the example in Figure 29.5a, we can employ the Pythagorean theorem to compute these squared Euclidean norms. Specifically, if we are interested in computing the squared norm from $\mathbf{r}$ to the detection vector $\mathbf{H}\hat{\mathbf{a}}$, where $\hat{\mathbf{a}} = \{-1, -1\}$, then one leg of a triangle corresponds to the squared distance from $\mathbf{r}$ to Φ and the other leg corresponds to the squared distance from Φ to $\mathbf{H}\hat{\mathbf{a}}$. Summing these squared distances via Pythagoras, i.e., $c^2 = a^2 + b^2$, yields the squared Euclidean distance $\|\mathbf{r} - \mathbf{H}\hat{\mathbf{a}}\|^2$ when $\hat{\mathbf{a}} = \{-1, -1\}$. Similarly the squared Euclidean distances to all other elements of $\mathbf{H}\mathcal{Z}$ can be computed.

Figure 29.5b depicts these same squared Euclidean distance calculations using a tree. The distance from $\mathbf{r}$ to Φ is the cost when the first detected symbol is -1 and is represented by the gray branch emanating from the root node of the detection tree. Similarly, the distance from $\mathbf{r}$ to Ω corresponds to the cost when the first detected symbol is $+1$ and is represented by the black branch emanating from the root node. Computing the squared Euclidean distances from Φ to $\mathbf{H}\hat{\mathbf{a}}$, where $\hat{a} = \{-1, -1\}$ and $\hat{\mathbf{a}} = \{-1, +1\}$, completes the left half of the tree and computing the squared Euclidean distances from Ω to $\mathbf{H}\hat{\mathbf{a}}$, where $\hat{\mathbf{a}} = \{+1, -1\}$ and $\hat{\mathbf{a}} = \{+1, +1\}$, completes the right half of the tree. Consequently, each of the nodes at the bottom of the detection tree, referred to as *leaf nodes*, corresponds to a unique decision from the set of all possible transmission vectors $\mathcal{Z}$.

In contrast to the geometric view, the algebraic view is based on a QR decomposition. Specifically, the squared distance for a candidate $\hat{\mathbf{a}}$ is

$$J(\hat{\mathbf{a}}) = \|\mathbf{r} - \mathbf{H}\hat{\mathbf{a}}\|^2 \tag{29.11}$$

$$= \|\mathbf{y} - \mathbf{R}\hat{\mathbf{a}}\|^2 \tag{29.12}$$

$$= \sum_{1 \le i \le N_t} \left| y_i - \sum_{1 \le v \le i} R_{iv}\hat{a}_v \right|^2, \tag{29.13}$$

where $\mathbf{H} = \mathbf{QR}$ is the QR decomposition of the channel matrix $\mathbf{H}$, where $\mathbf{R}$ is an $N_t \times N_t$ lower triangular matrix, where $\mathbf{Q}$ is an orthogonal matrix and where $\mathbf{y} = \mathbf{Q}^* \mathbf{r}$. The QR decomposition is an orthogonal and triangular decomposition of the channel matrix $\mathbf{H}$ that generalizes the projection process for the example in Figure 29.5a.

The cost function (Equation 29.13) can be interpreted as the sum of N_t *branch metrics*, one for each branch in a path from the root of the detection tree to a leaf node, where the metric for a branch in the ith stage of the detection tree with path history $\{a_1, a_2, \ldots, a_i\}$ is [4]

$$\left| y_i - \sum_{1 \le v \le i} R_{iv} a_v \right|^2. \tag{29.14}$$

We refer to the sum of the i branch metrics in the path from the root node to the node of interest in the detection tree as a *path metric*.

Figure 29.6 depicts a detection tree for a three-input MIMO system employing a 4-ary alphabet. There are three *layers* of the tree, one for each input, and there are $q^{N_t} = 4^3 = 64$ leaf nodes. There are $4 + 16 + 64 = 84$ total branches. Some of the branch metrics are shown; where the superscript for a decision $\hat{a}_i^{(\cdot)}$ denotes the index from the q-ary alphabet.

Keeping in mind the problem of list detection, the objective of tree-based detection is to find the $\ell = |\mathcal{L}|$ leaf nodes in the tree, corresponding to the valid elements from the set $\mathcal{Z}$, that yields an accurate solution to Equation 29.10. Using the minimum-distance list as our guide, we seek to find the ℓ leaf nodes in the tree with minimum cost.

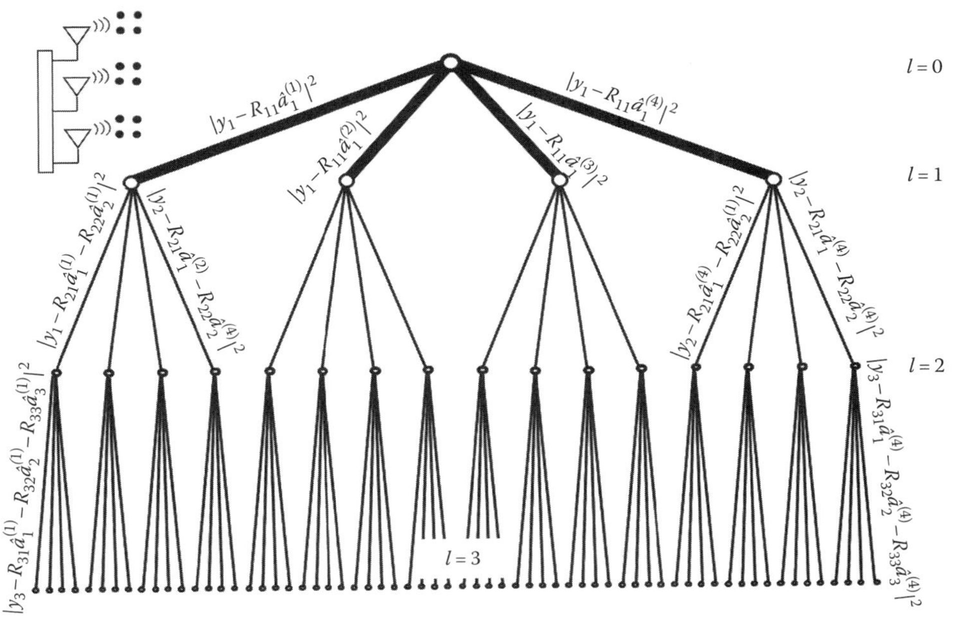

FIGURE 29.6 A 4-ary tree for $N_t = 3$.

29.4.1 Optimal Search

As described in the last section, the problem of soft-output MIMO detection boils down to the following:

> **Goal:** Find the ℓ leaf nodes in the tree with minimum cost.

A naive but effective solution would be to compute the path metrics for all the leaf nodes in the tree and then determine the ℓ leaf nodes with minimum cost. This brute force approach is undesirable because it visits all the nodes in the tree and thus has high complexity. For example, brute force detection of a four-layer tree with $q = 64$ would have over 16 million leaf nodes. Fortunately, there exist more efficient ways to achieve our goal.

29.4.1.1 The List Sphere Detector

The list sphere detector (LSD) [2] is a famous algorithm for achieving our goal. In this section, we begin with an intuitive description of the LSD. Following this, we provide a flowchart for implementing the LSD. Finally, we illustrate how it works via an example.

The LSD begins at the root node and advances through the tree in a greedy fashion, at each node selecting the child node with minimum weight. This process continues until we reach a leaf node or the cost of the node we are visiting exceeds a threshold. For the purposes of this discussion we initialize the threshold to ∞. Consequently, a leaf node will always be found at the start of our search. This leaf node becomes the first element of a list, although it might later be replaced.

The LSD then backtracks one layer and considers the inclusion of the "sibling" nodes of the leaf node it has just found. A Fincke and Pohst [12] enumeration would explore these siblings in a natural order, say from left to right, with no regard to their weights. Fewer nodes will be visited if a Schnorr–Euchner [13] enumeration is adopted, in which the siblings are explored in an order determined by their weights, with the best first. We assume Schnorr–Euchner enumeration in our discussion.

The process of adding sibling nodes to the first leaf node found continues until either there are no more sibling nodes to enumerate, or until the list consists of ℓ leaf nodes. In either case the algorithm backtracks an additional layer in the tree so that it is two layers removed from the leaf nodes. If the algorithm backtracks because the list is comprised of ℓ leaf nodes, then the threshold value must be updated to the weight of the highest-cost leaf node in the list.

Now two layers removed from the leaf nodes in the tree, the LSD enumerates the lowest-weight child node it has not yet explored and continues in a greedy fashion to either add leaf nodes to the list (if its cost is less than the threshold) or backtrack (if the cost exceeds the threshold). Anytime a leaf node is found with weight less than the threshold and the list size is already ℓ, the new leaf node replaces the leaf node in the list with highest cost. Upon a replacement event, the LSD updates the threshold to the cost of the largest-weight leaf node remaining in the list.

When there are no more nodes left to be explored at a given layer in the tree either because of exhaustive search or because all remaining nodes exceed the threshold, the algorithm backtracks up the tree to determine if there are any nodes left to be explored at higher layers in the tree. The LSD terminates when it can no longer backtrack.

The process just described for finding leaf nodes in the detection tree, where the search begins at the root node and proceeds as far as possible in the tree before backtracking, is known as *depth-first* search. Figure 29.7 summarizes the depth-first LSD using a flowchart. We assume that all branch weights are known so that the only input required by the algorithm is the list length ℓ. The output of the LSD is the list $\mathcal{L}$.

We now provide a concrete example of the LSD. To do so, we will use an example detection tree whose metrics are provided for us. Figure 29.8 depicts this example tree using the same parameters as the tree in Figure 29.5, i.e., $N_t = 3$ and $q = 4$. For our example, we will assume the desired list length is $\ell = 7$.

LSD example: Find the $\ell = 7$ leaf nodes with minimum cost for the tree in Figure 29.8 using the list sphere detector [2].

Considering the example tree in Figure 29.8 and implementing the algorithm just described, we begin in a greedy fashion by finding the leaf node $\{1 \rightsquigarrow 7 \rightsquigarrow 2\}$ and cost 10. We add the leaf node to the list, which up to this point was empty. We then backtrack one layer in the tree, extend the next best leaf node, i.e. $\{1 \rightsquigarrow 7 \rightsquigarrow 5\}$ with cost 13, and include it in the list. This process continues until all four leaf nodes are included in the list, i.e., we also include leaf nodes $\{1 \rightsquigarrow 7 \rightsquigarrow 6\}$ with cost 14 and $\{1 \rightsquigarrow 7 \rightsquigarrow 9\}$ with cost 17. At this point the list consists of four leaf nodes and the threshold still possesses its initial value of ∞. Note that each time a leaf node is added to the list, its associated weight is maintained as well.

The algorithm then backtracks two layers in the tree to the child node of the root with cost 1 at which point it once more advances through the tree in a greedy fashion until it reaches the leaf node $\{1 \rightsquigarrow 9 \rightsquigarrow 5\}$ with cost 15. This leaf node is added to the list and its associated cost maintained. The algorithm then backtracks one layer and proceeds to add the leaf node $\{1 \rightsquigarrow 9 \rightsquigarrow 6\}$ with cost 16. This is repeated to obtain the leaf node $\{1 \rightsquigarrow 9 \rightsquigarrow 8\}$ with cost 18. At this point the list consists of the desired amount of seven leaf nodes and it is time to update the threshold from ∞ to the cost of the worst-case leaf node in the list, i.e., 18. Because the costs of extending additional child nodes from the node $\{1 \rightsquigarrow 9\}$ would exceed the newly established threshold the algorithm backtracks to the child node of the root with cost 1. At this point, the algorithm considers extension to any further nodes but elects not to do so because all unexplored paths exceed the threshold, i.e., $1 + 18 > 18$.

Now the algorithm backtracks to the root node and enumerates the next best child node, i.e., the one with cost 2. We extend this best child node in a greedy fashion until either a leaf node is reached or the

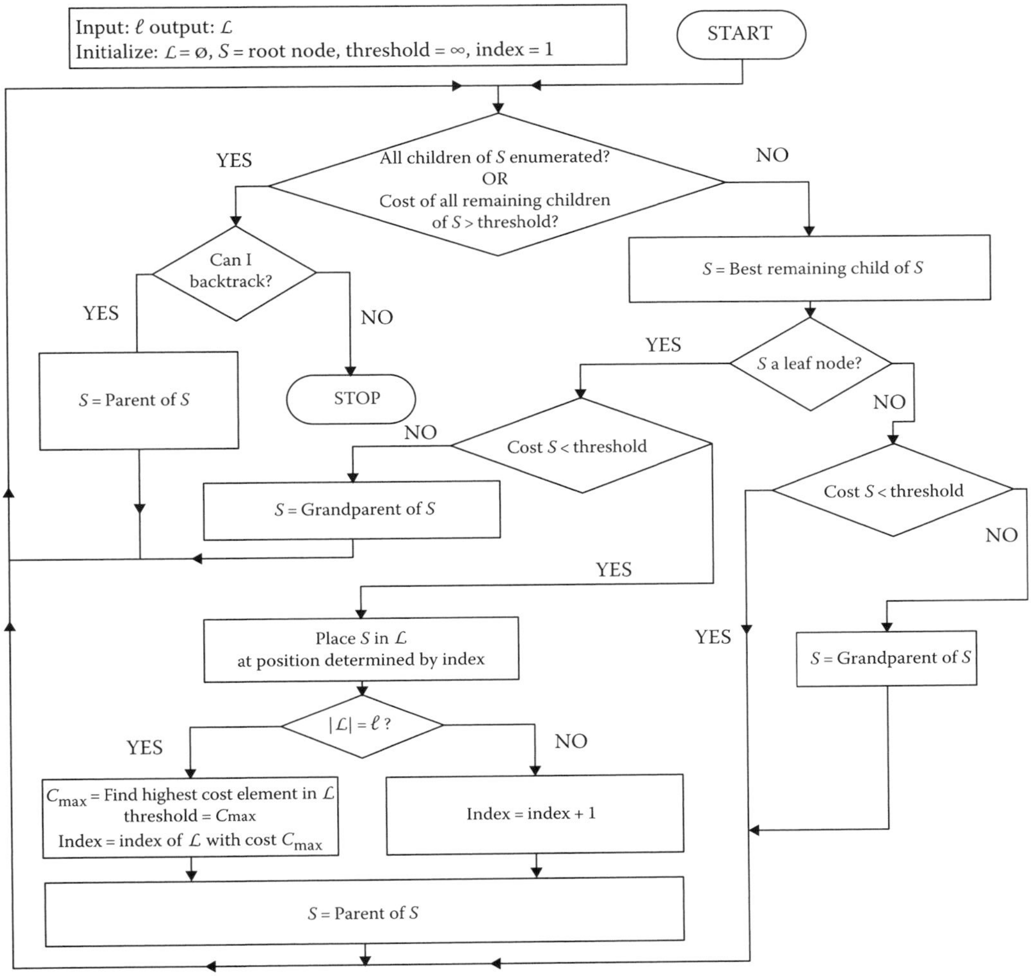

FIGURE 29.7 Flow chart for Schnorr–Euchner realization of list sphere detector.

node cost exceeds the threshold. For our example we reach the leaf node $\{2 \rightsquigarrow 1 \rightsquigarrow 1\}$ with cost 4. As this leaf node has lower cost then our threshold, and the list already has the desired size, we must update the list. This is accomplished by removing the highest-cost leaf node from the list, i.e., node $\{1 \rightsquigarrow 9 \rightsquigarrow 8\}$ with cost 18, and replacing it with the newly found leaf node with cost 4. Additionally, we decrease the threshold to the weight of the highest-cost remaining leaf node in the list, i.e., 17 corresponding to node $\{1 \rightsquigarrow 7 \rightsquigarrow 9\}$. The algorithm then backtracks one layer and extends the next best child node $\{2 \rightsquigarrow 1 \rightsquigarrow 2\}$ with cost 5. We add this leaf node to the list, replacing the one with cost 17, and decrease the threshold to 16. Next we add the leaf node $\{2 \rightsquigarrow 1 \rightsquigarrow 4\}$ with cost 7 to the list, replacing the leaf node with cost 16, and decrease the threshold to 15, corresponding to the highest-cost leaf node in the list, i.e., node $\{1 \rightsquigarrow 9 \rightsquigarrow 5\}$. We backtrack once more and add the leaf node $\{2 \rightsquigarrow 1 \rightsquigarrow 9\}$ with cost 12 to the list, replace the leaf node with cost 15, and decrease the threshold to the new highest cost of 14. Because no more leaf nodes can be enumerated from the parent node $\{2 \rightsquigarrow 1\}$, we backtrack in the tree to the child node of the root with cost 2. We then extend this node to its best child node and proceed in a greedy fashion to the best

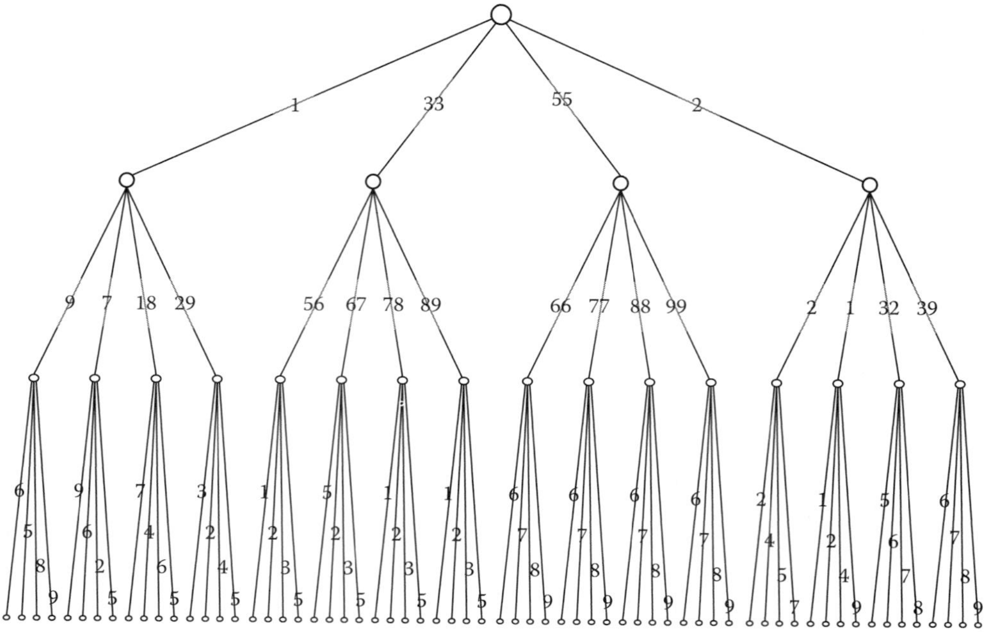

FIGURE 29.8 An example tree for $N_t = 3$ and $q = 4$.

leaf node hanging off this child node, i.e., the next leaf node we find is $\{2 \rightsquigarrow 2 \rightsquigarrow 2\}$ with cost 6. We then decrease the threshold to 13. Next, we add the leaf node $\{2 \rightsquigarrow 2 \rightsquigarrow 4\}$ with cost 8 to the list and decrease the threshold to 12. Then, we add the leaf node $\{2 \rightsquigarrow 2 \rightsquigarrow 5\}$ with cost 9 to the list and decrease the threshold to 10. This threshold value corresponds to the leaf node $\{1 \rightsquigarrow 7 \rightsquigarrow 2\}$. At this point enumerating further child nodes from the node $\{2 \rightsquigarrow 2\}$ would exceed our threshold because $2 + 2 + 7 > 10$. We then backtrack two layers to the child node hanging off the root node with cost 2.

At this point it is clear that no additional child nodes can be explored from the node $\{2\}$ because the threshold would be exceeded, i.e., $2 + 32 > 10$, so we backtrack to the root of the tree where we again realize that no additional nodes can be extended because 33 is greater than the current threshold of 10. Concluding that there are no remaining unexplored nodes, we terminate our search. The final list possesses exactly the seven minimum-weight leaf nodes $\{2 \rightsquigarrow 1 \rightsquigarrow 1\}$, $\{2 \rightsquigarrow 1 \rightsquigarrow 2\}$, $\{2 \rightsquigarrow 2 \rightsquigarrow 2\}$, $\{2 \rightsquigarrow 1 \rightsquigarrow 4\}$, $\{2 \rightsquigarrow 2 \rightsquigarrow 4\}$, $\{2 \rightsquigarrow 2 \rightsquigarrow 5\}$, and $\{1 \rightsquigarrow 7 \rightsquigarrow 2\}$ with associated costs 4, 5, 6, 7, 8, 9, and 10, respectively.

29.4.1.2 The List Sequential Detector

An alternative to the LSD for finding the ℓ minimum-cost leaf nodes in the tree is the list sequential (LISS) detector [14,15]. In contrast to the LSD which maintains only one node at a time in the detection tree, the LISS maintains multiple nodes in the tree simultaneously, where the nodes maintained need not be at the same layer in the tree. We call this type of search *metric-first* search [16].

The LISS algorithm is implemented using a stack to maintain the nodes currently under consideration. In this section, we begin with an intuitive description of the LISS. We follow up with a flowchart description. We also illustrate the LISS with an example.

With the stated objective of finding the ℓ leaf nodes with minimum weights in the tree, the LISS algorithm begins by initializing the stack to be the root node and its associated cost to be 0. After this initialization, we remove this minimum-cost node in the stack (at this point it is the only node in the

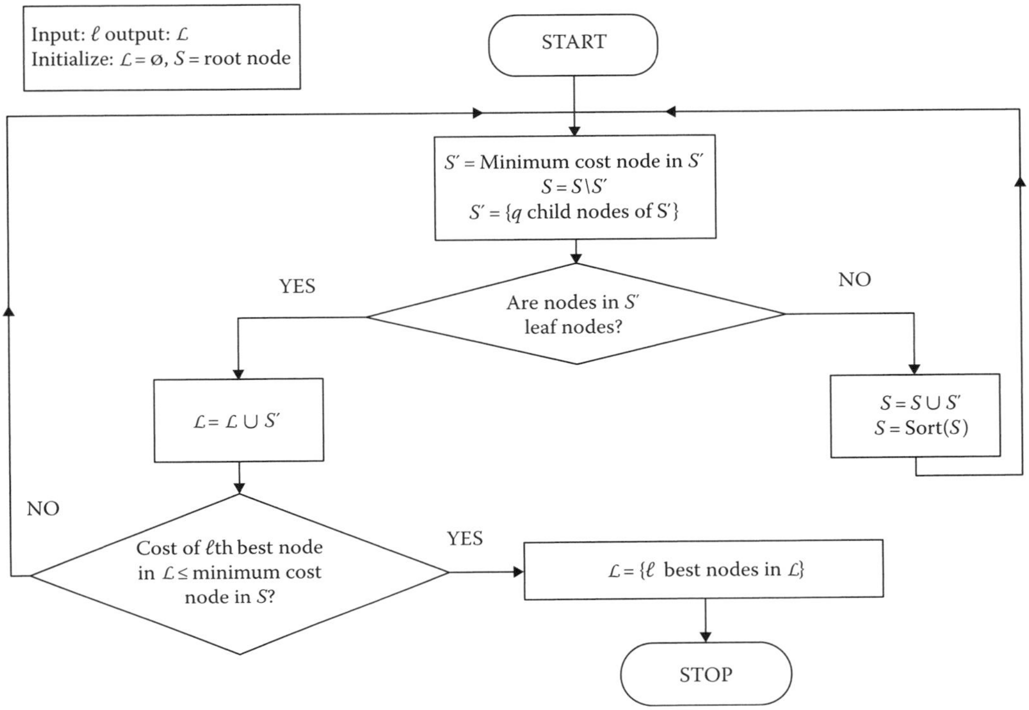

FIGURE 29.9 Flow chart for LISS detector.

stack) and replace it in the stack with all $q = |\mathcal{A}|$ of its child nodes. We then order the stack in terms of costs, placing the minimum-cost node at the top of the stack.

Here, the process begins to repeat itself. As before we remove the minimum-cost node from the stack and replace it with its q children. We reorder the stack once more and replace the minimum-cost node by its q children. Each time leaf nodes are reached we remove them from the stack and place them in a list.

The algorithm terminates when the node on the top of the ordered stack (i.e., the one with minimum cost) has weight greater than or equal to the cost of the ℓth minimum-weight leaf node in the list.* Upon termination we can truncate the list to the ℓ minimum-weight leaf nodes.

Figure 29.9 summarizes the LISS algorithm using a flow chart. As before, we assume that all branch weights are known so that the only input required by the algorithm is the list length ℓ. The output of the algorithm is the list $\mathcal{L}$. Unlimited memory is assumed to avoid a discussion about truncation of the stack.

LISS example: Find the $\ell = 7$ leaf nodes with minimum cost for the tree in Figure 29.8 using the LISS detector [14].

Returning to the example tree of Figure 29.8, we can utilize the approach just described to find the ℓ minimum-cost leaf nodes. After initializing the stack to be the root node and its associated cost to be

* This termination condition is slightly different than the one in [14], but is needed to ensure that we find the ℓ minimum weight leaf nodes.

zero, we extend the $q = 4$ child nodes from the root and place them in the stack. We then order these nodes based on their associated cost metrics (minimum cost to maximum cost). This yields the ordered stack of nodes {1}, {2}, {33}, and {55}. Next we take the best node, i.e., the one with cost 1, and replace it in the stack with its q child nodes, i.e., nodes $\{1 \rightsquigarrow 7\}$, $\{1 \rightsquigarrow 9\}$, $\{1 \rightsquigarrow 18\}$, and $\{1 \rightsquigarrow 29\}$ with associated cost metrics 8, 10, 19, and 30, respectively. Ordering the nodes in the stack, which now come from both the first and second layers in the tree, yields the ordered set of nodes: {2}, $\{1 \rightsquigarrow 7\}$, $\{1 \rightsquigarrow 9\}$, $\{1 \rightsquigarrow 18\}$, $\{1 \rightsquigarrow 29\}$, {33}, and {55}. Once more extending the best node, i.e., {2}, to its q children and ordering the stack yields the ordered set of nodes: $\{2 \rightsquigarrow 1\}$, $\{2 \rightsquigarrow 2\}$, $\{1 \rightsquigarrow 7\}$, $\{1 \rightsquigarrow 9\}$, $\{1 \rightsquigarrow 18\}$, $\{1 \rightsquigarrow 29\}$, {33}, $\{2 \rightsquigarrow 32\}$, $\{2 \rightsquigarrow 39\}$, and {55}.

At this point the stack consists of 10 nodes with eight nodes at the second layer in the tree and two at the first. The minimum-cost node in the stack is at the second layer in the tree, i.e. $\{2 \rightsquigarrow 1\}$ and so removing it from the stack and extending it to its q best children produces the leaf nodes $\{2 \rightsquigarrow 1 \rightsquigarrow 1\}$, $\{2 \rightsquigarrow 1 \rightsquigarrow 2\}$, $\{2 \rightsquigarrow 1 \rightsquigarrow 4\}$, and $\{2 \rightsquigarrow 1 \rightsquigarrow 9\}$. Rather than placing these leaf nodes in the stack, they become the first leaf nodes to be placed in the list for future consideration.

Having decreased its size by one, the stack possesses the ordered nodes $\{2 \rightsquigarrow 2\}$, $\{1 \rightsquigarrow 7\}$, $\{1 \rightsquigarrow 9\}$, $\{1 \rightsquigarrow 18\}$, $\{1 \rightsquigarrow 29\}$, {33}, $\{2 \rightsquigarrow 32\}$, $\{2 \rightsquigarrow 39\}$, and {55}. Again the minimum-cost node in the tree is a parent of leaf nodes and yields the following nodes for inclusion in the list: $\{2 \rightsquigarrow 2 \rightsquigarrow 2\}$, $\{2 \rightsquigarrow 2 \rightsquigarrow 4\}$, $\{2 \rightsquigarrow 2 \rightsquigarrow 5\}$, and $\{2 \rightsquigarrow 2 \rightsquigarrow 7\}$. At this point the associated costs for the leaf nodes in the list are 4, 5, 7, 12, 6, 8, 9, and 11.

Returning to the stack we are left with the nodes $\{1 \rightsquigarrow 7\}$, $\{1 \rightsquigarrow 9\}$, $\{1 \rightsquigarrow 18\}$, $\{1 \rightsquigarrow 29\}$, {33}, $\{2 \rightsquigarrow 32\}$, $\{2 \rightsquigarrow 39\}$, and {55}. We then remove the node $\{1 \rightsquigarrow 7\}$ from the stack and include its children $\{1 \rightsquigarrow 7 \rightsquigarrow 2\}$, $\{1 \rightsquigarrow 7 \rightsquigarrow 5\}$, $\{1 \rightsquigarrow 7 \rightsquigarrow 6\}$, and $\{1 \rightsquigarrow 7 \rightsquigarrow 9\}$ in the list. Now the associated costs for the leaf nodes in the list are 4, 5, 7, 12, 6, 8, 9, 11, 10, 13, 14, and 17 and the associated costs for the nodes in the stack are 10, 19, 30, 33, and 35.

Observe that at this point there are zero nodes in the stack whose cost is less than that of the seven best nodes in the list. Specifically, the minimum-cost node in the stack has a weight of 10 and the $\ell = 7$ best leaf nodes in the list have weights 4, 5, 6, 7, 8, 9, and 10. Upon making this observation we terminate the algorithm because it is impossible for the stack to produce a leaf node with a lower cost than the highest-cost nodes in $\mathcal{L}$. The list is truncated to the seven best leaf nodes and we are done.

29.4.2 Suboptimal Search

The LSD and LISS algorithms just described are efficient ways to achieve the goal of finding the ℓ leaf nodes in a tree with minimum weights. What happens if we modify our goal by relaxing the constraint that we must find the ℓ leaf nodes with *minimum* weights and instead search for ℓ leaf nodes with *small* weights? The advantage of such an approach would be that our search could visit fewer nodes in the tree, thereby reducing complexity. The obvious disadvantage would be a suboptimal solution to our problem.

For the LSD and LISS algorithms described in the previous section, we can solve the relaxed constraint problem through early termination. For the LSD this can be achieved by stopping once a certain number of nodes have been visited or by aggressively reducing the threshold value to search fewer nodes in the tree. For the LISS algorithm, we can terminate early once a certain number of leaf nodes are in the list or we could reduce the size of the stack to a predetermined fixed value such that it retains fewer nodes. A third option is to bias paths based on their layer in the tree to avoid extending nodes which are unlikely to produce a leaf node with small weight [17]. Additionally, to achieve our new goal of finding the ℓ leaf nodes with *small* weight, we cannot only modify the LSD or LISS algorithms, but we can also consider other ways to search the tree.

An efficient way to search a tree when a suboptimal list is allowed is to search the tree layer-by-layer and at each layer remove nodes which are unlikely to produce a leaf node with small weight. Algorithms that search the tree using a layer-by-layer approach are *breadth-first*. Breadth-first search algorithms have higher complexity than depth-first or metric-first approaches when the goal is to find the minimum-cost(s) leaf node(s) in the tree. They are, however, a viable alternative when suboptimal search is allowed. This is because, in contrast to depth-first or metric-first approaches, breadth-first algorithms have fixed complexity, meaning that they visit the same number of nodes in the tree independent of the branch weights on the tree. This is significant because it means that the algorithm has a regular structure which lends itself to practical implementation.

We now describe a breadth-first algorithm, which we call the soft M algorithm because of its close relation to the classical M algorithm [8]. The soft M algorithm begins by extending the $\beta \leq q$ minimum-weight nodes from the root of the tree using Schnorr–Euchner enumeration. Assuming the algorithm would like to keep all of these nodes, it can then extend the β best child nodes from each of the β parent nodes yielding β^2 nodes at the second layer in the tree. If β^2 is greater than some value, call it M, then the algorithm sorts the β^2 nodes and retains only the M best. This process of extending β nodes from each retained parent node and retaining only the M minimum-weight nodes (assuming M is less than the number of retained nodes) continues until we reach the final layer in the tree, where ℓ leaf nodes are found for inclusion in the list (assuming $\ell \leq M$).

Figure 29.10 summarizes the soft M algorithm using a flow chart. The inputs to the algorithm are the list length ℓ, the β parameter, which determines the number of nodes to extend from each retained parent node, and the M parameter, which is used to prune the tree when the number of nodes in the tree exceeds M. The output of the algorithm is the list $\mathcal{L}$.

> **Soft *M* algorithm example**: Find the leaf nodes resulting from searching the tree in Figure 29.8 using the soft-output M algorithm described in Figure 29.10 for $\beta = 2$ and $M = \ell = 7$.

From the root node in the tree we extend the $\beta = 2$ child nodes with minimum weight, i.e. {1} and {2}. This number is less than $M = 7$, so we maintain both the nodes. These maintained nodes become the parent nodes for the next layer in the tree, where each parent node produces $\beta = 2$ child nodes, yielding a total of $2 \times 2 = 4$ nodes. Again this number is less than $M = 7$ and so we maintain all of the nodes, i.e. $\{1 \rightsquigarrow 7\}$, $\{1 \rightsquigarrow 9\}$, $\{2 \rightsquigarrow 1\}$, and $\{2 \rightsquigarrow 2\}$. Once more we extend the $\beta = 2$ best nodes from each of the

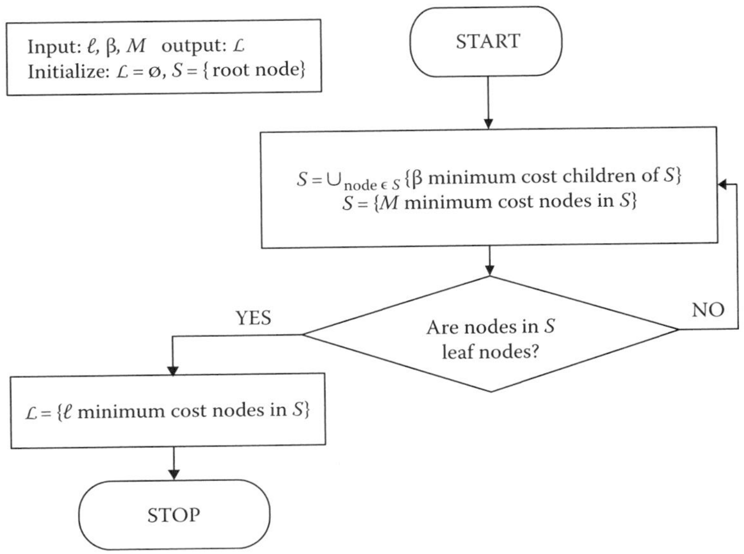

FIGURE 29.10 Flow chart for soft M algorithm.

retained nodes. This produces the eight leaf nodes $\{1 \rightsquigarrow 7 \rightsquigarrow 2\}$, $\{1 \rightsquigarrow 7 \rightsquigarrow 5\}$, $\{1 \rightsquigarrow 9 \rightsquigarrow 5\}$, $\{1 \rightsquigarrow 9 \rightsquigarrow 6\}$, $\{2 \rightsquigarrow 1 \rightsquigarrow 1\}$, $\{2 \rightsquigarrow 1 \rightsquigarrow 2\}$, $\{2 \rightsquigarrow 2 \rightsquigarrow 2\}$, and $\{2 \rightsquigarrow 2 \rightsquigarrow 4\}$ with associated costs 9, 13, 15, 16, 4, 5, 6, and 8, respectively. Maintaining the $\ell = 7$ minimum-cost nodes found using this approach, i.e., 4, 5, 6, 8, 9, 13, and 15, compares favorably with the seven true minimum-cost leaf nodes, i.e., 4, 5, 6, 7, 8, 9, and 10. This is because five of the seven list elements are the same, but far fewer tree nodes were visited to obtain the suboptimal list.

29.4.3 Candidate Adding

We now return to the problem statement at the end of Section 29.3, namely Equation 29.10. Ignoring for the moment the use of a list, one way to solve this expression exactly is to first find the $\hat{\mathbf{a}}^{\text{MAP}} \in \mathcal{Z}$ minimizing $\|\mathbf{r} - \mathbf{H}\hat{\mathbf{a}}\|^2$, where $\hat{\mathbf{a}}^{\text{MAP}}$ denotes the MAP solution. Then, perform a constrained search for all $j = \{1, \ldots bN_t\}$ for $\hat{\mathbf{a}}_j^{-\text{MAP}} \neq \hat{\mathbf{a}}^{\text{MAP}}$, where $\hat{\mathbf{a}}_j^{-\text{MAP}}$ denotes the constrained MAP solution subject to the constraint that its jth bit is the negation of the jth bit for $\hat{\mathbf{a}}^{\text{MAP}}$. This negation is a Boolean logic negation if we are talking about binary bits 1 and 0 and straightforward in the case of 1 and -1. The vectors $\hat{\mathbf{a}}^{\text{MAP}}$ and $\hat{\mathbf{a}}_j^{-\text{MAP}}$ are counterhypotheses of one another because the bit of interest is a different hypothesis (e.g., either 1 or -1) for each vector.

Smart candidate adding is the name given to approaches that search for candidate lists that either exactly or approximately include (1) $\hat{\mathbf{a}}^{\text{MAP}}$ and (2) $\hat{\mathbf{a}}^{-\text{MAP}}$ for each of the bN_t transmitted bits [18]. Because there are bN_t bits for each $\mathbf{a}$, the size of the list $\mathcal{L}$ optimally solving Equation 29.10, for a given $\mathbf{r}$ and $\mathbf{H}$, is at most $\ell = 1 + bN_t$, where the 1 term corresponds to $\hat{\mathbf{a}}^{\text{MAP}}$ and the bN_t terms stem from the fact that the constrained solutions for each of the bN_t bits might be a unique element from $\mathcal{Z}$.

Smart candidate adding can be incorporated into a tree search by not only considering the weights for the branches in the tree, but also their corresponding bit mappings. As with tree search in general, there are many ways to find a list using smart candidate adding. Some of these candidate adding methods, as we have just described, first find $\hat{\mathbf{a}}^{\text{MAP}}$ (or an approximation thereof) and then search the tree to find the counterhypotheses. Others first find not just a single candidate in their first pass through the tree, but a list of candidates, call it $\mathcal{L}_1$, and then find a second list $\mathcal{L}_2$ with counterhypotheses for bits without a counterhypothesis in $\mathcal{L}_1$. Still other candidate adding algorithms search the tree in a single pass and try to find, in parallel, $\hat{\mathbf{a}}^{\text{MAP}}$ and associated counterhypotheses as they progress down the layers of the tree. This single-pass parallel approach reduces complexity at the cost of a small performance loss [19].

Candidate adding example: Find the leaf nodes resulting from searching the tree in Figure 29.11a using the soft-output M algorithm described in Figure 29.10 for $\beta = 1$ and $M = \ell = 5$ by performing parallel smart candidate adding at each layer in the tree.

Figure 29.11a depicts the entire detection tree with all branch metrics provided. The path from the root of the tree with a branch metric of 2 at the first detection layer and 1 at the second detection layer corresponds to the MAP decision. Figure 29.11b begins the algorithm by enumerating the $\beta = 1$ best node from the root of the tree, i.e., the node corresponding to a branch metric of 2. Next, the algorithm considers the bit mapping information in Figure 29.11c. Observing that the branch corresponding to a branch metric of 2 employs the bit mapping of 00, the algorithm performs parallel smart candidate adding. The result is that the paths corresponding to the independent counterhypotheses of the first and second bits are added to the search, i.e., the paths corresponding to 10 and 01, respectively. The result of this parallel candidate adding is that the bit pattern 11 at the first detection layer is never used and can be removed from consideration for the rest of the search. Then, because $M = 5$ and only three nodes have

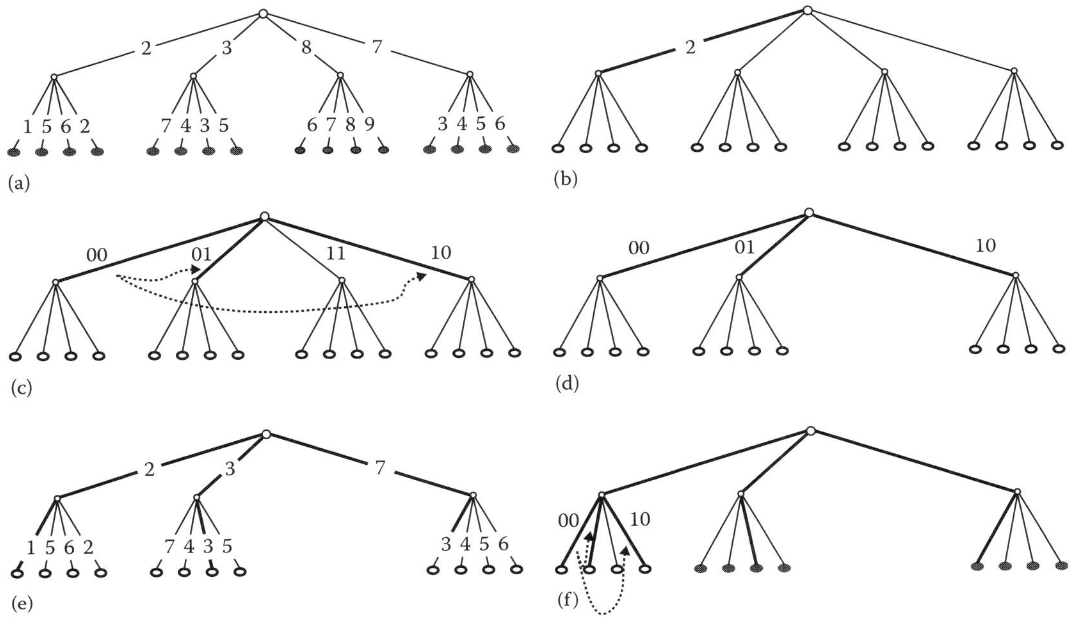

FIGURE 29.11 Example of candidate adding for a 4-ary tree with two layers.

been enumerated, no pruning occurs. These three remaining nodes serve as parent nodes for the next layer in the detection tree.

The process described in the previous paragraph is then repeated for the remaining layer in the tree. We continue by enumerating the $\beta = 1$ best child node from each of the retained parent nodes. This yields $3 \cdot 1 = 3$ leaf nodes in the detection tree. We then perform parallel candidate adding on the best estimate in the tree, i.e. $\{2 \rightsquigarrow 1\}$ as shown in Figure 29.11f yielding five leaf nodes. Because there are five leaf nodes and $M = \ell = 5$, we retain all the leaf nodes for our list and we are done. Observe that the bit patterns corresponding to the five nodes {2, 1}, {2, 2}, {2, 5}, {3, 3}, and {7, 3} are 0000, 0010, 0001, 0111, and 1000. As desired every bit has an associated counterhypothesis. Observe that for this example the node {2, 5}, with cost of 7 and bit pattern 0001, is unnecessary because node {3, 3}, with its lower cost of 6 and bit pattern 0111, serves as the counterhypothesis to the MAP node {2, 1} with bit pattern 0000 for both the second and fourth bits. For completeness we remark that node {2, 2} serves as the counterhypothesis for bit three and node {7, 3} serves as the counterhypothesis for bit one.

29.5 Discussion

In this section, we analyze the relative strengths and weakness of the list tree-based detectors presented in the chapter. To do so we use performance and computational complexity metrics. Before doing so, however, let us first step back to compare the universe of soft-output MIMO detection algorithms.

Table 29.1 classifies soft-output MIMO detectors into four quadrants based on whether or not a list $\mathcal{L} \subseteq \mathcal{Z}$ and/or a tree is used in the detection process. In this chapter, we elected to focus on the upper left quadrant of this grid, i.e., tree-based list detectors. We made the decision to focus on list detectors because, at the same SNR, these detectors are capable of significantly lower error rates than detectors that do not use a list. We made the decision to focus on tree-based approaches for finding the list because

TABLE 29.1 Classification of Soft-Output MIMO Detection Algorithms

	List	No List
Tree	List sphere detector [2] LISS detector [14,15] Soft *M* algorithm [8,20]	
No tree	Monte Carlo methods [21–23] Semidefinite programming [26,27] Soft sphere projection [30]	Soft ZF/MMSE [24,25] Soft interference cancellation [28,29]

these detectors efficiently solve the list detection problem and have a desirable performance-complexity trade-off.

In order to fairly assess the relative strengths and weaknesses of the algorithms in the grid, we first need to know what metrics to use. Clearly the receiver's error rate performance is a critical metric because the ultimate goal in communications is to reliably decode the transmitted message. However, the error rate performance is only one factor in establishing a preference for one detection algorithm over another. The receiver's computational complexity is critically important because it affects the chip size, execution time and power requirements for practical systems.

Analyzing computational complexity is often a difficult undertaking because there are many ways by which we measure complexity. Additionally, different wireless applications have different demands. Despite these obstacles there are a number of effective techniques for roughly analyzing computational complexity.

One measure of computational complexity that lends itself well to the presentation in this chapter is to compute the number of branch metrics which must be calculated as a search progresses through the detection tree. Unlike the examples in this chapter that assumed that all of the branch weights were known, in practice these weights must be computed as we progress through the tree. The disadvantages of this metric is that it does not tell the entire story. For example, the number of branch metric computations does not explicitly tell us about the time required to search through the tree or memory required to store nodes in a stack. The advantage of using the number of branch metric computations as our metric of computational complexity is that it is easy to calculate relative to other metrics, such as floating point operations or logic gates, and provides valuable insight into the overall system complexity.

Using our new complexity metric, we can analyze the algorithms presented in this chapter. We begin with the worst case. Specifically, the brute-force detector must compute all $(q^{N_t+1} - q)/(q - 1)$ branch metrics in the tree. By now we are well aware that we can do better.

The number of branch metric computations performed by the LSD and LISS is a random variable whose probability mass function is a strong function of the SNR. As the SNR increases the number of branch metric computations decreases. While the worst-case complexity of the list sphere detector can be very high (i.e., comparable to the worst-case complexity), the average complexity can be extremely small. We will now illustrate this using Example 1.

In contrast to these variable complexity approaches, breadth-first detection has fixed computational complexity in terms of the number of branch metric computations. Consequently, the number of branch metrics for these algorithms can be expressed deterministically. For example, a soft M algorithm with $\beta = 4$ and $M = 10$ for a tree with four layers and a 64-ary alphabet computes exactly $4 + 16 + 10 \cdot 4 + 10 \cdot 4 = 80$ branch metrics.

Example 1: LSD Branch Metric Example: Consider an 8-input, 8-output memoryless spatially and temporally i.i.d. fading channel in AWGN, and assume that the inputs are independent uncoded 64-QAM symbols. An exhaustive search would have to consider $64^8 = 2^{48} > 2.8 \times 10^{14}$ leaf nodes. Let N denote the number of branch metric computations performed by the list sphere detector, assuming the inputs are ordered according to the near-optimal MMSE sorted-QR decomposition [31]. The probability mass function for N is easy to estimate using simulation. Two examples are shown in Figure 29.12. When the SNR is 20 dB as in Figure 29.1a, the list sphere detector computes $\bar{N} = 550.7$ and $\bar{N} = 2010.4$ branch metrics on average for $\ell = 7$ and $\ell = bN_t + 1 = 49$, respectively. When the SNR is 30 dB, as shown in Figure 29.1b, the list sphere detector computes only $\bar{N} = 462.2$ and $\bar{N} = 1861.4$ branch metrics on average for $\ell = 7$ and $\ell = 49$. From these results, we observe that the average number of branch metric computations decreases with an increase in SNR and increases with the list length.

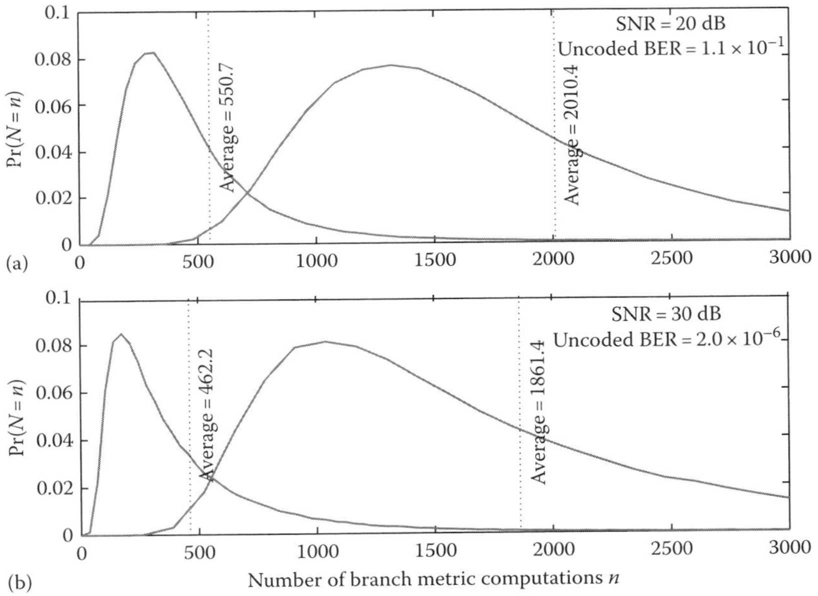

FIGURE 29.12 Estimated probability mass functions for N the number of branch metrics computed by the depth-first list sphere detector at (a) SNR $=$ 20 dB and (b) SNR $=$ 30 dB, assuming 8×8 Rayleigh-fading channel with 64-QAM inputs. These results were found by simulating the list sphere detector T times, with independent noise, channel, and symbol realizations for each trial, then estimating the pmf for N according to $\Pr[N = n] = I_n/T$, where I_n is the number of trials for which n nodes were visited. The number T of trials was 2×10^5 for (a) and (b).

29.5.1 Performance Comparison

We now briefly compare the performance of many of the detection methods described in previous sections. Our results were found using transmission over a spatially and temporally i.i.d. fading 5×5 MIMO channel using 16-QAM modulation alphabets. The information block size (including tail bits) was 9216 bits. Detection was performed based on the complex-valued system model. We used a coded setup based on the one in [2]: a rate 1/2 PCCC based on $(7,5)_{octal}$ convolutional codes using one internal iteration of log MAP decoding. Performance was measured in terms of the E_b/N_0 in dB to achieve a bit error rate of 10^{-5}. Additionally, an MMSE sorted-QR preprocessing is employed [31] and the LLR values were clipped to ± 8 for all algorithms shown.

Figure 29.13 provides performance versus complexity results for the LSD, the LISS, and the soft M algorithm with and without parallel smart candidate adding as outlined in Fig 29.13. Results for the LSD and the LISS are shown for list lengths of $\ell = \{1, 5, 10, 15, 20\}$ and the soft M with parameters $\beta = M = \ell = \{2, 3, 4\}$ and $\beta = \{2, 3, 4\}$ and $M = \ell = \{32, 243, 1024\}$, respectively, when no candidate adding is performed. Results for candidate adding are denoted by the open square with CA next to it. Here we set $\beta = 1$ and $M = \ell = 25$ so that $\omega = 4$ candidates were added to the list at each layer in the detection tree. For both the LSD and the LISS, results for the average computational complexity, as well as the 99.9th percentile computational complexity, are provided. Both the LSD and the LISS have variable computational complexity (in contrast to the soft M algorithms). The exact APP performance is given as a reference.

Figure 29.13 shows us that the LISS computes fewer branch metrics than the LSD when the desired list lengths $\ell = |\mathcal{L}|$ are the same for both the average and maximum number of branch metric computations. We note that although the LISS computes fewer branch metrics to find the minimum distance list, it does so at the cost of a potentially large memory requirement [32,33]. The soft M algorithm with $\beta = M = 4$ and no candidate adding has a less desirable performance-complexity trade-off relative to the average complexity for the LSD and LISS, but is more desirable relative to the 99.9th percentile complexity (often a more important metric of complexity). The soft M algorithm has fixed computational complexity, in contrast to the LSD and the LISS. Finally, we observe that the soft M algorithm with parallel smart candidate adding has a more desirable performance-complexity trade-off relative to all algorithms depicted due to its low and fixed computational complexity. This demonstrates the power of the candidate adding idea.

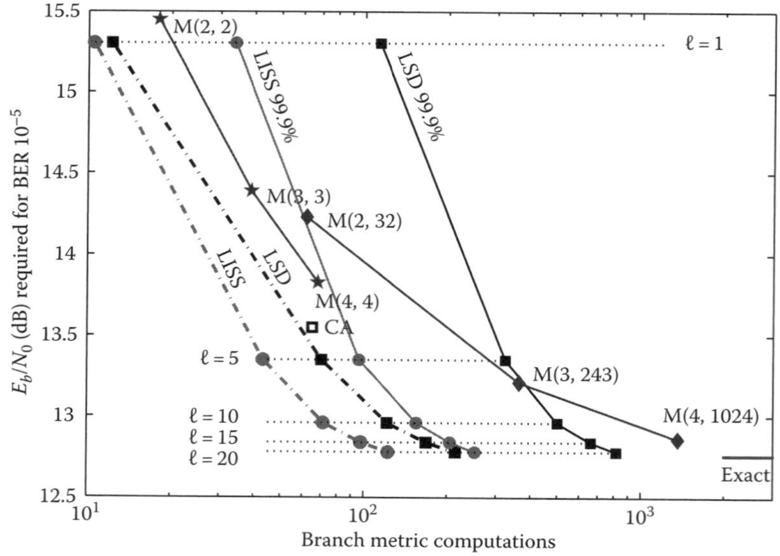

FIGURE 29.13 Performance versus complexity for soft-output 5 × 5 MIMO detection schemes using 16-QAM transmission in fast Rayleigh fading.

29.6 Further Information

Other Soft-Output MIMO Algorithms: Other soft-output tree-based MIMO detection algorithms include: iterative tree search [34], the list fixed-complexity sphere detector [35], the soft fixed-complexity sphere detector [36]. Additionally, as detailed in Table 29.1, there are many soft-output MIMO detection algorithms that fall outside the scope of this chapter. Some of these approaches include Monte Carlo methods [21–23], semidefinite programming [26,27], space-time Chase detection [37], and soft sphere projection [30]. Details on architectural issues pertaining to soft-output MIMO detection algorithms are provided in [38–40].

Search/Sort/Enumeration: Detailed treatment of tree search algorithms can be found in [8,9]. Recent advances relating to Schnorr–Euchner enumeration can be found in [41,42].

Preprocessing: It is often advantageous to detect the receiver vector signal in an order other than the natural ordering, i.e., a detection tree layer ordering differing from $a_1, \ldots, a_{N_t}$. An ordered QR decomposition is often used to achieve the desired ordering, i.e., $\mathbf{HP} = \mathbf{QR}$, where $\mathbf{P}$ is a permutation of the $N_t \times N_t$ identity matrix. Ordering can improve performance for suboptimal detection algorithms and reduces the search complexity of optimal algorithms such as the LSD. Some of these ordering algorithms include the BLAST ordering [43], sorted QR-decomposition [31], and related approaches used by the parallel detector [44], the B-Chase detector [45], and the fixed-complexity sphere decoder [46]. The use of a minimum mean-square error effective channel matrix when performing the ordered QR decomposition can further improve performance or reduce complexity [31]. See [47] for accurate path metric calculation when using MMSE preprocessing.

LLR Clipping: In the absence of a counterhypothesis in the list for a given bit, it is common to set the soft-output of the MIMO detector to a certain predefined value. Recent advances in this area for suboptimal list detectors can be found in [48,49].

References

1. T. Cui and C. Tellambura, An efficient generalized sphere decoder for rank-deficient MIMO systems, *IEEE Communications Letters*, 9, 423–425, May 2005.
2. B. Hochwald and S. ten Brink, Achieving near-capacity on a multiple-antenna channel, *IEEE Transactions on Communications*, 51, 389–399, Mar. 2003.
3. L. Hanzo, J. P. Woodard, and P. Robertson, Turbo decoding and detection for wireless applications, *Proceedings of the IEEE*, 95(6), 1178–1200, June 2007.
4. J. R. Barry, E. A. Lee, and D. G. Messerschmitt, *Digital Communication*, 3rd edn. Boston, MA: Kluwer Academic Publishers, 2004.
5. J. Hagenauer, P. Robertson, and L. Papke, Iterative (turbo) decoding of systematic convolutional codes with the MAP and SOVA algorithms, in *Proceedings of the ITG Conference on Source and Channel Coding* (*SCC*'94), Munich, Germany, 1994, pp. 21–29.
6. P. Robertson, E. Villebrun, and P. Hoeher, A comparison of optimal and sub-optimal MAP decoding algorithms operating in the log domain, in *IEEE International Conference on Communications* (*ICC*'95), vol. 2, Seattle, WA, June 1995, pp. 1009–1013.
7. E. Zimmermann and G. Fettweis, Generalized smart candidate adding for tree search based MIMO detection, in *ITG/IEEE Workshop on Smart Antennas* (*WSA*'07), Vienna, Austria, Feb. 2007.
8. J. Anderson and S. Mohan, Sequential coding algorithms: A survey and cost analysis, *IEEE Transactions on Communications*, 32, 169–176, Feb. 1984.
9. D. E. Knuth, *The Art of Computer Programming. Volume 3: Sorting and Searching*, 2nd edn. Reading, MA: Addison-Wesley Professional, 1998.
10. E. Zimmermann, Complexity Aspects in Near-Capacity MIMO Detection-Decoding, PhD dissertation, Technische Universität Dresden, Dresden, Germany, Aug. 2007.

11. K. Su and I. J. Wassel, Efficient MIMO detection by successive projection, in *IEEE International Symposium on Information Theory*, Adelaide, SA, Sep. 2005.
12. U. Fincke and M. Pohst, Improved methods for calculating vectors of short length in lattice, including a complexity analysis, *Mathematics of Computation*, 44, 463–471, Apr. 1985.
13. C. P. Schnorr and M. Euchner, Lattice basis reduction: Improved practical algorithms and solving subset sum problems, in *Mathematical Programming*, 66 (1–3), 181–199, Aug. 1994.
14. S. Bäro, J. Hagenauer, and M. Witzke, Iterative detection of MIMO transmission using a list-sequential (LISS) detector, in *IEEE International Conference on Communications* (*ICC'03*), Vol. 4, Anchorage, MA, Mar. 2003, pp. 2653–2657.
15. J. Hagenauer and C. Kuhn, The list-sequential (LISS) algorithm and its application, *IEEE Transactions on Communications*, 55 (5), 918–928, May 2007.
16. S. Mohan and J. Anderson, Computationally optimal metric-first code tree search algorithms, *IEEE Transactions on Communications*, 32, 710–717, June 1984.
17. J. Hagenauer and C. Kuhn, Turbo equalization for channels with high memory using a list-sequential (LISS) equalizer, in *Proceedings of the International Symposium on Turbo Codes and Related Topics* (*ISTC03*), Brest, France, Sep. 2003.
18. P. Marsch, E. Zimmermann, and G. Fettweis, Smart candidate adding: A new low-complexity approach towards near-capacity MIMO detection, in *13th European Signal Processing Conference* (*EUSIPCO'05*), Antalya, Turkey, 04–08 Sep. 2005.
19. E. Zimmermann, D. L. Milliner, G. Fettweis, and J. R. Barry, A parallel smart candidate adding algorithm for soft-output MIMO detection, in *Proceedings of the 7th International ITG Conference on Source and Channel Coding* (*SCC'08*), Ulm, Germany, Jan. 2008.
20. D. L. Milliner, E. Zimmermann, J. R. Barry, and G. Fettweis, A framework for fixed complexity single-stage breadth-first detection, in *10th International Symposium on Spread Spectrum Techniques and Applications* (*ISSSTA'08*), Bologna, Italy, pp. 25–28, Aug. 2008.
21. B. Dong, X. Wang, and A. Doucet, A new class of MIMO demodulation algorithms, *IEEE Transactions on Signal Processing*, 51, 2752–2763, Nov. 2003.
22. D. Guo and X. Wang, Blind detection in MIMO systems via sequential Monte Carlo, *IEEE Journal on Selected Areas in Communications*, 21, 453–464, Apr. 2003.
23. B. Farhang-Boroujeny, Z. Haidong, and S. Zhenning, Markov chain Monte Carlo algorithms for CDMA and MIMO communication systems, *IEEE Transactions on Signal Processing*, 54, 1896–1909, May 2006.
24. A. Dejonghe and L. Vandendorpe, Turbo equalization for multilevel modulation: An efficient low complexity scheme, in *Proceedings of the IEEE International Conference on Communications* (*ICC'02*), New York City, Apr. 2002 .
25. M. Butler and I. Collings, A zero-forcing approximate log-likelihood receiver for MIMO bit interleaved coded modulation, *IEEE Communications Letters*, 8, 105–107, Feb. 2004.
26. B. Steingrimsson, Z. Luo, and K. Wong, Soft quasi-maximum likelihood detection for multipleantenna wireless channels, *IEEE Transactions on Signal Processing*, 51, 2710–2718, Nov. 2003.
27. M. Nekuii and T. N. Davidson, List-based soft demodulation of MIMO QPSK via semidefinite relaxation, in *IEEE 8th Workshop on Signal Processing Advances in Wireless Communications*, Helsinki, Finland, June 2007, pp. 1–5.
28. W. J. Choi, K. W. Cheong, and J. M. Cioffi, Iterative soft interference cancellation for multiple antenna systems, in *Proceedings of the IEEE Wireless Communications and Networking Conference* (*WCNC00*), vol. 1, Chicago, IL, 2000, pp. 304–309.
29. S. Bittner, E. Zimmermann, and G. Fettweis, Low complexity soft interference cancellation for MIMO-systems, in *Proceedings of the IEEE Vehicular Technology Conference (VTC'06)*, Melbourne, Australia, May 2006.
30. D. Seethaler, G. Matz, and F. Hlawatsch, Efficient soft demodulation in MIMO-OFDM systems with BICM and constant modulus alphabets, in *IEEE International Conference on Acoustics, Speech and Signal Processing*, vol. 4, Toulouse, France, May 2006.

31. D. Wübben, R. Böhnke, V. Kühn, and K. D. Kammeyer, MMSE extension of V-BLAST based on sorted QR decomposition, in *Proceedings of the IEEE Semiannual Vehicular Technology Conference* (*VTC2003-Fall*), Orlando, FL, Oct. 2003.
32. S. Bittner, E. Zimmermann, W. Rave, and G. Fettweis, List sequential MIMO detection: Noise bias term and partial path augmentation, in *IEEE International Conference on Communications* (*ICC'06*), Istanbul, Turkey, June 11–15, 2006.
33. E. Zimmermann and G. Fettweis, On the efficiency of the sequential detector for solving soft-output detection problems, *IEEE Communications Letters*, 12, 840–842, Nov. 2008.
34. Y. L. C. de Jong and T. J. Willink, Iterative tree search detection for MIMO wireless systems, *IEEE Transactions on Communications*, 53, 930–935, June 2005.
35. L. G. Barbero and J. S. Thompson, Extending a fixed-complexity sphere decoder to obtain likelihood information for turbo-MIMO systems, *IEEE Transactions on Vehicular Technology*, Sep. 2008.
36. L. G. Barbero, T. Ratnarajah, and C. Cowan, A comparison of complex lattice reduction algorithms for MIMO detection, in *IEEE International Conference on Acoustics, Speech, and Signal Processing* (*ICASSP '08*), Las Vegas, NV, Apr. 2008.
37. D. J. Love, S. Hosur, A. Batra, and R. W. Heath, Jr., Space-time chase decoding, *IEEE Transactions on Wireless Communications*, 4, 2035–2039, Sep. 2005.
38. Z. Guo and P. Nilsson, Algorithm and implementation of the K-best sphere decoding for MIMO detection, *IEEE Journal on Selected Areas in Communications*, 24, 491–503, Mar. 2006.
39. S. Chen and T. Zhang, Low power soft-output signal detector design for wireless MIMO communication systems, in *International Symposium on Low Power Electronics and Design* (*ISLPED'07*), Troy, NY, Aug. 2007.
40. C. Studer, A. Burg, and H. Bölcskei, Soft-output sphere decoding: Algorithms and VLSI implementation, *IEEE Journal on Selected Areas in Communications*, 26, 290–300, Feb. 2008.
41. A. Burg, M. Borgmann, M. Wenk, M. Zellweger, W. Fichtner, and H. Bölcskei, VLSI implementation of MIMO detection using the sphere decoding algorithm, *IEEE Journal of Solid-State Cicuits*, 40, 1566–1577, July 2005.
42. B. Mennenga and G. Fettweis, Search sequence determination for tree search based detection algorithms, in *IEEE Sarnoff Symposium*, Princeton, NJ, Mar. 2009.
43. G. J. Foschini, G. D. Golden, R. A. Valenzuela, and P. W. Wolniansky, Simplified processing for wireless communication at high spectral efficiency, *IEEE Journal on Selected Areas in Communications*, 17, 1841–1852, Nov. 1999.
44. Y. Li and Z.-Q. Luo, Parallel detection for V-BLAST system, in *Proceedings of the IEEE International Conference on Communications* (*ICC'02*), vol. 1, New York, NY, Apr. 2002, pp. 340–344.
45. D. W. Waters and J. R. Barry, The chase family of detection algorithms for MIMO channels, *IEEE Transactions on Signal Processing*, 56, 739–747, Feb. 2008.
46. L. G. Barbero and J. S. Thompson, A fixed-complexity MIMO detector based on the complex sphere decoder, in *IEEE Workshop on Signal Processing Advances for Wireless Communications* (*SPAWC'06*), Cannes, France, July 2006.
47. E. Zimmermann and G. Fettweis, Unbiased MMSE tree search MIMO detection, in *International Symposium on Wireless Personal Multimedia Communications* (*WPMC'06*), San Diego, CA, Sep. 17–20, 2006.
48. D. L. Milliner, E. Zimmermann, G. Fettweis, and J. R. Barry, Channel state information based LLR clipping for list MIMO detection, in *IEEE International Symposium on Personal, Indoor and Mobile Radio Communications* (*PIMRC'08*), Cannes, France, Sep. 15–18, 2008.
49. L. Wang, L. Xu, S. Chen, and L. Hanzo, Apriori-LLR-threshold-assisted K-best sphere detection for MIMO channels, in *IEEE Vehicular Technology Conference* (*VTC'08*), Marina Bay, Singapore, May 2008, pp. 867–871.

30

Lattice Reduction–Aided Equalization for Wireless Applications

Wei Zhang
Georgia Institute of Technology

Xiaoli Ma
Georgia Institute of Technology

30.1 Introduction 30-2
30.2 Lattice Reduction 30-3
The LLL Algorithm • Seysen's Algorithm • Other LR Algorithms
30.3 Lattice Reduction-Aided Equalizers 30-9
LR-Aided LEs • LR-Aided DFEs • Dual LR-Aided Equalizers
30.4 Applications 30-12
MIMO Systems • LP-OFDM Systems
30.5 Conclusions 30-15
References 30-15

As wireless services spread and become integrated into people's daily life, the expectation of the performance and reliability of wireless devices naturally increases. As a result, system designers now face more challenges such as limited bandwidth, dynamic resource allocation, and particularly, channel fading effects introduced by the variability in the time, frequency, and space domains. Diversity techniques have been widely adopted to combat deleterious channel fading effects. To exploit the diversity embedded in the channel, both transmitter and receiver must be designed appropriately. To collect the full diversity enabled by the transmitter, maximum-likelihood equalizers (MLEs) or near-MLEs are usually adopted at the receiver. However, the high decoding complexity makes MLE or near-MLE infeasible in practical systems. To reduce the decoding complexity, low-complexity equalizers, such as linear equalizers (LEs) and decision feedback equalizers (DFEs), are often adopted. These methods, however, may not utilize the diversity enabled by the transmitter and as a result have degraded performance relative to the system with the MLE. Recently, lattice reduction (LR) techniques have been introduced to improve the performance of low-complexity equalizers without increasing the complexity significantly. In this chapter, we first present the development of the LR techniques by introducing the various LR algorithms. Then, the detailed procedures of the LR-aided LEs and the LR-aided DFEs are given for general linear systems. The performance and the complexity of LR-aided equalizers with different LR algorithms are compared. Finally, we demonstrate the effectiveness of LR-aided equalizer by two applications on multiantenna multi-input multi-output (MIMO) systems and precoded orthogonal frequency division multiplexing (OFDM) systems.

30.1 Introduction

Most wireless transmissions, such as the OFDM systems in IEEE 802.11a, multiantenna MIMO systems in IEEE 802.11n, and multiuser code division multiple access (CDMA) systems, can be modeled as a linear block transmission system. Consider a linear block transmission model depicted in Figure 30.1

$$\boldsymbol{y} = \boldsymbol{H}\boldsymbol{s} + \boldsymbol{w}, \tag{30.1}$$

where

- $\boldsymbol{H}$ is an $M \times N$ complex Gaussian channel matrix with zero mean, the $N \times 1$ vector $\boldsymbol{s}$ consists of the information symbols
- $\boldsymbol{y}$ is the $M \times 1$ received vector
- $\boldsymbol{w}$ is independent and identically distributed (i.i.d.) complex additive white Gaussian noise with variance σ_w^2

We assume that the channel matrix $\boldsymbol{H}$ is known at the receiver, but unknown at the transmitter. Note that the channel matrix $\boldsymbol{H}$ is general enough to represent a number of cases. The results in the following sections are based on this linear model, and thus can be easily applied to the specific wireless systems whose input/output (I/O) relationship can be expressed as the general system model in Equation 30.1.

Given the linear block transmission model in Equation 30.1, there are various ways to decode the transmitted symbol vector $\boldsymbol{s}$ from the observation $\boldsymbol{y}$. For the general linear model, here, we generalize the term "equalizer" as the one to equalize the channel effect (it does not have to be intersymbol interference channels). One often used and also optimal detector (if there is no prior information about the symbols and/or symbols are treated as deterministic parameters) is the MLE, which is based on an exhaustive search among all $N \times 1$ symbol vectors. MLE provides optimal error performance with high decoding complexity ($\mathcal{O}(|\mathcal{S}|^N)$). Some near-MLEs have also been proposed to reduce the complexity and achieve near-ML performance, e.g., the sphere-decoding (SD) method [7]. However, these near-MLEs exhibit either high average complexity or high complexity variance, especially when the size of the channel matrix and/or the constellation size is large. Furthermore, early termination and fixed memory considerations for hardware implementations may degrade the performance of near-MLEs.

There are some equalizers that are usually characterized and referred to as low-complexity equalizers: LEs and DFEs. Two often adopted LEs are the zero-forcing (ZF) equalizer and the linear minimum mean-square error (MMSE) equalizer. Given the model in Equation 30.1, the estimate of ZF-LE is given as

$$\hat{\boldsymbol{s}}_{zf} = \mathcal{Q}(\boldsymbol{H}^\dagger \boldsymbol{y}) = \mathcal{Q}(\boldsymbol{s} + \boldsymbol{H}^\dagger \boldsymbol{w}) = \mathcal{Q}(\boldsymbol{s} + \boldsymbol{\eta}), \tag{30.2}$$

where

- $\boldsymbol{H}^\dagger = (\boldsymbol{H}^{\mathcal{H}}\boldsymbol{H})^{-1}\boldsymbol{H}^{\mathcal{H}}$ denotes the Moore–Penrose pseudoinverse of the channel matrix $\boldsymbol{H}$
- $\mathcal{Q}(\cdot)$ denotes element-wise quantization to the nearest constellation point for a given modulation scheme
- $\boldsymbol{\eta}$ is the colored noise after equalization with covariance matrix $(\boldsymbol{H}^{\mathcal{H}}\boldsymbol{H})^{-1}$

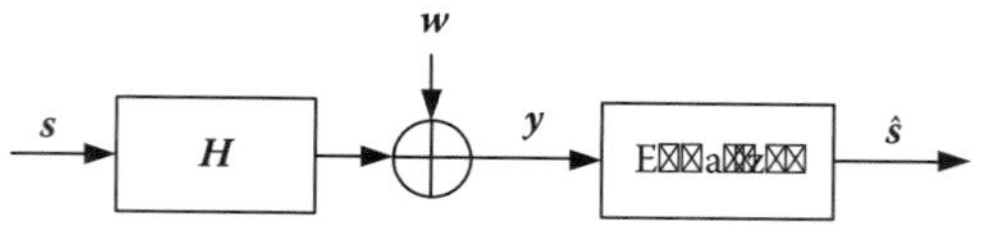

FIGURE 30.1 Block diagram of linear transmission system model.

Another often adopted LE, the linear MMSE-LE, is defined as

$$\boldsymbol{x}_{\text{mmse}} = \left(\boldsymbol{H}^{\mathcal{H}}\boldsymbol{H} + \sigma_w^2\boldsymbol{I}_N\right)^{-1}\boldsymbol{H}^{\mathcal{H}}\boldsymbol{y}, \tag{30.3}$$

followed by the quantization step. Here, we notice that, with the definition of an extended system (also shown in [8,24])

$$\bar{\boldsymbol{H}} = \begin{bmatrix} \boldsymbol{H} \\ \sigma_w\boldsymbol{I}_N \end{bmatrix} \quad \text{and} \quad \bar{\boldsymbol{y}} = \begin{bmatrix} \boldsymbol{y} \\ \boldsymbol{0}_{N\times 1} \end{bmatrix}, \tag{30.4}$$

the MMSE equalizer in Equation 30.3 can be rewritten as $\boldsymbol{x}_{\text{mmse}} = \bar{H}^{\dagger}\bar{y}$. Thus, the MMSE-LE has the same form as the ZF-LE in Equation 30.2 with respect to this extended system. The analysis based on the ZF equalizer in Equation 30.2 can be extended to MMSE equalizer accordingly. Since the MMSE equalizer further takes some noise information, MMSE-LE achieves better performance than ZF-LE in general, but requires an estimate of the noise variance at the receiver. The complexities of both equalizers are dominated by matrix inversion, which requires polynomial complexity $\mathcal{O}(N^3)$ via Gaussian elimination.

To improve the performance, successive-interference-cancellation (SIC) equalizers are proposed [6]. Since the DFEs have been shown to be equivalent to SICs in [5], we consider them as one category of low-complexity equalizers. The major difference between DFEs and LEs is the feedback of the detected symbols through a feedback matrix $\boldsymbol{B}$. According to the equalization method, DFEs are divided into two categories: ZF-DFE (ZF-SIC) and MMSE-DFE (MMSE-SIC). The specific designs of the feedforward matrix $\boldsymbol{F}$ and the feedback matrix $\boldsymbol{B}$ for both DFEs can be found in [5,10]. Different from LEs, matrix decompositions, e.g., QR decomposition, comprise the major part of the DFEs' complexity. Such algorithms are usually associated with the complexity of $\mathcal{O}(MN^2)$. Compared to LEs, the corresponding DFEs achieve better performance. However, the performance of DFEs is greatly affected by the decoding order and the error propagation. To improve the performance of DFEs and to mitigate the complexity overhead introduced by the feedback filter, optimum ordering is usually adopted in DFEs. For example, V-BLAST ordering optimizes the BER performance but the complexity is suboptimal [10].

30.2 Lattice Reduction

The main drawback of the aforementioned low-complexity equalizers is that these equalizers usually sacrifice the performance relative to (near-) MLEs. For example, the diversity order which is an important figure of merit to quantify the performance collected by LEs and DFEs is only $M-N+1$ for spatial multiplexing systems with i.i.d. channels, while (near-) MLE exploits diversity M [18]. The impact of the lack of diversity order becomes especially severe when the channel matrix is square, e.g., $M=N$. Furthermore, as shown in [10], optimal ordering cannot increase the diversity order collected by DFEs. Since (near-) MLEs exhibit either high average complexity or high complexity variance, the cubic order complexity makes LEs and DFEs widely adopted in practical systems. A natural question is whether the complexity reduction is worth the performance sacrifice. In the following, we introduce a technique which keeps the complexity low while improving the performance of low-complexity equalizers.

Given the system model in Equation 30.1, if the symbols $\boldsymbol{s}$ are drawn from Gaussian integer ring $\mathbb{Z}[j]$ (e.g., QAM, PAM constellations), then $\boldsymbol{H}\,\boldsymbol{s}$ belongs to a lattice spanned by the columns of $\boldsymbol{H}$ [1]. Hence, to decode s now becomes to find the nearest point to y on the lattice spanned by H. As we know, when the lattice basis H is orthogonal, i.e., $H^{\mathcal{H}}H$ is diagonal, the correlation matrix of the colored noise η after equalization in Equation 30.2 is diagonal, which means the performance of LEs is the same as that of MLE. Therefore, if we can find another basis $\tilde{H}$ in the same lattice which is more orthogonal than H, and apply LEs/DFEs based on $\tilde{H}$, intuitively the performance should be closer to that of MLE. The process to find a more orthogonal basis is called LR.

Theoretically, finding an optimal set of bases in a lattice is computationally expensive [1]. Therefore, for a given channel matrix H, the ultimate goal of LR is "relaxed" to find a matrix $\tilde{H} = HT$ that is closer to orthogonality than the original H. T is a unimodular matrix, which means that all the entries of T and T^{-1} are Gaussian integers and the determinant of T is ± 1 or $\pm j$. LR techniques have been studied by mathematicians for decades, and many LR algorithms have been proposed. Two widely adopted LR methods in wireless communications are the Lenstra–Lenstra–Lovász (LLL) algorithm [14] and Seysen's algorithm (SA) [22]. In this section, we focus on these two algorithms while we also briefly introduce the other LR algorithms for readers' interest.

30.2.1 The LLL Algorithm

So far, the most popular LR algorithm is the LLL algorithm [14]. The LLL algorithm does not guarantee to find the optimal basis, but it guarantees in polynomial average time to find a basis within a factor to the optimal one [9,14,16,25]. The real LLL (RLLL) algorithm is first applied to improve the performance of low-complexity equalizers by extending the complex transmission system in Equation 30.1 into an equivalent real model [24]. Furthermore, the complex LLL (CLLL) algorithm is proposed in [4] based on the Gram–Schmidt orthonormalization and in [18] based on the QR-decomposition. An $M \times N$ complex matrix $\tilde{H}$ is called an LLL-reduced basis of a lattice, if the QR decomposition $\tilde{H} = \tilde{Q}\tilde{R}$ satisfies the following two conditions:

$$\begin{aligned} &\left|\Re\left[\tilde{R}_{i,k}\right]\right| \leq \tfrac{1}{2}|\tilde{R}_{i,i}|, \ \left|\Im\left[\tilde{R}_{i,k}\right]\right| \leq \tfrac{1}{2}|\tilde{R}_{i,i}|, \quad \forall i < k, \\ &\delta|\tilde{R}_{i-1,i-1}|^2 \leq |\tilde{R}_{i,i}|^2 + |\tilde{R}_{i-1,i}|^2, \quad \forall i \in [2, N], \end{aligned} \tag{30.5}$$

where the parameter δ now is arbitrarily chosen from $\left(\frac{1}{2}, 1\right)$, and $\tilde{R}_{i,k}$ is the (i,k)th entry of $\tilde{R}$. The detailed pseudocode of the CLLL algorithm can be found in Table 30.1. The parameter δ controls the complexity and performance of the LLL algorithm and the bigger δ is, the higher is the complexity.

It has been shown that the CLLL algorithm reduces the complexity of the RLLL algorithm without sacrificing performance [4,18]. Though the RLLL algorithm only requires real operations, as shown in Table 30.2, it requires more basis updates than the CLLL algorithm. One basis update is defined as the process that updates the nth basis vector using the mth basis as $h_n \leftarrow h_n + a_{m,n}\, h_m$. Furthermore, the sorted QR decomposition (SQRD) in [24] is introduced into the LLL process to further reduce the complexity as shown in Table 30.2. The complexity of the LLL algorithm depends on the specific realization of the channel and thus is random. The results in [16] theoretically proved that the upper bound on the average complexity of LLL is $\mathcal{O}(N^3 \log N)$. Furthermore [9] reduces the upper bound on the average complexity to $\mathcal{O}\left(N^2 \log\left(\frac{N}{M-N+1}\right)\right)$. We need to note that the worst-case complexity of the LLL algorithm can be infinite, since we can build a special structure that needs exactly K iterations to finish the LLL algorithm [9,25]. However, the probability that a system meets such a special structure in practice or simulations is zero, not to mention the probability that we meet it many times. Thus, though theoretically the worst-case complexity of the LLL algorithm is not upper bounded, the complexity is still polynomial in practice and simulations.

TABLE 30.1 The Complex LLL Algorithm (Using MATLAB® Notation)

INPUT: $\boldsymbol{H}$; OUTPUT: $\tilde{\boldsymbol{Q}}, \tilde{\boldsymbol{R}}, \boldsymbol{T}$
(1) $[\tilde{\boldsymbol{Q}}, \tilde{\boldsymbol{R}}] = \text{QR Decomposition}(\boldsymbol{H})$;
(2) $\delta \in \left(\frac{1}{2}, 1\right)$;
(3) $m = \text{size}(\boldsymbol{H}, 2)$;
(4) $\boldsymbol{T} = \boldsymbol{I}_m$;
(5) $k = 2$;
(6) while $k \le m$
(7) for $n = k-1 : -1 : 1$
(8) $u = \text{round}((\tilde{\boldsymbol{R}}(n,k)/\tilde{\boldsymbol{R}}(n,n)))$;
(9) if $u \sim= 0$
(10) $\tilde{\boldsymbol{R}}(1:n,k) = \tilde{\boldsymbol{R}}(1:n,k) - u \cdot \tilde{\boldsymbol{R}}(1:n,n)$;
(11) $\boldsymbol{T}(:,k) = \boldsymbol{T}(:,k) - u \cdot \boldsymbol{T}(:,n)$;
(12) end
(13) end
(14) if $\delta\, \lvert\tilde{\boldsymbol{R}}(k-1,k-1)\rvert^2 > \lvert\tilde{\boldsymbol{R}}(k,k)\rvert^2 + \lvert\tilde{\boldsymbol{R}}(k-1,k)\rvert^2$
(15) Swap the $(k-1)$th and kth columns in $\tilde{\boldsymbol{R}}$ and $\boldsymbol{T}$
(16) $\boldsymbol{\Theta} = \begin{bmatrix} \alpha^* & \beta \\ -\beta & \alpha \end{bmatrix}$ where $\alpha = \dfrac{\tilde{\boldsymbol{R}}(k-1,k-1)}{\lVert\tilde{\boldsymbol{R}}(k-1:k,k-1)\rVert}$; $\beta = \dfrac{\tilde{\boldsymbol{R}}(k,k-1)}{\lVert\tilde{\boldsymbol{R}}(k-1:k,k-1)\rVert}$;
(17) $\tilde{\boldsymbol{R}}(k-1:k, k-1:m) = \boldsymbol{\Theta}\tilde{\boldsymbol{R}}(k-1:k, k-1:m)$;
(18) $\tilde{\boldsymbol{Q}}(:,k-1:k) = \tilde{\boldsymbol{Q}}(:,k-1:k)\boldsymbol{\Theta}^{\mathcal{H}}$;
(19) $k = \max(k-1, 2)$;
(20) else
(21) $k = k+1$;
(22) end
(23) end

Source: Ma, X. and Zhang, W., *IEEE Commun. Mag.*, 47(1), 56–62, 2008.

TABLE 30.2 Number of Basis Updates Needed for Different LR Algorithms for i.i.d. Channels

$M = N = n$		2	4	6	8	10
Greedy SA	Average	1.0733	5.4579	11.1725	16.9698	22.0766
	Std. deviation	0.6378	2.3032	4.4139	6.9265	9.3423
Real LLL	Average	3.5204	19.0711	46.5706	84.7531	132.36
	Std. deviation	2.7614	9.6056	21.5811	39.5851	61.8293
Complex LLL	Average	1.1151	6.6624	16.2276	29.0076	44.2684
	Std. deviation	0.6963	3.0824	7.0284	12.6450	19.4835
Complex LLL with SQRD	Average	1.0505	5.7555	13.4083	23.3189	35.2554
	Std. deviation	0.6208	2.6426	5.9486	10.3063	15.5275

Source: Zhang, W., Ma, X., Gestner, B., and Anderson, D.V., *IEEE Commun. Mag.*, 47(1), 56–62, 2009.

Following the CLLL algorithm, we find a "better" channel matrix $\tilde{\boldsymbol{H}} = \boldsymbol{HT}$ from the original channel matrix $\boldsymbol{H}$. To quantify the condition of the output matrix $\tilde{\boldsymbol{H}}$, we first introduce the following metric to quantify the distance to orthogonality for matrices

Definition 30.1: For an $M \times N$ matrix $\boldsymbol{H} = [\boldsymbol{h}_1, \boldsymbol{h}_2, \ldots, \boldsymbol{h}_N]$, with $\boldsymbol{h}_n$ being $\boldsymbol{H}$'s nth column, its orthogonality deficiency (od($\boldsymbol{H}$)) is defined as

$$\text{od}(\boldsymbol{H}) = 1 - \frac{\det(\boldsymbol{H}^{\mathcal{H}}\boldsymbol{H})}{\prod_{n=1}^{N} \|\boldsymbol{h}_n\|^2}, \tag{30.6}$$

where $\|\boldsymbol{h}_n\|$, $1 \leq n \leq N$ is the norm of the nth column of $\boldsymbol{H}$.*

Note that $0 \leq \text{od}(\boldsymbol{H}) \leq 1$, $\forall \boldsymbol{H}$. If $\boldsymbol{H}$ is singular, $\text{od}(\boldsymbol{H}) = 1$, and if the columns of $\boldsymbol{H}$ are orthogonal, $\text{od}(\boldsymbol{H}) = 0$. In general, when od($\boldsymbol{H}$) is smaller, we say that $\boldsymbol{H}$ is closer to an orthogonal matrix. Adopting od as the orthogonality metric, we quantify the condition of the output matrix $\tilde{\boldsymbol{H}}$ of the CLLL algorithm. Given a matrix $\boldsymbol{H} \in \mathbb{C}^{M \times N}$ with rank N, $\tilde{\boldsymbol{H}}$ is obtained after applying the CLLL algorithm in Table 30.1 for a given parameter $\delta \in \left(\frac{1}{2}, 1\right)$. Then, as shown in [18], the od of $\tilde{\boldsymbol{H}}$ satisfies:

$$\text{od}(\tilde{\boldsymbol{H}}) \leq 1 - 2^N \left(\frac{2}{2\delta - 1}\right)^{-\frac{N(N+1)}{2}}. \tag{30.7}$$

For real $\boldsymbol{H}$, the upper bound in Equation 30.7 is consistent with the result in [14, Proposition 1.8]. Here, we extend it to the complex field according to the CLLL algorithm in Table 30.1. If $\boldsymbol{H}$ is singular, i.e., rank ($\boldsymbol{H}$) $< N$, then the upper bound in Equation 30.7 does not hold true since $\boldsymbol{H}$ is not a basis anymore. In this case, we need to reduce the size of $\boldsymbol{H}$ and then apply the CLLL algorithm. From Equation 30.7, we can see that the CLLL algorithm does not guarantee to reduce the od for every realization of $\boldsymbol{H}$, but the new basis $\tilde{\boldsymbol{H}}$ now has an upper bound on od which is strictly less than one. In [18,20], we show that, thanks to this upper bound on od, the LLL-based equalizers collect the same diversity as MLEs.

30.2.2 Seysen's Algorithm

As an alternative to the LLL algorithm, SA is an iterative method to reduce the lattice based on the Seysen's metric [13,22]. The Seysen's metric that SA adopts to quantify the orthogonality of matrices is defined as follows (see e.g., [13,22]):

Definition 30.2: For an $M \times N$ matrix $\boldsymbol{H}$, the Seysen's metric ($S(\boldsymbol{H})$) is defined as

$$S(\boldsymbol{H}) = \sum_{n=1}^{N} \|\boldsymbol{h}_n\|^2 \|\boldsymbol{a}_n\|^2, \tag{30.8}$$

where $\boldsymbol{h}_n$ is the nth column of $\boldsymbol{H}$ and $\mathbf{a}_n^T$ is the nth row of $\boldsymbol{H}^{\dagger}$.

For any $\boldsymbol{H}$, $S(\boldsymbol{H}) \geq N$, with equality when $\boldsymbol{H}$ is a unitary matrix. In general, smaller $S(\boldsymbol{H})$ indicates that $\boldsymbol{H}$ is closer to being a unitary matrix. Compared with the definition of od in Equation 30.6, we can see that $S(\boldsymbol{H})$ optimizes the orthogonality of both $\boldsymbol{H}$ and $\boldsymbol{H}^{\mathcal{H}}$, while od($\boldsymbol{H}$) focuses on the orthogonality of $\boldsymbol{H}$ only.

The ultimate goal of SA is to find a set of bases $\tilde{\boldsymbol{H}}$, of which the Seysen's metric cannot be reduced any more. The lazy method and the greedy method are first proposed to implement SA [13], while a simplified greedy implementation is proposed in [21] to further reduce the complexity. The lazy implementation of SA guarantees to find the optimal bases that minimize $S(\tilde{\boldsymbol{H}})$ but requires extremely

* *od* has also been defined as $\frac{\prod_{n=1}^{N_t} \|\boldsymbol{h}_n\|}{|\det(\boldsymbol{H})|}$ in [11], which is equivalent to Equation 30.6 when $\boldsymbol{H}$ is square, but its value is unbounded.

high complexity. The greedy implementation requires much fewer operations, but the algorithm may stop at a certain set of bases $\tilde{\boldsymbol{H}}$ with suboptimal $S(\tilde{\boldsymbol{H}})$ (a local minimum).

SA adopts iterative basis updates to generate $\tilde{\boldsymbol{H}}$ whose $S(\boldsymbol{H})$ cannot be reduced any more. The new basis and the unimodular matrix are initialized as $\tilde{\boldsymbol{H}} = \boldsymbol{H}$ and $\boldsymbol{T} = \boldsymbol{I}_N$. In each basis update, an index pair (m,n) and an integer $\lambda_{m,n}$ are found to update $\tilde{\boldsymbol{H}}$ and $\tilde{\boldsymbol{H}}^{\dagger}$ as

$$\tilde{\boldsymbol{h}}_m \leftarrow \tilde{\boldsymbol{h}}_m + \lambda_{m,n}\tilde{\boldsymbol{h}}_n \quad \text{and} \quad \tilde{\boldsymbol{a}}_n^{\mathrm{T}} \leftarrow \tilde{\boldsymbol{a}}_n^{\mathrm{T}} - \lambda_{m,n}^{*}\,\tilde{\boldsymbol{a}}_m^{\mathrm{T}}, \tag{30.9}$$

where

$\tilde{\boldsymbol{h}}_m$ is the mth column of $\tilde{\boldsymbol{H}}$

$\tilde{\boldsymbol{a}}_n^T$ is the nth row of $\tilde{\boldsymbol{H}}^{\dagger}$

T, the unimodular matrix, is also updated correspondingly as $\boldsymbol{t}_m \leftarrow \boldsymbol{t}_m + \lambda_{m,n}\,\boldsymbol{t}_n$ [13,21]

According to [13,22], given the indices m and n, the integer $\lambda_{m,n}$ is chosen as

$$\lambda_{m,n} = \left\lfloor 0.5\left(\frac{\tilde{\boldsymbol{a}}_n^T\tilde{\boldsymbol{a}}_m^*}{\|\tilde{\boldsymbol{a}}_m\|^2} - \frac{\tilde{\boldsymbol{h}}_n^{\mathcal{H}}\tilde{\boldsymbol{h}}_m}{\|\tilde{\boldsymbol{h}}_n\|^2}\right)\right\rceil \tag{30.10}$$

to maximize the reduction of $S(\boldsymbol{H})$, where $\lfloor\cdot\rceil$ is the rounding operator that rounds the real and the imaginary parts to the nearest integers. The corresponding reduction of $S(\boldsymbol{H})$ is expressed as

$$\Delta_{m,n} = 2\|\tilde{\boldsymbol{h}}_m\|^2\|\tilde{\boldsymbol{a}}_n^{\mathrm{T}}\|^2\left(\Re\left(\lambda_{m,n}^*\left(\frac{\tilde{\boldsymbol{a}}_n^{\mathrm{T}}\tilde{\boldsymbol{a}}_m^*}{\|\tilde{\boldsymbol{a}}_m\|^2} - \frac{\tilde{\boldsymbol{h}}_n^{\mathcal{H}}\tilde{\boldsymbol{h}}_m}{\|\tilde{\boldsymbol{h}}_n\|^2}\right)\right) - |\lambda_{m,n}|^2\right). \tag{30.11}$$

The algorithm continues updating the bases as in Equation 30.9 until no more reduction can be made on $S(\boldsymbol{H})$, i.e., all the entries of the matrix $\boldsymbol{\Delta}$ defined in Equation 30.11 are zero. However, the algorithm may stop at $\tilde{\boldsymbol{H}}$ with suboptimal $S(\boldsymbol{H})$ (a local minimum). Only when the right index pairs are chosen in the right order can the global optimal $\tilde{\boldsymbol{H}}$ (i.e., $\tilde{\boldsymbol{H}}$ with minimum $S(\boldsymbol{H})$) be found. Thus, how to choose the index pairs to update the bases is the crucial problem of SA. As shown in [13,21], the index pair (m, n) is chosen randomly in the lazy implementation of SA, while in the greedy implementation, $(m, n) = \arg \max_{m,n=1,\ldots,N} \Delta_{m,n}$. In the following, we propose a tree-search algorithm to choose index pairs based on the corresponding reduction in $S(\boldsymbol{H})$.

We design our implementation of SA by formulating the choice of index pairs into a tree structure. The tree is composed of different versions of $\boldsymbol{H}$ as

1. The tree roots at the initialization $\tilde{\boldsymbol{H}} = \boldsymbol{H}$
2. The pth level of the tree denotes the pth basis update
3. Each leaf node on the pth level of the tree represents a candidate of $\tilde{\boldsymbol{H}}$ after basis updating with a selected index pair of its parent and is associated with the corresponding $S(\boldsymbol{H})$
4. Each end node with a path from the origin represents a candidate of $\tilde{\boldsymbol{H}}$ after a series of basis updates

Thus, finding the optimal $\tilde{\boldsymbol{H}}$ in terms of $S(\tilde{\boldsymbol{H}})$ is equivalent to finding the end node with the smallest $S(\tilde{\boldsymbol{H}})$ of this spanning tree. This is a regular tree where every interior node has $N\ (N-1)$ children. There are many existing search strategies to go through the tree and find the optimal end node, e.g., depth-first, breadth-first, and best-first tree-search algorithms. However, we need to note that this tree is built up by $\boldsymbol{H}$s and thus is random. This is different from the setup of other tree-search algorithms, e.g., the sphere-decoding method [7], where the spanning tree is based on the symbol constellation and thus fixed for every $\boldsymbol{H}$. Therefore, depth-first and best-first strategies are not computationally efficient for our problem here since we cannot even predict how many levels this tree has for a given $\boldsymbol{H}$. To simplify the tree-search process, we adopt the breadth-first approach.

In the pth basis update, instead of updating only one pair of basis indices as in the greedy implementation, we find the best K_c index pairs (K_c nodes with the smallest $S(\boldsymbol{H})$ at the pth level of the tree), which give us at most K_c different candidates of $\tilde{\boldsymbol{H}}$, $\{\tilde{\boldsymbol{H}}_k\}_{k=1}^{K_c}$. For each candidate $\tilde{\boldsymbol{H}}_k$, we check K_ℓ children nodes (the largest K_ℓ entries in the $\boldsymbol{\Delta}_{m,n}^{(k)}$ matrix associated with the current candidate). Then from these K_cK_ℓ children nodes, we again choose the best K_c nodes with the smallest $S(\boldsymbol{H})$. In other words, we select K_c nodes on each level of the tree, which are from the K_cK_ℓ children nodes of the K_c nodes we chose in the previous level. The program stops when none of the candidates has children nodes, i.e., $\boldsymbol{\Delta}$ matrix is zero. Then, we can choose the end node with the smallest $S(\boldsymbol{H})$ as the output $\tilde{\boldsymbol{H}}$. One major advantage of SA over the LLL algorithm is that it reduces the metric $S(\boldsymbol{H})$ by $\Delta_{m,n}$ in each basis update, whereas the LLL algorithm does not guarantee reduction of od($\boldsymbol{H}$).

Specifically, if $K_c = K_\ell = 1$, the tree-search SA becomes the greedy method [13,21]. If $K_c = \infty$ and $K_\ell = N\ (N-1)$, it is the lazy implementation of SA, which is also the implementation that achieves optimal $S(\boldsymbol{H})$ [13,21]. The detailed algorithm chart is given in Table 30.3. With this tree-search implementation of SA, we can adjust K_c and K_ℓ according to the complexity that we can afford. The larger K_c and K_ℓ are, the higher is the probability that the global optimal $\tilde{\boldsymbol{H}}$ ($\tilde{\boldsymbol{H}}$ with the smallest $S(\tilde{\boldsymbol{H}})$) is found, while the complexity is also higher.

Different from the LLL algorithm, SA needs a fixed number of arithmetic operations in each basis update, though the number of basis updates is still random. Another difference between the SA and the LLL algorithm is that the SA does not require QR decomposition but needs to compute the channel matrix inverse at the preprocessing stage, while the LLL algorithm needs to compute the matrix inverse after $\tilde{H}$ is found. As shown in Table 30.2, the number of basis updates needed by the simplified greedy SA in [21] is less than that needed by the CLLL algorithm and even the CLLL with SQRD, in both average

TABLE 30.3 The Tree-Search Implementation for SA (Using MATLAB Notation)

INPUT: $\boldsymbol{H}$, K_c, K_ℓ; OUTPUT: $\tilde{\boldsymbol{H}}$, $\boldsymbol{T}$
Initialization:
(1) $\tilde{\boldsymbol{H}} = \boldsymbol{H}$;
(2) $\boldsymbol{T} = \boldsymbol{I}_N$;
(3) Calculate $\tilde{\boldsymbol{H}}^\dagger$;
(4) Calculate $\boldsymbol{\lambda}_{m,n}$ and $\boldsymbol{\Delta}_{m,n}$ matrices as in Equations 30.10 and 30.11;
(5) Find K_c index pairs associated with the K_c largest value in $\boldsymbol{\Delta}$;
(6) for $k = 1$: K_c
(7) Update $\tilde{\boldsymbol{H}}_k$, $\tilde{\boldsymbol{H}}_k^\dagger$, and $\boldsymbol{T}_k$ as in Equation 30.9
(8) Update $\boldsymbol{\lambda}_{m,n}^{(k)}$ and $\boldsymbol{\Delta}_{m,n}^{(k)}$ matrices based on Equations 30.10 and 30.11;
(9) end
Iteration:
(10) while there exists $\boldsymbol{\lambda}_{m,n}^{(k)} \neq 0$
(11) for $k = 1$: K_c
(12) Find K_ℓ index pairs associated with the K_ℓ largest value in $\boldsymbol{\Delta}_{m,n}^{(k)}$;
(13) end
(14) Find K_c index pairs among K_cK_ℓ index pairs with the smallest $S(\boldsymbol{H})$;
(15) for $c = 1 : K_c$
(16) Update $\tilde{\boldsymbol{H}}_k$, $\tilde{\boldsymbol{H}}_k^\dagger$, and $\boldsymbol{T}_k$ as in Equation 30.9;
(17) Update $\boldsymbol{\lambda}_{m,n}^{(k)}$ and $\boldsymbol{\Delta}_{m,n}^{(k)}$ matrices based on Equations 30.10 and 30.11;
(18) end
(19) end
(20) Find $\tilde{\boldsymbol{H}}$ with the smallest $S(\tilde{\boldsymbol{H}})$ among K_c candidates $\{\tilde{\boldsymbol{H}}_k\}_{k=1}^{K_c}$

and standard deviation. However, the number of arithmetic operations needed by the SA in each basis update ($16M + 104N - 90$) is far more than that of the CLLL algorithm (at most ($28M + 46N + 6$) even if δ condition is violated), which leads to higher algorithm complexity. Another major drawback of SA is that it requires more memory storage during the updating process.

The analysis on the final condition ($\lambda_{mn} = 0$) reveals that for any two-dimensional (2-D) lattice, Seysen-reduced basis is the same as the Gaussian-reduced basis (up to signs) [26]. Furthermore, for a 2-D lattice, Seysen's metric of a Seysen-reduced basis is upper bounded by $\frac{8}{3}$ if the lattice is real, and by 4 if the lattice is complex [26]. Because the SA and the Gaussian reduction algorithm yield the same reduced basis, the same performance can be expected for applications (e.g., for 2×2 MIMO communication systems). Furthermore, the upper bounds on $S(\boldsymbol{H})$ guarantee the resulting basis is within a certain distance from orthogonality. In general, there is no theoretical result on whether the Seysen's metric is also upper bounded for N-D lattices with $N > 2$. However, the complementary cumulative distributive function (ccdf) of $S(\boldsymbol{H})$ of the output matrices of SA obtained in simulations shows the existence of a finite upper bound [26].

30.2.3 Other LR Algorithms

In addition to the LLL algorithm and the SA, many other LR algorithms have been proposed in the literature. The Gaussian reduction [3,26], Minkowski reduction, and Korkine–Zolotareff (KZ) reduction algorithms [12] find the optimal basis for a lattice based on the successive minimal criteria. However, the Gaussian reduction method is only for 2×2 systems and has been shown to be equivalent to SA [26]. The Minkowski and the KZ algorithms do not have polynomial time implementation and therefore infeasible for communications systems (see [25] and references therein). Furthermore, it has been shown that for LR-aided equalizers, the KZ algorithm only achieves performance similar to that of the LLL algorithm [23].

A simplified Brun's algorithm is proposed and implemented in [2] to reduce complexity but also sacrifices performance. Based on the approximation of eigenvectors of $\boldsymbol{H}$, the complexity of the simplified Brun's algorithm is even lower than the LLL algorithm. However, the performance of the simplified Brun's algorithm is much worse than that of the LLL algorithm as shown in [2]. Furthermore, the property of resulting matrices of the Brun's algorithm is not studied analytically. Thus, whether LR-aided equalizers based on the simplified Brun's algorithm can collect the same diversity as LLL-based equalizers is not clear. In Section 30.3, we provide the performance of LR-aided equalizers based on the simplified Brun's algorithm.

30.3 Lattice Reduction-Aided Equalizers

LR algorithms find a better basis $\tilde{\boldsymbol{H}} = \boldsymbol{H}\boldsymbol{T}$ that is more orthogonal than $\boldsymbol{H}$ based on different metrics. Note that the equivalence of the two lattices spanned by $\boldsymbol{H}$ and $\tilde{\boldsymbol{H}}$ is based on the assumption that all the entries of $\boldsymbol{s}$ belong to the complex integer set. In the following, we show how to adopt LR algorithms into the wireless communication decoding process.

With the new channel matrix $\tilde{\boldsymbol{H}}$ generated by applying LR algorithms onto the channel matrix, the system model in Equation 30.1 can be written as

$$\boldsymbol{y} = \boldsymbol{H}\boldsymbol{T}(\boldsymbol{T}^{-1}\boldsymbol{s}) + \boldsymbol{w} = \tilde{\boldsymbol{H}}\boldsymbol{z} + \boldsymbol{w}. \qquad (30.12)$$

Since all the entries of $\boldsymbol{T}^{-1}$ and the signal constellation belong to Gaussian integer ring, the entries of $\boldsymbol{z}$ are also Gaussian integers. Basically, LR-aided equalization is to apply traditional equalizers like LEs and DFEs onto the system model in Equation 30.12 to obtain the estimate of $\boldsymbol{z}$ and then the estimate of $\boldsymbol{s}$. We now provide the detailed process for LR-aided LEs and LR-aided DFEs (SICs), respectively.

30.3.1 LR-Aided LEs

Based on the general system model in Equation 30.1, we first introduce LR-aided LEs using LR-aided ZF-LE as an example. Since the MMSE-LE agrees with the ZF-LE with respect to the extended system in

Equation 30.4, to perform LR-aided MMSE-LE, it is equivalent to perform LR-aided ZF-LE on the extended system. For LR-aided ZF-LE, we first apply the ZF equalizer in Equation 30.2 onto the system in Equation 30.12 to obtain $\hat{z}$, the estimate of z, by taking the constellation of z as the whole Gaussian integer ring. After obtaining $\hat{z}$, we recover s by mapping $T\hat{z}$ to the original signal constellation. These two hard decoding steps consist of the LR-aided low-complexity equalizers for linear block transmission systems.

Note that the possible values of z are determined by the original signal constellation and the unimodular matrix T. Given a specific constellation of s, the actual constellation of z is random due to the randomness of H. To simplify the estimation of z, we assume that all the entries of z belong to Gaussian integer ring. Since T is random, the assumption is valid if the real and the imaginary parts of s belong to consecutive integer sets. However, for $\mathcal{M}$-QAM symbols, the real and the imaginary parts of each symbol are drawn from the set $\{-(\sqrt{\mathcal{M}}-1), \ldots, -1, 1, \ldots, \sqrt{\mathcal{M}}-1\}$, which is not a consecutive integer set. Therefore, we need to shift and/or scale the constellation to make sure that the real/imaginary part of the constellation belongs to a consecutive integer set. For example, by applying $(s-(1+j)\mathbf{1})/2$, we transfer the real and the imaginary parts of the $\mathcal{M}$ -QAM constellation to a consecutive integer set $\{-\sqrt{\mathcal{M}}/2, \ldots, -1, 0, 1, \ldots, (\sqrt{\mathcal{M}}-2)/2\}$, which makes the real and the imaginary parts of z as also consecutive integers. Therefore, the quantization to obtain $\hat{z}$ is now as simple as rounding to the nearest complex integer. Furthermore, the scaling and the shifting operation on signal constellations must be considered when obtaining the estimate of s. Using LR-aided ZF-LE as an example, the estimate $\hat{s}$ is expressed as

$$\hat{s} = 2\overline{\mathcal{Q}}\left(T\left\lfloor\frac{1}{2}(x - T^{-1}(1+j)\mathbf{1})\right\rceil\right) + (1+j)\mathbf{1}, \tag{30.13}$$

where

$x = \tilde{H}^{\dagger} y$,

$\overline{\mathcal{Q}}$ denotes the quantization operation that maps the real and the imaginary parts to the nearest integer in the set $\{-\sqrt{\mathcal{M}}/2, \ldots, -1, 0, 1, \ldots, (\sqrt{\mathcal{M}}-2)/2\}$

We summarize the main steps of the LR-aided LEs for QAM signals in Table 30.4. The inputs are y and H for LR-aided ZF-LE, and $\bar{y}$ and $\bar{H}$ as in Equation 30.4 for LR-aided MMSE-LE. In step (3), we manually shift and scale the original constellation to make sure that the constellation of z is in the whole complex integer set. Therefore, the estimation of z in step (4) is a simple rounding operation. The quantization operation $\overline{\mathcal{Q}}$ in step (5) is the same as the one in Equation 30.13. Other constellations like PSK that belong to the Gaussian integer ring and can be transferred to consecutive integer sets through scaling/shifting can also be adopted. The shifting and scaling operations in steps (3) and (5) need to be modified accordingly.

TABLE 30.4 LR-Aided LEs with QAM Constellations (Using MATLAB Notation)

INPUT: y, H for ZF-LE ($\bar{y} = [y; \mathbf{0}_{N\times 1}]$, $\bar{H} = [H; \sigma_\omega I_N]$ for MMSE-LE); OUTPUT: $\hat{s}$;
(1) $[\tilde{H}, T] = \text{LR}(H)$;
(2) $x = \tilde{H}^{\dagger} y$;
(3) $\hat{x} = (x - T^{-1}(1+j)\mathbf{1})/2$;
(4) $\hat{x} = \text{round}(\hat{x})$;
(5) $\hat{s} = 2\overline{\mathcal{Q}}(T\hat{z}) + (1+j)\mathbf{1}$;

30.3.2 LR-Aided DFEs

LR-aided DFEs (SICs) can be obtained by replacing the linear estimation of $\boldsymbol{z}$ in LR-aided LEs by the DFEs (SICs). Here, we introduce the detailed process of LR-aided ZF-DFE, while LR-aided MMSE-DFE can be obtained by applying LR-aided DFE onto the extended system in Equation 30.4. For the system model in Equation 30.12 obtained by applying LR algorithms onto the channel matrix $\boldsymbol{H}$, ZF-DFE with the feedforward matrix $\boldsymbol{F} = \tilde{\boldsymbol{Q}}^{\mathcal{H}}$ and the feedback matrix $\boldsymbol{B} = \tilde{\boldsymbol{R}}$ is applied to obtain the estimate of $\boldsymbol{z}$ by assuming the constellation of $\boldsymbol{z}$ to be the whole complex integer set. $\tilde{\boldsymbol{Q}}$ and $\tilde{\boldsymbol{R}}$ are the QR decomposition of the new channel matrix $\tilde{\boldsymbol{H}}$. For readers' convenience, we summarize the equalization process of the LR-aided ZF-DFE in Table 30.5. Note that the quantization in step (7) is the same as the mapping operation in Equation 30.13.

30.3.3 Dual LR-Aided Equalizers

The above LR-aided equalizers apply LR algorithms onto the channel matrix $\boldsymbol{H}$ to obtain a new matrix $\tilde{\boldsymbol{H}}$, which is closer to orthogonality. However, as shown in Equation 30.2, the covariance matrix of the colored noise $\boldsymbol{\eta}$ after equalization is $(\boldsymbol{H}^{\mathcal{H}}\boldsymbol{H})^{-1}$. Therefore, a more orthogonal $\boldsymbol{H}^{\dagger}$ ($(\boldsymbol{H}^{\mathcal{H}}\boldsymbol{H})^{-1}$ is more like a diagonal matrix) may lead to better performance than a more orthogonal $\boldsymbol{H}$. The dual LR-aided equalizers proposed in [15] aim to improve the orthogonality of $\boldsymbol{H}^{-1}$ and have been shown to achieve better performance than LR-aided equalizers.

Dual LR-aided equalizers apply LR algorithms onto the dual basis of the channel matrix $(\boldsymbol{H}^{\dagger})^{\mathcal{H}}$ to obtain a new matrix $\hat{\boldsymbol{H}} = (\boldsymbol{H}^{\dagger})^{\mathcal{H}}\boldsymbol{P}$, where $\boldsymbol{P}$ is also a unimodular matrix. Then, the linear system model in Equation 30.1 can be rewritten as

$$\boldsymbol{y} = \boldsymbol{H}(\boldsymbol{P}^{\mathcal{H}})^{-1}\boldsymbol{P}^{\mathcal{H}}\boldsymbol{s} + \boldsymbol{w} = \left(\hat{\boldsymbol{H}}^{\mathcal{H}}\right)^{\dagger}\boldsymbol{z} + \boldsymbol{w}. \tag{30.14}$$

Comparing to the system model of LR-aided equalizers in Equation 30.12, it is straightforward to see that LR-aided equalizers can be applied onto Equation 30.14 to obtain the estimate of $\boldsymbol{s}$ by replacing $\boldsymbol{T}$ with $(\boldsymbol{P}^{\mathcal{H}})^{-1}$ and $\tilde{\boldsymbol{H}}$ with $\left(\hat{\boldsymbol{H}}^{\mathcal{H}}\right)^{\dagger}$. As shown in [15], the dual LLL (DLLL)-aided equalizers achieve better performance than LLL-aided equalizers. Since the LLL algorithm focuses on the orthogonality of the input matrix, applying the LLL algorithm onto the dual basis makes $(\boldsymbol{H}^{\mathcal{H}}\boldsymbol{H})^{-1}$ more orthogonal than the one generated by applying LLL onto $\boldsymbol{H}$. However, this is not true for all LR algorithms. For example, applying SA onto the dual basis leads to the same performance as SA-aided equalizers, because SA balances the orthogonality between the original basis and the dual basis by adopting Seysen's metric as shown in Equation 30.8. In the following, we compare the performance of the aforementioned (dual) LR-aided equalizers.

TABLE 30.5 LR-Aided DFEs with QAM Constellations (Using MATLAB Notation)

INPUT: $\boldsymbol{y}$, $\boldsymbol{H}$ for ZF-DFE ($\bar{\boldsymbol{y}} = [\boldsymbol{y}; \boldsymbol{0}_{N\times 1}]$, $\bar{\boldsymbol{H}} = [\boldsymbol{H}; \sigma_w \boldsymbol{I}_N]$ for MMSE-DFE); OUTPUT: $\hat{\boldsymbol{s}}$;
(1) $[\tilde{\boldsymbol{Q}}, \tilde{\boldsymbol{R}}, \boldsymbol{T}] = \text{LR}(\boldsymbol{H})$;
(2) $\boldsymbol{x} = \tilde{\boldsymbol{Q}}^{\mathcal{H}}\boldsymbol{y}$;
(3) $\hat{\boldsymbol{x}} = (\boldsymbol{x} - \tilde{\boldsymbol{R}}\boldsymbol{T}^{-1}(1+j)\,\boldsymbol{1})/2$;
(4) for $n = N: (-1): 1$
(5) $\hat{z}(n) = \lfloor(\hat{x}(n) - \sum_{k=n+1}^{N} \tilde{\boldsymbol{R}}(n,k)\hat{z}(k))\tilde{\boldsymbol{R}}(n,n)\rceil$;
(6) end
(7) $\hat{\boldsymbol{s}} = 2\overline{\mathcal{Q}}(\boldsymbol{T}\hat{\boldsymbol{z}}) + (1+j)\boldsymbol{1}$;

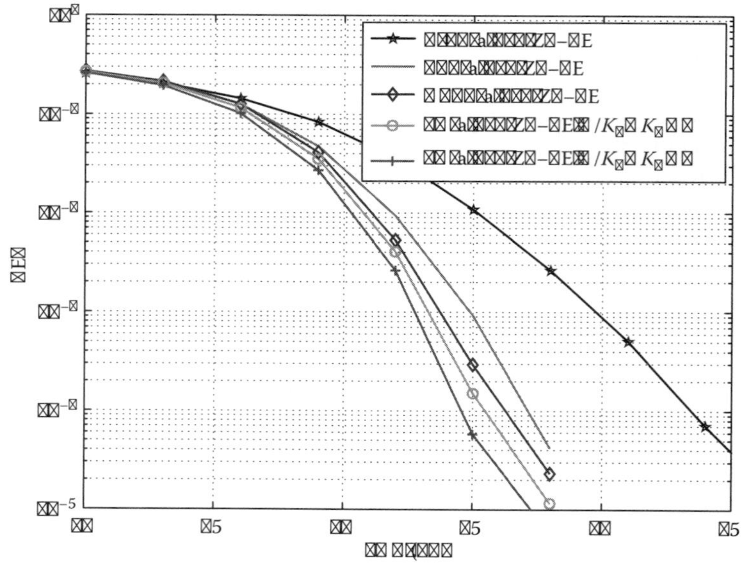

FIGURE 30.2 Performance comparisons for 8 × 8 systems with 64-QAM.

We compare the BER performance of LR-aided ZF-LE with three different LR algorithms: the complex LLL algorithm, SA, and Brun's algorithm. SA is adopted with $K_c = K_\ell = 1$ (greedy SA) and $K_c = K_\ell = 3$, respectively. Dual LR-aided ZF-LE based on the CLLL algorithm is also provided for comparisons. The performance for the linear system in Equation 30.1 with 8 × 8 i.i.d. channel matrices is plotted in Figure 30.2. A 64-QAM is adopted for the signal constellation. From the figure, we can see, Brun-aided equalizers have the worst performance while SA achieves the best performance on LR-aided ZF-LE. Increasing K_c and K_ℓ improves the performance of SA-aided equalizers. Furthermore, the DLLL-aided ZF-LE achieves better performance than the LLL-aided ZF-LE.

Comparing the traditional LEs and DFEs with LR-aided equalizers, it is straightforward to see that the major complexity overhead of LR-aided equalizers lies in the LR algorithms adopted to obtain $\tilde{\boldsymbol{H}}$. Since the average complexity of LR algorithms is polynomial, the complexity of LR-aided equalizers is still much less than that of (near-) MLEs. However, LR-aided equalizers consume higher complexity than the traditional low-complexity equalizers. Therefore, one may wonder the improvement on performance by adopting LR algorithms in the equalization process. In Section 30.4, we illustrate the performance of LR-aided equalizers by using two examples.

30.4 Applications

After introducing various LR algorithms and LR-aided equalizers, we now provide applications of LR-aided equalizers on wireless communications. Two specific transmission systems, the MIMO system with i.i.d. channels in [6] and the linear precoded (LP-) OFDM system in [17], are adopted as examples to show the performance of LR-aided equalizers. The performance of different equalizers based on the CLLL algorithm is provided for comparisons. Here, we adopt diversity gain (see e.g., [19] for definition) as a figure of merit for performance.

30.4.1 MIMO Systems

For MIMO systems, the system model is the same as the one in Equation 30.1. All the entries of the $M \times N$ channel matrix $\boldsymbol{H}$ are i.i.d. complex Gaussian distributed with zero mean and unit variance. It has been shown that for MIMO systems, the traditional low-complexity equalizers only collect diversity $M - N + 1$, while (near-) MLEs exploit diversity M [18]. For LR-aided equalizers, it has been proved in [18] that the diversity order collected by the LLL-aided equalizers is M, which is the same as that achieved by MLE.

We verify the performance improvement by plotting BER curves of different equalizers for i.i.d. channels with $M = N = 4$ and QPSK modulation in Figure 30.3. Five equalizers are adopted: ZF-LE, ZF-DFE, LR-aided ZF-LE, LR-aided ZF-DFE, and SD method. For LR-aided equalizers, the CLLL algorithm is adopted. We notice that the LEs only achieve diversity order $M - N + 1$, which is 1 in this case. The SD method collects diversity $M = 4$. As expected, the LR-aided ZF-LE and LR-aided ZF-DFE achieve the same diversity order as SD does. However, there still exist performance gaps between LR-aided equalizers and the near-MLE. The complexity of those five equalizers is compared by plotting the average number of arithmetic operations (including real additions and real multiplications) per channel realization for $M = N = 2$ to 16 in Figure 30.4. Here, the complexity of ZF-LE is obtained by adopting the Gaussian elimination for matrix inversion, while ZF-DFE adopts the QR decomposition implemented through householder transformation. The SD method is implemented as in [7]. From the figure, we can see that the complexity gap between LR-aided equalizers and traditional low-complexity equalizers is much smaller than that between SD and LEs/DFEs. The difference becomes larger as the dimension increases. Provided the significant performance improvement by the LR-aided equalizers, the complexity of LR-aided equalizers is really low. Furthermore, the complexity of LR-aided ZF-LE is slightly smaller than that of LR-aided ZF-DFE. Comparing the operations in Tables 30.4 and 30.5, we can see LR-aided ZF-DFE only requires an extra matrix–vector multiplication.

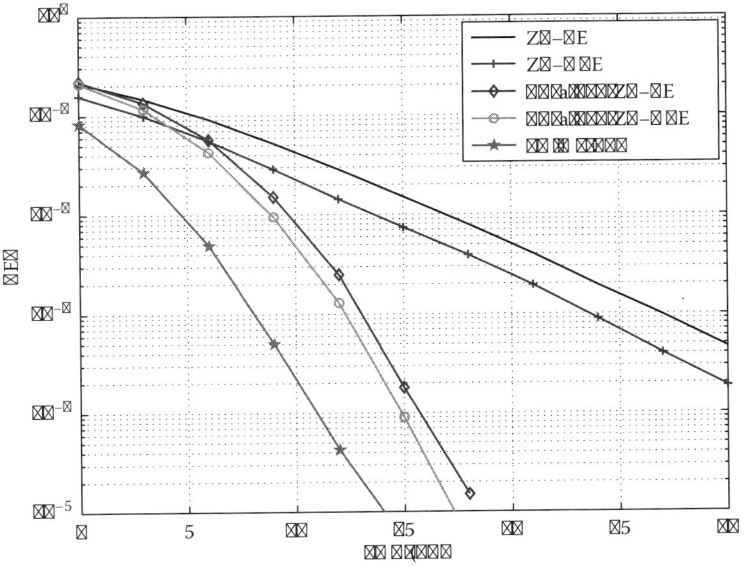

FIGURE 30.3 Performance of i.i.d. channels with $M = N = 4$ and QPSK modulation.

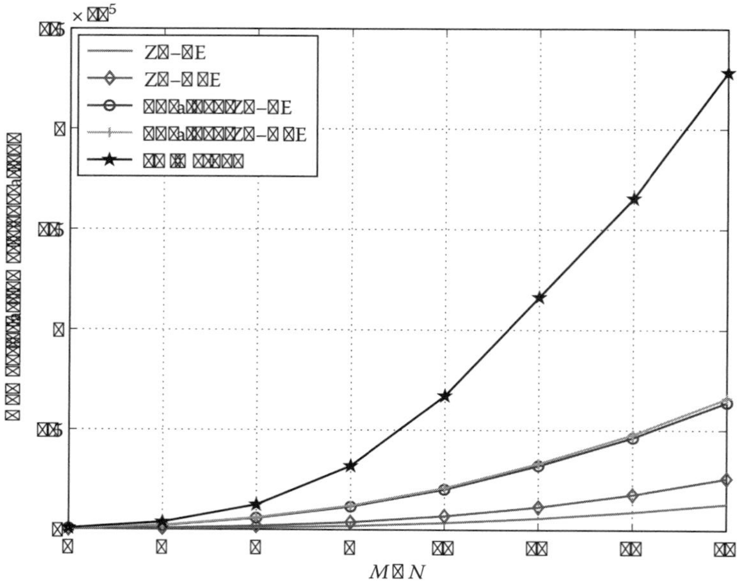

FIGURE 30.4 Complexity comparison of different equalizers with QPSK modulation.

30.4.2 LP-OFDM Systems

For LP-OFDM systems in [17], the equivalent channel matrix is a $K \times K$ matrix $\boldsymbol{H} = \boldsymbol{D}_H \boldsymbol{\Gamma}$, where $\boldsymbol{D}_H$ is a diagonal matrix with subcarrier response of the OFDM system as diagonal entries and $\boldsymbol{\Gamma}$ is a square unitary precoding matrix. The detailed design of $\boldsymbol{\Gamma}$ can be found in [17]. Assuming the rank of the covariance matrix for the frequency-selective channel taps is ρ_h, the diversity exploited by (near-) MLEs for LP-OFDM is min (K, ρ_h), while the diversity for low-complexity equalizers is always 1 [20]. Therefore, we can see that, for both systems, the traditional low-complexity equalizers fail to collect the diversity enabled at the transmitter.

The performance of the LLL-aided LEs for the LP-OFDM system in [17] is analyzed in [20]. It has been shown that the diversity order collected by the LR-aided LEs is min (K, ρ_h) which is the same as that obtained by the (near-) MLEs. To verify the performance for LP-OFDM systems, we fix the frequency-selective channel order as $L = 3$, and the number of subcarriers per group as $K = 4$ (for the detailed LP-OFDM design refer [17]). The channel taps are independent complex Gaussian random variables with zero mean and variance $\frac{1}{L+1}$, which means $\rho_h = L + 1 = 4$. Figure 30.5 shows the BER performance of ZF-LE, ZF-DFE, LR-aided ZF-LE, LR-aided ZF-DFE, and SD method. From this figure, we observe that traditional low-complexity equalizers only collect diversity 1, while LR-aided equalizers collect diversity order 4, the same as the MLE does, although there still exists a gap between the performance of the LR-aided equalizers and the MLE. The performance of the LR-aided ZF-DFE equalizer is better than that of the LR-aided ZF-LE.

Similar to the LLL algorithm, if the observation that SA upper bounds the output $S(\tilde{\boldsymbol{H}})$ by a finite constant can be theoretically proved, it can be easily proved that SA-aided low-complexity equalizers collect the same diversity as MLE does. For 2×2 systems, the result in [26] shows that SA upper bounds the Seysen's metric of the output matrix $\tilde{\boldsymbol{H}}$ by a finite number. For an N-D lattice, the simulation results show the existence of a finite upper bound, but it is not theoretically proved yet. However, according to the simulation results in [21] and [27], SA-aided equalizers collect the same diversity as LLL-aided equalizers for many transmission systems.

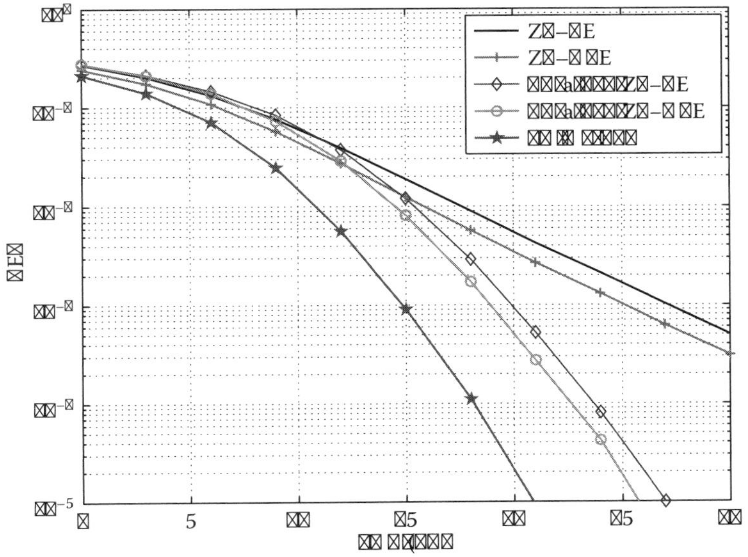

FIGURE 30.5 Performance for LP-OFDM with QPSK modulation.

30.5 Conclusions

Among different equalizers for linear block transmissions, traditional low-complexity equalizers such as LEs and DFEs are favored for their cubic-order polynomial complexity, but they often suffer from diversity loss. LR techniques have been adopted to improve the performance of these traditional low-complexity equalizers. We presented the development of LR techniques by introducing various LR algorithms. The details of the complex LLL algorithm and the Seysen's algorithm are provided. After introducing the detailed procedures of LR-aided LEs and LR-aided DFEs, the performance of LR-aided equalizers for wireless communication systems is compared. It has been shown that LR-aided equalizers collect the same diversity as that of (near-) MLEs with much lower complexity for two specific transmissions, MIMO and LP-OFDM. The encouraging results summarized in this chapter demonstrate that LR techniques are an attractive solution for future wireless receiver designs.

References

1. E. Agrell, T. Eriksson, A. Vardy, and K. Zeger, Closest point search in lattices, *IEEE Trans. Information Theory*, 48(8), 2201–2214, August 2002.
2. A. Burg, D. Seethaler, and G. Matz, VLSI implementation of a lattice-reduction algorithm for multi-antenna broadcast precoding, in *Proceedings of the IEEE International Symposium on Circuits and Systems*, New Orleans, LA, May 27–30, 2007, pp. 673–676.
3. H. Daudé, P. Flajolet, and B. Vallée, An analysis of the Gaussian algorithm for lattice reduction, in *Proceedings of the 1st International symposium on Algorithmic Number Theory*, Ithaca, NY, May 1994, pp. 144–158.
4. Y. H. Gan and W. H. Mow, Complex lattice reduction algorithms for low-complexity MIMO detection, in *Proceedings of the IEEE Global Telecommunications Conference*, 5, St. Louis, MO, November 28–December 2, 2005, pp. 2953–2957.
5. G. Ginis and J. M. Cioffi, On the relation between V-BLAST and the GDFE, *IEEE Communications Letters*, 5(9), 364–366, September 2001.

6. G. D. Golden, G. J. Foschini, R. A. Valenzuela, and P. W. Wolniansky, Detection algorithm and initial laboratory results using V-BLAST space-time communication architecture, *Electron. Lett.*, 35 (1), 14–16, January 7, 1999.
7. B. Hassibi and H. Vikalo, On the sphere-decoding algorithm I. Expected complexity, *IEEE Trans. Signal Process.*, 53(8), 2806–2818, August 2005.
8. B. Hassibi, An efficient square-root algorithm for BLAST, in *Proceedings on the IEEE International Conference on Acoustics, Speech, Signal Processing*, 2, Istanbul, Turkey, June 5–9, 2000, pp. 737–740.
9. J. Jaldén, D. Seethaler, and G. Matz, Worst- and average-case complexity of LLL lattice reduction in MIMO wireless systems, in *Proceedings of the IEEE International Conference on Acoustics, Speech and Signal Processing*, Las Vegas, NV, March 30–April 4, 2008.
10. Y. Jiang, X. Zheng, and J. Li, Asymptotic performance analysis of V-BLAST, in *Proceedings of the IEEE Global Telecommunications Conference*, 6, St. Louis, MO, November 28–December 2, 2005, pp. 3882–3886.
11. E. Kaltofen and G. Villard, Computing the sign or the value of the determinant of an integer matrix, a complexity survey, *J. Comput. Appl. Math.*, 162(1), 133–146, 2004.
12. A. Korkine and G. Zolotareff, Sur les formes quadratiques, (in French) *Math. Ann.*, 6, 366–389, 1873.
13. B. LaMacchia, Basis reduction algorithms and subset sum problems, Master's thesis, Massachusetts Institute of Technology, Cambridge, MA, May 1991.
14. A. K. Lenstra, H. W. Lenstra, and L. Lovász, Factoring polynomials with rational coefficients, *Math. Ann.*, 261, 515–534, 1982.
15. C. Ling, Approximate lattice decoding: Primal versus dual basis reduction, in *Proceedings of the IEEE International Symposium on Information Theory*, Seattle, WA, July 9–14 2006.
16. C. Ling and N. Howgrave-Graham, Effective LLL reduction for lattice decoding, in *Proceedings of the IEEE International Symposium on Information Theory*, Nice, France, June 2007, pp. 196–200.
17. Z. Liu, Y. Xin, and G. B. Giannakis, Linear constellation precoded OFDM with maximum multipath diversity and coding gains, *IEEE Trans. Commun.*, 51(3), 416–427, March 2003.
18. X. Ma and W. Zhang, Performance analysis for MIMO systems with linear equalization, *IEEE Trans. Commun.*, 56(2), 309–318, February 2008.
19. X. Ma and W. Zhang, Fundamental limits of linear equalizers: Diversity, capacity and complexity, *IEEE Trans. Info. Theory*, 54(8), 3442–3456, August 2008.
20. X. Ma, W. Zhang, and A. Swami, Lattice-reduction aided equalization for OFDM systems, *IEEE Trans. Wireless Commun.*, 8(2), February 2009.
21. D. Seethaler, G. Matz, and F. Hlawatsch, Low-complexity MIMO data detection using Seysen's lattice reduction algorithm, in *Proceedings of the IEEE International Conference on Acoustics, Speech and Signal Processing*, 3, Honolulu, HI, April 15–20 2007, pp. 53–56.
22. M. Seysen, Simultaneous reduction of a lattice basis and its reciprocal basis, *Combinatorica*, 13, 363–376, 1993.
23. C. Windpassinger and R. F. H. Fischer, Low-complexity near-maximum-likelihood detection and precoding for MIMO systems using lattice reduction, in *Proceedings of the Information Theory Workshop*, Munich, Germany, March 31–April 4 2003, pp. 345–348.
24. D. Wübben, R. Böhnke, V. Kühn, and K.-D. Kammeyer, Near-maximum-likelihood detection of MIMO systems using MMSE-based lattice reduction, in *Proceedings of the IEEE International Conference on Communications*, 2, Paris, France, June 20–24, 2004, pp. 798–802.
25. H. Yao, Efficient signal, code, and receiver designs for MIMO communication systems, Doctoral Thesis, Department of Electrical Engineering and Computer Science, Massachusetts Institute of Technology, Cambridge, MA, 2003.
26. W. Zhang, F. Arnold, and X. Ma, An analysis of Seysen's lattice reduction algorithm, *Signal Processing*, 88(10), 2573–2577, October 2008.
27. W. Zhang and X. Ma, Quantifying diversity for wireless systems with finite-bit representation, in *Proceedings of the IEEE International Conference on Acoustics, Speech and Signal Processing*, Las Vegas, NV, March 30–April 4, 2008.

31

Overview of Transmit Diversity Techniques for Multiple Antenna Systems

D. A. Zarbouti
National Technical University of Athens

D. A. Kateros
National Technical University of Athens

D. I. Kaklamani
National Technical University of Athens

G. N. Prezerakos
National Technical University of Athens

and

Technological Education Institute of Piraeus

31.1 Introduction **31**-1
31.2 System Model **31**-2
MIMO-OFDM
31.3 Antenna and Temporal Diversity Techniques for OFDM **31**-3
Trellis Codes for OFDM • Space–Time Block Codes for Flat Fading Channels • Space–Time Block Codes for Selective Fading Environments
31.4 Antenna and Frequency Diversity Techniques for OFDM **31**-10
Alamouti Technique • Space–Frequency Codes for Selective Fading Environments
31.5 Antenna, Time, and Frequency Diversity Techniques for OFDM **31**-13
References **31**-18

31.1 Introduction

High data rates are a clear demand of modern wireless communications. The recent advances in the field of multiple antenna systems have shown the great capacity and performance gains that can be achieved with multiple antenna technology and especially with multiple-input multiple-output (MIMO) systems. These systems take advantage of the spatial domain in order to provide with high data rates the communication links and overcome the bandwidth limitations.

Furthermore, orthogonal frequency division multiplexing (OFDM) has drawn much attention as a promising modulation scheme for the broadband wireless communications and it is a strong candidate for the next generation communication systems. The application of OFDM technique in a wideband and as a result frequency-selective wireless channel leads to the division of the initial signal bandwidth into several narrowband and as a result flat fading subchannels. Moreover, the basic principle of OFDM is to split a high-rate data stream into a number of lower rate substreams that are simultaneously transmitted over the OFDM carriers. This way the symbol duration of the substreams increases and the delay spread caused by multipath has minor effect on the received signal. Finally, the guard interval that it is introduced for every OFDM symbol decrease further the intersymbol interference while eliminated the intercarrier interference. A more thorough analysis concerning the OFDM basic principles can be found in [1].

The use of OFDM in conjunction with multiple antenna systems with the associating MIMO signaling strategies is investigated in the chapter. The discussion is primarily dedicated to the presentation of several antenna transmit diversity techniques that appear in the "References" section. Special interest is also given in schemes that concern selective fading channels since they dominate the wideband wireless technology.

31.2 System Model

31.2.1 MIMO-OFDM

The transceiver of a MIMO-OFDM system can be viewed in Figure 31.1. The transmitter is equipped with M_t antennas while the receiver is equipped with M_r antennas. Each transmitter chain involves an OFDM modulator which is fed by the output of the appropriate encoder that is used. The serial data stream leads N_s data symbols into the encoder which transforms them into a $N \cdot T \times M_t$ codematrix. In the last relation N are the available carriers provided by the inverse discrete Fourier transform (IDFT) module of the transmitter and T is the duration of each time slot. Let us consider $\mathbf{S} = [s_1, s_2, \ldots, s_{N_s}]$ the modulated data symbols that enter the encoder, as it is depicted in Figure 31.1, the produced codeword is given by

$$\mathbf{C} = \begin{pmatrix} \mathbf{C}_1 \\ \vdots \\ \mathbf{C}_T \end{pmatrix} \tag{31.1}$$

where C_i, $i = 1, \ldots, T$ is a $N \times M_t$ matrix that corresponds to the transmission codeword at each slot and given by

$$C_i = \begin{pmatrix} c_{11} & \cdots & c_{1M_t} \\ \vdots & \ddots & \vdots \\ c_{N1} & \cdots & c_{NM_t} \end{pmatrix} \tag{31.2}$$

the coded symbol $c_{j,k}$ of the above matrix is transmitted by the kth antenna with jth carrier during the ith time slot.

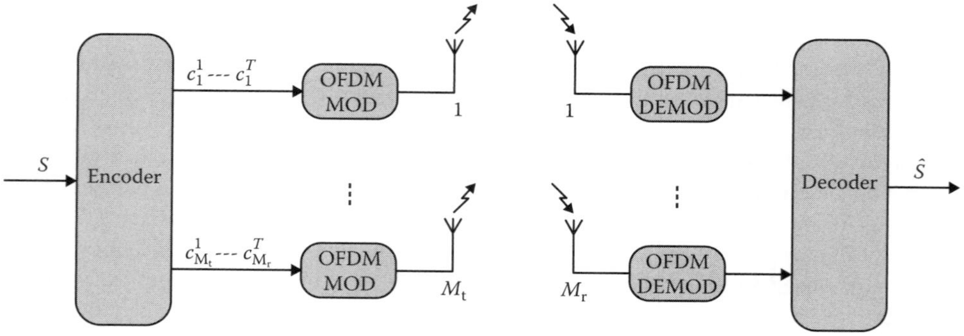

FIGURE 31.1 MIMO-OFDM transceiver.

In Sections 31.2.1.1 through 31.2.1.3 we present three basic performance parameters of such a system that are used for comparing encoding schemes.

31.2.1.1 Code Rate

A simple parameter of comparison for different MIMO-OFDM encoding techniques is the code rate that they achieve. Generally, the code rate of an encoding scheme is a metric that states the amount of useful information that is transmitted during each transmission burst. In the general case of the MIMO-OFDM system described above, the N_s useful data symbols are sent through $N \cdot T$ channels (transmission units). As a result the code rate in this case is

$$R = \frac{N_s}{N \cdot T} \tag{31.3}$$

31.2.1.2 Diversity Gain

Another crucial parameter in the investigation of the performance of a space diversity encoding technique is the achievable diversity gain [9,10]. In order to avoid the reception of poor quality signals in the receiver, wireless communication systems exploit the spatial, spectral and/or temporal resources in order to send replicas of the transmitted signal. In this way the receiver can improve its performance, since the probability of simultaneous fading is extremely low. In the case of flat fading MIMO channels the maximum diversity gain that can be reached is $M_t \cdot M_r$ while for selective fading the diversity gain can be increased to $M_t \cdot M_r \cdot L$, where L are the channel paths of independent fading. We must note at this point that when the antennas elements are too dense or the wireless channel is degenerated, i.e., because of a keyhole phenomenon, then the achievable diversity gain is obviously lower than the aforementioned upper bounds. A study on the achievable diversity gain of a MIMO channel can be found in [9] and mainly in [11].

31.2.1.3 Decoding Complexity

Finally, an encoding scheme is investigated under the decoding complexity that it demands. Obviously, special interest is given to schemes that involve low complexity receivers with fast decoding methods like maximum likelihood (ML) and maximum ratio combining (MRC). All the antenna diversity schemes that will be presented in the following will be commented towards the aforementioned parameters.

31.3 Antenna and Temporal Diversity Techniques for OFDM

In this section we present encoding techniques that offer spatial and temporal diversity gain in flat and selective fading MIMO channels. In the "References" section, two kinds of codes are investigated, the trellis and block encoding space–time (ST) schemes. However, because if the high decoding complexity induced by trellis codes, we conduct a more extensive analysis on block codes. Specifically, we mainly focus on orthogonal codes that offer full transmission rate (one) as they have attracted the attention of the research community to a larger extent.

31.3.1 Trellis Codes for OFDM

Space–time trellis codes (STTCs) [24] are a type of ST code used in multiple-antenna wireless communications. This scheme involves the transmission of multiple, redundant copies of a trellis code distributed over time and space (a number of antennas). These multiple copies of the data are used by the receiver to enhance its capability to reconstruct the actual transmitted data. The number of transmit antennas must necessarily be greater than one, but only a single receive antenna is required. However, multiple receive antennas are often used, when it is feasible, since this improves the performance of the system. In MIMO-OFDM systems STTCs are mainly considered in an OFDM framework where the incoming information symbols are trellis coded across both the OFDM subchannels and transmit antennas, in order to obtain the additional multipath diversity. In Section 31.3.2 we present examples of STTCs for MIMO-OFDM systems found in the "References" section.

In [24] the authors present STTCs for 4-PSK and 8-PSK constellations. Specifically, 4-PSK codes for transmission of 2 b/s/Hz using two transmit antennas and 8-PSK codes for transmission of 3 b/s/Hz using two transmit antennas as well. Assuming one receive antenna these codes provide a diversity advantage of two. In Figure 31.2 we include the 4-PSK codes assuming 8 and 16 states.

In [25] the authors make use of the STC encoder depicted in Figure 31.3. The encoder makes use of memory registers and the application of complex channel frequency responses for each subcarrier. The response $\mathbf{H}_i$ for subcarrier i is obtained assuming a tapped-delay line model. The encoder produces two copies of the input signal, which are fed to two 4-PSK mappers.

Lastly, in [26] the authors present codes that are shown to outperform the 2-antenna, 16-state code given in [24] and shown in Figure 31.2 above and the codes given in [26,27]. The suggested

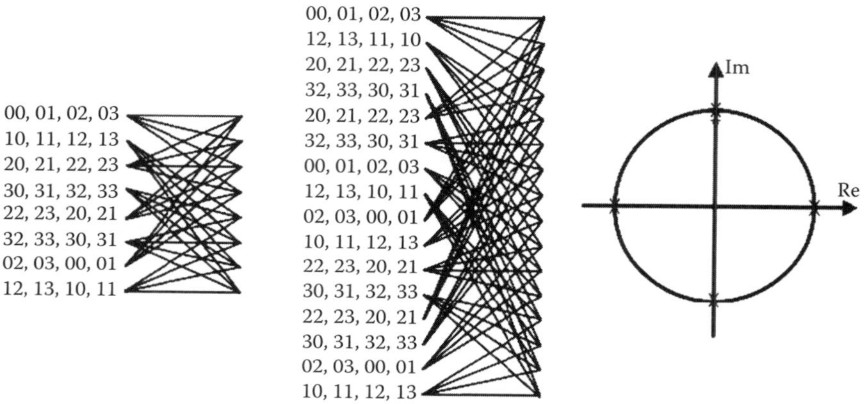

FIGURE 31.2 4-PSK, 8 and 16-state 2-space–time codes, 2 b/s/Hz.

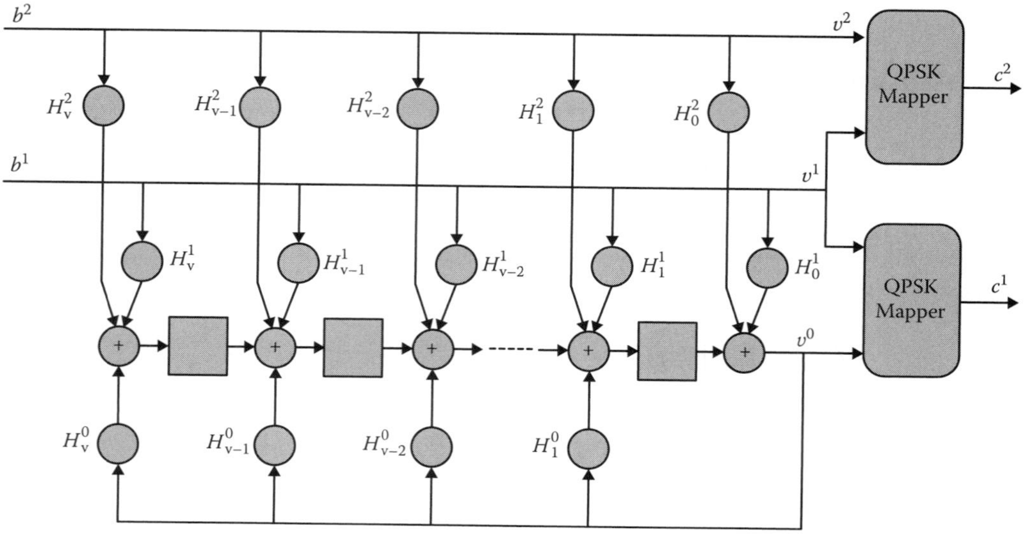

FIGURE 31.3 STC encoder structure of [25].

$$\begin{pmatrix} D+2D^2 & 1+2D^2 \\ 2D & 2 \end{pmatrix} \quad \begin{pmatrix} D+D^2 & 2+D \\ 2+D & 2+2D+2D^2 \end{pmatrix} \quad \begin{pmatrix} 2D^2 & 2+D+2D^2 \\ 2+D & 2D+2D^2 \end{pmatrix}$$

4-PSK, 8 states, 2 transmit antennas — 4-PSK, 8 states, 2 transmit antennas — 4-PSK, 8 states, 2 transmit antennas

$$\begin{pmatrix} (1+a)+D & a+(1+a)D & a+D & 1+(1+a)D \\ a+(1+a)D & a+D & 1+(1+a)D & 1+(1+a)D \\ a+D & 1+(1+a)D & 1+(1+a)D & (1+a)+aD \\ 1+(1+a)D & 1+(1+a)D & (1+a)+aD & 1+aD \end{pmatrix}$$

4-PSK, 16 states, 4 transmit antennas

$$\begin{pmatrix} (1+a)+(1+a)D+aD^2 & (1+a)D+aD^2 & 1+D^2 & 1+D^2 \\ (1+a)D+aD^2 & 1+D^2 & 1+(1+a)D+(1+a)D^2 & (1+a)+aD+(1+a)D^2 \\ 1+D^2 & 1+(1+a)D+(1+a)D^2 & (1+a)+aD+(1+a)D^2 & (1+a)+aD \\ 1+(1+a)D+(1+a)D^2 & (1+a)+aD+(1+a)D^2 & (1+a)+aD & D+D^2 \end{pmatrix}$$

4-PSK, 256 states, 4 transmit antennas

FIGURE 31.4 2-Space–time trellis codes, 4-PSK, 8 and 16 states, 2 b/s/Hz.

codes are 4-antenna and 16- and 256-state designed using an ad hoc approach. The generator matrices for the aforementioned codes are shown in Figure 31.4.

31.3.2 Space–Time Block Codes for Flat Fading Channels

31.3.2.1 Alamouti Technique

The breakthrough in recent ST coding was the Alamouti orthogonal ST block code (OSTBC) design [2]. The Alamouti design aims at flat fading channels and is a simple transmit diversity technique which considers two transmit antennas. The Alamouti technique achieves full diversity order without the need of channel state information at the transmitter. The scheme can be applied with single or multiple receive antennas. In Figure 31.5 the Alamouti transmitter is been depicted in case of one receive antenna.

The concept of Alamouti is rather simple, two coded symbols, c_0 and c_1 are being launched and transmitted simultaneously from the two transmit antennas during the first time slot, while the encoded symbols $-c_1^*$ and c_0^* are transmitted during the second time slot. In Table 31.1 the encoding scheme across time is presented. In order for a ST encoding scheme to be shortly described we use the code matrix. The code matrix of a general single carrier orthogonal design is a $T \times M_t$ matrix with T the number of the time slots and M_t the number of transmit antennas. When $T = M_t$ the produced schemes are of rate one while in the cases where $T > M_t$ the schemes sacrifice their transmission rate for orthogonality reasons achieving rates lower that one. Obviously, Alamouti scheme is an orthogonal scheme of rate one.

In the following we will give the simple procedures that are taken place in the receiver of the Alamouti scheme in order to verify the full diversity potential of this technique.

The channel gains for time slot t_0 for the two antennas are $h_0(t)$ and $h_1(t)$ while for the next time slot the channel gains are $h_0(t+T)$ and $h_1(t+T)$. According to Table 31.1 the received signals during both time slots are

$$y_0 = \sqrt{\frac{E_s}{2}} h_0 c_0 + \sqrt{\frac{E_s}{2}} h_1 c_1 + n_0 \tag{31.4}$$

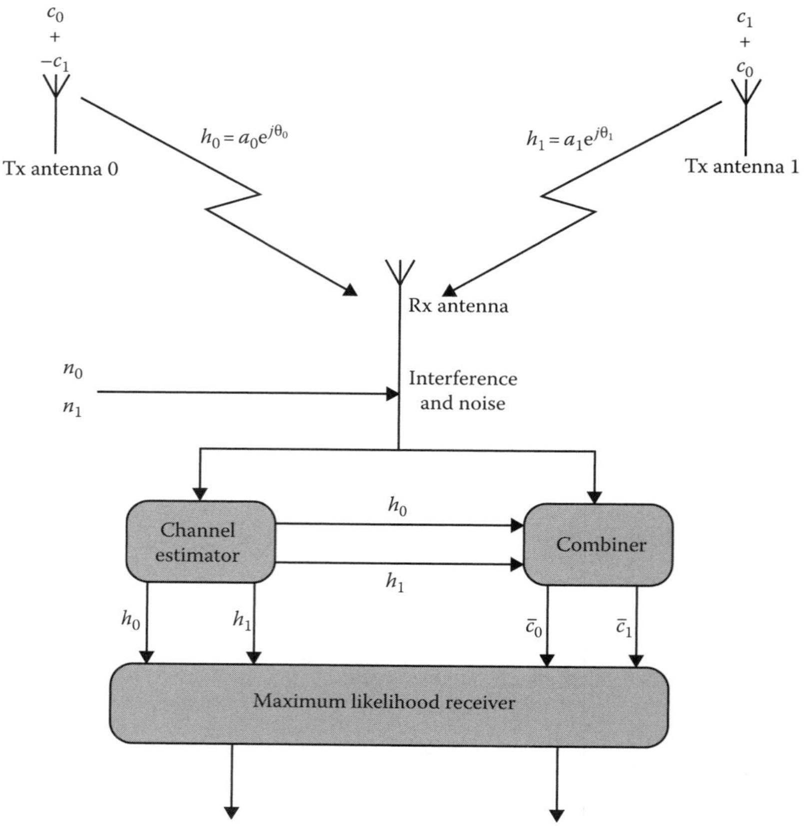

FIGURE 31.5 The Alamouti encoding technique. (From Alamouti, S.M., *IEEE J. Select. Areas Commun.*, 16, 1458, October, 1998.)

TABLE 31.1 Transmitted Coded Symbols in Alamouti Scheme

	Antenna 1	Antenna 2
T	$\mathbf{c}_1$	$\mathbf{c}_2$
$T+\tau$	$-\mathbf{c}_2^*$	$\mathbf{c}_1^*$

$$y_1 = -\sqrt{\frac{E_s}{2}} h_0 c_1^* + \sqrt{\frac{E_s}{2}} h_1 c_0^* + n_1 \tag{31.5}$$

We note that channel gains h_0 and h_1 are considered constant over the two time slots which follow the initial assumption of the slow and flat fading channel.

Equations 31.4 and 31.5 can be written as

$$y = \sqrt{\frac{E_s}{2}} \begin{bmatrix} h_0 & h_1 \\ h_1^* & -h_0^* \end{bmatrix} \begin{bmatrix} c_0 \\ c_1 \end{bmatrix} + \begin{bmatrix} n_0 \\ n_1^* \end{bmatrix} \tag{31.6}$$

We can observe that in Equation 31.6 the $\mathbf{H}_{\text{eff}} = \begin{bmatrix} h_1 & h_2 \\ h_2^\star & -h_1^\star \end{bmatrix}$ matrix which characterizes the channel is orthogonal. If instead of vector $\mathbf{y}$ we use the equivalent vector $\mathbf{z}$, given in Equation 31.7,

$$\mathbf{z} = \mathbf{H}_{\text{eff}}^{\text{H}}\mathbf{y} = \sqrt{\frac{E_s}{2}}\mathbf{H}_{\text{eff}}^{\text{H}}\mathbf{H}_{\text{eff}}\mathbf{c} + \mathbf{H}_{\text{eff}}^{\text{H}}\mathbf{n} \Rightarrow \mathbf{z} = \sqrt{\frac{E_s}{2}}\|h\|_{\text{F}}^2 I_2\mathbf{c} + \mathbf{H}_{\text{eff}}^{\text{H}}\mathbf{n} \tag{31.7}$$

then the output of a simple ML decoder is given by

$$\hat{\mathbf{C}} = \arg\min_{\mathbf{C}} \left\|\mathbf{Z} - \|h_{\text{eff}}\|_{\text{F}}^2\mathbf{C}\right\|_{\text{F}}^2 = \arg\min_{\mathbf{C}} \left\|\mathbf{Z} - \left(|h_1|^2 + |h_2|^2\right)\mathbf{C}\right\|_{\text{F}}^2 \tag{31.8}$$

The decoding procedure appears to be a linear process at the receiver while at the same time the $\mathbf{H}_{\text{eff}}$ matrix retains its orthogonality regardless of the $\mathbf{H}$ channel matrix. The last two observations have made the Alamouti concept for achieving diversity the dominant philosophy in STBC techniques.

The diversity order that this scheme can achieve is $2M_r$.

31.3.2.2 Extended Alamouti Schemes

The simple orthogonal scheme described above has been extended for more than two transmit antennas [3]. In fact, we can produce orthogonal designs for any number of transmit antennas. The designs appear in literature may be classified to two main categories, the real orthogonal designs, constructed out of real code matrices and the complex orthogonal designs constructed out of complex code matrices.

Real constellations (Pulse Amplitude Modulation, PAM) lead to real orthogonal designs. In a real orthogonal design the code matrix consist of the real entries $\pm c_1, \pm c_2, \ldots, \pm c_N$ and the problem of existence is a Hurwitz–Radon problem [4].

In case that a square code matrix is required the limitations are many, so only three full rate schemes can be produced. Specifically, real orthogonal designs of full rate and square code matrix can only be implemented for 2, 4, and 8 transmit antennas. However, generalized real orthogonal schemes of full rate and of $T \times M_t$ code matrix can also be produced, they are called delay–optimal designs and offer easily decoding schemes for $M_t \leq 8$.

In the case of complex orthogonal designs the code matrices consist of the entries $\pm c_1, \pm c_2, \ldots, \pm c_{M_t}$ and their conjugates $\pm c_1^\star, \pm c_2^\star, \ldots, \pm c_{M_t}^\star$. The Alamouti scheme described above belongs to this category. We note that the construction of orthogonal encoding schemes of any number of transmit antennas is possible, however the rate of such codes is lower than one. In addition, when the number of transmit antennas increases the transmission rate of OSTBC designs reduces.

At this point, the category of quasi-orthogonal ST block codes must also be discussed. While orthogonal schemes offer full diversity at the expense of reduced rate, the quasi-orthogonal STBCs (QOSTBCs) aim at designs of code rate one with partial diversity gain. These schemes are mostly built upon an original OSTBC design which they apply into a group of transmit antennas. In other words, the transmit antennas are divided into groups, and each group employs an OSTBC design.

In [5] for example, the author presents a quasi-orthogonal ST block design for $M_t = 4, 8$ and $M_r = 1$. The columns of the produced code matrix are not orthogonal to each other, but instead the matrix columns are divided into two orthogonal groups. This way, the columns of the same group do not have to be orthogonal to each other. The produced designs offer full transmission rate and a diversity order of 2. The author concludes that the proposed QOSTBC performs better than its orthogonal counterpart only in low SNRs since the orthogonal designs perform always better in high SNR regimes [6].

The next step in QOSTBC designs was realized through constellation rotation which resulted in improved performance in comparison to the original QOSTBC. Specifically, the authors in [6] suggest that if the original constellation set, $\mathcal{A}$, which is used for the production of the encoded symbols, is rotated around itself by a fixed angle φ, then there will be a significant gain in the received SNR depending on the modulation used. The authors present a design suitable for $M_t = 4$ and compare their results with those of [5]. Finally the same authors continue their work on QOSTBC with rotation constellation and in [7] they propose a class of linear QOSTBC designs that offer full rate and full diversity for an arbitrary number of transmit antennas.

31.3.3 Space–Time Block Codes for Selective Fading Environments

The discussion up to now concerned diversity encoding schemes for flat fading channels. Nevertheless, technologies beyond 3G must deal with frequency-selective channels so the construction of STBC for these kinds of channels are of great concern to this chapter. Frequency-selective channels destroy the orthogonality of those techniques. In the following we try to summarize diversity techniques for selective fading environments follow.

The simplest ST coding scheme for this kind of channels is presented in [8]. The OFDM technique is used in order to transform the frequency-selective channel to multiple flat fading ones, then the implementation of a typical ST code is straightforward.

The proposed transmitter consists of two antennas with an OFDM chain attached to each of them. The block diagram is shown in Figure 31.6.

Let us assume a discrete Fourier transform (DFT) module of N carriers. The serial to parallel converter produces $2N$ data symbols per time slot, so the $N \times 1$ data vectors $\mathbf{X_o}$ and $\mathbf{X_e}$ are constructed and transmitted by the antennas 1 and 2. During the next time slot the data vectors $-\mathbf{X_e^*}$ and $\mathbf{X_o^*}$ are transmitted by antennas 1 and 2 correspondingly. The scheme appears to be analogous to the simple Alamouti scheme described above.

The channel can be modeled as a diagonal matrix $N \times N$. The diagonal entries of the channel matrix are the channel gains ($h_1(n), h_2(n)$ for $n = 1, \ldots, N$) of each carrier which are subject to flat fading. Let $\mathbf{H_1}$ and $\mathbf{H_2}$ be the channel matrices and lets assume that the channel is subject to slow fading,

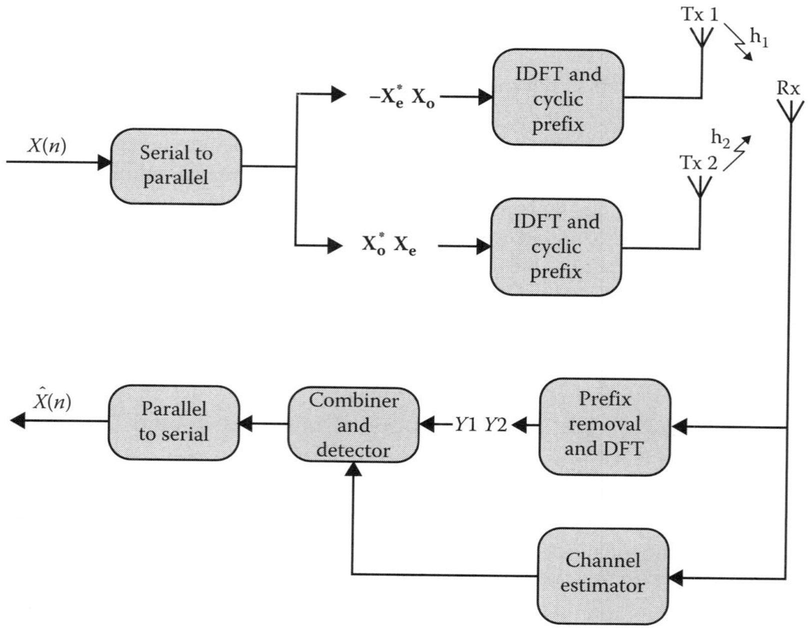

FIGURE 31.6 Block diagram of the Tx.

in that case $\mathbf{H}_1(T_0) = \mathbf{H}_1(T_0 + \tau)$ and $\mathbf{H}_2(T_0) = \mathbf{H}_2(T_0 + \tau)$. The vectors ($\mathbf{Y}_1$ and $\mathbf{Y}_2$) received during the two time slots are given by the equations

$$\mathbf{Y}_1 = \mathbf{H}_1\mathbf{X}_0 + \mathbf{H}_2\mathbf{X}_e + \mathbf{Z}_1 \tag{31.9}$$

$$\mathbf{Y}_1 = \mathbf{H}_1\mathbf{X}_0 + \mathbf{H}_2\mathbf{X}_e + \mathbf{Z}_1 \tag{31.10}$$

Assuming perfect channel knowledge at the receiver the following estimations of the transmitted vectors can be made (Equations 31.11 and 31.12). This way, the described scheme requires a simple decoding process.

$$\hat{\mathbf{X}}_0 = \mathbf{H}_1^*\mathbf{Y}_1 + \mathbf{H}_2\mathbf{Y}_2^* \tag{31.11}$$

$$\hat{\mathbf{X}}_e = \mathbf{H}_2^*\mathbf{Y}_1 - \mathbf{H}_1\mathbf{Y}_2^* \tag{31.12}$$

Obviously, the diversity order of this scheme is again $2M_R$ which indicates that this scheme does not exploit the multipath diversity that the frequency-selective channel offers. As it was proved in [9] the maximum diversity gain over a selective fading MIMO channel is $M_r \cdot M_t \cdot L$ where L is the number of delay taps of the channel impulse response.

The next ST block code that we describe achieves a diversity gain of $M_r \cdot M_t \cdot L$. The basic advantage of this algorithm is the fast decoding process that can be implemented. The code structure of this code is based on general OSTBC designing [3] applied to single carrier systems. Specifically, if $\mathbf{G}$ is the $T \times M_t$ code matrix of a general OSTBC code scheme, then Equation 31.13 applies

$$\mathbf{G}^H \cdot \mathbf{G} = \left(|c_1|^2 + |c_2|^2 + \cdots + |c_{N_s}|^2\right) \cdot \mathbf{I}_{M_t} \tag{31.13}$$

In Equation 31.13 $\mathbf{G}^H$ is the Hermitian of $\mathbf{G}$. The code structure that is proposed in [12] has the following code matrix:

$$\mathbf{C} = \sqrt{\gamma}\mathbf{G}' \otimes \mathbf{1}_{\Gamma\times 1} \tag{31.14}$$

where

- $\otimes$ is the symbol of the Kronecker product
- γ is a scalar that ensures that $\|\mathbf{C}\|_F^2 = T \cdot N \cdot M_t$ which corresponds to the power constraint
- $\mathbf{1}$ is a $\Gamma \times 1$ vector of all ones
- $\mathbf{G}$ is the aforementioned $\mathbf{G}$ code matrix but each entry c_i of the original $\mathbf{G}$ matrix is replaced by a symbol vector $\mathbf{c}_i' = \left[c_i(1) \cdots c_i\left(\frac{N}{\Gamma}\right)\right]^T$ of dimensions $\frac{N}{\Gamma} \times 1$

The elements $\mathbf{c}_i'(j)$, $j = 1, \ldots, \frac{N}{\Gamma}$ of the $\mathbf{c}_i'$ vector are members of any kind of modulation constellation and $\Gamma = 2^{\lceil \log_2 (L') \rceil}$ where $L' < L$. In [12] the verification of Equation 31.13 for the $\mathbf{C}$ code matrix is presented.

For simplicity and understanding purposes we cite the code matrix for two transmit antennas in Equation 31.15 and the transmission scheme in Table 31.2. Obviously, the simple ST encoding scheme presented in the previous paragraph is a special case of Equation 31.15 where $\Gamma = 1$.

$$\mathbf{C} = \begin{pmatrix} \mathbf{c}_1' & \mathbf{c}_2' \\ -\mathbf{c}_2'^* & \mathbf{c}_1'^* \end{pmatrix} \otimes \mathbf{1}_{\Gamma\times 1} = \begin{pmatrix} \mathbf{c}_1 & \mathbf{c}_2 \\ -\mathbf{c}_2^* & \mathbf{c}_1^* \end{pmatrix} \tag{31.15}$$

The OSTBC code of Equation 31.14 can offer a code rate that is Γ times less that the code rate of the code described with $\mathbf{G}$. However, it achieves a diversity gain of $M_t \cdot M_r \cdot L'$ and when $L' = L$ a full diversity gain is provided.

TABLE 31.2 Transmission Scheme for Selective Fading Channel

	Antenna 1	Antenna 2
First OFDM symbol	c'_1	c'_2
Second OFDM symbol	$-c'^*_2$	c'^*_1

31.4 Antenna and Frequency Diversity Techniques for OFDM

The simplest way to produce algorithms of this category is by applying the same ST coding schemes across space and frequency. However, for wideband MIMO systems the algorithms produced fail to exploit the frequency diversity introduced by those systems. In the following we present one scheme for flat fading channels while special interest is attributed to space–frequency (SF) schemes appropriate for selective fading environments. The design criteria of SF codes can be found in [13].

31.4.1 Alamouti Technique

The concept of this design is rather simple, the coded symbols are spread over two OFDM carriers instead of two time slots [10]. A clarifying depiction of this is shown in Figure 31.7.

Although simple, this technique does not achieve full diversity gain which is possible in the case of wideband technologies.

31.4.2 Space–Frequency Codes for Selective Fading Environments

In [14] a code design is presented that achieves full frequency and spatial diversity gain. The scheme concerns exclusively OFDM systems and is based on the fast Fourier transform (FFT) matrix usage. Specifically, the scheme is addressed to a system of M_t transmit antennas and M_r receive antennas, while the channel considered has L independent paths. The elements of $N \times N$ FFT matrix (**F**) are given in Equation 31.16:

$$[\mathbf{F}]_{m,n} = \frac{1}{\sqrt{N}} \exp\left(-j2\pi \frac{mn}{N}\right), \quad m, n = 0, 1, \ldots, N-1 \tag{31.16}$$

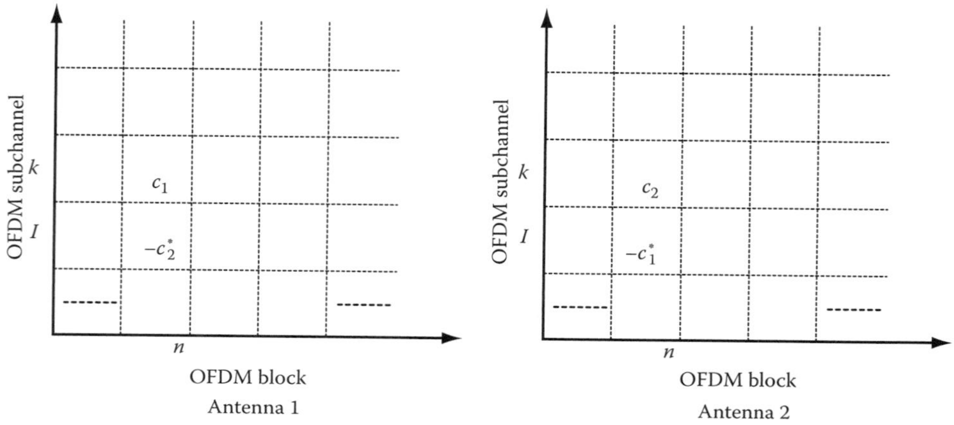

FIGURE 31.7 Alamouti scheme as a SF encoding technique.

As a result **F** is a unitary matrix with its columns being orthogonal to each other. The proposed $N \times M_t$ code matrix is provided by

$$\mathbf{C}^{\mathrm{T}} = [\mathbf{F}_1 \cdot \mathbf{c}\ \mathbf{F}_2 \cdot \mathbf{c} \cdots \mathbf{F}_{M_t} \cdot \mathbf{c}] \tag{31.17}$$

In Equation 31.17 **c** is a $K \times 1$ vector of modulated symbols extracted from the constellation alphabet used and $\mathbf{F}_i, i = 1, \ldots, M_t$ are $N \times K$ matrices containing vectors of the **F** matrix. The vectors that are used for the construction of the $\mathbf{F}_i$ matrices are selected in a way that the following criteria are satisfied:

- $\mathbf{F}_i \cdot \mathbf{F}_i^{\mathrm{H}} = \mathbf{I}_K, \quad i = 1, \ldots, M_t$ or none of the matrices $\mathbf{F}_i$ and $\mathbf{F}_j$ can have two identical columns.
- $\mathbf{F}_i \cdot \mathbf{F}_j^{\mathrm{H}} = \mathbf{0}_K \quad \text{for} \quad i \neq j$ or the $\mathbf{F}_i$ and $\mathbf{F}_j$ matrices should not share the same column.
- $\mathbf{F}_i^{\mathrm{H}} \cdot \mathbf{D}^{\mathrm{H}} \cdot \mathbf{F}_i = \mathbf{0}_K, \quad i = 1, \ldots, M_t$ or none of the $\mathbf{F}_i$ and $\mathbf{F}_j$ matrices should contain two neighboring columns of the **F** matrix.
- $\mathbf{F}_i^{\mathrm{H}} \cdot \mathbf{D}^{\mathrm{H}} \cdot \mathbf{F}_j = \mathbf{0}_K \quad \text{for} \quad i \neq j$ or the matrices $\mathbf{F}_i$ and $\mathbf{F}_j$ cannot share two neighboring columns of the **F** matrix.

The diversity gain of this scheme is $M_t \cdot M_r \cdot L$ but the achievable symbol rate remains under $\frac{1}{M_t \cdot L}$.

Another approach that achieves full diversity in selective fading MIMO environments can be found in [15]. The authors in this work prove that the coding schemes that provide full diversity gain in flat fading environments can be used to construct SF codes that provide full diversity gain in selective fading environments. In other words, the proposed schemes consist of a typical ST encoder, producing the $T \times M_t$ codeword, and a mapper that introduces the L channel taps. The proposed SF encoder concept is shown in Figure 31.8. We note that the encoder is suitable for STBC as well as STTC with the produced codeword to alter correspondingly.

Let $\mathbf{g} = [g_1\ g_2 \cdots g_{M_t}]$ be the $1 \times M_t$ vector of the encoded symbols generated by the ST encoder. The mapper of Figure 31.8 performs the following mapping:

$$(g_1\ g_2\ \ldots\ g_{M_t}) \rightarrow \mathbf{1}_{l \times 1}(g_1\ g_2\ \ldots\ g_{M_t}) \tag{31.18}$$

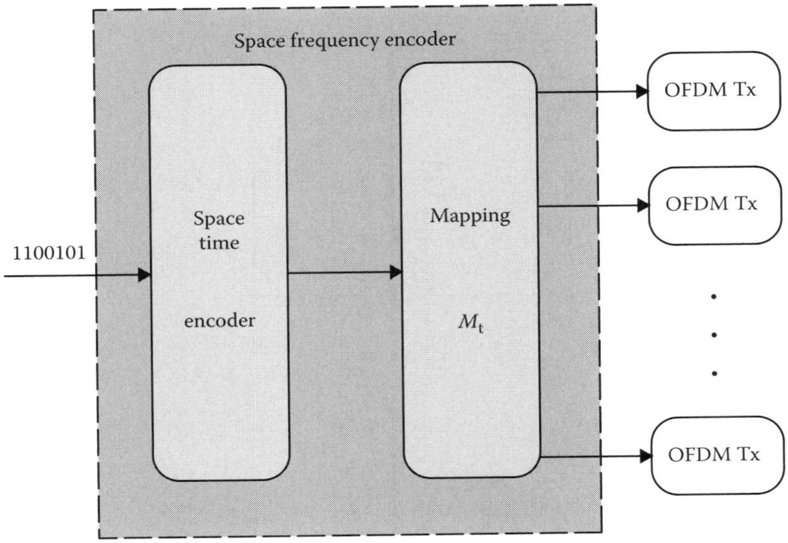

FIGURE 31.8 SF encoder proposed in [15].

where

$\mathbf{1}$ is $l \times 1$ vector of ones

l is a number that $1 \leq l \leq L$

Obviously, the output of this mapper is an $l \times M_t$ matrix. We choose the integer k so as $k = \left\lfloor \frac{N}{l \cdot M_t} \right\rfloor$ and we call $\mathbf{G}$ the ST encoder matrix of dimensions $k \cdot M_t \times M_t$. The proposed SF coding scheme constructs the following $N \times M_t$ code matrix.

$$\mathbf{C} = \begin{bmatrix} \mathbf{M}_l(\mathbf{G}) \\ \mathbf{0}_{(N-klM_t)\times M_t} \end{bmatrix} \tag{31.19}$$

where $\mathbf{M}_l(\mathbf{G}) = [\mathbf{I}_k\, M_t \otimes \mathbf{1}_{l\times 1}] \cdot \mathbf{G}$ and the symbol $\otimes$ stands for the Kronecher product. The output of the encoder is actually the $\mathbf{G}$ matrix repeated l times and adding zeros. The zero padding is necessary for retaining the $N \times M_t$ matrix dimensions of the codeword while it can also be used for ST trellis initial encoding in order to drive the trellis encoder to zero state. The diversity gain of this code is $M_r \cdot M_t \cdot l$.

A more recent work of the same authors [16] presents SF schemes that achieve full diversity gain as well as full symbol rate one. The authors proceed to the decomposition of the diversity gain as the product of the "intrinsic" and the "extrinsic" diversity products. The intrinsic diversity product depends on the signal constellations and the code design while the extrinsic diversity product depends on the applied permutation and the power delay profile of the channel. The application of an optimum permutation strategy leads to the maximization of the extrinsic diversity product.

Another effort for producing full diversity SF codes with high transmission rate, equal to the number of transmit antennas, is presented in [17]. The authors establish a set of more general design criteria for SF codes and based on them they define a new class of SF codes which structure is provided in [16].

The code structure is based on code matrices of the following form:

$$\mathbf{C} = \left[\mathbf{G}_1^T\ \mathbf{G}_2^T \cdots \mathbf{G}_P^T\ 0_{N-PK}^T\right] \tag{31.20}$$

In Equation 31.20 $K = l \cdot M_t$, $1 \leq l \leq L$, and $P = \left\lfloor \frac{N}{K} \right\rfloor$, the $\mathbf{0}$ matrix is used for zero padding when the numbers of subcarriers are not a multiple of K. The matrices $\mathbf{G}_P$ are of dimensions $K \times M_t$ and follow the structure of Equation 31.21.

$$G(\mathbf{X}_1, \mathbf{X}_2, \ldots, \mathbf{X}_{M_t}) = \begin{bmatrix} \mathbf{X}_{11} & \phi\mathbf{X}_{21} & \phi^2\mathbf{X}_{31} & \cdots & \phi^{M_t-1}\mathbf{X}_{M_t1} \\ \phi^{M_t-1}\mathbf{X}_{M_t2} & \mathbf{X}_{12} & \phi\mathbf{X}_{22} & \cdots & \phi^{M_t-2}\mathbf{X}_{(M_t-1)2} \\ \phi^{M_t-2}\mathbf{X}_{(M_t-1)3} & \phi^{M_t-2}\mathbf{X}_{M_t3} & \mathbf{X}_{13} & \cdots & \phi^{M_t-3}\mathbf{X}_{(M_t-2)3} \\ \vdots & \vdots & \vdots & \ddots & \vdots \\ \phi\mathbf{X}_{2M_t} & \phi^2\mathbf{X}_{3M_t} & \phi^3\mathbf{X}_{4M_t} & \cdots & \mathbf{X}_{1M_t} \end{bmatrix} \tag{31.21}$$

The $\mathbf{X}_i$ columns of Equation 31.21 are constructed according to

$$\mathbf{X}_i = \left[X_{i1}^T\ X_{i2}^T\ \cdots X_{iM_t}^T\right]^T \in \Re = \left\{\Theta s | s \in S^{K\times 1}\right\} \tag{31.22}$$

where

Θ is a $K \times K$ matrix

s is the signal constellation set used

The code rate of the proposed code is $\frac{PKM_t}{N}$ and in case N is a multiple of K the transmission rate is M_t.

The matrix $\mathbf{\Theta}$ is chosen so that the difference vector $\mathbf{X} - \bar{\mathbf{X}}$ has a Hamming weight equal to K for any $\mathbf{X} \neq \bar{\mathbf{X}} \in \Re$. The interested reader can look in [16] for a summary of such matrices. Apart from the $\mathbf{\Theta}$ matrix the complex number ϕ is another innovation of this algorithm. ϕ is chosen according to the delay profile and the matrix $\mathbf{\Theta}$. Obviously, this code requires channel knowledge at the transmitter. The authors in [17] give several examples of ϕ number construction taking under consideration different delay profiles and $\mathbf{\Theta}$ matrices.

Finally, as in the case of ST coding schemes, quasi-orthogonal designs have been developed for SF coding as well. In [18] the authors provide SF codes built upon quasi-orthogonal structures that achieve rate one and full diversity in a frequency-selective MIMO-OFDM channel.

31.5 Antenna, Time, and Frequency Diversity Techniques for OFDM

The schemes of this category use the three dimensions of space, time, and frequency (STF) to code across, resulting in an additional temporal diversity gain when compared to the coding schemes of the previous class (SF). In [19] it is proved that STF coding can achieve diversity gain equal to the product of the number of transmit antennas, the number of receive antennas, the channel taps, and the rank of the temporal correlation matrix of the channel. However, the full diversity gain employing STF encoding techniques are usually of prohibitive complexity. In the following a number of STF encoding schemes are thoroughly analyzed and the weaknesses of each are pointed out.

The STF code proposed in [20] is based on the concept of grouping the correlated OFDM carriers into groups. In this way, the system is divided into groups of STF (GTFM) subsystems within the proposed encoding STF scheme is applied. The authors in [20] prove that the proposed concept of grouping retains the maximum diversity gain while involves simplified encoding and decoding architectures.

Since the information exchanged between the transmitter and receiver in the case of STF coding schemes can be expressed as a point in a 3-D space, a system model suitable for this kind of scenarios is cited. The $\mathbf{H}$ channel matrix elements for each OFDM carrier are given by

$$H_{i,j}(p) = \sum_{l=0}^{L} h_{ij}(l) \cdot e^{-j(2\pi/N)\cdot l \cdot p} \tag{31.23}$$

Equation 31.23 provides the channel gain between the ith transmit and the jth receive antenna for the pth OFDM carrier, while Equation 31.24 shows the link level model for the MIMO-OFDM system.

$$y_n^j(p) = \sum_{\mu=1}^{M_t} H_{ij}(p) x_n^i(p) + w_n^j(p), \quad j = 1, \ldots, M_r, \quad p = 0, \ldots, N-1 \tag{31.24}$$

In Equation 31.24 $x_n^i(p)$ is the transmitted symbol from the ith antenna during the nth time slot on the pth subcarrier and $w_n^j(p)$ is the additive white noise. The symbol $x_n^i(p)$ is produced by the STF code. The codeword of such a code is expected to be 3-D with STF dimensions. Let M_t be the transmit antennas, N the carriers, and N_x the available time slots. The total coded symbols of the produced codeword are $M_t \cdot N\, N_x$ and the encoding scheme under investigation performs the mapping Ψ: $\mathbf{s} \rightarrow \mathbf{X}$, where $\mathbf{s}$ is the $N_s \times 1$ symbol vector that has already been defined in the previous paragraphs. The graphical representation of $\mathbf{X}$ codeword in STF is shown in Figure 31.9. Obviously, the dimensions of $\mathbf{X}$ are $M_t \times N \cdot N_x$.

Next, the carrier grouping process is described. First, a number of carriers N_c, multiple of the channel length is chosen, which represents the number of groups.

$$N_c = N_g \cdot (L+1) \tag{31.25}$$

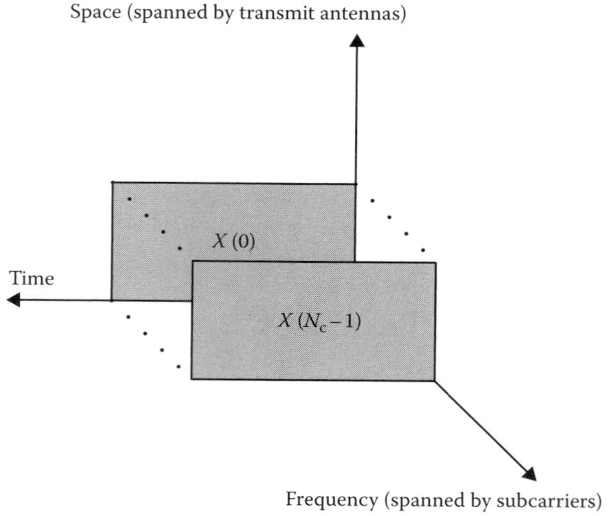

FIGURE 31.9 The spanned codeword in time and frequency.

The codeword **X** must be split into N_g subgroups and the codewords $\mathbf{X_g}$ of dimension $M_t \times N_x \cdot (L+1)$ must be produced. Specifically, the produced codewords are $\mathbf{X_g} = [X_g(0), X_g(1), \ldots, X_g(L)]$ for $g = 0, \ldots, N_g - 1$ and they must follow the rule

$$X_g(l) := X_g(N_g l + g) \tag{31.26}$$

The grouping of the carriers leads to the construction of a new simplified code mapping which is symbolized as Ψ_g: $\mathbf{s_g} \rightarrow \mathbf{X_g}$. Obviously, the symbol vector $\mathbf{s_g}$ is now a $N_s' \times 1$ vector, where $N_s' = \frac{N_s}{N_g}$.

Even the way the coding design criteria apply in this STF technique is not cited herein, some interesting remarks that are arisen in [20] are presented. The rank criterion is satisfied only when $N_x \geq M_t$. Under this constraint, it is obvious that the minimum dimension of the codeword $\mathbf{X_g}$ is $M_t \times M_t \cdot (L+1)$, while the minimum codeword size in the case of SF coding is no less than $M_t \times N$ with N the total number of carriers. Since the size of the codewords affects the design complexity, the proposed STF structure involves a simpler design process in comparison with the aforementioned SF tactics. Exploiting again the application of design criteria into the proposed encoding scheme it is deduced that the maximum diversity advantage of each GSTF subsystem is $M_t \cdot M_r \cdot (L+1)$, which equals to the diversity gain provided by the STF system without channel grouping. Finally, the authors in [20] point out that although the proposed GSTF does retain the diversity and code gain of the simple STF codes, the BER might be deteriorated.

The structure of the code can follow two steps: In the first one the constellation precoding takes place so as frequency diversity to be provided, while in the second step the ST encoding provides spatial diversity. This two-step encoding process is translated in a low-complexity, two-stage optimal decoding as it is well established in the same work of the authors.

According to [3] every generalized complex orthogonal design is characterized by an $M_t \times N_d$ matrix **O** with its nonzero entries taken from the set $\{d_i, d_i^*, i = 0, \ldots, N_t - 1\}$. The **O** matrix follows the rule

$$\mathbf{O}_{M_t}^{H} \mathbf{O}_{M_t} = a \sum_{i=0}^{N_s - 1} |d_i|^2 \mathbf{I}_{M_t} \tag{31.27}$$

In Equation 31.27 a is a positive constant. Every code design that is produced by this equation is of size (N_s, N_d). A generalized orthogonal complex design of size (N_s, N_t) exists only if the N_d and N_s depend on the transmit antenna M_t according to the following rule:

$$(N_s, N_d) = \begin{cases} (2,2), & \text{if } M_t = 2 \\ (3,4), & \text{if } M_t = 3,4 \\ (M_t, 2M_t), & \text{if } M_t > 4 \end{cases} \tag{31.28}$$

After this short reference to the generalized orthogonal designs we continue with the specific STF design. First, the parameter N_s' is chosen according to the equation $N_s' = N_t(L+1)$ and the original $\mathbf{s_g}$ symbol vector is demultiplexed into N_t subgroups in a way that

$$\mathbf{s}_g := \left[\mathbf{s}_{g,0}^{\mathrm{T}}, \ldots, \mathbf{s}_{g,N_t-1}^{\mathrm{T}}\right] \tag{31.29}$$

In Equation 31.29 each element $\mathbf{s}_{g,i}$ is a $(L+1) \times 1$ complex vector which must be submitted to the precoding process that will distribute the information symbols over multiple subcarriers. The precoding process leads to the precoded blocks $\tilde{S}_{g,i}$ that are $(L+1) \times (L+1)$ complex matrices produced by $\tilde{S}_{g,i} := \mathbf{\Theta}\mathbf{s}_{g,i}$. In the last equation, $\mathbf{\Theta}$ is a $(L+1) \times (L+1)$ complex matrix, which denotes the square constellation precoder. At the second stage the $\tilde{S}_{g,i}$ symbol matrix is transformed into the $\mathbf{X}_g$ codeword with the ST component.

In the precoding constellation stage the $\mathbf{\Theta}$ matrix is produced in relation to the modulation applied and the L value. The analytical construction of $\mathbf{\Theta}$ matrix is presented in [21], however we present here an example for easier understanding. In case of QPSK modulation and $L=1$, the $\mathbf{\Theta}$ matrix is given by

$$\mathbf{\Theta} = \frac{1}{\sqrt{2}} \begin{bmatrix} 1 & e^{j(\pi/4)} \\ 1 & e^{j(5\pi/4)} \end{bmatrix} \tag{31.30}$$

After having specified the precoding matrix $\mathbf{\Theta}$ and the symbol matrices $\tilde{\mathbf{s}}_{g,i} := \mathbf{\Theta}\mathbf{s}_{g,i}$, the codeword construction is as follows:

The matrix $\mathbf{O}_{M_t}$ that describes a general complex orthogonal design can be represented as

$$\mathbf{O}_{M_t} = \sum_{i=0}^{N_s-1} \mathbf{A}_i d_d + \mathbf{B}_i d_i^{\star} \tag{31.31}$$

In Equation 31.31 the real $N_d \times M_t$ $\mathbf{A}_i$ and $\mathbf{B}_i$ matrices satisfy the following equations:

$$A_i^{\mathrm{T}} A_{i'} + B_i^{\mathrm{T}} B_{i'} = a I_{M_t} \delta(i - i') \tag{31.32}$$

$$A_i^{\mathrm{T}} B_{i'} = 0 \tag{31.33}$$

The pairs $\{A_i, B_i\}_{i=0}^{N_s-1}$ are used in the construction of the codeword $\mathbf{X}_g$

$$X_g^{\mathrm{T}}(l) = \sum_{i=0}^{N_s-1} (A_i \tilde{s}_{g,i,l} + \mathbf{B}_i \tilde{s}_{g,i,l}^{\star}) \tag{31.34}$$

Equation 31.34 shows that the $\tilde{s}_{g,i,l}$ variables are to $\mathbf{X}_g$ what d_i variables are to $\mathbf{O}_{M_t}$.

The next STF code that is presented is analyzed in [23] where a systematic design of full diversity STF codes based on layered algebraic design is proposed. Most of the encoding schemes of this category consider a maximum diversity gain of $M_t \cdot M_r \cdot L$. However, the encoding scheme of [23] provides a maximum diversity gain of $M_r \cdot M_t \cdot L \cdot M_b$, where M_b is the number of the fading blocks of the frequency-selective fading MIMO channel. Contrary to the proposed SF encoding schemes, which are addressed to quasi-static fading channels, the authors suggest that in block fading channels the coding across multiple fading blocks can offer extra diversity advantage. Moreover, the proposed encoding scheme offers full rate, M_t under all circumstances, which makes this algorithm different from the already introduced ones when $M_b = 1$.

Before the code presentation a brief reference to the used system model is necessary. The $N \times M_t \cdot M_b$ codeword of the system is produced by mapping N_s data symbols:

$$\mathbf{C} = \left[\mathbf{C}^1\ \mathbf{C}^2\ \ldots\ \mathbf{C}^{M_b}\right] \tag{31.35}$$

In Equation 31.35 each element of $\mathbf{C}$ is given by $\mathbf{C}^{t} = \left[\mathbf{c}_1^t\ \mathbf{c}_2^t\ \ldots\ \mathbf{c}_{M_t}^t\right]$. Each of the column vectors of $\mathbf{C}^t$, $\mathbf{c}_j^t$, is sent as an OFDM block to the jth transmit antenna. The received signal after the cyclic prefix removal and the FFT module for the ith receive antenna for the t fading block is given by

$$\mathbf{Y}_i^t = \sum_{j=1}^{M_t} \mathrm{diag}\left(\mathbf{c}_j^t\right) \cdot \mathbf{H}_{i,j}^t \tag{31.36}$$

In Equation 31.36 $\mathbf{H}_{i,j}^t$ is the frequency response of the channel and is provided by

$$\mathbf{H}_{i,j}^t = \mathbf{F} \cdot \mathbf{h}_{i,j}^t \tag{31.37}$$

In Equation 31.37 $\mathbf{F} = [\mathbf{f}_0\ \mathbf{f}_1 \ldots \mathbf{f}_{L-1}]$ and each column vector of $\mathbf{F}$ is given by $\mathbf{f}_l = \left[1\ \omega_l\ \ldots\ \omega_l^{N-1}\right]^{\mathrm{T}}$, where $\omega_l = \exp\left(-j2\pi \frac{\tau_l}{T_s}\right)$ and T_s is the OFDM symbol duration. In the same equation, $\mathbf{h}_{i,j}^t$ is the $L \times 1$ impulse response vector for the channel with each element $h_{i,j,l}^t$, $l = 0, \ldots, L-1$ being the complex amplitude of the lth channel path.

In order to give a compact equation for the channel model, let $\mathbf{D}_l = \mathrm{diag}(\mathbf{f}_l)$, $l = 0, \ldots, L$ and $\mathbf{D}_l \mathbf{c}_j^t = \mathrm{diag}(\mathbf{c}_j^t) \cdot \mathbf{f}_l$. If we use Equation 31.37 in Equation 31.36 we find that

$$\mathbf{Y}_i^t = \sum_{j=1}^{M_t} \left[\mathbf{D}_0\mathbf{c}_j^t\ \mathbf{D}_1\mathbf{c}_j^t \cdots \mathbf{D}_{L-1}\mathbf{c}_j^t\right] \cdot \mathbf{h}_{i,j}^t \tag{31.38}$$

Equation 31.38 after some matrix permutations becomes

$$\mathbf{Y}_i^t = \sum_{l=1}^{L} \left[\mathbf{D}_l\mathbf{c}_1^t\ \mathbf{D}_l\mathbf{c}_2^t \cdots \mathbf{D}_l\mathbf{c}_{M_t}^t\right] \cdot \mathbf{h}_{i,l}^t = \sum_{l=1}^{L} \mathbf{D}_l \cdot \mathbf{C}^{t} \cdot \mathbf{h}_{i,l}^t \tag{31.39}$$

We note $\mathbf{X}_t = [\mathbf{D}_0\mathbf{C}^t\ \mathbf{D}_1\mathbf{C}^t \cdots \mathbf{D}_{L-1}\mathbf{C}^t]$ and $\mathbf{h}_i^t = \left[\mathbf{h}_{i,0}^{t\ \mathrm{T}}\ \mathbf{h}_{i,1}^{t\ \mathrm{T}} \cdots \mathbf{h}_{i,L-1}^{t\ \mathrm{T}}\right]^{\mathrm{T}}$ in order for Equation 31.39 to be written as

$$\mathbf{Y}_i^t = \mathbf{X}_t \cdot \mathbf{h}_i^t \tag{31.40}$$

for $t = 1, 2, \ldots, M_b$ and $i = 1, 2, \ldots, M_r$. Finally, the following notations are used in order to obtain the system model presented in Equation 31.41.

$$\begin{aligned}
\mathbf{Y} &= \left[\mathbf{Y}_1^{1^T} \cdots \mathbf{Y}_1^{M_b{}^T} \cdots \mathbf{Y}_{M_r}^{1\ T} \cdots \mathbf{Y}_{M_r}^{M_b{}^T}\right]^T \\
\mathbf{h} &= \left[\mathbf{h}_1^{1^T} \cdots \mathbf{h}_1^{M_b{}^T} \cdots \mathbf{h}_{M_r}^{1\ T} \cdots \mathbf{h}_{M_r}^{M_b{}^T}\right]^T \\
\mathbf{X} &= \mathbf{I}_{M_r} \otimes \mathrm{diag}(\mathbf{X}_1\ \mathbf{X}_2\ \cdots \mathbf{X}_{M_b}) \\
\mathbf{Y} &= \sqrt{\frac{\rho}{M_t}}\mathbf{X} \cdot \mathbf{h} + \mathbf{n}
\end{aligned} \tag{31.41}$$

In Equation 31.41 $\mathbf{Y}$ and $\mathbf{n}$ are complex $N \cdot M_b \cdot M_r \times 1$ vectors representing the received vector and the noise vector correspondingly, $\mathbf{X}$ is a $N \cdot M_b \cdot M_r \times M_t \cdot L \cdot M_b \cdot M_r$ complex matrix of the transmitted signal and $\mathbf{h}$ is the $M_t \cdot L \cdot M_b \cdot M_r \times 1$ vector of the channel impulse response.

In the following the code structure that the authors in [23] suggest is presented. N_s data symbols, form the $N \cdot M_t \cdot M_b \times 1$ symbol vector and they are parsed into $J = N/K$ subblocks, where $K = 2^{\lceil \log_2 (M_t \cdot L) \rceil}$. Each of the parsed data vectors $\mathbf{S}_i \in A^{KM_tM_b}(i = 1, 2, \ldots, J)$ are encoded into a STF coded matrix $\mathbf{B}_i \in C^{K \times M_tM_b}$. Let $K = N_p \cdot N_q$ so $N_p = 2^{\lceil \log_2 L \rceil}$ and $N_q = 2^{\lceil \log_2 M_t \rceil}$. If $L > 1$ and $M_t > 1$ then $K > 4$. Additionally, since N is a power of 2, according to the used FFT, J is always an integer.

Since the same STF encoding scheme is applied in every $\mathbf{B}_i$ matrix [23], the code structure will be presented only for the case of $\mathbf{B}_1$. The concept of the proposed scheme is also depicted in Figure 31.10, where the STF design along with its layered algebraic nature is shown.

$$\mathbf{B}_1 = \begin{pmatrix} \bar{\mathbf{X}}_1^1 & \bar{\mathbf{X}}_1^2 & \cdots & \bar{\mathbf{X}}_1^{M_b} \\ \bar{\mathbf{X}}_2^1 & \bar{\mathbf{X}}_2^2 & \cdots & \bar{\mathbf{X}}_2^{M_b} \\ \vdots & \vdots & \ddots & \vdots \\ \bar{\mathbf{X}}_{N_p}^1 & \bar{\mathbf{X}}_{N_p}^2 & \cdots & \bar{\mathbf{X}}_{N_p}^{M_b} \end{pmatrix} \tag{31.42}$$

Since $\mathbf{B}_i \in C^{K \times M_tM_b}$ each element of Equation 31.42 is a $N_q \times M_t$ complex matrix that is presented in Equation 31.43.

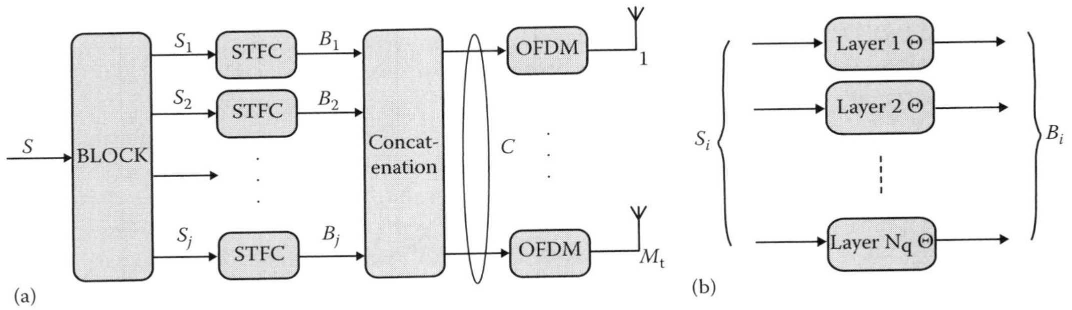

FIGURE 31.10 The STF code based on layered algebraic design.

$$\bar{\mathbf{X}}_m^t = \begin{pmatrix} X_{m,1}^t(1) & \phi X_{m,2}^t(1) & \cdots & \phi^{M_t-1} X_{m,M_t}^t(1) \\ \phi^{N_q-1} X_{m,N_q}^t\left(\left\lfloor \frac{M_t}{N_q} + 1 \right\rfloor\right) & X_{m,1}^t(2) & \cdots & \phi^{M_t-2} X_{m,M_t-1}^t(2) \\ \vdots & \vdots & \ddots & \vdots \\ \phi X_{m,2}^t(M_t) & \phi^2 X_{m,3}^t(M_t) & \cdots & \phi^{\left(1-\left\lfloor \frac{M_t}{N_q} \right\rfloor\right) M_t} X^t_{m,\left(1-\left\lfloor \frac{M_t}{N_q} \right\rfloor\right) M_t+1}(M_t) \end{pmatrix} \quad (31.43)$$

In Equation 31.43 ϕ is a complex that is introduced below. The diagonal layers of the matrix in Equation 31.43, and consequently the matrix as a whole, are constructed by the proposed design. For simplicity, we represent each diagonal layer of Equation 31.43 with the row vector of Equation 31.44.

$$\bar{\mathbf{X}}_{m,n}^t = \left[X_{m,n}^t(1)\ X_{m,n}^t(2) \cdots X_{m,n}^t(M_t)\right] \quad (31.44)$$

In Equation 31.44 $n = 1, \ldots, N_q$. We note that since $N_q = 2^{\lceil \log_2 M_t \rceil}$, obviously $N_q > M_t$. This way, the N_q diagonal layers represent the whole matrix of Equation 31.43. Since each layer given by Equation 31.44 is independent of m and t, it can be represented by Equation 31.45 and can be produced with the usage of $\mathbf{\Theta}$ matrix as is shown in Equation 31.46.

$$\bar{X}_n = \left[\bar{X}_{1,n}^1 \cdots \bar{X}_{N_p,n}^1\ \bar{X}_{1,n}^2 \cdots \bar{X}_{N_p,n}^2 \cdots \bar{X}_{1,n}^{M_b} \cdots \bar{X}_{N_p,n}^{M_b}\right] \quad (31.45)$$

$$\bar{\mathbf{X}}_n = \mathbf{\Theta} \cdot \mathbf{S}_n \quad (31.46)$$

If $N = N_p \cdot M_t \cdot M_b$, then $\mathbf{\Theta}$ is a unitary matrix which is given by Equation 31.47.

$$\mathbf{\Theta} = \mathbf{F}_N^H \text{diag}(1, \theta, \ldots, \theta^{N-1}) \quad (31.47)$$

In Equation 31.47 θ is algebraic over K with degree at least $N_p N_q M_b$, and K is the field extension of the original field Q, that contains all the entries of $\mathbf{\Theta}$, the signal alphabet and the $e^{-j2\pi\frac{\tau_l}{T_s}}$, $(l = 0, 1, \ldots, L-1)$. $\mathbf{F}_N$ is the $N \times N$ DFT matrix while $\phi = \theta^{1/N_q}$. The diversity achieved with the STF that is described above is $M_t M_r M_b L$ while the M_t rate is also achieved.

References

1. R. Prasad, *OFDM for Wireless Communications Systems*, Boston, MA, Artech House, 2004.
2. S.M. Alamouti, A simple transmit diversity technique for wireless communication, *IEEE J. Select. Areas Commun.*, 16, 1458, October 1998.
3. V. Tarokh, H. Jafarkhani, and A.R. Calderbank, Space-time block codes from orthogonal designs, *IEEE Trans. Inform. Theory*, 45, 1456–1467, July 1999.
4. A.V. Germita and J. Seberry, *Orthogonal Designs, Quadratic Forms and Hadamard Matrices, Lecture Notes in Pure and Applied Mathematics*, vol. 43. New York and Basel, Marcel Dekker, 1979.
5. H. Jafarkhani, A quasi-static space-time block code, *IEEE Trans. Commun.*, 49(1), January 2001.
6. N. Sharma and C.B. Papadias, Improved quasi-orthogonal codes through constellation rotation, *IEEE Trans. Commun.*, 51(3), 332–335, March 2003.
7. N. Sharma and C.B. Papadias, Full-rate full-diversity linear quasi-orthogonal space-time codes for any number of transmit antennas, *EURASIP J. Appl. Signal Process.*, 2004, (1), 1246–1256, January 2004.

8. K.F Lee and D.B. Williams, A space-time transmitter diversity technique for frequency selective fading channels, in *Proceedings of the IEEE Sensor Array and Multichannel Signal Processing Workshop*, Cambridge, MA, pp. 149–152, March 2000.
9. A. Paulraj, R. Nabar, and D. Gore, *Introduction to Space-Time Wireless Communications*, Cambridge University Press, Cambridge, U.K., 2003.
10. W. Zhang, X. Xia, and K. Letaief, Space-time/frequency coding for MIMO-OFDM in next generation broadband wireless systems, *IEEE Wireless Commun. Mag.*, 14, 32–43, June 2007.
11. G.J. Foschini and M.J. Gans, On limits of wireless communications in a fading environment when using multiple antennas, *Wireless Personal Commun.*, 6(3), 311–335, March 1998.
12. W. Zhang, X. Xia, and P.C. Ching, Full-diversity and fast ML decoding properties of general orthogonal space-time block codes for MIMO-OFDM systems, *IEEE Trans. Wireless Commun.*, 6(5), 1647–1653, May 2007.
13. H. Bolcskei and A.J. Paulraj, Space-frequency coded broadband OFDM systems, in *Proceedings of the IEEE WCNC-,62000*, Chicago, IL 1, 1–6, September 2000.
14. K.F. Lee and D.B. Williams, A space frequency transmitter diversity technique for OFDM systems, in *Proceedings of the IEEE Global Communication Conference*, San Francisco, CA, 3, pp. 1473–1477, November 27–December 1, 2000.
15. W. Su, Z. Safar, M. Olfat, and K.J.R. Liu, Obtaining full-diversity space-frequency codes from space-time codes via mapping, *IEEE Trans. Signal Process.*, 51, 2905–2916, November 2003.
16. W. Su, Z. Safar, M. Olfat, and K.J.R. Liu, Full-rate full-diversity space–frequency codes with optimum coding advantage, *IEEE Trans. Inform. Theory*, 51(1), 229–249, January 2005.
17. T. Kiran and B. Sundar Rajan, A systematic design of high-rate full-diversity space-frequency codes for MIMO-OFDM systems, in *Proceedings of the IEEE International Symposuim on Information Theory*, Adelaide, South Australia, pp. 2075–2079, September 2005.
18. F. Fazel and H. Jafarkhani, Quasi-orthogonal space-frequency and space-time-frequency block codes for MIMO OFDM channels, *IEEE Trans. Wireless Commun.*, 7(1), 184–192, January 2008.
19. W. Su, Z. Safar, and K.J.R. Liu, Towards maximum achievable diversity in space, time and frequency: Performance analysis and code design, *IEEE Trans. Wireless Commun.*, 4(4), 1847–1857, July 2005.
20. Z. Liu, Y. Xin, and G.B. Giannakis, Space-time-frequency coded OFDM over frequency-selective fading channels, *IEEE Trans. Signal Process.*, 50, 2465–2476, October 2002.
21. Y. Xin, Z. Wang, and G.B. Giannakis, Fellow, IEEE space-time diversity systems based on linear constellation precoding, *IEEE Trans. Wireless Commun.*, 2(2),294–309, March 2003.
22. W. Zhang, X.G. Xia, and P.C. Ching, High-rate full-diversity space-time-frequency codes for MIMO multipath block-fading channels, in *Proceedings of the IEEE Global Communication Conference (GLOBECOM2005)*, St. Louis, MO, 3, pp. 1587–1591, November 28–December 2, 2005.
23. W. Zhang, X.G. Xia, and P.C. Ching, High-rate full-diversity space-time-frequency codes for broadband MIMO block-fading channels, *IEEE Trans. Commun.*, 55, 25–34, January 2007.
24. V. Tarokh, N. Seshadri, and A.R. Calderbank, Space-time codes for high data rate wireless communication: Performance criterion and code construction, *IEEE Trans. Inform. Theory*, 44, 744–765, March 1998.
25. B. Lu and X. Wang, Space-time code design in OFDM systems, in *Proceedings of the IEEE Global Communication Conference*, San Francisco, CA, pp. 1000–1004, November 2000.
26. R.S. Blum, Y. Li, J.H. Winters, and Q. Yan, Improved space-time coding for MIMO-OFDM wireless communications, *IEEE Trans. Commun.*, 49, 1873–1878, November 2001.
27. Q. Yan and R.S. Blum, Optimum space–time convolutional codes, *Wireless Communication and Networking Conference*, Chicago, IL, 3, pp. 1351–1355, November 2000.

V

Radar Systems

Vijay K. Madisetti
Georgia Institute of Technology

32 **Radar Detection** *Bassem R. Mahafza and Atef Z. Elsherbeni* **32**-1
Detection in the Presence of Noise • Probability of False Alarm • Probability of Detection • Pulse Integration • Detection of Fluctuating Targets • Probability of Detection Calculation • The Radar Equation Revisited • Cumulative Probability of Detection • Constant False Alarm Rate • MATLAB® Program and Function Listings • References

33 **Radar Waveforms** *Bassem R. Mahafza and Atef Z. Elsherbeni* **33**-1
Low Pass, Band Pass Signals, and Quadrature Components • The Analytic Signal • CW and Pulsed Waveforms • Linear Frequency Modulation Waveforms • High Range Resolution • Stepped Frequency Waveforms • The Matched Filter • The Replica • Matched Filter Response to LFM Waveforms • Waveform Resolution and Ambiguity • "MyRadar" Design Case Study—Visit 3 • MATLAB® Program and Function Listings • Reference

34 **High Resolution Tactical Synthetic Aperture Radar** *Bassem R. Mahafza, Atef Z. Elsherbeni, and Brian J. Smith* **34**-1
Introduction • Side Looking SAR Geometry • SAR Design Considerations • SAR Radar Equation • SAR Signal Processing • Side Looking SAR Doppler Processing • SAR Imaging Using Doppler Processing • Range Walk • A Three-Dimensional SAR Imaging Technique • MATLAB® Programs and Functions • Reference

32

Radar Detection

32.1 Detection in the Presence of Noise 32-1
32.2 Probability of False Alarm 32-5
32.3 Probability of Detection 32-6
32.4 Pulse Integration 32-7
Coherent Integration • Noncoherent Integration • Mini Design Case Study 32.1
32.5 Detection of Fluctuating Targets 32-14
Threshold Selection
32.6 Probability of Detection Calculation 32-18
Detection of Swerling V Targets • Detection of Swerling I Targets • Detection of Swerling II Targets • Detection of Swerling III Targets • Detection of Swerling IV Targets
32.7 The Radar Equation Revisited 32-25
32.8 Cumulative Probability of Detection 32-27
Mini Design Case Study 32.2
32.9 Constant False Alarm Rate 32-29
Cell-Averaging CFAR (Single Pulse) • Cell-Averaging CFAR with Noncoherent Integration
32.10 MATLAB® Program and Function Listings 32-32
References 32-50

Bassem R. Mahafza
Deceibel Research, Inc.

Atef Z. Elsherbeni
University of Mississippi

32.1 Detection in the Presence of Noise

A simplified block diagram of a radar receiver that employs an envelope detector followed by a threshold decision is shown in Figure 32.1. The input signal to the receiver is composed of the radar echo signal $s(t)$ and additive zero mean white Gaussian noise $n(t)$, with variance ψ^2. The input noise is assumed to be spatially incoherent and uncorrelated with the signal.

The output of the bandpass IF filter is the signal $v(t)$, which can be written as

$$\begin{aligned} v(t) &= v_I(t)\cos\omega_0 t + v_Q(t)\sin\omega_0 t = r(t)\cos(\omega_0 t - \varphi(t)) \\ v_I(t) &= r(t)\cos\varphi(t) \\ v_Q(t) &= r(t)\sin\varphi(t) \end{aligned} \tag{32.1}$$

where

$\omega_0 = 2\pi f_0$ is the radar operating frequency

$r(t)$ is the envelope of $v(t)$

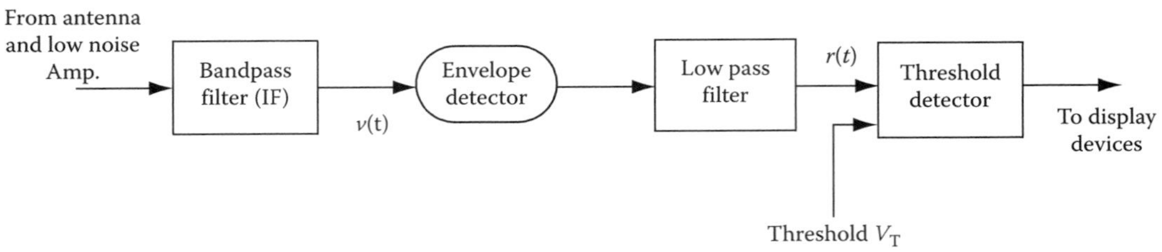

FIGURE 32.1 Simplified block diagram of an envelope detector and threshold receiver.

The phase is $\varphi(t) = \text{atan}(v_Q/v_I)$, and the subscripts I, Q, respectively, refer to the in-phase and quadrature components.

A target is detected when $r(t)$ exceeds the threshold value V_T, where the decision hypotheses are

$$\begin{aligned} &s(t) + n(t) > V_T \quad &\text{Detection} \\ &n(t) > V_T \quad &\text{False alarm} \end{aligned}$$

The case when the noise subtracts from the signal (while a target is present) to make $r(t)$ smaller than the threshold is called a miss. Radar designers seek to maximize the probability of detection for a given probability of false alarm.

The IF filter output is a complex random variable that is composed of either noise alone or noise plus target return signal (sine wave of amplitude A). The quadrature components corresponding to the first case are

$$\begin{aligned} v_I(t) &= n_I(t) \\ v_Q(t) &= n_Q(t) \end{aligned} \tag{32.2}$$

and for the second case,

$$\begin{aligned} v_I(t) &= A + n_I(t) = r(t)\cos\varphi(t) \quad \Rightarrow \quad n_I(t) = r(t)\cos\varphi(t) - A \\ v_Q(t) &= n_Q(t) = r(t)\sin\varphi(t) \end{aligned} \tag{32.3}$$

where the noise quadrature components $n_I(t)$ and $n_Q(t)$ are uncorrelated zero mean low pass Gaussian noise with equal variances, ψ^2. The joint probability density function (pdf) of the two random variables $n_I;n_Q$ is

$$\begin{aligned} f(n_I, n_Q) &= \frac{1}{2\pi\psi^2}\exp\left(-\frac{n_I^2 + n_Q^2}{2\psi^2}\right) \\ &= \frac{1}{2\pi\psi^2}\exp\left(-\frac{(r\cos\varphi - A)^2 + (r\sin\varphi)^2}{2\psi^2}\right) \end{aligned} \tag{32.4}$$

The pdfs of the random variables $r(t)$ and $\varphi(t)$, respectively, represent the modulus and phase of $v(t)$. The joint pdf for the two random variables $r(t);\varphi(t)$ is given by

$$f(r, \varphi) = f(n_I, n_Q)|J| \tag{32.5}$$

where $[J]$ is a matrix of derivatives defined by

$$[J] = \begin{bmatrix} \frac{\partial n_I}{\partial r} & \frac{\partial n_I}{\partial \varphi} \\ \frac{\partial n_Q}{\partial r} & \frac{\partial n_Q}{\partial \varphi} \end{bmatrix} = \begin{bmatrix} \cos\varphi - r\sin\varphi \\ \sin\varphi \; r\cos\varphi \end{bmatrix} \tag{32.6}$$

The determinant of the matrix of derivatives is called the Jacobian, and in this case it is equal to

$$|J| = r(t) \tag{32.7}$$

Substituting Equations 32.4 and 32.7 into Equation 32.5 and collecting terms yield

$$f(r,\varphi) = \frac{r}{2\pi\psi^2} \exp\left(-\frac{r^2 + A^2}{2\psi^2}\right) \exp\left(\frac{rA\cos\varphi}{\psi^2}\right) \tag{32.8}$$

The pdf for r alone is obtained by integrating Equation 32.8 over φ

$$f(r) = \int_0^{2\pi} f(r,\varphi)\mathrm{d}\varphi = \frac{r}{\psi^2} \exp\left(-\frac{r^2 + A^2}{2\psi^2}\right) \frac{1}{2\pi} \int_0^{2\pi} \exp\left(\frac{rA\cos\varphi}{\psi^2}\right) \mathrm{d}\varphi \tag{32.9}$$

where the integral inside Equation 32.9 is known as the modified Bessel function of zero order,

$$I_0(\beta) = \frac{1}{2\pi} \int_0^{2\pi} e^{\beta\cos\theta}\,\mathrm{d}\theta \tag{32.10}$$

Thus,

$$f(r) = \frac{r}{\psi^2} I_0\left(\frac{rA}{\psi^2}\right) \exp\left(-\frac{r^2 + A^2}{2\psi^2}\right) \tag{32.11}$$

which is the Rician pdf. If $A/\psi^2 = 0$ (noise alone), then Equation 32.11 becomes the Rayleigh pdf

$$f(r) = \frac{r}{\psi^2} \exp\left(-\frac{r^2}{2\psi^2}\right) \tag{32.12}$$

Also, when (A/ψ^2) is very large, Equation 32.11 becomes a Gaussian pdf of mean A and variance ψ^2:

$$f(r) \approx \frac{1}{\sqrt{2\pi\psi^2}} \exp\left(-\frac{(r - A)^2}{2\psi^2}\right) \tag{32.13}$$

Figure 32.2 shows plots for the Rayleigh and Gaussian densities. For this purpose, use MATLAB® program "*fig32_2.m*" given in Listing 32.1. This program uses MATLAB functions "*normpdf.m*" and "*raylpdf.m*". Both functions are part of the MATLAB Statistics toolbox. Their associated syntax is as follows

normpdf(x, mu, sigma)
raylpdf(x, sigma)

"*x*" is the variable, "*mu*" is the mean, and "*sigma*" is the standard deviation.

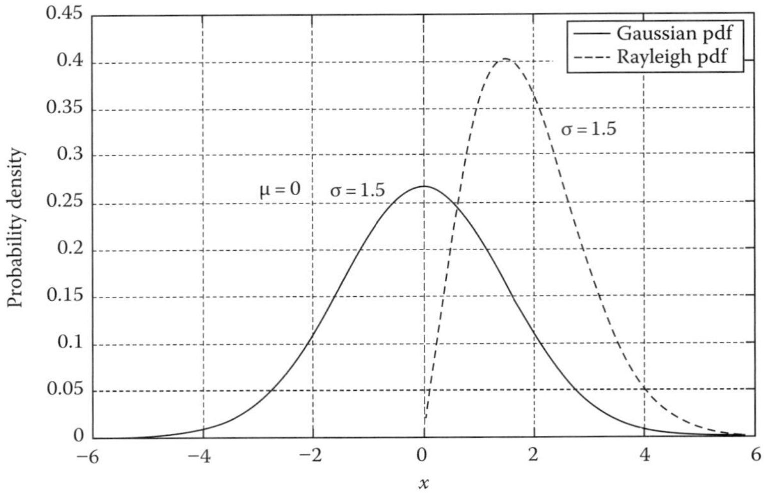

FIGURE 32.2 Gaussian and Rayleigh probability densities.

The density function for the random variable φ is obtained from

$$f(\varphi) = \int_0^r f(r, \varphi)\, dr \tag{32.14}$$

While the detailed derivation is left as an exercise, the result of Equation 32.14 is

$$f(\varphi) = \frac{1}{2} \exp\left(\frac{-A^2}{2\psi^2}\right) + \frac{A \cos \varphi}{\sqrt{2\pi\psi^2}} \exp\left(\frac{-(A \sin \varphi)^2}{2\psi^2}\right) F\left(\frac{A \cos \varphi}{\psi}\right) \tag{32.15}$$

where

$$F(x) = \int_{-\infty}^{x} \frac{1}{\sqrt{2\pi}} e^{-\xi^2/2}\, d\xi \tag{32.16}$$

The function $F(x)$ can be found tabulated in most mathematical formula reference books. Note that for the case of noise alone ($A = 0$), Equation 32.15 collapses to a uniform pdf over the interval $\{0, 2\pi\}$.

One excellent approximation for the function $F(x)$ is

$$F(x) = 1 - \left(\frac{1}{0.661x + 0.339\sqrt{x^2 + 5.51}}\right) \frac{1}{\sqrt{2\pi}} e^{-x^2/2} \quad x \geq 0 \tag{32.17}$$

and for negative values of x

$$F(-x) = 1 - F(x) \tag{32.18}$$

MATLAB Function "que_func.m"

The function "*que_func.m*" computes $F(x)$ using Equations 32.17 and 32.18 and is given in Listing 32.2. The syntax is as follows:

$$fofx = que\_func(x)$$

32.2 Probability of False Alarm

The probability of false alarm P_{fa} is defined as the probability that a sample R of the signal $r(t)$ will exceed the threshold voltage V_T when noise alone is present in the radar,

$$P_{fa} = \int_{V_T}^{\infty} \frac{r}{\psi^2} \exp\left(-\frac{r^2}{2\psi^2}\right) dr = \exp\left(\frac{-V_T^2}{2\psi^2}\right) \tag{32.19a}$$

$$V_T = \sqrt{2\psi^2 \ln\left(\frac{1}{P_{fa}}\right)} \tag{32.19b}$$

Figure 32.3 shows a plot of the normalized threshold versus the probability of false alarm. It is evident from this figure that P_{fa} is very sensitive to small changes in the threshold value. This figure can be reproduced using MATLAB program "*fig32_3.m*" given in Listing 32.3.

The false alarm time T_{fa} is related to the probability of false alarm by

$$T_{fa} = t_{int}/P_{fa} \tag{32.20}$$

where t_{int} represents the radar integration time, or the average time that the output of the envelope detector will pass the threshold voltage. Since the radar operating bandwidth B is the inverse of t_{int}, then by substituting Equation 32.19 into Equation 32.20 we can write T_{fa} as

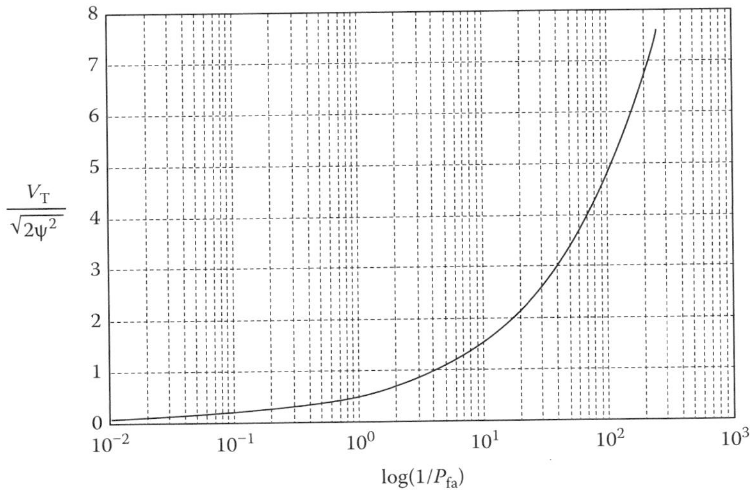

FIGURE 32.3 Normalized detection threshold versus probability of false alarm.

$$T_{fa} = \frac{1}{B} \exp\left(\frac{V_T^2}{2\psi^2}\right) \tag{32.21}$$

Minimizing T_{fa} means increasing the threshold value, and as a result the radar maximum detection range is decreased. Therefore, the choice of an acceptable value for T_{fa} becomes a compromise depending on the radar mode of operation.

Fehlner[1] defines the false alarm number as

$$n_{fa} = \frac{-\ln(2)}{\ln(1 - P_{fa})} \approx \frac{\ln(2)}{P_{fa}} \tag{32.22}$$

Other slightly different definitions for the false alarm number exist in the literature, causing a source of confusion for many nonexpert readers. Other than the definition in Equation 32.22, the most commonly used definition for the false alarm number is the one introduced by Marcum.[4] Marcum defines the false alarm number as the reciprocal of P_{fa}. In this text, the definition given in Equation 32.22 is always assumed. Hence, a clear distinction is made between Marcum's definition of the false alarm number and the definition in Equation 32.22.

32.3 Probability of Detection

The probability of detection P_D is the probability that a sample R of $r(t)$ will exceed the threshold voltage in the case of noise plus signal:

$$P_D = \int_{V_T}^{\infty} \frac{r}{\psi^2} I_0\left(\frac{rA}{\psi^2}\right) \exp\left(\frac{r^2 + A^2}{2\psi^2}\right) dr \tag{32.23}$$

If we assume that the radar signal is a sine waveform with amplitude A, then its power is $A/2$. Now, by using $\mathrm{SNR} = A^2/2\psi^2$ (single-pulse SNR) and $\left(V_T^2/2\psi^2\right) = \ln(1/P_{fa})$, then Equation 32.23 can be rewritten as

$$P_D = \int_{\sqrt{2\psi^2 \ln(1/P_{fa})}}^{\infty} \frac{r}{\psi^2} I_0\left(\frac{rA}{\psi^2}\right) \exp\left(-\frac{r^2 + A^2}{2\psi^2}\right) dr = Q\left[\sqrt{\frac{A^2}{\psi^2}}, \sqrt{2 \ln\left(\frac{1}{P_{fa}}\right)}\right] \tag{32.24}$$

$$Q[\alpha, \beta] = \int_{\beta}^{\infty} \zeta I_0(\alpha\zeta) e^{-(\zeta^2 + \alpha^2)/2} \, d\zeta \tag{32.25}$$

Q is called Marcum's Q-function. When P_{fa} is small and P_D is relatively large so that the threshold is also large, Equation 32.24 can be approximated by

$$P_D \approx F\left(\frac{A}{\psi} - \sqrt{2 \ln\left(\frac{1}{P_{fa}}\right)}\right) \tag{32.26}$$

where $F(x)$ is given by Equation 32.16. Many approximations for computing Equation 32.24 can be found throughout the literature. One very accurate approximation presented by North (see References) is given by

$$P_D \approx 0.5 \times erfc\left(\sqrt{-\ln P_{fa}} - \sqrt{SNR + 0.5}\right) \tag{32.27}$$

where the complementary error function is

$$erfc(z) = 1 - \frac{2}{\sqrt{\pi}} \int_0^z e^{-v^2} dv \tag{32.28}$$

MATLAB Function "marcumsq.m"

The integral given in Equation 32.24 is complicated and can be computed using numerical integration techniques. Pari[2] developed an excellent algorithm to numerically compute this integral. It is summarized as follows:

$$Q[a,b] = \left\{ \begin{array}{ll} \frac{\alpha_n}{2\beta_n} \exp\left(\frac{(a-b)^2}{2}\right) & a < b \\ 1 - \left(\frac{\alpha_n}{2\beta_n} \exp\left(\frac{(a-b)^2}{2}\right) \right. & \left. a \geq b \right) \end{array} \right\} \tag{32.29}$$

$$\alpha_n = d_n + \frac{2n}{ab}\alpha_{n-1} + \alpha_{n-2} \tag{32.30}$$

$$\beta_n = 1 + \frac{2n}{ab}\beta_{n-1} + \beta_{n-2} \tag{32.31}$$

$$d_{n+1} = d_n d_1 \tag{32.32}$$

$$\alpha_0 = \left\{ \begin{array}{ll} 1 & a < b \\ 0 & a \geq b \end{array} \right\} \tag{32.33}$$

$$d_1 = \left\{ \begin{array}{ll} a/b & a < b \\ b/a & a \geq b \end{array} \right\} \tag{32.34}$$

$\alpha_{-1} = 0.0$, $\beta_0 = 0.5$, and $\beta_{-1} = 0$. The recursive Equations 32.30 through 32.32 are computed continuously until $\beta_n > 10^p$ for values of $p \geq 3$. The accuracy of the algorithm is enhanced as the value of p is increased. The MATLAB function "*marcumsq.m*" given in Listing 32.4 implements Part's algorithm to calculate the probability of detection defined in Equation 32.24. The syntax is as follows:

$$Pd = marcumsq\ (alpha,\ beta)$$

where *alpha* and *beta* are from Equation 32.25. Figure 32.4 shows plots of the probability of detection, P_D, versus the single pulse SNR, with the P_{fa} as a parameter. This figure can be reproduced using the MATLAB program "*prob_snr1.m*" given in Listing 32.5.

32.4 Pulse Integration

In this section a more comprehensive analysis of this topic is introduced in the context of radar detection. The overall principles and conclusions presented earlier will not change; however, the mathematical formulation and specific numerical values will change. Coherent integration preserves the phase relationship between the received pulses, thus achieving a build up in the signal amplitude. Alternatively, pulse integration performed after the envelope detector (where the phase relation is destroyed) is called noncoherent or postdetection integration.

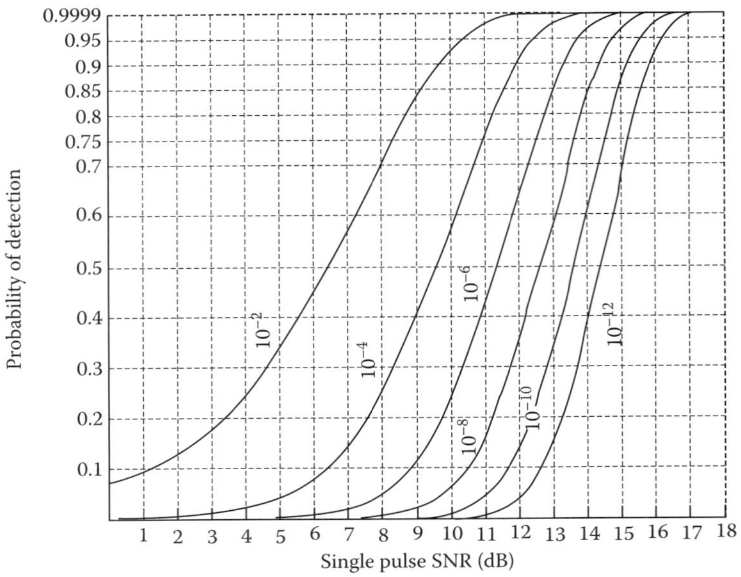

FIGURE 32.4 Probability of detection versus single pulse SNR, for several values of P_{fa}.

32.4.1 Coherent Integration

In coherent integration, if a perfect integrator is used (100% efficiency), then integrating n_P pulses would improve the SNR by the same factor. Otherwise, integration loss occurs which is always the case for noncoherent integration. In order to demonstrate this signal buildup, consider the case where the radar return signal contains both signal plus additive noise. The mth pulse is

$$y_m(t) = s(t) + n_m(t) \tag{32.35}$$

where

$s(t)$ is the radar return of interest
$n_m(t)$ is white uncorrelated additive noise signal

Coherent integration of n_P pulses yields

$$z(t) = \frac{1}{n_P}\sum_{m=1}^{n_P} y_m(t) = \sum_{m=1}^{n_P}\frac{1}{n_P}[s(t) + n_m(t)] = s(t) + \sum_{m=1}^{n_P}\frac{1}{n_P}n_m(t) \tag{32.36}$$

The total noise power in $z(t)$ is equal to the variance. More precisely,

$$\psi_{nz}^2 = E\left[\left(\sum_{m=1}^{n_P}\frac{1}{n_P}n_m(t)\right)\left(\sum_{l=1}^{n_P}\frac{1}{n_P}n_l(t)\right)^*\right] \tag{32.37}$$

where $E[\ ]$ is the expected value operator. It follows that

$$\psi_{nz}^2 = \frac{1}{n_P^2}\sum_{m,l=1}^{n_P} E[n_m(t)n_l^*(t)] = \frac{1}{n_P^2}\sum_{m,l=1}^{n_P}\psi_{ny}^2\delta_{ml} = \frac{1}{n_P}\psi_{ny}^2 \tag{32.38}$$

where

ψ_{ny}^2 is the single pulse noise power

δ_{ml} is equal to zero for $m \neq l$ and unity for $m = l$

Observation of Equations 32.36 and 32.38 shows that the desired signal power after coherent integration is unchanged, while the noise power is reduced by the factor $1/n_P$. Thus, the SNR after coherent integration is improved by n_P.

Denote the single pulse SNR required to produce a given probability of detection as $(\text{SNR})_1$. Also, denote $(\text{SNR})_{n_P}$ as the SNR required to produce the same probability of detection when n_P pulses are integrated. It follows that

$$(\text{SNR})_{n_P} = \frac{1}{n_P}(\text{SNR})_1 \tag{32.39}$$

The requirements of knowing the exact phase of each transmitted pulse as well as maintaining coherency during propagation is very costly and challenging to achieve. Thus, radar systems would not utilize coherent integration during search mode, since target microdynamics may not be available.

32.4.2 Noncoherent Integration

Noncoherent integration is often implemented after the envelope detector, also known as the quadratic detector. A block diagram of radar receiver utilizing a square law detector and noncoherent integration is illustrated in Figure 32.5. In practice, the square law detector is normally used as an approximation to the optimum receiver.

The pdf for the signal $r(t)$ was derived earlier and it is given in Equation 32.11. Define a new dimensionless variable y as

$$y_n = \frac{r_n}{\psi} \tag{32.40}$$

and also define

$$\Re_P = \frac{A^2}{\psi^2} = 2\,\text{SNR} \tag{32.41}$$

It follows that the pdf for the new variable is then given by

$$f(y_n) = f(r_n)\left|\frac{dr_n}{dy_n}\right| = y_n I_0\left(y_n\sqrt{\Re_P}\right)\exp\left(\frac{-(y_n^2 + \Re_P)}{2}\right) \tag{32.42}$$

The output of a square law detector for the nth pulse is proportional to the square of its input, which, after the change of variable in Equation 32.40, is proportional to y_n. Thus, it is convenient to define a new change variable,

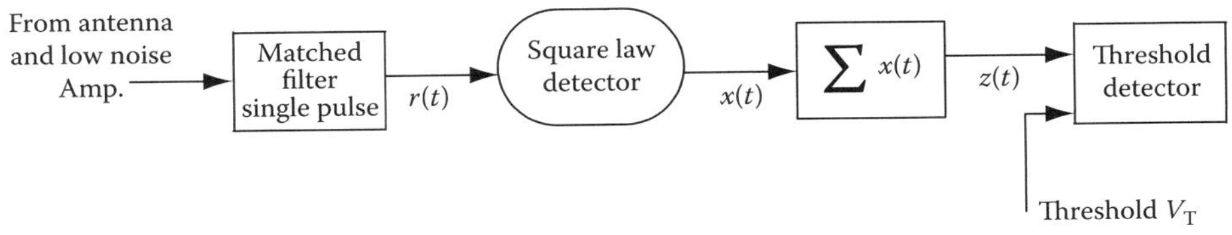

FIGURE 32.5 Simplified block diagram of a square law detector and noncoherent integration.

$$x_n = \frac{1}{2} y_n^2 \tag{32.43}$$

The pdf for the variable at the output of the square law detector is given by

$$f(x_n) = f(y_n)\left|\frac{dy_n}{dx_n}\right| = \exp\left(-\left(x_n + \frac{\Re_P}{2}\right)\right) I_0\left(\sqrt{2x_n\Re_p}\right) \tag{32.44}$$

Noncoherent integration of n_P pulses is implemented as

$$z = \sum_{n=1}^{n_P} x_n \tag{32.45}$$

Since the random variables x_n are independent, the pdf for the variable z is

$$f(z) = f(x_1) \bullet f(x_2) \bullet \cdots \bullet f(x_{n_P}) \tag{32.46}$$

The operator $\bullet$ symbolically indicates convolution. The characteristic functions for the individual pdfs can then be used to compute the joint pdf in Equation 32.46. The details of this development are left as an exercise. The result is

$$f(z) = \left(\frac{2z}{n_P\Re_p}\right)^{(n_P-1)/2} \exp\left(-z - \frac{1}{2}n_P\Re_p\right) I_{n_P-1}\left(\sqrt{2n_P z\Re_p}\right) \tag{32.47}$$

I_{n_p-1} is the modified Bessel function of order $n_P - 1$. Therefore, the probability of detection is obtained by integrating $f(z)$ from the threshold value to infinity. Alternatively, the probability of false alarm is obtained by letting $\Re_p$ be zero and integrating the pdf from the threshold value to infinity. Closed form solutions to these integrals are not easily available. Therefore, numerical techniques are often utilized to generate tables for the probability of detection.

32.4.2.1 Improvement Factor and Integration Loss

Denote the SNR that is required to achieve a specific P_D given a particular P_{fa} when n_P pulses are integrated noncoherently by $(SNR)_{NCI}$. And thus, the single pulse SNR $(SNR)_1$, is less than $(SNR)_{NCI}$. More precisely,

$$(SNR)_{NCI} = (SNR)_1 \times I(n_P) \tag{32.48}$$

where $I(n_P)$ is called the integration improvement factor. An empirically derived expression for the improvement factor that is accurate within 0.8 dB is reported in Peebles[3] as

$$\begin{aligned}[I(n_P)]_{dB} = 6.79(1 + 0.235P_D)\left(1 + \frac{\log(1/P_{fa})}{46.6}\right)\log(n_P) \\ (1 - 0.140\log(n_P) + 0.018310(\log n_P)^2)\end{aligned} \tag{32.49}$$

Figure 32.6a shows plots of the integration improvement factor as a function of the number of integrated pulses with P_D and P_{fa} as parameters, using Equation 32.49. This plot can be reproduced using the

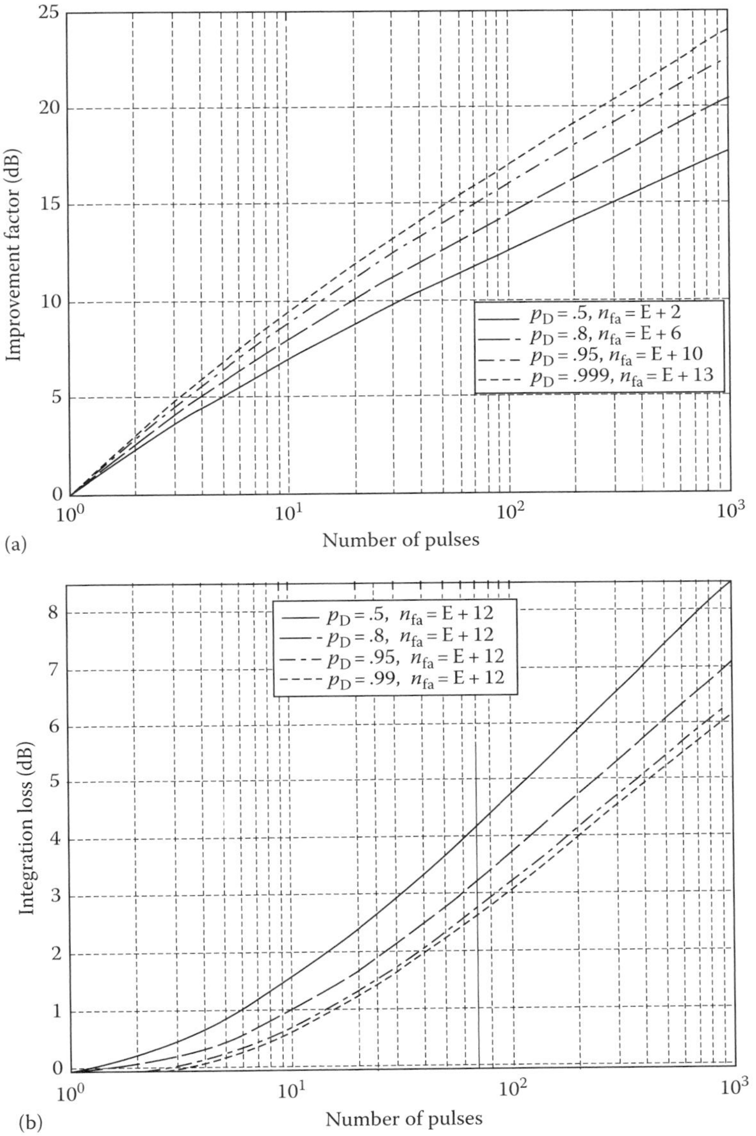

FIGURE 32.6 (a) Improvement factor versus number of noncoherently integrated pulses. (b) Integration loss versus number of noncoherently integrated pulses.

MATLAB program "*fig32_6a.m*" given in Listing 32.6. Note this program uses the MATLAB function "*improv_fac.m*," which is given in Listing 32.7.

MATLAB Function "improv_fac.m"

The function "*improv_fac.m*" calculates the improvement factor using Equation 32.49. It is given in Listing 32.7. The syntax is as follows:

$$[impr\_of\_np] = improv\_fac(np, pfa, pd)$$

where

Symbol	Description	Units	Status
np	Number of integrated pulses	None	Input
pfa	Probability of false alarm	None	Input
pd	Probability of detection	None	Input
impr_of_np	Improvement factor	Output	dB

The integration loss is defined as

$$L_{NCI} = n_P / I(n_P) \tag{32.50}$$

Figure 32.6b shows a plot of the integration loss versus n_P. This figure can be reproduced using MATLAB program "*fig32_6b.m*" given in Listing 32.8. It follows that, when noncoherent integration is utilized, the corresponding SNR required to achieve a certain P_D given a specific P_{fa} is now given by

$$(\mathrm{SNR})_{NCI} = (n_P \times (\mathrm{SNR})_1)/L_{NCI} \tag{32.51}$$

which is very similar to Equation 1.86 derived in Chapter 1.

32.4.3 Mini Design Case Study 32.1

An L-band radar has the following specifications: operating frequency $f_0 = 1.5$ GHz, operating bandwidth $B = 2$ MHz, noise figure $F = 8$ dB, system losses $L = 4$ dB, time of false alarm $T_{fa} = 12$ min, detection range $R = 12$ km, the minimum required SNR is SNR = 13.85 dB, antenna gain $G = 5000$, and target RCS $\sigma = 1\ \mathrm{m}^2$. (a) Determine the PRF f_r, the pulsewidth τ, the peak power P_t, the probability of false alarm P_{fa}, the corresponding P_D, and the minimum detectable signal level S_{min}. (b) How can you reduce the transmitter power to achieve the same performance when 10 pulses are integrated noncoherently? (c) If the radar operates at a shorter range in the single pulse mode, find the new probability of detection when the range decreases to 9 km.

32.4.3.1 A Solution

Assume that the maximum detection corresponds to the unambiguous range. From that the PRF is computed as

$$f_r = \frac{c}{2R_u} = \frac{3 \times 10^8}{2 \times 12{,}000} = 12.5\ \mathrm{kHz}$$

The pulsewidth is proportional to the inverse of the bandwidth,

$$\tau = \frac{1}{B} = \frac{1}{2 \times 10^6} = 0.5\ \mu\mathrm{s}$$

The probability of false alarm is

$$P_{fa} = \frac{1}{BT_{fa}} = \frac{1}{2 \times 10^6 \times 12 \times 60} = 6.94 \times 10^{-10}$$

It follows that by using MATLAB function "*marcumsq.m*" the probability of detection is calculated from

$$Q\left[\sqrt{\frac{A^2}{\psi^2}}, \sqrt{2\ \ln\left(\frac{1}{P_{fa}}\right)}\right]$$

with the following syntax

$$marcumsq(alpha,\ beta)$$

where

$$alpha = \sqrt{2} \times \sqrt{10^{13.85/10}} = 6.9665$$

$$beta = \sqrt{2\ \ln\left(\frac{1}{6.94 \times 10^{-10}}\right)} = 6.494$$

Remember that $(A^2/\psi^2) = 2\text{SNR}$. Thus, the detection probability is

$$P_D = marcumsq(6.9665,\ 6.944) = 0.508$$

Using the radar equation one can calculate the radar peak power. More precisely,

$$P_t = \text{SNR}\frac{(4\pi)^3 R^4 k T_0 BFL}{G^2 \lambda^2 \sigma} \Rightarrow$$

$$P_t = 10^{1.385}\frac{(4\pi)^3 \times 12{,}000^4 \times 1.38 \times 10^{-23} \times 290 \times 2 \times 10^6 \times 6.309 \times 2.511}{5000^2 \times 0.2^2 \times 1}$$

$$= 126.61\ \text{W}$$

And the minimum detectable signal is

$$S_{min} = \frac{P_t G^2 \lambda^2 \sigma}{(4\pi)^3 R^4 L} = \frac{126.61 \times 5000^2 \times 0.2^2 \times 1}{(4\pi)^3 \times 12{,}000^4 \times 2.511} = 1.2254 \times 10^{-12}\ \text{V}$$

when 10 pulses are integrated noncoherently, the corresponding improvement factor is calculated from the MATLAB function "*improv_fac.m*" using the following syntax

$$improv\_fac(10{,}1e\text{-}11{,}0.5)$$

which yields $I(10) = 6 \Rightarrow 7.78$ dB. Consequently, by keeping the probability of detection the same (with and without integration) the SNR can be reduced by a factor of almost 6 dB (13.85–7.78). The integration loss associated with a 10-pulse noncoherent integration is calculated from Equation 32.50 as

$$L_{NCI} = \frac{n_P}{I(10)} = \frac{10}{6} = 1.67 \Rightarrow 2.2\ \text{dB}$$

Thus the net single pulse SNR with 10-pulse noncoherent integration is

$$(\text{SNR})_{NCI} = 13.85 - 7.78 + 2.2 = 8.27\ \text{dB}$$

Finally, the improvement in the SNR due to decreasing the detection range to 9 km is

$$(\mathrm{SNR})_{9\,\mathrm{km}} = 10\log\left(\frac{12{,}000}{9000}\right)^4 + 13.85 = 18.85\,\mathrm{dB}$$

32.5 Detection of Fluctuating Targets

So far the probability of detection calculations assumed a constant target cross section (nonfluctuating target). This work was first analyzed by Marcum.[4] Swerling[5] extended Marcum's work to four distinct cases that account for variations in the target cross section. These cases have come to be known as Swerling models. They are: Swerling I, Swerling II, Swerling III, and Swerling IV. The constant RCS case analyzed by Marcum is widely known as Swerling 0 or equivalently Swerling V. Target fluctuation lowers the probability of detection, or equivalently reduces the SNR.

Swerling I targets have constant amplitude over one antenna scan; however, a Swerling I target amplitude varies independently from scan to scan according to a Chi-square pdf with two degrees of freedom. The amplitude of Swerling II targets fluctuates independently from pulse to pulse according to a Chi-square pdf with two degrees of freedom. Target fluctuation associated with a Swerling III model is similar to Swerling I, except in this case the target power fluctuates independently from pulse to pulse according to a Chi-square pdf with four degrees of freedom. Finally, the fluctuation of Swerling IV targets is from pulse to pulse according to a Chi-square pdf with four degrees of freedom. Swerling showed that the statistics associated with Swerling I and II models apply to targets consisting of many small scatterers of comparable RCS values, while the statistics associated with Swerling III and IV models apply to targets consisting of one large RCS scatterer and many small equal RCS scatterers. Noncoherent integration can be applied to all four Swerling models; however, coherent integration cannot be used when the target fluctuation is either Swerling II or Swerling IV. This is because the target amplitude decorrelates from pulse to pulse (fast fluctuation) for Swerling II and IV models, and thus phase coherency cannot be maintained.

The Chi-square pdf with $2N$ degrees of freedom can be written as

$$f(\sigma) = \frac{N}{(N-1)!\bar{\sigma}}\left(\frac{N\sigma}{\bar{\sigma}}\right)^{N-1}\exp\left(-\frac{N\sigma}{\bar{\sigma}}\right) \tag{32.52}$$

where $\bar{\sigma}$ is the average RCS value. Using this equation, the pdf associated with Swerling I and II targets can be obtained by letting $N=1$, which yields a Rayleigh pdf more precisely,

$$f(\sigma) = \frac{1}{\bar{\sigma}}\exp\left(-\frac{\sigma}{\bar{\sigma}}\right) \quad \sigma \geq 0 \tag{32.53}$$

Letting $N=2$ yields the pdf for Swerling III and IV type targets,

$$f(\sigma) = \frac{4\sigma}{\bar{\sigma}^2}\exp\left(-\frac{2\sigma}{\bar{\sigma}}\right) \quad \sigma \geq 0 \tag{32.54}$$

The probability of detection for a fluctuating target is computed in a similar fashion to Equation 32.23, except in this case $f(r)$ is replaced by the conditional pdf $f(r/\sigma)$. Performing the analysis for the general case (i.e., using Equation 32.47) yields

$$f(z/\sigma) = \left(\frac{2z}{n_P\sigma^2/\psi^2}\right)^{(n_P-1)/2}\exp\left(-z-\frac{1}{2}n_P\frac{\sigma^2}{\psi^2}\right)I_{n_P-1}\left(\sqrt{2n_P z\frac{\sigma^2}{\psi^2}}\right) \tag{32.55}$$

To obtain $f(z)$ use the relations

$$f(z,\sigma) = f(z/\sigma)f(\sigma) \tag{32.56}$$

$$f(z) = \int f(z,\sigma)\,\mathrm{d}\sigma \tag{32.57}$$

Finally, using Equation 32.56 in Equation 32.57 produces

$$f(z) = \int f(z/\sigma)f(\sigma)\,\mathrm{d}\sigma \tag{32.58}$$

where

$f(z/\sigma)$ is defined in Equation 32.55
$f(\sigma)$ is in either Equation 32.53 or 32.54

The probability of detection is obtained by integrating the pdf derived from Equation 32.58 from the threshold value to infinity. Performing the integration in Equation 32.58 leads to the incomplete Gamma function.

32.5.1 Threshold Selection

When only a single pulse is used, the detection threshold V_{T} is related to the probability of false alarm P_{fa} as defined in Equation 32.19. DiFranco and Rubin[6] derived a general form relating the threshold and P_{fa} for any number of pulses when noncoherent integration is used. It is

$$P_{\mathrm{fa}} = 1 - \Gamma_{\mathrm{I}}\left(\frac{V_{\mathrm{T}}}{\sqrt{n_{\mathrm{P}}}}, n_{\mathrm{P}} - 1\right) \tag{32.59}$$

where Γ_{I} is used to denote the incomplete Gamma function, and is defined by DiFranco and Rubin as

$$\Gamma_{\mathrm{I}}(u,s) = \int_0^{u\sqrt{s+1}} \frac{\mathrm{e}^{-\gamma}\gamma^s}{s!}\,\mathrm{d}\gamma \tag{32.60}$$

Note that the limiting values for the incomplete Gamma function are

$$\Gamma_{\mathrm{I}}(0,N) = 0 \quad \Gamma_{\mathrm{I}}(\infty,N) = 1 \tag{32.61}$$

For our purposes, the incomplete Gamma function can be approximated by

$$\Gamma_{\mathrm{I}}\left(\frac{V_{\mathrm{T}}}{\sqrt{n_{\mathrm{P}}}}, n_{\mathrm{P}} - 1\right) = 1 - \frac{V_{\mathrm{T}}^{n_{\mathrm{P}}-1}\mathrm{e}^{-V_{\mathrm{T}}}}{(n_{\mathrm{P}}-1)!}\left[1 + \frac{n_{\mathrm{P}}-1}{V_{\mathrm{T}}} + \frac{(n_{\mathrm{P}}-1)(n_{\mathrm{P}}-1)}{V_{\mathrm{T}}^2} + \cdots + \frac{(n_{\mathrm{P}}-1)!}{V_{\mathrm{T}}^{n_{\mathrm{P}}-1}}\right] \tag{32.62}$$

The threshold value V_{T} can then be approximated by the recursive formula used in the Newton–Raphson method. More precisely,

$$V_{\mathrm{T},m} = V_{\mathrm{T},m-1} - \frac{G(V_{\mathrm{T},m-1})}{G'(V_{\mathrm{T},m-1})}; \quad m = 1,2,3,\ldots \tag{32.63}$$

The iteration is terminated when $|V_{T,m} - V_{T,m-1}| < V_{T,m-1}/10000.0$. The functions G and G' are

$$G(V_{T,m}) = (0.5)^{n_P/n_{fa}} - \Gamma_I(V_T, n_P) \tag{32.64}$$

$$G'(V_{T,m}) = -\frac{e^{-V_T} V_T^{n_P-1}}{(n_P-1)!} \tag{32.65}$$

The initial value for the recursion is

$$V_{T,0} = n_P - \sqrt{n_P} + 2.3\sqrt{-\log P_{fa}}\left(\sqrt{-\log P_{fa}} + \sqrt{n_P} - 1\right) \tag{32.66}$$

MATLAB Function "incomplete_gamma.m"

In general, the incomplete Gamma function for some integer N is

$$\Gamma_I(x, N) = \int_0^x \frac{e^{-v} v^{N-1}}{(N-1)!} \, dv \tag{32.67}$$

The function "*incomplete_gamma.m*" implements Equation 32.67. It is given in Listing 32.9. Note that this function uses the MATLAB function "*factor.m*" which is given in Listing 32.10. The function "*factor.m*" calculates the factorial of an integer. Figure 32.7 shows the incomplete Gamma function for $N = 1, 3, 6, 10$. This figure can be reproduced using the MATLAB program "*fig32_7.m*" given in Listing 32.11. The syntax for this function is as follows:

$$[value] = incomplete\_gamma(x, N)$$

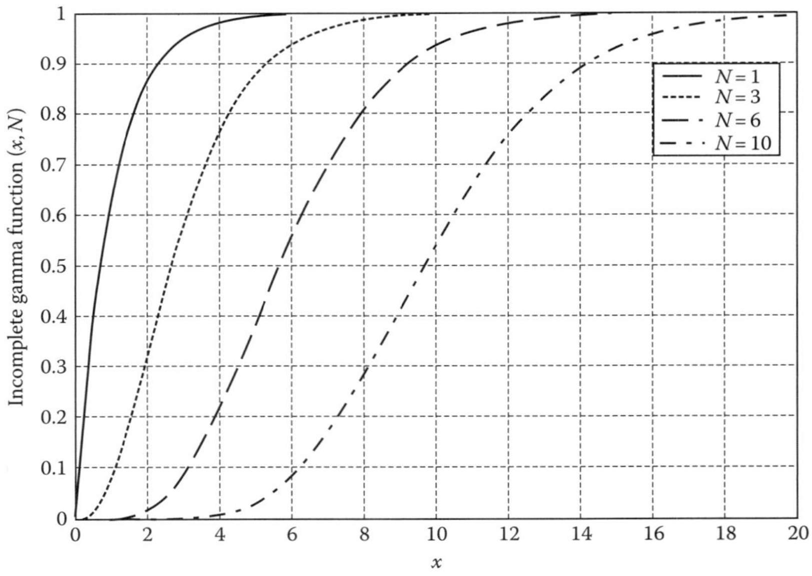

FIGURE 32.7 The incomplete Gamma function for four values of N.

where

Symbol	Description	Units	Status
X	Variable input to $\Gamma_1(x, N)$	Units of x	Input
N	Variable input to $\Gamma_1(x, N)$	None/integer	Input
value	$\Gamma_1(x, N)$	None	Output

MATLAB Function "threshold.m"

The function "*threshold.m*" calculates the threshold using the recursive formula used in the Newton–Raphson method. It is given in Listing 32.12. The syntax is as follows:

$$[pfa, vt] = threshold(nfa, np)$$

where

Symbol	Description	Units	Status
nfa	Marcum's false alarm number	None	Input
np	Number of integrated pulses	None	Input
pfa	Probability of false alarm	None	Output
vt	Threshold value	None	Output

Figure 32.8 shows plots of the threshold value versus the number of integrated pulses for several values of n_{fa}; remember that $P_{fa} \approx \ln(2)/n_{fa}$. This figure can be reproduced using MATLAB program "*fig32_8.m*" given in Listing 32.13. This program uses both "*threshold.m*" and "*incomplete_gamma*."

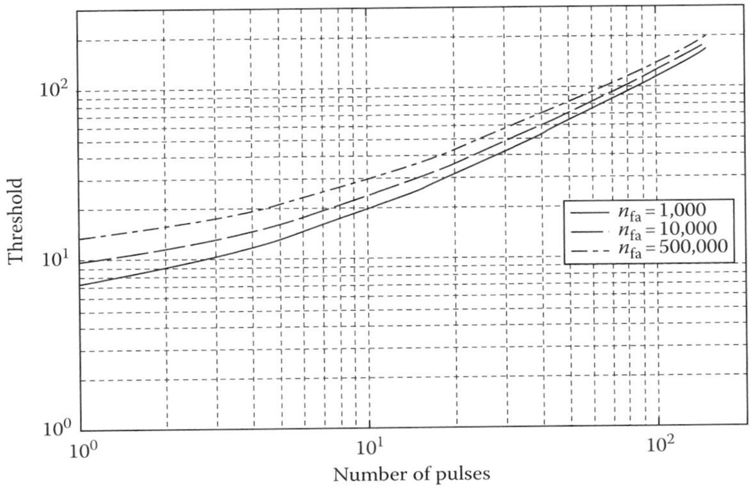

FIGURE 32.8 Threshold V_T versus n for several values of n_{fa}.

32.6 Probability of Detection Calculation

Marcum defined the probability of false alarm for the case when $n_P > 1$ as

$$P_{fa} \approx \ln(2)(n_P/n_{fa}) \tag{32.68}$$

The single pulse probability of detection for nonfluctuating targets is given in Equation 32.24. When $n_P > 1$, the probability of detection is computed using the Gram-Charlier series. In this case, the probability of detection is

$$P_D \cong \frac{erfc(V/\sqrt{2})}{2} - \frac{e^{-V^2/2}}{\sqrt{2\pi}}\left[C_3(V^2-1) + C_4V(3-V^2) - C_6V(V^4 - 10V^2 + 15)\right] \tag{32.69}$$

where the constants C_3, C_4, and C_6 are the Gram-Charlier series coefficients, and the variable V is

$$V = \frac{V_T - n_P(1 + \mathrm{SNR})}{\bar{\omega}} \tag{32.70}$$

In general, values for C_3, C_4, C_6, and $\bar{\omega}$ vary depending on the target fluctuation type.

32.6.1 Detection of Swerling V Targets

For Swerling V (Swerling 0) target fluctuations, the probability of detection is calculated using Equation 32.69. In this case, the Gram-Charlier series coefficients are

$$C_3 = -\frac{\mathrm{SNR} + 1/3}{\sqrt{n_P}(2\,\mathrm{SNR} + 1)^{1.5}} \tag{32.71}$$

$$C_4 = \frac{\mathrm{SNR} + 1/4}{n_P(2\,\mathrm{SNR} + 1)^2} \tag{32.72}$$

$$C_6 = C_3^2/2 \tag{32.73}$$

$$\bar{\omega} = \sqrt{n_P(2\,\mathrm{SNR} + 1)} \tag{32.74}$$

MATLAB Function "pd_swerling5.m"

The function "*pd_swerling5.m*" calculates the probability of detection for Swerling V targets. It is given in Listing 32.14. The syntax is as follows:

$$[pd] = pd\_swerling5\ (input\ 1, indicator, np, snr)$$

where

Symbol	Description	Units	Status
input1	P_{fa}, or n_{fa}	None	Input
indicator	When *input1* $= P_{fa}$	None	Input
	When *input1* $= n_{fa}$		
np	Number of integrated pulses	None	Input
snr	SNR	dB	Input
pd	Probability of detection	None	Output

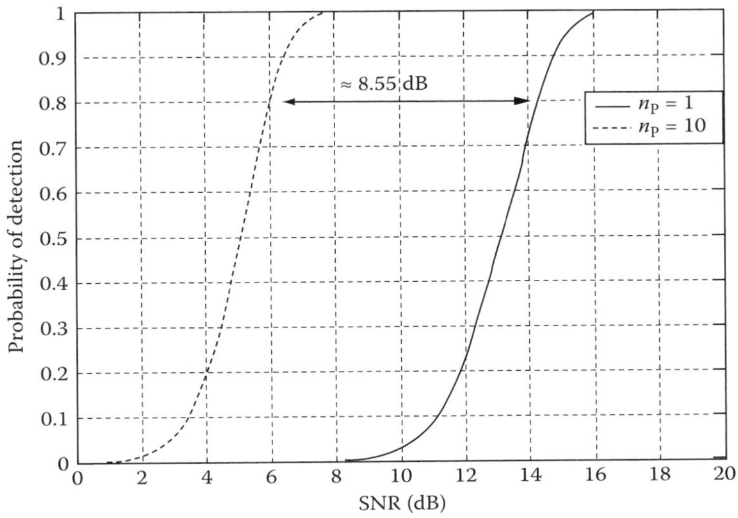

FIGURE 32.9 Probability of detection versus SNR, $P_{fa} = 10^{-9}$, and noncoherent integration.

Figure 32.9 shows a plot for the probability of detection versus SNR for cases $n_P = 1, 10$. This figure can be reproduced using the MATLAB program "*fig32_9.m*." It is given in Listing 32.15.

Note that it requires less SNR, with 10 pulses integrated noncoherently, to achieve the same probability of detection as in the case of a single pulse. Hence, for any given P_D the SNR improvement can be read from the plot. Equivalently, using the function "*improv_fac.m*" leads to about the same result. For example, when $P_D = 0.8$ the function "*improv_fac.m*" gives an SNR improvement factor of $I(10) \approx 8.55$ dB. Figure 32.9 shows that the 10 pulse SNR is about 6.03 dB. Therefore, the single pulse SNR is about (from Equation 32.49) 14.5 dB, which can be read from the figure.

32.6.2 Detection of Swerling I Targets

The exact formula for the probability of detection for Swerling I type targets was derived by Swerling. It is

$$P_D = e^{-V_T/(1+\mathrm{SNR})}; \quad n_P = 1 \tag{32.75}$$

$$P_D = 1 - \Gamma_I(V_T, n_P - 1) + \left(1 + \frac{1}{n_P\mathrm{SNR}}\right)^{n_P - 1} \Gamma_I\left(\frac{V_T}{1 + \frac{1}{n_P\mathrm{SNR}}}, n_P - 1\right) \times e^{-V_T/(1+n_P\mathrm{SNR})}; \quad n_P > 1 \tag{32.76}$$

MATLAB Function "pd_swerlingl.m"

The function "*pd_swerling1.m*" calculates the probability of detection for Swerling I type targets. It is given in Listing 32.16. The syntax is as follows:

$$[pd] = pd\_swerling1(nfa, np, snr)$$

where

Symbol	Description	Units	Status
nfa	Marcum's false alarm number	None	Input
np	Number of integrated pulses	None	Input
snr	SNR	dB	Input
pd	Probability of detection	None	Output

Figure 32.10 shows a plot of the probability of detection as a function of SNR for $n_P = 1$ and $P_{fa} = 10^{-9}$ for both Swerling I and V type fluctuations. Note that it requires more SNR, with fluctuation, to achieve the same P_D as in the case with no fluctuation. This figure can be reproduced using MATLAB program "*fig32_10.m*" given in Listing 32.17.

Figure 32.11a shows a plot of the probability of detection versus SNR for $n_P = 1$, 10, 50, 100, where $P_{fa} = 10^{-8}$. Figure 32.11b is similar to Figure 32.11a; in this case $P_{fa} = 10^{-11}$. These figures can be reproduced using MATLAB program "*fig32_11ab.m*" given in Listing 32.18.

32.6.3 Detection of Swerling II Targets

In the case of Swerling II targets, the probability of detection is given by

$$P_D = 1 - \Gamma_I\left(\frac{V_T}{(1 + \text{SNR})}, n_P\right); \quad n_P \leq 50 \tag{32.77}$$

For the case when $n_P > 50$ Equation 32.69 is used to compute the probability of detection. In this case,

$$C_3 = -\frac{1}{3\sqrt{n_P}}, \quad C_6 = \frac{C_3^2}{2} \tag{32.78}$$

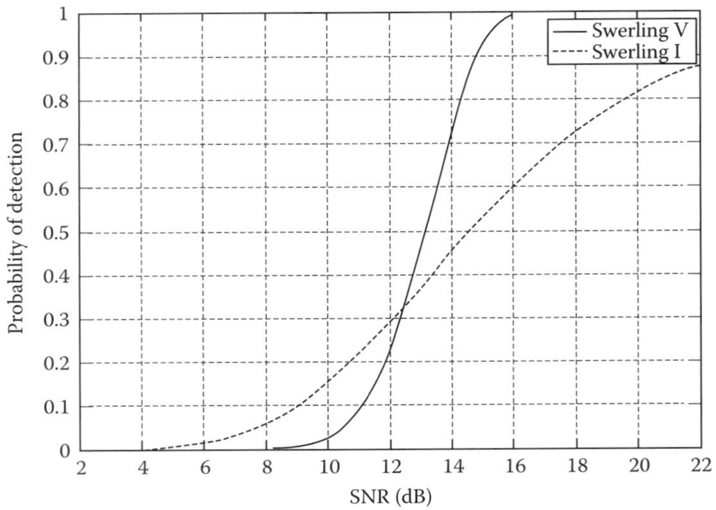

FIGURE 32.10 Probability of detection versus SNR, single pulse. $P_{fa} = 10^{-9}$.

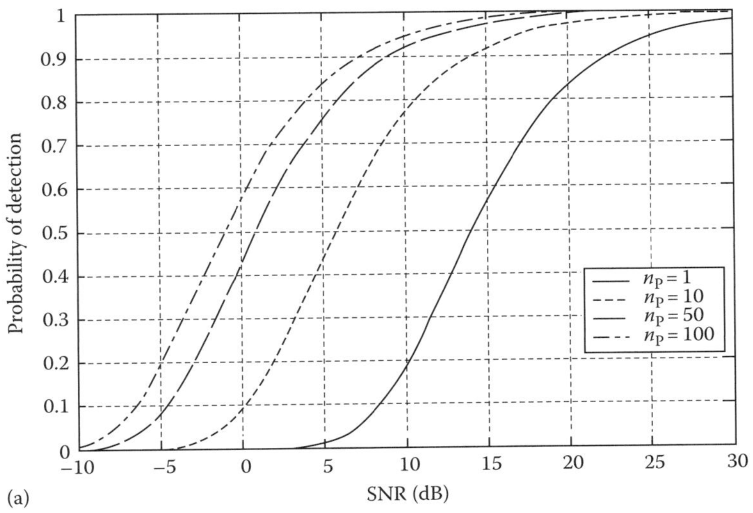

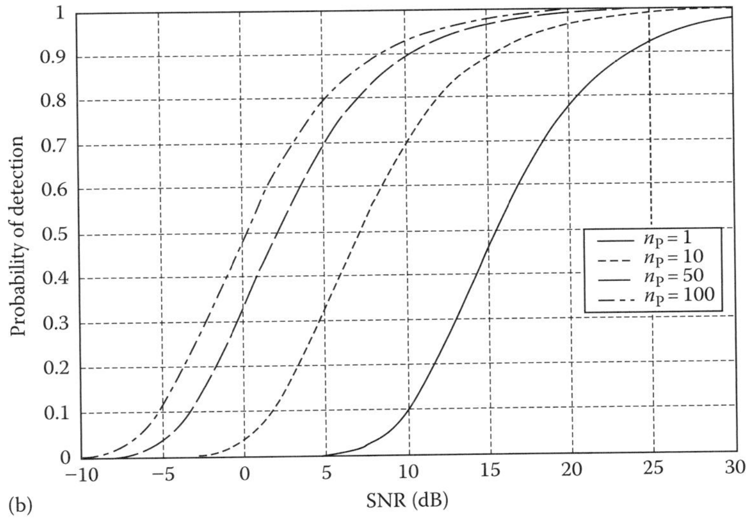

FIGURE 32.11 (a) Probability of detection versus SNR, swerling I ($P_{fa} = 10^{-8}$). (b) Probability of detection versus SNR, swerling I ($P_{fa} = 10^{-11}$).

$$C_4 = \frac{1}{4n_P} \tag{32.79}$$

$$\bar{\omega} = \sqrt{n_P}(1 + \text{SNR}) \tag{32.80}$$

MATLAB Function "pd_swerling2.m"

The function "*pd_swerling2.m*" calculates P_D for Swerling II type targets. It is given in Listing 32.19. The syntax is as follows:

$$[pd] = pd\_swerling2(nfa, np, snr)$$

where

Symbol	Description	Units	Status
nfa	Marcum's false alarm number	None	Input
np	Number of integrated pulses	None	Input
snr	SNR	dB	Input
pd	Probability of detection	None	Output

Figure 32.12 shows a plot of the probability of detection as a function of SNR for $n_P = 1, 10, 50, 100$, where $P_{fa} = 10^{-10}$. This figure can be reproduced using MATLAB program "*fig32_12.m*" given in Listing 32.20.

32.6.4 Detection of Swerling III Targets

The exact formulas, developed by Marcum, for the probability of detection for Swerling III type targets when $n_P = 1, 2$ is

$$\begin{aligned} P_D &= \exp\left(\frac{-V_T}{1 + n_P \mathrm{SNR}/2}\right)\left(1 + \frac{2}{n_P \mathrm{SNR}}\right)^{n_P - 2} \times K_0 \\ K_0 &= 1 + \frac{V_T}{1 + n_P \mathrm{SNR}/2} - \frac{2}{n_P \mathrm{SNR}}(n_P - 2) \end{aligned} \tag{32.81}$$

For $n_P > 2$ the expression is

$$\begin{aligned} P_D = {} & \frac{V_T^{n_P - 1} e^{-V_T}}{(1 + n_P \mathrm{SNR}/2)(n_P - 2)!} + 1 - \Gamma_I(V_T, n_P - 1) + K_0 \\ & \times \Gamma_I\left(\frac{V_T}{1 + 2/n_P \mathrm{SNR}}, n_P - 1\right) \end{aligned} \tag{32.82}$$

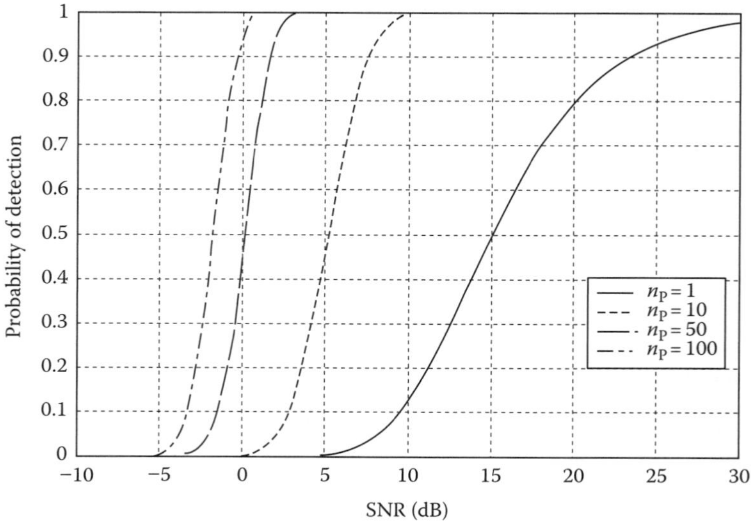

FIGURE 32.12 Probability of detection versus SNR. Swerling II. $P_{fa} = 10^{-10}$.

MATLAB Function "pd_swerling3.m"

The function "*pd_swerling3.m*" calculates P_D for Swerling III type targets. It is given in Listing 32.21. The syntax is as follows:

$$[pd] = pd\_swerling3(nfa,\ np,\ snr)$$

where

Symbol	Description	Units	Status
nfa	Marcum's false alarm number	None	Input
np	Number of integrated pulses	None	Input
snr	SNR	dB	Input
pd	Probability of detection	None	Output

Figure 32.13 shows a plot of the probability of detection as a function of SNR for $n_P = 1, 10, 50, 100$, where $P_{fa} = 10^{-9}$. This figure can be reproduced using MATLAB program "*fig32_13.m*" given in Listing 32.22.

32.6.5 Detection of Swerling IV Targets

The expression for the probability of detection for Swerling IV targets for $n_P < 50$ is

$$P_D = 1 - \left[\gamma_0 + \left(\frac{SNR}{2}\right) n_P \gamma_1 + \left(\frac{SNR}{2}\right)^2 \frac{n_P(n_P - 1)}{2!} \gamma_2 + \cdots + \left(\frac{SNR}{2}\right)^{n_P} \gamma_{n_P}\right] \left(1 + \frac{SNR}{2}\right)^{-n_P} \tag{32.83}$$

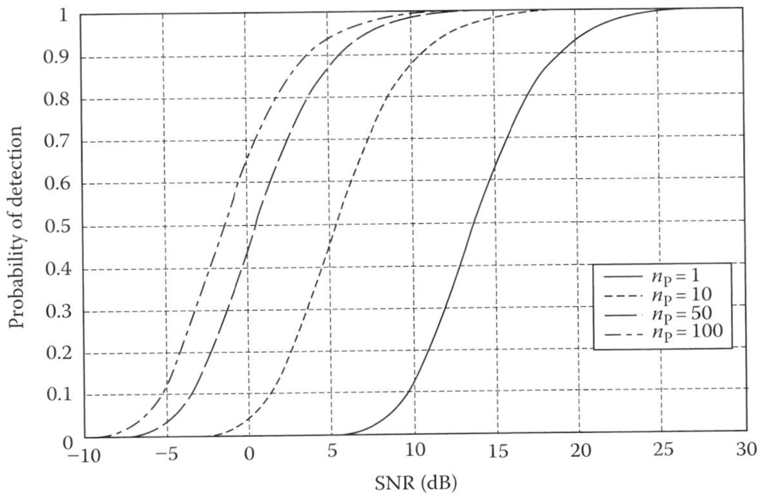

FIGURE 32.13 Probability of detection versus SNR. Swerling III. $P_{fa} = 10^{-9}$.

where

$$\gamma_i = \Gamma_I\left(\frac{V_T}{1 + (\text{SNR})/2}, n_P + i\right) \tag{32.84}$$

By using the recursive formula

$$\Gamma_I(x, i+1) = \Gamma_I(x, i) - \frac{x^i}{i! \exp(x)} \tag{32.85}$$

then only γ_0 needs to be calculated using Equation 32.84 and the rest of γ_i are calculated from the following recursion:

$$\gamma_i = \gamma_{i-1} - A_i; \quad i > 0 \tag{32.86}$$

$$A_i = \frac{V_T/(1 + (\text{SNR})/2)}{n_P + i - 1} A_{i-1}; \quad i > 1 \tag{32.87}$$

$$A_1 = \frac{(V_T/(1 + (\text{SNR})/2))^{n_P}}{n_P! \exp(V_T/(1 + (\text{SNR})/2))} \tag{32.88}$$

$$\gamma_0 = \Gamma_I\left(\frac{V_T}{(1 + (\text{SNR})/2)}, n_P\right) \tag{32.89}$$

For the case when $n_P > 50$, the Gram-Charlier series and Equation 32.69 can be used to calculate the probability of detection. In this case,

$$C_3 = \frac{1}{3\sqrt{n_P}} \frac{2\beta^3 - 1}{(2\beta^2 - 1)^{1.5}}; \quad C_6 = \frac{C_3^2}{2} \tag{32.90}$$

$$C_4 = \frac{1}{4n_P} \frac{2\beta^4 - 1}{(2\beta^2 - 1)^2} \tag{32.91}$$

$$\bar{\omega} = \sqrt{n_P(2\beta^2 - 1)} \tag{32.92}$$

$$\beta = 1 + \frac{\text{SNR}}{2} \tag{32.93}$$

MATLAB Function "pd_swerling4.m"

The function "*pd_swerling4.m*" calculates P_D for Swerling IV type targets. It is given in Listing 32.23. The syntax is as follows:

$$[pd] = pd\_swerling4(nfa, np, snr)$$

where

Symbol	Description	Units	Status
nfa	Marcum's false alarm number	None	Input
np	Number of integrated pulses	None	Input
snr	SNR	dB	Input
pd	Probability of detection	None	Output

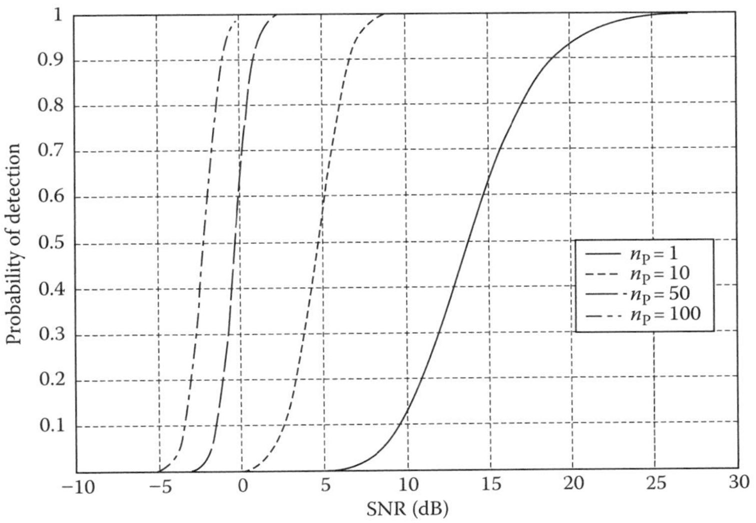

FIGURE 32.14 Probability of detection versus SNR. Swerling IV. $P_{fa} = 10^{-9}$.

Figure 32.14 shows a plot of the probability of detection as a function of SNR for $n_P = 1, 10, 50, 100$, where $P_{fa} = 10^{-9}$. This figure can be reproduced using MATLAB program "*fig32_14.m*" given in Listing 32.24.

32.7 The Radar Equation Revisited

In this section, a more comprehensive form of the radar equation is introduced. In this case, the radar equation is given by

$$R^4 = \frac{P_{av} G_t G_r \lambda^2 \sigma I(n_P)}{(4\pi)^3 k T_e F B \tau f_r L_t L_f (\mathrm{SNR})_1} \tag{32.94}$$

where

$P_{av} = P_t \tau f_r$ is the average transmitted power
P_t is the peak transmitted power
τ is pulsewidth
f_r is PRF
G_t is transmitting antenna gain
G_r is receiving antenna gain
λ is wavelength
σ is target cross section
$I(n_P)$ is improvement factor
n_P is the number of integrated pulses
k is the Boltzmann constant
T_e is effective noise temperature
F is the system noise figure
B is receiver bandwidth
L_t is total system losses including integration loss
L_f is loss due to target fluctuation
$(\mathrm{SNR})_1$ is the minimum single pulse SNR required for detection

The fluctuation loss, L_f, can be viewed as the amount of additional SNR required to compensate for the SNR loss due to target fluctuation, given a specific P_D value. This was demonstrated for a Swerling I fluctuation in Figure 32.10. Kanter[7] developed an exact analysis for calculating the fluctuation loss. In this text the authors will take advantage of the computational power of MATLAB and the MATLAB functions developed for this text to numerically calculate the amount of fluctuation loss with an accuracy of 0.005 dB or better. For this purpose the MATLAB function "*fluct_loss.m*" was developed. It is given in Listing 32.25. Its syntax is as follows:

$$[Lf, Pd\_Sw5] = fluct\_loss(pd, pfa, np, sw\_case)$$

where

Symbol	Description	Units	Status
pd	Desired probability of detection	None	Input
pfa	Probability of false alarm	None	Input
np	Number of pulses	None	Input
sw_case	1,2, 3, or 4 depending on the desired Swerling case	None	Input
Lf	Fluctuation loss	dB	Output
Pd_Sw5	Probability of detection corresponding to a Swerling V case	None	Output

For example, using the syntax

$$[Lf, Pd\_Sw5] = fluct\_loss(0.65, 1e-9, 10, 1)$$

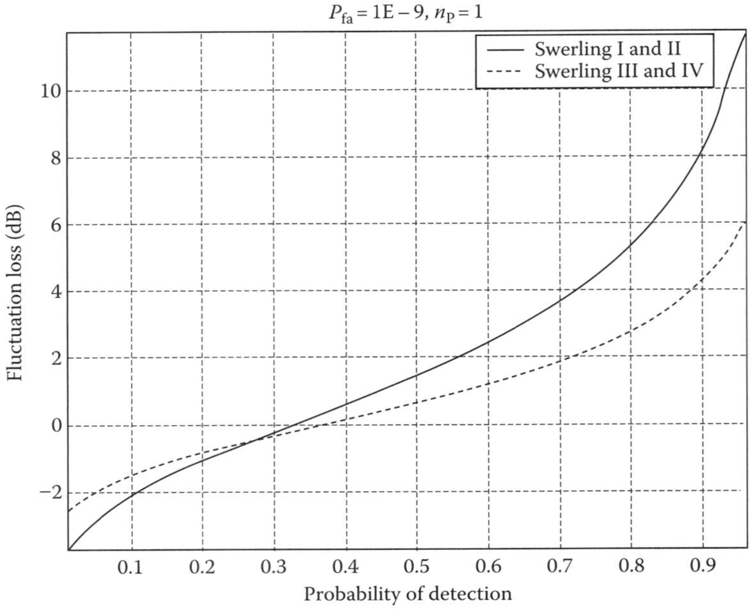

FIGURE 32.15 Fluctuation loss versus probability of detection.

will calculate the SNR corresponding to both Swerling V and Swerling I fluctuation when the desired probability of detection $P_D = 0.65$ and probability of false alarm $P_{fa} = 10^{-9}$ and 10 pulses of noncoherent integration. The following is a reprint of the output:

$$PD\_SW5 = 0.65096989459928$$
$$SNR\_SW5 = 5.52499999999990$$
$$PD\_SW1 = 0.65019653294095$$
$$SNR\_SW1 = 8.32999999999990$$
$$Lf = 2.80500000000000$$

Note that a negative value for L_f indicates a fluctuation SNR gain instead of loss. Finally, it must be noted that the function "*fluct_loss.m*" always assumes noncoherent integration. Figure 32.15 shows a plot for the additional SNR (or fluctuation loss) required to achieve a certain probability of detection. This figure can be reproduced using MATLAB program "*fig32_16.m*" given in Listing 32.26.

32.8 Cumulative Probability of Detection

Denote the range at which the single pulse SNR is unity (0 dB) as R_0, and refer to it as the reference range. Then, for a specific radar, the single pulse SNR at R_0 is defined by the radar equation and is given by

$$(\mathrm{SNR})_{R_0} = \frac{P_t G^2 \lambda^2 \sigma}{(4\pi)^3 k T_0 B F L R_0^4} = 1 \tag{32.95}$$

The single pulse SNR at any range R is

$$\mathrm{SNR} = \frac{P_t G^2 \lambda^2 \sigma}{(4\pi)^3 k T_0 B F L R^4} \tag{32.96}$$

Dividing Equation 32.96 by Equation 32.95 yields

$$\frac{\mathrm{SNR}}{(\mathrm{SNR})_{R_0}} = \left(\frac{R_0}{R}\right)^4 \tag{32.97}$$

Therefore, if the range R_0 is known then the SNR at any other range R is

$$(\mathrm{SNR})_{\mathrm{dB}} = 40 \log\left(\frac{R_0}{R}\right) \tag{32.98}$$

Also, define the range R_{50} as the range at which $P_D = 0.5 = P_{50}$. Normally, the radar unambiguous range R_u is set equal to $2P_{50}$.

The cumulative probability of detection refers to detecting the target at least once by the time it is at range R. More precisely, consider a target closing on a scanning radar, where the target is illuminated only during a scan (frame). As the target gets closer to the radar, its probability of detection increases since the SNR is increased. Suppose that the probability of detection during the nth frame is P_{D_n}; then, the cumulative probability of detecting the target at least once during the nth frame (see Figure 32.16) is given by

$$P_{C_n} = 1 - \prod_{i=1}^{n} (1 - P_{D_i}) \tag{32.99}$$

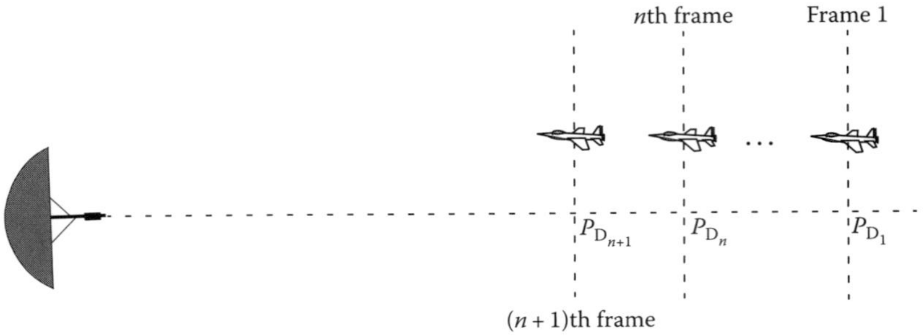

FIGURE 32.16 Detecting a target in many frames.

P_{D_1} is usually selected to be very small. Clearly, the probability of not detecting the target during the nth frame is $1 - P_{C_n}$. The probability of detection for the ith frame, P_{D_i}, is computed as discussed in the previous section.

32.8.1 Mini Design Case Study 32.2

A radar detects a closing target at $R = 10$ km, with probability of detection P_D equal to 0.5. Assume $P_{fa} = 10^{-7}$. Compute and sketch the single look probability of detection as a function of normalized range (with respect to $R = 10$ km), over the interval (2–20) km. If the range between two successive frames is 1 km, what is the cumulative probability of detection at $R = 8$ km?

32.8.1.1 A Solution

From the Junction "*marcumsq.m*" the SNR corresponding to $P_D = 0.5$ and $P_{fa} = 10^{-7}$ is approximately 12 dB. By using a similar analysis to that which led to Equation 32.98, we can express the SNR at any range R as

$$(\mathrm{SNR})_R = (\mathrm{SNR})_{10} + 40 \log \frac{10}{R} = 50 - 40 \log R$$

By using the function "*marcumsq.m*" we can construct the following table:

R, km	(SNR), dB	P_D
2	39.09	0.999
4	27.9	0.999
6	20.9	0.999
8	15.9	0.999
9	13.8	0.9
10	12.0	0.5
11	10.3	0.25
12	8.8	0.07
14	6.1	0.01
16	3.8	ε
20	0.01	ε

where ε is very small. A sketch of P_D versus normalized range is shown in Figure 32.17.

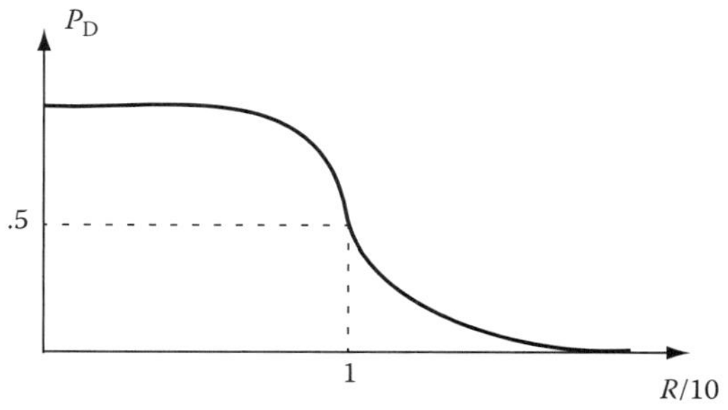

FIGURE 32.17 Cumulative probability of detection versus normalized range.

The cumulative probability of detection is given in Equation 32.95, where the probability of detection of the first frame is selected to be very small. Thus, we can arbitrarily choose frame 1 to be at $R = 16$ km. Note that selecting a different starting point for frame 1 would have a negligible effect on the cumulative probability (we only need P_{D_1} to be very small). Below is a range listing for frames 1 through 9, where frame 9 corresponds to $R = 8$ km. The cumulative probability of detection at 8 km is then

$$\begin{aligned} P_{C_9} &= 1 - (1 - 0.999)(1 - 0.9)(1 - 0.5)(1 - 0.25)(1 - 0.07)(1 - 0.01)(1 - \varepsilon)^2 \\ &\approx 0.9998 \end{aligned}$$

Frame	1	2	3	4	5	6	7	8	9
Range in km	16	15	14	13	12	11	10	9	8

32.9 Constant False Alarm Rate

The detection threshold is computed so that the radar receiver maintains a constant predetermined probability of false alarm. Equation 32.19b gives the relationship between the threshold value V_T and the probability of false alarm P_{fa}, and for convenience is repeated here as Equation 32.100:

$$V_T = \sqrt{2\Psi^2 \ln\left(\frac{1}{P_{fa}}\right)} \tag{32.100}$$

If the noise power ψ^2 is assumed to be constant, then a fixed threshold can satisfy Equation 32.100. However, due to many reasons this condition is rarely true. Thus, in order to maintain a constant probability of false alarm the threshold value must be continuously updated based on the estimates of the noise variance. The process of continuously changing the threshold value to maintain a constant probability of false alarm is known as Constant False Alarm Rate (CFAR).

Three different types of CFAR processors are primarily used. They are adaptive threshold CFAR, nonparametric CFAR, and nonlinear receiver techniques. Adaptive CFAR assumes that the interference distribution is known and approximates the unknown parameters associated with these distributions. Nonparametric CFAR processors tend to accommodate unknown interference distributions. Nonlinear receiver techniques attempt to normalize the root mean square amplitude of the interference. In this book only analog Cell-Averaging CFAR (CA-CFAR) technique is examined. The analysis presented in this section closely follows Urkowitz.[8]

32.9.1 Cell-Averaging CFAR (Single Pulse)

The CA-CFAR processor is shown in Figure 32.18. Cell averaging is performed on a series of range and/or Doppler bins (cells). The echo return for each pulse is detected by a square law detector. In analog implementation these cells are obtained from a tapped delay line. The Cell Under Test (CUT) is the central cell. The immediate neighbors of the CUT are excluded from the averaging process due to a possible spillover from the CUT. The output of M reference cells ($M/2$ on each side of the CUT) is averaged. The threshold value is obtained by multiplying the averaged estimate from all reference cells by a constant K_0 (used for scaling). A detection is declared in the CUT if

$$Y_1 \geq K_0 Z \qquad (32.101)$$

Cell-averaging CFAR assumes that the target of interest is in the CUT and all reference cells contain zero mean independent Gaussian noise of variance ψ^2. Therefore, the output of the reference cells, Z, represents a random variable with gamma pdf (special case of the Chi-square) with $2M$ degrees of freedom. In this case, the gamma pdf is

$$f(z) = \frac{z^{(M/2)-1} e^{(-z/2\psi^2)}}{2^{M/2}\psi^M \Gamma(M/2)}; \quad z > 0 \qquad (32.102)$$

The probability of false alarm corresponding to a fixed threshold was derived earlier. When CA-CFAR is implemented, then the probability of false alarm can be derived from the conditional false alarm probability, which is averaged over all possible values of the threshold in order to achieve an unconditional false alarm probability. The conditional probability of false alarm when $y = V_T$ can be written as

$$P_{fa}(V_T = y) = e^{-y/2\psi^2} \qquad (32.103)$$

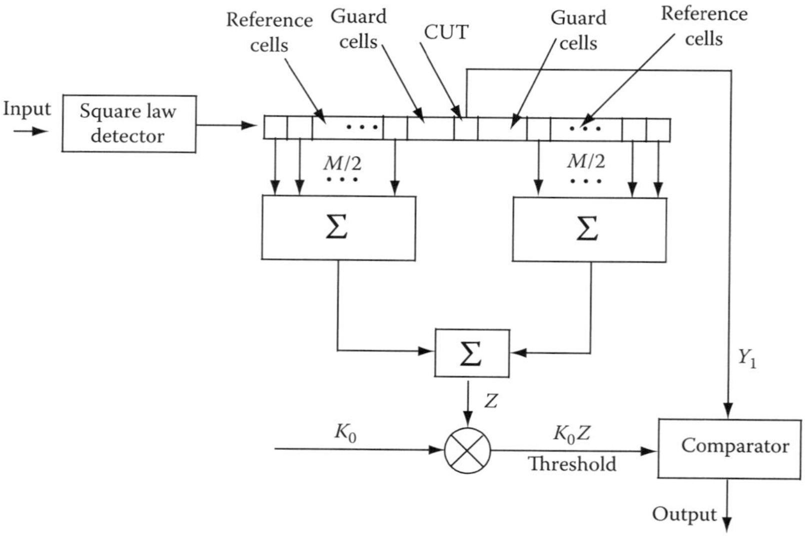

FIGURE 32.18 Conventional CA-CFAR.

It follows that the unconditional probability of false alarm is

$$P_{fa} = \int_0^\infty P_{fa}(V_T = y) f(y)\, dy \tag{32.104}$$

where $f(y)$ is the pdf of the threshold, which except for the constant K_0 is the same as that defined in Equation 32.102. Therefore,

$$f(y) = \frac{y^{M-1} e^{(-y/2K_0\psi^2)}}{(2K_0\psi^2)\Gamma(M)}; \quad y \geq 0 \tag{32.105}$$

Performing the integration in Equation 32.104 yields

$$P_{fa} = \frac{1}{(1 + K_0)^M} \tag{32.106}$$

Observation of Equation 32.106 shows that the probability of false alarm is now independent of the noise power, which is the objective of CFAR processing.

32.9.2 Cell-Averaging CFAR with Noncoherent Integration

In practice, CFAR averaging is often implemented after noncoherent integration, as illustrated in Figure 32.19. Now, the output of each reference cell is the sum of r_P squared envelopes. It follows that the total number of summed reference samples is Mn_P. The output Y_1 is also the sum of n_P squared envelopes. When noise alone is present in the CUT, Y_1 is a random variable whose pdf is a gamma distribution with $2n_P$ degrees of freedom. Additionally, the summed output of the reference cells is the sum of Mn_P squared envelopes. Thus, Z is also a random variable which has a gamma pdf with $2Mn_P$ degrees of freedom.

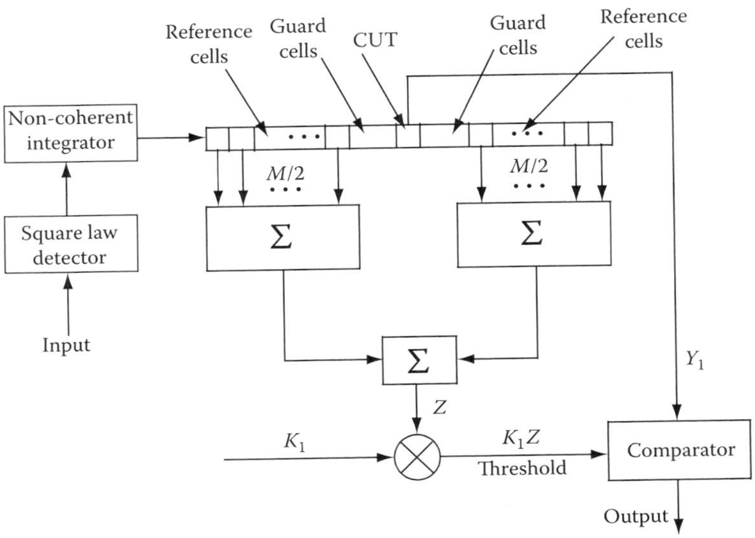

FIGURE 32.19 Conventional CA-CFAR with noncoherent integration.

The probability of false alarm is then equal to the probability that the ratio Y_1/Z exceeds the threshold. More precisely,

$$P_{fa} = Prob\{Y_1/Z > K_1\} \tag{32.107}$$

Equation 32.107 implies that one must first find the joint pdf for the ratio Y_1/Z. However, this can be avoided if P_{fa} is first computed for a fixed threshold value V_T, then averaged over all possible values of the threshold. Therefore, let the conditional probability of false alarm when $y = V_T$ be $P_{fa}(V_T = y)$. It follows that the unconditional false alarm probability is given by

$$P_{fa} = \int_0^\infty P_{fa}(V_T = y) f(y)\,dy \tag{32.108}$$

where $f(y)$ is the pdf of the threshold. In view of this, the pdf describing the random variable K_1Z is given by

$$f(y) = \frac{(y/K_1)^{Mn_P - 1} e^{(-y/2K_0\psi^2)}}{(2\psi^2)^{Mn_P} K_1 \Gamma(Mn_P)}; \quad y \geq 0 \tag{32.109}$$

It can be shown that in this case the probability of false alarm is independent of the noise power and is given by

$$P_{fa} = \frac{1}{(1+K_1)^{Mn_P}} \sum_{k=0}^{n_P-1} \frac{1}{k!} \frac{(Mn_P + k)}{\Gamma(Mn_P)} \left(\frac{K_1}{1+K_1}\right)^k \tag{32.110}$$

which is identical to Equation 32.106 when $K_1 = K_0$ and $n_P = 1$.

32.10 MATLAB® Program and Function Listings

This section presents listings for all MATLAB programs/functions used in this chapter. The user is advised to rerun these programs with different input parameters.

Listing 32.1. MATLAB Program "fig32_2.m"

```
% This program can be used to reproduce Figure 32.2 of the text
clear all
close all
xg=linspace(-6,6,1500); % random variable between -6 and 6
xr=linspace(0,6,1500); % random variable between 0 and 6
mu=0; % zero mean Gaussian pdf mean
sigma=1.5; % standard deviation (sqrt(variance))
ynorm=normpdf(xg,mu,sigma); % use MATLAB function normpdf
yray=raylpdf(xr,sigma); % use MATLAB function raylpdf
plot(xg, ynorm, 'k',xr,yray 'k-.');
grid
legend('Gaussian pdf', 'Rayleigh pdf')
xlabel('x')
ylabel('Probability density')
gtext('\mu=0; \sigma=1.5')
gtext('\sigma =1.5)
```

Listing 32.2. MATLAB Function "que_func.m"

```
function fofx=que_func(x)
% This function computes the value of the Q-function
% listed in Eq. (32.16). It uses the approximation in Eqs. (32.17) and (32.18)
if (x≥0)
  denom=0.661 * x+0.339 *sqrt(x^2+5.51);
  expo=exp(-x^2/2.0);
  fofx=1.0 - (1.0/sqrt(2.0 * pi)) * (1.0/denom) * expo;
else
  denom = 0.661 * x + 0.339 * sqrt(x^2 + 5.51);
  expo = exp(-x^2/2.0);
  value = 1.0- (1.0/sqrt(2.0 * pi)) * (1.0/denom) * expo;
  fofx = 1.0 - value;
end
```

Listing 32.3. MATLAB Program "fig32_3.m"

```
% This program generates Figure 32.3.
close all
clear all
logpfa=linspace(01,250,1000);
var=10.^ (logpfa./10.0);
vtnorm=sqrt(log (var));
semilogx(logpfa, vtnorm, 'k')
grid
```

Listing 32.4. MATLAB Function "marcumsq.m"

```
function Pd=marcumsq (a,b)
% This function uses Parl's method to compute PD
max_test_value=5000.;
if (a<b)
  alphan0=1.0;
  dn=a/b;
else
  alphan0=0.;
  dn=b/a;
end
alphan_1=0.;
betan0=0.5;
betan_1=0.;
D1=dn;
n=0;
ratio =2.0/(a*b);
r1=0.0;
betan=0.0;
alphan=0.0;
while betan<1000.,
  n=n+1;
  alphan=dn+ratio * n * alphan0+alphan;
```

```
  betan = 1.0 + ratio * η * betan0 + betan;
  alphan_1 = alphan0;
  alphan0 = alphan;
  betan_1 = betan0;
  betan0 = betan;
  dn = dn * D1;
end
PD = (alphan0/(2.0 * betan0)) * exp(- (a-b)∧ 2/2.0);
if (a>=b)
  PD = 1.0 - PD;
end
return
```

Listing 32.5. MATLAB Program "prob_snr1.m"

```
% This program is used to produce Fig. 32.4
close all
clear all
for nfa = 2:2:12
  b = sqrt(-2.0 * log(10^ (-nfa)));
  index = 0;
  hold on
  for snr = 0:.1:18
    index = index + 1;
    a = sqrt(2.0*10^(.1*snr));
    pro(index) = marcumsq(a,b);
  end
  x = 0:.1:18;
  set(gca, 'ytick',[.1 .2 .3 .4 .5 .6 .7 .75 .8 .85 .9 . . . .95 .9999])
  set(gca, 'xtick',[1 2 3 4 5 6 7 8 9 10 11 12 13 14 15 16 17 18])
  loglog(x, pro, 'k');
end
hold off
xlabel ('Single pulse SNR - dB')
ylabel ('Probability of detection')
grid
```

Listing 32.6. MATLAB program "fig32_6a.m"

```
% This program is used to produce Figure 32.6a
% It uses the function ''improv_fac''
clear all
close all
pfa1 = 1.0e-2;
pfa2 = 1.0e-6;
pfa3 = 1.0e-10;
pfa4 = 1.0e-13;
pd1 = .5;
pd2 = .8;
pd3 = .95;
pd4 = .999;
```

```
index= 0;
for np= 1:1:1000
  index= index+ 1;
  I1 (index) = improv_fac (np, pfa1, pd1);
  I2 (index) = improv_fac (np, pfa2, pd2);
  I3 (index) = improv_fac (np, pfa3, pd3);
  I4 (index) = improv_fac (np, pfa4, pd4);
end
  np= 1:1:1000;
  semilogx (np, I1, 'k', np, I2, 'k-', np, I3, 'k--.', np, I4, 'k:')
  xlabel ('Number of pulses');
  ylabel ('Improvement factor I - dB')
  legend ('pd= .5, nfa= e+2', 'pd= .8, nfa= e+6', 'pd= .95, nfa= e+10',
  'pd= .999, nfa= e+13');
   grid
```

Listing 32.7. MATLAB Function "improv_fac.m"

```
function impr_of_np= improv_fac (np, pfa, pd)
% This function computes the noncoherent integration improvement
% factor using the empirical formula defined in Eq. (32.49)
fact1= 1.0 + log10(1.0/pfa)/46.6;
fact2= 6.79 * (1.0+ 0.235 *pd);
fact3= 1.0-0.14 * log10(np) + 0.0183 * (log10(np))^ 2;
impr_of_np= fact1 *fact2 * fact3 * log10(np);
return
```

Listing 32.8. MATLAB Program "fig32_6b.m"

```
% This program is used to produce Fig.32.6b
% It uses the function ''improv_fac''.
clear all
close all
pfa1= 1.0e-12;
pfa2= 1.0e-12;
pfa3= 1.0e-12;
pfa4= 1.0e-12;
pd1= .5;
pd2= .8;
pd3= .95;
pd4= .99;
index= 0;
for np= 1:1:1000
  index= index + 1;
  I1= improv_fac (np, pfa1, pd1);
  I1= 10.^(0.1*I1);
  L1 (index) = -1*10*log10(i1./np);
  I2= improv_fac (np, pfa2, pd2);
  I2= 10.^ (0.1*I2);
  L2 (index) = -1*10*log10(i2./np);
  I3= improv_fac (np, pfa3, pd3);
```

```
  i3=10.^ (0.1*I3);
  L3(index)=-1*10*log10(i3./np);
  I4=improv_fac (np, pfa4, pd4);
  i4=10.^ (0.1*I4);
  L4 (index)=-1*10*log10(i4 ./np);
end
np=1:1:1000;
semilogx (np, L1, 'k', np, L2, 'k--', np, L3, 'k-.', np, L4, 'k:')
axis tight
xlabel ('Number of pulses');
ylabel ('Integration loss - dB')
legend ('pd= .5, nfa=e + 12', 'pd= .8, nfa=e + 12', 'pd= .95, nfa=e + 12'
'pd= .99, nfa=e + 12');
grid
```

Listing 32.9. MATLAB Function "incomplete_gamma.m"

```
function [value]=incomplete_gamma (vt, np)
% This function implements Eq. (2.67) to compute the Incomplete Gamma Func-
tion
% This function needs ''factor.m'' to run
format long
eps=1.000000001;
% Test to see if np=1
if(np=1)
  value1=vt* exp(-vt);
  value=1.0-exp(-vt);
  return
end
  sumold=1.0;
  sumnew =1.0;
  calc1=1.0;
  calc2=np;
xx=np* log(vt + 0.0000000001)-vt-factor (calc2);
temp1=exp(xx);
temp2=np/(vt + 0.0000000001);
diff =. 0;
ratio=1000.0;
if(vt >= np)
  while (ratio> = eps)
    diff=diff+1.0;
    calc1=calc1 * (calc2-diff)/vt;
    sumnew=sumold+calc1;
    ratio=sumnew/sumold;
    sumold=sumnew;
  end
  value=1.0-temp1 * sumnew * temp2;
  return
```

```
else
  diff=0.;
  sumold=1.;
  ratio=1000.;
  calc1=1.;
  while (ratio >= eps)
    diff=diff+1.0;
    calc1=calc1 *vt/(calc2+diff);
    sumnew=sumold+calc1;
    ratio=sumnew/sumold;
    sumold=sumnew;
    end
    value=temp1 * sumnew;
end
```

Listing 32.10. MATLAB Function "factor.m"

```
function [val]=factor (n)
% Compute the factorial of n using logarithms to avoid overflow.
format long
n=n+9.0;
n2=n * n;
temp= (n-1)  *  log(n) -n+log(sqrt(2.0  *  pi  *n)) . . .+  ((1.0-(1.0/30.+
(1.0/105)/n2)/n2)/12)/n;
val=temp-log((n-1)*(n-2)*(n-3)*(n-4)*(n-5)*(n-6) . . .*(n-7)*(n-8));
return
```

Listing 32.11. MATLAB Program "fig32_7. m"

```
% This program can be used to reproduce Fig. 32.7
close all
clear all
format long
ii=0;
for x=0:. 1:20
  ii=ii + 1;
  val1 (ii)=incomplete_gamma(x, 1);
  val2(ii)=incomplete_gamma(x, 3);
  val=incomplete_gamma(x, 6);
  val3(ii)=val;
  val=incomplete_gamma(x, 10);
  val4(ii)=val;
end
xx=0:. 1:20;
plot(xx,val1, 'k',xx,val2, 'k:',xx,val3, 'k--',xx,val4, 'k-.')
legend('N=1', 'N=3', 'N=6', 'N=10')
xlabel('x')
ylabel('Incomplete Gamma function (x,N)')
grid
```

Listing 32.12. MATLAB Function "threshold.m"

```
function [pfa, vt] = threshold (nfa, np)
% This function calculates the threshold value from nfa and np.
% The Newton-Raphson recursive formula is used (Eqs. (32-63) through (32-66))
% This function uses ''incomplete_gamma.m''.
delmax = .00001;
eps = 0.000000001;
delta = 10000.;
pfa = np * log(2)/nfa;
sqrtpfa = sqrt(-log10(pfa));
sqrtnp = sqrt(np);
vt0 = np - sqrtnp + 2.3 * sqrtpfa * (sqrtpfa + sqrtnp - 1.0);
vt = vt0;
while (abs(delta) >= vt0)
  igf = incomplete_gamma(vt0,np);
  num = 0.5^(np/nfa) - igf;
  temp = (np-1) * log(vt0 + eps) - vt0 - factor(np-1);
  deno = exp(temp);
  vt = vt0 + (num/(deno + eps));
  delta = abs(vt - vt0) * 10000.0;
  vt0 = vt;
end
```

Listing 32.13. MATLAB Program "fig32_8.m"

```
% Use this program to reproduce Fig. 32.8 of text
clear all
forn = 1: 1: 150
  [pfal y1(n)] = threshold(1000,n);
  [pfa2 y3(n)] = threshold(10000,n);
  [pfa3 y4(n)] = threshold(500000,n);
end
n = 1:1:150;
loglog(n,y1, 'k',n,y3, 'k--',n,y4, 'k-.');
axis([0 200 1 300])
xlabel ('Number of pulses');
ylabel('Threshold')
legend('nfa = 1000', 'nfa = 10000', 'nfa = 500000')
grid
```

Listing 32.14. MATLAB Function "pd_swerling5.m"

```
function pd = pd_swerling5 (input1, indicator, np, snrbar)
% This function is used to calculate the probability of
% for Swerling 5 or 0 targets for np > 1.
if(np == 1)
  'Stop, np must be greater than 1'
  return
end
format long
```

```
snrbar=10.0.^(snrbar./10.);
eps=0.00000001;
delmax=.00001;
delta=10000.;
% Calculate the threshold Vt
if (indicator ~=1)
  nfa=input1;
  pfa=np * log(2)/nfa;
else
  pfa=input1;
  nfa=np * log(2)/pfa;
end
sqrtpfa=sqrt(-log10(pfa));
sqrtnp=sqrt(np);
vt0=np-sqrtnp+2.3 * sqrtpfa * (sqrtpfa+sqrtnp-1.0);
vt=vt0;
while (abs(delta) >= vt0)
  igf=incomplete_gamma(vt0,np);
  num=0.5^(np/nfa) - igf;
  temp= (np-1) * log(vt0 + eps) - vt0 - factor (np-1);
  deno=exp(temp);
  vt=vt0+ (num/(deno+eps));
  delta=abs(vt - vt0) * 10000.0;
  vt0=vt;
end
% Calculate the Gram-Charlier coefficients
temp1=2.0 .*snrbar+1.0;
omegabar=sqrt(np .* temp1);
c3=-(snrbar+1.0/3.0)./(sqrt(np) .* templ.^1.5);
c4= (snrbar+0.25)./(np .* templ.^2.);
c6=c3.*c3./2.0;
V= (vt - np .* (1.0+snrbar))./omegabar;
Vsqr=V. *V;
val1=exp(-Vsqr ./2.0)./sqrt(2.0 * pi);
val2=c3 .* (V.^2-1.0)+c4 .*V.* (3.0-V^2) -...c6.*V * (V.^4-10..
*V.^2+15.0);
q=0.5 .* erfc (V./sqrt(2.0));
pd=q-val1. * val2;
```

Listing 32.15. MATLAB Program “fig32_9.m”

```
% This program is used to produce Fig. 32.9
close all
clear all
pfa=1e-9;
nfa=log(2)/pfa;
b=sqrt(-2.0*log(pfa));
index=0;
for snr=0:.1:20
  index=index+1;
```

```
  a=sqrt(2.0 * 10^(.1 *snr));
  pro(index)=marcumsq(a,b);
  prob205(index)=pd_swerling5 (pfa, 1, 10, snr);
end
x= 0:. 1:20;
plot(x, pro, 'k',x,prob205, 'k:');
axis([0 20 0 1])
xlabel ('SNR - dB')
ylabel ('Probability of detection')
legend('np=1', 'np=10')
grid
```

Listing 32.16. MATLAB Function "pd_swerling1.m"

```
function pd=pd_swerling1 (nfa, np, snrbar)
% This function is used to calculate the probability of detection
% for Swerling 1 targets.
format long
snrbar=10.0^(snrbar/10.);
eps=0.00000001;
delmax=.00001;
delta=10000.;
% Calculate the threshold Vt
pfa=np * log(2)/nfa;
sqrtpfa=sqrt(-log10(pfa));
sqrtnp=sqrt(np);
vt0=np-sqrtnp+2.3 * sqrtpfa * (sqrtpfa+sqrtnp-1.0);
vt=vt0;
while (abs(delta) ≥ vt0)
  igf=incomplete_gamma(vt0,np);
  num=0.5^(np/nfa) - igf;
  temp= (np-1) * log(vt0 + eps) - vt0 - factor (np-1);
  deno=exp(temp);
  vt=vt0+ (num/(deno+eps));
  delta=abs(vt - vt0) * 10000.0;
  vt0=vt;
end
if(np=1)
  temp=-vt/(1.0+snrbar);
  pd=exp(temp);
  return
end
  temp1=1.0+np * snrbar;
  temp2=1.0/(np *snrbar);
  temp=1.0+temp 2;
  val1=temp^(np-1.);
  igf1=incomplete_gamma(vt,np-1);
  igf2=incomplete_gamma(vt/temp,np-1);
  pd=1.0-igf1+val1 * igf2 * exp(-vt/temp1);
```

Listing 32.17. MATLAB Program "fig32_10.m"

```
% This program is used to reproduce Fig. 32.10
close all
clear all
pfa=1e-9;
nfa=log(2)/pfa;
b=sqrt(-2.0*log(pfa));
index=0;
for snr=0:.1:22
  index=index+1;
  a=sqrt(2.0 * 10^(.1 *snr));
  pro(index)=marcumsq(a,b);
  prob(index)=pd_swerling1 (nfa, 1, snr);
end
x=0:.1:22;
plot(x, pro, 'k',x,prob, 'k:');
axis([2 22 0 1])
xlabel ('SNR - dB')
ylabel ('Probability of detection')
legend('Sterling V', 'Sterling I')
grid
```

Listing 32.18. MATLAB Program "fig32_11ab.m"

```
% This program is used to produce Fig.32.11a&b
clear all
pfa=1e-11;
nfa=log(2)/pfa;
index=0;
for snr=-10:.5:30
  index=index +1;
  prob1(index)=pd_swerling1 (nfa, 1, snr);
  prob10(index)=pd_swerling1 (nfa, 10, snr);
  prob50(index)=pd_swerling1 (nfa, 50, snr);
  prob 100(index)=pd_swerling1 (nfa, 100, snr);
end
x=-10:.5:30;
plot(x, prob1, 'k',x,prob10, 'k:',x,prob50, 'k--', ...x,prob100, 'k-.');
axis([-10 30 0 1])
xlabel ('SNR - dB')
ylabel ('Probability of detection')
legend('np=1', 'np=10', 'np=50', 'np=100')
grid
```

Listing 32.19. MATLAB Function "pd_swerling2.m"

```
function pd=pd_swerling2 (nfa, np, snrbar)
% This function is used to calculate the probability of detection
% for Swerling 2 targets.
format long
```

```
snrbar=10.0^(snrbar/10.);
eps=0.00000001;
delmax=.00001;
delta=10000.;
% Calculate the threshold Vt
pfa=np * log(2)/nfa;
sqrtpfa=sqrt(-log10(pfa));
sqrtnp=sqrt(np);
vt0=np-sqrtnp+2.3 * sqrtpfa * (sqrtpfa+sqrtnp-1.0);
vt=vt0;
while (abs(delta) >= vt0)
  igf=incomplete_gamma(vt0,np);
  num=0.5^(np/nfa) - igf;
  temp=(np-1) * log(vt0 + eps) - vt0 - factor (np-1);
  deno=exp(temp);
  vt=vt0+ (num/(deno + eps));
  delta=abs(vt-vt0) * 10000.0;
  vt0=vt;
end
if(np<=50)
  temp=vt/(1.0+snrbar);
  pd=1.0-incomplete_gamma(temp,np);
  return
else
  temp1=snrbar+1.0;
  omegabar=sqrt(np) * temp1;
  c3 =-1.0/sqrt(9.0*np);
  c4=0.25/np;
  c6=c3 *c3/2.0;
  V=(vt - np * temp1)/omegabar;
  Vsqr=V *V;
  val1=exp(-Vsqr/2.0)/sqrt(2.0 *pi);
  val2=c3* (V^2 - 1.0) + c4*V* (3.0 - V^2) - ... c6*V* (V^4 - 10. *V^2+15.0);
  q=0.5 * erfc (V/sqrt(2.0));
  pd=q - val1 * val2;
end
```

Listing 32.20. MATLAB Program "fig32_12.m"

```
% This program is used to produce Fig. 32.12
clear all
pfa=1e-10;
nfa=log(2)/pfa;
index=0;
for snr=-10:.5:30
  index=index +1;
  prob1(index)=pd_swerling2 (nfa, 1, snr);
  prob 10(index)=pd_swerling2 (nfa, 10, snr);
```

```
  prob50(index)=pd_swerling2 (nfa, 50, snr);
  prob 100(index)=pd_swerling2 (nfa, 100, snr);
end
  x=-10:.5:30;
  plot(x, prob1, 'k',x,prob10, 'k:',x,prob50, 'k--', . . .x, prob100, 'k-.');
  axis([-10 30 0 1])
  xlabel ('SNR - dB')
  ylabel ('Probability of detection')
  legend('np=1', 'np=10', 'np=50', 'np=100')
  grid
```

Listing 32.21. MATLAB Function "pd_swerling3.m"

```
function pd=pd_swerling3 (nfa, np, snrbar)
% This function is used to calculate the probability of detection
% for Swerling 3 targets.
format long
snrbar=10.0^(snrbar/10.);
eps=0.00000001;
delmax=.00001;
delta=10000.;
% Calculate the threshold Vt
pfa=np * log(2)/nfa;
sqrtpfa=sqrt(-log10(pfa));
sqrtnp=sqrt(np);
vt0=np-sqrtnp+2.3 * sqrtpfa * (sqrtpfa+sqrtnp-1.0);
vt=vt0;
while (abs(delta)≥vt0)
  igf=incomplete_gamma(vt0,np);
  num=0.5∧(np/nfa) - igf;
  temp=(np-1) * log(vt0 + eps) - vt0 - factor (np-1);
  deno=exp(temp);
  vt=vt0+ (num/(deno+eps));
  delta=abs(vt - vt0) * 10000.0;
  vt0=vt;
end
temp1 =vt/(1.0+0.5 * np *snrbar);
temp2=1.0+2.0/(np * snrbar);
temp3 =2.0* (np-2.0)/(np * snrbar);
ko=exp(-temp1) * temp2^(np-2.) * (1.0+temp1-temp3);
if(np<=2)
  pd=ko;
  return
else
  temp4=vt^(np-1.) * exp(-vt)/(temp1 * exp(factor(np-2.)));
  temp5=vt/(1.0+2.0/(np *snrbar));
  pd=temp4+1.0-incomplete_gamma(vt,np-1.)+ko  * . . . incomplete_gamma
   (temp5, np-1.);
end
```

Listing 32.22. MATLAB Program "fig32_13.m"

```
% This program is used to produce Fig. 32.13
clear all
pfa=1e-9;
nfa=log(2)/pfa;
index=0;
for snr=-10:.5:30
  index=index+1;
  prob1(index)=pd_swerling3 (nfa, 1, snr);
  prob10(index)=pd_swerling3 (nfa, 10, snr);
  prob50(index)=pd_swerling3(nfa, 50, snr);
  prob100(index)=pd_swerling3 (nfa, 100, snr);
end
x=-10:.5:30;
plot(x, prob1, 'k',x,prob10, 'k:',x,prob50, 'k-', ... x,prob100, 'k-.');
axis([-10 30 0 1])
xlabel ('SNR - dB')
ylabel ('Probability of detection')
legend('np=1', 'np=10', 'np=50', 'np=100')
grid
```

Listing 32.23. MATLAB Function "pd_swerling4.m"

```
function pd=pd_swerling4 (nfa, np, snrbar)
% This function is used to calculate the probability of detection
% for Swerling 4 targets.
format long
snrbar=10.0^(snrbar/10.);
eps=0.00000001;
delmax=.00001;
delta=10000.;
% Calculate the threshold Vt
pfa=np * log(2)/nfa;
sqrtpfa=sqrt(-log10(pfa));
sqrtnp=sqrt(np);
vt0=np - sqrtnp+2.3 * sqrtpfa * (sqrtpfa+sqrtnp - 1.0);
vt=vt0;
while (abs(delta) >= vt0)
  igf=incomplete_gamma(vt0,np);
  num=0.5^(np/nfa) - igf;
  temp=(np-1) * log(vt0 + eps) - vt0 - factor (np-1);
  deno=exp(temp);
  vt=vt0+(num/(deno + eps));
  delta=abs(vt - vt0) * 10000.0;
  vt0=vt;
end
h8=snrbar/2.0;
beta=1.0+h8;
beta2=2.0 * beta^2 - 1.0;
beta3=2.0*beta^3;
```

```
if (np>=50)
  temp1=2.0 *beta -1;
  omegabar=sqrt(np * temp1);
  c3=(beta3-1.)/3.0/beta2/omegabar;
  c4=(beta3 * beta3-1.0)/4./np/beta2/beta2;
  c6=c3 * c3/2.0;
  V=(vt-np * (1.0+snrbar))/omegabar;
  Vsqr=V * V;
  val1=exp(-Vsqr/2.0)/sqrt(2.0 *pi);
  val2=c3* (V^2-1.0) + c4*V* (3.0-V^2) - . . . c6*V* (V^4-10. *V^2+15.0);
  q=0.5* erfc (V/sqrt(2.0));
  pd=q-val1 * val2;
  return
else
  snr=1.0;
  gamma0=incomplete_gamma(vt/beta,np);
  a1=(vt/beta)^np/(exp(factor(np)) * exp(vt/beta));
  sum=gamma0;
  for i=1:1:np
    temp1=1;
    if (i ==1)
      ai=a1;
    else
      ai=(vt/beta) * a1/(np+i-1);
    end
    a1=ai;
    gammai=gamma0-ai;
    gamma0=gammai;
    a1=ai;
    for ii=1:1: i
      temp1=temp1 * (np+1-ii);
    end
    term=(snrbar/2.0) ^i * gammai * temp1/exp(factor(i));
    sum=sum+term;
  end
  pd=1.0-sum/beta^np;
end
pd=max(pd, 0.);
```

Listing 32.24. MATLAB Program "fig32_14.m"

```
% This program is used to produce Fig. 32.14
clear all
pfa=1e-9;
nfa=log(2)/pfa;
index=0;
for snr=-10:. 5:30
  index=index +1;
  prob1(index)=pd_swerling4 (nfa, 1, snr);
  prob10(index)=pd_swerling4 (nfa, 10, snr);
```

```
  prob50(index) = pd_swerling4(nfa, 50, snr);
  prob100(index) = pd_swerling4 (nfa, 100, snr);
end
x = -10:.5:30;
plot(x, prob1, 'k',x,prob10, 'k:',x,prob50, 'k--', ...x,prob100, 'k-.');
axis([-10 30 0 1.1])
xlabel ('SNR - dB')
ylabel ('Probability of detection')
legend('np = 1', 'np = 10', 'np = 50', 'np = 100')
grid
axis tight
```

Listing 32.25. MATLAB Function "fluct_loss.m"

```
function [Lf,Pd_Sw5] = fluct_loss(pd, pfa, np, sw_case)
% This function calculates the SNR fluctuation loss for Swerling models
% A negative Lf value indicates SNR gain instead of loss
format long
% compute the false alarm number
nfa = log(2)/pfa;
% *************** Swerling 5 case ***************
% check to make sure that np > 1
if(np == 1)
  b = sqrt(-2.0 * log(pfa));
  Pd_Sw5 = 0.001;
  snr_inc = 0.1 - 0.005;
  while(Pd_Sw5 <= pd)
    snr_inc = snr_inc + 0.005;
    a = sqrt(2.0 * 10^(.1*snr_jnc));
    Pd_Sw 5 = marcumsq(a,b);
  end
  PD_SW5 = Pd_Sw5
  SNR_SW5 = snr_inc
else
  % np > 1 use MATLAB function pd_swerling5.m
  Snr_inc = 0.1 - 0.005;
  Pd_Sw5 = 0.001;
  while(Pd_Sw5 <= pd)
    snr_inc = snr_inc + 0.005;
    Pd_Sw5 = pd_swerling5(pfa, 1, np, snr_inc);
  end
  PD_SW5 = Pd_Sw5
  SNR_SW5 = snr_inc
End
If sw_case == 5
  Lf = 0.
  Return
end
% *************** End Swerling 5 case ***************
```

```
% *************** Swerling 1 case ***************
if (sw_case = = 1)
  Pd_Sw1=0.001;
  snr_inc=0.1-0.005;
  while(Pd_Sw1< = pd)
    snr_inc=snr_inc+0.005;
    Pd_Sw1=pd_swerling1 (nfa, np, snr_inc);
  end
  PD_SW1=Pd_Sw1
  SNR_SW1=snr_inc
  Lf=SNR_SW1-SNR_SW5
end
% *************** End Swerling 1 case ***************
% *************** Swerling 2 case ***************
if (sw_case = =2)
  Pd_Sw2=0.001;
  Snr_inc=0.1-0.005;
  while(Pd_Sw2< = pd)
    snr_inc=snr_inc+0.005;
    Pd_Sw2=pd_swerling2(nfa, np, snr_inc);
  end
  PD_SW2=Pd_Sw2
  SNR_SW2=snr_inc
  Lf=SNR_SW2-SNR_SW5
end
% *************** End Swerling 2 case ***************
% *************** Swerling 3 case ***************
if (sw case = =3)
  Pd_Sw3=0.001;
  snr_inc=0.1-0.005;
  while(Pd_Sw3< = pd)
    snr_inc=snr_inc+0.005;
    Pd_Sw3=pd_swerling3(nfa, np, snr_inc);
  end
  PD_SW3=Pd_Sw3
  SNR_SW3=snr_inc
  Lf=SNR_SW3-SNR_SW5
end
% *************** End Swerling 3 case ***************
% *************** Swerling 4 case ***************
if (sw_case = =4)
  Pd_Sw4=0.001;
  snr_inc=0.1-0.005;
  while(Pd_Sw4< = pd)
    snr_inc=snr_inc+0.005;
    Pd_Sw4=pd_swerling4(nfa, np, snr_inc);
  end
  PD_SW4=Pd_Sw4
  SNR_SW4=snr_inc
```

```
  Lf = SNR_SW4 - SNR_SW5
  end
  % *************** End Swerling 4 case ***************
return
```

Listing 32.26. MATLAB Program "fig32_15.m"

```
% Use this program to reproduce Fig. 32.15 of text
clear all
close all
index= 0.;
for pd= 0.01 :.05:1
  index= index+ 1;
  [LfPd_Sw5] = fluct_loss(pd, 1e-9,1,1);
  Lf1 (index) = Lf;
  [Lf,Pd_Sw5] = fluct_loss(pd, 1e-9,1,4);
  Lf4(index) = Lf;
end
pd= 0.01:.05:1;
figure (2)
plot(pd, Lf1, 'k',pd, Lf4, 'K:')
xlabel('Probability of detection')
ylabel('Fluctuation loss - dB')
legend('Swerling I & II', 'Swerling III & IV')
title('Pfa= 1e-9,np= 1')
grid
```

Listing 32.27. MATLAB Program "myradar_visit32_1.m"

```
% Myradar design case study visit 32_1
close all
clear all
pfa= 1e-7;
pd= 0.995;
np= 7;
pt= 165.8e3; % peak power in Watts
freq= 3e + 9; % radar operating frequency in Hz
g= 34.5139; % antenna gain in dB
sigmam= 0.5; % missile RCS m squared
sigmaa= 4; % aircraft RCS m squared
te= 290.0; % effective noise temperature in Kelvins
b= 1.0e + 6; % radar operating bandwidth in Hz
nf= 6.0; %noise figure in dB
loss= 8.0; % radar losses in dB
% compute the improvement factor due to 7-pulse noncoherent integration
Improv= improv_fac (np, pfa, pd);
% calculate the integration loss
lossnci= 10*log10(np) - Improv;
% calculate net gain in SNR due to integration
SNR_net= Improv- lossnci;
loss_total= loss+ lossnci;
```

```
rangem=55e3;
rangea=90e3;
SNR_single_pulse missile = radar_eq(pt, freq, g, sigmam, te, b, nf, loss,
rangem)
SNR_7_pulse_NCI_missile=SNR_single_pulse_missile+SNR_net
SNR_single_pulse_aircraft = radar_eq(pt, freq, g, sigmaa, te, b, nf, loss,
rangea) SNR_7_pulse_NCI_aircraft=SNR_single_pulse_aircraft+SNR_net
```

Listing 32.28. MATLAB Program "myradar_visit2_2.m"

```
%clear all
% close all
% swid=3;
%pfa=1e-7;
%np=1;
% R_1st_frcane=61 e3; % Range for first frame
% R0=55e3; % range to last frame
%SNRO=9;% SNR at R0
% frame=0.3e3; % frame size
nfa=log(2)/pfa;
range_frame=R_1st_frame:-frame:R0; % Range to each frame
% implement Eq. (32.98)
SNRi=SNR0+40* log 10 ( (R0./range_frame));
% calculate the Swerling 5 Pd at each frame
b=sqrt(-2.0 * log(pfa));
if np ==1
  for frame=1:1:size(SNRi,2)
    a=sqrt(2.0 * 10^(.1*SNRi(frame)));
    pd5(frame) =marcumsq(a,b);
  end
else
  [pd5]=pd_swerling5(pfa, 1, np, SNRi);
end
% compute additional SNR needed due to fluctuation
for frame=1:1:size(SNRi,2)
  [Lf(frame), Pd_Sw5]=fluct_loss (pd5 (frame), pfa, np, swid);
end
% adjust SNR at each frame
SNRi=SNRi-Lf;
% compute the frame Pd
for frame=1:1:size(SNRi,2)
  if(swid==1)
    Pdi(frame)=pd_swerling1 (nfa, np, SNRi (frame));
  end
  if(swid==2)
    Pdi(frame)=pd_swerling2 (nfa, np, SNRi(frame));
  end
  if(swid==3)
    Pdi(frame)=pd_swerling3 (nfa, np, SNRi(frame));
  end
```

```
    if(swid= =4)
      Pdi(frame)=pd_swerling4 (nfa, np, SNRi(frame));
    end
    if(swid= =5)
      Pdi(frame)=pd5(frame);
    End
end
Pdc(1 : size(SNRi, 2))=0;
Pdc(1)=1-Pdi(1);
% compute the cumulative Pd
for frame=2:1:size (SNRi, 2)
  Pdc(frame)= (1-Pdi(frame)) * Pdc(frame-1);
end
PDC=1-Pdc(21)
```

References

1. Fehlner, L. F., *Marcum's and Swerling's Data on Target Detection by a Pulsed Radar*, Johns Hopkins University, Applied Physics Lab. Rpt. TG451, July 2, 1962, and Rpt. TG451A, September 1964.
2. Parl, S., A new method of calculating the generalized Q function, *IEEE Transactions on Information Theory*, IT-26(1), 121–124, January 1980.
3. Peebles, R. Z., Jr. *Radar Principles,* John Wiley & Sons, Inc., New York, 1998.
4. Marcum, J. I., A statistical theory of target detection by pulsed radar, *IRE Transactions on Information Theory*, IT-6, 59–267, April 1960.
5. Swerling, P., Probability of detection for fluctuating targets, *IRE Transactions on Information Theory*, IT-6, 269–308, April 1960.
6. DiFranco, J. V. and Rubin, W. L., *Radar Detection,* Artech House, Norwood, MA, 1980.
7. Kanter, I., Exact detection probability for partially correlated Rayleigh targets, *IEEE Transactions*, AES-22, 184–196, March 1986.
8. Urkowitz, H., *Decision and Detection Theory*, unpublished lecture notes. Lockheed Martin Co., Moorestown, NJ.

33

Radar Waveforms

Bassem R. Mahafza
Deceibel Research, Inc.

Atef Z. Elsherbeni
University of Mississippi

Choosing a particular waveform type and a signal processing technique in a radar system depends heavily on the radar's specific mission and role. The cost and complexity associated with a certain type of waveform hardware and software implementation constitute a major factor in the decision process. Radar systems can use continuous waveforms (CW) or pulsed waveforms with or without modulation. Modulation techniques can be either analog or digital. Range and Doppler resolutions are directly related to the specific waveform frequency characteristics. Thus, knowledge of the power spectrum density (PSD) of a waveform is very critical. In general, signals or waveforms can be analyzed using time domain or frequency domain techniques. This chapter introduces many of the most commonly used radar waveforms. Relevant uses of a specific waveform will be addressed in the context of its time and frequency domain characteristics. In this book, the terms waveform and signal are used interchangeably to mean the same thing.

33.1 Low Pass, Band Pass Signals, and Quadrature Components

Signals that contain significant frequency composition at a low frequency band including DC are called low pass (LP) signals. Signals that have significant frequency composition around some frequency away from the origin are called band pass (BP) signals. A real BP signal $x(t)$ can be represented mathematically by

$$x(t) = r(t)\cos\left(2\pi f_0 t + \phi_x(t)\right) \tag{33.1}$$

where

$r(t)$ is the amplitude modulation or envelope

$\phi_x(t)$ is the phase modulation

f_0 is the carrier frequency

Both $r(t)$ and $\phi_x(t)$ have frequency components significantly smaller than f_0. The frequency modulation is

$$f_{\text{m}}(t) = \frac{1}{2\pi}\frac{\text{d}}{\text{d}t}\phi_x(t) \tag{33.2}$$

and the instantaneous frequency is

$$f_i(t) = \frac{1}{2\pi}\frac{\text{d}}{\text{d}t}(2\pi f_0 t + \phi_x(t)) = f_0 + f_m(t) \tag{33.3}$$

If the signal bandwidth is B, and if f_0 is very large compared to B, the signal $x(t)$ is referred to as a narrow band pass signal.

Band pass signals can also be represented by two low pass signals known as the quadrature components; in this case Equation 33.1 can be rewritten as

$$x(t) = x_I(t)\cos 2\pi f_0 t - x_Q(t)\sin 2\pi f_0 t \tag{33.4}$$

where $x_I(t)$ and $x_Q(t)$ are real LP signals referred to as the quadrature components and are given, respectively, by

$$\begin{aligned} x_I(t) &= r(t)\cos\phi_x(t) \\ x_Q(t) &= r(t)\sin\phi_x(t) \end{aligned} \tag{33.5}$$

Figure 33.1 shows how the quadrature components are extracted.

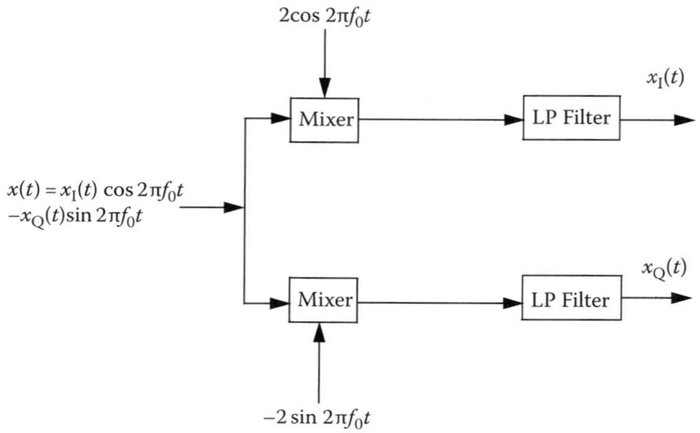

FIGURE 33.1 Extraction of quadrature components.

33.2 The Analytic Signal

The sinusoidal signal $x(t)$ defined in Equation 33.1 can be written as the real part of the complex signal $\psi(t)$. More precisely,

$$x(t) = \mathrm{Re}\{\psi(t)\} = \mathrm{Re}\{r(t)e^{j\phi_x(t)}e^{j2\pi f_0 t}\} \tag{33.6}$$

Define the "analytic signal" as

$$\psi(t) = v(t)e^{j2\pi f_0 t} \tag{33.7}$$

where

$$v(t) = r(t)e^{j\phi_x(t)} \tag{33.8}$$

and

$$\Psi(\omega) = \begin{bmatrix} 2X(\omega) \\ 0 \end{bmatrix} \quad \begin{matrix} \omega \geq 0 \\ \omega < 0 \end{matrix} \tag{33.9}$$

where
$\Psi(\omega)$ is the Fourier transform (FT) of $\psi(t)$
$X(\omega)$ is the FT of $x(t)$

Equation 33.9 can be written as

$$\Psi(\omega) = 2U(\omega)X(\omega) \tag{33.10}$$

where $U(\omega)$ is the step function in the frequency domain. Thus, it can be shown that $\psi(t)$ is

$$\psi(t) = x(t) + j\tilde{x}(t) \tag{33.11}$$

$\tilde{x}(t)$ is the Hubert transform of $x(t)$.

Using Equations 33.6 and 33.11, one can then write (shown here without proof)

$$x(t) = u_{0I}(t)\cos\omega_0 t - u_{0Q}(t)\sin\omega_0 t \tag{33.12}$$

which is similar to Equation 33.4 with $\omega_0 = 2\pi f_0$.

Using Parseval's theorem it can be shown that the energy associated with the signal $x(t)$ is

$$E_x = \frac{1}{2}\int_{-\infty}^{\infty} x^2(t)\mathrm{d}t = \frac{1}{2}\int_{-\infty}^{\infty} u^2(t)\mathrm{d}t = \frac{1}{2}E_\psi \tag{33.13}$$

33.3 CW and Pulsed Waveforms

The spectrum of a given signal describes the spread of its energy in the frequency domain. An energy signal (finite energy) can be characterized by its energy spectrum density (ESD) function, while a power signal (finite power) is characterized by the PSD function. The units of the ESD are Joules per Hertz and the PSD has units Watts per Hertz.

The signal bandwidth is the range of frequency over which the signal has a nonzero spectrum. In general, any signal can be defined using its duration (time domain) and bandwidth (frequency domain). A signal is said to be band-limited if it has finite bandwidth. Signals that have finite durations (time-limited) will have infinite bandwidths, while band-limited signals have infinite durations. The extreme case is a continuous sine wave, whose bandwidth is infinitesimal.

A time domain signal $f(t)$ has a FT $F(\omega)$ given by

$$F(\omega) = \int_{-\infty}^{\infty} f(t)e^{-j\omega t}dt \tag{33.14}$$

where the inverse Fourier transform (IFT) is

$$f(t) = \frac{1}{2\pi}\int_{-\infty}^{\infty} F(\omega)e^{j\omega t}d\omega \tag{33.15}$$

The signal autocorrelation function $R_f(\tau)$ is

$$R_f(\tau) = \int_{-\infty}^{\infty} f^*(t)f(t+\tau)dt \tag{33.16}$$

The asterisk indicates the complex conjugate. The signal amplitude spectrum is $|F(\omega)|$. If $f(t)$ were an energy signal, then its ESD is $|F(\omega)|^2$; and if it were a power signal, then its PSD is $\bar{S}_f(\omega)$ which is the FT of the autocorrelation function

$$\bar{S}_f(\omega) \int_{-\infty}^{\infty} \bar{R}_f(\tau)e^{-j\omega\tau}d\tau \tag{33.17}$$

First, consider a CW waveform given by

$$f_1(t) = A\cos\omega_0 t \tag{33.18}$$

The FT of $f_1(t)$ is

$$F_1(\omega) = A\pi[\delta(\omega-\omega_0)+\delta(\omega-\omega_0)] \tag{33.19}$$

where

$\delta(\cdot)$ is the Dirac delta function

$\omega_0 = 2\pi f_0$

As indicated by the amplitude spectrum shown in Figure 33.2, the signal $f_1(t)$ has infinitesimal bandwidth, located at $\pm f_0$.

Next consider the time domain signal $f_2(t)$ given by

$$f_2(t) = A\mathrm{Rect}\left(\frac{t}{\tau}\right) = \begin{Bmatrix} A & -\frac{\tau}{2} \le t \le \frac{\tau}{2} \\ 0 & \text{otherwise} \end{Bmatrix} \tag{33.20}$$

It follows that the FT is

$$F_2(\omega) = A\tau\mathrm{Sinc}\left(\frac{\omega\tau}{2}\right) \tag{33.21}$$

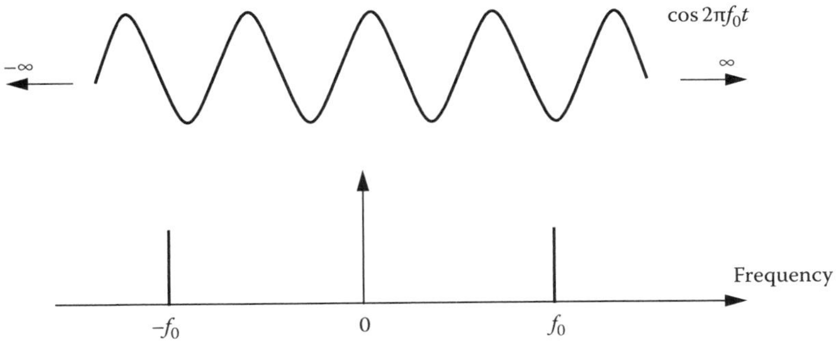

FIGURE 33.2 Amplitude spectrum for a continuous sine wave.

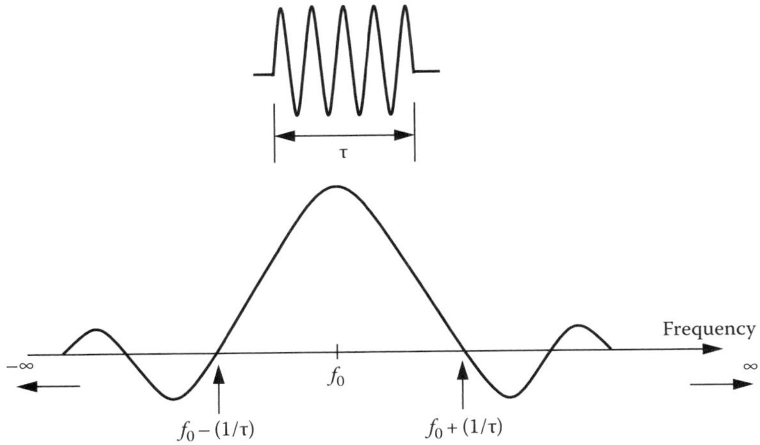

FIGURE 33.3 Amplitude spectrum for a single pulse, or a train of non-coherent pulses.

where

$$\text{Sinc}(x) = \frac{\sin(\pi x)}{\pi x} \tag{33.22}$$

The amplitude spectrum of $f_2(t)$ is shown in Figure 33.3. In this case, the bandwidth is infinite. Since infinite bandwidths cannot be physically implemented, the signal bandwidth is approximated by $2\pi/\tau$ rad/s or $1/\tau$ Hz. In practice, this approximation is widely accepted since it accounts for most of the signal energy.

Now consider the coherent gated CW waveform $f_3(t)$ given by

$$f_3(t) = \sum_{n=-\infty}^{\infty} f_2(t - nT) \tag{33.23}$$

Clearly $f_3(t)$ is periodic, where T is the period (recall that $f_r = 1/T$ is the PRF). Using the complex exponential Fourier series we can rewrite $f_3(t)$ as

$$f_3(t) = \sum_{n=-\infty}^{\infty} F_n e^{\frac{j2\pi nt}{T}} \tag{33.24}$$

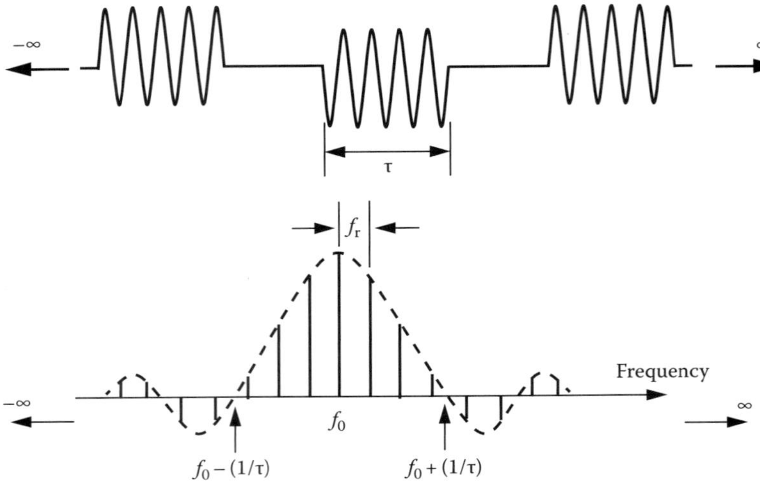

FIGURE 33.4 Amplitude spectrum for a coherent pulse train of infinite length.

where the Fourier series coefficients F_n are given by

$$F_n = \frac{A\tau}{T}\,\mathrm{Sinc}\left(\frac{n\tau\pi}{T}\right) \tag{33.25}$$

It follows that the FT of $f_3(t)$ is

$$F_3(\omega) = 2\pi \sum_{n=-\infty}^{\infty} F_n \delta(\omega - 2n\pi f_r) \tag{33.26}$$

The amplitude spectrum of $f_3(t)$ is shown in Figure 33.4. In this case, the spectrum has a sin x/x envelope that corresponds to F_n. The spacing between the spectral lines is equal to the radar PRF, f_r.

Finally, define the function $f_4(t)$ as

$$f_4(t) = \sum_{n=0}^{N} f_2(t - nT) \tag{33.27}$$

Note that $f_4(t)$ is a limited duration of $f_3(t)$. The FT of $f_4(t)$ is

$$F_4(\omega) = AN\tau\left(\mathrm{Sinc}\left(\omega\frac{NT}{2}\right) \bullet \sum_{n=-\infty}^{\infty} \mathrm{Sinc}(n\pi\tau f_r)\delta(\omega - 2n\pi f_r)\right) \tag{33.28}$$

where the operator (•) indicates convolution. The spectrum in this case is shown in Figure 33.5. The envelope is still a sin x/x which corresponds to the pulsewidth. But the spectral lines are replaced by sin x/x spectra that correspond to the duration NT.

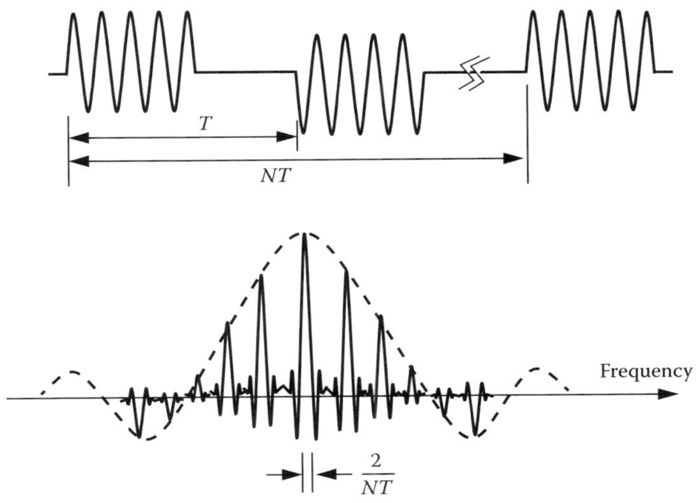

FIGURE 33.5 Amplitude spectrum for a coherent pulse train of finite length.

33.4 Linear Frequency Modulation Waveforms

Frequency or phase modulated waveforms can be used to achieve much wider operating bandwidths. Linear frequency modulation (LFM) is commonly used. In this case, the frequency is swept linearly across the pulsewidth, either upward (up-chirp) or downward (down-chirp). The matched filter bandwidth is proportional to the sweep bandwidth, and is independent of the pulsewidth. Figure 33.6 shows a typical example of an LFM waveform. The pulsewidth is τ, and the bandwidth is B.

The LFM up-chirp instantaneous phase can be expressed by

$$\psi(t) = 2\pi\left(f_0 t + \frac{\mu}{2}t^2\right) \quad -\frac{\tau}{2} \leq t \leq \frac{\tau}{2} \tag{33.29}$$

where

f_0 is the radar center frequency

$\mu = (2\pi B)/\tau$ is the LFM coefficient

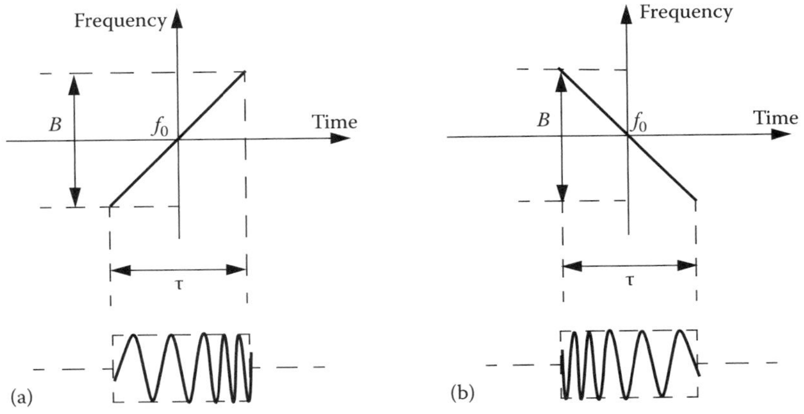

FIGURE 33.6 Typical LFM waveforms: (a) up-chirp; (b) down-chirp.

Thus, the instantaneous frequency is

$$f(t) = \frac{1}{2\pi}\frac{\mathrm{d}}{\mathrm{d}t}\psi(t) = f_0 + \mu t \qquad -\frac{\tau}{2} \le t \le \frac{\tau}{2} \tag{33.30}$$

Similarly, the down-chirp instantaneous phase and frequency are given, respectively, by

$$\psi(t) = 2\pi\left(f_0 t - \frac{\mu}{2}t^2\right) \qquad -\frac{\tau}{2} \le t \le \frac{\tau}{2} \tag{33.31}$$

$$f(t) = \frac{1}{2\pi}\frac{\mathrm{d}}{\mathrm{d}t}\psi(t) = f_0 - \mu t \qquad -\frac{\tau}{2} \le t \le \frac{\tau}{2} \tag{33.32}$$

A typical LFM waveform can be expressed by

$$s_1(t) = \mathrm{Rect}\left(\frac{t}{\tau}\right)\mathrm{e}^{j2\pi\left(f_0 t + \frac{\mu}{2}t^2\right)} \tag{33.33}$$

where
 Rect(t/τ) denotes a rectangular pulse of width τ

Equation 33.33 is then written as

$$s_1(t) = \mathrm{e}^{j2\pi f_0 t}s(t) \tag{33.34}$$

where

$$s(t) = \mathrm{Rect}\left(\frac{t}{\tau}\right)\mathrm{e}^{j\pi\mu t^2} \tag{33.35}$$

is the complex envelope of $s_1(t)$.

The spectrum of the signal $s_1(t)$ is determined from its complex envelope $s(t)$. The complex exponential term in Equation 33.34 introduces a frequency shift about the center frequency f_0. Taking the FT of $s(t)$ yields

$$S(\omega) = \int_{-\infty}^{\infty} \mathrm{Rect}\left(\frac{t}{\tau}\right)\mathrm{e}^{j\pi\mu t^2}\mathrm{e}^{-j\omega t}\mathrm{d}t = \int_{\frac{\tau}{2}}^{\frac{\tau}{2}} \exp\left(\frac{j2\pi\mu t^2}{2}\right)\mathrm{e}^{-j\omega t}\mathrm{d}t \tag{33.36}$$

Let $\mu' = 2\pi\mu = 2\pi B/\tau$, and perform the change of variable

$$x = \sqrt{\frac{\mu'}{\pi}}\left(t - \frac{\omega}{\mu'}\right); \quad \mathrm{d}x = \sqrt{\frac{\mu'}{\pi}}\mathrm{d}t \tag{33.37}$$

Thus, Equation 33.36 can be written as

$$S(\omega) = \sqrt{\frac{\pi}{\mu'}}\mathrm{e}^{-j\omega^2/2\mu'}\int_{-x_1}^{x_2} \mathrm{e}^{j\pi x^2/2}\mathrm{d}x \tag{33.38}$$

$$S(\omega) = \sqrt{\frac{\pi}{\mu'}}\mathrm{e}^{-j\omega^2/2\mu'}\left\{\int_0^{x_2} \mathrm{e}^{j\pi x^2/2}\mathrm{d}x - \int_0^{-x_1} \mathrm{e}^{j\pi x^2/2}\mathrm{d}x\right\} \tag{33.39}$$

where

$$x_1 = \sqrt{\frac{\mu'}{\pi}}\left(\frac{\tau}{2} + \frac{\omega}{\mu'}\right) = \sqrt{\frac{B\tau}{2}}\left(1 + \frac{f}{B/2}\right) \quad (33.40)$$

$$x_2 = \sqrt{\frac{\mu'}{\pi}}\left(\frac{\tau}{2} + \frac{\omega}{\mu'}\right) = \sqrt{\frac{B\tau}{2}}\left(1 + \frac{f}{B/2}\right) \quad (33.41)$$

The Fresnel integrals, denoted by $C(x)$ and $S(x)$, are defined by

$$C(x) = \int_0^x \cos\left(\frac{\pi \upsilon^2}{2}\right) d\upsilon \quad (33.42)$$

$$S(x) = \int_0^x \sin\left(\frac{\pi \upsilon^2}{2}\right) d\upsilon \quad (33.43)$$

Fresnel integrals are approximated by

$$C(x) \approx \frac{1}{2} + \frac{1}{\pi x} \sin\left(\frac{\pi}{2}x^2\right); \quad x \gg 1 \quad (33.44)$$

$$S(x) \approx \frac{1}{2} + \frac{1}{\pi x} \cos\left(\frac{\pi}{2}x^2\right); \quad x \gg 1 \quad (33.45)$$

Note that $C(-x) = -C(x)$ and $S(-x) = -S(x)$. Figure 33.7 shows a plot of both $C(x)$ and $S(x)$ for $0 \leq x \leq 4.0$. This figure can be reproduced using MATLAB® program *"fig33_7.m"* given in Listing 33.1.

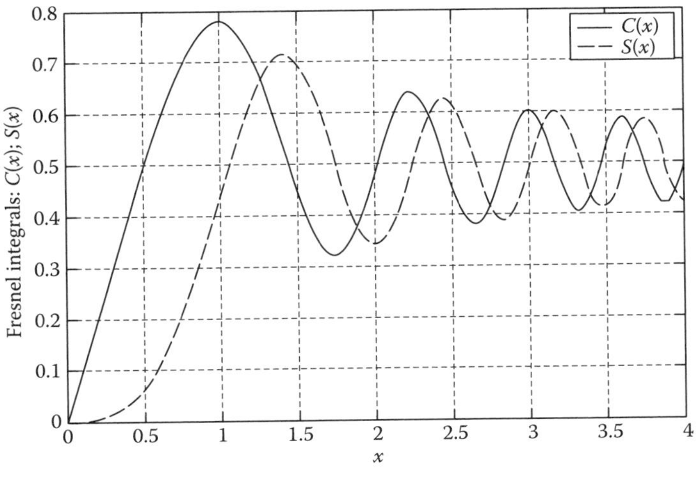

FIGURE 33.7 Fresnel integrals.

Using Equations 33.42 and 33.43 into Equation 33.39 and performing the integration yield

$$S(\omega) = \tau\sqrt{\frac{1}{B\tau}}e^{-j\omega^2/(4\pi B)}\left\{\frac{[C(x_2) + C(x_1)] + j[S(x_2) + S(x_1)]}{\sqrt{2}}\right\} \tag{33.46}$$

Figure 33.8 shows typical plots for the LFM real part, imaginary part, and amplitude spectrum. The square-like spectrum shown in Figure 33.8c is widely known as the Fresnel spectrum. This figure can be reproduced using MATLAB program *"fig33_8.m,"* given in Listing 33.2.

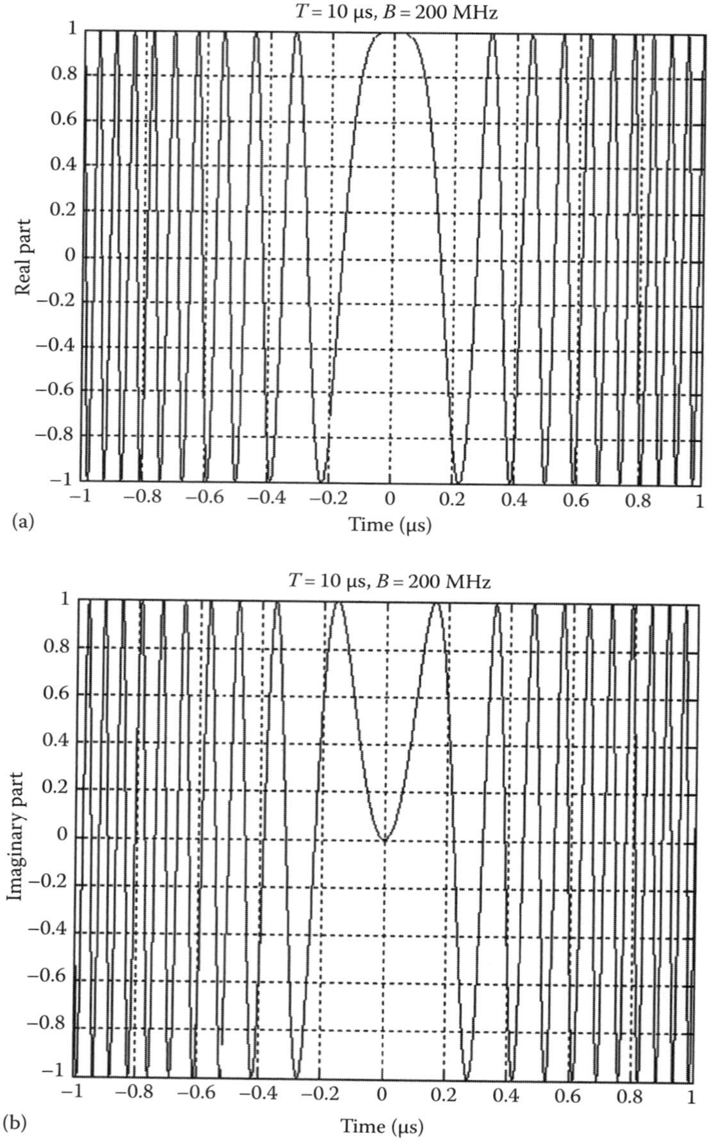

FIGURE 33.8 (a) Typical LFM waveform, real part. (b) Typical LFM waveform, imaginary part.

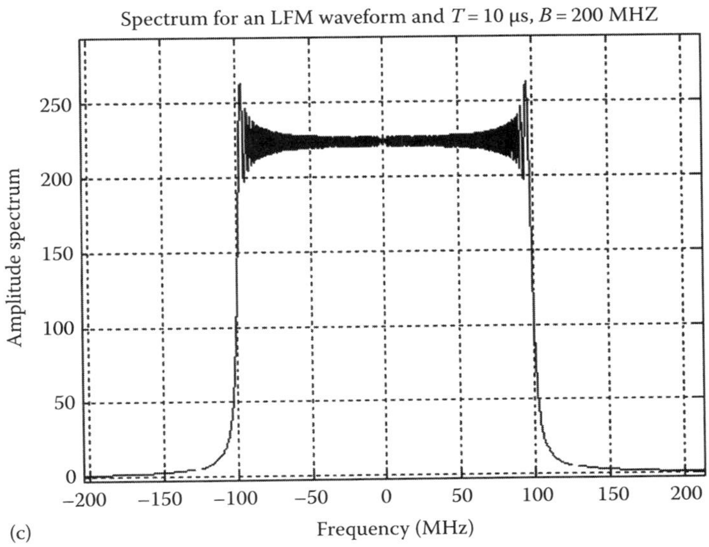

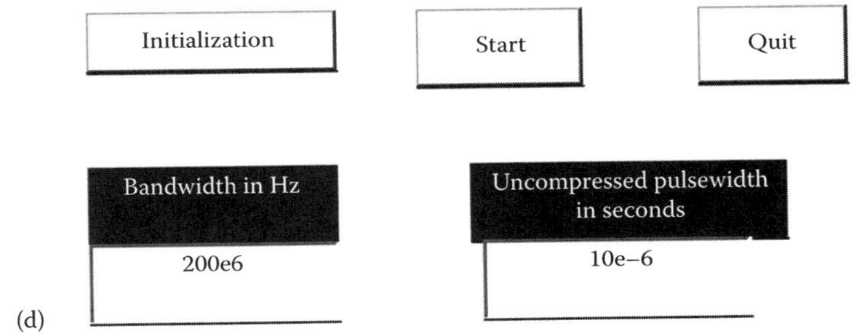

FIGURE 33.8 (continued) (c) Typical spectrum for an LFM waveform. (d) GUI workspace *"LFM_gui.m."*

A MATLAB GUI (see Figure 33.8d) was developed to input LFM data and display outputs as shown in Figure 33.8. It is called *"LFM_gui.m."* Its inputs are the uncompressed pulsewidth and the chirp bandwidth.

33.5 High Range Resolution

When pulse compression is not used, the instantaneous bandwidth B of radar receiver is normally matched to the pulse bandwidth, and in most radar applications this is done by setting $B = 1/\tau$. Therefore, range resolution is given by

$$\Delta R = (c\tau)/2 = c/(2B) \qquad (33.47)$$

Radar users and designers alike seek to accomplish high range resolution (HRR) by minimizing ΔR. However, as suggested by Equation 33.47 in order to achieve HRR one must use very short pulses and consequently reduce the average transmitted power and impose severe operating bandwidth requirements. Achieving fine range resolution while maintaining adequate average transmitted power can be accomplished by using pulse compression techniques. By means of frequency or phase modulation, pulse compression allows us to achieve the average transmitted power of a relatively long pulse, while obtaining the range resolution corresponding to a very short pulse. As an example, consider an LFM waveform whose bandwidth is B and uncompressed pulsewidth (transmitted) is τ. After pulse compression the compressed pulsewidth is denoted by τ', where $\tau' \ll \tau$, and the HRR is

$$\Delta R = \frac{c\tau'}{2} \ll \frac{c\tau}{2} \tag{33.48}$$

LFM and frequency-modulated (FM) CW waveforms are commonly used to achieve HRR. HRR can also be synthesized using a class of waveforms known as the "Stepped Frequency Waveforms" (SFW). Stepped frequency waveforms require more complex hardware implementation as compared to LFM or FM-CW; however, the radar operating bandwidth requirements are less restrictive. This is true because the receiver instantaneous bandwidth is matched to the SFW subpulse bandwidth which is much smaller than the LFM or FM-CW bandwidth. A brief discussion of SFW waveforms is presented in the following section.

33.6 Stepped Frequency Waveforms

SFW produce Synthetic HRR target profiles because the target range profile is computed by means of inverse discrete Fourier ransformation (IDFT) of frequency domain samples of the actual target range profile. The process of generating a synthetic HRR profile is described in Wehner [1]. It is summarized as follows:

1. A series of n narrow-band pulses are transmitted. The frequency from pulse to pulse is stepped by a fixed frequency step Δf. Each group of n pulses is referred to as a burst.
2. The received signal is sampled at a rate that coincides with the center of each pulse.
3. The quadrature components for each burst are collected and stored.
4. Spectral weighting (to reduce the range sidelobe levels) is applied to the quadrature components. Corrections for target velocity, phase, and amplitude variations are applied.
5. The IDFT of the weighted quadrature components of each burst is calculated to synthesize a range profile for that burst. The process is repeated for N bursts to obtain consecutive synthetic HRR profiles.

Figure 33.9 shows a typical SFW burst. The pulse repetition interval (PRI) is T, and the pulsewidth is τ'. Each pulse can have its own LFM, or other type of modulation; in this book LFM is assumed. The center frequency for the ith step is

$$f_i = f_0 + i\Delta f; \quad i = 0, n-1 \tag{33.49}$$

Within a burst, the transmitted waveform for the ith step can be described as

$$s_i(t) = \begin{pmatrix} C_i \cos 2\pi f_i t + \theta_i; & iT \le t \le iT + \tau' \\ 0 & \text{elsewhere} \end{pmatrix} \tag{33.50}$$

where

θ_i are the relative phases

C_i are constants

The received signal from a target located at range R_0 at time $t = 0$ is then given by

$$s_{ri}(t) = C_i' \cos(2\pi f_i(t - \tau(t)) + \theta_i); \quad iT + \tau(t) \le t \le iT + \tau' + \tau(t) \tag{33.51}$$

where C'_i are constant and the round trip delay $\tau(t)$ is given by

$$\tau(t) = \frac{R_0 - vt}{c/2} \tag{33.52}$$

c is the speed of light and v is the target radial velocity.

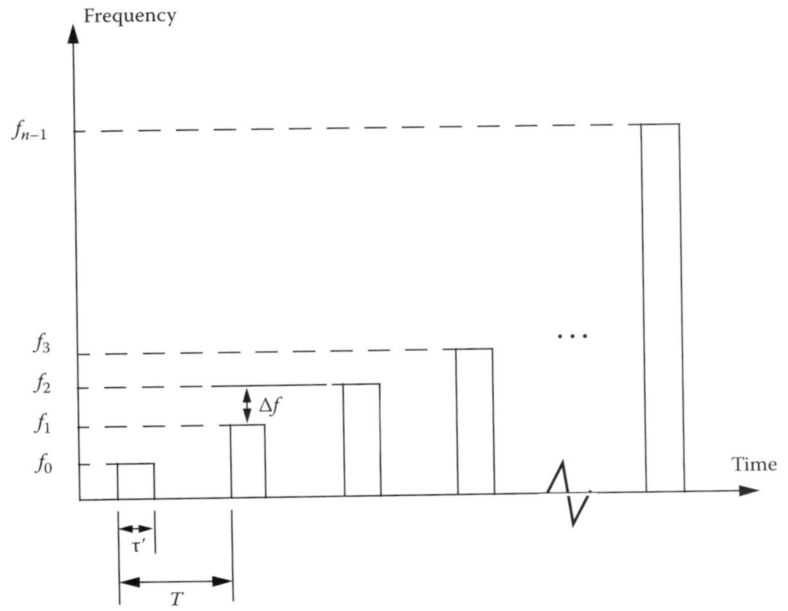

FIGURE 33.9 Stepped frequency waveform burst.

The received signal is down-converted to base-band in order to extract the quadrature components. More precisely, $s_{ri}(t)$ is mixed with the signal

$$y_i(t) = C\cos(2\pi f_i t + \theta_i); \quad iT \leq t \leq iT + \tau' \tag{33.53}$$

After low pass filtering, the quadrature components are given by

$$\begin{pmatrix} x_I(t) \\ x_Q(t) \end{pmatrix} = \begin{pmatrix} A_i \cos\psi_i(t) \\ A_i \sin\psi_i(t) \end{pmatrix} \tag{33.54}$$

where A_i are constants, and

$$\psi_i(t) = -2\pi f_i\left(\frac{2R_0}{c} - \frac{2vt}{c}\right) \tag{33.55}$$

where now $f_i = \Delta f$. For each pulse, the quadrature components are then sampled at

$$t_i = iT + \frac{\tau_r}{2} + \frac{2R_0}{c} \tag{33.56}$$

τ_r is the time delay associated with the range that corresponds to the start of the range profile.

The quadrature components can then be expressed in complex form as

$$X_i = A_i e^{j\psi_i} \tag{33.57}$$

Equation 33.57 represents samples of the target reflectivity, due to a single burst, in the frequency domain. This information can then be transformed into a series of range delay reflectivity (i.e., range profile) values by using the IDFT. It follows that

$$H_l = \frac{1}{n}\sum_{i=0}^{n-1} X_i \exp\left(j\frac{2\pi li}{n}\right); \quad 0 \le l \le n-1 \tag{33.58}$$

Substituting Equations 33.57 and 33.55 into Equation 33.58 and collecting terms yield

$$H_l = \frac{1}{n}\sum_{i=0}^{n-1} A_i \exp\left\{j\left(\frac{2\pi li}{n} - 2\pi f_i\left(\frac{2R_0}{c} - \frac{2vt_i}{c}\right)\right)\right\} \tag{33.59}$$

By normalizing with respect to n and by assuming that $A_i = 1$ and that the target is stationary (i.e., $v = 0$), then Equation 33.59 can be written as

$$H_l = \sum_{i=0}^{n-1} \exp\left\{j\left(\frac{2\pi li}{n} - 2\pi f_i \frac{2R_0}{c}\right)\right\} \tag{33.60}$$

Using $f_i = i\Delta f$ inside Equation 33.60 yields

$$H_l = \sum_{i=0}^{n-1} \exp\left\{j\frac{2\pi i}{n}\left(-\frac{2nR_0\Delta f}{c} + l\right)\right\} \tag{33.61}$$

which can be simplified to

$$H_l = \frac{\sin \pi\chi}{\sin \frac{\pi\chi}{n}} \exp\left(j\frac{n-1}{2}\frac{2\pi\chi}{n}\right) \tag{33.62}$$

where

$$\chi = \frac{-2nR_0\Delta f}{c} + l \tag{33.63}$$

Finally, the synthesized range profile is

$$|H_l| = \left|\frac{\sin \pi\chi}{\sin \frac{\pi\chi}{n}}\right| \tag{33.64}$$

33.6.1 Range Resolution and Range Ambiguity in SFW

As usual, range resolution is determined from the overall system bandwidth. Assuming a SFW with n steps, and step size Δf, then the corresponding range resolution is equal to

$$\Delta R = \frac{c}{2n\Delta f} \tag{33.65}$$

Range ambiguity associated with a SFW can be determined by examining the phase term that corresponds to a point scatterer located at range R_0. More precisely,

$$\psi_i(t) = 2\pi f_i \frac{2R_0}{c} \tag{33.66}$$

It follows that

$$\frac{\Delta\psi}{\Delta f} = \frac{4\pi(f_{i+1} - f_i)}{(f_{i+1} - f_i)} \frac{R_0}{c} = \frac{4\pi R_0}{c} \tag{33.67}$$

or equivalently,

$$R_0 = \frac{\Delta\psi}{\Delta f} \frac{c}{4\pi} \tag{33.68}$$

It is clear from Equation 33.68 that range ambiguity exists for $\Delta\psi = \Delta\psi + 2n\pi$. Therefore,

$$R_0 = \frac{\Delta\psi + 2n\pi}{\Delta f} \frac{c}{4\pi} = R_0 + n\left(\frac{c}{2\Delta f}\right) \tag{33.69}$$

and the unambiguous range window is

$$R_u = \frac{c}{2\Delta f} \tag{33.70}$$

Hence, a range profile synthesized using a particular SFW represents the relative range reflectivity for all scatterers within the unambiguous range window, with respect to the absolute range that corresponds to the burst time delay. Additionally, if a specific target extent is larger than R_u, then all scatterers falling outside the unambiguous range window will fold over and appear in the synthesized profile. This fold-over problem is identical to the spectral fold-over that occurs when using a fast Fourier transform (FFT) to resolve certain signal frequency contents. For example, consider an FFT with frequency resolution $\Delta f = 50$ Hz, and size NFFT = 64. In this case, this FFT can resolve frequency tones between −1600 Hz and 1600 Hz. When this FFT is used to resolve the frequency content of a sine-wave tone equal to 1800 Hz, fold-over occurs and a spectral line at the fourth FFT bin (i.e., 200 Hz) appears. Therefore, in order to avoid fold-over in the synthesized range profile, the frequency step Δf must be

$$\Delta f \leq c/2E \tag{33.71}$$

where E is the target extent in meters.

Additionally, the pulsewidth must also be large enough to contain the whole target extent. Thus,

$$\Delta f \leq 1/\tau' \tag{33.72}$$

and, in practice,

$$\Delta f \leq 1/2\tau' \tag{33.73}$$

This is necessary in order to reduce the amount of contamination of the synthesized range profile caused by the clutter surrounding the target under consideration.

MATLAB Function "hrr_profile.m"

The function *"hrr_profile.m"* computes and plots the synthetic HRR profile for a specific SFW. It is given in Listing 33.3. This function utilizes an inverse fast Fourier transform (IFFT) of a size equal to twice the number of steps. Hamming window of the same size is also assumed. The syntax is as follows:

[hl] = hrr_profile (nscat, scat_range, scat_rcs, n, deltaf, prf, v, r0, winid) where

Symbol	Description	Units	Status
nscat	Number of scatterers that make up the target	None	Input
scat_range	Vector containing range to individual scatterers	m	Input
scat_rcs	Vector containing RCS of individual scatterers	m^2	Input
n	Number of steps	None	Input
deltaf	Frequency step	Hz	Input
prf	PRF of SFW	Hz	Input
v	Target velocity	m/s	Input
r0	Profile starting range	m	Input
winid	Number > 0 for Hamming window number < 0 for no window	None	Input
hl	Range profile	dB	Output

For example, assume that the range profile starts at $R_0 = 900$ m and that

nscat	tau	N	deltaf	Prf	V
3	100 μ s	64	10 MHz	10 KHz	0.0

In this case,

$$\Delta R = \frac{3 \times 10^8}{2 \times 64 \times 10 \times 10^6} = 0.235 \text{ m}$$

$$R_u = \frac{3 \times 10^8}{2 \times 10 \times 10^6} = 15 \text{ m}$$

Thus, scatterers that are more than 0.235 m apart will appear as distinct peaks in the synthesized range profile. Assume two cases; in the first case,

[scat range] = [908, 910, 912] meters, and in the second case, *[scat range] = [908, 910, 910.2] meters.* In both cases, let *[scat rcs] = [100, 10, 1] meters squared.*

Figure 33.10 shows the synthesized range profiles generated using the function *"hrr_profile.m"* and the first case when the Hamming window is not used. Figure 33.11 is similar to Figure 33.10, except in this case the Hamming window is used. Figure 33.12 shows the synthesized range profile that corresponds to the second case (Hamming window is used). Note that all three scatterers were resolved in Figures 33.10 and 33.11; however, the last two scatterers are not resolved in Figure 33.12, since they are separated by less than ΔR.

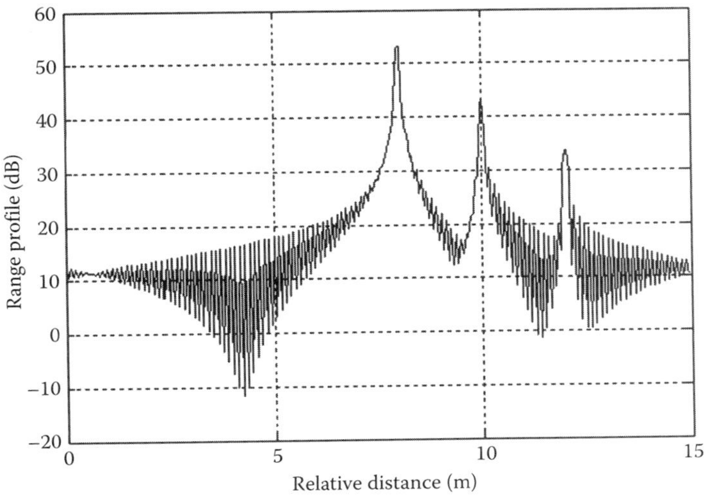

FIGURE 33.10 Synthetic range profile for three resolved scatterers. No window.

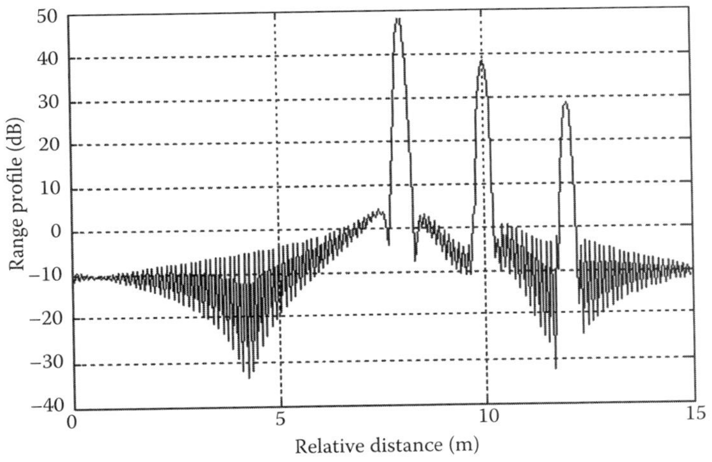

FIGURE 33.11 Synthetic range profile for three scatterers. Hamming window.

Next, consider another case where *[scat_range] = [908, 912, 916] meters*. Figure 33.13 shows the corresponding range profile. In this case, fold-over occurs, and the last scatterer appears at the lower portion of the synthesized range profile. Also, consider the case where

$$[scat\_range] = [908, 910, 923]\ meters$$

Figure 33.14 shows the corresponding range profile. In this case, ambiguity is associated with the first and third scatterers since they are separated by 15 m. Both appear at the same range bin.

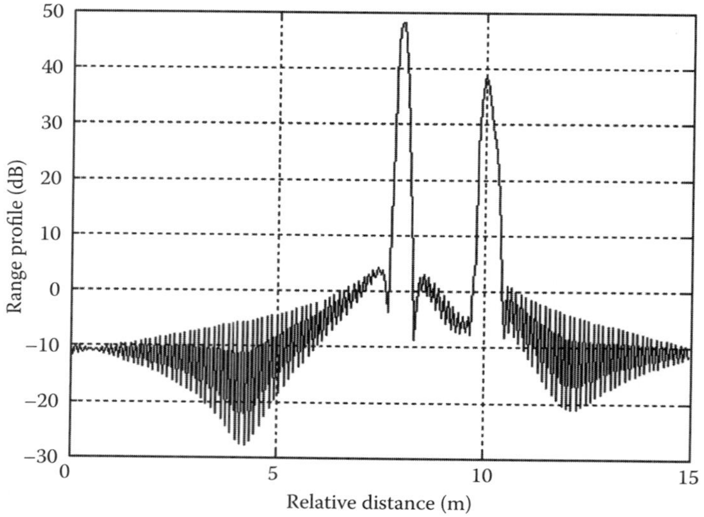

FIGURE 33.12 Synthetic range profile for three scatterers. Two are unresolved.

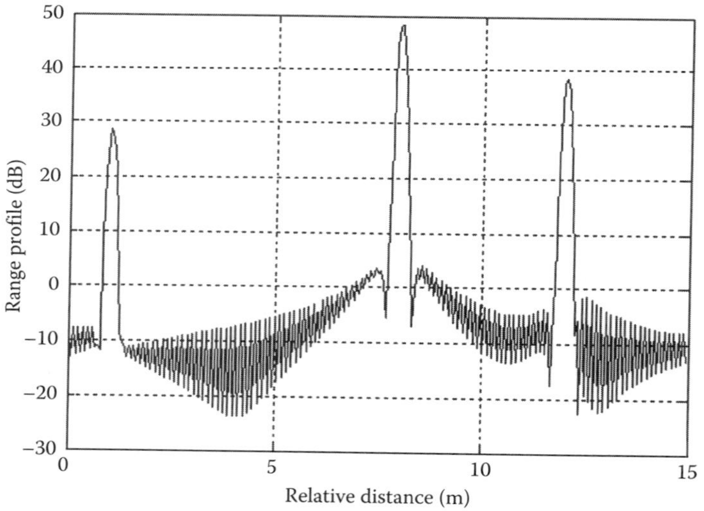

FIGURE 33.13 Synthetic range profile for three scatterers. Third scatterer folds over.

33.6.2 Effect of Target Velocity

The range profile defined in Equation 33.64 is obtained by assuming that the target under examination is stationary. The effect of target velocity on the synthesized range profile can be determined by substituting Equations 33.55 and 33.56 into Equation 33.58, which yields

$$H_l = \sum_{i=0}^{n-1} A_i \exp\left\{ j\frac{2\pi li}{n} - j2\pi f_i \left[\frac{2R}{c} - \frac{2v}{c}\left(iT + \frac{\tau_1}{2} + \frac{2R}{c} \right) \right] \right\} \tag{33.74}$$

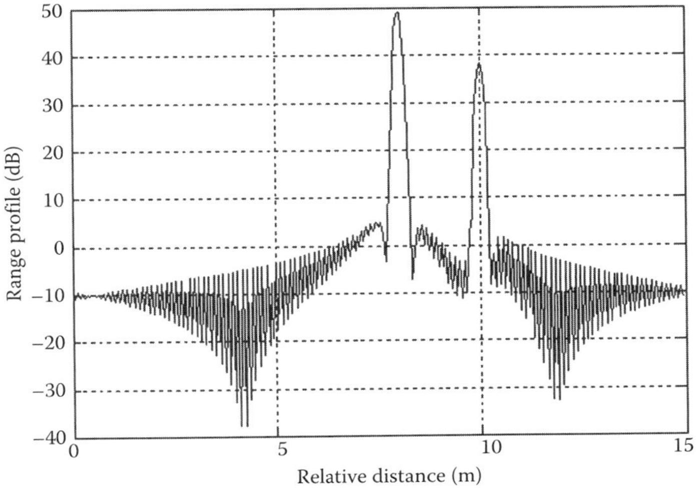

FIGURE 33.14 Synthetic range profile for three scatterers. The first and third scatterers appear in the same FFT bin.

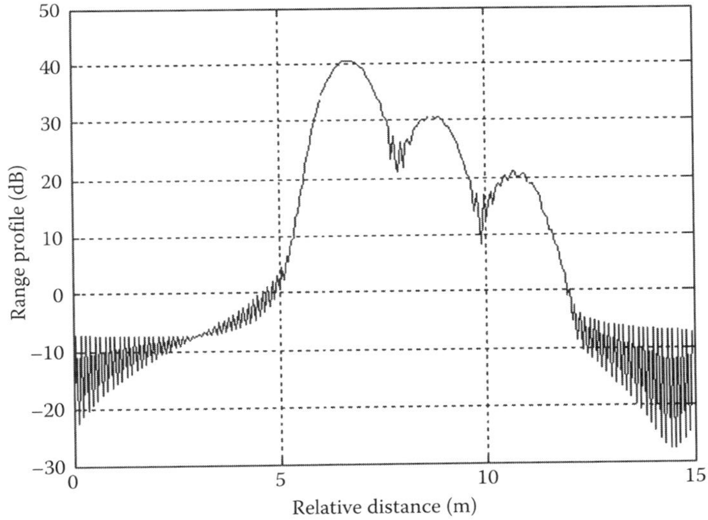

FIGURE 33.15 Illustration of range profile distortion due to target velocity.

The additional phase term present in Equation 33.74 distorts the synthesized range profile. In order to illustrate this distortion, consider the SFW described in Section 33.6.1, and assume the three scatterers of the first case. Also, assume that $v = 100$ m/s. Figure 33.15 shows the synthesized range profile for this case. Comparisons of Figures 33.11 and 33.15 clearly show the distortion effects caused by the uncompensated target velocity. Figure 33.16 is similar to Figure 33.15 except in this case, $v = -100$ m/s. Note in either case, the targets have moved from their expected positions (to the left or right) by $Disp = 2 \times n \times v/PRF$ *(1.28 m)*.

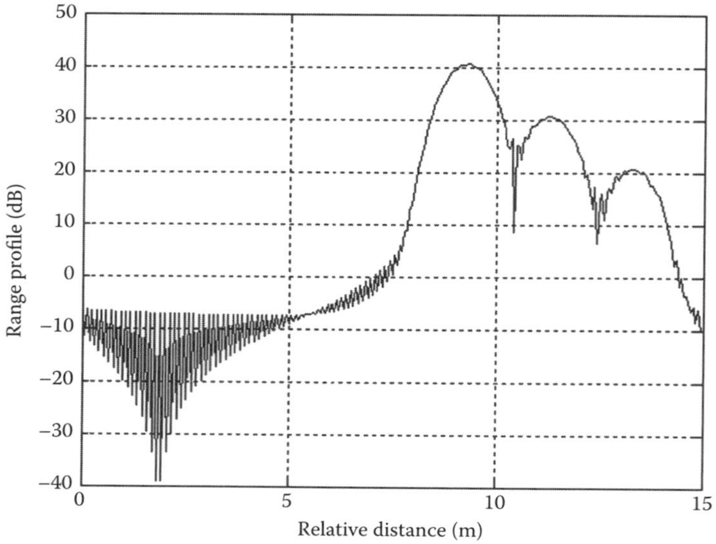

FIGURE 33.16 Illustration of range profile distortion due to target velocity.

This distortion can be eliminated by multiplying the complex received data at each pulse by the phase term

$$\Phi = \exp\left(-j2\pi f_i\left[\frac{2\underset{\sim}{v}}{c}\left(iT + \frac{\tau_1}{2} + \frac{2\underset{\sim}{R}}{c}\right)\right]\right) \tag{33.75}$$

$\underset{\sim}{v}$ and $\underset{\sim}{R}$ are, respectively, estimates of the target velocity and range. This process of modifying the phase of the quadrature components is often referred to as "phase rotation." In practice, when good estimates of $\underset{\sim}{v}$ and $\underset{\sim}{R}$ are not available, then the effects of target velocity are reduced by using frequency hopping

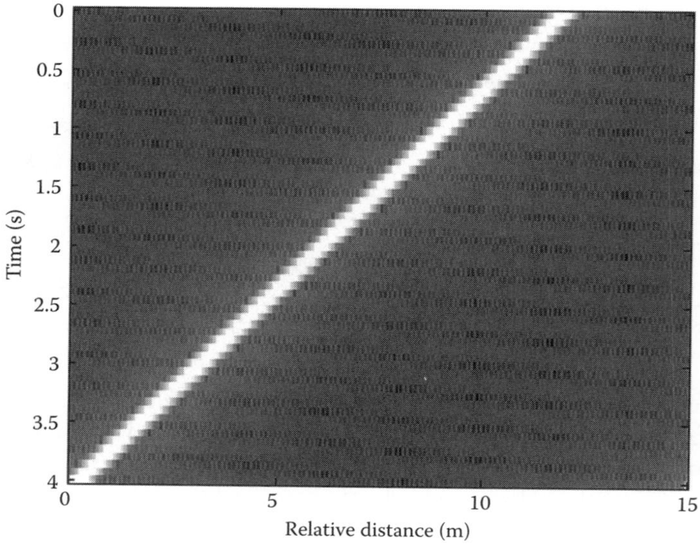

FIGURE 33.17 Synthesized range profile for a moving target (4 s long).

between the consecutive pulses within the SFW. In this case, the frequency of each individual pulse is chosen according to a predetermined code. Waveforms of this type are often called frequency coded waveforms (FCW). Costas waveforms or signals are a good example of this type of waveform.

Figure 33.17 shows a synthesized range profile for a moving target whose RCS is $\sigma = 10\ \text{m}^2$ and $v = 15$ m/s. The initial target range is at $R = 912$ m. All other parameters are as before. This figure can be reproduced using the MATLAB program "*fig33_17.m*" given in Listing 33.4.

33.7 The Matched Filter

The most unique characteristic of the matched filter is that it produces the maximum achievable instantaneous SNR at its output when a signal plus additive white noise is present at the input. The noise does not need to be Gaussian. The peak instantaneous SNR at the receiver output can be achieved by matching the radar receiver transfer function to the received signal. We will show that the peak instantaneous signal power divided by the average noise power at the output of a matched filter is equal to twice the input signal energy divided by the input noise power, regardless of the waveform used by the radar. This is the reason why matched filters are often referred to as optimum filters in the SNR sense. Note that the peak power used in the derivation of the radar equation (SNR) represents the average signal power over the duration of the pulse, not the peak instantaneous signal power as in the case of a matched filter. In practice, it is sometimes difficult to achieve perfect matched filtering. In such cases, suboptimum filters may be used. Due to this mismatch, degradation in the output SNR occurs.

Consider a radar system that uses a finite duration energy signal $s_i(t)$. Denote the pulsewidth as τ', and assume that a matched filter receiver is utilized. The main question that we need to answer is: What is the impulse, or frequency, response of the filter that maximizes the instantaneous SNR at the output of the receiver when a delayed version of the signal $s_i(t)$ plus additive white noise is at the input?

The matched filter input signal can then be represented by

$$x(t) = Cs_i(t - t_1) + n_i(t) \tag{33.76}$$

where

C is a constant

t_1 is an unknown time delay proportional to the target range

$n_i(t)$ is input white noise

Since the input noise is white, its corresponding autocorrelation and power spectral density (PSD) functions are given, respectively, by

$$\bar{R}_{n_i}(t) = \frac{N_0}{2}\delta(t) \tag{33.77}$$

$$\bar{S}_{n_i}(\omega) = \frac{N_0}{2} \tag{33.78}$$

where N_0 is a constant.

Denote $s_o(t)$ and $n_o(t)$ as the signal and noise filter outputs, respectively. More precisely, we can define

$$y(t) = Cs_o(t - t_1) + n_o(t) \tag{33.79}$$

where

$$s_o(t) = s_i(t) \bullet h(t) \tag{33.80}$$

$$n_o(t) = n_i(t) \bullet h(t) \tag{33.81}$$

The operator ($\bullet$) indicates convolution, and $h(t)$ is the filter impulse response (the filter is assumed to be linear time invariant).

Let $R_h(t)$ denote the filter autocorrelation function. It follows that the output noise autocorrelation and PSD functions are

$$\bar{R}_{n_o}(t) = \bar{R}_{n_i}(t) \bullet R_h(t) = \frac{N_0}{2}\delta(t) \bullet R_h(t) = \frac{N_0}{2}R_h(t) \tag{33.82}$$

$$\bar{S}_{n_o}(\omega) = \bar{S}_{n_i}(\omega)|H(\omega)|^2 = \frac{N_0}{2}|H(\omega)|^2 \tag{33.83}$$

where $H(\omega)$ is the FT for the filter impulse response, $h(t)$. The total average output noise power is equal to $\bar{R}_{n_o}(t)$ evaluated at $t = 0$. More precisely,

$$\bar{R}_{n_o}(0) = \frac{N_0}{2}\int_{-\infty}^{\infty}|h(u)|^2 du \tag{33.84}$$

The output signal power evaluated at time t is $|Cs_o(t - t_1)|^2$, and by using Equation 33.80 we get

$$s_o(t - t_1) = \int_{-\infty}^{\infty} s_i(t - t_1 - u)h(u)du \tag{33.85}$$

A general expression for the output SNR at time t can be written as

$$\text{SNR}(t) = \frac{|Cs_o(t - t_1)|^2}{\bar{R}_{n_o}(0)} \tag{33.86}$$

Substituting Equations 33.84 and 33.85 into Equation 33.86 yields

$$\text{SNR}(t) = \frac{C^2\left|\int_{-\infty}^{\infty} s_i(t - t_1 - u)h(u)du\right|^2}{\frac{N_0}{2}\int_{-\infty}^{\infty}|h(u)|^2 du} \tag{33.87}$$

The Schwartz inequality states that

$$\left|\int_{-\infty}^{\infty} P(x)Q(x)dx\right|^2 \leq \int_{-\infty}^{\infty}|P(x)|^2 dx \int_{-\infty}^{\infty}|Q(x)|^2 dx \tag{33.88}$$

where the equality applies only when $P = kQ^*$, where k is a constant and can be assumed to be unity. Then by applying Equation 33.88 on the numerator of Equation 33.87, we get

$$\mathrm{SNR}(t) \leq \frac{C^2 \int_{-\infty}^{\infty} |s_i(t - t_1 - u)|^2 du \int_{-\infty}^{\infty} |h(u)|^2 du}{\frac{N_0}{2} \int_{-\infty}^{\infty} |h(u)|^2 du} = \frac{2C^2 \int_{-\infty}^{\infty} |s_i(t - t_1 - u)|^2 du}{N_0} \tag{33.89}$$

Equation 33.89 tells us that the peak instantaneous SNR occurs when equality is achieved (i.e., from Equation 33.88 $h = ks_i^*$). More precisely, if we assume that equality occurs at $t = t_0$, and that $k = 1$, then

$$h(u) = s_i^*(t_0 - t_1 - u) \tag{33.90}$$

and the maximum instantaneous SNR is

$$\mathrm{SNR}(t_0) = \frac{2C^2 \int_{-\infty}^{\infty} |s_i(t_0 - t_1 - u)|^2 du}{N_0} \tag{33.91}$$

Equation 33.91 can be simplified using Parseval's theorem,

$$E = C^2 \int_{-\infty}^{\infty} |s_i(t_0 - t_1 - u)|^2 du \tag{33.92}$$

where E denotes the energy of the input signal; consequently we can write the output peak instantaneous SNR as

$$\mathrm{SNR}(t_0) = \frac{2E}{N_0} \tag{33.93}$$

Thus, we can draw the conclusion that the peak instantaneous SNR depends only on the signal energy and input noise power, and is independent of the waveform utilized by the radar.

Finally, we can define the impulse response for the matched filter from Equation 33.90. If we desire the peak to occur at $t_0 = t_1$, we get the noncausal matched filter impulse response,

$$h_{nc}(t) = s_i^*(-t) \tag{33.94}$$

Alternatively, the causal impulse response is

$$h_c(t) = s_i^*(\tau - t) \tag{33.95}$$

where, in this case, the peak occurs at $t_0 = t_1 + \tau$. It follows that the FTs of $h_{nc}(t)$ and $h_c(t)$ are given, respectively, by

$$H_{nc}(\omega) = s_i^*(\omega) \tag{33.96}$$

$$H_c(\omega) = s_i^*(\omega)e^{-j\omega\tau} \tag{33.97}$$

where $S_i(\omega)$ is the FT of $s_i(t)$. Thus, the moduli of $H(\omega)$ and $S_i(\omega)$ are identical; however, the phase responses are opposite of each other.

Example

Compute the maximum instantaneous SNR at the output of a linear filter whose impulse response is matched to the signal $x(t) = \exp(-t^2/2T)$.

Solution

The signal energy is

$$E = \int_{-\infty}^{\infty} |x(t)|^2 dt = \int_{-\infty}^{\infty} e^{(-t^2)/T} dt = \sqrt{\pi T}\ \text{J}$$

It follows that the maximum instantaneous SNR is

$$\text{SNR} = \frac{\sqrt{\pi T}}{N_0/2}$$

where $N_0/2$ is the input noise PSD.

33.8 The Replica

Again, consider a radar system that uses a finite duration energy signal $s_i(t)$, and assume that a matched filter receiver is utilized. The input signal is given in Equation 33.76 and is repeated here as Equation 33.98,

$$x(t) = Cs_i(t - t_1) + n_i(t) \tag{33.98}$$

The matched filter output $y(t)$ can be expressed by the convolution integral between the filter's impulse response and $x(t)$,

$$y(t) = \int_{-\infty}^{\infty} x(u)h(t - u)du \tag{33.99}$$

Substituting Equation 33.95 into Equation 33.99 yields

$$y(t) = \int_{-\infty}^{\infty} x(u)s_i^*(\tau - t + u)du = \bar{R}_{xs_i}(t - \tau) \tag{33.100}$$

where $\bar{R}_{xs_i}(t - \tau)$ is a cross-correlation between $x(t)$ and $s_i(\tau - t)$. Therefore, the matched filter output can be computed from the cross-correlation between the radar received signal and a delayed replica of the transmitted waveform. If the input signal is the same as the transmitted signal, the output of the matched filter would be the autocorrelation function of the received (or transmitted) signal. In practice, replicas of the transmitted waveforms are normally computed and stored in memory for use by the radar signal processor when needed.

33.9 Matched Filter Response to LFM Waveforms

In order to develop a general expression for the matched filter output when an LFM waveform is utilized, we will consider the case when the radar is tracking a closing target with velocity v. The transmitted signal is

$$s_1(t) = \text{Rect}\left(\frac{t}{\tau'}\right) e^{j2\pi\left(f_0 t + \frac{\mu}{2}t^2\right)} \tag{33.101}$$

The received signal is then given by

$$s_{r_1}(t) = s_1(t - \Delta(t)) \tag{33.102}$$

$$\Delta(t) = t_0 - \frac{2v}{c}(t - t_0) \tag{33.103}$$

where

t_0 is the time corresponding to the target initial detection range

c is the speed of light

Using Equation 33.103 we can rewrite Equation 33.102 as

$$s_{r_1}(t) = s_1\left(t - t_0 + \frac{2v}{c}(t - t_0)\right) = s_1(\gamma(t - t_0)) \tag{33.104}$$

and

$$\gamma = 1 + 2\frac{v}{c} \tag{33.105}$$

is the scaling coefficient. Substituting Equation 33.101 into Equation 33.104 yields

$$s_{r_1}(t) = Rect\left(\frac{\gamma(t - t_0)}{\tau'}\right) e^{j2\pi f_0 \gamma(t-t_0)} e^{j\pi\mu\gamma^2(t-t_0)^2} \tag{33.106}$$

which is the analytical signal representation for $s_{r_1}(t)$. The complex envelope of the signal $s_{r_1}(t)$ is obtained by multiplying Equation 33.106 by $\exp(-j2\pi f_0 t)$. Denote the complex envelope by $s_r(t)$; then after some manipulation we get

$$s_r(t) = e^{-j2\pi f_0 t_0} Rect\left(\frac{\gamma(t - t_0)}{\tau'}\right) e^{j2\pi f_0(\gamma-1)(t-t_0)} e^{j\pi\mu\gamma^2(t-t_0)^2} \tag{33.107}$$

The Doppler shift due to the target motion is

$$f_d = \frac{2v}{c} f_0 \tag{33.108}$$

and since $\gamma - 1 = 2v/c$, we get

$$f_d = (\gamma - 1) f_0 \tag{33.109}$$

Using the approximation $\gamma \approx 1$ and Equation 33.109, Equation 33.107 is rewritten as

$$s_r(t) \approx e^{j2\pi f_d(t-t_0)} s(t - t_0) \tag{33.110}$$

where

$$s(t - t_0) = e^{-j2\pi f_0 t} s_1(t - t_0) \tag{33.111}$$

$s_1(t)$ is given in Equation 33.101. The matched filter response is given by the convolution integral

$$s_o(t) = \int_{-\infty}^{\infty} h(u) s_r(t - u) du \tag{33.112}$$

For a noncausal matched filter the impulse response $h(u)$ is equal to $s^*(-t)$; it follows that

$$s_o(t) = \int_{-\infty}^{\infty} s^*(-u)s_r(t-u)du \tag{33.113}$$

Substituting Equation 33.111 into Equation 33.113, and performing some algebraic manipulations, we get

$$s_o(t) = \int_{-\infty}^{\infty} s^*(u)e^{j2\pi f_d(t+u-t_0)}s(t+u-t_0)du \tag{33.114}$$

Finally, making the change of variable $t' = t + u$ yields

$$s_o(t) = \int_{-\infty}^{\infty} s^*(t'-t)s(t'-t_0)e^{j2\pi f_d(t'-t_0)}dt' \tag{33.115}$$

It is customary to set $t_0 = 0$. It follows that

$$s_o(t{:}f_d) = \int_{-\infty}^{\infty} s(t')s^*(t'-t)e^{j2\pi f_d t'}dt' \tag{33.116}$$

where we used the notation $s_o(t;f_d)$ to indicate that the output is a function of both time and Doppler frequency.

33.10 Waveform Resolution and Ambiguity

As indicated by Equation 33.93 the radar sensitivity (in the case of white additive noise) depends only on the total energy of the received signal and is independent of the shape of the specific waveform. This leads us to ask the following question: If the radar sensitivity is independent of the waveform, then what is the best choice for a transmitted waveform? The answer depends on many factors; however, the most important consideration lies in the waveform's range and Doppler resolution characteristics.

As discussed in Chapter 32, range resolution implies separation between distinct targets in range. Alternatively, Doppler resolution implies separation between distinct targets in frequency. Thus, ambiguity and accuracy of this separation are closely associated terms.

33.10.1 Range Resolution

Consider radar returns from two stationary targets (zero Doppler) separated in range by distance ΔR. What is the smallest value of ΔR so that the returned signal is interpreted by the radar as two distinct targets? In order to answer this question, assume that the radar transmitted pulse is denoted by $s(t)$,

$$s(t) = A(t)\cos(2\pi f_0 t + \phi(t)) \tag{33.117}$$

where

f_0 is the carrier frequency
$A(t)$ is the amplitude modulation
$\phi(t)$ is the phase modulation

The signal $s(t)$ can then be expressed as the real part of the complex signal $\psi(t)$, where

$$\psi(t) = A(t)e^{j(\omega_0 t - \phi(t))} = u(t)e^{j\omega_0 t} \tag{33.118}$$

and

$$u(t) = A(t)e^{-j\phi(t)} \tag{33.119}$$

It follows that

$$s(t) = \mathrm{Re}\{\psi(t)\} \tag{33.120}$$

The returns from both targets are respectively given by

$$s_{r1}(t) = \psi(t - \tau_0) \tag{33.121}$$

$$s_{r2}(t) = \psi(t - \tau_0 - \tau) \tag{33.122}$$

where τ is the difference in delay between the two returns. One can assume that the reference time is τ_0, and thus without any loss of generality one may set $\tau_0 = 0$. It follows that the two targets are distinguishable by how large or small the delay τ can be.

In order to measure the difference in range between the two targets consider the integral square error between $\psi(t)$ and $\psi(t-\tau)$. Denoting this error as ε_R^2, it follows that

$$\varepsilon_R^2 = \int_{-\infty}^{\infty} |\psi(t) - \psi(t-\tau)|^2 dt \tag{33.123}$$

Equation 33.123 can be written as

$$\varepsilon_R^2 = \int_{-\infty}^{\infty} |\psi(t)|^2 dt + \int_{-\infty}^{\infty} |\psi(t-\tau)|^2 dt - \int_{-\infty}^{\infty} \{(\psi(t)\psi^*(t-\tau) + \psi^*(t)\psi(t-\tau))dt\} \tag{33.124}$$

Using Equation 33.118 into Equation 33.124 yields

$$\begin{aligned}\varepsilon_R^2 &= 2\int_{-\infty}^{\infty} |u(t)|^2 dt - 2\mathrm{Re}\left\{\int_{-\infty}^{\infty} \psi^*(t)\psi(t-\tau)dt\right\} \\ &= 2\int |u(t)|^2 dt - 2\mathrm{Re}\left\{e^{-j\omega_0\tau}\int u^*(t)u(t-\tau)dt\right\}\end{aligned} \tag{33.125}$$

The first term in the right hand side of Equation 33.125 represents the signal energy, and is assumed to be constant. The second term is a varying function of τ with its fluctuation tied to the carrier frequency. The integral inside the right-most side of this equation is defined as the "range ambiguity function,"

$$\chi_R(\tau) = \int_{-\infty}^{\infty} u^*(t)u(t-\tau)dt \tag{33.126}$$

The maximum value of $\chi_R(\tau)$ is at $\tau = 0$. Target resolvability in range is measured by the squared magnitude $|\chi_R(\tau)|^2$. It follows that if $|\chi_R(\tau)| = \chi_R(0)$ for some nonzero value of τ, then the two targets are indistinguishable. Alternatively, if $|\chi_R(\tau)| \neq \chi_R(0)$ for some nonzero value of τ, then the two targets may be distinguishable (resolvable). As a consequence, the most desirable shape for $\chi_R(\tau)$ is a very sharp peak (thumb tack shape) centered at $\tau = 0$ and falling very quickly away from the peak.

The time delay resolution is

$$\Delta\tau = \frac{\int_{-\infty}^{\infty} |\chi_R(\tau)|^2 d\tau}{\chi_R^2(0)} \tag{33.127}$$

Using Parseval's theorem, Equation 33.127 can be written as

$$\Delta\tau = 2\pi \frac{\int_{-\infty}^{\infty} |U(\omega)|^4 d\omega}{\left[\int_{-\infty}^{\infty} |U(\omega)|^2 d\omega\right]^2} \tag{33.128}$$

The minimum range resolution corresponding to $\Delta\tau$ is

$$\Delta R = c\Delta\tau/2 \tag{33.129}$$

However, since the signal effective bandwidth is

$$B = \frac{\left[\int_{-\infty}^{\infty} |U(\omega)|^2 d\omega\right]}{2\pi \int_{-\infty}^{\infty} |U(\omega)|^4 d\omega} \tag{33.130}$$

the range resolution is expressed as a function of the waveform's bandwidth as

$$\Delta R = c/(2B) \tag{33.131}$$

The comparison between Equations 33.116 and 33.126 indicates that the output of the matched filter and the range ambiguity function have the same envelope (in this case the Doppler shift f_d is set to zero). This indicates that the matched filter, in addition to providing the maximum instantaneous SNR at its output, also preserves the signal range resolution properties.

33.10.2 Doppler Resolution

It was shown in Chapter 32 that the Doppler shift corresponding to the target radial velocity is

$$f_d = \frac{2v}{\lambda} = \frac{2vf_0}{c} \tag{33.132}$$

where

v is the target radial velocity
λ is the wavelength
f_0 is the frequency
c is the speed of light

Let

$$\Psi(f) = \int_{-\infty}^{\infty} \psi(t) e^{-j2\pi f t} dt \tag{33.133}$$

Due to the Doppler shift associated with the target, the received signal spectrum will be shifted by f_d. In other words the received spectrum can be represented by $\Psi(f - f_d)$. In order to distinguish between the two targets located at the same range but having different velocities, one may use the integral square error. More precisely,

$$\varepsilon_f^2 = \int_{-\infty}^{\infty} |\Psi(f) - \Psi(f - f_d)|^2 df \tag{33.134}$$

Using similar analysis as that which led to Equation 33.125, one should minimize

$$2\mathrm{Re}\left\{ \int_{-\infty}^{\infty} \Psi^*(f)\Psi(f - f_d) df \right\} \tag{33.135}$$

By using the analytic signal in Equation 33.118 it can be shown that

$$\Psi(f) = U(2\pi f - 2\pi f_0) \tag{33.136}$$

Thus, Equation 33.135 becomes

$$\int_{-\infty}^{\infty} U^*(2\pi f) U(2\pi f - 2\pi f_d) df = \int_{-\infty}^{\infty} U^*(2\pi f - 2\pi f_0) U(2\pi f - 2\pi f_0 - 2\pi f_d) df \tag{33.137}$$

The complex frequency correlation function is then defined as

$$\chi_f(f_d) = \int_{-\infty}^{\infty} U^*(2\pi f) U(2\pi f - 2\pi f_d) df = \int_{-\infty}^{\infty} |u(t)|^{2j2\pi f_d t} dt \tag{33.138}$$

and the Doppler resolution constant Δf_d is

$$\Delta f_d = \frac{\int_{-\infty}^{\infty} \left|\chi_f(f_d)\right|^2 df_d}{\chi_f^2(0)} = \frac{\int_{-\infty}^{\infty} |u(t)|^4 dt}{\left[\int_{-\infty}^{\infty} |u(t)|^4 dt\right]^2} = \frac{1}{\tau'} \tag{33.139}$$

where τ' is pulsewidth.

Finally, one can define the corresponding velocity resolution as

$$\Delta v = \frac{c \Delta f_d}{2 f_0} = \frac{c}{2 f_0 \tau'} \tag{33.140}$$

Again observation of Equations 33.138 and 33.116 indicate that the output of the matched filter and the ambiguity function (when $\tau = 0$) are similar to each other. Consequently, one concludes that the matched filter preserves the waveform Doppler resolution properties as well.

33.10.3 Combined Range and Doppler Resolution

In this general case, one needs to use a two-dimensional function in the pair of variables (τ, f_d). For this purpose, assume that the complex envelope of the transmitted waveform is

$$\psi(t) = u(t)e^{j2\pi f_0 t} \tag{33.141}$$

Then the delayed and Doppler-shifted signal is

$$\psi'(t - \tau) = u(t - \tau)e^{j2\pi(f_0 - f_d)(t-\tau)} \tag{33.142}$$

Computing the integral square error between Equations 33.142 and 33.141 yields

$$\varepsilon^2 = \int_{-\infty}^{\infty} |\psi(t) - \psi'(t - \tau)|^2 dt = 2\int_{-\infty}^{\infty} |\psi(t)|^2 dt - 2\mathrm{Re}\left\{\int_{-\infty}^{\infty} \psi^*(t) - \psi'(t - \tau)dt\right\} \tag{33.143}$$

which can be written as

$$\varepsilon^2 = 2\int_{-\infty}^{\infty} |u(t)|^2 dt - 2\mathrm{Re}\left\{e^{j2\pi(f_0 - f_d)\tau}\int_{-\infty}^{\infty} u(t)u^*(t - \tau)e^{j2\pi f_d t}dt\right\} \tag{33.144}$$

Again, in order to maximize this squared error for $\tau \neq 0$ one must minimize the last term of Equation 33.144.

Define the combined range and Doppler correlation function as

$$\chi(\tau, f_d) = \int_{-\infty}^{\infty} u(t)u^*(t - \tau)e^{j2\pi f_d t}dt \tag{33.145}$$

In order to achieve the most range and Doppler resolution, the modulus square of this function must be minimized for $\tau \neq 0$ and $f_d \neq 0$. Note that the output of the matched filter in Equation 33.116 is identical to that given in Equation 33.145. This means that the output of the matched filter exhibits maximum instantaneous SNR as well as the most achievable range and Doppler resolutions.

33.11 "MyRadar" Design Case Study—Visit 3

33.11.1 Problem Statement

Assuming a matched filter receiver, select a set of waveforms that can meet the design requirements as stated in the previous two chapters. Assume linear frequency modulation. Do not use more than a total of five waveforms. Modify the design so that the range resolution $\Delta R = 30$ m during the search mode, and $\Delta R = 7.5$ m during tracking.

33.11.2 A Design

The major characteristics of radar waveforms include the waveform's energy, range resolution, and Doppler (or velocity) resolution. The pulse (waveform) energy is

$$E = P_t\tau \tag{33.146}$$

where
P_t is the peak transmitted power
τ is the pulsewidth

Range resolution is defined in Equation 33.131, while the velocity resolution is in Equation 33.140.

Close attention should be paid to the selection process of the pulsewidth. In this design we will assume that the pulse energy is the same as that computed in Chapter 32. The radar operating bandwidth during search and track are calculated from Equation 33.131 as

$$\begin{Bmatrix} B_{\text{search}} \\ B_{\text{track}} \end{Bmatrix} = \begin{Bmatrix} 3 \times 10^8/(2 \times 30) = 5\text{MHz} \\ 3 \times 10^8/(2 \times 7.5) = 20\text{MHz} \end{Bmatrix} \tag{33.147}$$

Since the design calls for a pulsed radar, then for each pulse transmitted (one PRI) the radar should not be allowed to receive any signal until that pulse has been completely transmitted. This limits the radar to a minimum operating range defined by

$$R_{\min} = \frac{c\tau}{2} \tag{33.148}$$

In this design choose $R_{\min} \geq 15$ km. It follows that the minimum acceptable pulsewidth is $\tau_{\max} \leq 100$ μs (Table 33.1).

For this design select five waveforms, one for search and four for track. Typically search waveforms are longer than track waveforms; alternatively, tracking waveforms require wider bandwidths than search waveforms. However, in the context of range, more energy is required at longer ranges (for both track and search waveforms), since one would expect the SNR to get larger as range becomes smaller. This was depicted in the example shown in Figure 32.13 in Chapter 32.

Assume that during search and initial detection the single pulse peak power is to be kept under 10 kW (i.e., $P_t \leq 20$ kW). Then by using the single pulse energy, one can compute the minimum required pulsewidth as

$$\tau_{\min} \geq \frac{0.1147}{20 \times 10^3} = 5.735\mu\text{s} \tag{33.149}$$

Choose $\tau_{\text{search}} = 20$ μs, with bandwidth $B = 5$ MHz and use LFM modulation. Figure 33.18 shows plots of the real part, imaginary part, and the spectrum of this search waveform. This figure was produced using the GUI workspace "*LFM_gui.m.*" As far as the track waveforms, choose four waveforms of the same bandwidth ($B_{\text{track}} = 20$ MHz) and with the following pulsewidths.

TABLE 33.1 "MyRadar" Design Case Study Track Waveforms

Pulsewidth	Range Window
$\tau_{t1} = 20$ μs	$R_{\max} \rightarrow 0.75\ R_{\max}$
$\tau_{t2} = 17.5$ μs	$0.75\ R_{\max} \rightarrow 0.5\ R_{\max}$
$\tau_{t3} = 15$ μs	$0.5\ R_{\max} \rightarrow 0.25\ R_{\max}$
$\tau_{t4} = 12.5$ μs	$R \leq 0.25\ R_{\max}$

Note that R_{max} refers to the initial range at which track has been initiated. Figure 33.19 is similar to Figure 33.18 except it is for τ_{t3}.

For the waveform set selected in this design option, the radar duty cycle varies from 1.25%–2.0%. The PRF is calculated as $f_r = 1$ KHz; thus the PRI is $T = 1$ ms.

At this point of the design, one must verify that the selected waveforms provide the radar with the desired SNR. In other words, one must now rerun these calculations and verify that the SNR has not been degraded.

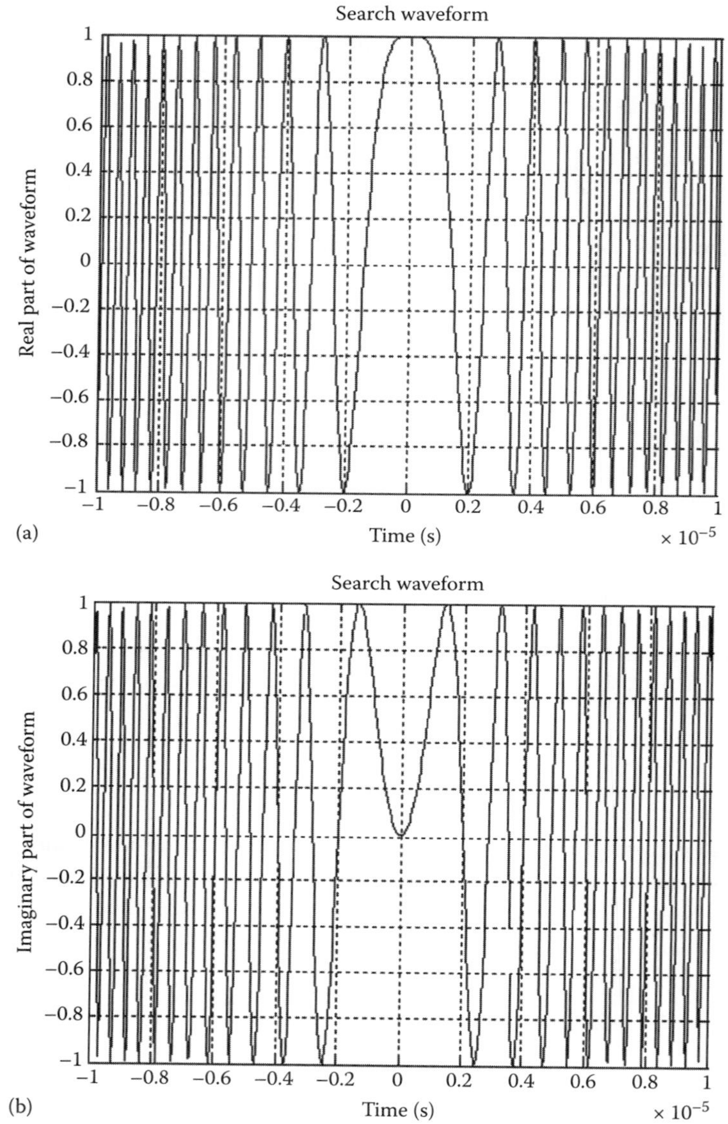

FIGURE 33.18 (a) Real part of search waveform. (b) Imaginary part of search waveform.

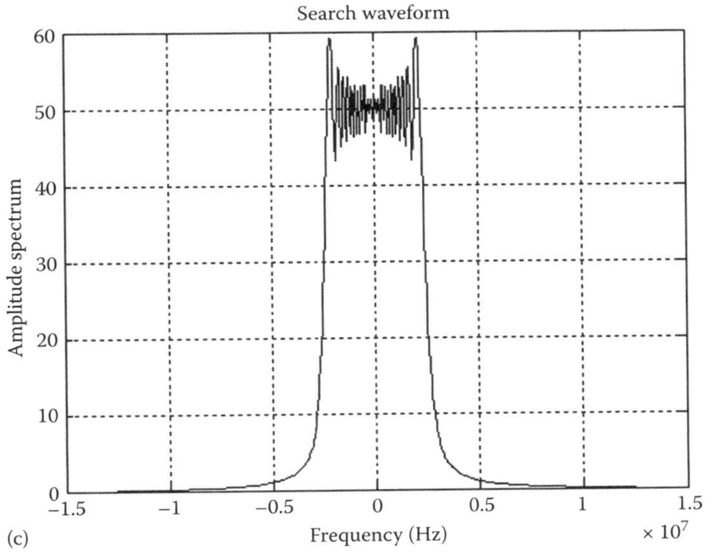

FIGURE 33.18 (continued) (c) Amplitude spectrum.

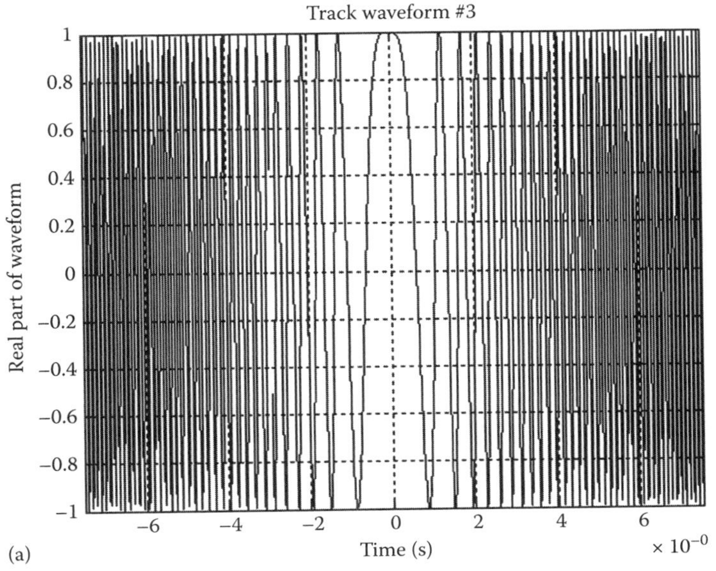

FIGURE 33.19 (a) Real part of waveform.

(continued)

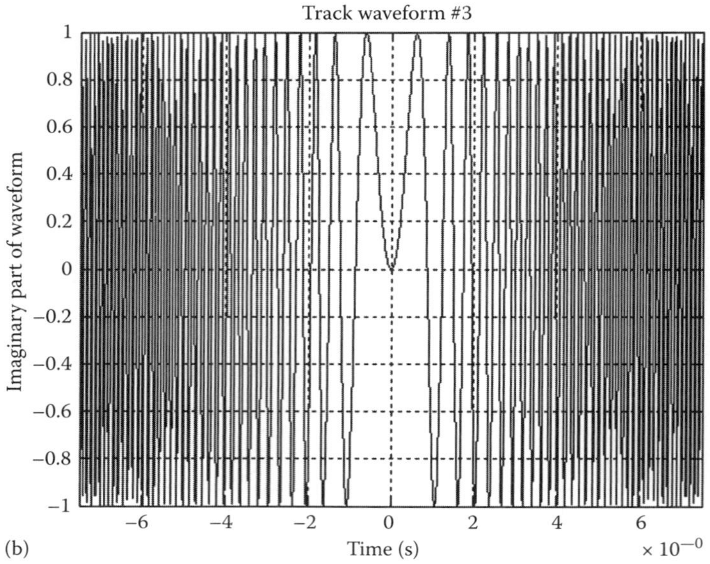

(b)

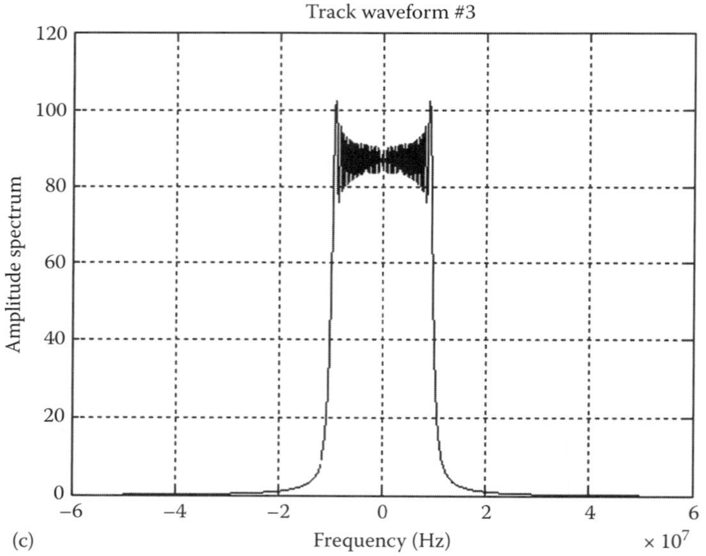

(c)

FIGURE 33.19 (continued) (b) Imaginary part of waveform. (c) Amplitude spectrum.

33.12 MATLAB® Program and Function Listings

This section presents listings for all MATLAB programs/functions used in this chapter.

Listing 33.1. MATLAB Program "fig33_7.m"

```
% Use this program to reproduce Fig 33.7 from text
clear all
close all
```

```
n=0;
for x=0:.05:4
        n=n+1;
        sx(n)=quadl('fresnels'.0,x);
        cx(n)=quadl('fresnelc',.0,x);
        end
        plot(cx)
        x=0:.05:4;
        plot(x,cx, 'k',x,sx, 'k--') grid
        xlabel ('x')
        ylabel ('Fresnel integrals: C(x); S(x)')
        legend('C(x)','S(x)')
```

Listing 33.2. MATLAB Program "fig33_8.m"

```
% Use this program to reproduce Fig. 33.8 of text
close all
clear all
eps=0.000001;
%Enter pulsewidth and bandwidth
B=200.0e6; %200 MHZ bandwidth
T=10.e-6; %10 micro second pulse;
% Compute alpha
mu=2. * pi * B/T;
% Determine sampling times
delt=linspace(-T/2., T/2., 10001); % 1 nano second sampling interval
% Compute the complex LFM representation
Ichannal=cos(mu. * delt.^2/2.); % Real part
Qchannal=sin(mu. * delt.^2/2.); % Imaginary Part
LFM=Ichannal+sqrt(-1). * Qchannal; % complex signal
%Compute the FFT of the LFM waveform
LFMFFT =fftshift(fft(LFM));
% Plot the real and Imaginary parts and the spectrum
freqlimit=0.5 / 1.e-9;% the sampling interval 1 nano-second
freq=linspace(-freqlimit/1.e6,freqlimit/1.e6,10001);
figure(1)
plot(delt*1e6,Ichannal, 'k');
axis([-11 -11])
grid
xlabel('Time - microsecs')
ylabel('Real part')
title('T=10 Microsecond, B=200 MHz')
figure(2)
plot(delt*1e6,Qchannal, 'k');
axis([-11 -11])
grid
xlabel('Time - microsecs')
ylabel('Imaginary part')
title('T=10 Microsecond, B=200 MHz')
figure(3)
```

```
plot(freq, abs(LFMFFT), 'k');
%axis tight
grid
xlabel('Frequency - MHz')
ylabel('Amplitude spectrum')
title('Spectrum for an LFM waveform and T=10 Microsecond, . . .
B=200 MHZ')
```

Listing 33.3. MATLAB Function "hrr_profile.m"

```
function [hl]=hrr_profile (nscat, scat_range, scat_rcs, n, deltaf, prf, v,
rnote,winid)
% Range or Time domain Profile
% Range_Profile returns the Range or Time domain plot of a simulated
% HRR SFWF returning from a predetermined number of targets with a pre-
determined
% RCS for each target.
c=3.0e8; % speed of light (m/s)
num_pulses=n;
SNR dB=40;
nfft=256;
%carrier_freq=9.5e9; %Hz (10GHz)
freq_step=deltaf; %Hz (10MHz)
V= v; % radial velocity (m/s) -- (+) =towards radar (-) =away
PRI=1./prf; % (s)
if (nfft>2*num_pulses)
num_pulses=nfft/2;
end
Inphase=zeros((2*num_pulses),1);
Quadrature=zeros((2*num_pulses),1);
Inphase tgt=zeros(num_pulses,1);
Quadrature_tgt=zeros(num_pulses,1);
IQ_freq_domain=zeros((2*num_pulses),1);
Weighted_I_freq_domain=zeros((num_pulses), 1);
Weighted_Q_freq_domain=zeros((num_pulses), 1);
Weighted_IQ_time_domain=zeros((2*num_pulses),1);
Weighted_IQ_freq_domain=zeros((2*num_pulses),1);
abs_Weighted_IQ_time_domain=zeros((2*num_pulses),1);
dB_abs_Weighted_IQ_time_domain=zeros((2*num_pulses),1);
taur=2. * rnote / c;
for jscat=1:nscat
ii=0;
for i=1:num_pulses
ii=ii+1;
rec_freq= ((i-1)*freq_step);
Inphase_tgt(ii)=Inphase_tgt(ii)+sqrt(scat_rcs(jscat)) * cos(-
                    2*pi*rec_freq* . . .
               (2.*scat_range(jscat)/c - 2*(V/c)*((i-1)*PRI+taur/2+
                    2 * scat_range (¡scat)/c)));
```

```
            Quadrature_tgt(ii)=Quadrature_tgt(ii)+sqrt(scat_rcs
            (jscat))*sin(-2*pi*rec_freq*...
            (2*scat_range(jscat)/c - 2*(V/c)*((i-1)*PRI+taur/2+
                2 * scat_range (jscat)/c)));
end
end
if(winid >= 0)
        window(1 :num_pulses)=hamming(num_pulses);
else
        window(1 :num_pulses)=1;
end
Inphase=Inphase_tgt;
Quadrature=Quadrature_tgt;
Weighted_I_freq_domain(1:num_pulses)=Inphase(1:num_pulses). * window';
Weighted_Q_freq_domain(1:num_pulses)=Quadrature(1:num_pulses).*window';
Weighted_IQ_freq_domain(1 :num_pulses)=Weighted_I_freq_domain+...
Weighted_Q_freq_domain*j;
Weighted_IQ_freq_domain (num_pulses: 2*num_pulses) =0.+0. i;
Weighted_IQ_time_domain= (ifft(Weighted_IQ_freq_domain));
abs_Weighted_IQ_time_domain= (abs(Weighted_IQ_time_domain));
dB_abs_Weighted_IQ_time_domain =
            20.0*log10(abs_Weighted_IQ_time_domain)+SNR_dB;
% calculate the unambiguous range window size
Ru=c/2/deltaf;
hl=dB_abs_Weighted_IQ_time_domain;
numb=2*num_pulses;
delx_meter=Ru /numb;
xmeter= 0:delx_meter:Ru-delx_meter;
plot(xmeter, dB_abs_Weighted_IQ_time_domain, 'k')
xlabel ('relative distance - meters')
ylabel ('Range profile - dB')
grid
```

Listing 33.4. MATLAB Program "fig33_1 7.m"

```
% use this program to reproduce Fig. 33.17 of text
clear all
close all
nscat=1;
scat_range =912;
scat_rcs=10;
n =64;
deltaf= 10e6;
prf= 10e3;
v=15;
rnote=900,
winid=1;
count=0;
for time= 0:.05:3
```

```
        count=count +1;
        hl=hrr_profile (nscat, scat_range, scat_rcs, n, deltaf, prf, v,
        rnote,winid);
        array (count,:)=transpose(hl);
        hl(1: end)=0;
        scat_range=scat_range - 2 *n *v / prf;
end
figure (1)
numb=2*256;% this number matches that used in hrr_profile.
Delx_meter=15 /numb;
xmeter=0:delx_meter: 15-delx_meter;
imagesc(xmeter, 0:0.05:4, array)
colormap(gray)
ylabel ('Time in seconds')
xlabel('Relative distance in meters')
```

Reference

1. Wehner, D. R., *High Resolution Radar*, 2nd edition, Artech House, Norwood, MA, 1993.

34

High Resolution Tactical Synthetic Aperture Radar

Bassem R. Mahafza
Deceibel Research, Inc.

Atef Z. Elsherbeni
University of Mississippi

Brian J. Smith
U.S. Army Aviation and Missile Command

This chapter provides an introduction to Tactical Synthetic Aperture Radar (TSAR). The purpose of this chapter is to further develop the readers' understanding of synthetic aperture radar (SAR) by taking a closer look at high resolution spotlight SAR image formation algorithms, motion compensation techniques, autofocus algorithms, and performance metrics.

34.1 Introduction

Modern airborne radar systems are designed to perform a large number of functions which range from detection and discrimination of targets to mapping large areas of ground terrain. This mapping can be performed by the SAR. Through illuminating the ground with coherent radiation and measuring the echo signals, SAR can produce high resolution two-dimensional (2-D) (and in some cases three-dimensional [3-D]) imagery of the ground surface. The quality of ground maps generated by SAR is determined by the size of the resolution cell. A resolution cell is specified by both range and azimuth resolutions of the system. Other factors affecting the size of the resolution cells are (1) size of the processed map and the amount of signal processing involved; (2) cost consideration; and (3) size of the objects that need to be resolved in the map. For example, mapping gross features of cities and coastlines does not require as much resolution when compared to resolving houses, vehicles, and streets.

SAR systems can produce maps of reflectivity versus range and Doppler (cross range). Range resolution is accomplished through range gating. Fine range resolution can be accomplished by using pulse compression techniques. The azimuth resolution depends on antenna size and radar wavelength. Fine azimuth resolution is enhanced by taking advantage of the radar motion in order to synthesize a larger antenna aperture. Let N_r denote the number of range bins and let N_a denote the number of azimuth cells. It follows that the total number of resolution cells in the map is N_rN_a. SAR systems that are generally concerned with improving azimuth resolution are often referred to as Doppler beam-sharpening (DBS) SARs. In this case, each range bin is processed to resolve targets in Doppler which correspond to azimuth. This chapter is presented in the context of DBS.

Due to the large amount of signal processing required in SAR imagery, the early SAR designs implemented optical processing techniques. Although such optical processors can produce high quality radar images, they have several shortcomings. They can be very costly and are, in general, limited to making strip maps. Motion compensation is not easy to implement for radars that utilize optical processors. With the recent advances in solid state electronics and very large scale integration (VLSI) technologies, digital signal processing in real time has been made possible in SAR systems.

34.2 Side Looking SAR Geometry

Figure 34.1 shows the geometry of the standard side looking SAR. We will assume that the platform carrying the radar maintains both fixed altitude h and velocity v. The antenna 3 dB beamwidth is θ, and the elevation angle (measured from the z-axis to the antenna axis) is β. The intersection of the antenna beam with the ground defines a footprint. As the platform moves, the footprint scans a swath on the ground.

The radar position with respect to the absolute origin $\vec{O} = (0,0,0)$, at any time, is the vector $\vec{a}(t)$. The velocity vector $\vec{a}'(t)$ is

$$\vec{a}'(t) = 0 \times \hat{a}_x + v \times \hat{a}_y + 0 \times \hat{a}_z \tag{34.1}$$

The line of sight (LOS) for the current footprint centered at $\vec{q}(t_c)$ is defined by the vector $\vec{R}(t_c)$, where t_c denotes the central time of the observation interval T_{ob} (coherent integration interval). More precisely,

$$(t = t_a + t_c); \quad -\frac{T_{ob}}{2} \leq t \leq \frac{T_{ob}}{2} \tag{34.2}$$

where t_a and t are the absolute and relative times, respectively. The vector $\vec{m}_g$ defines the ground projection of the antenna at central time. The minimum slant range to the swath is R_{min}, and the maximum range is denoted R_{max}, as illustrated by Figure 34.2. It follows that

$$\begin{aligned} R_{min} &= h/\cos(\beta - \theta/2) \\ R_{max} &= h/\cos(\beta + \theta/2) \\ |\vec{R}(t_c)| &= h/\cos\beta \end{aligned} \tag{34.3}$$

Notice that the elevation angle β is equal to

$$\beta = 90 - \psi_g \tag{34.4}$$

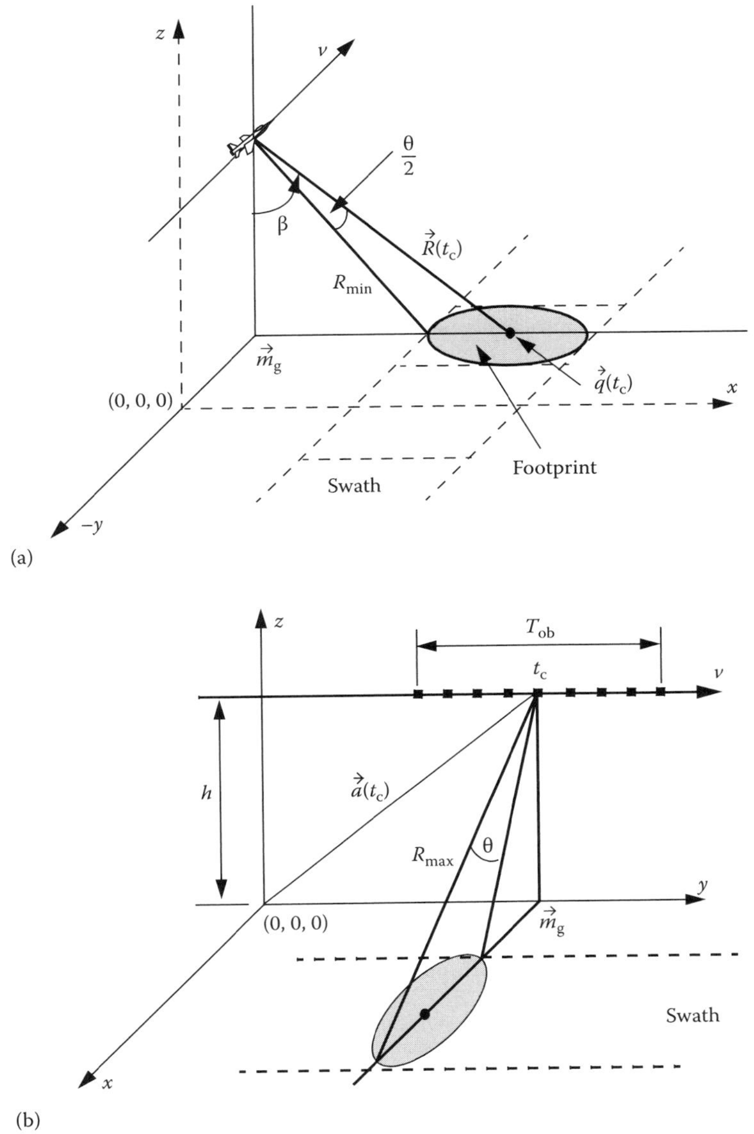

FIGURE 34.1 Side looking SAR geometry.

where ψ_g is the grazing angle. The size of the footprint is a function of the grazing angle and the antenna beamwidth, as illustrated in Figure 34.3. The SAR geometry described in this section is referred to as SAR "strip mode" of operation. Another SAR mode of operation, which will not be discussed in this chapter, is called "spot-light mode," where the antenna is steered (mechanically or electronically) to continuously illuminate one spot (footprint) on the ground. In this case, one high resolution image of the current footprint is generated during an observation interval.

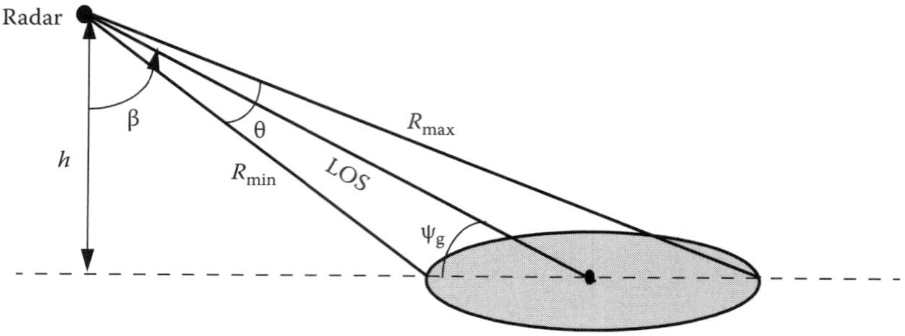

FIGURE 34.2 Definition of minimum and maximum range.

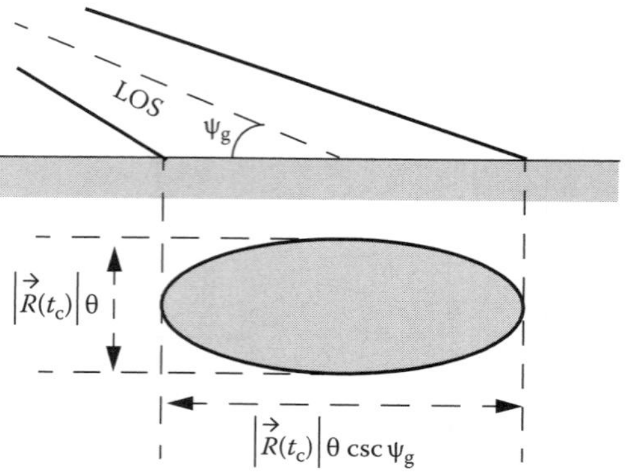

FIGURE 34.3 Footprint definition.

34.3 SAR Design Considerations

The quality of SAR images is heavily dependent on the size of the map resolution cell shown in Figure 34.4. The range resolution, ΔR, is computed on the beam LOS, and is given by

$$\Delta R = (c\tau)/2 \tag{34.5}$$

where τ is the pulsewidth. From the geometry in Figure 34.5 the extent of the range cell ground projection ΔR_g is computed as

$$\Delta R_g = \frac{c\tau}{2} \sec\psi_g \tag{34.6}$$

The azimuth or cross range resolution for a real antenna with a 3 dB beamwidth θ (radians) at range R is

$$\Delta A = \theta R \tag{34.7}$$

However, the antenna beamwidth is proportional to the aperture size,

$$\theta \approx \frac{\lambda}{L} \tag{34.8}$$

where

λ is the wavelength

L is the aperture length

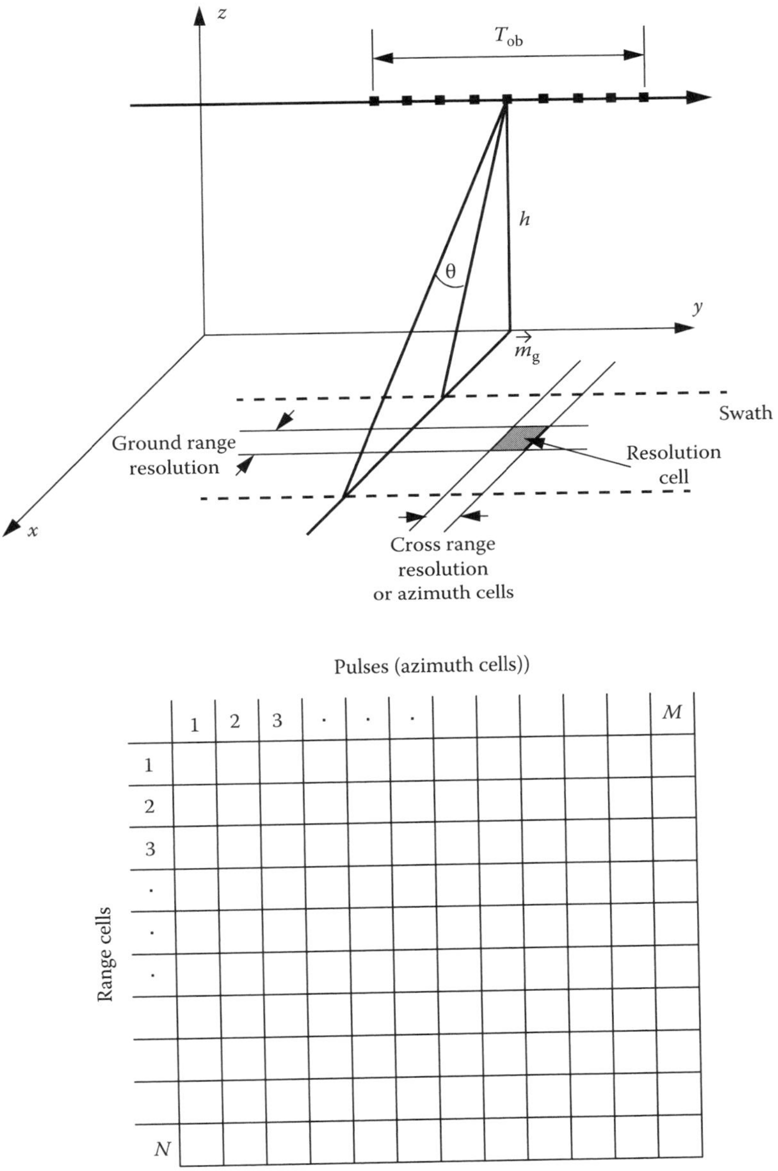

FIGURE 34.4 Definition of a resolution cell.

It follows that

$$\Delta A = \frac{\lambda R}{L} \tag{34.9}$$

And since the effective synthetic aperture size is twice that of a real array, the azimuth resolution for a synthetic array is then given by

$$\Delta A = \frac{\lambda R}{2L} \tag{34.10}$$

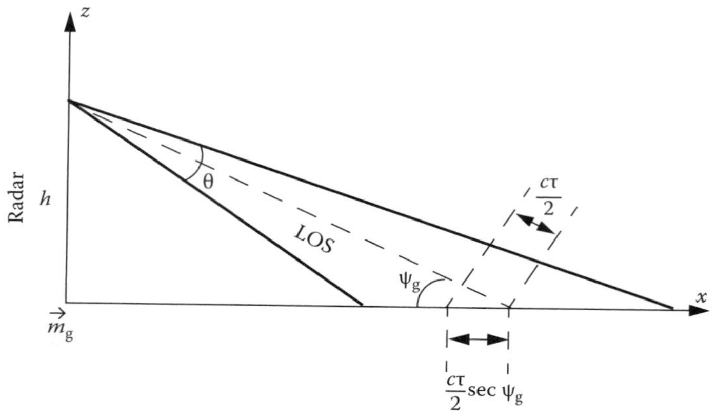

FIGURE 34.5 Definition of a range cell on the ground.

Furthermore, since the synthetic aperture length L is equal to vT_{ob}, Equation 34.10 can be rewritten as

$$\Delta A = \frac{\lambda R}{2vT_{ob}} \tag{34.11}$$

The azimuth resolution can be greatly improved by taking advantage of the Doppler variation within a footprint (or a beam). As the radar travels along its flight path the radial velocity to a ground scatterer (point target) within a footprint varies as a function of the radar radial velocity in the direction of that scatterer. The variation of Doppler frequency for a certain scatterer is called the "Doppler history."

Let $R(t)$ denote the range to a scatterer at time t, and v_r be the corresponding radial velocity; thus the Doppler shift is

$$f_d = -\frac{2R'(t)}{\lambda} = \frac{2v_r}{\lambda} \tag{34.12}$$

where $R'(t)$ is the range rate to the scatterer. Let t_1 and t_2 be the times when the scatterer enters and leaves the radar beam, respectively, and t_c be the time that corresponds to minimum range. Figure 34.6 shows a sketch of the corresponding $R(t)$. Since the radial velocity can be computed as the derivative of $R(t)$ with respect to time, one can clearly see that Doppler frequency is maximum at t_1, zero at t_c, and minimum at t_2, as illustrated in Figure 34.7.

In general, the radar maximum PRF, $f_{r_{max}}$, must be low enough to avoid range ambiguity. Alternatively, the minimum PRF, $f_{r_{min}}$, must be high enough to avoid Doppler ambiguity. SAR unambiguous range

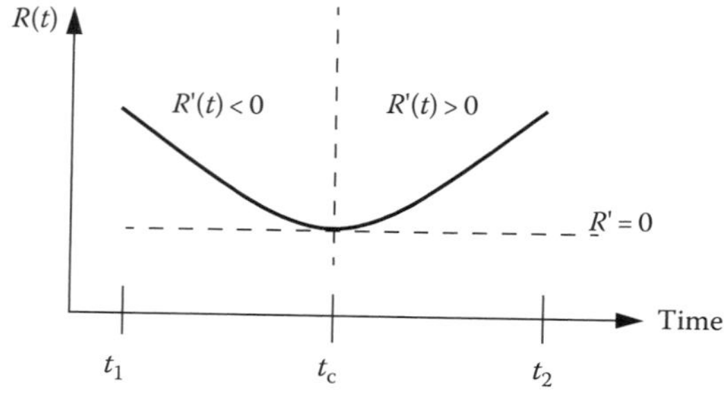

FIGURE 34.6 Sketch of range versus time for a scatterer.

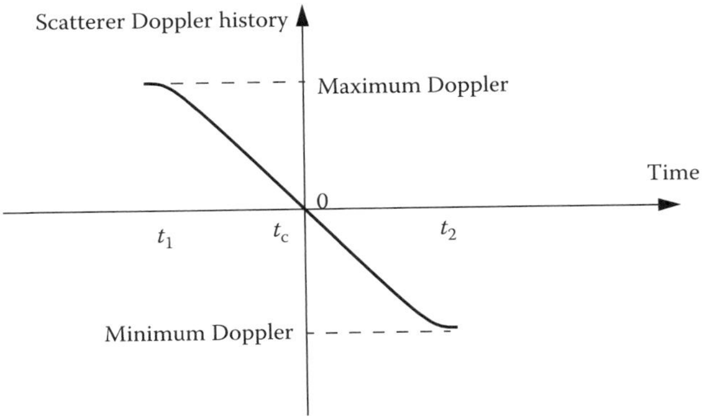

FIGURE 34.7 Point scatterer Doppler history.

must be at least as wide as the extent of a footprint. More precisely, since target returns from maximum range due to the current pulse must be received by the radar before the next pulse is transmitted, it follows that SAR unambiguous range is given by

$$R_u = R_{max} - R_{min} \tag{34.13}$$

An expression for unambiguous range was derived in Chapter 32, and is repeated here as Equation 34.14,

$$R_u = \frac{c}{2f_r} \tag{34.14}$$

Combining Equations 34.14 and 34.13 yields

$$f_{r_{max}} \le \frac{c}{2(R_{max} - R_{min})} \tag{34.15}$$

SAR minimum PRF, $f_{r_{min}}$, is selected so that Doppler ambiguity is avoided. In other words, $f_{r_{min}}$ must be greater than the maximum expected Doppler spread within a footprint. From the geometry of Figure 34.8, the maximum and minimum Doppler frequencies are, respectively, given by

$$f_{d_{max}} = \frac{2v}{\lambda}\sin\left(\frac{\theta}{2}\right)\sin\beta;\ \text{at } t_1 \tag{34.16}$$

$$f_{d_{min}} = -\frac{2v}{\lambda}\sin\left(\frac{\theta}{2}\right)\sin\beta;\ \text{at } t_2 \tag{34.17}$$

It follows that the maximum Doppler spread is

$$\Delta f_d = f_{d_{max}} - f_{d_{min}} \tag{34.18}$$

Substituting Equations 34.16 and 34.17 into Equation 34.18 and applying the proper trigonometric identities yield

$$\Delta f_d = \frac{4v}{\lambda}\sin\frac{\theta}{2}\sin\beta \tag{34.19}$$

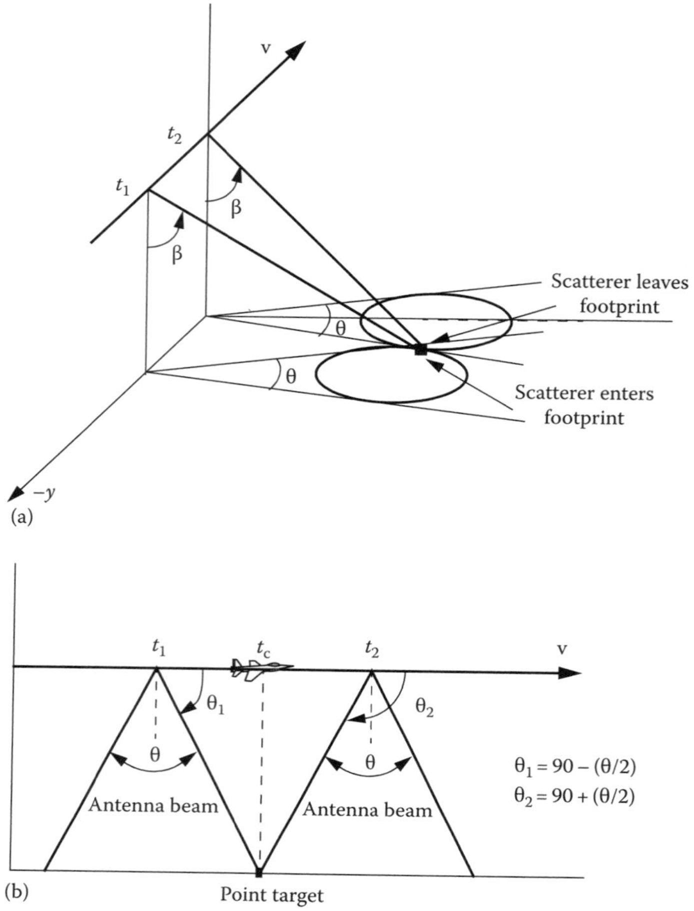

FIGURE 34.8 Doppler history computation. (a) Full view; (b) top view.

Finally, by using the small angle approximation we get

$$\Delta f_{\mathrm{d}} \approx \frac{4v}{\lambda}\frac{\theta}{2}\sin\beta = \frac{2v}{\lambda}\theta\sin\beta \tag{34.20}$$

Therefore, the minimum PRF is

$$f_{\mathrm{r_{min}}} \geq \frac{2v}{\lambda}\theta\sin\beta \tag{34.21}$$

Combining Equations 34.15 and 34.21 we get

$$\frac{c}{2(R_{\max} - R_{\min})} \geq f_{\mathrm{r}} \geq \frac{2v}{\lambda}\theta\sin\beta \tag{34.22}$$

It is possible to resolve adjacent scatterers at the same range within a footprint based only on the difference of their Doppler histories. For this purpose, assume that the two scatterers are within the kth range bin.

Denote their angular displacement as $\Delta\theta$, and let $\Delta f_{d_{min}}$ be the minimum Doppler spread between the two scatterers such that they will appear in two distinct Doppler filters. Using the same methodology that led to Equation 34.20 we get

$$\Delta f_{d_{min}} = \frac{2v}{\lambda} \Delta\theta \sin \beta_k \tag{34.23}$$

where β_k is the elevation angle corresponding to the kth range bin.

The bandwidth of the individual Doppler filters must be equal to the inverse of the coherent integration interval T_{ob} (i.e., $\Delta f_{d_{min}} = 1/T_{ob}$). It follows that

$$\Delta\theta = \frac{\lambda}{2vT_{ob} \sin \beta_k} \tag{34.24}$$

Substituting L for vT_{ob} yields

$$\Delta\theta = \frac{\lambda}{2L \sin \beta_k} \tag{34.25}$$

Therefore, the SAR azimuth resolution (within the kth range bin) is

$$\Delta A_g = \Delta\theta R_k = R_k \frac{\lambda}{2L \sin \beta_k} \tag{34.26}$$

Note that when $\beta_k = 90°$, Equation 34.26 is identical to Equation 34.10.

34.4 SAR Radar Equation

The single pulse radar equation is written as follows:

$$\text{SNR} = \frac{P_t G^2 \lambda^2 \sigma}{(4\pi)^3 R_k^4 k T_0 B L_{Loss}} \tag{34.27}$$

where

- P_t is peak power
- G is antenna gain
- λ is wavelength
- σ is radar cross section
- R_k is radar slant range to the kth range bin
- k is the Boltzmann constant
- T_0 is receiver noise temperature
- B is receiver bandwidth
- L_{Loss} is radar losses

The radar cross section is a function of the radar resolution cell and terrain reflectivity. More precisely,

$$\sigma = \sigma^0 \Delta R_g \Delta A_g = \sigma^0 \Delta A_g \frac{c\tau}{2} \sec \psi_g \tag{34.28}$$

where

- σ^0 is the clutter scattering coefficient
- ΔA_g is the azimuth resolution

Equation 34.6 was used to replace the ground range resolution. The number of coherently integrated pulses within an observation interval is

$$n = f_{\mathrm{r}} T_{\mathrm{ob}} = \frac{f_{\mathrm{r}} L}{v} \tag{34.29}$$

where L is the synthetic aperture size. Using Equation 34.26 in Equation 34.29 and rearranging terms yield

$$n = \frac{\lambda R f_{\mathrm{r}}}{2 \Delta A_{\mathrm{g}} v} \csc \beta_k \tag{34.30}$$

The radar average power over the observation interval is

$$P_{\mathrm{av}} = (P_{\mathrm{t}}/B) f_{\mathrm{r}} \tag{34.31}$$

The SNR for n coherently integrated pulses is then

$$(\mathrm{SNR})_n = n\mathrm{SNR} = n \frac{P_{\mathrm{t}} G^2 \lambda^2 \sigma}{(4\pi)^3 R_k^4 k T_0 B L_{\mathrm{Loss}}} \tag{34.32}$$

Substituting Equations 34.31, 34.30, and 34.28 into Equation 34.32 and performing some algebraic manipulations give the SAR radar equation,

$$(\mathrm{SNR})_n = \frac{P_{\mathrm{av}} G^2 \lambda^3 \sigma^0}{(4\pi)^3 R_k^3 k T_0 L_{\mathrm{Loss}}} \frac{\Delta R_{\mathrm{g}}}{2v} \csc \beta_k \tag{34.33}$$

Equation 34.33 leads to the conclusion that in SAR systems the SNR is (1) inversely proportional to the third power of range; (2) independent of azimuth resolution; (3) function of the ground range resolution; (4) inversely proportional to the velocity v; and (5) proportional to the third power of wavelength.

34.5 SAR Signal Processing

There are two signal processing techniques to sequentially produce a SAR map or image; they are line-by-line processing and Doppler processing. The concept of SAR line-by-line processing is as follows: Through the radar linear motion a synthetic array is formed, where the elements of the current synthetic array correspond to the position of the antenna transmissions during the last observation interval. Azimuth resolution is obtained by forming narrow synthetic beams through combinations of the last observation interval returns. Fine range resolution is accomplished in real time by utilizing range gating and pulse compression. For each range bin and each of the transmitted pulses during the last observation interval, the returns are recorded in a 2-D array of data that is updated for every pulse. Denote the 2-D array of data as MAP.

To further illustrate the concept of line-by-line processing, consider the case where a map of size $N_\alpha \times N_{\mathrm{r}}$ is to be produced, where N_{a} is the number of azimuth cells and N_{r} is the number of range bins. Hence, MAP is of size $N_{\mathrm{a}} \times N_{\mathrm{r}}$, where the columns refer to range bins, and the rows refer to azimuth cells. For each transmitted pulse, the echoes from consecutive range bins are recorded sequentially in the first row of MAP. Once the first row is completely filled (i.e., returns from all range bins have been received), all data (in all rows) are shifted downward one row before the next pulse is transmitted. Thus, one row of MAP is generated for every transmitted pulse. Consequently, for the current observation interval, returns from the first transmitted pulse will be located in the bottom row of MAP, and returns from the last transmitted pulse will be in the first row of MAP.

In SAR Doppler processing, the array MAP is updated once every N pulses so that a block of N columns is generated simultaneously. In this case, N refers to the number of transmissions during an observation interval (i.e., size of the synthetic array). From an antenna point of view, this is equivalent to having N adjacent synthetic beams formed in parallel through electronic steering.

34.6 Side Looking SAR Doppler Processing

Consider the geometry shown in Figure 34.9, and assume that the scatterer C_i is located within the kth range bin. The scatterer azimuth and elevation angles are μ_i, and β_i, respectively. The scatterer elevation angle β_i, is assumed to be equal to β_k, the range bin elevation angle. This assumption is true if the ground range resolution, ΔR_g, is small; otherwise, $\beta_i = \beta_k + \varepsilon_i$ for some small ε_i; in this chapter $\varepsilon_i = 0$.

The normalized transmitted signal can be represented by

$$s(t) = \cos(2\pi f_0 t - \xi_0) \tag{34.34}$$

where
f_0 is the radar operating frequency
ξ_0 denotes the transmitter phase

The returned radar signal from C_i is then equal to

$$s_i(t, \mu_i) = A_i \cos[2\pi f_0 (t - \tau_i(t, \mu_i)) - \xi_0] \tag{34.35}$$

where
$\tau_i(t, \mu_i)$ is the round-trip delay to the scatterer
A_i includes scatterer strength, range attenuation, and antenna gain

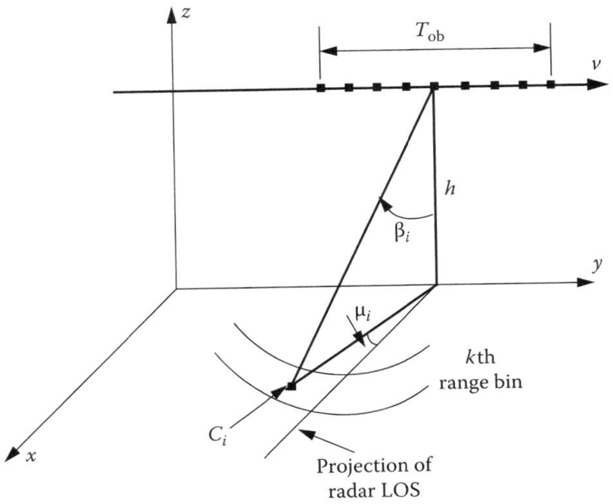

FIGURE 34.9 A scatterer C_i within the kth range bin.

The round-trip delay is

$$\tau_i(t, \mu_i) = \frac{2r_i(t, \mu_i)}{c} \tag{34.36}$$

where
 c is the speed of light
 $r_i(t, \mu_i)$ is the scatterer slant range

From the geometry in Figure 34.9, one can write the expression for the slant range to the ith scatterer within the kth range bin as

$$r_i(t, \mu_i) = \frac{h}{\cos\beta_i}\sqrt{1 - \frac{2vt}{h}\cos\beta_i\cos\mu_i\sin\beta_i + \left(\frac{vt}{h}\cos\beta_i\right)^2} \tag{34.37}$$

And by using Equation 34.36 the round-trip delay can be written as

$$\tau_i(t, \mu_i) = \frac{2}{c}\frac{h}{\cos\beta_i}\sqrt{1 - \frac{2vt}{h}\cos\beta_i\cos\mu_i\sin\beta_i + \left(\frac{vt}{h}\cos\beta_i\right)^2} \tag{34.38}$$

The round-trip delay can be approximated using a 2-D second-order Taylor series expansion about the reference state $(t, \mu) = (0, 0)$. Performing this Taylor series expansion yields

$$\tau_i(t, \mu_i) \approx \bar{\tau} + \bar{\tau}_{t\mu} \quad \mu_i t + \bar{\tau}_{tt}\frac{t^2}{2} \tag{34.39}$$

where the overbar indicates evaluation at the state (0, 0), and the subscripts denote partial derivatives. For example, $\bar{t}_{t\mu}$ means

$$\bar{\tau}_{t\mu} = \frac{\partial^2}{\partial t \partial \mu}\tau_i(t, \mu_i)|_{(t,\mu)=(0,0)} \tag{34.40}$$

The Taylor series coefficients are

$$\bar{\tau} = \left(\frac{2h}{c}\right)\frac{1}{\cos\beta_i} \tag{34.41}$$

$$\bar{\tau}_{t\mu} = \left(\frac{2v}{c}\right)\sin\beta_i \tag{34.42}$$

$$\bar{\tau}_{tt} = \left(\frac{2v^2}{hc}\right)\cos\beta_i \tag{34.43}$$

Note that other Taylor series coefficients are either zeros or very small. Hence, they are neglected. Finally, we can rewrite the returned radar signal as

$$\begin{aligned} s_i(t, \mu_i) &= A_i \cos\left[\hat{\psi}_i(t, \mu_i) - \xi_0\right] \\ \hat{\psi}_i(t, \mu_i) &= 2\pi f_0\left[(1 - \bar{\tau}_{t\mu}\mu_i)t - \bar{\tau} - \bar{\tau}_{tt}\frac{t^2}{2}\right] \end{aligned} \tag{34.44}$$

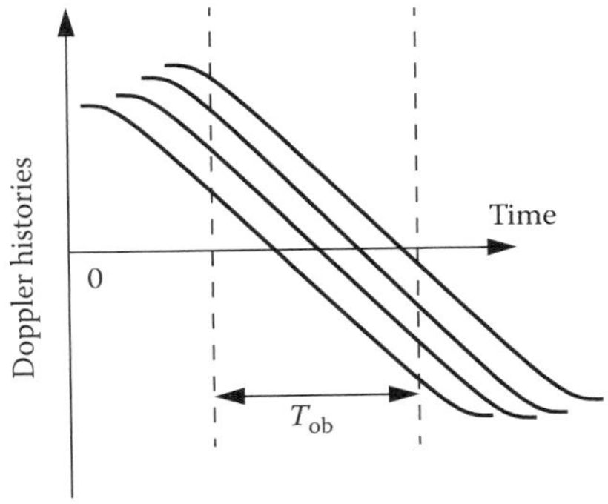

FIGURE 34.10 Doppler histories for several scatterers within the same range bin.

Observation of Equation 34.44 indicates that the instantaneous frequency for the ith scatterer varies as a linear function of time due to the second-order phase term $2\pi f_0(\bar{\tau}_{tt}t^2/2)$ (this confirms the result we concluded about a scatterer Doppler history). Furthermore, since this phase term is range-bin dependent and not scatterer dependent, all scatterers within the same range bin produce this exact second-order phase term. It follows that scatterers within a range bin have identical Doppler histories. These Doppler histories are separated by the time delay required to fly between them, as illustrated in Figure 34.10.

Suppose that there are I scatterers within the kth range bin. In this case, the combined returns for this cell are the sum of the individual returns due to each scatterer as defined by Equation 34.44. In other words, superposition holds, and the overall echo signal is

$$s_r(t) = \sum_{i=1}^{I} s_i(t, \mu_i) \tag{34.45}$$

A signal processing block diagram for the kth range bin is illustrated in Figure 34.11. It consists of the following steps. First, heterodyning with the carrier frequency is performed to extract the quadrature components.

This is followed by LP filtering and A/D conversion. Next, deramping or focusing to remove the second-order phase term of the quadrature components is carried out using a phase rotation matrix. The last stage of the processing includes windowing, performing an fast Fourier transform (FFT) on the windowed quadrature components, and scaling the amplitude spectrum to account for range attenuation and antenna gain.

The discrete quadrature components are

$$\begin{aligned} \tilde{x}_I(t_n) &= \tilde{x}_I(n) = A_i \cos\left[\tilde{\psi}_i(t_n, \mu_i) - \xi_0\right] \\ \tilde{x}_Q(t_n) &= \tilde{x}_Q(n) = A_i \sin\left[\tilde{\psi}_i(t_n, \mu_i) - \xi_0\right] \end{aligned} \tag{34.46}$$

$$\tilde{\psi}_i(t_n, \mu_i) = \psi_i(t_n, \mu_i) - 2\pi f_0 t_n \tag{34.47}$$

and t_n denotes the nth sampling time (remember that $-T_{ob}/2 \le t_n \le T_{ob}/2$). The quadrature components after deramping (i.e., removal of the phase $\psi = -\pi f_0 \bar{\tau}_{tt} t_n^2$ are given by

$$\begin{bmatrix} x_I(n) \\ x_Q(n) \end{bmatrix} = \begin{bmatrix} \cos\psi & -\sin\psi \\ \sin\psi & \cos\psi \end{bmatrix} \begin{bmatrix} \tilde{x}_I(n) \\ \tilde{x}_Q(n) \end{bmatrix} \tag{34.48}$$

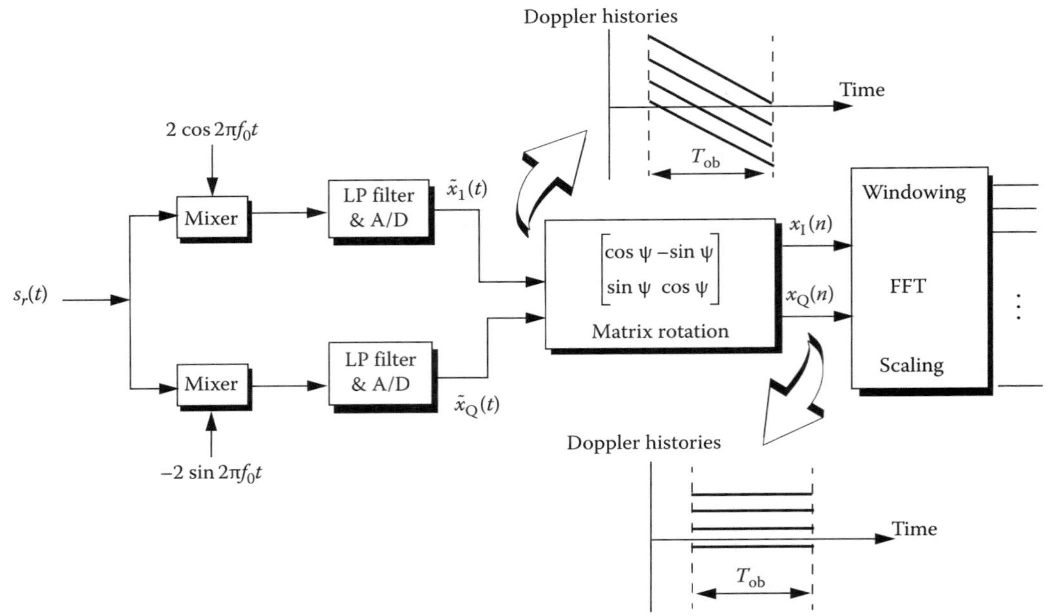

FIGURE 34.11 Signal processing block diagram for the kth range bin.

34.7 SAR Imaging Using Doppler Processing

It was mentioned earlier that SAR imaging is performed using two orthogonal dimensions (range and azimuth). Range resolution is controlled by the receiver bandwidth and pulse compression. Azimuth resolution is limited by the antenna beamwidth. A one-to-one correspondence between the FFT bins and the azimuth resolution cells can be established by utilizing the signal model described in the previous section. Therefore, the problem of target detection is transformed into a spectral analysis problem, where detection is based on the amplitude spectrum of the returned signal. The FFT frequency resolution Δf is equal to the inverse of the observation interval T_{ob}. It follows that a peak in the amplitude spectrum at $k_1\Delta f$ indicates the presence of a scatterer at frequency $f_{d1} = k_1\Delta f$.

For an example, consider the scatterer C_i within the kth range bin. The instantaneous frequency f_{di} corresponding to this scatterer is

$$f_{di} = \frac{1}{2\pi}\frac{d\psi}{dt} = f_0\bar{\tau}_{t\mu}\mu_i = \frac{2v}{\lambda}\sin\beta_i\mu_i \tag{34.49}$$

This is the same result derived in Equation 34.23, with $\mu_i = \Delta\theta$. Therefore, the scatterers separated in Doppler by more than Δf can then be resolved.

Figure 34.12 shows a 2-D SAR image for three point scatterers located 10 km downrange. In this case, the azimuth and range resolutions are equal to 1 m and the operating frequency is 35 GHz. Figure 34.13 is similar to Figure 34.12, except in this case the resolution cell is equal to 6 in. One can clearly see the blurring that occurs in the image. Figures 34.12 and 34.13 can be reproduced using the program "*fig34_34_13.m*" given in Listing 34.1 in Section 34.10.

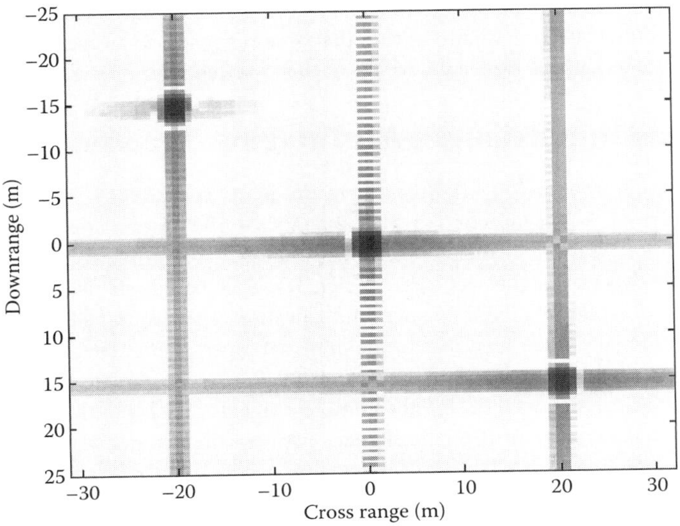

FIGURE 34.12 Three point scatterer image. Resolution cell is 1 m^2.

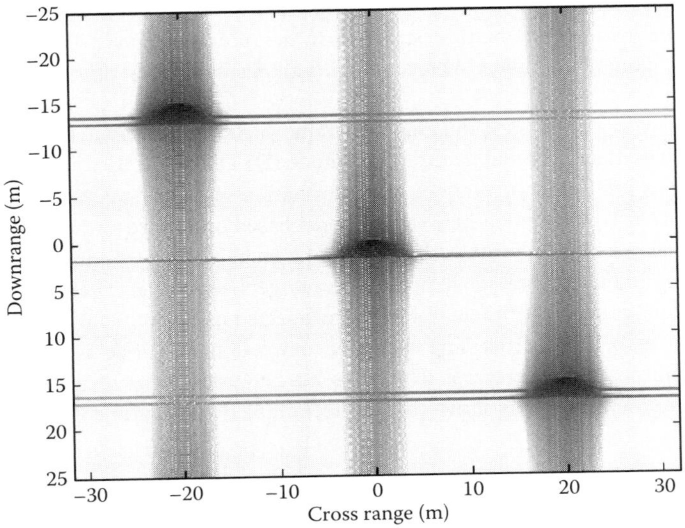

FIGURE 34.13 Three point scatterer image. Resolution cell is squared inches.

34.8 Range Walk

As shown earlier, SAR Doppler processing is achieved in two steps: first, range gating and second, azimuth compression within each bin at the end of the observation interval. For this purpose, azimuth compression assumes that each scatterer remains within the same range bin during the observation interval. However, since the range gates are defined with respect to a radar that is moving, the range gate grid is also moving relative to the ground. As a result a scatterer appears to be moving within its range bin. This phenomenon is known as range walk. A small amount of range walk does not bother Doppler processing as long as the scatterer remains within the same range bin. However, range walk over several range bins can constitute serious problems, where in this case Doppler processing is meaningless.

34.9 A Three-Dimensional SAR Imaging Technique

This section presents a new 3-D SAR imaging technique.* It utilizes a linear array in transverse motion to synthesize a 2-D synthetic array. Elements of the linear array are fired sequentially (one element at a time), while all elements receive in parallel. A 2-D information sequence is computed from the equiphase two-way signal returns. A signal model based on a third-order Taylor series expansion about incremental relative time, azimuth, elevation, and target height is used. Scatterers are detected as peaks in the amplitude spectrum of the information sequence. Detection is performed in two stages. First, all scatterers within a footprint are detected using an incomplete signal model where target height is set to zero. Then, processing using the complete signal model is performed only on range bins containing significant scatterer returns. The difference between the two images is used to measure target height. Computer simulation shows that this technique is accurate and virtually impulse invariant.

34.9.1 Background

Standard SAR imaging systems are generally used to generate high resolution 2-D images of ground terrain. Range gating determines resolution along the first dimension. Pulse compression techniques are usually used to achieve fine range resolution. Such techniques require the use of wideband receiver and display devices in order to resolve the time structure in the returned signals. The width of azimuth cells provides resolution along the other dimension. Azimuth resolution is limited by the duration of the observation interval.

This section presents a 3-D SAR imaging technique based on discrete Fourier transform (DFT) processing of equiphase data collected in sequential mode (DFTSQM). It uses a linear array in transverse motion to synthesize a 2-D synthetic array. A 2-D information sequence is computed from the equiphase two-way signal returns. To this end, a new signal model based on a third-order Taylor series expansion about incremental relative time, azimuth, elevation, and target height is introduced. Standard SAR imaging can be achieved using an incomplete signal model where target height is set to zero. Detection is performed in two stages. First, all scatterers within a footprint are detected using an incomplete signal model, where target height is set to zero. Then, processing using the complete signal model is performed only on range bins containing significant scatterer returns. The difference between the two images is used as an indication of target height. Computer simulation shows that this technique is accurate and virtually impulse invariant.

34.9.2 DFTSQM Operation and Signal Processing

34.9.2.1 Linear Arrays

Consider a linear array of size N, uniform element spacing d, and wavelength λ. Assume a far field scatterer P located at direction-sine $\sin\beta_l$. DFTSQM operation for this array can be described as follows. The elements are fired sequentially, one at a time, while all elements receive in parallel. The echoes are collected and integrated coherently on the basis of equal phase to compute a complex information sequence $\{b(m); m=0, 2N-1\}$. The x-coordinates, in d-units, of the x_n^{th} element with respect to the center of the array is

$$x_n = \left(-\frac{N-1}{2} + n\right); \quad n = 0, \ldots N-1 \tag{34.50}$$

* This section is extracted from Mahafza et al. [1]

The electric field received by the x_2^{th} element due to the firing of the x_1^{th}, and reflection by the lth far field scatterer P, is

$$E(x_1, x_2; s_l) = G^2(s_l)\left(\frac{R_0}{R}\right)^4 \sqrt{\sigma_l}\exp\left(j\phi(x_1, x_2; s_l)\right) \tag{34.51}$$

$$\phi(x_1, x_2; s_l) = \frac{2\pi}{\lambda}(x_1 + x_2)(s_l) \tag{34.52}$$

$$s_l = \sin\beta_l \tag{34.53}$$

where

$\sqrt{\sigma_l}$ is the target cross section
$G^2(s_l)$ is the two-way element gain
$(R_0/R)^4$ is the range attenuation with respect to reference range R_0

The scatterer phase is assumed to be zero; however it could be easily included. Assuming multiple scatterers in the array's FOV, the cumulative electric field in the path $x_1 \Rightarrow x_2$ due to reflections from all scatterers is

$$E(x_1, x_2) = \sum_{\text{all } l}\left[E_I(x_1, x_2; s_l) + jE_Q(x_1, x_2; s_l)\right] \tag{34.54}$$

where the subscripts (I, Q) denote the quadrature components. Note that the variable part of the phase given in Equation 34.52 is proportional to the integers resulting from the sums $\{(x_{nl} + x_{n2});\ (n1, n2) = 0, \ldots N-1\}$. In the far field operation there are a total of ($2N$–1 distinct $(x_{nl} + x_{n2})$ sums. Therefore, the electric fields with paths of the same $(x_{nl} + x_{n2})$ sums can be collected coherently. In this manner the information sequence $\{b(m);\ m = 0,\ 2\ N-1\}$ is computed, where $b(2N-1)$ is set to zero. At the same time one forms the sequence $\{c(m);\ m = 0, \ldots 2N-2\}$ which keeps track of the number of returns that have the same $(x_{n1} + x_{n2})$ sum. More precisely, for $m = n1 + n2;\ (n1, n2) = \ldots 0, N-1$

$$b(m) = b(m) + E(x_{n1}, x_{n2}) \tag{34.55}$$

$$c(m) = c(m) + 1 \tag{34.56}$$

It follows that

$$\{c(m); m = 0, \ldots 2N-2\} = \left\{\begin{array}{lllll} m+1 & ; & m=0 & \ldots & N-2 \\ N & ; & m & = & N-1 \\ 2N-1 & -m & m=N, & \ldots & 2N-2 \end{array}\right\} \tag{34.57}$$

which is a triangular shape sequence.

The processing of the sequence $\{b(m)\}$ is performed as follows: (1) the weighting takes the sequence $\{c(m)\}$ into account; (2) the complex sequence $\{b(m)\}$ is extended to size N_F, a power integer of two, by zero padding; (3) the DFT of the extended sequence $\{b'(m);\ m = 0,\ N_F - 1\}$ is computed,

$$B(q) = \sum_{m=0} b'(m) \cdot \exp\left(-j\frac{2\pi qm}{N_F}\right); \quad q = 0, \ldots N_F - 1 \tag{34.58}$$

and, (4) after compensation for antenna gain and range attenuation, scatterers are detected as peaks in the amplitude spectrum $|B(q)|$. Note that step (4) is true only when

$$\sin \beta_q = \frac{\lambda q}{2Nd}; \quad q = 0, \ldots 2N - 1 \tag{34.59}$$

where

$\sin \beta_q$ denotes the direction-sine of the qth scatterer
$N_F = 2N$ is implied in Equation 34.59

The classical approach to multiple target detection is to use a phased array antenna with phase shifting and tapering hardware. The array beamwidth is proportional to (λ/Nd), and the first sidelobe is at about -13 dB. On the other hand, multiple target detection using DFTSQM provides a beamwidth proportional to $(\lambda/2Nd)$ as indicated by (Equation 34.59), which has the effect of doubling the array's resolution. The first sidelobe is at about -27 dB due to the triangular sequence $\{c(m)\}$. Additionally, no phase shifting hardware is required for detection of targets within a single element's field of view.

34.9.2.2 Rectangular Arrays

DFTSQM operation and signal processing for 2-D arrays can be described as follows. Consider an $N_x \times N_y$ rectangular array. All N_xN_y elements are fired sequentially, one at a time. After each firing, all the N_xN_y array elements receive in parallel. Thus, N_xN_y samples of the quadrature components are collected after each firing, and a total of $(N_xN_y)^2$ samples will be collected. However, in the far field operation, there are only $(2N_x - 1) \times (2N_y - 1)$ distinct equiphase returns. Therefore, the collected data can be added coherently to form a 2-D information array of size $(2N_x - 1) \times (2N_y - 1)$. The two-way radiation pattern is computed as the modulus of the 2-D amplitude spectrum of the information array. The processing includes 2-D windowing, 2-D DFT, antenna gain, and range attenuation compensation. The field of view of the 2-D array is determined by the 3 dB pattern of a single element. All the scatterers within this field will be detected simultaneously as peaks in the amplitude spectrum.

Consider a rectangular array of size $N \times N$, with uniform element spacing $d_x = d_y = d$, and wavelength λ. The coordinates of the nth element, in d-units, are

$$x_n = \left(-\frac{N-1}{2} + n\right); \quad n = 0, \ldots N - 1 \tag{34.60}$$

$$y_n = \left(-\frac{N-1}{2} + n\right); \quad n = 0, \ldots N - 1 \tag{34.61}$$

Assume a far field point P defined by the azimuth and elevation angles (α, β). In this case, the one-way geometric phase for an element is

$$\varphi'(\mathrm{x}, \mathrm{y}) = \frac{2\pi}{\lambda}[x \sin \beta \cos \alpha + y \sin \beta \sin \alpha] \tag{34.62}$$

Therefore, the two-way geometric phase between the (x_1, y_1) and (x_2, y_2) elements is

$$\varphi(x_1, y_1, x_2, y_2,) = \frac{2\pi}{\lambda} \sin \beta[(x_1 + x_2) \cos \alpha + (y_1 + y_2) \sin \alpha] \tag{34.63}$$

The two-way electric field for the lth scatterer at (α_l, β_l) is

$$E(x_1, x_2, y_1, y_2; \alpha_l, \beta_l) = G^2(\beta_l)\left(\frac{R_0}{R}\right)^4 \sqrt{\sigma_l}\exp[j(\phi(x_1, y_1, x_2, y_2))] \tag{34.64}$$

Assuming multiple scatterers within the array's FOV, then the cumulative electric field for the two-way path $(x_1, y_1) \Rightarrow (x_2, y_2)$ is given by

$$E(x_1, x_2, y_1, y_2) = \sum_{\text{all scatteres}} E(x_1, x_2, y_1, y_2; \alpha_l, \beta_l) \tag{34.65}$$

All formulas for the 2-D case reduce to those of a linear array case by setting $N_y = 1$ and $\alpha = 0$.

The variable part of the phase given in Equation 34.63 is proportional to the integers $(x_1 + x_2)$ and $(y_1 + y_2)$. Therefore, after completion of the sequential firing, electric fields with paths of the same (i, j) sums, where

$$\{i = x_{n1} + x_{n2}; \quad i = -(N-1), \ldots (N-1)\} \tag{34.66}$$

$$\{j = y_{n1} + y_{n2}; \quad j = -(N-1), \ldots (N-1)\} \tag{34.67}$$

can be collected coherently. In this manner the 2-D information array $\{b\{m_x, m_y); \{m_x, m_y) = 0, \ldots 2N-1\}$ is computed. The coefficient sequence $\{c(m_x, m_y); \{m_x, m_y) = 0, \ldots 2N-2\}$ is also computed. More precisely,

$$\begin{gathered} \text{for } m_x = n1 + n2 \quad \text{and} \quad m_y = n1 + n2 \\ n1 = 0, \ldots N-1, \quad \text{and} \quad n2 = 0, \ldots N-1 \end{gathered} \tag{34.68}$$

$$b(m_x, m_y) = b(m_x, m_y) + E(x_{n1}, y_{n1}, x_{n2}, y_{n2}) \tag{34.69}$$

It follows that

$$c(m_x, m_y) = (N_x - |m_x - (N_x - 1)|) \times (N_y - |m_y - (N_y - 1)|) \tag{34.70}$$

The processing of the complex 2-D information array $\{b(m_x, m_y)\}$ is similar to that of the linear case with the exception that one should use a 2-D DFT. After antenna gain and range attenuation compensation, scatterers are detected as peaks in the 2-D amplitude spectrum of the information array. A scatterer located at angles (α_l, β_l) will produce a peak in the amplitude spectrum at DFT indexes (p_l, q_l), where

$$\alpha_l = a\tan\left(\frac{q_l}{p_l}\right) \tag{34.71}$$

$$\sin\beta_l = \frac{\lambda p_l}{2Nd\cos\alpha_l} = \frac{\lambda q_l}{2Nd\sin\alpha_l} \tag{34.72}$$

Derivation of Equation 34.71 is in Section 34.9.7.

34.9.3 Geometry for DFTSQM SAR Imaging

Figure 34.14 shows the geometry of the DFTSQM SAR imaging system. In this case, t_c denotes the central time of the observation interval, D_{ob}. The aircraft maintains both constant velocity v and height h.

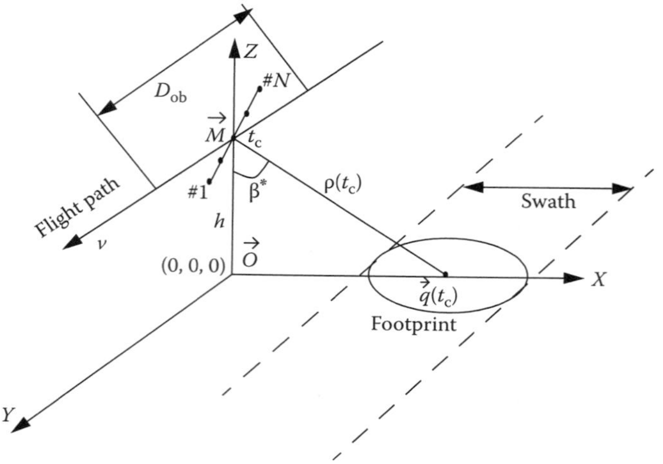

FIGURE 34.14 Geometry for DFTSQM imaging system.

The origin for the relative system of coordinates is denoted as $\vec{O}$. The vector $\vec{O}M$ defines the radar location at time t_c. The transmitting antenna consists of a linear real array operating in the sequential mode. The real array is of size N, element spacing d, and the radiators are circular dishes of diameter $D = d$. Assuming that the aircraft scans M transmitting locations along the flight path, then a rectangular array of size $N \times M$ is synthesized, as illustrated in Figure 34.15.

The vector $\vec{q}(t_c)$ defines the center of the 3 dB footprint at time t_c. The center of the array coincides with the flight path, and it is assumed to be perpendicular to both the flight path and the LOS $\rho(t_c)$. The unit vector $\vec{a}$ along the real array is

$$\vec{a} = \cos\beta^* \vec{a}_x + \sin\beta^* \vec{a}_Z \tag{34.73}$$

where β^* is the elevation angle, or the complement of the depression angle, for the center of the footprint at central time t_c.

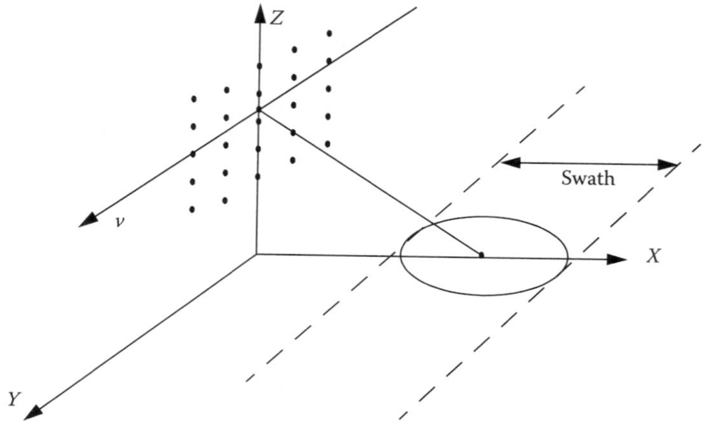

FIGURE 34.15 Synthesized 2-D array.

34.9.4 Slant Range Equation

Consider the geometry shown in Figure 34.16 and assume that there is a scatterer $\vec{C}_i$ within the kth range cell. This scatterer is defined by

$$\{\text{ampltiude, phase, elevation, azimuth, height}\} = \{a_i, \phi_i, \beta_i, \mu_i, \tilde{h}_i\} \tag{34.74}$$

The scatterer $\vec{C}_i$ (assuming rectangular coordinates) is given by

$$\vec{C}_i = h \tan \beta_i \cos \mu_i \vec{a}_x + h \tan \beta_i \sin \mu_i \vec{a}_y + \tilde{h}_i \vec{a}_Z \tag{34.75}$$

$$\beta_i = \beta_k + \varepsilon \tag{34.76}$$

where
β_k denotes the elevation angle for the kth range cell at the center of the observation interval
ε is an incremental angle

Let $\vec{O}e_n$ refer to the vector between the nth array element and the point $\vec{O}$, then

$$\vec{O}e_n = D_n \cos \beta^\star \vec{a}_x + vt\vec{a}_y + (D_n \sin \beta^\star + h)\vec{a}_z \tag{34.77}$$

$$D_n = \left(\frac{1-N}{2} + n\right)d; \quad n = 0, \ldots N-1 \tag{34.78}$$

The range between a scatterer $\vec{C}$ within the kth range cell and the nth element of the real array is

$$\begin{aligned} r_n^2(t, \varepsilon, \mu, \tilde{h}; D_n) = {} & D_n^2 + v^2t^2 + (h - \tilde{h})^2 + 2D_n \sin \beta^\star(h - \tilde{h}) \\ & + h \tan(\beta_k + \varepsilon)[h \tan(\beta_k + \varepsilon) - 2D_n \cos \beta^\star \cos \mu - 2vt \sin \mu] \end{aligned} \tag{34.79}$$

It is more practical to use the scatterer's elevation and azimuth direction-sines rather than the corresponding increments. Therefore, define the scatterer's azimuth and elevation direction-sines as

$$s = \sin \mu \tag{34.80}$$

$$u = \sin \varepsilon \tag{34.81}$$

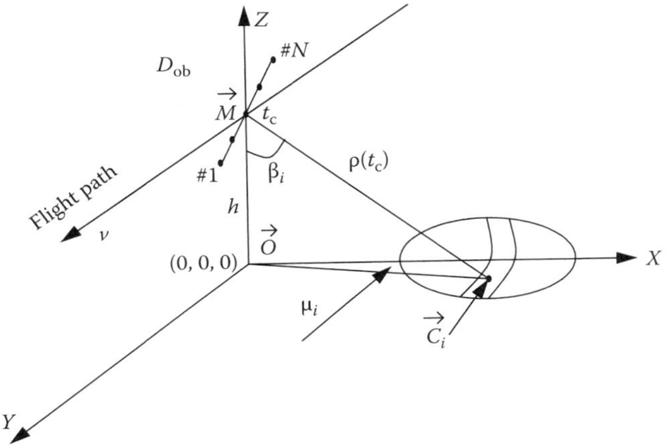

FIGURE 34.16 Scatterer $\vec{C}_i$ within a range cell.

Then, one can rewrite Equation 34.79 as

$$r_n^2(t,s,u,\tilde{h};\ D_n) = D_n^2 + v^2t^2 + (h-\tilde{h})^2 + h^2f^2(u) + 2D_n \sin\beta^*(h-\tilde{h}) - \left(2D_nh\cos\beta^* f(u)\sqrt{1-s^2} - 2vhtf(u)s\right) \quad (34.82)$$

$$f(u) = \tan(\beta_k + a\sin u) \quad (34.83)$$

Expanding r_n as a third-order Taylor series expansion about incremental (t,s,u,h) yields

$$\begin{aligned} r(t,s,u,\tilde{h};D_n) &= \bar{r} + \bar{r}_{\tilde{h}}\tilde{h} + \bar{r}_u u + \bar{r}_{\tilde{h}\tilde{h}}\frac{\tilde{h}^2}{2} + \bar{r}_{\tilde{h}u}\tilde{h}u + \bar{r}_{ss}\frac{s^2}{2} + \bar{r}_{st}st \\ &\quad + \bar{r}_{tt}\frac{t^2}{2} + \bar{r}_{uu}\frac{u^2}{2} + \bar{r}_{\tilde{h}\tilde{h}\tilde{h}}\frac{\tilde{h}^3}{6} + \bar{r}_{\tilde{h}\tilde{h}u}\frac{\tilde{h}^2u}{2} + \bar{r}_{\tilde{h}st}\tilde{h}st + \bar{r}_{\tilde{h}uu}\frac{\tilde{h}u^2}{2} \\ &\quad + \bar{r}_{\tilde{h}ss}\frac{\tilde{h}s^2}{2} + \bar{r}_{uss}\frac{us^2}{2} + \bar{r}_{stu}stu + \bar{r}_{suu}\frac{su^2}{2} + \bar{r}_{t\tilde{h}\tilde{h}}\frac{t\tilde{h}^2}{2} + \bar{r}_{utt}\frac{ut^2}{2} + \bar{r}_{uuu}\frac{u^3}{6} \end{aligned} \quad (34.84)$$

where subscripts denote partial derivations, and the over-bar indicates evaluation at the state $(t,s,u,\tilde{h}) = (0,0,0,0)$. Note that

$$\left\{\bar{r}_s = \bar{r}_t = \bar{r}_{\tilde{h}x} = \bar{r}_{\tilde{h}t} = \bar{r}_{su} = \bar{r}_{tu} = \bar{r}_{\tilde{h}\tilde{h}t} = \bar{r}_{\tilde{h}su} = \bar{r}_{\tilde{h}tu} = \bar{r}_{sss} = \bar{r}_{sst} = \bar{r}_{stt} = \bar{r}_{ttt} = \bar{r}_{tsu} = 0\right\} \quad (34.85)$$

Section 34.9.8 has detailed expressions of all nonzero Taylor series coefficients for the kth range cell. Even at the maximum increments t_{mx}, s_{mx}, u_{mx}, $\tilde{h}_{mx}$, the terms:

$$\left\{\bar{r}_{\tilde{h}\tilde{h}\tilde{h}}\frac{\tilde{h}^3}{6},\ \bar{r}_{\tilde{h}\tilde{h}u}\frac{\tilde{h}^2u}{2},\ \bar{r}_{\tilde{h}uu}\frac{\tilde{h}u^2}{2},\ \bar{r}_{\tilde{h}xx}\frac{\tilde{h}s^2}{2},\ \bar{r}_{uss}\frac{us^2}{2},\ \bar{r}_{stu}stu,\ \bar{r}_{suu}\frac{su^2}{2},\ \bar{r}_{t\tilde{h}\tilde{h}}\frac{t\tilde{h}^2}{2},\ \bar{r}_{utt}\frac{ut^2}{2},\ \bar{r}_{uuu}\frac{u^3}{6}\right\} \quad (34.86)$$

are small and can be neglected. Thus, the range r_n is approximated by

$$r(t,s,u,\tilde{h};D_n) = \bar{r} + \bar{r}_{\tilde{h}}\tilde{h} + \bar{r}_u u + \bar{r}_{\tilde{h}\tilde{h}}\frac{\tilde{h}^2}{2} + \bar{r}_{\tilde{h}u} + \bar{r}_{ss}\frac{s^2}{2} + \bar{r}_{st}st + \bar{r}_{tt}\frac{t^2}{2} + \bar{r}_{uu}\frac{u^2}{2} + \bar{r}_{\tilde{h}st}\tilde{h}st \quad (34.87)$$

Consider the following two-way path: the n_1^{th} element transmitting, scatterer $\vec{C}_i$ reflecting, and the n_2^{th} element receiving. It follows that the round-trip delay corresponding to this two-way path is

$$\tau_{n_1n_2} = \frac{1}{c}\left(r_{n_1}(t,s,u,\tilde{h};D_{n_1}) + r_{n_2}(t,s,u,\tilde{h};D_{n_2})\right) \quad (34.88)$$

where c is the speed of light.

34.9.5 Signal Synthesis

The observation interval is divided into M subintervals of width $\Delta t = (D_{ob} \div M)$. During each subinterval, the real array is operated in sequential mode, and an array length of $2N$ is synthesized.

The number of subintervals M is computed such that Δt is large enough to allow sequential transmission for the real array without causing range ambiguities. In other words, if the maximum range is denoted as R_{mx} then

$$\Delta t > N\frac{2R_{mx}}{c} \tag{34.89}$$

Each subinterval is then partitioned into N sampling subintervals of width $2R_{mx}/c$. The location t_{mn} represents the sampling time at which the nth element is transmitting during the mth subinterval.

The normalized transmitted signal during the mth subinterval for the nth element is defined as

$$s_n(t_{mn}) = \cos(2\pi f_o t_{mn} + \zeta) \tag{34.90}$$

where

ζ denotes the transmitter phase
f_o is the system operating frequency

Assume that there is only one scatterer, $\vec{C}_i$, within the kth range cell defined by $(a_i, \phi_i, s_i, u_i, \tilde{h}_i)$. The returned signal at the n_2^{th} element due to firing from the n_1^{th} element and reflection from the $\vec{C}_i$ scatterer is

$$s_i(n_1, n_2; t_{mn_1}) = a_i G^2(\sin\beta_i)(\rho_k(t_c)/\rho(t_c))^4 \cos[2\pi f_o(t_{mn_1} - \tau_{n_1 n_2}) + \zeta - \phi_i] \tag{34.91}$$

where

G^2 represents the two-way antenna gain
$(\rho_k(t_c)/\rho(t_c))^4$ denotes the range attenuation at the kth range cell

The analysis in this chapter will assume here on that ζ and ϕ_i, are both equal to zeroes.

Suppose that there are N_o scatterers within the kth range cell, with angular locations given by

$$\{(a_i, \phi_i, s_i, u_i, \tilde{h}_i); \quad i = 1, \ldots N_o\} \tag{34.92}$$

The composite returned signal at time t_{mn_1} within this range cell due to the path ($n_1 \Rightarrow$ all $\vec{C}_i \Rightarrow n_2$) is

$$s(n_1, n_2; t_{mn_1}) = \sum_{i=1} s_i(n_1, n_2; t_{mn_1}) \tag{34.93}$$

The platform motion synthesizes a rectangular array of size $N \times M$, where only one column of N elements exists at a time. However, if $M = 2N$ and the real array is operated in the sequential mode, a square planar array of size $2N \times 2N$ is synthesized. The element spacing along the flight path is $d_y = \nu D_{ob}/M$.

Consider the kth range bin. The corresponding 2-D information sequence $\{b_k(n, m);\ (n, m) = 0, \ldots 2N - 2\}$ consists of $2N$ similar vectors. The mth vector represents the returns due to the sequential firing of all N elements during the mth subinterval. Each vector has $(2N-1)$ rows, and it is extended, by adding zeroes, to the next power of two. For example, consider the mth subinterval, and let $M=2N=4$. Then, the elements of the extended column $\{b_k(n,m)\}$ are

i

$$\begin{aligned}\{b_k(0,m), b_k\ (1,m), b_k\ (2,m), b_k(3,m), b_k(4,m), b_k(5,m), b_k(6,m), b_k(7,m)\}\\ = \{s(0,0;t_{mn_0}),\ s(0,1;t_{mn_0}) + s(1,0;t_{mn_1}),\ s(0,2;t_{mn_0}) + s(1,1;t_{m,n_1}) + s(2,0;t_{mn_2}),\ s(0,3;t_{mn_0})\\ + s(1,2;t_{mn_1}) + s(2,1;t_{mn_2}) + s(3,0;t_{mn_3}),\ s(1,3;t_{mn_1}) + s(2,2;t_{mn_2}) + s(3,1;t_{mn_3}),\ s(2,3;t_{mn_2})\\ + s(3,2;t_{mn_3}), s(3,3;t_{mn_3}), 0\}\end{aligned} \tag{34.94}$$

34.9.6 Electronic Processing

Consider again the kth range cell during the mth subinterval, and the two-way path: n_1^{th} element transmitting and n_2^{th} element receiving. The analog quadrature components corresponding to this two-way path are

$$s_I^{\perp}(n_1,n_2;t) = B\cos\psi^{\perp} \tag{34.95}$$

$$s_Q^{\perp}(n_1,n_2;t) = B\sin\psi^{\perp} \tag{34.96}$$

$$\begin{aligned}\psi^{\perp} = 2\pi f_0\Bigg\{t - \frac{1}{c}\Bigg[&2\bar{r} + (\bar{r}_{\tilde{h}}(D_{n_1}) + \bar{r}_{\tilde{h}}(D_{n_2}))\tilde{h} + (\bar{r}_u(D_{n_1}) + \bar{r}_u(D_{n_2}))u\\ &+ (\bar{r}_{\tilde{h}\tilde{h}}(D_{n_1}) + \bar{r}_{\tilde{h}\tilde{h}}(D_{n_2}))\frac{\tilde{h}^2}{2} + (\bar{r}_{\tilde{h}u}(D_{n_1}) + \bar{r}_{\tilde{h}u}(D_{n_2}))\tilde{h}u\\ &+ (\bar{r}_{ss}(D_{n_1}) + \bar{r}_{ss}(D_{n_2}))\frac{s^2}{2} + 2\bar{r}_{st}st + 2\bar{r}_{tt}\frac{t^2}{2}\\ &+ (\bar{r}_{uu}(D_{n_1}) + \bar{r}_{uu}(D_{n_2}))\frac{u^2}{2} + (\bar{r}_{\tilde{h}st}(D_{n_1}) + \bar{r}_{\tilde{h}st}(D_{n_2}))\tilde{h}st\Bigg]\Bigg\}\end{aligned} \tag{34.97}$$

where B denotes antenna gain, range attenuation, and scatterers' strengths. The subscripts for t have been dropped for notation simplicity. Rearranging Equation 34.97 and collecting terms yields

$$\begin{aligned}\psi^{\perp} = \frac{2\pi f_0}{c}\Bigg\{&\Big\{tc - [2\bar{r}_{st}s + (\bar{r}_{\tilde{h}st}(D_{n_1}) + \bar{r}_{\tilde{h}st}(D_{n_2}))\tilde{h}s]t - \bar{r}_{tt}t^2\Big\}\\ &- [2\bar{r} + (\bar{r}_{\tilde{h}}(D_{n_1}) + \bar{r}_{\tilde{h}}(D_{n_2}))\tilde{h} + (\bar{r}_u(D_{n_1}) + \bar{r}_u(D_{n_2}))u\\ &+ (\bar{r}_{uu}(D_{n_1}) + \bar{r}_{uu}(D_{n_2}))\frac{u^2}{2} + (\bar{r}_{\tilde{h}\tilde{h}}(D_{n_1}) + \bar{r}_{\tilde{h}\tilde{h}}(D_{n_2}))\frac{\tilde{h}^2}{2}\\ &+ (\bar{r}_{\tilde{h}\tilde{h}}(D_{n_1}) + \bar{r}_{\tilde{h}u}(D_{n_2}))\tilde{h}u + (\bar{r}_{ss}(D_{n_1}) + \bar{r}_{ss}(D_{n_2}))\frac{s^2}{2}\Bigg]\Bigg\}\end{aligned} \tag{34.98}$$

After analog to digital (A/D) conversion, deramping of the quadrature components to cancel the quadratic phase $(-2\pi f_0 \bar{r}_{tt}t^2/c)$ is performed. Then, the digital quadrature components are

$$s_I(n_1,n_2;t) = B\cos\psi \tag{34.99}$$

$$s_Q(n_1,n_2;t) = B\sin\psi \tag{34.100}$$

$$\psi = \psi^{\perp} - 2\pi f_0 t + 2\pi f_0 \bar{r}_{tt} \frac{t^2}{c} \tag{34.101}$$

The instantaneous frequency for the ith scatterer within the kth range cell is computed as

$$f_{di} = \frac{1}{2\pi}\frac{d\psi}{dt} = -\frac{f_0}{c}[2\bar{r}_{st}s + (\bar{r}_{\tilde{h}st}(D_{n_1}) + \bar{r}_{\tilde{h}st}(D_{n2}))\tilde{h}s] \tag{34.102}$$

Substituting the actual values for $\bar{r}_{st}$, $\bar{r}_{\tilde{h}st}(D_{n_1})$, $\bar{r}_{\tilde{h}st}(D_{n_2})$ and collecting terms yields

$$f_{di} = -\left(\frac{2v \sin\beta_k}{\lambda}\right)\left(\frac{\tilde{h}s}{\rho_k^2(t_c)}(h + (D_{n_1} + D_{n_2})\sin\beta^*) - s\right) \tag{34.103}$$

Note that if $\tilde{h} = 0$, then

$$f_{di} = \frac{2v}{\lambda}\sin\beta_k \sin\mu \tag{34.104}$$

which is the Doppler value corresponding to a ground patch (see Equation 34.49).

The last stage of the processing consists of three steps: (1) 2-D windowing; (2) performing a 2-D DFT on the windowed quadrature components; and (3) scaling to compensate for antenna gain and range attenuation.

34.9.7 Derivation of Equation 34.71

Consider a rectangular array of size $N \times N$, with uniform element spacing $d_x = d_y = d$, and wavelength λ. Assume sequential mode operation where elements are fired sequentially, one at a time, while all elements receive in parallel. Assume far field observation defined by azimuth and elevation angles (α, β). The unit vector $\vec{\mu}$ on the LOS, with respect to $\vec{O}$, is given by

$$\vec{u} = \sin\beta\cos\alpha\ \vec{a}_x + \sin\beta\sin\alpha\vec{a}_y + \cos\beta\vec{a}_z \tag{34.105}$$

The (n_x, n_y)th element of the array can be defined by the vector

$$\vec{e}(n_x, n_y) = \left(n_x - \frac{N-1}{2}\right)d\vec{a}_x + \left(n_x - \frac{N-1}{2}\right)d\vec{a}_y \tag{34.106}$$

where $(n_x, n_y = 0, \ldots N-1)$. The one-way geometric phase for this element is

$$\varphi'(n_x, n_y) = k(\vec{u} \bullet \vec{e}(n_x, n_y)) \tag{34.107}$$

where
- $k = 2\pi/\lambda$ is the wave number
- the operator ($\bullet$) indicates dot product

Therefore, the two-way geometric phase between the (n_{x1}, n_{y1}) and (n_{x2}, n_{y2}) elements is

$$\varphi(n_{x1}, n_{y1}, n_{x2}, n_{y2}) = k\left[\vec{u} \bullet \left\{\vec{e}(n_{x1}, n_{y1}) + \vec{e}(n_{x2}, n_{y2})\right\}\right] \tag{34.108}$$

The cumulative two-way normalized electric field due to all transmissions is

$$E(\vec{u}) = E_t(\vec{u})E_r(\vec{u}) \tag{34.109}$$

where the subscripts t and r, respectively, refer to the transmitted and received electric fields. More precisely,

$$E_t(\vec{u}) = \sum_{n_{xt}=0}^{N-1}\sum_{n_{yt}=0}^{N-1} w(n_{xt}, n_{yt}) \exp\left[jk\left\{\vec{u} \bullet \vec{e}(n_{xt}, n_{yt})\right\}\right] \tag{34.110}$$

$$E_r(\vec{u}) = \sum_{n_{xr}=0}^{N-1}\sum_{n_{yr}=0}^{N-1} w(n_{xr}, n_{yr}) \exp\left[jk\left\{\vec{u} \bullet \vec{e}(n_{xr}, n_{yr})\right\}\right] \tag{34.111}$$

In this case, $w(n_x, n_y)$ denotes the tapering sequence. Substituting Equations 34.108, 34.110, and 34.111 into Equation 34.109 and grouping all fields with the same two-way geometric phase yields

$$E(\vec{u}) = e^{j\delta} \sum_{m=0}^{N_a-1}\sum_{n=0}^{N_a-1} w'(m, n) \exp[jk \sin\beta(m\cos\alpha + n\sin\alpha)] \tag{34.112}$$

$$N_a = 2N - 1 \tag{34.113}$$

$$m = n_{xt} + n_{xr}; \quad m = 0, \ldots 2N - 2 \tag{34.114}$$

$$n = n_{yt} + n_{yr}; \quad m = 0, \ldots 2N - 2 \tag{34.115}$$

$$\delta = \left(\frac{-d\sin\beta}{2}\right)(N - 1)(\cos\alpha + \sin\alpha) \tag{34.116}$$

The two-way array pattern is then computed as

$$|E(\vec{u})| = \left|\sum_{m=0}^{N_a-1}\sum_{n=0}^{N_a-1} w'(m, n) \exp[jkd \sin\beta(m\cos\alpha + n\sin\alpha)]\right| \tag{34.117}$$

Consider the 2-D DFT transform, $W(p, q)$, of the array $w'(n_x, n_y)$

$$W'(p, q) = \sum_{m=0}^{N_a-1}\sum_{n=0}^{N_a-1} w'(m, n) \exp\left(-j\frac{2\pi}{N_a}(pm + qn)\right); \quad p, q = 0, \ldots N_a - 1 \tag{34.118}$$

Comparison of Equations 34.117 and 34.118 indicates that $|E(\vec{u})|$ is equal to $|W'(p, q)|$ if

$$-\left(\frac{2\pi}{N_a}\right)p = \frac{2\pi}{\lambda} d\sin\beta\cos\alpha \tag{34.119}$$

$$-\left(\frac{2\pi}{N_a}\right)q = \frac{2\pi}{\lambda} d\sin\beta\cos\alpha \tag{34.120}$$

It follows that

$$\alpha = \tan^{-1}\left(\frac{q}{p}\right) \tag{34.121}$$

34.9.8 Nonzero Taylor Series Coefficients for the kth Range Cell

$$\bar{r} = \sqrt{D_n^2 + h^2(1+\tan\beta_k) + 2hD_n\sin\beta^* - 2hD_n\cos\beta^*\tan\beta_k} = \rho_k(t_c) \quad (34.122)$$

$$\bar{r}_{\tilde{h}} = \left(\frac{-1}{\bar{r}}\right)(h + D_n\sin\beta^*) \quad (34.123)$$

$$\bar{r}_u = \left(\frac{h}{\bar{r}\cos^2\beta_k}\right)(h\tan\beta_k - D_n\cos\beta^*) \quad (34.124)$$

$$\bar{r}_{\tilde{h}\tilde{h}} = \left(\frac{1}{\bar{r}}\right) - \left(\frac{1}{\bar{r}^3}\right)(h + D_n\sin\beta^*) \quad (34.125)$$

$$\bar{r}_{\tilde{h}u} = \left(\frac{1}{\bar{r}^3}\right)\left(\frac{h}{\cos^2\beta_k}\right)(h + D_n\tan\beta^*)(h\tan\beta_k - D_n\cos\beta^*) \quad (34.126)$$

$$\bar{r}_{ss} = \left(\frac{-1}{4\bar{r}^3}\right) + \left(\frac{1}{\bar{r}}\right)(h\tan\beta_k - D_n\cos\beta^*) \quad (34.127)$$

$$\bar{r}_{st} = \left(\frac{-1}{\bar{r}}\right)hv\tan\beta_k \quad (34.128)$$

$$\bar{r}_{tt} = \frac{v^2}{\bar{r}} \quad (34.129)$$

$$\bar{r}_{uu} = \left(\frac{h}{\bar{r}\cos^3\beta_k}\right)\left\{\left(\frac{h}{\bar{r}^2\cos\beta_k}\right)(h\tan\beta_k - D_n\cos\beta^*)\right.$$
$$\left. + h\left(\left(\frac{1}{\cos\beta_k}\right) + 2\tan\beta_k\sin\beta_k\right) - 2\sin\beta_k D_n\cos\beta^*\right\} \quad (34.130)$$

$$\bar{r}_{\tilde{h}\tilde{h}\tilde{h}} = \left(\frac{3}{\bar{r}^3}\right)(h + D_n\sin\beta^*)\left[\left(\frac{1}{\bar{r}^2}\right)(h + D_n\sin\beta^*)^2 - 1\right] \quad (34.131)$$

$$\bar{r}_{\tilde{h}\tilde{h}u} = \left(\frac{h}{\bar{r}^3\cos^2\beta_k}\right)(h\tan\beta_k - D_n\cos\beta^*)\left[\left(\frac{-3}{\bar{r}^2}\right)(h + D_n\sin\beta^*)^2 + 1\right] \quad (34.132)$$

$$\bar{r}_{\tilde{h}st} = \left(\frac{hv\tan\beta_k}{\bar{r}^3}\right)(h + D_n\sin\beta^*) \quad (34.133)$$

$$\bar{r}_{\tilde{h}uu} = \left(\frac{-3}{\bar{r}^5}\right)\left(\frac{h^2}{\cos^4\beta_k}\right)(h + D_n\sin\beta^*)(h\tan\beta_k - D_n\cos\beta^*) \quad (34.134)$$

$$\bar{r}_{\tilde{h}ss} = \left(\frac{-1}{\bar{r}^3}\right)(h\tan\beta_k - D_n\cos\beta^*)(h + D_n\sin\beta^*) \quad (34.135)$$

$$\bar{r}_{uss} = \left(\frac{h}{\bar{r}\cos^2\beta_k}\right)(D_n\cos\beta^*)\left[\left(\frac{1}{\bar{r}^2}\right)(h\tan\beta_k - D_n\cos\beta^*)(h\tan\beta_k) + 1\right] \quad (34.136)$$

$$\bar{r}_{stu} = \left(\frac{-h\tan\beta_k}{\bar{r}^3\cos^2\beta_k}\right)(h\tan\beta_k - D_n\cos\beta^*) \quad (34.137)$$

$$\bar{r}_{suu} = \left(\frac{hD_n\cos\beta_k}{\bar{r}\cos^2\beta_k}\right)\left[\left(\frac{h\tan\beta_k}{\bar{r}^2}\right)(h\tan\beta_k - D_n\cos\beta^*) + 1\right] \quad (34.138)$$

$$\bar{r}_{\tilde{h}tt} = \left(\frac{v^2h}{\bar{r}^3\cos^2\beta_k}\right)(h\tan\beta_k - D_n\cos\beta^*) \quad (34.139)$$

$$\bar{r}_{uuu} = \left(\frac{h}{\bar{r}\cos^4\beta_k}\right)\left[8h\tan\beta_k + \sin^2\beta_k(h - D_n\cos\beta^*) - 2D_n\cos\beta^*\right]$$
$$+ \left(\frac{3h^2}{\bar{r}^3\cos^5\beta_k}\right)(h\tan\beta_k - D_n\cos\beta^*) + \left[\left(\frac{3h^2}{\bar{r}^3\cos^5\beta_k}\right)(h\tan\beta_k - D_n\cos\beta^*)\right.$$
$$\left.\times\frac{1}{2\cos\beta_k} + (h\tan\beta_k - D_n\cos\beta^*)\right] + \left(\frac{3h^3}{\bar{r}^5\cos^6\beta_k}\right)(h\tan\beta_k - D_n\cos\beta^*) \quad (34.140)$$

34.10 MATLAB® Programs and Functions

Listing 34.1. MATLAB Program "fig34_34-13.m"

```
%                    Figures 34.12 and 34.13
% Program to do Spotlight SAR using the rectangular format and
%       HRR for range compression.
%                       13 June 2003
%                       Dr. Brian J. Smith
clear all;
%%%%%%%%% SAR Image Resolution %%%%
dr = .50;
da = .10;
%dr= 6*2.54/100
% da= 6*2.54/100;
%%%%%%%%%Scatter Locations%%%%%%%
xn= [10000 10015 9985]; % Scatter Location, x-axis
yn= [0 -20 20];              % Scatter Location, y-axis
Num_Scatter=3;   % Number of Scatters
Rnom= 10000;
%%%%%%%%% Radar Parameters %%%%%%%%
f_0 =   35.0e9;          % Lowest Freq. in the HRR Waveform
df=   3. 0e6;            % Freq. step size for HRR, Hz
c =   3e8;               % Speed of light, m/s
Kr= 1.33;
Num_Pulse= 2^(round(log2(Kr*c/(2*dr*df))));
Lambda= c/( f_0+Num_Pulse*df/2);
%%%%%%%%% Synthetic Array Parameters %%%%%%%
du= 0.2;
L= round(Kr*Lambda*Rnom/(2*da));
U= -(L/2):du:(L/2);
Num_du= length(U);
%%%%%%%%% This section generates the target returns %%%%%%
Num_U= round(LZdu);
```

```
I_Temp = 0
Q_Temp = 0;
for I = 1:Num_U
  for J = 1:Num Pulse
    for K = 1:Num_Scatter
      Yr = yn(K) - ((I-1)*du - (L/2));
      Rt = sqrt(xn(K)^2 + Yr^2);
      F_ci = f_0 + (J - 1)*df;
      PHI = -4*pi*Rt*F_ci/c;
      I_Temp = cos(PHI) + I_Temp;
      Q_Temp = sin(PHI) + Q_Temp;
    end;
    IQ_Raw(J,I) = I_Temp + i*Q_Temp;
    I_Temp = 0.0;
    Q_Temp = 0.0;
  end;
end;
%%%%%%%%%% End target return section %%%%%
%%%%%%%%%% Range Compression %%%%%%%%%%%%
Num_RB = 2*Num Pulse;
WR = hamming(Num_Pulse);
for I = 1:Num_U
  Range_Compressed(:,I) = fftshift(ifft(IQ_Raw(:,I).*WR,Num_RB));
end;
%%%%%%%%%% Focus Range Compressed Data %%%%
dn = (1:Num_U)*du - L/2;
PHI_Focus = -2*pi*(dn.^2)/(Lambda*xn(1));
for I = 1:Num_RB
   Temp = angle(Range_Compressed(I,:)) - PHI_Focus;
   Focused(I,:) = abs(Range_Compressed(I,:)).*exp(i*Temp);
end;
%Focused = Range Compressed;
%%%%%%%%%% Azimuth Compression %%%%%%%%%%%
WA = hamming(Num_U);
for I = 1:Num_RB
   AZ_Compressed(I,:) = fftshift(ifft(Focused(I,:).*WA'));
end;
   SAR_Map = 10*log10(abs(AZ Compressed));
   Y_Temp = (1:Num_RB)*(c/(2*Num_RB*df);
   Y = YTemp - max(Y_Temp)/2;
```

```
X_Temp= (1:length(IQ_Raw))*(Lambda*xn(1)/(2*L));
X=XTemp - max(X_Temp)/2;
image(X,Y,20-SAR_Map); %
%image(X,Y,5-SAR_Map); %
axis([-25 25 -25 25]); axis equal; colormap(gray(64));
xlabel('Cross Range (m)'); ylabel('Down Range (m)');
grid
%print -djpeg.Jpg
```

Reference

1. Mahafza, B. R. and Sajjadi, M., Three-dimensional SAR imaging using a linear array in transverse motion, *IEEE AES Trans.*, 32(1), January 1996, 499–510.

VI

Advanced Topics in Video and Image Processing

Vijay K. Madisetti
Georgia Institute of Technology

35

3D Image Processing

André Redert
Philips Research Europe

Emile A. Hendriks
Delft University of Technology

35.1 Introduction

Images are the projection of three-dimensional (3D) real-world scenes onto a two-dimensional (2D) plane. As a result of this planar projection, images do not contain explicit depth information of objects in the scene. Only by indirect cues such as relative object size, sharpness and interposition (objects partly occluding each other), shadows, perspective factors, and object motion can we recover this depth information.

In a wide variety of image processing applications, explicit depth information is required in addition to the scene's gray value information [Dhon89, Hang95, Horn86, Marr82, Teka95]. Examples of such applications are found in 3D vision (robot vision, photogrammetry, remote sensing systems); medical imaging (computer tomography [CT], magnetic resonance imaging [MRI], microsurgery); remote handling of objects, for instance in inaccessible industrial plants or in space exploration; and visual communications aiming at entertainment (TV, gaming, cell phones), signage (advertisement), and virtual presence (conferencing, education, virtual travel and shopping, virtual reality). In these applications, depth information is essential for accurate image analysis or enhancing efficiency, impact, or realism.

The primary role of image processing in the above applications is to convert 3D information from an available representation, e.g., from sensors (cameras) or a transmission channel, into a representation required by, e.g., a storage system, display, or interaction system. For example, a stereoscopic camera pair records a pair of images, which are stored in compressed form, decompressed, and extrapolated to a multiview set of images to be shown on a multiview 3D display (note that although stereoscopic means spatial, or "3D," it usually refers to image pairs in the context of 3D imaging).

In this chapter, we discuss 3D image processing focusing on the application of visual communication, see Figure 35.1. This involves a wide variety of representations and conversion techniques also relevant in other applications. Necessarily though, in the fast-evolving area of 3D sensors, displays, and processing,

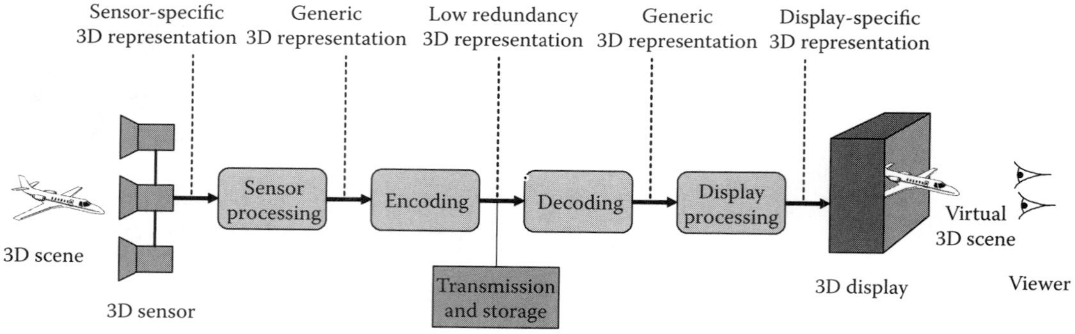

FIGURE 35.1 Systems for 3D visual communication involve a wide variety of representations for 3D information. Subsequently, many 3D image processing techniques exist to convert these representations (implemented in sensor/display processing and encoding/decoding).

an overview is incomplete by definition. We consider specialized or embryonic techniques (e.g., holography, specific data compression schemes, specific object sensors) beyond the current scope.

Next, we treat 3D human perception shortly (Section 35.2), followed by overviews of 3D sensors (Section 35.3), displays (Section 35.4), 3D representations (Section 35.5), and finally 3D image processing techniques (Section 35.6).

35.2 3D Perception

In this section, we discuss 3D human perception in short, which is sufficient for the 3D visual communication system. More about this topic can be found in [Past91, Seku02, Lamb07].

In the real world, objects emit or reflect light rays in different directions, with different intensities and colors. Viewers only observe those light rays that pass through the pupils of their eyes. On the retinas, 2D images of the scene are formed containing photometric information.

To extract 3D geometry information, the brain uses several depth cues that can be divided into two categories, namely psychological and physiological cues. Psychological cues involve knowledge about the scene, such as known shapes of certain objects, the way shading works, occlusion (objects that move in front of each other), relative size between objects, perspective, and color. Current 2D visual communication systems already provide all visual information to extract these cues.

The physiological cues are the cues that we would like to introduce by 3D visual communication systems since they are absent in 2D systems. These cues are more related to direct measurable physical properties, such as the accommodation of the eye lens and the convergence of the eyeballs toward an object of interest. We categorize them as follows by scale of viewer position (large), intereye distance (medium), and pupil size (small), see Figure 35.2.

35.2.1 Viewer Position: Viewer Motion Parallax Cue

When viewers walk or move their head, the selection of light rays that they observe changes over time. In specific, for a horizontally translating viewer, nearby objects seem to move faster than objects that are far away. This is called the viewer motion parallax cue. Figure 35.2 shows a viewer moving in a train. Within the 2D images on both his retinas, the nearby trees move faster than the distant trees.

35.2.2 Intereye Distance: Binocular Disparity Cue

Viewers have two eyes separated by some moderate distance ($\sim$6.5 cm). This provides two slightly different viewpoints, enabling to determine the distances to objects via the so-called binocular disparity

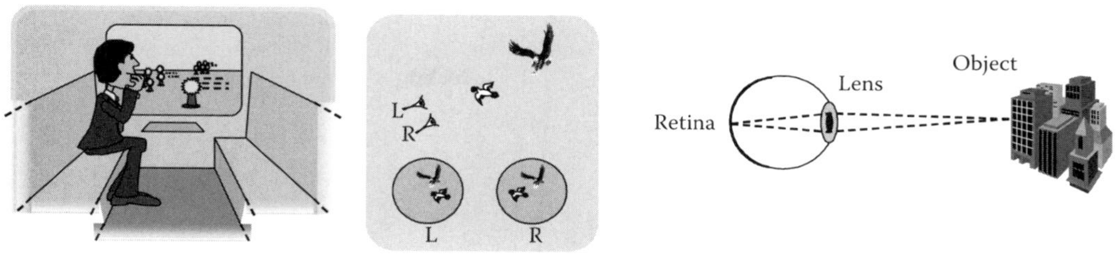

FIGURE 35.2 Visual cues to be introduced by 3D communication systems, Left: viewer motion parallax, Middle: binocular disparity, and Right: accommodation.

cue. Figure 35.2 shows the difference (disparity) in relative positions of the two birds on the viewer's retinas, a cue for their difference in distance.

35.2.3 Pupil Size: Accommodation Cue

The pupils of the eyes have finite size and thus, from each point of the scene, several light rays forming a cone enter each pupil. The eye lens diffracts all these light rays, which then form a second cone within the eyeball. If the lens is accommodated to the correct distance to the object of interest, the apex of the cone is at the retina, producing a sharp image of that object. The required accommodation provides a cue for the distance to the object. Other objects will be blurred on the retina according to their distance relative to the object of interest.

35.2.4 All Three Cues

The three physiological cues are quite alike differing only by the scale in which the human eyes select light rays from all rays the scene emits. Despite this similarity, many 3D displays provide only one or two of these cues. The possible relation between cue mismatch and viewer/eye fatigue is an active area of research [Lamb07].

35.3 3D Sensors

Figure 35.3 shows a taxonomy including most basic and well-known 3D sensor types. Next we discuss these briefly, more details can be found in [Dist95, Ohm98, 3dvs, Rede02, Kana97, Mats04, Wasc05].

First, we make the distinction between photometric and geometric sensors. The photometric sensors are most well-known: these are ordinary video cameras providing images containing the scene's texture, by capturing light rays passively. Even a single camera is a 3D sensor, as its images provide humans with the psychological 3D cues. A pair of cameras, often called a stereo camera, has typically been used to provide the binocular disparity cue, e.g., in 3D cinemas and experimental 3DTV [Dist95]. Current research is focusing on handling multicamera systems [Kana97, Mats04, Wasc05].

Geometric 3D sensors provide explicit depth measurements by actively scanning material points of the scene. Volume sensors provide information about the entire 3D space scanned, and are typically found in the medical area, such as computed tomography (CT)/magnetic resonance imaging (MRI)/positron emission tomography (PET) and ultrasound scanners, while infrasound is also used for underground scanning. These sensors include highly specialized processing, including the so-called radon transforms and the (audio-) holographic techniques, which are beyond the scope of this chapter (see, e.g. [Lanz98] for the radon transform). Surface sensors provide information only from the object surface closest to the sensor. Within material research, the tunneling microscope is based on actually touching and following the surface.

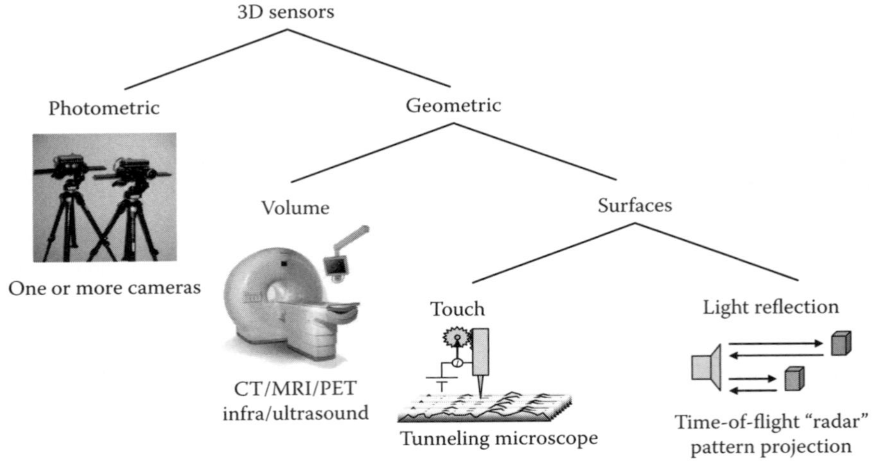

FIGURE 35.3 3D sensors.

Several geometric surface sensors are available for more general applications in the visual domain, based on light reflection. Radar-like sensors are based on time-of-flight measurements: the scene is illuminated with a very short, intense pulse of (invisible) light. The time it takes for the pulse to hit the scene and be reflected back into the sensor is a direct measure of the scene distance. Laser-based systems scan the scene point-by-point, and are therefore slow but very accurate. The so-called depth or range cameras have the pulse to illuminate the entire scene at once and capture the reflection by a pixel-matrix sensor [3dvs]. They enable real-time video capture but at a lower accuracy.

Pattern projection (also called structured light) sensors operate by projecting a certain light pattern on the scene, see, e.g. [Grie04]. A conventional beamer often suffices for the projection. A camera located at some distance to the beamer records the pattern, which will be spatially distorted according to the scene depth. This type of sensor is equivalent with the stereo camera, with the beamer playing the role of one of the cameras. The advantage is that the scene texture is actively imposed by the pattern instead of passively observed. The pattern is chosen to maximize the accuracy of subsequent stereoscopic image processing which relies on scene texture (see Section 35.6). A typical challenge for the pattern projector is to maximize 3D accuracy while also recording the scene's texture undisturbed and in case of humans in the scene, ensuring their visual comfort and eye safety.

35.4 3D Displays

Figure 35.4 shows a taxonomy including most basic and well-known 3D display types. Next we discuss these briefly; more details can be found in [Blun02, Born00, Javi06, Mcal02, Sext99, Trav97].

On top of the taxonomy, we make the distinction between autostereoscopic displays enabling free viewing, and displays that need viewers to wear (additional) optics. Examples of the latter are virtual reality helmets (head-mounted displays), and goggle-based displays ranging from television-based, to immersive (e.g., the CAVE concept is a room-sized box [Cruz93]), and to 3D cinema. These displays provide two distinct images (a stereo image pair), one for each eye. With this, they provide the binocular disparity cue, but not the accommodation and the viewer motion parallax cues. The parallax cue can be included by active viewer tracking and subsequent synthesis of appropriate stereo images. This is typically done in virtual reality applications.

In autostereoscopic displays, we make a distinction between geometric and photometric displays. Geometric displays create real images in the optical sense, that is, light that emanates from some virtual

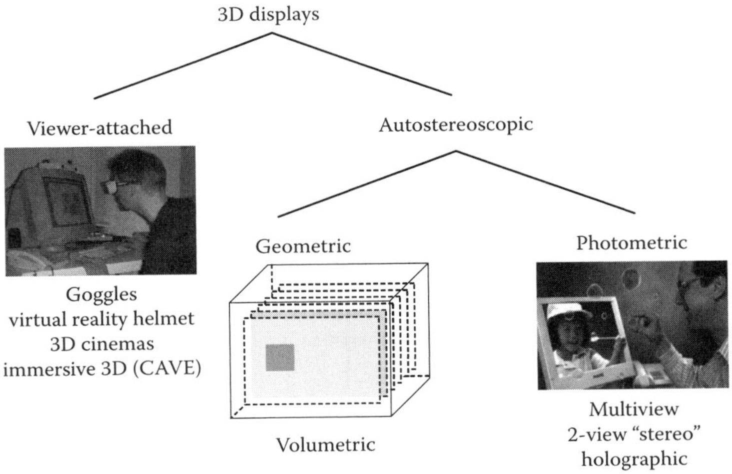

FIGURE 35.4 3D displays.

scene point is actually created by matter at that location. Photometric displays create virtual images by creating light rays within the display and focus of these toward virtual scene points possibly outside the display.

Volumetric displays are geometric 3D displays. Since any 3D object is visualized within the physical display housing, volumetric displays are voluminous to achieve considerable depth. The volumetric display shown in Figure 35.4 is based on projecting several different 2D images onto a stack at different depths. Other volumetric display techniques include tiling a 3D volume with individually addressable light emitters such as LEDs, or exciting a substrate volume in such a way to have it emit light locally (e.g., by focused excitation with several invisible laser beams), and projecting a sequence of 2D images onto a fast rotating 2D display plane, with the 2D image being dependent on the instantaneous rotation angle in such a way to provide a 3D image.

Since volumetric displays create object points in space that are real in the optical sense, all three 3D cues are provided: accommodation, binocular, and viewer motion parallax. However, most volumetric displays can visualize objects only transparently, disabling the important occlusion cue: opaque foreground objects should occlude background objects, but they do not. This cue is in fact correctly provided by any conventional 2D display; so this is an important challenge for volumetric 3D displays.

Photometric 3D displays are able to visualize 3D objects outside their physical housing. These displays come in several types. The hologram is the most well-known, yet also most different from other display types: its native resolution is orders of magnitude higher, as pixel dimensions must be in the order of light wavelengths (~500 nm). Real-time 3D video holograms with color, sufficient size, and viewing angle are still extremely challenging due to bandwidth and laser illumination issues.

More feasible photometric 3D displays are multiview displays, based on conventional technology as LCDs with additional optics. The principle is to have pixels emit light independently in several directional beams instead of one beam across the full viewing angle. Several pixels can create a virtual 3D point outside the display plane by illuminating beams that focus on the desired point, see Figure 35.5. Within each pixel, the beams can be created in several ways. Figure 35.5 shows a popular method also used in the well-known 3D postcards: a pixel-width lens focuses the light of N ordinary subpixels (ordinary in the sense of emitting light uniformly, without beams) into N beams, where each subpixel has a width of $1/N$ times the lens/pixel width.

Current multiview displays have in the order of $N = 5$–20 pixels cooperate. Effectively, these displays project N full 2D images, or "views," in different directions toward the viewer, who observes two of them, or two different combinations of views.

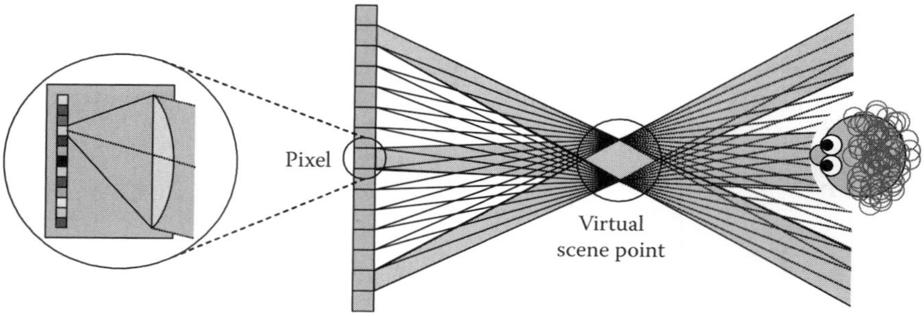

FIGURE 35.5 Multiview display principle: N pixels cooperate to produce one virtual 3D scene point (here $N = 13$).

As each beam's width increases linearly with depth, so does the size of any visualized 3D scene point. The display's resolution thus drops with depth. Objects visualized within the display plane have a spatial resolution equal to pixels, but become progressively more blurred outside the display plane. Effectively, the display has a so-called depth-of-field (DOF) property, similar to photography where objects sufficiently far from the focus plane are out of focus. A depth range can be arbitrarily defined, e.g., a range in which object resolution is better than halved, meaning scene point size is between 1 and 2 pixels (a kind of −3 dB point). The product of the overall viewing angle and the depth range scales linearly with N and the pixel size, leading to typical depth ranges of 5%–20% times the display size.

Multiview displays have properties similar to holograms, which also have a DOF limit to depth. With this, they provide all three 3D cues, and binocular disparity and viewer motion parallax are easily observed. In contrast, the accommodation cue is less obvious: when a viewer accommodates to some object of interest, the blurring imposed by the eye on other objects within the display depth range is typically below the display's spatial resolution. The effect, if present at all, of this upon a viewer is unclear at the moment. Furthermore, most multiview displays have N pixels cooperate along the horizontal axis only, enabling 3D cues only in the horizontal direction. These displays will not provide binocular disparity for viewers with 90° rotated heads (eyes vertically separated) or viewer motion parallax when moving vertically, but few applications require this.

A special type of multiview display is the two-view stereo display. Instead of having pixels focus their beams on virtual points (which would just be an $N = 2$ display), their beams are all focused on either the left or the right eye of the viewer. In principle, the total viewing angle is very small and viewers are forced to sit still in a very narrow so-called sweet-spot. However, adaptive systems can track the viewer and redirect the sweet-spot to the actual viewer position, enabling freedom of movement, see Figure 35.6. This requires highly accurate, low latency viewer tracking, and subsequent sweet-spot adaptation.

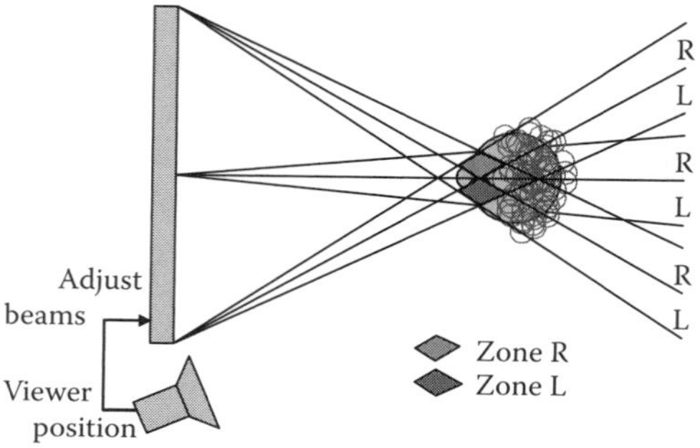

FIGURE 35.6 Tracked 2-view stereo display (not equal to an $N = 2$ multiview display).

The advantage of focusing the beams to the viewer's eyes is that two fully independent images can be shown to the left and the right eye, enabling substantial depth visualization without DOF effects. This display only provides the binocular disparity cue. The viewer motion parallax cue can be added by using the measured viewer position to adaptively generate appropriate stereo images [Ohm98].

35.5 3D Representations

Figure 35.7 shows a taxonomy including most basic and well-known 3D data representation types. Next we discuss these briefly, more details can be found in [Carp84, Shad98, Taub98].

First, we make a distinction between photometric and geometric 3D representations. Photometric representations consist of one or more conventional 2D images. Even a single 2D image represents 3D information, as it provides humans with the psychological 3D cues. An image pair, often called a stereo pair, has typically been used to provide the binocular disparity cue, e.g., in experimental 3DTV [Dist95] and 3D cinemas. Current research is focusing on handling multicamera images [Kana97, Mats04, Wasc05], also called image arrays or light fields [Levo96, Magn00].

The primary advantage of photometric 3D representations is that they are very well suited to capture and display visual 3D phenomena: there exist direct sensors (cameras) and 3D displays (2D, stereo-, and multiview displays) for it, and these correspond closely to the needs of the human visual system. If the sensor and the display representations are the same, a 3D scene can be communicated in arbitrary detail without 3D image processing. The drawback of such a system is its heavy demand on the number of cameras, and the high data bandwidth required for the many images. Especially if the application requires some geometrical flexibility, e.g., depth effect adjustment in 3D displays, all images possibly needed must be available in the representation.

Geometric 3D representations contain explicit depth information of objects, rather than via implicit visual depth cues. This is useful in applications where interaction with the virtual scene is required such as 3D computer vision and medical imaging (making measurements), 3D gaming, 3D video editing, haptic interfaces (touch control), 3D displays (flexible visualization by geometric adjustments as depth scaling), etc. Furthermore, geometric surface representations (see below) typically are bandwidth efficient, containing far less redundant information than photometric representations.

A huge challenge for geometric 3D representations is to correctly represent phenomena that do not relate easily to objects and depth. These include indeterminate object presence (dirty windows,

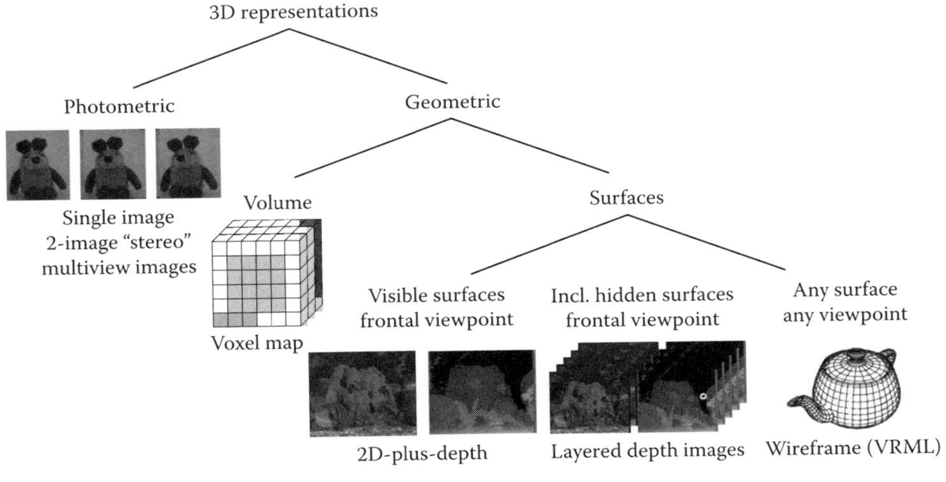

FIGURE 35.7 3D data representations.

rain, fog, water), visual effects unrelated to real objects (specular reflections from metals, mirrors), and objects with complex geometry/topology (trees with many leaves, sand in the air). Despite all these drawbacks, geometric representations play an essential role in the handling of photometric representations, see Section 35.6.

Geometric representations come in volume and surface type representations. The voxel map is a well-known volume representation, where every voxel represents a small volume element of a scene. Each voxel has several attributes such as transparency (matter presence), color, or other. Medical CT/MRI/PET scanners typically provide voxels with a radiation absorption attribute (indicative for different tissue types), while volumetric displays can visualize voxels with color attributes.

Geometric surface representations do not describe the inner parts of the objects, but only their surfaces. Several surface representations exist, differing in which surfaces can be described and which cannot. Wireframes (also called meshes, or VRML models) are capable of representing any surface at any location in a scene, see [Taub98]. These are typically used for synthetic 3D objects; in CAD/CAM and 3D computer games they represent objects or even an entire virtual world in high detail.

Image-based surface representations are popular for representing natural 3D scenes. They combine the advantages of photometric and geometric representations, representing depth explicitly, while being an image-like format. The so-called 2D-plus-depth format has been proposed as a general 3D video format for 3DTV [Rede07]. It consists of a conventional 2D image enabling to drive 2D displays directly, plus a second image containing gray values indicating scene depth per pixel. In this way, every small piece of scene surface visible in the image is represented and assigned a depth value. Obviously, this representation includes only those parts of the scene that are visible from one particular viewpoint. This incompleteness adds to the compactness and simplicity of the representation, but also to its drawback for 3D visualization: if the represented scene is observed from a viewpoint slightly different from the original, it will become clear that some background parts are missing, namely those just occluded in the original image. Use of the 2D-plus-depth representation is therefore always accompanied by processing to complete the background, see Section 35.6.3.

Several approaches exist to enable more complete scenes to be represented explicitly. Typically, the "A buffer" method [Carp84] and the layered-depth-image (LDI) method [Shad98] add more 2D-plus-depth layers. All are related to the same viewpoint, but each next layer represents the first surface hidden from view that was not already in the representation. In Figure 35.7, the first extra layer will thus contain those parts of the mountain that lie behind the plants and fishes, and those parts of the dark background that are behind the mountain, while the second extra layer will contain the background behind the mountain behind the fishes, etc.

Not shown in the taxonomy are compressed or low-redundancy representations, as they exist for practically every representation. For example, photometric and image-based geometric 3D representations can be compressed within existing image compression formats as MPEG2, MPEG4, and H264 [Brul07]. The schemes can be applied "as is" on the multiple images and gray-scale depth maps, or through standardized extensions. A considerable number of compression schemes have been developed for specific representations, e.g., image arrays [Levo96, Magn00] and wireframes [Peng05, Taub98].

35.6 3D Image Processing: Representation Conversion

In any system, representation conversions are necessary when sensor, storage/transmission, and display/application 3D representations are not equal. Early stereoscopic television systems did use one representation throughout the system: the two-image or stereo image representation, enabling systems without 3D processing [Dist95]. Current 3DTV system approaches aim at decoupling sensor, display, and transmission representations, requiring conversion processing while enabling more flexibility [Rede02].

Figure 35.8 shows many typical 3D image processing techniques, involving a conversion from one representation into the other. We omitted compression schemes, as most of these are based on 2D

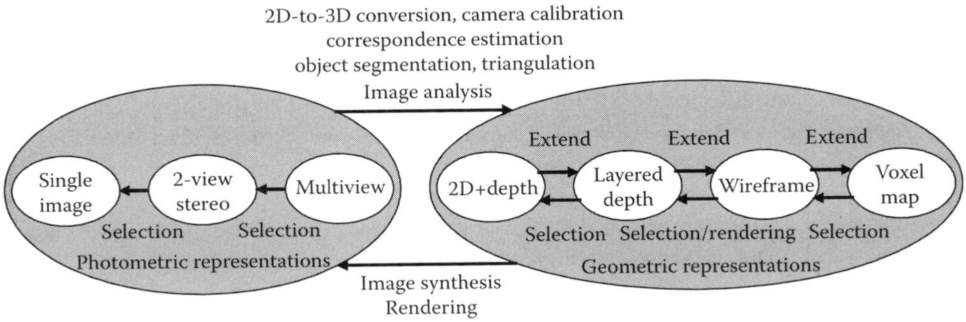

FIGURE 35.8 3D image processing is about converting 3D representations. The challenge is in the image analysis algorithms that convert photometric (light-ray-based) representations into geometric (matter-based) representations.

techniques, or are very specific for one representation (e.g., wireframes). Next we discuss the conversions as in Figure 35.8, grouped in being across geometric and photometric representations, or within one of them toward simpler representations (down-conversion) or more complex (up-conversion).

35.6.1 Down-Converting: Selection

When the output representation is a subset of the input, e.g., down-converting a stereo image into a single image, a simple selection process suffices (e.g., taking either the left or the right image). Only certain geometric down-conversions, from nonimage-based to image-based (e.g., wireframes to 2D-plus-depth), require some processing in the form of rendering, see Section 35.6.4.

35.6.2 Up-Converting Photometry: To Geometry and Back

It is impossible to up-convert a photometric representation with any reasonable quality by a process as simple as selection. In fact, the only general way is to convert to a geometric representation and back, requiring considerable processing discussed in Sections 35.6.4 and 35.6.5.

35.6.3 Up-Converting Geometry: Completing a 3D Representation

Geometric to geometric up-conversions are simple in the sense that the output representation easily encompasses all data from the input representation. However, to actually make use of the better representational capabilities of the output representation, one has to extend the represented scene by making it more complete, that is, adding surfaces or volume elements.

Many 2D image restoration techniques such as hole filling or inpainting [Bert03, Wang03] can be used to extend or complete image-based 3D representations such as 2D-plus-depth and LDIs. Obviously, these techniques are limited as objects cannot be recovered from scratch. When objects are at least partially present, a small part of their nonrepresented parts can be estimated by extrapolating the visible object texture near the occluded object edge. Although such extrapolation is limited (to say 1–10 pixels for arbitrary objects in image-based representations), it may suffice for the purpose of 3D visualization without artifacts.

Many techniques are available to repair or complete geometric representations, e.g., wireframes [Noor03].

35.6.4 Converting Geometry into Photometry: Rendering

One of the most important and widespread 3D image processing techniques is called rendering [Fole95]. It converts a geometric representation into a photometric representation by synthesizing images from some set of desired viewpoints. The rendering process is a simulation of the physical process of light rays emanating from a scene moving through the air into a camera arriving at its imaging plane. This physical process is very well understood; research in rendering mainly focuses on efficiency improvements enabling more and more light rays to be simulated, thereby increasing image quality. Often, however, research in rendering encompasses the topic of the 3D representations itself, as the latter highly determines the final quality of the rendered images.

Current PC graphics cards are equipped with a dedicated 3D rendering processor capable of tracing billions of light rays per second for rendering images from wireframes. Natural 3D scenes are mainly rendered from image-based 3D representations, see e.g. [Shum00].

35.6.5 Converting Photometry into Geometry: Image Analysis

Finally, the conversion of a photometric representation into a geometric representation is a very diverse processing task involving 3D image analysis. In case the input is only one image, we have the most challenging task of 2D-to-3D conversion: obtaining depth information via 2D computer vision and heuristics by extracting the psychological cues mimicking the human visual system. Whenever a photometric 3D representation contains two or more images, conversion to a geometric representation can be performed in a more analytic way using traditional methods applied also in fields other than 3D image processing. Next we discuss approaches for one, two, and more input images.

35.6.5.1 Single-Image Analysis (Conversion of 2D-to-3D): Computer Vision

Single-image analysis, or 2D-to-3D conversion, is part of 3D computer vision. The conversion quality increases steadily, along with growing knowledge about the human visual system and newly discovered heuristics. Still, however, the current state-of-the-art is nowhere near the capabilities of human operators in converting arbitrary 2D images into 3D. For dedicated applications, however, viable 2D-to-3D algorithms exist (e.g., in 3DTV where pleasing 3D rather than correct 3D is required).

Figure 35.9 shows an example of the automated way of extracting and using the occlusion (interposition) cue [Rede07]. Other methods are, e.g., using shading [Zhan99], object defocus (DOF) [Ens93], perspective lines [Batt04], and supervised learning [Saxe05].

The only well-known and mature 2D-to-3D conversion method is called structure from motion (SFM) [Jeba99, Poll99]. It applies for monoscopic video with a static scene and moving camera. Subsequent video frames are equivalent with images taken at once by multiple cameras located around the scene, for which the stereo- and multicamera approaches discussed next apply.

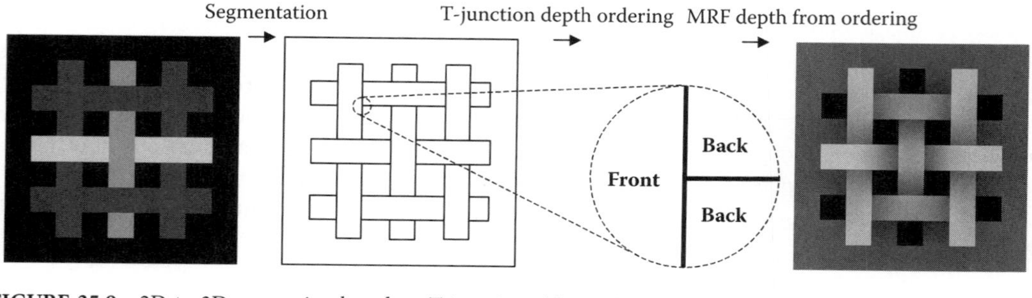

FIGURE 35.9 2D-to-3D conversion based on T-junctions (three-object points). The sparse depth ordering information is made dense by a MRF approach utilizing a piecewise smoothness constraint. (From Redert, P.A. et al., *SPIE*, vol. 6508, San Jose, CA, 2007.)

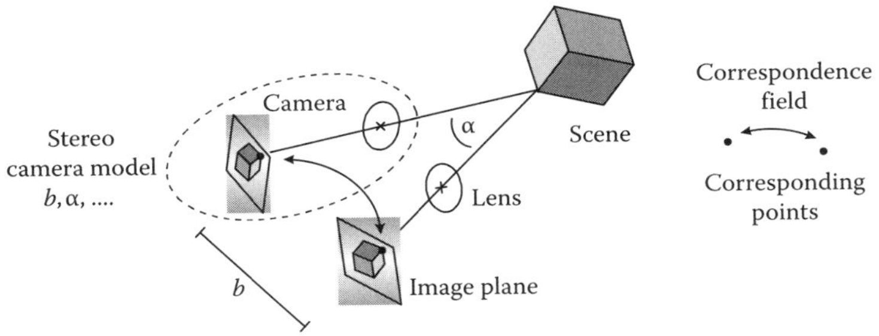

FIGURE 35.10 Conversion of a stereo image into a geometric 3D representation involves traditional 3D image processing tasks of camera calibration and correspondence estimation.

35.6.5.2 Stereo Image Analysis

Figure 35.10 shows an example of stereo (binocular) image pair analysis, with conversion into a geometric 3D representation using several traditional 3D image processing tasks. By correspondence estimation, pixels in one image are paired with pixels in the other that originate from the same scene point. By camera calibration, a geometrical model for the cameras (position and orientation of image planes and lenses, lens distortion) is found. Together, for each pixel pair two light rays can be reconstructed. Finally, by triangulation we find the 3D coordinates of the scene point at the rays' intersection.

While triangulation is a simple and fully understood process, similar to rendering (tracing light rays), camera calibration [Tsai87, Truc98, Pede99, Lei05] and correspondence estimation [Rede99, Scha02] are still active and challenging research topics. Next, we discuss the latter two topics, followed by the concept of epipolar geometry that relates to both.

35.6.5.2.1 Camera Calibration

Two main types of camera calibration methods exist. The so-called fixed calibration relies on capturing images from a 3D calibration object with accurately known shape and dimensions; see Figure 35.11.

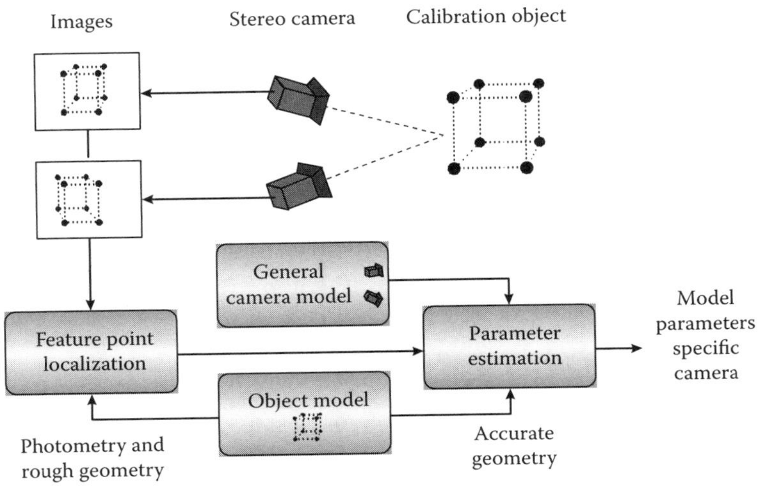

FIGURE 35.11 Fixed camera calibration is a manual method using a calibration object with accurately known shape and color.

Feature points of this object are detected and localized in the two images by a dedicated algorithm applying knowledge of the object. The set of localized feature points provides many corresponding point pairs. Triangulation of these pairs with some set of camera parameters yields some 3D reconstruction of the calibration, with some error compared to the real calibration object.

Typical calibration algorithms use a mix of direct analytical methods and numerical optimizations to obtain the best set of camera parameters, defined by minimizing the calibration objects' reconstruction error. As long as (nonlinear) lens distortions are not taken into account in the camera parameters, typical camera models can be described by matrix algebra (the so-called projective geometry) [Truc98] enabling analytic solutions.

The camera calibration parameters found enable accurate reconstruction of objects of interest other than the calibration object, at least, within the space originally occupied by the calibration object. However, manufacturing and maintaining a large and accurate 3D calibration object is hard. In practice, often a smaller and flat calibration object is used to scan (by hand) the volume required for the objects of interest. With this, the calibration algorithm also has to estimate and take into account the several positions and orientations of the calibration object. Figure 35.12 shows a typical flat calibration object. Feature (dot) localization accuracies are in the subpixel range of 0.01–1 pixel [Rede00].

The second method of camera calibration is the so-called self-calibration. It can be applied when only images of the objects of interest are at hand, without the need for any calibration object. Although less accurate and robust, self-calibration is the only option when recording the calibration object is impossible, e.g., in handling existing stereo video material.

In camera self-calibration, corresponding pixel pairs are found by applying correspondence estimation. These pairs are triangulated to lead to some reconstruction of the scene, whatever it may be, using some camera parameters. Since there is no reference scene to define a reconstruction error, parameter optimization must be guided by a different criterion. The only available option is that the triangulation process leads to intersections. Clearly, many sets of wrong camera parameters lead to light rays being simulated for a corresponding pixel pair that do not intersect (in Figure 35.10 the rays do intersect).

Conceptually, the reconstruction error for a corresponding pixel pair can be defined as the distance between the two closest points on the two light rays. However, this is not scale-independent leading to wrong camera parameters biased to very small 3D reconstructed objects. The solution to this is to compute the distance in length units related to pixels, that is, via projecting the two closest points back to one (or both) of the images.

FIGURE 35.12 Typical flat calibration object with dots as contrast-rich and well-localized features.

The accuracy and robustness of self-calibration differ from fixed calibration in several ways. First of all, using self-calibration, the overall stereo image conversion process yields a geometrical 3D representation given in length units other than meter, as there is no reference whatsoever to the meter unit. Some unit must be chosen, however, and it can be related to the captured objects (some kind of object size normalization), or related to the stereo camera (e.g., the baseline unit equaling the distance between the two cameras), or any arbitrary choice.

Further, self-calibration accuracy and robustness depend on the chosen camera model and the accuracy of the pixel correspondences found. Since the latter originate from the correspondence estimation, the accuracy is generally lower than in fixed calibration, where correspondences are found by dedicated algorithms using an accurate calibration object model. In terms of the camera model, it has been theoretically proven that self-calibration methods can at most measure seven free parameters [Faug93, Truc98]. This suffices to include the position and orientation of the two cameras relative to one another (six parameters minus one that is normalized to a fixed value to define the length unit = 5 in total), and two parameters, e.g., the zoom factors (distance between lens and image plane) of each camera. This proof applies to camera models without any lens distortion. The inclusion of lens distortion makes the camera model more accurate, but the mathematics is less tractable. Experimental evidence suggests that more than seven parameters can be measured when lens distortion is included [Rede00].

35.6.5.2.2 *Correspondence Estimation*

In correspondence estimation, the goal is to pair pixels in one image with pixels in the other that originate from the same scene point. The typical output of the algorithm is a correspondence field: to each pixel x_1, y_1 in the first image, a vector is assigned $[x_2 - x_1, y_2 - y_1]$ "pointing" to the corresponding pixel x_2, y_2 in the second image. This kind of two-image analysis is commonly used in monoscopic video processing under the name of motion estimation [Teka95, Stil97], with the two images being captured by the same camera at different time instants t_1, t_2.

The basic estimation principle is simple: pair up pixels with similar gray-level or color. To be invariant to photometric differences among the two images, derived features as spatial color derivatives or local texture are used as well. Despite this simple principle, correspondence estimation is a major task: a pair of images with 10^6 pixels each gives rise to $10^6!$ possible correspondence fields, considering all 1:1 pixel pairings. This number goes up further for subpixel pairing. Apart from this, correspondence estimation is a typical "ill-posed" problem [Bert88]; mostly, several scene points share the same color (or local texture), so the stereo image pair does not provide sufficient information to find a correspondence field unambiguously.

Besides similarity in color for correspondences, additional constraints are essential to find a correspondence field reliably. Two main constraints are generally applied. First, it is often assumed that the two images are obtained by cameras relatively close to one another, and in more or less the same orientation. Then, the correspondence field is limited in value, typically in a range $-5 \cdots +5$ to $-100 \cdots +100$ pixels (for both horizontal and vertical components), limiting the number of possible correspondence fields.

The second typical constraint is the so-called smoothing constraint: the final obtained 3D geometry should be composed of smooth planes, in reasonable accordance with most scenes and objects in the real world. In terms of the correspondence field, it means that the spatial derivative of the field has a low value around zero (for both x and y components). This is equivalent with two pixels neighboring in one image being paired to pixels also neighboring in the other image, at least approximately. A good example is block-based pairing, where a block of say 8×8 neighboring pixels are paired to an 8×8 block of neighboring pixels in the other, see Figure 35.13. Although the blocks enable reliable correspondence estimation, the spatial resolution goes down to block level.

Many techniques exist to apply smoothing without sacrificing spatial resolution. With post-processing methods, the spurious block edges in Figure 35.13 can be realigned with original texture

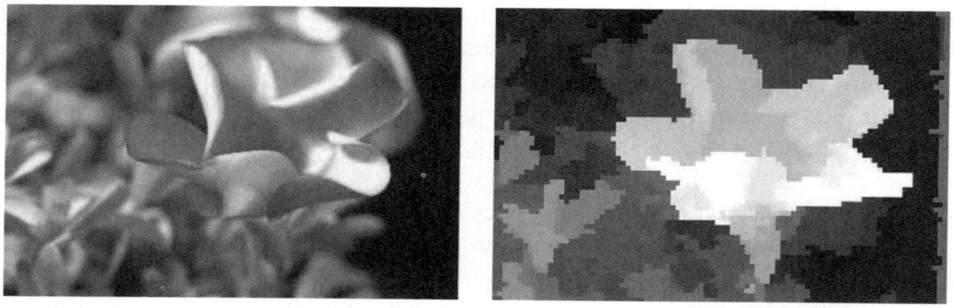

FIGURE 35.13 2D-plus-depth obtained from a stereo image pair via block-based correspondence estimation.

FIGURE 35.14 2D-plus-depth obtained from a stereo image pair via MRF method applying object-based smoothing.

edges. In the so-called Markov Random Field (MRF) methods [Gema84, Rede99, Stil97] the challenge is to apply smoothing only within objects or object segments, rather than across different objects (see also Figure 35.9). The latter would undesirably result in object boundaries being blurred in the correspondence field, and subsequently objects being "glued" together in the geometric 3D representation obtained.

Although image segmentation into semantically meaningful objects remains a very challenging research topic, the requirements in correspondence estimation are more relaxed. Grouping pixels into segments smaller than objects but large enough to allow for reliable matching is sufficient and currently within reach, see Figure 35.14.

In correspondence estimation, there are many exceptions to 1:1 pixel correspondences. First, a pixel location in one image may correspond to a subpixel location of the other image, requiring subpixel estimation. Further, one-to-many (or some-to-more) correspondences occur whenever the apparent size of an object is not the same in the two images. This is the case in general, unless the object is a flat surface parallel to the baseline (line through the camera centers). Finally, some (parts of) objects will be visible in one image, but not at all in the other as something else is in front of it. For such so-called occlusions, there are no valid correspondences to be found and any attempt to do so will result in errors or inaccuracy. Depending on application and required accuracy, occlusion handling is or is not required. The result in Figure 35.14 was obtained without occlusion handling.

35.6.5.2.3 Epipolar Geometry

Epipolar geometry relates to correspondence estimation and camera calibration. Figure 35.15 introduces all relevant concepts. First, the line through the camera lens centers is called baseline (baseline also refers to the distance between the cameras). Any plane that encompasses the baseline is called

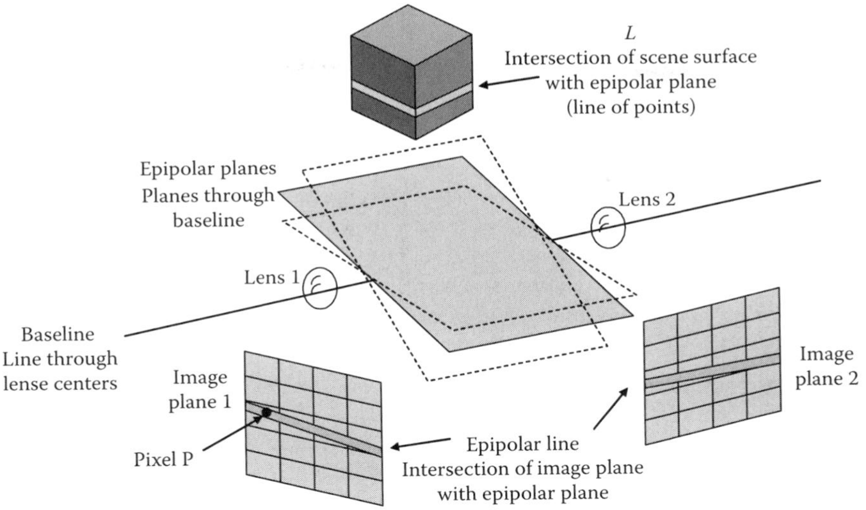

FIGURE 35.15 Epipolar geometry relates correspondence estimation and camera calibration (see text).

epipolar plane; all epipolar planes can be parameterized by a single rotation angle. Every epipolar plane intersects the image plane of both cameras somewhere, resulting in epipolar lines. The intersection with the scene is some plane, of which the parts visible in the images encompass only a line L-shaped to the scene.

When camera calibration is performed before correspondence estimation (necessarily by fixed calibration), we know the location of the camera lenses and the image planes. Given some arbitrary pixel P, we know in which epipolar plane P lies, and thus which part of the scene possibly corresponds to P (one of the points in L), and thus where the pairing pixel of P can be found: somewhere on the epipolar line in the other image. The latter is regardless of L, the scene shape. With that, the search for possible correspondences is reduced from all pixels in the other image to only a single line of pixels. This is called the epipolar constraint, providing a major reduction in complexity.

A more mathematical and thorough treatment on epipolar geometry can be found in [Faug93, Truc98]. For a wide overview of correspondence estimation algorithms applying the epipolar constraint we refer to [Scha02].

35.6.5.3 Multiple Image Analysis

The advantage of using more than two images is obviously higher accuracy, less occlusions (objects or parts visible by only one camera), and more complete 3D representations, since information can be fused from more sources observing the scene from more angles [Kana97, Mats04, Wasc05].

Conversion to geometric 3D representations can be performed basically along the same lines as for stereo camera pairs. The tasks of camera calibration, correspondence estimation, and triangulation can be scaled to more than two images, leading to, e.g., "trifocal tensors" (three-camera models) and multi-camera bundle adjustment (triangulation of several light rays). Increasingly popular are modular system approaches, where many two/three-camera modules are each equipped with all processing to generate geometric 3D representations of parts of the scene, after which these parts are fused by techniques such as depth/range map fusion into representations of the full scene [Oish05, Xian04].

The so-called shape-from-silhouette methods can add robustness by segmenting specific objects (e.g., humans in front of a blue screen) within individual images, determining their outline, and combining all outlines via the so-called visual hull approaches [Lamb07, Laur94].

References

[Batt04] S. Battiato, S. Curti, M. La Cascia, M. Tortora, and E. Scordato, Depth map generation by image classification, *SPIE*, Vol. 5302, San Jose, CA, 2004.

[Beec01] M. Op de Beeck and P.A. Redert, Three dimensional video for the home, *Proceedings of the EUROIMAGE ICAV3D 2001 International Conference on Augmented, Virtual Environments and Three-Dimensional Imaging*, Mykonos, Greece, 2001, pp. 188–191.

[Bert88] M. Bertero, T. Poggio, and V. Torre, Ill-posed problems in early vision, *Proceedings of the IEEE*, 76(8), 869–889, 1988.

[Bert03] M. Bertalmio, L. Vese, G. Sapiro, and S. Osher, Simultaneous structure and texture image inpainting, *IEEE Transactions on Image Processing*, 12(8), 882–889, 2003.

[Blun02] B. Blundell and A.J. Schwarz, The classification of volumetric display systems: Characteristics and predictability of the image space, *IEEE Transactions on Visualization and Computer Graphics*, 8(1), 66–76, 2002.

[Born00] R. Börner, B. Duckstein, O. Machui, R. Röder, T. Sinning, and T. Sikora, A family of single-user autostereoscopic displays with head-tracking capabilities, *IEEE Transactions on Circuits and Systems for Video Technology*, 10(2), 234–243, 2000.

[Brul07] W.H.A. Bruls, C. Varekamp, A. Bourge, J. van der Meer, and R. Klein Gunnewiek, Enabling introduction of stereoscopic (3D) video: Compression standards & its impact on display rendering, *International Conference on Consumer Electronics*, Paper 6.3–4, Las Vegas, NV, 2007.

[Carp84] L. Carpenter, The A-buffer, an antialiased hidden surface method, *ACM SIGGRAPH Conference on CGIT*, New York, 1984.

[Cruz93] C. Cruz-Neira, D.J. Sandlin, and T.A. DeFanti, Surround-screen projection-based virtual reality: The design and implementation of the CAVE, *SIGGRAPH*, Chicago, IL, 1993.

[Dist95] European RACE 2045 project DISTIMA, 1992–1995.

[Dhon89] U.R. Dhond and J.K. Aggerwal, Structure from stereo, *IEEE Transactions on System, Man and Cybernetics*, 19(6), 1489–1509, 1989.

[Ens93] J. Ens and P. Lawrence, An investigation of methods for determining depth from focus, *IEEE Transactions on PAMI*, 15(2), 97–108, 1993.

[Faug93] O. Faugeras, *Three-Dimensional Computer Vision, A Geometric Viewpoint*, MIT Press, Cambridge, MA, 1993.

[Fole95] J.D. Foley, A. van Dam, S.K. Feiner, and J.F. Hughes, *Computer Graphics: Principles and Practice in C*, 2nd edn., Addison-Wesley, MA, 1995.

[Gema84] S. Geman and D. Geman, Stochastic relaxation, Gibbs distributions, and the Bayesian restoration of images, *IEEE Transactions on PAMI*, 6(6), 721–741, 1984.

[Grie04] A. Griesser, T.P. Koninckx, and L. Van Gool, Adaptive real-time 3D acquisition and contour tracking within a multiple structured light system, *12th Pacific Conference on Computer Graphics and Applications*, Seoul, Korea, pp. 361–370, 2004.

[Hang95] H.M. Hang and J.W. Woods, *Handbook of Visual Communications*, Academic Press, San Diego, CA, 1995.

[Horn86] B.K.P. Horn, *Robot Vision*, MIT Press, Cambridge, 1986.

[Javi06] B. Javidi and F. Okano (editors), Special issue on 3-D technologies for imaging and display, *Proceedings of the IEEE*, 94(3), 487–489, 2006.

[Jaya84] N.S. Jayant and P. Noll, *Digital Coding of Waveforms*, Prentice-Hall, London, 1984.

[Jeba99] T. Jebara, A. Azarbayejani, and A. Pentland, 3D structure from 2D motion, in *IEEE Signal Processing Magazine, Special Issue on 3D and Stereoscopic Visual Communication*, 16(3), 66–84, 1999.

[Kana97] T. Kanade, P. Rander, and P.J. Narayanan, Virtualized reality: Constructing virtual worlds from real scenes, *IEEE Multimedia, Immersive Telepresence*, 4(1), 34–47, 1997.

[Lamb05] E. Lamboray, S. Wurmlin, and M. Gross, Data streaming in telepresence environments, *IEEE Transactions on Visualization and Computer Graphics*, 11(6), 637–648, 2005.

[Lamb07] M.T.M. Lambooij, W.A. IJsselsteijn, and I. Heynderickx, Visual discomfort in stereoscopic displays: A review, *SPIE*, Vol. 6490, San Jose, CA, 2007.

[Lanz98] S. Lanzavecchia and P.L. Bellon, Fast computation of 3D radon transform via a direct Fourier method, *Bioinformatics*, 14(2), 212–216, 1998.

[Laur94] A. Laurentini, The visual hull concept for silhouette-based image understanding, *IEEE Transactions on PAMI*, 16(2), 150–162, 1994.

[Lei05] B.J. Lei, E.A. Hendriks, and A.K. Katsaggelos, Camera calibration, Chapter 3 in N. Sarris and M.G. Strintzis (editors), *3D Modeling & Animation: Synthesis and Analysis Techniques for the Human Body*, IRM Press, Hershey, PA, 2005.

[Levo96] M. Levoy and P. Hanrahan, Light field rendering, *ACM SIGGRAPH*, Los Angeles, CA, pp. 31–42, 1996.

[Magn00] M. Magnor and B. Girod, Data compression for light-field rendering, *IEEE Transactions on Circuits and Systems for Video Technology*, 10(3), 338–343, 2000.

[Marr82] D. Marr, *Vision*, Freeman, San Francisco, CA, 1982.

[Mats04] T. Matsuyama, X. Wu, T. Takai, and T. Wada, Real-time dynamic 3-D object shape reconstruction and high-fidelity texture mapping for 3-D video, *IEEE Transactions on Circuits and Systems for Video Technology*, 14(3), 357–369, 2004.

[Mcal02] D.F. McAllister, 3D displays, *Wiley Encyclopedia in Imaging*, Wiley, New York, pp. 1327–1344, 2002 (http://research.csc.ncsu.edu/stereographics/wiley.pdf).

[Noor03] F.S. Nooruddin and G. Turk, Simplification and repair of polygonal models using volumetric techniques, *IEEE Transactions on Visualization and Computer Graphics*, 9(2), 191–205, 2003.

[Ohm98] J.R. Ohm, K. Grueneberg, E.A. Hendriks, M.E. Izquierdo, D. Kalivas, M. Karland, D. Papadimatos, and P.A. Redert, A real-time hardware system for stereoscopic videoconferencing with viewpoint adaptation, *Signal Processing: Image Communication*, 14(1–2), 147–171, 1998.

[Oish05] T. Oishi, A. Nakazawa, R. Kurazume, and K. Ikeuchi, Fast simultaneous alignment of multiple range images using index images, *5th International Conference on 3-D Digital Imaging and Modeling (3DIM)*, Ottawa, Ontario, Canada, pp. 476–483, 2005.

[Past91] S. Pastoor, 3-D television: A survey of recent research results on subjective requirements, *Signal Processing: Image Communications*, 4(1), 21–32, 1991.

[Pede99] F. Pedersini, A. Sarti, and S. Tubaro, Multi-camera systems, *IEEE Signal Processing Magazine*, 16(3), 55–65, 1999.

[Peng05] J. Peng, C.S. Kim, and C.C. Jay Kuo, Technologies for 3D mesh compression: A survey, *Elsevier Journal of Visual Communication and Image Representation*, 16(6), 688–733, 2005.

[Perk92] M.G. Perkins, Data compression of stereopairs, *IEEE Transaction Communication*, 40(4), 684–696, 1992.

[Poll99] M. Pollefeys, R. Koch, M. Vergauwen, and L. Van Gool, Hand-held acquisition of 3D models with a video camera, *Proc. 3DIM'99 (Second International Conference on 3-D Digital Imaging and Modeling)*, Ottawa, Canada, pp. 14–23, 1999.

[Rede99] P.A. Redert, E.A. Hendriks, and J. Biemond, Correspondence estimation in image pairs, *IEEE Signal Processing Magazine, Special Issue on 3D Imaging*, 16(3), 29–46, 1999.

[Rede00] P.A. Redert, Multi-viewpoint systems for 3D visual communication, PhD thesis, advisors J. Biemond and E.A. Hendriks, TU Delft, ISBN 90-901-3985-0, 2000 (pdf at http://ict.ewi.tudelft.nl).

[Rede02] P.A. Redert, M. Op de Beeck, C. Fehn, W. IJsselsteijn, M. Pollefeys, L. Van Gool, E. Ofek, I. Sexton, and P. Surman, ATTEST: Advanced three-dimensional television system technologies, *Proc. of 1st International Symposium on 3D Processing, Visualization, Transmission (3DPVT)*, Padova, Italy, pp. 313–319, 2002.

[Rede07] P.A. Redert, R.P. Berretty, C. Varekamp, B.W.D. van Geest, J. Bruijns, R. Braspenning, and Q. Wei, Challenges in 3DTV image processing, *SPIE*, Vol. 6508, San Jose, CA, 2007.

[Saxe05] A. Saxena, S.H. Chung, and A.Y. Ng, Learning depth from single monocular images, *Proceedings of the 19-th Annual Conference on Neural Information Processing Systems*, Stanford, CA, 2005.

[Scha02] D. Scharstein and R. Szeliski, A taxonomy and evaluation of dense two-frame stereo correspondence algorithms, *IJCV*, 47(1/2/3), 7–42, 2002.

[Seku02] R. Sekuler and R. Blake, *Perception*, 4th edn., McGraw-Hill, New York, 2002.

[Sext99] I. Sexton and P. Surman, Stereoscopic and autostereoscopic display systems, *IEEE Signal Processing Magazine*, 16(3), 85–99, May 1999.

[Shad98] J. Shade, S. Gortler, L.-W. He, and R. Szeliski, Layered depth images, *ACM SIGGRAPH*, Orlando, FL, pp. 231–242, 1998.

[Shum00] H.Y. Shum and S.B. Kang, A review of image-based rendering techniques, *Conf. on Visual communications and Image Processing* (*VCIP*), Perth, Australia, 2000.

[Stil97] C. Stiller, Object-based estimation of dense motion fields, *IEEE Transactions on Image Processing*, 6(2), 234–250, 1997.

[Taub98] G. Taubin, W.P. Horn, F. Lazarus, and J. Rossignac, Geometry coding and VRML, *Proceedings of the IEEE*, 86(6), 1228–1243, 1998.

[Teka95] A.M. Tekalp, *Digital Video Processing*, Prentice-Hall, Upper Saddle River, NJ, 1995.

[Trav97] A.R.L. Travis, The display of three-dimensional video images, *Proceedings of the IEEE*, 85, 1817–1832, 1997.

[Truc98] E. Trucco and A. Verri, *Introductory Techniques for 3-D Computer Vision*, Prentice Hall, New York, 1998.

[Tsai87] R.Y. Tsai, A versatile camera calibration technique for high-accuracy 3D machine vision metrology using off-the-shelf TV cameras and lenses, *IEEE Journal of Robotics and Automation*, 3(4), 323–344, 1987.

[Wang03] J. Wang and M.M. Oliveira, A hole-filling strategy for reconstruction of smooth surfaces in range images, *Computer Graphics and Image Processing* (*SIBGRAPI*), São Carlos, Brazil, pp. 11–18, 2003.

[Wasc05] M. Waschbüsch, S. Würmlin, D. Cotting, F. Sadlo, and M. Gross, Scalable 3D video of dynamic scenes, *Proceedings of the Pacific Graphics*, Switzerland, 2005.

[Xian04] J. Xiangyang, T. Boyling, J.P. Siebert, N. McFarlane, W. Jiahua, and R. Tillett, Integration of range images in a multi-view stereo system, *Proceedings of the 17th International Conference on Pattern Recognition*, Vol 4, U.K., pp. 280–283, 2004.

[Zhan99] R. Zhang, P.S. Tsai, J. Cryer, and M. Shah, Shape from shading: A survey, *IEEE Transactions on PAMI*, 21(8), 690–706, 1999.

[3dvs] Range camera, www.3dvsystems.com.

Index

D

E

F

G

H

I

J

K

L

P

Q

R

S

T